# Algebra 1

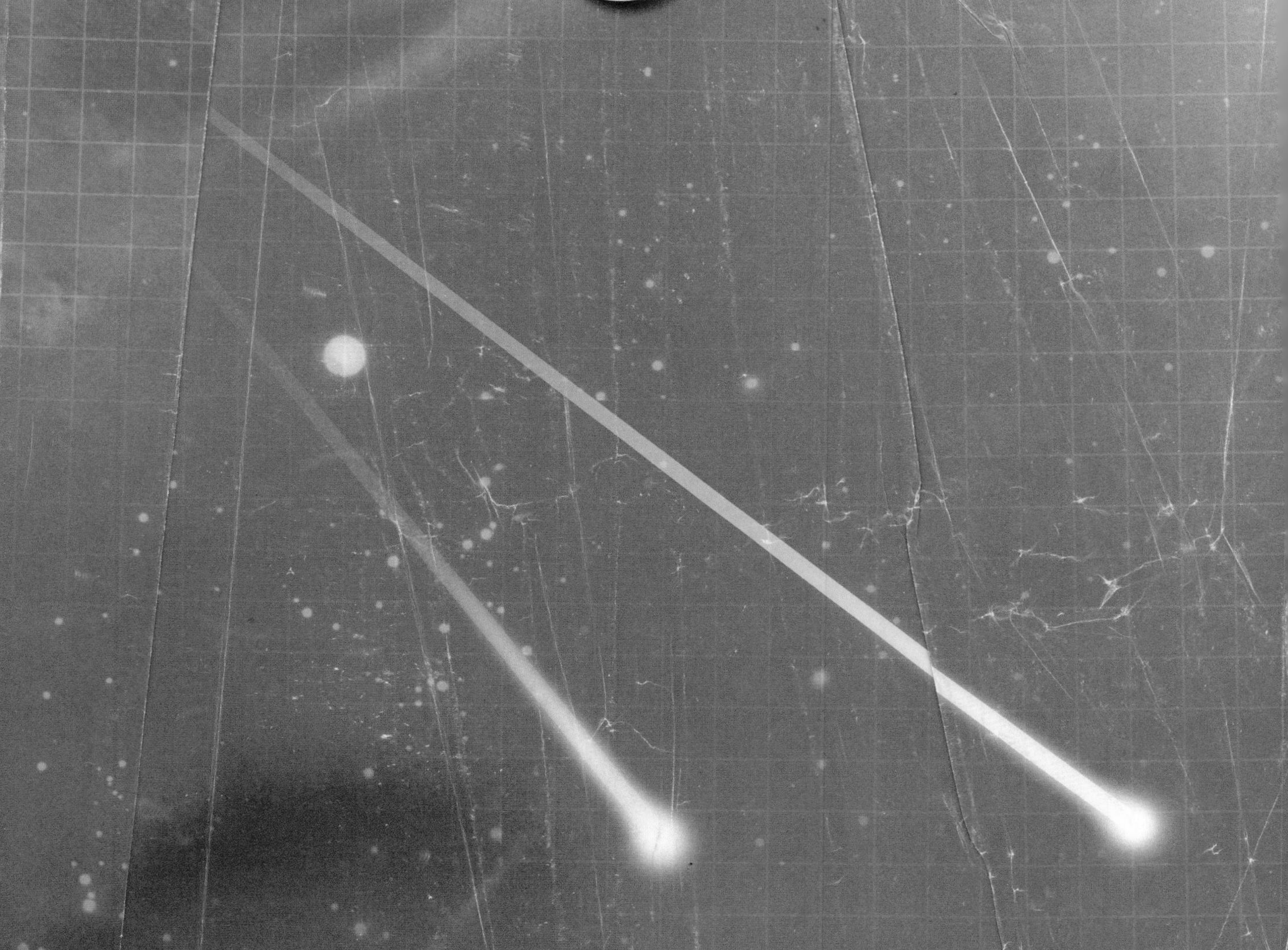

HOLT, RINEHART AND WINSTON

A Harcourt Classroom Education Company

Austin • New York • Orlando • Atlanta • San Francisco • Boston • Dallas • Toronto • London

## AUTHORS

**James F**ultz, ***Senior Series Author***
Dr. Sch has over 30 years of experience teaching at the high school a levels is the Robert L. Morton Professor of Mathematics Education Univ ty. He helped to establish standards for mathematics instruction auth f the NCTM *Curriculum and Evaluation Standards for School Ma* and A re *Curriculum: Making Mathematics Count for Everyone.*

**Paul Kennedy, *Senior Author***
A pr ssor in the Department of Mathematics at Southwest Texas State U Dr. nnedy is a leader in mathematics education reform. His research foc dev ping algebraic thinking by using multiple representations and techn ha een the author of numerous publications and he is often invited to sp co uct workshops on the teaching of secondary mathematics.

**de Ellis, Jr.**
fessor Ellis has co-authored numerous books and articles on how to inte hnology realistically and meaningfully into the mathematics curriculum. y contributor to the landmark study *Everybody Counts: A Report to the Na e Future of Mathematics Education.*

**Kathleen A. Hollowell**
Dr. Hollowell is an experienced high school mathematics and computer scien teacher who currently serves as Director of the Mathematics & Science Educa Resource Center, University of Delaware. Dr. Hollowell is particularly well ver the special challenge of motivating students and making the classroom a more dynamic place to learn.



(Acknowledgments appear on pages 867–868, which are extensions of the copyright page.)

Printed in the United States of America

ISBN: 0-03-066051-3

1 2 3 4 5 6 7 8 9 032 05 04 03 02 01

## CONTRIBUTING AUTHOR

**Irene "Sam" Jovell**
An award winning teacher at Niskayuna High School, Niskayuna, New York, Ms. Jovell served on the writing team for the *New York State Mathematics, Science, and Technology Framework*. A popular presenter at state and national conferences, her workshops focus on technology-based innovative education. Currently she is serving as president of the Association of Mathematics Teachers of New York State.

## CONTENT CONSULTANT

**Rosemary A. Garmann**

Rosemary Garmann is a graduate of Marian College (B.A.) and the University of Notre Dame (M.S. in Mathematics) and has completed post-graduate studies in mathematics and education at the University of Cincinnati, Xavier University, University of Akron, and Butler University. Ms. Garmann has taught elementary, middle school, and high school classes and is currently employed at Miami University in Oxford, Ohio, as a mathematics instructor. Formerly, she was a mathematics supervisor, building administrator, and mathematics teacher in the Cincinnati City Schools, as well as a mathematics consultant for the Hamilton County, Ohio, school districts.

## REVIEWERS

**Kenneth Adiekweh**
LAUSD-Fremont
Los Angeles, California

**Anthony Antonelli**
St. Ignatius High School
Cleveland, Ohio

**Catherine Barker**
LAUSD-Carson
Carson, California

**Judy B. Basara**
St. Hubert High School
Philadelphia, Pennsylvania

**Jeffery M. Blount**
San Diego Unified School District
San Diego, California

**Charene Borden**
LAUSD-Fran Polytech-North Hollywood
Sun Valley, California

**Kathi Bowers**
Palo Alto Unified School District
Palo Alto, California

**Paula Brown**
San Diego Unified School District
La Jolla, California

**Blanche Brownley**
District of Columbia Public Schools
Washington, D.C.

**Jeanette Burds**
LAUSD-Chatsworth-Granada Hills
Northridge, California

**Lynn Carlucci**
Scottsdale Unified School District
Scottsdale, Arizona

**David Casoli**
Aliquippa School District
Aliquippa, Pennsylvania

**Felisha Cheatem**
Akron Public Schools
Akron, Ohio

**Portia Clinton**
Elk Grove Unified School District
Elk Grove, California

**Elden Cozort**
Scottsdale Unified School District
Scottsdale, Arizona

**Nancy Crouse**
San Diego Unified School District
San Diego, California

**David Depner**
Montour School District
Coraopolis, Pennsylvania

**Guy Gnash DeRosa-Reed, Jr.**
LAUSD-Crenshaw-Dorsey
Los Angeles, California

**Rekha Desai**
LAUSD-Belmont
Los Angeles, California

**Mak Family**
LAUSD-Belmont
Los Angeles, California

**Thomas Fitzgerald**
Brevard County Schools
Melbourne, Florida

**Beverly Fleming**
Fairfax Public Schools
Fairfax, Virginia

**Mark Freathy**
Elk Grove Unified School District
Elk Grove, California

**Paul Fulkerson**
Round Rock Independent School District
Round Rock, Texas

**Linda Fulmore**
Phoenix Union High School District
Phoenix, Arizona

**Rick Hartley**
San Diego Unified School District
San Diego, California

**Stan Heeb**
San Ramon Valley Unified School District
San Ramon, California

**John Holmes**
West Contra Costa Unified School District
Richmond, California

**David Honda**
San Diego Unified School District
San Diego, California

**Terry Mahaffey**
San Diego Unified School District
San Diego, California

**Sally Maloney**
Worchester Public Schools
Worchester, Massachusetts

**Pam Mason**
LAUSD-Chatsworth-Granada Hills
Granada Hills, California

**Amy Mikolajczyk**
South Western City School District
Grove City, Ohio

**Cheryl Mockel**
Mead School District
Spokane, Washington

**Derek Moriuchi**
LAUSD-Garfield
Los Angeles, California

**Judy Muhlethaler**
San Dieguito Union High School District
Encinitas, California

**Adrian Ortega**
LAUSD-Venice-Westchester
Los Angeles, California

**Martin Paco**
Albuquerque Public Schools
Albuquerque, New Mexico

**Curt Perry**
Plymouth-Canton Community Schools
Plymouth, Michigan

**Mai Pham**
LAUSD-Southgate
South Gate, California

**Mary Prater**
LAUSD-Fairfax-LA-Hollywood
Los Angeles, California

**John Remensky**
South Park School District
Library, Pennsylvania

**Elaine Riding**
McGuffey School District
Claysville, Pennsylvania

**Dolores Salvo**
Mt. Lebanon School District
Pittsburgh, Pennsylvania

**Mike Shakelford**
San Ramon Valley Unified School District
San Ramon, California

**Karen Smith**
Sumner County School District
Gallatin, Tennessee

**Kathleen Yamashita**
San Ramon Valley Unified School District
San Ramon, California

**Diane Young**
Austin Independent School District
Austin, Texas

# Table of Contents

### MATH CONNECTIONS



### APPLICATIONS

## 5 LINEAR FUNCTIONS 216



## 6 INEQUALITIES AND ABSOLUTE VALUE 274



### MATH CONNECTIONS



### APPLICATIONS

## 7 SYSTEMS OF EQUATIONS AND INEQUALITIES 316



## 8 EXPONENTS AND EXPONENTIAL FUNCTIONS 368

### MATH CONNECTIONS



### APPLICATIONS

## 11 RATIONAL FUNCTIONS 524

## 12 RADICALS, FUNCTIONS & COORDINATE GEOMETRY 574

### MATH CONNECTIONS



### APPLICATIONS

## INFO BANK 736



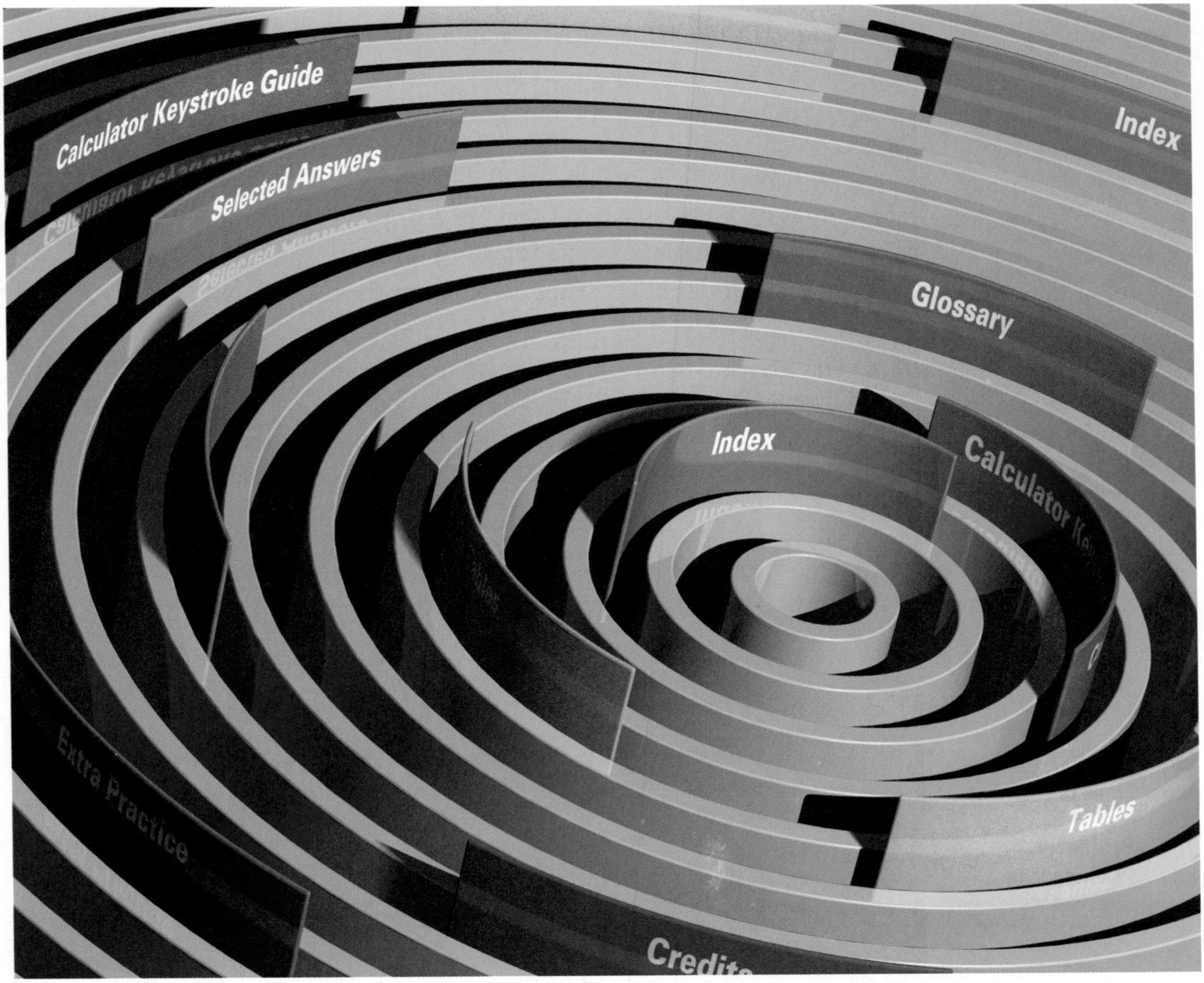

# From Patterns to Algebra

THROUGHOUT HISTORY, PEOPLE FROM ALL PARTS of the world have been fascinated with patterns. The early native tribes of the American Southwest were aware of many patterns that influenced their lives. Many of their artifacts show a deep appreciation for the beauty of patterns.

In the technological world of modern society, patterns provide a basis for discoveries in science and mathematics. Scientists use patterns to study, understand, and predict nature. Mathematicians look for regular patterns when investigating systems of numbers. Patterns can provide a powerful tool for solving problems.

## Lessons

*The rug, wedding basket, and pot show the rich use of patterns in Navaho artifacts.*

*Geologists use seismographs to record patterns in earthquake tremors.*

CHAPTER PORTFOLIO ACTIVITIES PROJECT

## About the Chapter Project

Recognizing patterns in data is useful for solving problems. Algebra provides tools that help you to generalize and make predictions based on patterns. In the Chapter Project, *Working With Patterns in Data* on page 44, you will learn how to recognize mathematical patterns in golden rectangles, in a set of circular objects in your classroom, and in a simulation of a wave cheer. After completing the Chapter Project, you will be able to do the following:

- Analyze patterns in data by using scatter plots and tables.

## About the Portfolio Activities

Throughout the chapter, you will be given opportunities to complete Portfolio Activities that are designed to support your work on the Chapter Project.

- Building staircases with cubes to represent triangular numbers is the topic of the Portfolio Activity on page 10.
- Representing a number sequence by using triangles constructed with toothpicks is the topic of the Portfolio Activity on page 17.
- A famous sequence of numbers named for the mathematician Fibonacci is the topic of the Portfolio Activity on page 29.
- Correlations in real-world sets of data are the topic of the Portfolio Activity on page 43.

1.1

# Using Differences to Identify Patterns

*Lightning heats the air, which expands rapidly and creates thunder.*

**Why** ***Organizing data and using differences to help identify patterns provide a way to better understand relationships between things like the time you hear thunder and the time you see lightning.***

## Objectives

- Use differences to identify patterns in number sequences.
- Make predictions by using patterns in number sequences.

There are many patterns in nature. Analyzing these patterns helps us to understand nature better. Mathematics provides tools for studying the patterns around us.

A **number sequence** is a string of numbers, or terms, in a certain order. Often there will be three dots after the last given number. The three dots (...) indicate that there are more terms that are not listed.

If the difference from one term to the next in a number sequence is always the same, the difference is called a **constant difference.**

**EXAMPLE 1** **Find the next three terms of each sequence by using constant differences.**

**a.** 1, 3, 5, 7, 9, ...  **b.** 80, 73, 66, 59, 52, ...

**SOLUTION**

**a.** Find the constant differences.

**1** 1+2=3 **3** 3+2=5 **5** 5+2=7 **7** 7+2=9 **9**

2 2 2 2 *Notice that the terms increase by 2.*

Add 2 to the previous term to find the next term.

$$9 + 2 = 11 \qquad 11 + 2 = 13 \qquad 13 + 2 = 15$$

The next three terms of the sequence are 11, 13, and 15.

**b.** Find the constant differences.

**80** 80 − 7=73 **73** 73 − 7=66 **66** 66 − 7=59 **59** 59 − 7=52 **52**

−7 −7 −7 −7

*Notice that the terms decrease by 7.*

Subtract 7 from the previous term to find each new term.

$$52 - 7 = 45 \qquad 45 - 7 = 38 \qquad 38 - 7 = 31$$

The next three terms are 45, 38, and 31.

**TRY THIS** Find the next three terms of each sequence by using constant differences.

**a.** 1, 4, 7, 10, 13, …

**b.** 30, 25, 20, 15, 10, …

**EXAMPLE 2** **Find the next three terms of each sequence by using constant differences.**

**a.** 1, 4, 9, 16, 25, …

**b.** 37, 41, 48, 58, 71, …

**SOLUTION**

**a.** Find the differences.

1 1+3=4 4 4+5=9 9 9+7=16 16 16+9=25 25

3 5 7 9

Since the first differences are not constant, find the second differences.

1 4 9 16 25

3 3+2=5 5 5+2=7 7 7+2=9 9

2 2 2

The second differences are a constant 2. Use the constant second difference to find the next three first differences.

1 4 9 16 25

3 5 7 9 9+2=11 11 11+2=13 13 13+2=15 15

2 2 2 2 2 2

Use these three first differences to find the next three terms.

1 4 9 16 25 25+11=36 36 36+13=49 49 49+15=64 64

3 5 7 9 11 13 15

2 2 2 2 2 2

The next three terms of the sequence are 36, 49, and 64.

**b.** Find the first differences.

37 41 48 58 71

4 7 10 13

Since the first differences are not constant, find the second differences.

37 41 48 58 71

4 7 10 13

3 3 3

The second differences are a constant 3. Use the constant second difference to find the next three first differences.

37 41 48 58 71

4 7 10 13 16 19 22

3 3 3 3 3 3

Use these three first differences to find the next three terms.

37 41 48 58 71 87 106 128

4 7 10 13 16 19 22

3 3 3 3 3 3

The next three terms of the sequence are 87, 106, and 128.

**TRY THIS** Find the next three terms of each sequence by using constant differences.

**a.** 2, 6, 12, 20, 30, ...

**b.** 8, 20, 30, 38, 44, ...

## Patterns in Number Sequences

When mathematicians study numbers for patterns, they often make conjectures.

A **conjecture** is a statement about observations that is believed to be true.

When mathematicians make a conjecture, they try to either prove that the conjecture is true or find a counterexample to show that the conjecture is not true.

You may make conjectures when studying a pattern such as a number sequence.

**EXAMPLE 3**

**APPLICATION PHYSICS**

The table below shows the relationship between temperatures in Celsius and temperatures in Fahrenheit.

| Celsius | 0 | 10 | 20 | 30 | 40 |
|---|---|---|---|---|---|
| Fahrenheit | 32 | 50 | 68 | 86 | 104 |

Use the method of constant differences to find the Fahrenheit temperatures that correspond to the Celsius temperatures of 50, 60, and 70.

**SOLUTION**

Look at the differences to discover the pattern.

Find the constant differences of the Fahrenheit temperatures.

| Celsius | 0 | 10 | 20 | 30 | 40 | 50 | 60 | 70 |
|---|---|---|---|---|---|---|---|---|
| Fahrenheit | 32 | 50 | 68 | 86 | 104 | ? | ? | ? |

18 18 18 18

Since the differences are constant, the Fahrenheit temperatues increase by 18 for every Celsius increase by 10.

To find the next three Fahrenheit temperatures add 18 to the previous temperatue.

$104 + 18 = 122 \qquad 122 + 18 = 140 \qquad 140 + 18 = 158$

The three temperatures that corresponds to the Celsius temperatures of 50, 60, and 70 are the Fahrenheit temperatures of 122, 140, and 158.

**TRY THIS** Find the Fahrenheit temperature that corresponds to the Celsius temperature of 100. What does this temperature represent?

## Using Problem-Solving Strategies

CONNECTION
GEOMETRY

Some sequences can also be studied with diagrams. For example, the sequence 2, 6, 12, 20, 30, … is found by counting the number of dots in the pattern below.

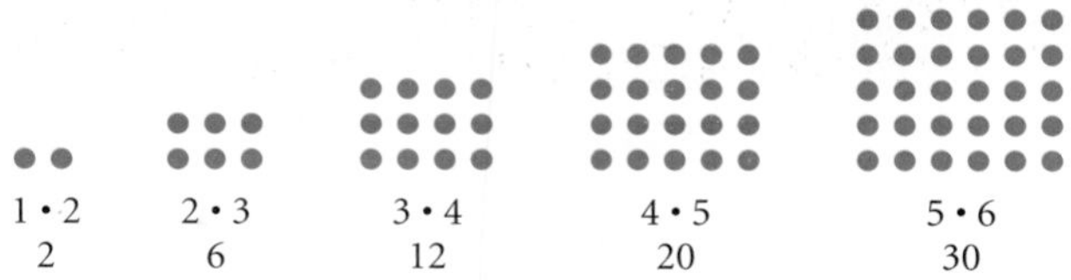


The next three terms are $6 \cdot 7 = 42$, $7 \cdot 8 = 56$, and $8 \cdot 9 = 72$.

**TRY THIS** Why do you think the numbers in the sequence 2, 6, 12, 20, 30, … are called rectangular numbers?

Drawing diagrams is just one problem solving strategy that can be helpful in extending sequences. The next example uses several strategies.

Problem solving strategies include **solving a simpler problem**, **making a table or chart**, and **looking for a pattern.**

**EXAMPLE 4** Suppose that 10 friends have just returned to school. Each friend has exactly one conversation with each of the other friends to talk about what they did during summer break. Use problem-solving strategies to determine how many conversations there will be.

**SOLUTION**

Begin with a smaller number of people. With just one person, no conversations are possible. With two people, one conversation is possible. The dot patterns below represent the number of conversations for up to four people.

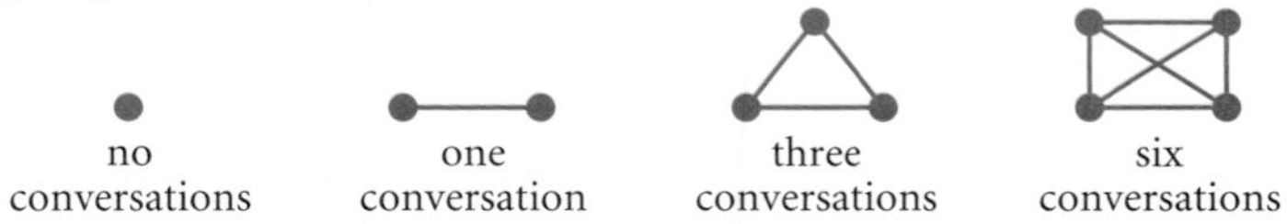

Arrange the information from the simpler problems in a table.

| **People** | 1 | 2 | 3 | 4 | 5 | 6 |
|---|---|---|---|---|---|---|
| **Conversations** | 0 | 1 | 3 | 6 | ? | ? |

Use differences to determine how the number of conversations is increasing. Then extend the pattern to 10 people.

| **People** | 1 | 2 | 3 | 4 | 5 | 6 | 7 | 8 | 9 | 10 |
|---|---|---|---|---|---|---|---|---|---|---|
| **Conversations** | 0 | 1 | 3 | 6 | 10 | 15 | 21 | 28 | 36 | 45 |

Differences: 1 2 3 4 5 6 7 8 9

With 10 friends, it takes 45 conversations for each person to have exactly one conversation with each other person.

APPLICATION
METEOROLOGY

*Activity*

## Exploring Differences

The table below shows the approximate distance from a lightning strike to an observer based on the time it takes the observer to hear the thunder.

| Seconds | 2 | 4 | 6 | 8 | 10 |
|---|---|---|---|---|---|
| Miles | 0.4 | 0.8 | 1.2 | ? | ? |

1. Copy and complete the table.
2. For each 2-second increase in time, what is the increase in miles?
3. Use this rate of change to calculate the distance in miles for 12, 14, and 16 seconds.

CHECKPOINT ✔

4. The sound of thunder travels 4.4 miles in 22 seconds. Find the distance that the sound of thunder travels in 23 seconds.

# Exercises

## Communicate

1. Explain how to find the first and second differences for the following sequence: 11, 14, 19, 26, 35, 46, …
2. Describe the method for predicting the next two terms in the following sequence: 88, 76, 64, 52, 40, …
3. Explain how to work backward from a constant second difference to extend the following sequence: 1, 6, 16, 31, 51, 76, …

## Guided Skills Practice

**Find the next three terms of each sequence by using constant differences.** *(EXAMPLES 1 AND 2)*

4. 1, 3, 5, 7, 9, …
5. 10, 12, 14, 16, 18, …
6. 87, 78, 69, 60, 51, …
7. 1, 4, 9, 16, 25, …
8. 10, 17, 26, 37, 50, …
9. 3, 13, 27, 45, 67, …

## Practice and Apply

internet connect
**Homework Help Online**
Go To: **go.hrw.com**
Keyword:
**MA1 Homework Help**
for Exercises 10-19

**Find the next three terms of each sequence.**

10. 18, 32, 46, 60, 74, …
11. 33, 49, 65, 81, 97, …
12. 100, 94, 88, 82, 76, …
13. 44, 41, 38, 35, 32, …
14. 20, 21, 26, 35, 48, …
15. 30, 31, 35, 42, 52, …
16. 1, 8, 18, 31, 47, …
17. 4, 7, 15, 28, 46, …
18. 1, 2.5, 4, 5.5, 7.0, …
19. 0.5, 2, 4.5, 8, 12.5, …

## CHALLENGE

**20.** If the first differences of a sequence are a constant 4 and the second term is 8, find the first 5 terms of the sequence.

**21.** If the first differences of a sequence are a constant −7 and the third term is 22, find the first 5 terms of the sequence.

**22.** The second and third terms of a sequence are 29 and 42. If the first differences are constant, find the first 5 terms of the sequence.

**23.** The third and fourth terms of a sequence are 101 and 83. If the first differences are constant, find the first 5 terms of the sequence.

**24.** If the second differences of a sequence are a constant 3, the first of the first differences is 7, and the first term is 2, find the first 5 terms of the sequence.

**25.** For each sequence, determine which differences are constant and what the constant difference is.

**a.** 1, 2, 3, 4, 5, … **b.** $1^2, 2^2, 3^2, 4^2, 5^2, \ldots$ **c.** $1^3, 2^3, 3^3, 4^3, 5^3, \ldots$

**d.** Based on **a**, **b**, and **c**, determine how many differences you have to compute until you find a constant difference for the sequence $1^4, 2^4, 3^4, 4^4, 5^4, \ldots$

## CONNECTIONS

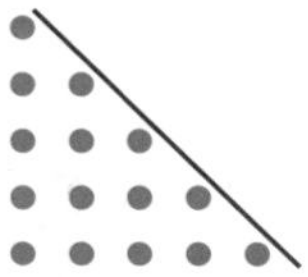

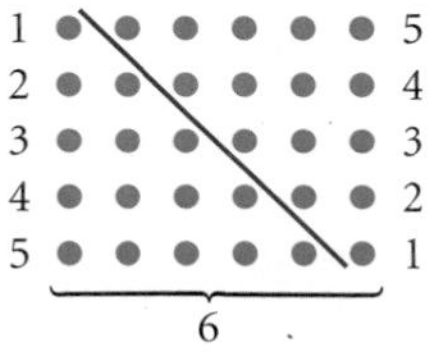


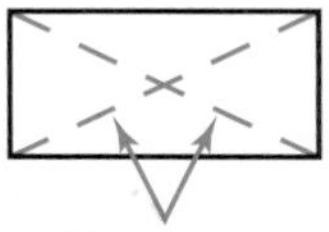

diagonals

**GEOMETRY** The sum of the counting numbers from 1 to 5 can be represented by drawing a triangle of dots as shown at left. A rectangle can be formed by copying the triangle and rotating it. The sum of the counting numbers from 1 to 5 is half of the number of dots in the rectangle, or $\frac{5 \cdot 6}{2}$. For Exercises 26–28, sketch similar geometric dot patterns and find each sum.

**26.** 1 + 2 + 3 + 4 + 5 + 6 + 7 **27.** 1 + 2 + 3 + 4 + 5 + 6 + 7 + 8 + 9 + 10

**28.** 1 + 2 + 3 + 4 + 5 + 6 + 7 + 8 + 9 + 10 + 11 + 12

**GEOMETRY** Find each sum below. Think of a geometric dot pattern, but do not draw a sketch.

**29.** 1 + 2 + · · · + 40 **30.** 1 + 2 + · · · + 60 **31.** 1 + 2 + · · · + 80

**32. GEOMETRY** The table below shows the number of diagonals for various types of polygons. Use constant differences to fill in the missing values in the table.

| | Triangle | Rectangle | Pentagon | Hexagon | Heptagon | Octagon | Nonagon | Decagon |
|---|---|---|---|---|---|---|---|---|
| **Sides** | 3 | 4 | 5 | 6 | 7 | 8 | 9 | 10 |
| **Diagonals** | 0 | 2 | 5 | 9 | 14 | ? | ? | ? |

## APPLICATION

**33. PHYSICS** As the altitude increases, the boiling point of water decreases, as shown in the table below. For example, in Las Vegas, Nevada, at an altitude of 2180 feet, the boiling point of water is approximately 208.4°F.

| **Altitude (in thousands of feet)** | **0** | **1** | **2** | **3** | **4** | **5** |
|---|---|---|---|---|---|---|
| **Boiling point (in °F)** | **212** | **210.2** | **208.4** | **206.6** | **204.8** | **203** |

**a.** Estimate the boiling point of water in Colorado Springs, Colorado, which is at an altitude of 6170 feet.

**b.** According to the pattern in the table, what is the boiling point of water at an altitude of 18,000 feet?

APPLICATIONS

**34.** RECREATION If each of 11 football teams in a college athletic conference played each of the other teams just once in a season, how many games would there be? As part of your answer, make a table similar to the one in Example 4 on page 7. How many games would there be for 12 teams?

**35.** TELECOMMUNICATIONS Suppose that 8 cities are to be linked by fiber optic cables, with 2 cables between each pair of cities. How many cables will there be?

## Look Back

**Multiply.**

**36.** $2.7 \cdot 3.1$ **37.** $4.31 \cdot 2$ **38.** $\frac{1}{2} \cdot \frac{2}{3}$ **39.** $2\frac{2}{3} \cdot 1\frac{1}{4}$

**Divide.**

**40.** $3.64 \div 2$ **41.** $0.52 \div 1.3$ **42.** $\frac{1}{2} \div \frac{1}{2}$ **43.** $2\frac{3}{8} \div 2$

## Look Beyond

internet connect
**Portfolio Extension**
Go To: **go.hrw.com**
Keyword:
**MA1 Triangular**

**44.** The method of finding differences works only for certain kinds of number sequences. Try it on the following sequences and tell what happens. Predict the next three terms of each sequence without using differences. Explain your method.

**a.** 1, 2, 4, 8, 16, 32, … **b.** 1, 10, 100, 1000, 10,000, …

PORTFOLIO ACTIVITY

Include this activity in your portfolio.

You can use cubes to build the sequence of stairways shown above.

**1.** Sketch a picture of the fourth stairway on graph paper. How many cubes does it take to build the fourth stairway? Build a table of values for the number of cubes in each stairway, as shown below.

| Number of steps | 1 | 2 | 3 | 4 | 5 | 6 | 7 | 8 | 9 | 10 |
|---|---|---|---|---|---|---|---|---|---|---|
| Cubes in the stairway | 1 | 3 | 6 | ? | ? | ? | ? | ? | ? | ? |

**2.** Sketch the stairways up to the 10th stairway.

**3.** What is the pattern in the sequence of stairways? How does the 10th stairway compare with your answer in the table you made in Step 1?

The sequence of numbers that you have been working with, 1, 3, 6, 10, …, is called the sequence of triangular numbers.

**4.** Show visually that the square number 16 is the sum of two triangular numbers and state the two triangular numbers.

**5.** Show visually that 100 is the sum of two triangular numbers and state the two triangular numbers.

1.2

# Variables, Expressions, and Equations

**Objectives**

- Use variables to represent unknown quantities.
- Represent real-world situations with equations and solve by guess-and-check.

**Why** *Using variables and expressions to form equations provides a systematic way of representing and solving many types of problems.*

Suppose that a rental shop charges an \$8 fee plus \$3 per hour to rent a bike. The amount charged for different times is shown in the following table.

| Number of hours | Calculation | Cost (\$) |
|---|---|---|
| 1 | $8 + 3$<br>$8 + 3(1)$ | 11 |
| 2 | $8 + 3 + 3$<br>$8 + 3(2)$ | 14 |
| 3 | $8 + 3 + 3 + 3$<br>$8 + 3(3)$ | 17 |

There is a pattern in how the cost is calculated.

**The bike rental cost is 8 plus 3 times the number of hours.**

**Variables** are letters that are used to represent quantities in algebra. For example, the letter $h$ can be used to represent the number of hours a bike will be rented.

The cost of renting a bike can be written as an *algebraic expression.*

$$8 + 3(h) \qquad \text{or} \qquad 8 + 3h$$

An **algebraic expression** is formed by combining numbers and variables using the operations of addition, subtraction, multiplication, division, powers, and roots. The algebraic expression $8 + 3h$ represents the cost of renting a bike.

## Values of Expressions

The cost of renting a bike for 5 hours can be found by replacing $h$ with 5 and evaluating the expression. Let $h = 5$ in $8 + 3h$.

$$8 + 3(5)$$
$$8 + 15$$
$$23$$

The cost of renting a bike for 5 hours is \$23.

**EXAMPLE 1** **Make a table to show the value of the expression $8 + 3h$ when $h$ is 6, 8, and 12 hours.**

**SOLUTION**

Make a table as shown and substitute to find the value of the expression for each value of $h$.

| $h$ | 6 | 8 | 12 |
|---|---|---|---|
| $8 + 3h$ | $8 + 3(6)$<br>$8 + 18$ | $8 + 3(8)$<br>$8 + 24$ | $8 + 3(12)$<br>$8 + 36$ |
| Value | 26 | 32 | 44 |

## Activity: Modeling Data With a Table and an Equation

Suppose that the average speed of a mountain biker in a race is 20 miles per hour. Build a table for time and distance.

**Table for 20 miles per hour**

| Time (hours) | 0 | 0.5 | 1.0 | 1.5 | 2.0 | 2.5 | 3.0 | 3.5 | 4.0 |
|---|---|---|---|---|---|---|---|---|---|
| Distance (miles) | 0 | ? | 20 | ? | ? | ? | ? | ? | ? |

1. Copy and complete the table. Describe your method for filling in the distances. How could you find the distance cycled in 7 hours? in 8 hours?

CHECKPOINT ✔

2. How do you find the distance for a specific time? Rewrite your answer with variables instead of words. Use $d$ for distance and $t$ for time. Use symbols for "equals" and "multiplied by." Your result will be an *equation* with variables.

### Equations

Two algebraic expressions separated by an equal sign form an **equation.** When two expressions represent the same quantity, an equation can be written.

**EXAMPLE 2** At City Park you can rent in-line skates and protective gear for $2 per hour plus a $4 fee. **Write an expression that represents the cost of renting skates and then write an equation that relates the expression to a total cost of $10.**

**SOLUTION**

| Number of Hours | Expression for cost |
|---|---|
| 1 | $4 + 2(1)$ |
| 2 | $4 + 2(2)$ |
| 3 | $4 + 2(3)$ |

The rental cost is 4 plus 2 times the number of hours.

$$4 + 2h$$

If the total cost of renting skates is $10, the equation relating the expression and the actual cost is

$$4 + 2h = 10$$

## Solutions to Equations

When one side of an equation has the same value as the other side, the equation is said to be true. To determine whether an equation is true, simplify both sides.

$$4 + 2(3) \stackrel{?}{=} 6 + 2(2)$$ *Is this equation true?*

$$4 + 6 \stackrel{?}{=} 6 + 4$$ *Simplify.*

$$10 = 10 \checkmark$$ *Simplify. The statement is true.*

> The left and right sides of the equation have the same value. Therefore, the equation is true. → $10 = 10$

If an equation has a variable in it, you can substitute a value for the variable. If the substitution gives you a true equation, then the value you substituted is a *solution* of the equation. In the equation $2h = 15$, 7.5 is a solution.

$$2h = 15$$ *The given equation.*

$$2(7.5) \stackrel{?}{=} 15$$ *Substitute 7.5 for the variable h. Ask: Is the result true?*

$$15 = 15 \checkmark$$ *Simplify. The statement is true.*

Since $h = 7.5$ gives a true equation, 7.5 is a solution of the original equation.

The **solutions** to an equation are all values of the variables that give a true equation when substituted for the variables. In many equations, such as the equation in Example 3, there is just one solution.

Finding solutions to equations is a major focus of algebra. In this book you will learn methods for finding solutions. As a beginning to finding solutions, you can use trial and error ("guess and check") to identify values that give a true equation when substituted.

**EXAMPLE 3** Jessica places 3 photos on the cover of her picture album. She places 4 photos on each of the other pages. She has 155 photos in all. **How many pages will she need if she uses all of her photos?**

**SOLUTION**

PROBLEM SOLVING

Write an equation and use a **guess-and-check** strategy.

Let $p$ equal the number of pages in Jessica's picture album. To answer the question, solve for $p$ in the equation $4p + 3 = 155$. Guess the number of pages, substitute the number you guessed for $p$, and check your guess.

| **Guess 1:** Try $p = 20$. | **Guess 2:** Try $p = 40$. | **Guess 3:** Try $p = 38$. |
|---|---|---|
| $4p + 3 = 155$ | $4p + 3 = 155$ | $4p + 3 = 155$ |
| $4(20) + 3 = 83$ | $4(40) + 3 = 163$ | $4(38) + 3 = 155$ |
| 83 is too small. Try a larger number for $p$. | 163 is too large. Try a smaller number for $p$. | Correct. |

The solution to the problem is 38 pages.

**CRITICAL THINKING** In Example 3, if Jessica had 160 photos, there would be no whole-number solution to the equation—notice that $4(39) + 3 = 159$ and $4(40) + 3 = 163$. How many pages would she need for 160 photos? Explain.

**TRY THIS** Marge's class is going on a field trip to an aquarium. The class rate is \$12 per student plus \$25. How many students can go if the class has \$335 to spend?

You can also solve a problem by using a calculator to model or *simulate* the problem situation. In Example 4 the loss of forest acreage over a period of time is modeled by repeated subtractions using a calculator.

## EXAMPLE 4

**APPLICATION**
**ECOLOGY**

In 1990, the United Nations reported a deforestation figure of around 114,364 acres of trees per day. The area of Olympic National Park is about 922,654 acres. **How long would it take to lose an area of forest equal to the area of Olympic National Park if 114,364 acres are destroyed each day worldwide?**

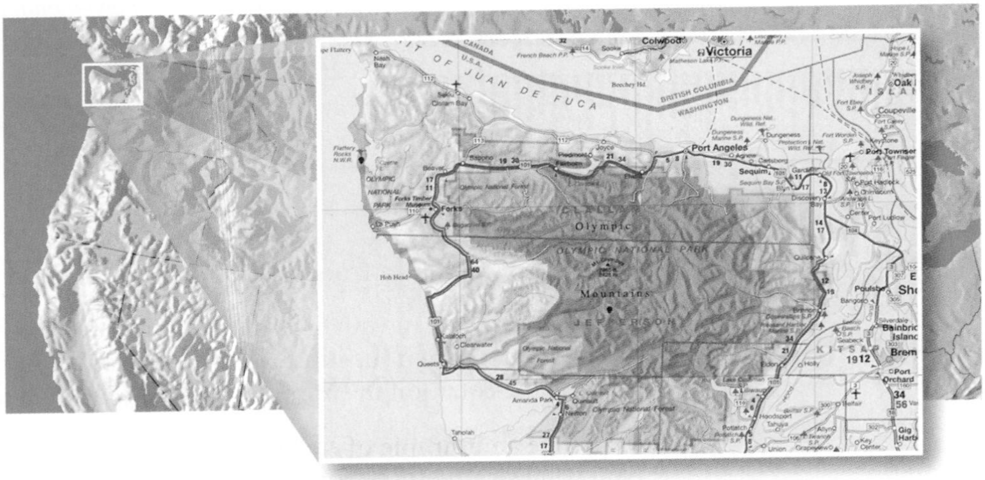


**a.** Solve the problem by using the constant feature of a calculator.

**b.** Express the situation as an equation and use the equation to find the number of acres left at the end of the third day.

### SOLUTION

**TECHNOLOGY**
**GRAPHICS CALCULATOR**

**a.** On a calculator with a constant feature that repeats an operation, enter the area of Olympic National Park. Subtract 114,364 repeatedly, counting the number of times you subtract.

As you continue to subtract, you can see how quickly the forest area is diminished. Continue repeating the subtraction until the entire forest area is depleted.

After subtracting 114,364 eight times, only 7742 acres remain. Therefore, it would take 9 days to lose 922,654 acres of forest land.

**b.** To express the situation as an equation, you need to interpret and relate the information in the problem.

The acreage left is 922,654 minus 114,364 acres for each day that passes.

If $A$ represents the number of acres left and $D$ the number of days, the situation can be expressed by the following equation:

$$A = 922{,}654 - 114{,}364D$$

You can use the equation to find the amount of forest left on any given day. To find the number of acres left at the end of the third day, substitute 3 for the variable $D$.

$$\begin{aligned} A &= 922{,}654 - 114{,}364(3) \\ &= 922{,}654 - 343{,}092 \\ &= 579{,}562 \end{aligned}$$

There are 579,562 acres left at the end of the third day.

# Exercises

## Communicate

**1.** Give an example of each: *variable*, *expression*, and *equation*.

**2.** Describe how you would set up the equation for the following problem: If 12 pencils cost $1.92, find the cost of 1 pencil.

**3.** Describe how you would set up the equation for the following problem: If tickets for a concert cost $10 each, how many can you buy with $35?

**4.** Explain how you would use the guess-and-check strategy to solve the equation $10x + 3 = 513$.

## Guided Skills Practice

**5.** Make a table to show the value of the expression $5c + 7$ when $c$ is 1, 2, 3, and 4. ***(EXAMPLE 1)***

**APPLICATIONS**

**6.** **RECREATION** At the roller rink, skating costs $2.50 per hour plus $7. Complete the table of values for this situation. Write an equation to represent the situation. ***(EXAMPLE 2)***

| **Hours** | 1 | 2 | 3 | 4 |
|---|---|---|---|---|
| **Cost ($)** | ? | ? | ? | ? |

**7.** **RECREATION** Write and solve an equation for the following problem: Bob has $32. How many hours can he skate if skating costs $2.50 per hour plus $7? ***(EXAMPLE 3)***

8. Write an equation and use the table feature on a graphics calculator to determine the number of the term that has a value of 0 in the sequence below. *(EXAMPLE 4)*

13,500, 13,000, 12,500, 12,000, ...

9. Write an equation that describes the linear pattern in the data below. Use $t$ for time and $d$ for distance. *(EXAMPLE 4)*

| Time (hours) | 0 | 1 | 2 | 3 | 4 | 5 |
|---|---|---|---|---|---|---|
| Distance (miles) | 0 | 24 | 48 | 72 | 96 | 120 |

## Practice and Apply

**Make a table showing the value of each expression when the value of the variable is 1, 2, 3, 4, and 5.**

**10.** $4x$ **11.** $5y$ **12.** $7s + 4$ **13.** $3n$

**14.** $5p - 3$ **15.** $2f - 1$ **16.** $6q + 9$ **17.** $9g - 4$

**18.** $8x$ **19.** $3r + 2$ **20.** $9y + 3$ **21.** $3z - 3$

**Use guess-and-check to solve the following equations.**

**22.** $4x + 3 = 47$ **23.** $19x + 13 = 51$ **24.** $3x + 17 = 56$

**25.** $7x - 12 = 44$ **26.** $13x - 3 = 62$ **27.** $21x - 17 = 193$

**28.** $5p - 10 = 50$ **29.** $3s + 2 = 53$ **30.** $11a - 2 = 64$

APPLICATIONS

**CONSUMER ECONOMICS** If pencils cost 20 cents each and pens cost 30 cents each, find the cost of the following items.

**31.** 0 pencils **32.** 4 pencils **33.** 10 pencils

**34.** $p$ pencils **35.** $r$ pens **36.** $r$ pens and $p$ pencils

internet connect
**Homework Help Online**
Go To: **go.hrw.com**
Keyword:
**MA1 Homework Help**
for Exercises 37-38

**37. ENTERTAINMENT** If tickets for a concert cost \$9 each, how many tickets can you buy if you have \$135? Write an equation and solve by guess-and-check.

**38. ENTERTAINMENT** If tickets for a concert cost \$11 each, how many tickets can you buy if you have \$132? Write an equation and solve by guess-and-check.

**39. HOME IMPROVEMENT** Write an equation that describes the linear pattern in the data below. Use $g$ for the number of gallons of paint and $c$ for the total cost.

| Gallons of paint | 0 | 1 | 2 | 3 | 4 | 5 |
|---|---|---|---|---|---|---|
| Total cost | 0 | 16 | 32 | 48 | 64 | 80 |

**TECHNOLOGY** **Tell how you would generate each sequence below on a calculator. Find the 20th term of each sequence.**

**40.** 6, 8, 10, 12, 14, ... **41.** 15, 25, 35, 45, 55, ...

**42.** 100, 90, 80, 70, 60, ... **43.** 52, 48, 44, 40, 36, ...

## Look Back

**44.** The sum $1 + 2 + 3 + 4 + 5$ is $\frac{5 \cdot 6}{2}$. Examine the next two cases—the sum of the first six integers and the sum of the first seven integers. Make a conjecture about the sum of the first $n$ integers. ***(LESSON 1.1)***

**45.** Find the first and second differences for the sequence 2, 8, 24, 51, 90, … ***(LESSON 1.1)***

**46.** Find the next two terms in the sequence 11, 22, 33, 44, 55, 66, … ***(LESSON 1.1)***

**47.** The first differences of a sequence are 7, 9, 11, and 13. The first term of the sequence is 5. What are the first five terms of the sequence? ***(LESSON 1.1)***

**48.** The first term of a sequence is 11. The first difference is 3, and the second differences are a constant 2. What are the first five terms of the sequence? ***(LESSON 1.1)***

**internet connect**

**Activities Online**
Go To: **go.hrw.com**
Keyword:
**MA1 Mars Weight**

## Look Beyond

**49.** The expression $x^2$ means "$x$ times $x$." Use guess-and-check to solve for $x$ in the equation $x^2 = 256$.

**PORTFOLIO ACTIVITY**

Include this activity in your portfolio.

$3 = 2(1) + 1$ $5 = 2(2) + 1$ $7 = 2(3) + 1$

**1.** Study the sequence of toothpick triangles. Sketch the next two sets in the sequence. How many toothpicks does it take to build each set? Build a table for the sets of triangles and compute the first differences.

| **Set number, $x$** | 0 | 1 | 2 | 3 | 4 | 5 |
|---|---|---|---|---|---|---|
| **Segments, $y$** | ? | 3 | 5 | ? | ? | ? |

First differences: 2 2 ? ?

**2.** How many toothpicks does it take to draw the 100th set? How could you use differences to work back to the number of toothpicks in set 0?

The differences can be used to write an equation, $y = 2x + 1$. If you separate the toothpicks, it is easy to visualize the equation.

**WORKING ON THE CHAPTER PROJECT**

You should now be able to complete Activities 1 and 2 of the Chapter Project on page 44.

1.3

# The Algebraic Order of Operations

**Objective**

- Apply the algebraic order of operations.

**Why** ***Calculating how much floor covering is needed for a particular room can involve a series of operations such as addition, subtraction, multiplication, and division. Knowing the correct algebraic order of operations is important.***

**APPLICATION**
**HOME IMPROVEMENT**

Mary plans to buy floor covering for two rooms. One room is 12 feet by 15 feet, and the other room is 20 feet by 14 feet. To find the number of square feet, she uses a simple four-function calculator and enters the following:

12 [×] 15 [+] 20 [×] 14 [=]

Much to Mary's surprise, her calculator's answer is [2800]. Mary knows that 2800 square feet is too much because her entire house is less than 2000 square feet. Why did her calculator give a wrong answer?

To avoid misunderstandings and errors, mathematicians have agreed on certain rules for computation called the **algebraic order of operations**.

As you will see, Mary's calculator was not designed to follow these rules.

### The Algebraic Order of Operations

1. Perform all operations within grouping symbols from innermost outward.
2. Perform all operations with exponents.
3. Perform all multiplications and divisions in order from left to right.
4. Perform all additions and subtractions in order from left to right.

Symbols like parentheses, ( ), brackets, [ ], braces, { }, and the fraction bar, —, are called **grouping symbols** or **inclusion symbols**. Treat any grouped numbers and variables as a single quantity. Operations should always be done within the innermost grouping symbols first. Then work outward.

**EXAMPLE 1** **Use the algebraic order of operations to evaluate each expression.**

**a.** $3 - 12 \div 4$ **b.** $6^2 \div 4 \cdot 3$ **c.** $(8 \cdot 7 + 2) + (3 \cdot 2 - 2 \cdot 2)$

**SOLUTION**

**a.** Divide first and then subtract: $3 - 12 \div 4 = 3 - 3 = 0$

**b.** Evaluate the exponent and then divide and multiply in order:
$6^2 \div 4 \cdot 3 = 36 \div 4 \cdot 3 = 9 \cdot 3 = 27$

**c.** Perform all operations within parentheses first:
$(8 \cdot 7 + 2) + (3 \cdot 2 - 2 \cdot 2) = (56 + 2) + (6 - 4) = 58 + 2 = 60$

**TRY THIS** Use the algebraic order of operations to evaluate each expression.

**a.** $13 + 2 \cdot 5$ **b.** $6^2 - 4 \cdot 3 + 7^2$ **c.** $(3 \cdot 4 + 7) + (2 \cdot 4 + 3 \cdot 4)$

In the activity below, you will investigate the rules of operations that different calculators follow.

## *Activity*

## Exploring the Order of Operations

**You will need:** a graphics calculator or a scientific calculator

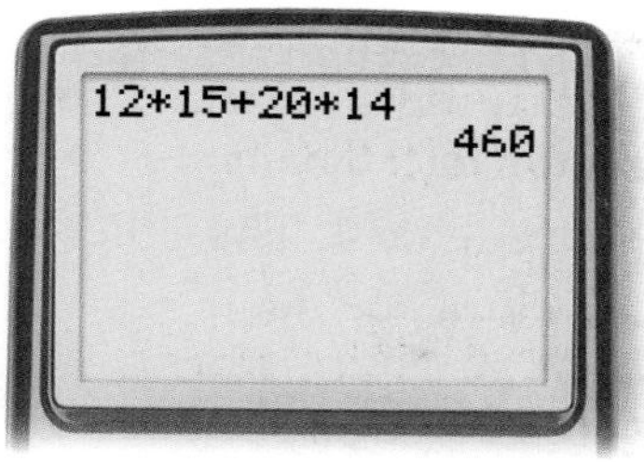


1. What steps did Mary's simple calculator use to get 2800? (See page 18.)
2. Unlike Mary's calculator, graphics and scientific calculators follow the algebraic order of operations. The solution to Mary's problem is shown at left on a graphics calculator. What order did this calculator follow?
3. Predict the value of each expression below. Then use a calculator that follows the algebraic order of operations to verify your prediction.
   **a.** $2 + 3 \cdot 5$ **b.** $2^3 + 3 \cdot 5$ **c.** $15 \cdot 2 - 3^2$ **d.** $2 + 3 \cdot 5 - 4$
   **e.** $6 \div 2 \cdot 3$ **f.** $(2 + 3) \cdot 5$ **g.** $(2^3 + 3) \cdot 5$ **h.** $(15 \cdot 2 - 3)^2$

**CHECKPOINT ✓**

4. Write a paragraph to describe the algebraic order of operations that the graphics calculator follows.

**EXAMPLE 2** **Insert grouping symbols to make $30 + 4 \div 2 - 1 = 16$ true.**

**SOLUTION**

Use parentheses to group $30 + 4$ before dividing by 2.

$$[(30 + 4) \div 2] - 1 \stackrel{?}{=} 16$$
$$(34 \div 2) - 1 \stackrel{?}{=} 16$$
$$17 - 1 \stackrel{?}{=} 16$$
$$16 = 16 \quad \textit{True}$$

**TRY THIS** Insert grouping symbols to make $80 \div 2 + 6 \cdot 5 = 50$ true.

**EXAMPLE 3** **Use the algebraic order of operations to evaluate $5x^2 + 7y$ when $x$ is 3 and $y$ is 2.**

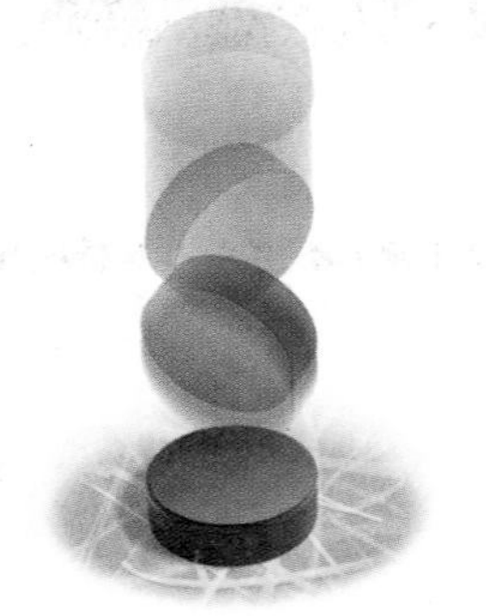

**SOLUTION**

| | |
|---|---|
| | $5x^2 + 7y$ |
| Replace $x$ with 3 and $y$ with 2. | $5 \cdot 3^2 + 7 \cdot 2$ |
| First square the 3. | $5 \cdot 9 + 7 \cdot 2$ |
| Perform all multiplications. | $45 + 14$ |
| Finally, add the results. | $59$ |

**TRY THIS** The height, in feet, of a falling object dropped from a height of 1000 feet after $s$ seconds is given by $1000 - 16s^2$. What is its height after 5 seconds?

**EXAMPLE 4** **Use a graphics or scientific calculator to evaluate $\frac{57 + 95}{16} + \frac{220}{88 + 104}$. Describe the keystrokes.**

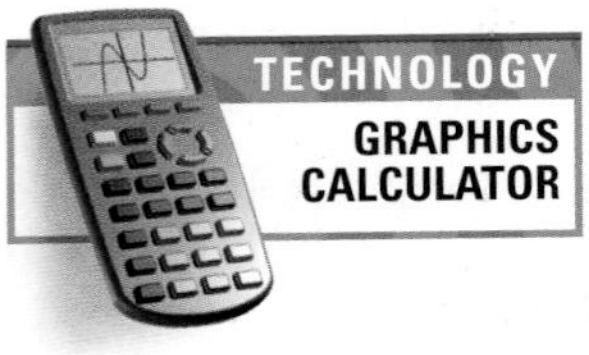


**SOLUTION**

Since the entire quantity $57 + 95$ is divided by 16, place parentheses around $57 + 95$. Do the same for $88 + 104$.

$$(57 + 95) \div 16 + 220 \div (88 + 104)$$

The keystrokes are shown below.

[ ( ] 57 [ + ] 95 [ ) ] [ ÷ ] 16 [ + ] 220 [ ÷ ] [ ( ] 88 [ + ] 104 [ ) ] [ ENTER ]

The answer is 10.646 to three decimal places. Is this reasonable? The first fraction is about $\frac{160}{16}$, or 10. The second fraction is about $\frac{200}{200}$, or 1. Thus, the estimated sum is about $10 + 1$, or 11. The answer 10.646 is reasonable.

**TRY THIS** Use a graphics or scientific calculator to evaluate $\frac{17 + 97}{2} - \frac{42 - 6}{12}$. Describe the keystrokes.

**EXAMPLE 5** **Evaluate $2\{2[2(3) + 1] + 1\} + 1$.**

**SOLUTION**

When you evaluate an expression that contains several pairs of grouping symbols, begin with the innermost pair and work outward.

$$2\{2[2(3) + 1] + 1\} + 1$$
$$2[2(7) + 1] + 1$$
$$2(15) + 1$$
$$31$$

If you use a calculator or computer, the expression will usually contain only parentheses. The expression would be entered as $2(2(2(3) + 1) + 1) + 1$. This does not affect the way you simplify the expression.

**TRY THIS** Evaluate $3\{4[5(6) - 3(2)] - 2\} + 7$.

# Exercises

## Communicate

1. Explain how to evaluate $3 + 2 \cdot 4$.
2. What two possible answers might you get for $20 - 2 \cdot 5$? Which answer is correct and why?
3. Describe the steps for evaluating $\{[3(8-4)]^2 - 6\} \div (4-2)$.
4. Explain why the rules called the *algebraic order of operations* are necessary for computation.
5. Explain why 4 is a reasonable estimate for $\frac{173+223}{151-47}$.

## Guided Skills Practice

**Use the algebraic order of operations to evaluate each expression.** ***(EXAMPLE 1)***

6. $(7 + 3^2) - 2 \cdot 4$
7. $4 \div 2 + 70 \div 2$
8. $12 \div 2 \cdot 3 - 4^2$

**Insert grouping symbols to make each equation true.** ***(EXAMPLE 2)***

9. $14 \div 5 + 2 + 8 = 10$
10. $4 + 7 \cdot 3 + 1 = 34$
11. $8 + 25 \cdot 2 \div 6 = 11$

**Evaluate each expression for $a = 4$ and $b = 6$.** ***(EXAMPLE 3)***

12. $a^2 + b^2$
13. $\frac{a+b}{2} - \frac{b-a}{2}$
14. $a^2 - (b^2 - a^2 - a)$
15. Evaluate $\frac{3+7}{5^2} + 2$. Describe the scientific or graphics calculator keystrokes that are used to evaluate the expression. ***(EXAMPLE 4)***

**Evaluate.** ***(EXAMPLE 5)***

16. $2\{2[3(2) + 4]\} + 6$
17. $4\{5[13 - 2(6)]\} - 19$

## Practice and Apply

internet connect
**Homework Help Online**
Go To: **go.hrw.com**
Keyword:
**MA1 Homework Help** for Exercises 18-41

**Evaluate each expression.**

18. $57 \cdot 29 + 89$
19. $72(98) + 12$
20. $89 + 57 \cdot 29$
21. $3(15) + 9$
22. $43 \cdot 32 + 91 \cdot 67$
23. $45(75) + 9(24)$
24. $157 - 29 + 23 \cdot 9$
25. $91 \div 7 + 6$
26. $187 - 34 \div 17$
27. $2(5 + 4) \div 9$
28. $12 - 7 \cdot 3 + 9^2$
29. $3 - 1 + 24 \div 6$
30. $7 + 6 \div 2 \cdot 10$
31. $3 \cdot 2 + 7 \cdot 6$
32. $100 - (3)(6)(4)$
33. $3^2 + 7 \cdot 2 - 8 \cdot 2$
34. $80 \div 4 \cdot 2 - 2 \cdot 2$
35. $2 \cdot 6^2 - 100 \div 50$
36. $3(2 + 7 - 8) + 16$
37. $4^2 \cdot 2 + [7 - (3^2 - 5)]$
38. $[15(10) - 12(10)] \div 10$
39. $4[(3 + 2 \cdot 3) - 5] + 7$
40. $(8 - 4) \cdot (12 - 3) \div (2 + 1 \cdot 2)$
41. $8^2 - 7^2 + 6^2$

**Place grouping symbols to make each equation true.**

**42.** $28 - 2 \cdot 0 = 0$ **43.** $59 - 4 \cdot 6 - 4 = 51$ **44.** $25 - 15 \cdot 6 + 4 = 64$

**45.** $22 - 4 \cdot 3 = 54$ **46.** $81 \div 9 + 18 = 3$ **47.** $108 - 17 \cdot 2 + 3 = 23$

**Given $a = 5$, $b = 3$, and $c = 4$, evaluate each expression.**

**48.** $a + b - c$ **49.** $a^2 + b^2$ **50.** $a^2 - b^2$ **51.** $(a + b) \cdot c$

**52.** $a^2 - b - c$ **53.** $a^2 - (b + c)$ **54.** $a^2 + b^2 + c^2$ **55.** $b^2 + (c^2 - a)$

**Evaluate each expression. Round to the nearest thousandth if necessary. Describe the graphics or scientific calculator keystrokes used to evaluate each expression.**

**56.** $\frac{28 + 59}{97 - 68}$ **57.** $\frac{97 - 17}{72 + 8}$ **58.** $\frac{40 \cdot 90}{8 \cdot 25}$

**59.** $\frac{28 + 59}{97 - 17}$ **60.** $\frac{97 - 17}{72 + 7}$ **61.** $\frac{43 \cdot 91}{8 \cdot 25}$

**Evaluate each expression.**

**62.** $3\{4[5(4 + 7)] - 2\}$ **63.** $10 + 2\{44 - [3(2 + 7)]\}$ **64.** $2\{2[2(2 + 2)]\} - 2$

CHALLENGE

**65.** Place grouping symbols to make the equation below true.

$$100 - 99 \cdot 3 + 7 \cdot 2 + 1 \div 24 = 1$$

CONNECTION

**66.** **STATISTICS** The table shows the results of an algebra quiz from a teacher's grade book. Find the average for the class and describe your method.

| Score | 100 | 90 | 80 | 70 | 60 |
|---|---|---|---|---|---|
| Number of students | 4 | 12 | 7 | 0 | 1 |

APPLICATION

**67.** **HOME IMPROVEMENT** Mr. Hilbert owns two rental properties, house A and house B. He wants to carpet two rooms in house A and three rooms in house B. He finds a store selling carpet for \$12.99 per square yard. Use the dimensions indicated in the picture. (Hint: $1 \text{ yd}^2 = 9 \text{ ft}^2$)

**a.** Find the total number of square feet of carpet needed for house A.
**b.** Find the total number of square feet of carpet needed for house B.
**c.** Find the cost of carpeting the rooms in house A.
**d.** Find the cost of carpeting the rooms in house B.

APPLICATION

**68. METEOROLOGY** The temperature-humidity index, or THI, is a number that measures the degree of discomfort that a person may feel because of the amount of humidity in the air. The higher the index, the greater the discomfort. You can find the THI by using the formula $\text{THI} = 0.4(t + s) + 15$, where $t$ is the air temperature and $s$ is the temperature from a thermometer with a wet cloth on its bulb. Find the THI for each pair of Fahrenheit temperatures.

**a.** $t = 80, s = 65$ **b.** $t = 86, s = 74$
**c.** $t = 100, s = 81$

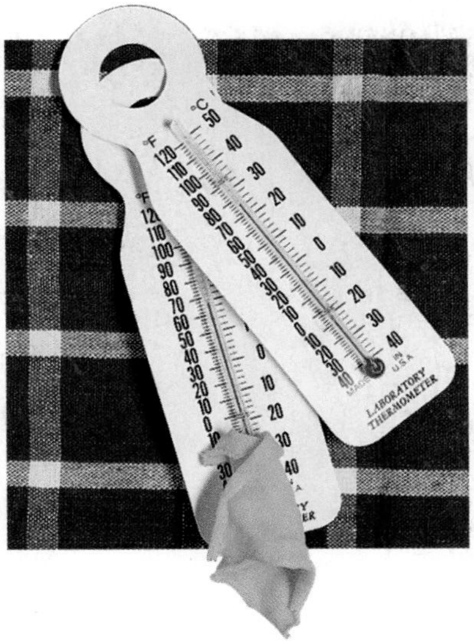

## Look Back

**69. CULTURAL CONNECTION: AFRICA** Thousands of years ago Hypatia, the earliest known female mathematician, investigated the relationship between square numbers and triangular numbers.

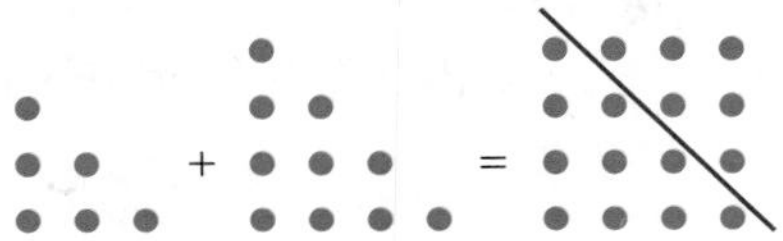

**a.** Find at least three more examples of how a square number can be represented as the sum of two triangular numbers. ***(LESSON 1.1)***
**b.** Are there any square numbers that cannot be represented as the sum of two triangular numbers? ***(LESSON 1.1)***

**Find the next two terms of each sequence.** ***(LESSON 1.1)***

**70.** 2, 5, 8, 11, 14, … **71.** 59, 54, 49, 44, 39, … **72.** 3, 6, 9, 12, 15, …

**Write an equation for each situation and solve it by using guess-and-check.** ***(LESSON 1.2)***

**73.** Jean has \$45. How many hours can Jean ice skate if skating costs \$5 plus \$3 per hour?

**74.** How many oranges can you buy with \$3 if oranges cost 13¢ each?

CONNECTION

**75. GEOMETRY** The perimeter of a rectangle is the number of units in the distance around the border, or the sum of twice the length and twice the width. The area is the product of the length and width. For what length and width will the perimeter and the area both be 16?

## Look Beyond

**An exponent indicates repeated multiplication. Tell whether each statement is true or false.**

**76.** $(3 + 4)^2 \stackrel{?}{=} 3 + 4^2$ **77.** $(3 + 4)^2 \stackrel{?}{=} (3 + 4)(3 + 4)$

**78.** $(3 + 4)^2 \stackrel{?}{=} 3^2 + 2(3)(4) + 4^2$ **79.** $(3 + 4)^2 \stackrel{?}{=} 3^2 + 4^2$

# 1.4 Graphing With Coordinates

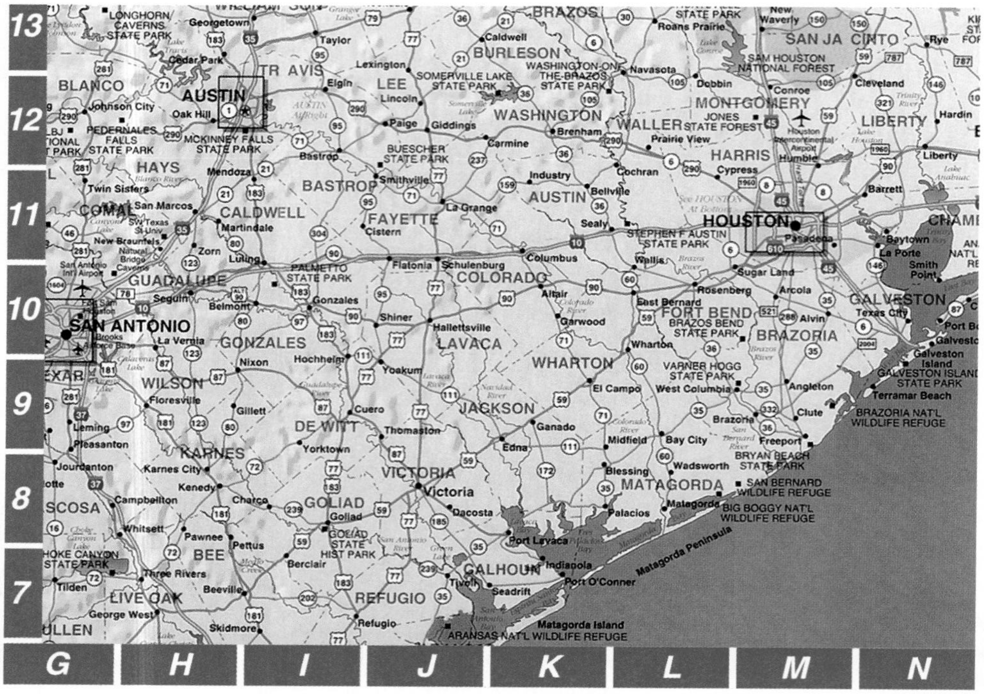


## Objective

- Plot points and lines on a coordinate plane.

**Why** *Just as locations can be related to each other with a map, graphing provides a picture of how two variables relate to one another.*

The letter and number grid on many road maps is an example of the use of a coordinate system to help locate a specific point. For example, Houston, Texas is located at (*M, 11*) on the map. Mathematicians use coordinates to represent points on a graph.

One basic form of a mathematical graph is a number line.

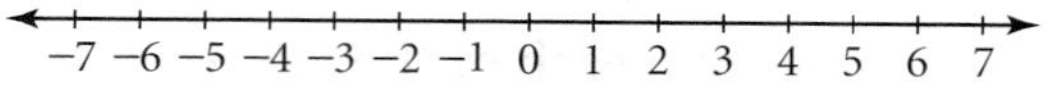


The arrows indicate that the line and the numbers extend infinitely in both directions.

A horizontal and a vertical numer line are used to divide a plane into four regions, called **quadrants.**

A plane set up according to this system is known as a **coordinate plane**.

The horizontal line is called the ***x*-axis**, and numbers on this axis are called ***x*-coordinates.**

The vertical line is called the ***y*-axis**, and numbers on this axis are called ***y*-coordinates.**

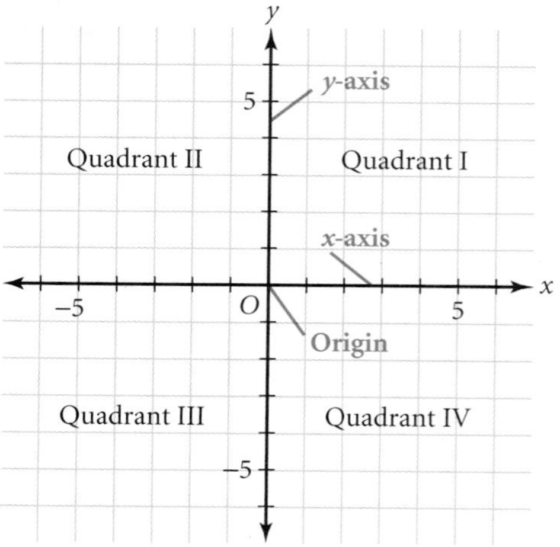


The intersection of the *x*-axis and the *y*-axis is a point called the **origin.**

Letters other than $x$ and $y$ can represent the axes. For example, you might see the $x$-axis shown as the $t$-axis when $t$ refers to time in a problem.

The order in which you name the coordinates is important. This means that a point identified by (10, 5) is different from a point identified by (5, 10). For this reason, the coordinates of a point on a coordinate plane are called an **ordered pair**.

The origin has coordinates (0, 0).

**EXAMPLE 1** **Determine the quadrant and coordinates for points *A*, *B*, and *C* on the graph.**

CONNECTION
COORDINATE GEOMETRY

**SOLUTION**

Point $A$ is 4 units to the right and 4 units up. It is in quadrant I and the coordinates are (4, 4).

Point $B$ is 1 unit to the left and 3 units up. It is in quadrant II and the coordinates are (−1, 3).

Point $C$ is 6 units to the right and 2 units down. It is in quadrant IV and the coordinates are (6, −2).

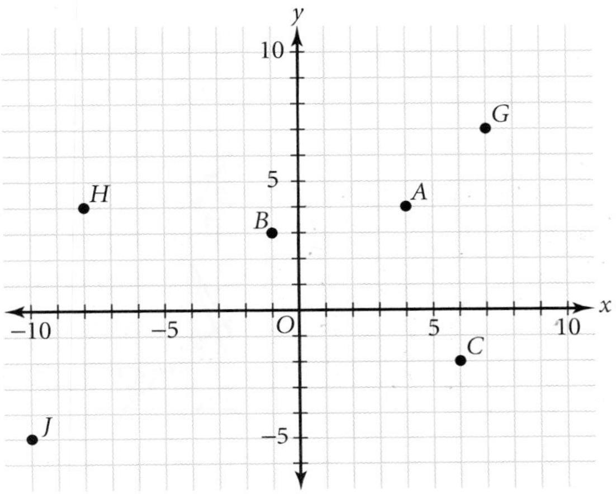


**TRY THIS** Determine the coordinates for points $G$, $H$, and $J$ on the graph in Example 1.

**EXAMPLE 2** **Graph each of the following points: $D(2, 6)$, $E(10, -3)$, $F(-7, -4)$.**

**SOLUTION**

Begin at the origin, (0, 0). Move 2 units to the right and 6 units up. Label the point $D$.

Begin at the origin, (0, 0). Move 10 units to the right and 3 units down. Label the point $E$.

Begin at the origin, (0, 0). Move 7 units to the left and 4 units down. Label the point $F$.

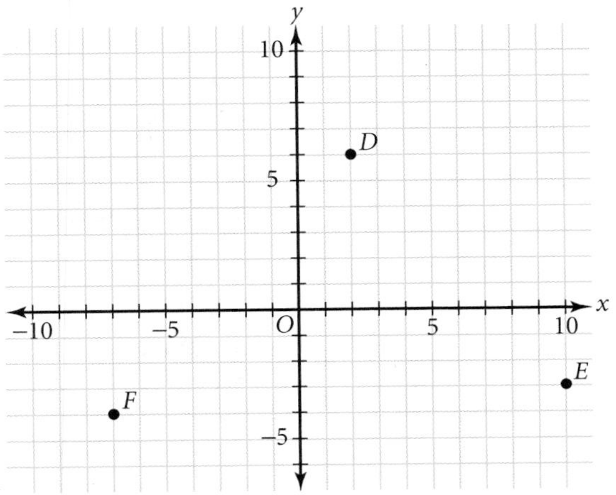


**TRY THIS** Graph each of the following points: $K(-3, 8)$, $L(-5, -7)$, and $M(8, -2)$.

## *Activity* Graphing Data

APPLICATION
RECREATION

**You will need:** graph paper and a pencil

You can use the data in the table below to build a graph that shows the distance a bicyclist travels in a given period of time, not counting rest periods.

| **Time in hours** | 0.5 | 1.0 | 1.5 | 2.0 | 2.5 |
|---|---|---|---|---|---|
| **Distance in miles** | 11.5 | 23.0 | ? | ? | ? |

1. Copy and complete the table. Assume that the bicyclist maintains a steady speed.
2. List the five ordered pairs in your table.
3. Graph the ordered pairs on a coordinate plane and connect the points with a line. The resulting graph models the data for the bicyclist.
4. If the bicyclist continues at the same speed, what distance is covered in 10 hours? How far does the bicyclist travel in 5 hours? in 9 hours?

CHECKPOINT ✔

5. Write an equation for the distance, $d$, the bicyclist travels in terms of time, $t$.

EXAMPLE 3

CONNECTION
COORDINATE GEOMETRY

**Make a table for the equation $y = 2x$, and find the values for $y$ by substituting 1, 2, 3, 4, and 5 for $x$. Graph the ordered pairs on a coordinate plane, and connect the points to make a line.**

**SOLUTION**

Make a table to show the values of $x$ and $y$.

| $x$ | 1 | 2 | 3 | 4 | 5 |
|---|---|---|---|---|---|
| $y$ | 2 | 4 | 6 | 8 | 10 |

Graph the ordered pairs on a coordinate plane, and connect the points to form a line.

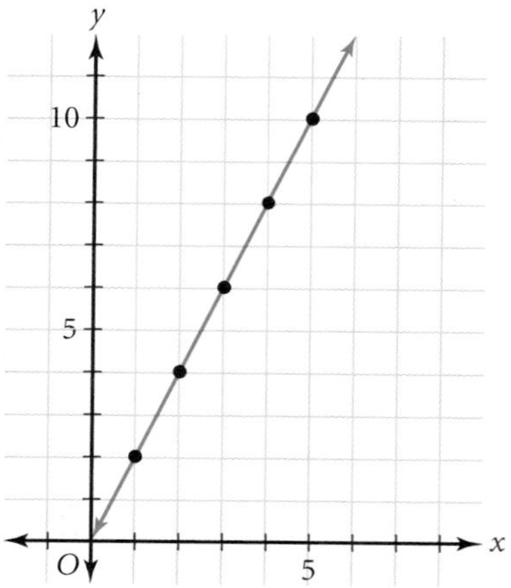


**TRY THIS** Make a table for the equation $y = 3x$, and find the values for $y$ by substituting 1, 2, 3, 4, and 5 for $x$. Graph the ordered pairs on a coordinate plane and connect the points to make a line.

# Exercises

## Communicate

**1.** Explain how you identify the coordinates of a given point on a graph.

**2.** Is (6, 7) the same point as (7, 6)? Explain.

**3.** Describe the steps for plotting the point with the coordinates (7, 3).

**4.** Discuss the advantages and disadvantages of displaying data in a table and in a graph.

**5.** Describe how to make a table of values for $x$ and $y$ for the equation $y = 2x + 5$.

## Guided Skills Practice

**Determine the quadrant and coordinates of the given points.** *(EXAMPLE 1)*

**6.** $E$ **7.** $F$

**8.** $G$ **9.** $H$

**10.** $I$ **11.** $J$

**12.** $K$ **13.** $L$

**14.** $M$ **15.** $N$

**16.** $R$ **17.** $P$

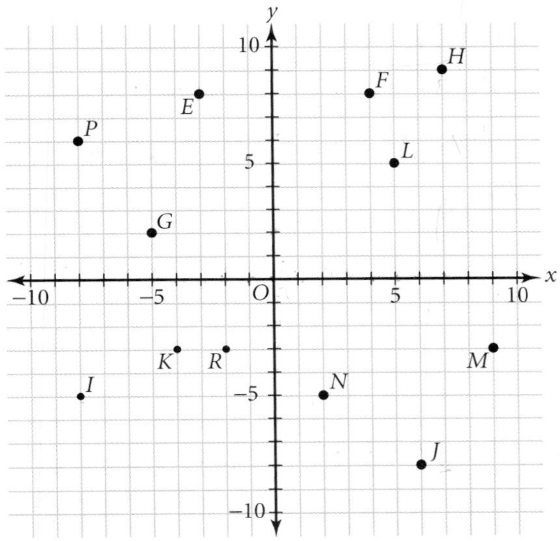


**Graph each point on a coordinate plane.** *(EXAMPLE 2)*

**18.** $L(4, 3)$ **19.** $M(1, 3)$ **20.** $N(5, 1)$

**Make a table for each equation and find values for *y* by substituting 1, 2, 3, 4, and 5 for *x*. Graph the ordered pairs, and connect the points to make a line.** *(EXAMPLE 3)*

**21.** $y = 3x$ **22.** $y = 2x + 1$ **23.** $y = 3x - 2$

## Practice and Apply

**Graph each list of ordered pairs. State whether they lie on a straight line.**

**24.** $(1, -3), (2, -6), (3, -9)$

**25.** $(1, 5), (2, -4), (3, 1)$

**26.** $(1, 10), (2, 7), (3, 2)$

**27.** $(0, 3), (2, 5), (4, 7)$

**28.** $(5, 2), (7, 2), (9, 2)$

**29.** $(-4, 1), (-4, 5), (-4, 9)$

**internet connect** go.hrw.com

**Homework Help Online**
Go To: **go.hrw.com**
Keyword:
**MA1 Homework Help**
for Exercises 30-43

**Determine the quadrant and coordinates of the given points.**

**30.** $A$ **31.** $B$

**32.** $C$ **33.** $D$

**34.** $E$ **35.** $F$

**36.** $G$ **37.** $H$

**38.** $R$ **39.** $S$

**40.** $T$ **41.** $U$

**42.** $V$ **43.** $W$

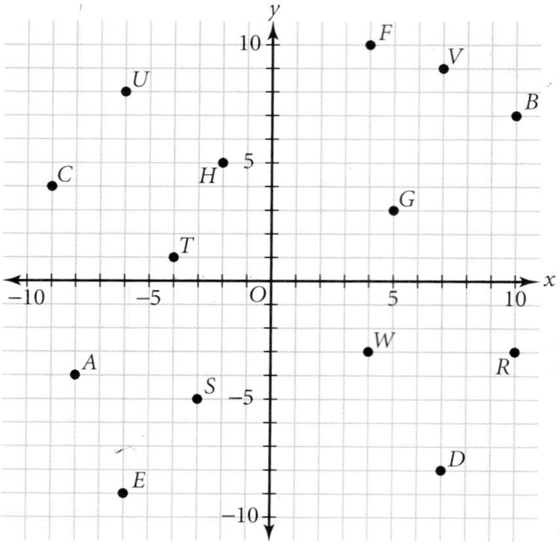


**Make a table for each equation, and find the values for $y$ by substituting 1, 2, 3, 4, and 5 for $x$.**

**44.** $y = x + 3$ **45.** $y = x - 4$ **46.** $y = 2x$

**47.** $y = 2x + 5$ **48.** $y = 4x - 3$ **49.** $y = 5x + 2$

**50.** $y = x + 1$ **51.** $y = 7x - 5$ **52.** $y = \frac{1}{2}x + \frac{3}{4}$

**CONNECTIONS**

**COORDINATE GEOMETRY** Graph and compare the following two equations by making a table and plotting points. Connect the points for each equation to make a line. What effect does the operation before the $x$ have on the graph?

**53.** $y = 3 + x$ **54.** $y = 3 - x$

**COORDINATE GEOMETRY** Graph each of the following equations on the same set of coordinate axes by making tables, plotting points, and drawing lines. Then compare the graphs. Explain your conclusions.

**55.** $y = x + 7$ **56.** $y = x - 7$

**57.** $y = 7 - x$ **58.** $y = -7 - x$

**APPLICATIONS**

**59. BUSINESS** Suppose that a mail-order T-shirt company charges $8 per T-shirt plus a $3 handling charge per order.

**a.** How much does an order of 2 shirts cost?

**b.** How much does an order of 5 shirts cost?

**c.** Make a set of ordered pairs from the given information, and plot them as points on a graph. Do the points lie on a straight line?

**d.** Use your graph to find how many shirts you could order with $75.

**60. RECREATION** Suppose that a bicyclist is riding at a steady speed. The table below represents the distance she travels in a given period of time. Copy and complete the table. List the ordered pairs from the table, and graph them on a coordinate plane. Use the graph to determine how far the bicyclist would travel in 8 hours.

| Time (hours) | 1 | 2 | 3 | 4 | 5 |
|---|---|---|---|---|---|
| Distance (miles) | 24.5 | 49 | ? | ? | ? |

## Look Back

*Karl Friedrich Gauss (1777–1855) was one of the greatest of all mathematicians.*

**61.** Find the sum $1 + 2 + 3 + 4 + \cdots + 18$. ***(LESSON 1.1)***

**62.** Find the next three numbers in the following pattern: 43, 49, 55, 61, 67, … ***(LESSON 1.1)***

**CULTURAL CONNECTION: EUROPE** When Gauss was very young, a teacher told him to find the sum of all of the numbers from 1 to 100, thinking this would keep him very busy for a long time. But Gauss very cleverly paired the numbers and observed that $1 + 99 = 100$, $2 + 98 = 100$, $3 + 97 = 100$, and so on. ***(LESSON 1.1)***

**63.** How many pairs of numbers could be formed this way without repeating any numbers?

**64.** Which whole numbers from 1 to 100 would not appear in any of the pairs?

**65.** Based on the previous two exercises, what is the sum of the numbers from 1 to 100?

## Look Beyond

internet connect
**Portfolio Extension**
Go To: **go.hrw.com**
Keyword:
**MA1 Fibonacci**

**66.** Plot the data points from the table below on a coordinate plane. Use a straightedge to draw the line that best fits the points on the scatter plot.

| $x$ | 0 | 3 | 6 | 9 | 12 |
|---|---|---|---|---|---|
| $y$ | 14 | 20 | 26 | 32 | 38 |

**67.** What is the relationship of the points to the line?

## PORTFOLIO ACTIVITY

Include this activity in your portfolio.

Leonard Fibonacci was a famous mathematician who lived in the Middle Ages. He is famous for a special sequence of numbers now called the Fibonacci sequence. The sequence starts as follows:

1, 1, 2, 3, 5, 8, …

This sequence of numbers appears in many places in nature. Pine cones usually have 5 spiral arms in one direction and 8 in the other. The sneezewort plant has 1 main stem with 2 branches, 3 branches, 5 branches, and so on. In the sunflower at right, spirals of 21, 34, and 55 (in different directions) can be seen.

Find the next four Fibonacci numbers in the sequence. Copy and complete the table below.

| **Term number** | 1 | 2 | 3 | 4 | 5 | 6 | 7 | 8 | 9 | 10 |
|---|---|---|---|---|---|---|---|---|---|---|
| **Fibonacci number** | 1 | 1 | 2 | 3 | 5 | 8 | ? | ? | ? | ? |

Describe the pattern that you discover in the sequence.

1.5

# Representing Linear Patterns

**Objectives**

- Represent linear patterns with equations.
- Represent linear equations with graphs.

**Why** *Many situations, like the cost of participation in a buyers club, follow linear patterns. Linear equations can help you calculate solutions for problems that involve these situations.*

## Generalizing Data Patterns With Equations

A DVD buyers club charges a \$20 membership fee and \$15 per DVD purchased. The table below represents this situation.

| **Number of DVDs purchased** | 0 | 1 | 2 | 3 | 4 | 5 |
|---|---|---|---|---|---|---|
| **Total cost (\$)** | 20 | 35 | 50 | 65 | 80 | 95 |

15 15 15 15 15

*First differences*

Since the data for total cost show constant first differences, the pattern is linear.

The equation $c = 15d + 20$ represents this situation, where $c$ represents the total cost and $d$ represents the number of DVDs purchased. Notice that the number 15 in the equation is the same as the first differences, and represents the price per DVD. The number 20 in the equation represents the initial cost to join the club.

**CRITICAL THINKING** Explain how the first differences are used to write an equation for total cost.

**EXAMPLE 1**

**APPLICATION**
**CONSUMER ECONOMICS**

The costs associated with being a member of DVD Club are presented in the table below. **Find the first differences and write an equation to represent the data pattern.**

| Number of DVDs purchased | 0 | 1 | 2 | 3 | 4 | 5 |
|---|---|---|---|---|---|---|
| Total cost ($) | 26 | 39 | 52 | 65 | 78 | 91 |

**SOLUTION**

26   39   52   65   78   91
  13   13   13   13   13

The first differences are 13. Therefore, the club charges $13 per DVD.

The cost for 0 DVDs is $26. Therefore, the membership fee is $26.
The relationship is *total cost* = *13* • *number of DVDs* + 26.

Use $c$ for total cost and $d$ for the number of DVDs to write the equation.

$$c = 13d + 26$$

**TRY THIS**

The table shows the costs associated with an audio CD club that charges a membership fee. Write an equation to represent the pattern in the data.

| Number of CDs purchased | 0 | 1 | 2 | 3 | 4 | 5 |
|---|---|---|---|---|---|---|
| Total cost ($) | 21 | 31 | 41 | 51 | 61 | 71 |

## *Activity*
## Modeling an Equation With a Graph

**APPLICATION**
**TRAVEL**

**You will need:** a graphics calculator or graph paper

A passenger ferry from Seattle, Washington, to Victoria, British Columbia, travels at a rate of 53 miles per hour. The distance the ferry travels in a given period of time can be represented by the following:

$$\text{distance} = \text{rate} \cdot \text{time}$$
$$d = 53t$$

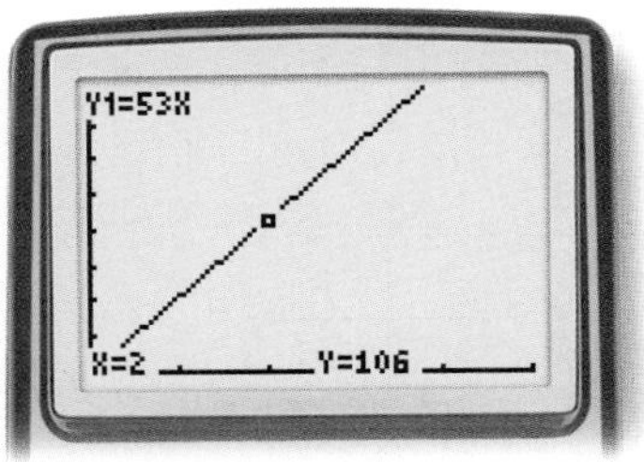


*Use the graph and the trace feature on a graphics calculator.*

1. Complete the table of values for the equation $d = 53t$.

| Time in hours | 0.5 | 1.0 | 1.5 | 2.0 | 2.5 |
|---|---|---|---|---|---|
| Distance in miles | 26.5 | 53.0 | ? | ? | ? |

2. Graph each ordered pair from the table on a coordinate plane with time, $t$, on the horizontal axis and distance, $d$, on the vertical axis.
3. Connect the points. Describe the shape that is formed by connecting the points.

**CHECKPOINT ✓**

4. Use the graph to predict how far the ferry would travel in 3 hours and in 4 hours.

## Linear Equations and Graphs

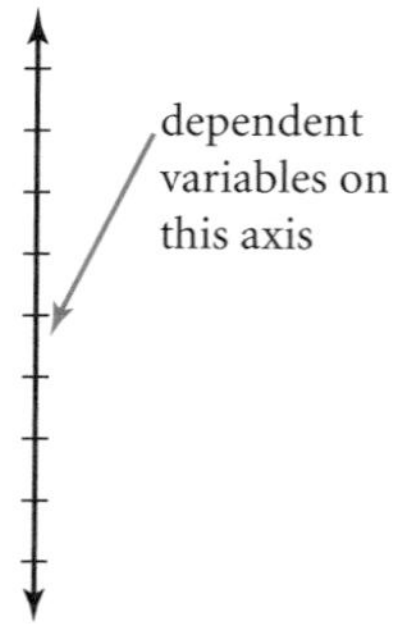


When you connect the points plotted in the previous activity, the graph of the equation $d = 53t$ is a straight line. For this reason, the equation is called a **linear equation**.

Drawing a line to connect a series of points does not automatically make every point on the line a solution to the problem. For example, it would make no sense to solve a distance problem for a negative time value, although you could certainly extend the line to include some negative values. As you solve problems, you must decide which kinds of answers are reasonable for the situation you are representing.

In the equation $d = 53t$, $t$ is called the **independent variable**, and $d$ is called the **dependent variable**. The value of $d$ depends on the value chosen for $t$.

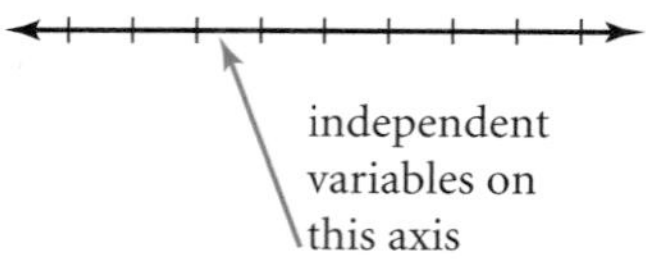


On a graph, the independent variable is represented on the horizontal axis and the dependent variable is represented on the vertical axis.

**EXAMPLE 2**

**APPLICATION**
**TRAVEL**

A jet airplane is flying at a rate of 280 miles per hour. **Write a linear equation to express the distance traveled in a given amount of time. Graph the equation for times from 0 to 10 hours.**

**SOLUTION**

The distance traveled can be represented by the following relationship:

$$\text{distance} = \text{rate} \cdot \text{time}$$

The rate is 280 miles per hour. Let $d$ represent the distance in miles and $t$ represent the time in hours. The linear equation that expresses the distance traveled is as follows:

$$d = 280t$$

To find ordered pairs for the graph, make a table. Since $t$ is the independent variable, choose values of $t$. Use the equation to determine the resulting values of $d$.

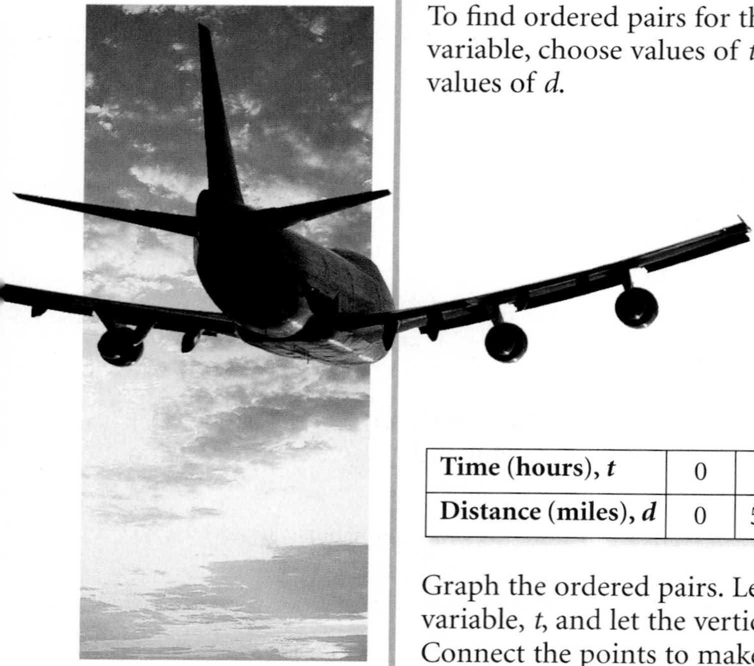

| Time (hours), $t$ | 0 | 2 | 5 | 10 |
|---|---|---|---|---|
| **Distance (miles), $d$** | 0 | 560 | 1400 | 2800 |

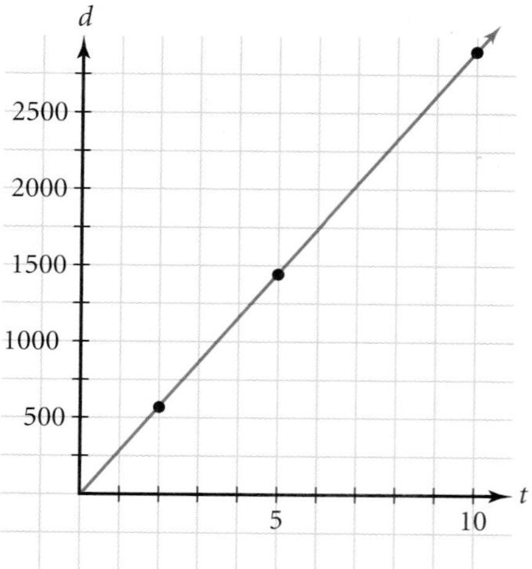


Graph the ordered pairs. Let the horizontal axis represent the independent variable, $t$, and let the vertical axis represent the dependent variable, $d$. Connect the points to make a line.

You can also graph the equation on a graphics calculator. Remember that some calculators use **Y1** to represent the dependent variable, distance, and **X** to represent the independent variable, time.

Enter **280X** after **Y1=**.

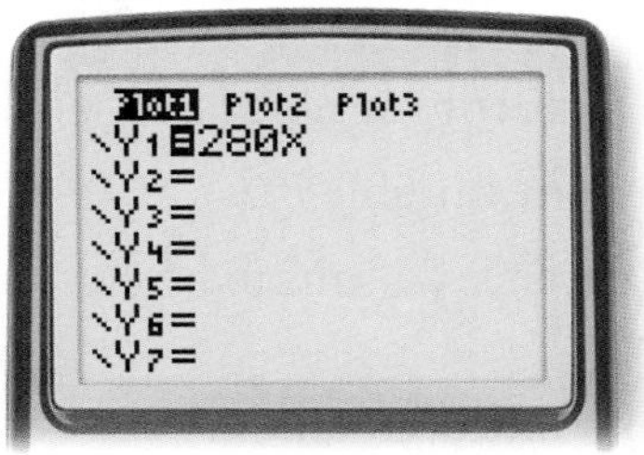


Set the window of the calculator as shown.

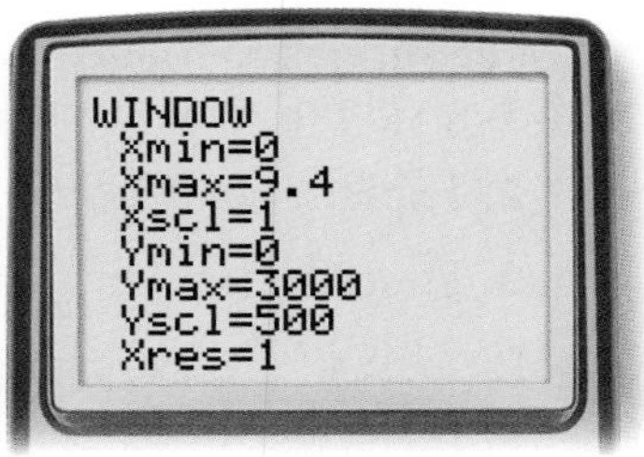


Press **GRAPH**.

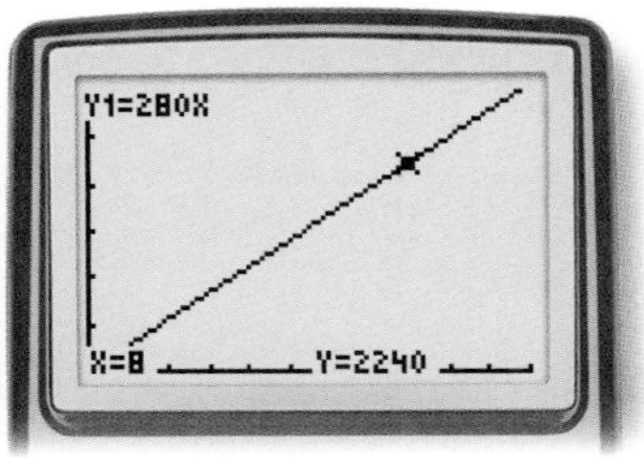


**TRY THIS** A passenger train averages 54 mph. Write a linear equation that represents the distance traveled in a given time. Graph the equation for times from 0 to 15 hours.

## EXAMPLE 3

**APPLICATION**
**HEALTH**

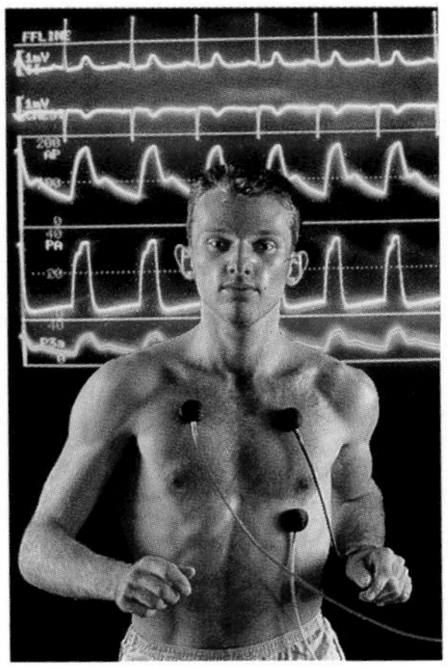

Exercise physiologists suggest that a reasonable estimate for the maximum heart rate during exercise is no more than 220 beats per minute minus the person's age. **Represent the maximum heart rate by *r* and the age by *a*. Write a linear equation to express the maximum heart rate, *r*, in terms of age, *a*. Graph the equation for ages from 10 to 50.**

### SOLUTION

Since the rate, $r$, equals 220 beats per minute minus the age, $a$, write the equation $r = 220 - a$. This equation is a linear equation. The independent variable is $a$ and the dependent variable is $r$.

Use the equation $r = 220 - a$ to create a table for the following values of $a$: 10, 20, 30, 40, and 50.

| $a$ | 10 | 20 | 30 | 40 | 50 |
|---|---|---|---|---|---|
| $r$ | 210 | 200 | 190 | 180 | 170 |

Let the horizontal axis represent age, $a$, and the vertical axis represent heart rate, $r$. Graph each ordered pair, and connect the points to draw a line.

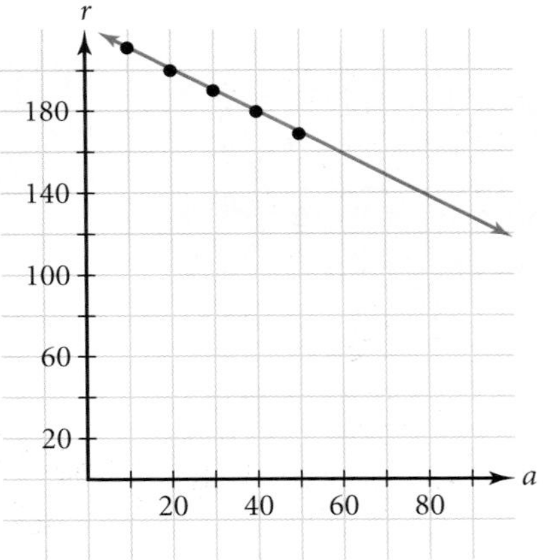


To graph the equation on a graphics calculator, enter **220 – X** after **Y1=** on a graphics calculator. Set the limits of the domain and range in the range window. Then press **GRAPH**.

Note: For some people the proper level of exercise should be lower. You should consult a doctor if there is any doubt about your exercise routine.

**CRITICAL THINKING** In Example 3, what values of $a$ are reasonable for the problem?

# Exercises

## Communicate

1. Explain what an independent variable is.
2. Explain what a dependent variable is.
3. Describe how first differences of number sequences relate to linear equations.
4. Describe the graph of a linear equation.

## Guided Skills Practice

APPLICATIONS

5. **TELECOMMUNICATIONS** The costs associated with a particular cellular phone plan are shown below. Find the first differences and write an equation to represent the data pattern. ***(EXAMPLE 1)***

| Hours per month | 0 | 1 | 2 | 3 | 4 | 5 |
|---|---|---|---|---|---|---|
| Total cost ($) | 25 | 37 | 49 | 61 | 73 | 85 |

*You can cross the English Channel by hovercraft in under 30 minutes.*

6. **TRANSPORTATION** A hovercraft ferry crossing the English Channel averages 47 miles per hour. Make a table, draw a graph, and write an equation to show the distance the hovercraft travels for times from 0 to 3 hours. ***(EXAMPLE 2)***

7. **CONTESTS** The prize for a baking contest is $200. If a contestant spends $6 per batch on ingredients to practice a new recipe, write an equation to describe the profit, $p$, the contestant will make if he wins in terms of the number of practice batches, $b$, used to perfect the recipe. ***(EXAMPLE 3)***

## Practice and Apply

**Find the first differences for each data set, and write an equation to represent the data pattern.**

8.

| | | | | | | |
|---|---|---|---|---|---|---|
| 0 | 1 | 2 | 3 | 4 | 5 | 6 |
| 17 | 23 | 29 | 35 | 41 | 47 | 53 |

9.

| | | | | | | |
|---|---|---|---|---|---|---|
| 0 | 1 | 2 | 3 | 4 | 5 | 6 |
| 25 | 31 | 37 | 43 | 49 | 55 | 61 |

10.

| | | | | | | |
|---|---|---|---|---|---|---|
| 0 | 1 | 2 | 3 | 4 | 5 | 6 |
| 60 | 180 | 300 | 420 | 540 | 660 | 780 |

11.

| | | | | | | |
|---|---|---|---|---|---|---|
| 0 | 1 | 2 | 3 | 4 | 5 | 6 |
| 0 | 3 | 6 | 9 | 12 | 15 | 18 |

12.

| | | | | | | |
|---|---|---|---|---|---|---|
| 0 | 1 | 2 | 3 | 4 | 5 | 6 |
| 0 | 110 | 220 | 330 | 440 | 550 | 660 |

13.

| | | | | | | |
|---|---|---|---|---|---|---|
| 0 | 1 | 2 | 3 | 4 | 5 | 6 |
| 500 | 455 | 410 | 365 | 320 | 275 | 230 |

14.

| | | | | | | |
|---|---|---|---|---|---|---|
| 0 | 1 | 2 | 3 | 4 | 5 | 6 |
| 250 | 240 | 230 | 220 | 210 | 200 | 190 |

15.

| | | | | | | |
|---|---|---|---|---|---|---|
| 0 | 1 | 2 | 3 | 4 | 5 | 6 |
| 5 | 26 | 47 | 68 | 89 | 110 | 131 |

16.

| | | | | | | |
|---|---|---|---|---|---|---|
| 0 | 1 | 2 | 3 | 4 | 5 | 6 |
| 201 | 181 | 161 | 141 | 121 | 101 | 81 |

**internet connect**
**Homework Help Online**
Go To: **go.hrw.com**
Keyword:
**MA1 Homework Help** for Exercises 17-28

**Make a table of values for each equation, using 1, 2, 3, 4, and 5 as values for *x*. Draw a graph for each equation by plotting points from your data set.**

**17.** $y = 12 + 7x$ **18.** $y = 220 - 7x$ **19.** $y = 30x$ **20.** $y = 12x$

**21.** $y = 300 - 2x$ **22.** $y = 23 + 5x$ **23.** $y = 15 + 7x$ **24.** $y = 425 - 5x$

**25.** $y = 120 - 10x$ **26.** $y = 65x$ **27.** $y = 110x$ **28.** $y = 53 + 11x$

## CONNECTIONS

**29.** **COORDINATE GEOMETRY** Graph the data from the DVD buyers club (page 30) and the DVD Club (Example 1, page 31) on the same coordinate grid. Connect the points with a line for each club. From the graph, find the point of intersection. What does this point represent?

**30.** **STATISTICS** Suppose that the average salary for a particular career begins at $30,000 for someone with no experience and increases by $2500 per year of experience.

**a.** Complete a table that shows the average salary through 5 years of experience. Graph each ordered pair from the table.

**b.** Use the graph to predict the average salary in the field after 10 years of experience.

## APPLICATIONS

**31.** **TRAVEL** A paddleboat traveling on the Colorado River, in Austin, Texas, averages 14 miles per hour. Write an equation, make a table, and draw a graph to show the distance that the paddleboat travels for times from 0 to 5 hours.

**32.** **TRAVEL** A family plans a summer vacation to several national parks. They plan to average 58 miles per hour when they are traveling between the parks. Write an equation, make a table, and draw a graph to show the distance the family can travel for times from 0 to 8 hours. Use your graph to determine how far they could travel if they drove for 10 hours.

33. **CONSUMER ECONOMICS** Frank enjoys listening to music at work. The CD club that he is a member of charges \$11.25 for each CD plus a \$26.00 membership fee. Make a table of the costs for buying 1, 2, 3, 4, and 5 CDs from the club. Draw a graph and write an equation to represent the situation.

34. **RECREATION** Don walks at a rate of 3 miles per hour. You can determine the distance he walks by multiplying the rate by the number of hours he walks.
   a. Represent hours by $h$, and write an equation for the distance, $d$.
   b. Make a table and graph the equation for the distance that Don travels based on the number of hours that he walks.

35. **HEALTH** Using the equation given in Example 3, $r = 220 - a$, what should be the maximum heart rate for a 36-year-old person?

36. **ENTERTAINMENT** How many \$5 movie tickets can you buy with \$32?

37. **FUND-RAISING** How many \$5 raffle tickets do you have to sell to raise \$32?

38. **GIFTS** If 5 people equally split the cost of a \$32 birthday gift, how much should each pay?

**CHALLENGE**

39. Explain why the reasonable answers to Exercises 36, 37, and 38 are all different.

## Look Back

40. What are the next three terms of the sequence 1, 1, 4, 10, 19, …? ***(LESSON 1.1)***

41. How many differences are needed to reach a constant difference for the sequence 1, 2, 6, 15, 31, …? What is the constant difference for the sequence? ***(LESSON 1.1)***

42. Make a table to solve the equation $7x + 5 = 40$. ***(LESSON 1.2)***

**Evaluate each expression.** ***(LESSON 1.3)***

43. $2 \cdot 14 \div 2 + 5$

44. $6 + 12 \div 6 - 4$

45. $4[(12 - 3) \cdot 2] \div 12$

46. $[3(4) - 6] - [(15 - 7) \div 4]$

## Look Beyond

**For each equation, make a table, plot the points, and tell whether the graph is linear.**

47. $y = \frac{x}{3}$

48. $y = \frac{3}{x}$

49. $x + y = 12$

50. $x - y = 12$

51. The graph of the path of a rocket is a curve called a parabola. The highest point of this curve is called the vertex. Graph the set of points below, and determine which point is the vertex.

| $x$ | 0 | 1 | 2 | 3 | 4 | 5 | 6 | 7 | 8 | 9 | 10 |
|---|---|---|---|---|---|---|---|---|---|---|---|
| $y$ | 0 | 9 | 16 | 21 | 24 | 25 | 24 | 21 | 16 | 9 | 0 |

1.6

# Scatter Plots and Lines of Best Fit

## Objectives

- Interpret data in a scatter plot.
- Find a line of best fit on a scatter plot by inspection.

**Why** ***Scatter plots provide a picture of the relationship between two sets of data, like those related to weather. Both scatter plots and lines of best fit can be used to predict relationships and future occurrences, such as tornado frequency.***

| Year | 1950 | 1955 | 1960 | 1965 | 1970 | 1975 | 1980 | 1985 | 1990 | 1995 |
|---|---|---|---|---|---|---|---|---|---|---|
| # of tornadoes | 201 | 593 | 616 | 897 | 654 | 919 | 866 | 684 | 1133 | 1234 |

Frequency Table

APPLICATION
**METEOROLOGY**

Scientists, such as meteorologists, often look for patterns and relationships between sets of numerical data. The frequency of tornadoes in a certain geographic area or during a specific period of time might show a pattern.

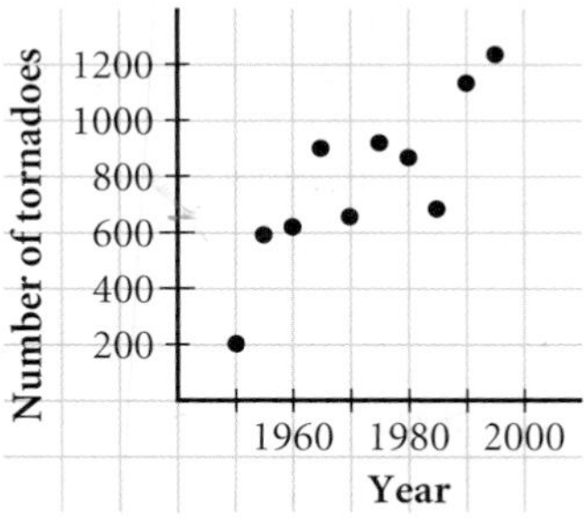


Scatter Plot

An effective way to see a relationship in data is to display the information as a scatter plot. A **scatter plot** shows how two variables relate to each other by showing how closely the data points cluster to a line.

## Correlation

Scatter plots provide a convenient way to determine whether a relationship exists between two variables. The relationship is called a **correlation**.

**EXAMPLE 1** Compare the scatter plots of data related to characteristics of children from 0 to 18 years old with the statements below. **Match scatter plots a, b, and c with statements 1, 2, and 3.**

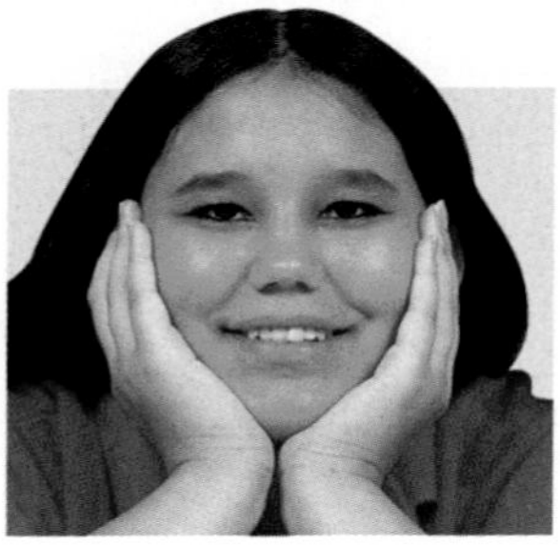

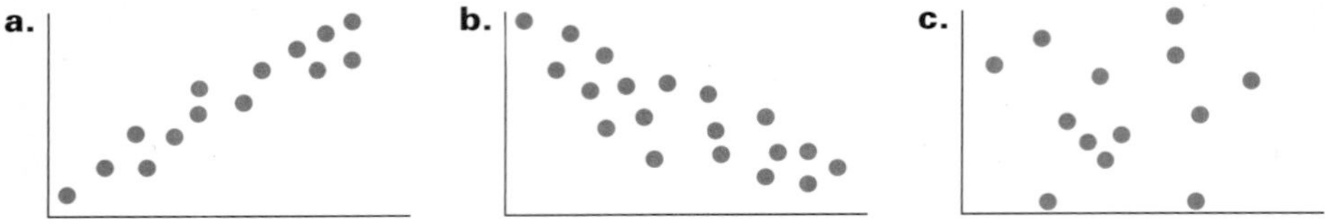


1. The variables studied are *age* and *eye color.*
2. The variables studied are *age* and *the time needed to run a fixed distance.*
3. The variables studied are *age* and *height.*

**SOLUTION**

Scatter plot **a** shows a strong positive correlation. A **positive correlation** occurs when both variables increase. The correlation is strong because the data points are close to a straight line. As a child's age increases, his or her height increases, so this scatter plot would fit statement **3**.

Scatter plot **b** shows a strong negative correlation. A **negative correlation** occurs when one variable increases and the other variable decreases. As a child gets older and stronger, the time needed to run a fixed distance decreases, so this scatter plot would fit statement **2**.

Scatter plot **c** shows little or no correlation because the data points are randomly scattered. Eye color has no relation to age, so this scatter plot would fit statement **1**.

**TRY THIS** Think of two examples of related data. Describe the kind of correlation you expect from the data and why.

*Activity*

## Working With Trends in Data

APPLICATION
ECOLOGY

TECHNOLOGY
GRAPHICS CALCULATOR

**You will need:** a graphics calculator, or graph paper and a pencil

Suppose that a certain wildlife reserve has an increase in its buffalo population. The data is shown below.

| Year | 0 | 2 | 4 | 6 | 8 | 10 | 12 | 14 | 16 |
|---|---|---|---|---|---|---|---|---|---|
| Population | 50 | 75 | 90 | 115 | 145 | 165 | 180 | 200 | 230 |
| Differences | | 25 | 15 | 25 | ? | ? | ? | ? | ? |

1. Use a graphics calculator or draw a scatter plot of the data.
2. Copy the table and complete the sequence of first differences.

3. A line of best fit is a line drawn on a scatter plot that shows the trend in the data. Looking at the pattern of differences, you might think that the population has increased by about 20 buffalo every 2 years or by about 10 every year. Begin trying to draw a line of best fit by using $y = 10x + 50$ as your equation.
4. Since that line is not the line of best fit, adjust the equation to better fit the data by using a guess-and-check strategy.

CHECKPOINT ✓

5. Use your equation to predict the population in the 10th year ($x = 10$). How close is it to the value that was given in the table? Use it to predict the population in the 20th year. You can also make a prediction by using the TABLE or TRACE feature of your graphics calculator.

## Finding Lines of Best Fit

Scatter plots can be used to show trends in data. When the points in a scatter plot are represented by a trend line or a **line of best fit**, you can study the line to see how the data behave. You may have a basis to predict what the data might be for values not given.

**EXAMPLE 2**

**APPLICATION**
**METEOROLOGY**

Use the data from the Storm Prediction Center in Norman, Oklahoma, to find the line of best fit for the scatter plot of the number of storms in given years. **Describe the correlation shown in the scatter plot.**

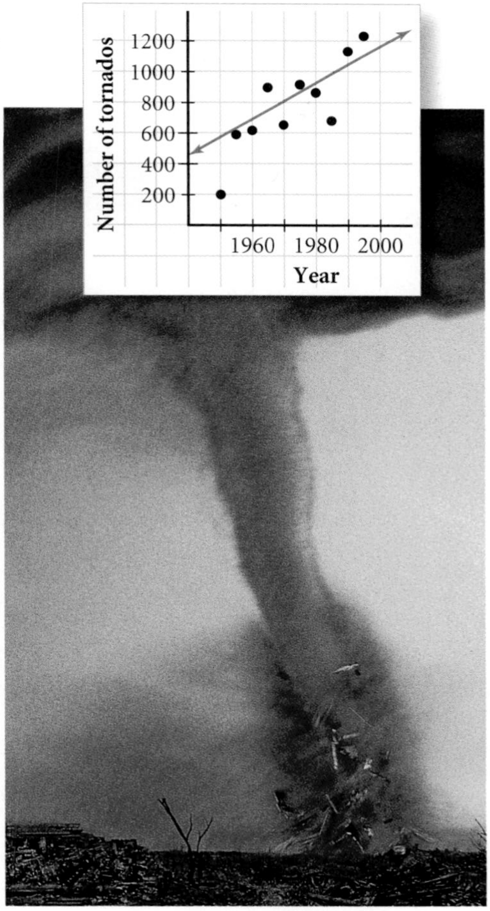


**SOLUTION**

Plot the data as coordinates, (year, number of tornadoes), on a grid with the years on the horizontal axis and the number of tornadoes on the vertical axis.

Use the following steps to help you find the line of best fit:

- Use a straightedge, such as a clear ruler or a piece of uncooked spaghetti, to model the line.
- To fit the line to the points, choose your line so that it best matches the overall trend. The line does not have to pass through any of the points.

Since the data for both variables in the scatter plot are increasing, the correlation is positive.

**TRY THIS** Use the line of best fit to predict how many tornadoes may be reported in the United States in 2015 if the trend continues.

There is usually only one line that *best* fits a set of data. A line of best fit should take all of the points into consideration.

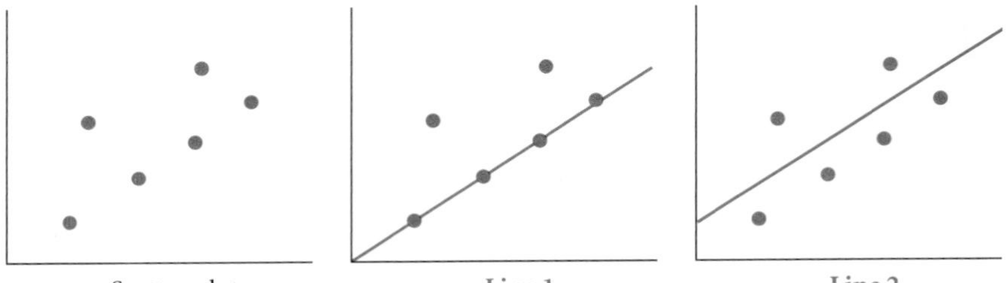


Line 1 goes through several of the points, but it completely ignores the others.

Line 2 is closer to four points than it is to the other two, but it takes all of the points into account. Line 2 is a better fit, even though it doesn't go through any of the points.

**CRITICAL THINKING** Explain how the scales used for the axes of a scatter plot could affect the interpretation of the line of best fit.

# Exercises

## Communicate

**1.** Refer to the scatter plot in Example 2. Explain how to use the line of best fit to make a prediction.

**2.** Describe two data sets that have a strong negative correlation.

**3.** Describe two data sets that have a strong positive correlation.

**4.** Describe the correlation for the following scatter plots as strong positive, strong negative, or little to none. Explain the reason for your answer.

**a.**

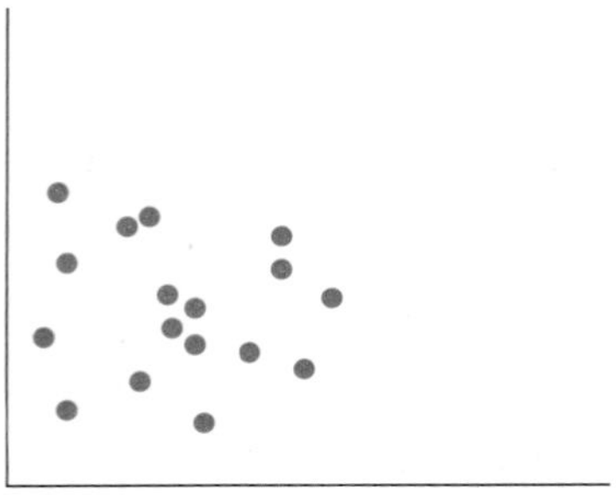

**b.**

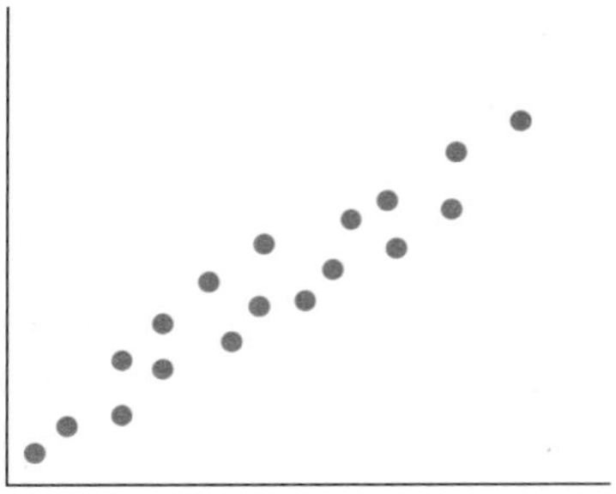

## Guided Skills Practice

APPLICATION

**ECONOMICS** The table below shows the price of a half-gallon of whole milk from 1989 to 1995.

| Year | 1989 | 1990 | 1991 | 1992 | 1993 | 1994 | 1995 |
|---|---|---|---|---|---|---|---|
| Cost | 1.37 | 1.39 | 1.40 | 1.39 | 1.43 | 1.44 | 1.48 |

*Source: Statistical Abstract of the U.S.*

**5.** Plot the data in a scatter plot and draw a line of best fit. ***(EXAMPLES 1 AND 2)***

**6.** What kind of correlation exists? ***(EXAMPLE 1)***

**7.** Use the line of best fit you drew in Exercise 5 to estimate the cost of a half-gallon of whole milk in 1999. ***(EXAMPLE 2)***

## Practice and Apply

**For each scatter plot, describe the correlation as strong positive, strong negative, or little to none.**

**8.**

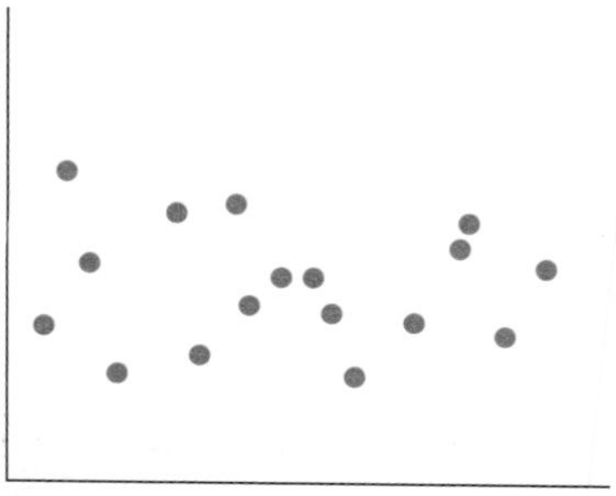

**9.**

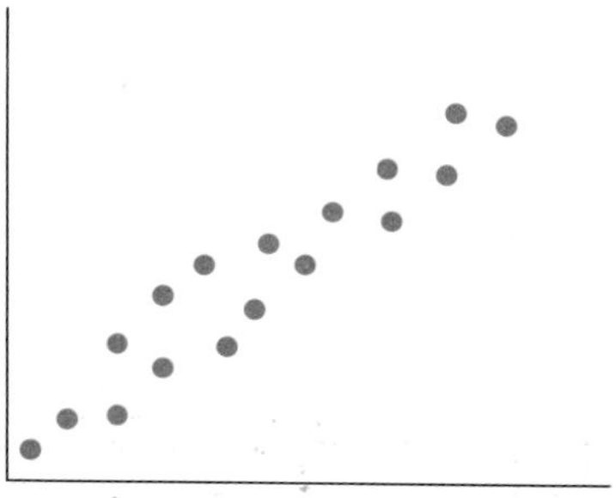

**10.** Draw the two coordinate axes shown below on your graph paper. Plot the same points, (1, 2), (2, 3), and (3, 4), on each set of axes. Connect the points with a line.

**Graph 1**

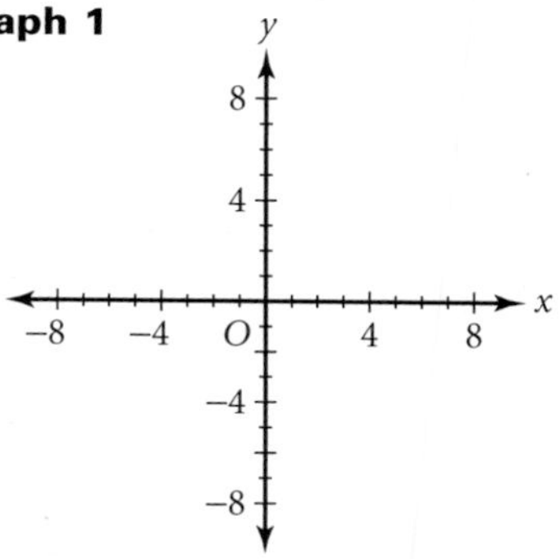


**Graph 2**

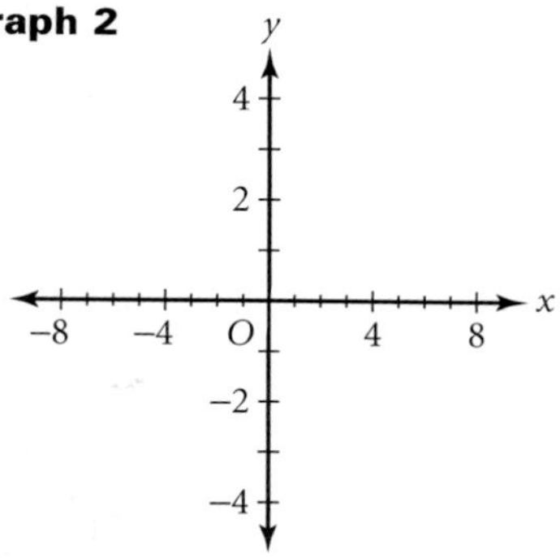


**a.** Which graph is steeper, graph 1 or graph 2?

**b.** What caused one graph to be steeper, even though the points are the same?

**c.** How does a change in scale on the axes affect the steepness of the line? (*Scale* refers to the distance between the units on an axis.)

**d.** Does the scale of the graph affect the correlation? Explain your answer.

**internet connect**

**Homework Help Online**

Go To: **go.hrw.com**
Keyword:
**MA1 Homework Help**
for Exercises 11-15

**11.** **DEMOGRAPHICS** Make a scatter plot for the data below on the number of people working on farms in various years, and draw a line of best fit. Describe the correlation as strong positive, strong negative, or little to none.

| Year | 1940 | 1950 | 1960 | 1970 | 1980 | 1990 |
|---|---|---|---|---|---|---|
| Number of farm workers in thousands | 8995 | 6858 | 4132 | 2881 | 2818 | 2864 |

*Source: The World Almanac, 1997*

**GEOGRAPHY** Use the following data to answer Exercises 12–14:

| City | North latitude | Elevation (in feet) | Maximum normal temp. for Jan. (in °F) |
|---|---|---|---|
| Miami, FL | 26° | 7 | 75 |
| Charleston, SC | 33° | 40 | 50 |
| Washington, DC | 39° | 10 | 43 |
| Boston, MA | 42° | 15 | 36 |
| Portland, ME | 44° | 43 | 31 |

**12.** Graph latitude and temperature, and draw a line of best fit. Describe the correlation.

**13.** Graph latitude and elevation, and draw a line of best fit. Describe the correlation.

**14.** Graph elevation and temperature, and draw a line of best fit. Describe the correlation.

**15.** **SPORTS** Make a scatter plot of the following data for the women's 400-meter run in the Olympics:

| Year | 1964 | 1968 | 1972 | 1976 | 1980 | 1984 | 1988 | 1992 |
|---|---|---|---|---|---|---|---|---|
| Seconds | 52.00 | 52.00 | 51.08 | 49.29 | 48.88 | 48.83 | 48.65 | 48.83 |

*Source: The World Almanac, 1997*

**16.** Describe the correlation for the scatter plot you made in Exercise 15.

**APPLICATION**

**MUSIC** As the length of a string such as a guitar string changes, the frequency, measured in vibrations per second (vps), changes according to data in the following table:

| Length of string (cm) | 40 | 60 | 80 | 100 | 120 |
|---|---|---|---|---|---|
| Frequency (vps) | 660 | 440 | 330 | 264 | 220 |

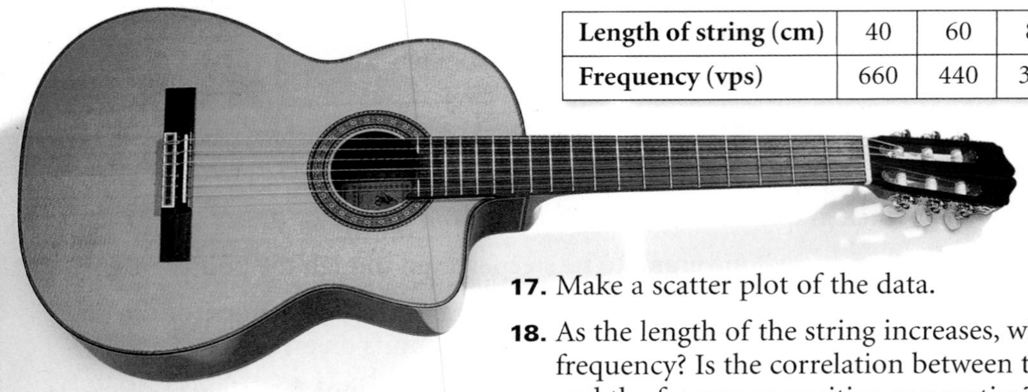

**17.** Make a scatter plot of the data.

**18.** As the length of the string increases, what happens to the frequency? Is the correlation between the length of a string and the frequency positive or negative?

## Look Back

internet connect
**Activities Online**
Go To: **go.hrw.com**
Keyword:
**MA1 Nose Count**

**Perform the following calculations:**

**19.** $\sqrt{144}$ **20.** $\sqrt{7+2}$ **21.** $\sqrt{170-1}$ **22.** $2\sqrt{16}$

**23.** What is the greatest common factor for 30, 60, and 15?

**24.** Pair the equivalent expressions, and explain the relationship between them.

**a.** $2 \cdot 2 \cdot 2 \cdot 2$ **b.** $4(2)$ **c.** $2 + 2 + 2 + 2$ **d.** $2^4$

## Look Beyond

**25.** What values of $x$ will make $2x^2 + 4 = 12$ true?

**26.** What values of $n$ will make $42 - 2n^2 = 10$ true?

## PORTFOLIO ACTIVITY

Include this activity in your portfolio.

| | | | | | | | | | |
|---|---|---|---|---|---|---|---|---|---|
| **Math test scores** | 60 | 40 | 80 | 20 | 65 | 55 | 100 | 90 | 85 |
| **Science test scores** | 70 | 35 | 90 | 50 | 65 | 40 | 95 | 85 | 90 |

*The graph shows a median-fit line for a scatter plot that compares students' math and science test scores. The dashed red lines divide the data into three groups. The median points and median-fit line are indicated in blue.*

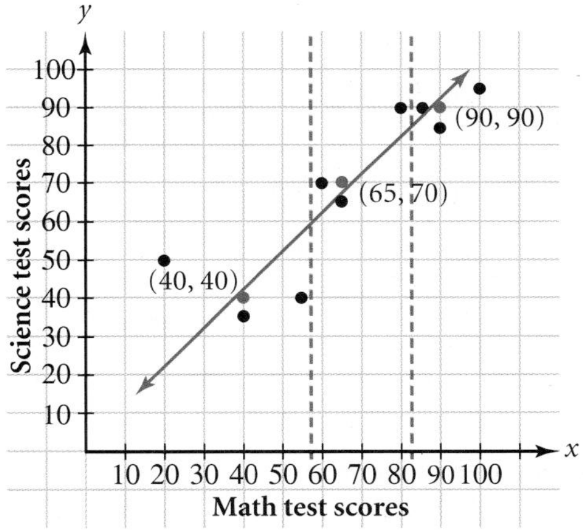


**1.** Find a set of real-world data similar to the tornado data in this lesson. You might look for information on the amount of precipitation in a particular region or on the occurrence of hurricanes, for example. Be sure that you have at least 15 points to plot.

**2.** Make a table and a scatter plot of your data. Describe the correlation.

**3.** Use two vertical lines to separate the data points on your scatter plot into 3 equal groups.

**4.** Within each group of data points, find the median of the $x$-coordinates and the median of the $y$-coordinates. The median is found by listing the numbers from least to greatest and picking the middle value. For example, if 1, 3, 5, 8, and 9 are the $y$-coordinates in one region, the median value is 5.

**5.** Use the median coordinates to create 3 ordered pairs, one for each region. For example, if the median $x$-coordinate for the first region is 5 and the median $y$-coordinate for the first region is 7, the ordered pair is (5, 7). Plot each of these points on your scatter plot.

**6.** Align a ruler with the outer two median points. Keeping the ruler parallel to the original position, slide the ruler one-third of the distance toward the inner median point. Draw the line. The result is a **median-fit line**, which is one type of *best-fit line*.

**7.** Make a prediction for a future year based on your data and the median-fit line.

**WORKING ON THE CHAPTER PROJECT**

You should now be able to complete Activities 2 and 3 of the Chapter Project on page 44.

CHAPTER PROJECT ONE

# WORKING WITH Patterns in Data

*The growth pattern of the chambered nautilus forms an equiangular spiral that is based on the golden ratio.*

**People have been working with patterns to solve problems from the earliest evidence of mathematical thought. In this project, you will work with patterns to solve a variety of problems and to gain insight into the world of algebra.**

## Activity 1

The ancient Greeks were interested in constructing golden rectangles, which were considered pleasing to look at. The ratio of the length to the width of these rectangles (approximately 1.618) was called the golden ratio. Psychologists have found that consumers subconsciously prefer rectangular products that are close to golden rectangles.

You can use consecutive Fibonacci numbers to build rectangles that are close to golden rectangles. The rectangle at right is built by using the consecutive Fibonacci numbers 3 and 5.

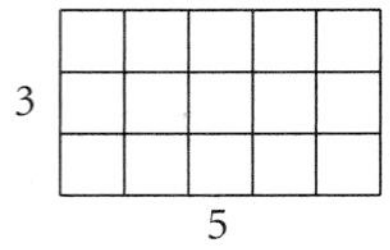


**1.** Use a calculator to write the ratio $\frac{5}{3}$ as a decimal by dividing, $5 \div 3$. How close is the ratio $\frac{5}{3}$ to the golden ratio?

| Length | Width | Ratio |
|---|---|---|
| 5 | 3 | $\frac{5}{3} = 1.667$ |
| 8 | 5 | |
| 13 | 8 | |
| ? | 13 | |
| ? | ? | |
| | | |

Use consecutive pairs of Fibonacci numbers to construct "almost" golden rectangles.

- Complete the table by using consecutive Fibonacci numbers for width and length through a length of 55.
- Sketch a rectangle with the length and width in centimeters or millimeters.
- Compute the ratio $\frac{\text{length}}{\text{width}}$ as a decimal and compare it with the golden ratio.

**2.** Which of the following closely resemble golden rectangles?

**a.** $3 \times 5$ index cards
**b.** $5 \times 7$ photograph
**c.** $5 \times 8$ photograph
**d.** $8 \times 10$ photograph

**3.** Do you think that golden rectangles are the most pleasing to the eye?

**4.** Find five rectangles in the real world that have ratios close to the golden ratio.

## Activity 2

1. Work with your class to find 20 circular objects. Measure their circumference, *c*, and diameter, *d*, and build a table of values.
2. For each circular object, compute the ratio $\frac{c}{d}$. Average the ratios.
3. Make a scatter plot of the data.
4. How do the number $\pi$ and the equation $c = \pi d$ relate to this activity?

## Activity 3

Simulate a wave cheer with members of your class.

1. To simulate the cheer, each person stands and sits in succession. Record the time it takes from start to finish and enter it in the table. (If there are not enough people for a wave, go back to the first person and repeat the wave until you reach the required number.)

| **People** | 4 | 8 | 12 | 16 | 20 | *P* |
|---|---|---|---|---|---|---|
| **Time** | ? | ? | ? | ? | ? | $T = ?$ |

2. Make a scatter plot of the data on graph paper. Draw a line of best fit for the data.
3. Use the line of best fit to answer these questions.
   a. How long would it take 60 people to complete the wave?
   b. How many people would be needed to complete the wave in 2 minutes?

1

# Chapter Review and Assessment

## VOCABULARY



## Key Skills & Exercises

### LESSON 1.1

#### Key Skills

**Extend a sequence by working backward from the differences.**

Apply the method of differences until you get a constant. Then use the differences to find the next two terms.

10, 13, 19, 28, 40, ...

10 13 19 28 40 ? ?
3 6 9 12 ? ?
3 3 3 3 3

Work backward to find the next two terms.

$40 + 15 = 55$ $55 + 18 = 73$

The next two terms are 55 and 73.

#### Exercises

**Find the next two terms of each sequence.**

**1.** 1, 5, 11, 19, 29, ...

**2.** 1, 1, 6, 16, 31, ...

**3.** 90, 70, 54, 42, 34, ...

**4.** 1, 4, 7, 10, ...

**5.** 8, 16, 26, 38, ...

**6.** 100, 99, 97, 94, ...

**7.** The fourth and fifth terms of a sequence are 21 and 32. If the second differences are a constant 3, what are the first five terms of the sequence?

### LESSON 1.2

#### Key Skills

**Evaluate expressions for given values of the variable.**

Evaluate $2x + 3$ for $x = 2$ and $x = 3$.

| | |
|---|---|
| $2(2) + 3$ | $2(3) + 3$ |
| $4 + 3$ | $6 + 3$ |
| $7$ | $9$ |

**Write an equation to represent a problem situation.**

Write an equation for the cost of renting a car, using the variables *c* and *d*.

The cost is \$100 plus \$20 per day.

$c = 100 + 20d$

#### Exercises

**Evaluate each expression when *x* is 1, 2, 3, and 4.**

**8.** $3x$ **9.** $5x + 2$ **10.** $10 - x$

**11.** $2x + 4$ **12.** $x + 7$ **13.** $4x - 5$

**14.** Mr. O'Neil is counting algebra books in the school bookroom. He counts 50 crates plus 7 additional books. Write an equation for this situation in terms of the total number of books, *t*, and the number of books in a crate, *c*.

## LESSON 1.3

### Key Skills

**Use the algebraic order of operations to find the value of expressions.**

To evaluate the expression $5(7 - 4) - 6^2 \div 3 + 1$, follow the algebraic order of operations: grouping symbols, exponents, multiplication/division, and addition/subtraction.

$$5(7 - 4) - 6^2 \div 3 + 1$$
$$5(3) - 6^2 \div 3 + 1$$
$$5(3) - 36 \div 3 + 1$$
$$15 - 12 + 1$$
$$4$$

### Exercises

**Evaluate.**

**15.** $17 - 4 \cdot 3$

**16.** $32 - 24 \div 6 - 4$

**17.** $3 \cdot 4^2 - [24 \div (6 - 4)]$

**18.** $3(4 + 7 \cdot 2) - 100 \div 50$

**19.** $2[3(4 \cdot 5 + 2) - 1] - 4$

**20.** $(3^2 + 2^2) - 2(3^2 - 2^2)$

## LESSON 1.4

### Key Skills

**Graph an equation by plotting points.**

Make a table for the equation $y = 2x$ by substituting the values 1, 2, 3, 4, and 5 for $x$ and finding the values for $y$.

| $x$ | 1 | 2 | 3 | 4 | 5 |
|---|---|---|---|---|---|
| $y$ | 2 | 4 | 6 | 8 | 10 |

Graph the ordered pairs on a coordinate plane and connect the points to make a line.

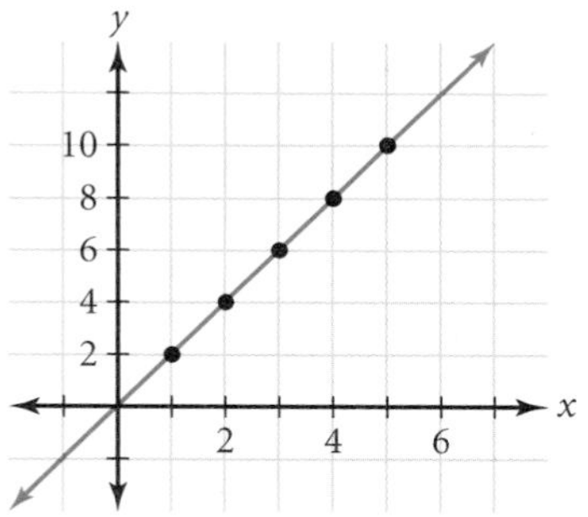


### Exercises

**Graph each list of ordered pairs. Tell whether the points lie on a straight line.**

**21.** (1, 2), (2, 3), (3, 4)

**22.** (2, 5), (4, 2), (5, 2)

**23.** (1, 3), (1, 4), (3, 5)

**Graph each equation by plotting points and connecting the points to make a line.**

**24.** $y = x + 2$

**25.** $y = 3x - 1$

**26.** $y = x + 2$

**27.** $y = x$

## LESSON 1.5

### Key Skills

**Write and graph linear equations representing problem situations.**

The cost for membership in a CD club is shown in the table below.

| Number of CDs purchased ($n$) | 0 | 1 | 2 | 3 | 4 |
|---|---|---|---|---|---|
| Cost in dollars ($c$) | 10 | 21 | 32 | 43 | 54 |

11 11 11 11

### Exercises

**28.** A book club charges a membership fee in addition to the cost of each book purchased. Write an equation to represent the data shown in the table below, and then graph.

| Number of books purchased ($n$) | 0 | 1 | 2 | 3 | 4 |
|---|---|---|---|---|---|
| Cost in dollars ($c$) | 15 | 27 | 39 | 51 | 63 |

Since the data show constant first differences, the pattern is linear. The linear equation $c = 11n + 10$ represents the situation. Graph the equation by plotting points from the table and connecting them to form a line.

**29.** Marcia rides her bike at a rate of 5 miles per hour. Write an equation, make a table, and draw a graph to show the distance Marcia travels for times from 0 to 3 hours.

LESSON 1.6

## Key Skills

**Draw a scatter plot for data, and determine the correlation.**

The table shows the change in a population over a period of 80 years.

| Year | 1900 | 1920 | 1940 | 1960 | 1980 |
|---|---|---|---|---|---|
| People per square mile | 21.5 | 29.9 | 37.2 | 50.6 | 64 |

Plot the ordered pairs from the table on a coordinate plane to form a scatter plot.

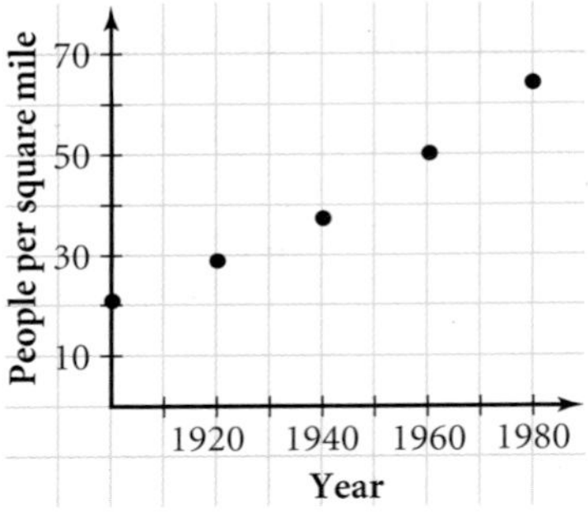


There is a strong positive correlation because both variables increase.

## Exercises

The table below shows how the cost of food items that were \$1 in 1982 has increased over the years.

| Year | 1982 | 1985 | 1988 | 1991 |
|---|---|---|---|---|
| Cost | \$1 | \$1.06 | \$1.18 | \$1.36 |

**30.** Make a scatter plot for the data.

**31.** Describe the correlation in the scatter plot from Exercise 30.

The table below shows the average daily high temperature at certain points on a mountain.

| Elev. (ft) | 3000 | 6000 | 9000 | 12,000 |
|---|---|---|---|---|
| Temp. (°F) | 75 | 65 | 52 | 40 |

**32.** Make a scatter plot for the data.

**33.** Describe the correlation in the scatter plot from Exercise 32.

## Applications

**34.** ENTERTAINMENT If it costs \$13 per person to enter an amusement park, how many people can get in for \$98?

**35.** ENTERTAINMENT Suppose that the cost to order concert tickets is \$22 per ticket, plus a \$5 handling charge per order (regardless of how many tickets are ordered).

**a.** How much does an order of 4 tickets cost?
**b.** How much does an order of 8 tickets cost?
**c.** Let $t$ represent the number of tickets, and write an equation for the cost, $c$, of an order of tickets.

# 1 Chapter Test

**Find the next two terms of each sequence.**

1. 3, 7, 11, 15, 19, …
2. 2, 13, 24, 35, 46, …
3. 101, 96, 91, 86, 81, …
4. −3, 0, 5, 12, 21, …
5. The second difference of a sequence is a constant 5. The differences between the first term and the second term, 12, is 8. Find the first five terms.

**Evaluate each expression when the value of the variables is 1, 3, and 6.**

6. $s + 19$
7. $2x + 5$
8. $15p - 9$
9. $12 + 3y$
10. $2 + 7r$
11. $4q - 7$
12. If theater tickets are $7 each, how many can you buy for $112? Write an equation and solve by guess and check.

**Evaluate each expression.**

13. $2 + 5 \cdot 3$
14. $3(5 - 2) + 7$
15. $2 + 8 - 3^2$
16. $3 \cdot 42 - 25 \div 5$
17. $[12 + (4 - 3) \cdot 2] + 12 \div 3$
18. $5[(3 \cdot 8 \div 2 + 6) - 5] + 9$

**Place grouping symbols to make each equation true.**

19. $17 + 3 \div 10 = 2$
20. $24 \div 8 + 4 + 16 \div 2 = 10$

**Given *a* = 6, *b* = 3, and *c* = 5, evaluate each expression.**

21. $a^2 + b + c$
22. $a^2 - c^2$

**Graph each list of ordered pairs. Tell whether the points lie on a straight line.**

23. (1, 0), (3, 2), (7, 6)
24. (−1, −5), (2, 0), (3, 3)
25. (0, −4), (2, 2), (3, 5)

**Make a table for each equation, and find the values for *y* by substituting 1, 2, 3, 4, and 5 for *x*. Then graph each equation.**

26. $y = x - 4$
27. $y = 3x - 8$
28. $y = 5x$
29. $y = 10 - 2x$

**Find the first differences for each data set, and write an equation to represent the data pattern.**

30.

| | | | | | | |
|---|---|---|---|---|---|---|
| 0 | 1 | 2 | 3 | 4 | 5 | 6 |
| 7 | 11 | 15 | 19 | 23 | 27 | 31 |

31. TRAVEL A tour guide plans a trip to historical sites in the Southwest. He plans to travel between sites at an average speed of 47 miles per hour. Write an equation, make a table, and draw a graph to show the distance the tour guide can travel for times from 0 to 6 hours.

**Describe each correlation as positive, negative, or little to none.**

32.
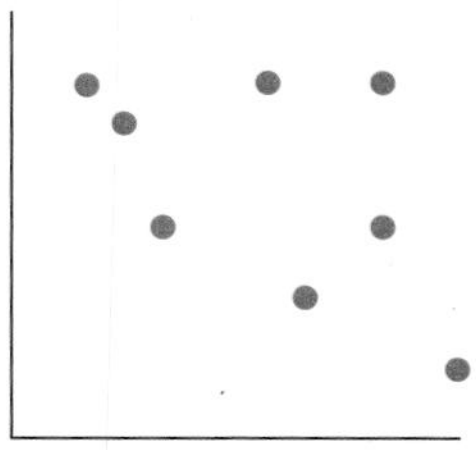

33.
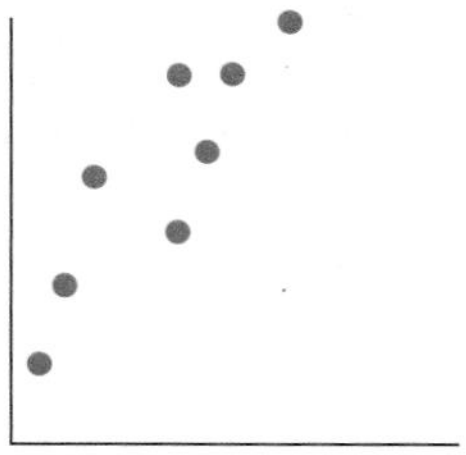

34. **Trace the scatter plot below, and draw on your paper the line of best fit.**

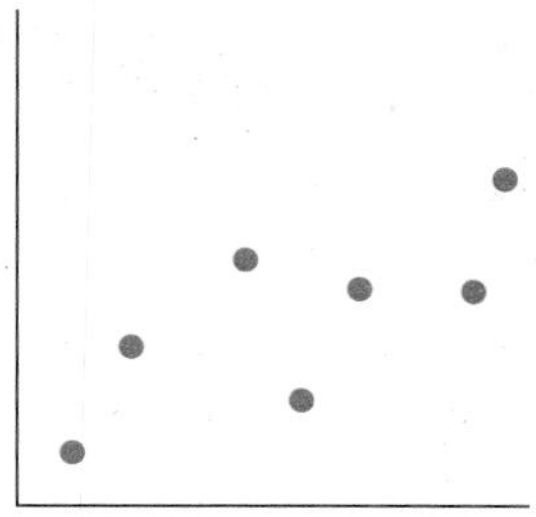

# College Entrance Exam Practice

internet connect
**Standardized Test Prep Online**
Go To: **go.hrw.com**
Keyword: **MM1 Test Prep.**

**QUANTITATIVE COMPARISON** For Items 1–4, write
A if the quantity in Column A is greater than the quantity in Column B;
B if the quantity in Column B is greater than the quantity in Column A;
C if the two quantities are equal; or
D if the relationship cannot be determined from the information given.

| | Column A | Column B | Answers |
|---|---|---|---|
| **1.** | $\frac{1}{2}+\frac{3}{5}$ | $\frac{1}{2}\cdot\frac{3}{5}$ | Ⓐ Ⓑ Ⓒ Ⓓ |
| **2.** | The next term in the sequence 7, 16, 25, 34, 43, . . . | The next term in the sequence 2, 4, 8, 16, 32, . . . | Ⓐ Ⓑ Ⓒ Ⓓ [Lesson 1.1] |
| **3.** | $\sqrt{225}$ | $3^2 + 2(7-4)$ | Ⓐ Ⓑ Ⓒ Ⓓ [Lesson 1.3] |
| **4.** | $\frac{6}{10}$ | $\left(\frac{6}{10}\right)^2$ | Ⓐ Ⓑ Ⓒ Ⓓ |

**5.** If Shelly drives at an average speed of 50 miles per hour, how many hours will it take her to drive 500 miles? ***(LESSON 1.2)***
**a.** 6 hours  **b.** 8 hours
**c.** 10 hours  **d.** more than 10 hours

**6.** Simplify the following expression:
$(90 \div 3^2) - 9 + 5^2$ ***(LESSON 1.3)***
**a.** 24  **b.** 26
**c.** 98  **d.** 126

**7.** What number is missing from this pattern?
232, 343, ______, 565, 676, 787 ***(LESSON 1.1)***
**a.** 424  **b.** 403
**c.** 454  **d.** 554

**If notebooks cost 59 cents, find the cost of the items below.** ***(LESSON 1.2)***

**8.** 2 notebooks  **9.** 5 notebooks

**10.** 12 notebooks

**11.** Write an equation to model this situation: How many notebooks can you buy for $14.75?

**Alice sells memberships for an exercise club. Her weekly pay is $250 plus $20 for each membership sold.** ***(LESSONS 1.4 AND 1.5)***

**12.** Write an equation for Alice's weekly pay, $P$, in terms of the number of memberships she sells, $m$.

**13.** Make a table to show Alice's weekly pay for 1, 2, 3, 4, and 5 memberships sold.

**14.** Suppose that she earns $330 one week. How many memberships did she sell?

**15.** Make a graph by using the data from your table.

**16.** Use your graph to find how many memberships Alice needs to sell in order to earn at least $380 in one week.

**17.** Make a table of the values and evaluate the expression $5x + 7$ by substituting 1, 2, 3, 4, and 5 for $x$. ***(LESSON 1.2)***

**18.** Use the table of values in Exercise 17 to solve the equation $5x + 7 = 22$.

**Write a list of factors for each number. Circle the prime factors.**

**19.** 20 **20.** 42 **21.** 76

**Determine whether each number is prime or composite.**

**22.** 19 **23.** 30 **24.** 31

**Evaluate.** ***(LESSON 1.3)***

**25.** $\frac{3}{4} - \frac{3}{5}$

**26.** $1\frac{2}{3} - \frac{7}{8}$

**27.** $\frac{7}{15} \cdot \frac{3}{28}$

**28.** $\frac{1}{2} \div \frac{3}{5}$

**29.** $3 + 27 \div 3^2 - (7 - 5)$

**30.** $(9 + 37) + 11$

**31.** $20 \cdot (5 \cdot 19)$

**Graph each list of ordered pairs and tell whether the points lie on a straight line.** ***(LESSON 1.4)***

**32.** (0, 1), (4, 9), (2, 5)

**33.** (1, 9), (8, 4), (4, 5)

**Describe the correlation for each scatter plot below as strong positive, strong negative, or little to none.** ***(LESSON 1.6)***

**34.**

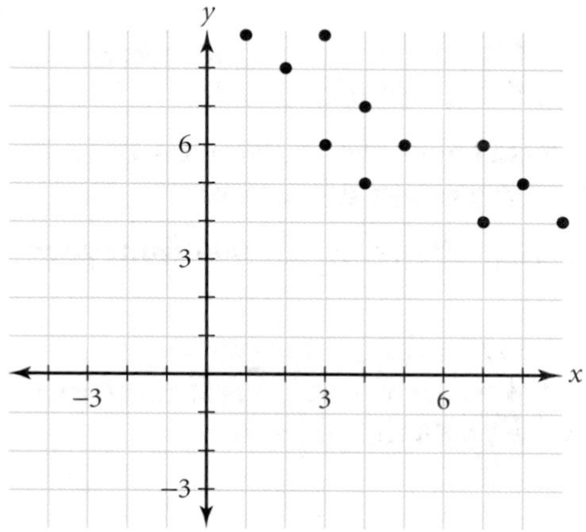


**35.**

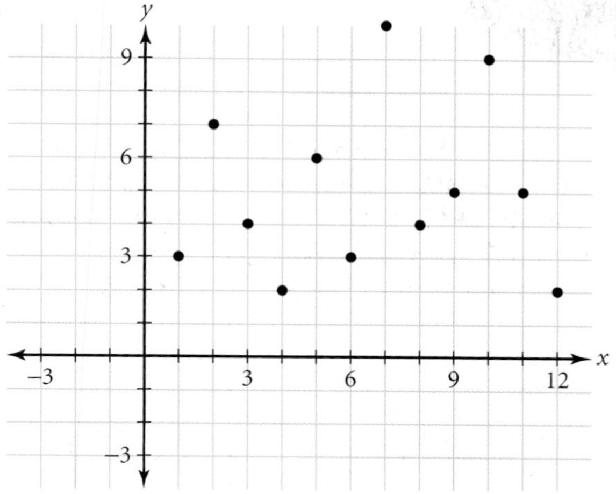


**36.** Make a scatter plot of this data for the amount of time spent studying and the score on a test the next day. ***(LESSON 1.6)***

| | | | | | | | | | | |
|---|---|---|---|---|---|---|---|---|---|---|
| **Time studying (in minutes)** | 15 | 12 | 20 | 30 | 25 | 40 | 60 | 50 | 55 | 45 |
| **Test score** | 65 | 68 | 75 | 80 | 79 | 85 | 97 | 80 | 88 | 73 |

**37.** Use a straightedge to estimate the line of best fit for the scatter plot from Exercise 36. ***(LESSON 1.6)***

**FREE-RESPONSE GRID**

**The following questions may be answered by using a free-response grid such as that commonly used by standardized test services.**

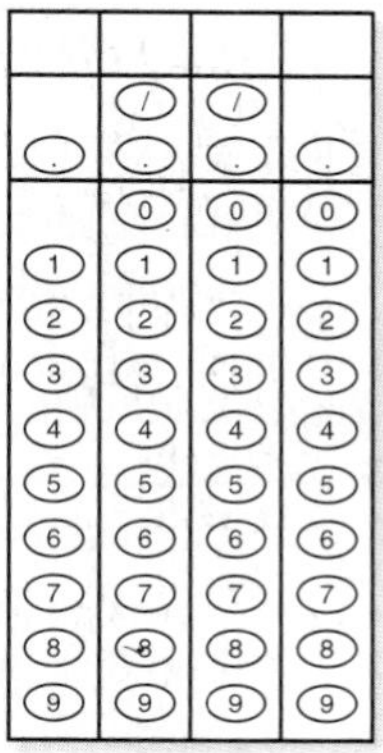


**Refer to the sequence 1, 2, 5, 10, 17, 26, ...** ***(LESSON 1.1)***

**38.** Find the second differences.

**39.** Find the next term.

**40.** If the constant second difference of a sequence is 2, the first of the first differences is 5, and the first term is 12, find the fifth term of the sequence. ***(LESSON 1.1)***

**Find the next term of each sequence.** ***(LESSON 1.1)***

**41.** 1, 3, 7, 13, ...

**42.** 99, 93, 86, 78, ...

# 2 Operations in Algebra

One of the first books about algebra was written in Arabic by a ninth-century scientist named Muhammad ibn Muas al-Khwarizmi. The title of the book was shortened to *al-jabr,* now spelled *algebra.* The word *algebra* comes from part of the Arab title that means "equals can be added to both sides of an equation." Al-Khwarizmi used his *al-jabr* to help him in his scientific work in geography and astronomy.

*An astrolabe is a device that was used by early astronomers to study the positions of stars.*

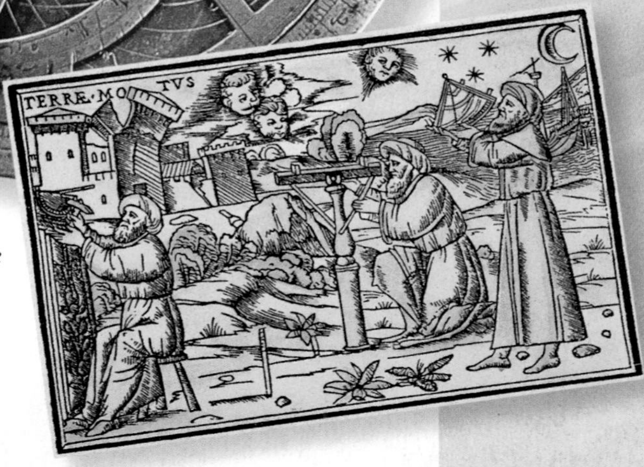


## Lessons

- 2.1 • The Real Numbers and Absolute Value
- 2.2 • Adding Real Numbers
- 2.3 • Subtracting Real Numbers
- 2.4 • Multiplying and Dividing Real Numbers
- 2.5 • Properties and Mental Computation
- 2.6 • Adding and Subtracting Expressions
- 2.7 • Multiplying and Dividing Expressions

Project Using a Spreadsheet

*Pages from an Arabic manuscript of al-Khwarizmi's algebra book*

CHAPTER PORTFOLIO ACTIVITIES PROJECT

## About the Chapter Project

A spreadsheet is a tool for organizing and analyzing numeric data that uses positive and negative numbers, expressions, and equations. The Chapter Project, *Using a Spreadsheet*, on page 104, is an introduction to the use of spreadsheets and provides an opportunity to apply the algebraic concepts covered in the chapter. After completing the Chapter Project, you will be able to do the following:

- Use a computer spreadsheet to analyze simple data sets.

## About the Portfolio Activities

Throughout the chapter, you will be given opportunities to complete Portfolio Activities that are designed to support your work on the Chapter Project.

- The Portfolio Activity on page 66 describes how profit and loss are indicated on a spreadsheet.
- Average monthly temperature ranges for Calgary, Alberta, Canada, are the topic of the Portfolio Activity on page 79.
- In the Portfolio Activity on page 88, a spreadsheet is used to help describe windchill data.
- The Portfolio Activity on page 103 presents a bicycle-rental problem.

# 2.1 The Real Numbers and Absolute Value

**Venus' flytrap** *Dionaea muscipula* | **House cat** *Felis domesticus* | **Indonsian gliding frog** *Rhacophorus rheinwardti*

**Objectives**

- Compare real numbers.
- Simplify expressions involving opposites and absolute value.

**Why** *Taxonomy deals with the naming and classification of plants and animals. Another system helps to classify numbers in mathematics. Understanding the different types of numbers and comparing numbers are important skills in many real-world situations.*

**APPLICATION**
**BIOLOGY**

Scientists have classified millions of plants and animals by using a system called taxonomy that was developed in the mid-1700s by Swedish naturalist Carolus Linnaeus. This system groups living things by common characteristics in increasingly smaller and more specific categories. In mathematics numbers are classified in a similar, though much less complicated, system.

## Sets of Numbers

The Venn diagram below shows how all of the sets of numbers that will be discussed relate to one another.

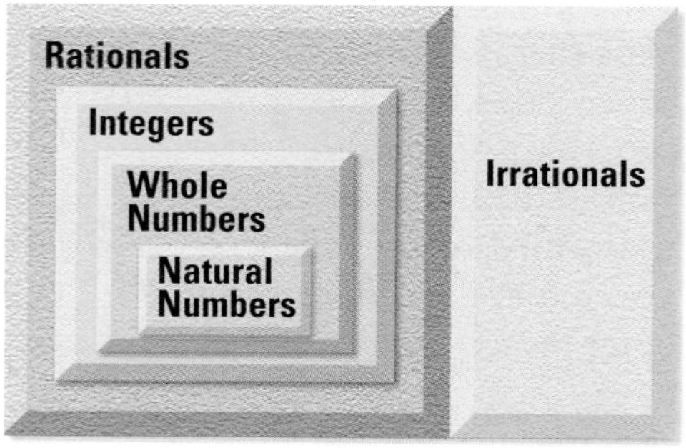


*The real numbers*

Different sets of numbers can be represented by lists of their members. For example:

**Natural** or counting numbers: $N = 1, 2, 3, 4, \ldots$
**Whole numbers:** $W = 0, 1, 2, 3, 4, \ldots$
**Integers:** $I = \ldots -3, -2, -1, 0, 1, 2, 3, \ldots$

The sets $N$, $W$, and $I$ can also be shown on a number line. Remember that 0 is known as the origin and is neither positive nor negative. Numbers to the right of 0 are **positive**, and those to the left of 0 are **negative**.

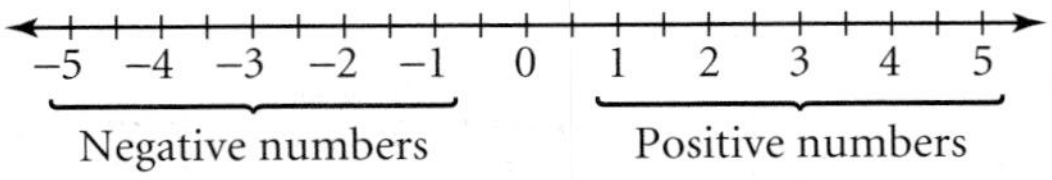


Although the labeled numbers are typically integers, there are many numbers on a number line that are not integers. Numbers like $\frac{2}{3}$, $-\frac{4}{15}$, and $\frac{23}{5}$ are included on a number line and are called **rational numbers.**

**Definition of Rational Number**

Any number that can be expressed in the form $\frac{a}{b}$, where $a$ and $b$ are integers and $b \neq 0$, is a rational number.

The set of rational numbers cannot be listed as the integers can be, but this set does include the integers since any integer, $a$, can be expressed as $\frac{a}{1}$.

The set of rational numbers has an important characteristic that distinguishes it from other types of numbers. Any rational number can be written as either a terminating decimal or a repeating decimal. Repeating decimals do not end: at some point they continuously repeat a pattern of digits to the right of the decimal. Terminating decimals simply end at a certain point.

Use a bar above the repeating digits to indicate a repeating decimal. For example, $\frac{2}{11}$ is equal to $0.1818\ldots$, which is written $0.\overline{18}$.

**EXAMPLE 1** **Use division to write each rational number as a decimal and indicate whether the number is a terminating decimal or a repeating decimal.**

**a.** $\frac{4}{5}$ **b.** $\frac{4}{11}$ **c.** $\frac{5}{8}$ **d.** $\frac{7}{9}$ **e.** $\frac{5}{12}$ **f.** $\frac{3}{4}$

**SOLUTION**

**a.** $\frac{4}{5} = 0.8$; terminating decimal

**b.** $\frac{4}{11} = 0.\overline{36}$; repeating decimal

**c.** $\frac{5}{8} = 0.625$; terminating decimal

**d.** $\frac{7}{9} = 0.\overline{7}$; repeating decimal

**e.** $\frac{5}{12} = 0.41\overline{6}$; repeating decimal

**f.** $\frac{3}{4} = 0.75$; terminating decimal

Numbers such as $\sqrt{7}$ and $\sqrt{11}$ are the square roots of numbers that are not perfect squares. You will study square roots, or radicals, in a later chapter. When evaluating these square roots, the results are nonrepeating decimals.

$$\sqrt{7} = 2.64575131\ldots$$
$$\sqrt{11} = 3.3166247\ldots$$

These numbers continue without repeating a pattern of digits. Since they are neither terminating nor repeating decimals, these numbers are not rational and cannot be expressed in the form $\frac{a}{b}$, where $a$ and $b$ are integers and $b \neq 0$. They are called **irrational numbers.**

The set of rational numbers and the set of irrational numbers together form the set of **real numbers.**

## Ordering Numbers

The number line represents the entire set of real numbers. Each point on the number line represents exactly one number. Each number in the set of real numbers corresponds with exactly one point on the number line. This allows the numbers on the number line to be ordered. Numbers on the number line increase in value from left to right.

The symbols used for ordering are as follows:

| | |
|---|---|
| $<$ less than | $\leq$ less than or equal to |
| $>$ greater than | $\geq$ greater than or equal to |
| $=$ equal to | |

**CRITICAL THINKING** Since each point on the number line represents exactly one number, how can two numbers be equal to one another? That is, why do we need the equal sign for ordering numbers?

**EXAMPLE 2** **Insert an ordering symbol to make each statement true.**

**a.** $5$ ___ $-7$ **b.** $\frac{1}{2}$ ___ $\frac{8}{16}$ **c.** $-4.8$ ___ $-4.7$ **d.** $4\frac{2}{3}$ ___ $6$

**SOLUTION**

**a.** 5 is to the right of $-7$ on a number line, so $5 > -7$.

**b.** $\frac{1}{2}$ and $\frac{8}{16}$ represent the same number, so $\frac{1}{2} = \frac{8}{16}$.

**c.** $-4.8$ is to the left of $-4.7$ on a number line, so $-4.8 < -4.7$.

**d.** $4\frac{2}{3}$ is to the left of 6 on a number line, so $4\frac{2}{3} < 6$.

**TRY THIS** Insert an ordering symbol to make each statement true.

**a.** $-9$ ___ $-12$ **b.** $\frac{1}{2}$ ___ $\frac{3}{5}$

**c.** $-6.3$ ___ $5$ **d.** $\frac{4}{12}$ ___ $\frac{7}{21}$

## Opposites and Absolute Value

On the number line, two numbers that lie on opposite sides of 0 and are the same distance from 0 are **opposites.** The opposite of 2 is negative 2, written as –2. The opposite of a positive number is always a negative number.

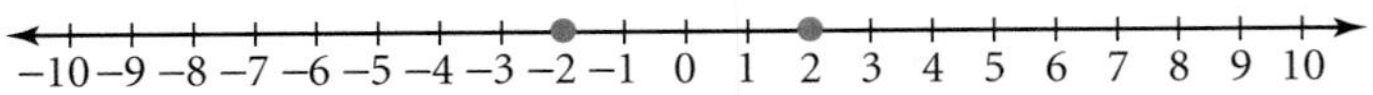


As you can see from the number line, the opposite of –2 is +2. In fact, the opposite of a negative number is always a positive number. This is represented with symbols as follows:

$$-(-2) = 2$$

The opposite of zero is zero itself.

$$-0 = 0$$

**EXAMPLE 3** **Evaluate each expression.**

**a.** $-\left(-\frac{2}{3}\right)$ **b.** $-(0)$ **c.** $-(7.4 - 3.2)$

**SOLUTION**

**a.** The opposite of negative $\frac{2}{3}$ is $\frac{2}{3}$.

**b.** The opposite of 0 is 0.

**c.** The opposite of the difference between 7.4 and 3.2 is –4.2.

**TRY THIS** Evaluate.

**a.** $-\left(-\frac{4}{7}\right)$ **b.** $-(-0)$ **c.** $-(6.2 - 2.7)$

Opposite numbers are helpful in understanding other important concepts in algebra, such as absolute value.

### Definition of Absolute Value

The absolute value of a real number $x$ is the distance from $x$ to 0 on a number line. The symbol $|x|$ means the absolute value of $x$.

Notice that a number and its opposite have the same absolute value. The absolute values of 2 and –2 are the same, 2. Each is 2 units from 0.

The absolute value of a positive number or 0 is the same as the number itself. The absolute value of a negative number is the opposite of the number.

**EXAMPLE 4** **Simplify.**

**a.** $|7|$ **b.** $|-14.2|$
**c.** $|0|$ **d.** $|9 + 4|$

**SOLUTION**

**a.** $|7| = 7$ **b.** $|-14.2| = 14.2$

**c.** $|0| = 0$ **d.** $|9 + 4| = |13| = 13$

**TRY THIS** Evaluate.

**a.** $|13.2|$ **b.** $|17 - 12|$

# Exercises

## Communicate

**1.** How are the numbers $\frac{2}{3}$ and $\frac{3}{5}$ different in terms of their decimal equivalents?

**2.** Explain why an integer is also a rational number.

**3.** What does absolute value tell you about a number? Why is the absolute value of 6 the same as the absolute value of –6?

## Guided Skills Practice

**Using division, label each number as a terminating decimal or a repeating decimal.** *(EXAMPLE 1)*

**4.** $\frac{7}{8}$ **5.** $\frac{19}{22}$ **6.** $\frac{18}{55}$ **7.** $\frac{5}{16}$

**Insert <, >, or = to make each statement true.** *(EXAMPLE 2)*

**8.** 7 ___ 3 **9.** $2\frac{1}{2}$ ___ –3 **10.** –0.2 ___ –0.1 **11.** $\frac{1}{2}$ ___ $\frac{1}{3}$

**Find the opposite of each expression.** *(EXAMPLE 3)*

**12.** 2 **13.** $-\frac{4}{5}$ **14.** 0.5 **15.** 0

**Evaluate.** *(EXAMPLE 4)*

**16.** $|15|$ **17.** $\left|-12\frac{1}{2}\right|$ **18.** $|2 + 6|$ **19.** $|6 - 7|$

## Practice and Apply

**Insert <, >, or = to make each statement true.**

**20.** 0 ___ 5 **21.** $-4\frac{1}{4}$ ___ –4 **22.** –0.6 ___ 0.6 **23.** 5.4 ___ $5\frac{1}{2}$

**24.** –2 ___ 1 **25.** –7 ___ –9 **26.** $\frac{2}{3}$ ___ $-1\frac{1}{3}$ **27.** 6.7 ___ $6\frac{3}{4}$

**28.** –4.62 ___ –4.6 **29.** 7.94 ___ 7.95 **30.** $-\frac{1}{5}$ ___ $-\frac{1}{4}$ **31.** –8.2 ___ $-8\frac{1}{4}$

**internet connect**

**Homework Help Online**
Go To: **go.hrw.com**
Keyword:
**MA1 Homework Help**
for Exercises 32-47

**Find the opposite of each number.**

**32.** 17 **33.** $-17$ **34.** 0 **35.** (7.8)

**36.** $x$ **37.** $-1.2$ **38.** $\frac{4}{5}$ **39.** $-\frac{13}{20}$

**Find the absolute value of each number.**

**40.** $-6$ **41.** 18 **42.** 0 **43.** $-10.3$

**44.** $\frac{9}{19}$ **45.** $-3.8$ **46.** 17.1 **47.** $-\frac{3}{4}$

**48. Name a value of *x* such that**

**a.** $-x$ is negative.

**b.** $-x$ is positive.

**c.** $-x$ is neither positive nor negative.

**Simplify each expression.**

**49.** $-(-10)$ **50.** $-(6.8 - 4.9)$ **51.** $-(3 \cdot 8)$ **52.** $-(17 - 17)$

**53.** $|3|$ **54.** $|-20|$ **55.** $\left|\frac{20}{5}\right|$ **56.** $|-(-3)|$

**57.** $|6| + |-6|$ **58.** $\left|-2\frac{1}{3}\right|$ **59.** $|-1| \cdot |8|$ **60.** $-|16 - 10|$

**Determine whether each statement is true or false.**

**61.** On a number line, $x$ and $-x$ are the same distance from 0.

**62.** 0 is neither a whole number nor an integer.

**63.** If $x$ is a positive integer, then $|x| = -x$.

**64.** There are two real numbers whose absolute value is 6.

CONNECTION

**65.** COORDINATE GEOMETRY Determine the final point on a number line if you start at $-7$, move 3 spaces to the left, and then move $4\frac{1}{2}$ spaces to the right. Write an equation that models this situation.

## Look Back

**Find the next two terms in each number sequence.** *(LESSON 1.1)*

**66.** 6, 13, 22, 33, . . . **67.** 1, 7, 23, 50, 89, . . .

APPLICATION

CONSUMER ECONOMICS A truck rental firm charges \$25 plus 30 cents per mile to rent a moving truck. *(LESSON 1.2)*

**68.** Write an equation that shows the cost, $c$, of renting the truck in terms of the number of miles, $m$.

**69.** Make a table and find the rental cost for 25, 50, 75, 100, and 125 miles.

**70.** How many miles can you drive the truck if you have budgeted \$58 for the rental?

## Look Beyond

**71.** Use a graphics calculator to plot the function $y = \text{ABS}(x)$. Describe the graph.

2.2

# Adding Real Numbers

**Objectives**

- Use algebra tiles to model addition.
- Add numbers with like signs.
- Add numbers with unlike signs.

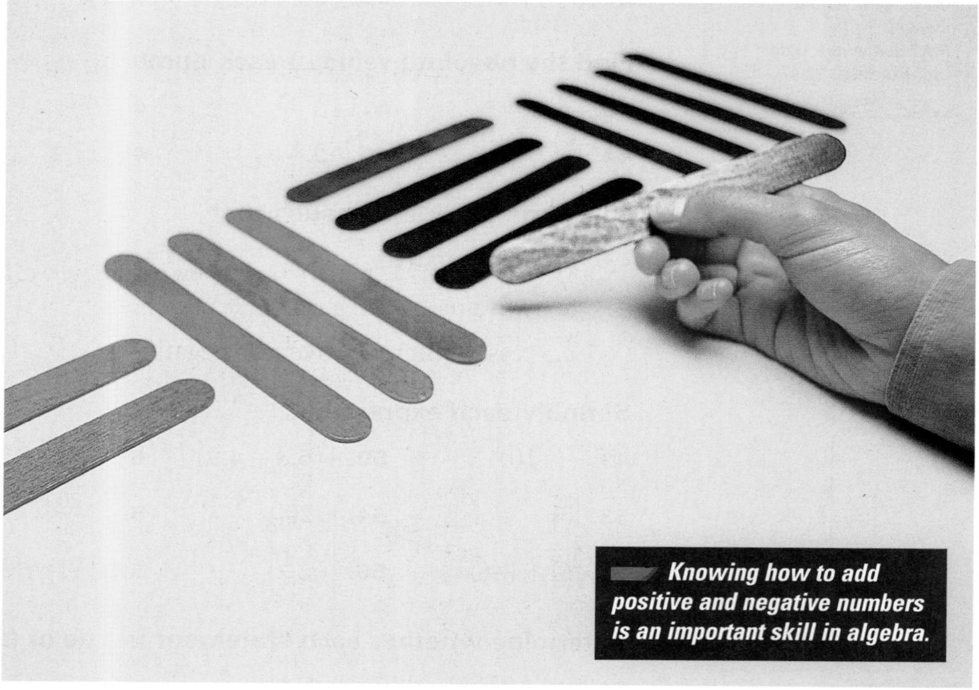


*You can use sticks of different colors to practice an ancient Chinese method of addition.*

APPLICATION
BUSINESS

**CULTURAL CONNECTION: ASIA** More than 2000 years ago, the Chinese used rods of different colors to represent positive and negative numbers. Consider the following problem:

> A business transaction involved a loss of 23 monetary units (−23) and a gain of 54 monetary units (+54). What was the overall gain or loss?

To solve a problem like this, vertical rods were used to represent units or ones, and horizontal rods were used to represent tens. Rods of one color were positive (gains), and rods of another color were negative (losses).

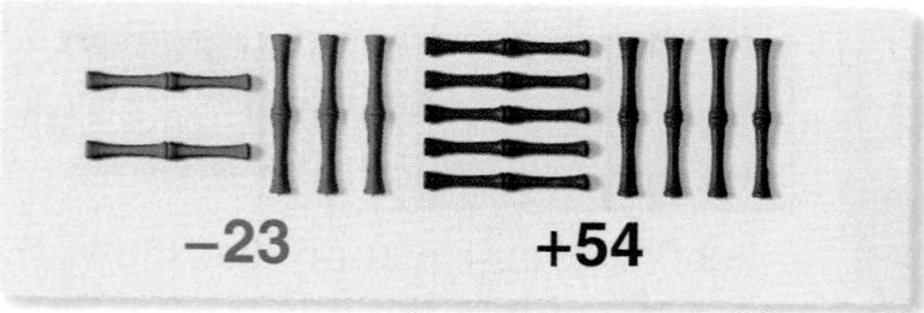


*A gain and a loss of equal amounts do not change the total.* Therefore, pairs of red and black rods can be eliminated from the layout because they have no effect on the total.

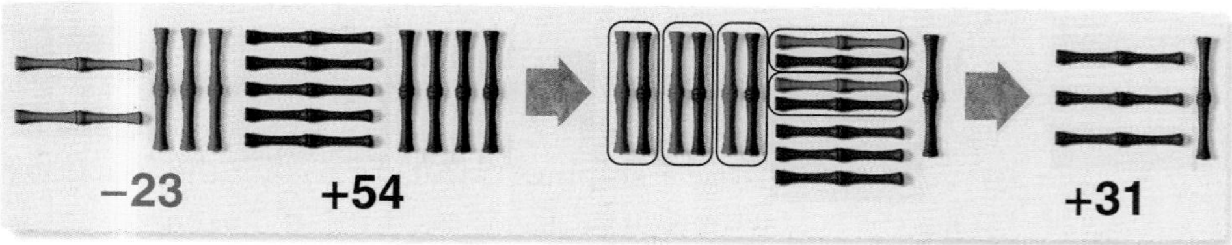

## Adding Integers

Like the rods in the Chinese mathematical example, algebra tiles can be used to model integer addition. One color of tile can represent the positive integers, and another color of tile can represent the negative integers.

To model the addition of two integers with like signs (both positive or both negative), represent each number with appropriately colored tiles. Then combine the tiles. Count the combined tiles and use the same sign (positive or negative) as the numbers you added.

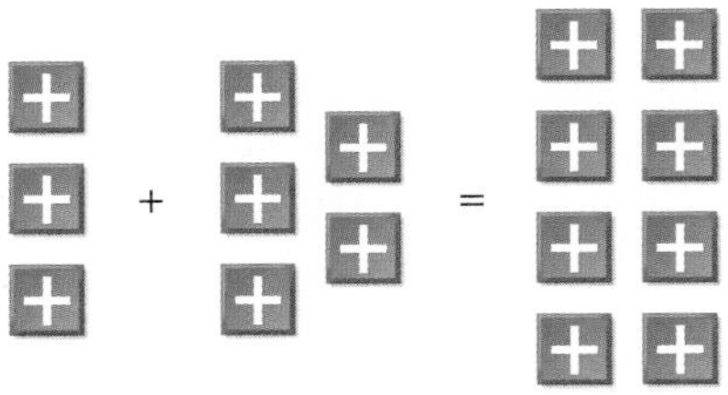

3 + 5 = +8

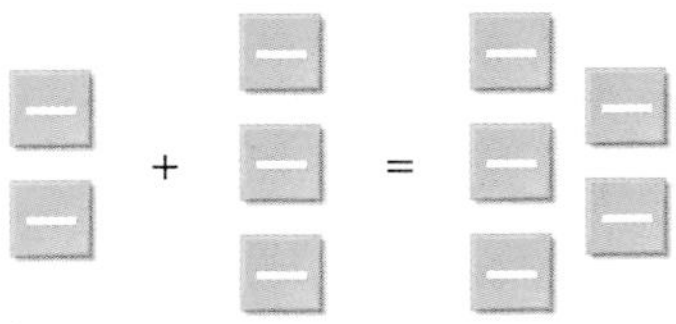

−2 + (−3) = −5

Modeling the addition of two integers with unlike signs (one positive and one negative) involves neutral pairs of tiles. A **neutral pair** consist of one positive and one negative tile. Since the value of a neutral pair is zero, they do not affect the total value of a group of tiles.

The value of a neutral pair is 0.

To model the addition of two integers with unlike signs, represent each number with appropriately colored tiles. Then combine the tiles. Remove all neutral pairs. The remaining tiles represent the sum. The color of the tiles indicates the sign of the sum.

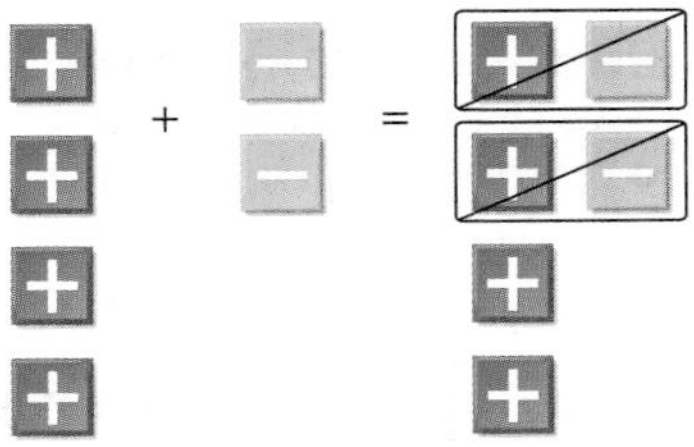

4 + (−2) = +2

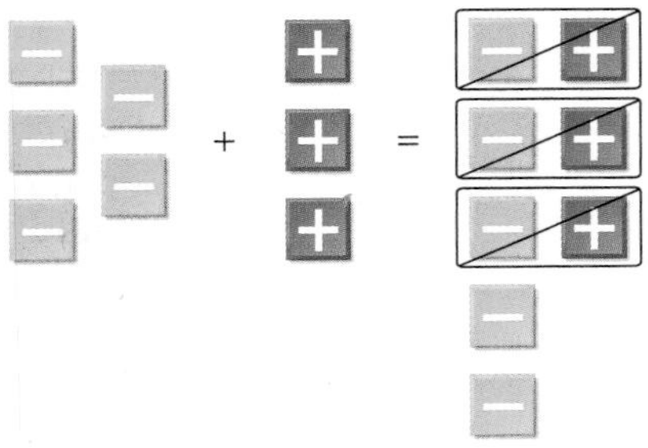

−5 + 3 = −2

## Activity
## Modeling Addition—Unlike Signs

**You will need:** positive and negative tiles

1. Arrange 5 positive tiles and 8 negative tiles as shown.

2. Rearrange the tiles to form as many neutral pairs as you can. (One neutral pair has been circled.)

3. Remove all of the neutral pairs. How many tiles remain? Are they positive or negative? What is the answer to $5 + (-8)$?

4. Experiment by modeling other sums, such as $-10 + 6$ and $10 + (-3)$.
5. Look for a pattern in your results. For example, when the number of negative tiles is greater than the number of positive tiles, what is the sign of the answer?

CHECKPOINT ✓

6. How can you add numbers with unlike signs without using algebra tiles? Write your own rule. (Hint: Use absolute value.)

## Using Rules for Addition

By now, you may have already discovered the rules for addition in algebra. The rules use the definition of absolute value (Lesson 2.1).

### Rules for Adding Two Signed Numbers

**Addition of two numbers with like signs**
1. Find the *sum* of the absolute values.
2. Use the sign common to both numbers.

**Addition of two numbers with unlike signs**
1. Find the *difference* of the absolute values.
2. Use the sign of the number with the greater absolute value.

## EXAMPLE 1 Find each sum.

**a.** $-2 + (-8)$ **b.** $15 + (-23)$ **c.** $14 + (-5)$

**SOLUTION**

**a.** Since the numbers have like signs (both negative), find the sum of their absolute values. Use the sign common to both numbers.

$|-2| = 2$ $|-8| = 8$ $2 + 8 = 10$ $-2 + (-8) = -10$

**b.** Since the numbers have unlike signs, find the difference of their absolute values. Use the sign of the number with the greater absolute value.

$|15| = 15$ $|-23| = 23$ $23 - 15 = 8$ $15 + (-23) = -8$

**c.** Since the numbers have unlike signs, find the difference of their absolute values. Use the sign of the number with the greater absolute value.

$|14| = 14$ $|-5| = 5$ $14 - 5 = 9$ $14 + (-5) = 9$

**TRY THIS** Find each sum.

**a.** $-3 + (-12)$ **b.** $17 + (-32)$ **c.** $-15 + 18$

The rules of algebraic addition apply to all real numbers.

## EXAMPLE 2 Find each sum.

**a.** $-1.3 + 2.5$ **b.** $-\frac{1}{4} + \left(-1\frac{1}{4}\right)$

**c.** $-3.7 + (-8.6)$ **d.** $-4\frac{2}{3} + 5\frac{1}{3}$

**SOLUTION**

**a.** Since the numbers have unlike signs, find the difference of their absolute values. Use the sign of the number with the greater absolute value.

$|-1.3| = 1.3$ $|2.5| = 2.5$ $2.5 - 1.3 = 1.2$ $-1.3 + 2.5 = 1.2$

**b.** Since the numbers have like signs (both negative), find the sum of their absolute values. Use the sign common to both numbers.

$\left|-\frac{1}{4}\right| = \frac{1}{4}$ $\left|-1\frac{1}{4}\right| = 1\frac{1}{4}$ $\frac{1}{4} + 1\frac{1}{4} = 1\frac{2}{4} = 1\frac{1}{2}$ $-\frac{1}{4} + \left(-1\frac{1}{4}\right) = -1\frac{1}{2}$

**c.** Since the numbers have like signs, find the sum of their absolute values. Use the sign common to both numbers.

$|-3.7| = 3.7$ $|-8.6| = 8.6$ $3.7 + 8.6 = 12.3$ $-3.7 + (-8.6) = -12.3$

**d.** Since the numbers have unlike signs, find the difference of their absolute values. Use the sign of the number with the greater absolute value.

$\left|-4\frac{2}{3}\right| = 4\frac{2}{3}$ $\left|5\frac{1}{3}\right| = 5\frac{1}{3}$ $5\frac{1}{3} - 4\frac{2}{3} = \frac{2}{3}$ $-4\frac{2}{3} + 5\frac{1}{3} = \frac{2}{3}$

**TRY THIS** Find each sum.

**a.** $-9 + 3.7$ **b.** $7 + \left(-1\frac{3}{4}\right)$ **c.** $-6.3 + (-2.5)$ **d.** $4\frac{1}{5} + \left(-7\frac{2}{3}\right)$

## The Additive Identity and Additive Inverse

The sum of any number and 0 is equal to the number. Therefore, 0 is called the **identity element for addition**.

**Identity Property for Addition**

For all real numbers $a$, $a + 0 = a$ and $0 + a = a$.

Related to the identity element, 0, are **additive inverses**. Another name for additive inverse is *opposite*. The sum of two additive inverses is 0, the identity element.

$-7 + 7 = 0$ $\quad$ $4 + (-4) = 0$ $\quad$ $-3 + 3 = 0$

**Additive Inverse Property**

For every real number $a$, there is exactly one real number $-a$ such that $a + (-a) = 0$ and $-a + a = 0$.

# Exercises

## Communicate

1. Describe how to add two integers that have the same sign.
2. Describe how to add two integers that have different signs.
3. Explain how to use algebra tiles to model the following sums:
   **a.** $-4 + (-5)$ **b.** $-3 + 6$
4. Explain how the Identity Property for Addition and the Additive Inverse Property are useful when modeling addition with algebra tiles.

## Guided Skills Practice

**Find each sum.** *(EXAMPLE 1)*

**5.** $-3 + (-15)$ **6.** $12 + (-7)$ **7.** $-24 + 37$

**Find each sum.** *(EXAMPLE 2)*

**8.** $-2.35 + (-1.76)$ **9.** $-5.77 + 12.5$ **10.** $-5\frac{3}{4} + 1\frac{1}{4}$

## Practice and Apply

internet connect
**Homework Help Online**
Go To: **go.hrw.com**
Keyword:
**MA1 Homework Help**
for Exercises 11-54

**Use algebra tiles to find the following sums:**

**11.** $-5 + (-2)$ **12.** $-3 + 3$ **13.** $2 + (-6)$ **14.** $-2 + 9$

**15.** $7 + (-3)$ **16.** $-2 + (-6)$ **17.** $8 + (-5)$ **18.** $6 + (-6)$

**Find each sum.**

**19.** $-7 + (-15)$ **20.** $-10 + 7$ **21.** $23 + (-15)$

**22.** $28 + (-50)$ **23.** $5.7 + (-3.2)$ **24.** $-17 + (-34)$

**25.** $\frac{4}{17} + \left(-\frac{12}{17}\right)$ **26.** $84 + (-18)$ **27.** $-29 + 14$

**28.** $-43 + (-82)$ **29.** $-308 + (-80)$ **30.** $-56 + 107$

**31.** $-5 + (-5)$ **32.** $-5 + 5$ **33.** $-13 + 20$

**34.** $13 + (-20)$ **35.** $2.85 + 4.96$ **36.** $-5.3 + 1.4$

**37.** $6.98 + (-0.2)$ **38.** $0.17 + (-3.45)$ **39.** $-0.5 + 0.25$

**40.** $0.26 + (-1.5)$ **41.** $3.7 + (-0.7)$ **42.** $5.6 + (-3.2)$

**43.** $-8.27 + (-3.2)$ **44.** $5.23 + (-2.7)$ **45.** $-3.4 + 7$

**46.** $-14 + 2.5$ **47.** $-\frac{4}{15} + \left(-\frac{7}{15}\right)$ **48.** $\frac{11}{25} + \left(-\frac{13}{25}\right)$

**49.** $-\frac{1}{2} + \frac{1}{3}$ **50.** $-\frac{3}{4} + \frac{1}{5}$ **51.** $-1\frac{2}{5} + 3\frac{1}{4}$

**52.** $2\frac{1}{4} + \left(-5\frac{2}{5}\right)$ **53.** $\frac{11}{3} + \left(-\frac{11}{3}\right)$ **54.** $-\frac{3}{2} + \left(-\frac{10}{7}\right)$

**Let $a = 2$, $b = 3$, and $c = -5$. Evaluate each expression.**

**55.** $a + b$ **56.** $a + (-b)$ **57.** $-a + b$ **58.** $-a + (-b)$

**59.** $a + c$ **60.** $a + (-c)$ **61.** $-b + c$ **62.** $-b + (-c)$

**63.** $a + b + c$ **64.** $(a + b) + (-a) + (-b)$ **65.** $-c + (-b) + (a) + (-b)$

CHALLENGE

**66.** $2(a^2 - b) + 3c - 5ab$

CONNECTION

**67. STATISTICS** Suppose that a quarterback had the following plays in a football game: a pass for a gain of 7 yards, a sack for a loss of 2 yards, a sneak for a gain of 6 yards, and a sack for a loss of 13 yards.
**a.** What was the total yardage gained or lost on these four plays?
**b.** Suppose that the next two plays were a gain of 24 yards and then a loss of 17 yards. What was the total yardage for all six plays?

APPLICATIONS

**68. BANKING** Jeff opened a savings account with \$57. Later, he deposited another \$32. Finally, he made a withdrawal of \$45 in order to buy some new CDs. What was the remaining balance in his savings account?

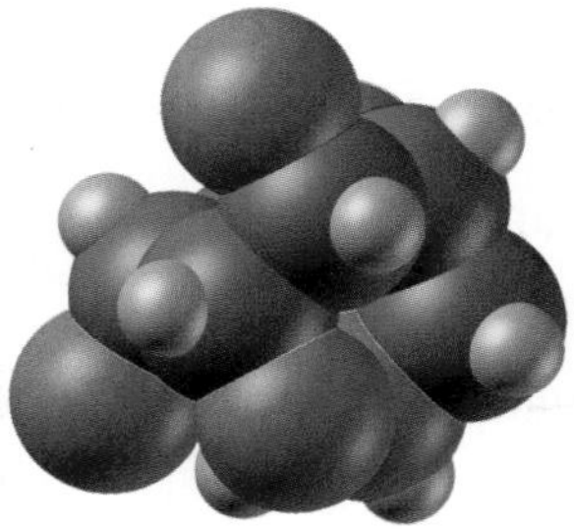
Glucose, $C_6H_{12}O_6$

**69. CHEMISTRY** Protons have a charge of +1 and electrons have a charge of −1. The total charge of a particle is the sum of its proton charges and electron charges. Find the total charge of a particle with the following numbers of protons and electrons:
**a.** 12 protons, 10 electrons
**b.** 16 protons, 18 electrons
**c.** 10 protons, 10 electrons

## Look Back

**70.** Find the sum $1 + 2 + 3 + \cdots + 40$. ***(LESSON 1.1)***

**Find the next two terms in each number sequence.** ***(LESSON 1.1)***

**71.** 40, 37, 34, 31, . . .  **72.** −1, 3, 7, . . .  **73.** 1, 3, 6, 10, . . .

**74.** 5, 11, 18, 26, . . .  **75.** 2, 4, 8, 16, . . .  **76.** 2, 4, 7, 11, . . .

**Evaluate.** ***(LESSON 1.3)***

**77.** $15 \cdot 2 - 7$  **78.** $[3(4 - 2)^2] + 7$  **79.** $12 + 3^2 + (9 - 6)$

**80.** $5^2 + 2^2 + 4^2$  **81.** $(7 + 3 \cdot 2) - 2^2$  **82.** $100 - 99 \div 33 \div 3$

**Plot each point on the same coordinate plane.** ***(LESSON 1.4)***

**83.** $A(4, 4)$  **84.** $B(5, 2)$  **85.** $C(3, 4)$  **86.** $D(6, 3)$

**Find the absolute value of each number.** ***(LESSON 2.1)***

**87.** 17  **88.** −1.2  **89.** $-\frac{13}{20}$  **90.** $(12.8 - 5)$

## Look Beyond

**91.** You may recall from arithmetic that adding the same number repeatedly can be represented by multiplication and vice versa.

$$2 + 2 + 2 + 2 + 2 = 5(2) = 10$$

With this in mind, represent $7(-2)$ as an addition problem, and then find the value of $7(-2)$.

**PORTFOLIO ACTIVITY**

Include this activity in your portfolio.

The table below comes from the quarterly spreadsheet that a business prints out. Study the spreadsheet and answer the questions below.

| | Revenue | Expenses | Profit |
|---|---|---|---|
| **January** | 25, 626 | 12, 810 | 12, 816 |
| **February** | 10, 887 | 14, 098 | −3211 |
| **March** | 12, 978 | 12, 995 | −17 |
| **TOTAL** | 49, 491 | 39, 903 | 9588 |

**1.** Explain how the numbers in the profit column are calculated.

**2.** Why are some of the numbers in the profit column negative?

**3.** Explain how 9588 in the profit column can be calculated in two different ways.

# 2.3 Subtracting Real Numbers

## Objectives

- Use algebra tiles to model subtraction.
- Define subtraction in terms of addition.
- Subtract numbers with like and unlike signs.

**Why** *In algebra, the properties of the basic operations are defined in terms of addition and multiplication. You must understand how to convert subtraction to addition in order to use the properties when you solve equations.*

**APPLICATION**
**PERSONAL FINANCE**

When Mark uses his debit card, the bank subtracts, or debits, his account by the amount of his purchase. Suppose that he starts with \$50 in his account and uses his debit card for a \$20 purchase. This transaction can be represented in two ways.

Subtraction: $50 - 20$ Addition: $50 + (-20)$

According to either method, Mark will have \$30 left.

$50 - 20 = 30$ $50 + (-20) = 30$

In algebra, it is often useful to think of subtraction as *adding the opposite* of the number that is being subtracted, as shown above. Let's first consider the more familiar idea of subtraction as "taking away."

Using algebra tiles, you can model a subtraction problem such as $6 - 2$ by beginning with 6 tiles and taking away 2 tiles.

How could you model $2 - 6$ by taking away tiles? The answer involves neutral pairs.

**EXAMPLE 1** **Use algebra tiles to model the subtraction $2 - 6$.**

**SOLUTION**

Start with 2 positive tiles.

Since you want to subtract 6, you will need 4 more positive tiles than you already have. Add 4 neutral pairs (4 positive and 4 negative tiles). Remember that neutral pairs do not change the value.

Now you can subtract, or *take away*, 6 positive tiles. There will be 4 negative tiles left.

$$2 - 6 = -4$$

**TRY THIS** Use algebra tiles to model the subtraction $4 - 10$.

## Subtraction as Adding the Opposite

The bank account example on page 67 showed that a subtraction problem can be represented by an addition problem. The addition problem $50 + (-20)$ has the same answer as the subtraction problem $50 - 20$.

**Definition of Subtraction**

For all real numbers $a$ and $b$:

$$a - b = a + (-b)$$

*Activity*

### Modeling the Definition of Subtraction

**You will need:** algebra tiles

1. Suppose that you want to *subtract* 3 positive tiles from some number that is represented by another group of tiles (not shown). No matter what the other number is, you can make sure that there will be enough tiles to subtract if you add 3 neutral pairs to the group.

2. After you have added the 3 neutral pairs, what will the next step be?
3. What is the overall effect of adding 3 neutral pairs and removing 3 positive tiles? **CHECKPOINT ✔**
4. State how the steps in this activity show that subtracting a given number is the same as adding the opposite of that number.

**EXAMPLE 2** **Use the definition of subtraction to find each difference.**

**a.** $3 - (-4)$ **b.** $-5.6 - (-1.4)$

**c.** $-1\frac{2}{3} - 4\frac{1}{3}$ **d.** $5 - 2.3$

**SOLUTION**

To subtract a number, find its opposite, and then add.

**a.** The opposite of $-4$ is 4.

$3 - (-4) = 3 + 4 = 7$

**b.** The opposite of $-1.4$ is 1.4.

$-5.6 - (-1.4) = -5.6 + 1.4 = -4.2$

**c.** The opposite of $4\frac{1}{3}$ is $-4\frac{1}{3}$.

$-1\frac{2}{3} - 4\frac{1}{3} = -1\frac{2}{3} + \left(-4\frac{1}{3}\right) = -6$

**d.** The opposite of 2.3 is $-2.3$.

$5 - 2.3 = 5 + (-2.3) = 2.7$

**TRY THIS** Use the definition of subtraction to find each difference.

**a.** $-11 - \left(-5\frac{1}{2}\right)$ **b.** $-2.7 - 5.3$ **c.** $2 - 5$ **d.** $8.2 - 6$

## Measuring Distance on a Number Line

Sometimes you will need to determine the distance between points on a number line. One method of doing this uses absolute value.

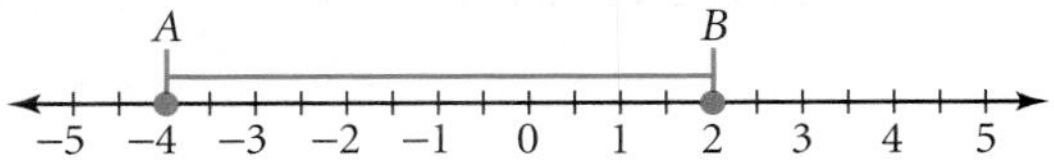


To find the distance between points *A* and *B*, subtract the values, and then find the absolute value of the difference.

$$\text{The distance between } -4 \text{ and } 2 \text{ is } |2 - (-4)| = |6| = 6,$$
$$\text{or } |(-4) - 2| = |-6| = 6.$$

In general, if you let *a* represent the value for point *A* and let *b* represent the value for point *B*, the distance between points *A* and *B* on a number line is $|b - a|$ or $|a - b|$.

**EXAMPLE 3** **What is the distance between $-5$ and 3 on a number line?**

**SOLUTION**

The distance between $-5$ and 3 is

$$|-5 - 3| = |-8| = 8,$$
$$\text{or } |3 - (-5)| = |8| = 8.$$

**TRY THIS**

**a.** Find the distance between $-2$ and 13 on a number line.

**b.** Find the distance between $-11\frac{3}{4}$ and $-5\frac{1}{4}$ on a number line.

## Closure

The sum of any two real numbers is a real number. To mathematicians, this means that the set of real numbers is *closed under addition*. The set of real numbers is also closed under subtraction. As a comparison, consider the whole numbers. Remember that the whole numbers consist of only the positive integers and zero. Are the whole numbers closed under addition and subtraction?

Adding two positive integers, adding a positive integer and zero, and adding zero and zero all result in a whole number. Therefore, the whole numbers are closed under addition.

$2 + 4 = 6$ $\quad$ $3 + 0 = 3$ $\quad$ $0 + 0 = 0$

*The sums are all whole numbers.*

What about subtraction? You cannot determine the value of a subtraction expression such as $4 - 8$ without using negative integers.

$4 - 8 = -4$ $\quad$ *Not a whole number*

Because the whole numbers do not include the negative integers, the whole numbers are not closed under subtraction.

# Exercises

### Communicate

1. How does the addition of neutral pairs of algebra tiles model a subtraction problem?

APPLICATION

2. **METEOROLOGY** The temperature is 5°F. Explain how to find the temperature if it dropped by 25°F.
3. Describe how you would solve the problem $-4 - 7$ by using algebra tiles.
4. What is meant by "adding the opposite"?
5. Explain why the real numbers are closed under subtraction.

### Guided Skills Practice

**Find each difference.** ***(EXAMPLES 1 AND 2)***

**6.** $-8 - 5$ **7.** $-9 - (-11)$ **8.** $-5 - 14$ **9.** $6 - (-8)$

APPLICATION

**10.** **NAVIGATION** A submarine at a depth of 98 feet descends another 89 feet. What is the new depth of the submarine? ***(EXAMPLES 1 AND 2)***

**Find the distance between each pair of points on a number line.** ***(EXAMPLE 3)***

**11.** 3, −7 **12.** −5, −6 **13.** 13, −13 **14.** −3, 13 **15.** 8, −6

## Practice and Apply

**internet connect**
**Homework Help Online**
Go To: **go.hrw.com**
Keyword:
**MA1 Homework Help** for Exercises 16-47

**Find each difference.**

**16.** $12 - 7$ **17.** $-9 - 4$ **18.** $-15 - (-12)$ **19.** $11 - (-3)$

**20.** $-10 - 7$ **21.** $24 - (-11)$ **22.** $-86 - (-92)$ **23.** $-57 - 14$

**24.** $-117 - 82$ **25.** $84.2 - (-12.5)$ **26.** $-65.4 - 32.8$ **27.** $94 - (-16)$

**Find each sum or difference.**

**28.** $67 - 3$ **29.** $42 - (-9)$ **30.** $-10 - (-21)$ **31.** $-35 - 17$

**32.** $33 - (-33)$ **33.** $-78 + (-45)$ **34.** $990 - (-155)$ **35.** $-97 - 88$

**36.** $-43 + (-15)$ **37.** $-77 - (-77)$ **38.** $-108 + 118$ **39.** $85 - (-12)$

**40.** $-2.05 - 30.4$ **41.** $1.08 - (-6.79)$ **42.** $-0.012 - 0.65$ **43.** $64.5 - 65$

**44.** $\frac{1}{5} - \frac{4}{5}$ **45.** $\frac{3}{16} - \left(-\frac{7}{16}\right)$ **46.** $-\frac{1}{2} - \frac{1}{3}$ **47.** $\frac{3}{8} - \left(-\frac{3}{4}\right)$

**Let $x = 5$, $y = -3$, and $z = -10$. Evaluate each expression.**

**48.** $x - y$ **49.** $-x - y$ **50.** $-x - z$ **51.** $x - (-x)$

**52.** $y - z$ **53.** $-z - y$ **54.** $-z - (-y)$ **55.** $x - (-y)$

**Find the distance between each pair of points on a number line.**

**56.** 4, 9 **57.** −6, 15 **58.** −47, −23 **59.** −12, 74

**60.** −17, 5 **61.** 12, 39 **62.** −3, 52 **63.** −86, −23

**CONNECTIONS**

**64. STATISTICS** Susan, a marathon athlete, wants to improve her time in the Boston Marathon by 12 minutes this year.

**a.** If her finish time last year was 2 hours and 7 minutes, what is her goal for this year?

**b.** Suppose that at the race Susan manages to improve her time by 14 minutes over the previous year. What is her new finish time?

**65. PATTERNS IN DATA** Use constant differences to find the next three terms in each sequence.

**a.** −10, −7, −4, −1, . . . **b.** 60, 45, 30, 15, . . . **c.** −20, −11, −3, 4, . . .

APPLICATIONS

**66.** INVENTORY The school cafeteria had 27 cases of juice before lunch. After lunch, there were 14 cases of juice left. How many cases of juice were used during lunch?

**67.** TEMPERATURE Theresa noticed that the thermostat in her freezer was set at 5°F. She lowered the thermostat by 7°F. Later, she lowered the thermostat another 2°F. At what temperature was the thermostat finally set?

**68.** BANKING Millie's checking account balance is \$205 on the 13th of the month. She doesn't get paid again until the 1st of the following month. Millie has to pay a cable bill of \$42.50 and a phone bill of \$37.23.

**a.** Once these two bills are paid, how much money is left for Millie to spend this month?

CHALLENGE

**b.** If Millie usually withdraws \$20 for spending money every few days, how many times during the rest of the month can she do this if she has no other bills to pay? If she has withdrawn spending money twice and then withdraws \$50 for a special occasion on the 21st of the month, how much money will Millie have left?

## Look Back

internet connect

**Activities Online**

Go To: **go.hrw.com**
Keyword:
**MA1 Stocks**

**Evaluate.** *(LESSON 1.3)*

**69.** $36 - 12 \div 3 - 20$

**70.** $3 \cdot 5 + 7 \div 2$

**71.** $28 \div 2 \cdot 7 + 4$

**72.** $100 - 99 \div 3$

**73.** $(8^2 + 3 \cdot 2) - 16 \cdot 2$

**74.** $100 \div [5 - (2 \cdot 3 - 6)]$

**Plot each set of ordered pairs on a coordinate plane and tell whether the points lie on a straight line.** *(LESSON 1.4)*

**75.** (4, 11), (1, 5), (3, 9) **76.** (1, −3), (7, 2), (5, 6) **77.** (4, 9), (2, 8), (8, 11)

**Find each sum.** *(LESSON 2.2)*

**78.** $-18 + (-23)$

**79.** $-95 + 78$

**80.** $-31 + (-27)$

**81.** $53 + (-29)$

**82.** $-3.2 + (-2.7)$

**83.** $5 + (-3.6)$

**84.** $-10.11 + (-3.9)$

**85.** $3.7 + (-5.9)$

## Look Beyond

**86.** Let $x$ represent an unknown point on a number line. If the distance between $x$ and 4 on the number line is 7, what number(s) does $x$ represent? (Hint: Draw a number line, and try different values of $x$.)

**87.** What number(s) does $x$ represent if the distance between $x$ and 4 is 10?

2.4

# Multiplying and Dividing Real Numbers

**Why** *In business, measures of gain and loss are reported with positive and negative numbers. Calculations with this data often involve multiplying and dividing. When calculating with positive and negative numbers, it is important to know what sign the result will have.*

## Objectives

- Multiply and divide positive and negative numbers.
- Define the Properties of Zero.

**APPLICATION**
**BANKING**

A college student is planning her budget based on her bank statements. She is studying the role that her monthly sources of income and her expenditures play in her final balance. Examine what would happen if she made the following changes:

| Change | Effect | Representation | Result |
|---|---|---|---|
| Begin new part-time job | Add 2 deposits of \$100 | $(2)(100) = 200$ | Increase of \$200 |
| Take bus to new job | Add 15 withdrawals of \$1 | $(15)(-1) = -15$ | Decrease of \$15 |
| Quit old part-time job | Remove 2 deposits of \$50 | $(-2)(50) = -100$ | Decrease of \$100 |
| Eat out less often | Remove 2 withdrawals of \$15 | $(-2)(-15) = 30$ | Increase of \$30 |

## *Activity*
## Multiplication and Division Patterns

### Part I

1. Complete each pattern.

| a. | b. | c. |
|---|---|---|
| $(2)(3) = 6$ | $(3)(3) = 9$ | $(3)(-3) = -9$ |
| $(2)(2) = 4$ | $(2)(3) = 6$ | $(2)(-3) = -6$ |
| $(2)(1) = 2$ | $(1)(3) = 3$ | $(1)(-3) = -3$ |
| $(2)(0) = 0$ | $(0)(3) = 0$ | $(0)(-3) = 0$ |
| $(2)(-1) = \underline{?}$ | $(-1)(3) = \underline{?}$ | $(-1)(-3) = \underline{?}$ |
| $(2)(-2) = \underline{?}$ | $(-2)(3) = \underline{?}$ | $(-2)(-3) = \underline{?}$ |
| $(2)(-3) = \underline{?}$ | $(-3)(3) = \underline{?}$ | $(-3)(-3) = \underline{?}$ |

**CHECKPOINT ✔**

2. Complete each statement.

**a.** $(+) \cdot (+) = \underline{\quad + \quad}$  **b.** $(+) \cdot (-) = \underline{\quad ? \quad}$

**c.** $(-) \cdot (+) = \underline{\quad ? \quad}$  **d.** $(-) \cdot (-) = \underline{\quad ? \quad}$

**Part II**

1. Find each product. Examine all of the signs.
   a. $(8)(7) = \underline{\quad ? \quad}$  b. $(6)(-5) = \underline{\quad ? \quad}$
   c. $(-4)(2) = \underline{\quad ? \quad}$  d. $(-8)(-1) = \underline{\quad ? \quad}$
2. Multiplication is related to division. You can write related division facts for any multiplication fact. For example, the related division facts for $6 \cdot 5 = 30$ are $30 \div 5 = 6$ and $30 \div 6 = 5$. Write the related division facts for each of the multiplication facts in Step 1. Compare the signs in the related division and multiplication facts.

CHECKPOINT ✓

3. Complete each statement.
   a. $(+) \div (+) = \underline{\quad ? \quad}$  b. $(+) \div (-) = \underline{\quad ? \quad}$
   c. $(-) \div (+) = \underline{\quad ? \quad}$  d. $(-) \div (-) = \underline{\quad ? \quad}$

When multiplying or dividing signed numbers, follow the rules below.

### Rules for Multiplying Two Signed Numbers

**Multiplying Two Numbers With Like Signs**

1. Find the product of the absolute values of the numbers.
2. Write the product as a positive number.

$(+) \cdot (+) = (+) \qquad (-) \cdot (-) = (+)$

**Multiplying Two Numbers With Unlike Signs**

1. Find the product of the absolute values of the numbers.
2. Write the product as a negative number.

$(+) \cdot (-) = (-) \qquad (-) \cdot (+) = (-)$

### Rules for Dividing Two Signed Numbers

**Dividing Two Numbers With Like Signs**

1. Find the quotient of the absolute values of the numbers.
2. Write the quotient as a positive number.

$(+) \div (+) = (+) \qquad (-) \div (-) = (+)$

**Dividing Two Numbers With Unlike Signs**

1. Find the quotient of the absolute values of the numbers.
2. Write the quotient as a negative number.

$(+) \div (-) = (-) \qquad (-) \div (+) = (-)$

**EXAMPLE 1** **Find each product.**

**a.** $(9)(-4)$ **b.** $(-4.5)(-2)$ **c.** $(-5)(1.43)$

**SOLUTION**

**a.** $(9)(-4) = -36$ *The answer is negative.*
**b.** $(-4.5)(-2) = 9$ *The answer is positive.*
**c.** $(-5)(1.43) = -7.15$ *The answer is negative.*

**TRY THIS** Find each product.

**a.** $(8)(-6)$ **b.** $(-4)(-5.3)$ **c.** $(-7)(2.6)$

**EXAMPLE 2** **Find each quotient.**

**a.** $\frac{9}{-3}$ **b.** $(-4.5) \div (-1.5)$ **c.** $\frac{-50}{25}$

**SOLUTION**

**a.** $(9) \div (-3) = -3$ *The answer is negative.*
**b.** $(-4.5) \div (-1.5) = 3$ *The answer is positive.*
**c.** $(-50) \div (25) = -2$ *The answer is negative.*

**TRY THIS** Find each quotient.

**a.** $\frac{16}{-4}$ **b.** $(-3.2) \div (-0.8)$ **c.** $\frac{-25}{5}$

## The Multiplicative Identity and Reciprocals

The product of any number and 1 is equal to the number. Therefore, 1 is called the **identity element for multiplication**.

### Identity Property for Multiplication

For all real numbers $a$, $a \cdot 1 = a$ and $1 \cdot a = a$.

Related to the identity element, 1, are **multiplicative inverses**, or **reciprocals**. The product of two reciprocals is 1.

$7 \cdot \frac{1}{7} = 1$ $\frac{2}{3} \cdot \frac{3}{2} = 1$ $-\frac{4}{3} \cdot \left(-\frac{3}{4}\right) = 1$

### Multiplicative Inverse Property

For every nonzero real number $a$, there is exactly one number $\frac{1}{a}$ such that

$$a \cdot \frac{1}{a} = 1 \text{ and } \frac{1}{a} \cdot a = 1.$$

The number $\frac{1}{a}$ is called the reciprocal or multiplicative inverse of $a$.

Another way to divide numbers is to *multiply by reciprocals.* The quotient $12 \div (-3)$ can also be written as the product $12 \cdot \frac{1}{-3}$. The result is the same: $\frac{12}{-3}$, or $-4$.

**EXAMPLE 3** **Find each quotient by using reciprocals.**

**a.** $24 \div (-8)$ **b.** $-4 \div \frac{8}{10}$ **c.** $\frac{4}{5} \div \frac{3}{10}$

**SOLUTION**

**a.** The reciprocal of $-8$ is $\frac{1}{-8}$.

$$24 \div (-8) = 24 \cdot \frac{1}{-8} = \frac{24}{-8} = -3$$

**b.** The reciprocal of $\frac{8}{10}$ is $\frac{10}{8}$.

$$-4 \div \frac{8}{10} = -4 \cdot \frac{10}{8} = \frac{-40}{8} = -5$$

**c.** The reciprocal of $\frac{3}{10}$ is $\frac{10}{3}$.

$$\frac{4}{5} \div \frac{3}{10} = \frac{4}{5} \cdot \frac{10}{3} = \frac{40}{15} = 2\frac{2}{3}$$

**TRY THIS** Find each quotient by using reciprocals.

**a.** $-49 \div -7$ **b.** $\frac{5}{9} \div (-3)$ **c.** $\frac{2}{3} \div \left(-\frac{4}{5}\right)$

## Properties of Zero

Zero is the number that falls between positive numbers and negative numbers on a number line. However, *zero is neither positive nor negative.* There are some special rules for multiplying and dividing with zero.

### Properties of Zero

Let $a$ represent any real number.

1. The product of any real number and zero is zero.

   $a \cdot 0 = 0$ and $0 \cdot a = 0$

2. Zero divided by any nonzero real number is zero.

   $\frac{0}{a} = 0$, where $a \neq 0$

3. Division by zero is undefined. (Division by zero is not possible.)

Zero does not have a reciprocal because any number multiplied by 0 is 0.

**EXAMPLE 4** **Find each product or quotient.**

**a.** $(0)(-4)$ **b.** $\frac{0}{-4}$ **c.** $4.5 \div 0$

**SOLUTION**

**a.** $(0)(-4) = 0$ — *The product of any number and zero is zero.*

**b.** $\frac{0}{-4} = 0$ — *Zero divided by any nonzero number is zero.*

**c.** $4.5 \div 0$ is undefined. — *Division by zero is undefined.*

**TRY THIS** Find each product or quotient.

**a.** $\frac{0}{-8}$ **b.** $\frac{0}{27}$ **c.** $-45 \cdot 0$ **d.** $\frac{17}{0}$

**CRITICAL THINKING** Which one of the Properties of Zero applies to $0 \div 0$?

**TECHNOLOGY GRAPHICS CALCULATOR**

Calculators will indicate an error when you try to divide by zero.

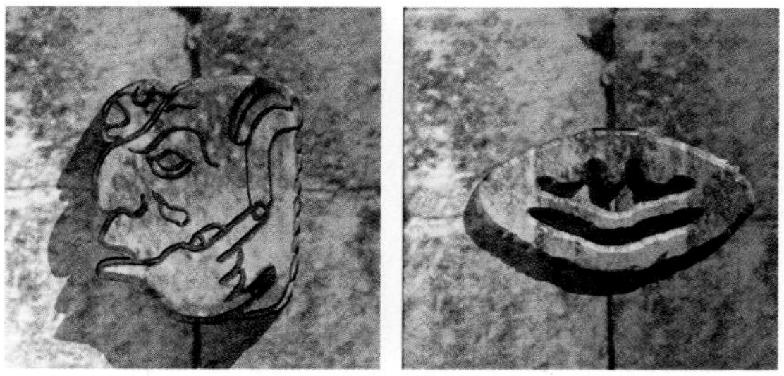

*These are two symbols the Maya used for zero.*

# Exercises

## Communicate

1. Explain the rule for multiplying 7 and $-8$.
2. Explain the rule for dividing $-36$ by $-9$.
3. Explain what a reciprocal is.
4. The multiplication equation for $\frac{6}{0} = n$ would be $0 \cdot n = 6$ or "zero times the number $n$ is 6." What is the value of $n$? Why?
5. Explain how multiplying by the reciprocal in order to divide is similar to adding the opposite in order to subtract.

## Guided Skills Practice

**Find each product.** ***(EXAMPLE 1)***

**6.** $(-3)(-8)$ **7.** $(-5)(-6)$ **8.** $(-15)(9)$ **9.** $(-3)(17)$

**Find each quotient.** ***(EXAMPLES 2 AND 3)***

**10.** $\frac{15}{-3}$ **11.** $-42 \div 6$ **12.** $\frac{-18}{-6}$ **13.** $\frac{3}{7} \div \frac{6}{14}$

**Find each product or quotient.** ***(EXAMPLE 4)***

**14.** $7 \cdot 0$ **15.** $0 \cdot (-9)$ **16.** $\frac{-3}{0}$ **17.** $\frac{0}{3}$

## Practice and Apply

internet connect
**Homework Help Online**
Go To: **go.hrw.com**
Keyword:
**MA1 Homework Help**
for Exercises 18-39

**Find each product or quotient.**

**18.** $(-4)(6)$ **19.** $(5)(-7)$ **20.** $(-14) \div (-2)$ **21.** $(12)(-11)$

**22.** $27 \div (-3)$ **23.** $(-17)(-3)$ **24.** $(-21)(-6)$ **25.** $(-17)(6)$

**26.** $115 \div (-5)$ **27.** $(-84) \div 7$ **28.** $(-1.2)(4.8)$ **29.** $(-8)(-8)$

**30.** $(10.7)(-0.5)$ **31.** $8.8 \div (-1.1)$ **32.** $-2.7 \div (-0.3)$ **33.** $(-5.2)(-3.6)$

**34.** $(-23)(56)$ **35.** $(-0.591)(3.6)$ **36.** $(-220.345) \div (-34.7)$

**37.** $\left(\frac{7}{8}\right)\left(-\frac{23}{12}\right)$ **38.** $\frac{412.59}{-17}$ **39.** $\left(\frac{10}{3}\right) \div \left(-\frac{3}{10}\right)$

**Write the reciprocal of each number.**

**40.** $-12$ **41.** $17$ **42.** $\frac{6}{25}$ **43.** $\frac{4}{-9}$

**Evaluate.**

**44.** $\frac{8}{-2}$ **45.** $-22 \div (-1)$ **46.** $(-12)6$ **47.** $-1.2 \div (-6)$

**48.** $\frac{3.4}{-17}$ **49.** $(-5)(6)(-6)$ **50.** $-(-8)(-6)$ **51.** $(-4)(-7)$

**52.** $\frac{-8+10}{-2}$ **53.** $\frac{-4+7}{-6}$ **54.** $\frac{-225}{-5}$ **55.** $(-47)(23)$

**56.** $\frac{-1208}{24}$ **57.** $\frac{(7)(-1)}{(-7)}$ **58.** $\frac{(-6)(-12)}{3}$ **59.** $96 \div \left(-\frac{2}{3}\right)$

**Kammi says, "Two negatives always make a positive." State whether this is true for the following operations:**

**60.** addition **61.** subtraction

**62.** multiplication **63.** division

CONNECTION

**64.** STATISTICS Suppose that a football team scored 27 touchdowns in a 9-game season and made the two-point conversion each time.
- **a.** How many points came from touchdowns? from two-point conversions?
- **b.** How many total points did they score?
- **c.** If they scored the same number of points per game, how many points were scored in one game?

APPLICATION

**65.** BANKING Juan opened a savings account with a $20 deposit. He made 4 more deposits of $20 each and 5 withdrawals of $10 each.
- **a.** What was the amount of Juan's deposits?
- **b.** What was the amount of his withdrawals?
- **c.** If Juan wants to save another $200, how many $25 deposits must he make?

**66.** BUSINESS A pet store owner is dividing a 50-pound sack of birdseed into smaller sacks to sell individually.
- **a.** How many $3\frac{1}{3}$-pound bags are there in one large sack?
- **b.** How many $1\frac{2}{3}$-pound bags are there in one large sack?

## Look Back

**Find the second differences for each sequence.** *(LESSON 1.1)*

**67.** 1, 3, 7, 13, 21, . . . **68.** 17, 16, 14, 11, . . . **69.** $-24, -20, -12, 0, \ldots$

**Evaluate.** *(LESSON 1.3)*

**70.** $14 \cdot 2 - 7$ **71.** $15 \div 3 - 2 \cdot 2$ **72.** $3^2 \cdot 2 + 14 \div 7$

**Find each sum or difference.** *(LESSONS 2.2 AND 2.3)*

**73.** $-23 + (-51)$ **74.** $-17 - (-17)$ **75.** $-18 + (-83)$ **76.** $-32 - (-5)$

**Let $q = 8$, $r = -2$, and $s = -5$. Evaluate each expression.** *(LESSON 2.3)*

**77.** $q - r$ **78.** $-q - r$ **79.** $-r - (-s)$ **80.** $s - (-r)$

## Look Beyond

**Rewrite each expression in a simpler form.**

**81.** $\frac{3w + 24}{-3}$ **82.** $\frac{-5y - 45}{5}$ **83.** $\frac{6x + 12}{2}$

**internet connect**

**Portfolio Extension**

Go To: **go.hrw.com**
Keyword:
**MA1 TEMP Extremes**

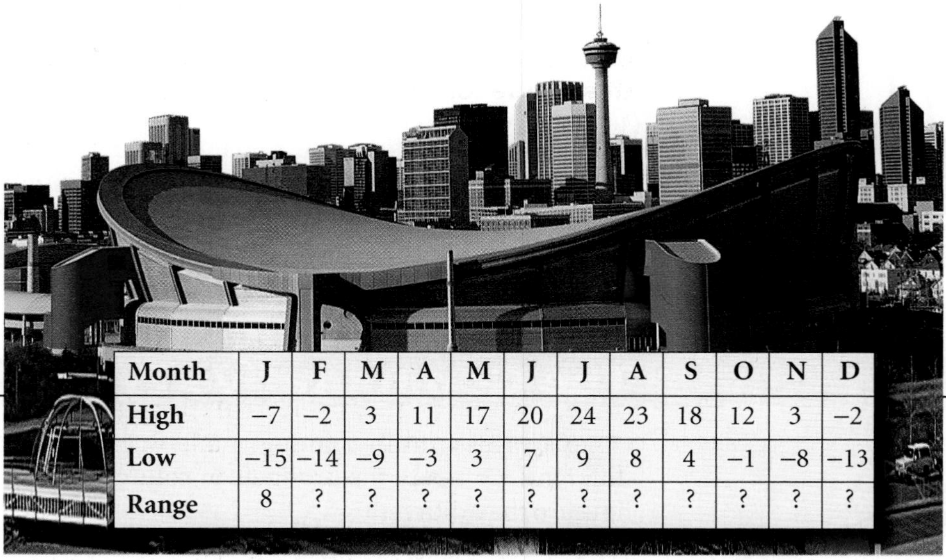

| Month | J | F | M | A | M | J | J | A | S | O | N | D |
|---|---|---|---|---|---|---|---|---|---|---|---|---|
| High | −7 | −2 | 3 | 11 | 17 | 20 | 24 | 23 | 18 | 12 | 3 | −2 |
| Low | −15 | −14 | −9 | −3 | 3 | 7 | 9 | 8 | 4 | −1 | −8 | −13 |
| Range | 8 | ? | ? | ? | ? | ? | ? | ? | ? | ? | ? | ? |

**PORTFOLIO ACTIVITY**

Include this activity in your portfolio.

Average high and low temperatures are part of the data recorded about a region's climate. The table above shows the average high and low temperatures in Calgary, Alberta, Canada, in degrees Celsius.

1. The range is the difference between the average high and the average low for a given period. Copy and complete the table by filling in the range data.
2. Use points on a number line or thermometer to show that the smallest range is in January.
3. Display the data as a double line graph with the average highs on one line and the average lows on the other line. (Note: The data do not represent linear equations.)

**WORKING ON THE CHAPTER PROJECT**

You should now be able to complete Activities 1 and 2 of the Chapter Project on page 104.

# 2.5 Properties and Mental Computation

**Objectives**

- State and apply the Commutative, Associative, and Distributive Properties.
- Use the Commutative, Associative, and Distributive Properties to perform mental computations.

**Why** *In algebra, it is necessary to rearrange and rewrite expressions in order to solve equations. The Commutative, Associative, and Distributive Properties enable you to do this. The properties are also useful when performing mental computations.*

**APPLICATION**
**INVENTORY**

Felicia works at a river-tube rental company. Each morning she checks the inventory of tubes stored in three sheds. She counts 44, 27, and 56 tubes and quickly calculates the total, 127, mentally. How do you think Felicia used mental computation to calculate the sum 44 + 27 + 56?

## The Commutative Properties

Compare the sums in columns A and B. How are they alike and how are they different? Compare the products in columns C and D. How are they alike and how are they different?

**Sums**

| Column A | Column B |
|---|---|
| $56 + 11 = 67$ | $11 + 56 = 67$ |
| $28 + 32 = 60$ | $32 + 28 = 60$ |
| $13 + 87 = 100$ | $87 + 13 = 100$ |

**Products**

| Column C | Column D |
|---|---|
| $6 \cdot 11 = 66$ | $11 \cdot 6 = 66$ |
| $8 \cdot 9 = 72$ | $9 \cdot 8 = 72$ |
| $10 \cdot 54 = 540$ | $54 \cdot 10 = 540$ |

Columns A and B illustrate the **Commutative Property of Addition**. Changing the order of the numbers in an addition expression does not change the sum.

Columns C and D illustrate the **Commutative Property of Multiplication**. Changing the order of the numbers in a multiplication expression does not change the product.

Changing the order of two terms in an expression is known as *commuting the terms.*

**Commutative Property of Addition**

For all real numbers $a$ and $b$:

$$a + b = b + a$$

**Commutative Property of Multiplication**

For all real numbers $a$ and $b$:

$$a \cdot b = b \cdot a$$

**CRITICAL THINKING** Do you think that any two numbers can be divided in either order without changing the quotient? Explain.

## The Associative Properties

Compare the sums in columns A and B. How are they alike and how are they different? Compare the products in columns C and D. How are they alike and how are they different?

**Sums**

| Column A | Column B |
|---|---|
| $(12 + 8) + 15 = 35$ | $12 + (8 + 15) = 35$ |
| $(39 + 4) + 16 = 59$ | $39 + (4 + 16) = 59$ |
| $(24 + 36) + 7 = 67$ | $24 + (36 + 7) = 67$ |

**Products**

| Column C | Column D |
|---|---|
| $(2 \cdot 5) \cdot 16 = 160$ | $2 \cdot (5 \cdot 16) = 160$ |
| $(20 \cdot 5) \cdot 13 = 1300$ | $20 \cdot (5 \cdot 13) = 1300$ |
| $(3 \cdot 11) \cdot 10 = 330$ | $3 \cdot (11 \cdot 10) = 330$ |

Columns A and B illustrate the **Associative Property of Addition.** Changing the grouping of the numbers in an addition expression does not change the sum.

Columns C and D illustrate the **Associative Property of Multiplication**. Changing the grouping of the numbers in a multiplication expression does not change the product.

**The Associative Property of Addition**

For all real numbers $a$, $b$, and $c$:

$$(a + b) + c = a + (b + c)$$

**The Associative Property of Multiplication**

For all real numbers $a$, $b$, and $c$:

$$(a \cdot b) \cdot c = a \cdot (b \cdot c)$$

**EXAMPLE 1** At the beginning of this lesson, Felicia used mental computation to calculate the number of tubes stored in three sheds. **Show how the Commutative and Associative Properties are used to regroup the numbers to make the sum easier to compute mentally.**

**SOLUTION**

$$\begin{aligned} &(44 + 27) + 56 \\ &= (27 + 44) + 56 && \textit{Commutative Property of Addition} \\ &= 27 + (44 + 56) && \textit{Associative Property of Addition} \\ &= 27 + 100 \\ &= 127 \end{aligned}$$

## The Distributive Property

The **Distributive Property** involves addition, subtraction, and multiplication. An important property of numbers is that multiplication can be *distributed* over addition and subtraction.

The Distributive Property works in two directions. To find the product $3(5 + 2)$, multiplication is distributed over addition.

$$3(5 + 2) = 3 \cdot 5 + 3 \cdot 2$$

When used in reverse, a common factor can be removed from the terms.

$$3 \cdot 5 + 3 \cdot 2 = 3(5 + 2)$$

The Distributive Property works similarly for multiplication over subtraction. For example:

$$4(7 - 3) = 4 \cdot 7 - 4 \cdot 3$$

And, in reverse:

$$4 \cdot 7 - 4 \cdot 3 = 4(7 - 3)$$

**The Distributive Property of Multiplication Over Addition and Subtraction**

For all real numbers $a$, $b$, and $c$:

$a(b + c) = ab + ac$ and $(b + c)a = ba + ca$

and

$a(b - c) = ab - ac$ and $(b - c)a = ba - ca$

You can use the Commutative, Associative, and Distributive Properties to rearrange expressions for quick mental computation.

## *Activity*
## Exploring Mental Computation

1. Examine the table below, reading each row from left to right. Copy and complete the table.

| Column A | Column B | Column C | Column D |
|---|---|---|---|
| 4(15) | 4(10 + 5) | 4 • 10 + 4 • 5 | 40 + 20 = 60 |
| 8(17) | 8(10 + 7) | 8 • 10 + 8 • 7 | ? |
| 5(22) | 5(10 + 10 + 2) | ? | ? |
| 12(25) | (10 + 2)25 | 10 • 25 + 2 • 25 | ? |
| 7(21) | 7(20 + 1) | ? | ? |
| 6(128) | ? | ? | ? |
| 8(452) | ? | ? | ? |
| 11(3651) | ? | ? | ? |

2. Describe the process shown in each row from left to right.

CHECKPOINT ✔

3. Based on the process that you described in Step 2, explain how the Distributive Property can be used in the mental computation of products.

**CRITICAL THINKING**

Explain how knowing multiplication facts, such as $3 \times 8 = 24$, by memory helps you to do mental computation. Describe the advantages of using mental computation instead of relying on a calculator.

**EXAMPLE 2**

**APPLICATION**
**RECREATION**

John rents a river tube for 8 hours at $3.50 per hour. **Use the Distributive Property to find the total cost.**

**SOLUTION**

The total cost is modeled by the expression 8(3.5). Use the Distributive Property to find the sum of two products.

$$\begin{aligned} 8(3.5) &= 8(3) + 8(0.5) \\ &= 24 + 4 \\ &= 28 \end{aligned}$$

The total cost is $28.

**TRY THIS**

Use the Distributive Property to find the cost of renting a river tube for 8 hours at $4.75 per hour. Find the cost of renting a river tube for 7 hours at $5.50 per hour.

**CRITICAL THINKING**

Use the Distributive Property to explain why $-(4 - 7) = -4 + 7$ and $-(-5 + 6) = 5 - 6$.

## Opposites of Numbers and Expressions

You can find the opposite of a number by multiplying the number by $-1$. For example, $-1 \cdot 8 = -8$, and $-1 \cdot -8 = 8$. You can find the opposite of an addition or subtraction expression in a similar way by using the Distributive Property.

**EXAMPLE 3** **Find the opposite of each expression.**

**a.** $a + b$ **b.** $x - 8$

**SOLUTION**

Multiply each expression by $-1$.

**a.** $-1(a + b) = (-1)(a) + (-1)(b) = -a + (-b) = -a - b$

**b.** $-1(x - 8) = (-1)(x) - (-1)(8) = -x - (-8) = -x + 8$

**TRY THIS** Find the opposite of each expression.

**a.** $a - b$ **b.** $-x + 8$

## Proving Statements

The table below summarizes the important properties of the real numbers discussed in this chapter.

| THE PROPERTIES BELOW ARE TRUE FOR ALL REAL NUMBERS *a, b,* AND *c.* | Addition | Multiplication |
|---|---|---|
| **Commutative** | $a + b = b + a$ | $ab = ba$ |
| **Associative** | $(a + b) + c = a + (b + c)$ | $(ab)c = a(bc)$ |
| **Identity** | 0 is the identity. $a + 0 = 0 + a = a$ | 1 is the identity. $a \cdot 1 = 1 \cdot a = a$ |
| **Inverse** | $-a$ is the additive inverse of $a$. | For $a \neq 0$, $\frac{1}{a}$ is the multiplicative inverse of $a$. |
| **Distributive** | $a(b + c) = ab + ac$ and $(b + c)a = ba + ca$ | |

The following properties of equality are also true:

**Properties of Equality**

For all real numbers $a$, $b$, and $c$:

**Reflexive Property** $a = a$ (A number is equal to itself.)

**Symmetric Property** If $a = b$, then $b = a$.

**Transitive Property** If $a = b$ and $b = c$, then $a = c$.

**Substitution Property** If $a = b$, then $a$ can be replaced by $b$ and $b$ can be replaced by $a$.

All of the properties of real numbers can be used to prove mathematical statements. In mathematics, a statement that can be proved true is called a **theorem**. In statements of theorems, variables represent real numbers.

**EXAMPLE 4** **Prove: $(a + b) + (-a) = b$. Justify each step.**

**SOLUTION**

**Proof**

| Statement | Reason |
|---|---|
| **1.** $(a + b) + (-a) = (b + a) + (-a)$ | **1.** Commutative Prop. of Add. |
| **2.** $= b + [a + (-a)]$ | **2.** Associative Prop. of Add. |
| **3.** $= b + 0$ | **3.** Additive Inverse Prop. |
| **4.** $= b$ | **4.** Identity Prop. Add. |
| **5.** $(a + b) + (-a) = b$ | **5.** Transitive Prop. of Equality |

**TRY THIS** Prove: $a + (b + c) = (b + a) + c$.

**EXAMPLE 5** **Prove: If $a = b$, then $ac = bc$. Justify each step.**

**SOLUTION**

**Proof**

| Statement | Reason |
|---|---|
| **1.** $a = b$ | **1.** Given |
| **2.** $ac = ac$ | **2.** Reflexive Prop. of Equality |
| **3.** $ac = bc$ | **3.** Substitution Prop. of Equality |

**TRY THIS** Prove: If $a = b$, then $ca = cb$.

# Exercises

## Communicate

**1.** Explain why it is helpful to know the properties discussed in this lesson.

**2.** Give examples of how you can use the Commutative and Associative Properties to help you compute mentally.

**3.** Explain how to use mental computation and the Distributive Property to find each product.

**a.** 8(21) **b.** 11(35) **c.** 14(22)

**4.** River tubes rent for $3.50 per hour. One weekend Bill and Tracy rent a tube for 5 hours on Saturday and 7 hours on Sunday. Explain how to use the Distributive Property to compute the total rental fee for the weekend.

## Guided Skills Practice

**Complete each step, and name the property used.** ***(EXAMPLE 1)***

**5.** $(36 + 15) + 64$
$= (15 + \underline{?}) + 64$ Comm. Prop.
$= 15 + (36 + \underline{?})$ $\underline{?}$ Prop.
$= 15 + \underline{?}$
$= \underline{?}$

**6.** $20 \cdot (72 \cdot 5)$
$= 20 \cdot (\underline{?} \cdot 72)$ $\underline{?}$ Prop.
$= (\underline{?} \cdot 5) \cdot 72$ $\underline{?}$ Prop.
$= \underline{?} \cdot \underline{?}$
$= \underline{?}$

**APPLICATION**

**7. INCOME** Jarrel works 9 hours on Saturday and 6.5 hours on Sunday. He earns $12 per hour. Use the Distributive Property to compute Jarrel's total wages for the weekend. Remember to use the algebraic order of operations and to check your answer. ***(EXAMPLE 2)***

**Find the opposite of each expression.** ***(EXAMPLE 3)***

**8.** $x + y$ **9.** $-a - b$ **10.** $x - y$

**11.** Prove: $a + (b + c) = (a + c) + b$. ***(EXAMPLES 4 AND 5)***

## Practice and Apply

**Complete each step, and name the property used.**

**12.** $(24 + 27) + 56$
$= (27 + \underline{?}) + 56$ Comm. Prop.
$= 27 + (24 + \underline{?})$ $\underline{?}$ Prop.
$= 27 + \underline{?}$
$= \underline{?}$

**13.** $25(7 + 4)$
$= 25 \cdot \underline{?} + \underline{?} \cdot 4$ $\underline{?}$ Prop.
$= \underline{?} + \underline{?}$
$= \underline{?}$

**Use the Associative and Commutative Properties to find each sum or product. Show your work and name the properties used.**

**14.** $(27 + 98) + 73$ **15.** $(45 \cdot 32) \cdot 0$

**16.** $(87 \cdot 5) \cdot 2$ **17.** $50 \cdot (118 \cdot 20)$

**Find the opposite of each expression.**

**18.** $3 + 4x$ **19.** $2y - 9$

**20.** $15x + y - 7$ **21.** $-a + 2b + 8$

**internet connect**
**Homework Help Online**
Go To: **go.hrw.com**
Keyword:
**MA1 Homework Help** for Exercises 22-31

**Use the Distributive Property and mental computation to calculate each product.**

**22.** $4 \cdot 28 = 4(20 + 8)$
$= 4 \cdot 20 + 4 \cdot 8$
$= 80 + 32$
$= \underline{?}$

**23.** $40 \cdot 18 = 40(10 + 8)$
$= 40 \cdot \underline{?} + 40 \cdot \underline{?}$
$= \underline{?} + \underline{?}$
$= \underline{?}$

**24.** $9 \cdot 680 = 9(600 + 80)$
$= \underline{?} \cdot 600 + \underline{?} \cdot 80$
$= \underline{?} + \underline{?}$
$= \underline{?}$

**25.** $70 \cdot 540 = 70(500 + 40)$
$= \underline{?} \cdot \underline{?} + \underline{?} \cdot \underline{?}$
$= \underline{?} + \underline{?}$
$= \underline{?}$

**Use the Distributive Property to rewrite each expression.**

**26.** $9xy - 21xyz$ **27.** $rs + rq$ **28.** $xy - wy$

**29.** $4pq + pr$ **30.** $3de - 15df$ **31.** $35st + 20rs$

**Name the property illustrated. Be specific.**

**32.** $32 + 17 = 17 + 32$ **33.** $13 \cdot 21 - 13 \cdot 9 = 13(21 - 9)$

**34.** $6(4.7 - 2) = 6(4.7) - 6(2)$ **35.** $4(5x) = (4 \cdot 5)x$

**36.** $-8.2(2 + 5.3) = (2 + 5.3)(-8.2)$ **37.** $(6 - 3)5 = 6 \cdot 5 - 3 \cdot 5$

**38.** Prove: $(a + b) + c = (c + b) + a$. Justify each step.

**39.** Prove: $(ab)c = a(cb)$. Justify each step.

**40.** Prove: If $a = b$, then $a + c = b + c$. Justify each step.

**41.** Prove: If $a = b$, then $c + a = c + b$. Justify each step.

CONNECTION

**42.** **GEOMETRY** Use the Distributive Property to find the perimeter of the following rectangles:

**a.** a rectangle with a length of 57 and a width of 43

**b.** the total perimeter of 3 rectangles with a length of 167 and a width of 133 plus 3 rectangles with a length of 82 and a width of 68

APPLICATIONS

**43.** **INCOME** Marcia earns \$15 per hour tutoring math students. She tutored for 4 hours on Friday and for 6 hours on Saturday. Using two different methods, find the total amount Marcia earned for tutoring on Friday and Saturday.

**44.** **INCOME** Janie works 5 hours on Saturday and 4.5 hours on Sunday. She earns \$12 per hour. Use the Distributive Property to compute Janie's total earnings for the weekend. Remember to use the algebraic order of operations and to check your answer.

**45.** **ACCOUNTING** At a music store, 876 CDs are sold for \$12 each during a sale. Use the Distributive Property to compute the total value of the CDs sold during the sale. Show your work.

**46.** **INVENTORY** Sally owns a bakery. Each morning she has to count all of the baked goods that are delivered. How can Sally use mental computation to count the baked goods shown below? What is the total?

## Look Back

**Make a table of values for $x$ and $y$ by substituting 1, 2, 3, 4, 5, and 10 for $x$ in the equations below. Plot the resulting points on a coordinate plane.** *(LESSON 1.5)*

**47.** $y = 3x$

**48.** $y = 2x - 1$

APPLICATION

**49.** ENTERTAINMENT A recording studio charges a fee of \$100 plus \$45 per hour of recording. Write an equation for the total charge, $c$, based on the number of hours, $h$, of recording. Make a table of charges for 1 through 8 hours. *(LESSON 1.5)*

**Evaluate.** *(LESSONS 2.2, 2.3, AND 2.4)*

**50.** $-5 + (-3)$

**51.** $8 - (-17)$

**52.** $14 \cdot (-3)$

## Look Beyond

Use the Distributive Property to complete each exercise.

**53.**
$$\begin{aligned} 25 \cdot 76 &= (20 + 5) \cdot (70 + 6) \\ &= 20 \cdot (70 + 6) + 5 \cdot (70 + 6) \\ &= 20 \cdot 70 + 20 \cdot 6 + 5 \cdot 70 + 5 \cdot 6 \\ &= \underline{?} + \underline{?} + \underline{?} + \underline{?} \\ &= \underline{?} \end{aligned}$$

**54.**
$$\begin{aligned} 26 \cdot 34 &= (30 - 4) \cdot (30 + 4) \\ &= (30 - 4) \cdot 30 + (30 - 4) \cdot 4 \\ &= 30 \cdot 30 - 4 \cdot 30 + 30 \cdot 4 - 4 \cdot 4 \\ &= \underline{?} - \underline{?} + \underline{?} - \underline{?} \\ &= \underline{?} \end{aligned}$$

internet connect

**Portfolio Extension**

Go To: **go.hrw.com**
Keyword:
**MA1 Windchill**

## PORTFOLIO ACTIVITY

Include this activity in your portfolio.

Windchill is a measure of the heat loss from exposed skin due to a combination of wind and low air temperature. The table below shows the windchill at an actual temperature of 25°F.

**Windchill Temperatures at 25°F**

| Wind speed (mph) | 15 | 20 | 25 | 30 | 35 | 40 | 45 |
|---|---|---|---|---|---|---|---|
| Windchill (°F) | 2 | −3 | −7 | −10 | ? | ? | ? |

1. Use a computer spreadsheet to compute the differences in the sequence of windchill numbers. Is the sequence linear?
2. Use differences to extend the sequence in the table. What do you observe?
3. Make a prediction for the windchill at wind speeds of 35, 40, and 45 mph.

# 2.6 Adding and Subtracting Expressions

**Objectives**

- Use the Distributive Property to combine like terms.
- Simplify expressions with several variables.

**Why** *In algebra you work with expressions that contain numbers and variables. When you add and subtract expressions, you need to write the sums and differences in simplified form.*

## Combining Like Terms

In the illustration above, the lumber shipment on the truck is represented by the expression $3x + 4$ (3 bundles of lumber plus 4 loose boards). The shipment on the ground is represented by $5x + 2$ (5 bundles plus 2 loose boards). The sum of the two expressions—a new expression—might be written as follows:

$$(3x + 4) + (5x + 2) \qquad \textit{sum of the shipments}$$

In an expression, parts that are added are called **terms**. The terms $3x$ and $5x$ are **like terms** because they contain *the same form of the variable x*. The numerical part of an $x$-term, in this case 3 or 5, is called the **coefficient** of the $x$-term. The numbers 4 and 2, which are also like terms, are called **constants**.

The expression for the sum can be *simplified* by combining like terms. The terms for the bundles of lumber ($3x$ and $5x$) and the terms for the loose boards (4 and 2) are combined.

$$8x + 6 \qquad \textit{simplified expression}$$

An expression is **simplified** when all like terms have been combined and all parentheses have been removed.

## *Activity*
## Exploring Combining Like Terms

**You will need:** algebra tiles

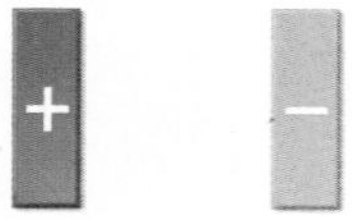

$+x$-tile $-x$-tile

1. Add $3x + 4$ and $2x + 1$ by using algebra tiles.

$3x + 4 \quad + \quad 2x + 1 \quad = \quad 5x + 5$

2. Add $4x + 3$ and $3x - 1$ by using algebra tiles. Recall the technique of removing neutral pairs.

$4x + 3 \quad + \quad 3x - 1 \quad = \quad 7x + 2$

3. Subtract $2x + 1$ from $3x + 4$. You can remove two $x$-tiles and 1 unit tile, or you can change the tiles representing $2x + 1$ to their opposites and add.
$(3x + 4) - (2x + 1) = (3x + 4) + (-2x - 1)$

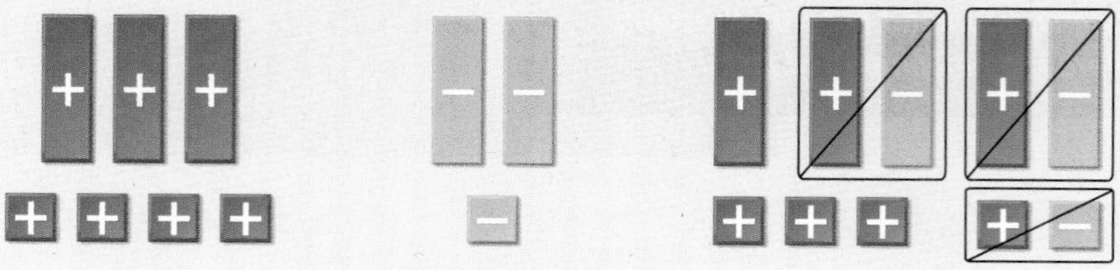

$3x + 4 \quad + \quad (-2x - 1) \quad = \quad x + 3$

**CHECKPOINT ✓** 4. Subtract $4x - 1$ from $2x + 3$. You may want to add neutral pairs of $x$-tiles to the original expression.

Combining like terms is an application of the Distributive Property.

**EXAMPLE 1** **Use the Distributive Property to show that the following are true statements:**

**a.** $2x + 3x = 5x$ **b.** $4x - 2x = 2x$

**SOLUTION**

**a.** $2x + 3x = (2 + 3)x$ *Dist. Prop.*
$(2 + 3)x = 5x$ *Simplify.*

**b.** $4x - 2x = (4 - 2)x$ *Dist. Prop.*
$(4 - 2)x = 2x$ *Simplify.*

**TRY THIS** Use the Distributive Property to show that the following statements are true:

**a.** $7x + 2x = 9x$ **b.** $9x - 3x = 6x$

## Adding and Subtracting Expressions

To add two expressions, use the Commutative and Associative Properties to rearrange and group like terms, and simplify by combining like terms.

| | |
|---|---|
| $(3x + 2) + (4x + 3)$ | *Given* |
| $3x + 2 + 4x + 3$ | |
| $(3x + 4x) + (2 + 3)$ | *Rearrange and group like terms.* |
| $7x + 5$ | *Combine like terms.* |

You can subtract one number from another by adding its opposite. This is also true for expressions. To subtract one expression from another, change the subtracted expression to its opposite by changing the sign of each term in the expression. Then add the expressions.

| | |
|---|---|
| $(2x + 3) - (4x - 1)$ | *Given* |
| $(2x + 3) + (-4x + 1)$ | *Definition of subtraction* |
| $2x + 3 - 4x + 1$ | *(Note: + (–4x) becomes –4x by the definition of subtraction.)* |
| $(2x - 4x) + (3 + 1)$ | *Rearrange and group like terms.)* |
| $-2x + 4$ | *Combine like terms.* |

**EXAMPLE 2** **Simplify.**

**a.** $(3x + 4) + (2x - 1)$
**b.** $(4d - 2) - (5d - 3)$
**c.** $(8x - 2y) + (5x + 6y)$
**d.** $(5a + 3b) - (2a + 5b)$

**SOLUTION**

| | |
|---|---|
| **a.** $(3x + 4) + (2x - 1)$ | *Given* |
| $(3x + 2x) + (4 - 1)$ | *Rearrange and group like terms.* |
| $5x + 3$ | *Combine like terms.* |
| **b.** $(4d - 2) - (5d - 3)$ | *Given* |
| $(4d - 2) + (-5d + 3)$ | *Definition of subtraction* |
| $(4d - 5d) + (-2 + 3)$ | *Rearrange and group like terms.* |
| $-d + 1$ | *Combine like terms.* |
| **c.** $(8x - 2y) + (5x + 6y)$ | *Given* |
| $(8x + 5x) + (-2y + 6y)$ | *Rearrange and group like terms.* |
| $13x + 4y$ | *Combine like terms.* |
| **d.** $(5a + 3b) - (2a + 5b)$ | *Given* |
| $(5a + 3b) + (-2a - 5b)$ | *Definition of subtraction* |
| $(5a - 2a) + (3b - 5b)$ | *Rearrange and group like terms.* |
| $3a - 2b$ | *Combine like terms.* |

**TRY THIS** Simplify.

**a.** $(3x + 6) + (-4x + 5)$
**b.** $(-4m + 3) - (-6m + 3)$
**c.** $(4x - 3y) + (2x + 2y)$
**d.** $(3m + 17n) - (5m + 15n)$

# Exercises

## Communicate

1. Define *like terms*. Give three examples of like terms.
2. Describe the steps involved in using the Distributive Property to add $3x$ and $8x$.
3. Explain what is meant by "adding the opposite" when subtracting expressions.
4. Explain how to simplify the following subtraction problem: $(9x+3)-(4x-1)$.

## Guided Skills Practice

**Use the Distributive Property to show that the statements below are true.** *(EXAMPLE 1)*

**5.** $-7x+10x=3x$ **6.** $5h-9h=-4h$ **7.** $3d+(-2d)=d$

**Simplify.** *(EXAMPLE 2)*

**8.** $(2x+5)+(8x-9)$ **9.** $(6c+2)-(-3c+7)$

**10.** $(5f-g)-(3f+g)$ **11.** $(8x+2y)+(8x-2y)$

## Practice and Apply

**Use the Distributive Property to show that the following are true statements:**

**12.** $3x+4x=7x$ **13.** $4m-10m=-6m$ **14.** $5z+(-2z)=3z$

**15.** $\frac{1}{3}p+\frac{2}{3}p=p$ **16.** $4.7d-6.7d=-2d$ **17.** $11f-15f=-4f$

**18.** $0.3r+4.6r=4.9r$ **19.** $-8x-4x=-12x$ **20.** $\frac{1}{4}y-\frac{1}{2}y=-\frac{1}{4}y$

internet connect
**Homework Help Online**
Go To: **go.hrw.com**
Keyword:
**MA1 Homework Help** for Exercises 21-48

**Simplify the following expressions:**

**21.** $9x-3x$ **22.** $8y-2y$

**23.** $5c-(3-2c)$ **24.** $7d-(1-d)$

**25.** $(7r+2s)+(6r+3s)$ **26.** $(9k+2j)+(11k-2j)$

**27.** $(2a-1)-(-5a-5)$ **28.** $(4m-2n)+(-3m-4n)$

**29.** $(5x+7)-(2x-7)$ **30.** $(-x+2y)+(2x+3y)$

**31.** $(3r-7)+(17r-6)$ **32.** $(-7f+2)-(6f+3)$

**33.** $(5x-1)+(-3x-1)$ **34.** $(-4x-2y)-(4x+2y)$

**35.** $(3.7x+2)-(1.7x+3)$ **36.** $(-5x+2y)-(3x+2y)$

**37.** $(9v-8w)-(8v-9w)$ **38.** $(2q+3)-(-4q+5)+(6q-7)$

**39.** $(9+4y)-(-1+8y)+(7-y)$ **40.** $(1.1a+1.2b)+(2a-0.8b)$

**41.** $(5x + 5y) - (5x + 7y)$

**42.** $(-x - y) + (-x - y)$

**43.** $(3.5m - 2.5n) + (2.7m - 3.7n)$

**44.** $(5p - 6r) - (p + r)$

**45.** $(2x + 7y) + (2x + 7y)$

**46.** $(515x + 755y) - (350x + 250y)$

**CHALLENGE**

**47.** $\left(\frac{x}{2} + 1\right) - \left(\frac{x}{3} - 1\right)$

**48.** $\left(\frac{2m}{5} + \frac{1}{2}\right) + \left(\frac{m}{10} + \frac{5}{2}\right)$

**CONNECTIONS**

**49.** **GEOMETRY** Write an expression for the area of the whole rectangle. Explain how the Distributive Property can be used to find this area.

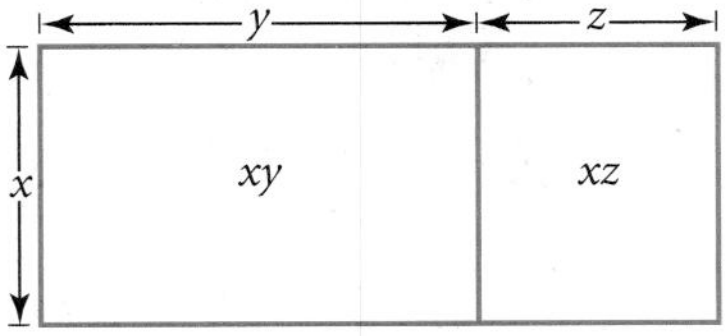


**GEOMETRY** Give an expression in simplified form for the perimeter of each figure.

**50.**

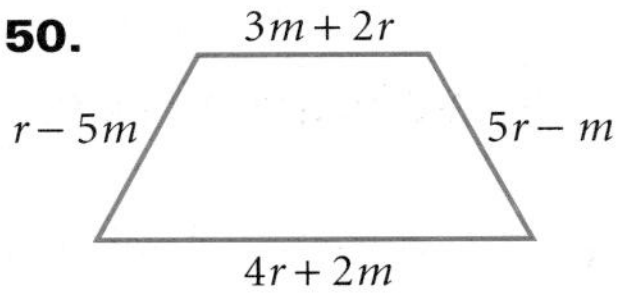


**51.**

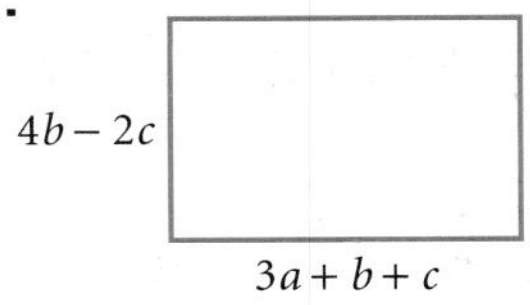


**52.**

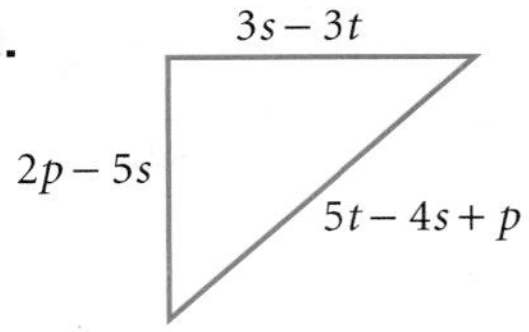


**53.** **MANUFACTURING** A piece of fabric is cut into 12 equal lengths with 3 inches left over. Define a variable, and write an expression for the length of the original piece.

**54.** **INVENTORY** The school kitchen had 11 cases of juice plus 3 extra cans of juice. After lunch they had 6 cases of juice and no extra cans. How many cans of juice were sold during lunch? (Note: A case contains 24 cans.)

## Look Back

**55.** Find the next three terms in the sequence below. ***(LESSON 1.1)***

10, 20, 40, 80, 160, . . .

**Place parentheses to make each equation true.** ***(LESSON 1.3)***

**56.** $28 \div 2 - 4 \cdot 1 = 10$

**57.** $16 \div 5 + 3 \div 2 = 1$

**58.** $40 \cdot 2 + 10 \cdot 4 = 1680$

**Evaluate.** ***(LESSONS 2.2 AND 2.3)***

**59.** $-2 + (-3)$

**60.** $-7 + 6 \quad -1$

**61.** $-5 - 5(-2)$

**62.** $-10 - 5$

**Compute mentally.** ***(LESSON 2.5)***

**63.** $67 + 25 + 33$

**64.** $5 \cdot 11 \cdot 2$

## Look Beyond

**CONNECTION**

**65.** **GEOMETRY** Draw a diagram to model the area represented by $(x + y)(x + y)$.

# EYEWITNESS MATH

VOL. LXI. ..NO. 19,900 NEW YORK, ... APRIL 16, 1912— TWENTY-FOUR PAGES, ONE CENT

## ABANDON SHIP!

### *Titanic*, Giant White Star Liner, Sinks After Collision With Iceberg on Her Maiden Voyage...

*Richmond Times-Dispatch*

| | |
|---|---|
| chives | G13 |
| es | A2 |
| | F10 |
| | D7 |
| e | G1 |
| page | A26 |
| | A32 |
| | F1 |
| | B23 |

NEW YORK, APRIL 15, 1912 — The greatest marine disaster in the history of the world occurred last night, when the *Titanic*, of the White Star Line, the biggest and finest of steamships, crammed herself against an iceberg and sank . . . in less than four hours.

Swinging from the westerly steamship lane at the Grand Banks of Newfoundland to take the direct run to this port, she hurled her giant bulk against an iceberg that rose from an immense field, drifted unseasonably from the Arctic. Running at high speed into that grim and silent enemy of seafarers, the shock crushed her bow . . . Through rent plates and timbers, water rushed . . .

When the *Titanic* plunged headlong against a wall of ice . . . her fate established that no modern steamship is unsinkable.

Ms. ... ill conti... rganizati... with spec...

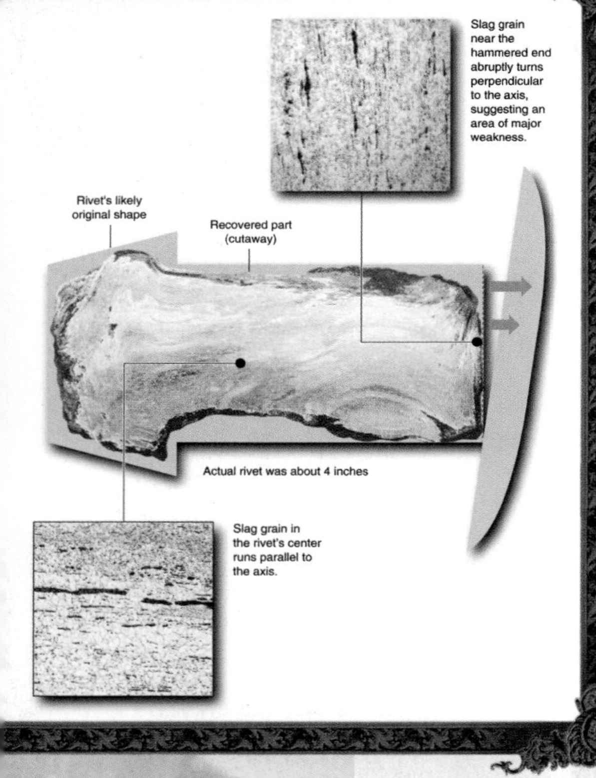


**When the passenger ship Titanic sank only four days into its maiden voyage, it set the stage for years of debate and speculation. The design of the ship was thought to be so safe that it would be virtually impossible for it to sink. Since the night of April 14, 1912, survivors, scholars, and theorists have presented various explanations for what happened and how the "unsinkable" ship sank so quickly. Until the wreckage was discovered in 1985, there was no evidence to support or refute any of the theories**

## Faulty Rivets Emerge as Clues to *Titanic* Disaster

January 27, 1998 — Ever since the *Titanic* was discovered in the depths of the North Atlantic . . . her steel plates melting into rivers of rust, expeditions have repeatedly probed the hulk. And investigators armed with a growing body of evidence have been working to solve riddles posed by the opulent liner's sinking.

Now, after years of analysis and any number of false leads, experts say they have preliminary evidence suggesting that the *Titanic*, the biggest ship of her day, a dream of luxury come to life, may have been done in by structural weaknesses in some of her smallest and least glamorous parts: the rivets.

Experts, diving down nearly two and a half miles to peer through thick mud with sound waves, discovered that the *Titanic*'s bow had been pierced by six thin wounds.

The finding laid to rest the myth that the iceberg had sliced open a 300-foot gash in the ship's side and strengthened interest in the possibility of rivet failure.

the... fro... D... per... the...

The 46,000-ton *Titanic* was made of steel held together with some three million rivets. They secured both beams and plates. Suspicions of rivet failure have long haunted the famous disaster . . . But mysteries remained because the area in the lower bow damaged by the iceberg remained hidden under thick layers of mud.

Now new evidence suggests that they were defective and that the collision with the iceberg, rather than cutting a gash in the hull, caused the rivets to pop. An analysis shows that they had high concentrations of slag, a glassy material, making them brittle and prone to fractures. The hammered ends, or tails, of the analyzed rivets have vanished. They presumably popped off during the disaster.

Testimony given by witnesses to the tragedy was analyzed during the British Inquiry in 1912 by naval architect Edward Wilding. The evidence implied damage to the first through fifth and possibly the sixth of the ship's "watertight" compartments. This meant a possible total of at least 249 feet for the length of the damage.

Based on testimony regarding the times at which water reached various decks within the hull, Wilding concluded that the *Titanic* took on about 16,000 cubic feet of water in the first 40 minutes. Using that rate of entry, he calculated that a total area of 12 square feet must have been opened in the ship's hull by the iceberg.

*by William J. Broad, New York Times*

## Cooperative Learning

1. If 1 cubic foot is equivalent to 7.48 gallons, calculate the number of gallons of water that the *Titanic* took on during the first 40 minutes after colliding with the iceberg.
2. If one gallon of water weighs 8.33 pounds, find the total added weight of the water that entered Titanic during the first 40 minutes.
3. Speculation about the length of the damage ranged as high as 300 feet. Assuming that the damage to the hull actually covered 200 feet in length, calculate the average width of the hole based on Wilding's estimate for the total area of the damage. Use the formula *area* = *length* • *width.*
4. Do you think that it is possible for a gash of that length to have the width that you calculated? How likely do you think it is that a continuous gash of this size would be created by an iceberg?
5. Given the information about the rivets and the six thin wounds from the article, what do you think is the most likely explanation for the damage to the *Titanic*?
6. **Project** Imagine that you are planning a new expedition to the wreckage of the *Titanic.* Write a brief report describing the purpose of your expedition and what information you would look for. Would you try to assess the damage more accurately through observations, collect pieces of evidence, or perhaps bring back artifacts and souvenirs? How would you justify the purpose of your mission in relation to its cost?

# 2.7 Multiplying and Dividing Expressions

### Objectives

- Multiply expressions containing variables.
- Divide expressions containing variables.

**Why** *When solving equations, you often need to multiply or divide not only single numbers, but also larger expressions that may include variables.*

*The number of tiles in a rectangular area can be found by multiplying the number of tiles along the length by the number of tiles along the width.*

## Modeling Multiplication With Algebra Tiles

To model multiplication with algebra tiles, you can arrange the tiles in rectangular patterns (including squares). For example, the arrangement below models the multiplication 2 • 3.

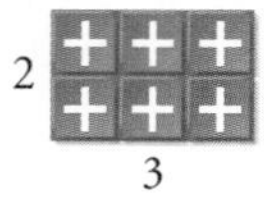


The area of the rectangle is 2 • 3, or 6 unit tiles.
2 • 3 = 6

You can also model more complicated multiplications, such as (2 + 3) • (4 + 2).

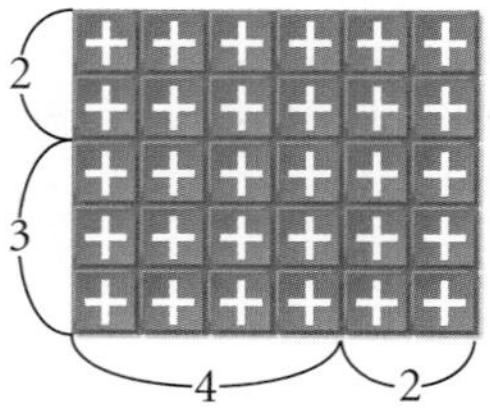


The area of the rectangle is (2 + 3) • (4 + 2), or 30 unit tiles.
(2 + 3) • (4 + 2) = 30

When *one of the factors* of a product contains a variable such as $x$, you will need to include $x$-tiles. The arrangement below models the product $2 \cdot 3x$.

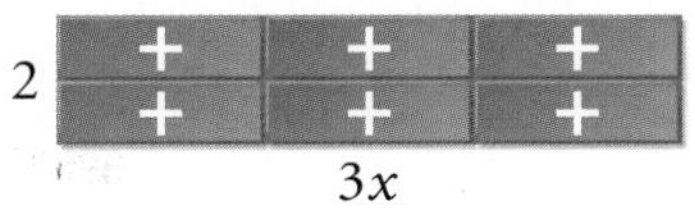


The area of the rectangle is
$2 \cdot 3x$, or 6 $x$-tiles.
$2 \cdot 3x = 6x$

When *both of the factors* of a product contain a variable such as $x$, a new tile is needed to represent the product. Think of a tile that has the same height and length as the long side of an $x$-tile. The area of the tile is $x \cdot x$, which is written as $x^2$ and read as "$x$ squared."

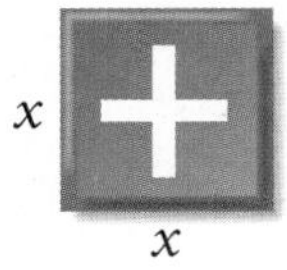


The area of the new tile is
$x \cdot x$, or $x^2$.

To model $2x \cdot 3x$, use $x^2$-tiles. Think of $2x \cdot 3x$ as the area of a rectangle with dimensions of $2x$ by $3x$.

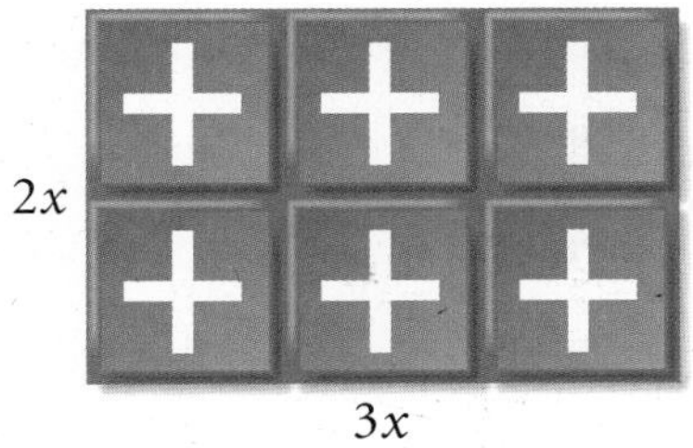


The area of the rectangle is
$2x \cdot 3x$, or 6 $x^2$-tiles.
$2x \cdot 3x = 6x^2$

## Exponents

In later work you will use variables that occur as factors more than twice, such as $x \cdot x \cdot x = x^3$ (read as "$x$ cubed").

An expression such as $x^2$ or $x^3$ is called a **power** (in this case, a power of $x$). Other powers are illustrated in the table below.

| Symbol | Read as | Mathematical meaning |
|---|---|---|
| $3^1$ | 3 to the first power | 3 |
| $3^2$ | 3 to the second power, or 3 squared | $3 \cdot 3$ |
| $3^3$ | 3 to the third power, or 3 cubed | $3 \cdot 3 \cdot 3$ |
| $3^4$ | 3 to the fourth power | $3 \cdot 3 \cdot 3 \cdot 3$ |
| $3^5$ | 3 to the fifth power | $3 \cdot 3 \cdot 3 \cdot 3 \cdot 3$ |
| $y^4$ | $y$ to the fourth power | $y \cdot y \cdot y \cdot y$ |
| $5y^4$ | 5 multiplied by $y$ to the fourth power | $5 \cdot y \cdot y \cdot y \cdot y$ |
| $y^n$ | $y$ to the $n$th power | $y \cdot y \cdot y \cdot \ldots \cdot y$ |

In an expression such as $x^n$, $x$ is called the **base** and $n$ is called the exponent. The **exponent** is the number of times the base appears as a factor.

**CRITICAL THINKING** Describe the shape of the geometric object that could be used to represent $x^3$.

## Multiplying and Dividing Expressions

To multiply expressions, you can use the definition of exponents along with the Commutative, Associative, and Distributive Properties.

**EXAMPLE 1** **Simplify.** **a.** $2x(3x - 4)$ **b.** $(-2)(5a - 4)$

**SOLUTION**

**a.** $2x(3x - 4) = 2x(3x) - 2x(4)$
$= 6x^2 - 8x$

*Use the Distributive Property.*

**b.** $(-2)(5a - 4) = (-2)(5a) - (-2)(4)$
$= -10a + 8$

**TRY THIS** Simplify. **a.** $3x(-5x - 4)$ **b.** $(-4)(3x + 4)$

**EXAMPLE 2** **Simplify: $(5x + 3y - 7) - 3(2x - y)$.**

**SOLUTION**

Use the definition of subtraction to rewrite the expression as an addition problem.

$(5x + 3y - 7) - 3(2x - y) = (5x + 3y - 7) + (-3)(2x - y)$

Distribute $-3$ over $(2x - y)$.

$$\begin{aligned}(5x + 3y - 7) + (-3)(2x - y) &= 5x + 3y - 7 + (-3)(2x) - (-3)(y)\\ &= 5x + 3y - 7 - 6x + 3y\\ &= (5x - 6x) + (3y + 3y) - 7\\ &= -x + 6y - 7\end{aligned}$$

**TRY THIS** Simplify: $(2b - 5c) - 4(3b + c)$.

**EXAMPLE 3** **Model the division problem $\frac{2x+6}{2}$ with algebra tiles.**

**SOLUTION**

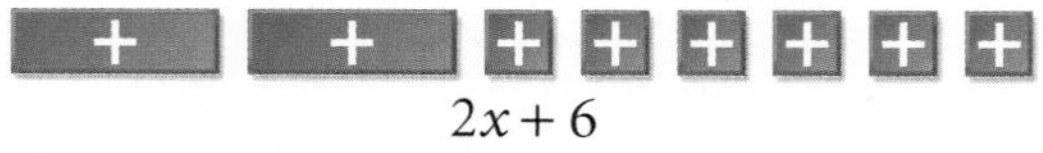


Because you are dividing by 2, divide the tiles into two equal groups.

Since each set contains 1 $x$-tile and 3 unit-tiles, $\frac{2x+6}{2} = x + 3$.

**TRY THIS** Use algebra tiles to model the division problem $\frac{8x-4}{4}$.

As the algebra-tile model suggests, when you divide an expression by a number, each term in the expression is divided by that number.

**Dividing an Expression**

For all real numbers $a$, $b$, and $c$, where $c \neq 0$:

$$\frac{a+b}{c} = \frac{a}{c} + \frac{b}{c} \text{ and } \frac{a-b}{c} = \frac{a}{c} - \frac{b}{c}$$

**CRITICAL THINKING** The division expression $\frac{a+b}{c}$ can be written as $\frac{1}{c}(a+b)$. Show how $\frac{1}{c}(a+b)$ gives the rule for dividing a sum.

**EXAMPLE 4** **Simplify $\frac{2x^2+6}{2}$.**

**SOLUTION**

There are two ways of working this problem. One way is to divide each term in the numerator by the divisor:

$$\frac{2x^2+6}{2} = \frac{2x^2}{2} + \frac{6}{2}$$
$$= x^2 + 3$$

Another way of solving the problem is to represent it as a multiplication problem and to use the Distributive Property.

$$\frac{2x^2+6}{2} = \frac{1}{2}(2x^2+6)$$
$$= \frac{1}{2}(2x^2) + \frac{1}{2}(6)$$
$$= x^2 + 3$$

**EXAMPLE 5**

**APPLICATION**
**RECREATION**

George is in charge of supplies for a mountain-climbing guide company. He is trying to locate carabiners for four climbing teams. George's preferred brand of carabiners are available at three outfitters as follows:

Outfitter A has 4 boxes plus 6 loose carabiners. Outfitter B has 6 boxes, but one box is missing 3 carabiners. Outfitter C has 10 boxes plus 5 loose carabiners.

**Assuming each box contains an equal number of carabiners, write an expression describing the total number of carabiners available.**

**Write another expression describing the number of carabiners available for each team if the carabiners are divided equally among the teams.**

*Carabiners are used in mountain climbing to attach the climber to a rope.*

**SOLUTION**

Using $c$ to represent the number of carabiners in one box, the number available from each outfitter can be written as follows:

Outfitter A: $4c + 6$
Outfitter B: $6c - 3$
Outfitter C: $10c + 5$

The total available carabiners can be represented by adding the three expressions.

$$\begin{aligned}(4c + 6) + (6c - 3) + (10c + 5) &= 4c + 6 + 6c - 3 + 10c + 5 \\ &= (4c + 6c + 10c) + (6 - 3 + 5) \\ &= 20c + 8\end{aligned}$$

To represent the number of carabiners available for each team, divide the $20c + 8$ by 4.

$$\begin{aligned}\frac{20c + 8}{4} &= \frac{20c}{4} + \frac{8}{4} \\ &= 5c + 2\end{aligned}$$

**EXAMPLE 6** Simplify $\frac{8x - 4}{-4}$.

**SOLUTION**

Remember that because the quantity $8x - 4$ is to be divided by $-4$, each term in the quantity must be divided by $-4$.

$$\begin{aligned}\frac{8x - 4}{-4} &= \frac{8x}{-4} - \frac{4}{-4} \\ &= -2x - (-1) \\ &= -2x + 1\end{aligned}$$

Using the Distributive Property, multiply each term in the numerator by $\frac{1}{-4}$.

$$\begin{aligned}\frac{8x - 4}{-4} &= \frac{1}{-4}(8x - 4) \\ &= \left(\frac{1}{-4}\right)(8x) - \left(\frac{1}{-4}\right)(4) \\ &= -2x + 1\end{aligned}$$

**TRY THIS** Simplify each expression by using two different methods.

**a.** $\frac{6x^2 - 9}{3}$ **b.** $\frac{10x^2 - 15}{-5}$

# Exercises

## Communicate

**1.** Explain how to write an expression that shows Jan's income if she earns \$6 per hour and works
**a.** 4 hours. **b.** 2.5 hours. **c.** $h$ hours.

**2.** Explain how to simplify $-2(4m + 5)$.

**3.** Describe how to simplify the expression $\frac{5p - 15}{-5}$.

**4.** Are all of the expressions below equivalent? If not, explain how they are different.
**a.** $(4x - 6) \div 2$ **b.** $(4x - 6)(0.5)$ **c.** $\frac{4x - 6}{2}$ **d.** $\frac{-6 + 4x}{2}$

**5.** Explain why $\frac{2x + 6}{2}$ is not equivalent to $x + 6$.

## Guided Skills Practice

**Simplify.** *(EXAMPLE 1)*

**6.** $3(4x - 2)$ **7.** $-3(5x + 2)$ **8.** $-8(4 - x)$

**Simplify.** *(EXAMPLE 2)*

**9.** $(2x + 3y - 6) + 3(2x + y)$ **10.** $5(b + c) - 4(3b - 2c + 1)$

**Use algebra tiles to model each division problem.** *(EXAMPLE 3)*

**11.** $\frac{2x + 8}{2}$ **12.** $\frac{4x^2 + 10}{2}$ **13.** $\frac{5y^2 + 20}{5}$

**Simplify.** *(EXAMPLES 4, 5, AND 6)*

**14.** $\frac{-6x + 10}{-2}$ **15.** $\frac{12x - 18}{-6}$ **16.** $\frac{3x^2 - 21}{3}$

## Practice and Apply

**internet connect**
**Homework Help Online**
Go To: **go.hrw.com**
Keyword:
**MA1 Homework Help** for Exercises 17-37

**Simplify the following expressions. Use the Distributive Property if needed.**

**17.** $2 \cdot 6x$ **18.** $-6x \cdot 2$ **19.** $6x \cdot 2x$
**20.** $12x \cdot 3x$ **21.** $-66x \div 2$ **22.** $-12x \div 3$
**23.** $-1.2x \cdot 3x$ **24.** $-2(6x + 3)$ **25.** $7x^2 - (3 - x^2)$
**26.** $2x^2 - (4 - x^2)$ **27.** $x^2 - 2(3 - x^2)$ **28.** $3x \cdot 5 + 2x \cdot 2$
**29.** $-2 \cdot 8x$ **30.** $-2(4x - 1)$ **31.** $8x \cdot 2x$
**32.** $-8x \div 2$ **33.** $-3(7x - 3)$ **34.** $-21x \cdot 3x$
**35.** $21x \cdot 3x$ **36.** $-21x \div 3$ **37.** $8x^2 - (2 - 5x^2)$

38. $8x^2 - 10(2 - 5x^2)$

39. $\dfrac{11 - 33y}{11}$

40. $\dfrac{-10x + 35}{5}$

41. $\dfrac{-90x + 2.7}{-9}$

42. $\dfrac{5w + 15}{-5}$

43. $\dfrac{8m + 10}{2}$

44. $\dfrac{30x + 20}{-5}$

45. $\dfrac{4y^2 + 4}{4}$

46. $\dfrac{27x - 18}{-9}$

**CHALLENGE**

47. $\dfrac{30x^2 + 20x - 8}{-5}$

48. $\dfrac{36y - 30}{12}$

49. $\dfrac{49.5 + 10x}{10x}$

**50.** Which of the expressions below are equivalent?
**a.** $5(3x - 7)$ **b.** $5 \cdot (3x - 7)$ **c.** $(5)(7 - 3x)$ **d.** $5(3x + 7)$

**51.** Which of the expressions below are equivalent?
**a.** $(6 - 12x) \div 6$ **b.** $(6 - 12x) \cdot \frac{1}{6}$ **c.** $\frac{6 - 12x}{6}$ **d.** $1 - 12x$

**CONNECTIONS**

**GEOMETRY** The formula for the area of a rectangle is $A = lw$. Find the area for each rectangle by evaluating the formula for the given length and width.

**52.**

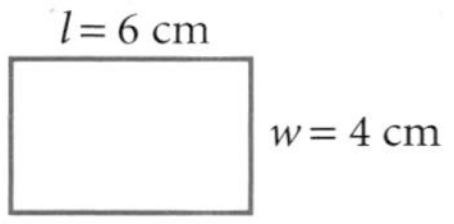


**53.**

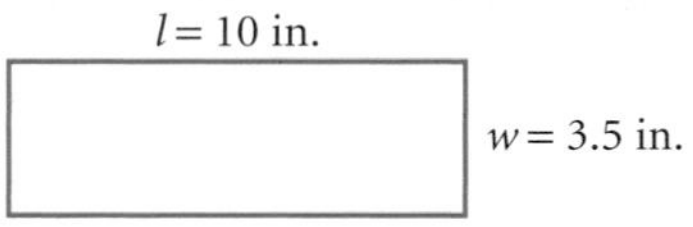


**54.**

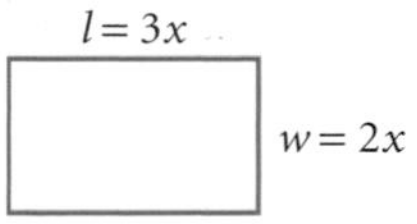


**GEOMETRY** The formula for the volume of a rectangular prism is $V = lwh$. Find the area for each rectanglular prism by evaluating the formula for the given length, width, and height.

**55.**

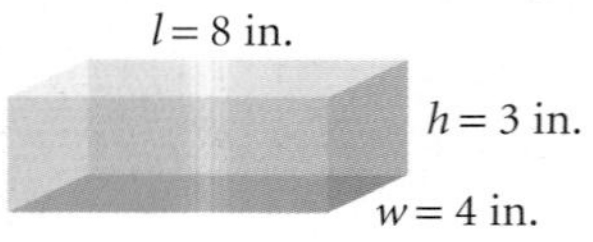


**56.**

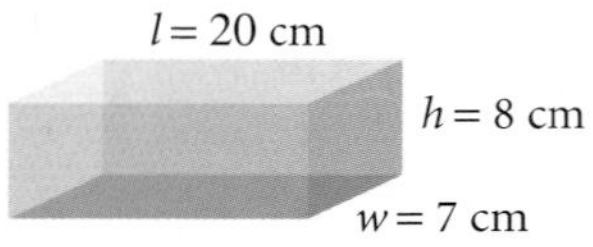


**57.**

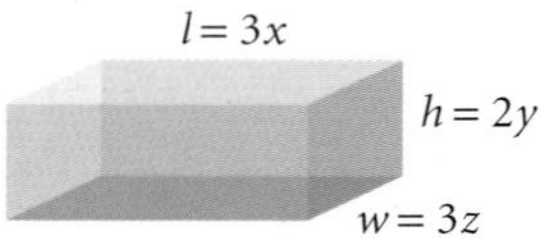


**APPLICATIONS**

**58. INCOME** Nicole has a part-time job three days after school and on Saturdays. If she earns \$5.25 per hour, find Nicole's earnings for each of the following four days:
**a.** 4 hours on Friday **b.** 7 hours on Saturday
**c.** $h$ hours on Monday **d.** $(h + 1)$ hours on Tuesday

**CHALLENGE**

**59.** Write an expression in simplified form that shows the total number of hours that Nicole worked during the four days described in Exercise 58.

**60. BUSINESS** A plumber charges \$35 an hour plus a fixed service fee of \$20 per job. Write an expression and determine how much the plumber would charge for a job that takes
**a.** 1 hour. **b.** 3 hours. **c.** 7.5 hours.

**61. BUSINESS** On Sundays, the plumber described in Exercise 60 doubles his hourly rate. Write an expression and determine how much the plumber would charge on Sundays for a job that takes
**a.** 1 hour. **b.** 3 hours. **c.** 7.5 hours.

## Look Back

**Evaluate.** ***(LESSONS 2.2, 2.3, AND 2.4)***

**62.** $3 + (-2)$

**63.** $1.4 - (-5)$

**64.** $-10 \cdot 21$

**65.** $4.7 \cdot (-2)$

**66.** $3.6 - (-3.6)$

**67.** $2.7 + (-5)$

**68.** $1\frac{1}{2} - 2$

**69.** $87 \div (-3)$

**70.** $\frac{-8}{4}$

**71.** $-150 \div (-15)$

**72.** $-15 \cdot 2.1$

**73.** $3.7 - (-2.6)$

**74.** $-8.5 \div 5$

**75.** $0.1 - 2.7$

**76.** $15 \cdot (-3)$

**77.** $86 + (-92)$

**78.** $19 + (-2)$

**79.** $3\frac{1}{3} + \left(-1\frac{1}{2}\right)$

**80.** $5\frac{1}{2} - (-5)$

**81.** $-3.25 \div (-2.5)$

**Simplify.** ***(LESSON 2.6)***

**82.** $(-3x + 7) + (2x - 1)$

**83.** $(5m - 2) + (5m - 2)$

**84.** $(-4x - 2y) - (-3x - 4y)$

**85.** $(3x + 2y) + (3x - 2y) - (3x - 2y)$

**86.** $(x^2 + 2y + 4) - (2x - 3y - 2)$

## Look Beyond

**87.** Simplify: $\frac{2x^2 + 4x}{4x}$.

**88.** Solve $3x = 9a - 6b$ for $x$.

### PORTFOLIO ACTIVITY

Include this activity in your portfolio.

Visitors to a city park can rent bicycles at the following rates:

Adult: $2 per hour plus $4
Child: $1 per hour plus $3

**1.** Write an expression for the cost of renting an adult bicycle with $H$ as the number of hours.

**2.** Write an expression for the cost of renting a child's bicycle with $H$ as the number of hours.

**3.** Write two different expressions for renting 1 adult's bicycle and 1 child's bicycle for $H$ hours.

**4.** Write two different expressions for renting 2 adult's bicycles and 3 children's bicycles for $H$ hours.

**WORKING ON THE CHAPTER PROJECT**

You should now be able to complete Activity 3 of the Chapter Project on page 104.

CHAPTER PROJECT TWO

# Using a Spreadsheet

Computer spreadsheets can perform many computations quickly. Business people use spreadsheets to organize and analyze data.

The cells are identified by the column letter and row number, as shown.

| | A | B | C |
|---|---|---|---|
| 1 | A1 | B1 | C1 |
| 2 | A2 | B2 | C2 |
| 3 | A3 | B3 | C3 |

## Activity 1

Below is an example of a simple spreadsheet.

| | A | B | C |
|---|---|---|---|
| 1 | 1 | -4 | 3 |
| 2 | 2 | 0 | -2 |
| 3 | -3 | 4 | -1 |

1. What is the number in cell B2?
2. What is the number in cell C1?
3. What is the number in cell A3?
4. Use mental math to find the sum of each row, column, and diagonal.
5. What did you discover about all of the sums?
6. If you are using a spreadsheet on a computer, use the SUM command to find all of the row, column, and diagonal sums.
7. Adjust the values to create other number arrays in which the sums of the rows, columns, and diagonals are all the same (these are called “magic squares”).

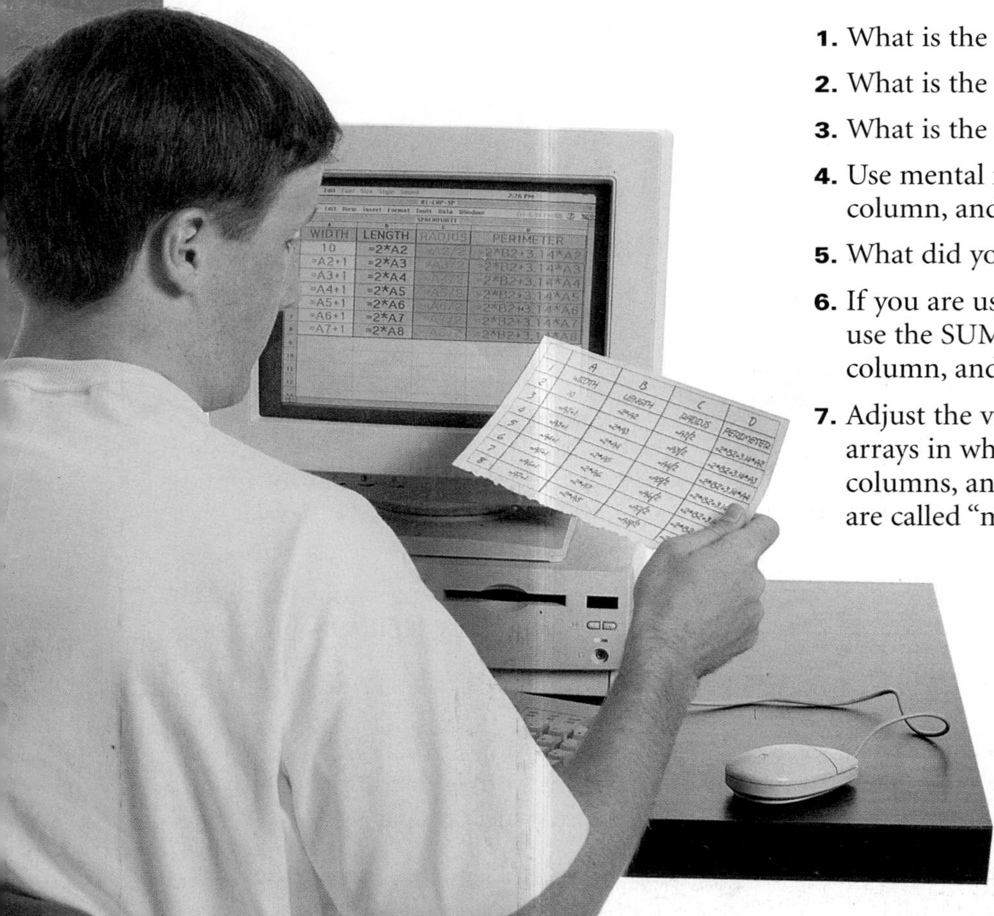

## Activity 2

The table below shows the average high and low temperature in Minneapolis, Minnesota, in degrees Fahrenheit.

| | A | B | C | D | E | F | G | H | I | J | K | L | M |
|---|---|---|---|---|---|---|---|---|---|---|---|---|---|
| 1 | Month | J | F | M | A | M | J | J | A | S | O | N | D |
| 2 | Average high | 21 | 27 | 39 | 57 | 69 | 79 | 84 | 81 | 71 | 59 | 41 | 26 |
| 3 | Average low | 3 | 9 | 23 | 36 | 48 | 58 | 63 | 60 | 50 | 39 | 25 | 10 |
| 4 | Average range | ? | ? | ? | ? | ? | ? | ? | ? | ? | ? | ? | ? |

1. Put the data into a spreadsheet. Compute the range for each month.
2. Use the spreadsheet to build a double line graph for the data. How does the double line graph display the monthly ranges?
3. Assume that the months are all the same length, and compute the average annual high, the average annual low, and the average annual range.
4. Write a paragraph describing the differences in the averages found in Step 3.

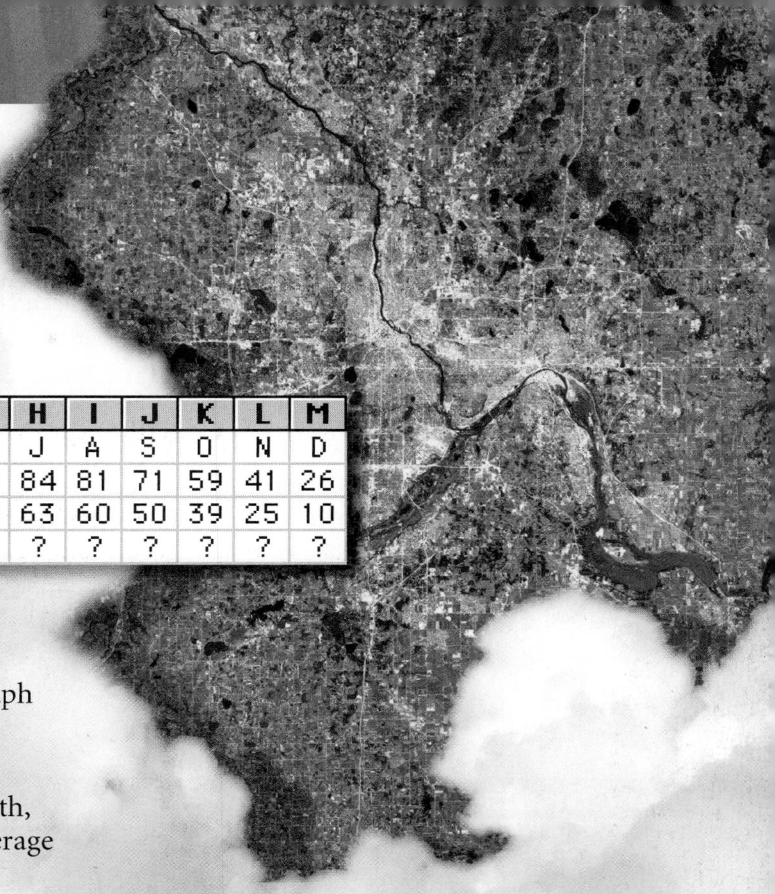

*Satellite photo of Minneapolis-St. Paul, Minnesota*

## Activity 3

Visitors to a national park can rent equipment at the following rates:

| | |
|---|---|
| Canoe | \$6 per hour plus \$14 |
| Kayak | \$5 per hour plus \$12 |

Build a spreadsheet to determine the cost of renting the following combinations of canoes and kayaks for 1 through 10 hours:

| | A | B | C | D | E | F | G | H | I |
|---|---|---|---|---|---|---|---|---|---|
| 1 | Combinations | | | | | | | | |
| 2 | Number of canoes | 1 | 1 | 1 | 1 | 2 | 2 | 2 | 2 |
| 3 | Number of kayaks | 1 | 2 | 3 | 4 | 1 | 2 | 3 | 4 |

# 2 Chapter Review and Assessment

## VOCABULARY



## Key Skills & Exercises

### LESSON 2.1

#### Key Skills

**Compare values of real numbers.**

Insert an ordering symbol to make the following statements true:

**a.** 8 ___ 5 **b.** −2 ___ 7

**a.** 8 is to the right of 5 on a number line, so $8 > 5$.

**b.** 7 is to the right of −2 on a number line, so $7 > -2$.

**Find the absolute value of a number.**

To find the absolute value of −7, find the distance it is from 0 on the number line.

$$|-7| = 7$$

#### Exercises

**Insert an ordering symbol to make each statement true.**

**1.** 12 ___ 15
**2.** −19 ___ 0.2
**3.** 23.4 ___ 23
**4.** $-1\frac{2}{3}$ ___ $\frac{23}{47}$

**Find the absolute value of each number.**

**5.** −73
**6.** 12
**7.** −15.2
**8.** $-\frac{16}{53}$
**9.** $\frac{23}{54}$
**10.** 0

### LESSON 2.2

#### Key Skills

**Add two or more real numbers.**

When numbers have like signs, find the sum of their absolute values, and use the common sign. When numbers have unlike signs, find the difference of their absolute values and use the sign of the number with the greater absolute value.

$-4 + (-7) = -11$ $26 + (-9) = 17$ $-26 + 9 = -17$

#### Exercises

**Find each sum.**

**11.** $-17 + 6$
**12.** $48 + (-15)$
**13.** $-23 + (-25) + 3$
**14.** $-39 + 68$
**15.** $33 + (-55)$
**16.** $-214 + 214$
**17.** $6 + (-7) + (-9)$
**18.** $-8 + 8 + (-12)$

LESSON 2.3

**Key Skills**

**Subtract real numbers.**

To subtract –8, add its opposite.
$-16 - (-8) = -16 + 8 = -8$

To subtract 19, add its opposite.
$-38 - 19 = -38 + (-19) = -57$

**Exercises**

**Find each difference.**

**19.** $9 - (-15)$ **20.** $48 - (-48)$

**21.** $-13 - 28$ **22.** $39 - (-18)$

**23.** $-67 - (-42)$ **24.** $-23 - (-72)$

LESSON 2.4

**Key Skills**

**Multiply and divide real numbers.**

When multiplying or dividing numbers with like signs, the result is positive. When multiplying or dividing numbers with unlike signs, the result is negative.

**a.** $(-14)(-3) = 42$
**b.** $(-72) \div (24) = -3$
**c.** $(-8)[(-7) + 7] = (-8)(0) = 0$

**Exercises**

**Find each product or quotient.**

**25.** $(-12)(-5)$ **26.** $(-0.9)(3)$

**27.** $(54) \div (-18)$ **28.** $(-121) \div (-11)$

**29.** $(-2)(-2)(-2)(-2)$ **30.** $(-6)(-3) \div (-9)$

**31.** $45[8 + (-8)]$ **32.** $(-4)(4) \div (-1)$

**33.** $(-5)(-5)(-1)(1)$ **34.** $(-3)(4 + 7)$

LESSON 2.5

**Key Skills**

**Identify number properties. Find a sum or product by using mental computation.**

**Commutative Property**

$a + b = b + a \quad a \cdot b = b \cdot a$

To simplify $25 + 23.9 + 75$:
$25 + 23.9 + 75 = 25 + 75 + 23.9$
$= 100 + 23.9 = 123.9$

**Associative Property**

$(a + b) + c = a + (b + c)$
$(a \cdot b) \cdot c = a \cdot (b \cdot c)$

To simplify $(17 \cdot 5) \cdot 2$:
$(17 \cdot 5) \cdot 2 = 17 \cdot (5 \cdot 2)$
$= 17 \cdot 10 = 170$

**Distributive Property**

$a(b + c) = ab + ac$

To simplify $7(31)$:
$7(31) = 7(30 + 1)$
$= 7(30) + 7(1)$
$= 210 + 7 = 217$

**Exercises**

**Use mental math to find each sum or product. Identify the property or properties that you use.**

**35.** $(27 + 8) + 12$ **36.** $(25 \cdot 87) \cdot 4$

**37.** $(6.2 + 7.1) + 3.8$ **38.** $(63 \cdot 20) \cdot 5$

**39.** $(3 + 10) + 7$ **40.** $(3 \cdot 5) \cdot 20$

**41.** $(8.7 + 21.1) + 1.3$ **42.** $(7 + 15) + 13$

**Complete each step, and name the property used.**

**43.**

| | |
|---|---|
| $(7 + 48) + 23$ | Given |
| $= (48 + \underline{?}) + 23$ | Commutative Property |
| $= 48 + (7 + \underline{?})$ | $\underline{?}$ Property |
| $= 48 + \underline{?}$ | |
| $= \underline{?}$ | |

**44.**

| | |
|---|---|
| $25 \cdot (18 \cdot 4)$ | Given |
| $= 25 \cdot (\underline{?} \cdot 18)$ | $\underline{?}$ Property |
| $= (\underline{?} \cdot 4) \cdot 18$ | $\underline{?}$ Property |
| $= \underline{?} \cdot \underline{?}$ | |
| $= \underline{?}$ | |

LESSON 2.6

**Key Skills**

**Add and subtract expressions, and combine like terms.**

When an expression in parentheses has a plus sign in front of it, drop the parentheses and combine like terms. When an expression in parentheses has a minus sign in front of it, change the minus sign to a plus sign and multiply the terms inside the parentheses by −1.

| | |
|---|---|
| $(7x + 4y) - (6x - 5y)$ | *Given* |
| $(7x + 4y) + (-6x + 5y)$ | *Definition of subtraction* |
| $7x + (-6x) + 4y + 5y$ | *Rearrange terms.* |
| $x + 9y$ | *Combine like terms.* |

**Exercises**

**Simplify.**

**45.** $(6a - 1) + (5a - 4)$

**46.** $(7 - t) + (3t + 4)$

**47.** $(8n - 4) - (6n - 3)$

**48.** $(6d + 3a) - (4d - 7a)$

**49.** $\left(\frac{x}{3} - 2\right) + \left(\frac{x}{2} + 4\right)$

**50.** $(4a - 3b - c) - (4b - 7) + (9a + 5)$

**51.** $(3x + 2) - (2x - 1)$

**52.** $(7y - 3) + (6y + 2) - (3y - 7)$

LESSON 2.7

**Key Skills**

**Multiply and divide expressions.**

**a.**

| | |
|---|---|
| $-4(5x - 6)$ | *Given* |
| $= (-4)(5x) - (-4)(6)$ | *Distributive Property* |
| $= -20x + 24$ | *Multiply.* |

**b.**

| | |
|---|---|
| $(2a - 3b + 1) - 4(a + 6b - 3)$ | *Given* |
| $= 2a - 3b + 1 - 4a - 24b + 12$ | *Dist. Prop.* |
| $= -2a - 27b + 13$ | *Combine like terms.* |

**c.** $\frac{12 - 8y}{-4} = \frac{12}{-4} + \left(\frac{-8y}{-4}\right) = -3 + 2y$

**Exercises**

**Simplify.**

**53.** $3 \cdot 9x$

**54.** $-33d \div 3$

**55.** $-2(7b - 2)$

**56.** $-2.4f \cdot 2f$

**57.** $\frac{-30v + 3.6}{-3}$

**58.** $9r^2 - 8(4 - 3r^2)$

**59.** $\frac{6(x + 1)}{3}$

**60.** $(5a^2 + 2a)(6m^2)$

**61.** $\frac{-90w + 24w}{-3}$

**62.** $\frac{3x^2 + 30}{3}$

## Applications

**GEOMETRY** The formula for the perimeter, $P$, of a triangle with sides of length $a$, $b$, and $c$ is $P = a + b + c$.

**63.** Use the formula to find $P$ for $a = (x + 7)$, $b = (y - 8)$, and $c = (3x + 2y)$.

**64.** Use the formula to find $P$ for $a = (x + 3)$, $b = (x + 7)$, and $c = (x - 10)$.

# 2 Chapter Test

**Insert <, >, or = to make each statement true.**

1. 23 ___ 32
2. $-5$ ___ $-15$
3. 19.9 ___ 19.7
4. $5\frac{3}{5}$ ___ $5\frac{2}{3}$

**Simplify each expression.**

5. $|3 \cdot (-4)|$
6. $-|7-3|$
7. If $x$ is a negative number, then what is the absolute value of $x$?

**Find each sum.**

8. $729 + (-201)$
9. $-54 + (-103)$
10. $-3.14 + 2.72$
11. $-\frac{5}{4} + \left(-\frac{7}{8}\right)$

**Let $a = 3$, $b = 5$, and $c = -4$. Evaluate each expression.**

12. $2c + 7a$
13. $c + (-3b)$
14. CHEMISTRY Protons have a charge of +1 and electrons have a charge of $-1$. The total charge on an ion is the sum of its electron and proton charges. An ion of calcium has 20 protons and 18 electrons. What is its charge?

**Find each difference.**

15. $2 - 5$
16. $5 - (-2)$
17. $12\frac{1}{2} - 10\frac{2}{5}$
18. $1.62 - (-7.29)$

**Let $x = -4$, $y = 2$, and $z = -3$. Evaluate each expression.**

19. $x - y$
20. $-y - (-z)$

21. WEATHER At 8 A.M., the temperature was 5°F. By noon, the temperature had risen 6 degrees. From noon to 8 P.M., the temperature dropped 14 degrees. What was the temperature at 8 P.M.?

**Find each product or quotient.**

22. $(3)(-6)$
23. $14 \div (-7)$
24. $(-25)(-8)$
25. $(-11.61) \div (-0.9)$
26. $-[(-3)(5)(-6)]$
27. $\frac{3 + (-18)}{5}$

**Use the Distributive Property to rewrite each.**

28. $12a - 9b$
29. $4a - 10ac$

**Name the property illustrated.**

30. $6 + (4 + 9) = (6 + 4) + 9$
31. $3 + 9 + 7 = 3 + 7 + 9$
32. $45xy - 35x = 5x(9y - 7)$
33. INCOME Khoi earns $45 per hour teaching piano. On Monday he teaches for 5 hours, and on Wednesday he teaches for 6.5 hours. Use the Distributive Property to compute Khoi's earnings for the two days.

**Simplify the following expressions.**

34. $12r - 7r$
35. $2.1c - 3.5c$
36. $(2a - 3b) - (5a - 7b)$
37. $(3r + 4s) + (2s - 5t) + (-5r - 3s)$
38. $3x \cdot 8x$
39. $-7y \cdot 8$
40. $5x^2 - 2(7x - 3x^2)$
41. $-2q^2 + 3(5 + 4q^2)$
42. $\frac{18x - 27}{-3}$
43. $\frac{24t^2 + 48t - 16}{8}$

# College Entrance Exam Practice

internet connect
**Standardized Test Prep Online**
Go To: go.hrw.com
Keyword: MM1 Test Prep.

**QUANTITATIVE COMPARISON** For Items 1–4, write
A if the quantity in Column A is greater than the quantity in Column B;
B if the quantity in Column B is greater than the quantity in Column A;
C if the two quantities are equal; or
D if the relationship cannot be determined from the information given.

| | Column A | Column B | Answers |
|---|---|---|---|
| **1.** | $23 \cdot 43 + 23 \cdot 59$ | $23 \cdot (43 + 59)$ | Ⓐ Ⓑ Ⓒ Ⓓ [Lesson 2.5] |
| **2.** | $-17 - (-46)$ | $-17 - 46$ | Ⓐ Ⓑ Ⓒ Ⓓ [Lesson 2.3] |
| **3.** | $-113 \div 48$ | $113 \div (-48)$ | Ⓐ Ⓑ Ⓒ Ⓓ [Lesson 2.4] |
| **4.** | $\frac{2}{3}$ | $\frac{5}{6}$ | Ⓐ Ⓑ Ⓒ Ⓓ |

**5.** Evaluate the expression $a + (-b)$ for $a = 2$ and $b = -5$. ***(LESSON 2.2)***
**a.** $-3$ **b.** 7
**c.** 3 **d.** none of these

**6.** Apply the method of finding differences to the sequence 4, 19, 44, 79, 124, . . . What is the value of the constant difference that you get? ***(LESSON 1.1)***
**a.** 5 **b.** 10
**c.** 15 **d.** 25

**7.** Which of the following is equivalent to $(8x - 4z) - (5z + x)$? ***(LESSON 2.6)***
**a.** $8x - 4z - 5z + x$ **b.** $8x + x - 4z - 5z$
**c.** $8x - 4z - 5z - x$ **d.** none of these

**8.** The following equation is an example of which property? ***(LESSON 2.5)***
$$7 + (11 + 4) = (7 + 11) + 4$$
**a.** Associative Property
**b.** Commutative Property
**c.** Distributive Property
**d.** Addition Property

**9.** Which number is the opposite of $-7$? ***(LESSON 2.1)***
**a.** 0 **b.** $+7$ **c.** $-14$ **d.** $-1$

**10.** Find $|-6|$. ***(LESSON 2.1)***
**a.** $+6$ **b.** $-6$ **c.** 0 **d.** 12

**11.** Multiply: $2x(3x - 1)$. ***(LESSON 2.7)***

**Simplify.** ***(LESSON 2.6)***

**12.** $(4x + 5y) + (x + 7y)$

**13.** $(x + y + z) + (2w + 3y + 7)$

**14.** $(2a - 3) - (a - 5)$

**15.** $(4y + 9) - (7y - 2) + (13 - y)$

**16.** $(3x + 7) - (2x - 2)$

**17.** $(8x + 2y) + (3x + 5y)$

**18.** $(17x - 2y + z) + (3x - y) + (2y - z)$

**19.** $(8x - 3y) + (8x - 3y) - (2x + 2y)$

**What are the coordinates of the given points? *(LESSON 1.4)***

**20.** *A*

**21.** *B*

**22.** *C*

**23.** *D*

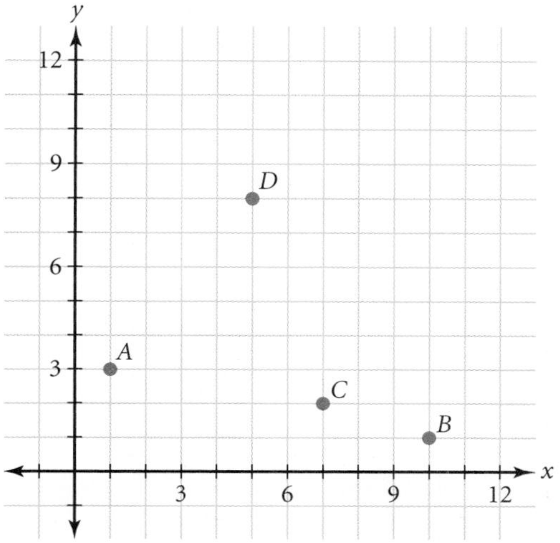


**Find each sum or difference. *(LESSONS 2.2 AND 2.3)***

**24.** $-39 + 68$

**25.** $8 - 14 - 27$

**26.** $-23 - (-72)$

**27.** $-8 + 8 + (-12)$

**Make a table of values for each equation, using 1, 2, 3, 4, and 5 as values for *x*. Draw a graph for each equation by plotting points from your table. *(LESSON 1.5)***

**28.** $y = 2x + 3$

**29.** $y = x - 7$

**30.** The chart below shows how the cost of food items that were \$3.00 in 1982 has increased over the years. Make a scatter plot, and use a straightedge to estimate the line of best fit. ***(LESSON 1.6)***

| Year | 1982 | 1985 | 1988 | 1991 | 1994 | 1997 |
|---|---|---|---|---|---|---|
| Cost | 3.00 | 3.18 | 3.54 | 3.54 | 4.08 | 4.95 |

**31.** Simplify: $(3b + 109) - (2b + 53)$. ***(LESSON 2.6)***

**32.** Simplify: $\frac{4x - 12}{2}$. ***(LESSON 2.7)***

**33.** Simplify: $(8x + 3y - 2z) + (4x - y + 5z)$. ***(LESSON 2.6)***

**34.** Evaluate: $(5 + 2^3) - (-3)(2 + 4)$. ***(LESSON 1.3)***

**35.** Find the next three terms in the sequence: 1, 5, 11, 19, . . . ***(LESSON 1.1)***

**FREE-RESPONSE GRID**

**The following questions may be answered by using a free-response grid such as that commonly used by standardized test services.**

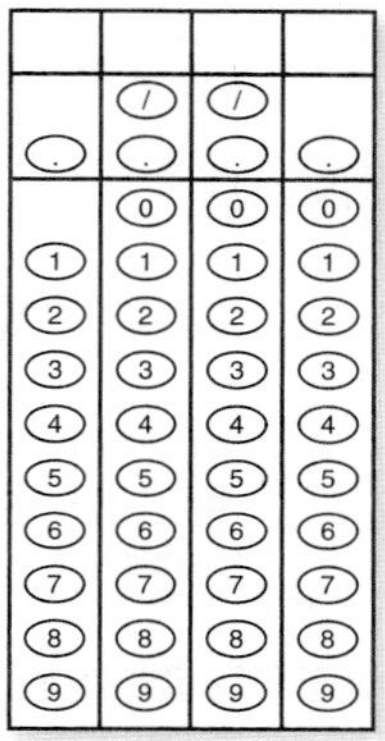


**36.** Find the next term in the sequence: 7, 16, 25, 34, 43, . . . ***(LESSON 1.1)***

**37.** How many \$6 movie tickets can you buy with \$42? ***(LESSON 1.2)***

**38.** Evaluate: $(6 + 3^2) - 5(4 - 7)$. ***(LESSON 1.3)***

**39.** Find the opposite of $-5$. ***(LESSON 2.1)***

**40.** Evaluate: $8 + (-3.4)$. ***(LESSON 2.2)***

**41.** Evaluate: $-14 \div \left(-\frac{1}{2}\right)$. ***(LESSON 2.4)***

**42.** Let $q = 6$, $r = -3$, and $s = -2$.
Evaluate: $q + r - s$. ***(LESSON 2.3)***

**Find the value of each expression when the variable is 4. *(LESSON 1.2)***

**43.** $3p$

**44.** $6y - 7$

**45.** $2m + 13$

**46.** $-2s + 9$

**47.** Find the second difference for the sequence 1, 3, 8, 16, 27, . . . ***(LESSON 1.1)***

**48.** Evaluate: $|-7|$. ***(LESSON 2.1)***

**49.** Evaluate: $\frac{(-3)(-7)}{42}$. ***(LESSON 2.4)***

# Equations

SOLVING EQUATIONS IS OFTEN NECESSARY WHEN working with problems that require a number for an answer. For example, in business, equations are used to determine bank interest and sales commissions. In sports, batting averages and other statistics are often determined by using equations. You can use an equation to determine how many miles per gallon of gasoline your car can be driven.

## Lessons

CHAPTER PORTFOLIO ACTIVITIES PROJECT

## About the Chapter Project

Throughout Chapter 3 you will be introduced to algebraic procedures that will expand your mathematical problem-solving skills. These skills include recognizing a pattern in a numeric relationship that can be modeled by an equation. In the Chapter Project, *Track Team Schemes*, on page 154 you will apply these skills to a fund-raising scenario. After completing the Chapter Project, you will be able to do the following:

- Write and solve multistep equations.

## About the Portfolio Activities

Throughout the chapter, you will be given opportunities to complete Portfolio Activities that are designed to support your work on the Chapter Project.

- Egyptian equation-solving techniques are the topic of the Portfolio Activity on page 121.
- The Portfolio Activity on page 128 introduces a problem concerning the sale of T-shirts by a high school track team.
- The Portfolio Activity on page 140 extends the T-shirt sale problem from page 128.
- The Portfolio Activity on page 153 presents a famous problem about Greek mathematician Diophantus.

3.1

# Solving Equations by Adding and Subtracting

**Objective**

- Solve equations by using addition and subtraction.

**Why** *Many real-world situations, such as purchasing problems, can be represented by equations involving addition and subtraction.*

## Solving Equations by Using Subtraction

**APPLICATION**
**PERSONAL FINANCE**

Think about the simple problem below. How can it be modeled by a variable and an equation?

> Riva has \$20 and wants to purchase a pair of boots that cost \$32. How much more money does she need to buy the boots?

The situation can be modeled by the following equation, where $x$ represents the additional amount in dollars that Riva needs:

$$20 + x = 32, \text{ or } x + 20 = 32$$

Tiles can be used to represent the steps in solving the equation. Suppose that the $x$-tile represents an unknown number of unit tiles. The $x$-tile and the unit tiles represent the equation $x + 20 = 32$.

*The oblong tile represents x.*

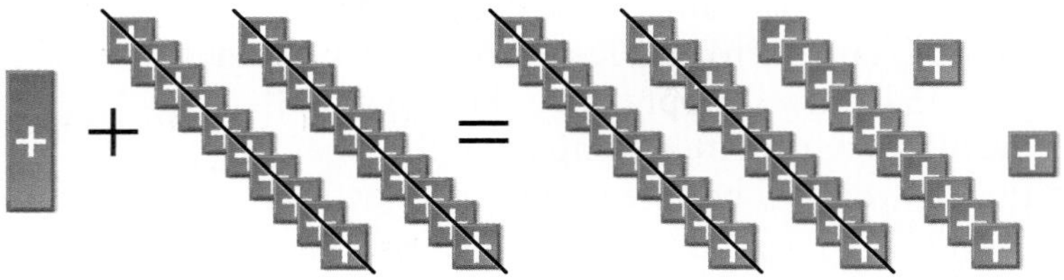

Take away 20 unit tiles from each side of the equation. The quantities on either side of the equal sign remain equal. The $x$-tile is now isolated. Count the remaining tiles on the other side of the equal sign to find the value of $x$.

Subtracting the same number of tiles from each side of an equation models the Subtraction Property of Equality.

### Subtraction Property of Equality

For all real numbers $a$, $b$, and $c$, if $a = b$, then $a - c = b - c$.

Subtracting equal amounts from each side of an equation results in an equivalent equation.

**EXAMPLE 1** **Use the Subtraction Property of Equality to solve each equation. Check the solution.**

**a.** $x + 7 = 9$ **b.** $8 + a = 2$ **c.** $10.7 = 6.2 + m$

**SOLUTION**

**a.** Subtract 7 from each side of the equation.

$$x + 7 = 9$$
$$x + 7 - 7 = 9 - 7$$
$$x = 2$$

To check the solution, substitute 2 for $x$ in the original equation.

$$2 + 7 \stackrel{?}{=} 9$$
$$9 = 9 \quad \textit{True}$$

**b.** Subtract 8 from each side of the equation.

$$8 + a = 2$$
$$8 + a - 8 = 2 - 8$$
$$a = -6$$

To check the solution, substitute $-6$ for $a$ in the original equation.

$$8 + (-6) \stackrel{?}{=} 2$$
$$2 = 2 \quad \textit{True}$$

**c.** Subtract 6.2 from each side of the equation.

$$10.7 = 6.2 + m$$
$$10.7 - 6.2 = 6.2 + m - 6.2$$
$$4.5 = m$$
$$m = 4.5 \quad \textit{Sym. Prop.}$$

To check the solution, substitute 4.5 for $m$ in the original equation.

$$10.7 \stackrel{?}{=} 6.2 + 4.5$$
$$10.7 = 10.7 \quad \textit{True}$$

**TRY THIS** Solve each equation. Check the solution.

**a.** $a + 5 = 4$ **b.** $\frac{1}{2} + n = 11\frac{1}{2}$ **c.** $27.2 = h + 45.6$

**EXAMPLE 2** A school band needs new uniforms. So far they have raised \$2344.10 toward a goal of \$5000. **Write and solve an equation to find how much more money the band must raise.**

**PROBLEM SOLVING**

**SOLUTION**

**Organize the information.**
Unknown: the amount still needed
Known: the amount already raised

**Write a sentence with words and operation symbols.**
the amount still needed + the amount already raised = the goal

**Write an equation.** Let $x$ represent the amount still needed in dollars.
$x + 2344.10 = 5000$

**Solve the equation.**

$$x + 2344.10 - 2344.10 = 5000 - 2344.10 \quad \textit{Subtr. Prop. of Eq.}$$
$$x = 2655.90$$

The band needs to raise \$2655.90.

## Solving Equations by Using Addition

### Activity
### Modeling the Addition Property of Equality

**You will need:** algebra tiles

Use algebra tiles to model the process of solving the equation $x - 2 = 8$.

1. Place 1 $x$-tile and 2 negative unit tiles on the left side of the work area and 8 unit tiles on the right side of the work area. How many positive unit tiles must be added to each side to isolate the $x$-tile?
2. Simplify the model by removing the neutral pairs. What is the solution?

CHECKPOINT ✔

3. Fill in the blanks to show the process for arriving at the solution.

$$x - 2 + \_\_\_\_ = 8 + \_\_\_\_$$
$$x = \_\_\_\_$$

Equations involving subtraction are solved by using addition.

**Addition Property of Equality**

For all real numbers $a$, $b$, and $c$, if $a = b$, then $a + c = b + c$.

Adding equal amounts to each side of an equation results in an equivalent equation.

**EXAMPLE 3** **Use the Addition Property of Equality to solve each equation. Check the solution.**

**a.** $x - 3 = 2$ **b.** $-12 = a - 8$ **c.** $m - 3.2 = 5.6$

**SOLUTION**

**a.** Add 3 to each side of the equation.

$$x - 3 = 2$$
$$x - 3 + 3 = 2 + 3$$
$$x = 5$$

To check the solution, substitute 5 for $x$ in the original equation.

$$5 - 3 \stackrel{?}{=} 2$$
$$2 = 2 \quad \textit{True}$$

**b.** Add 8 to each side of the equation.

$$-12 = a - 8$$
$$-12 + 8 = a - 8 + 8$$
$$-4 = a$$
$$a = -4 \quad \textit{Sym. Prop.}$$

To check the solution, substitute $-4$ for $a$ in the original equation.

$$-12 \stackrel{?}{=} -4 - 8$$
$$-12 = -12 \quad \textit{True}$$

**c.** Add 3.2 to each side of the equation.

$$m - 3.2 = 5.6$$
$$m - 3.2 + 3.2 = 5.6 + 3.2$$
$$m = 8.8$$

To check the solution, substitute 8.8 for $m$ in the original equation.

$$8.8 - 3.2 \stackrel{?}{=} 5.6$$
$$5.6 = 5.6 \quad \textit{True}$$

**TRY THIS** Solve each equation. Check the solution.

**a.** $a - 5 = 7$ **b.** $-8 = n - 3$ **c.** $h - 4.6 = 2.2$

**EXAMPLE 4** **Solve each equation. Check the solution.**

**a.** $5 - x = 7$ **b.** $13 - m = 7$

**SOLUTION**

**a.**

$$5 - x = 7$$
$$5 - x - 5 = 7 - 5 \quad \textit{Subtraction Property of Equality}$$
$$-x = 2$$
$$x = -2 \quad \textit{The opposite of x is 2; therefore, x is −2.}$$

**CHECK**

$$5 - (-2) \stackrel{?}{=} 7$$
$$7 = 7 \quad \textit{True}$$

**b.**

$$13 - m = 7$$
$$13 - m - 13 = 7 - 13 \quad \textit{Subtraction Property of Equality}$$
$$-m = -6$$
$$m = 6 \quad \textit{The opposite of m is −6; therefore, m is 6.}$$

**CHECK**

$$13 - 6 \stackrel{?}{=} 7$$
$$7 = 7 \quad \textit{True}$$

**EXAMPLE 5** In a dart game, the range of scores was 47 points, and the lowest score was 52. **Write and solve an equation to find the highest score in this dart game.**

CONNECTION
STATISTICS

PROBLEM SOLVING

**SOLUTION**

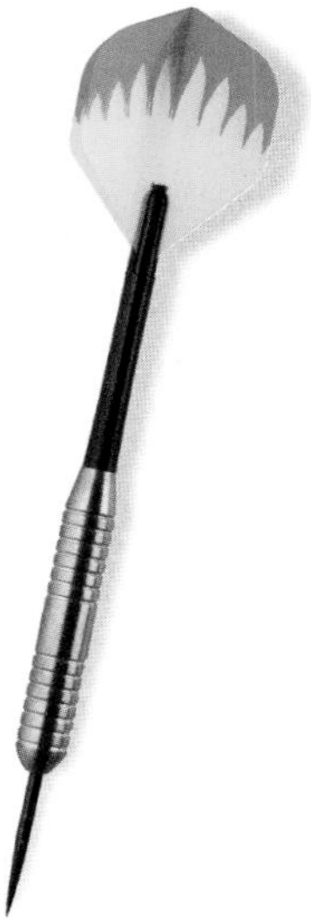

**ORGANIZE**

The range is the difference between the highest and lowest scores.

range = highest score − lowest score

Because the highest score is unknown, let $h$ represent this value.

**WRITE**

range = highest score − lowest score

$$47 = h - 52$$

**SOLVE**

$$47 = h - 52$$
$$47 + 52 = h - 52 + 52 \quad \textit{Add. Prop. of Eq.}$$
$$99 = h$$
$$h = 99 \quad \textit{Symmetric Property of Equality}$$

**CHECK**

Substitute 99 for $h$ in the equation $47 = h - 52$.
Since $47 = 99 - 52$, the highest score was 99.

**TECHNOLOGY**
**GRAPHICS CALCULATOR**

The equation $x + 7 = 10$ is the linear equation $x + 7 = y$ when $y = 10$. You can find the solution to the equation by graphing the lines $y = x + 7$ and $y = 10$ and finding the $x$-value at the intersection point.

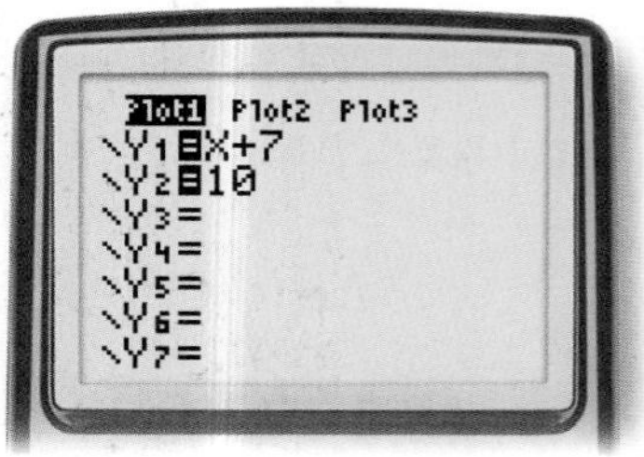


*Enter the equations in the* **Y=** *window.*

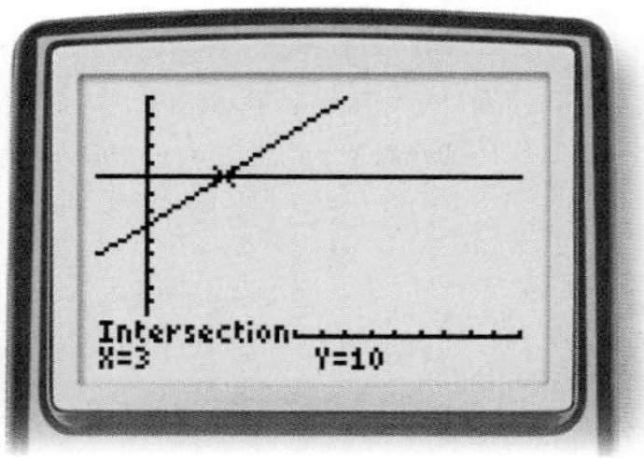


*Find the point of intersection: $x = 3$ when $y = 10$.*

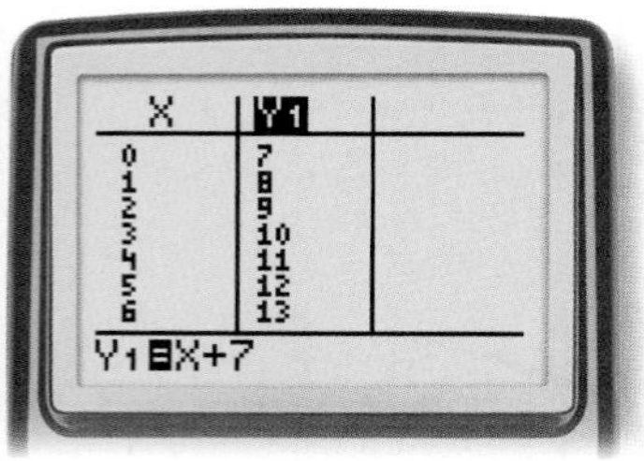


*You can also observe values for $y = x + 7$ in the table window to confirm that $x = 3$ when $y = 10$.*

# Exercises

## Communicate

**1.** Compare the Addition Property of Equality with the Subtraction Property of Equality. Explain how to decide which one to use to solve an equation.

**2.** Describe situations that can be modeled by the following equations:
**a.** $x + 20 = 50$ **b.** $x - 20 = 50$

**3.** Explain how to solve each equation by using algebra tiles.
**a.** $x + 6 = 10$ **b.** $x - 6 = 10$

**4.** Explain how to solve $10 = x + 7$ algebraically.

## Guided Skills Practice

**Solve each equation by using the Subtraction Property of Equality. Check the solution.** *(EXAMPLE 1)*

**5.** $x + 7 = 31$ **6.** $-7.2 = x + 3.5$ **7.** $3\frac{1}{4} + m = 5\frac{3}{8}$

**8.** Jill has saved \$137 toward the purchase of a mountain bike. The bike costs \$547. Write and solve an equation to find out how much more money she needs. *(EXAMPLE 2)*

**Solve each equation by using the Addition Property of Equality. Check the solution.** *(EXAMPLE 3)*

**9.** $y - 7.5 = -2.7$ **10.** $32 = m - 3$ **11.** $r - 4\frac{1}{3} = 1\frac{1}{2}$

**12.** $x - 7 = 10$ **13.** $p - 10 = -15$ **14.** $m - 3.5 = 7.5$

**Solve each equation. Check the solution.** ***(EXAMPLE 4)***

**15.** $4 - m = 7$ **16.** $10 - x = 2$ **17.** $13 = 7 - x$

**18.** The difference between the normal level and the 50-year flood level of a lake is 29 feet. If the normal level is 690 feet, what is the 50-year flood level? ***(EXAMPLE 5)***

## Practice and Apply

**Solve each equation. Check the solution.**

**internet connect**
**Homework Help Online**
Go To: **go.hrw.com**
Keyword:
**MA1 Homework Help**
for Exercises 19-63

**19.** $a - 16 = 15$ **20.** $a - 4 = -10$ **21.** $11 = t + 29$

**22.** $m + 54 = 36$ **23.** $r - 10 = -80$ **24.** $g - 27 = 148$

**25.** $b - 109 = 58$ **26.** $24 = h - 53$ **27.** $y + 37 = -110$

**28.** $396 = z + 256$ **29.** $819 = g + 75$ **30.** $1\frac{1}{4} = x + \frac{3}{4}$

**31.** $x + 9 = 6$ **32.** $x - 10 = -4$ **33.** $x + 6 = -4$

**34.** $y + \frac{2}{3} = \frac{7}{9}$ **35.** $3 = g - 1.2$ **36.** $b - 4.4 = 7$

**37.** $c + 2.1 = 3.5$ **38.** $d + 8.7 = 11.9$ **39.** $k - 2.1 = -6.3$

**40.** $j - 6.8 = -12.4$ **41.** $g - \frac{2}{6} = \frac{5}{6}$ **42.** $1\frac{2}{7} = x + \frac{4}{5}$

**43.** $y - 7.4 = 8.3$ **44.** $-3.7 = y - 1.2$ **45.** $x + 4.2 = -6.1$

**46.** $5 + m = 10$ **47.** $-8 + x = 4$ **48.** $9 - x = 3$

**49.** $-3 - p = 2$ **50.** $6 - y = -3$ **51.** $-4 - m = -3$

**52.** $x + 6.2 = -5.3$ **53.** $456 = a - 529$ **54.** $d + 904 = -759$

**55.** $t + \frac{6}{7} = \frac{2}{3}$ **56.** $t - \frac{3}{5} = \frac{1}{2}$ **57.** $k + 5.8 = -3.2$

**58.** $h + 9.1 = -5.3$ **59.** $n - \frac{7}{9} = \frac{1}{6}$ **60.** $-\frac{1}{3} = j + \frac{1}{5}$

**61.** $m + 2.3 = 7$ **62.** $15 = x - 4.6$ **63.** $x + \frac{1}{2} = 1$

**In Exercises 64–76, write an equation for each situation. Then solve the equation.**

**CONNECTIONS**

**64.** **STATISTICS** The range of a set of scores is 28. If the highest score is 47, what is the lowest score?

**65.** **GEOMETRY** Supplementary angles are pairs of angles whose measures total 180°. Determine the measure of an angle that is supplementary to an angle with a measure of 92°.

**66.** **GEOMETRY** The measures of the interior angles of a triangle total 180°. Two angles of a triangle have a measure of 50° each. Find the measure of the third angle.

**CONNECTIONS**

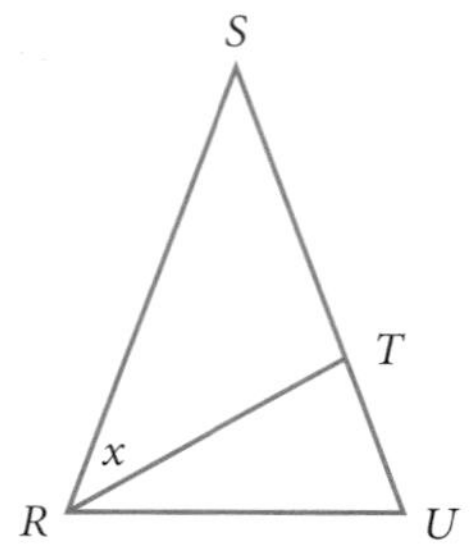


67. **GEOMETRY** Angle *SRU* of triangle *RSU* measures 70°. Angle *TRU* measures 30°. Write and solve an equation to find the measure of angle *SRT*.

68. **GEOMETRY** Angle *SRU* of triangle *RSU* measures 70°. Angle *TRU* measures 18°. Write and solve an equation to find the measure of angle *SRT*.

**APPLICATIONS**

69. **PERSONAL FINANCE** Lisa wants to buy a $53 computer program and has $67. How much will she have left after she buys the program?

70. **PERSONAL FINANCE** Robert has saved $23 more than he needs to buy a tennis racket that costs $49. How much has he saved?

71. **PERSONAL FINANCE** Sandy has saved $23 less than she needs to buy a tennis racket that costs $49. How much has she saved?

72. **SPORTS** The first- and second-string running backs on a football team ran for a total of 94 yards. If the first-string back ran 89 yards, how many yards did the second-string back run?

73. **TRAVEL** If the odometer on Sarah's car registered 23,580 when she finished a vacation trip of 149 miles, what did it register when she started the trip?

74. **TRAVEL** If the odometer on Sarah's car registered 23,580 before a vacation trip of 149 miles, what did it register when she finished her trip?

75. **CALENDAR** Carl sees a calendar which indicates that his birthday, December 21, is the 355th day of the year and that the current day, October 15, is the 288th day of the year. How many days away is his birthday?

76. **CONSUMER ECONOMICS** Sam bought two items that cost $12.98 and $14.95, plus tax. He got $0.39 in change from $30. How much was the tax?

## Look Back

**77.** Find the sum $1 + 2 + 3 + \cdots + 12$. ***(LESSON 1.1)***

**78.** State the pattern and find the next three numbers in the sequence 6, 0, 12, 6, 18, 12, ... ***(LESSON 1.1)***

**79.** Suppose that a rocket took 23 seconds from takeoff to its return to Earth. Find how many seconds after takeoff the rocket reached its maximum height. ***(LESSON 1.1)***

**80.** Find the perimeter of a square with sides that are 2.5 inches long. ***(LESSON 2.2)***

**81.** If poster board costs \$0.39 a piece, how many pieces of poster board can Shannon buy for \$2.50? ***(LESSON 2.4)***

**Perform the indicated operations.** ***(LESSON 2.6)***

**82.** $(3x - 7) - (x - 4)$ **83.** $(8 - 7y) + 2y$ **84.** $(n - 2) - (-3n + 4)$

internet connect

**Portfolio Extension**

Go To: **go.hrw.com**
Keyword:
**MA1 Egyptian**

## Look Beyond

**Show how to solve these equations by using algebra tiles.**

**85.** $3x = 18$ **86.** $2x = 12$ **87.** $2x + 5 = 11$

PORTFOLIO ACTIVITY

Include this activity in your portfolio.

**Egyptian Equation Solving**
About 4000 years ago, Egyptian mathematicians wrote textbooks to prepare students for government and business jobs. In 1858, a Scotsman named Rhind bought one of these textbooks, which had been copied on papyrus by a scribe named Ahmes. Part of this papyrus is shown in the photo at right.

Problem 70 of the Rhind papyrus asks, "What is the amount of meal in each loaf of bread if 2520 *ro* of meal is made into 100 loaves?" In the form of an equation, this is written as $100m = 2520$.

The Egyptians solved the equation by using the table method and by guess-and-check.

**100*m* = 2520**

| $m$ | 1 | 10 | 20 | 5 | $\frac{1}{5}$ |
|---|---|---|---|---|---|
| **100*m*** | 100 | 1000 | 2000 | 500 | 20 |

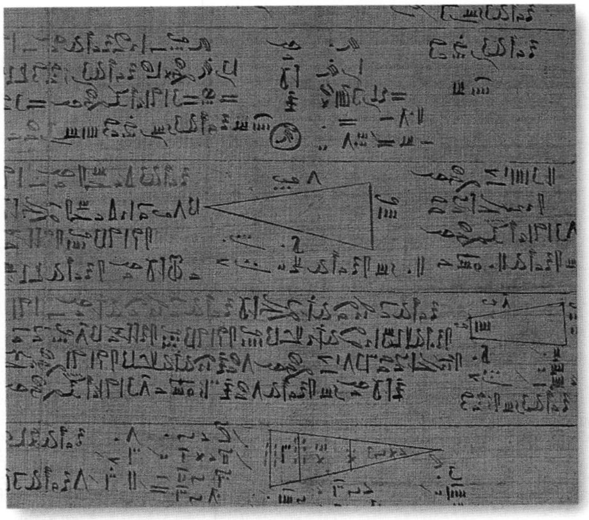

For the equation $100m = 2520$, they needed to multiply $100(20) + 100(5) + 100\left(\frac{1}{5}\right)$ on the left side to get 2520 on the right side. They found that $m$ was $20 + 5 + \frac{1}{5}$, or $25\frac{1}{5}$ *ro*.

Solve these equations as the Egyptians might have.

**1.** $40m = 4800$ **2.** $60j = 930$

**3.** $24p = 252$ **4.** $10w = 2432$

# 3.2 Solving Equations by Multiplying and Dividing

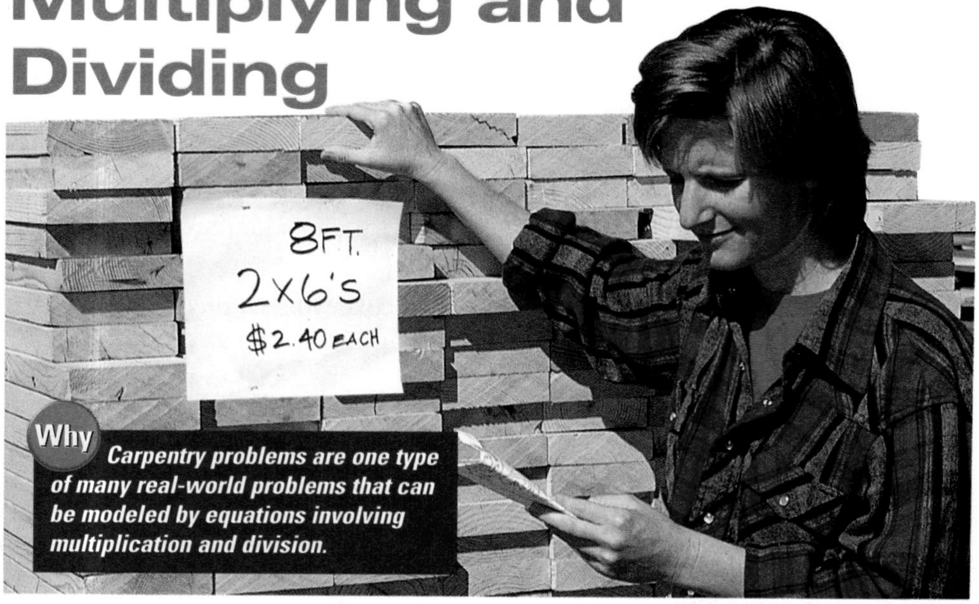


**Why** *Carpentry problems are one type of many real-world problems that can be modeled by equations involving multiplication and division.*

**Objective**

- Solve equations by using multiplication and division.

**APPLICATION**
**CARPENTRY**

Alicia needs to purchase twenty 8-foot boards to complete the framing of a house. In order to compare costs, she needs to know the cost per foot. At one store she finds 8-foot boards for $2.40 each. What is the cost per foot?

## Solving Equations by Using Division

You can use the Division Property of Equality to solve problems such as the one presented above. (See Example 3.)

**The Division Property of Equality**

For all real numbers $a$, $b$, and $c$, if $a = b$ and $c \neq 0$, then $\frac{a}{c} = \frac{b}{c}$.

Dividing each side of an equation by equal amounts results in an equivalent equation.

**EXAMPLE 1** **Solve the equation $3x = -9$. Check the solution.**

**SOLUTION**

Divide each side of the equation by 3.

$$3x = -9$$
$$\frac{3x}{3} = \frac{-9}{3}$$
$$x = -3$$

To check the solution, substitute −3 for $x$ in the original equation.

$$3(-3) \stackrel{?}{=} -9$$
$$-9 = -9 \quad \textit{True}$$

**TRY THIS** Solve the equation $6x = 72$. Check the solution.

**EXAMPLE 2** **Solve the equation $2 = 7a$. Check the solution.**

**SOLUTION**

Divide each side of the equation by 7.

$$2 = 7a$$
$$\frac{2}{7} = \frac{7a}{7}$$
$$\frac{2}{7} = a$$
$$a = \frac{2}{7} \quad \textit{Sym. Prop.}$$

To check the solution, substitute $\frac{2}{7}$ for $a$ in the original equation.

$$2 \stackrel{?}{=} 7\left(\frac{2}{7}\right)$$
$$2 = 2 \quad \textit{True}$$

**TRY THIS** Solve the equation $-3 = 15b$. Check the solution.

**EXAMPLE 3** **A store sells 8-foot boards for \$2.40 each. What is the cost per foot?**

**APPLICATION**
**CARPENTRY**

**SOLUTION**

Let $x$ be the cost in dollars per foot. Then the equation $8x = 2.40$ models this situation.

Divide each side of the equation by 8.

$$8x = 2.40$$
$$\frac{8x}{8} = \frac{2.40}{8}$$
$$x = 0.30$$

To check the solution, substitute 0.30 for $x$ in the original equation.

$$8(0.30) \stackrel{?}{=} 2.40$$
$$2.40 = 2.40 \quad \textit{True}$$

The boards cost \$0.30 per foot.

**TRY THIS** A store sells 12-foot boards for \$4.80 each. What is the cost per foot?

**EXAMPLE 4** In a regular pentagon, all of the interior angles are equal in measure. **The sum of the angle measures is 540°. Write and solve an equation to find the measure of each angle.**

**CONNECTION**
**GEOMETRY**

**SOLUTION**

The total number of angles is 5. The sum of the measures of the angles is 540°. Let $x$ be the measure of each angle. Then $5x$ is the sum of the measures of all 5 angles and $5x = 540$.

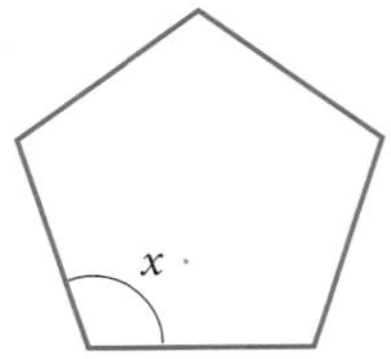


**Solve for $x$.**

$$5x = 540$$
$$\frac{5x}{5} = \frac{540}{5} \quad \textit{Div. Prop. of Equality}$$
$$x = 108 \quad \textit{Simplify.}$$

**CHECK**

$$5(108) \stackrel{?}{=} 540$$
$$540 = 540 \quad \textit{True}$$

Each angle has a measure of 108°.

## Solving Equations by Using Multiplication

**APPLICATION**
**SPORTS**

To compute a batting average, the number of hits is divided by the number of times at bat.

$$\frac{\text{Hits}}{\text{Times at bat}} = \text{Average}$$

Suppose that Alex batted 150 times during his varsity baseball career. If his final batting average is 0.300, how many hits did he get?

Let $H$ represent the number of hits. The problem can be modeled by the following division equation:

$$\frac{H}{150} = 0.300$$

In algebraic notation, $\frac{H}{150}$ is the same as $\frac{1}{150}(H)$. To solve the equation, multiply both sides of the equation by 150 because 150 is the reciprocal of $\frac{1}{150}$, the coefficient of $H$.

$$150\left(\frac{H}{150}\right) = 150(0.300)$$
$$H = 45$$

Alex had 45 hits in his varsity baseball career.

### The Multiplication Property of Equality

For all real numbers $a$, $b$, and $c$, if $a = b$, then $ac = bc$.

Multiplying each side of an equation by equal amounts results in an equivalent equation.

**EXAMPLE 5** **Solve each equation. Check the solution.**

**a.** $\frac{m}{5} = 1.3$ **b.** $-3 = \frac{x}{-4}$ **c.** $\frac{w}{5} = -10$

**SOLUTION**

**a.** Multiply each side of the equation by 5.

$$\frac{m}{5} = 1.3$$
$$(5)\frac{m}{5} = (5)1.3$$
$$m = 6.5$$

To check the solution, substitute 6.5 for $m$ in in the original equation.

$$\frac{6.5}{5} \stackrel{?}{=} 1.3$$
$1.3 = 1.3$ *True*

**b.** Multiply each side of the equation by −4.

$$-3 = \frac{x}{-4}$$
$$(-4)(-3) = (-4)\frac{x}{-4}$$
$$12 = x$$
$$x = 12$$

To check the solution, substitute 12 for $x$ in the original equation.

$$-3 \stackrel{?}{=} \frac{12}{-4}$$
$-3 = -3$ *True*

**c.** Multiply each side of the equation by 5.

$$\frac{w}{5} = -10$$
$$(5)\frac{w}{5} = (5)(-10)$$
$$w = -50$$

To check the solution, substitute −50 for $w$ the original equation.

$$\frac{-50}{5} \stackrel{?}{=} -10$$
$-10 = -10$ *True*

**TRY THIS** Solve each equation. Check the solution.

**a.** $\frac{t}{-3.2} = 5$ **b.** $\frac{d}{3} = -5.4$ **c.** $3 = \frac{m}{1.5}$

## Using Reciprocals

Recall that in order to divide by a nonzero number, you multiply by the reciprocal of that number. For example, $10 \div 5$ can be rewritten as $10 \times \frac{1}{5}$. The value of both expressions is the same, 2.

Note the following:

$$\begin{aligned} \frac{a}{b} \bullet \frac{b}{a} &= \frac{ab}{ba} \\ &= \frac{ab}{ab} \\ &= 1 \end{aligned}$$

This tells you that the reciprocal of a rational number, $\frac{a}{b}$, is $\frac{b}{a}$. Reciprocals can be used to solve multiplication and division equations.

**EXAMPLE 6** **Solve each equation by using reciprocals. Check each solution.**

**a.** $\frac{3}{5}y = 10$ **b.** $-\frac{1}{3} = 5x$ **c.** $\frac{2b}{3} = -10$

**SOLUTION**

**a.** The reciprocal of the coefficient of $y$ is $\frac{5}{3}$. Multiply each side of the equation by $\frac{5}{3}$.

$$\left(\frac{5}{3}\right)\frac{3}{5}y = \left(\frac{5}{3}\right)10$$

$$y = \frac{50}{3}, \text{ or } 16\frac{2}{3}$$

To check the solution, substitute $\frac{50}{3}$ for $y$ in the original equation.

$$\frac{3}{5}\left(\frac{50}{3}\right) \stackrel{?}{=} 10$$

$$\frac{50}{5} \stackrel{?}{=} 10$$

$$10 = 10 \quad \textit{True}$$

**b.** The reciprocal of the coefficient of $x$ is $\frac{1}{5}$. Multiply each side of the equation by $\frac{1}{5}$.

$$\left(\frac{1}{5}\right)\left(-\frac{1}{3}\right) = \left(\frac{1}{5}\right)5x$$

$$-\frac{1}{15} = x$$

$$x = -\frac{1}{15} \quad \textit{Sym. Prop.}$$

To check the solution, substitute $-\frac{1}{15}$ for $x$ in the original equation.

$$-\frac{1}{3} \stackrel{?}{=} 5\left(-\frac{1}{15}\right)$$

$$-\frac{1}{3} \stackrel{?}{=} -\frac{5}{15}$$

$$-\frac{1}{3} = -\frac{1}{3} \quad \textit{True}$$

**c.** The reciprocal of the coefficient of $b$ is $\frac{3}{2}$. Multiply each side of the equation by $\frac{3}{2}$.

$$\left(\frac{3}{2}\right)\frac{2b}{3} = \left(\frac{3}{2}\right)(-10)$$

$$b = -\frac{30}{2}, \text{ or } -15$$

To check the solution, substitute $-15$ for $b$ in the original equation.

$$\frac{2(-15)}{3} \stackrel{?}{=} -10$$

$$-\frac{30}{3} \stackrel{?}{=} -10$$

$$-10 = -10 \quad \textit{True}$$

**TRY THIS** Solve by using reciprocals. Check the solution.

**a.** $4m = -\frac{3}{8}$ **b.** $\frac{2d}{5} = \frac{3}{2}$ **c.** $-\frac{4}{5} = 3x$

**CRITICAL THINKING** Explain how multiplying by a reciprocal instead of dividing is similar to adding the opposite instead of subtracting.

**EXAMPLE 7** An apple pie is cut into 6 pieces. Each piece contains a total of 18 grams of fat. **Write and solve an equation to determine how many grams of fat are in the whole pie.**

APPLICATION
FOOD

PROBLEM SOLVING

**SOLUTION**

**Write an equation.** Let $x$ be the number of grams of fat in the whole pie. Since the pie is cut into 6 equal pieces, the equation is as follows:

$$\frac{1}{6}x = 18$$

Solve for $x$.

$$\frac{1}{6}x = 18$$
$$(6)\left(\frac{1}{6}x\right) = (6)(18)$$
$$x = 108$$

The whole pie contains 108 grams of fat.

## Exercises

### Communicate

**Explain how to solve the following equations for $x$:**

**1.** $592x = 812$ **2.** $5x = \frac{1}{10}$ **3.** $-21 = \frac{x}{12}$ **4.** $\frac{4x}{5} = \frac{1}{2}$

CONNECTION

**5.** GEOMETRY From the formula for the circumference of a circle, explain how to find the diameter, $d$, in terms of the circumference, $C$, and $\pi$.

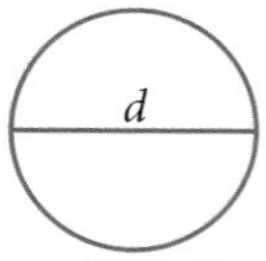


### Guided Skills Practice

**Solve each equation. Check the solution.** ***(EXAMPLES 1, 2, AND 3)***

**6.** $-14x = 28$ **7.** $0.5x = -10$ **8.** $4 = 16x$ **9.** $-3m = 15$

CONNECTION

**10.** GEOMETRY The sum of the measures of the interior angles of a regular octagon is 940°. Write and solve an equation to find the measure of each angle. ***(EXAMPLE 4)***

**Solve each equation. Check the solution.** ***(EXAMPLE 5)***

**11.** $\frac{x}{2} = -3$ **12.** $\frac{x}{-5} = 1.2$ **13.** $4 = \frac{x}{2.1}$ **14.** $\frac{m}{-3} = -2.1$

**Use reciprocals to solve each equation. Check your solution.**
***(EXAMPLES 6 AND 7)***

**15.** $\frac{3x}{2} = -18$ **16.** $5x = \frac{1}{4}$ **17.** $\frac{x}{4} = \frac{3}{4}$ **18.** $\frac{2}{3}m = -\frac{3}{2}$

**19.** **FOOD** A chocolate cake is cut into 8 pieces. Each piece contains a total of 14 grams of fat. Write and solve an equation to determine how many grams of fat are in the whole cake.

## *Practice and Apply*

**internet connect**
**Homework Help Online**
Go To: **go.hrw.com**
Keyword:
**MA1 Homework Help** for Exercises 20-52

**Solve each equation and check your solution.**

| | | |
|---|---|---|
| **20.** $7x = 56$ | **21.** $-7m = -14$ | **22.** $-3x = 9$ |
| **23.** $5.6v = 7$ | **24.** $-13 = \frac{y}{3}$ | **25.** $\frac{b}{-9} = 6$ |
| **26.** $\frac{x}{27} = -26$ | **27.** $\frac{p}{-9} = 0.9$ | **28.** $84 = -12y$ |
| **29.** $111x = -888$ | **30.** $-3x = -4215$ | **31.** $-4x = -3228$ |
| **32.** $\frac{x}{-7} = -1.4$ | **33.** $6 = \frac{x}{0.5}$ | **34.** $7 = -56w$ |
| **35.** $-3f = 15$ | **36.** $888x = 111$ | **37.** $0.505 = 0.505x$ |
| **38.** $4b = -15$ | **39.** $2a = 1$ | **40.** $\frac{p}{111} = -10$ |
| **41.** $2a = 13$ | **42.** $\frac{m}{-9} = 0$ | **43.** $\frac{b}{15} = 1$ |
| **44.** $\frac{2x}{3} = \frac{3}{4}$ | **45.** $-\frac{7x}{5} = \frac{3}{10}$ | **46.** $-\frac{1}{4} = \frac{x}{2}$ |
| **47.** $\frac{-m}{4} = 2$ | **48.** $-4p = 16$ | **49.** $\frac{x}{-7} = 5$ |
| **50.** $-2w = 13$ | **51.** $\frac{w}{-5} = 10$ | **52.** $\frac{-x}{3} = 10$ |

**APPLICATIONS**

**BUSINESS** Max is apprenticing as a carpenter. Write and solve an equation to describe each situation in Exercises 53–57.

**53.** If one roll of masking tape costs \$1.15, how many rolls can Max buy with \$5.75?

**54.** Max finds masking tape in packages of 4 rolls for \$4.32. What is the cost of one roll in the package? Is the cost of a single roll of masking tape (from Exercise 53) more than the cost of one roll in a package of 4.

**55.** Max's employer has \$18.00 to spend on tape measures for his crew. How many can he buy with that amount if each tape measure costs \$4.50?

**56.** What is the price of one extension cord if Max can buy a package of 6 for \$7.26?

**57.** Max buys a package of 4 AA-type batteries for \$2.52. What is the price for each battery?

**58.** **TRAVEL** Natalie's family wants to make a 400-mile drive in 8 hours. Write and solve an equation to determine what speed they must average for the trip.

**59.** **TRAVEL** Maria wants to drive 320 miles in 8 hours. Write and solve an equation to determine what speed she must average for the trip.

## Look Back

**60.** Insert inclusion symbols to make $30 \cdot 7 - 4 \cdot 10 \div 2 \div 2 = 225$ true. ***(LESSON 1.3)***

**Evaluate each absolute value.** ***(LESSON 2.1)***

**61.** $|0|$ **62.** $|-5|$ **63.** $\left|-\frac{1}{6}\right|$ **64.** $|9|$

**Evaluate each expression for $a = -3$, $b = 2$, and $c = 0$.** ***(LESSON 2.4)***

**65.** $\frac{a + b \cdot c}{-3}$ **66.** $\frac{a \cdot b}{b + c}$ **67.** $\frac{a + b}{b + c}$

**68.** Shawna started at ground level and walked down 13 steps. She noticed that she had forgotten her purse, so she walked 47 steps back up to her apartment. She then walked down 27 steps to meet her friend. How many steps from ground level is she? Show the numerical expression that models this situation. ***(LESSONS 2.2 AND 2.3)***

**internet connect**
**Activities Online**
Go To: **go.hrw.com**
Keyword:
**MA1 Oscar at Bat**

**Solve each equation. Check the solution.** ***(LESSON 3.1)***

**69.** $x + 5 = 7$ **70.** $x - 2.4 = 5.7$ **71.** $x + \frac{7}{5} = \frac{2}{3}$

## Look Beyond

**Solve each equation. Check the solution.**

**72.** $3y = 10 + 5y$ **73.** $4x + 0.5 = 5x$ **74.** $x + 4 = 3x - 2$

## PORTFOLIO ACTIVITY

Include this activity in your portfolio.

The track team at Hale High School plans to sell T-shirts to raise money for new equipment. In order for the track team to make a profit, the income from selling the T-shirts must be greater than the cost of making the T-shirts. At Ted's Tees, printing costs are \$0.80 per shirt and the cost for each T-shirt is \$3.75. The shop also charges a \$125 fee per order for the silk-screen design.

**1.** Based on the prices at Ted's Tees, determine the cost for ordering 100, 250, 500, and 750 T-shirts.

**2.** Determine the cost of a single printed T-shirt for orders of 100, 250, 500, and 750 T-shirts.

**3.** If the track team orders 250 T-shirts and wishes to make a \$500 profit, how much should they charge for each T-shirt?

**WORKING ON THE CHAPTER PROJECT**

You should now be able to complete Activity 1 of the Chapter Project on page 154.

3.3

# Solving Two-Step Equations

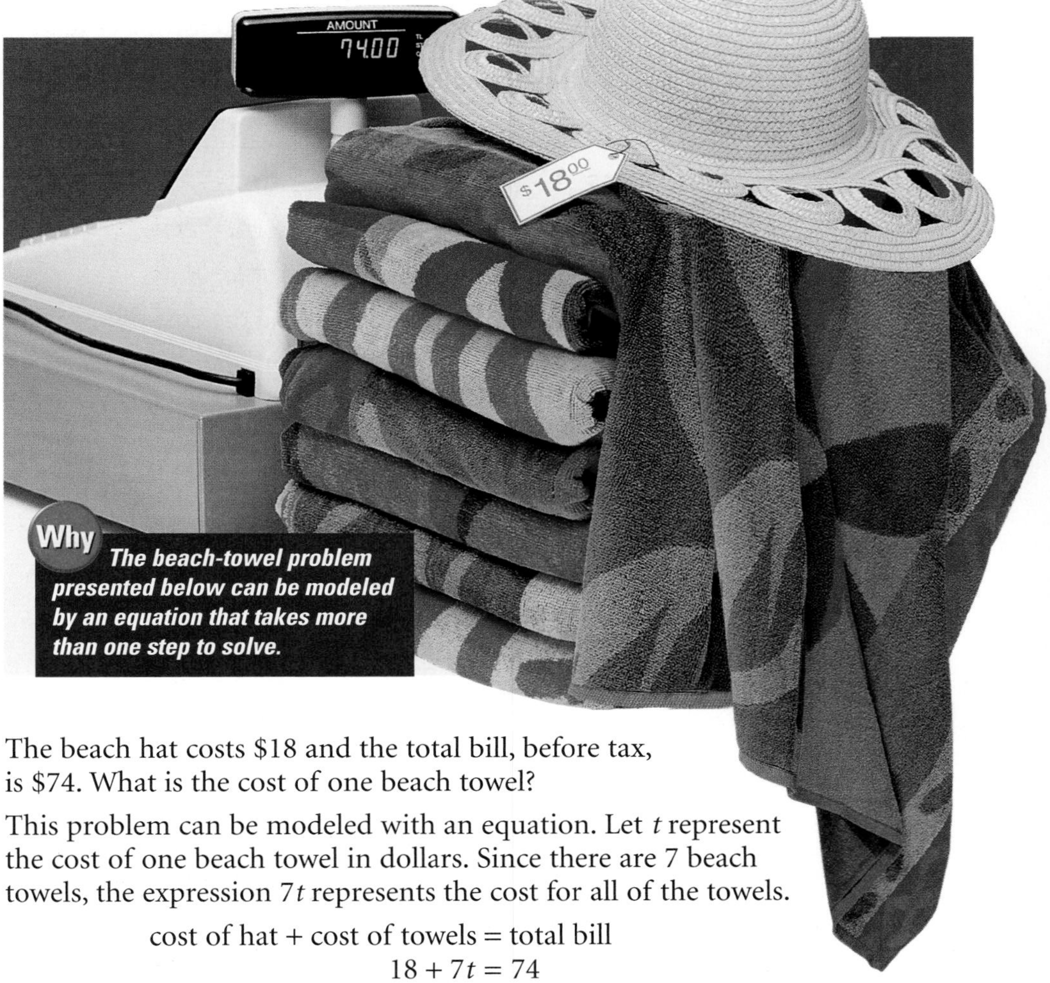


## Objectives

- Write equations that represent real-world situations.
- Solve two-step equations.

**Why** *The beach-towel problem presented below can be modeled by an equation that takes more than one step to solve.*

**APPLICATION**
**CONSUMER ECONOMICS**

The beach hat costs $18 and the total bill, before tax, is $74. What is the cost of one beach towel?

This problem can be modeled with an equation. Let $t$ represent the cost of one beach towel in dollars. Since there are 7 beach towels, the expression $7t$ represents the cost for all of the towels.

$$\text{cost of hat} + \text{cost of towels} = \text{total bill}$$
$$18 + 7t = 74$$

Solving the equation fot $t$ requires both a subtraction and a division step.

**EXAMPLE 1** **Solve $18 + 7t = 74$ for $t$ in order to find the cost of one beach towel.**

**SOLUTION**

| | | |
|---|---|---|
| $18 + 7t = 74$ | | |
| $18 + 7t - 18 = 74 - 18$ | *Subtr. Prop. Eq.* | |
| $7t = 56$ | *Simplify.* | |
| $\frac{7t}{7} = \frac{56}{7}$ | *Div. Prop. Eq.* | |
| $t = 8$ | | |

**CHECK**

$18 + 7(8) \stackrel{?}{=} 74$

$18 + 56 \stackrel{?}{=} 74$

$74 = 74$ *True*

One beach towel costs $8.

You can also approach the beach towel problem by using a table, as in the following Activity.

## Activity
## Using a Table

Solve the equation for the cost, $t$, of one beach towel by using a table. Complete the table through \$10 for the cost of a single towel.

| Cost of single towel (\$) | Cost of hat + cost of towels, $18 + 7t$ | Total (\$) |
|---|---|---|
| 0 | $18 + 7(0)$ | 18 |
| 1 | $18 + 7(1)$ | 25 |
| 2 | ? | ? |

1. Copy and complete the table above.

CHECKPOINT ✔

2. Explain how to determine the cost of a single towel when the total bill is \$74.
3. Describe other ways to solve the equation $18 + 7t = 74$ for $t$.

**EXAMPLE 2** **Solve.**

**a.** $10x + 3 = 33$

**b.** $-47 = 3x - 50$

**SOLUTION**

**a.**

$$10x + 3 = 33$$
$$10x + 3 - 3 = 33 - 3 \quad \textit{Subtr. Prop. Eq.}$$
$$10x = 30 \quad \textit{Simplify.}$$
$$\frac{10x}{10} = \frac{30}{10} \quad \textit{Div. Prop. Eq.}$$
$$x = 3$$

CHECK

$$10(3) + 3 \stackrel{?}{=} 33$$
$$30 + 3 \stackrel{?}{=} 33$$
$$33 = 33 \quad \textit{True}$$

**b.**

$$-47 = 3x - 50$$
$$-47 + 50 = 3x - 50 + 50 \quad \textit{Add. Prop. Eq}$$
$$3 = 3x \quad \textit{Simplify.}$$
$$\frac{3}{3} = \frac{3x}{3} \quad \textit{Div. Prop. Eq.}$$
$$1 = x$$

CHECK

$$-47 \stackrel{?}{=} 3(1) - 50$$
$$-47 \stackrel{?}{=} 3 - 50$$
$$-47 = -47 \quad \textit{True}$$

**TRY THIS** **Solve.**

**a.** $15y + 31 = 61$

**b.** $-35 = 2p + 10$

**EXAMPLE 3** **Solve.**

**a.** $\frac{m}{-5} + 3 = -2.5$

**b.** $5 = \frac{x}{3} - 14$

**SOLUTION**

**a.**

$$\frac{m}{-5} + 3 = -2.5$$
$$\frac{m}{-5} + 3 - 3 = -2.5 - 3 \quad \textit{Subtr. Prop. Eq.}$$
$$\frac{m}{-5} = -5.5 \quad \textit{Simplify.}$$
$$-5\left(\frac{m}{-5}\right) = -5(-5.5) \quad \textit{Mult. Prop. Eq.}$$
$$m = 27.5$$

CHECK

$$\frac{27.5}{-5} + 3 \stackrel{?}{=} -2.5$$
$$-5.5 + 3 \stackrel{?}{=} -2.5$$
$$-2.5 = -2.5 \quad \textit{True}$$

**b.**

$$5 = \frac{x}{3} - 14$$
$$5 + 14 = \frac{x}{3} - 14 + 14 \qquad \textit{Add. Prop. Eq.}$$
$$19 = \frac{x}{3} \qquad \textit{Simplify.}$$
$$3(19) = 3\left(\frac{x}{3}\right) \qquad \textit{Mult. Prop. Eq.}$$
$$x = 57$$

**CHECK**

$$5 \stackrel{?}{=} \frac{57}{3} - 14$$
$$5 \stackrel{?}{=} 19 - 14$$
$$5 = 5 \qquad \textit{True}$$

**TRY THIS**

**Solve.**

**a.** $-4.2 = \frac{m}{2} + 1.8$

**b.** $\frac{x}{-4} - 3 = 10$

**EXAMPLE 4**

**APPLICATION**
**CONSUMER ECONOMICS**

Jim bought 6 glasses and 6 mugs. Each mug cost \$5.98 and each glass cost \$2.99. The salesperson tells Jim that his total bill, before tax, is \$60.60. Jim estimates that 1 mug and 1 glass would cost between \$8 and \$9, so 6 mugs and 6 glasses should cost between \$48 and \$54. Jim thinks that \$60.60 is too high. Jim noticed that the correct price was scanned for the mugs. **What price did the scanner give for each of the glasses?**

**SOLUTION**

The cost for 6 mugs is \$5.98 • 6, or \$35.88.

Let $g$ represent the cost of one glass. The cost of 6 glasses is represented by the expression $6g$.

**PROBLEM SOLVING**

**Write an equation** that models the problem.

cost of mugs + cost of glasses = total bill

$$35.88 \quad + \quad 6g \quad = \quad 60.60$$

Solve for $g$.

$$35.88 + 6g = 60.60 \qquad \textit{Given}$$
$$35.88 - 35.88 + 6g = 60.60 - 35.88 \qquad \textit{Subtraction Property of Equality}$$
$$6g = 24.72 \qquad \textit{Simplify.}$$
$$\frac{6g}{6} = \frac{24.72}{6} \qquad \textit{Division Property of Equality}$$
$$g = 4.12 \qquad \textit{Simplify.}$$

Jim was charged \$4.12 instead of \$2.99 for each glass.

**TECHNOLOGY**
**GRAPHICS CALCULATOR**

The solution to the beach towels and hat problem presented at the beginning of the lesson (page 129) can be represented by the intersection of two lines. The equations for the lines are based on the two sides of the original equation, $18 + 7t = 74$.

$$y = 18 + 7x \qquad \text{Replace } t \text{ with } x.$$
$$y = 74$$

Enter the equations in a graphics calculator and graph the lines. The solution for $x$ is the value of the $x$-coordinate at the point of intersection. Use the trace or intersection feature to find the point of intersection.

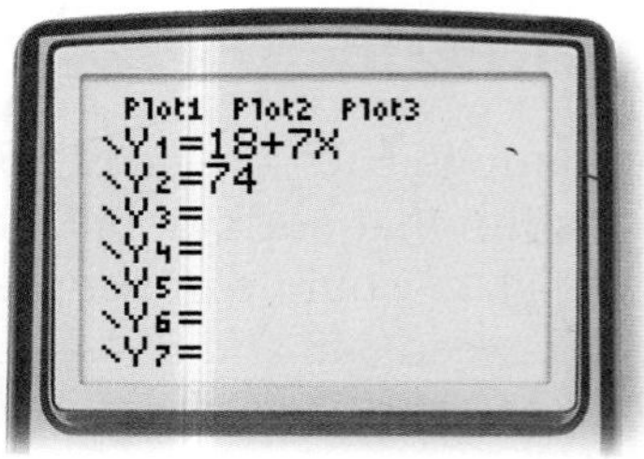


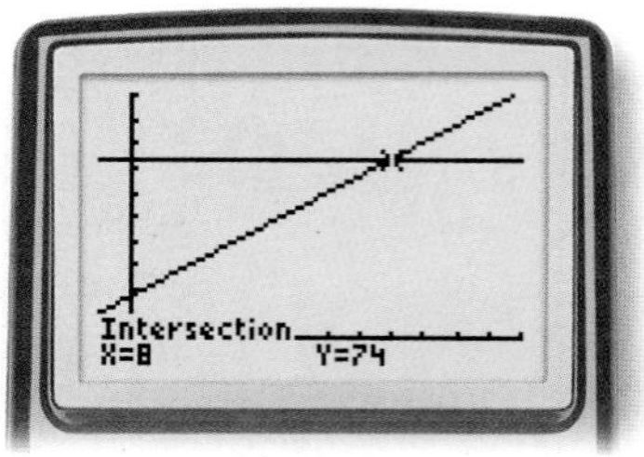


*The value of the x-coordinate at the point of intersection is 8.*

# Exercises

## Communicate

**Explain the steps needed to solve the following equations:**

**1.** $2x + 1 = 7$

**2.** $15 - 4m = 10$

**3.** $\frac{x}{24} + 33 = 84$

**4.** Explain how to use a table to solve an equation.

**5.** Explain how to use a graph to solve an equation.

## Guided Skills Practice

**Solve each equation.** ***(EXAMPLES 1 AND 2)***

**6.** $4p - 13 = 27$ **7.** $-2 = 6x + 4$ **8.** $15 + 2.5m = 70$

**Solve each equation.** ***(EXAMPLE 3)***

**9.** $\frac{x}{4} - 3 = 21$ **10.** $-40 = \frac{g}{-3} - 20$ **11.** $\frac{m}{-4} + 5.5 = -3.2$

**12.** Leo bought 6 glasses and 6 mugs. Leo knows that he was charged \$3.99 for each mug and that the total bill before tax was \$65.88. Write and solve an equation to determine how much Leo was charged for each glass. ***(EXAMPLE 4)***

## Practice and Apply

internet connect

**Homework Help Online**

Go To: **go.hrw.com**
Keyword:
**MA1 Homework Help** for Exercises 13-51

**Solve each equation.**

**13.** $5x + 9 = 39$

**14.** $9p + 11 = -7$

**15.** $6 - 2d = 42$

**16.** $9 - c = -13$

**17.** $2m + 5 = 17$

**18.** $9p + 20 = -7$

**19.** $5x + 9 = 54$

**20.** $3 + 2x = 21$

**21.** $6 - 8d = -42$

**22.** $9 - 14z = 51$

**23.** $12 = 9x - 6$

**24.** $16 = 5w - 9$

**25.** $-4 - 11w = 18$

**26.** $-7 - 13y = 32$

**27.** $36 = -3y + 12$

**28.** $4y + 3 = 13$

**29.** $10m + 3.4 = 7$

**30.** $-3.7 = 2m + 5.1$

**31.** $127 = 2x + 17$

**32.** $4m + 3 = 15$

**33.** $-8m - 12 = 20$

**34.** $47 = 2f - 3$

**35.** $5.2 + 1.3x = -1.3$

**36.** $8p - 15 = 87$

**37.** $\frac{x}{5} - 2 = 3.7$

**38.** $4 - \frac{m}{2} = 10$

**39.** $15 = \frac{a}{3} - 2$

**40.** $10 = \frac{x}{4} + 5$

**41.** $\frac{p}{3} - 2 = -56$

**42.** $45 + \frac{x}{3} = -20$

**43.** $\frac{z}{5} - 22 = -20$

**44.** $\frac{v}{4} + 8 = 11$

**45.** $\frac{d}{10} - 1 = -31$

**46.** $\frac{h}{12} - 5 = -17$

**47.** $-4 = \frac{x}{2} - 5$

**48.** $-45 = \frac{m}{-2} + 4$

**49.** $\frac{x}{-5} + 3.2 = 1.4$

**50.** $1\frac{1}{2} + \frac{m}{-2} = 3\frac{1}{2}$

**51.** $\frac{x}{2} - 4\frac{1}{3} = -2\frac{1}{4}$

### APPLICATIONS

**52. ECOLOGY** The number of trees planted by the Alpine Nursery in April was 3 more than twice the number of trees planted by the nursery in March. If 71 trees were planted in April, write and solve an equation to find how many trees were planted in March.

**53. CONSUMER ECONOMICS** The $347 selling price for a stereo is $35 more than 3 times the wholesale cost. Write and solve an equation to find the wholesale cost.

**54. COMMUNICATION** A long-distance phone company charges $4.99 per month plus 10¢ per minute. Write and solve an equation to determine how many minutes were used in a month if the monthly bill is $16.99.

**55. CONSUMER ECONOMICS** An online music club sells compact discs for $15.95 each plus $2.95 shipping and handling per order. If Kara's total bill is $98.65, write and solve an equation to determine the number of compact discs Kara purchased.

**APPLICATION**

**56. CONSUMER ECONOMICS** Mike bought 8 pairs of socks and a T-shirt. His bill was \$31.50. Write and solve an equation to determine the price for each pair of socks if the cost of the T-shirt was \$7.50.

**57. CULTURAL CONNECTION: AFRICA** Write and solve an equation to find the answer to this problem from ancient Egypt, written about 3800 years ago. Fill a large basket $1\frac{1}{2}$ times. Then add 4 *hekats* (a *hekat* is about half of a bushel). The total is 10 *hekats*. How many *hekats* does the basket hold?

## Look Back

**Find each product.** ***(LESSON 2.4)***

**58.** $-4(-9)$ **59.** $-7(-8)$ **60.** $-13(7)$ **61.** $6(-5)$

**62.** $(-5)(-0.5)$ **63.** $(4.2)(-3)$ **64.** $(-1.5)(-1.5)$ **65.** $(45)(-0.5)$

**Find each quotient.** ***(LESSON 2.4)***

**66.** $\frac{27}{-3}$ **67.** $-45 \div 9$ **68.** $\frac{-42}{-6}$ **69.** $-96 \div 12$

**70.** $\frac{-30}{-6}$ **71.** $\frac{40}{-5}$ **72.** $-10 \div 2$ **73.** $\frac{-55}{11}$

**Solve each equation and check the solution.** ***(LESSON 3.1)***

**74.** $d + 23 = 54$ **75.** $g - 18 = -16$ **76.** $x + 73 = -853$ **77.** $h - \frac{7}{12} = \frac{5}{12}$

**78.** $x + 14 = 24$ **79.** $5 - m = 2$ **80.** $m + 17 = -3$ **81.** $p - 13 = -3.2$

**Solve each equation and check the solution.** ***(LESSON 3.2)***

**82.** $3x = 18$ **83.** $\frac{2}{3}x = 8$ **84.** $\frac{x}{5} = -17$ **85.** $-17y = -51$

**86.** $-5x = 20$ **87.** $\frac{m}{-2} = 30$ **88.** $\frac{n}{5} = -15$ **89.** $-13p = 26$

## Look Beyond

**Solve each equation.**

**90.** $2x + 3 = 4x - 5$ **91.** $7(x - 2) = 4x$ **92.** $7x - 2 = x + 16$

**93.** $3(x - 3) + 10 = 4x - 3x + 8$

3.4

# Solving Multistep Equations

**Objective**

- Write and solve multistep equations.

**APPLICATION**
**BUSINESS**

Mateo orders office supplies for a small company. Two suppliers offer pens at the same price per box. Supplier A gives a \$2 discount on 4 boxes of pens. *For the same total cost*, Supplier B sells 3 boxes of pens and charges \$4 shipping. What is the price of a box of pens?

Let $x$ equal the price of a box of pens. $4x - 2$ models the cost of a purchase from Supplier A, and $3x + 4$ models the cost from Supplier B. To find the price of a box of pens, set the expressions equal to each other and solve for $x$.

$$4x - 2 = 3x + 4$$

**EXAMPLE 1** **Solve $4x - 2 = 3x + 4$ to find the cost of one box of pens. Check the solution.**

**APPLICATION**
**BUSINESS**

**SOLUTION**

Begin by subtracting $3x$ to get the variable on just one side of the equation.

| | |
|---|---|
| $4x - 2 = 3x + 4$ | *Given* |
| $4x - 2 - 3x = 3x + 4 - 3x$ | *Subtraction Property of Equality* |
| $x - 2 = 4$ | *Simplify.* |
| $x - 2 + 2 = 4 + 2$ | *Addition Property of Equality* |
| $x = 6$ | *Simplify.* |

**CHECK**

$4(6) - 2 \stackrel{?}{=} 3(6) + 4$

$24 - 2 \stackrel{?}{=} 18 + 4$

$22 = 22$ *True*

The cost of one box of pens is \$6.

## EXAMPLE 2 Solve $5x - 6 = -6x + 16$. Check the solution.

**SOLUTION**

Begin by adding $6x$ to each side.

| | |
|---|---|
| $5x - 6 = -6x + 16$ | *Given* |
| $5x - 6 + 6x = -6x + 16 + 6x$ | *Addition Property of Equality* |
| $11x - 6 = 16$ | *Simplify.* |
| $11x - 6 + 6 = 16 + 6$ | *Addition Property of Equality* |
| $11x = 22$ | *Simplify.* |
| $\frac{11x}{11} = \frac{22}{11}$ | *Division Property of Equality* |
| $x = 2$ | |

**CHECK**

$5(2) - 6 \stackrel{?}{=} -6(2) + 16$

$4 = 4$ *True*

**TRY THIS** Solve. Check the solution.

**a.** $5x - 7 = 4x + 3$ **b.** $-3m + 14 = 5m + 38$

## EXAMPLE 3 Solve $5x + 2 = 3x$. Check the solution.

**SOLUTION**

Begin by subtracting $3x$ from each side.

| | |
|---|---|
| $5x + 2 = 3x$ | *Given* |
| $5x + 2 - 3x = 3x - 3x$ | *Subtraction Property of Equality* |
| $2x + 2 = 0$ | *Simplify.* |
| $2x + 2 - 2 = 0 - 2$ | *Subtraction Property of Equality* |
| $2x = -2$ | *Simplify.* |
| $\frac{2x}{2} = \frac{-2}{2}$ | *Division Property of Equality* |
| $x = -1$ | |

**CHECK**

$5(-1) + 2 \stackrel{?}{=} 3(-1)$

$-3 = -3$ *True*

**TRY THIS** Solve. Check the solution.

**a.** $5x - 7 = 4x$ **b.** $-3m + 14 = 5m$

**TECHNOLOGY**
**GRAPHICS CALCULATOR**

The solution to the equation in Example 2 can be represented by the intersection of two lines. The solution for $x$ is the value of the $x$-coordinate at the point of intersection. The two lines are represented by the following equations:

$$y = 5x - 6$$
$$y = -6x + 16$$

Enter the equations in a graphics calculator and graph the lines. Use the trace or intersection feature to find the point of intersection. The value of the $x$-coordinate at the point of intersection is 2.

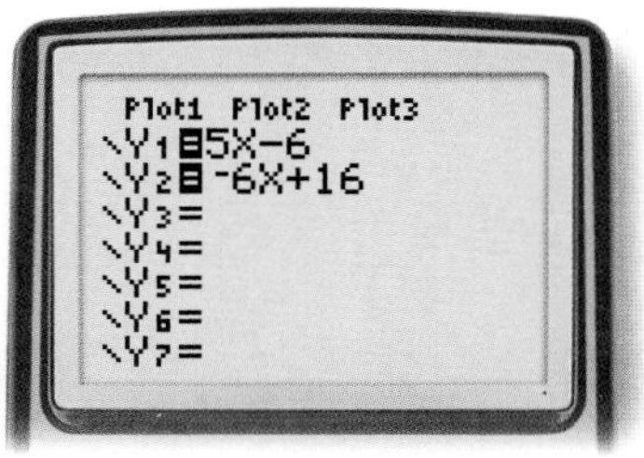


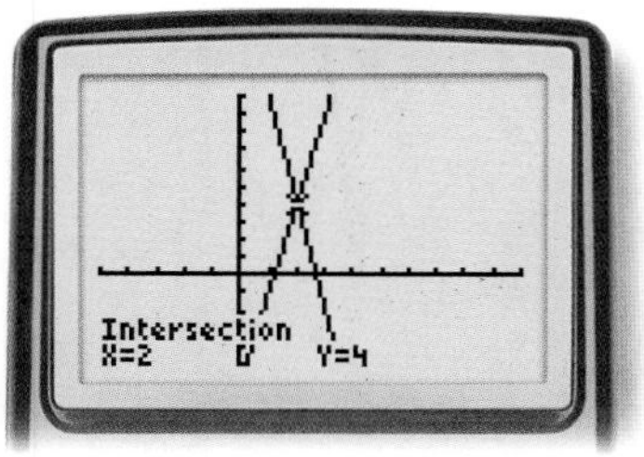


**EXAMPLE 4** Solve $x - \frac{1}{4} = \frac{x}{3} + 7\frac{3}{4}$.

**SOLUTION**

Clear the equation of fractions by changing the mixed number to a fraction and then multiplying both sides by the least common denominator, 12. Then solve for $x$.

| | |
|---|---|
| $x - \frac{1}{4} = \frac{x}{3} + \frac{31}{4}$ | *Given* |
| $12\left(x - \frac{1}{4}\right) = 12\left(\frac{x}{3} + \frac{31}{4}\right)$ | *Multiplication Property of Equality* |
| $12x - 12\left(\frac{1}{4}\right) = 12\left(\frac{x}{3}\right) + 12\left(\frac{31}{4}\right)$ | *Distributive Property* |
| $12x - 3 = 4x + 93$ | *Simplify.* |
| $12x - 3 - 4x = 4x + 93 - 4x$ | *Subtraction Property of Equality* |
| $8x - 3 = 93$ | *Simplify.* |
| $8x - 3 + 3 = 93 + 3$ | *Addition Property of Equality* |
| $8x = 96$ | *Simplify.* |
| $\frac{8x}{8} = \frac{96}{8}$ | *Division Property of Equality* |
| $x = 12$ | *Simplify.* |

Check the result by substituting 12 for $x$ in the original equation.

$$12 - \frac{1}{4} \stackrel{?}{=} \frac{12}{3} + 7\frac{3}{4}$$

$11\frac{3}{4} = 11\frac{3}{4}$ *True*

**TRY THIS** Solve $x - 2 = \frac{x}{5} + 6$.

**EXAMPLE 5** Heather is studying for her third test in history. She scored 79 and 88 on her first two tests. **What score does Heather need on her third test so that her average will be exactly 85?**

**SOLUTION**

Let $s$ represent the score Heather needs on her third test. The average of her three tests can be modeled by $\frac{79 + 88 + s}{3}$.

| | | |
|---|---|---|
| $\frac{79 + 88 + s}{3} = 85$ | | *Given* |
| $\frac{167 + s}{3} = 85$ | | *Simplify.* |
| $3\left(\frac{167 + s}{3}\right) = 3(85)$ | | *Multiplication Property of Equality* |
| $167 + s = 255$ | | *Simplify.* |
| $167 + s - 167 = 255 - 167$ | | *Subtraction Property of Equality* |
| $s = 88$ | | *Simplify.* |

Heather needs to score 88 on her third test to have an average of exactly 85.

**CHECK**

$\frac{79 + 88 + 88}{3} \stackrel{?}{=} 85$

$85 = 85$ *True*

**TRY THIS** Solve $\frac{90 + 80 + x}{3} = 70$.

# Exercises

## Communicate

**Explain the steps needed to solve the following equations:**

**1.** $23p + 57 = p + 984$ **2.** $5z - 5 = -2z + 19$ **3.** $3 - 2x = 4x + 15$

**4.** Explain how to solve $3x + 6 = 2x + 4$ by using a graph.

CONNECTION

**5.** STATISTICS Brian scored 85 on each of his first two algebra tests. Explain how to determine the score that Brian needs on his third test in order to have an average of exactly 90.

## Guided Skills Practice

**Solve each equation and check the solution.** *(EXAMPLES 1, 2, AND 3)*

**6.** $2b + 4 = -8$ **7.** $3 - 7h = 10$ **8.** $f = -6 + 2f$ **9.** $4x - 1 = 2x + 5$

**Solve each equation and check the solution.** *(EXAMPLE 4)*

**10.** $\frac{2}{3} - 3x = \frac{1}{3} + 2x$ **11.** $\frac{1}{6} + 3n = \frac{7}{24} + n$ **12.** $\frac{3a}{4} + \frac{7}{6} = -a + \frac{7}{24}$

CONNECTION

**13.** **STATISTICS** Steve is about to take his fifth math test. His scores on the first four tests were 84, 92, 71, and 94. What score does Steve need on his fifth test in order to have an average of exactly 85? ***(EXAMPLE 5)***

## Practice and Apply

internet connect

**Homework Help Online**
Go To: **go.hrw.com**
Keyword:
**MA1 Homework Help** for Exercises 14-31

**Solve and check each equation.**

**14.** $2x - 2 = 4x + 6$

**15.** $3x + 5 = 2x + 2$

**16.** $4x + 3 = 5x - 4$

**17.** $2x - 5 = 4x - 1$

**18.** $5x + 24 = 2x + 15$

**19.** $5y - 10 = 14 - 3y$

**20.** $12 - 6z = 10 - 5z$

**21.** $5m - 7 = -6m - 29$

**22.** $-10x + 3 = -3x + 12 - 4x$

**23.** $6p - 12 = -4p + 18$

**24.** $1.8x + 2.8 = 2.5x + 2.1$

**25.** $2.6h + 18 = 2.4h + 22$

**26.** $5h - 7 = 2h + 2$

**27.** $4n + 1 = 12 + 5n$

**28.** $1 - 3x = 2x + 8$

**29.** $3a - 8 = \frac{a}{2} + 2$

**30.** $\frac{w}{2} + 7 = \frac{w}{3} + 9$

**31.** $6 - \frac{t}{4} = 8 + \frac{t}{2}$

CHALLENGE

**32.** $x + \frac{5}{8} + \frac{3x}{4} = \frac{2}{3} + 5x$

CONNECTIONS

**33.** **GEOMETRY** The area of a rectangle increased by 5 is the same as 3 less than twice the area of the rectangle. Write and solve an equation to determine the area of the rectangle.

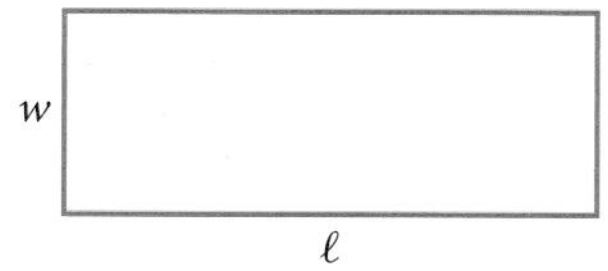


**34.** **GEOMETRY** The perimeter of a triangle increased by 3 is the same as 35 decreased by 7 times the perimeter. Write and solve an equation to determine the perimeter of the triangle.

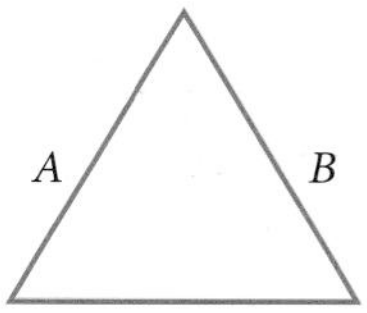

**35.** **STATISTICS** In her first five basketball games, Rachel scored 27, 18, 27, 32, and 21 points. Write and solve an equation to determine how many points she must score in the sixth game in order to have an average of exactly 25 points per game.

APPLICATION

**36.** **HOME IMPROVEMENT** Hannah wants to hire a painter to paint her house. Painters Plus charges \$360 plus \$12 per hour. Davis & Sons charges \$279 plus \$15 per hour. Write and solve an equation to determine the number of hours for which the two costs would be the same.

**APPLICATIONS**

**37.** **TEMPERATURE** If a temperature is increased by 80 degrees, it is the same as 6 times the temperature. Write and solve an equation to determine the temperature.

**38.** **GARDENING** Jake and Sergio are planting tulip bulbs. Jake has planted 60 bulbs and is planting at a rate of 44 bulbs per hour. Sergio has planted 96 bulbs and is planting at a rate of 32 bulbs per hour. In how many hours will Jake and Sergio have planted the same number of bulbs? What number of bulbs will this be? Write and solve an equation to determine the answer.

## Look Back

**Solve each equation and check the solution.** *(LESSON 3.1)*

**39.** $x - 3 = 4$ **40.** $y + 17 = 2$ **41.** $-3 = m + 12$ **42.** $5 - x = 10$

**Solve each equation and check the solution.** *(LESSON 3.2)*

**43.** $3m = 24$ **44.** $\frac{x}{-3} = 5$ **45.** $35 = -3.5x$ **46.** $-8.2 = \frac{m}{-2}$

**Solve each equation and check the solution.** *(LESSON 3.3)*

**47.** $4x + 3 = 20$ **48.** $\frac{2}{3}x - 4 = 12$ **49.** $-2.3 = 3y + 0.7$ **50.** $15 = \frac{m}{8} + 12$

## Look Beyond

**Solve $\frac{1}{x} = \frac{2}{x-2}$ by using each method.**

**51.** graphing **52.** algebra

**Solve $\frac{6}{x+2} = \frac{3}{x-2}$ by using each method.**

**53.** graphing **54.** algebra

**PORTFOLIO ACTIVITY**

Include this activity in your portfolio.

Refer to the T-shirt problem in the Portfolio Activity on page 128. Suppose that the track team sells the T-shirts for \$7.50.

1. If $t$ is the number of T-shirts sold, explain why the equation $125 + 4.55t = 7.50t$ represents the point at which the income from the sales is equal to the cost of making the T-shirts.
2. Solve the equation in Step 1.
3. Suppose that the T-shirts sell for \$10. What is the new value of $t$?

**WORKING ON THE CHAPTER PROJECT**

You should now be able to complete Activity 2 of the Chapter Project on page 154.

3.5

# Using the Distributive Property

## Objectives

- Use the Distributive Property to solve equations.
- Solve real-world problems by using multistep equations.

**APPLICATION**
**CONSUMER ECONOMICS**

Paul buys 3 booklets of bus passes at the student discount of $4 off each booklet for a total of $48. How can you find the original price for each of the booklets?

To find the original price of the bus-pass booklet, write an equation. Let $x$ represent the original price of the booklet in dollars. Then $x-4$ represents the discounted price that Paul paid for each booklet.

Since Paul bought 3 booklets, the following equation models the situation:

$$3(x-4)=48$$

## *Activity*
## Using a Table to Solve for *x*

You can solve the problem presented above by building a table of values for the equation.

| Original price, $x$ | $3(x-4)$ | Total ($) |
|---|---|---|
| 5 | $3(5-4)$ | 3 |
| 10 | $3(10-4)$ | ? |
| 15 | ? | ? |
| 20 | ? | ? |
| 25 | ? | ? |

1. Copy and complete the table.
2. Explain how to determine the original cost of each booklet when the total is $48.

**CHECKPOINT ✔**

3. Describe how to solve $3(x-4)=48$ by using the Distributive Property.

You can solve the problem presented on page 141 algebraically. Start by using the Distributive Property.

**EXAMPLE 1** **Solve the equation from page 141, $3(x - 4) = 48$, to find the original price of each of the booklets that Paul bought.**

**SOLUTION**

To solve an equation that has a variable inside parentheses, you can use the Distributive Property to remove the parentheses.

| | |
|---|---|
| $3(x - 4) = 48$ | *Given* |
| $3x - 12 = 48$ | *Distributive Property* |
| $3x - 12 + 12 = 48 + 12$ | *Addition Property of Equality* |
| $3x = 60$ | *Simplify.* |
| $\frac{3x}{3} = \frac{60}{3}$ | *Division Property of Equality* |
| $x = 20$ | *Simplify.* |

**CHECK**

| | |
|---|---|
| $3(20 - 4) \stackrel{?}{=} 48$ | |
| $3(16) \stackrel{?}{=} 48$ | |
| $48 = 48$ | *True* |

The original price of each booklet is $20.

**CRITICAL THINKING** Explain how the equation in Example 1 could be solved algebraically without using the Distributive Property.

**EXAMPLE 2** **Solve each equation. Check the solution.**

**a.** $3y - 8 - y = 6$ **b.** $2(x + 5) = -16$ **c.** $4x - 8(x + 1) = 8$

**SOLUTION**

**a.**

| | |
|---|---|
| $3y - 8 - y = 6$ | *Given* |
| $2y - 8 = 6$ | *Combine like terms.* |
| $2y - 8 + 8 = 6 + 8$ | *Addition Property of Equality* |
| $2y = 14$ | *Simplify.* |
| $\frac{2y}{2} = \frac{14}{2}$ | *Division Property of Equality* |
| $y = 7$ | *Simplify.* |

**CHECK**

| | |
|---|---|
| $3(7) - 8 - 7 \stackrel{?}{=} 6$ | |
| $6 = 6$ | *True* |

**b.**

| | |
|---|---|
| $2(x + 5) = -16$ | *Given* |
| $2x + 10 = -16$ | *Distributive Property* |
| $2x + 10 - 10 = -16 - 10$ | *Subtraction Property of Equality* |
| $2x = -26$ | *Simplify.* |
| $\frac{2x}{2} = \frac{-26}{2}$ | *Division Property of Equality* |
| $x = -13$ | *Simplify.* |

**CHECK**

$$2(-13+5) \stackrel{?}{=} -16$$
$$-16 = -16 \qquad \textit{True}$$

**c.**

| | |
|---|---|
| $4x - 8(x+1) = 8$ | *Given* |
| $4x - 8x - 8 = 8$ | *Distributive Property* |
| $-4x - 8 = 8$ | *Combine like terms.* |
| $-4x - 8 + 8 = 8 + 8$ | *Addition Property of Equality* |
| $-4x = 16$ | *Simplify.* |
| $\frac{-4x}{-4} = \frac{16}{-4}$ | *Division Property of Equality* |
| $x = -4$ | *Simplify.* |

**CHECK**

$$4(-4) - 8(-4+1) \stackrel{?}{=} 8$$
$$8 = 8 \qquad \textit{True}$$

**TRY THIS** Solve each equation.

**a.** $4t + 7 - t = 19$ **b.** $5(p-2) = -15$ **c.** $3m + 2(4m - 6) = 10$

Solving an equation sometimes results in a special situation, as illustrated in Example 3.

**EXAMPLE 3** **Solve each equation.**

**a.** $2(x+4) - 5 = 2x + 3$ **b.** $4m + 1 - m = 3m + 9$

**SOLUTION**

**a.**

| | |
|---|---|
| $2(x+4) - 5 = 2x + 3$ | |
| $2x + 8 - 5 = 2x + 3$ | *Distributive Property* |
| $2x + 3 = 2x + 3$ | *Simplify.* |
| $2x + 3 - 2x = 2x + 3 - 2x$ | *Subtraction Property of Equality* |
| $3 = 3$ | *True* |

All real numbers are solutions to this equation because the simplified form is always true.

**b.**

| | |
|---|---|
| $4m + 1 - m = 3m + 9$ | |
| $3m + 1 = 3m + 9$ | *Combine like terms.* |
| $3m + 1 - 3m = 3m + 9 - 3m$ | *Subtraction Property of Equality* |
| $1 = 9$ | *False* |

There are no solutions to this equation because the simplified form is always false.

**TRY THIS** Solve each equation.

**a.** $3(y-2) + 4 = 3y - 2$ **b.** $7s + 3 - 2s = 5s + 6$

**EXAMPLE 4** **Solve $4x - 3(2x + 4) = 8x - 25$.**

**SOLUTION**

| | |
|---|---|
| $4x - 3(2x + 4) = 8x - 25$ | *Given* |
| $4x - 3(2x) + (-3)(4) = 8x - 25$ | *Distributive Property* |
| $4x - 6x - 12 = 8x - 25$ | *Simplify.* |
| $-2x - 12 = 8x - 25$ | *Simplify.* |
| $-2x - 12 - 8x = 8x - 25 - 8x$ | *Subtraction Property of Equality* |
| $-10x - 12 = -25$ | *Simplify.* |
| $-10x - 12 + 12 = -25 + 12$ | *Addition Property of Equality* |
| $-10x = -13$ | *Simplify.* |
| $\frac{-10x}{-10} = \frac{-13}{-10}$ | *Division Property of Equality* |
| $x = 1.3$ | *Simplify.* |

Check by substituting 1.3 for $x$ in the original equation.

$$4x - 3(2x + 4) = 8x - 25$$
$$4(1.3) - 3[2(1.3) + 4] \stackrel{?}{=} 8(1.3) - 25$$
$$5.2 - 3(6.6) \stackrel{?}{=} 10.4 - 25$$
$$5.2 - 19.8 \stackrel{?}{=} 10.4 - 25$$
$$-14.6 = -14.6 \quad \textit{True}$$

**TRY THIS** Solve $4y - 7(y + 6) = 5y - 2$.

# Exercises

## Communicate

**Explain the steps needed to solve the following equations:**

**1.** $2(x - 4) = 11$ **2.** $3(x + 1) + x = 5x$ **3.** $4(x - 4) + 3 = 2x - 1$

**4.** Describe how to solve $4(x + 4) = 108$ by using a table.

## Guided Skills Practice

**Solve each equation.** ***(EXAMPLES 1 AND 2)***

**5.** $5s + 7 - 2s = 16$ **6.** $4(f + 2) = 20$ **7.** $5r - 3(r + 6) = 24$

**Solve each equation.** ***(EXAMPLE 3)***

**8.** $4(m + 3) + 7 = 4m + 19$ **9.** $3p - 7 + 2p = 5p + 12$

**Solve each equation.** *(EXAMPLE 4)*

**10.** $3y + 7(2y - 4) = y - 12$

**11.** $5(y + 4) - 3y = 4y + 12$

## Practice and Apply

**Solve each equation.**

**internet connect**
**Homework Help Online**
Go To: **go.hrw.com**
Keyword:
**MA1 Homework Help**
for Exercises 12-43

**12.** $4n - 2 + 7n = 20$

**13.** $3(r - 4) = 9$

**14.** $6x - 4(2x + 1) = 12$

**15.** $6(t + 7) - 20 = 6t$

**16.** $4x + 3 - 2x = 2x + 7$

**17.** $3y - 2(3y + 2) = 8$

**18.** $3(x + 1) = 2x + 7$

**19.** $3w - 1 - 4w = 4 - 2w$

**20.** $8x - 5 = 4x + 4 - 2x$

**21.** $15 - 3y = y + 13 + y$

**22.** $4a - 4 = -2a + 14$

**23.** $4m - 5 = 3m + 7$

**24.** $2(x + 1) = 3x - 3$

**25.** $2m - 4 = 2(6 - 7m)$

**26.** $8y - 3 = 5(2y + 1)$

**27.** $9y - 8 + 4y = 7y + 16$

**28.** $0.3w - 4 = 0.8 - 0.2w$

**29.** $5x + 32 = 8 - x$

**30.** $4y + 2 = 3(6 - 4y)$

**31.** $m - 12 = 3m + 4$

**32.** $2(8y - 7) = 2(3 + 8y)$

**33.** $2(2x + 3) = 8x + 5$

**34.** $3 - 2(y - 1) = 2 + 4y$

**35.** $3(2r - 1) + 5 = 5(r + 1)$

**36.** $5(x + 1) + 2 = 2x + 11$

**37.** $4z - (z + 6) = 3z - 4$

**38.** $3x - 2(x + 6) = 4x - (x - 10)$

**39.** $8y - 4 + 3(y + 7) = 6y - 3(y - 3)$

**40.** $1.4m - 0.6(m - 2) = 2.4m$

**41.** $16x - 3(4x + 7) = 6x - (2x + 21)$

**42.** $6.3y = 5.2y - 1.1y + 12.1$

**43.** $(w - 3) - 5(w + 7) = 10(w + 3) - (7w + 5)$

**CONNECTIONS**

**44. GEOMETRY** The perimeters of the two rectangles at right are equal. Write and solve an equation to find the value of $x$ and the common perimeter.

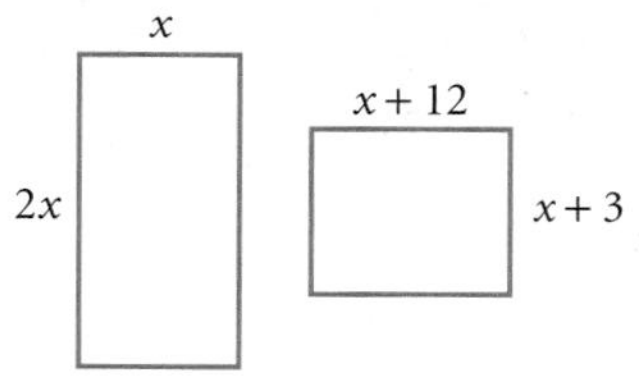


**45. GEOMETRY** Write and solve an equation to determine the value of $x$ for which the perimeter of the rectangle at right will be equal to the perimeter of the square.

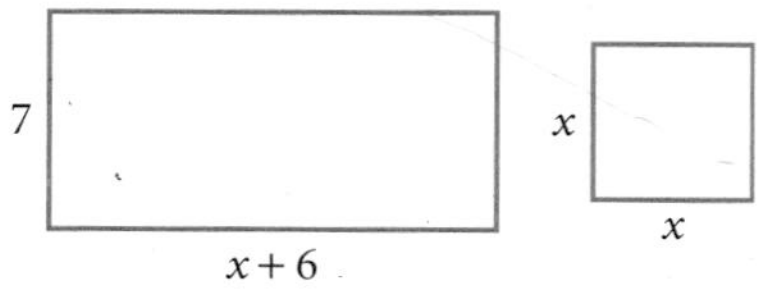


**46. GEOMETRY** The sum of the angle measures in a triangle is 180°. In triangle $ABC$ at right, $m\angle A = 4x + 34$, $m\angle B = 7x - 10$, and $m\angle C = 5x + 12$. Write and solve an equation to find the value of $x$ and the measures of angles $A$, $B$, and $C$.

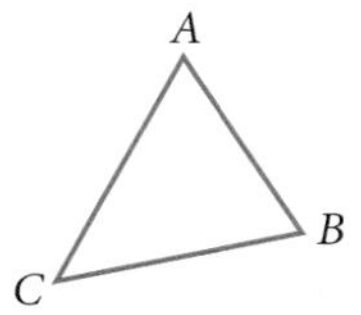

APPLICATIONS

**47.** **STUDENT GOVERNMENT** In the last student-council elections, Jill got $x$ number of votes. Morgan got twice as many votes as Jill. Trisha won the election with 7 more votes than Morgan. If 327 students voted in the election, write and solve an equation to determine how many votes each candidate got.

**48.** **SMALL BUSINESS** To reduce inventory, Millennium Games offers single-game CDs at a \$5 discount. Roger buys 4 of these single-game CDs for \$80. Write and solve an equation to determine the original price of a single-game CD.

## Look Back

**Determine the next three terms of each sequence.** ***(LESSON 1.1)***

**49.** $1, 3, 7, 13, \ldots$ **50.** $50, 41, 32, 23, \ldots$ **51.** $1, \frac{1}{2}, \frac{1}{4}, \frac{1}{8}, \ldots$

**Add.** ***(LESSON 2.2)***

**52.** $-3 + (-7)$ **53.** $4.6 + (-2.7)$ **54.** $-1\frac{1}{2} + \left(-1\frac{1}{2}\right)$ **55.** $3 + (-2.1)$

**Subtract.** ***(LESSON 2.3)***

**56.** $3 - (-7)$ **57.** $3 - 4$ **58.** $-1\frac{2}{3} - \left(-3\frac{1}{3}\right)$ **59.** $0 - 3$

**Multiply.** ***(LESSON 2.4)***

**60.** $(3)(-2)$ **61.** $(-2.7)(-2.3)$ **62.** $(5)(-0.3)$ **63.** $\left(-\frac{1}{2}\right)\left(-\frac{1}{2}\right)$

**Divide.** ***(LESSON 2.4)***

**64.** $\frac{-30}{5}$ **65.** $-\frac{1}{5} \div \frac{1}{5}$ **66.** $-5.2 \div (-1.3)$ **67.** $300 \div (-2.5)$

**Solve.** ***(LESSONS 3.1 AND 3.2)***

**68.** $3x = 9$ **69.** $\frac{x}{5} = 2$ **70.** $4 = x - 7.2$ **71.** $4 - x = -2.1$

**72.** $5m = -20$ **73.** $\frac{r}{-2} = 10$ **74.** $m + 4.7 = -2.5$ **75.** $-15 = x + 2$

**76.** $27p = 81$ **77.** $100x = 50$ **78.** $3 - x = 2$ **79.** $-7 = 4 - x$

**80.** $-5p = 35$ **81.** $202 = -101m$ **82.** $m + 7.6 = 14$ **83.** $p - 3.1 = -2.7$

## Look Beyond

**Solve each inequality and graph the solution on a number line.**

**84.** $3x > 4$ **85.** $-4x < 32$ **86.** $x - 3 \leq 5$ **87.** $2y - 7 \geq 11$

3.6

# Using Formulas and Literal Equations

## Objectives

- Solve literal equations for a specific variable.
- Use formulas to solve problems.

**Why** *Formulas are algebraic generalizations that can be applied to a variety of problems.*

An equation that involves two or more variables is called a **literal equation.** A **formula** is a literal equation that states a rule for a relationship among quantities. For example, the formula $C = \frac{5}{9}(F - 32)$ relates degrees Fahrenheit to degrees Celsius. The variables in a formula are often related to the names of the quantities involved, such as $F$ for Fahrenheit.

When working with formulas, the values for one or more of the variables are sometimes known, leaving the value of one variable to be determined. You can use the temperature conversion formula given above, for example, to easily convert from degrees Fahrenheit to degrees Celsius.

**EXAMPLE 1**

**APPLICATION**
**TEMPERATURE**

**Convert each temperature from degrees Fahrenheit to degrees Celsius by using the formula $C = \frac{5}{9}(F - 32)$.**

**a.** 68°F **b.** 32°F **c.** 23°F

**SOLUTION**

Substitute the given Fahrenheit temperature for $F$ in the formula.

**a.** 68°F

$C = \frac{5}{9}(68 - 32)$

$C = \frac{5}{9}(36)$

$C = 20$

$68°F = 20°C$

**b.** 32°F

$C = \frac{5}{9}(32 - 32)$

$C = \frac{5}{9}(0)$

$C = 0$

$32°F = 0°C$

**c.** 23°F

$C = \frac{5}{9}(23 - 32)$

$C = \frac{5}{9}(-9)$

$C = -5$

$23°F = -5°C$

**TRY THIS** Convert each temperature from degrees Fahrenheit to degrees Celsius.

**a.** 50°F **b.** 212°F **c.** 104°F

## Solving for a Variable

With some literal equations and formulas you will need to solve the equation for one variable in terms of the other variables.

**EXAMPLE 2**

APPLICATION
INVESTMENTS

The formula $A = P + I$ shows that the total amount of money, $A$, received from an investment equals the principal, $P$ (the original amount of money invested), plus the interest, $I$. **Write a formula for the interest, $I$, in terms of the principal, $P$, and the total amount, $A$. In other words, solve the formula for $I$.**

**SOLUTION**

To isolate $I$, subtract $P$ from both sides of the equation.

$$A = P + I$$

$$A - P = P + I - P \quad \textit{Subtraction Property of Equality}$$

$$A - P = I$$

$$I = A - P \quad \textit{Symmetric Property of Equality}$$

**EXAMPLE 3**

APPLICATION
PHYSICS

*Density* is the mass of a material divided by its volume. The formula for density is $D = \frac{m}{V}$. **Find a formula for the mass, $m$, in terms of the density, $D$, and the volume, $V$.**

**SOLUTION**

Multiply both sides of the equation by $V$.

$$D = \frac{m}{V}$$

$$V(D) = V\left(\frac{m}{V}\right) \quad \textit{Multiplication Property of Equality}$$

$$VD = m \quad \textit{Simplify.}$$

$$m = VD \quad \textit{Symmetric Property of Equality}$$

**TRY THIS** Solve each formula for the indicated variable.

**a.** $A = P + I$, for $P$ **b.** $C = 2\pi r$, for $r$

**EXAMPLE 4**

**Solve each equation for the indicated variable.**

**a.** $y = mx + b$, for $b$ **b.** $3x + 4y = 12$, for $y$

**SOLUTION**

**a.**

$$y = mx + b$$

$$y - mx = mx + b - mx \quad \textit{Subtraction Property of Equality}$$

$$y - mx = b \quad \textit{Simplify.}$$

$$b = y - mx \quad \textit{Symmetric Property of Equality}$$

**b.**

$$3x + 4y = 12$$

$$3x + 4y - 3x = 12 - 3x \quad \textit{Subtraction Property of Equality}$$

$$4y = 12 - 3x \quad \textit{Simplify.}$$

$$\frac{4y}{4} = \frac{12 - 3x}{4} \quad \textit{Division Property of Equality}$$

$$y = 3 - \frac{3}{4}x \quad \textit{Simplify.}$$

$$y = 3 + \left(-\frac{3}{4}x\right) \quad \textit{Definition of subtraction}$$

$$y = -\frac{3}{4}x + 3 \quad \textit{Commutative Property}$$

*This is called the $y = mx + b$ form and will be useful in the study of linear equations.*

## Working With Formulas

**EXAMPLE 5**

CONNECTION
**GEOMETRY**

Jayne has 396 linear feet of chicken wire and is putting a fence up on her small farm to fence in her chickens. She wants the length of the pen to be twice the width. **Find the dimensions of the pen.**

**SOLUTION**

PROBLEM SOLVING

**Draw a diagram** of the pen. Since the length is twice the width, label the sides $w$ and $2w$.

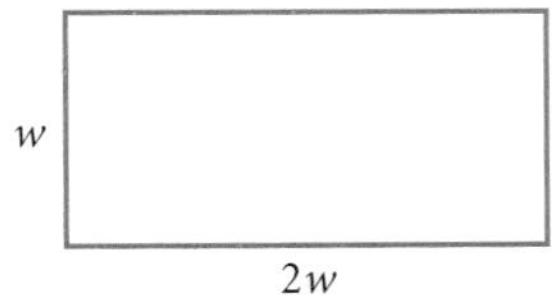


The length of fencing that Jayne has equals the perimeter of the pen. Use the formula for the perimeter of a rectangle, $P = 2l + 2w$, to write an equation for a perimeter of 396 feet.

$$P = 2l + 2w \quad \textit{Perimeter formula}$$

$$396 = 2(2w) + 2w \quad \textit{Substitute 2w for l.}$$

Solve the equation to find the dimensions.

$$396 = 2(2w) + 2w$$

$$396 = 4w + 2w$$

$$396 = 6w$$

$$\frac{396}{6} = \frac{6w}{6} \quad \textit{Division Property of Equality}$$

$$66 = w \quad \textit{Simplify.}$$

$$w = 66 \quad \textit{Symmetric Property of Equality}$$

The width is 66 feet, so the length $2w$, is $2(66)$, or 132 feet.

**EXAMPLE 6**

**CONNECTION GEOMETRY**

Jeffrey is building a tabletop in the shape of a trapezoid. The area of the tabletop is to be 18 square feet. The longer base of the trapezoid is 7 feet and the height of the trapezoid is 3.5 feet. **Use the formula for the area of a trapezoid, $A = \frac{1}{2}h(b_1 + b_2)$, to find the length of the shorter base.**

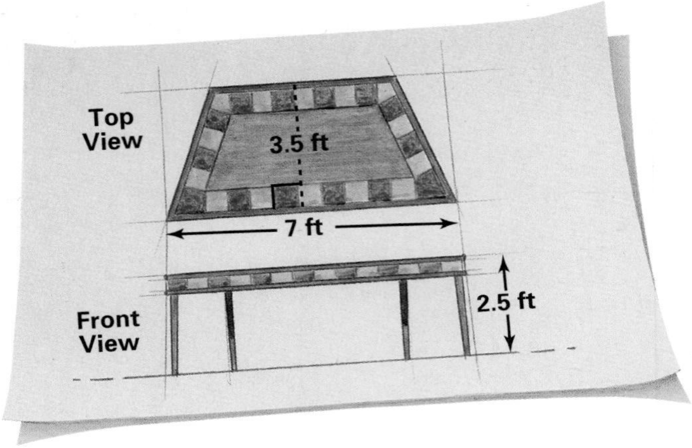


**SOLUTION**

Substitute 18 for $A$, 3.5 for $h$, and 7 for $b_1$ in the formula and solve for $b_2$, the length of the second base.

| | |
|---|---|
| $A = \frac{1}{2}h(b_1 + b_2)$ | |
| $18 = \frac{1}{2}(3.5)(7 + b_2)$ | *Substitute.* |
| $18 = (1.75)(7 + b_2)$ | *Simplify.* |
| $18 = (1.75)(7) + (1.75)(b_2)$ | *Distributive Property* |
| $18 = 12.25 + 1.75b_2$ | *Simplify.* |
| $18 - 12.25 = 12.25 + 1.75b_2 - 12.25$ | *Subtraction Property of Equality* |
| $5.75 = 1.75b_2$ | *Simplify.* |
| $\frac{5.75}{1.75} = \frac{1.75b_2}{1.75}$ | *Division Property of Equality* |
| $3.2857 \approx b_2$ | *Simplify.* |
| $b_2 \approx 3.2857$ | The shorter base should be approximately 3.29 feet. |

**EXAMPLE 7**

**CONNECTION GEOMETRY**

The circumference of a circle $C$, can be determined from the formula $C = 2\pi r$, where $r$ is the radius of the circle. **Use the formula to find the value of $r$ for each circumference below. Use 3.14 as an approximation for $\pi$, and round your answers to the nearest hundredth.**

**a.** $C = 25$

**b.** $C = 12$

**SOLUTION**

**a.**

| | |
|---|---|
| $C = 2\pi r$ | |
| $25 \approx 2(3.14)r$ | *Substitute.* |
| $25 \approx 6.28r$ | *Simplify.* |
| $\frac{25}{6.28} \approx \frac{6.28r}{6.28}$ | *Div. Prop. of Eq.* |
| $3.98 \approx r$ | *Simplify.* |
| $r \approx 3.98$ | |

**b.**

| | |
|---|---|
| $C = 2\pi r$ | |
| $12 \approx 2(3.14)r$ | *Substitute.* |
| $12 \approx 6.28r$ | *Simplify.* |
| $\frac{12}{6.28} \approx \frac{6.28r}{6.28}$ | *Div. Prop. of Eq.* |
| $1.91 \approx r$ | *Simplify* |
| $r \approx 1.91$ | |

**TRY THIS** Use the formula $C = 2\pi r$ to find the value of $r$ for each circumference below. Use 3.14 as an approximation for $\pi$, and round your answers to the nearest hundredth.

**a.** $C = 34$

**b.** $C = 7.0$

# Exercises

## Communicate

1. Explain what a literal equation is.
2. Describe how formulas can be helpful in mathematics.
3. Explain how to solve each equation for the indicated variable.
   **a.** $C = \pi d$, for $d$ **b.** $A + B = C$, for $B$ **c.** $3m + 7 = y$, for $m$
4. Explain how to find the length of a rectangle when you know the perimeter and the width.

## Guided Skills Practice

**Convert each temperature from degrees Fahrenheit to degrees Celsius by using the formula $C = \frac{5}{9}(F - 32)$.** *(EXAMPLE 1)*

**5.** 59°F **6.** 14°F **7.** 149°F

**Solve for the indicated variable.** *(EXAMPLES 2, 3, AND 4)*

**8.** $A - P = I$, for $A$ **9.** $y = mx$, for $m$ **10.** $a + bc = d$, for $b$

**11.** A rectangle has a length that is 3 times its width. Use the formula $P = 2l + 2w$ to determine the length and width if the perimeter is 32. *(EXAMPLE 5)*

**12.** A trapezoid has an area of 22 square inches. The longer base measures 6 inches and the height is 4 inches. Use the formula $A = \frac{1}{2}h(b_1 + b_2)$ to find the length of the shorter base. *(EXAMPLE 6)*

**13.** The circumference of a circle is 30. Use the formula $C = 2\pi r$ to find the radius, $r$. Use 3.14 as an approximation for $\pi$, and round your answer to the nearest hundredth. *(EXAMPLE 7)*

## Practice and Apply

**Use the formula $C = \frac{5}{9}(F - 32)$ to convert each temperature from degrees Fahrenheit to degrees Celsius.**

**14.** 23°F **15.** 86°F **16.** 18°F **17.** 125°F

**Solve each equation for the indicated variable.**

**18.** $T + M = R$, for $T$ **19.** $M = T - R$, for $T$ **20.** $R + T = M$, for $T$

**21.** $a + b = c$, for $b$ **22.** $c = a - b$, for $b$ **23.** $a + b = -c$, for $a$

**24.** $a - b = c$, for $a$ **25.** $C = \pi d$, for $d$ **26.** $y = mx + 6$, for $m$

**27.** $A = lw$, for $w$ **28.** $3t = r$, for $t$ **29.** $5p + 9c = p$, for $c$

**30.** $ma = q$, for $a$ **31.** $ax + r = 7$, for $x$ **32.** $2a + 2b = c$, for $b$

**33.** $4y + 3x = 5$, for $x$ **34.** $3x + 7y = 2$, for $y$ **35.** $y = 3x + 3b$, for $b$

internet connect
go.hrw.com
**Homework Help Online**
Go To: **go.hrw.com**
Keyword:
**MA1 Homework Help** for Exercises 18-35

**CONNECTIONS**

**GEOMETRY** Use the formula $C = 2\pi r$ to find the radius, $r$, for each circumference below. Use 3.14 as an approximation for $\pi$, and round your answers to the nearest hundredth.

**36.** $C = 30$ inches
**37.** $C = 5$ centimeters
**38.** $C = 100$ meters
**39.** $C = 15$ meters
**40.** $C = 71$ inches
**41.** $C = 52$ feet
**42.** $C = 3$ feet
**43.** $C = 11$ inches
**44.** $C = 200$ meters

**GEOMETRY** Use the formula $A = \frac{1}{2}h(b_1 + b_2)$ to find the length of the missing base for trapezoids with the given dimensions. Round your answers to the nearest hundredth.

**45.** $A = 200, b_1 = 24, h = 10$
**46.** $A = 150, b_1 = 20, h = 9$
**47.** $A = 28, b_1 = 8, h = 4$
**48.** $A = 50, b_1 = 10, h = 6$
**49.** $A = 120, b_1 = 20, h = 10$
**50.** $A = 200, b_1 = 50, h = 5$

**APPLICATIONS**

**51.** **LANDSCAPING** Fred is putting a decorative border around a rectangular flower garden. The total perimeter of the garden is 200 feet and the length is 4 times the width. Use the formula $P = 2l + 2w$ to find the length and the width.

**52.** **TRAVEL** You can estimate the time, $t$, in hours that it takes to fly a distance, $d$, in miles by using the formula $t = \frac{d}{500} + \frac{1}{2}$. For example, a flight of 1000 miles takes about 2.5 hours.

**a.** Use the formula to estimate the time that it takes to fly 1300 miles.
**b.** Solve the formula for $d$.
**c.** Use the formula from **b** to find how many miles you can fly in 4 hours.

**53.** **THEATRE** Alicia is painting a rectangular box as a part of the set for a school play. The box has a length of 3 feet, a width of 2 feet, and a height of 1 foot. Use the formula $A = 2(lw + hw + lh)$, where $A$ is the total surface area, $l$ is the length, $w$ is the width, and $h$ is the height, to find the total surface area of the box.

## Look Back

**Evaluate.** ***(LESSON 2.3)***

**54.** $89 - (-14)$ **55.** $400 - (-111)$ **56.** $-674 - (-900)$

**Evaluate.** ***(LESSON 2.4)***

**57.** $(-3)(-3)(-1)(-1)$ **58.** $(1)(-1)(-1)(-1)$ **59.** $\frac{-22}{2}$ **60.** $\frac{-16}{-4}$

**Use the Distributive Property to simplify each expression.** ***(LESSON 2.5)***

**61.** $(3x - 2y + 1) - 3(x + 2y - 1)$ **62.** $3(a + b) - 2(a - b)$

**Solve each equation.** ***(LESSON 3.3)***

**63.** $3x - 2 = 19$ **64.** $-4x + 15 = -45$ **65.** $18x + 53 = 7x - 68$

## Look Beyond

**Sometimes inequalities are used to indicate values that make sense in an equation, such as in these examples.**

If $y = \frac{1}{x}$, then $x \neq 0$. If $y = \sqrt{x}$, then $x \geq 0$.

**Complete each of the following:**

**66.** If $y = \frac{1}{x-2}$, then $x \neq \underline{?}$. **67.** If $y = \sqrt{x+3}$, then $x \geq \underline{?}$.

## PORTFOLIO ACTIVITY

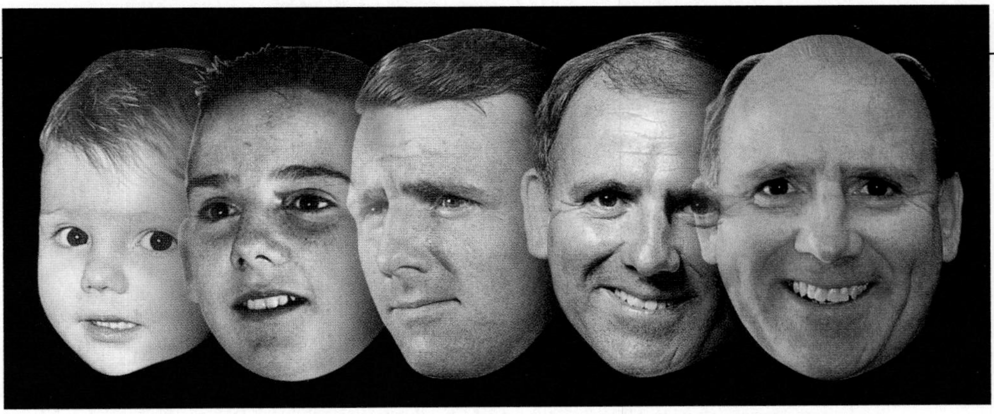

*This series of photos shows the age progression for one man.*

Include this activity in your portfolio.

Diophantus of Alexandria was a mathematician who lived in the third century C.E. He has been called the father of algebra. The following problem was written about Diophantus:

> Diophantus lived $\frac{1}{6}$ of his life as a child,
> $\frac{1}{12}$ more as a youth,
> $\frac{1}{7}$ more before he married, and
> 5 years more before his son was born.
> His son was alive for only $\frac{1}{2}$ Diophantus's life.
> Diophantus then found solace in his studies for 4 years more, before he too died.

Write and solve an equation to determine how long Diophantus lived.

**WORKING ON THE CHAPTER PROJECT**

You should now be able to complete Activity 3 of the Chapter Project on page 154.

CHAPTER PROJECT THREE

# TRACK TEAM *Schemes*

**The school board for Hale High School has approved funding for a new track. The Hale High School track team is raising money to buy new equipment for the new track and the track program at the school.**

## Activity 1

The Hale High School track team is selling T-shirts to raise money. The Portfolio Activity on page 136 included the following information about the cost of the T-shirts:

> Ted's Tees charges \$3.75 per shirt, \$0.80 to print each shirt with the design, and \$125 per order for the silk-screen design.

1. Suppose that the track team wishes to make a profit of \$1000 from the sale of T-shirts. Explore several pricing schemes that result in this level of profit by adjusting the amount that the track team charges for the shirts and the number of shirts that need to be sold.
2. Optional: Create a spreadsheet that shows the cost, income, and profit for each of the pricing schemes from Step 1.

## Activity 2

1. For each pricing scheme that you explored in Activity 1, write an equation that represents how many T-shirts need to be sold to reach the break-even point, the point at which the cost and income are the same.
2. For each pricing scheme that you explored in Activity 1, write an equation that represents how many T-shirts need to be sold to make a profit of \$1000.

## Activity 3

The Hale High School track coaches need to make a preliminary survey of the school property in order to determine the best site for the new track.

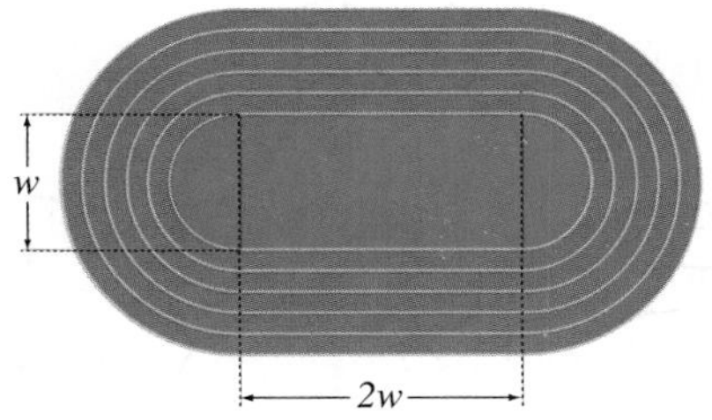


*The inside lane of the track will enclose an area that is the shape of a rectangle with a semicircle at each end. The length of the rectangular part of the track should be twice the width.*

The International Amateur Athletic Federation (IAAF) is the governing body of track and field. The IAAF rules specify that an outdoor running track should measure no less than 400 meters around the inside lane (perimeter). If available, a spreadsheet program is a useful for tool for completing this activity.

1. Write an equation for the perimeter of the track, using the information in the diagram at left.
2. Find the value of $w$ which ensures that the inside lane of the track will produce a lap length of no less than 400 meters.
3. IAAF rules state that each lane should measure about 1.25 meters in width. Find the perimeter of the outer edge of the inside lane.
4. Suppose that the track is to have 6 lanes. Find the perimeter of the outer edge of each lane.
5. Assume that the inner edge of the inside lane is 400 meters. Explain how you could stagger the starting point for each lane to account for the differences in the lane perimeters.

# 3 Chapter Review and Assessment

## VOCABULARY



## Key Skills & Exercises

### LESSON 3.1

**Key Skills**

**Solve equations by using addition and subtraction.**

Solve $x + 15 = 11$.

| | |
|---|---|
| $x + 15 = 11$ | |
| $x + 15 - 15 = 11 - 15$ | *Subtr. Prop. of Equality* |
| $x = -4$ | *Simplify.* |
| Check: $-4 + 15 = 11$ | *True* |

Solve $-8 = y - 14$.

| | |
|---|---|
| $-8 = y - 14$ | |
| $-8 + 14 = y - 14 + 14$ | *Add. Prop. of Equality* |
| $6 = y$ | *Simplify.* |
| $y = 6$ | |
| Check: $-8 = 6 - 14$ | *True* |

**Exercises**

**Solve each equation. Check the solution.**

**1.** $r + 26 = 16$

**2.** $a + 1.5 = 3.6$

**3.** $t + 7 = -5$

**4.** $y - 13 = 12$

**5.** $24 = x - 19$

**6.** $m + \frac{1}{2} = \frac{5}{6}$

**7.** $h - \frac{1}{6} = \frac{2}{3}$

**8.** $k + 5 = 7$

**9.** $m - 3 = 15$

**10.** $x + 17 = 2$

### LESSON 3.2

**Key Skills**

**Solve equations by using multiplication and division.**

Solve $-9k = 108$.

| | |
|---|---|
| $-9k = 108$ | *Given* |
| $\frac{-9k}{-9} = \frac{108}{-9}$ | *Division Property of Equality* |
| $k = -12$ | *Simplify.* |

Solve $\frac{w}{-5} = -2.2$.

| | |
|---|---|
| $\frac{w}{-5} = -2.2$ | *Given* |
| $-5\left(\frac{w}{-5}\right) = -5(-2.2)$ | *Mult. Prop. of Equality* |
| $w = 11$ | *Simplify.* |

**Exercises**

**Solve each equation.**

**11.** $17x = -85$

**12.** $-4g = -56$

**13.** $-2.2h = 33$

**14.** $24f = 150$

**15.** $-8w = 0.5$

**16.** $\frac{y}{-2.4} = -10$

**17.** $\frac{t}{5} = -24$

**18.** $\frac{-c}{3} = 12$

**19.** $\frac{w}{8} = -19$

**20.** $\frac{x}{-3} = 0$

## LESSON 3.3

### Key Skills

**Solve two-step equations.**

Solve $4x + 5 = 15$.

$$4x + 5 = 15$$
$$4x + 5 - 5 = 15 - 5 \quad \textit{Sub. Prop. of Equal.}$$
$$4x = 10 \quad \textit{Simplify.}$$
$$\frac{4x}{4} = \frac{10}{4} \quad \textit{Div. Prop. of Equal.}$$
$$x = 2.5 \quad \textit{Simplify.}$$

Solve $-5 - \frac{x}{3} = 2$.

$$-5 - \frac{x}{3} = 2$$
$$-5 - \frac{x}{3} + 5 = 2 + 5 \quad \textit{Add. Prop. of Equal.}$$
$$-\frac{x}{3} = 7 \quad \textit{Simplify.}$$
$$-3\left(-\frac{x}{3}\right) = -3(7) \quad \textit{Mult. Prop. of Equal.}$$
$$x = -21 \quad \textit{Simplify.}$$

### Exercises

**Solve.**

**21.** $3a + 7 = 31$

**22.** $-2x + 10 = -4$

**23.** $\frac{y}{5} + 4 = 40.5$

**24.** $5x + 3 = -17$

**25.** $-7 = 4 + \frac{x}{3}$

**26.** $-8 = \frac{x}{2} - 5$

**27.** $-3x - 2 = -4$

**28.** $5x + 17 = -17$

**29.** $3 - 2x = 3$

**30.** $\frac{x}{4} - 5 = 5$

**31.** $4x + 3 = 23$

**32.** $45 = 2w - 32$

**33.** $60 = 5 - \frac{m}{7}$

## LESSON 3.4

### Key Skills

**Solve multistep equations.**

Solve $5x + 7 = 2x - 2$.

$$5x + 7 = 2x - 2$$
$$5x + 7 - 2x = 2x - 2 - 2x \quad \textit{Sub. Prop. of Equal.}$$
$$3x + 7 = -2 \quad \textit{Simplify.}$$
$$3x + 7 - 7 = -2 - 7 \quad \textit{Sub. Prop. of Equal.}$$
$$3x = -9 \quad \textit{Simplify.}$$
$$\frac{3x}{3} = \frac{-9}{3} \quad \textit{Div. Prop. of Equal.}$$
$$x = -3 \quad \textit{Simplify.}$$

### Exercises

**Solve.**

**34.** $n - 6 = 2n - 14$

**35.** $2m + 5 = \frac{m}{5} - 4$

**36.** $4x + 18 = x$

**37.** $5m - 3 = 2m - 9$

**38.** $15p + 4 = 5p$

**39.** $\frac{m}{2} + 3 = \frac{m}{5} + 5$

**40.** $8x - 1.5 = 2x$

**41.** $3f - 2 = 7$

**42.** $2x + 3 = 3x + 5$

**43.** $5m - 3 = 2$

**44.** $-14x = 3x + 17$

**45.** $-2x + 3 = x - 6$

**46.** $8f + 7 = 15$

**47.** $11m + 2 = 9m - 18$

## LESSON 3.5

### Key Skills

**Use the Distributive Property to solve multistep equations.**

Solve $5x + 3(x - 3) + 7 = 2x - 1$.

| | |
|---|---|
| $5x + 3(x - 3) + 7 = 2x - 1$ | |
| $5x + 3x - 9 + 7 = 2x - 1$ | *Distributive Prop.* |
| $8x - 2 = 2x - 1$ | *Combine like terms.* |
| $8x - 2 - 2x = 2x - 1 - 2x$ | *Subtr. Prop. of Equal.* |
| $6x - 2 = -1$ | *Simplify.* |
| $6x - 2 + 2 = -1 + 2$ | *Add. Prop. of Equal.* |
| $6x = 1$ | *Simplify.* |
| $\frac{6x}{6} = \frac{1}{6}$ | *Div. Prop. Equal.* |
| $x = \frac{1}{6}$ | |

### Exercises

**Solve.**

**48.** $7x - 2(x + 6) = -2$

**49.** $3x + 4(x - 2) = 7 + x$

**50.** $5(x - 3) = 10$

**51.** $6(2x + 7) = 102$

**52.** $3 - 4x = 5(x + 6)$

**53.** $x + 7 = 2(x - 1)$

**54.** $5m - 3(m - 3) = 2(m + 3)$

**55.** $6(x + 2x + 1) = 12x$

## LESSON 3.6

### Key Skills

**Manipulate literal equations.**

Solve $4m + s = 3p$ for $m$.

| | |
|---|---|
| $4m + s = 3p$ | |
| $4m + s - s = 3p - s$ | *Sub. Prop. of Equal.* |
| $4m = 3p - s$ | *Simplify.* |
| $\frac{4m}{4} = \frac{3p - s}{4}$ | *Div. Prop. of Equal.* |
| $m = \frac{3p - s}{4}$ | *Simplify.* |

**Use formulas to solve problems.**

Use the formula $A = lw$ to find $l$ if $A$ is 40 and $w$ is 4.

| | |
|---|---|
| $A = lw$ | |
| $40 = 4l$ | *Substitution* |
| $\frac{40}{4} = \frac{4l}{4}$ | *Div. Prop. of Equal.* |
| $10 = l$ | *Simplify.* |
| $l = 10$ | *Sym. Prop.* |

### Exercises

**Solve for the indicated variable.**

**56.** $A + B = C$, for $A$

**57.** $y = mx + b$, for $m$

**58.** $9x + 3y = 10$, for $y$

**59.** $ax + b = c$, for $b$

**60.** $T + \frac{S}{3} = R$, for $S$

**Use the formula $C = 2\pi r$ to find $r$ for each circumference below. Use 3.14 as an approximation for $\pi$, and round your answers to the nearest hundredth.**

**61.** $C = 100$ meters

**62.** $C = 10$ centimeters

**63.** $C = 15$ feet

**64.** $C = 25$ centimeters

**65.** $C = 28$ inches

## Applications

**66.** **CONSUMER ECONOMICS** Marsha needs to buy painting supplies. Paintbrushes cost \$1.75 each and paint costs \$2.45 per jar. Write an equation that shows how much will be spent, *s*, if she buys *b* paintbrushes and *j* jars of paint. If she has \$125 to spend and she buys 22 paintbrushes, how many jars of paint can she buy?

# 3 Chapter Test

**Solve each equation.**

1. $g + 17 = 35$
2. $13 = a - 6$
3. $8 - x = 15$
4. $g + \frac{2}{3} = \frac{7}{9}$
5. $-1.7 + d = -3.2$

**Write an equation then solve.**

6. SPORTS Amy and Jenna played 27 games of tennis in their match. Amy won 12 of the games. How many did Jenna win?

**Solve each equation.**

7. $3x = 9$
8. $-5p = -110$
9. $\frac{h}{5} = -12$
10. $\frac{-v}{3} = \frac{3}{8}$

**Write an equation then solve.**

11. CONSUMER ECONOMICS A package of 12 VHS tapes costs \$41.88. What is the price of an individual VHS tape?

**Solve each equation.**

12. $2x + 5 = 9$
13. $5p - 9 = 11$
14. $\frac{h}{3} - 7 = 40$
15. $12 + 3y = -21$
16. $5s + 1.9 = 16.9$
17. $5y + 7 = \frac{3}{2}$
18. $2.3 - 7r = -25.7$
19. CONSUMER ECONOMICS It costs \$102.50 to repair the CD player in your stereo system. If you pay \$15 for new parts and \$25 per hour for labor, write and solve an equation to determine how much time the repair took.

**Solve each equation.**

20. $3x + 4 = 5x - 2$
21. $8 - 3z = 2z + 9$
22. $\frac{d}{3} + 5 = \frac{d}{4} - 7$
23. $8k - 4.7 = 3k + 0.3$
24. $3\frac{1}{5} + \frac{v}{-5} = 18$
25. CONSUMER ECONOMICS Three new ink cartridges cost \$9 less than 5 new ink cartridges. Write and solve an equation to find the cost of each ink cartridge.

**Solve each equation.**

26. $5x + 3(x + 6) = 66$
27. $4(m - 3) = 10m - 6(m - 7)$

**Solve for the indicated variable.**

28. $t + u = v$, for $u$
29. $ax - r = 12$, for $x$
30. Use the formula $C = \frac{5}{9}(F - 32)$ to convert $15°C$ to Fahrenheit degrees.

# College Entrance Exam Practice



**QUANTITATIVE COMPARISON** For Items 1–4, write
A if the quantity in Column A is greater than the quantity in Column B;
B if the quantity in Column B is greater than the quantity in Column A;
C if the two quantities are equal; or
D if the relationship cannot be determined from the information given.

| | Column A | Column B | Answers |
|---|---|---|---|
| **1.** | $x = 6$<br>$2x + 1$ | $4x - 3$ | Ⓐ Ⓑ Ⓒ Ⓓ [Lesson 1.3] |
| **2.** | $(-11) \cdot (121)$ | $(121) \div (-11)$ | Ⓐ Ⓑ Ⓒ Ⓓ [Lesson 2.4] |
| **3.** | $(3y + 2) + (2y - 5)$ | $(2y + 6) - (y + 7)$ | Ⓐ Ⓑ Ⓒ Ⓓ [Lesson 2.6] |
| **4.** | The solution to the equation $-\frac{3}{4}x = 3$ | The solution to the equation $-\frac{4}{3}x = -4$ | Ⓐ Ⓑ Ⓒ Ⓓ [Lesson 3.2] |

**5.** Which expression is equal to 19? ***(LESSON 1.3)***
**a.** $15 \cdot (6 + 4) - 57$ **b.** $2^3 - 6 \cdot (3 + 2)$
**c.** $4^2 - 5^2 \div 5 + 2(0)$ **d.** $9 \div 3 \cdot 2^3 - 5$

**6.** What is the value of $-5\frac{5}{6} \cdot \frac{3}{5}$? ***(LESSON 2.4)***
**a.** $-5\frac{1}{2}$ **b.** $2\frac{1}{3}$
**c.** $-3\frac{1}{2}$ **d.** none of these

**7.** Which expression is *not* equivalent to $\frac{32 - 8x}{4}$? ***(LESSON 2.7)***
**a.** $8 - 2x$ **b.** $\frac{16 - 4x}{2}$
**c.** $(32 - 8x) \div 4$ **d.** none of these

**8.** Which expression is not equal to the others? ***(LESSON 2.6)***
**a.** $3x - 2(x - 3)$ **b.** $3x + 2(3 - x)$
**c.** $3x - 2(x + 3)$ **d.** $3x - 2(-3 + x)$

**9.** What is the solution to the equation $\frac{q}{139.2} = -58$? ***(LESSON 3.2)***
**a.** $-2.4$ **b.** $-8073.6$
**c.** $-0.417$ **d.** $-7551.6$

**Find each sum.** ***(LESSON 2.2)***

**10.** $-25 + (-4)$

**11.** $-36 + 6 + (-6)$

**12.** $452 + (-452)$

**13.** Solve the equation $-3w + 5 = -16$. ***(LESSON 3.3)***

**14.** The perimeter of a rectangle is 52 and its width is 6 less than its length. Find the length and width of the rectangle. ***(LESSON 3.6)***

**15.** Solve: $6 - (3 + 2x) = -3x + (2x - 4)$. ***(LESSON 3.5)***

**16.** Solve: $y + 4.5 = 6$. ***(LESSON 3.1)***

**17.** Solve: $2(x + 7) = 15$. ***(LESSON 3.5)***

**18.** Solve: $3p - 2 + 2p = 8$. ***(LESSON 3.5)***

**Solve each equation.** ***(LESSON 3.1)***

**19.** $x + 7 = 6$

**20.** $t - 9 = 11$

**21.** $z - 8.4 = 1.25$

**22.** $82 = 21 + r$

**Determine the next three terms of each sequence.** ***(LESSON 1.1)***

**23.** 1, 4, 8, 13, 19, . . .

**24.** 2, 4, 8, 14, 22, . . .

**25.** 10, 5, 0, –5, . . .

**26.** 3, 6, 12, 24, . . .

**27.** Use the graph of Ellen's earnings below to determine how much she would earn if she worked 12 hours. ***(LESSON 1.5)***

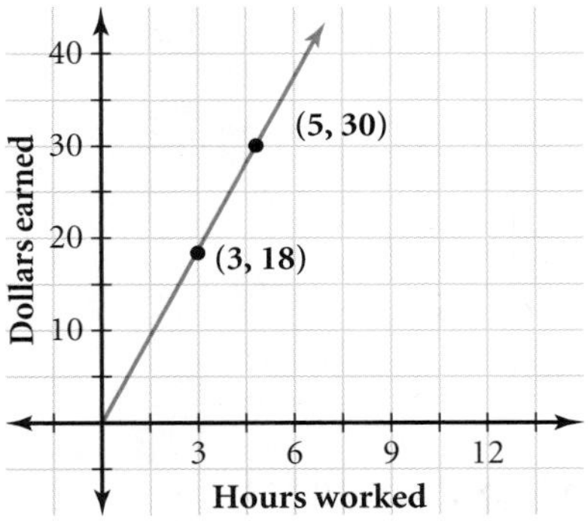


**Solve each equation for the variable indicated.** ***(LESSONS 3.1, 3.2, 3.3 AND 3.6)***

**28.** $a + 7.5 = 3.2$, for $a$

**29.** $-6.5 = x + 1.7$, for $x$

**30.** $T = I + P$, for $P$

**31.** $d = rt$, for $r$

**32.** $3x + 7 = 2x - 6$, for $x$

**33.** $\frac{x}{4} + 2 = 3x + 35$, for $x$

**34.** $-5 + 2p = 3p$, for $p$

FREE-RESPONSE GRID

**The following questions may be answered by using a free-response grid such as that commonly used by standardized test services.**

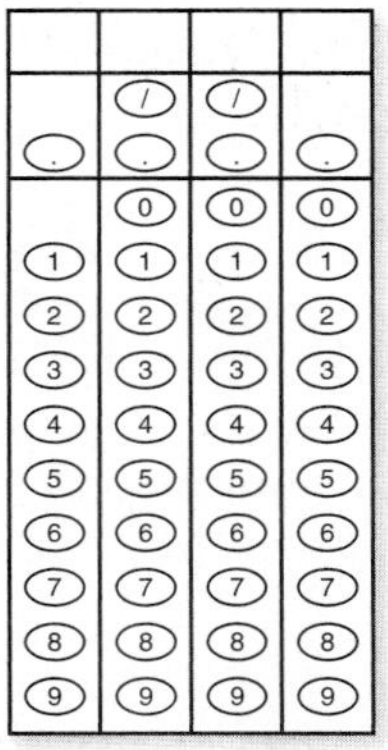

**35.** Misty's team is in a league with a total of 12 teams. If each team plays each of the other team once, how many games will be played? ***(LESSON 1.1)***

**36.** Evaluate: $(-4)(4) \div (-1)$. ***(LESSON 2.4)***

**37.** Use the method of differences to find the next term of the sequence 45, 42, 37, 30, 21, . . . ***(LESSON 1.1)***

**38.** Use mental math to simplify $(45 + 23) + 107$. ***(LESSON 2.5)***

**Determine the *x*-coordinate of each point listed below.** ***(LESSON 1.4)***

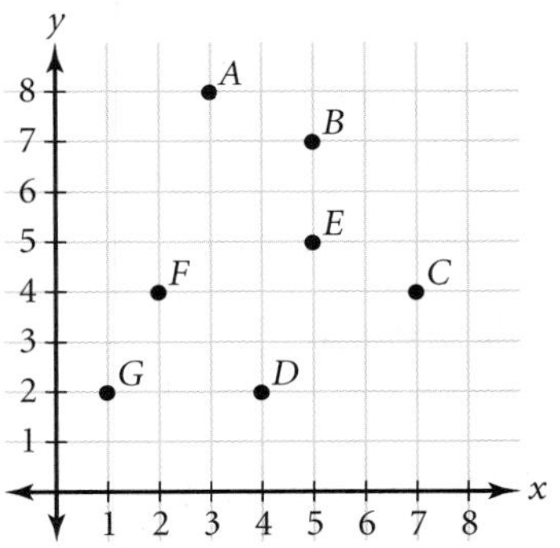


**39.** point $A$

**40.** point $D$

**41.** point $F$

**42.** point $G$

4

# Proportional Reasoning and Statistics

## Lessons



The need for data collection goes as far back as ancient times, when political and religious leaders gathered information about people and property. The word *statistic* comes from the German word *statistik*, which was commonly used in German universities in the 1700s to describe the comparison of data about different nations.

The field of statistics was important to the United States war effort during World War II. As a result, the use of statistics grew to cover a great variety of areas after the war. Today statisticians can find careers in many fields, including law, education, science, business, sports, and journalism.

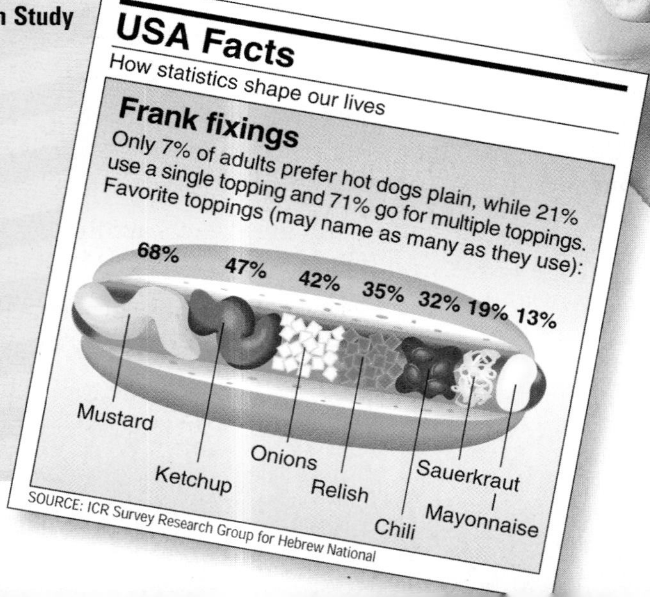

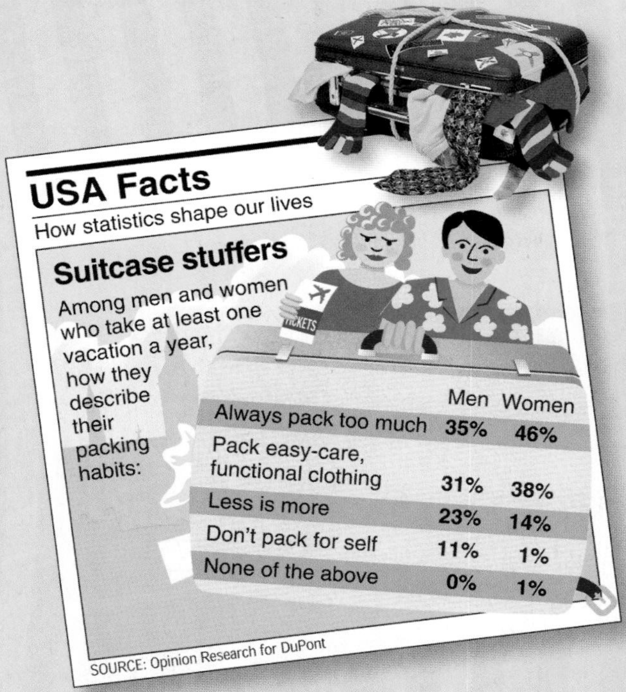

*Central Park,*
*New York City*

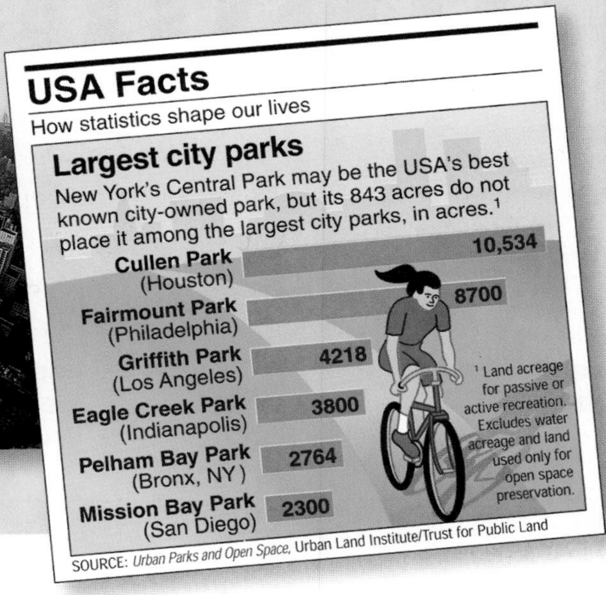


CHAPTER PORTFOLIO ACTIVITIES PROJECT

## About the Chapter Project

The Chapter Project, *Designing a Research Study*, on page 208, provides an opportunity for you to design your own research study, compile the results, and report on the results. After completing the Chapter Project, you will be able to do the following:

- Analyze data by using graphs, plots, and measures of central tendency.

## About the Portfolio Activities

Throughout the chapter, you will be given opportunities to complete Portfolio Activities that are designed to support your work on the Chapter Project.

- Soft-drink preferences are the topic of the Portfolio Activity on page 170.
- Simulating a probability experiment is the topic of the Portfolio Activity on page 185.
- Graphing test scores is the topic of the Portfolio Activity on page 207.

4.1

# Using Proportional Reasoning

### Objectives

- Identify the means and extremes of a proportion.
- Use proportions to solve problems.

**Why** ***Proportional reasoning is an important tool in many professions, including science, engineering, and business. Proportions are also common in everyday activities such as cooking and gardening.***

## Proportions

**APPLICATION**
**SPORTS**

In the first round of a timed free-throw contest, Ben made 28 out of 40 shots and Alicia made 21 out of 30 shots. Who was more accurate?

| **Ben** | **Alicia** |
|---|---|
| $\frac{28}{40} = \frac{7}{10}$ | $\frac{21}{30} = \frac{7}{10}$ |

Because both fractions simplify to $\frac{7}{10}$, $\frac{28}{40}$ is equal to $\frac{21}{30}$. The contestants were equally accurate. Fractions, such as $\frac{7}{10}$ and $\frac{15}{3}$ are *ratios.* A **ratio** is a comparison of two quantities that uses division.

The equation $\frac{28}{40} = \frac{21}{30}$ is called a **proportion**. Any statement that two ratios are equal, $\frac{a}{b} = \frac{c}{d}$, is a proportion. You can use proportions to solve many types of problems.

## Means and Extremes

A proportion such as $\frac{a}{b} = \frac{c}{d}$ can also be written in the form $a\!:\!b = c\!:\!d$. The inside terms, $b$ and $c$, are called the **means**. The outside terms, $a$ and $d$, are called the **extremes.** The product of the means and the product of the extremes can be used to solve proportion problems.

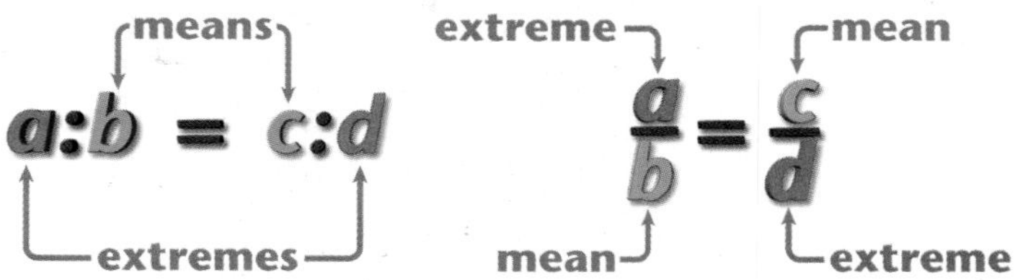


*Activity*

### Exploring Relationships in Proportions

1. For each proportion, identify the means and the extremes.
   **a.** $\frac{1}{2} = \frac{5}{10}$ **b.** $\frac{3}{4} = \frac{9}{12}$
   **c.** $\frac{15}{25} = \frac{30}{50}$ **d.** $\frac{4}{6} = \frac{6}{9}$
2. For each proportion in Step 1, find the product of the means and the product of the extremes. (For example, in part **a**, find $2 \cdot 5$ and $1 \cdot 10$.)
3. Describe what you observe about the relationship between the product of the means and the product of the extremes.
4. For any proportion $\frac{a}{b} = \frac{c}{d}$, where $b \neq 0$ and $d \neq 0$, $ad$ and $bc$ are called cross products. Explain how cross products and products of the means and extremes are related.

CHECKPOINT ✔

5. Explain how you can use the cross products to solve the following proportion for $x$:
$$\frac{x}{4} = \frac{42}{24}$$

In a true proportion, the product of the means and the product of the extremes, also called **cross products**, are equal. Because of this relationship, it is possible to use cross products to determine whether a proportion is true and to solve for a missing value in a proportion.

**Cross Products Property**

For all real numbers $a$, $b$, $c$, and $d$, where $b \neq 0$ and $d \neq 0$:

If $\frac{a}{b} = \frac{c}{d}$, then $a \cdot d = b \cdot c$.

**EXAMPLE 1** **Use cross products to determine whether each proportion is true.**

**a.** $\frac{2}{3} \stackrel{?}{=} \frac{3}{4}$

**b.** $\frac{3}{5} \stackrel{?}{=} \frac{15}{25}$

**SOLUTION**

**a.** Find the cross products.

$2 \cdot 4 = 8 \quad 3 \cdot 3 = 9 \quad 8 \neq 9$

The proportion is not true.

$\frac{2}{3} \neq \frac{3}{4}$

**b.** Find the cross products.

$3 \cdot 25 = 75 \quad 5 \cdot 15 = 75 \quad 75 = 75$

The proportion is true.

$\frac{3}{5} = \frac{15}{25}$

When one term of a proportion is unknown, such as in $\frac{3}{7} = \frac{x}{14}$, you can use cross products to find the value of the variable that makes a true proportion. First find the cross products and set them equal to each other.

$$7 \cdot x = 3 \cdot 14$$
$$7x = 42$$

Then solve for the variable.

$$7x = 42$$
$$\frac{7x}{7} = \frac{42}{7}$$
$$x = 6$$

**EXAMPLE 2** **Solve each proportion.**

**a.** $\frac{15}{m} = \frac{10}{9}$

**b.** $\frac{5}{60} = \frac{3}{x}$

**c.** $\frac{x}{8} = \frac{7.5}{20}$

**d.** $\frac{9}{3} = \frac{m}{5}$

**SOLUTION**

**a.**
$$\frac{15}{m} = \frac{10}{9}$$
$$15 \cdot 9 = 10m$$
$$135 = 10m$$
$$\frac{135}{10} = \frac{10m}{10}$$
$$13.5 = m$$
$$m = 13.5$$

**b.**
$$\frac{5}{60} = \frac{3}{x}$$
$$5x = 180$$
$$\frac{5x}{5} = \frac{180}{5}$$
$$x = 36$$

**c.**
$$\frac{x}{8} = \frac{7.5}{20}$$
$$20x = 60$$
$$\frac{20x}{20} = \frac{60}{20}$$
$$x = 3$$

**d.**
$$\frac{9}{3} = \frac{m}{5}$$
$$45 = 3m$$
$$\frac{45}{3} = \frac{3m}{3}$$
$$15 = m$$
$$m = 15$$

**TRY THIS** Solve each proportion. **a.** $\frac{x}{3.5} = \frac{5}{2.5}$ **b.** $\frac{13}{2} = \frac{65}{n}$

**CRITICAL THINKING** What steps would be required to solve each proportion in Example 2 without using cross products?

You may recall that similar polygons have corresponding sides that are proportional. The symbol ~ is used to indicate that two figures are similar.

**EXAMPLE 3**

CONNECTION
GEOMETRY

Given: rectangle $ABCD \sim$ rectangle $WXYZ$. **Find *s*.**

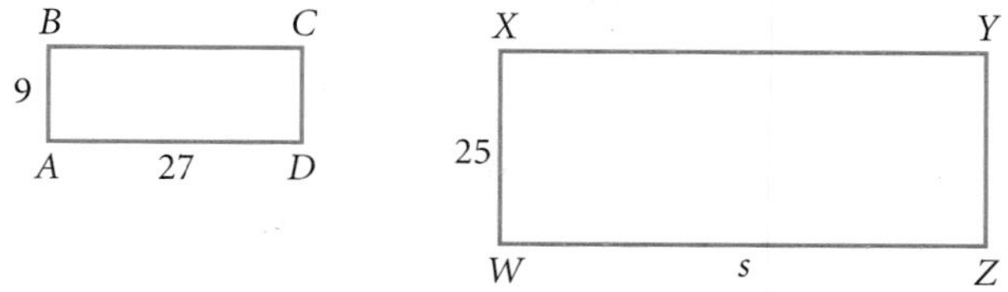


**SOLUTION**

Because the rectangles are similar, the corresponding sides are proportional.

Set up a proportion. $\frac{9}{25} = \frac{27}{s}$

Solve the proportion for $s$. $675 = 9s$

$75 = s$

The length is 75.

**TRY THIS**

Given: triangle $MNO \sim$ triangle $UVW$. Find $x$.

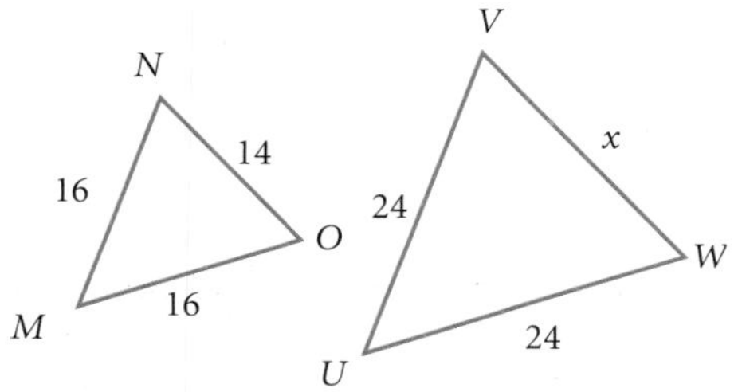


**EXAMPLE 4**

APPLICATION
INVENTORY

The goal of the Bartlesville City Library is to have 2 books in stock for every 5 residents. The most recent census shows that the population is 12,455. **Find the total number of books that the library should have in order to meet the goal.**

**SOLUTION**

PROBLEM SOLVING

First **make a table** to organize the information.

| Number of books | 2 | $n$ |
|---|---|---|
| Number of residents | 5 | 12,455 |

Write the proportion shown in the table. $\frac{2}{5} = \frac{n}{12{,}455}$

Then use cross products to solve the proportion.

$$5n = 2 \cdot 12{,}455$$
$$5n = 24{,}910$$
$$\frac{5n}{5} = \frac{24{,}910}{5}$$
$$n = 4982$$

The library should have 4982 books in order to meet its goal.

# Exercises

## Communicate

**1.** What is a proportion? Give an example.

**2.** Explain how to find the means and the extremes of a proportion.

**3.** Make a statement about the relationship between the means and the extremes of a proportion.

**4.** Describe two ways to solve for $n$ in the proportion $\frac{n}{36} = \frac{2}{3}$.

## Guided Skills Practice

**Determine whether each proportion is true.** ***(EXAMPLE 1)***

**5.** $\frac{3}{4} \stackrel{?}{=} \frac{6}{8}$ **6.** $\frac{12}{19} \stackrel{?}{=} \frac{5}{7}$ **7.** $\frac{4}{21} \stackrel{?}{=} \frac{16}{84}$

**Solve each proportion.** ***(EXAMPLE 2)***

**8.** $\frac{5}{7} = \frac{x}{21}$ **9.** $\frac{10}{p} = \frac{85}{153}$ **10.** $\frac{12}{14} = \frac{84}{m}$

CONNECTION

**11.** **GEOMETRY** Rectangle *QRST* and rectangle *LMNO* are similar. Write and solve a proportion to find the missing length, $b$. ***(EXAMPLE 3)***

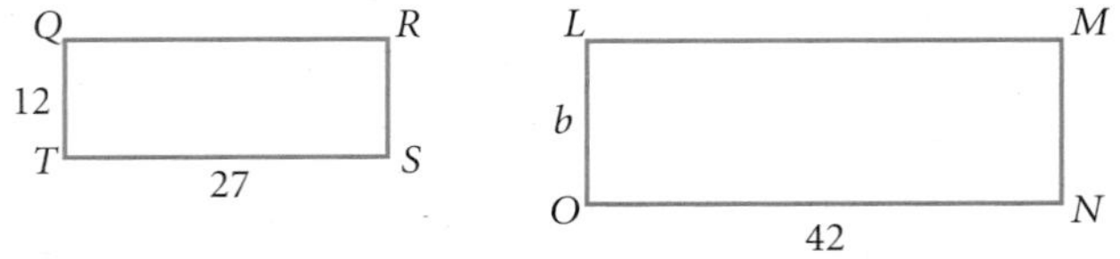


APPLICATION

**12.** **BIOLOGY** To keep the proper balance between fish species A and fish species B in an experimental fish pond, a wildlife biologist needs to keep the ratio of species A to species B at 2 to 3. Write and solve a proportion to find how many fish of species B are needed if there are 256 fish of species A. ***(EXAMPLE 4)***

## Practice and Apply

internet connect

**Homework Help Online**

Go To: **go.hrw.com**
Keyword:
**MA1 Homework Help** for Exercises 13-24

**Determine whether each proportion is true.**

**13.** $\frac{15}{9} \stackrel{?}{=} \frac{35}{21}$ **14.** $\frac{12}{9} \stackrel{?}{=} \frac{18}{12}$ **15.** $\frac{56}{24} \stackrel{?}{=} \frac{49}{21}$

**16.** $\frac{27}{21} \stackrel{?}{=} \frac{35}{28}$ **17.** $\frac{18}{8} \stackrel{?}{=} \frac{108}{48}$ **18.** $\frac{3}{5} \stackrel{?}{=} \frac{81}{135}$

**19.** $\frac{3}{13} \stackrel{?}{=} \frac{10}{65}$ **20.** $\frac{12}{20} \stackrel{?}{=} \frac{27}{45}$ **21.** $\frac{8}{10} \stackrel{?}{=} \frac{24}{30}$

**22.** $\frac{12}{8} \stackrel{?}{=} \frac{48}{34}$ **23.** $\frac{24}{3} \stackrel{?}{=} \frac{72}{12}$ **24.** $\frac{12}{16} \stackrel{?}{=} \frac{60}{80}$

**Solve each proportion. Round to the nearest hundredth if necessary.**

**25.** $\frac{27}{18} = \frac{42}{n}$ **26.** $\frac{38}{19} = \frac{n}{20}$ **27.** $\frac{42}{28} = \frac{36}{n}$

**28.** $\frac{n}{48} = \frac{72}{96}$ **29.** $\frac{21.5}{x} = \frac{64.5}{18}$ **30.** $\frac{x}{37.2} = \frac{16}{24.8}$

**31.** $\frac{30.8}{112} = \frac{y}{10}$ **32.** $\frac{t}{25} = \frac{471}{15}$ **33.** $\frac{1.2}{1.5} = \frac{8}{m}$

**34.** $\frac{72}{34} = \frac{9}{g}$ **35.** $\frac{7}{15} = \frac{f}{48}$ **36.** $\frac{4}{m} = \frac{1.6}{22}$

**37.** $\frac{21}{36} = \frac{4.8}{d}$ **38.** $\frac{21}{56} = \frac{x}{7.2}$ **39.** $\frac{p}{12} = \frac{21}{63}$

**40.** $\frac{45}{18} = \frac{3}{k}$ **41.** $\frac{22}{36} = \frac{x}{198}$ **42.** $\frac{60}{21} = \frac{20}{s}$

**43.** $\frac{52}{13} = \frac{8}{q}$ **44.** $\frac{y}{16} = \frac{12}{4}$ **45.** $\frac{74}{p} = \frac{92}{46}$

**46.** Suppose that a 6-foot tall person casts a shadow that is 10 feet long. A nearby statue casts a 16-foot shadow. How tall is the statue?

**47.** A 52-foot building casts a shadow that is 48 feet long. At the same time, the shadow cast by a nearby statue is 18 feet long. How tall is the statue?

CONNECTION

**48.** **GEOMETRY** Triangle *ABC* is similar to triangle *MNP*. Find *s*.

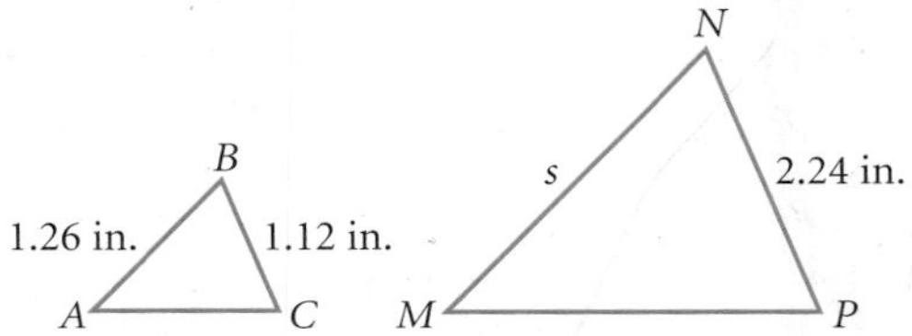


APPLICATIONS

**49.** **CONSUMER ECONOMICS** A fertilizer manufacturer recommends 4 pounds of fertilizer for every 1500 square feet of lawn. How many pounds of fertilizer are needed to cover 2400 square feet of lawn?

**50.** **TRAVEL** Matthew drove his car 425 miles on 18.2 gallons of gasoline. At this rate, how many gallons of gasoline will Matthew need to travel 640 miles? Give your answer to the nearest tenth of a gallon.

**51.** **COOKING** A recipe for chili for 12 people calls for 3 pounds of ground beef, 8 tomatoes, and various spices.

**a.** How many pounds of ground beef are needed to make enough chili for 50 people?

**b.** How many pounds of ground beef are needed to make enough chili for 65 people?

## Look Back

**52.** Subtract: $\frac{1}{12} - \frac{2}{30}$. ***(LESSON 2.3)***

**Simplify each expression.** ***(LESSONS 2.6 AND 2.7)***

**53.** $3 \cdot 5x$ **54.** $-2y \cdot 3y$ **55.** $7(x-3)-(5-4x)$

**56.** $4s+7-2s-5$ **57.** $4t \cdot (-3t)$ **58.** $\frac{4y+2}{2}$

**Solve each equation.** ***(LESSON 3.2)***

**59.** $\frac{w}{-12} = 13$ **60.** $-18 = \frac{x}{12}$ **61.** $100x = 216$

**62.** $3p = -18$ **63.** $-4y = 124$ **64.** $5t = 123$

## Look Beyond

**CONNECTION**

**65.** **GEOMETRY** The ratio of both the lengths and the widths of the rectangles is 1:3. Find the ratio of the areas.

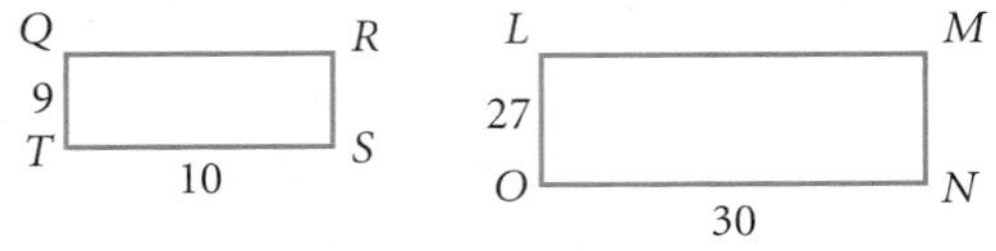


**66.** For the rectangles in Exercise 65, does the ratio of the areas equal the ratio of the lengths and the widths? Why or why not?

**PORTFOLIO ACTIVITY**

Include this activity in your portfolio.

Horacha used a computer to build a bar graph for the soft-drink preferences reported in a survey of her science class.

**1.** Which soft drink was the most popular?

**2.** How many students responded to the survey?

**3.** Do students prefer colas over lemon-lime? Explain your reasoning.

**4.** Can you conclude that over 50% of the class prefers some kind of cola? Explain your reasoning.

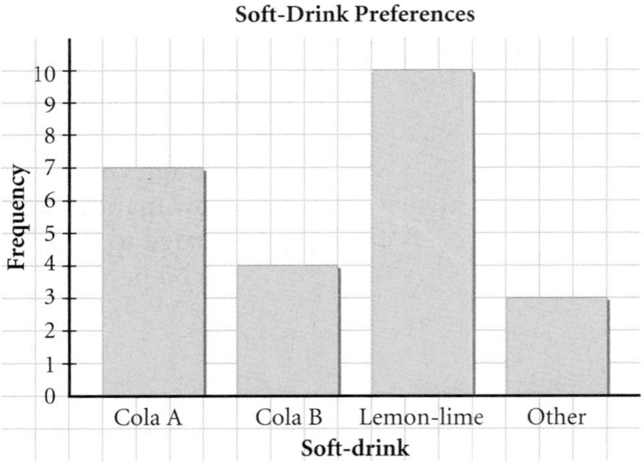


**5.** How many soft-drink varieties do you think are included in the "other" category? Why?

**WORKING ON THE CHAPTER PROJECT**

You should now be able to complete Activities 1 and 2 of the Chapter Project on page 208.

# 4.2 Percent Problems

**Objectives**

- Find equivalent fractions, decimals, and percents.
- Solve problems involving percent.

**Why** *Percents are often found in newspapers, magazines, and other media. They are used in tables, charts, graphs, and advertisements to describe situations like discounts, taxes, and weather.*

## Fractions, Decimals, and Percents

A **percent** is a *ratio* that compares a number with 100. For example, 50% is the same as $\frac{50}{100}$, or $\frac{1}{2}$, and 100% is the same as $\frac{100}{100}$, or 1.

**APPLICATION**
**METEOROLOGY**

Relative humidity is a common example of percent. For example, at 30°C, 1 cubic meter of air can hold no more than 26 grams of water. At this amount, the relative humidity is 100%. At other amounts, ratios are used to determine the relative humidity. If 1 cubic meter of air at 30°C contains 6.5 grams of water, the relative humidity is 25% because $\frac{6.5}{26} = \frac{1}{4}$, or 25%.

When solving these types of problems, it is helpful to know how to write percents as decimals and as fractions. Sometimes you may wish to write the fractions in simplest terms.

$$60\% = \frac{60}{100} = 0.60 \qquad \frac{60}{100} = \frac{3}{5}$$

**EXAMPLE 1** **Write each percent as a decimal and as a fraction or mixed number in simplest terms.**

**a.** 75% **b.** 110% **c.** 4.4%

**SOLUTION**

**a.** $75\% = \frac{75}{100} = 0.75$ $\qquad \frac{75}{100} = \frac{3}{4}$

**b.** $110\% = \frac{110}{100} = 1.10$ $\qquad \frac{110}{100} = \frac{11}{10}$, or $1\frac{1}{10}$

**c.** $4.4\% = \frac{4.4}{100} = \frac{44}{1000} = 0.044$ $\qquad \frac{4.4}{100} = \frac{44}{1000} = \frac{11}{250}$

**TRY THIS** Write each percent as a decimal and as a fraction or mixed number in simplest terms.

**a.** 40% **b.** 175% **c.** 2.5%

In a percent statement such as "25% of 40 is 10," 25 is the **percent rate**, 40 is known as the **base**, and 10 is the **percentage**. If you know two of the parts in a percent problem, there are different methods to find the third part.

### The Proportion Method

You can use proportions to solve percent problems. A percent bar can help you to visualize the proportion.

**1.** The percentage is unknown. 25% of 40 is $x$.

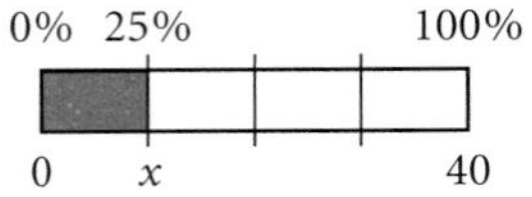


$$\frac{25}{100} = \frac{x}{40} \rightarrow 100x = 1000 \rightarrow x = 10$$

**2.** The base is unknown. 25% of $x$ is 10.

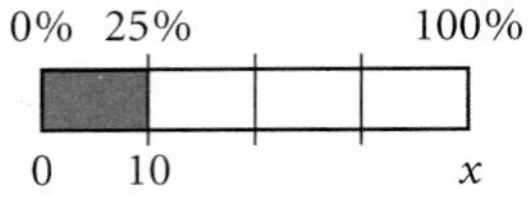


$$\frac{25}{100} = \frac{10}{x} \rightarrow 25x = 1000 \rightarrow x = 40$$

**3.** The percent rate is unknown. $x$% of 40 is 10.

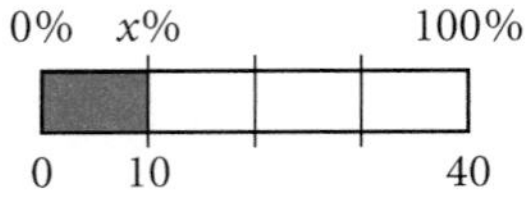


$$\frac{x}{100} = \frac{10}{40} \rightarrow 40x = 1000 \rightarrow x = 25$$

### The Equation Method

You can set up and solve an equation for the unknown part of a percent problem if the other two parts are known. Write percents as decimals when using them in this type of equation.

**1.** The percentage is unknown.

25% of 40 is $x$.

$0.25 \cdot 40 = x$

$10 = x$

**2.** The base is unknown.

25% of $x$ is 10.

$0.25 \cdot x = 10$

$x = 40$

**3.** The percent rate is unknown.

$x$ of 40 is 10.

$x \cdot 40 = 10$

$x = 0.25$, or 25%

**EXAMPLE 2** The sophomore class is sponsoring a trip to see a play. Student tickets normally cost $8. If at least 20 people buy tickets as a group, there is a 30% discount. **How much will each ticket cost at the discounted price?**

**APPLICATION**
**DISCOUNTS**

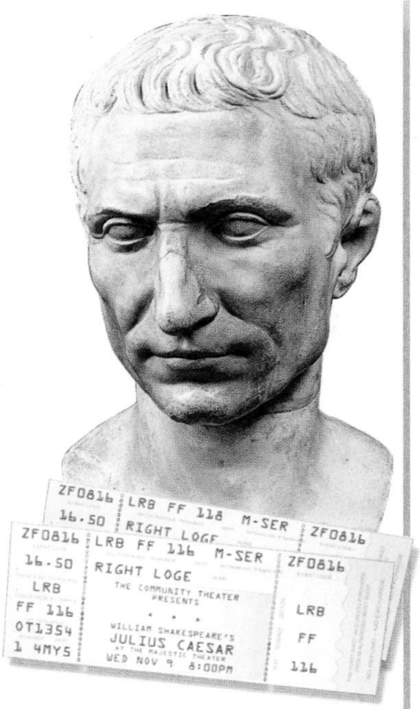


*The play "Julius Caesar" was written by William Shakespeare between 1599 and 1600.*

**SOLUTION**

### Proportion Method

You can visualize the problem by using a percent bar. The base, or regular cost, $8, is the length of the bar. Because the tickets are 30% off, the discounted price is 70% of the original price (100% − 30% = 70%). Shade 70% of the bar, and use $x$ to represent the percentage, or discounted price.

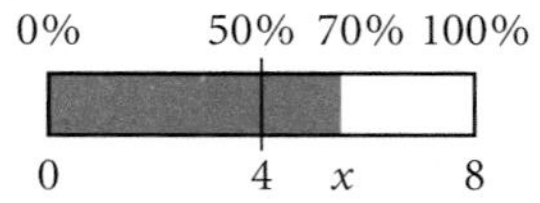


Write the proportion $\frac{70}{100} = \frac{x}{8}$ from the percent bar.

Solve the proportion. $\frac{70}{100} = \frac{x}{8} \rightarrow 100x = 560 \rightarrow x = 5.60$

The discounted price is $5.60.

### Equation Method

As in the proportion method, use $x$ to represent the discounted price.

The discounted price, $x$, is 70% of the full price, $8.

Model this statement with an equation and solve for $x$.

$x = 0.70 \cdot 8$

$x = 5.60$

The discounted price is $5.60.

**TRY THIS** Find each answer by using the proportion method or the equation method.

**a.** What percent of 80 is 15?

**b.** Find 115% of 200.

**c.** 45 is 40% of what number?

**EXAMPLE 3** A VCR with a sale price of $239.40 is advertised as 40% off. **What was the original price?**

**APPLICATION**
**DISCOUNTS**

**SOLUTION**

Let $w$ represent the original price, or base, and note that $239.40 is 60% of the original price (100% − 40% = 60%). Model this information with an equation.

$$239.40 = 0.60w$$

Apply the Division Property of Equality and simplify.

$$\frac{239.40}{0.60} = \frac{0.60w}{0.60}$$

$$399 = w$$

The original price was $399.

**EXAMPLE 4** Most states have a sales tax for certain items. Linda paid \$47.74 for an item priced at \$45.25. **Find the sales tax rate to the nearest tenth of a percent.**

**APPLICATION**
**TAXES**

**SOLUTION**

First find the amount of sales tax. Subtract the price of the item from the total amount paid.

$$\$47.74 - \$45.25 = \$2.49$$

The amount of sales tax was \$2.49. You can use a percent bar to show the proportion. Let $p$ represent the percent rate.

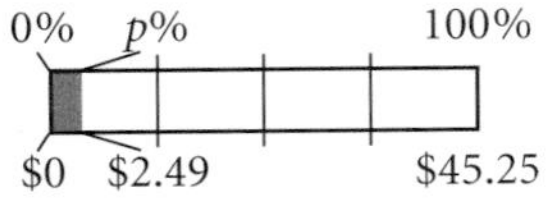


Write the proportion from the percent bar.

$$\frac{p}{100} = \frac{2.49}{45.25}$$

Solve the proportion.

$p \approx 5.5$ The sales tax rate is 5.5%.

**TRY THIS** Find the sales tax rate to the nearest tenth of a percent if the total paid for an item priced at \$50.75 was \$54.05.

Many types of problems involve finding a percent of increase or percent of decrease after a change has occurred. It is important to remember when working these types of problems that the original amount is always used as the base.

**EXAMPLE 5** A family is adding additional rooms to their house. The house originally had 1500 square feet of floor space. After the additions are completed, the house will have 1800 square feet of floor space. **Find the percent of increase in floor space.**

**SOLUTION**

First find the amount of the increase in floor space. Find the difference between the amounts of floor space in square feet before and after the additions.

$$1800 - 1500 = 300$$

The amount of increase in floor space is 300 square feet. Thus, some percent of 1500 is 300. Let $p$ represent the percent rate in decimal form. You can model the problem with an equation.

$$p \cdot 1500 = 300, \text{ or } 1500p = 300$$

Solve for $p$ and convert the decimal to a percent.

$$\frac{1500p}{1500} = \frac{300}{1500} \rightarrow p = 0.20$$

The percent of increase in floor space is 20%.

**TRY THIS** Suppose that the amount of floor space in Example 5 had been increased from 1500 square feet to 1875 square feet. Find the percent of increase.

**EXAMPLE 6** The price of a car that originally sold for $17,000 has been reduced to $14,450. **Find the percent of decrease in the price of the car.**

**SOLUTION**

First find the amount of the decrease in the price. Find the difference between the original price and the reduced price.

$$17{,}000 - 14{,}450 = 2550$$

The amount of decrease in the price is $2550. Thus, some percent of 17,000 is 2550. Let $p$ represent the percent rate in decimal form. You can model the problem with an equation.

$$p \cdot 17{,}000 = 2550, \text{ or } 17{,}000p = 2550$$

Solve for $p$ and convert the decimal to a percent.

$$\frac{17{,}000p}{17{,}000} = \frac{2550}{17{,}000} \rightarrow p = 0.15$$

The percent of decrease is 15%.

**TRY THIS** Suppose that the price of the car in Example 6 had been reduced from $17,000 to $13,430. Find the percent of decrease in the price of the car.

# Exercises

## Communicate

internet connect
**Activities Online**
Go To: **go.hrw.com**
Keyword:
**MA1 Rural Exodus**

1. Explain the procedure for changing a percent to a decimal.
2. Explain how a percent bar helps to model a percent problem.
3. What are the advantages of using the proportion method to solve percent problems?
4. What are the advantages of using the equation method to solve percent problems?

## Guided Skills Practice

**Write each percent as a decimal and as a fraction or mixed number in simplest terms.** *(EXAMPLE 1)*

5. 34%
6. 20%
7. 130%
8. 7.5%

**Find each answer.** *(EXAMPLE 2)*

9. 70% of 150 is what number?
10. 14% of 50 is what number?

**APPLICATION**

11. **HEALTH** Tim has reduced his cholesterol level by 20% by dieting and exercising. If his original level was 238, what is his new cholesterol level? *(EXAMPLE 2)*

**Find each answer.** ***(EXAMPLE 3)***

**12.** 17 is 50% of what number?

**13.** 25 is 40% of what number?

**Find each answer.** ***(EXAMPLE 4)***

**14.** What percent of 50 is 5?

**15.** What percent of 350 is 70?

**APPLICATION**

**16.** **TAXES** Jim paid $25.85 to purchase a shirt that cost $23.50. What was the tax rate, rounded to the nearest tenth of a percent? ***(EXAMPLE 4)***

**A yard originally had a perimeter of 200 meters. Find the percent of increase for each new perimeter below.** ***(EXAMPLE 5)***

**17.** 240 meters

**18.** 230 meters

**19.** 280 meters

**A car originally cost $30,000. Find the percent of decrease for each new price below.** ***(EXAMPLE 6)***

**20.** $24,000

**21.** $27,500

**22.** $16,500

## Practice and Apply

**Write each percent as a decimal.**

**23.** 55%
**24.** 1.2%
**25.** 8%
**26.** 45.3%
**27.** 0.5%
**28.** 73%
**29.** 186%
**30.** 24%
**31.** 4.5%
**32.** 17.2%

**Write each percent as a fraction or mixed number in lowest terms.**

**33.** 47%
**34.** 18%
**35.** 3.5%
**36.** 0.01%
**37.** 125%
**38.** 52%
**39.** 38%
**40.** 0.8%
**41.** 210%
**42.** 7.3%

**internet connect**
**Homework Help Online**
Go To: **go.hrw.com**
Keyword:
**MA1 Homework Help** for Exercises 43-62

**Draw a percent bar for each problem, and use a proportion to find the solution.**

**43.** Find 35% of 80.

**44.** 18 is 20% of what number?

**45.** What percent of 40 is 60?

**46.** What percent of 20 is 50?

**47.** 48 is 40% of what number?

**48.** Find 45% of 120.

**Find each answer.**

**49.** Find 40% of 50.

**50.** Find 200% of 50.

**51.** 2 is what percent of 100?

**52.** Find 45% of 80.

**53.** 30 is 60% of what number?

**54.** What percent of 80 is 10?

**55.** Find 180% of 40.

**56.** 8 is 20% of what number?

**57.** Find 30% of 120.

**58.** Find 50% of 124.

**59.** What percent of 80 is 60?

**60.** Find 175% of 92.

**61.** 27 is 60% of what number?

**62.** What percent of 35 is 105?

**A house originally had 1700 square feet of floor space. Find the percent of increase for each floor space given below.**

**63.** 1955 square feet

**64.** 1870 square feet

**65.** 1785 square feet

**66.** 2057 square feet

**A car originally cost $15,000. Find the percent of decrease for each new price.**

**67.** $14,250 **68.** $13,050 **69.** $13,650 **70.** $12,300

**71.** A teacher says that 40% of your grade will be based on homework. If there is a total of 500 points possible toward your grade, how many points are possible for homework?

CONNECTION

**72.** STATISTICS The 1996 presidential election had the following results:

| Candidate | Popular vote | Electoral vote |
|---|---|---|
| Clinton (Democrat) | 45,590,703 | 379 |
| Dole (Republican) | 37,816,307 | 159 |
| Perot (Independent) | 7,866,284 | 0 |

**a.** What percent of the popular vote did each candidate get?

**b.** What percent of the electoral vote did each candidate get?

CHALLENGE

**c.** If 12,300,000 Clinton voters and 8,100,000 Dole voters voted for Perot, what would each candidate's percentage of the popular vote have been?

APPLICATIONS

**73.** DISCOUNTS A book is marked down from an original price of $20 to $15. By what percent has the book been marked down?

**74.** DISCOUNTS A coat is marked for sale at $160, which is 40% off the original price. What was the original price to the nearest dollar?

**75.** TAXES If there is a 5% sales tax on the sale price of the coat in Exercise 74, what is the total amount that you would have to pay?

**76.** POLITICAL SCIENCE The newspaper reports that only 42% of the registered voters in a small town voted in an election. If 11,960 people voted, how many registered voters are there?

## Look Back

**Evaluate.** *(LESSONS 2.2, 2.3, AND 2.4)*

**77.** $-3 + 7$ **78.** $(-3)(7)$ **79.** $-3 - (-7)$ **80.** $(-3)(-7)$

**Simplify.** *(LESSON 2.6)*

**81.** $(4x + 7) + (9x - 9)$ **82.** $(3x^2 + 2x) + (3x^2 - 2x)$

**83.** $(-6c + 5) - (-3c + 8)$ **84.** $(-y + 2z) + (-3y + 3z)$

**Simplify.** *(LESSON 2.7)*

**85.** $\frac{4x + 18x^2}{2}$ **86.** $\frac{15f^2 + 35}{-5}$ **87.** $\frac{3x^2 - 21}{3}$

## Look Beyond

**88.** Suppose that a $1000 investment grows by 8% each year. Complete the table below.

| Years | 0 | 1 | 2 | 3 | 4 |
|---|---|---|---|---|---|
| Value ($) | 1000 | 1080 | ? | ? | ? |

# EYEWITNESS MATH

# BARELY ENOUGH GRIZZLIES?

## Counting Big Bears

### Government Wants Them Off the Endangered List; How Many is Enough?

*By Marj Charlier–Staff Reporter of the Wall Street Journal*

KALISPELL, Mont.—The solitary monarch of the wild, the grizzly rules the rugged high country. Scientists call it *Ursus arctus horribilus*, and the name suggests the terror it inspires. Fearless and unpredictable, the bear can kill with a single rake of a claw.

Man is the grizzly's only natural enemy, but the clash almost proved fatal to the species. Unrestricted hunting and development of the animal's habitat decimated the bear population over the last century. By 1975 when the grizzly was placed on the federal government's Endangered Species List, only 1,000 were thought to be alive.

Now, federal officials say the grizzly population has rebounded, though they concede there is no hard evidence to support that conclusion, and they are proposing to take the bear off the list.

**A Death Sentence?**

Many environmentalists, however, believe that even if the number of bears has increased, the grizzly population is still too small to strip the animal of federal protection. Such a plan, they say, would be a death sentence for the species. "If the grizzly bear is delisted, it will be strangled by development and go extinct quite promptly," says Lance Olsen, president of the nonprofit Great Bear Foundation.

**Population Unknown**

Complicating the issue is the difficulty of determining how many grizzlies roam the Rockies and their foothills. Everyone concedes that the 1,000 figure for 1975 was only a guess. Attempts to take a bear census in the years since then have proved difficult, and no one knows how many grizzlies there are now.

Richard Mace, a wildlife biologist for the state of Montana, knows how tough it is to count grizzlies. Displaying a battered steel box with a jagged rip down one side, he explains that the box once held a tree-mounted camera that was rigged to snap photos when it sensed body heat from a large animal. Biologists had placed some 43 cameras in a section of Montana's Swan Mountains, hoping to get a more accurate bear count.

Apparently, Mr. Mace says, when the camera flashed, the grizzly slashed. Its two-inch long claws ripped through the steel casing, which is a bit thicker than a coffee can. The bear dug out the camera and chewed it to pieces.

## Cooperative Learning

Counting animals in the wild is not easy. How do you know if you have counted some more than once? How can you possibly find them all? Wildlife managers sometimes use a technique called *tag and recapture.* To see how the tag-and-recapture method works, you can try it—with objects instead of animals.

**You will need:**

- a batch of several hundred objects (macaroni, beans, crumpled pieces of paper) in a paper bag or other container
- a marker (or some other way to *tag*, or mark, the objects)

**1.** Follow steps **a–e** to estimate the number of objects in your batch.

**a.** Remove a handful of objects from the batch and mark them. Record how many you have marked.

**b.** Return the marked objects to the batch. Mix thoroughly.

**c.** Take 30 objects from the batch at random. Count the number of marked objects in your sample. Copy the chart and fill in the first row. Repeat for each group member.

| Sample | Total number of objects in sample ($N_s$) | Number of marked objects in sample ($M_s$) |
|---|---|---|
| 1. | 30 | |
| 2. | 30 | |

**d.** How do you think the ratio of marked objects to total number of objects in your samples compares with the ratio of marked objects to total number of objects in the whole batch?

**e.** Use the equation below to estimate the size of the whole batch. (Average your data for $M_s$ and $N_s$.)

$$\frac{M_b}{N_b} = \frac{M_s}{N_s}$$

$M_b$ is the number of **marked objects** in the batch.
$N_b$ is the **total number of objects** in the batch.
$M_s$ is the number of **marked objects** in the sample.
$N_s$ is the **total number of objects** in the sample.

**f.** Check your estimate by counting the actual number of objects in the batch. Within what percent of the actual number was your estimate?

**2.** Describe how the tag-and-recapture method could be used to estimate an animal population.

**3.** Do you think the tag-and-recapture method would work for a grizzly bear population? Why or why not?

# 4.3 Introduction to Probability

**Objective**

- Find the experimental probability that an event will occur.

**Why** ***Experimental probability is the basis for many conclusions and predictions, including the daily weather forecast and which sports teams are expected to win. The experiments that you perform in this lesson will help you understand more about experimental probability.***

The **probability** of an event is the likelihood that the event will occur.

If the weather forecaster says that the probability of rain is 0%, this means that she is certain that it will not rain. If she says that the probability of rain is 100%, this means that she is certain that it will rain.

Probability is measured on a scale from 0% to 100%, or from 0 to 1.

Impossible — Less likely — More likely — Certain

0% ———————————————— 100%

How can you find the probability that something will happen? One way is by *observing* the way that things happen in the real world. For example, meteorologists study weather data such as temperature, pressure, and humidity. Information about past weather patterns is compared with current patterns to predict the chance of rain each day.

You can conduct a simpler experiment. Consider the following question: If you roll a pair of number cubes repeatedly, how often would you expect the pair to show a sum of 7? One way to answer the question is to make observations about trials of your own—as in the following activity.

## *Activity*
## Two Probability Experiments

**You will need:** a coin and two number cubes

You may want to work in groups of two or more.

**Part I**

1. Roll a pair of number cubes 20 times and record the number of times that a sum of 7 occurs.
2. Compare your results with the results of other groups in your class.
3. Find a percent for the entire class by dividing the total number of sums of 7 that occurred by the total number of rolls for the whole class.

CHECKPOINT ✔

4. Are your group and class results what you expected? Explain.

**Part II**

1. Toss a coin 20 times and record how many times it comes up heads. Divide this number by 20 to find the percent of heads that occurred.

2. Compare your results with the results of other groups in your class.
3. Find a percent for the entire class by dividing the total number of heads that came up by the total number of tosses for the whole class.

CHECKPOINT ✔

4. Are your group and class results what you expected? Explain.

## Experimental Probability

A probability experiment consists of a number of *trials*, such as the tossing of a coin or the rolling of a pair of number cubes. For any given trial, the result, or outcome, may or may not be favorable. That is, it may or may not be the event that you are interested in studying. For example, if you are interested in the probability that a tossed coin comes up heads, then heads is a **favorable outcome**. In some instances you may be interested in counting unfavorable outcomes to find the probability that an event does not occur.

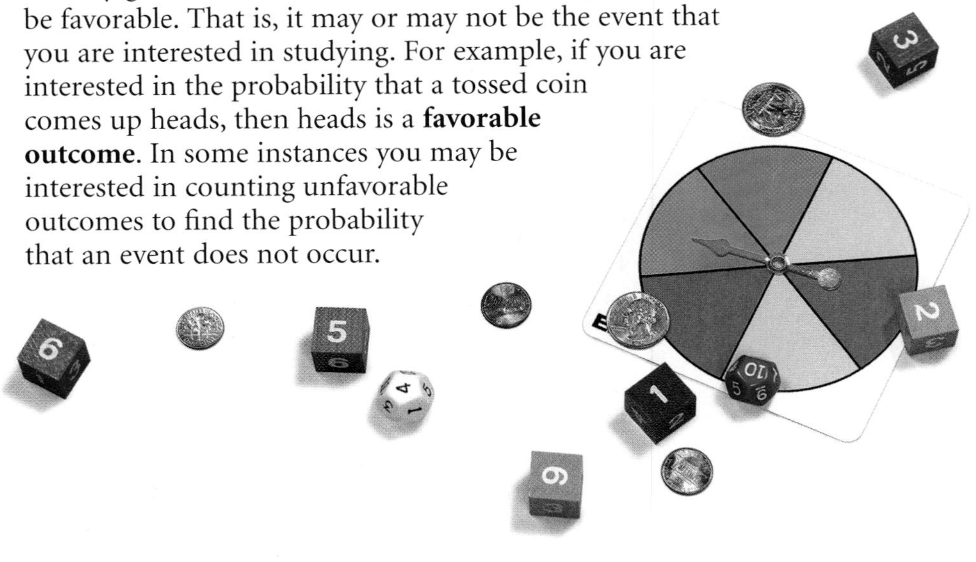

The number of favorable outcomes divided by the total number of trials is called the **experimental probability** of the event.

**Experimental Probability**

For a sufficient number of trials, the experimental probability of an event, $P(E)$, is given by the following formula:

$$P(E) = \frac{\text{(number of favorable outcomes)}}{\text{(total number of trials)}}$$

Assume that all examples and exercises in this section refer only to fair number cubes and coins.

**EXAMPLE 1** Two number cubes were rolled 100 times. An even number appeared as a sum 47 times. **Based on this experiment, what is the experimental probability that two rolled number cubes will show a sum that is an even number?**

**SOLUTION**

There were 100 trials and 47 successful outcomes. Thus, in this case, the experimental probability of a successful outcome (showing a sum that is an even number) is as follows:

$\frac{47}{100} = 0.47$, or 47%

**TRY THIS** Two number cubes were rolled 100 times. A sum greater than 8 appeared 29 times. Based on this experiment, what is the experimental probability that two rolled number cubes will show a sum greater than 8?

**EXAMPLE 2** Two coins were tossed 200 times. Two heads appeared 48 times. **Based on this experiment, what is the experimental probability that two flipped coins will show two heads?**

**SOLUTION**

There were 200 trials and 48 successful outcomes. Thus, in this case, the experimental probability of a successful outcome (showing two heads) is as follows:

$\frac{48}{200} = 0.24$, or 24%

**TRY THIS** Two coins were tossed 200 times. Two tails appeared 52 times. Based on this experiment, what is the experimental probability of two flipped coins showing two tails?

# Exercises

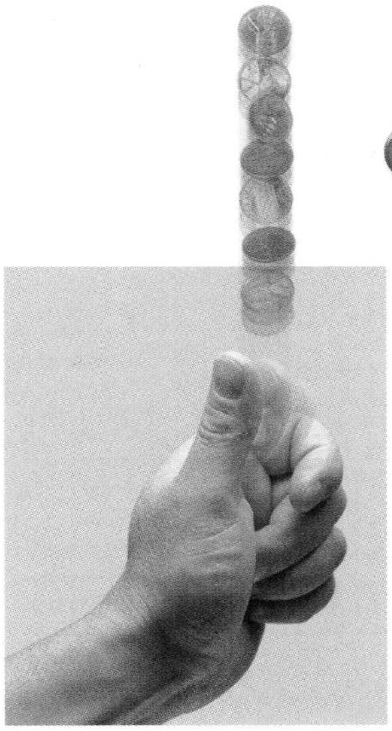

## Communicate

1. Explain what is meant by *experimental probability.*
2. Describe an experiment that could be used to find the experimental probability of getting at least 3 heads in 4 tosses of a coin.
3. Is it possible for two separate groups to conduct the same experiment to determine an experimental probability and get different results? Explain or give an example.
4. Is it possible for someone to conduct the same probability experiment twice and get different results? Explain or give an example.
5. Explain why the number of trials performed in a probability experiment can affect the outcome.

## Guided Skills Practice

**Two number cubes were rolled 100 times. Based on the results below, find the experimental probability of each outcome.** *(EXAMPLE 1)*

6. An odd number appeared as a sum 53 times.
7. A sum of less than 4 appeared 7 times.
8. At least one 6 appeared 29 times.

**internet connect**
**Activities Online**
Go To: **go.hrw.com**
Keyword:
**MA1 Simpson's Paradox**

**Two coins were flipped 200 times. Based on the results below, find the experimental probability of each outcome.** *(EXAMPLE 2)*

9. Two heads appeared 46 times.
10. Two tails appeared 51 times.
11. One heads and one tails appeared 106 times.

## Practice and Apply

**internet connect**
**Homework Help Online**
Go To: **go.hrw.com**
Keyword:
**MA1 Homework Help**
for Exercises 12-23

**Two number cubes were rolled 100 times. Based on the results below, find the experimental probability of each outcome.**

12. A sum of less than 5 appeared 18 times.
13. An even number appeared on at least one cube 76 times.
14. A sum of 9 appeared 13 times.
15. A sum of 7 or 11 appeared 24 times.
16. An even number appeared on both cubes 21 times.
17. Two 6s appeared 3 times.

**Three coins were tossed 200 times. Based on the results below, find the experimental probability of each outcome.**

**18.** At least one coin came up tails 178 times.

**19.** At least one coin came up heads 173 times.

**20.** Exactly two coins came up heads 84 times.

**21.** Exactly two coins came up tails 67 times.

**22.** All three coins came up tails 27 times.

**23.** All three coins came up heads 22 times.

**Two coins were tossed 20 times with the following results (H represents heads and T represents tails):**

| Trial | 1 | 2 | 3 | 4 | 5 | 6 | 7 | 8 | 9 | 10 | 11 | 12 | 13 | 14 | 15 | 16 | 17 | 18 | 19 | 20 |
|---|---|---|---|---|---|---|---|---|---|---|---|---|---|---|---|---|---|---|---|---|
| **Coin 1** | H | H | H | H | T | H | T | T | H | H | H | T | T | H | T | T | T | H | H | H |
| **Coin 2** | T | H | H | T | T | T | T | H | T | T | T | H | H | H | H | H | H | H | T | H |

**Based on the data above, find the following experimental probabilities:**

**24.** Both coins came up the same.

**25.** Both coins are heads.

**26.** At least one coin is heads.

**27.** Neither coin is heads.

**Conduct 20 trials of your own to find the following experimental probabilities. Report your results in a table.**

**28.** The sum of two rolls of a number cube is 9.

**29.** Three tosses of a coin will result in at least 2 heads.

**CHALLENGE**

**30.** To determine an experimental probability, Fred conducts 15 trials and Ted conducts 16 trials. Is it possible for them to arrive at the same experimental probability? Explain.

**CONNECTION**

**PROBABILITY** The result of a school experiment to find the probability of tossing three heads in a row was 11.5%. Twenty groups participated in the experiment.

**31.** Suppose each group performed 30 trials. How many favorable outcomes occurred? unfavorable outcomes?

**32.** Suppose there were 46 favorable outcomes total. How many trials did the twenty groups perform?

**APPLICATIONS**

**33. GAMES** Many board games offer an extra turn if a player rolls doubles (both cubes showing the same number). Conduct several experiments with two number cubes to determine the experimental probability of rolling doubles for different numbers of trials. Complete a table for 5, 10, 15, 20, 25, and 30 trials. Show the number of doubles and the experimental probability for each number of trials. Compare your results with others in your class.

**34. GAMES** Combine all of your trials from Exercise 33, and count all of the doubles. From those totals, find the overall experimental probability. Do you think this is a reliable estimate of the probability of rolling doubles during a game? Explain.

## Look Back

**Use guess-and-check to solve each equation.** *(LESSON 1.2)*

**35.** $-1 = 4 - x$ **36.** $2x + 1 = 1$

**Simplify.** *(LESSONS 2.2, 2.3, AND 2.4)*

**37.** $-17 + (-12)$ **38.** $(-3) \cdot (-13) + (-41)$ **39.** $\frac{-4}{12} - \left(\frac{3}{-6}\right)$

**Solve each equation for *t*.** *(LESSON 3.5)*

**40.** $7(t - r) = 2r - 3(4r + t)$ **41.** $\frac{3(4t - 2)}{2} = \frac{3}{2} - (5 - t)$

## Look Beyond

**internet connect**

**Portfolio Extension**

Go To: **go.hrw.com**
Keyword:
**MA1 Coin Toss**

**42.** How many different pairs of numbers can be generated by one red cube and one green cube if the faces of each cube are numbered from 1 to 6?

**43.** How many of the pairs from Exercise 42 have a sum of 3?

## PORTFOLIO ACTIVITY

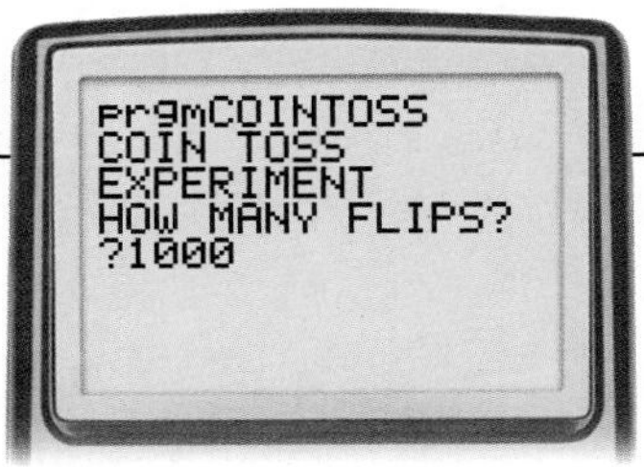


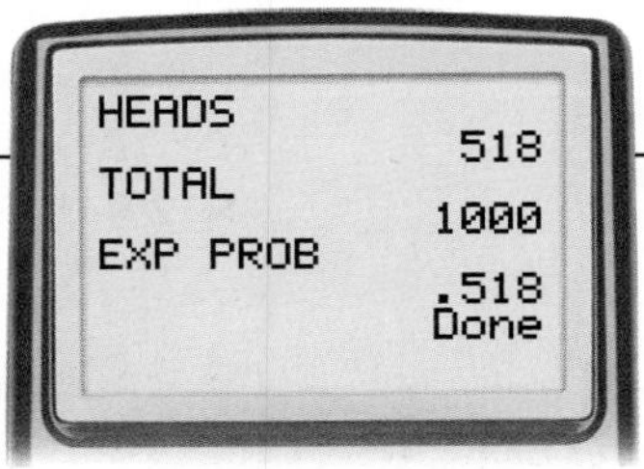


Include this activity in your portfolio.

You can use a graphics calculator to simulate a probability experiment. With a calculator, you can quickly conduct experiments with many more trials than you could by hand. The following program uses a random-number generator to simulate tossing a coin a given number of times:

```
Disp "COIN TOSS":Disp "EXPERIMENT"
Disp "HOW MANY FLIPS?"
Input N
0 → H
For(X,1,N)
int(2rand) → T
If T=0:H+1 → H
End
Disp "HEADS",H,"TOTAL",N
Disp "EXP PROB",H/N
```

**1.** What experimental probability is shown on the display for a 1000-trial experiment above?

**2.** Use the program and a graphics calculator to complete the table.

| Number of trials | Heads | Experimental probability |
|---|---|---|
| 4 | | |
| 10 | | |
| 20 | | |
| 30 | | |

**3.** Compare your results with those of others in your class.

**4.** Combine all of the trials and find the total number of heads. What is the overall experimental probability?

**WORKING ON THE CHAPTER PROJECT**

You should now be able to complete Activities 3 and 4 of the Chapter Project on page 208.

4.4

# Measures of Central Tendency

**Objectives**

- Find the mean, median, mode, and range of a data set.
- Represent data with frequency tables.

**Why** *Sometimes sets of numbers can be confusing. Statistics are used to present all kinds of information in a simplified form. It is important to understand different types of statistics and what they tell you about a particular set of data.*

## Descriptive Statistics

Statistics can help you understand and describe raw data. For example, the *mean*, *median*, and *mode* of a data set each tell you something about the *central tendency* of the data—that is, how the data tend to cluster around a value. The *range* provides information about the *spread* of the data.

**Definitions: Mean, Median, Mode, and Range**

The **mean** of a data set is the quotient when the sum of all of the elements is divided by the total number of elements.

The **median** of a data set is the middle number in the set when the elements are placed in numerical order. If there are an even number of elements, the median is the average of the two middle numbers.

The **mode** of a data set is the element, if any, that occurs most often. There may be no mode, one mode, or several modes.

The **range** of a data set is the difference between the greatest and least values in the set.

EXAMPLE 1 **Find the mean, median, mode, and range of the batting averages on page 186 for five National League players: 0.372, 0.366, 0.362, 0.333, and 0.327.**

APPLICATION
SPORTS

SOLUTION

To find the *mean*, add the 5 batting averages, and then divide the sum by 5.
$0.372 + 0.366 + 0.362 + 0.333 + 0.327 = 1.76$
$1.76 \div 5 = 0.352$
The mean is **0.352**.

To find the *median*, arrange the five batting averages in ascending order. Because the number of averages is odd, take the middle value.
0.327, 0.333, 0.362, 0.366, 0.372
The median is **0.362**, the middle value.

To find the *mode*, look for a value that occurs more often than any other. Because each batting average occurs just once, there is no mode for the set.

To find the *range*, subtract the lowest value from the highest value in the set.
$0.372 - 0.327 = 0.045$
The range is **0.045**.

EXAMPLE 2 **Find the mean, median, mode, and range for each set of data.**

**a.** 3, 5, 8, 5, 8, 7, 16, 10, 2, 7, 1, 5, 6, 1
**b.** 160, 100, 150, 125, 125, 150, 114

SOLUTION

**a.** To find the mean, add the numbers, and then divide by 14 (the number of numbers in the set of data).

$$3 + 5 + 8 + 5 + 8 + 7 + 16 + 10 + 2 + 7 + 1 + 5 + 6 + 1 = 84$$

$84 \div 14 = 6$ The mean is 6.

To find the median and mode, arrange the data in ascending order.

1, 1, 2, 3, 5, 5, **5, 6,** 7, 7, 8, 8, 10, 16

The median is the middle value. Since there is an even number of numbers, 14, the middle value falls between the seventh and eighth numbers, 5 and 6. Add and divide by 2 in order to find the median.

$5 + 6 = 11$ $11 \div 2 = 5.5$ The median is 5.5.

The mode is the value that occurs most frequently. There are three 5s. The mode is 5.

To find the range, subtract the lowest from the highest value in a set.

$16 - 1 = 15$ The range is 15.

**b.** To find the mean, add the numbers, and then divide by 7 (the number of numbers in the set of data).

$$160 + 100 + 150 + 125 + 125 + 150 + 114$$

$924 \div 7 = 132$ The mean is 132.

To find the median and the mode, arrange the data in ascending order.

100, 114, 125, 125, 150, 150, 160

The median is the middle value. Since there is an odd number of numbers, 7, the middle value is the fourth number.

The median is 125.

The mode is the value that occurs most frequently. There are two 125s and two 150s. Therefore, this set of data has two modes, 125 and 150.
To find the range, subtract the lowest from the highest value in the set.

$160 - 100 = 60$ The range is 60.

**TRY THIS** Give the mean, median, mode, and range of the batting averages on page 186 for five American League players.

**EXAMPLE 3** Suppose that Dale has test scores of 75, 95, 82, and 77. He has one test remaining and his goal is to have a test average of exactly 85. **What must Dale score on the final test in order to meet his goal?**

**SOLUTION**

Valuable information about the data set can be found without any calculations. First, order the data.

$$\{75, 77, 82, 95\}$$

The highest and lowest scores are both 10 points from 85, so their average is 85, and they can be disregarded. The other two scores are both lower than 85, which means that the new score must to be higher than 85 in order to balance these two scores.

The exact score can be calculated algebraically by solving a related equation.

$$\frac{75 + 77 + 82 + 95 + x}{5} = 85$$

$$329 + x = 425$$

$$x = 96$$

Dale must earn a 96 on his final test in order to have an average of 85 for his test grades.

## Frequency Tables

A **frequency table** is a method of displaying data in which tally marks are used to show how often each element of the set occurs. Instead of listing each occurrence of a number, a tally mark is placed next to the number each time it occurs. The following frequency table shows the scores on a 100-point quiz. How many scores are represented in the table?

| **Score** | 55 | 60 | 65 | 70 | 75 | 80 | 85 | 90 | 95 | 100 |
|---|---|---|---|---|---|---|---|---|---|---|
| **Frequency** | \| | \| | \|\| | \|\|\| | \|\|\|\| | \|\|\| | \|\|\|\| | \|\|\|\| | \|\| | \| |

*Activity*

## Using Frequency Tables

Use the frequency table on page 188 to complete this activity.

1. What is the mode of the scores? Explain how a frequency table makes it easy to find the mode.
2. What is the median of the scores? Explain how to find the median from the frequency table.
3. Without computing, estimate the mean. Then list all of the scores and compute the mean.

CHECKPOINT ✔

4. How can a frequency table make it easier to compute the mean, median, and mode of a set of data?

**EXAMPLE 4**

APPLICATION
SPORTS

Cathy has been swimming in the 50-meter butterfly event since she started high school. During that time, she has kept track of what place she finished in the event at each meet she attended. **Find the mode, median, mean, and range of the places finished, using the data in the table below.**

| **Place** | 1st | 2nd | 3rd | 4th | 5th | 6th | 7th | 8th | 9th | 10th |
|---|---|---|---|---|---|---|---|---|---|---|
| **Frequency** | \|\|\|\| | \|\|\| | \|\|\|\| | ~~\|\|\|\|~~ | \|\| | \|\|\|\| | \|\|\|\| | \|\|\| | \|\|\| | \| |

**SOLUTION**

To find the mode, find the value in the frequency table with the most tally marks. Fourth place has five marks, so the mode of the places finished is 4th.

To find the median, first count the tally marks in the frequency table. There are 33 marks in this table. Next count to find the location of the center tally mark, in this case the 17th mark. That mark is found at 5th place, so the median of the places finished is 5th.

To find the mean, first multiply each place finished by the number of tally marks below it. Then add these products and divide by the total number of tally marks, 33.

$4 + 6 + 12 + 20 + 10 + 24 + 28 + 24 + 27 + 10 = 165$

$165 \div 33 = 5$

The mean of the places finished is 5th.

The range is the difference between the highest and lowest places finished, in this case, $10 - 1 = 9$. Thus, the range of the places finished is 9.

CRITICAL THINKING

Mean, median, and mode each tell you something different about a set of data. Which statistic do you think is the best average for the places that Cathy finished in the 50-meter butterfly event? Which statistic do you think should be used to find the average of your grades? the average number of people in a family?

# Exercises

## Communicate

**1.** Explain the difference between the mean and the median of a set of data.

**2.** Describe how you would find the median of the following set of data:
6, 8, 4, 9, 3, 4, 2, 5, 4, 0, 7

CONNECTION

**3.** STATISTICS A census report says that the average American family has 1.5 children. What measure of central tendency was most likely used? What problems do you think occur with statistics like this?

**4.** Explain how to determine the mean of the data set displayed in the frequency table below.

| Grade | 70 | 75 | 80 | 85 | 90 | 95 | 100 |
|---|---|---|---|---|---|---|---|
| Frequency | ∣ | ∣∣ | ∣∣∣ | ∣∣∣∣ | ∣∣∣ | ∣∣∣∣ | ∣∣∣∣ |

## Guided Skills Practice

**Find the mean, median, mode, and range for each set of data.**
***(EXAMPLES 1 AND 2)***

**5.** 2, 5, 8, 4, 2, 8, 3, 7, 6

**6.** 12, 29, 14, 8, 36, 26, 15

**7.** 5, 9, 8, 10, 7, 3

**8.** 0.25, 0.91, 0.65, 0.63, 0.82, 0.73, 0.21

**9.** 8.3, 2.5, 1.7, 7.6, 9, 5.2, 3.1, 4

**10.** 2, 5, 3, 6, 12, 9, 1, 4, 2, 4

APPLICATION

SPORTS Use the frequency table below, which compares the number of runs per game scored by a softball team, for Exercises 11–13.

| Total runs scored | 0 | 1 | 2 | 3 | 4 | 5 | 6 | 7 | 8 |
|---|---|---|---|---|---|---|---|---|---|
| Frequency | ∣∣ | ∣∣∣∣ | ∣∣∣ | 卌 | ∣∣∣ | ∣∣∣ | ∣∣ | ∣∣∣∣ | ∣ |

**11.** How many games are represented in the frequency table? ***(EXAMPLE 2)***

**12.** Find the mean number of runs scored by the softball team. ***(EXAMPLE 2)***

**13.** Find the median, mode, and range for the runs scored per game. ***(EXAMPLE 2)***

## Practice and Apply

**Find the mean of each set of data.**

**14.** 18, 15, 12, 14, 16

**15.** 5, 10, 12, 13, 10

**16.** 18, 24, 16, 13, 27, 22

**17.** 87, 86, 92, 95, 75

**18.** 45, 54, 53, 47, 38, 63

**19.** 91, 46, 85, 96, 78, 84

internet connect

**Homework Help Online**
Go To: **go.hrw.com**
Keyword:
**MA1 Homework Help**
for Exercises 14-53

**20.** 1.4, 2.6, 3.2, 1.8, 4.2, 3.3

**21.** 123, 114, 98, 102, 115

**22.** 0.25, 0.68, 0.12, 0.45, 0.5

**23.** 45, 57, 78, 23, 59, 43, 52

**24–33.** Find the median of each set of data in Exercises 14–23.

**34–43.** Find the range for each set of data in Exercises 14–23.

**Find the mode, if any, of each set of data.**

**44.** 5, 2, 6, 8, 2, 3, 1, 3, 2, 4, 7, 9

**45.** 16, 23, 19, 6, 15, 17, 18, 12, 19, 23, 19

**46.** 50, 48, 50, 55, 48, 57, 56

**47.** 97, 101, 98, 103, 101, 98, 97, 107, 98

**48.** 91, 87, 98, 97, 91, 100, 94

**49–53.** Find the mean, median, and range of each data set in Exercises 44–48.

CONNECTION

**54. STATISTICS** A class had the following test scores:
85, 100, 60, 81, 80, 75, 70, 90, 85, 65, 85, 75, 100, 75, 70, 95, 80, 60, 90, 70, 60, 85, 55, 60, 65, 65, 65, 70, 73, 75, 80, 81, 85, 90, 95

**a.** Make a frequency table.
**b.** Find the mean.
**c.** Find the median.
**d.** Find the mode.

**55.** Suppose that Frank has test grades of 84, 92, 89, and 93. He has one test left, and his goal is to have a test average of 90. What must Frank score on the final test in order to meet his goal?

APPLICATIONS

**56. SPORTS** The basketball coach at the high school made the following list of the points scored by each member of the freshman basketball team: 26, 18, 4, 18, 18, 5, 6, 11, 4, 9, and 8.

**a.** Make a frequency table for the data.
**b.** Find the median number of points scored.
**c.** Find the mean of the data.
**d.** What is the mode?
**e.** Which measure of central tendency would you use to give the best impression of the players' abilities? Explain your choice.

**57. BIOLOGY** A biologist studying red-tailed hawks counts the eggs in their nests. Four nests have 1 egg, 65 nests have 2 eggs, and 5 nests have 3 eggs. He concludes that red-tailed hawks usually have 2 eggs at a time in the nest. Which measure of central tendency did the biologist most likely use?

*The red-tailed hawk, found in North America, frequently preys on rodents.*

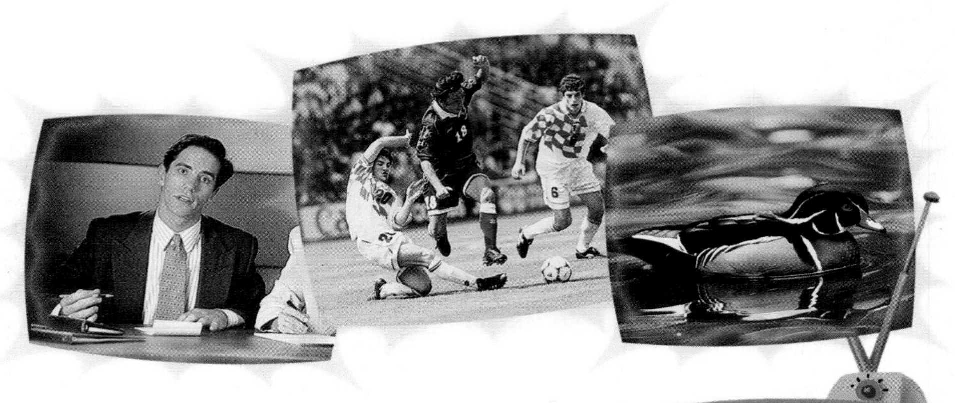

APPLICATION

**58.** **RECREATION** The table at right shows the average number of hours spent watching television for certain age groups.

**a.** Find the mean number of hours.

**b.** Find the median number of hours.

**c.** What is the mode of the data?

**d.** What is the range of the data?

CHALLENGE

**e.** Suppose that the number of hours per week spent watching television for females aged 25–54 is 32. Would this change the mean, median, mode, or range? Explain.

| Age group (in years) | Number of hours per week spent watching TV |
|---|---|
| Children 2–5 | 24 |
| Children 6–12 | 20 |
| Female 13–17 | 21 |
| Female 18–24 | 25 |
| Female 25–54 | 30 |
| Female 55+ | 44 |
| Male 13–17 | 21 |
| Male 18–24 | 22 |
| Male 25–54 | 28 |
| Male 55+ | 38 |

## Look Back

**59.** Write an equation to describe the following situation and solve it by using guess-and-check: If tickets for a movie cost \$7 each, how many can you buy with \$91? ***(LESSON 1.2)***

**internet connect**
**Activities Online**
Go To: **go.hrw.com**
Keyword:
**MA1 Voting Schemes**

**Find each sum.** ***(LESSON 2.1)***

**60.** $-12 + (-7)$ **61.** $23.4 + 142.7$ **62.** $-302 + (-128) + 23$

**Solve each equation. Check your solution.** ***(LESSONS 3.2 AND 3.3)***

**63.** $2a + 3 = 27$ **64.** $413y = 1239$ **65.** $\frac{x}{-16} = 23$

**66.** $3x - 14 = 88$ **67.** $4.3b + 12.7 = 29.9$

**Solve for the variable in each proportion.** ***(LESSON 4.1)***

**68.** $\frac{5}{7} = \frac{d}{56}$ **69.** $\frac{20}{q} = \frac{45}{36}$

## Look Beyond

**70. Use the frequency table you created in Exercise 54.**

**a.** What percent of the students scored 60 or below? from 61 to 70? from 71 to 80? from 81 to 90? from 91 to 100?

**b.** Use your answer from part **a** to make a circle graph, or pie chart, for the data.

# 4.5 Graphing Data

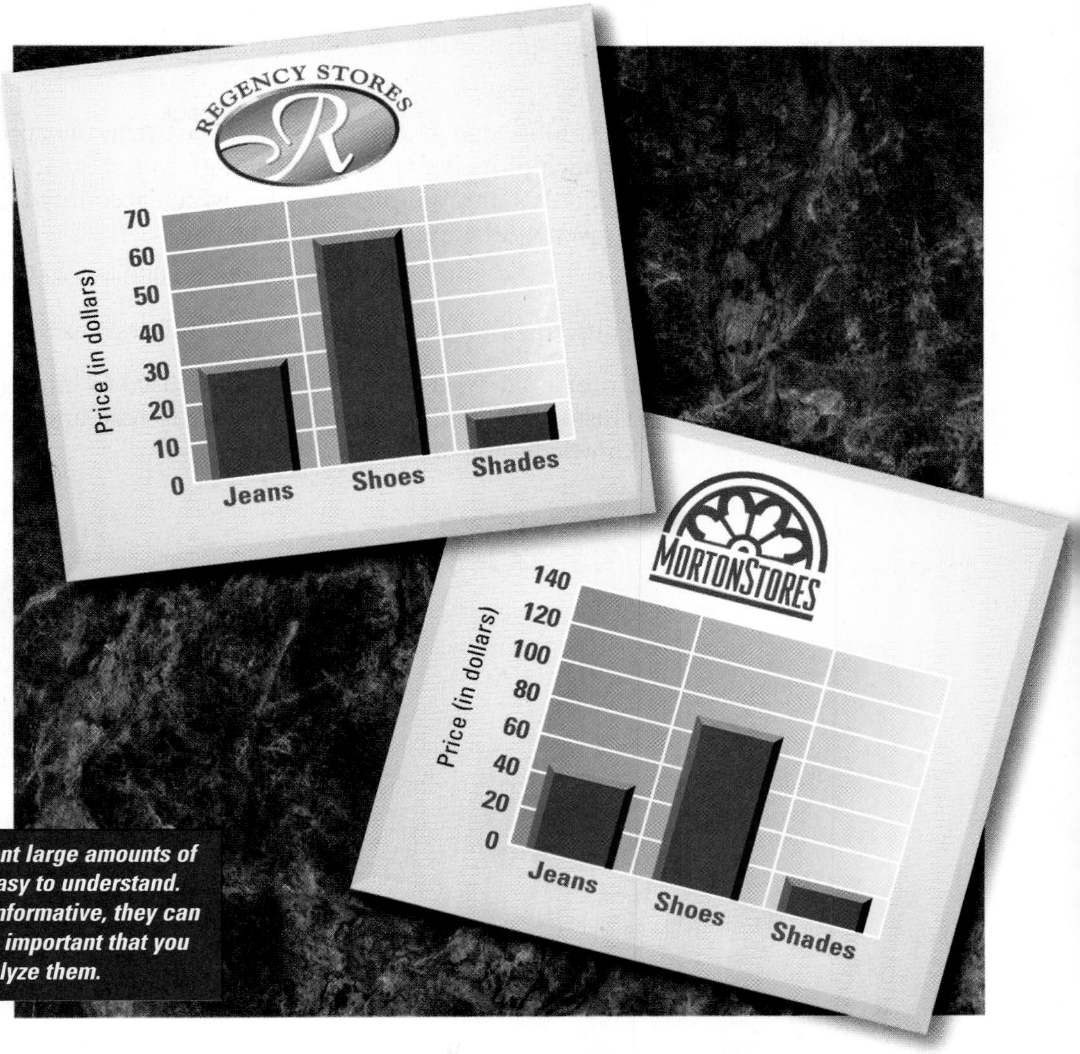


### Objectives

- Interpret line graphs, bar graphs, and circle graphs.
- Represent data with circle graphs.
- Analyze graphs in order to find a misleading presentation of data.

**Why** ***Graphs are used to present large amounts of information in ways that are easy to understand. Although graphs can be very informative, they can sometimes be misleading. It is important that you learn how to interpret and analyze them.***

## Common Types of Graphs

Graphs are very useful for interpreting large amounts of data at a glance. Three common forms of graphs are bar graphs, line graphs, and circle graphs. A circle graph is sometimes called a pie chart because of its resemblance to a pie that has been sliced.

**Bar graphs** use rectangular bars or objects to represent data. They can be used to compare individual years or other distinct items, as shown for the stores above.

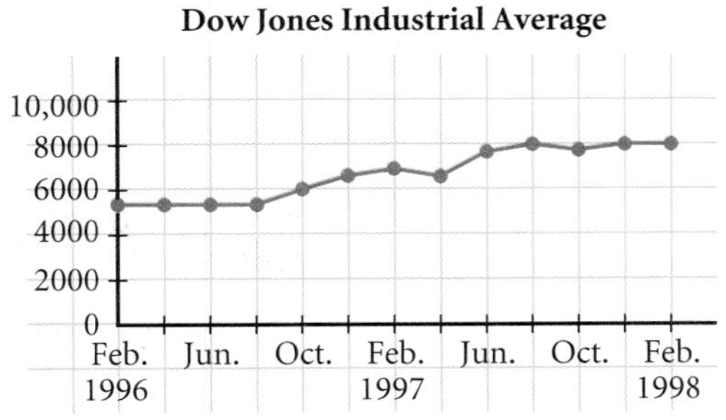


A **line graph** is a type of graph that uses line segments between known data points to show changes that have occurred over time. An example of a line graph is shown above. By looking at the general trend, or pattern, of this graph, can you make a reasonable guess about the stock market in the future?

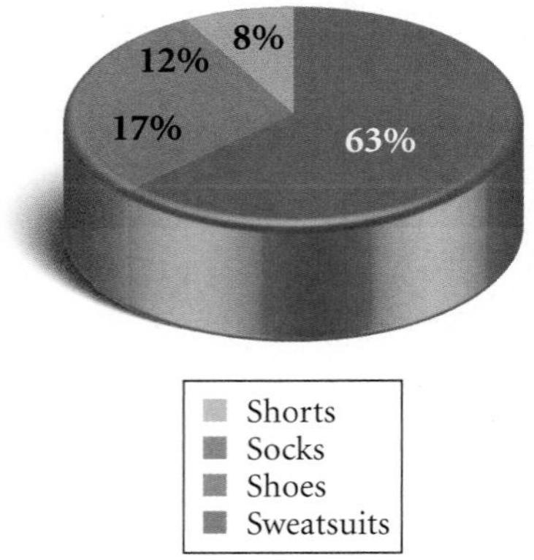


In a **circle graph**, or pie chart, data are represented as parts of a circle. This is most effective for looking at parts or percents of a whole. For example, a sporting goods store that needs to know what percent of its sales are from its shoes, shorts, socks, and sweatsuits, might use a circle graph like the one at left.

If the store had \$1,000,000 in sales last year, the approximate dollar amount for each of its items is the percent given for the item multiplied by \$1,000,000. The dollar amount of shoe sales, which accounted for 17% of the store's sales, is given by the following:

$$0.17 \cdot 1{,}000{,}000 = 170{,}000$$

Thus, the total amount of sales from shoes was \$170,000.

Unfortunately, graphs are sometimes used to intentionally mislead consumers. This is why it is important to be able to read and interpret graphs in a knowledgeable manner.

## *Activity*
## Misleading Graphs

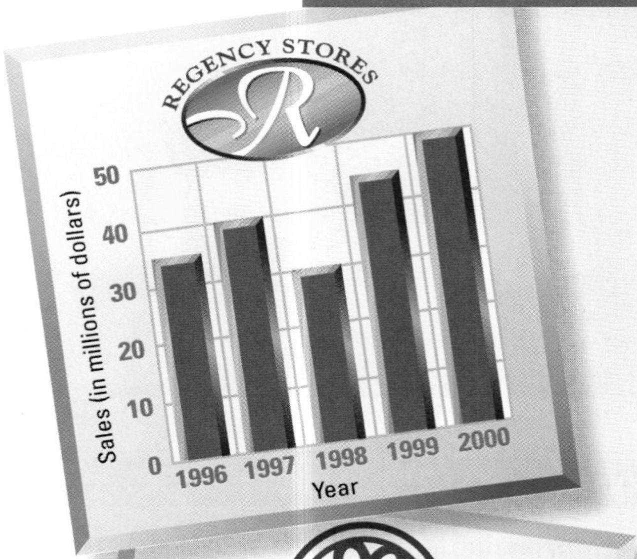


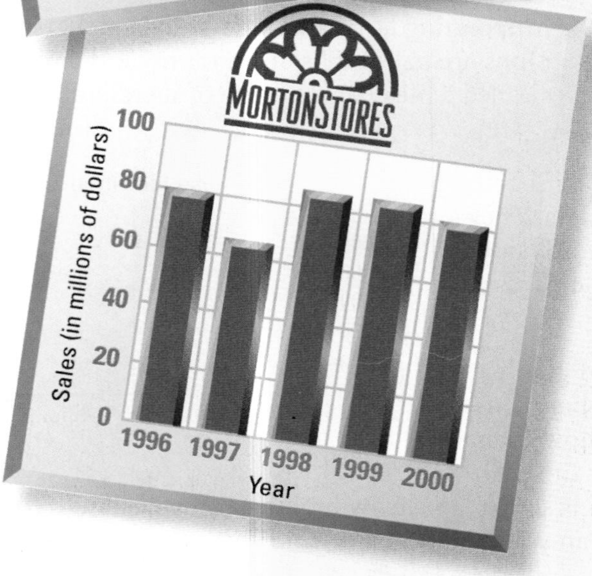


The bar graphs at left show the sales (in millions of dollars) of two retail companies for five consecutive years.

1. What is the approximate amount of sales for the Regency stores in 1997? What is the approximate amount of sales for the Morton stores in 1997? Which company's bar graph has a longer bar to represent sales in 1997? Which company actually had greater sales in 1997?

2. What is the approximate amount of sales for the Regency stores in 2000? What is the approximate amount of sales for the Morton stores in 2000? Which company's bar graph has a shorter bar to represent sales in 2000? Which company actually had lower sales in 2000?

3. What features of the graphs make it possible to misinterpret the information? How might the information be presented to make a more accurate visual comparison?

4. Redraw the graphs to present a more accurate visual comparison.

## Working With Graphs

**APPLICATION**
**ECONOMICS**

A line graph can be used to make predictions based on current trends.

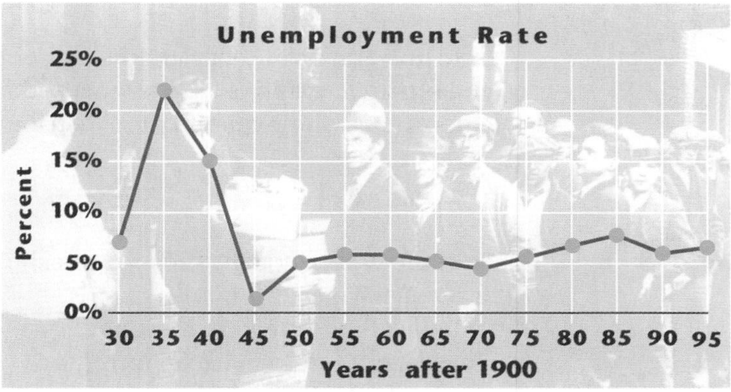


From the graph above, you can tell in what year the unemployment rate was the greatest. You can also find the year of the lowest unemployment rate, 1945, and several years when the rate was the same, 1955, 1960, 1975, and 1990. The graph also clearly shows the trend of increasing unemployment from 1930 to 1935 and sharply decreasing unemployment from 1935 to 1945.

**CRITICAL THINKING**

What historical event might account for the rise in the unemployment rate from 1930 to 1935? What event would account for the decline in the unemployment rate from 1935 to 1945?

**EXAMPLE 1**

**APPLICATION**
**INVESTMENTS**

Don invested $100 each in four companies in 1995. The line graph at right shows the value of the stock over several years. **Find the approximate values of Don's MRI and MMart stock in 1999. Construct a bar graph that shows the approximate value of Don's WMB stock from 1995 to 1999.**

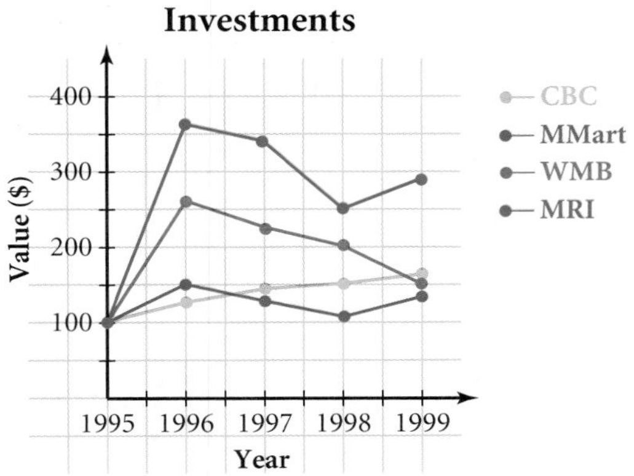


**SOLUTION**

**MRI:** The graph shows that the value of Don's MRI stock was about $290 in 1999.

**MMart:** The graph shows that the value of Don's MMart stock was about $140 in 1999.

**WMB:** The approximate value of Don's WMB stock was $100 in 1995, $260 in 1996, $220 in 1997, $200 in 1998, and $140 in 1999. The bar graph for this information is shown at right.

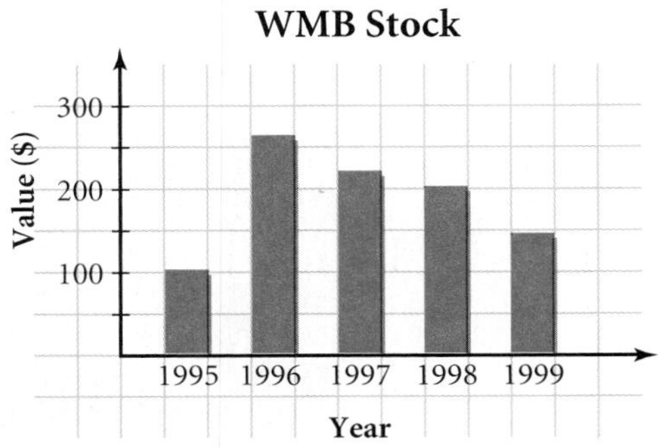

**TRY THIS** Find the approximate value of Don's MRI stock in 1997 and of Don's CBC stock in 1999. Construct a bar graph that shows the approximate value of Don's CBC stock from 1995 to 1999.

While line graphs and bar graphs are useful for presenting data and analyzing trends, a circle graph is a better method for representing how different elements in a whole quantity are distributed.

**EXAMPLE 2**

**CONNECTION**
**STATISTICS**

The circle graph at right shows how the average American family spends its money. **Find how much money, on average, a family with an annual income of $32,000 spends on housing, transportation, and food.**

**American Family Spending**

**SOLUTION**

**Housing:** Multiply 31% by $32,000.

$$0.31 \cdot 32{,}000 = 9920$$

The average amount that an American family with an annual income of $32,000 spends on housing is $9920.

**Transportation:** Multiply 18% by $32,000.

$$0.18 \cdot 32{,}000 = 5760$$

The average amount that an American family with an annual income of $32,000 spends on transportation is $5760.

**Food:** Multiply 15% by $32,000.

$$0.15 \cdot 32{,}000 = 4800$$

The average amount that an American family with an annual income of $32,000 spends on food is $4800.

**TRY THIS** Find how much money, on average, a family with an annual income of $32,000 spends on investments, clothing, and miscellaneous expenses.

Reading circle graphs is a useful skill. There may be times when you need to create a circle graph. In some cases, you will need to use a protractor to construct your graph.

**EXAMPLE 3** **Use the information in the table to make a circle graph for the number of cats in each breed that are entered in the Feline Fanciers cat show.**

**APPLICATION**
**CATS**

| **Breed** | Persian | Abyssinian | Balinese | Manx | Siamese | Total |
|---|---|---|---|---|---|---|
| **Number of cats** | 99 | 30 | 66 | 41 | 106 | 342 |

**SOLUTION**

Find the percent for each category.

There are 99 Persian cats in the show. The total number of cats entered in the show is 342.

$$\frac{99}{342} \approx 0.29, \text{ or approximately } 29\%$$

The Persian breed accounts for 29% of the cats entered in the show.

Find the number of degrees that should be used to represent this category in the graph. The total number of degrees in a circle graph is 360.

$$0.29 \cdot 360° = 104.4°$$

Continue with the remaining categories. The original table can be extended to include columns for the percent and the number of degrees.

| **Breed** | **Number of cats** | **Percent** | **Number of degrees** |
|---|---|---|---|
| Persian | 99 | $\frac{99}{342} \approx 0.29 \approx 29\%$ | $0.29 \cdot 360° \approx 104.4°$ |
| Abyssinian | 30 | $\frac{30}{342} \approx 0.09 \approx 9\%$ | $0.09 \cdot 360° \approx 32.4°$ |
| Balinese | 66 | $\frac{66}{342} \approx 0.19 \approx 19\%$ | $0.19 \cdot 360° \approx 68.4°$ |
| Manx | 41 | $\frac{41}{342} \approx 0.12 \approx 12\%$ | $0.12 \cdot 360° \approx 43.2°$ |
| Siamese | 106 | $\frac{106}{342} \approx 0.31 \approx 31\%$ | $0.31 \cdot 360° \approx 111.6°$ |
| Total | 342 | 100% | 360° |

Use a compass to construct the circle, and mark the center of the circle with a point. Then use a protractor to draw angles with the indicated measures. It may be necessary to round your angle measures to the nearest whole number.

**Cat Show Entries**

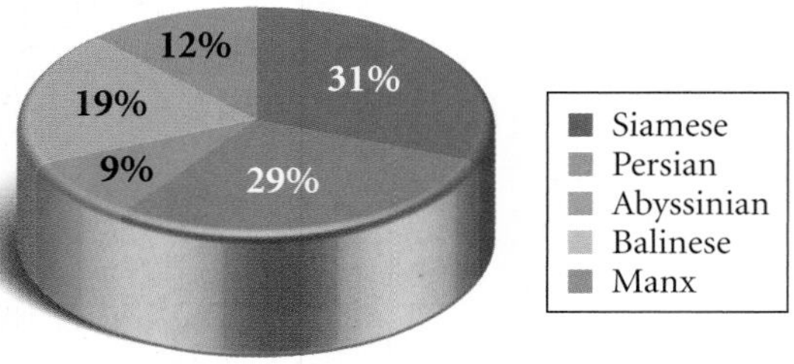

**TRY THIS** A survey of 500 Americans was conducted to find the position in which they fall asleep. Make a circle graph that shows the percent of people in each category.

| **Sleeping position** | Side | Back | Stomach | Varies | Not sure |
|---|---|---|---|---|---|
| **Number** | 280 | 25 | 55 | 55 | 85 |

# Exercises

## Communicate

1. Why are graphs important statistical tools?
2. Explain how graphs can be misleading.
3. Name and describe three different types of graphs that can be used to display data.
4. Explain the difference between a bar graph and a line graph.

## Guided Skills Practice

**The graphs below are based on the 1990 United States census.**

CONNECTION

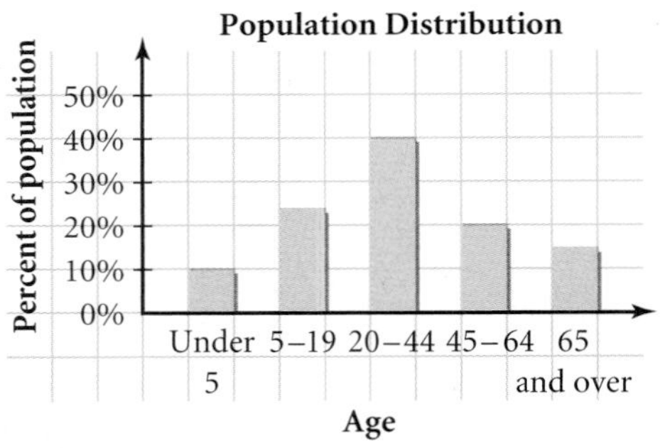


5. **STATISTICS** Based on the bar graph, what is the smallest population group in the United States? What is the second smallest group? ***(EXAMPLE 1)***
6. Based on the circle graph, how many people lived in each region listed if the total population in the United States in 1990 was approximately 253,000,000? ***(EXAMPLE 2)***

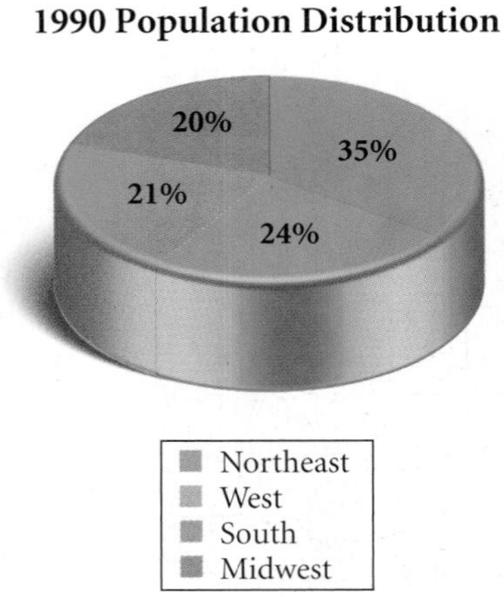


7. How many people would you expect to have been living in the Northeast in 1995 if the distribution stayed the same and the total population that year was estimated at 265,000,000? ***(EXAMPLE 2)***
8. Suppose that you asked 90 people whether or not they liked a certain movie. Thirty people told you that they did not like the movie, 45 people said that they did like it, and 15 people were unsure whether they liked it. Make a circle graph that displays the data. ***(EXAMPLE 3)***

## Practice and Apply

**CONNECTION**

**STATISTICS** These graphs show the price, in dollars, for the stock of two companies. Use the graphs for Exercises 9–11.

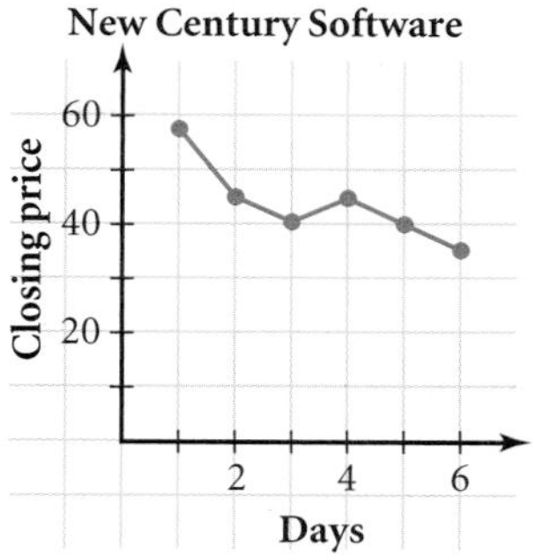


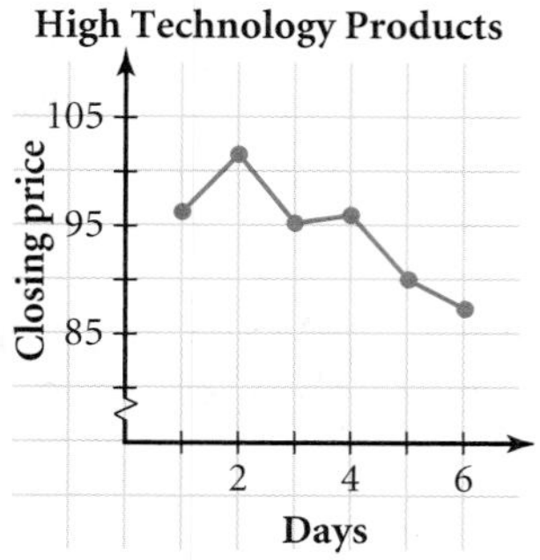


**9.** Which company had the highest one-day stock price for the week?

**10.** Which company had the lowest one-day stock price during the week?

**11.** Why might it be misleading to display these two graphs together?

**APPLICATIONS**

**GEOGRAPHY** Use the graph of average high temperatures for Exercises 12–14.

**12.** Which city had the highest temperature? In which month did this temperature occur?

**13.** Which city had the lowest temperature? In which month did this temperature occur?

**14.** Which city had the greatest increase in temperature from January to July? What was the approximate amount of the increase?

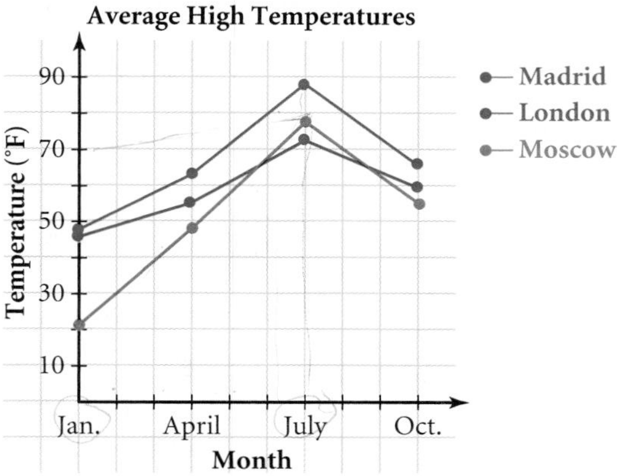


**15.** **DOGS** Use the information below to make a circle graph for the number of dogs in each breed that are in a Canine Companions program.

| **Breed** | Golden Retriever | Yellow Lab | Collie | Cocker Spaniel | Dalmatian | Total |
|---|---|---|---|---|---|---|
| **Number of dogs** | 105 | 73 | 24 | 47 | 91 | 340 |

**FIRE PREVENTION** Use the circle graph about the causes of house fires for Exercises 16–19.

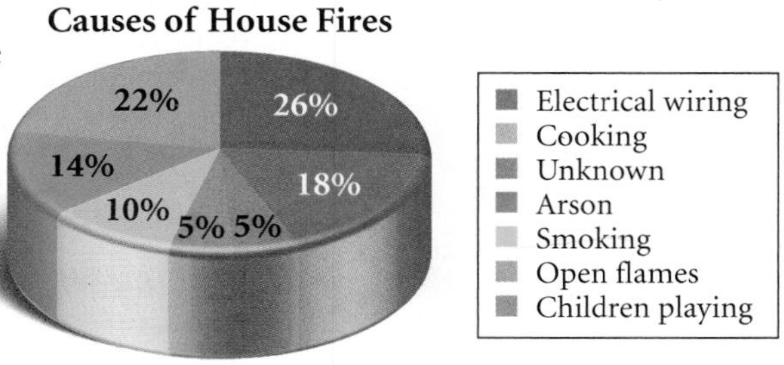


**16.** If there were 957,000 fires in one year, how many of them were caused by smoking?

**17.** If there were 957,000 fires in one year, how many were caused by faulty electrical wiring?

**18.** According to this graph, what is the most likely cause of a house fire?

**19.** Based on these statistics, what steps could you take to reduce fire hazards in your home?

**APPLICATION**

**internet connect**

**Homework Help Online**

Go To: **go.hrw.com**

Keyword: **MA1 Homework Help** for Exercises 20-24

**HEALTH** The bar graph shown here is in the annual report for a rehabilitation hospital. Use the graph for Exercises 20–24.

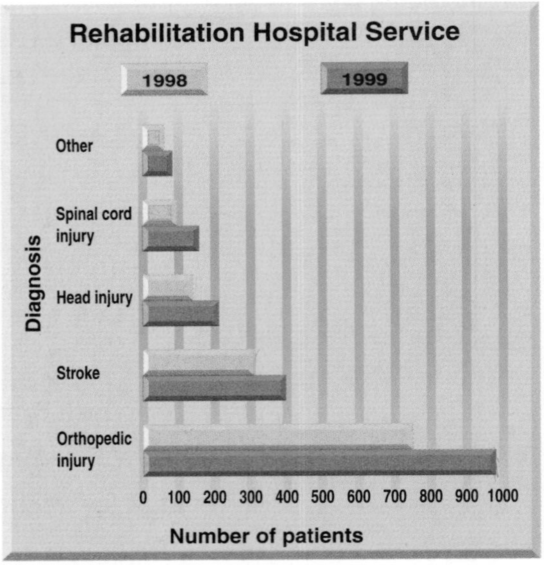


**20.** Which category had the greatest increase in the number of patients from 1998 to 1999?

**21.** Which category had the smallest increase in the number of patients from 1998 to 1999?

**22.** How many patients in 1999 were diagnosed with an orthopedic injury? a spinal cord injury?

**23.** What was the increase in the number of patients diagnosed with a stroke from 1998 to 1999?

**CHALLENGE**

**24.** Was the percent increase in the number of patients from 1998 to 1999 the same for each type of injury diagnosed at the rehabilitation hospital? Explain your reasoning.

## *Look Back*

**Simplify.** ***(LESSONS 2.5 AND 2.6)***

**25.** $(3x - 7) + (5x + 2)$

**26.** $4(7s - 3) - 8(s + 2)$

**Solve each equation. Check your answers.** ***(LESSONS 3.4 AND 3.5)***

**27.** $4t + 7 = 23 + 2t$

**28.** $3(5h - 12) = 9$

**29.** $2x + 3(x + 7) = 41$

**30.** $4y + 17 = 3y - 26$

**31.** $4(m + 7) + 3m - 5 = 2m + 23$

**32.** $3(x + 5) = 5(x + 3)$

**Find each answer.** ***(LESSON 4.2)***

**33.** What is 35% of 160?

**34.** 14 is what percent of 200?

**Find the mean, median, mode, and range for each set of data.** ***(LESSON 4.4)***

**35.** 4, 5, 18, 12, 6, 4, 10, 15

**36.** 100, 87, 92, 94, 90, 100, 88

## *Look Beyond*

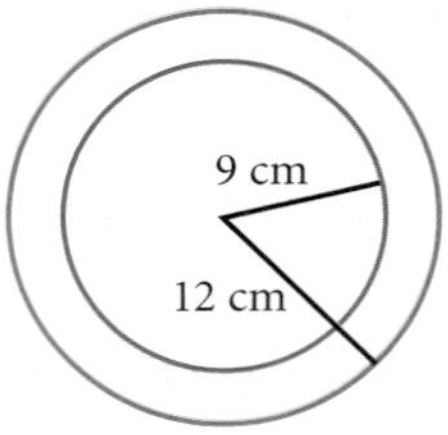


**37.** The formula for the circumference of a circle is $C = 2\pi r$, where $r$ is the radius of the circle. Use this formula to find the sum of the circumferences of the two circles shown at left.

4.6

# Other Data Displays

*Gray whales, which live in the northern Pacific Ocean, were recently taken off the endangered species list. Some scientists collect data such as the length of adult gray whales.*

**Why** ***Displays such as stem-and-leaf plots, histograms, and box-and-whisker plots provide a way to study large sets of data more easily and arrive at conclusions more quickly. This process is helpful in fields like science, medicine, and technology.***

### Objectives

- Interpret stem-and-leaf plots, histograms, and box-and-whisker plots.
- Represent data with stem-and-leaf plots, histograms, and box-and-whisker plots.

## Stem-and-Leaf Plots

| Length of Adult Gray Whales (in meters) | | | |
|---|---|---|---|
| 12.7 | 13.5 | 14.0 | 14.2 |
| 12.8 | 13.5 | 14.0 | 14.3 |
| 12.8 | 13.6 | 14.0 | 14.3 |
| 12.8 | 13.6 | 14.1 | 14.3 |
| 12.9 | 13.6 | 14.1 | 14.4 |
| 13.0 | 13.7 | 14.1 | 14.4 |
| 13.1 | 13.7 | 14.1 | 14.4 |
| 13.1 | 13.7 | 14.1 | 14.5 |
| 13.2 | 13.8 | 14.1 | 14.5 |
| 13.3 | 13.8 | 14.1 | 14.5 |
| 13.4 | 13.9 | 14.2 | 14.6 |
| 13.4 | 13.9 | 14.2 | 15.2 |

A **stem-and-leaf plot** displays data by dividing each element into two parts. The stem-and-leaf plot below gives the lengths of several adult whales. The numbers to the left of the vertical line are called **stems** and represent the first part of each element. The numbers to the right of the line are called **leaves** and frequently represent the last digit of each element.

| Stem-and-Leaf Plot | |
|---|---|
| **Stems** | **Leaves** |
| 12 | 7, 8, 8, 8, 9 |
| 13 | 0, 1, 1, 2, 3, 4, 4, 5, 5, 6, 6, 6, 7, 7, 7, 8, 8, 9, 9 |
| 14 | 0, 0, 0, 1, 1, 1, 1, 1, 1, 1, 2, 2, 2, 3, 3, 3, 4, 4, 4, 5, 5, 5, 6 |
| 15 | 2 |

To record data in a stem-and-leaf plot, first examine the data set to determine what the stems and leaves will be. In the stem-and-leaf plot above, the stems are whole numbers, and the leaves are tenths.

If the data set consists of numbers from 15 to 55, the stems can represent the tens digits, and the leaves can represent the ones digits.

**EXAMPLE 1** Gray whales breathe every 5 to 15 minutes. They can, however, go as long as 40 minutes without breathing. A scientist studying a group of whales recorded the whales' breathing intervals (in minutes). **Construct a stem-and-leaf plot of the data in the table at left.**

| Breathing Intervals (in minutes) | | | |
|---|---|---|---|
| 5.6 | 8.6 | 10.3 | 11.3 |
| 14.8 | 14.7 | 13.6 | 12.2 |
| 6.5 | 15.5 | 8.6 | 13.8 |
| 8.4 | 11.0 | 13.5 | 15.8 |
| 15.6 | 12.6 | 10.5 | 6.7 |
| 7.9 | 9.6 | 10.8 | 11.7 |
| 15.6 | 11.4 | 9.2 | 6.8 |
| 7.1 | 9.2 | 10.5 | 11.4 |
| 16.2 | 14.0 | 12.9 | 11.5 |
| 12.3 | 13.8 | 10.1 | 14.8 |
| 13.7 | 14.2 | 12.6 | 8.5 |
| 10.1 | 12.7 | 12.2 | 11.4 |
| 12.6 | 13.9 | 13.8 | 14.7 |
| 12.2 | 13.7 | 14.2 | 11.4 |
| 10.7 | 13.9 | 12.7 | 16.7 |
| 13.9 | 15.6 | 7.3 | |

**SOLUTION**

To construct the stem-and-leaf plot, first determine the stems and leaves. The numbers range from 5.6 to 16.7, so let the stems be whole numbers, from 5 to 16, and let the leaves be tenths. Then record the leaves next to the appropriate stems. Arrange the leaves in each row from least to greatest.

| Breathing Intervals (in minutes) | |
|---|---|
| **Stems** | **Leaves** |
| 5 | 6 |
| 6 | 5, 7, 8 |
| 7 | 1, 3, 9 |
| 8 | 4, 5, 6, 6 |
| 9 | 2, 2, 6 |
| 10 | 1, 1, 3, 5, 5, 7, 8 |
| 11 | 0, 3, 4, 4, 4, 4, 5, 7 |
| 12 | 2, 2, 2, 3, 6, 6, 6, 7, 7, 9 |
| 13 | 5, 6, 7, 7, 8, 8, 8, 9, 9, 9 |
| 14 | 0, 2, 2, 7, 7, 8, 8 |
| 15 | 5, 6, 6, 6, 8 |
| 16 | 2, 7 |

**TRY THIS** Make a stem-and-leaf plot for the following data:
97, 86, 64,79, 91, 81, 80, 99, 80, 96, 79, 76.

## Histograms

A **histogram** is a bar graph that has no space between the bars and that shows the frequency of data. You can construct a histogram very easily from the data in a stem-and-leaf plot. Refer to the stem-and-leaf plot for the lengths of gray whales on page 201.

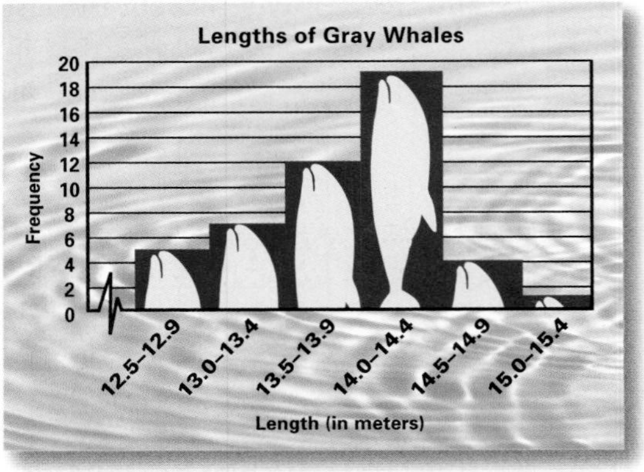


The lengths are listed along the horizontal axis of the histogram, and the frequencies of the numbers in the data set are listed along the vertical axis. Notice that the lengths are listed on the horizontal axis in equal intervals, such as from 12.5 to 12.9, instead of by each individual measurement.

# Box-and-Whisker Plots

**Box-and-whisker plots,** also called box plots, show how data are distributed by using the median, the upper and lower quartiles, and the greatest and least values of the data. Another value, called the **interquartile range,** is a useful measure of the spread of data.

*Activity*

## Interpreting a Box-and-Whisker Plot

Student grades for two of Ms. Garth's classes are listed below.

| | Student Grades | | | | | | | | | | | | |
|---|---|---|---|---|---|---|---|---|---|---|---|---|---|
| **Class 1** | 60 | 80 | 65 | 85 | 80 | 75 | 75 | 100 | 70 | 90 | 90 | 80 | 75 |
| **Class 2** | 70 | 85 | 75 | 85 | 70 | 85 | 85 | 90 | 100 | 90 | 95 | 85 | 85 |

1. Order the grades for Class 1 from lowest to highest, and find the median grade.
2. Find the median grade for the 6 lowest grades of Class 1.
3. Find the median grade for the 6 highest grades of Class 1.
4. The data for Class 1 are displayed below in a box-and-whisker plot. What value is represented by the vertical line in the middle of the box?

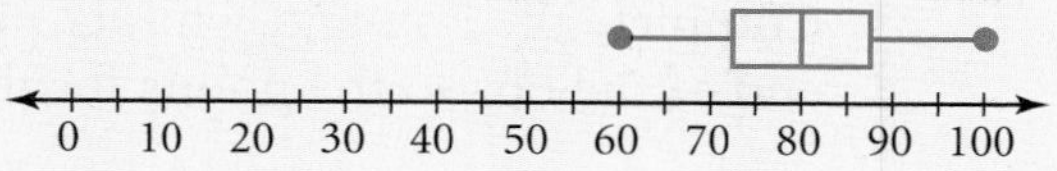


5. The vertical line at 72.5 represents the *lower quartile* and the vertical line at 87.5 represents the *upper quartile.* How do these values compare with the lower and upper medians you found in Steps 2 and 3?
6. Explain in your own words what the dots at the far left and right of the horizontal line represent.

CHECKPOINT ✔

7. Sketch a box-and-whisker plot of the data for Class 2.
8. The *interquartile range* is found by subtracting the lower quartile from the upper quartile. Find the interquartile range for Classes 1 and 2. How does a box-and-whisker plot show this value?

**EXAMPLE 2** **Use the stem-and-leaf plot of the length of gray whales (in meters) to construct a box-and-whisker plot. Find the interquartile range of the data.**

APPLICATION
BIOLOGY

| Length of Gray Whales (in meters) | |
|---|---|
| **Stems** | **Leaves** |
| 12 | 7, 8, 8, 8, 9 |
| 13 | 0, 1, 1, 2, 3, 4, 4, 5, 5, 6, 6, 6, 7, 7, 7, 8, 8, 9, 9 |
| 14 | 0, 0, 0, 1, 1, 1, 1, 1, 1, 1, 2, 2, 2, 3, 3, 3, 4, 4, 4, 5, 5, 5, 6 |
| 15 | 2 |

**SOLUTION**

First draw a number line. Use points to mark the least value, 12.7, and greatest value, 15.2. Identify the median with a vertical mark above the number line.

The median is 13.95, the mean of the 24th and 25th ordered values.

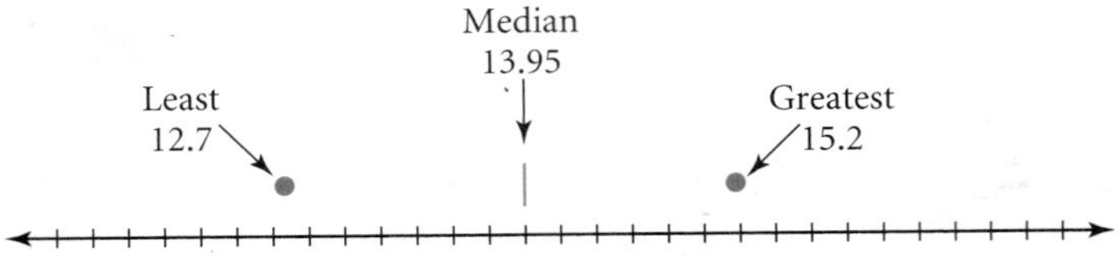


Find the median of all the measurements in the lower half, *below* 13.95. This measurement, called the **lower quartile**, is 13.45.

Find the median of all the measurements in the upper half, *above* 13.95. This measurement, called the **upper quartile**, is 14.2.

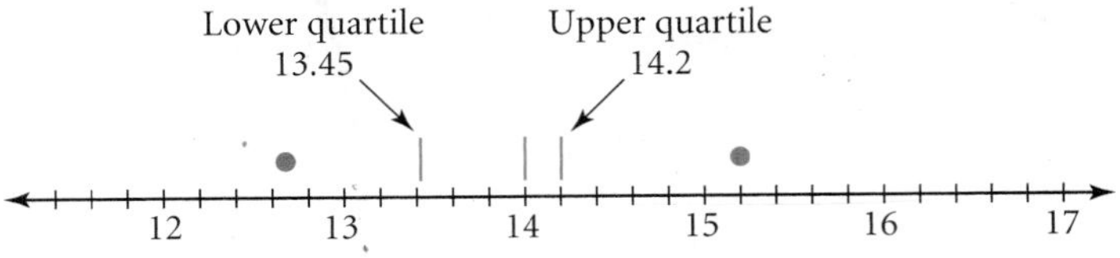


Identify the upper and lower quartiles with vertical marks above the number line as you did for the median.

Draw a rectangular box from the lower quartile to the upper quartile. Complete the box-and-whisker plot by drawing the line segments from each end of the box to the points for the least and greatest values.

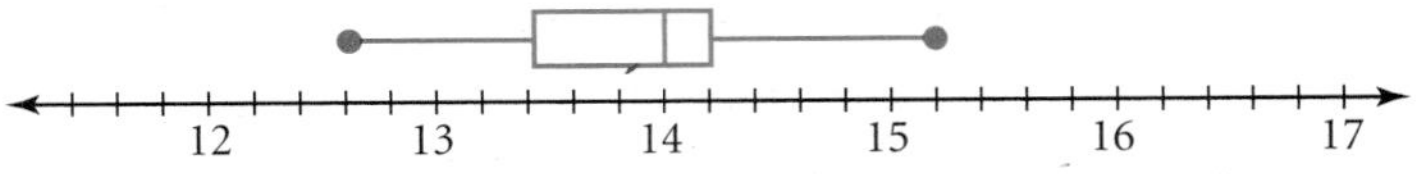


Find the difference between the upper quartile and the lower quartile. This measurement, called the **interquartile range,** is 0.75. Notice that this value is the width of the box in the box-and-whisker plot.

**TRY THIS**

For the data set represented by the box-and-whisker plot below, give the values of the following:

**a.** the lowest score
**b.** the highest score
**c.** the upper quartile
**d.** the lower quartile
**e.** the median
**f.** the interquartile range

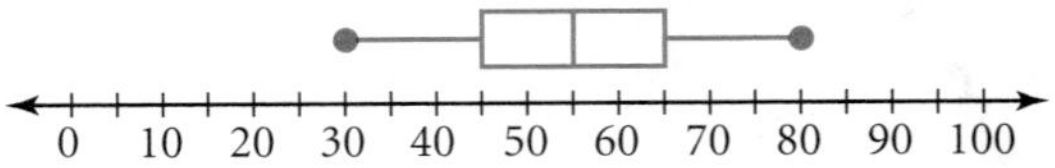

**TECHNOLOGY**
**GRAPHICS CALCULATOR**

Once data are entered into a graphics calculator, they can easily be represented in a number of ways. The box-and-whisker plot from Example 2 is shown below on a graphics calculator display.

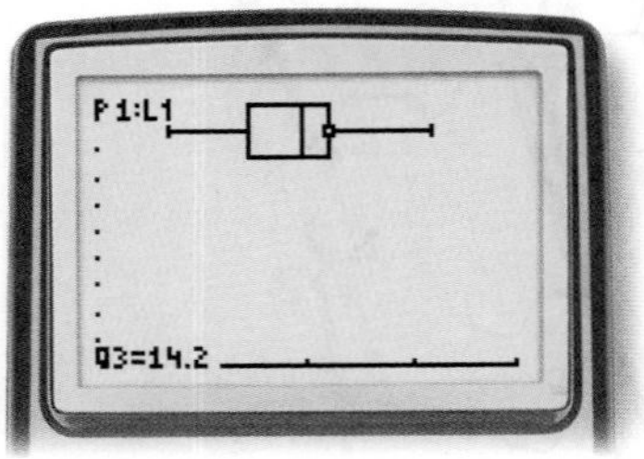

# Exercises

## Communicate

1. Explain how to construct a stem-and-leaf plot.
2. Explain how to make a histogram from a stem-and-leaf plot.
3. Explain how to determine the median and the range from a box-and-whisker plot.
4. Explain how to find the upper and lower quartiles and the interquartile range from a box-and-whisker plot.

## Guided Skills Practice

5. Construct a stem-and-leaf plot of the following data: 12.1, 10.3, 12.6, 15.2, 11.5, 14.4, 14.3, 13.9, 17.3, 14.1, 13.7, 16.6, 15.5, 13.8, 10.7, 16.7, 15.2, 13.8, 12.1, 15.2, 12.2, 13.4. ***(EXAMPLE 1)***
6. Construct a box-and-whisker plot of the data from Exercise 5. What is the interquartile range? ***(EXAMPLE 2)***

## Practice and Apply

**internet connect**
**Homework Help Online**
Go To: **go.hrw.com**
Keyword:
**MA1 Homework Help** for Exercises 17-26

**Construct a stem-and-leaf plot for each set of data.**

7. 14.8, 12.2, 20.3, 18.7, 16.4, 15.7, 17.2, 15.9, 12.5, 18.9, 16.6, 16.3, 15.1, 19.1, 13.3, 16.4
8. 47, 42, 39, 55, 68, 61, 14, 53, 61, 36, 53, 48, 59, 23, 44, 37, 62, 11, 36, 41, 44, 48, 53
9. 144, 179, 136, 183, 153, 147, 127, 142, 136, 176, 148, 153, 125, 127, 128, 133, 135, 138, 139, 141, 153, 154, 159, 162, 175, 184
10. 17, 12, 22, 38, 84, 76, 37, 28, 11, 46, 83, 78, 65, 32, 48, 56, 13, 82, 11, 13, 14, 23, 32, 41, 47, 49, 73
11. 1.2, 5.6, 8.4, 6.5, 9.7, 6.3, 2.5, 4.6, 4.8, 6.8, 6.7, 2.5, 3.4, 7.8, 9.1, 5.2, 8.4, 9.2, 2.7
12. 13, 25, 29, 38, 27, 26, 12, 24, 25, 27, 26, 28, 16, 18, 32, 24, 34, 29, 21, 20, 19, 37
13. 7, 12, 18, 9, 5, 23, 5, 16, 19, 18, 4, 8, 26, 29, 5, 6, 7, 19, 10, 12, 17
14. 12, 11, 16, 18, 20, 21, 28, 34, 19, 16, 14, 17, 12, 18, 14, 16, 24
15. 26.4, 26.1, 27.6, 29.5, 29.4, 23.1, 25.8, 29.6, 24.7, 23.6, 21.4, 20.8, 29.4, 21.6, 23.9
16. 328, 349, 336, 307, 378, 355, 322, 374, 371, 337, 314, 407, 303, 335, 339, 359, 366, 393

**17–26.** Construct a box-and-whisker plot for each set of data in Exercises 7–16. What is the interquartile range for each set?

**STATISTICS** The stem-and-leaf plot below shows the number of runs credited to the top players in a baseball league. The stems are tens digits, and the leaves are ones digits.

| Number of Runs | |
|---|---|
| **Stems** | **Leaves** |
| 3 | 5, 5 |
| 4 | 0, 2, 2 |
| 5 | 2, 2, 5, 6, 7, 9, 9, 9 |
| 6 | 0, 0, 1, 2, 3, 5 |
| 7 | 4, 6, 7, 9, 9 |
| 8 | 1, 1, 8, 9 |
| 9 | — |
| 10 | 4 |

**27.** Another player had 60 runs. If his data were added to the stem-and-leaf plot, would the range, mean, median, or mode be affected? If so, how?

**28.** What percent of the players scored more than 80 runs? Round your answer to the nearest tenth of a percent.

**29.** What percent of the players scored fewer than 60 runs? Round your answer to the nearest tenth of a percent.

**30.** Construct a histogram of the data.

**31.** Construct a box-and-whisker plot of the data. What is the interquartile range of the data?

## APPLICATIONS

**BIOLOGY** A scientist recorded the duration of humpback whale songs. Use the data below for Exercises 32–38.

| Duration of Humpack Whale Songs (in minutes) | | | |
|---|---|---|---|
| 16.6 | 19.9 | 17.2 | 18.9 |
| 14.5 | 18.1 | 15.6 | 17.8 |
| 15.3 | 16.3 | 17.2 | 16.2 |
| 19.7 | 14.1 | 15.5 | 19.3 |
| 20.0 | 17.4 | 18.2 | 16.1 |
| 19.3 | 15.0 | 14.9 | 15.2 |

**32.** Use the data to make a stem-and-leaf plot.

**33.** What is the range of the data?

**34.** Find the mean and median of the data.

**35.** Is there a mode? If so, what is it?

**36.** What measure of central tendency do you think best answers this question? Why?

**37.** Construct a histogram of the data.

**38.** Construct a box-and-whisker plot of the data. What is the interquartile range of the data?

**BIOLOGY** One well-known type of dolphin is the bottle-nosed dolphin, which can weigh up to 600 pounds. Use the data below for Exercises 39–44.

**39.** Use the data to make a stem-and-leaf plot.

**40.** What is the range of the data?

**41.** What is the median of the data?

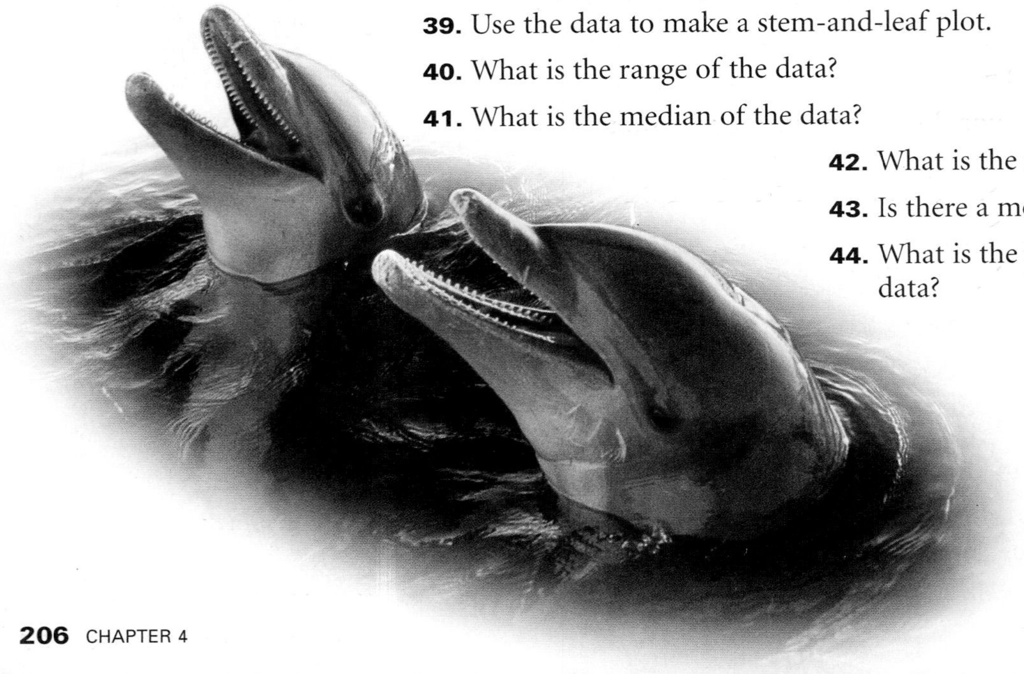

**42.** What is the mean of the data?

**43.** Is there a mode? If so, what is it?

**44.** What is the interquartile range of the data?

| Weight of Bottle-nose Dolphins (in pounds) | | | | |
|---|---|---|---|---|
| 345 | 574 | 249 | 497 | 588 |
| 567 | 586 | 383 | 484 | 491 |
| 275 | 499 | 478 | 343 | 549 |
| 399 | 512 | 589 | 576 | 497 |

## Look Back

**Find each product.** *(LESSON 2.4)*

**45.** $8 \cdot 1\frac{1}{4}$

**46.** $\frac{2}{3} \cdot 2\frac{3}{5}$

**47.** $6\frac{1}{5} \cdot 3\frac{3}{10}$

**Simplify.** *(LESSONS 2.6 AND 2.7)*

**48.** $2(3x + 7) - x$

**49.** $5y - 7 + 2y + 3$

**50.** $\frac{3x + 18}{x + 6} - 2x + 4$

**51.** Solve $4y - 3 = 6y - 2$ for $y$. *(LESSON 3.4)*

**Write each percent as a fraction or mixed number in simplest terms.** *(LESSON 4.2)*

**52.** 80%

**53.** 38%

**54.** 3.5%

## Look Beyond

APPLICATION

**TRAVEL** Mary leaves the city in her car at a steady rate of 40 miles per hour. The distance that she drives away from the city depends on the number of hours that she drives. Her distance from the city equals the rate times the number of hours.

**55.** Write an equation to describe the distance that Mary drives in terms of the number of hours that she drives.

**56.** Find the distance she would travel in 1, 2, 3, 4, and 5 hours and plot the points on a graph. Connect the points.

**57.** Describe the graph. How far up does the graph move for every hour that it moves over?

internet connect
**Portfolio Extension**
Go To: **go.hrw.com**
Keyword:
**MA1 Endangered**

PORTFOLIO ACTIVITY

Include this activity in your portfolio.

The grades on a test in a history class are listed in the table below.

| Student | 1 | 2 | 3 | 4 | 5 | 6 | 7 | 8 | 9 | 10 | 11 | 12 | 13 |
|---|---|---|---|---|---|---|---|---|---|---|---|---|---|
| Grade | 60 | 80 | 65 | 85 | 80 | 75 | 75 | 100 | 70 | 90 | 90 | 80 | 75 |

1. Represent the data in three different types of displays.
2. Explain why you chose each of the three different types of displays.
3. Which display is the best way to show this data and why?

**WORKING ON THE CHAPTER PROJECT**

You should now be able to complete Activities 5 and 6 of the Chapter Project on page 208.

CHAPTER PROJECT FOUR

# Designing a Research Study

Many types of questions can be investigated through research studies. In this project, you will choose a question and answer it experimentally by collecting and analyzing data.

**Activity 1**

### Choose a Topic

Decide on a topic that you would like to research. State the question that your study will be designed to answer.

**Activity 2**

### Make a Hypothesis

Write a statement to describe the answer you think you will find in your research. Explain why you think this will be the outcome.

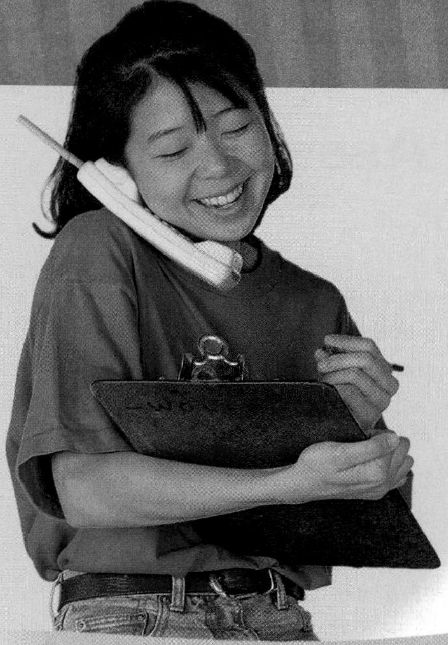

## Activity 3

### Design the Study

Write the questions that you will use in your study or describe the data that you will collect, and explain why each question or observation is necessary. Read through your completed study, and write a justification of why you think it is reasonable.

**Where do teenagers usually do their homework?**

## Activity 4

### Conduct the Study

Collect as large a sample for your study as you can manage. Be sure that you collect enough data (at least 10 values) so that you will be able to accurately test your hypothesis.

## Activity 5

### Display the Data

Summarize the results of your study in a table. Then use one of the methods you studied in this chapter (bar graph, line graph, circle graph, histogram, stem-and-leaf plot, or box-and-whisker plot) to display your data. You might also want to use a scatter plot.

**Is a person's height related to his or her running speed?**

## Activity 6

### Analyze the Results

Explain how the data you gathered supports or contradicts the hypothesis you made in Activity 2.

# 4 Chapter Review and Assessment

## VOCABULARY



## Key Skills & Exercises

### LESSON 4.1

#### Key Skills

**Solve a proportion.**

Solve $\frac{n}{32} = \frac{3}{8}$.

**Method A**

$$\frac{n}{32} = \frac{3}{8}$$
$$32 \cdot \frac{n}{32} = 32 \cdot \frac{3}{8}$$
$$n = \frac{96}{8} = 12$$

**Method B**

$$\frac{n}{32} = \frac{3}{8}$$
$$32 \cdot 3 = 8 \cdot n$$
$$96 = 8n$$
$$12 = n$$

#### Exercises

**Solve each proportion.**

1. $\frac{n}{6} = \frac{12}{9}$
2. $\frac{13}{n} = \frac{39}{27}$
3. $\frac{11.4}{8} = \frac{45.6}{x}$
4. $\frac{16}{41} = \frac{x}{820}$
5. $\frac{x}{5} = \frac{3}{15}$
6. $\frac{n}{2} = \frac{18}{3}$
7. $\frac{4}{5} = \frac{3}{x}$
8. $\frac{15}{3} = \frac{m}{30}$

### LESSON 4.2

#### Key Skills

**Solve percent problems.**

Find 30% of 15.

Use the proportion method.

$$\frac{30}{100} = \frac{x}{15}$$
$$100x = 450$$
$$\frac{100x}{100} = \frac{450}{100}$$
$$x = 4.5$$

30% of 15 is 4.5.

What percent of 50 is 75?

Use the equation method.

$$x \cdot 50 = 75$$
$$\frac{x \cdot 50}{50} = \frac{75}{50}$$
$$x = 1.5 = 150\%$$

75 is 150% of 50.

#### Exercises

**Find each answer.**

9. What is 55% of 60?
10. 28 is 70% of what number?
11. What is 200% of 40?
12. What is 35% of 140?
13. What percent of 90 is 40.5?
14. 4 is what percent of 50?
15. 10 is what percent of 40?
16. 47% of 250 is what number?
17. 31 is 50% of what number?
18. 2 is 200% of what number?
19. 87% of 1500 is what number?
20. What is 14% of 350?

## LESSON 4.3

### Key Skills

**Find the experimental probability for a particular event.**

Two number cubes were tossed 100 times. A sum greater than 8 appeared 34 times. Find the experimental probability of rolling a sum greater than 8.

There were 100 trials and 34 successful outcomes. Thus, in this case, the experimental probability of a successful outcome (rolling a sum that is greater than 8) is as follows:

$$\frac{34}{100} = 0.34 = 34\%$$

### Exercises

**Two number cubes were tossed 150 times. Based on the results given below, find the experimental probability for each outcome.**

**21.** A sum greater than 6 appeared 67 times.

**22.** A pair of 3s appeared 4 times.

**23.** A sum of 11 appeared 11 times.

**24.** At least one 4 appeared 23 times.

## LESSON 4.4

### Key Skills

**Find the mean, median, mode, and range for a set of data.**

Lucy received the following grades on her mathematics tests during a one-month period: 80, 75, 80, 95, and 100.

**mean:** $\frac{80 + 75 + 80 + 95 + 100}{5} = 86$

**median:** Arrange the grades in order: 75, 80, 80, 95, 100. The median is the middle grade, which is 80.

**mode:** The grade that appears most often is 80.

**range:** The range is 100 – 75, or 25.

### Exercises

**Find the mean, median, mode, and range for each set of data.**

**25.** 6, 8, 8, 11, 7

**26.** 5, 7, 10, 13, 21, 3

**27.** 55, 85, 96, 102, 135, 85, 55, 96, 55, 206

**28.** 87, 65, 32, 103, 59

**29.** 0.352, 0.376, 0.385, 0.368, 0.395

**30.** 1.5, 1.6, 1.8, 1.1, 1.4

**31.** 8, 10, 12, 24, 36, 47, 8, 12, 8

## LESSON 4.5

### Key Skills

**Interpret and construct bar graphs, line graphs, and circle graphs.**

The circle graph below shows annual operating expenses for a car. The total expenses were $2148, and $720 was spent on gasoline. The angle of the section for gasoline is calculated as follows:

$$\frac{720}{\text{total cost}} \cdot 360° = \frac{720}{2148} \cdot 360°$$
$$\approx 120.7°$$

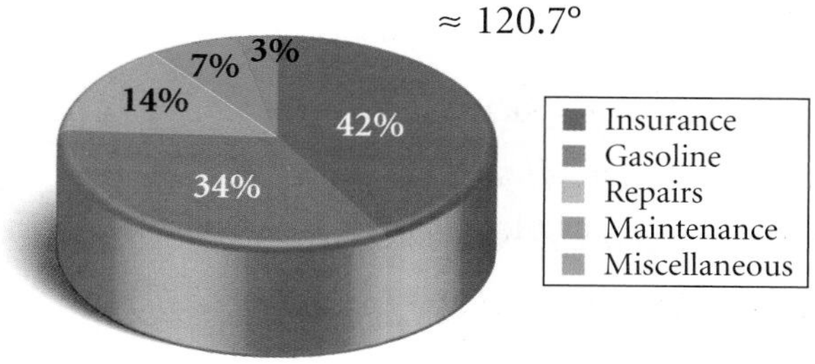


### Exercises

**32.** If the total amount of the operating expenses were $2500, what amount would have been spent on insurance? Use the circle graph at left.

**33.** Brenda used her car-expense record to budget her expenses for the next year. Make a circle graph of the budget shown below.

| | |
|---|---|
| **Gasoline** | $800 |
| **Maintenance** | $200 |
| **Repairs** | $200 |
| **Insurance** | $900 |
| **Miscellaneous** | $100 |

LESSON 4.6

## Key Skills

**Use data to construct stem-and-leaf plots, histograms, and box-and-whisker plots.**

Employees in a company are surveyed to find out their ages. A stem-and-leaf plot, histogram, and box-and-whisker plot are each made from the data: 23, 18, 46, 23, 51, 27, 34, 39, 31, and 56.

| Stems | Leaves |
|---|---|
| 1 | 8 |
| 2 | 3, 3, 7 |
| 3 | 1, 4, 9 |
| 4 | 6 |
| 5 | 1, 6 |

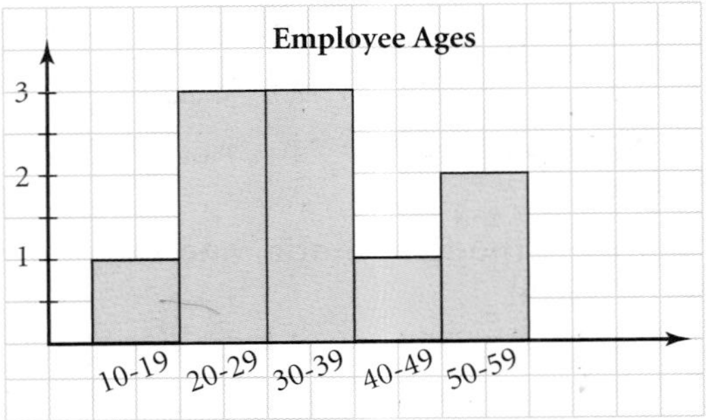


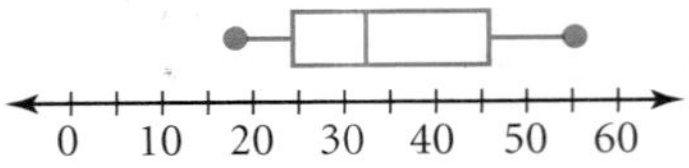


## Exercises

**For the data set at left, identify**

**34.** the lower quartile.

**35.** the upper quartile.

**36.** the median.

**37.** the range.

**38.** the interquartile range.

**Display the following data for test scores as indicated below:**

76, 78, 92, 65, 98, 80, 92, 60, 85, 85, 99, 50, 74, 75, 81, 93, 58, 79, 84, 68, 88

**39.** in a stem-and-leaf plot

**40.** in a histogram

**41.** in a box-and-whisker plot

## Applications

**42.** **EDUCATION** The tuition costs at five universities are $11,800; $14,500; $12,500; $17,000; and $18,600. Find the mean, median, mode, and range of the tuition costs.

**43.** **CONSUMER ECONOMICS** In late December, the McGroty family budgeted the following amounts for essential expenses during the new year: food, $6500; housing, $10,300; clothing, $2500; transportation, $7000; health and personal care, $1800; and miscellaneous, $2000. Draw a circle graph to display the data.

**44.** **PATTERNS IN DATA** Typing speeds on an exam for Ms. Huxley's Typing I class were: 24, 32, 19, 22, 35, 25, 42, 36, 21, 32, 24, 34, 31, 21, 18, 15, 28, 38, 22, 36, and 27. Ms Huxley's Typing II class had speeds of 32, 45, 44, 42, 53, 62, 27, 36, 44, 52, 38, 47, 26, 38, and 48. Construct a box-and whisker plot for each of these data sets, and compare the results.

# Chapter Test

**Solve each proportion.**

**1.** $\frac{x}{5} = \frac{9}{15}$ **2.** $\frac{12}{r} = \frac{48}{16}$ **3.** $\frac{2}{3} = \frac{k}{18}$

**4.** $\frac{6}{8} = \frac{27}{m}$ **5.** $\frac{1.8}{3} = \frac{2.4}{p}$ **6.** $\frac{c}{1.6} = \frac{0.8}{6.4}$

**7.** CONSUMER ECONOMICS It takes 4 pounds of grass seed for 3 acres of lawn. How many pounds of grass seed would be needed to cover 5 acres?

**Write each percent as a fraction or mixed number in lowest terms.**

**8.** 5% **9.** 34% **10.** 230%

**Find each answer.**

**11.** What is 30% of 18?

**12.** 54 is 60% of what number?

**13.** What percent of 84 is 28?

**14.** BUSINESS A certain piece of machinery depreciates from \$9600 to \$8160 in one year. What percent decrease is this?

**15.** GAMES At a carnival, Erin observed a spin-the-wheel game. She noticed that in 30 spins, a prize was won 7 times. What is the experimental probability of winning a prize?

**Find the mean, median, mode, and range of each set of data.**

**16.** 4, 5, 1, 7, 6, 5, 3, 5, 4, 5

**17.** 3.5, 3.3, 4.0, 3.8, 3.4, 3.6

**18.** STATISTICS The mean of six data points is 85 and the mode is 87. If four of the data points are 78, 83, 86, and 87, what are the other two points?

**19.** Based on the bar graph, which class has the smallest population

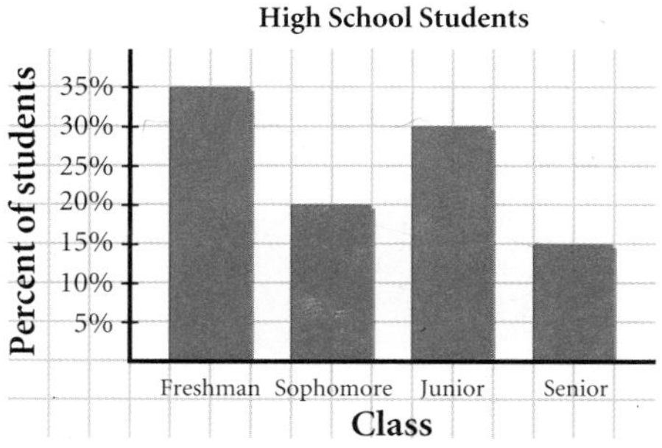


**20.** If there are 860 students enrolled at the high school (see Number 19), how many are juniors?

**21.** EDUCATION Peggy is registering for classes at the local university. She needs 50 credits to satisfy her degree requirements. Use the circle graph to determine how many math credits she needs.

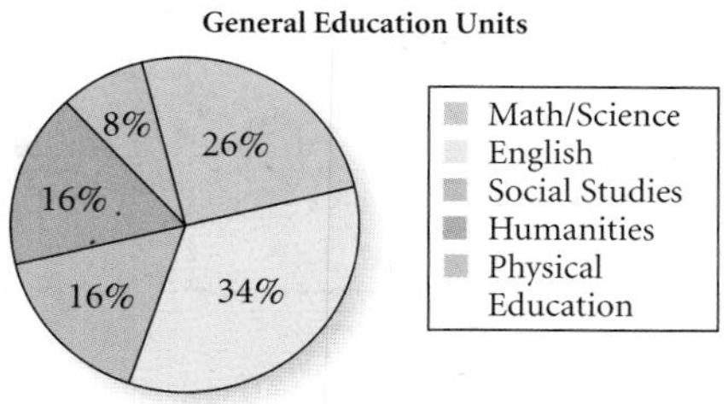


**22.** Construct a stem-and-leaf plot for the following data.

| | | | | |
|---|---|---|---|---|
| 62 | 55 | 66 | 73 | 51 |
| 57 | 71 | 54 | 65 | 71 |
| 62 | 57 | 69 | 71 | 74 |

**23.** For the data set represented by the box-and-whisker plot below give the values of the following:

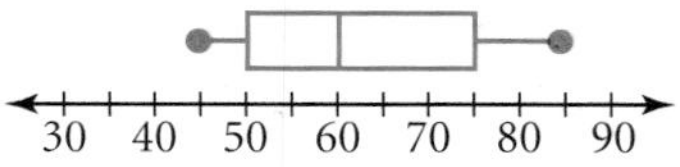


**a.** the lowest score **b.** the highest score

**c.** the median **d.** the lower quartile

**24.** HEALTH A nurse recorded the following pulse rates in beats per minute. Construct a histogram of the data.

| | | | | | | |
|---|---|---|---|---|---|---|
| 46 | 49 | 55 | 56 | 58 | 62 | 65 |
| 66 | 68 | 68 | 69 | 70 | 72 | 72 |
| 75 | 75 | 75 | 78 | 80 | 82 | 96 |

# College Entrance Exam Practice

internet connect
**Standardized Test Prep Online**
Go To: **go.hrw.com**
Keyword: **MM1 Test Prep.**

**QUANTITATIVE COMPARISON** For Questions 1–6, write
A if the quantity in Column A is greater than the quantity in Column B;
B if the quantity in Column B is greater than the quantity in Column A;
C if the two quantities are equal; or
D if the relationship cannot be determined from the information given.

| | Column A | Column B | Answers |
|---|---|---|---|
| **1.** | The number of \$14 CDs you can buy for \$86 | The number of \$12 CDs you can buy for \$75 | (A) (B) (C) (D) [Lesson 1.2] |
| **2.** | $\|-12.4\|$ | $\|12.4\|$ | (A) (B) (C) (D) [Lesson 2.1] |
| **3.** | $15 \cdot (6 + 4) - 7$ | $2^3 - 6 \cdot (3 + 2)$ | (A) (B) (C) (D) [Lesson 1.3] |
| **4.** | $-18 + (-3)$ | $-17 + (-3)$ | (A) (B) (C) (D) [Lessons 2.1 and 2.2] |
| **5.** | The solution to $v + 10 = 7$ | The solution to $v - 6 = 5$ | (A) (B) (C) (D) [Lesson 3.1] |
| **6.** | $(-15)(-4) \div (10)$ | $(-15)(-4) \div (-10)$ | (A) (B) (C) (D) [Lesson 2.4] |

**7.** The correlation between two variables is described as strong positive. What is the relationship between the two variables? ***(LESSON 1.6)***

**a.** As one variable increases, the other also tends to increase.
**b.** As one variable increases, you cannot tell if the other tends to increase or decrease.
**c.** As one variable increases, the other tends to decrease.

**8.** Which sequence shows a linear relationship? ***(LESSON 1.5)***

**a.** 60, 30, 15, 7.5, 3.75, …
**b.** 6, 12, 24, 48, 96, …
**c.** 6, 9, 13.5, 20.25, 30.375, …
**d.** 64, 58, 52, 46, …

**9.** What is the value of $\frac{5^2 + 11}{9 \cdot 3 + 9}$? ***(LESSON 2.4)***

**a.** 36 **b.** 0
**c.** 1 **d.** $\frac{4}{9}$

**10.** What percent is equivalent to 0.5? ***(LESSON 4.2)***

**a.** 0.5% **b.** 50%
**c.** 5% **d.** 0.05%

**11.** Jamie made 80% of his free throws in his game last night. If he attempted 15 free throws, how many free throws did he make? ***(LESSON 4.2)***

**12.** What is the solution to the equation $\frac{q}{139.2} = -58$? ***(LESSON 3.2)***

**a.** $-2.4$ **b.** $-8073.6$

**c.** $-0.417$ **d.** $-7551.6$

**13.** Construct a stem-and-leaf plot for the following data: 12, 18, 26, 24, 29, 32, 37, 16, 14, 12, 29, 27, 34, 30, and 11. ***(LESSON 4.6)***

**14.** Is $18t$ a variable, an expression, or an equation? Explain your answer. ***(LESSON 1.2)***

**What are the coordinates of the given points? *(LESSON 1.4)***

**15.** $X$

**16.** $Y$

**17.** $Z$

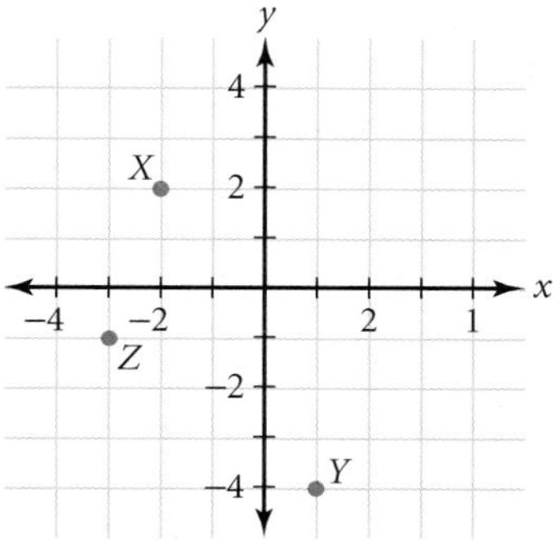


**18.** Solve the proportion $\frac{g}{12} = \frac{-3}{20}$. ***(LESSON 4.1)***

**19.** What is the reciprocal of $-\frac{1}{9}$? ***(LESSON 2.4)***

**Simplify each expression. *(LESSONS 2.6 AND 2.7)***

**20.** $(9w - 5) - (3w - 4)$

**21.** $(-54s) \div (6s)$

**22.** $4x^2 - 4(3x^2 + 7)$

**23.** $16x^2 - 8x^2$

**24.** Solve the proportion $\frac{15}{n} = \frac{20}{32}$. ***(LESSON 4.1)***

**25.** Solve $2y + 4.5 = 6y - 1.5$. ***(LESSON 3.4)***

**26.** What is the solution to the proportion $\frac{c}{16} = \frac{3}{8}$? ***(LESSON 4.1)***

**Solve each equation.** ***(LESSONS 3.1, 3.2, AND 3.3)***

**27.** $x - 3 = 14$

**28.** $y + 4 = 23$

**29.** $3x = -39$

**30.** $6p = 24$

**31.** $3d - 15 = 30$

**32.** Evaluate $\frac{(-5)(-10)}{(-2)(5)}$. ***(LESSON 2.4)***

**33.** Find the opposite of the expression $3x - 7$. ***(LESSON 2.5)***

**34.** Complete each step, and name the property used. ***(LESSON 2.5)***

$(74 + 23) + 26$

$= (23 + ?) + 26$ ____________

$= 23 + (74 + ?)$ ____________

$= 23 + ?$ ____________

$= ?$ ____________

**FREE-RESPONSE GRID**

**The following questions may be answered by using a free-response grid such as that commonly used by standardized test services.**

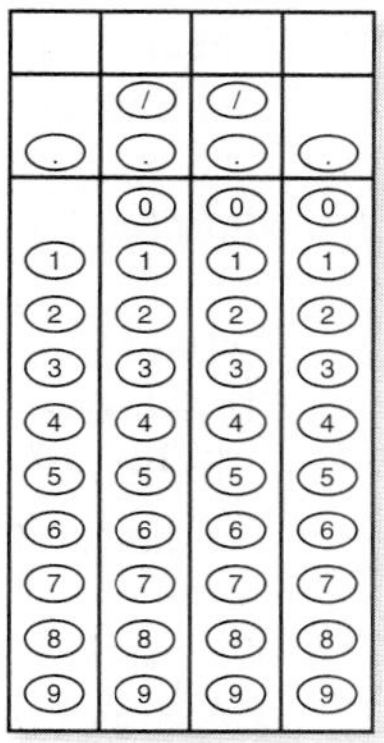

**35.** What is the solution to the equation $3r + 2(r - 4) = 12$? ***(LESSON 3.5)***

**36.** Find the mean of the following grades: 96, 83, 92, 79, 94, and 87. ***(LESSON 4.3)***

**37.** Hannah got 84% of the questions correct on a test. If there were 50 questions on the test, how many did she answer correctly? ***(LESSON 4.2)***

**38.** Two number cubes were rolled 100 times. A sum greater than 7 appeared 29 times. Based on this experiment, what is the experimental probability, expressed as a decimal, that a sum greater than 7 will appear on two rolled cubes? ***(LESSON 4.4)***

# Linear Functions

5

THE GREAT PYRAMID OF GIZEH IN EGYPT WAS built over a 30-year period around 2900 B.C.E. The builders used blocks averaging 2.5 tons, but some weighed as much as 54 tons. The blocks were transported from a quarry many miles away. The pyramid has a square base, and the error in constructing the right angles of the base was only 1 part in 14,000.

In order to build the pyramids, the builders needed to solve problems about slope. One problem from the Rhind papyrus asks how to find the *seked* (a form of slope) of a pyramid with a base of 2520 *palms* (618 feet) on each side and a height of 250 *cubits* (429 feet).

## Lessons

*The pyramid at the entrance to the Louvre in Paris, France, provides a contrast to the older architecture of the main building.*

*The strong lines in these high-rise buildings show the importance of linear functions in architecture.*

*Pyramids, Gizeh, Egypt*

## About the Chapter Project

Algebra provides the power to model real-world data. Beyond merely representing data on a graph or chart, algebra makes it possible to use *functions* to model data. In the Chapter Project, *Modeling Trends in Data*, on page 264, you will learn how to model data with linear equations by using linear regression. After completing the Chapter Project, you will be able to do the following:

- Draw a scatter plot of real-world data and determine the equation of the line of best fit or regression line.
- Make predictions about trends in data based on regression lines.

## About the Portfolio Activities

Throughout the chapter, you will be given opportunities to complete Portfolio Activities that are designed to support your work on the Chapter Project.

- Relating measures of diameter and measures of circumference is the topic of the Portfolio Activity on page 233.
- Recognizing real-world data that can be modeled with a graph is the topic of the Portfolio Activity on page 243.
- Correlations between sets of data are the topic of the Portfolio Activity on page 263.

# 5.1 Linear Functions and Graphs

## Objectives

- Determine whether a relation is a function.
- Describe the domain and range of a function.

**Why** *Linear functions are the most basic algebraic functions. Many real-world problems can be modeled by linear functions.*

*Many people keep track of their height at different ages.*

Sets of paired numbers have a special meaning in algebra. Consider the following tables:

APPLICATION HEALTH

**Table 1**

| | Age (years) | Height (inches) |
|---|---|---|
| **Chad** | 14 | 68 |
| **Twan** | 11 | 64 |
| **Joshua** | 13 | 65 |
| **Sam** | 16 | 72 |
| **Sarah** | 15 | 64 |
| **Paul** | 14 | 67 |
| **Mike** | 15 | 66 |
| **Crystal** | 12 | 62 |

*This table shows a pairing of ages with heights.*

**Table 2**

| Joshua's age (years) | Joshua's height (inches) |
|---|---|
| 11 | 63 |
| 12 | 64 |
| 13 | 65 |
| 14 | 70 |
| 15 | 72 |
| 16 | 72 |

*This table shows the age and height data for one student.*

Each table presents two sets of paired numbers. A **relation** is a pairing between two sets of numbers. The two sets of numbers in Table 2 represent a special relation called a *function.*

## *Activity*
## Exploring Relations and Functions

1. Write the ordered pairs from each table on page 218. The first few have been done for you.

   Table 1: (14, 68), (11, 64), ...   Table 2: (11, 63), (12, 64), ...

2. Study the first coordinates of the set of ordered pairs from Table 1. Are any of the first coordinates repeated?
3. Study the first coordinates of the set of ordered pairs from Table 2. Are any of the first coordinates repeated?
4. Study the second coordinates from Table 1. Are any of the second coordinates repeated?
5. Study the second coordinates from Table 2. Are any of the second coordinates repeated?

CHECKPOINT ✓

6. Both tables represent relations. The second table represents a function. Based on your observations from Steps 2–5, what makes a relation a function?

## Relations and Functions

A convenient way to represent a relation is as a set of ordered pairs, using braces to enclose the set. A set of ordered pairs is {(14, 68), (11, 64), (13, 65)}.

A relation is a pairing between two sets of numbers. A function is a relation that meets a special condition.

### Definition of a Function

A **function** is a pairing between two sets of numbers in which each element of the first set is paired with exactly one element of the second set.

Table 2 on page 218 represents a function. The set of ordered pairs from Table 2 is {(11, 63), (12, 64), (13, 65), (14, 70), (15, 72), (16, 72)}.

Notice that no two ordered pairs have the same first coordinate. In a function, some of the second coordinates may be the same, but all of the first coordinates must be different.

The first coordinates in the set of ordered pairs are the **domain** of the relation, and the second coordinates are the **range** of the relation. The domain of the relation in Table 1 is {11, 12, 13, 14, 15, 16}. The range of the relation is {62, 64, 65, 66, 67, 68, 72}.

A function has a domain and a range just as any relation does. The domain of the function from Table 2 on page 218 is {11, 12, 13, 14, 15, 16}. The range is {63, 64, 65, 70, 72}.

**EXAMPLE 1** **Determine whether each set of ordered pairs is a function. Describe the domain and range for each.**

**a.** $\{(3,4), (3, 6), (5, 14), (7, 14)\}$ **b.** $\{(4, 12), (5, 18), (7, 12), (8, 19)\}$

**SOLUTION**

**a.** The ordered pairs (3, 4) and (3, 6) have the same first coordinate. This relation is not a function. The domain is $\{3, 5, 7\}$, and the range is $\{4, 6, 14\}$.

**b.** Since none of the ordered pairs have the same first coordinate, this relation is a function. The domain is $\{4, 5, 7, 8\}$, and the range is $\{12, 18, 19\}$.

**TRY THIS** Write a set of four ordered pairs that is a function and a set of four ordered pairs that is not a function.

## Linear Functions

Three common methods are used to describe functions—tables that list ordered pairs, function rules, and graphs.

Since a function is a special pairing between two sets of numbers, a convenient way to provide a rule for a function is to use an equation in two variables such as those shown below.

$y = 5x$ $\qquad d = 5t$ $\qquad 3x + 2y = 10$ $\qquad -10p = 2q + 1$

An equation in two variables can have infinitely many solutions which can be represented by ordered pairs of numbers. If an equation *describes* a function, then the set of all ordered pairs that are solutions to the equation *define* the function. Functions and equations in two variables can be linear, as presented in this lesson, or not linear, as presented in later lessons.

**EXAMPLE 2** **Complete each ordered pair so that it is a solution to $2x + y = 9$.**

**a.** $(1, ?)$ **b.** $\left(\frac{5}{2}, ?\right)$ **c.** $(?, 9)$ **d.** $(4, ?)$

**SOLUTION**

**a.** Substitute 1 for $x$ and solve for $y$.

$2(1) + y = 9 \rightarrow 2 + y = 9 \rightarrow y = 7 \qquad (1, 7)$

**b.** Substitute $\frac{5}{2}$ for $x$ and solve for $y$.

$2\left(\frac{5}{2}\right) + y = 9 \rightarrow 5 + y = 9 \rightarrow y = 4 \qquad \left(\frac{5}{2}, 4\right)$

**c.** Substitute 9 for $y$ and solve for $x$.

$2x + 9 = 9 \rightarrow 2x = 0 \rightarrow x = 0 \qquad (0, 9)$

**d.** Substitute 4 for $x$ and solve for $y$.

$2(4) + y = 9 \rightarrow 8 + y = 9 \rightarrow y = 1 \qquad (4, 1)$

**TRY THIS** Complete each ordered pair so that it is a solution to $3x + y = 15$.

**a.** (3, ?) **b.** (?, 3) **c.** (0, ?) **d.** (5, ?)

A table of values can help you to determine a rule for a linear function.

**EXAMPLE 3**

**APPLICATION**
**RECREATION**

The table below shows the amount that a company charges for a raft rental for up to 10 hours. An initial deposit is included in the amounts shown. **Write an equation for the function. Write a set of ordered pairs for the function. Identify the domain and range values.**

| **Time in hours, $x$** | 1 | 2 | 3 | 4 | 5 | 6 | 7 | 8 | 9 | 10 |
|---|---|---|---|---|---|---|---|---|---|---|
| **Rental cost in dollars, $y$** | 15 | 20 | 25 | 30 | 35 | 40 | 45 | 50 | 55 | 60 |

**SOLUTION**

**PROBLEM SOLVING**

Use the differences in the rental cost to **determine a pattern**. For each hour, the cost increases by \$5. The cost for 1 hour is \$15, which includes the deposit. If the rental fee is \$5 per hour, then the deposit is \$10. Renting a raft costs \$10 plus \$5 per hour. The equation is as follows:

$y = 5x + 10$

The set of ordered pairs is {(1, 15), (2, 20), (3, 25), (4, 30), (5, 35), (6, 40), (7, 45), (8, 50), (9, 55), (10, 60)}.

The domain is {1, 2, 3, 4, 5, 6, 7, 8, 9, 10}.

The range is {15, 20, 25, 30, 35, 40, 45, 50, 55, 60}.

The domain and range do not include amounts that are not whole numbers because the rental fee is based on hourly rates and the domain of the function is restricted to hours from 1 to 10 only.

**TECHNOLOGY**
**GRAPHICS CALCULATOR**

On some calculators you can list the ordered pairs and plot the corresponding points to show the graph of the function in Example 3.

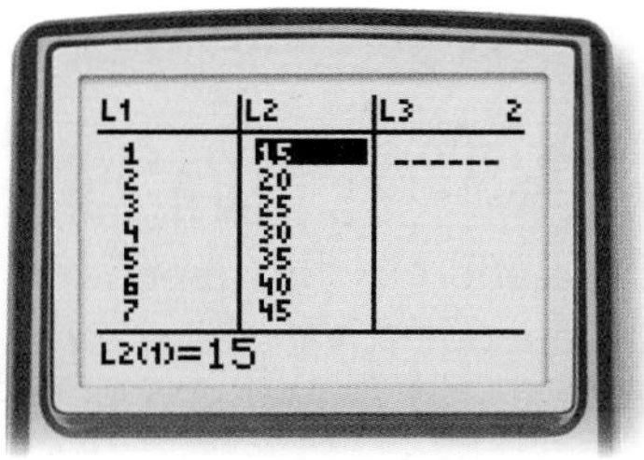


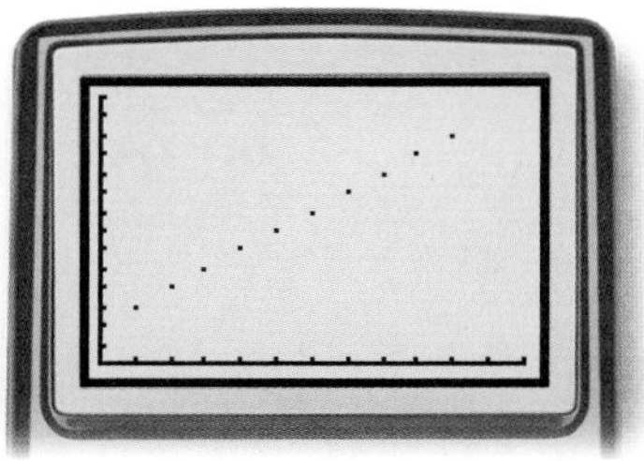

**CRITICAL THINKING** Does a charge of \$10 for 0 hours make sense for the problem in Example 3? Why or why not?

**TRY THIS** The table below shows the amount that a company charges for a raft rental for up to 10 hours. An initial deposit is included in the amounts shown. Write an equation for the function. Write a set of ordered pairs for the function. Identify the domain and range values.

| **Time in hours, $x$** | 1 | 2 | 3 | 4 | 5 | 6 | 7 | 8 | 9 | 10 |
|---|---|---|---|---|---|---|---|---|---|---|
| **Rental cost in dollars, $y$** | 18.50 | 22 | 25.50 | 29 | 32.50 | 36 | 39.50 | 43 | 46.50 | 50 |

Graphs can also be used to represent functions. This method is especially helpful for functions with infinite domains that are impossible to list in a table.

You can read some of the ordered pairs from the graph of a linear function and use these ordered pairs to write an equation.

## EXAMPLE 4

APPLICATION
**TRANSPORTATION**

The graph shows the distance that a plane travels over time. **Identify the dependent and independent variables, and describe the domain and range. Write an equation for the linear function.**

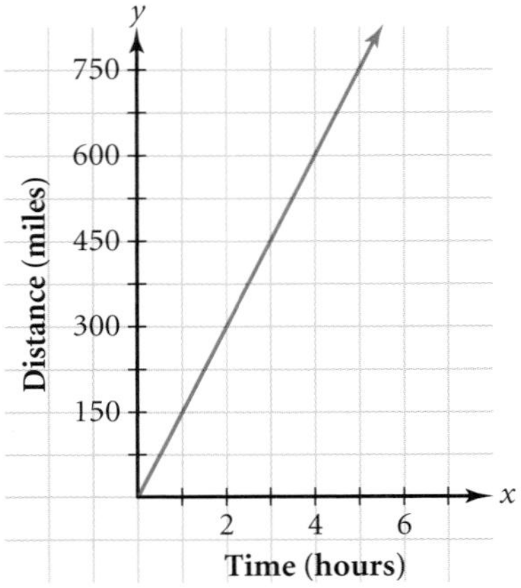

**SOLUTION**

The dependent variable is distance which is represented by $y$. The independent variable is time which is represented by $x$. (See page 32.) Both the domain values and the range values are the set of all real numbers greater than or equal to 0.

Build a table of values to find the equation.

| **Time in hours, $x$** | 0 | 1 | 2 | 3 | 4 | 5 |
|---|---|---|---|---|---|---|
| **Distance in miles, $y$** | 0 | 150 | 300 | 450 | 600 | 750 |

Use the differences between the $y$-values to determine a pattern. The $y$-values start with 0 and increase consistently by 150. An equation that represents this function is $y = 150x$.

**TRY THIS** The graph shows the distance that a tour bus is from its home base over time. Identify the dependent and independent variables and describe the domain and range. Write an equation for the linear function.

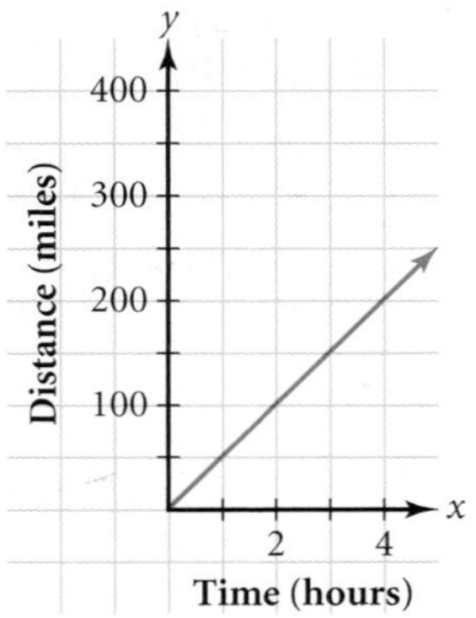

Another way to express the equation for the function in Example 4 is to use **function notation**, $f(x) = 150x$. This is read "$f$ of $x$ equals $150x$." Notice that "$f(x)$" replaces "$y$." Any letters may be used to represent the variables, but the independent variable must appear within the parentheses and in the expression on the right side of the equation. You will study the use of function notation further in Chapter 14.

# Exercises

## Communicate

**1.** Study each set of ordered pairs and determine whether each is a function. Explain your reasoning. Describe the domain and range for each.
**a.** $\{(4, 5), (5, 4), (3, 4), (1, 5)\}$
**b.** $\{(-2, 3), (2, -3), (4, 2), (2, -4)\}$
**c.** $\{(1, 3), (2, 4), (3, 5), (4, 6), (5, 7)\}$
**d.** $\{(1, 4), (1, 5), (1, 6), (1, 7)\}$

**2.** Explain what makes a function different from any other relation.

**3.** How do you know that a point lies on a line given the equation of the line?

**4.** Describe how to find an equation from the graph of a linear function.

## Guided Skills Practice

**Describe the domain and range of each relation below. Determine whether each relation is a function.** *(EXAMPLE 1)*

**5.** $\{(9, -5), (4, 5), (3, -2)\}$
**6.** $\{(1, 1), (2, 1), (3, 1), (4, 1)\}$

**Complete each ordered pair so that it is a solution to $4x + y = 21$.** *(EXAMPLE 2)*

**7.** $(5, ?)$
**8.** $(?, 9)$
**9.** $(0, ?)$
**10.** $(?, 0)$

**11.** The table shows per minute rates for a cellular phone call, including a \$2 fixed fee per call. Use the table to write an equation for this function. Identify the domain and the range shown in the table. *(EXAMPLE 3)*

| Minutes of use | 1 | 2 | 3 | 4 | 5 | 6 |
|---|---|---|---|---|---|---|
| Rate in dollars | 2.50 | 3.00 | 3.50 | 4.00 | 4.50 | 5.00 |

**12.** The graph shows a linear function. Make a table of values, and determine an equation. *(EXAMPLE 4)*

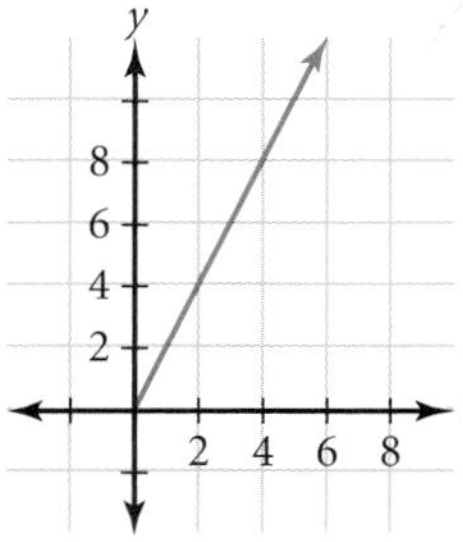

## Practice and Apply

internet connect
**Homework Help Online**
Go To: go.hrw.com
Keyword:
**MA1 Homework Help** for Exercises 13-17

**For each relation, describe the domain and range. Determine whether the relation is a function. Explain your reasoning.**

**13.** $\{(9.2, 3.4), (5.7, -3.4), (-2, 4)\}$

**14.** $\{(0, 3), (0, 4), (4, 0), (3, 0)\}$

**15.** $\{(-1, -2), (1, -2), (3, -1), (-3, -2), (0, 4)\}$

**16.** $\{(2, 1), (2, 1000)\}$

**17.** $\{(9, 9), (3, 3), (4, 4), (5, 5), (12, 12)\}$

**Complete each ordered pair so that it is a solution to $2x - y = 14$.**

**18.** (8, ?) **19.** (10, ?) **20.** (0, ?) **21.** (?, 0)

**22.** (5, ?) **23.** (−5, ?) **24.** (3, ?) **25.** (?, 3)

**26.** (?, 6) **27.** (?, −4) **28.** (?, −7) **29.** (?, 10)

CONNECTIONS

**30. COORDINATE GEOMETRY** Game Menagerie offers a deal for frequent video-game players. The graph shows the cost of playing from 0 to 30 video games. Make a table of values and write an equation for the situation.

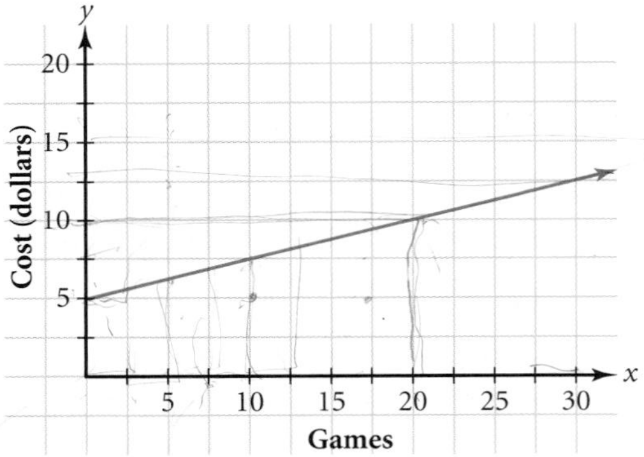


**31. COORDINATE GEOMETRY** The graph shows how far a hiker is from the starting point of a 30-mile trail. Make a table of values, write an equation for the graph, and describe the domain and range.

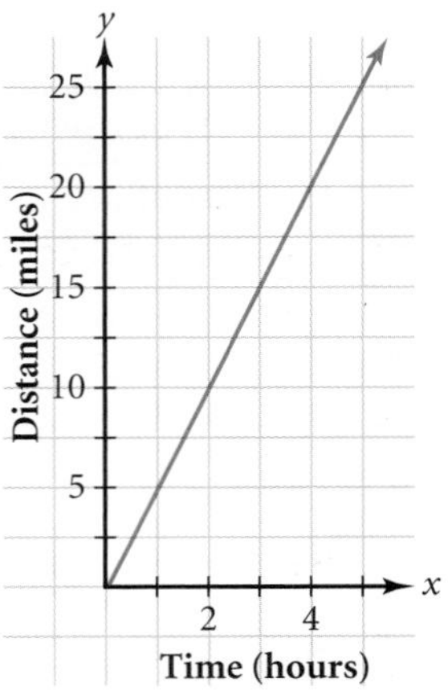

APPLICATIONS

**32. ENTERTAINMENT** The table shows the total costs associated with purchases from an audio CD club, including an initial membership fee of $25. Write an equation for this situation.

| CDs purchased | 0 | 2 | 4 | 6 | 8 | 10 | 12 | 14 |
|---|---|---|---|---|---|---|---|---|
| Cost | $25 | $45 | $65 | $85 | $105 | $125 | $145 | $165 |

**33. ENTERTAINMENT** The table shows the total costs associated with purchases from another audio CD club, including an initial membership fee of $35. Write an equation for this situation.

| CDs purchased | 0 | 2 | 4 | 6 | 8 | 10 | 12 | 14 |
|---|---|---|---|---|---|---|---|---|
| Cost | $35 | $51 | $67 | $83 | $99 | $115 | $131 | $147 |

**34. TECHNOLOGY** Use a graphics calculator to graph the equations from Exercises 32 and 33. Compare the two options. Which CD club offers the better deal and why?

## Look Back

**For Exercises 35 and 36, build a table of values by substituting 1, 2, 3, 4, 5, and 10 for *x*. Sketch a graph on graph paper.** ***(LESSON 1.5)***

**35.** $y = 2x + 1$

**36.** $y = 5x - 1$

APPLICATION

**DEMOGRAPHICS** In a small community, the percent of adults with college degrees has steadily increased, as shown in the table. ***(LESSON 1.6)***

| Year | 1955 | 1960 | 1965 | 1970 | 1975 | 1980 | 1985 | 1990 | 1995 | 2000 |
|---|---|---|---|---|---|---|---|---|---|---|
| Percent | 12 | 15 | 22 | 25 | 35 | 48 | 52 | 52 | 55 | 60 |

**37.** Make a scatter plot of the data.

**38.** Draw a line of best fit to predict the percent in the years 2005 and 2010.

**39.** Write an equation for your line of best fit.

**40.** Is the line a good predictor over the next 50 years? Why or why not?

**Use mental computation to evaluate each expression.** ***(LESSON 2.5)***

**41.** $300 - 196$

**42.** $10 \cdot 30$

**43.** $\frac{480}{16}$

**44.** $1000 \cdot 1000$

**45.** Find the area of a square with a side length of 7 centimeters.

## Look Beyond

| $x$ | 1 | 2 | 3 | 4 | 5 | 6 | 7 | 8 |
|---|---|---|---|---|---|---|---|---|
| $y$ | 1 | 4 | 9 | 16 | 25 | 36 | 49 | 64 |

**46.** Study the table above. Is the relation linear?

**47.** Write an equation for the data in the table above. Graph the equation, and describe the shape of the graph.

# 5.2 Defining Slope

**Objectives**

- Calculate the slope of a line by using the rise and run.
- Calculate the slope of a line from the ratio of the differences in *x*- and *y*-coordinates.

**Why** ***By using the information in the diagram, you can find the slope, or steepness, of the hill for each of the hikers.***

## Finding Slope From Rise and Run

The hikers' strides uphill can be represented by triangles that show the vertical **rise** and horizontal **run** of each stride. Placing the triangles on a coordinate plane provides a way to determine the *slopes* of the lines that represent the hikers' paths up the hill.

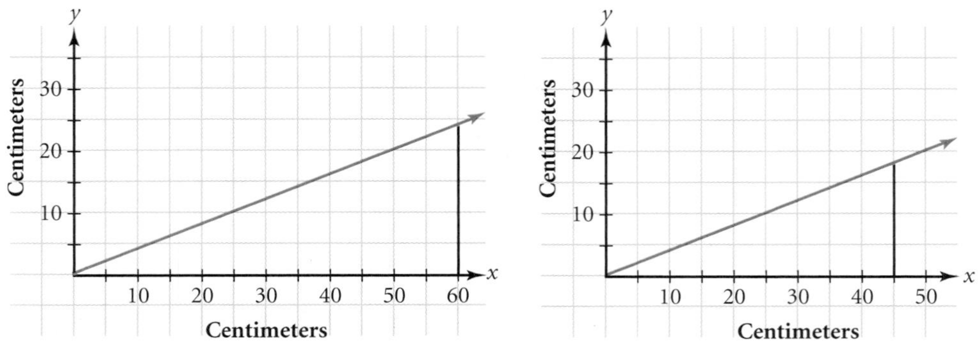

**Slope** is the ratio of vertical rise to horizontal run. Although the two hikers shown on page 226 take different-sized steps, the slope of the hill is the same for both of them.

**EXAMPLE 1** **Find the slopes of the lines that represent each hiker's path up the hill.**

APPLICATION
RECREATION

**SOLUTION**

To calculate the slope, find the ratio of rise to run for each hiker. The horizontal distance of Rich's step (the run) is 60 centimeters, and the vertical distance (the rise) is 24 centimeters. The run for Kara's step is 45 centimeters and the rise is 18 centimeters.

**Rich**

$$\text{slope} = \frac{\text{rise}}{\text{run}} = \frac{24}{60} = \frac{2}{5}, \text{ or } 0.4$$

**Kara**

$$\text{slope} = \frac{\text{rise}}{\text{run}} = \frac{18}{45} = \frac{2}{5}, \text{ or } 0.4$$

The slope is $\frac{2}{5}$, or 0.4.

**TRY THIS** Find the slope of a line that represents the uphill path of a hiker if the rise is 12 inches and the run is 24 inches.

A graph that represents hiking downhill shows a line that is falling from left to right. The rise is negative in this case.

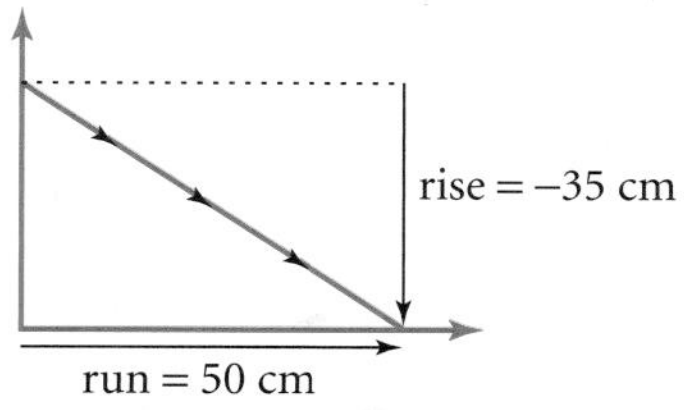


$$\text{slope} = \frac{\text{rise}}{\text{run}} = \frac{-35}{50} = -\frac{7}{10} = -0.7$$

On a coordinate plane, lines with **positive slope** go up as you move from left to right. Lines with **negative slope** go down as you move from left to right.

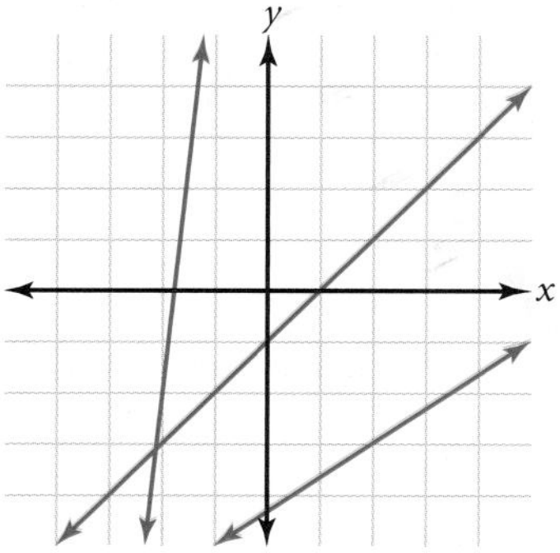


Lines with positive slope

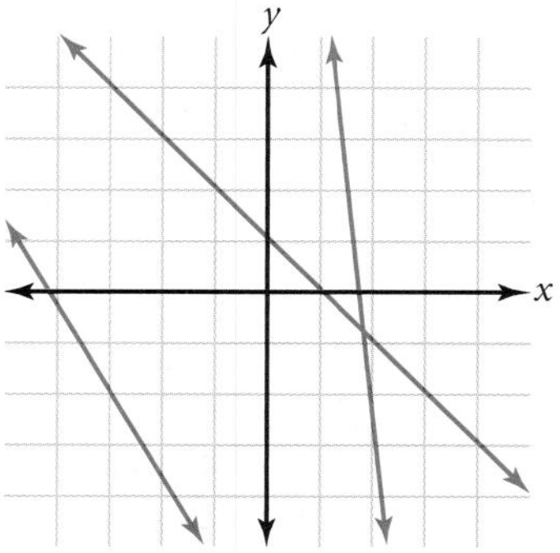


Lines with negative slope

In algebra, slope is defined in terms of the coordinate plane. Real-world information must be translated in order to be studied accurately.

**EXAMPLE 2**

**APPLICATION**
**PHYSICS**

Maya and her class are studying the effects of friction. In an experiment, they lean a meter stick against a wall. The object of the experiment is to find the steepness of the meter stick when it first starts to slide. They observe that it does not slide when the top of the meter stick is 80 centimeters from the tabletop and the bottom is 60 centimeters from the wall. When the steepness is any less, the meter stick slides.

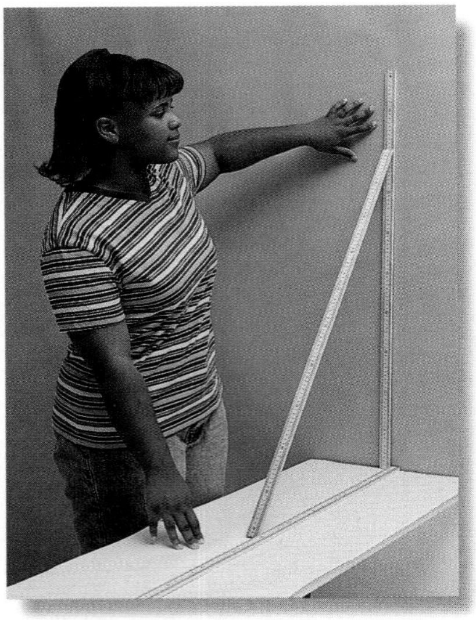

The students recorded this information in a graph. The $y$-axis represents the wall, and the $x$-axis represents the tabletop. **From Maya's perspective, what is the slope of the meter stick just before it begins to slide?**

**SOLUTION**

To find the slope from Maya's perspective, use the points where the line representing the meter stick touches the $x$- and $y$-axes. The run is the number of units to the right (60), and the rise is the number of units downward (–80).

$$\text{slope} = \frac{\text{rise}}{\text{run}} = \frac{-80}{60} = -\frac{4}{3}$$

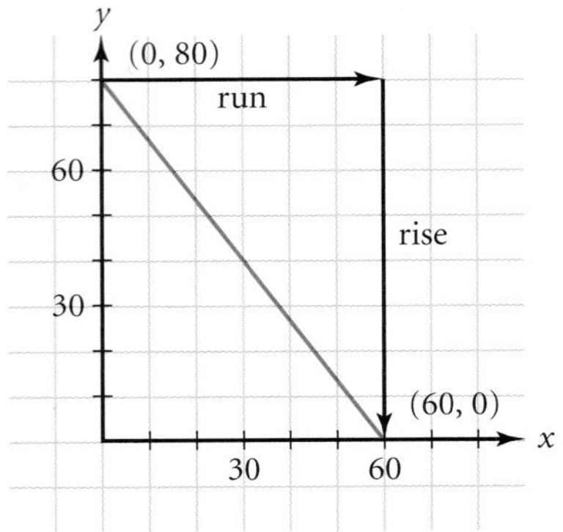


You can find the slope of a line if you know the coordinates of two points on the line.

**Definition of Slope**

Given two points with coordinates $(x_1, y_1)$ and $(x_2, y_2)$, the formula for the slope, $m$, of the line containing the points is

$$m = \frac{y_2 - y_1}{x_2 - x_1}.$$

**EXAMPLE 3** **Find the slope of the line containing the points $M(-2, -6)$ and $N(3, 5)$.**

**SOLUTION**

$$\text{slope} = \frac{y_2 - y_1}{x_2 - x_1}$$

Choose $M$ as the first point, with coordinates $(x_1, y_1)$. Let $N$ be the second point, with coordinates $(x_2, y_2)$. Subtract the $x$-coordinates in the same order as the $y$-coordinates.

$$\text{slope} = \frac{5 - (-6)}{3 - (-2)} = \frac{11}{5}$$

**TRY THIS** Find the slope of the line containing the points $P(2, -3)$ and $Q(-2, 15)$.

**CRITICAL THINKING** What is the effect on the value found for the slope if the $x$-coordinates are not subtracted in the same order as the $y$-coordinates?

**EXAMPLE 4** **Graph the line containing the point (1, 2) and having slope $m = \frac{3}{4}$.**

**SOLUTION**

Graph the point (1, 2).

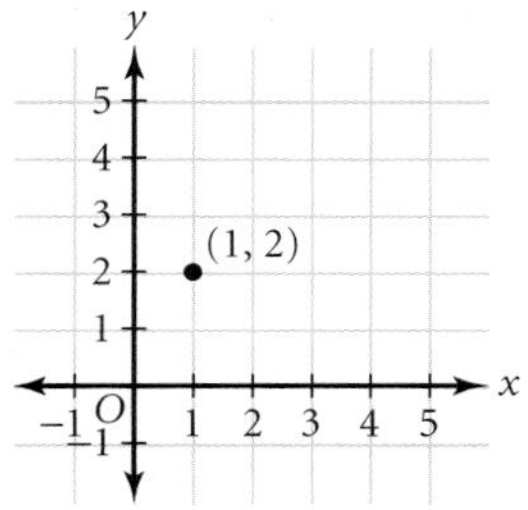


Use the **rise** and then the **run** of the slope to find a second point.

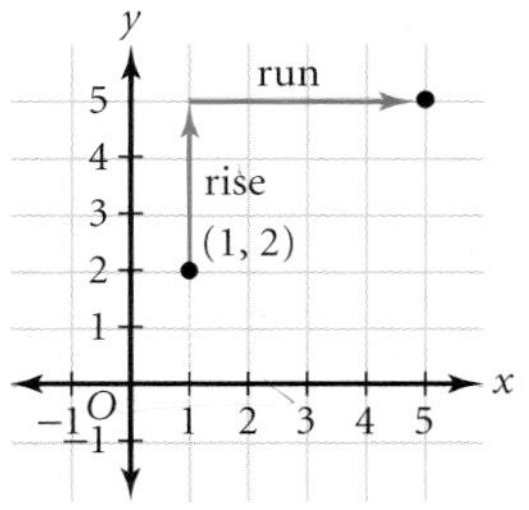


Draw the line containing the two points

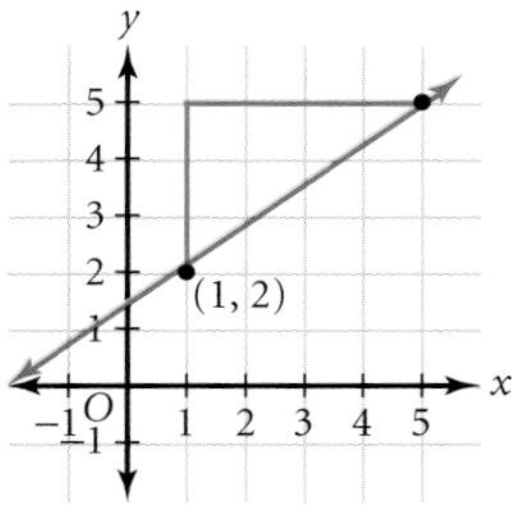


**TRY THIS** Graph the line containing the point (2, 3) and having slope $m = -\frac{5}{2}$.

## Horizontal and Vertical Lines

Horizontal and vertical lines are special cases that are important to the study of slope.

**EXAMPLE 5** **Find the slope of the horizontal line shown at right.**

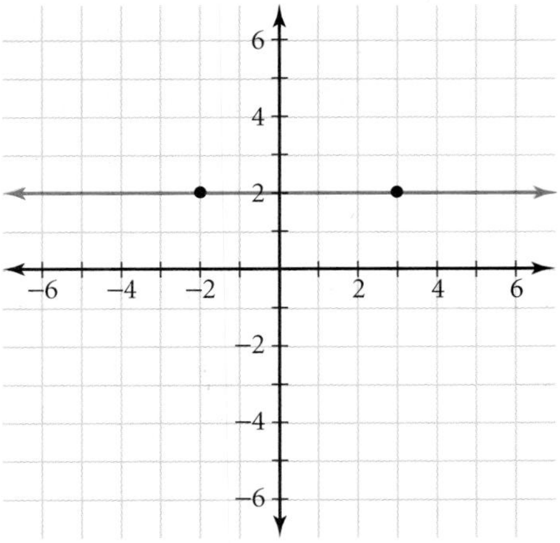

**SOLUTION**

For a horizontal line, the $y$-coordinate of every point on the line is the same.

Choose two points on the line, and use the slope formula.

For the points (3, 2) and (−2, 2):

$$\text{slope} = \frac{2-2}{-2-3} = \frac{0}{-5} = 0$$

The slope of the horizontal line is 0.

**EXAMPLE 6** **Find the slope of the vertical line shown at right.**

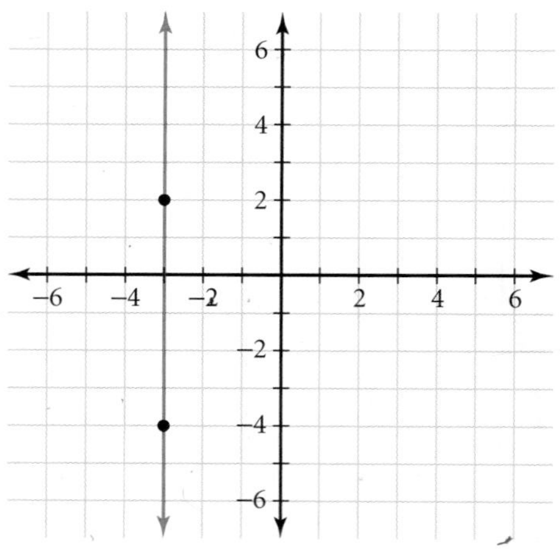


**SOLUTION**

For a vertical line, the *x*-coordinate of every point on the line is the same.

Choose two points on the line, and use the slope formula.

For the points (–3, –4) and (–3, 2):

$$\text{slope} = \frac{2-(-4)}{-3-(-3)} = \frac{6}{0}$$ *Dividing by 0 is not possible.*

The slope of the vertical line is *undefined* because division by 0 is impossible. Therefore, the slope formula cannot be applied to vertical lines.

**The Slopes of Horizontal and Vertical Lines**

The slope of a **horizontal line** is 0.

The slope of a **vertical line** is undefined.

**TRY THIS** Plot each pair of points and draw the line that contains them. Find the slope of each line.

**a.** (5, 6) and (5, –1) **b.** (–1, 4) and (5, 4)

## Exercises

### Communicate

**1.** Explain how to draw a line that has a rise of 4 and a run of 3 and that passes through the origin.

**2.** Describe the slope of a line with a negative rise and a positive run.

**3.** Why is the slope of a vertical line undefined?

**4.** Suppose that *s* is positive. Compare a line with a slope of *s* to a line with a slope of –*s*. Describe how the lines are different.

**5.** Points *R*(5, –3) and *N*(9, –4) are on line *k*. Explain how to find the slope of line *k*.

## Guided Skills Practice

**What is the slope of each line?** ***(EXAMPLES 1 AND 2)***

**6.**

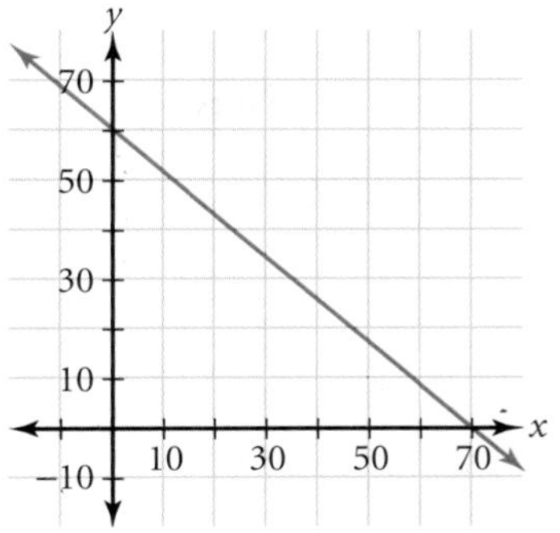


**7.**

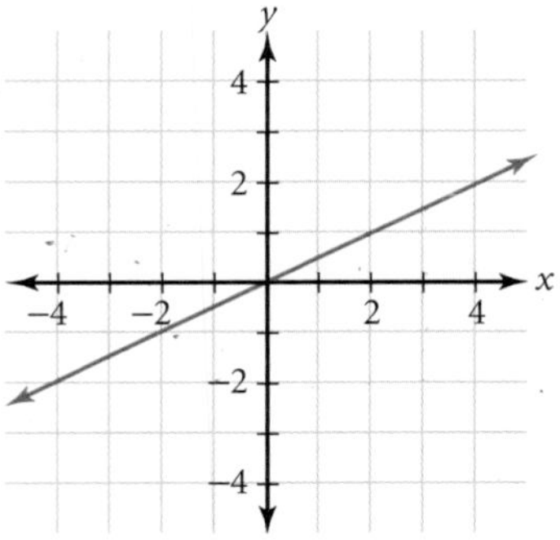


**internet connect**

**Activities Online**

Go To: **go.hrw.com**
Keyword:
**MA1 Marathon**

**8.** Find the slope of a line containing the points (–1, 7) and (8, –3). ***(EXAMPLE 3)***

**9.** Draw the line that has slope $m = \frac{4}{3}$ and that contains the point (–2, –1). ***(EXAMPLE 4)***

**10.** Plot the points and draw the line that contains them. Describe the line. What is its slope?
**a.** (–3, 3) and (6, 3) ***(EXAMPLE 5)*** **b.** (4, 7) and (4, 3) ***(EXAMPLE 6)***

## Practice and Apply

**Find the slope of the lines graphed below.**

**11.**

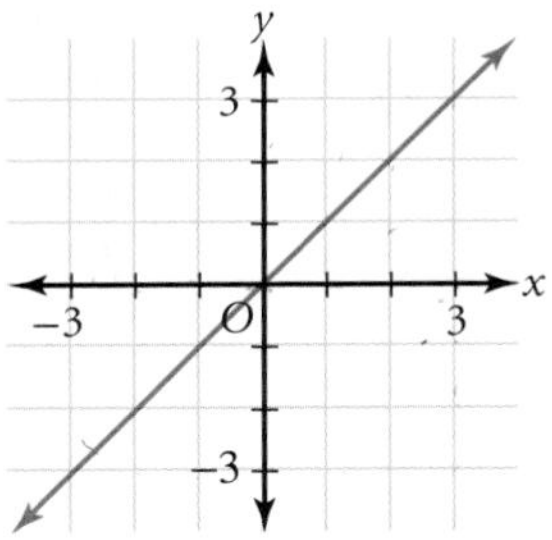


**12.**

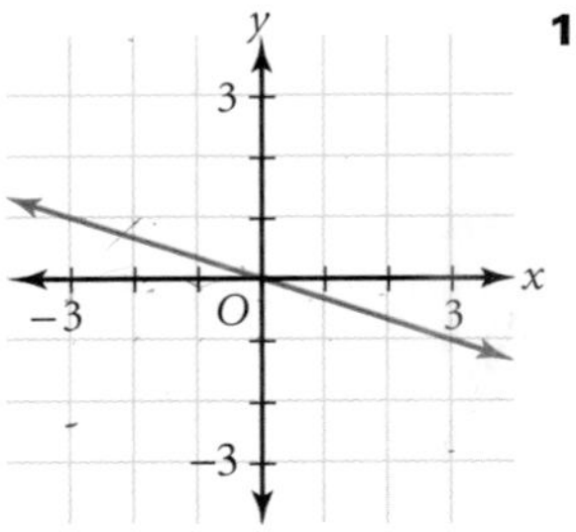


**13.**

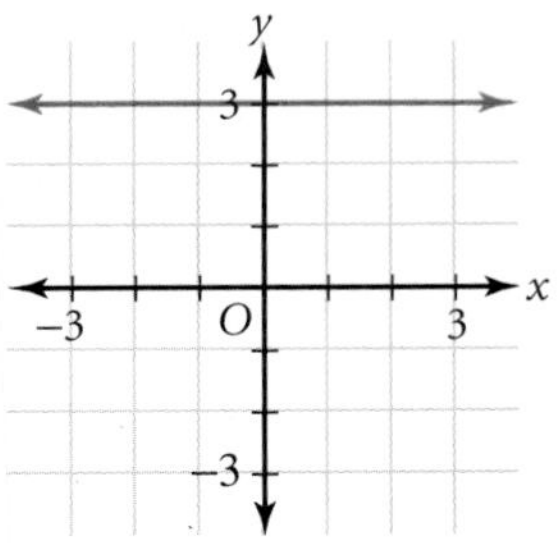


**14.**

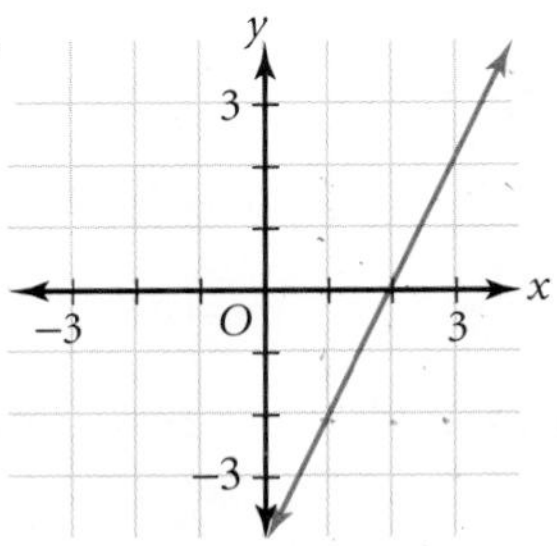


**15.**

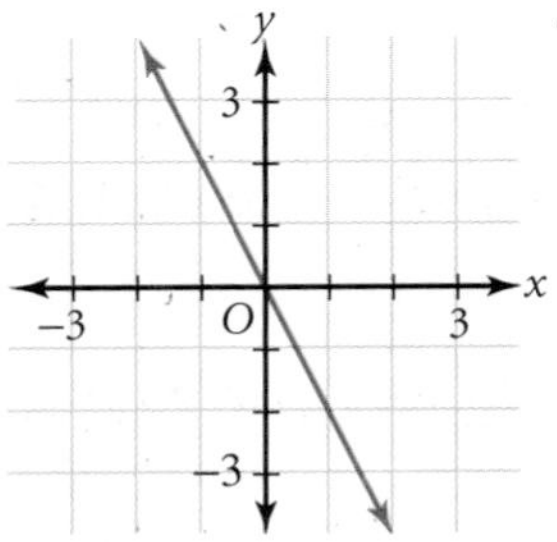


**16.**

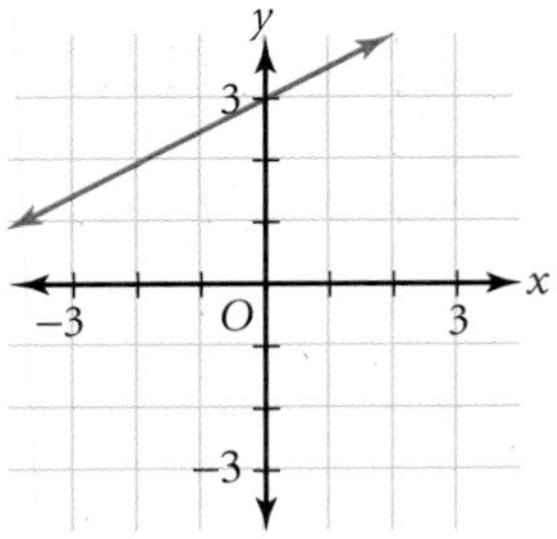


**17.**

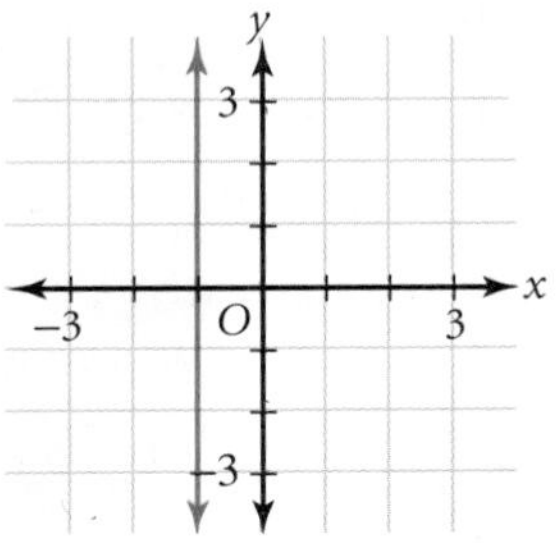


**18.**

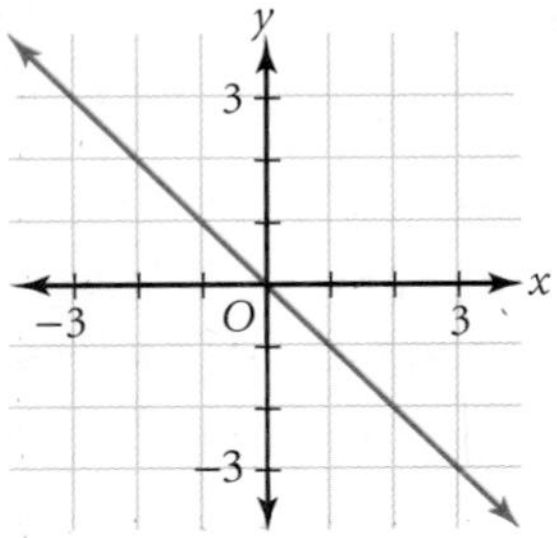


**19.**

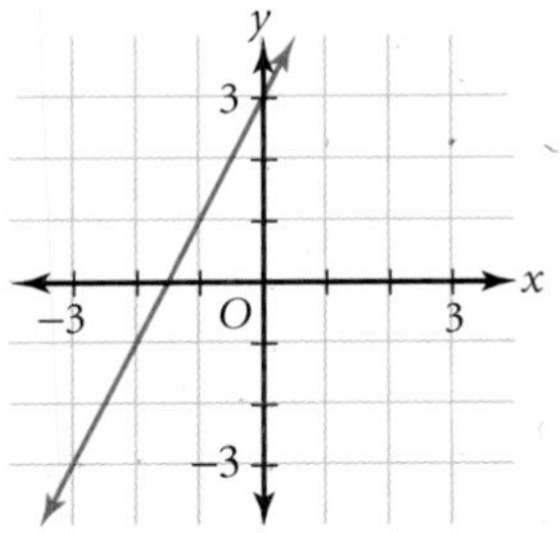

**Find the slope for each rise and run.**

**20.** rise: 1, run: 5

**21.** rise: 3, run: 10

**22.** rise: –1, run: 7

**23.** rise: –6, run: 3

**24.** rise: 5, run: –3

**25.** rise: 5, run: –2

**26.** rise: –9, run: –5

**27.** rise: –4, run: –3

**Find the slope of the line that contains each pair of points.**

internet connect

**Homework Help Online**

Go To: **go.hrw.com**
Keyword:
**MA1 Homework Help** for Exercises 28-35

**28.** $M(9, 6), N(1, 4)$

**29.** $M(23, 1), N(2, 6)$

**30.** $M(2, 24), N(2, 9)$

**31.** $M(8, 10), N(8, 7)$

**32.** $M(-2, 2), N(4, -4)$

**33.** $M(10, 4), N(7, 4)$

**34.** $M(-9, 16), N(-11, 16)$

**35.** $M(2, 7), N(0, 0)$

**Draw the graph of the line that has the given slope and that contains the given point.**

**36.** $(5, 4); m = \frac{2}{3}$

**37.** $(-5, -3); m = \frac{3}{4}$

**38.** $(-3, 4); m = -\frac{1}{2}$

**39.** $(0, 0); m = \frac{1}{2}$

**40.** $(-1, -5); m = 0$

**41.** $(3, -4)$; undefined slope

**42.** $(3, 1); m = -2$

**43.** $(-1, -2); m = -3$

**CHALLENGE**

**44.** A line passes through the origin and the point $(c, d)$. What is the slope of the line?

**45.** A line passes through the point $(a, b)$ and the point $(c, d)$. What is the slope of the line?

**CONNECTION**

**46.** **GEOMETRY** Find or draw at least four pictures of houses or buildings that have roofs with different slopes. Use a ruler to find the slope from the rise and run.

**APPLICATIONS**

**47.** **PHYSICS** If the temperature of a beverage changes from 200°F at a time of 0 minutes to 150°F at 10 minutes, what is the rate of change? In other words, what is the slope of the line connecting the points (0, 200) and (10, 150)?

**48.** **GEOGRAPHY** Denver's elevation is 1 mile (5280 feet). The elevation of Loveland Pass, 70 miles west of Denver, is 11,990 feet. What is the average slope of the land from Denver to Loveland Pass? (Use 70 miles to approximate the run)

*There are 91 steps in each of the 4 stairways on the Mayan-Toltec pyramid at Chichén Itzá, in the Yucatán.*

**CULTURAL CONNECTION: AMERICAS** Mexican pyramids are somewhat different in design from Egyptian pyramids. Mexican pyramids have stairways on each side.

**49.** If the height (rise) of the stairway is 27 meters and the length (run) is 25 meters, what is the slope of the stairway?

**50.** The average rise of a step on the pyramid is 29.7 centimeters, and the average run is 27.5 centimeters. Find the slope of a step, and compare it with the slope of the stairway.

## Look Back

**Evaluate.** *(LESSON 2.1)*

**51.** $|-7|$ **52.** $|50|$ **53.** $-|29|$ **54.** $-|-99|$

**55.** If there are 8 people contributing to a charity, how much would each person have to contribute to raise $98? *(LESSON 2.4)*

**Solve for *x*.** *(LESSON 3.3, 3.4, AND 3.6)*

**56.** $3x + 16 = 19$ **57.** $28 = -4 + 4x$ **58.** $4x + 5 = 2x + 5$

**59.** $-2x + 10 = 5x$ **60.** $a + x = 3x + b$ **61.** $2x - 7 = a + 8$

## Look Beyond

**internet connect**
**Portfolio Extension**
Go To: **go.hrw.com**
Keyword:
**MA1 Golden Ratio**

**62.** Graph $y = 4x$, $y = 4x - 2$, and $y = 4x + 3$ on the same axes. Find the slope of each line. What do you notice about the lines?

**63.** Graph $y = 3x - 2$ and $y = -\frac{1}{3}x - 2$ on the same axes. Find the slope of each line. What do you notice about the lines?

**PORTFOLIO ACTIVITY**

Include this activity in your portfolio.

Use a strip of paper or cloth tape measure to measure the diameter and circumference of several round objects. Record the measurements in a table. Then graph the data with the diameters on the *x*-axis and the circumferences on the *y*-axis. Fit the data with a straight line. How well does the line fit the data? Find the slope of the line. From this data and graph, what conjectures can you make about the relationship between diameter and circumference?

**WORKING ON THE CHAPTER PROJECT**

You should now be able to complete Activity 1 of the Chapter Project on page 264.

# EYEWITNESS MATH

# Watch Your Step!

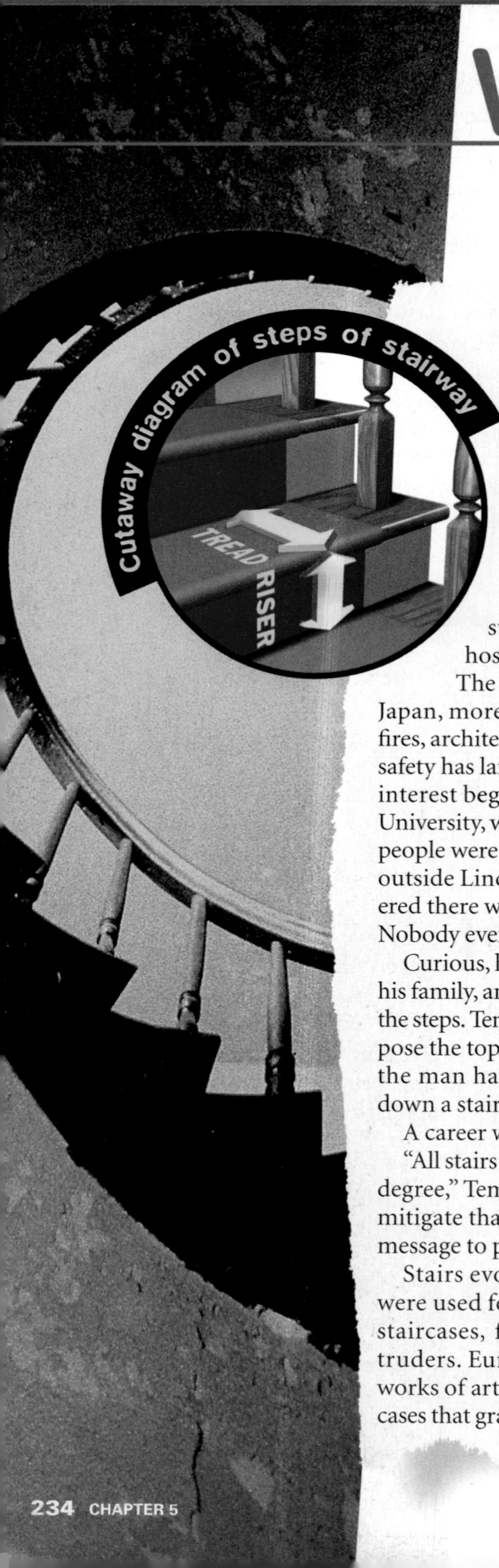


## Atlanta Architect Steps Up Quest for Safe Staircases

### Numerous Accidents, Deaths Push Researcher into Career

ATLANTA— . . . Every year, 1 million Americans seek medical treatment for falls on staircases. About 50,000 are hospitalized and 4,000 die.

The United States isn't unique. In Japan, more people die on stairs than in fires, architect John Templer said. Yet stair safety has largely been ignored. Templer's interest began as a student at Columbia University, when someone asked him why people were always falling down the steps outside Lincoln Center. Templer discovered there was no research on stair safety. Nobody even counted falls.

Curious, he visited Lincoln Center with his family, and his sister-in-law tripped on the steps. Templer called his mentor to propose the topic as a thesis and learned that the man had just broken his leg falling down a stairwell of a subway station.

A career was born.

"All stairs are dangerous; it's a matter of degree," Templer said. "There are ways to mitigate that danger, if we could get that message to people."

Stairs evolved from ladders and first were used for defense. Narrow winding staircases, for instance, hampered intruders. Europeans wrought stairs into works of art, building grand palace staircases that gradually steepened, forcing visitors into a slow, stately pace as they approached royalty.

Most stairs now have 9-inch treads and $8\frac{1}{2}$-inch risers, a size determined around 1850, Templer said. But people today have bigger feet that hang over the edges of stairs, throwing them off balance, he said.

Stairs also are too high, Templer concluded after experiments in which he forced volunteers to trip on collapsible stairs. They were harnessed so they didn't tumble all the way to the floor, but Templer used videos to simulate how they would have landed.

He wants building codes revised for stairs with 11-inch treads and 7-inch risers. His proposal prompted a lobbying blitz from the National Association of Home Builders, which contends that larger stairs would add at least 150 square feet and $1,500 in costs to a house.

NAHB's Richard Meyer dismissed Templer's work, saying people fall when stairs are improperly lighted, have loose carpeting, or have objects placed in the way.

That's true, too, Templer said. But he said his experiments, funded by the National Science Foundation, prove stair shape is a large problem.

*By Lauran Neergaard*
*—THE ASSOCIATED PRESS*

## Count Your Way to Stair Success

The riser height (rise) and the tread width (run) determine how comfortable the stairs will be to use. If the rise and run are too great, the stairs will strain your legs and be hard to climb. If the rise and run are too small, you may whack your toe on the back of each step.

Over the years, carpenters have determined that tread width times riser height should equal somewhere between 72 and 75 when the measurements are in inches.

On the main stair, the maximum rise should be no more than $8\frac{1}{4}$ inches and the minimum run should be no less than 9 inches.

To determine how many steps , or treads you need, measure from the top of the finished floor on the lower level to the top of the finished floor on the upper level.

To figure the rise and run in a house with 8-foot ceilings, for instance, start by figuring the total vertical rise. By the time you add floor joists, subfloor, and finish floor, the total is about 105 inches.

A standard number of treads in a stair between first and second floors is 14. One hundred-five divided by 14 equals $7\frac{1}{2}$. That means the distance from the top of each step to the top of the next step will be $7\frac{1}{2}$ inches.

With a riser height of $7\frac{1}{2}$ inches, tread width (run) should be at least 9 inches. Ten inches is a more comfortable run; when you multiply $7\frac{1}{2}$ inches by 10 inches, you get 75—within the conventional *ratio* of 72 to 75. With fourteen 10-inch treads, the total run of the stairs will be 140 inches. In other words, the entire stair will be 105 inches tall and 140 inches deep.

You can alter the rise and run to some extent. If you used 15 risers instead of 14, for instance, the rise would be 7 inches, and the tread width would be $10\frac{1}{2}$ inches ($7$ times $10\frac{1}{2}$ equals 73.5, within the rise and run guidelines).

*By Karol V. Menzie*
*and Randy Johnson*
*—BALTIMORE SUN*

### Cooperative Learning

1. In the news article above, is the term *ratio* used correctly? Explain.
2. Find the angle of each of the following staircases. Draw the triangle and measure the angle.
   a. The 105-inch-by-140-inch staircase described in the article above.
   b. The same staircase changed as described in the article to include 15 risers instead of 14.
   c. The same staircase changed to meet Templer's preference for 7-inch risers and 11-inch treads.
3. Which of the three staircases in Step 2 would take up the most room? Explain.
4. Suppose that you did not want the building code for stairs to change. Write an argument (with numbers and a diagram) to show that larger stairs would add at least 150 square feet to a typical house. Make up your own estimates of the size of a typical house.
5. Measure the rise and run of stairs in your school or home. Do they match the information in the two news articles about standard stairs?

5.3

# Rate of Change and Direct Variation

*When aloft, commercial planes maintain a fairly constant speed. This makes it possible to model their motion during a flight with a linear graph.*

## Objectives

- Find the rate of change from a graph.
- Solve and graph direct-variation equations.

**Why** *Rate of change and direct variation are two concepts directly related to slope, which has many real-world applications.*

## Rate of Change

APPLICATION
**AVIATION**

The graph at right shows the relationship between the variables for a plane moving at a constant rate.

The **rate of change**, or speed, of the plane can be determined from the graph. After 20 minutes, the plane has traveled approximately 200 miles. Use a formula for rate of change as follows:

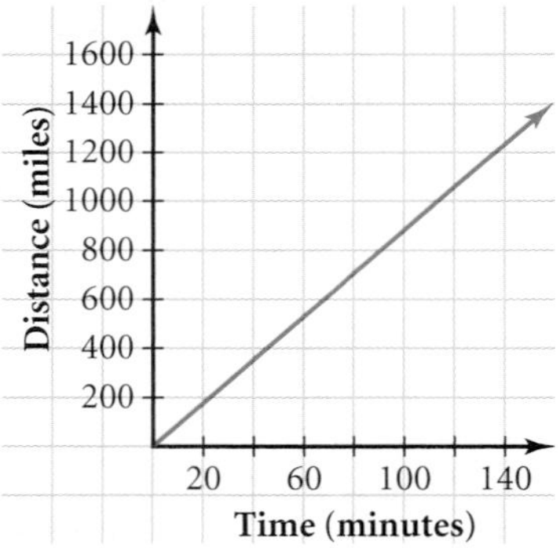


$$\text{rate of change} = \frac{\text{change in distance}}{\text{change in time}} \approx \frac{200}{20} \approx \frac{10}{1}$$

The plane travels approximately 10 miles for every 1 minute that it travels, or approximately 600 miles per hour.

Notice that the rate of change was determined by calculating the rise, 200 miles, over the run, 20 minutes.

In fact, the slope of a linear graph is the same as the rate of change of the linear function that it represents.

**EXAMPLE 1**

**APPLICATION**
**TRANSPORTATION**

The graph shows the distance a car travels at a constant speed. **What is the speed of the car?**

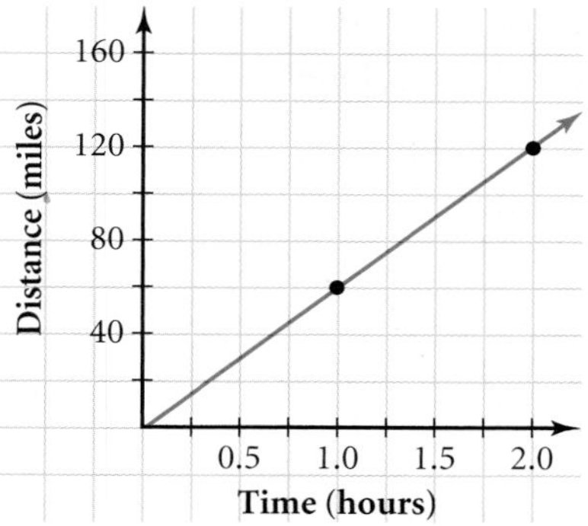


**SOLUTION**

Select two points, such as the ones shown on the graph. Use the slope formula.

Let $(x_1, y_1)$ be (1.0, 60).
Let $(x_2, y_2)$ be (2.0, 120).

$$\frac{\text{change in distance}}{\text{change in time}} = \frac{(y_2 - y_1)}{(x_2 - x_1)} = \frac{120 - 60}{2.0 - 1.0} = \frac{60 \text{ miles}}{1.0 \text{ hour}}$$

The car travels 60 miles for every 1 hour it travels.

**TRY THIS**

Find the rate of change of the ramp represented in the drawing at right. How many inches does the ramp rise for every foot of length?

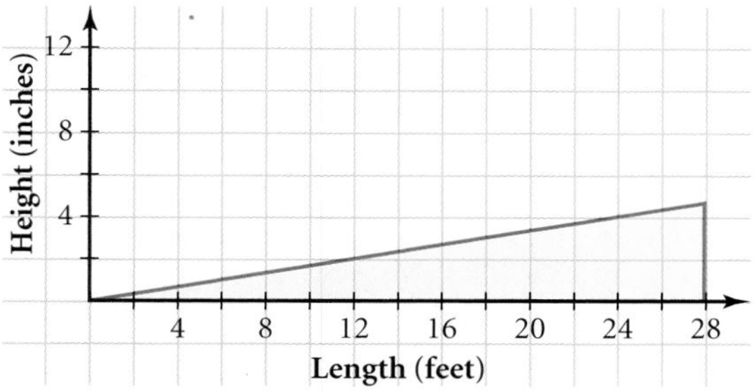


**EXAMPLE 2**

**TECHNOLOGY**
**GRAPHICS CALCULATOR**

The graph shown at right was made from data collected by a motion detector. In the experiment, a student walked in a straight line away from the motion detector. **Describe the movement of the student by identifying the rates of change shown by the graph.**

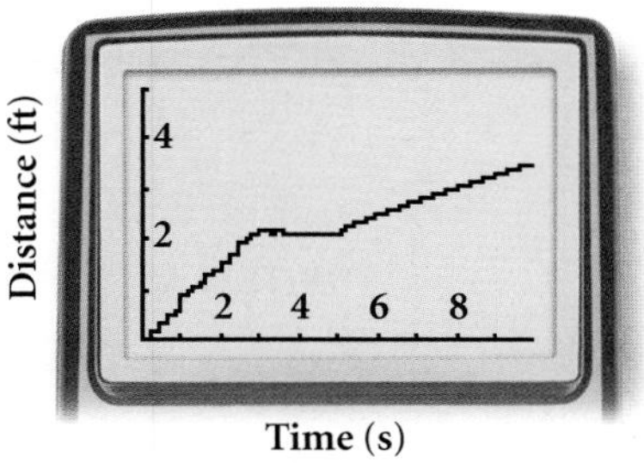


*Time in seconds is indicated on the horizontal axis and distance in feet is indicated on the vertical axis.*

**SOLUTION**

The graph of the data can be divided into three segments. Using the formula rate of change $= \frac{\text{change in distance}}{\text{change in time}}$, the rate of change for the first segment is $\frac{2}{3}$ feet per second. The middle segment is a horizontal line, so the rate of change is 0 feet per second. The rate of change for the last segment in the graph is approximately $\frac{1}{3}$ foot per second. Thus, the student walked away from the motion detector at $\frac{2}{3}$ feet per second for 3 seconds. Then the student stood still for 2 seconds. Finally, the student walked away from the motion detector at $\frac{1}{3}$ foot per second for 4 seconds.

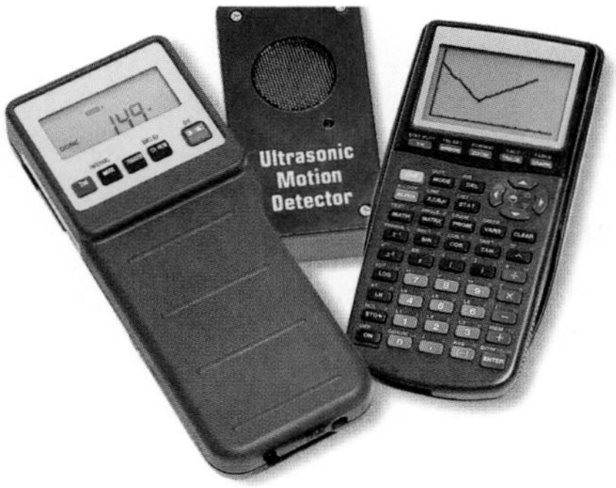


*Some calculators can be used with equipment like a motion detector to collect data for analysis.*

## Direct Variation

*A Japanese high-speed train passes near Mount Fuji.*

One type of rate of change is direct variation. An example of a rate of change that is a direct variation is average speed. For example, using $t$ for time in minutes and $d$ for distance in miles, the equation for a high-speed train traveling at 3 miles per minute can be written as follows:

$$d = 3t$$

Notice that 3 is a constant. In terms of direct variation, you can state the equation as follows:

"The distance traveled by a high-speed train moving at a constant speed of 3 miles per minute varies directly as the time traveled."

**Direct Variation**

If $y$ varies directly as $x$, then $y = kx$, or $\frac{y}{x} = k$, where $k$ is the **constant of variation.**

**EXAMPLE 3** **If $y$ varies directly as $x$ and $y = 4$ when $x = 0.5$, find the constant of variation and write an equation of direct variation.**

**SOLUTION**

Since $\frac{y}{x} = k$, substitute 4 for $y$ and 0.5 for $x$ to find the constant of variation.

$$\frac{4}{0.5} = k \qquad 8 = k \qquad k = 8$$

Use the constant of variation, $k$, to write an equation of direct variation. Substitute 8 for $k$.

$$y = 8x$$

**TRY THIS** If $y$ varies directly as $x$ and $y = 10$ when $x = 2$, find the constant of variation and write an equation of direct variation.

**EXAMPLE 4** **If $y$ varies directly as $x$ and $y = 27$ when $x = 6$, find $x$ when $y = 45$.**

**SOLUTION**

In a direct variation, all values of $\frac{y}{x}$ must equal some constant, $k$. Therefore, all values of $\frac{y}{x}$ must be equal. You can set up and solve a proportion to find $x$.

$$\frac{27}{6} = \frac{45}{x}$$
$$27x = 270$$
$$x = 10$$

**TRY THIS** If $y$ varies directly as $x$ and $y = 35$ when $x = 7$, find $y$ when $x = 84$.

**EXAMPLE 5** **Hooke's law** relates the distance a spring stretches to the amount of force applied to the spring. Hooke's law is expressed mathematically as follows:

$$F = kd$$

**APPLICATION**
**PHYSICS**

In the equation, the variable $d$ is the distance of the stretch in meters and the variable $F$ is the amount of the force in Newtons. (A force of one Newton is equivalent to the weight of one medium-sized apple.) The letter $k$ is a constant, meaning that $k$ has a definite value that does not change or vary in the equation for a particular spring. The actual value of $k$ for a given spring is determined by experiment.

A force of 20 Newtons stretches a spring a distance of 0.5 meter. **Find the constant, *k*, for the spring.**

**SOLUTION**

Substitute the values for $F$, the force applied, and $d$, the distance of the stretch, into the equation for Hooke's law. Then solve for $k$.

$$F = kd$$
$$20 = k \cdot (0.5)$$
$$k = 20 \div 0.5$$
$$k = 40$$

The spring constant is 40.

**TRY THIS** A spring with a spring constant, $k$, of 28 is stretched a distance of 0.25 meter by a lead weight. Find the force in Newtons exerted by the lead weight.

## *Activity*
## Exploring the Graphs of Direct Variations

**You will need:** graph paper

1. Complete the table of values below for the equation $y = 0.5x$.

| $x$ | 0 | 1 | 2 | 3 | 4 | 5 |
|---|---|---|---|---|---|---|
| $y$ | ? | ? | ? | ? | ? | ? |

2. Use the points in the table to draw a graph of the direct variation.
3. What is the shape of the graph? Does it pass through the origin?
4. Select two points on the graph and use them to find the slope of the graph.

**CHECKPOINT ✔**

5. Repeat Steps 1–4 with several different constants for the direct variation. Make conjectures about the shape of the graph, whether it passes through the origin, and its slope.

The graph of a direct variation is a straight line that passes through the origin. The constant of variation is the slope of the graph.

# Exercises

## Communicate

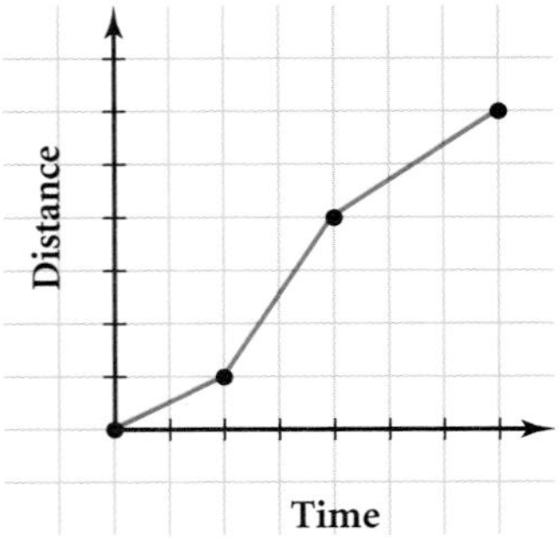


1. Allen is riding his mountain bike over three different types of terrain. Examine the graph. When is he riding uphill? downhill? on level ground?
2. Describe the slope of a hill in terms of a rate of change.
3. For a direct-variation equation, explain how the slope of the graph and the constant of variation are related.
4. **FUNDRAISING** At the homecoming football game, the senior class makes \$1.50 for every program sold. Describe the direct variation for the amount of money the senior class will make from selling football programs.

## Guided Skills Practice

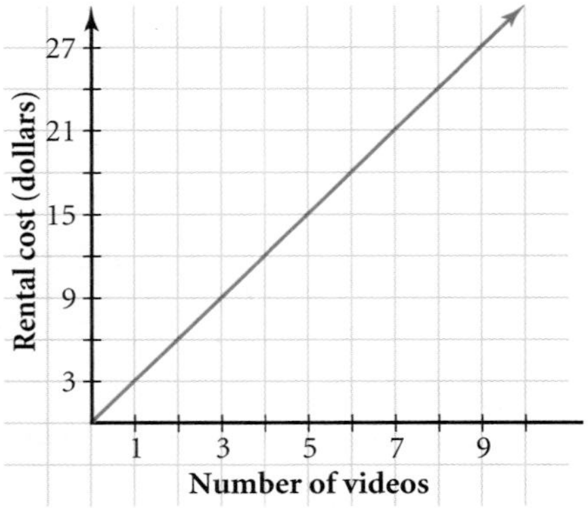


5. The graph at right shows the cost of renting videos. What is the rate of change, or cost per video? ***(EXAMPLE 1)***

**APPLICATIONS**

**INCOME** The graph below represents an employee's daily wages as a function of the number of hours that he works in a day. ***(EXAMPLE 2)***

Daily Wages

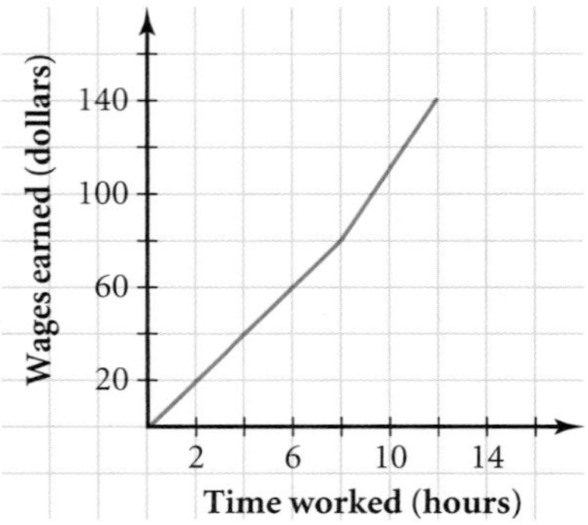


6. Use the graph to find the employee's hourly rate of pay if he works no more than 8 hours per day.
7. Use the graph to find his hourly rate of pay after he has worked 8 hours in a given day (overtime pay).
8. If $y$ varies directly as $x$ and $y = 7$ when $x = 28$, find the constant of variation and write an equation of direct variation. ***(EXAMPLE 3)***
9. If $y$ varies directly as $x$ and $y = 30$ when $x = 15$, find $x$ when $y = 50$. ***(EXAMPLE 4)***
10. **PHYSICS** A force of 8 Newtons stretches a spring a distance of 0.25 meter. Find the constant, $k$, for the spring. ***(EXAMPLE 5)***

internet connect
**Activities Online**
Go To: **go.hrw.com**
Keyword:
**MA1 Import Export**

## Practice and Apply

**APPLICATIONS**

**11. RENTALS** The graph at right shows the monthly cost of renting videos, including an initial membership fee. Find the rate of change, or cost per video.

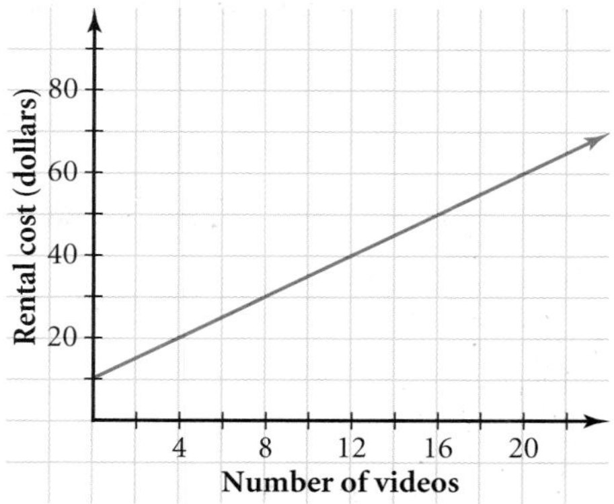


**12. INCOME** The graph at right shows Amy's wages as a function of the number hours she works in a week. Use the graph to find Amy's hourly pay rate if she works no more than 40 hours in a week and after she has worked more than 40 hours in a given week (overtime pay).

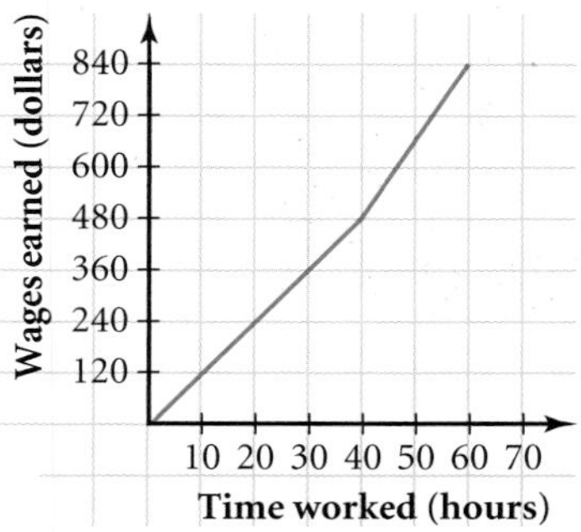


**For the following values of *x* and *y*, *y* varies directly as *x*. Find the constant of variation, and write an equation of direct variation.**

**13.** $y$ is 14 when $x$ is 2.

**14.** $y$ is 2.6 when $x$ is 13.

**15.** $y$ is 5 when $x$ is 8.

**16.** $y$ is 3 when $x$ is 5.

**17.** $y$ is 132 when $x$ is 12.

**18.** $y$ is –2 when $x$ is 5.

**19.** $y$ is 4.5 when $x$ is 15.

**20.** $y$ is 84 when $x$ is 21.

**21.** $y$ is 84 when $x$ is 12.

**22.** $y$ is 2 when $x$ is 8.

internet connect

**Homework Help Online**

Go To: **go.hrw.com**
Keyword:
**MA1 Homework Help** for Exercises 23-26

**For the following values of *x* and *y*, *y* varies directly as *x*. Find the missing value.**

**23.** $y$ is 14 when $x$ is 2. Find $x$ when $y$ is 21.

**24.** $y$ is 5 when $x$ is 8. Find $y$ when $x$ is 28.

**25.** $y$ is 56 when $x$ is 7. Find $x$ when $y$ is 8.

**26.** $y$ is 27 when $x$ is 3. Find $x$ when $y$ is 4.5.

**CONNECTIONS**

**27. GEOMETRY** The perimeter, $p$, of a square varies directly as the length, $l$, of a side. Write an equation of direct variation that relates the two variables.

**28. GEOMETRY** If the base area of a cylindrical container is kept constant, the volume of the container varies directly as the height. The table shows the volume and height for various cylinders with the same base area. Find the constant of variation. Then find the volume if the height is 12 centimeters.

| Height (cm) | 3 | 4 | 5 | 6 |
|---|---|---|---|---|
| Volume ($cm^3$) | 27 | 36 | 45 | 54 |

**APPLICATIONS**

**29. INCOME** Suppose that the hourly wage at a job is \$8. Let $x$ represent the number of hours, and let $y$ represent the total amount earned. Write an equation of direct variation, and graph the equation.

**30. AVIATION** An airplane ascends at a rate of 2000 feet per minute. Write an equation of direct variation, and graph the equation.

**31. HEALTH** Suppose that you are walking at 4 miles per hour. Let $x$ represent the time you spend walking, and let $y$ represent the distance you walk. Write an equation of direct variation, and graph the equation.

**RECREATION** On the second day of a hike, two hikers start out from different camps at the same time. The graph shows the hikers' progress on day two.

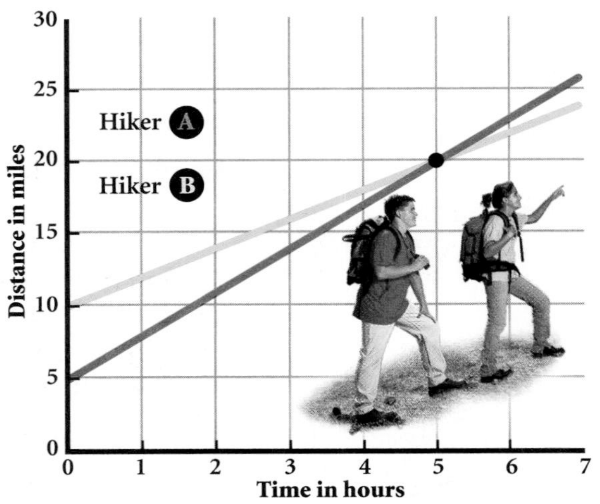


**32.** Find the rate of travel on day two for hiker A.

**33.** Find the rate of travel on day two for hiker B.

**34.** After how many hours on day two will the hikers have traveled the same total distance?

**CHALLENGE**

**35.** After how many hours on day two will hiker A be 3 miles ahead of hiker B in total distance?

**APPLICATION**

**36. PHYSICS** Find a sturdy rubber band or spring and suspend it so that you can measure its length without any weight on it. Find several weights that the rubber band or spring will hold. Hang the weights on the rubber band or spring, and measure the distance of the stretch. Make a table of the weights and the distances of the stretches. Make a scatter plot, and find the line of best fit. Use a graphics calculator, if available, to find the equation of the line of best fit. Write a description of your findings.

## Look Back

**Graph each point on a coordinate plane.** *(LESSON 1.4)*

**37.** $A(-2, 3)$ **38.** $B(0, 8)$ **39.** $C(-5, -1)$

**Make a table for each equation and find values of $y$ by substituting 1, 2, 3, 4, and 5 for $x$.** *(LESSON 1.5)*

**40.** $y = 7x$ **41.** $y = \frac{1}{2}x - 3$

**42.** A ski boat crosses a lake at an average speed of 35 mph. Write an equation, make a table, and draw a graph for the distance the boat travels for times from 0 to 3 hours. *(LESSON 1.5)*

**Use the Distributive Property to show that the equations below are true.** *(LESSON 2.5)*

**43.** $-7x + 11x = 4x$ **44.** $8h - 9h = -h$ **45.** $3d + 5d = 8d$

**Simplify.** *(LESSON 2.5)*

**46.** $(3x + 7) + (4x - 9)$

**47.** $(9c + 3) - (-8c + 2)$

**48.** $(-3d - 12) - (-2d + 4)$

**49.** $(6h - 2) - (4h + 3)$

**50.** $(-5t + 2) + (t - 12)$

**51.** $(f + 3) - (6f - 45)$

**52.** $(2a + b - c) + (a - 3b + 5c)$

**53.** $(b + 9f - 2g) - (5b - 4f + 12g)$

## Look Beyond

**internet connect**
**Portfolio Extension**
Go To: **go.hrw.com**
Keyword:
**MA1 Ordered Pairs**

**54.** Graph each function below on a coordinate plane. Let the horizontal axis represent $x$ and the vertical axis represent $f(x)$. Describe each graph. Are the graphs linear?

**a.** $f(x) = 2x^2$  **b.** $f(x) = \frac{2}{x},\ x \neq 0$  **c.** $f(x) = \sqrt{x},\ x \geq 0$

**PORTFOLIO ACTIVITY**

*A rain gauge is used to determine rainfall amounts.*

Include this activity in your portfolio.

1. Describe three sets of real-world data that can be modeled as ordered pairs on a coordinate plane (for example, regional rainfall amounts per month).
2. Pick one of your sets of data from Step 1 and draw a graph. Connect the data points on your graph. Describe the shape of the line connecting the points. Is it straight, curved, or jagged? (Recall that drawing a line to connect points is a tool to help visualize the relationship and does not mean that the relationship holds true for every point on the line.)
3. Using the data you graphed in Step 2, redraw the graph without connecting the data points—this creates a scatter plot of your data.
4. Use a clear ruler to draw a line of best fit for your data.
5. Make a prediction for a data point that is not on your graph. For example, if you chose to model monthly rainfall totals, predict the rainfall for the next month.
6. Explain why you are or are not able to make a prediction from your graph. Do you have confidence in your prediction? Why or why not?

**WORKING ON THE CHAPTER PROJECT**

You should now be able to complete Activity 2 of the Chapter Project on page 264.

# 5.4 The Slope-Intercept Form

## Objectives

- Define and explain the components of the slope-intercept form of a linear equation.
- Use the slope-intercept form of a linear equation.

**Why** *The slope-intercept form allows you to write an equation for a line when you can identify only the slope and y-intercept.*

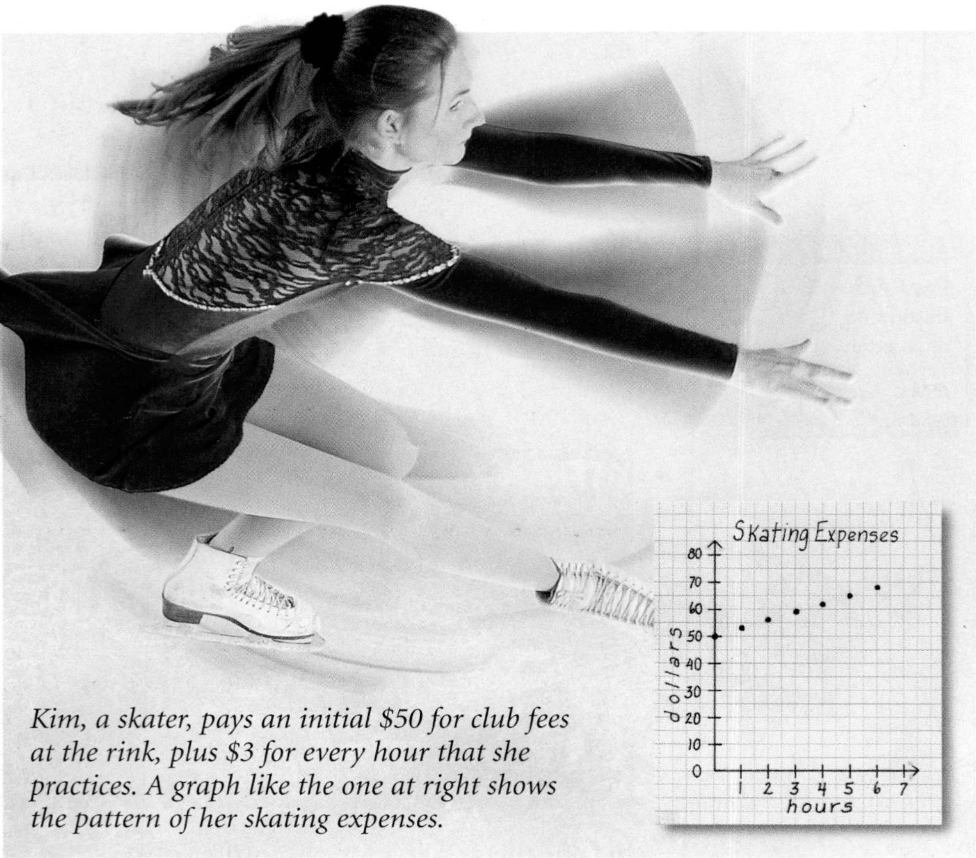


*Kim, a skater, pays an initial $50 for club fees at the rink, plus $3 for every hour that she practices. A graph like the one at right shows the pattern of her skating expenses.*

APPLICATION
SPORTS

A graph representing what Kim must pay for her practice time does *not* pass through the origin. This is because she has to pay an initial $50 in club fees each month before she can practice.

PROBLEM SOLVING

Examine the data. Let $x$ be the number of hours Kim practices. Let $y$ be the total cost for her skating expenses for the month. **Make a table** of $x$- and $y$-values from the data.

| **Time in hours, $x$** | 0 | 1 | 2 | 3 | 4 | 5 | 6 | 7 | 8 | 9 | 10 |
|---|---|---|---|---|---|---|---|---|---|---|---|
| **Cost in dollars, $y$** | 50 | 53 | 56 | 59 | 62 | 65 | 68 | 71 | 74 | 77 | 80 |

Differences: 3 3 3 3 3 3 3 3 3 3

Notice that the first differences are constant. Constant first differences indicate a linear pattern. Therefore, an equation that represents Kim's practice expenses is linear. In this case, the total cost is a function of the number of hours that Kim practices. She uses a time card that keeps precise track of her practice times so that the cost can also be calculated for fractional parts of hours.

To find the slope of the graph that models Kim's expenses, pick any two points from the table. For any points that you choose, the following is true:

$$\text{slope} = m = \frac{\text{change in } y}{\text{change in } x} = \frac{3}{1} = 3$$

The slope represents the fact that Kim must pay \$3 for every hour she uses the skating rink.

The table also shows that when $x$ is 0, $y$ is 50. Thus, (0, 50) is on the graph. The $y$-value of the point where the line crosses the $y$-axis is called the ***y*-intercept**. In this case, the $y$-intercept of 50 represents the initial \$50 fee that Kim must pay.

In the study of linear equations, the $y$-intercept is usually represented by $b$ and the slope by $m$. The graph of an equation of the form $y = mx + b$ has a slope of $m$ and crosses the $y$-axis at $b$. If you substitute 3 for $m$ and 50 for $b$, an equation that models the cost of Kim's practice becomes $y = 3x + 50$.

**Slope-Intercept Form**

The slope-intercept form for a line with a slope of $m$ and a $y$-intercept of $b$ is

$$y = mx + b.$$

The ordered pairs $(x, y)$ that are solutions to the equation $y = mx + b$ represent a function. For any such function, the set of independent-variable values, in this case $x$, is the domain. The set of dependent-variable values, in this case $y$, is the range. You will have to consider what values are reasonable for the domain and range of any particular problem.

**EXAMPLE 1**

**APPLICATION**
**SPORTS**

The equation of the line that represents Kim's monthly skating costs is $y = 3x + 50$. **Use the slope-intercept form to construct the graph of the equation.**

**SOLUTION**

The form is $y = mx + b$. Since $b$ is 50, measure 50 units up from the origin on the $y$-axis and graph a point.

From that point, measure the run and then the rise of the slope to locate a second point.

The slope, $m$, is 3. You can use any equivalent of $\frac{3}{1}$ to measure the rise and run. The $y$-axis is marked in intervals of 10 units, so use $\frac{30}{10}$. From the initial point (0, 50), move 10 units to the right and then 30 units up. Graph the second point. Draw a line through the points.

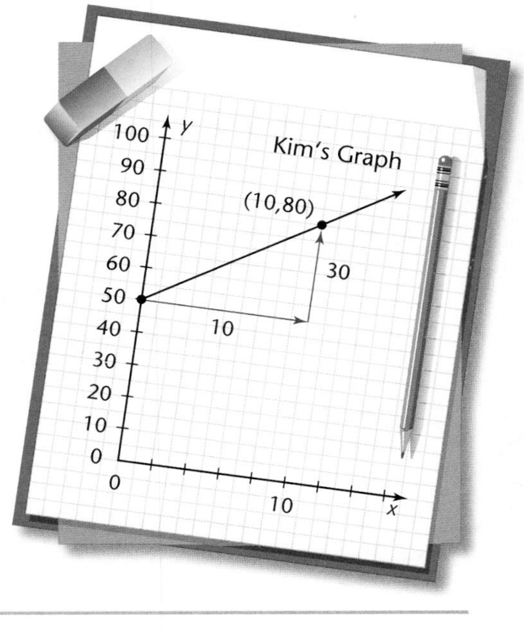

**TRY THIS** Use the slope-intercept form to construct the graph of the equation $y = 2x - 8$.

**CRITICAL THINKING** Why do you think the line in Kim's graph on page 245 does not extend past the $y$-axis? What do you think is a realistic maximum value for the domain? In other words, what is a realistic maximum number of hours of skating practice for one month?

*Activity*

## Exploring Graphs of Linear Functions

TECHNOLOGY
GRAPHICS CALCULATOR

**You will need:** graph paper or a graphics calculator

Graph the following linear functions on the same axes:

$y = 2x$ $\quad y = 2x + 3$ $\quad y = 2x + 5$ $\quad y = 2x - 3$

1. How are the graphs of these lines alike?
2. How are the graphs of these lines different?
3. Predict what the graph of $y = 2x - 4$ looks like.
4. Check your prediction by graphing.

CHECKPOINT ✔

5. Explain how changing the value of $b$ in an equation of the form $y = mx + b$ affects the graph of the function.

## From Two Points to an Equation

When you know two points on a line, you can determine the equation for that line. First calculate the slope, $m$, by using the slope formula. Then calculate $b$ by using the slope-intercept form and one of the points.

Recall from Lesson 5.2 that for two points with coordinates $(x_1, y_1)$ and $(x_2, y_2)$, the formula for the slope is as follows:

$$m = \frac{\text{change in } y}{\text{change in } x} = \frac{\text{difference in } y}{\text{difference in } x} = \frac{y_2 - y_1}{x_2 - x_1}$$

**EXAMPLE 2**

APPLICATION
SPORTS

A graph shows that after 3 hours of skating the total cost is \$57. It also shows that after 5 hours of skating the total cost is \$65. **Write an equation in slope-intercept form for the line that models this situation.**

**SOLUTION**

You can represent the data as points by writing the time and cost as ordered pairs, (5, 65) and (3, 57). To write the equation of the line from these two points, substitute the values into the slope formula.

$$m = \frac{y_2 - y_1}{x_2 - x_1}$$

$$m = \frac{57 - 65}{3 - 5} = \frac{-8}{-2} = 4$$

Substitute 4 for $m$ in $y = mx + b$.

$$y = 4x + b$$

Then choose either point, and substitute the values for $x$ and $y$ into the equation. If you use the point (5, 65), substitute 5 for $x$ and 65 for $y$. Then solve for $b$.

$$\begin{aligned} y &= 4x + b \\ 65 &= 4(5) + b \\ 65 &= 20 + b \\ 45 &= b \end{aligned}$$

Now substitute 4 for $m$ and 45 for $b$ in $y = mx + b$. The equation in slope-intercept form for the line is $y = 4x + 45$.

**TRY THIS** Write an equation in slope-intercept form for the line containing the points (3, 3) and (5, 7).

**CRITICAL THINKING** Explain why the slope-intercept form is a helpful way to write a linear equation. Is $3x - 4y = 7$ a linear equation? Why or why not?

## Finding Intercepts

The slope-intercept form makes it very easy to find the $y$-intercept since it is given by $b$ in the equation $y = mx + b$. You can also use this form to find the **$x$-intercept**, which is the $x$-coordinate of the point where the line crosses the $x$-axis.

**EXAMPLE 3** **Identify the $x$- and $y$-intercepts, and use them to graph the following line.**

$y = -3x + 12$

**SOLUTION**

To find the $x$-intercept, substitute 0 for $y$ in the equation. Then solve.

$$\begin{aligned} y &= -3x + 12 \\ 0 &= -3x + 12 \\ 3x &= 12 \\ x &= 4 \end{aligned}$$

The $x$-intercept is 4.

The $y$-intercept is 12.

Use the $x$- and $y$-intercepts to graph the line.

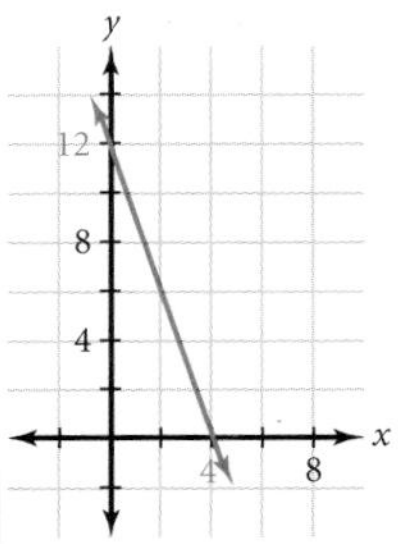


**TRY THIS** Identify the $x$- and $y$-intercepts, and use them to graph each line.

**a.** $y = -4x + 8$ **b.** $y = 3x - 15$

## Horizontal and Vertical Lines

The intercepts make it very easy to write equations for both horizontal and vertical lines. Since horizontal lines have a slope of 0, their equations are of the form $y = b$. Vertical lines have undefined slope (Lesson 5.2), but an equation can still be written by using the $x$-intercept. If the $x$-intercept is $a$, then the equation for a vertical line is $x = a$.

### Equations of Horizontal and Vertical Lines

The equation of a horizontal line is $y = b$, where $b$ is the $y$-intercept.

The equation of a vertical line is $x = a$, where $a$ is the $x$-intercept.

**EXAMPLE 4** Write an equation for each line.

**a.**

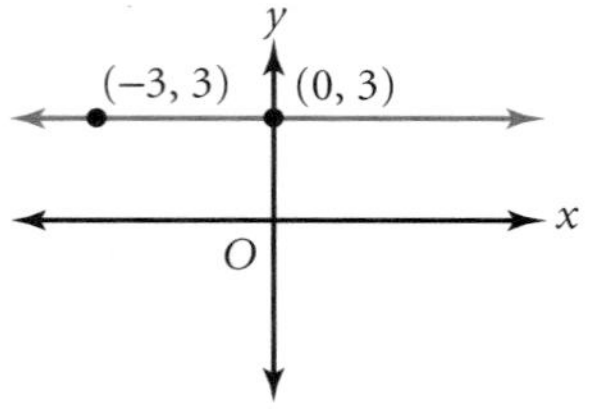


**b.**

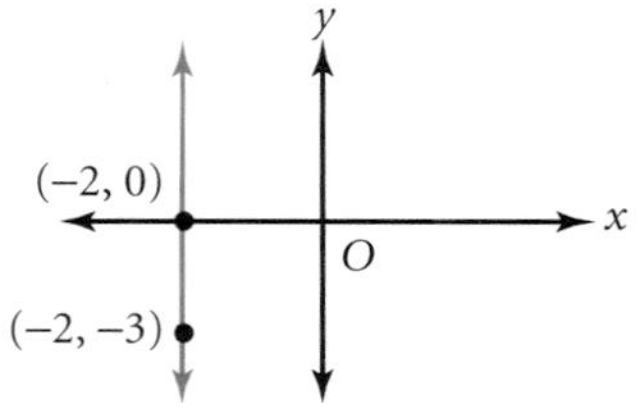


**SOLUTION**

**a.** Notice that the line is horizontal.
The $y$-intercept is 3.
Substitute into the equation for a horizontal line.
$y = 3$

**b.** Notice that the line is vertical.
The $x$-intercept is −2.
Substitute into the equation for a vertical line.
$x = -2$

**TRY THIS**

Write an equation for each line.

**a.**

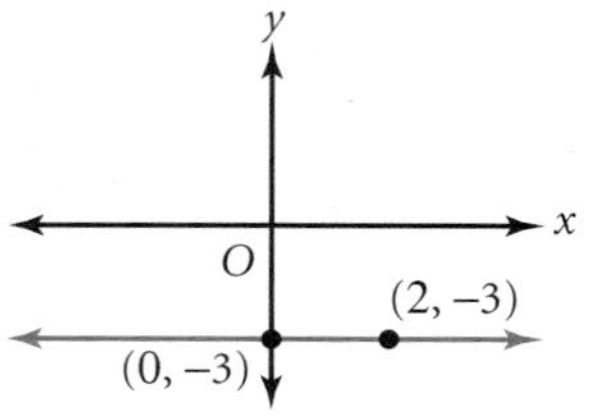


**b.**

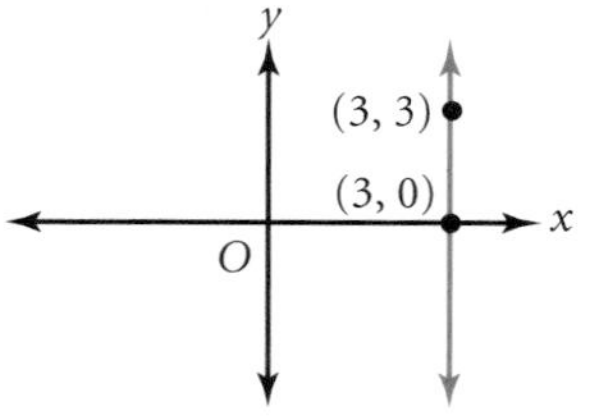

# Exercises

## Communicate

**1.** Line $l$ passes through the origin and the point $A(3, 6)$. Explain how to write an equation for line $l$.

**2.** If the slope of a line is $h$ and it passes through the origin, what is an equation of the line?

**3.** Describe how you can determine an equation of the function graphed at right.

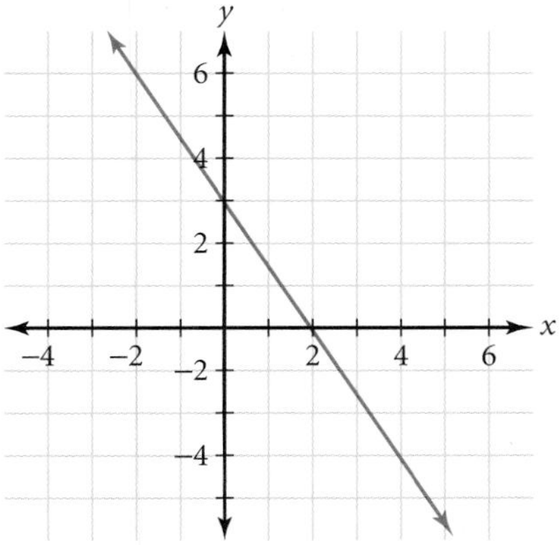


**4.** How does changing the value of $b$ affect the graph of $y = mx + b$?

**5.** How does changing the value of $m$ affect the graph of $y = mx + b$?

## Guided Skills Practice

**Graph each equation below by using the slope and $y$-intercept.** ***(EXAMPLE 1)***

**6.** $y = -4x - 1$

**7.** $y = \frac{3}{5}x - 2$

**8.** Write an equation for a line passing through the points $(4, 5)$ and $(6, -2)$. ***(EXAMPLE 2)***

**9.** Identify the $x$- and $y$-intercepts, and use them to graph the line $y = -x + 7$. ***(EXAMPLE 3)***

**Write an equation for each line.** ***(EXAMPLE 4)***

**10.** a horizontal line that passes through the point $(6, 0)$

**11.** a vertical line that passes through the point $(0, 0)$

## Practice and Apply

**Identify the $x$- and $y$-intercepts, and use them to graph each line.**

**12.** $y = 4x + 5$

**13.** $y = 8x - 1$

**14.** $y = -3x + 5$

**15.** $y = -2x + 13$

**16.** $y = 17x - 4$

**17.** $y = -5x - 9$

**Graph each equation by using the slope and $y$-intercept.**

**18.** $y = -5x + 7$

**19.** $y = \frac{4}{5}x - 3$

**20.** $y = -\frac{3}{2}x + 8$

**21.** $y = 0.5x - 4$

**22.** $y = -3x + 2.5$

**23.** $y = -x - 15$

**internet connect**

**Homework Help Online**
Go To: **go.hrw.com**
Keyword:
**MA1 Homework Help**
for Exercises 24-32

**Write an equation in slope-intercept form for the line that contains each pair of points.**

**24.** $(0, 6), (5, 0)$ **25.** $(3, 4), (-1, -2)$ **26.** $(1, 7), (5, 3)$

**27.** $(7, 2), (-4, -2)$ **28.** $(-5, 0), (8, -4)$ **29.** $(7, -7), (-4, -3)$

**30.** $(-4, -3), (-2, -6)$ **31.** $(6, 6), (-2, -2)$ **32.** $(-1, 1), (5, -7)$

**Write an equation in slope-intercept form for each line graphed below.**

**33.**

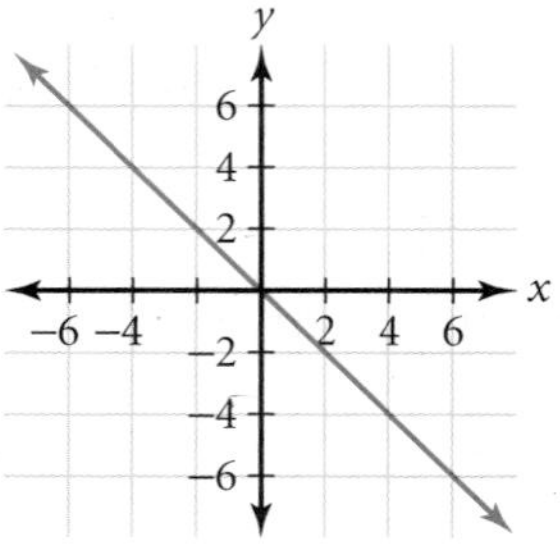


**34.**

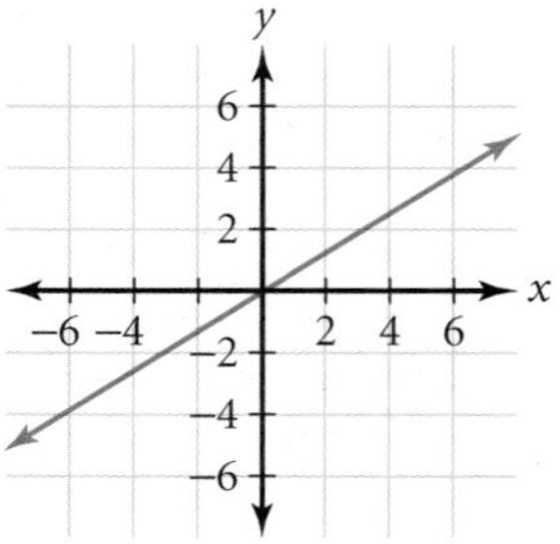


**35.**

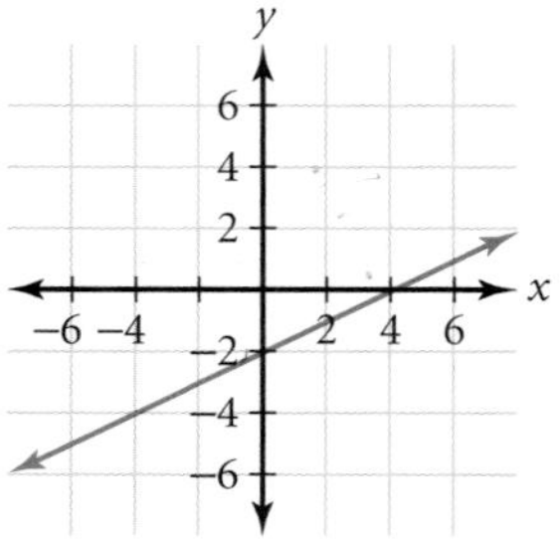


**36.**

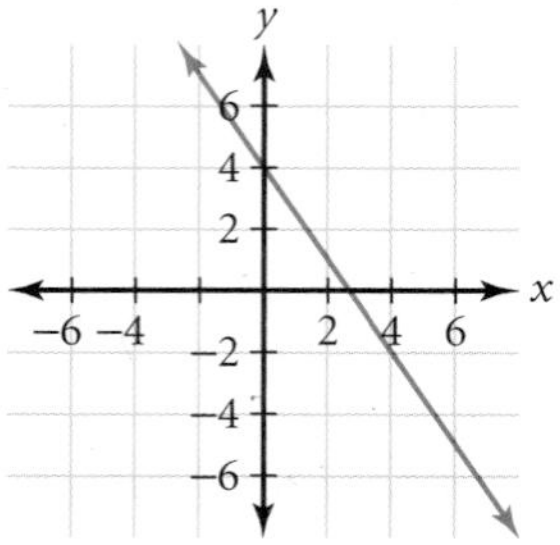


**37.**

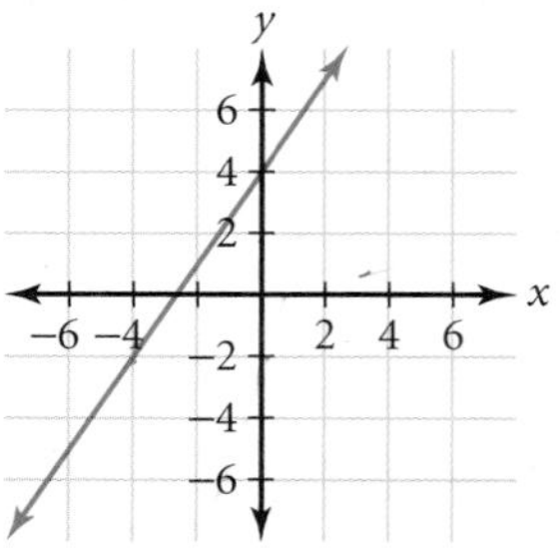


**38.**

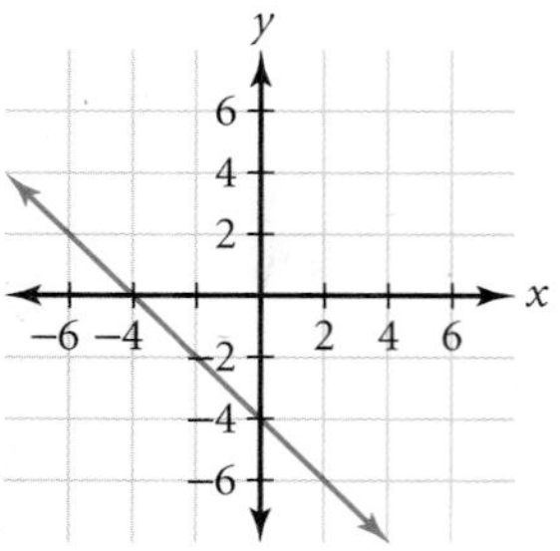


**Write an equation for the line that contains each pair of points.**

**39.** $(3, 4), (3, -3)$ **40.** $(-4, 1), (6, 1)$

**41.** $(2, -5), (10, -5)$ **42.** $(5, -3), (5, 1)$

**Write an equation in slope-intercept form for the line that fits each description below.**

**43.** contains the origin and has a slope of $-2$

**44.** contains the origin and has a slope of $\frac{2}{3}$

**45.** crosses the $y$-axis at $-1$ and has a slope of 5

**46.** contains the origin and $(4, 5)$

**47.** contains $(-3, 2)$ and $(6, 6)$

**48.** contains $(0, -2)$ and $(7, 1)$

**49.** contains $(6, 8)$ and $(-3, 4)$

**Without graphing, describe what the graph of each equation looks like. Include information about the slope and $y$-intercept.**

**50.** $y = -5x$ **51.** $y = -5x + 3$ **52.** $y = 7$

**53.** $x = 7$ **54.** $y = -x - 3$ **55.** $y = 2x - 8$

**56.** $y = \frac{1}{4}x + 10$ **57.** $y = \frac{4}{5}x + 9$ **58.** $y = 7 - x$

**59.** $y = 3 + \frac{1}{8}x$ **60.** $y = \frac{-2x}{3}$ **61.** $y = \frac{6x}{2} - 1$

CONNECTION

**COORDINATE GEOMETRY** Match each equation with the appropriate description.

**62.** $x + y = 9$ **a.** a horizontal line 9 units above the origin

**63.** $xy = 9$ **b.** a vertical line 9 units to the right of the origin

**64.** $x = 9$ **c.** a line through the origin with a slope of 9

**65.** $y = 9$ **d.** a line with a slope of $-1$ and a $y$-intercept of 9

**66.** $y = 9x$ **e.** something other than a straight line

APPLICATIONS

**67. ECOLOGY** Suppose that the water level of a river is 34 feet and that it is receding at a rate of 0.5 foot per day. Write an equation for the water level, $L$, after $d$ days. Graph the equation. In how many days will the water level be 26 feet?

CHALLENGE

**68. BUDGET** Saul's father is thinking of buying his son a six-month movie pass for $40. With the pass matinees cost $1.00. If matinees are normally $3.50 each, how many times must Saul attend in order for it to benefit his father to buy the pass?

## Look Back

APPLICATION

**FUND-RAISING** For the following numbers of people, how much will each person have to contribute to reach a goal of $250? *(LESSON 3.2)*

**69.** 25 people **70.** 15 people **71.** 20 people **72.** 150 people

**73.** Solve $3a + 9d = 7$ for $d$. *(LESSON 3.6)*

**74.** Dan earns 15% on each $20 box of candies he sells. How many boxes will he have to sell to earn at least $100? *(LESSON 4.2)*

## Look Beyond

**75. CULTURAL CONNECTION: AFRICA** The Rhind papyrus that describes the mathematics of building the ancient pyramids of Egypt refers to a *seked.* A seked can be stated mathematically as follows:

$$\text{seked} = \frac{\text{horizontal distance in palms}}{\text{vertical distance in cubits}}$$

Consider a pyramid with a base of 2520 palms on each side and a height of 250 cubits. Scribes took half of the length of the base as the run and the height as the rise. They calculated the seked to be $\frac{1260}{250}$. Compare the calculation of the seked with the calculation of the slope for the same pyramid. Explain the relationship between the seked and slope.

# 5.5 The Standard and Point-Slope Forms

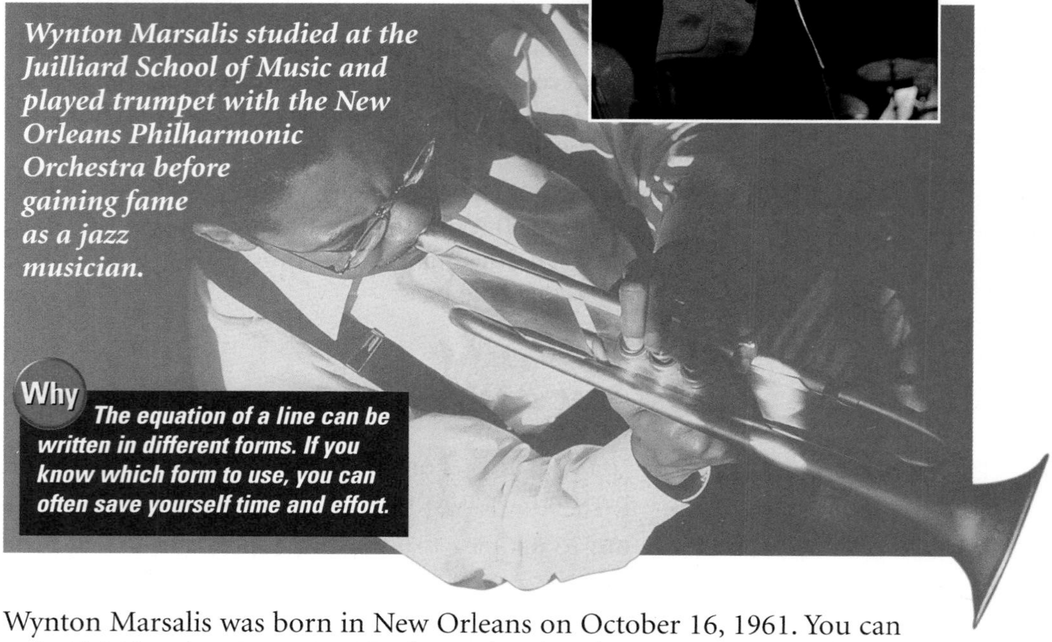

*Wynton Marsalis studied at the Juilliard School of Music and played trumpet with the New Orleans Philharmonic Orchestra before gaining fame as a jazz musician.*

**Objectives**

- Define and use the standard form for a linear equation.
- Define and use the point-slope form for a linear equation.

**Why** *The equation of a line can be written in different forms. If you know which form to use, you can often save yourself time and effort.*

Wynton Marsalis was born in New Orleans on October 16, 1961. You can write this date in many ways. Each way has its advantages. The long form is easy to read. The short form is quick and easy to write. The computer form is easy to sort by year, month, or day.

| October 16, 1961 | 10-16-61 | 19611016 |
|---|---|---|
| **long form** | **short form** | **computer form** |

You have been using the slope-intercept form of the equation for a line, $y = mx + b$, to solve problems. As for dates, there are other forms for the equation of a line. Each form has advantages.

## Standard Form

Consider an equation written in the form $Ax + By = C$. After values for $A$, $B$, and $C$ are determined, a solution to this equation is any ordered pair of numbers $(x, y)$ that makes the equation true. When these ordered pairs are graphed, they form a straight line.

**Standard Form of a Linear Equation**

A linear equation in the form $Ax + By = C$ is in standard form when
- $A$, $B$, and $C$ are real numbers, and
- $A$ and $B$ are not both zero.

**EXAMPLE 1** **Write each linear equation in standard form.**

**a.** $y = -2x + 3$ **b.** $x = -13y + 4$ **c.** $\frac{3}{4}x - 2 = 3y$

**SOLUTION**

**a.**
$$y = -2x + 3$$
$$2x + y = 3$$

**b.**
$$x = -13y + 4$$
$$x + 13y = 4$$

**c.**
$$\frac{3}{4}x - 2 = 3y$$
$$\frac{3}{4}x - 2 - 3y = 0$$
$$\frac{3}{4}x - 3y = 2$$

*This is acceptable as standard form. It could also be written as* $\frac{3}{4}x + (-3y) = 2$.

**TRY THIS** Write each linear equation in standard form.

**a.** $y = -4x - 12$ **b.** $2y - 5 = 3x$ **c.** $5.7x - 5.3 = 4.3y$

**EXAMPLE 2**

**APPLICATION**
**FUND-RAISING**

Jackie is in charge of selling tickets for the school jazz concert at \$2.00 for student tickets and \$4.00 for adult tickets. She hopes that the total ticket sales will be about \$600 in order to cover expenses and make a modest profit. **Write a linear equation in standard form to model this situation, and then graph the equation.**

**SOLUTION**

Let $x$ represent the number of \$4.00 adult tickets.
Let $y$ represent the number of \$2.00 student tickets.

An equation in standard form that models the problem is $4x + 2y = 600$.

To graph an equation in standard form, substitute 0 for $x$, and solve for $y$. This gives the $y$-intercept—the $y$-coordinate of the point where the line crosses the $y$-axis. Then substitute 0 for y and solve for $x$. This gives the $x$-intercept—the $x$-coordinate of the point where the line crosses the $x$-axis.

Start with the equation $4x + 2y = 600$.

First let $x = 0$.

$$4x + 2y = 600$$
$$0 + 2y = 600$$
$$y = 300$$

The $y$-intercept is 300.

Then let $y = 0$.

$$4x + 2y = 600$$
$$4x + 0 = 600$$
$$x = 150$$

The $x$-intercept is 150.

Draw the line that connects the two intercepts. The coordinates of any point on this line will solve the equation. Recall though that only whole numbers make sense for the domain and range in this situation. Why?

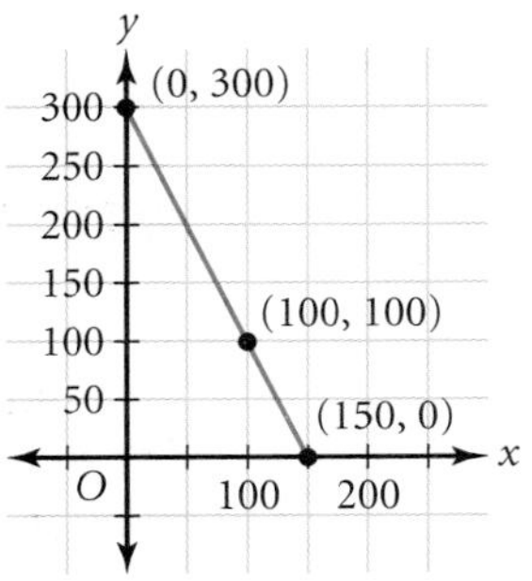

**TECHNOLOGY**
**GRAPHICS CALCULATOR**

To enter the equation from Example 2 in a graphics calculator, first rewrite the equation in slope-intercept form, $y = mx + b$.

$$\begin{aligned} 4x + 2y &= 600 \\ 2y &= 600 - 4x \\ y &= 300 - 2x \\ y &= -2x + 300 \end{aligned}$$

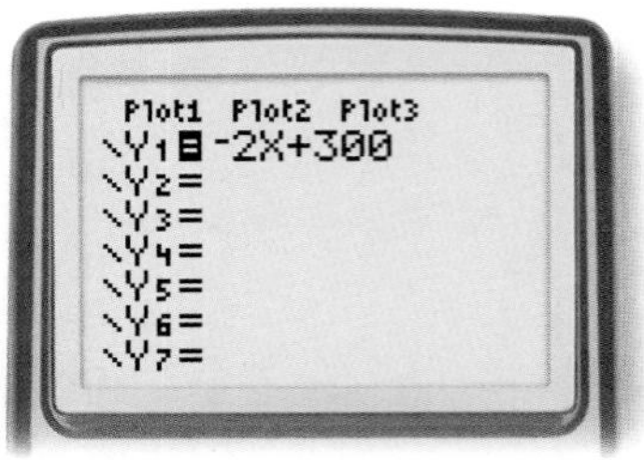


The equation can now be entered into the calculator by using the **Y=** button.

**CRITICAL THINKING** If the equation of a line is written in standard form, $Ax + By = C$, the slope is $-A/B$ and the $y$-intercept is $C/B$. Explain why.

## Point-Slope Form

If you know the slope of a line and the coordinates of one of its points, you can write a third form of an equation for a line called the **point-slope form.**

The point-slope form is derived from the slope formula, $m = \frac{y_2 - y_1}{x_2 - x_1}$.

$$\begin{aligned} m &= \frac{y_2 - y_1}{x_2 - x_1} \\ m(x_2 - x_1) &= \frac{y_2 - y_1}{x_2 - x_1}(x_2 - x_1) \\ m(x_2 - x_1) &= y_2 - y_1 \end{aligned}$$

If you replace the point $(x_2, y_2)$ in the equation with the notation for any point on the line, $(x, y)$, the result is as follows:

$$y - y_1 = m(x - x_1)$$

**Point-Slope Form**

The form $y - y_1 = m(x - x_1)$ is the point-slope form for the equation of a line. The coordinates $x_1$ and $y_1$ are taken from a given point, $(x_1, y_1)$, and $m$ is the slope.

**EXAMPLE 3** A line with a slope of 3 contains the point $(2, -7)$. **Write an equation in point-slope form for the line.**

**SOLUTION**

Let $m = 3$, $x_1 = 2$, and $y_1 = -7$. Substitute the given values into the point-slope form of the equation.

$$\begin{aligned} y - y_1 &= m(x - x_1) \\ y - (-7) &= 3(x - 2) \end{aligned}$$

**TRY THIS** A line with a slope of $-4$ contains the point $(4, 2)$. Write an equation in point-slope form for the line.

**EXAMPLE 4** **Write $y - 149 = 9(x - 16)$ in each form.**

**a.** slope-intercept form  **b.** standard form

**SOLUTION**

**a.** Begin with the original equation and solve for $y$.

$$y - 149 = 9(x - 16)$$
$$y - 149 = 9x - 144$$
$$y = 9x + 5$$

**b.** Then change the slope-intercept form to standard form, $Ax + By = C$.

$$y = 9x + 5$$
$$y - 9x = 5$$
$$-9x + y = 5$$
$$9x - y = -5$$

**TRY THIS** Write $y - 23 = 5(x - 4)$ in each form.

**a.** slope-intercept form  **b.** standard form

**EXAMPLE 5** **Use the points (−1, 10) and (5, 8) to write the equation of a line in point-slope form, and then write the equation in slope-intercept form.**

**SOLUTION**

The point-slope form simplifies the way this problem is solved. Let $(x_1, y_1)$ be (−1, 10) and $(x_2, y_2)$ be (5, 8). Find the slope.

$$\text{slope} = \frac{y_2 - y_1}{x_2 - x_1} = \frac{8 - 10}{5 - (-1)} = \frac{-2}{6} = -\frac{1}{3}$$

Use the point (−1, 10) for $(x_1, y_1)$. Substitute the slope and the coordinates of the point into the equation $y - y_1 = m(x - x_1)$.

$$y - 10 = -\tfrac{1}{3}[x - (-1)] \quad \textit{point-slope form}$$

Solve for $y$: $$y - 10 = -\tfrac{1}{3}(x + 1)$$
$$y - 10 = -\tfrac{1}{3}x - \tfrac{1}{3}$$
$$y = -\tfrac{1}{3}x + 9\tfrac{2}{3} \quad \textit{slope-intercept form}$$

**TRY THIS** Use the points (5, 65) and (7, 71) to write the equation of a line in point-slope form, and then write the equation in slope-intercept form.

**SUMMARY OF THE FORMS FOR THE EQUATION OF A LINE**

| Name | Form | Example |
|---|---|---|
| Slope-intercept | $y = mx + b$ | $y = -3x + 5$ |
| Standard | $Ax + By = C$ | $3x + y = 5$ |
| Point-slope | $y - y_1 = m(x - x_1)$ | $y - 2 = -3(x - 1)$ |

# Exercises

## Communicate

1. Explain how to write the equation $5y - 2 = -3x$ in standard form.
2. Tell how you would identify the intercepts for the graph of the equation $3x + 6y = 18$.
3. Explain how to rewrite $x - 3y = 9$ in slope-intercept form.
4. How would you use the point-slope formula to write the equation of a line containing the points $(-2, 4)$ and $(4, -8)$?

## Guided Skills Practice

**Write each equation in standard form.** ***(EXAMPLES 1 AND 2)***

5. $y = 3x + 7$
6. $2y = 3x - 4$
7. $3x = -7y - 17$

**Write an equation in point-slope form for the line with each slope that contains the given point.** ***(EXAMPLE 3)***

8. slope of 2; (3, 4)
9. slope of –2; (–3, 4)
10. slope of $\frac{1}{3}$; (3, – 4)

**Write the equations below in slope-intercept form and in standard form.** ***(EXAMPLE 4)***

11. $y - 50 = 8(x - 4)$
12. $3y = 9x + 15$
13. $y = 10(-4x + 3)$

**Use each pair of points to write the equation of a line in point-slope form, and then write the equation in slope-intercept form.** ***(EXAMPLE 5)***

14. (5, –2), (–2, 5)
15. (–3, 3), (–4, 4)
16. (3, 2), (–3, –2)

## Practice and Apply

internet connect
**Homework Help Online**
Go To: **go.hrw.com**
Keyword:
**MA1 Homework Help** for Exercises 17-28

**Write each equation in standard form.**

17. $4x = -3y + 24$
18. $7y = -5x - 35$
19. $6x + 4y + 12 = 0$
20. $2x = 4y$
21. $6x - 8 = 2y + 6$
22. $x = \frac{2}{3}y + 6$
23. $7x + 14y = 3x - 10$
24. $5 = y - x$
25. $7x + 2y + 14 = 0$
26. $2x - 4 = 3y + 6$
27. $6 = x - y$
28. $3 + 9y = -x - 12$

**Write an equation in point-slope form for the line that has the given slope and that contains the given point, and then change the equation to slope-intercept form.**

29. slope 3, (2, 6)
30. slope –4, (–3, 1)
31. slope $\frac{1}{5}$, (–4, –2)

**Find the *x*- and *y*-intercepts, and use them to graph each equation.**

32. $x + y = 10$
33. $3x - 2y = 12$
34. $5x + 4y = 12$
35. $4x - 5y = 20$
36. $2x - 7y = 14$
37. $9x + y = 18$

**Write an equation for the line that contains each pair of points.**

**38.** $(7, -2), (7, 8)$ **39.** $(0, 3), (-2, -1)$ **40.** $(-7, 5), (16, 5)$

**41.** The equation of a given line is $6x + 2y = 40$. What is the slope of the line?

CHALLENGE

**42.** Compare the graphs of $4a + 2s = 588$ and $2a + s = 294$. What do you find?

CONNECTION

**COORDINATE GEOMETRY** The equation $x = 4$ cannot be written in slope-intercept form because the slope is undefined. It can, however, be written in standard form as $1x + 0y = 4$. Copy and complete the table below. Write the equivalent forms of each equation or "undefined slope" when appropriate.

| | Given | Slope-intercept form | Standard form |
|---|---|---|---|
| **43.** | $x = 1$ | ? | ? |
| **44.** | $y = 4$ | ? | ? |
| **45.** | $x + y = 5$ | ? | ? |
| **46.** | $y = 4x$ | ? | ? |
| **47.** | $x = 4y$ | ? | ? |

APPLICATION

*Anne Frank 1929–1945*

**48. FUND-RAISING** Tickets to a school play are \$3 for students and \$5 for adults. Write an equation in standard form which shows that the total sales for $s$ student tickets and $a$ adult tickets is \$700. Graph the equation with adult tickets on the horizontal axis and student tickets on the vertical axis.

## Look Back

**49.** Describe a scatter plot with a strong negative correlation. *(LESSON 1.6)*

**Evaluate each expression for $x = 1$, $y = 1$, and $z = 2$.** *(LESSONS 2.2 AND 2.3)*

**50.** $x^2 + y + z^2$ **51.** $x - y + z$ **52.** $x + y - z$ **53.** $-(x + y + z)$

**Solve each equation.** *(LESSON 3.2)*

**54.** $-5y = 30$ **55.** $3x = 420$ **56.** $\frac{y}{9} = 36$ **57.** $\frac{x}{2} = 108$

**internet connect**

**Activities Online**

Go To: **go.hrw.com**
Keyword:
**MA1 Check Digits**

**Find the slope of the line containing the origin and the given point. Then give the equation of the line containing the origin and the given point.** *(LESSON 5.2)*

**58.** $A(3, 6)$ **59.** $B(2, 8)$ **60.** $C(6, 3)$ **61.** $D(-5, -7)$

## Look Beyond

**62.** Will two distinct vertical lines ever intersect?

**63.** What kind of angle is formed when a horizontal and a vertical line intersect?

# 5.6 Parallel and Perpendicular Lines

## Objectives

- Identify parallel lines and perpendicular lines by comparing their slopes.
- Write equations of lines that are parallel and perpendicular to given lines.

**Why** ***Identifying parallel and perpendicular lines by comparing slopes is an important step toward recognizing relationships between lines without graphing.***

| | °F | °C | K |
|---|---|---|---|
| **Water boils** | 212 | 100 | 373 |
| **Water freezes** | 32 | 0 | 273 |
| **Absolute zero** | −459 | −273 | 0 |

*For a given pressure, the physical state of water depends on its temperature—it can be a liquid, a solid, or a gas.*

The table above compares temperatures for three scales—Fahrenheit, Celsius, and Kelvin.

**APPLICATION**
**PHYSICS**

The formulas for converting from Celsius and Kelvin to Fahrenheit are $F = \frac{9}{5}C + 32$ (Celsius to Fahrenheit) and $F = \frac{9}{5}K - 460$ (Kelvin to Fahrenheit). The two equations can be rewritten by using $y$ for $F$ and $x$ for $C$ and $K$; then they can be graphed on the same coordinate plane.

$$y = \frac{9}{5}x + 32$$

$$y = \frac{9}{5}x - 460$$

Notice that the two lines are parallel. Both lines have the same slope, $\frac{9}{5}$.

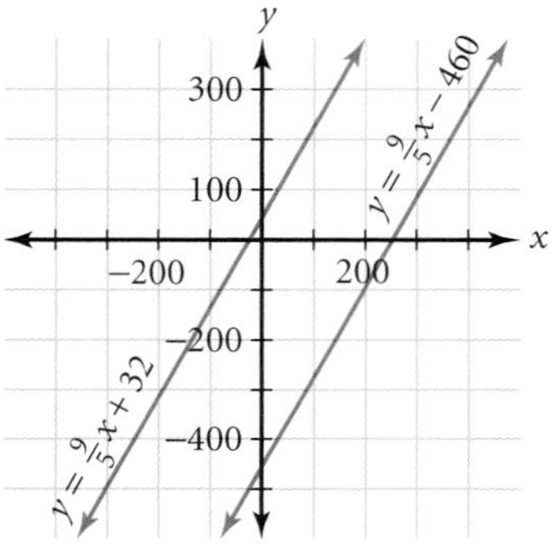

**Parallel Lines**

If two different lines have the same slope, the lines are **parallel**. If two nonvertical lines are parallel, they have the same slope. Two parallel, vertical lines have undefined slopes.

EXAMPLE 1 **Write an equation in slope-intercept form of a line parallel to $y = 3x - 7$ with a $y$-intercept of 4.**

SOLUTION

The slope of $y = 3x - 7$ is 3; a line parallel to this line has a slope of 3.

The $y$-intercept will be 4, so the equation of the new line is $y = 3x + 4$.

**TRY THIS** Write an equation in slope-intercept form of a line parallel to $y = 0.5x + 5$ with a $y$-intercept of –2.

CONNECTION GEOMETRY

In geometry, **perpendicular lines** form right angles. In the next activity, you will discover the algebraic relationship between the slopes of two lines that are perpendicular.

*Activity*

## Exploring Slopes and Perpendicular Lines

**You will need:** graph paper and a sheet of clear plastic or tracing paper

On a piece of graph paper with the $x$- and $y$-axes marked, draw a line that has a run of 4 units and a rise of 3 units, as shown below. Make a copy on the plastic or tracing paper. Rotate the copy of the graph 90 degrees clockwise, as shown. When one graph is placed on top of the other, you can see the relationship between the two lines.

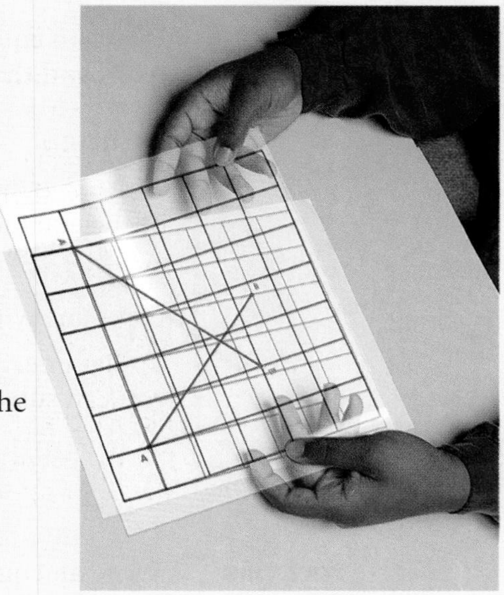

The lines are perpendicular because the line on the second graph was rotated 90 degrees from the line on the first graph.

1. What is the slope of the original line?
2. Find the slope of the rotated line. Count the squares to find the rise and the run.
3. What is the sign for the slope of the rotated line?

CHECKPOINT ✔

4. What is the relationship between the slope of the rotated line and the slope of the original line?

Consider the equations $y = 2x + 3$ and $y = -\frac{1}{2}x - 1$. The graph shows that the two lines are perpendicular to each other.

Notice that the slopes of the two lines are *negative reciprocals* of each other.

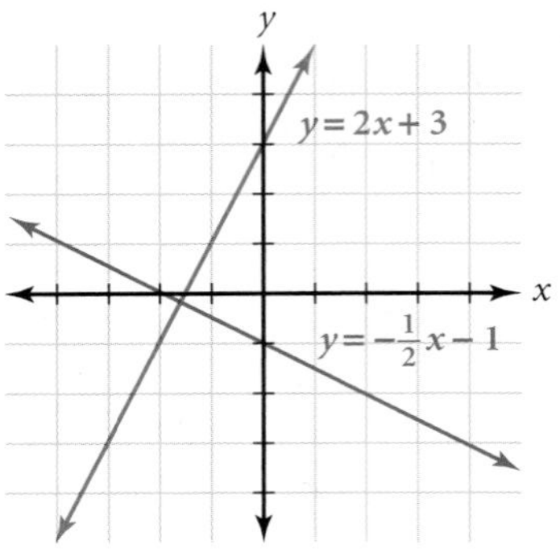


**Perpendicular Lines**

If the slopes of two lines are $m$ and $-\frac{1}{m}$, the lines are perpendicular.

**EXAMPLE 2** **Write an equation in slope-intercept form of a line that has a $y$-intercept of 4 and that is perpendicular to the line $y = 3x + 2$.**

**SOLUTION**

The slope of the perpendicular line is the negative reciprocal of 3, or $-\frac{1}{3}$, and the $y$-intercept is 4. The slope-intercept form of the equation is $y = -\frac{1}{3}x + 4$.

**TRY THIS** Write an equation in slope-intercept form of a line that has a $y$-intercept of 6 and that is perpendicular to $y = 4x + 2$.

**EXAMPLE 3** **Write an equation in point-slope form for the line that contains the point (4, 5) and that is perpendicular to the line $2x + 3y = 7$.**

**SOLUTION**

First solve the given equation for $y$ in order to find the slope-intercept form of the equation.

$$y = -\frac{2}{3}x + \frac{7}{3}$$

For a line with a slope of $-\frac{2}{3}$, any line with a slope of $\frac{3}{2}$ is perpendicular to that line. Since the perpendicular line also contains the point (4, 5), you can substitute the coordinates and slope into the point-slope form, $y - y_1 = m(x - x_1)$.

$$y - 5 = \frac{3}{2}(x - 4)$$

**TRY THIS** Write an equation in point-slope form for the line that contains the point (3, −2) and that is perpendicular to the line $4x - 2y = -6$.

# Exercises

## Communicate

**1.** Explain how to write an equation for a line parallel to $y = 4x + 3$.

**2.** The slope of a line is $\frac{2}{3}$. Explain how to find the slope of a line that is perpendicular to that line.

**3.** How would you find the slope of a line perpendicular to the line $y = \frac{1}{3}x + 2$?

**4.** Describe how to find an equation for a line perpendicular to $y = 4x + 3$.

## Guided Skills Practice

**Write an equation in slope-intercept form for the line parallel to each line below and with a *y*-intercept of 5.** *(EXAMPLE 1)*

**5.** $y = 2x + 3$ **6.** $y = -3x$ **7.** $4y = x$ **8.** $-y = -6x + 2$

**Write an equation in slope-intercept form for a line perpendicular to each line below and with a *y*-intercept of 4.** *(EXAMPLE 2)*

**9.** $y = 3x - 3$ **10.** $y = -3x$ **11.** $5y = x$ **12.** $-6y = x$

**Write an equation in point-slope form for the line that contains the point (4, 5) and is perpendicular to the graph of the equation.** *(EXAMPLE 3)*

**13.** $2x + 3y = 4$ **14.** $x - 3y = 8$ **15.** $-2x - 8y = 16$

## Practice and Apply

**internet connect**
**Homework Help Online**
Go To: **go.hrw.com**
Keyword:
**MA1 Homework Help**
for Exercises 16-27

**Find the slope of each line.**

**16.** $y = 4x + 10$ **17.** $3x + y = 7$ **18.** $10 = -5x + 2y$

**19.** $4x - 3y = 12$ **20.** $y = \frac{1}{3}x - 3$ **21.** $3x - y = 7$

**22.** $2x - y = 14$ **23.** $3x + 2y = 51$ **24.** $13 = 20x - 5y$

**25.** $3y = -4x + 2$ **26.** $\frac{2}{3}x + 6y = 1$ **27.** $4x - \frac{1}{4}y = 8$

**Find the slope of a line that is perpendicular to the following lines:**

**28.** $y = -\frac{1}{3}x + 10$ **29.** $-\frac{1}{2}x - y = 20$ **30.** $13 = -x + y$

**31.** $3x + 12y = 12$ **32.** $y = 5x + 10$ **33.** $3x + y = 2$

**34.** $20 = -5x + 2y$ **35.** $4x - 4y = 12$ **36.** $2y = 5x + 11$

**37.** $-4x + 8y = 17$ **38.** $12x - 3y = 10$ **39.** $4y = 20x - 3$

**Write an equation in point-slope form for a line containing the point (2, 3) and parallel to each of the following lines:**

**40.** $x + y = 1$ **41.** $3x = 7y + 2$ **42.** $y = 2x - 3$

**43.** $3y = 2x$ **44.** $7x - 2y = 10$ **45.** $11 = 3y + 2x$

**Write an equation in slope-intercept form for each line according to the given information.**

| | Contains | Is parallel to |
|---|---|---|
| **46.** | (3, −5) | $5x - 2y = 10$ |
| **47.** | (−2, 7) | $y = 3x - 4$ |
| **48.** | (2, 4) | $y = 7$ |
| **49.** | (2, −4) | $y = 3x - 4$ |
| **50.** | (−1, 4) | $y = 2x + 5$ |

| | Contains | Is perpendicular to |
|---|---|---|
| **51.** | (3, −5) | $5x + 2y = 10$ |
| **52.** | (2, 7) | $y = 3x - 4$ |
| **53.** | (2, −4) | $y = 7$ |
| **54.** | (2, −4) | $3x + y = 5$ |
| **55.** | (−1, 4) | $y = 2x - 5$ |

**Graph the line $y = 5x$.**

**56.** Sketch a line that is parallel to $y = 5x$. Write its equation.

**57.** Sketch a line that is perpendicular to $y = 5x$. Write its equation.

**What is the slope of a line that is**

**58.** parallel to a horizontal line? **59.** perpendicular to a horizontal line?

**60.** parallel to a vertical line? **61.** perpendicular to a vertical line?

CONNECTIONS

**GEOMETRY** Write the equations of four lines that intersect to form a square whose sides are

**62.** *parallel* to the axes. **63.** *not parallel* to the axes.

**64. GEOMETRY** If one side of a square is contained by the line $y = \frac{3}{4}x + 5$, write equations of lines that could contain the other 3 sides.

APPLICATION

**65. PHYSICS** Write the formulas for converting from Fahrenheit, $F$, to Celsius, $C$, and to Kelvin, $K$. Rewrite them by using $x$ for $F$ and $y$ for $K$ and $C$. Graph the two equations on the same coordinate plane.

**66.** What is the relationship between the two lines in Exercise 65? Find the slope of these two lines.

CHALLENGE

**67.** What relationship do these two lines have to the lines formed by the equations for converting from Celsius and Kelvin to Fahrenheit?

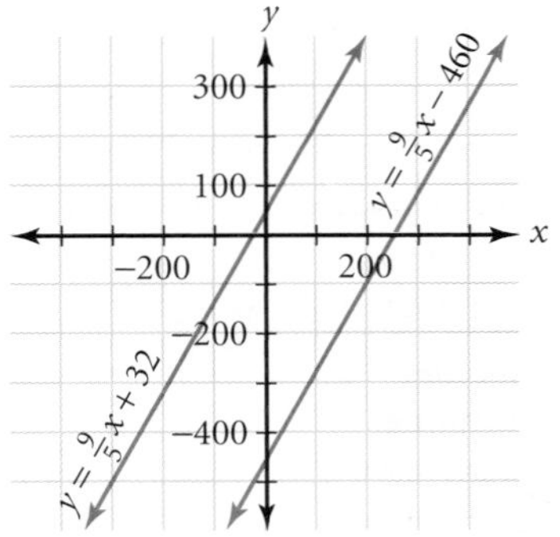


*The equation $y = \frac{9}{5}x + 32$ converts Celsius to Fahrenheit. $y = \frac{9}{5}x - 460$ converts Kelvin to Fahrenheit.*

## Look Back

**68.** Place parentheses to make $2 \cdot 7 + 35 \div 7 - 10 = 2$ true. ***(LESSON 1.3)***

**Evaluate each expression.** ***(LESSONS 2.2 AND 2.3)***

**69.** $-4 + (-3) + 1$

**70.** $-2 + 3 + (-7) + 3$

**71.** $-12 + 4 - (-4)$

**72.** $-3 + (-11) - (-8)$

**73.** $-5 + (-2) - 13$

**74.** $8 - (-5) + 23$

**75.** $15 + (-11) - 6$

**76.** $26 + (-16) - (-8)$

**Simplify.** ***(LESSON 2.6)***

**77.** $2x^2 + 3y + 4y + 3x^2$

**78.** $3x + 2 + 4y + 2 + 3y$

**79.** $2x + 3xy + 5x^2 + 7xy$

**80.** $3x + 4y + 2x + 5 + 6y$

**81.** $4x^2 + 5x + 8 + 11x^2 + 3x$

**82.** $3xy + 2x + 4y + xy$

**83.** $9x^2 + 5xy + 2x + 4x^2$

**84.** $4y^2 + 12y + 6xy + 3y$

## Look Beyond

**How many ordered pairs will satisfy two linear equations simultaneously if the equations have graphs**

**85.** that are parallel?

**86.** that are perpendicular?

### PORTFOLIO ACTIVITY

Include this activity in your portfolio.

One type of scientific research involves determining correlations between sets of data. For example, for many years researchers have studied the correlation between high speeds and automobile accident rates. Use current newspapers or other media to find reports that claim a correlation between sets of data. For each report, answer the following questions in your portfolio:

- Are you convinced that the claimed correlation actually exists? Why or why not?
- What evidence is used to support the claimed correlation?
- Does the report use graphs to represent the correlation? If so, describe whether or not the graph accurately represents the claimed correlation. If a graph is not used, explain how a graph might help or hinder the report's claim.

**WORKING ON THE CHAPTER PROJECT**

You should now be able to complete Activity 3 of the Chapter Project on page 264.

CHAPTER PROJECT FIVE

# Modeling Trends IN DATA

*The scoreboard shows the average speed at the Indianapolis 500 from 1915 to 1995 in 5-year intervals.*

*Over the last 100 years, technological improvements have significantly increased the top speeds for automobiles. Speeds at races, such as the Indianapolis 500, have steadily increased. This increase in speed over time can be approximately modeled by a linear equation. In this project, you will use your knowledge of linear equations to model real-world data.*

## Activity 1

1. Use the data from the scoreboard above to create a scatter plot with years on the horizontal axis and speeds on the vertical axis.
2. Use a clear ruler to draw a line of best fit on your scatter plot.
3. Find the slope and $y$-intercept of your line of best fit, and write the equation for the line.
4. Based on your line of best fit, estimate the speeds for 2000, 2005, and the year of the most recent Indianapolis 500. Use an almanac or other information source to find the actual speeds for the years that you estimated. What might explain the difference between your estimates and the actual speeds?

Linear equations can be useful for representing real-world data even when the data does not fit exactly on a line. The line of best fit you drew for Activity 1 is an example of a line used to model a trend in data. The line of best fit is also called a regression line (the data tends to "regress" to the line). Regression lines can help you to analyze trends in data.

## Activity 2 (Optional, requires the use of a graphics calculator)

Graphics calculator technology allows you to be more precise with the scatter plot and regression line. The calculator will automatically calculate the slope and the $y$-intercept of the regression line.

1. Select the **Stat** key of your calculator. Place the information from the table into a two-variable $(x, y)$ data table.

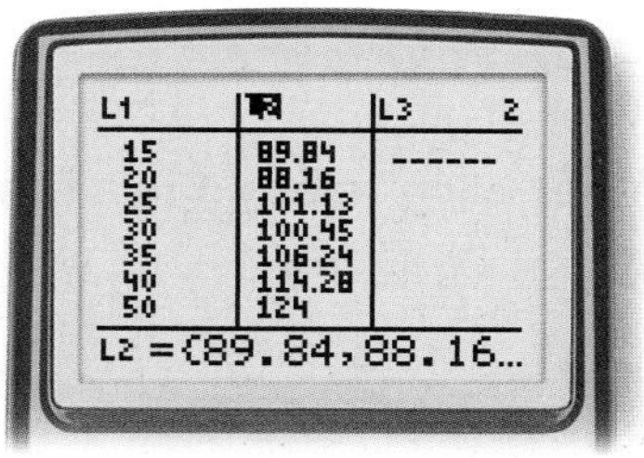


2. Locate and select **Reg** or **LinReg** for the linear regression. Here, the slope for the line of best fit is $a$, and the $y$-intercept is $b$.

3. The slope, represented by $a$ on the calculator screen, is the rate of increase in the average speed, about 1.03. Use the slope and the $y$-intercept (represented by $b$) to write the equation for the regression line (line of best fit). Graph the regression line by entering the equation after **Y=** on your calculator.

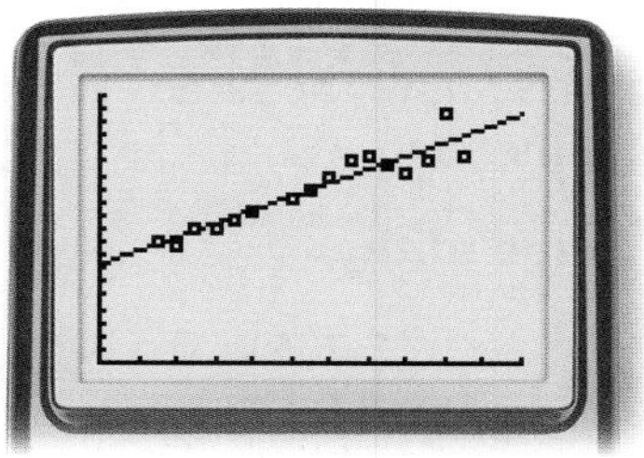

## Activity 3

Choose one of the topics listed at right or one of your own that can be modeled by linear regression. You will need to do research in the library, on the Internet, or by conducting your own experiment in order to find data for your topic. Write a report about your data. In your report, include a scatter plot of your data showing the regression line and the equation of the regression line. Discuss the type of correlation represented by your data. Make predictions for data points that are not shown on your scatter plot, and discuss why the actual data may vary. Explain why linear regression is or is not a good way to model your data.

**Suggested topics:**

- air temperature and the number of cricket chirps per minute
- foot length and hat size
- commuter travel time and the distance from the destination
- number of advertisements and sales
- years of experience and salary

# 5 Chapter Review and Assessment

## VOCABULARY

| | | | | | |
|---|---|---|---|---|---|
| constant of variation | 238 | parallel | 259 | run | 226 |
| direct variation | 238 | perpendicular | 259 | slope | 228 |
| domain | 219 | point-slope form | 254 | slope-intercept form | 245 |
| function | 219 | range | 219 | standard form | 252 |
| function notation | 223 | rate of change | 236 | vertical line | 230 |
| Hooke's law | 239 | relation | 218 | *x*-intercept | 247 |
| horizontal line | 230 | rise | 226 | *y*-intercept | 245 |

## Key Skills & Exercises

### LESSON 5.1

#### Key Skills

**Determine whether a given relation is a function.**

Determine whether the set of ordered pairs below is a function.

$$\{(4, 5), (5, 7), (7, 12), (4, 9)\}$$

The ordered pairs (4, 5) and (4, 9) have the same first coordinate. This is not a function.

**Describe the domain and range of the following relations:**

$$\{(2, 3), (7, 2), (8, 4), (9, 13), (6, 12)\}$$

The domain is {2, 6, 7, 8, 9}.

The range is {2, 3, 4, 12, 13}.

#### Exercises

**Determine whether each relation is a function.**

1. {(7, 12), (6, 9), (7, 8), (4, 11)}
2. {(5, 5), (10, 7), (7, 15), (4, 9)}
3. {(8, 5), (5, 3), (8, 12), (1, 5)}
4. {(6, 5), (3, 7), (7, 9), (8, 9)}

**Describe the domain and range for each relation.**

5. {(8, 5), (2, 5), (3, 12), (6, 5)}
6. {(5, 5), (7, 6), (12, 7), (13, 24)}
7. {(2, 5), (4, 7), (1, 9), (8, 5)}
8. {(34, 4), (4, 15), (10, 4), (9, 17)}

### LESSON 5.2

#### Key Skills

**Find the slope of a line from the rise and run.**

If the rise between two points on a line is 6 and the run between the same two points is 2, find the slope of the line.

$$\text{slope} = m = \frac{6}{2} = 3$$

#### Exercises

**Find the slope of a line with the given rise and run.**

9. rise of 4
   run of 8
10. rise of 16
    run of 4
11. rise of 12
    run of 18
12. rise of 8
    run of 12

**Find the slope of a line from two points.**

Find the slope of a line containing the points $A(-2, 4)$ and $B(3, 5)$.

$$\text{slope} = m = \frac{\text{difference in } y\text{-values}}{\text{difference in } x\text{-values}} = \frac{5-4}{3-(-2)} = \frac{1}{5}$$

**Find the slope of the line containing the given points.**

**13.** $A(-3, 2)$, $B(2, 3)$

**14.** $A(2, 5)$, $B(4, 1)$

**15.** $A(-5, 4)$, $B(1, 4)$

**LESSON 5.3**

## Key Skills

**Find the rate of change from a graph.**

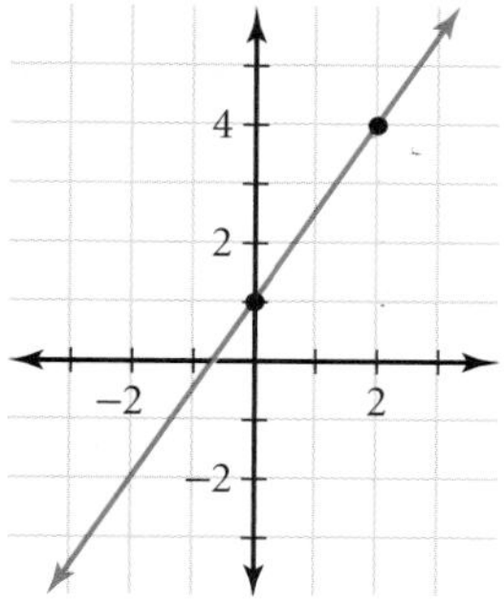


To find the rate of change from the linear graph, find the slope of the line.

Use the slope formula and points $(0, 1)$ and $(2, 4)$.

$$m = \frac{y_2 - y_1}{x_2 - x_1} = \frac{4-1}{2-0} = \frac{3}{2}$$

The slope of a line is referred to as the rate of change because the slope is the rate that the $y$-values change compared to the rate that the $x$-values change.

## Exercises

**Find the rate of change for each linear graph.**

**16.**

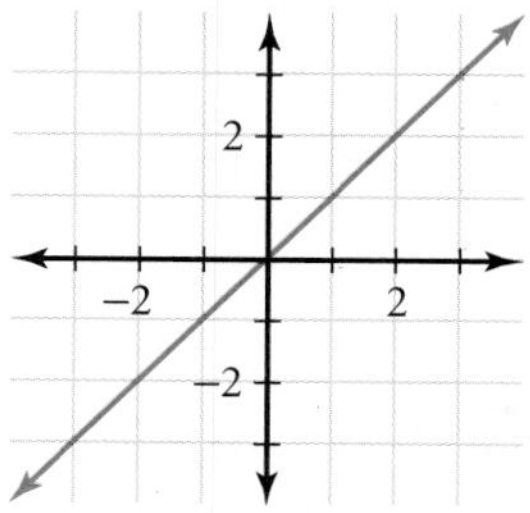

**17.**

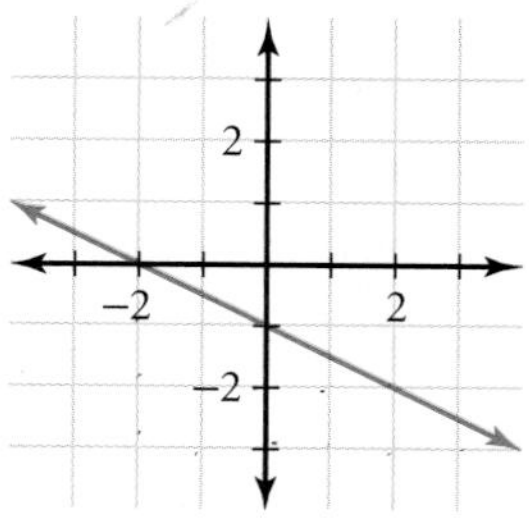

**18.**

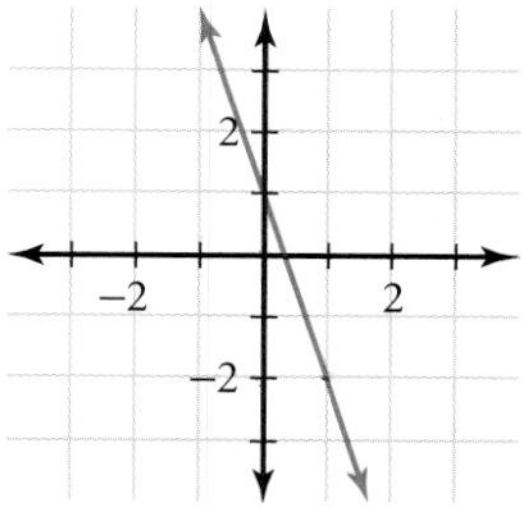

**19.**

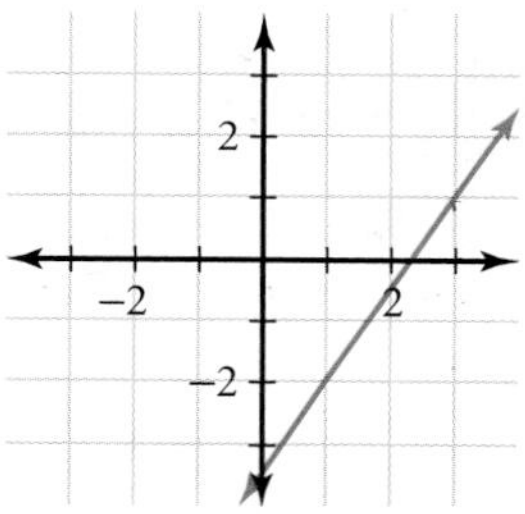

**20.**

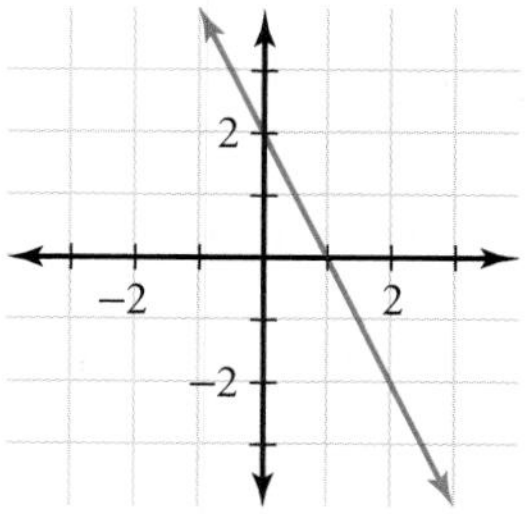

**21.**

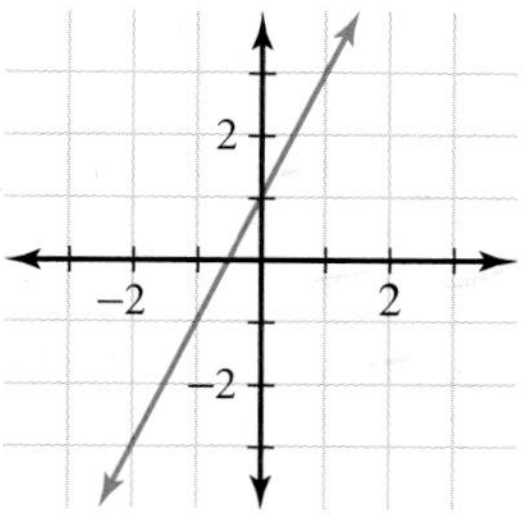

**22.**

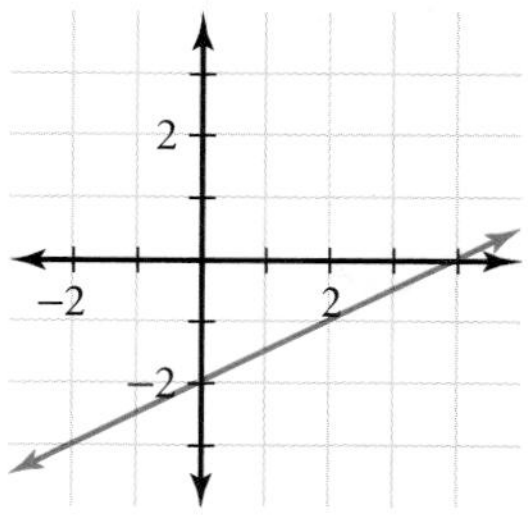

**23.**

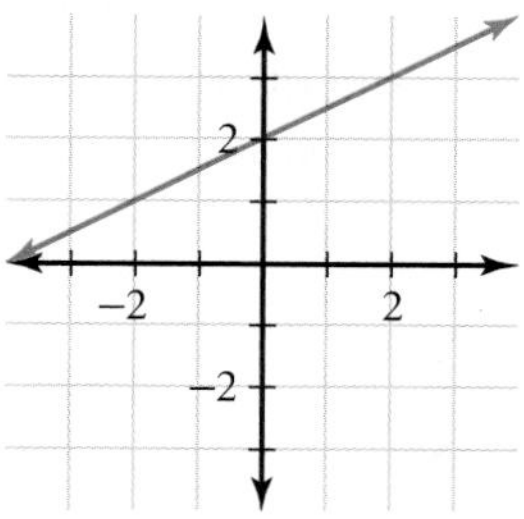

**Solve problems of direct variation.**

To find the constant of variation for a spring by using Hooke's law, substitute values for the force and distance and solve for the constant, $k$. A force of 8 Newtons stretches a spring 3 meters.

$$F = kd$$
$$8 = k(3)$$
$$k = \frac{8}{3}$$
$$k \approx 2.7$$

If $y$ varies directly as $x$ and $y = 24$ when $x = 6$, find the constant of variation, $k$, and write an equation of direct variation.

Substitute 24 for $y$ and 6 for $x$ in $\frac{y}{x} = k$ to find the constant of variation, $k$.

$$\frac{24}{6} = k \qquad k = 4$$

Use the constant of variation, $k$, to write an equation of direct variation. Substitute 4 for $k$.

$$\frac{y}{x} = k$$
$$\frac{y}{x} = 4$$
$$y = 4x$$

**Use Hooke's law, $F = kd$, to find the constant, $k$, for each spring described below.**

**24.** force: 2 Newtons; stretch: 0.6 meter

**25.** force: 10 Newtons; stretch: 0.4 meter

**26.** force: 10 Newtons; stretch: 1.0 meter

**27.** force: 5 Newtons; stretch: 0.1 meter

**28.** force: 20 Newtons; stretch: 0.8 meter

**If $y$ varies directly as $x$ for the values of $x$ and $y$ below, find the constant of variation, $k$, and write an equation of direct variation.**

**29.** $y = 4, x = 10$

**30.** $y = 16, x = 8$

**31.** $y = -5, x = 15$

**32.** $y = -12, x = 18$

**33.** $y = -4, x = -16$

**34.** $y = 2, x = 2$

**35.** $y = 4, x = -12$

**36.** $y = 6, x = 9$

LESSON 5.4

## Key Skills

**Find an equation in slope-intercept form for a line.**

Write an equation for the line
**a.** with a slope of 3 and $y$-intercept of 5.
**b.** passing through $(-2, 3)$ and $(-1, 5)$.

**a.** Substitute 3 for $m$ and 5 for $b$.

$$y = mx + b$$
$$y = 3x + 5$$

**b.** $m = \frac{5 - 3}{-1 - (-2)} = 2$

Substitute 2 for $m$, $-2$ for $x$, and 3 for $y$ in the equation $y = mx + b$. Then solve for $b$.

$$3 = 2(-2) + b$$
$$3 = -4 + b$$
$$7 = b$$

## Exercises

**Write an equation in slope-intercept form for each line described below.**

**37.** slope of 2 and $y$-intercept of 1

**38.** slope of $-\frac{1}{2}$ and $y$-intercept of 5

**39.** containing $(0, -4)$ with a slope of 8

**40.** containing $(4, 2)$ and $(5, -4)$

**41.** containing $(2, -1)$ and $(-3, -1)$

**42.** containing $(-6, 0)$ and $(0, -2)$

**43.** slope of $-3$ and $y$-intercept of 6

**44.** containing $(4, 6)$ and $(-2, 0)$

**45.** slope of 5 and $y$-intercept of $-2$

**46.** containing $(2, 5)$ and $(-6, 9)$

Substitute 2 for $m$ and 7 for $b$.

$$y = mx + b$$
$$y = 2x + 7$$

**Graph an equation in slope-intercept form.**

Graph the equation $y = \frac{1}{2}x + 5$.

The slope is $\frac{1}{2}$ and the $y$-intercept is 5.

Locate (0, 5) and move 2 units to the right and 1 unit up from that point. Graph this point. Draw the line containing the points.

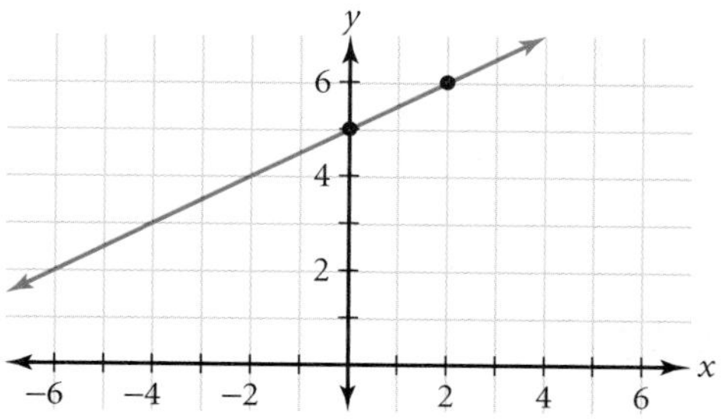


**Graph each equation.**

**47.** $y = -\frac{1}{2}x - 2$

**48.** $y = -x + 6$

**49.** $y = \frac{3}{4}x + 4$

**50.** $y = 4x - 1$

**51.** $y = \frac{2}{3}x + 5$

**52.** $y = -\frac{5}{4}x + 5$

**53.** $y = \frac{1}{2}x + 6$

**54.** $y = x + 4$

**55.** $y = -2x - 3$

**56.** $y = -4x + 1$

**LESSON 5.5**

## Key Skills

**Write equations in standard form.**

Write $y + 10 = -8x - 22$ in standard form.

$$y + 10 = -8x - 22$$
$$8x + y + 10 = -22$$
$$8x + y = -32$$

**Write equations in point-slope form.**

Write an equation for the line containing the points (4, 5) and (2, 1).

$m = \frac{1 - 5}{2 - 4} = \frac{-4}{-2} = 2$

Use the point (4, 5) for $(x_1, y_1)$.

$y - 5 = 2(x - 4)$

## Exercises

**Write the following equations in standard form:**

**57.** $y + 9 = 4x - 8$

**58.** $y - 4 = -x + 1$

**59.** $y - 13 = 2x + 4$

**60.** $3x + y + 6 = 9$

**Write an equation in point-slope form for the following lines:**

**61.** with a slope of 2 and a $y$-intercept of −1

**62.** containing (0, 4) and (1, 2)

**63.** containing (3, 5) with a slope of 4

**64.** crossing the $x$-axis at $x = -2$ and the $y$-axis at $y = 1$

**65.** with a slope of 3 and $y$-intercept of −3

**66.** containing (−2, 4) with a slope of −2

**67.** crossing the $x$-axis at $x = 3$ and the $y$-axis at $y = 6$

**68.** with a slope of −4 and $y$-intercept of −1

LESSON 5.6

## Key Skills

**Recongnize and write equations of lines parallel and perpendicular to other lines.**

Write an equation in slope-intercept form for the line containing the point (0, 4) and

**a.** parallel to the line $2x + 3y = 5$.
**b.** perpendicular to the line $2x + 3y = 5$.
Find the slope of the line $2x + 3y = 5$ by writing the equation in slope-intercept form.

$$y = -\frac{2}{3}x + \frac{5}{3}$$

**a.** The slope of the parallel line is $-\frac{2}{3}$. The $y$-intercept is 4. The equation of the parallel line is $y = -\frac{2}{3}x + 4$.

**b.** The slope of the perpendicular line is the negative reciprocal of $-\frac{2}{3}$, or $\frac{3}{2}$. The equation is $y = \frac{3}{2}x + 4$.

## Exercises

**Find the slopes of lines that are parallel and perpendicular to the following lines:**

**69.** $y = \frac{2}{3}x + 4$

**70.** $y = -7x - 1$

**Write an equation for the line that contains the point (1, 5) and is**

**71.** parallel to $2x + y = -1$.

**72.** perpendicular to $2x + y = -1$.

## Applications

**73.** **GEOGRAPHY** The elevation of Houston is 120 feet. The elevation of El Paso is 4000 feet. Suppose that the distance between the two cities is a straight line of 760 miles. Traveling from Houston to El Paso, what is the average slope of the land?

**TRAVEL** Marie had driven 110 miles when she stopped at a gas station. After she left the station, she kept track of the hours she drove. Based on an average of 50 miles per hour for the rest of the trip, write an equation to find the total distance she drove. Determine the total distance traveled if she drove the following number of hours after leaving the gas station:

**74.** 1.6 hours
**75.** 2.9 hours
**76.** 3.3 hours

**77.** **METEOROLOGY** If the air temperature is 76°F at 8 A.M. and 92°F at 4 P.M., find the rate of change of temperature per hour.

**78.** **METEOROLOGY** Would the rate of change in Exercise 77 be different if you used the Celsius temperature scale and converted hours to minutes? Write a paragraph explaining your answer.

**79.** **INCOME** A person's rate is the amount that he or she earns per hour. If you earn $500 for a 40-hour work week, what is your pay rate?

# Chapter Test

**Determine whether each relation is a function and give the domain and range of the relation.**

1. $\{(2, 4), (3, 9), (4, 4), (5, 9)\}$
2. $\{(3, 6), (5, 6), (7, 6), (9, 6)\}$
3. $\{(8, 5), (8, 3), (8, -3), (8, -5)\}$
4. $\{(3, 1), (4, 1), (5, 9), (2, 6)\}$

**Complete each ordered pair so that it is a solution to the equation $3x + y = 7$**

5. $(1, \underline{?})$ 6. $(-2, \underline{?})$ 7. $(\underline{?}, 2)$

**Find the slope of each line under the given conditions.**

8. rise of 7 and run of 3
9. containing the points $(1, 5)$, and $(4, 8)$
10. passing through the origin and $(c, d)$
11. horizontal line through $(5, 7)$
12.

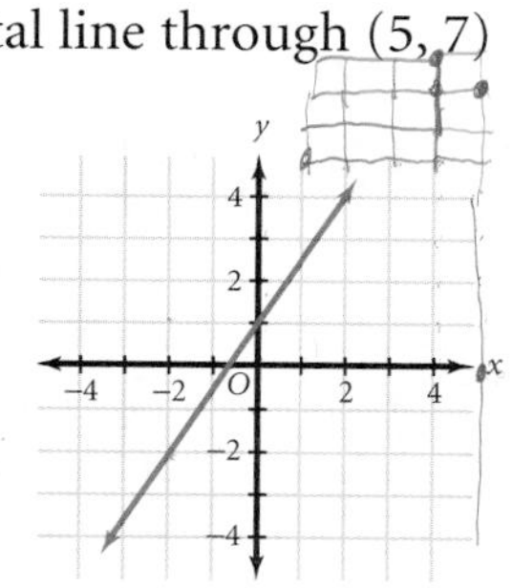


13. SPORTS A ski hill has a vertical drop of 1500 feet over a distance of 4200 feet. What is the slope of the ski hill?
14. INCOME The graph below represents an employee's daily earnings as a function of the number of hours worked. Use the graph to find the employee's rate of pay.

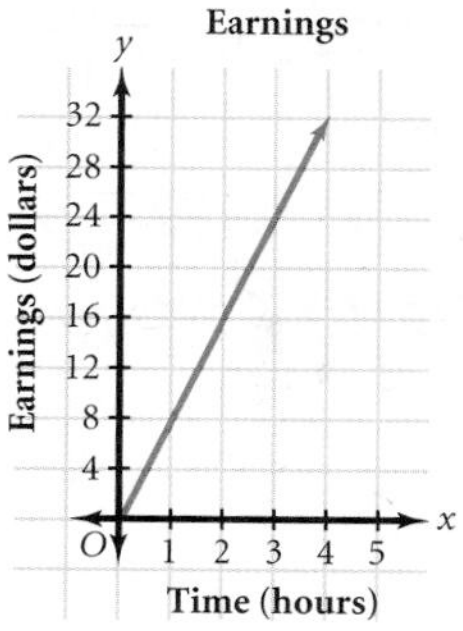


**For the following values, $y$ varies directly as $x$. Find the missing value.**

15. $y$ is 21 when $x$ is 3. Find $x$ when $y$ is 28.
16. $y$ is 48 when $x$ is 6. Find $y$ when $x$ is 15.
17. PHYSICS A force of 25 Newtons stretches a spring a distance of 0.8 m. How far will a force of 40 Newtons stretch the spring?
18. Describe the graph of $y = \frac{2}{3}x - 7$. Include information about slope and $y$-intercept.

**Write each equation in standard form.**

19. $y - 8 = 3x + 4$ 20. $3 - 2x = 5 - 7y$

**Write an equation in slope-intercept form for each line described below.**

21. slope of 5 and $y$-intercept of $-3$
22. containing $(2, 9)$ with a slope of $\frac{5}{8}$
23. containing $(9, -3)$ and $(0, 5)$
24. passing through $(3, 4)$ and $(5, 8)$
25. containing $(-5, 9)$ and $(-5, 6)$
26. $x$-intercept of 5 and $y$-intercept of 3
27. ECOLOGY The water level of a reservoir is at 27 feet. During a drought the water level falls by $1\frac{1}{2}$ feet per week. Write an equation for the water level $L$ after $w$ weeks. Graph the equation. In how many weeks will the water level be at 18 feet?

**Write an equation in slope-intercept form for each line, according to the given information.**

28. parallel to $y = 4x + 5$, contains $(2, -1)$
29. perpendicular to $y = \frac{2}{3}x - 8$, contains $(-1, 2)$

# College Entrance Exam Practice

**internet connect** go.hrw.com
**Standardized Test Prep Online**
Go To: **go.hrw.com**
Keyword: **MM1 Test Prep.**

**QUANTITATIVE COMPARISON** For Questions 1–6, write
A if the quantity in Column A is greater than the quantity in Column B;
B if the quantity in Column B is greater than the quantity in Column A;
C if the two quantities are equal; or
D if the relationship cannot be determined from the information given.

| | Column A | Column B | Answers |
|---|---|---|---|
| **1.** | The solution to the equation $4x + 7 = -13$ | The solution to the equation $-3x + 11 = 29$ | (A) (B) (C) (D) [Lesson 3.3] |
| **2.** | The slope of the line containing (7, 2) and (8, 7) | The slope of the line containing (−1, −3) and (2, 9) | (A) (B) (C) (D) [Lesson 5.2] |
| **3.** | (−21)(−4) | (−7)(12) | (A) (B) (C) (D) [Lesson 2.4] |
| **4.** | First differences of the sequence 2, 7, 12, 17, . . . | First differences of the sequence 1, 5, 14, 28, . . . | (A) (B) (C) (D) [Lesson 1.1] |
| **5.** | 20% of 120 | 35% of 70 | (A) (B) (C) (D) [Lesson 4.2] |
| **6.** | The slope of a horizontal line | The slope of a line containing (0, 1) and (9, 1) | (A) (B) (C) (D) [Lesson 5.2] |

**7.** What is the solution to the equation $3p - 173 = 220$? ***(LESSON 3.3)***
**a.** 78 **b.** 131
**c.** 15.6 **d.** none of these

**8.** What number is missing from the following sequence? ***(LESSON 1.1)***
12, 36, __, 84, 108, 132
**a.** 52 **b.** 70
**c.** 60 **d.** 44

**9.** Evaluate the expression $-s + 2t$ for $s = -3$ and $t = 7$. ***(LESSON 2.4)***
**a.** −7 **b.** 17
**c.** 11 **d.** none of these

**Solve each proportion.** ***(LESSON 4.1)***

**10.** $\frac{x}{7} = \frac{52}{13}$ **11.** $\frac{1.2}{y} = \frac{2.3}{13.8}$ **12.** $\frac{63}{102} = \frac{p}{64}$

**Write an equation for the line containing the origin and the points below.** ***(LESSON 5.2)***

**13.** (2, 1) **14.** (−3, 6)

**15.** (4, −12) **16.** (−1, −13)

**17.** Solve $-4w + 17 = 65$ for $w$. ***(LESSON 3.3)***

**Simplify.** ***(LESSON 1.3)***

**18.** $4 \cdot 3 + 12 \div 2$

**19.** $[14(8 \div 4) + 7] \div 5$

**Find each answer.** ***(LESSON 4.2)***

**20.** What is 15% of 110?

**21.** 105 is 70% of what number?

**22.** 12 is what percent of 50?

**If pencils cost 29 cents, find the cost of the following:** ***(LESSON 1.2)***

**23.** 3 pencils

**24.** 5 pencils

**25.** 14 pencils

**26.** Write an equation to model this situation: How many pencils can you buy for $8.70? ***(LESSON 1.2)***

**27.** Write an equation in standard form for the line graphed below. ***(LESSON 5.5)***

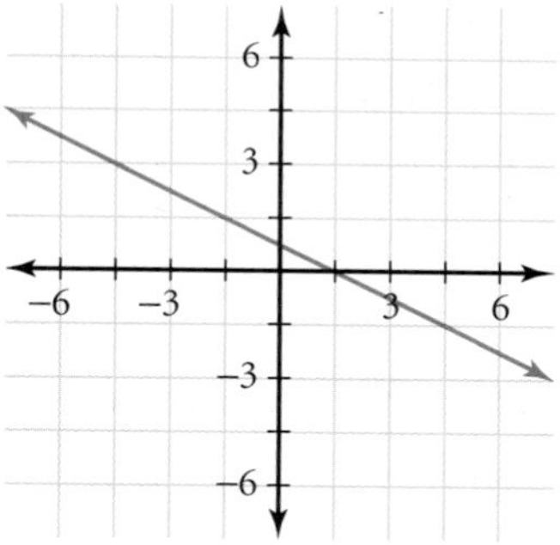


**Write an equation in slope-intercept form for each line described below.** ***(LESSON 5.5)***

**28.** slope = −3, $y$-intercept = 4

**29.** slope = 7, containing (0, −2)

**30.** containing (0, 2) and (2, 5)

**31.** Solve the equation $y - 7 = 24$. ***(LESSON 3.1)***

**Find the mean, median, mode, and range for each set of data.** ***(LESSON 4.4)***

**32.** {8, 11, 7, 9, 5}

**33.** {98, 91, 87, 100, 88}

**34.** {0.21, 0.17, 0.28, 0.14}

**35.** Determine the slope of the line represented by $2x + 4y = 1$. ***(LESSON 5.5)***

**36.** Solve: $3x - 5 = 13$. ***(LESSON 3.3)***

**37.** Solve: $4 + x = 6$. ***(LESSON 3.1)***

**38.** Solve: $2x - 5 = 7$. ***(LESSON 3.3)***

**39.** Solve: $x - 5 = 13$. ***(LESSON 3.1)***

**40.** Solve: $\frac{38}{14} = \frac{a}{7}$. ***(LESSON 4.1)***

**41.** Find 60% of 180. ***(LESSON 4.2)***

**42.** Determine the next three terms of the following sequence: 1, 4, 8, 13, 19, … ***(LESSON 1.1)***

**43.** Evaluate: $2 - [(3 \cdot 4 - 6 \cdot 7) \cdot 3 - 4]$. ***(LESSON 1.3)***

**44.** Find 30% of 180. ***(LESSON 4.2)***

FREE-RESPONSE GRID

**The following questions may be answered by using a free-response grid such as that commonly used by standardized test services.**

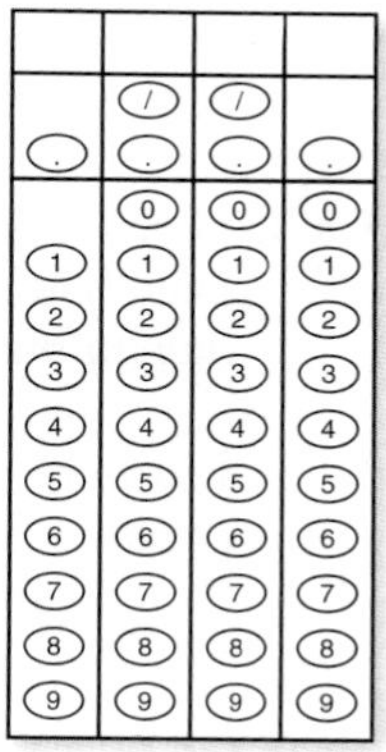


**45.** Find the slope of a line passing through $A(4, 5)$ and $B(-2, 2)$. ***(LESSON 5.2)***

**46.** Simplify: $5(7 + 13) - 40 \div 8$. ***(LESSON 1.3)***

**47.** What is 55% of 20? ***(LESSON 4.2)***

**48.** Solve the equation $\frac{t}{3.5} = 14.2$. ***(LESSON 3.2)***

6

# Inequalities and Absolute Value

## Lessons



THE CALIPERS, MICROMETER, AND SPHYGMOmanometer (blood-pressure gauge) shown on these pages are just three of the many measuring instruments used daily in work environments. But measurements are never absolute. Whatever unit is chosen as being accurate enough for the work at hand, a smaller unit is always possible. For example, centimeters or millimeters are the smallest units found on a common ruler, but physicists frequently use nanometers (one-billionth of a meter) or picometers (one-trillionth of a meter).

Because of the impossibility of making absolute measurements, there is always a *margin of error* in the size of manufactured products. Quality control standards are determined on the basis of how large a difference from the standard can be allowed without affecting the performance of the product. Similarly, statistical probability standards determine the margin of error that is acceptable in opinion polls and scientific studies.

Inequalities and absolute value can be used to express margins of error, and in this chapter, you will learn the meaning of these concepts and methods for applying them to a variety of situations.

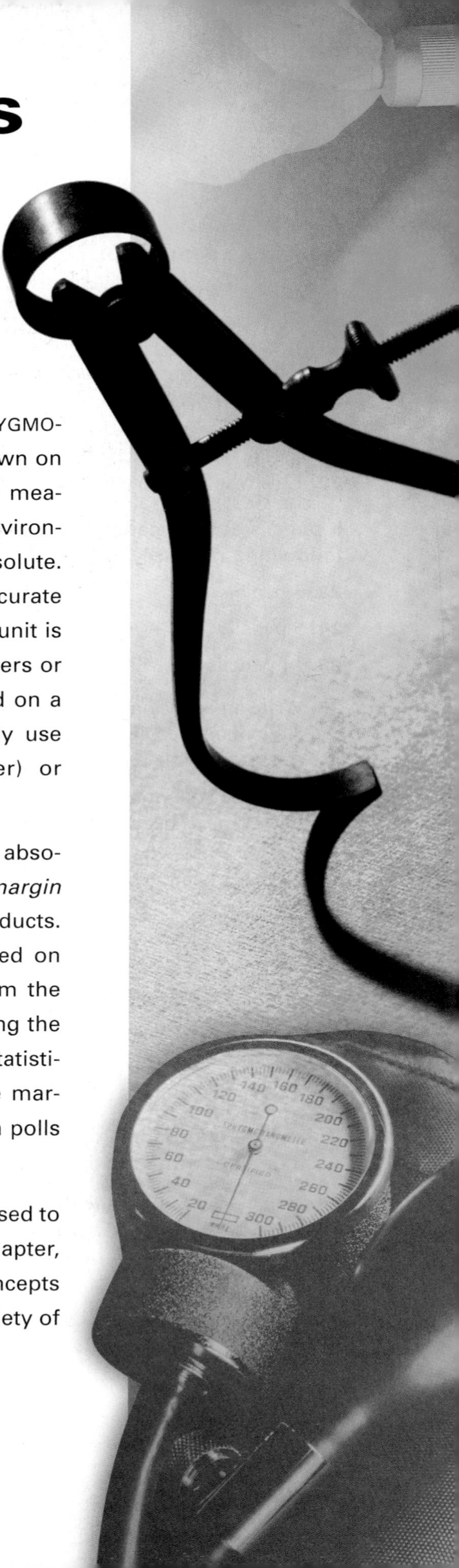

*The micrometer can be used to measure relatively small distances such as the diameter of a human hair.*

*A sphygmomanometer is used to measure blood pressure.*

*Precision instruments, such as assembly line robots, need to work within a very narrow margin of error.*

CHAPTER PROJECT
PORTFOLIO ACTIVITIES

## About the Chapter Project

Different algebraic functions can be used to model different situations. In the Chapter Project, *Planning a Snowboarding Trip*, on page 308, you will investigate the use of a *step function* to model a problem situation. After completing the Chapter Project, you will be able to do the following:

- Recognize the relationship between pairs of numbers as a step function.
- Graph a step function.
- Recognize, write, and graph a *greatest integer* function (an important type of step function).

## About the Portfolio Activities

Throughout the chapter, you will be given opportunities to complete Portfolio Activities that are designed to support your work on the Chapter Project.

- Piecewise functions are the topic of the Portfolio Activity on page 281.
- Absolute-value functions are the topic of the Portfolio Activity on page 288.

6.1

# Solving Inequalities

**Objectives**

- State and use symbols of inequality.
- Solve inequalities that involve addition and subtraction.

**Why** ***Many real-world problems, such as the price of a lunch, involve finding a range of possibilities instead of one specific answer. Inequalities can be used to model this type of problem.***

## Inequalities

When you are operating on a budget you need to keep different expenses under certain levels. For example, you may need to keep the price of a lunch below $5.

An **inequality** is a statement that two expressions are not equal. The sentence "Lunch will cost less than $5" can be represented by the inequality $L < 5$. The solution to an inequality is the set of numbers that make the inequality true. In this case, the range of values in the solution includes any amount less than $5.

Several inequalities are used in algebra.

| Statements of Inequality | |
|---|---|
| $a$ is less than $b$. | $a < b$ |
| $a$ is greater than $b$. | $a > b$ |
| $a$ is less than or equal to $b$ (or $a$ is at most $b$). | $a \leq b$ |
| $a$ is greater than or equal to $b$ (or $a$ is at least $b$). | $a \geq b$ |
| $a$ is greater than $b$ and less than $c$. | $b < a < c$ |
| $a$ is greater than or equal to $b$ and less than or equal to $c$. | $b \leq a \leq c$ |
| $a$ is not equal to $b$. | $a \neq b$ |

**EXAMPLE 1** Michael can spend at most \$3.10 for lunch. He buys a hamburger and a drink for \$2.15. **Write an inequality that models how much Michael can spend on dessert and stay within his spending limit.**

**APPLICATION**
**CONSUMER ECONOMICS**

**SOLUTION**

Let $d$ be the amount Michael can spend on dessert. The total amount he can spend is $d + 2.15$. Michael's situation can be modeled by the inequality $d + 2.15 \le 3.10$, which is read as "$d$ plus 2.15 is less than or equal to 3.10."

**TRY THIS** Trish has only \$3.75 to spend for lunch. She buys a cheeseburger for \$2.50. Write an inequality that models how much Trish can spend on dessert and stay within her spending limit.

## Properties of Inequality

The inequality $d + 2.15 \le 3.10$ can be thought of as the combination of the following two statements:

$$d + 2.15 < 3.10 \text{ or } d + 2.15 = 3.10$$

The equation $d + 2.15 = 3.10$ is true when $d = 0.95$. Notice that the inequality $d + 2.15 < 3.10$ is true when $d < 0.95$. This suggests that properties used to solve inequalities are similar to those used to solve equations.

### Addition and Subtraction Properties of Inequality

Let $a$, $b$, and $c$ be real numbers.

If $a < b$, then $a + c < b + c$.
If $a > b$, then $a + c > b + c$.

If $a < b$, then $a - c < b - c$.
If $a > b$, then $a - c > b - c$.

**EXAMPLE 2** **Solve the inequality $m - 8 \geq 2$. Check the solution.**

**SOLUTION**

$m - 8 \geq 2$

$m - 8 + 8 \geq 2 + 8$ — *Addition Property of Inequality*

$m \geq 10$ — *Simplify.*

To check, substitute values near 10 into the original inequality to see if the result is true.

Substitute 10 for $m$.

$10 - 8 \overset{?}{\geq} 2$

$2 \geq 2$ *True*

Substitute 11 for $m$.

$11 - 8 \overset{?}{\geq} 72$

$3 \geq 2$ *True*

**TRY THIS** Solve the inequality $c + 7 < -11$.

## Number Lines and Inequalities

The solution to an inequality is the set of all numbers that make the statement true. The solution set might be listed, described, or graphed.

An inequality is graphed as an interval, a ray, or a line. Included endpoints are shown as dots, •, while endpoints that are not included are shown as open circles, ∘.

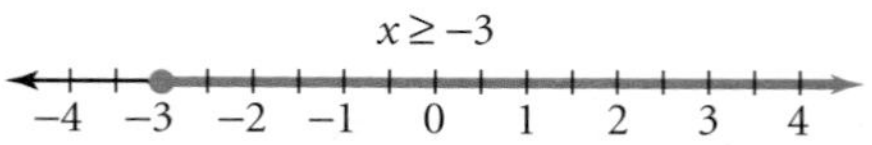


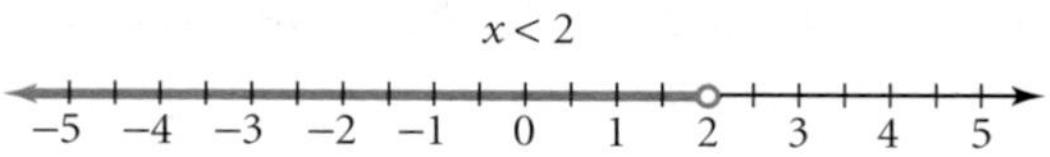


**EXAMPLE 3** **Graph the solution to $5 > x - 3$.**

**SOLUTION**

First solve the inequality for $x$.

$5 > x - 3$ — *Given*

$5 + 3 > x - 3 + 3$ — *Addition Property of Inequality*

$8 > x$ — *Simplify.*

$x < 8$ — *It is customary to put the variable to the left of the inequality sign in the final answer.*

To graph the result on a number line, shade all points to the left of 8 on the number line. This includes all points whose coordinates are less than 8. To indicate that 8 is *not included*, put an *open circle* at 8.

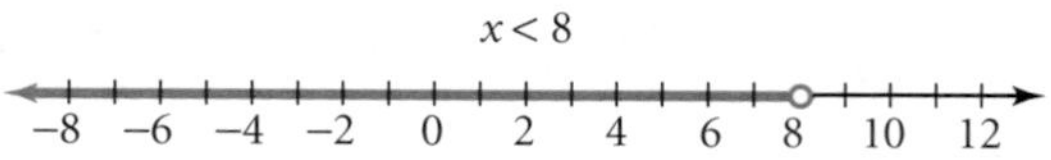


**TRY THIS** Graph the solution to $t + 7 < -3$.

**EXAMPLE 4** **Write an inequality that describes the points graphed on each number line.**

**a.** 

**b.** 

**SOLUTION**

**a.** This number line shows a ray that includes all real numbers greater than –4. Since the endpoint at –4 is an open circle, –4 is not included in the solution set. The inequality that describes this graph is $x > -4$.

**b.** This number line shows a ray that includes all real numbers less than or equal to 3. Since the endpoint at 3 is shaded, 3 is included in the solution set. The inequality that describes this graph is $x \leq 3$.

**TRY THIS** Write an inequality that describes the points graphed on each number line.

**a.** 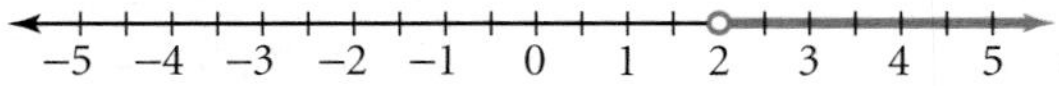


**b.** 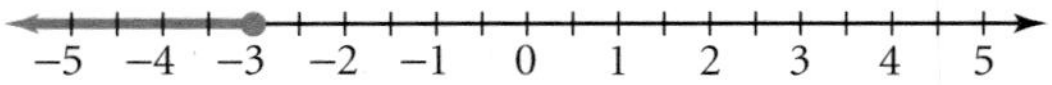


# Exercises

## Communicate

**1.** How are the Addition and Subtraction Properties of Inequality similar to the Addition and Subtraction Properties of Equality?

**2.** What steps are necessary to solve the inequality $x - 4 \leq 1$? Name the property you would use for each step.

**3.** Explain how to graph the solution to $x + 3 < 7$ on a number line.

**4.** How would you write an inequality that describes the points graphed on the number line below?

## Guided Skills Practice

**APPLICATION**

**5.** **CONSUMER ECONOMICS** Jerry has only \$27 for school supplies. He spends \$18 on a backpack. Write an inequality that models how much more Jerry can spend on school supplies and stay within his limit. ***(EXAMPLE 1)***

**Solve each inequality and graph each solution on a number line.** ***(EXAMPLES 2 AND 3)***

**6.** $b + 1 \leq -11$ **7.** $x - 9 > 5$ **8.** $m - 14 \geq -8$ **9.** $-2 < t + 3$

**10.** $x - 3 \geq 4$ **11.** $y + 7 < 2$ **12.** $3 \leq x - 17$ **13.** $-5 > 4 + x$

**14.** Write an inequality that describes the points graphed on the number line below. ***(EXAMPLE 4)***

## Practice and Apply

**internet connect**
**Homework Help Online**
Go To: **go.hrw.com**
Keyword:
**MA1 Homework Help**
for Exercises 15-35

**Solve each inequality and graph each solution on a number line.**

**15.** $x + 8 > -1$ **16.** $x - 6 \leq 7$ **17.** $18 < x - 7$

**18.** $b + 4 \leq 18$ **19.** $67 \geq y + 28$ **20.** $a + 18 > -3$

**21.** $m - 9 \leq 40$ **22.** $a - 7 > 2.3$ **23.** $x + 0.04 > 0.6$

**24.** $8 > x - 0.1$ **25.** $x + \frac{3}{4} \geq \frac{1}{2}$ **26.** $1 > x + \frac{3}{4}$

**27.** $v - 3 \leq 17$ **28.** $17 > b - 5$ **29.** $t - 76 \leq 50$

**30.** $h + 15 \geq 5$ **31.** $v + 6.2 \geq 8.1$ **32.** $m - 2.2 < -12.2$

**33.** $x + \frac{2}{3} \leq \frac{5}{9}$ **34.** $t - \frac{7}{4} \leq \frac{3}{4}$ **35.** $c - \frac{3}{7} \geq \frac{6}{13}$

**Write an inequality that describes the set of points graphed on each number line.**

**36.**
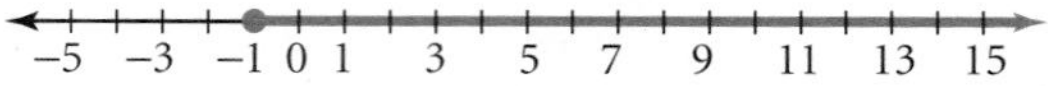


**37.**
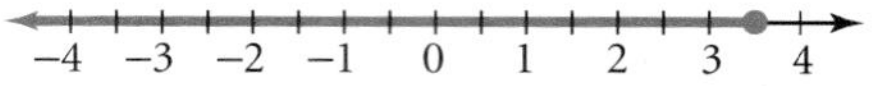


**38.**
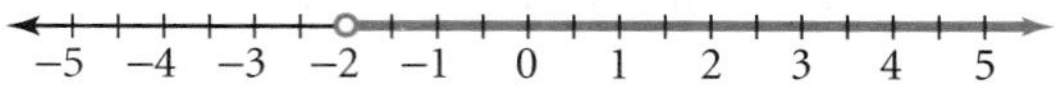


**CHALLENGES**

**39.** If 8 more than twice a number is less than 6 more than the number, what can the number be?

**40.** If 3 less than 4 times a number is greater than or equal to 3 less than 7 times the number, what is the largest the number can be?

**APPLICATIONS**

**41. AUDITORIUM SEATING** A school auditorium can seat 450 people for graduation. The graduates will use 74 seats. Write and solve an inequality to describe the number of additional people who can be seated in the auditorium.

**42. INVESTMENTS** A bank promises that the interest rate on a certain savings account will never drop below 3.5%. Write an inequality to represent this situation.

**APPLICATION**

**TEMPERATURE** Write an inequality or an equation in terms of $t$ to represent each situation.

**43.** a temperature of at least 40°F

**44.** a temperature of no more than 54°F

**45.** a temperature above 34°F

**46.** a temperature of 78°F

**47.** Let $t$ represent an initial temperature reading. After a rise of 5°F, the temperature was between 70°F and 80°F inclusive. What inequality represents this situation?

## Look Back

**48.** The points on a scatter plot lie near a line that rises from left to right. Name the type of correlation. *(LESSON 1.6)*

**49.** The points on a scatter plot lie near a line that falls from left to right. Name the type of correlation. *(LESSON 1.6)*

**Evaluate.** *(LESSON 2.1)*

**50.** $|-4.5|$ **51.** $|-9|$ **52.** $|-3 + 4|$

**Solve.** *(LESSONS 3.1 AND 3.3)*

**53.** $x + 7 = 10$ **54.** $15 - x = 25$ **55.** $3 - 2x = 17$

## Look Beyond

**Solve for *x*.**

**56.** $bx + c = a$ **57.** $tx - 7 = r$ **58.** $Ax + Bx = C$

**Find three values of *x* that make each inequality true and three values of *x* that make each inequality false.**

**59.** $2x < 8$ **60.** $4x + 5 \leq 16$ **61.** $8x - 3 > 33$

**PORTFOLIO ACTIVITY**

Include this activity in your portfolio.

Some functions have graphs that can be assembled from sections of other graphs. This type of function is called a *piecewise function.* The following function is a piecewise function:

$$y = \begin{cases} -x \text{ if } x < 0 \\ x \text{ if } x \geq 0 \end{cases}$$

The table below has been completed for $x$-values of $-2$ and 0.5. Because $x < 0$ for $x = -2$, the equation $y = -x$ is used to determine $y$, and $y = -(-2) = 2$. Because $x \geq 0$ for $x = 0.5$, the equation $y = x$ is used to determine $y$, and $y = 0.5$. Complete the table for the remaining $x$-values.

| $x$ | −4 | −3 | −2 | −1 | −0.5 | 0 | 0.5 | 1 | 2 | 3 | 4 |
|---|---|---|---|---|---|---|---|---|---|---|---|
| $y$ | ? | ? | 2 | ? | ? | ? | 0.5 | ? | ? | ? | ? |

# 6.2 Multistep Inequalities

*By sending a signal to the lander, scientists were able to pinpoint* Pathfinder*'s landing site on Mars to within 5 meters of its actual location. Thus the lander's position was a point within a range of values.*

## Objectives

- State and apply the Multiplication and Division Properties of Inequality.
- Solve multistep inequalities in one variable.

**Why** ***Many real-world situations are solved not by finding a single answer, but by finding a minimum, a maximum, or a range of answers that satisfy the conditions. For problems of this type, you often need to solve an inequality.***

The activity and the list of properties that follow will help you to understand how to use multiplication and division to solve inequalities.

*Activity*

## Multiplying and Dividing Inequalities

1. Copy and complete the table. Notice that in each row of the table $A < B$.

| $A$ | $B$ | $2A$ | $2B$ | $-2A$ | $-2B$ | $\frac{A}{2}$ | $\frac{B}{2}$ | $\frac{A}{-2}$ | $\frac{B}{-2}$ |
|---|---|---|---|---|---|---|---|---|---|
| 2 | 4 | 4 | 8 | −4 | −8 | 1 | 2 | −1 | −2 |
| −2 | 2 | ? | ? | ? | ? | ? | ? | ? | ? |
| −4 | 6 | ? | ? | ? | ? | ? | ? | ? | ? |
| −6 | −2 | ? | ? | ? | ? | ? | ? | ? | ? |
| −8 | −6 | ? | ? | ? | ? | ? | ? | ? | ? |

2. **a.** Compare $2A$ and $2B$ in each row. Write your findings as an inequality.
**b.** Compare $-2A$ and $-2B$ in each row. Write your findings as an inequality.

3. **a.** Compare $\frac{A}{2}$ and $\frac{B}{2}$ in each row. Write your findings as an inequality.
**b.** Compare $\frac{A}{-2}$ and $\frac{B}{-2}$ in each row. Write your findings as an inequality.

**CHECKPOINT ✔**

4. **a.** Write a rule to explain what you must do to an inequality when you multiply or divide each side by the same positive number.
**b.** Write a rule to explain what you must do to an inequality when you multiply or divide each side by the same negative number.

### Multiplication and Division Properties of Inequalities

Let $a$, $b$, and $c$ be nonzero real numbers.

For $c > 0$:

If $a < b$, then $ac < bc$ and $\frac{a}{c} < \frac{b}{c}$.

If $a > b$, then $ac > bc$ and $\frac{a}{c} > \frac{b}{c}$.

For $c < 0$:

If $a < b$, then $ac > bc$ and $\frac{a}{c} > \frac{b}{c}$.

If $a > b$, then $ac < bc$ and $\frac{a}{c} < \frac{b}{c}$.

These properties are the same for inequalities involving $\leq$ and $\geq$.

Example 1 illustrates the use of multiplication and division by positive numbers in solving inequalities.

**EXAMPLE 1** **Solve and graph each inequality.**

**a.** $3x > 6$

**b.** $16 \leq 2y$

**c.** $-25 \geq 5x$

**d.** $\frac{a}{3} \geq 4$

**SOLUTION**

**a.** $3x > 6$
$$\frac{3x}{3} > \frac{6}{3}$$
$$x > 2$$

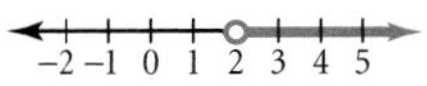


**b.** $16 \leq 2y$
$$\frac{16}{2} \leq \frac{2y}{2}$$
$$8 \leq y$$
$$y \geq 8$$

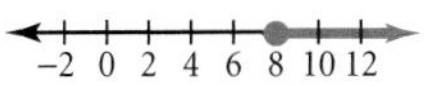


**c.** $-25 \geq 5x$
$$\frac{-25}{5} \geq \frac{5x}{5}$$
$$-5 \geq x$$
$$x \leq -5$$

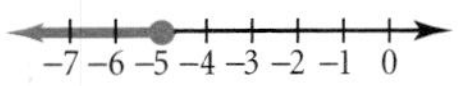


**d.** $\frac{a}{3} \geq 4$
$$3\left(\frac{a}{3}\right) \geq 3(4)$$
$$a \geq 12$$

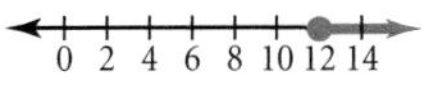


**TRY THIS** Solve and graph.

**a.** $6y > 12$

**b.** $18 < 3x$

**c.** $5m \leq -35$

**d.** $\frac{x}{7} \geq -6$

Example 2 illustrates the use of multiplication or division by negative numbers in solving inequalities.

**EXAMPLE 2** **Solve each inequality.**

**a.** $-x > 9$ **b.** $-2m < 10$

**c.** $15 \leq -5x$ **d.** $-\frac{r}{7} \geq -3$

**SOLUTION**

*Either multiplying or dividing by –1 will change –x to x.*

**a.**
$$-x > 9$$
$$-1(-x) < -1(9)$$
$$x < -9$$

**b.**
$$-2m < 10$$
$$\frac{-2m}{-2} > \frac{10}{-2}$$
$$m > -5$$

**c.**
$$15 \leq -5x$$
$$\frac{15}{-5} \geq \frac{-5x}{-5}$$
$$-3 \geq x$$
$$x \leq -3$$

**d.**
$$-\frac{r}{7} \geq -3$$
$$-7\left(-\frac{r}{7}\right) \leq -7(-3)$$
$$r \leq 21$$

**TRY THIS** Solve.

**a.** $-x < 12$ **b.** $-4x > 24$ **c.** $-16x \leq 32$ **d.** $\frac{x}{-9} \leq -108$

Example 3 illustrates multi-step inequalities, where multiplication (or division) and addition (or subtraction) are both required in the solutions.

**EXAMPLE 3** **Solve each inequality.**

**a.** $3x - (2 + 6x) \leq 28$

**b.** $2x - 6 \geq 3x - 2$

**SOLUTION**

**a.**

| | |
|---|---|
| $3x - (2 + 6x) \leq 28$ | *Given* |
| $3x - 2 - 6x \leq 28$ | *Distributive Property* |
| $-3x - 2 \leq 28$ | *Simplify.* |
| $-3x - 2 + 2 \leq 28 + 2$ | *Addition Property of Inequality* |
| $-3x \leq 30$ | *Simplify.* |
| $\frac{-3x}{-3} \geq \frac{30}{-3}$ | *Division Property of Inequality* |
| $x \geq -10$ | |

**b.** When the variable appears on both sides of the inequality, you can often avoid inequality-sign changes by choosing the operation(s) which will result in a positive coefficient for the variable.

$$2x - 6 \geq 3x - 2$$
$$2x - 2x - 6 \geq 3x - 2x - 2$$
$$-6 \geq x - 2$$
$$-6 + 2 \geq x$$
$$-4 \geq x$$
$$x \leq -4$$

*By choosing to subtract 2x in the second step, x has a positive coefficient in the the third step.*

**TRY THIS** Solve each inequality.

**a.** $2m + 7(m - 1) \geq 26$

**b.** $3p + 2 \leq 7p - 5$

**EXAMPLE 4** Eastern Cellular offers two monthly phone plans. **For what range of minutes would it cost less to use plan 1?**

APPLICATION
**CONSUMER ECONOMICS**

**SOLUTION**

**Method A**
First write an inequality to represent the problem.

Plan 1: $25 + 0.5m$
Plan 2: $0.7m$

| plan 1 | is less than | plan 2 |
|---|---|---|
| ↓ | ↓ | ↓ |
| $25 + 0.5m$ | $<$ | $0.7m$ |

Solve the inequality.

$$25 + 0.5m < 0.7m$$

$$25 + 0.5m - 0.5m < 0.7m - 0.5m \qquad \textit{Subtraction Property of Inequality}$$

$$25 < 0.2m \qquad \textit{Simplify.}$$

$$\frac{25}{0.2} < \frac{0.2m}{0.2} \qquad \textit{Division Property of Inequality}$$

$$125 < m \qquad \textit{Simplify.}$$

$$m > 125$$

Plan 1 will cost less when the phone is used for more than 125 minutes per month.

TECHNOLOGY
**GRAPHICS CALCULATOR**

**Method B**
You can use the table feature on a graphics calculator. Enter the expression for plan 1 as **Y1** and the expression for plan 2 as **Y2**. Then key in the table feature.

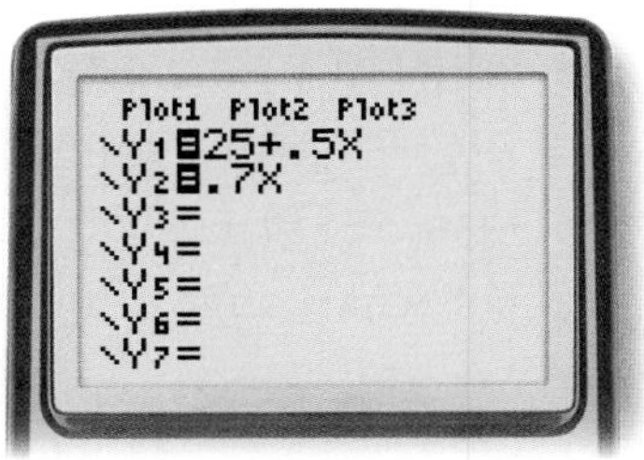


The table shows that the costs are equal at 125 minutes. Plan 1 costs less when the number of minutes is greater than 125.

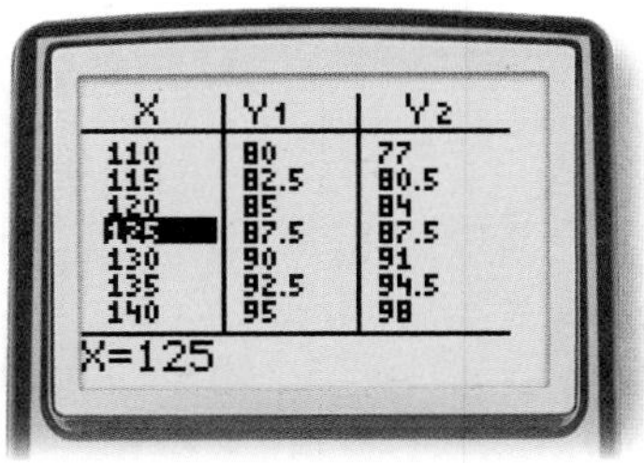


**TRY THIS** Jennifer, the manager of a fast-food restaurant, can choose one of two salary options.

Option 1: $2000 a month
Option 2: $500 a month plus $10 dollars per hour

For what range of hours would option 2 pay more?

# Exercises

## Communicate

**Describe the steps needed to solve each inequality.**

**1.** $x + 1 > 4$ **2.** $x - 3 \leq 13$ **3.** $-3p < 12$ **4.** $-4x - 2 \geq 2x + 3$

**5.** How do the Multiplication and Division Properties of Inequality differ from the Multiplication and Division Properties of Equality?

**6.** Why is $x < -2$ not the solution to the inequality $-3x < 6$?

**7.** Demonstrate why the inequality symbol must be reversed when both sides of $a < b$ are multiplied by $-1$. Use both a numerical example and a number line to support your reasoning.

## Guided Skills Practice

**Solve and graph each inequality.** ***(EXAMPLE 1)***

**8.** $8x \leq 48$ **9.** $3x < 48$ **10.** $\frac{x}{5} \geq -4$ **11.** $\frac{d}{7} \geq -9$

**Solve each inequality.** ***(EXAMPLE 2)***

**12.** $-3b > 27$ **13.** $-4x < -16$ **14.** $-\frac{x}{3} \geq 12$ **15.** $-\frac{x}{7} < -3$

**Solve each inequality.** ***(EXAMPLE 3)***

**16.** $2x - (4x + 2) \leq 38$ **17.** $-18y - 4 > 50$

**APPLICATION**

**18.** **CONSUMER ECONOMICS** A digital telephone service provider offers its customers two monthly service plans.

Plan A: \$30 + \$0.50 per minute Plan B: \$10 + \$0.75 per minute

For what range of minutes does plan B cost less? ***(EXAMPLE 4)***

## Practice and Apply

**internet connect**
**Homework Help Online**
Go To: **go.hrw.com**
Keyword:
**MA1 Homework Help** for Exercises 19-66

**Solve and graph each inequality.**

**19.** $2h > 32$ **20.** $8x \leq -56$ **21.** $5y \leq 22$

**22.** $-12b > 3$ **23.** $\frac{y}{6} > 17$ **24.** $\frac{z}{4} \leq -8$

**25.** $\frac{x}{8} < 1$ **26.** $\frac{u}{-3} \geq 21$ **27.** $-3y \geq 48$

**28.** $-6z \leq 96$ **29.** $-\frac{1}{5}m \geq 8$ **30.** $-\frac{g}{14} \leq 7$

**Solve each inequality.**

**31.** ~~$2x + 8 \geq 11$~~ **32.** $4x - 11 < 20$ **33.** $3x - 7 > 18$

**34.** $2.4z + 9.6 \geq 21.3$ **35.** ~~$8 - h > 9$~~ **36.** $6 - x > -1$

**37.** $-5b + 4 \leq 18$ **38.** $-3a - 7 > 2.3$ **39.** $6y + 7 \geq 2y + 28$

**40.** $1.3b - 9.2 < 7.7$ **41.** $4m + 1 < 5m - 7$ **42.** $10.8 \leq 2c - 11.3$

**43.** $38.25 > -2q + 29.5$ **44.** $\frac{-d}{3} + 4 < 10$ **45.** $\frac{-f}{5} - 1 < 3$

**Solve each inequality.**

**46.** $15 - \frac{t}{4} > 10$

**47.** $\frac{x}{4} - 2 \leq 13$

**48.** $\frac{5}{7}x + 27 \leq 12$

**49.** $2x + 5 \leq -3$

**50.** $-3t > 8$

**51.** $5c - 9 > 20$

**52.** $3(m + 7) \leq 28$

**53.** $4p + 7 - p \leq 31$

**54.** $7m - 2(m - 1) > -48$

**55.** $3p + (2 - 3p) > 28$

**56.** $y + 5 + 5y \geq -7$

**57.** $2(m - 3) + 8m \leq 11$

**58.** $-3b + 12 \leq 16$

**59.** $4y + 5 \geq y + 26$

**60.** $-6r - 10 < 3r - 8$

**61.** $3.3q - 8.4 < 10.8$

**62.** $-5.6h + 8.6 > 2.3$

**63.** $\frac{p}{7} - \frac{1}{14} \leq \frac{1}{2}$

**64.** $-\frac{2}{3}n + 12 < 21$

**65.** $\frac{1}{4}d - \frac{3}{9} > 2$

**66.** $-\frac{7}{8}m - 9 \leq 40$

CONNECTION

**67.** **GEOMETRY** The length, $l$, of a rectangle is at least 5 centimeters more than the width, $w$.

**a.** Express the statement above as an inequality.

**b.** Write an inequality for the statement above when the width is 20 centimeters.

APPLICATIONS

**68.** **SMALL BUSINESS** Brian charges \$15 plus \$5.50 per hour to mow yards. Greg charges \$12 plus \$6.25 per hour.

**a.** Write two equations to show how much Brian and Greg charge, $c$, in terms of the number of hours, $h$, that they work.

**b.** When is Brian's charge greater than Greg's charge?

**c.** When are Brian's and Greg's charges equal?

**69.** **TRAVEL** Jeff's car averages 18 miles per gallon of gasoline. What is the greatest number of gallons of gasoline that he will need if he travels no more than 450 miles?

**70.** **THEATER** Members of the drama club are putting on a play. They need to spend \$560 for costumes, sets, and programs. They plan to sell tickets for \$7 each. How many tickets must be sold in order for the drama club to make a profit?

**71.** **SMALL BUSINESS** A catering business specializes in catering wedding receptions. They charge \$550 for setting up the buffet and an additional \$6.50 per guest.

**a.** Mr. and Mrs. Hiroshige want to spend no more than \$1200 on the catering for their daughter's wedding reception. Write an inequality in terms of the number of guests, $g$, that they can invite to the reception.

**b.** Solve the inequality that you wrote for part **a**. What is the maximum number of guests that can be invited to the reception?

## Look Back

**Evaluate.** *(LESSON 2.3)*

**72.** $76 - (-43)$ **73.** $-111 - 400$ **74.** $12 - 546$

**75.** $80 - (-31)$ **76.** $-1 - 99$ **77.** $-31 - (-50)$

**Evaluate.** *(LESSON 2.4)*

**78.** $(-3)(4)(-3)$ **79.** $(-2)(-2)(-2)(-2)$ **80.** $(5)(-3)(-3)(5)$

**81.** $(-4)(-4)(-4)$ **82.** $(-2)(5)(-3)$ **83.** $(-25)(2)(-2)(-25)$

**Simplify each expression.** *(LESSONS 2.6 AND 2.7)*

**84.** $3x + 2y + 5x + 7y$ **85.** $3(x + 2y) + 4(2x + y)$

**Solve each equation.** *(LESSON 3.3)*

**86.** $2x + 7 = 15$ **87.** $65 = 3x - 7$ **88.** $6y + 17 = 32$

**89.** $3m - 10 = 23$ **90.** $130 = 4p - 30$ **91.** $8x + 7 = -17$

## Look Beyond

**internet connect**
**Portfolio Extension**
Go To: **go.hrw.com**
Keyword:
**MA1 Puzzles**

**Solve each inequality for *y*. Then state whether the point (0, 0) satisfies the inequality.**

**92.** $5(y + 2x) \leq 15x + 1$ **93.** $y - 6x - \frac{1}{2}(y - 3x) + 1 \leq \frac{3}{2}(y - 3x)$

**94.** $4x - 2(y + 1) > -30x$ **95.** $y - 2x + 1 < 1 - x$

**PORTFOLIO ACTIVITY**

Include this activity in your portfolio.

Refer to the table of values for the piecewise function in the Portfolio Activity on page 281.

1. Plot the points from the table on a coordinate plane. Sketch the graph of the function by connecting the points.
2. Describe the shape of the graph. What is the equation of the line to the left of the *y*-axis? What is the equation of the line to the right of the *y*-axis?

The function you graphed is a well-known and useful function in mathematics. It is called the *absolute-value function.* The notation $\begin{cases} -x \text{ if } x < 0 \\ x \text{ if } x \geq 0 \end{cases}$ is generally replaced by the symbol $|x|$. You will learn more about the absolute-value function in Lesson 6.4.

**WORKING ON THE CHAPTER PROJECT**

You should now be able to complete Activities 1 and 2 of the Chapter Project on page 308.

# 6.3 Compound Inequalities

## Objectives

- Graph the solution sets of compound inequalities.
- Solve compound inequalities.

*Swimming pools require maintenance to ensure that they stay clean and free of bacteria. Adding chlorine to a pool is one way to ensure that bacteria do not have a chance to grow.*

**Why** ***Many real-world situations have a range of values that are acceptable. Compound inequalities describe these types of limited ranges.***

When two inequalities are combined into one statement, the result is called a **compound inequality**. The words *and* and *or* are used to describe the relationship between the two parts of the inequality.

**EXAMPLE 1**

**APPLICATION**
**POOL MAINTENANCE**

The United States standard for safe levels of chlorine in swimming pools is at least 1 part per million and no greater than 3 parts per million. **Write and graph a compound inequality to describe this situation.**

**SOLUTION**

Let $c$ be the level of chlorine in parts per million.

The inequality representing at least 1 part per million is $c \geq 1$. The inequality representing no more than 3 parts per million is $c \leq 3$.

Both $c \geq 1$ AND $c \leq 3$ must be true. The two inequalities can be combined into one statement: $1 \leq c \leq 3$.

The graph of the compound inequality is the *intersection* region of the two parts of the inequality.

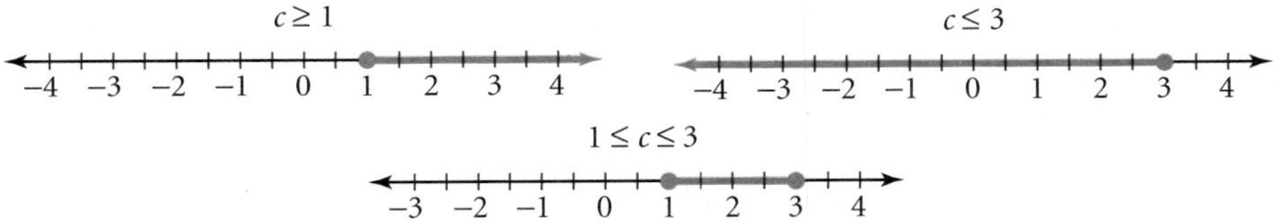


The graph of the solution set shows the set of values that make both inequalities true.

Notice that in Example 1 that $c \geq 1$ AND $c \geq 3$. A compound inequality involving AND has a solution region that represents an **intersection**, or overlapping, of the solution regions for the separate parts of the inequality. This type of compound inequality is called a **conjunction**.

$1 \leq c \leq 3$

Both $c \geq 1$ AND $c \leq 3$ must be true to make the conjunction true.

Example 2 illustrates the other type of compound inequality, a *disjunction.*

**EXAMPLE 2**

**APPLICATION**
**AIR TRAVEL**

In a certain restricted air space, airplanes must fly below 10,000 feet or above 15,000 feet. **Write and graph a compound inequality to describe this situation.**

**SOLUTION**

Let $a$ be the altitude in feet.

The inequality that represents flying below 10,000 feet is $a < 10{,}000$. The inequality that represents flying above 15,000 feet is $a > 15{,}000$.

*Either* $a < 10{,}000$ OR $a > 15{,}000$ can be true. The two inequalities can be combined into one statement using the word OR: $a < 10{,}000$ OR $a > 15{,}000$.

The graph of the compound inequality includes the regions on the opposite sides of the two endpoints determined by the individual inequalities.

$a < 10{,}000$

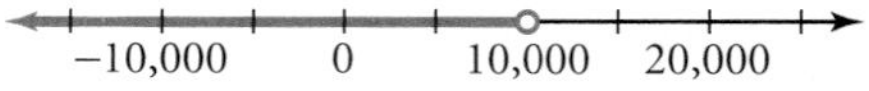


$a > 15{,}000$

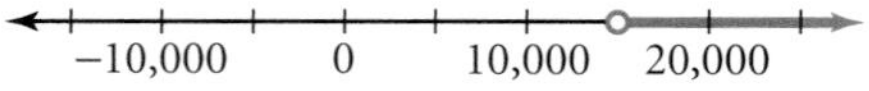


$a < 10{,}000$ OR $a > 15{,}000$

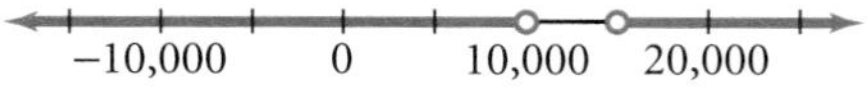


The graph of the solution set shows the set of values that make either one inequality or the other true.

In Example 2, either $a < 10{,}000$ OR $a > 15{,}000$. A compound inequality involving OR has solution regions that are the **union**, or the total, of the solution regions of the separate parts of the inequality. This type of compound inequality is a **disjunction**.

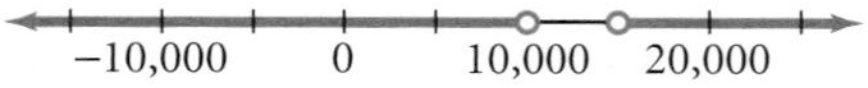


If either $a < 10{,}000$ OR $a > 15{,}000$ is true, then the disjunction is true.

## EXAMPLE 3 Graph each compound inequality.

**a.** $-4 < x$ AND $x < 2$ **b.** $x < 2$ OR $x \geq 5$ **c.** $x > 4$ AND $x < 3$

### SOLUTION

**a.** Notice that the acceptable values of $x$ must be both greater than $-4$ AND less than 2. This conjuction may also be written as $-4 < x < 2$.

$-4 < x$

$x < 2$

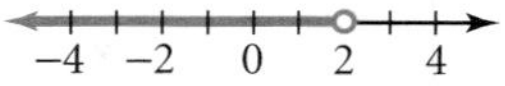


Therefore, the solution is the set of values between $-4$ and 2.

$-4 < x < 2$

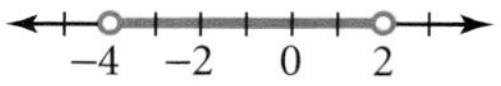


**b.** The acceptable values for this compound inequality may be either less than 2 OR greater than or equal to 5.

$x < 2$

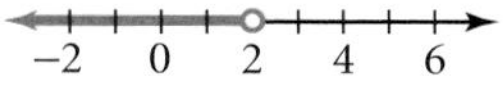


$x \geq 5$

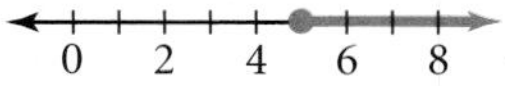


In this case the solution consists of the values that lie on opposite sides of the endpoints and that include the endpoint at 5.

$x < 2$ OR $x \geq 5$

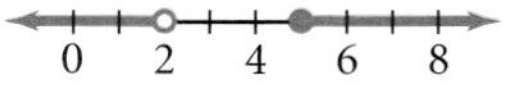


**c.** The acceptable values for this compound inequality must be both greater than 4 AND less than 3. There is *no solution,* because there are no numbers that satisfy both conditions.

$x > 4$

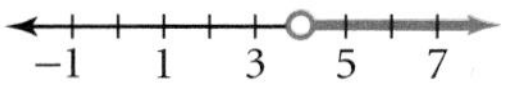


$x < 3$

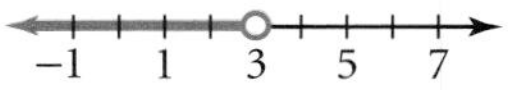


**TRY THIS** Graph each compound inequality.

**a.** $q < 5$ OR $q > 16$ **b.** $x > -2$ AND $x < -5$ **c.** $-3 \leq y \leq 6$

## EXAMPLE 4 Solve and graph $-5 \leq 3x + 7 \leq 10$.

### SOLUTION

Solve each inequality separately and then combine the results to form a compound inequality.

| | |
|---|---|
| $-5 \leq 3x + 7$ | $3x + 7 \leq 10$ |
| $-12 \leq 3x$ | $3x \leq 3$ |
| $-4 \leq x$ | $x \leq 1$ |

$x \geq -4$

$x \leq 1$

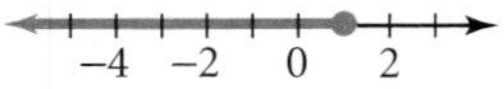


The solution to the compound inequality is region where both inequalities are true and where the graphs overlap.

$-4 \leq x \leq 1$

The solution is $-4 \leq x \leq 1$.

**TRY THIS** Solve and graph $-6 < 2x + 4 \leq 10$.

## EXAMPLE 5 Solve and graph $3x \leq -12$ OR $2x + 7 \geq 5$.

**SOLUTION**

Simply solve each inequality separately, keeping the word OR between them, and then graph the resulting statements.

$$3x \leq -12 \qquad \text{OR} \qquad 2x + 7 \geq 5$$
$$x \leq -4 \qquad\qquad\qquad 2x \geq -2$$
$$x \geq -1$$

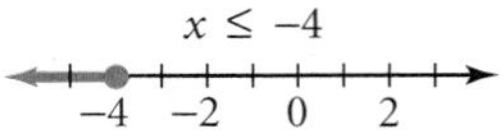


OR

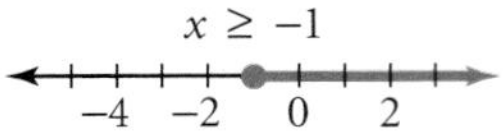


The resulting graph is constructed as follows:

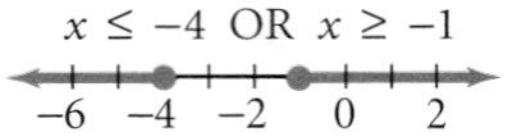


**TRY THIS** Solve and graph $5x \leq 15$ OR $3x - 7 > 23$.

## EXAMPLE 6 Solve and graph $2x + 1 \leq 13$ OR $x - 5 \geq -5$.

**SOLUTION**

$$2x + 1 \leq 13 \qquad \text{OR} \qquad x - 5 \geq -5$$
$$2x \leq 12 \qquad\qquad\qquad x \geq 0$$
$$x \leq 6$$

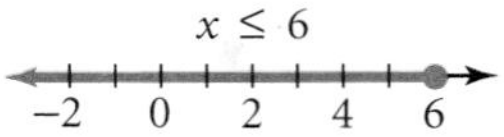


OR

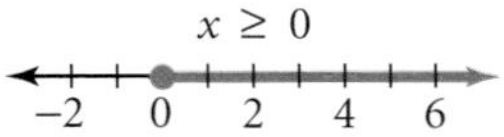


Since OR is used, the solution region includes all real numbers and is not confined to the intersection of the two individual solution sets.

The graph of the solution set is constructed as follows:

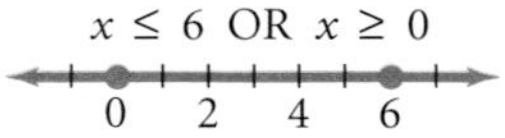


# Exercises

## Communicate

internet connect
**Activities Online**
Go To: **go.hrw.com**
Keyword:
**MA1 Hurricane Force**

1. Describe how to write $y > -6$ AND $y < 4$ as a compound inequality.
2. Describe how to solve the compound inequality $-2 < 2x < 4$.
3. **a.** Give a real-world situation that can be described by a compound inequality.
   **b.** Give the compound inequality that describes the situation from Exercise 3.
4. Describe the differences between the following compound inequalities:
   **a.** $-4 < x \leq 5$  **b.** $-4 < x$ OR $x \leq 5$

## Guided Skills Practice

**Graph each compound inequality.** ***(EXAMPLES 1, 2, AND 3)***

**5.** $-7 < x < 5$ **6.** $-2 \leq y \leq 4$ **7.** $q < 5$ OR $q \geq 7$

**8.** $45 < z \leq 47$ **9.** $v < 4$ OR $v \geq 12$ **10.** $-5 \leq x \leq 5$

**Solve and graph each compound inequality.** ***(EXAMPLE 4)***

**11.** $-7 \leq 3x + 2 \leq 8$ **12.** $-2 \leq 4x + 6 \leq 22$ **13.** $8 < 3x - 1 \leq 11$

**Solve and graph each compound inequality.** ***(EXAMPLE 5)***

**14.** $2x \leq 6$ OR $3x > 12$ **15.** $7x > 21$ OR $2x < -2$ **16.** $2x > 7$ OR $3x < 7$

**Solve and graph each compound inequality.** ***(EXAMPLE 6)***

**17.** $3x - 4 \leq 11$ OR $2x + 5 > 5$ **18.** $2x - 4 > -2$ OR $2x - 6 < -2$

## Practice and Apply

**internet connect**
**Homework Help Online**
Go To: **go.hrw.com**
Keyword:
**MA1 Homework Help**
for Exercises 19-30

**Graph each compound inequality.**

**19.** $2 < x < 5$ **20.** $x > 4$ AND $x < 2$ **21.** $-1 \leq x < 7$

**22.** $z < 4$ OR $z \geq 7$ **23.** $-4 < n \leq 6$ **24.** $x > -14$ OR $x < -18$

**25.** $-6 < v \leq 3$ **26.** $x < 4$ AND $x \geq 6$ **27.** $3.5 < t \leq 7.5$

**28.** $y > -4.5$ OR $y < -6.2$ **29.** $t < 6$ OR $t \geq 8.4$ **30.** $-4.3 < t \leq 6.9$

**Solve and graph the following compound inequalities:**

**31.** $-3 \leq 2x + 15 < 21$ **32.** $5 \leq z - 8 < 7$ **33.** $2x < 14$ OR $3x \geq 18$

**34.** $-14 < 2x - 6 \leq 12$ **35.** $-5 < 2x - 7 < 7$ **36.** $-2x > 16$ OR $-3x \leq 18$

**37.** $-5 < 2x < 7$ **38.** $-2x < 14$ OR $3x < -12$

**APPLICATIONS**

**39.** **WEATHER** The high temperature on a certain day was 98°F and the low was 78°F. If $t$ represents the temperature, write a compound inequality that describes the range of temperatures on this day.

**40.** **SALES** If a store makes a profit of \$3 on each T-shirt that it sells, how many T-shirts must it sell to make a profit of at least \$400?

## Look Back

**41.** Find the sum $1 + 2 + 3 + 4 + 5 + 6$ by using a geometric dot pattern. ***(LESSON 1.1)***

**42.** If each of 6 teams played each of the other teams just once in a tournament, how many games would there be? ***(LESSON 1.1)***

**43.** Evaluate: $4^2 \cdot 16[2 + 3(2 \cdot 4 + 7)] - 26$. ***(LESSON 1.3)***

**44.** Simplify: $(7 - 3y) - (8 + 2y)$. ***(LESSON 2.6)***

## Look Beyond

**45.** Solve $2x + 3y > 18$ for $y$.

**46.** When $x$ is 4, find the values of $y$ that make the inequality in Exercise 45 true.

6.4

# Absolute-Value Functions

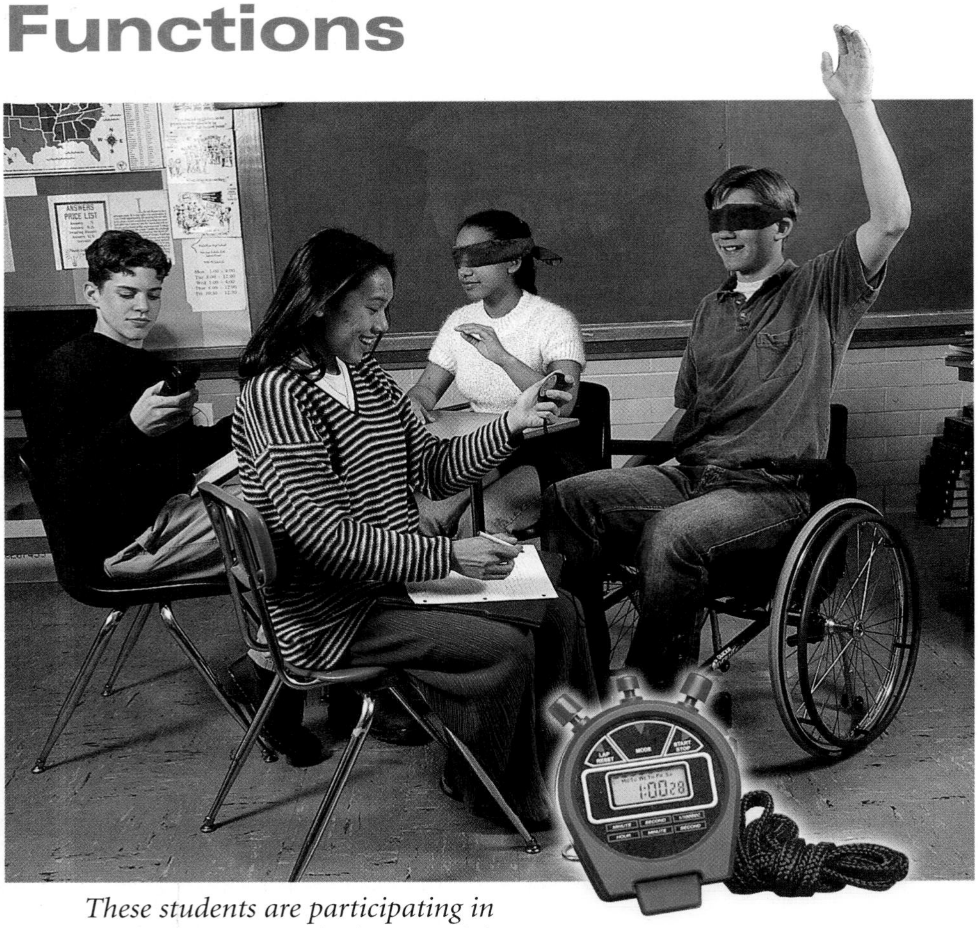

These students are participating in an experiment in which some of the students estimate elapsed time.

### Objectives

- Explore features of the absolute-value function.
- Explore basic transformations of the absolute-value function.

**Why** ***The absolute-value function allows you to consider the magnitude of a number without regard for its sign. This is useful in various kinds of measurements, including statistics.***

CONNECTION
STATISTICS

The students in the experiment shown above work in pairs. Half of the students are blindfolded and then told to raise a hand. They are told to put their hand down when they think that 1 minute has passed. The other half of the students act as timers. Later they switch roles.

TECHNOLOGY
SPREADSHEET

Data from the first group is entered in a spreadsheet with the students' names in column A and their estimates of the time elapsed in column B.

Compute the error *in seconds* in column C by subtracting 60 from each student's time.

In column C, a negative number means the estimate was *under* 1 minute. The absolute error in column D shows only the *amount* of the error.

| | A | B | C | D |
|---|---|---|---|---|
| 1 | Student | Time | Error | Abs. Error |
| 2 | Tricia | 49 | -11 | 11 |
| 3 | Keira | 59 | -1 | 1 |
| 4 | Tom | 51 | -9 | 9 |
| 5 | Louise | 65 | 5 | 5 |
| 6 | James | 68 | 8 | 8 |
| 7 | Sakeenah | 77 | 17 | 17 |
| 8 | Hong | 66 | 6 | 6 |
| 9 | Louis | 54 | -6 | 6 |
| 10 | Mary | 67 | 7 | 7 |
| 11 | Maria | 46 | -14 | 14 |
| 12 | Marcus | 62 | 2 | 2 |
| 13 | Shamar | 73 | 13 | 13 |
| 14 | Lois | 61 | 1 | 1 |
| 15 | Dianne | 53 | -7 | 7 |
| 16 | Suzanne | 64 | 4 | 4 |

For example, row 10 shows that Mary thought 1 minute had elapsed after 67 seconds. Her error was $67 - 60 = 7$, indicating that her answer was 7 seconds *over* 1 minute. Row 15 shows that Diane's time was 53 seconds, so her error was $53 - 60 = -7$, or 7 seconds *under* 1 minute.

Look carefully at column D in the spreadsheet. Compare the numbers in column C with those in column D. What is the sign of each number in column D? How does it compare with the sign of the corresponding number in column C?

The **absolute value** of a number can be shown on a number line, as you learned in Lesson 2.1. Recall that the absolute value of a number is its distance from zero.

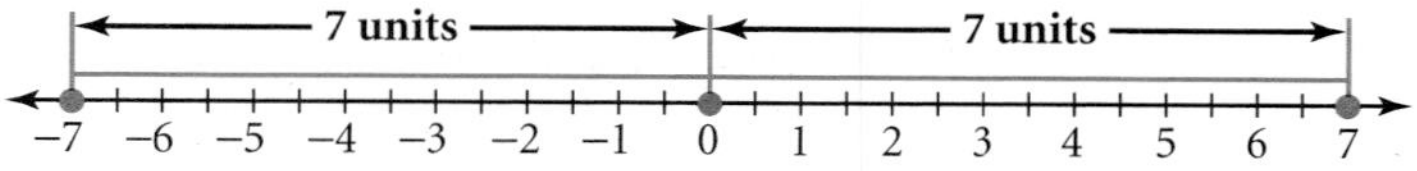


7 is 7 units from 0, so the absolute value of 7 is 7.

Thus, $|7| = 7$.

−7 is 7 units from 0, so the absolute value of −7 is 7.

Thus, $|-7| = 7$.

The geometric illustration leads to the following algebraic interpretation: The absolute value of a number is the number itself when the number is positive or zero. The absolute value of a number is the opposite of the number when the number is negative.

### Algebraic Definition of Absolute Value

For all real numbers $a$:

$$|a| = a, \text{ for } a \geq 0$$
$$|a| = -a, \text{ for } a < 0$$

**EXAMPLE 1** **Find each of the following absolute values:**

**a.** $|8 - 2|$ **b.** $|2 - 8|$ **c.** $|3 - 3|$

**SOLUTION**

**a.** $|8 - 2| = |6|$
$= 6$

**b.** $|2 - 8| = |-6|$
$= 6$

**c.** $|3 - 3| = |0|$
$= 0$

**TRY THIS** Find each of the following absolute values:

**a.** $|5 - 14|$ **b.** $|12 - 2|$

A function written in the form $y = |x|$ or $y = \text{ABS}(x)$ is an example of an **absolute-value function.**

**EXAMPLE 2** **a. Find the domain and the range of the absolute-value function $y = |x|$.**
**b. Graph the function $y = |x|$ for $x$-values from −3 to 3.**

**SOLUTION**

**a.** Since you can find the absolute value of any number, the domain is all real numbers. Since $|0| = 0$ and $|x|$ is positive otherwise, the range is all non-negative real numbers, or $y \geq 0$.

**b.** Make a table of ordered pairs. For every value of $x$ in the table, write the absolute value of $x$ as the $y$-value. For example, if $x$ is −2, then $y$ is $|-2|$ or 2.

| $x$ | −3 | −2 | −1 | 0 | 1 | 2 | 3 |
|---|---|---|---|---|---|---|---|
| $y$ | 3 | 2 | 1 | 0 | 1 | 2 | 3 |

CONNECTION
COORDINATE GEOMETRY

Plot the ordered pairs, and then connect the points because the domain includes all real numbers.

The graph of the absolute-value function has a distinctive **V** shape. The tip of the **V** is at the origin, (0, 0).

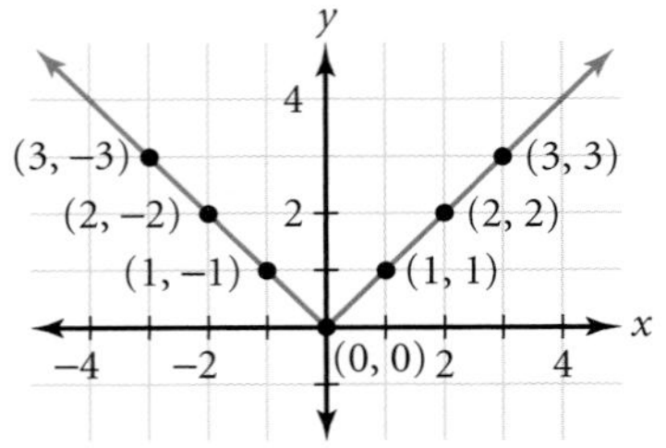


**TRY THIS** Find the domain and range of each function. Then graph each function.
**a.** $y = 2|x|$ **b.** $y = |x - 2|$

## *Activity* Manipulating the Absolute-Value Function

TECHNOLOGY
GRAPHICS CALCULATOR

**You will need:** a graphics calculator or graph paper

1. Graph the functions $y = |x|$ and $y = |x - 3|$ together on a same coordinate plane.
2. What effect does the 3 have on the graph? **CHECKPOINT ✔**
3. Graph the functions $y = |x|$ and $y = |x| - 3$ on the same coordinate plane.
4. What effect does the 3 have on the graph? **CHECKPOINT ✔**
5. Graph the functions $y = |x|$ and $y = -|x|$ together on a coordinate plane.
6. What effect does the negative sign have on the graph? **CHECKPOINT ✔**

The function $y = |x|$ is called the absolute-value **parent function** because it is the most basic absolute-value function.

The graph of a parent function changes if the function rule is changed in some way. There are certain kinds of changes called **transformations.** (Transformations will be presented in detail in Chapter 14.) Two types of transformations are illustrated in Example 3.

A **translation** shifts the parent-function graph horizontally or vertically.

A **reflection** creates a mirror image of the parent-function graph across the **line of reflection.** If the line of reflection is the $x$-axis, then each point $(a, b)$ on the original graph becomes $(a, -b)$ on the reflection.

**EXAMPLE 3** **Graph $y = |x|$ and the given transformation function on the same coordinate plane. Identify the type of transformation.**

**a.** $y = |x + 7|$ **b.** $y = |x| + 4$ **c.** $y = -|x|$

**SOLUTION**

**a.** Substituting $x + 7$ for the $x$ in the parent function causes its graph to shift 7 units to the left. This shift is a (horizontal) translation.

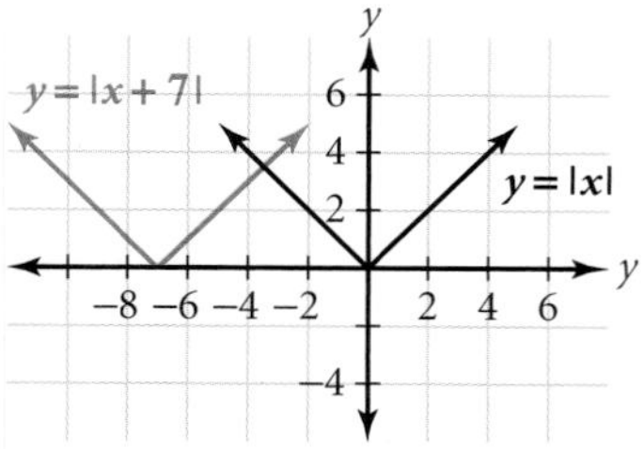


**b.** Adding 4 to the parent function causes its graph to shift 4 units up. This shift is a (vertical) translation.

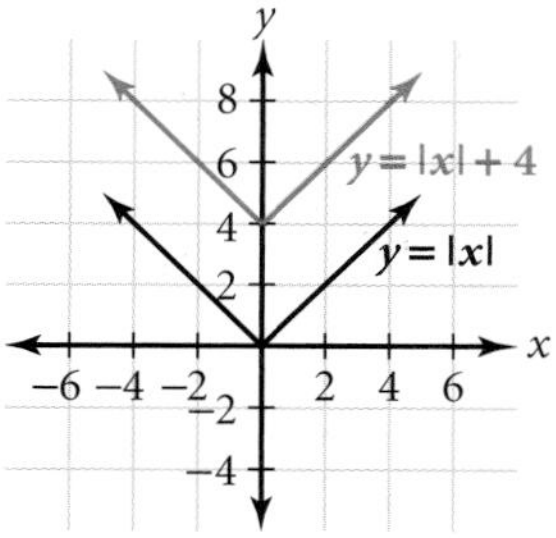


**c.** Multiplying the parent function by −1 causes the parent-function graph to reflect across the $x$-axis. Therefore, this transformation is a reflection.

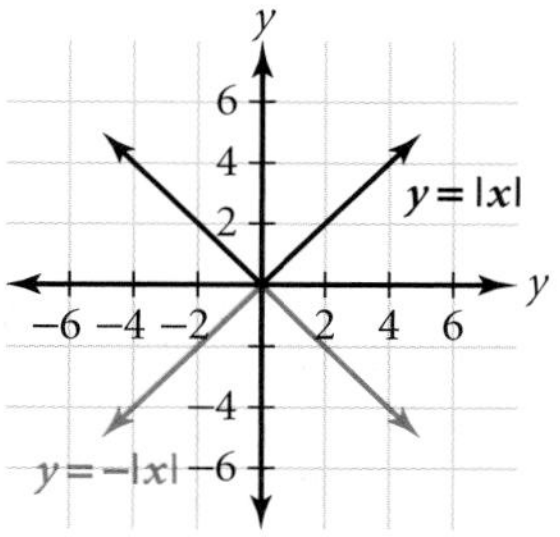


**CRITICAL THINKING** Normally, a reflection of the parent graph across the $y$-axis is caused by substituting $-x$ for $x$ in the parent function. What is the result of doing this in the absolute-value function? Explain.

# Exercises

## Communicate

1. When guessing how long it takes for a minute to elapse, Andy guessed 51 seconds and Abi guessed 68 seconds. Explain how to compare the amount of their errors.
2. Give an example of a value of $a$ for which $|a| = -a$, if possible. Explain.
3. Describe the graph of an absolute-value function.
4. Can the value of $|x|$ ever be negative? Explain.
5. Can the value of $|x|$ ever be zero? Explain.

## Guided Skills Practice

**Find each absolute-value.** ***(EXAMPLE 1)***

**6.** $|5 - 12|$ **7.** $|13 - 12|$ **8.** $|-3 - 3|$ **9.** $|4 - 4|$

**10.** Find the domain and range of each function, and then graph the function. ***(EXAMPLE 2)***

**a.** $y = |x + 1|$ **b.** $y = |x| + 1$

**11.** Graph $y = |x|$ and each of the following functions on the same coordinate plane. Identify the type(s) of transformation(s) of the graph of $y = |x|$ that occurs with each of the new functions. ***(EXAMPLE 3)***

**a.** $y = |x + 1|$ **b.** $y = |x| + 1$ **c.** $y = -|x| + 1$

## Practice and Apply

**Find the absolute value (ABS) of each number.**

**12.** 17 **13.** $-33$ **14.** $-7.11$ **15.** 8.67

**16.** $\frac{4}{3}$ **17.** $-2.5$ **18.** $-3\frac{5}{11}$ **19.** 79.2

**Evaluate.**

**20.** $|13 - 24|$ **21.** $|0 - 3|$ **22.** $|4 - 12|$ **23.** $|1 - 11|$

**24.** $|0 - (-3)|$ **25.** $|1 - 27|$ **26.** $|11 - 3|$ **27.** $|-14 - (-14)|$

**28.** $|-13 + 13|$ **29.** $|-5 - 2|$ **30.** $|5 - (-3)|$ **31.** $|-11 - 11|$

**32.** $|-5 + (-5)|$ **33.** $|0 - 5|$ **34.** $|5 - 10|$ **35.** $|-5 - 10|$

**Find the domain and range of each function.**

**36.** $y = |x + 4|$ **37.** $y = |x - 5|$ **38.** $y = |x| + 2$ **39.** $y = |x| - 4$

**40.** $y = -|x + 4|$ **41.** $y = -|x - 5|$ **42.** $y = -|x| + 2$ **43.** $y = -|x| - 4$

**44.** $y = 4|x|$ **45.** $y = \frac{1}{2}|x|$ **46.** $y = 4|x| - 1$ **47.** $y = 4|x - 1|$

**internet connect**

**Homework Help Online**

Go To: **go.hrw.com**
Keyword:
**MA1 Homework Help** for Exercises 48-55

**48–55.** For Exercises 48–55, refer to the functions given in Exercises 36–43. Graph $y = |x|$ and the given function on the same coordinate plane. Identify the transformations of the absolute-value parent function.

**Find $|a - b|$ and $|b - a|$ for each of the following:**

**56.** $a = 5, b = 3$ **57.** $a = 5, b = -3$ **58.** $a = -5, b = 3$ **59.** $a = 3, b = 5$

**60.** Based on your answers for Exercises 56–59, what conclusions can you make about $|a - b|$ and $|b - a|$?

**CHALLENGE**

**61.** **TRANSFORMATIONS** Write a function for each transformation of $y = |x|$ described below.

**a.** The graph of $y = |x|$ is translated 5 units to the left and then 2 units down.

**b.** The graph of $y = |x|$ is translated 5 units up and then reflected across the $x$-axis.

**c.** The graph of $y = |x|$ is reflected across the $x$-axis and then translated 2 units up.

APPLICATIONS

62. **CHEMISTRY** In chemistry lab, different students reported their findings about how much sodium was in a salt compound. If the actual amount was 8.4 milligrams, find the error and the absolute error for each student.
   a. Alisha reported 8.2 milligrams.
   b. Najee reported 9.0 milligrams.
   c. Nell reported 8.1 milligrams.
   d. Paula reported 8.4 milligrams.

**TRAVEL** An express train leaves from Houston traveling north to St. Louis at an average speed of 80 miles per hour. After about 3 hours the train passes through Dallas. If $t$ is the travel time in hours from Houston, the distance between the train and Dallas is given by $y = 80|3 - t|$.

63. After 2 hours, how far has the train traveled? Where is the train at this time in relation to Dallas?

64. After 4 hours, how far has the train traveled? Where is the train at this time in relation to Dallas?

## Look Back

65. Identify the next three numbers in the sequence below. *(LESSON 1.1)*
   2, 6, 10, 14, 18, . . .

66. If Miguel has \$2.00, how many bookmarks can he buy if the bookmarks cost \$0.38 each? *(LESSON 2.4)*

**Evaluate.** *(LESSONS 2.2, 2.3, AND 2.4)*

67. $-3 + 4$
68. $-3(0.3)$
69. $-15 - (-15)$
70. $60 \div -3$
71. $(-10)\left(\frac{1}{2}\right)$
72. $-1.4 - (-3)$
73. $4\left(-1\frac{1}{2}\right)$
74. $(-3.2) \div 4$

75. Graph the functions below on the same coordinate plane and describe the similarities and differences. *(LESSON 5.4)*
   a. $y = x + 2$
   b. $y = \frac{2}{3}x - 1$
   c. $y = x$

**Write each linear equation in slope-intercept form.** *(LESSON 5.4)*

76. $3x + 2y = 1$
77. $4x = 2y$
78. $4y = 0$
79. $2x - 2y = 17$

**Solve each inequality and graph the solution on a number line.** *(LESSON 6.1)*

80. $x + 7 \leq 3$
81. $x - 3 \geq 2$
82. $x + 15 \leq -1$
83. $x - 3 > 4$

## Look Beyond

84. On a number line, $\sqrt{x^2}$ can be thought of as the distance from the origin, 0, to the point $x$. In the coordinate plane, $\sqrt{x^2 + y^2}$ can be thought of as the distance from the origin (0, 0) to the point $(x, y)$. Use $\sqrt{x^2 + y^2}$ to find the distance from the origin to each of the following points:
   a. (3, 4)
   b. (12, 5)
   c. (−8, 6)
   d. (−15, −20)

6.5

# Absolute-Value Equations and Inequalities

*A company manufactures a special gear for a car. If the gear is made too large or too small, it will not turn smoothly and may wear out quickly.*

**Objectives**

- Solve absolute-value equations.
- Solve absolute-value inequalities, and express the solution as a range of values on a number line.

**Why** *Absolute-value expressions are often used to describe differences from a given value, such as tolerances in manufacturing.*

## Error and Absolute Value

**APPLICATION**
**MANUFACTURING**

A gear is designed with a specification of 3.50 centimeters for its diameter. Its actual diameter must be within 0.01 centimeter of 3.50 centimeters in order for the gear to work properly. This can be expressed as $3.50 \pm 0.01$.

The difference between the actual measure and the specified measure is called the **error.** For example, a gear with an actual measure of 3.48 centimeters would have the following error:

$$3.48 - 3.50 = -0.02 \qquad \textit{Error}$$

The **absolute error** is the absolute value of the difference between the actual measure and the specified measure.

$$|3.48 - 3.50| = |-0.02| = 0.02 \qquad \textit{Absolute error}$$

The greatest and least acceptable values for the diameter of the gear are the values for which the absolute error is 0.01. Letting $d$ represent the actual diameter of the gear, you can use the following **absolute-value equation** to find the greatest and least acceptable values:

$$|d - 3.50| = 0.01$$

Because of the definition of absolute value, *the equation will have two solutions.*

Recall the definition of the absolute value of a number that you learned in Lesson 6.4: $|a| = a$ for $a \geq 0$ and $|a| = -a$ for $a < 0$. If there is an expression inside the absolute-value sign, it is usually helpful to restate the definition as follows:

$$\text{For } a \geq 0,\ |x| = a \text{ is equivalent to } x = a \text{ or } x = -a.$$

**EXAMPLE 1** **Solve the absolute-value equation for the gear problem.**

$$|d - 3.50| = 0.01$$

CONNECTION
MAXIMUM/MINIMUM

**SOLUTION**

To solve an absolute-value equation, you must consider two cases.

**CASE 1**

Consider the quantity within the absolute-value symbols to be positive.

$$|d - 3.50| = 0.01$$
$$d - 3.50 = 0.01$$
$$d = 3.51$$

$|3.51 - 3.50| = 0.01$ *True*

**CASE 2**

Consider the quantity within the absolute-value symbols to be negative.

$$|d - 3.50| = 0.01$$
$$d - 3.50 = -0.01$$
$$d = -0.01 + 3.50$$
$$d = 3.49$$

$|3.49 - 3.50| = |-0.01| = 0.01$ *True*

The solution shows that the greatest and least allowable lengths for the gear diameter are 3.51 centimeters and 3.49 centimeters, respectively.

**TRY THIS** Solve $|x - 10| = 4.5$.

**EXAMPLE 2** **Solve $|3x - 2| = 10$.**

**SOLUTION**

**CASE 1**

Consider the quantity within the absolute-value symbols to be positive.

$$|3x - 2| = 10$$
$$3x - 2 = 10$$
$$3x = 12$$
$$x = 4$$

$$|3(4) - 2| \stackrel{?}{=} 10$$
$$|12 - 2| \stackrel{?}{=} 10$$
$|10| = 10$ *True*

**CASE 2**

Consider the quantity within the absolute-value symbols to be negative.

$$|3x - 2| = 10$$
$$3x - 2 = -10$$
$$3x = -8$$
$$x = -\frac{8}{3}$$

$$\left|3\left(-\frac{8}{3}\right) - 2\right| \stackrel{?}{=} 10$$
$$|-8 - 2| \stackrel{?}{=} 10$$
$|-10| = 10$ *True*

**TRY THIS** Solve $|2x - 4| = 8$.

**CRITICAL THINKING**

**a.** How can you tell immediately that $|5x - 2| = -12$ has no solution?
**b.** What apparent solutions do you get if you use the method in Example 2?

## Absolute Value and Inequalities

In the gear problem, the acceptable values for $d$ fall within the boundary values of 3.49 centimeters and 3.51 centimeters. To find the acceptable values, you can use an **absolute-value inequality**. In this case, the inequality $|d - 3.50| \leq 0.01$ expresses the requirement that the actual diameter of a gear must fall within 0.01 centimeter of 3.50 centimeters. A similar inequality is solved in Example 3.

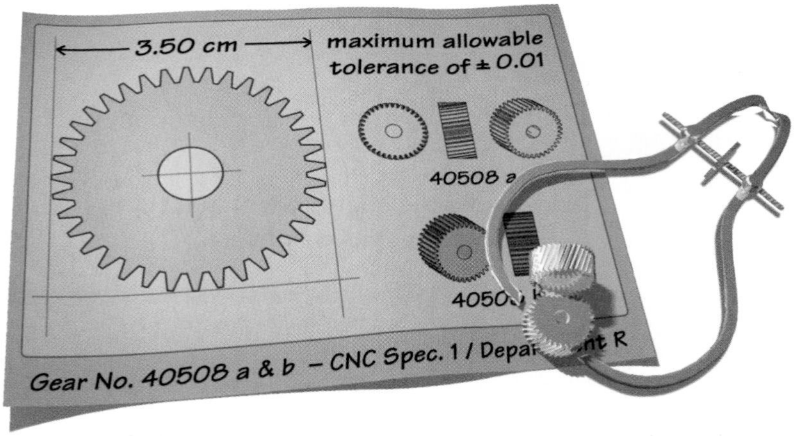


**EXAMPLE 3**

**APPLICATION**
**CRAFTS**

The square metal frame for a piece of stained glass will fit properly into the space in a block if the error in the actual length of the sides is within 0.15 centimeter of 48 centimeters. **Use an absolute-value inequality to find the boundary values.**

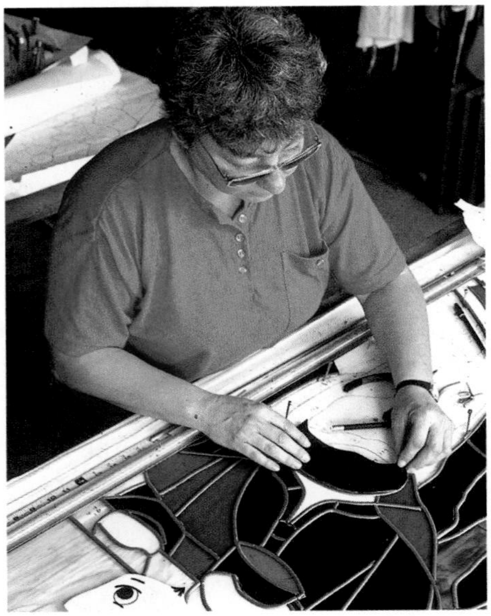

**SOLUTION**

The absolute error can be expressed as $|x - 48.00|$, where $x$ is the actual measurement of the square frame. The allowable absolute error is no greater than 0.15 centimeter. Since the absolute error must be less than or equal to 0.15 centimeter, use an inequality.

Solve $|x - 48.00| \leq 0.15$.

**CASE 1**

Find the upper boundary.

$$|x - 48.00| \leq 0.15$$
$$x - 48.00 \leq 0.15$$
$$x \leq 48.15$$

**CASE 2**

Find the lower boundary.

$$|x - 48.00| \leq 0.15$$
$$x - 48.00 \geq -0.15$$
$$x \geq 47.85$$

The allowable lengths for the sides of the square frame are all of the measures between 47.85 and 48.15 centimeters inclusive, or $47.85 \leq x \leq 48.15$.

## EXAMPLE 4 Solve $|x - (-5)| \leq 2$.

**SOLUTION**

Solve $-2 \leq x - (-5) \leq 2$ by solving each part of the inequality seperately.

**CASE 1**

$-2 \leq x - (-5)$
$-2 \leq x + 5$
$-7 \leq x$

**CASE 2**

$x - (-5) \leq 2$
$x + 5 \leq 2$
$x \leq -3$

Then combine the solutions in the form of a conjunction.

$$-7 \leq x \leq -3$$

Notice that when an absolute-value expression is in the form $|x + 5|$, $+5$ is a shortened form of $-(-5)$. Therefore, when you graph the solution of the inequality, $-5$ will be the midpoint of the solution region.

**TRY THIS** Solve $|x + 6| \leq 4$.

**CRITICAL THINKING** What is the total length of the solution interval for $|x - a| \leq b$?

## EXAMPLE 5 Solve $|x - 6| > 2$.

**SOLUTION**

Consider two cases.

**CASE 1**

The quantity $x - 6$ is positive.

$|x - 6| > 2$
$x - 6 > 2$
$x > 8$

**CASE 2**

The quantity $x - 6$ is negative.

$|x - 6| > 2$
$-(x - 6) > 2$
$-x + 6 > 2$
$-x > -4$
$x < 4$

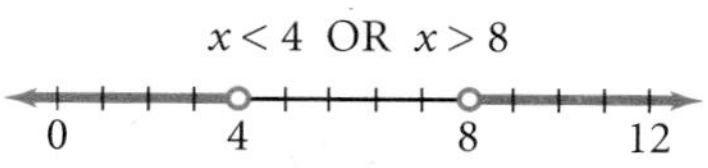


Check by substituting numbers from the solution region into the original inequality.

**TRY THIS** Solve and graph each absolute-value inequality.

**a.** $|x + 7| > 3$ **b.** $|x - 5| \geq 2$

**CRITICAL THINKING** How can you tell immediately that there is no solution for $|x - 5| < -1$?

## Graphing Absolute-Value Equations

CONNECTION
COORDINATE GEOMETRY

You can use graphs of absolute-value and linear equations in two variables to illustrate one-variable absolute-value equations and inequalities. A graphics calculator will also give you solutions using this method, as you can discover in the following activity.

### *Activity*
### Graphing to Find Solutions

TECHNOLOGY
GRAPHICS CALCULATOR

**You will need:** a graphics calculator or graph paper

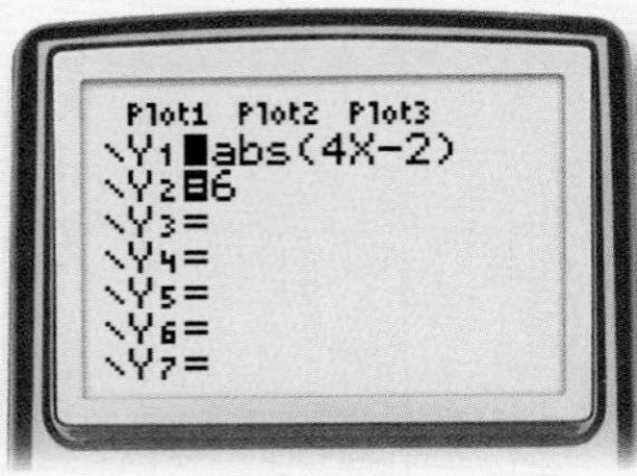


1. Graph $y = |4x - 2|$ and $y = 6$ on the same coordinate plane.

   If you are using a graphics calculator, press **graph** to see the graph on the calculator screen.

   *To set the window, use Xmin = −10, Xmax = 10, Ymin = −5, and Ymax = 10.*

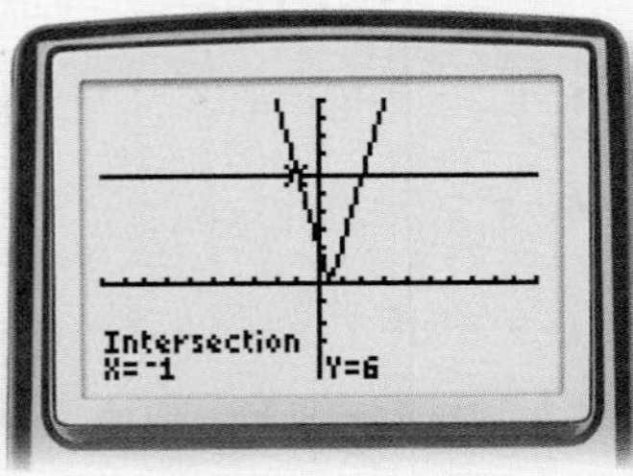


2. Explain how you can use the graph to find the solutions to the equation $|4x - 2| = 6$.
3. Explain how you can use the graph to find the solution to the inequalities $|4x - 2| < 6$ and $|4x - 2| > 6$.

CHECKPOINT ✓

4. Use a graphics calculator to solve each of the following:

   **a.** $|2x + 1| = 5$ **b.** $|2x + 1| < 5$ **c.** $|2x + 1| > 5$

**EXAMPLE 6** **Use graphing to solve:** **a.** $|x + 2| = 6$ **b.** $|x + 2| > 6$

**SOLUTION**

**Step 1**
Graph the absolute-value function $y = |x + 2|$.

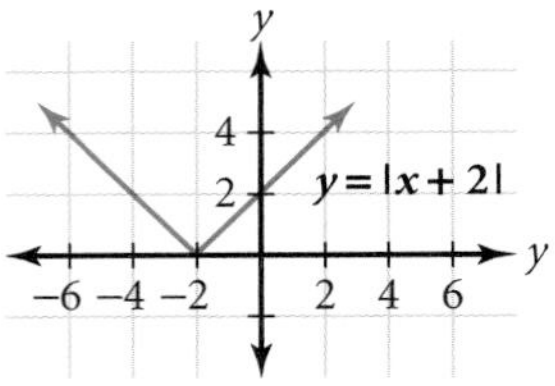


**Step 2**
Graph the constant function $y = 6$ on the same coordinate plane.

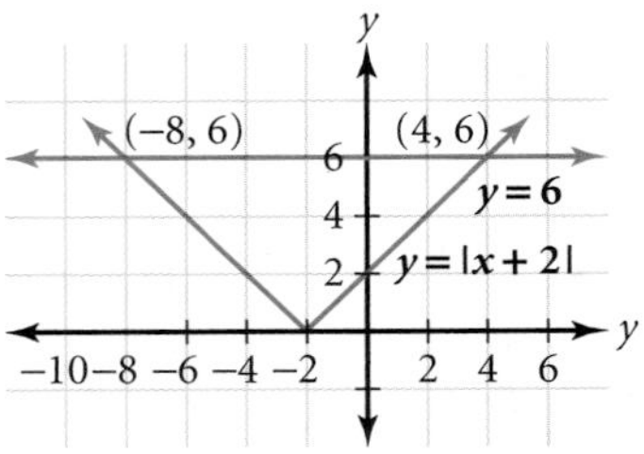


**a.** The $x$-values of the two intersection points of the graphs will be the solutions to $|x + 2| = 6$, $x = 4$ or $x = -8$.

**b.** All of the $x$-values of the linear graph on the outside of each intersection point will be the solutions to $|x + 2| > 6$, $x > 4$ or $x < -8$.

# Exercises

## Communicate

**1.** Compare the solutions of $4 = |x + 5|$ with the solution of $4 \geq |x + 5|$.

**2. a.** Explain the meaning of the specification $45 \pm 0.001$ centimeters.

**b.** Explain how to write an absolute-value inequality to represent $45 \pm 0.001$ centimeters.

**3.** Why must you consider two cases when solving absolute-value equations and inequalities?

**4.** Describe how to use an absolute-value inequality to represent all of the values on a number line that are within 3 units of −7.

## Guided Skills Practice

**Solve each equation.** *(EXAMPLE 1)*

**5.** $|x - 5.5| = 0.5$ **6.** $|x - 2| = 5$ **7.** $|a - 3.3| = 7$

**Solve each equation.** *(EXAMPLE 2)*

**8.** $|5x - 14.5| = 2.5$ **9.** $|3z + 5| = 16$ **10.** $|2v + 6| = 8$

**11.** What numbers are less than or equal to 4 units from 10 on a number line? Write and solve an inequality that represents this situation. *(EXAMPLE 3)*

**Solve each inequality.** *(EXAMPLE 4)*

**12.** $|5x - 14.5| \leq 2.5$ **13.** $|3x + 2.5| \leq 23.5$ **14.** $|4x + 5.5| \leq 22.5$

**15.** Solve $|x - 10| > 3$. *(EXAMPLE 5)*

**Solve by graphing on a coordinate plane.** *(EXAMPLE 6)*

**16.** $|8x + 4| \geq 20$ **17.** $|3x - 6| < 3$

## Practice and Apply

**Solve each equation if possible. Check your answers.**

**18.** $|x - 5| = 3$ **19.** $|x - 1| = 6$ **20.** $|x - 2| = 4$

**21.** $|x - 8| = 5$ **22.** $|5x - 1| = 4$ **23.** $|2x + 4| = 7$

**24.** $|4x + 5| = -1$ **25.** $|-1 + x| = -3$ **26.** $|7x + 4| = 18$

**Solve and graph each inequality. Check your answers.**

**27.** $|x - 3| < 7$ **28.** $|x - 12| < 6$ **29.** $|x + 4| > 8$

**30.** $|x + 9| > 1$ **31.** $|x - 8| \leq -4$ **32.** $|x - 5| \leq -2$

**33.** $|x - 2| > 6$ **34.** $|x - 2| \leq 10$ **35.** $|x - 16| \leq 2$

**36.** $|x + 1| < 5$ **37.** $|x + 4| > 2$ **38.** $|x + 5| \leq 35$

internet connect

**Homework Help Online**

Go To: **go.hrw.com**
Keyword:
**MA1 Homework Help** for Exercises 27-38

**Solve. Round your answers, if necessary, to the nearest thousandth.**

**39.** $|3d + 7| > 5$

**40.** $|2b - 5| > 8$

**41.** $|4 - 7y| \leq 2$

**42.** $|7 - 9w| \leq 6$

**43.** $\left|\frac{3}{4} - x\right| \geq \frac{7}{8}$

**44.** $\left|\frac{4}{9} - t\right| \geq \frac{2}{3}$

**45.** $\left|\frac{2}{3}x + \frac{3}{4}\right| < 4$

**46.** $\left|\frac{7}{9}h + \frac{2}{3}\right| < \frac{1}{2}$

**47.** $\left|\frac{1}{2}x - \frac{4}{5}\right| < 0.6$

**48.** $\left|\frac{1}{3}d + \frac{5}{6}\right| \leq 0.3$

**49.** $|0.5y - 0.75| > 0.42$

**50.** $|0.002x - 0.14| > 0.0065$

**51.** $|0.001 - 5.2x| > 2.1$

**52.** $|0.005 - 0.3x| > 0.002$

**53.** $|5.2 - 0.001x| > 2.1$

**54.** $|0.3 - 0.005x| > 0.002$

**55.** The distance between $x$ and $-2$ on a number line is 7.

**a.** Draw a number-line diagram to illustrate the given statement.

**b.** Translate the given statement into an absolute-value equation.

**Refer to the following inequality for Exercises 56–59: $|x - 3| < 4$.**

**56.** Translate the absolute-value inequality into a sentence that begins with "The distance between."

**57.** Graph the solution to the inequality on a number line.

**58.** Describe the solution in words.

**59.** Give five specific numbers that satisfy the inequality.

CHALLENGE

**60.** Write all the ways in which an absolute-value inequality can be written to represent all of the values within (and including) 3 units of $-7$. (Hint: Remember that $|a - b| = |b - a|$.)

APPLICATION

**61.** **HEALTH** According to a height and weight chart, Mark's ideal wrestling weight is 168 pounds. Mark will be satisfied if he is no more than 3 pounds from his ideal weight.

**a.** Draw a number-line diagram to illustrate the given situation. Identify the boundary values and acceptable weights.

**b.** Write an absolute-value equation to describe the boundary values. Show the algebraic solution.

**c.** Write an absolute-value inequality to describe the acceptable weights. Show the algebraic solution.

APPLICATIONS

62. **MANUFACTURING** A valve for the space shuttle must be within 0.001 millimeter of 5 millimeters in diameter.

a. Write an absolute-value equation to describe the boundary values for the given situation. Solve your equation.

b. What are the acceptable diameters for the valve?

c. Describe these acceptable values with an absolute-value inequality.

63. **GOVERNMENT** A polling organization reports that 52% of registered voters preferred candidate $A$. The polling technique used has a **margin of error** of ±3%, meaning that the results are considered to be accurate within a range of 3% on either side of the reported figure.

a. What are the upper and lower boundaries for the actual percent of voters who support candidate $A$?

b. Show how to represent the upper and lower boundaries, $B$, using an absolute-value equation.

c. Is it possible that candidate $A$ is not actually preferred by the majority of voters? Explain.

## Look Back

**Evaluate each expression for $p = 4$, $q = -1$, and $r = -2$.** *(LESSON 2.1)*

64. $pqr - q$ 65. $\frac{pq}{r}$ 66. $\frac{pqr}{q} + pqr$

**Simplify.** *(LESSON 2.5)*

67. $-5(8c + 3)$ 68. $9(7b + 2)$ 69. $-4(-5k + 8)$

**Solve each equation.** *(LESSONS 3.1, 3.2, AND 3.4)*

70. $x + 7 = 4$ 71. $2x + 3 = 3x + 1$ 72. $m - 7 = 2m + 3$

73. $\frac{y}{7} = 2$ 74. $3x = -21$ 75. $-14p = -28$

**Solve each inequality.** *(LESSON 6.2)*

76. $4x + 5 < 25$ 77. $6y - 10 > 5$ 78. $9m - 8 < 4 + 8m$

## Look Beyond

79. **Solve for $x$.**

a. $|x| = |-5|$ b. $|x| = |a|$ c. $|4x + 1| = |3x + 2|$

# CHAPTER PROJECT SIX

# Planning a Snowboarding Trip

Mr. Line coaches a snowboarding team and teaches algebra in Bellevue, Washington. He is preparing to take his snowboarding team to a halfpipe competition and asks his algebra class to determine the number of vans the team will need for the trip and the amount of money the team will need to pay the entry fee for the competition. In this project, you will imagine that you are in Mr. Line's class and use step functions to determine the number of vans and the amount money the team will need.

## Activity 1

### The Vans

The snowboarding team normally has 17 members, but several members have been playing too many video games after school and may have to stay home to catch up on their homework.

1. Let $n$ represent the number of team members going to the competition and let $v$ represent the number of vans the team will need. Given that each van seats 6 team members, complete the table at right.
2. Plot the pairs of numbers from the table on a coordinate plane. The independent variable is $n$ and the dependent variable is $v$. You can see that the value of $v$ stays the same until the value of $n$ reaches a certain number. Then the value of $v$ *jumps*. Because the graph looks like a flight of steps, this type of function is known as a step function.
3. Suppose that 4 students have to stay home and do their homework. Describe how Mr. Line could use the graph that you created to determine the number of vans the team will need.

| $n$ | $v$ |
|---|---|
| 1 | ? |
| 2 | ? |
| 3 | ? |
| 4 | ? |
| 5 | ? |
| 6 | ? |
| 7 | ? |
| 8 | ? |
| 9 | ? |
| 10 | ? |
| 11 | ? |
| 12 | ? |
| 13 | ? |
| 14 | ? |
| 15 | ? |
| 16 | ? |
| 17 | ? |

**Activity 2**

### The Entry Fee

The greatest-integer function is a step function that can be used to model many situations. The greatest-integer function rounds a given number down to the nearest integer. For example, $\text{INT}(2) = 2$, $\text{INT}(2.9) = 2$, $\text{INT}\left(\frac{7}{3}\right) = 2$, and $\text{INT}(-2.1) = -3$. It is important to note that the greatest-integer function always rounds *down*, and not *up*, to the nearest integer.

The entry fee for the competition depends on the number of team members competing. Let $n$ represent the number of team members competing and let $f$ represent the entry fee in dollars. The entry fee is given by the equation $f = 10\left[\text{INT}\left(\frac{n}{5}\right) + 1\right]$.

1. Make a table of values and graph the given equation above for integer values of $n$ such that $1 \le n \le 17$.
2. Refer to the equation above and your graph to describe the pricing structure for the entry fee.

# 6 Chapter Review and Assessment

## VOCABULARY



## Key Skills & Exercises

### LESSON 6.1

#### Key Skills

**Solve inequalities, and graph the solution on the number line.**

For inequalities involving addition and subtraction, solve as you would an equation. Then graph the solution on a number line.

$$5t - 6 \leq 4t + 3$$
$$5t - 4t - 6 \leq 4t - 4t + 3$$
$$t - 6 \leq 3$$
$$t - 6 + 6 \leq 3 + 6$$
$$t \leq 9$$

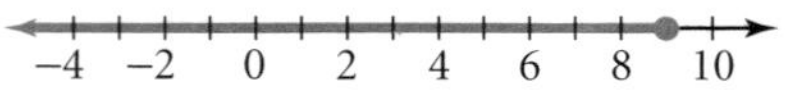

#### Exercises

**Solve each inequality, and graph the solution on a number line.**

1. $x + 5 > 10$
2. $n - 15 \leq 3$
3. $g + 6 < 11$
4. $11 \geq b + 4$
5. $y + 0.09 < 3.09$
6. $d - \frac{2}{3} \geq \frac{3}{9}$
7. $\frac{4}{5} \leq x - \frac{1}{5}$
8. $x - 3.2 \leq 2.7$
9. $x + 3.5 \leq 7$
10. $17 \leq x - 2.6$
11. $m + 3 > 0$
12. $m - 3.6 < 6$

### LESSON 6.2

#### Key Skills

**Solve multistep inequalities.**

Solve $14 - x \leq -6$.

| | |
|---|---|
| $14 - x \leq -6$ | *Given* |
| $14 - 14 - x \leq -6 - 14$ | *Subtr. Prop. of Inequality* |
| $-x \leq -20$ | *Simplify.* |
| $-1(-x) \geq -1(-20)$ | *Mult. Prop. of Inequality* |
| $x \geq 20$ | *Simplify.* |

Remember that if each side of an inequality is multiplied or divided by a negative number, the inequality symbol is reversed.

#### Exercises

**Solve.**

13. $2x + 4 < 6$
14. $8 - 3y \geq 2$
15. $5r - 12 > 48$
16. $\frac{-p}{8} \leq -3$
17. $7 < \frac{-b}{4}$
18. $t + 3 \geq 9 - t$
19. $11m - 17 > 13$
20. $5y + 2 > 7$
21. $8x + 7 \leq 2x + 2$
22. $4x + 1 > 7x - 14$
23. $k + \frac{1}{2} < \frac{k}{4} + 2$
24. $\frac{3}{5}k - 4 \leq \frac{1}{5}k + 7$
25. $-\frac{3}{5}k + 4 \leq -\frac{1}{5}k$

## LESSON 6.3

### Key Skills

**Solve and graph compound inequalities.**

Solve each inequality separately and then combine the result into a compound inequality.

Conjuction: Solve and graph $-4 \leq 2x - 6 < 12$.

| | | |
|---|---|---|
| $-4 \leq 2x - 6$ | AND | $2x - 6 < 12$ |
| $1 \leq x$ | AND | $x < 9$ |

$$1 \leq x < 9$$

Combine the graphs of the two inequalities in order to graph the compound inequality.

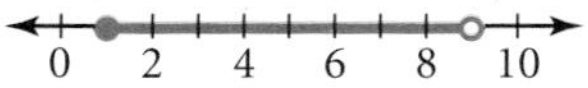

Disjunction: Solve $3x < -6$ OR $5x + 2 > -3$. Then graph the solution on a number line.

| | | |
|---|---|---|
| $3x < -6$ | | $5x + 2 > -3$ |
| | | $5x > -5$ |
| $x < -2$ | OR | $x > -1$ |

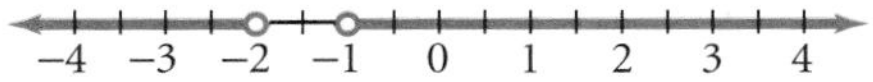

### Exercises

**Solve each inequality, if possible. Then graph the solution on a number line.**

**26.** $-5 \leq 4x \leq 12$

**27.** $3 < 2x + 1 \leq 9$

**28.** $-13 < 3x + 5 \leq 26$

**29.** $12 \leq -4y + 8 < 16$

**30.** $-5 \geq x + 2 \geq -7$

**31.** $-5 \leq 2 - x \leq 7$

**32.** $3x + 2 > 11$ OR $-(3x - 2) > 4$

**33.** $2x < 12$ OR $3x \geq 27$

**34.** $-3x \geq 12$ OR $2x > 6$

**35.** $3a + 2 \geq -2$ OR $-4a + 1 \geq -3$

**36.** $x + 3 \geq 4$ OR $3x + 7 \leq 22$

## LESSON 6.4

### Key Skills

**Find the absolute value of an expression.**

Find each of the following:

**a.** $|7 - 3|$ **b.** $|3 - 7|$

**a.** $|7 - 3| = |4|$
$= 4$

**b.** $|3 - 7| = |-4|$
$= 4$

**Describe the domain and range of an absolute-value function.**

Describe the domain and range of $y = 3|x|$.

Since you can find the absolute value of any number, the domain is all real numbers. When $x = 0$, $3|x| = 0$. When $x$ has any other value, the value of $3|x|$ will be a positive number. Therefore, the range is all non-negative numbers, or $y \geq 0$.

### Exercises

**Evalutate each of the following:**

**37.** $|4 - 6|$ **38.** $|3 - 10|$

**39.** $|14 - 18|$ **40.** $|18 - 14|$

**41.** $|3 - 20|$ **42.** $|13 - 6|$

**43.** $|17 - 3|$ **44.** $|3 - 17|$

**Describe the domain and range of each absolute-value function.**

**45.** $y = 5|x|$ **46.** $y = \frac{1}{2}|x|$

**47.** $y = |x - 2|$ **48.** $y = |x - 3|$

**49.** $y = |x| - 2$ **50.** $y = |x| + 3$

**51.** $y = -|x|$ **52.** $y = |x| - 4$

**53.** $y = 4|x| + 3$ **54.** $y = -4|x| + 3$

LESSON 6.5

**Key Skills**

**Solve absolute-value equations.**

Solve $|x-2| = 3$.

| Case 1 | Case 2 |
|---|---|
| $x-2=3$ | $x-2=-3$ |
| $x=5$ | $x=-1$ |

The solution is $x = 5$ or $x = -1$.

**Solve absolute-value inequalities.**

Conjuction: Solve $|x+5| < 7$.

| Case 1 | Case 2 |
|---|---|
| $x+5>-7$ | $x+5<7$ |
| $x>-12$ | $x<2$ |

The solution is $-12 < x < 2$.

Disconjuction: Solve $|3x+5| > 7$.

| Case 1 | Case 2 |
|---|---|
| $3x+5>7$ | $3x+5<-7$ |
| $3x>2$ | $3x<-12$ |
| $x>\frac{2}{3}$ | $x<-4$ |

The solution is $x < -4$ or $x > \frac{2}{3}$.

**Exercises**

**Solve, if possible.**

**55.** $|n-4| = 8$

**56.** $|3t+2| = 6$

**57.** $|2d-5| = 15$

**58.** $|x+7| = 3$

**59.** $|3x+2| = 4$

**60.** $|y-7| = 16$

**61.** $|k-3| \leq 1$

**62.** $|h-6| < 2$

**63.** $|2-m| \leq 2$

**64.** $|7-x| \leq 4$

**65.** $|x+2| \leq -1$

**66.** $|3x+3| < 0$

**67.** $|2r+4| \geq 6$

**68.** $|p-13| \geq 3$

**69.** $5 \leq |3+2m|$

**70.** $6 \leq |6x+3|$

**71.** $|x-7| \geq 0$

**72.** $|x-7| > -5$

## Applications

**73.** **CONSUMER ECONOMICS** Tamara paid $5.45 for a lunch. She had less than $5 left after paying her lunch bill. Write and solve an inequality representing the amount of money that Tamara could have had before she paid her lunch bill.

**74.** **BANKING** Randy had some money in his savings account. After he deposited $50, there was at least $180 in his account. Write and solve an inequality to determine the least amount that could have been in the account before his deposit.

**75.** **SAVINGS** The cost of a VCR is at least $280. Rhonda has saved $90. Write and solve an inequality to determine the least amount that Rhonda still needs to save in order to buy a VCR.

**76.** **HOBBIES** Rita has $15 to spend on new baseball cards. She decides to buy a card that costs $3. Write and solve an inequality to find the greatest amount that she can spend on other baseball cards.

# 6 Chapter Test

**Solve each inequality.**

1. $x + 4 < 7$
2. $m - 3 \geq 5$
3. $y - 2.72 \leq 1.62$
4. $\frac{1}{2} + f < 2\frac{1}{8}$

**Write the inequality that describes the set of points graphed on the number line.**

5. 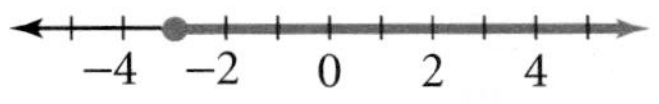

6. 

**Solve each inequality.**

7. $4r < 16$
8. $8 > \frac{g}{3}$
9. $7 - 3x \geq 22$
10. $4 - 5q < -24$
11. $2(x - 3) \leq -26$
12. $-5x + 4 > -2x + 31$
13. GEOMETRY The perimeter of a square is at least 15 centimeters more than the length of a side. Express the statement as an inequality, and solve for the possible length of the side.

**Graph each compound inequality.**

14. $-3 \leq x \leq 1$
15. $x < -2$ or $x \geq 0$

**Solve and graph each compound inequality.**

16. $-8 \leq 3x \leq 12$
17. $4 < 5u - 6 < 29$
18. $3p \leq 12$ or $7p > 35$
19. The high test score on a Geometry test was 97 and the low was 64. Write a compound inequality that describes the range of test scores, where $t$ represents the test scores.
20. SPORTS Three people on the volleyball team scored the same number of points. Together, their scores added up to a number between 18 and 33. Write and solve an inequality that represents the range of possibilities for their individual point totals.

**Evaluate.**

21. $|3 - 5|$
22. $|-10 - 12|$
23. $|18 - (-23)|$

**Describe the domain and range of each absolute value function.**

24. $y = 3|x|$
25. $y = \frac{2}{3}|x|$
26. $y = -|x + 1|$
27. $y = |x| - 6$

**Solve if possible and round your answer, if necessary, to the nearest thousandth.**

28. $|x + 3| = 7$
29. $|x - 4| > 3$
30. $|8 - x| = 21$
31. $|-2 + x| = -9$
32. $|x + 4| - 2 > 10$
33. $|5x + 3| \leq -2$
34. $|8 - 6x| < 5$
35. The distance between a number and 7 on a number line is less than 5. Write and solve an absolute value inequality to represent this statement.
36. ENGINEERING A bearing for a new sports car must be manufactured within 0.001 mm of its circumference of 2.35 mm. Write and solve an absolute value inequality to describe the largest and the smallest possible circumference for the bearing.

# College Entrance Exam Practice

internet connect
**Standardized Test Prep Online**
Go To: **go.hrw.com**
Keyword: **MM1 Test Prep.**

**QUANTITATIVE COMPARISON** For Questions 1–4 write
A if the quantity in Column A is greater than the quantity in Column B;
B if the quantity in Column B is greater than the quantity in Column A
C if the two quantities are equal; or
D if the relationship cannot be determined from the information given

| | Column A | Column B | Answers |
|---|---|---|---|
| **1.** | $-18+4$ | $-12+(-13)$ | Ⓐ Ⓑ Ⓒ Ⓓ [Lesson 2.2] |
| **2.** | 0.45 | $\frac{45}{100}$ | Ⓐ Ⓑ Ⓒ Ⓓ [Lesson 4.2] |
| **3.** | The smallest integer solution to $x+5 \geq 3$ | The greatest integer solution to $x-7<-4$ | Ⓐ Ⓑ Ⓒ Ⓓ [Lesson 6.1] |
| **4.** | The slope of the line passing through (2, 3) and (4, 7) | The slope of the line $3x+5y=12$ | Ⓐ Ⓑ Ⓒ Ⓓ [Lesson 5.5] |

**5.** Which expression is equal to 2? ***(LESSON 1.3)***
**a.** $3 \cdot 4 + 2 \div 5$ **b.** $2^7 \div 2^6$
**c.** $4(3-2) \div 0$ **d.** $2^4 - 2^3$

**6.** What is the solution to the equation $23y = 115$? ***(LESSON 3.2)***
**a.** $y = 45$ **b.** $y = 5$
**c.** $y = 192$ **d.** $y = 15$

**7.** What is the solution to the inequality $-2x + 3 \leq 7$? ***(LESSON 6.2)***
**a.** $x \leq 2$ **b.** $x \leq 5$
**c.** $x \geq -2$ **d.** $x > -2$

**8.** Which of the following is the standard form of the linear equation $y + 4 = 2x - 7$? ***(LESSON 5.5)***
**a.** $y = 2x - 11$ **b.** $2x + y = -11$
**c.** $-2x + y = 11$ **d.** $2x - y = 11$

**9.** What is the solution to the equation $\frac{m}{2.3} = 5$? ***(LESSON 3.2)***
**a.** $m = 1.5$ **b.** $m = -1.25$
**c.** $m = 1.15$ **d.** $m = 11.5$

**Find each sum or difference.**
***(LESSONS 2.2 AND 2.3)***

**10.** $-25 + 35$

**11.** $-36 + 7 + (-23)$

**12.** $323 + (-233)$

**13.** $690 - (-235)$

**14.** $-34 - (-34)$

**15.** $45 + (-23)$

**16.** $-3.4 + (-2.34)$

**17.** $7.8 - 19.2$

**18.** Solve: $y + 6.5 = 7$. ***(LESSON 3.1)***

**19.** Solve: $3x + 7 = x - 4$. ***(LESSON 3.4)***

**20.** Solve: $3(y + 4) - 23 = 5y$. ***(LESSON 3.5)***

**21.** Solve: $y + 7 < 3$. ***(LESSON 6.1)***

**22.** Solve: $-2x - 3 \geq 4x + 7$. ***(LESSON 6.2)***

**23.** What percent of 40 is 10? ***(LESSON 4.2)***

**24.** 24 is what percent of 72? ***(LESSON 4.2)***

**25.** 30% of 150 is what number? ***(LESSON 4.2)***

**26.** Find the next three terms of the sequence 5, 10, 20, 35, . . . ***(LESSON 1.1)***

**27.** Simplify $(7x + 2) - (3x - 2y + 2)$. ***(LESSON 2.6)***

**Find the slope of the line passing through the given points.** ***(LESSON 5.2)***

**28.** (3, 5) and (4, 6)

**29.** (9, –10) and (3, –2)

**30.** (5, 5) and (–5, 5)

**31.** (3, –4) and (4, –50)

**32.** (9, 0) and (0, 8)

**Write each linear equation in slope-intercept form.** ***(LESSON 5.4)***

**33.** $3y - 2x = 14$

**34.** $y - 3x = 2$

**35.** $4x = 3y + 2$

**Write an equation in standard form for a line passing through the given points.** ***(LESSON 5.5)***

**36.** (4, 6) and (3, 7)

**37.** (–2, –3) and (3, 2)

**38.** (–3, –3) and (3, 3)

**39.** (–2, 0) and (–3, –1)

**40.** (3, –5) and (7, –3)

**Find the mean, median, and mode for each set of numbers.** ***(LESSON 4.4)***

**41.** {1, 3, 3, 4, 5, 6, 6, 7, 8, 9, 10}

**42.** {300, 320, 120, 125, 126, 129}

**43.** {12, 12, 14, 14, 15, 11, 10, 10, 10, 15, 12}

**44.** {1, 2, 3, 4, 5, 5, 6, 5, 4, 3, 10}

**Simplify each expression.** ***(LESSON 2.6)***

**45.** $(9w - 5) - (3w - 4)$

**46.** $4x^2 - 4(3x^2 + 7)$

**47.** $(9m - 5n) + (-3m - 4n)$

**48.** $(3x + 7y) - (4x + 5y)$

**Solve each proportion.** ***(LESSON 4.1)***

**49.** $\frac{x}{5} = \frac{10}{2}$

**50.** $\frac{3}{m} = \frac{27}{45}$

**51.** $\frac{2}{5} = \frac{m}{27}$

### FREE-RESPONSE GRID

**The following questions may be answered by using a free-response grid such as that commonly used by standardized test services.**

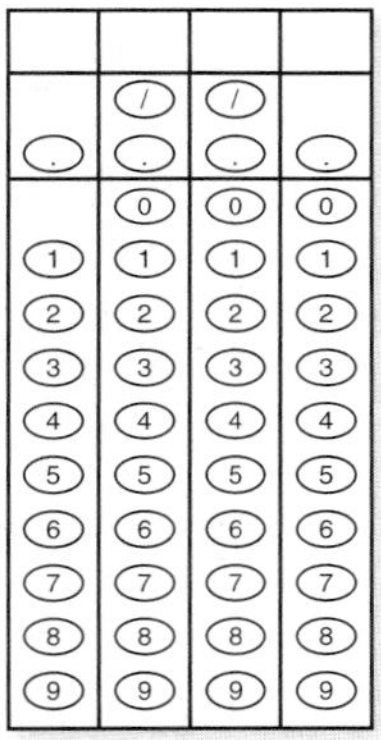

**52.** Simplify $4^2 \div 8 + 5(8 - 2) \cdot 2$. ***(LESSON 1.3)***

**53.** Solve: $\frac{5}{8} - x = \frac{1}{2}$. ***(LESSON 3.3)***

**54.** Maurice has scores of 88, 90, and 80 on three tests. What score must he make on his next test to have a mean score of exactly 85? ***(LESSON 4.4)***

**55.** Find the slope of a line that passes through the origin and (4, 2). ***(LESSON 5.2)***

7

# Systems of Equations and Inequalities



SOLVING REAL-WORLD PROBLEMS OFTEN INVOLVES determining the values of two or more variables. For example, organizers of music concerts need to make decisions about variables such as space, costs, and profits. In order to predict the weather, meteorologists must consider variables such as air pressure, wind direction, and the level of moisture in the atmosphere. Helicopter pilots need to consider variables such as weight, thrust, and wind speed when making decisions about taking off. Algebra, often performed by a computer, is used to solve these problems. Applying appropriate systems of equations makes it possible to solve for many variables at once.

In Chapter 7, you will learn how to solve systems of linear equations and inequalities in two variables and learn how to apply systems of equations to real-world situations.

*Weight, thrust, and wind speed are important variables to a pilot.*

*Concert organizers consider variables such as cost and space.*

*Meteorologists use systems of equations to make sense out of weather data.*

## About the Chapter Project

Writing and solving equations and systems of equations enable you to solve some very complex problems—graphically and algebraically. In the Chapter Project, *Minimum Cost/Maximum Profit,* on page 360 you will learn how a manager of a camping supply store can maximize profit when selling trail mix to backpackers. After completing the Chapter Project, you will be able to do the following:

- Write and solve systems of equations and inequalities.
- Solve an optimization problem.

## About the Portfolio Activities

Throughout the chapter, you will be given opportunities to complete Portfolio Activities that are designed to support your work on the Chapter Project.

- Baseball and basketball players' salaries are the topic of the Portfolio Activity on page 325.
- Finding the correct amount to charge for tickets to a magic show is the topic of the Portfolio Activity on page 337.
- Deciding the number of color pages for a high-school yearbook is the topic of the Portfolio Activity on page 352.

7.1

# Graphing Systems of Equations

*Newspaper advertisements provide the Smiths with information about costs.*

**Objectives**

- Graph systems of equations.
- Solve a system of equations by inspecting a graph.

**Why** ***Graphing systems of equations in two unknowns makes it possible to solve for more than one variable at the same time. You may encounter situations in which the graph of a system of equations can be used to help make the decision that is best for you.***

## Modeling Options With Equations

**APPLICATION**
**CONSUMER ECONOMICS**

The Smith family is trying to decide between installing cable TV or purchasing a VCR and renting tapes. The Smiths need to compare the cost of each option over several months in order to make an appropriate decision.

Cable TV: One-time installation charge of \$45, and \$34 per month
VCR: Purchase a VCR for \$185 and rent 8 tapes per month at \$2.50 per tape, which is \$20 per month

One way to compare the two options is to create a graph. The graph will show when both options will cost the same. First model each option by using an equation.

Let $m$ represent the number of months.
Let $c$ represent the total costs for an option.

Cable TV cost = cost for $m$ months + installation

$$c = 34m + 45$$

VCR and tape cost = tape rental for $m$ months + VCR

$$c = 20m + 185$$

Graph each equation on the same coordinate plane, with months on the horizontal axis and cost on the vertical axis.

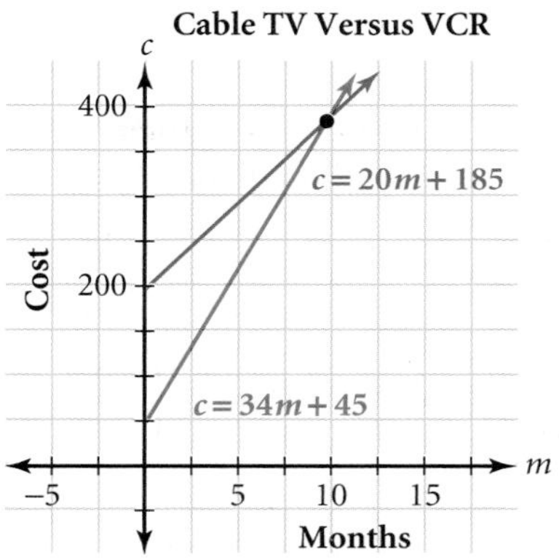


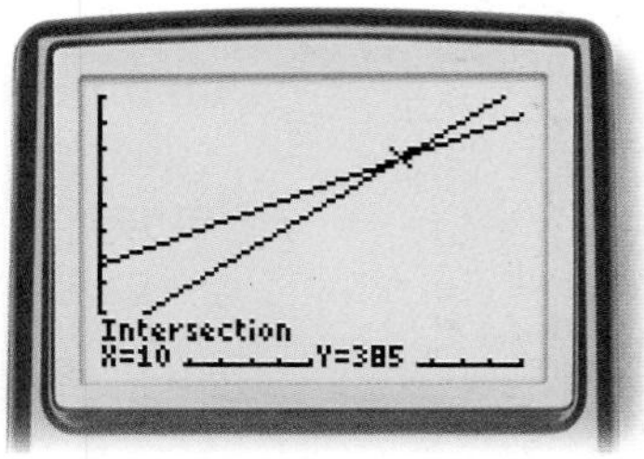


*On a graphics calculator, use the intersection feature to find the point of intersection.*

The lines representing the two options intersect at (10, 385). This means that at 10 months the cost of each option is \$385. The graph shows that each cost continues to increase, but the cost of cable TV increases at a faster rate. After 10 months, the VCR is a better choice.

Two equations in two variables are called a **system of equations.** A **solution** to a system of equations in two variables is an ordered pair of numbers that is a solution to each equation in the system. The solution to the system in the problem about the Smith family is (10, 385). The solution to a system of linear equations can be found by graphing.

**EXAMPLE 1** **Solve by graphing. Check your answer.**

$$\begin{cases} y = 3x + 1 \\ y = -x + 5 \end{cases}$$

*The brace { is used to indicate that the equations form a system.*

**TECHNOLOGY GRAPHICS CALCULATOR**

**SOLUTION**

Use a graphics calculator or graph paper to graph both equations on the same coordinate plane.

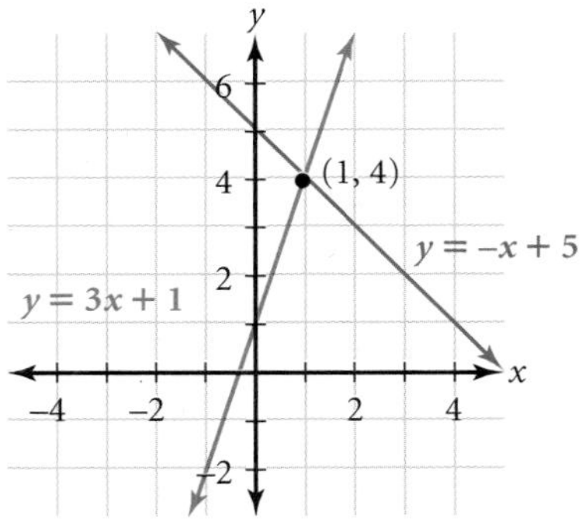


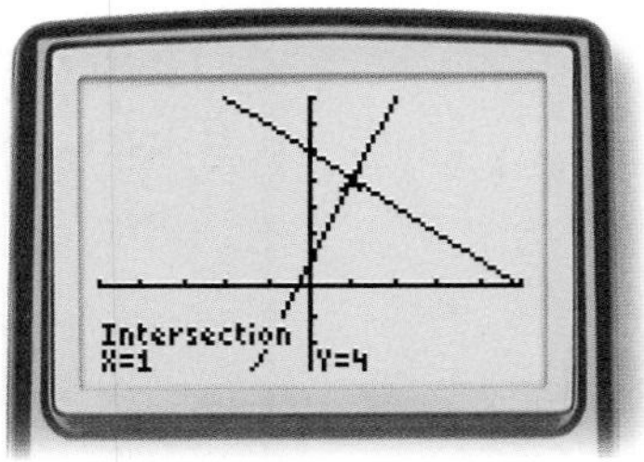


From the graph, the solution is (1, 4).

Check the solution by substituting 1 for $x$ and 4 for $y$ in each equation.

| | | | |
|---|---|---|---|
| $y = 3x + 1$ | | $y = -x + 5$ | |
| $4 \stackrel{?}{=} 3(1) + 1$ | | $4 \stackrel{?}{=} -(1) + 5$ | |
| $4 = 4$ | *True* | $4 = 4$ | *True* |

Because the substitution is true for each equation, (1, 4) is the solution.

**TRY THIS** Solve the system by graphing. Check your answer. $\begin{cases} y = 2x + 2 \\ y = -x - 1 \end{cases}$

In order to graph a system of equations, you may need to change each equation to slope-intercept form. Whether you are using a graphics calculator or graph paper, the slope-intercept form is often easier to use.

## EXAMPLE 2 Solve by graphing. Check your answer. $\begin{cases} 3x + y = 4 \\ x - 2y = 6 \end{cases}$

### SOLUTION

Change each equation to slope-intercept form, $y = mx + b$, by solving for $y$ in terms of $x$.

$$\begin{aligned} 3x + y &= 4 \\ y &= -3x + 4 \end{aligned} \qquad \begin{aligned} x - 2y &= 6 \\ -2y &= -x + 6 \\ y &= \frac{1}{2}x - 3 \end{aligned}$$

TECHNOLOGY
GRAPHICS CALCULATOR

Graph both equations on the same coordinate plane to find a common solution.

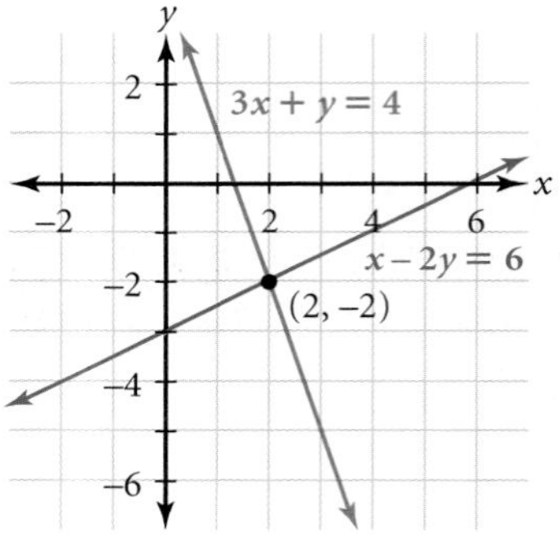


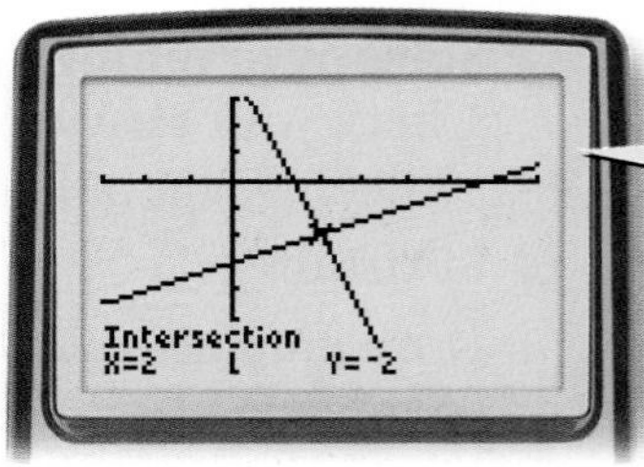


*To make your graph look this way, set your window to x: [−3, 7] and y: [−7, 3].*

The solution is $(2, -2)$.

Check the solution by substituting 2 for $x$ and $-2$ for $y$ in each equation.

$$\begin{aligned} 3x + y &= 4 \\ 3(2) + (-2) &\stackrel{?}{=} 4 \\ 4 &= 4 \quad \textit{True} \end{aligned} \qquad \begin{aligned} x - 2y &= 6 \\ 2 - 2(-2) &\stackrel{?}{=} 6 \\ 6 &= 6 \quad \textit{True} \end{aligned}$$

**TRY THIS** Solve each system below by graphing. Check your answers.

**a.** $\begin{cases} y + 2x = 2 \\ y + x = 1 \end{cases}$ **b.** $\begin{cases} y = 3x \\ y = -4x - 7 \end{cases}$

**CRITICAL THINKING** Consider the following system: $\begin{cases} y = -2x + 3 \\ y = -2x + 10 \end{cases}$

Does the graph of this system show a common solution? Why or why not?

*Activity*

## Exploring Approximate Solutions

**You will need:** graph paper and a graphics calculator

1. Graph this system on graph paper. $\begin{cases} 4x + 3y = 6 \\ 2y = x + 2 \end{cases}$
2. Estimate the solution from your graph. Check your estimated solution by substituting your values for $x$ and $y$ into the equations. How accurate are your estimates?

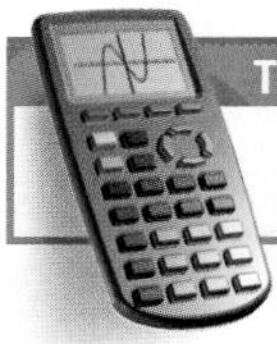

TECHNOLOGY
GRAPHICS CALCULATOR

3. Graph the system from Step 1 on a graphics calculator. Use the trace feature or intersection feature to find the solution to the system.

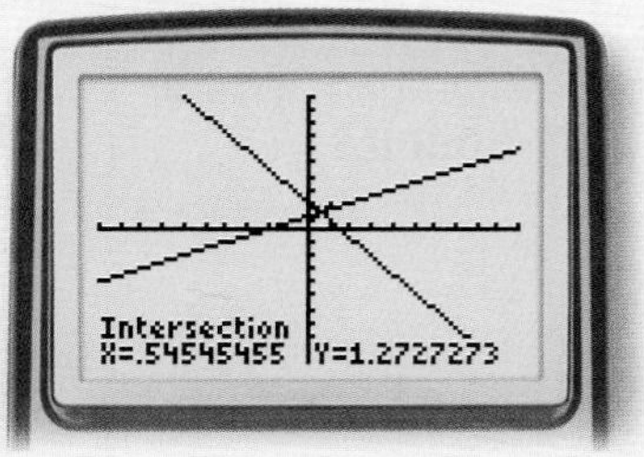


4. Test the solution from the calculator by substituting the values for $x$ and $y$ into the equations. Note that you can use the **Frac** feature on some calculators to find fractions for decimal values.

CHECKPOINT ✔

5. Discuss the advantages and disadvantages of the paper-and-pencil and calculator methods for solving systems of equations.

## Approximate Solution

It is sometimes difficult to find an exact solution from a hand-drawn graph of a system of equations. A reasonable estimate for a point of intersection is an **approximate solution**.

**EXAMPLE 3** **Find an approximate solution by graphing. Check your answer.** $\begin{cases} y = -0.2x + 1 \\ y = 2x - 7 \end{cases}$

**SOLUTION**

An approximate solution for this system is (3.7, 0.3).

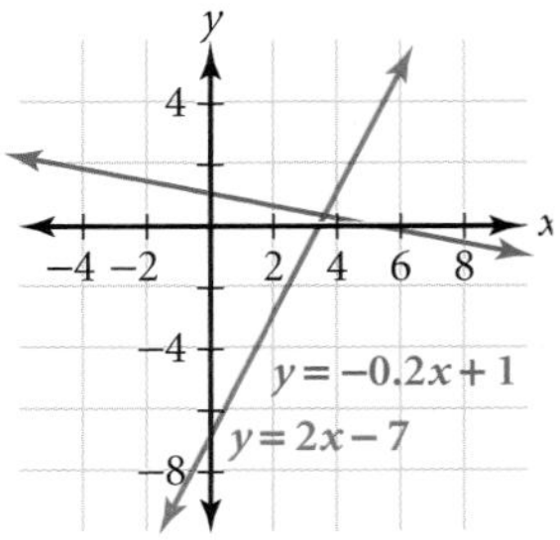


Check the solution by substituting 3.7 for $x$ and 0.3 for $y$ in the original equations.

| | |
|---|---|
| $y = -0.2x + 1$ | $y = 2x - 7$ |
| $0.3 \stackrel{?}{=} -0.2(3.7) + 1$ | $0.3 \stackrel{?}{=} 2(3.7) - 7$ |
| $0.3 \stackrel{?}{=} -0.74 + 1$ | $0.3 \stackrel{?}{=} 7.4 - 7$ |
| $0.3 \approx 0.26$ | $0.3 \approx 0.4$ |

A check shows that the approximate solution is reasonable to the nearest tenth.

You can also check your approximate solution with a graphics calculator.

**TRY THIS** Find an approximate solution by graphing. Check your answer. $\begin{cases} y = 4.6x + 3 \\ y = 1.5x - 7 \end{cases}$

## Using Systems to Solve Problems

**EXAMPLE 4**

**APPLICATION**
**HOBBIES**

A birding club holds an annual photography contest among its members. After a set time limit in a particular park, contestants receive 4 points for photos of songbirds and 20 points for photos of birds of prey. Last year's winner had a total of 200 points from 38 photos of individual birds. **How many of each type of bird did the winner photograph?**

**SOLUTION**

**PROBLEM SOLVING**

First define the two variables. Then write equations to model the problem.

**ORGANIZE**

Let $x$ represent the number of 4-point songbird photos.
Let $y$ represent the number of 20-point bird-of-prey photos.

**WRITE**

points for songbirds + points for birds of prey = 200 points

$4x + 20y = 200$

number of 4-point photos + number of 20-point photos = 38 photos

$x + y = 38$

**SOLVE**

Change the equations to slope-intercept form.

$$4x + 20y = 200$$
$$20y = -4x + 200$$
$$y = -\frac{1}{5}x + 10$$

$$x + y = 38$$
$$y = -x + 38$$

Graph the two equations on the same coordinate plane.

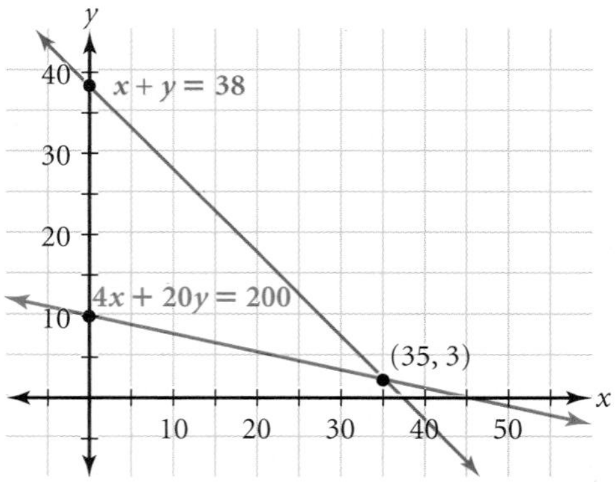


The winner photographed 35 songbirds and 3 birds of prey.

**CHECK**

$35 + 3 \stackrel{?}{=} 38$
$38 = 38$ *True*

$4(35) + 20(3) \stackrel{?}{=} 200$
$140 + 60 \stackrel{?}{=} 200$
$200 = 200$ *True*

# Exercises

## Communicate

**Exercises 1 and 2 refer to the following system of equations:** $\begin{cases} 2x - 3y = 4 \\ x + 4y = -9 \end{cases}$

**1.** How do you write each equation in slope-intercept form? Why is this form used to find values for the variables?

**2.** Explain how to find the solution for the system from a graph.

**3.** How do you graph the system at right? Explain how to make a reasonable estimate of the solution by inspecting the graph. Why is it important to check your estimate? $\begin{cases} x + y = 3 \\ x - y = 4 \end{cases}$

## Guided Skills Practice

**Solve by graphing. Check your answers.** ***(EXAMPLES 1 AND 2)***

**4.** $\begin{cases} -2x + y = 1 \\ y = -x + 4 \end{cases}$ **5.** $\begin{cases} y + 2x = 0 \\ 2y = -x - 6 \end{cases}$ **6.** $\begin{cases} 2x + 3y = -12 \\ 4x - 4y = 4 \end{cases}$

**Graph each system and find an approximate solution for each system by examining the graphs. Round your approximations to the nearest tenth. Check your answers.** ***(EXAMPLE 3)***

**7.** $\begin{cases} \frac{1}{2}x - y = 2 \\ y = -\frac{2}{3}x \end{cases}$ **8.** $\begin{cases} 2y - x = 6 \\ 3x + y = -5 \end{cases}$

**9.** Larry has \$4 in nickels and quarters. If there are 36 coins in all, how many quarters does Larry have? ***(EXAMPLE 4)***

## Practice and Apply

internet connect

**Homework Help Online**
Go To: **go.hrw.com**
Keyword:
**MA1 Homework Help**
for Exercises 10-27

**Solve by graphing. Round approximate solutions to the nearest tenth. Check by substituting the approximate solutions into the original equations.**

**10.** $\begin{cases} y = -\frac{1}{3}x + 2 \\ y = 2x + 12 \end{cases}$ **11.** $\begin{cases} y = \frac{3}{4}x + 3 \\ y = \frac{2}{3}x + 1 \end{cases}$ **12.** $\begin{cases} y = x \\ y = -x \end{cases}$

**13.** $\begin{cases} x + 7 = y \\ -4x + 2 = y \end{cases}$ **14.** $\begin{cases} a = 400 - 2b \\ a = b + 100 \end{cases}$ **15.** $\begin{cases} 3y = 4x - 2 \\ y = 2x - 2 \end{cases}$

**16.** $\begin{cases} x - 2y = 10 \\ 3y = 30 - 2x \end{cases}$ **17.** $\begin{cases} x = 2 \\ 2y = 4x + 2 \end{cases}$ **18.** $\begin{cases} -7m + 14n = -21 \\ 3m + 15 = 2n \end{cases}$

**19.** $\begin{cases} 9x - y = -7 \\ 6y - 2x = 15 \end{cases}$ **20.** $\begin{cases} x + 2 = y + 13 \\ 5 - x = y - 4 \end{cases}$ **21.** $\begin{cases} x - 3 = y - 3 \\ 2x + 3y = 10 \end{cases}$

22. $\begin{cases} 25y - 0.4x = 0.8 \\ 0.5x + 0.75y = 1 \end{cases}$

23. $\begin{cases} 3c + 2d = -6 \\ -3c + 2d = 6 \end{cases}$

24. $\begin{cases} 2x + 10y = -5 \\ 6x + 4y = 2 \end{cases}$

25. $\begin{cases} 5x + 6y = 14 \\ 3x + 5y = 7 \end{cases}$

26. $\begin{cases} 3x + 5y = 12 \\ 7x - 5y = 8 \end{cases}$

27. $\begin{cases} 7x - 9y = 13 \\ \frac{3}{4}y = \frac{1}{2}x + \frac{4}{3} \end{cases}$

**Determine whether the point (2, 10) is a solution for each system of equations.**

28. $\begin{cases} y = 2x - 4 \\ y = x + 8 \end{cases}$

29. $\begin{cases} y = -x + 12 \\ x = -y + 16 \end{cases}$

30. $\begin{cases} y = x + 8 \\ y = -3x + 16 \end{cases}$

31. $\begin{cases} x + 3y = 6 \\ -6 = 2x + 12 \end{cases}$

32. Graph the equations in Exercises 28–31 to check your answers.

33. Find the point of intersection of $y = x - 8$ and $y = -2x + 6$.

34. Use a graph to find the approximate point of intersection of $y = -x + 12$ and $y = 2x - 4$, giving values to the nearest tenth. Estimate the fractional value of these decimals. Check your estimate by substituting into the original equations.

**Refer to points *A*(3,5), *B*(4, – 1), and *C*(9,3). Lines *AB* and *AC* meet at point *A*.**

35. Write an equation for line *AB*.

36. Write an equation for line *AC*.

37. Check your equations by graphing.

38. Use the coordinates of the point of intersection to check your equations.

CONNECTIONS

**STATISTICS** Use the following data to solve Exercises 39–42:

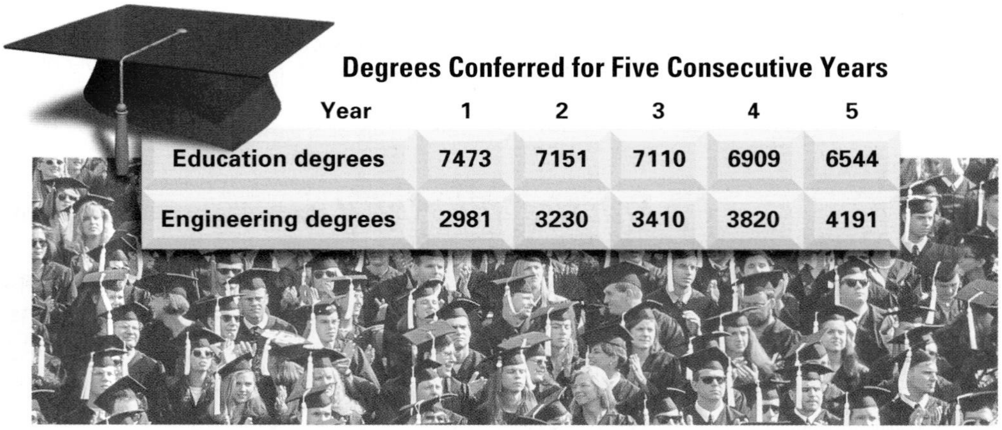

**Degrees Conferred for Five Consecutive Years**

| Year | 1 | 2 | 3 | 4 | 5 |
|---|---|---|---|---|---|
| Education degrees | 7473 | 7151 | 7110 | 6909 | 6544 |
| Engineering degrees | 2981 | 3230 | 3410 | 3820 | 4191 |

39. **TECHNOLOGY** Use a graphics calculator to enter the data for education degrees as ordered pairs (years, degrees) in a table. Use your calculator to show a scatter plot of the data.

40. **TECHNOLOGY** Use a graphics calculator to enter the data for engineering degrees as ordered pairs (years, degrees) in a table. Use your calculator to show a scatter plot of the data.

41. **TECHNOLOGY** Use your graphics calculator to find the equation for the line of best fit for each scatter plot in Exercises 39 and 40.

**42.** Find the approximate point of intersection for the two lines you found in Exercise 41. Based on the data in the table, about how many years beyond the given data will it be before the number of education degrees and the number of engineering degrees conferred are the same?

**APPLICATION**

**43.** **AVIATION** A plane initially at 28,000 feet is descending at a rate of 1800 feet per minute, and a second plane initially at 1500 feet is climbing at a rate of 2300 feet per minute. Write and graph a system of equations to determine how many minutes it will take for the planes to be at the same altitude.

## Look Back

**44.** Place parentheses and brackets to make $21 \div 3 - 4 + 0 + 6 \cdot (-3) = -39$ true. ***(LESSON 1.3)***

**45.** If data points are randomly scattered, what type of correlation exists? ***(LESSON 1.6)***

**Evaluate.** ***(LESSONS 2.2 AND 2.3)***

**46.** $-3 + \left(-\frac{1}{4}\right)$ **47.** $1.5 - (-1.5)$ **48.** $3\frac{1}{2} - 5\frac{2}{3}$

**Evaluate.** ***(LESSON 2.4)***

**49.** $(-3)(2)$ **50.** $(-1.7)(-2.5)$ **51.** $4\left(-2\frac{1}{3}\right)$

**Simplify each expression.** ***(LESSON 2.6)***

**52.** $-4x + 2x$ **53.** $3y + (-2x) - 3y$ **54.** $x - y + x$

**55.** Solve: $3(5 - 2x) - (8 - 6x) = -9 + 2(3x + 4) - 10$. ***(LESSON 3.5)***

**internet connect**
**Portfolio Extension**
Go To: **go.hrw.com**
Keyword:
**MA1 Athletes**

## Look Beyond

**Substitute each value of *x* into the equation $-2x + 4y = 12$, and solve for *y*.**

**56.** $x = -6$ **57.** $x = 8y$ **58.** $x = y - 3$ **59.** $x = 3y + 1$

## PORTFOLIO ACTIVITY

Include this activity in your portfolio.

From 1980 to 1990, there was a shift in salaries for basketball players and baseball players. The graph reflects the average salaries for the players.

- Estimate the year that the salaries were equal and estimate the salary.
- Estimate the $y$-intercept of each graph and use it with the point of intersection to determine the equation of each line.
- Extend the graph to predict the salaries for 2000.
- Find the current salary data and the data for 2000 in an almanac or on-line, and compare these data with your results.
- Find data for football salaries in an almanac or on-line, and compare them with the baseball and basketball salaries.

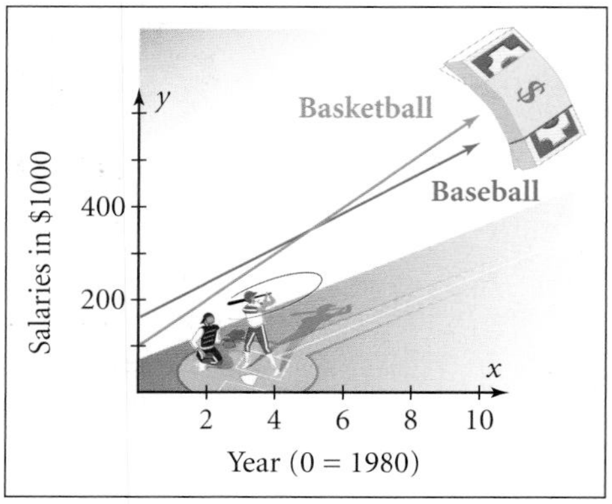

7.2

# The Substitution Method

**Objective**

- Find an exact solution to a system of linear equations by using the substitution method.

*The 12 Hours at Sebring automobile race is an endurance test. Two or more drivers for each racing team take turns driving in the race, which covers more than 1800 miles.*

**Why** ***Systems of equations often require exact solutions. In order to make a profit, the owner of a small business at the auto race track must know the exact amount to charge for each of her two products.***

You can find an exact solution to a system of linear equations without graphing. One method of doing this is to use substitution.

## Activity
## Exploring Substitution

APPLICATION
AUTO RACING

Two drivers on a team at the 12 Hours at Sebring automobile race drive for a total of 157 laps. Driver two drives 21 laps less than driver one. How many laps did each driver drive?

1. First use equations to model the problem.
   Let $x$ represent the number of laps for driver one.
   Let $y$ represent the number of laps for driver two.

   Then $\begin{cases} x + y = 157 \\ y = x - 21 \end{cases}$ is the system that models the problem.

PROBLEM SOLVING

2. Use the **guess-and-check** method to find the values for $x$ and $y$ that solve the system.
3. Look at the second equation in the system, $y = x - 21$. How can this information about $y$ be used in the first equation?
4. Because $y = x - 21$, substitute $x - 21$ in place of $y$ in the first equation. Solve the new equation for $x$.
5. Substitute this value of $x$ into the second equation to find $y$.

CHECKPOINT ✔

6. Compare the values for $x$ and $y$ with the answer you found by using guess-and-check. Are they the same?

If you know the value of one variable in a system of equations, you can find the solution for the system by substituting the known value of the variable into one of the equations. This method is called **substitution.**

**EXAMPLE 1** **Solve by using substitution.** $\begin{cases} 8x + 2y = 19 \\ x = 3 \end{cases}$

**SOLUTION**

Since $x = 3$, you can substitute the value of $x$ into the first equation.

$$8(3) + 2y = 19$$

Solve the resulting equation for $y$.

$$\begin{aligned} 24 + 2y &= 19 \\ 2y &= -5 \\ y &= -2.5 \end{aligned}$$

The solution is $(3, -2.5)$.

Check the solution by substituting the values for $x$ and $y$ into the first equation.

$$\begin{aligned} 8(3) + 2(-2.5) &\stackrel{?}{=} 19 \\ 24 + (-5) &\stackrel{?}{=} 19 \\ 19 &= 19 \quad \textit{True} \end{aligned}$$

**EXAMPLE 2** **Solve by using substitution.** $\begin{cases} 15x - 5y = 30 \\ y = 2x + 3 \end{cases}$

**SOLUTION**

Substitute $2x + 3$ for $y$ into the first equation, and solve for $x$.

$$\begin{aligned} 15x - 5(2x + 3) &= 30 \\ 15x - 10x - 15 &= 30 \\ 5x - 15 &= 30 \\ 5x &= 45 \\ x &= 9 \end{aligned}$$

Substitute 9 for $x$ into the equation $y = 2x + 3$, and solve for $y$.

$$\begin{aligned} y &= 2(9) + 3 \\ y &= 18 + 3 \\ y &= 21 \end{aligned}$$

The solution is $(9, 21)$.

Check the solution by substituting the values for $x$ and $y$ into the original equations.

$$\begin{aligned} 15(9) - 5(21) &\stackrel{?}{=} 30 \\ 135 - 105 &\stackrel{?}{=} 30 \\ 30 &= 30 \quad \textit{True} \end{aligned} \qquad \begin{aligned} 21 &\stackrel{?}{=} 2(9) + 3 \\ 21 &\stackrel{?}{=} 18 + 3 \\ 21 &= 21 \quad \textit{True} \end{aligned}$$

**TRY THIS** Solve each system by using substitution.

**a.** $\begin{cases} 2x + 5y = 14 \\ y = 5 \end{cases}$ **b.** $\begin{cases} -3x + 2y = 31 \\ x = 0.5y + 6 \end{cases}$

**EXAMPLE 3** **Solve by using substitution.** $\begin{cases} 3x + y = 4 \\ 5x - 7y = 11 \end{cases}$

**SOLUTION**

To use substitution in this example, solve the first equation for $y$.

$$3x + y = 4$$
$$y = 4 - 3x$$

*Choose the equation that is easier to solve.*

Substitute $4 - 3x$ for $y$ in the second equation and solve for $x$.

$$\begin{aligned} 5x - 7y &= 11 \\ 5x - 7(4 - 3x) &= 11 \\ 5x - 28 + 21x &= 11 \\ 26x - 28 &= 11 \\ 26x &= 39 \\ x &= 1.5 \end{aligned}$$

Substitute $1.5$ for $x$ in the first equation and solve for $y$.

$$\begin{aligned} 3(1.5) + y &= 4 \\ 4.5 + y &= 4 \\ y &= -0.5 \end{aligned}$$

The solution is $(1.5, -0.5)$. Check the solution by substituting the values for $x$ and $y$ into the original equations.

**TRY THIS** Solve by using substitution. $\begin{cases} 6x - 2y = 11 \\ x + 3y = 4 \end{cases}$

**CRITICAL THINKING** In Example 3, what is the advantage in choosing to solve for $y$ in the first equation?

**EXAMPLE 4**

**APPLICATION**
**SMALL BUSINESS**

Pam sells racing team caps at her concession stand at an automobile race track. She has 100 of last year's caps and 300 of this year's caps in stock. Pam knows from past experience that her customers will pay $7 more for this year's cap than for last year's. She has calculated that she needs to make $5300 in sales of the caps. **Assuming that she will sell all of the caps, how much should Pam charge for each type of cap?**

**SOLUTION**

Define the variables.
Let $m$ represent the cost of this year's cap.
Let $n$ represent the cost of last year's cap.

Write a system of linear equations. $\begin{cases} 300m + 100n = 5300 \\ m = n + 7 \end{cases}$

Substitute $n + 7$ for $m$ in the first equation, and solve for $n$.

$$\begin{aligned} 300(n + 7) + 100n &= 5300 \\ 300n + 2100 + 100n &= 5300 \\ 400n + 2100 &= 5300 \\ 400n &= 3200 \\ n &= 8 \end{aligned}$$

Substitute $8$ for $n$ in the second equation, and solve for $m$.

$$\begin{aligned} m &= 8 + 7 \\ m &= 15 \end{aligned}$$

Pam should charge $15 for this year's caps and $8 for last year's caps.

# Exercises

## Communicate

**1.** If you know $y = 42$, explain how to use substitution to solve $y = 2x + 8$ for $x$.

**2.** Given the equations $-4x + y = 2$ and $2x + 3y = 34$, tell how to find an expression to substitute and how to find the solution.

**3.** Explain how to use substitution to solve this system. $\begin{cases} x - 2y = 8 \\ 2x + 3y = 23 \end{cases}$

## Guided Skills Practice

**Solve by using substitution. Check your answers.** *(EXAMPLES 1, 2, AND 3)*

**4.** $\begin{cases} 5x = 3y + 12 \\ x = 5 \end{cases}$ **5.** $\begin{cases} 3x - 2y = 2 \\ y = 2x + 8 \end{cases}$ **6.** $\begin{cases} 5x - y = 1 \\ 3x + y = 1 \end{cases}$ **7.** $\begin{cases} 2x + y = 1 \\ 10x = 4y + 2 \end{cases}$

**Write and solve a system of equations for the problem below. Check your answer.** *(EXAMPLE 4)*

**8.** The sum of two numbers is 27. The larger number is 3 more than the smaller number. Find the two numbers.

## Practice and Apply

**internet connect**

**Homework Help Online**
Go To: **go.hrw.com**
Keyword:
**MA1 Homework Help** for Exercises 9-28

**Solve by using substitution, and check your answers.**

**9.** $\begin{cases} 2x + 8y = 1 \\ x = 2y \end{cases}$ **10.** $\begin{cases} x = 7 \\ 2x + y = 5 \end{cases}$ **11.** $\begin{cases} 3x + y = 5 \\ 2x - y = 10 \end{cases}$

**12.** $\begin{cases} y = 5 - x \\ 1 = 4x + 3y \end{cases}$ **13.** $\begin{cases} 2x + y = -92 \\ 2x + 2y = -98 \end{cases}$ **14.** $\begin{cases} 4x + 3y = 13 \\ x + y = 4 \end{cases}$

**15.** $\begin{cases} 6y = x + 18 \\ 2y - x = 6 \end{cases}$ **16.** $\begin{cases} 2x + y = 1 \\ 10x - 4y = 2 \end{cases}$ **17.** $\begin{cases} 2x + 3y = 7 \\ x + 4y = 9 \end{cases}$

**18.** $\begin{cases} 4x - y = 15 \\ -2x + 3y = 12 \end{cases}$ **19.** $\begin{cases} 2y + x = 4 \\ y - x = -7 \end{cases}$ **20.** $\begin{cases} 4y - x = 4 \\ y + x = 6 \end{cases}$

**21.** $\begin{cases} y = x - 3 \\ x + y = 5 \end{cases}$ **22.** $\begin{cases} 2x + 3y = 21 \\ -3x - 6y = -24 \end{cases}$ **23.** $\begin{cases} 5x - 7y = 31 \\ -4x + 2y = -14 \end{cases}$

**24.** $\begin{cases} 3x + y = 21 \\ 10x + 5y = 65 \end{cases}$ **25.** $\begin{cases} -3y = 9x + 24 \\ 6y + 2x = 32 \end{cases}$ **26.** $\begin{cases} 12x + 4y = 22 \\ 3x - 8y = -10 \end{cases}$

**27.** $\begin{cases} 11x + 4y = -17 \\ -6x + y = 22 \end{cases}$ **28.** $\begin{cases} -5x + 7y = -41 \\ 7x + y = 25 \end{cases}$

**CONNECTION**

**29. GEOMETRY** Find the dimensions of the rectangle shown at right if its length is twice its width.

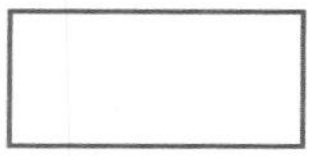

*Perimeter* = 208

CONNECTION

30. **GEOMETRY** In the triangle shown, the sum of the measures of angle $A$ and angle $B$ is 90°. If angle $A$ is 30° less than twice angle $B$, then find the degree measure for each angle.

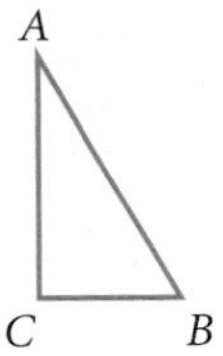


31. **NUMBER THEORY** One number is 4 less than 3 times a second number. If 3 more than twice the first number is decreased by twice the second, the result is 11. Find both numbers.

**Write and solve a system of equations for each problem.**

APPLICATIONS

32. **RECREATION** A hot-air balloon is rising from the ground at a rate of 4 feet every second. Another balloon at 756 feet is descending at a rate of 3 feet every second. In how many seconds will the two balloons be at the same altitude?

33. **FUND-RAISING** At the Boy Scout "all you can eat" spaghetti dinner, the troop served 210 people and raised $935. Adult meals cost $6.00 and child meals cost $3.50.
  - **a.** Write an equation for the *total amount raised* from the adult and child dinners served.
  - **b.** Write an equation for the *total number* of adult and child dinners served.
  - **c.** Solve the system of equations from parts **a** and **b**. How many adult and child dinners were served?

34. **CULTURAL CONNECTION: ASIA** A Chinese problem states that several people combined their money to buy a farm tool to share. If each person paid 8 coins, they had 3 coins too many. If each paid 7 coins, they were 4 coins short. How many people were there, and what was the price in coins of the farm tool?

## Look Back

APPLICATION

35. **RECREATION** In a foot race, Sam finished 20 feet in front of Joe. Joe was 5 feet behind Mark, and Mark was 10 feet behind Rob. Tom was 15 feet ahead of Rob. In what order did they finish? *(LESSON 2.2)*

internet connect
**Activities Online**
Go To: **go.hrw.com**
Keyword:
**MA1 House Gender**

**Solve each equation for $x$.** *(LESSON 3.2)*

**36.** $\frac{x}{15} = 3$ **37.** $\frac{3}{x} = 15$ **38.** $\frac{15}{x} = 3$ **39.** $\frac{x}{3} = 15$

**40.** The number 12.6 is 42% of what number? *(LESSON 4.2)*

**Tell whether the following equations represent lines that are parallel, perpendicular, or neither:** *(LESSON 5.6)*

**41.** $\begin{cases} 3y = 2x - 15 \\ 3x + 2y = 24 \end{cases}$ **42.** $\begin{cases} 2y = x - 12 \\ 2y - x = 12 \end{cases}$ **43.** $\begin{cases} y = x - 1 \\ y = -x + 3 \end{cases}$

## Look Beyond

**Examine the following systems. Use substitution to solve for $x$ and $y$.**

**44.** $\begin{cases} x + 2y + 3z = 8 \\ y + 2z = 3 \\ z = 2 \end{cases}$ **45.** $\begin{cases} 2x + 3y + 5z = 44 \\ 2y - 6z = 4 \\ z = 4 \end{cases}$

# 7.3 The Elimination Method

## Objectives

- Use the elimination method to solve a system of equations.
- Choose an appropriate method to solve a system of equations.

**Why** ***The elimination method is another technique that can provide a quick solution to complex systems of equations such as the one presented in the car-rental problem.***

*Jason drove 125 miles on the 2-day trip. He drove 350 miles on the 4-day trip.*

**APPLICATION**
**RENTAL FEES**

To find the daily fee and the per-mile cost at Airport Rent-A-Car, you can set up and solve a system of equations.

First define the variables.
Let $d$ represent the cost per day.
Let $m$ represent the cost per mile.

Use the information from the rental bills to set up a system of equations.

$$\begin{cases} 2d + 125m = 95.75 & \text{represents Bill No. 001} \\ 4d + 350m = 226.50 & \text{represents Bill No. 002} \end{cases}$$

Both the graphing method and the substitution method provide options for finding the solution to this system. However, neither of these methods would be easy to apply to this system. In this lesson you will develop a method for solving a system like this one.

*Activity*

## Exploring Using Opposites

Consider this system of equations: $\begin{cases} 3x + 2y = 7 \\ 5x - 2y = 9 \end{cases}$

1. Which terms of the equations are opposites?
2. Use the Addition Property of Equality to combine the two equations into one equation (add $3x$ and $5x$, $2y$ and $-2y$, and 7 and 9). How many variables does the resulting equation have?
3. Solve the resulting equation for $x$, and then solve for $y$ by substituting the value for $x$ into one of the original equations.
4. Check the solution in both equations of the system.

CHECKPOINT ✓

5. Explain how opposites can be used to solve a system of equations.

## The Elimination Method

In the activity, you used the **elimination method** to solve a system of equations. This method uses opposites to eliminate one of the variables.

**EXAMPLE 1** **Solve by elimination.** $\begin{cases} 3x + 4y = 7 \\ 2x - 4y = 13 \end{cases}$

**SOLUTION**

Use the Addition Property of Equality to combine the equations, and solve the resulting equation for $x$.

$$\begin{aligned} 3x + 4y &= 7 \\ \underline{2x - 4y} &\underline{= 13} \\ 5x + 0 &= 20 \\ 5x &= 20 \\ x &= 4 \end{aligned}$$

*4y and −4y are opposites.*

To find $y$, substitute 4 for $x$ in one of the original equations.

$$\begin{aligned} 3x + 4y &= 7 \\ 3(4) + 4y &= 7 \\ 12 + 4y &= 7 \\ 4y &= -5 \\ y &= -1.25 \end{aligned}$$

The solution is $(4, -1.25)$.

Check by substituting 4 for $x$ and $-1.25$ for $y$ in the original equations.

$$\begin{aligned} 3(4) + 4(-1.25) &\stackrel{?}{=} 7 \\ 12 + (-5) &\stackrel{?}{=} 7 \\ 7 &= 7 \quad \textit{True} \end{aligned}$$

$$\begin{aligned} 2(4) - 4(-1.25) &\stackrel{?}{=} 13 \\ 8 - (-5) &\stackrel{?}{=} 13 \\ 13 &= 13 \quad \textit{True} \end{aligned}$$

In the system of equations, notice that the $y$-terms are opposites. These types of systems are the easiest to solve by using the elimination method.

Sometimes it is necessary to multiply one or both equations by a number in order to produce opposites that will eliminate each other. This process is easiest when a variable in one of the equations has a coefficient of 1. However, it can be applied to more complicated systems such as the one solved in Example 2.

**EXAMPLE 2** **Solve by elimination.** $\begin{cases} 2x + 3y = 1 \\ 5x + 7y = 3 \end{cases}$

**SOLUTION**

Multiply the first equation by 5 and the second equation by −2 in order to produce opposites that will eliminate each other.

$$\begin{cases} (5)2x + (5)3y = (5)1 \\ (-2)5x + (-2)7y = (-2)3 \end{cases} \Rightarrow \begin{cases} 10x + 15y = 5 \\ -10x - 14y = -6 \end{cases}$$

Use the Addition Property of Equality to combine the equations, and solve the resulting equation for $y$.

$$\begin{aligned} 10x + 15y &= 5 \\ -10x - 14y &= -6 \\ \hline y &= -1 \end{aligned}$$

To find $x$, substitute −1 for $y$ in one of the original equations.

$$\begin{aligned} 2x + 3y &= 1 \\ 2x + 3(-1) &= 1 \\ 2x - 3 &= 1 \\ 2x &= 4 \\ x &= 2 \end{aligned}$$

The solution is (2, −1).

Check by substituting 2 for $x$ and −1 for $y$ in the original equations.

$$\begin{aligned} 2(2) + 3(-1) &\stackrel{?}{=} 1 \\ 4 + (-3) &\stackrel{?}{=} 1 \\ 1 &= 1 \quad \textit{True} \end{aligned} \qquad \begin{aligned} 5(2) + 7(-1) &\stackrel{?}{=} 3 \\ 10 + (-7) &\stackrel{?}{=} 3 \\ 3 &= 3 \quad \textit{True} \end{aligned}$$

**EXAMPLE 3** **Use elimination to solve the car-rental problem from page 331.**

$$\begin{cases} 2d + 125m = 95.75 \\ 4d + 350m = 226.50 \end{cases}$$

**APPLICATION**

**RENTAL FEES**

**SOLUTION**

Multiply each term of the first equation by −2.

$$\begin{cases} (-2)2d + (-2)125m = (-2)95.75 \\ 4d + 350m = 226.50 \end{cases}$$

Use the Addition Property of Equality to combine the equations, and solve the resulting equation for $m$.

$$\begin{aligned} -4d + (-250m) &= -191.50 \\ 4d + \quad 350m &= \quad 226.50 \\ \hline 100m &= 35 \\ m &= 0.35 \end{aligned}$$

Then substitute 0.35 for $m$ in the first equation to find $d$.

$$\begin{aligned} 2d + 125(0.35) &= 95.75 \\ 2d + 43.75 &= 95.75 \\ 2d &= 52 \\ d &= 26 \end{aligned}$$

The solution is (26, 0.35). A check shows that the solution is correct. The car was rented for $26 a day plus $0.35 per mile.

**TRY THIS** Solve by elimination. **a.** $\begin{cases} 2x - y = 7 \\ 5x + 4y = 11 \end{cases}$ **b.** $\begin{cases} 3a - 2b = 6 \\ 5a + 7b = 41 \end{cases}$

## Choosing a Method

You have learned several methods for solving systems of linear equations. What method should you choose? There may be more than one good choice for solving a system.

**SUMMARY OF METHODS FOR SOLVING SYSTEMS**

| Example | Suggested method | Why |
|---|---|---|
| $\begin{cases} 6x + y = 10 \\ y = 5 \end{cases}$ | Substitution | The value of $y$ is known and can be easily substituted into the other equation. |
| $\begin{cases} 2x - 5y = -20 \\ 4x + 5y = 14 \end{cases}$ | Elimination | $5y$ and $-5y$ are opposites and are easily eliminated. |
| $\begin{cases} 9a - 2b = -11 \\ 8a - 4b = 25 \end{cases}$ | Elimination | $b$ can be easily eliminated by multiplying the first equation by $-2$. |
| $\begin{cases} 324p + 456t = 225 \\ 178p - 245t = 150 \end{cases}$ | Graph on a graphics calculator | The coefficients are large numbers, so other methods may be cumbersome. |

# Exercises

## Communicate

internet connect
**Activities Online**
Go To: **go.hrw.com**
Keyword:
**MA1 Sports Opera**

**In the following systems, which terms are opposites? Explain how you would solve each system.**

**1.** $\begin{cases} x + 7y = 13 \\ x - 7y = 5 \end{cases}$ **2.** $\begin{cases} 2x - 3y = 8 \\ 5x + 3y = 20 \end{cases}$ **3.** $\begin{cases} 2a + b = 6 \\ -2a - 3b = 8 \end{cases}$

**Describe the steps necessary to solve each system by using the elimination method.**

**4.** $\begin{cases} 2x + 3y = 9 \\ 3x + 6y = 7 \end{cases}$ **5.** $\begin{cases} 2x - 5y = 1 \\ 3x - 4y = -2 \end{cases}$ **6.** $\begin{cases} 9a + 2b = 2 \\ 21a + 6b = 4 \end{cases}$

**Choose a method for solving each system, and explain your decision.**

**7.** $\begin{cases} 3x + 2y = 5 \\ 5x - 2y = 7 \end{cases}$ **8.** $\begin{cases} 4x + 3y = 15 \\ 3x + y = 10 \end{cases}$ **9.** $\begin{cases} 5.6x + 8.77y = 31.17 \\ 7.3x - 2.34y = 26.86 \end{cases}$

## Guided Skills Practice

**Solve each system by elimination, and check your solution.**

**10.** $\begin{cases} 3x + 2y = 5 \\ 5x - 2y = 7 \end{cases}$ ***(EXAMPLE 1)***

**11.** $\begin{cases} 4m + 3n = 13 \\ 2m - 4n = 1 \end{cases}$ ***(EXAMPLE 2)***

**12.** $\begin{cases} 2j - 2k = 4 \\ 3j + 5k = -10 \end{cases}$ ***(EXAMPLE 3)***

**13.** $\begin{cases} 2x + 3y = 1 \\ -3x - 4y = 0 \end{cases}$ ***(EXAMPLE 3)***

## Practice and Apply

**internet connect**
**Homework Help Online**
Go To: **go.hrw.com**
Keyword:
**MA1 Homework Help**
for Exercises 14-34

**Solve each system of equations by elimination, and check your solution.**

**14.** $\begin{cases} -x + 2y = 12 \\ x + 6y = 20 \end{cases}$

**15.** $\begin{cases} 2p + 3q = 18 \\ 5p - q = 11 \end{cases}$

**16.** $\begin{cases} -4x + 3y = -1 \\ 8x + 6y = 10 \end{cases}$

**17.** $\begin{cases} 2m - 3n = 5 \\ 5m - 3n = 11 \end{cases}$

**18.** $\begin{cases} 6x - 5y = 3 \\ -12x + 8y = 5 \end{cases}$

**19.** $\begin{cases} 4s + 3t = 6 \\ -2s + 6t = 7 \end{cases}$

**Solve each system of equations by using any method.**

**20.** $\begin{cases} 2x + 4y = 40 \\ 7x - 3y = 4 \end{cases}$

**21.** $\begin{cases} 13x - 3y = -50 \\ 12x + 5y = 16 \end{cases}$

**22.** $\begin{cases} 2x - 7y = 20 \\ 5x + 8y = -1 \end{cases}$

**23.** $\begin{cases} 3x - 2y = 2 \\ 4x - 7y = 33 \end{cases}$

**24.** $\begin{cases} -7x - 3y = 10 \\ 2x + 2y = -8 \end{cases}$

**25.** $\begin{cases} 2x - 5y = -14 \\ -7x + 4y = -5 \end{cases}$

**26.** $\begin{cases} \frac{3}{2}x + 3y = 2 \\ 3x - 4y = -16 \end{cases}$

**27.** $\begin{cases} 11x + 2y = -8 \\ 8x + 3y = 5 \end{cases}$

**28.** $\begin{cases} 3x - 2y = -0.3 \\ 5x + 3y = 4.25 \end{cases}$

**29.** $\begin{cases} -x - 7 = 3y \\ 6y = 2x - 14 \end{cases}$

**30.** $\begin{cases} 2m = 2 - 9n \\ 21n = 4 - 6m \end{cases}$

**31.** $\begin{cases} 2x = 3y - 12 \\ \frac{1}{3}x = 4y + 5 \end{cases}$

**32.** $\begin{cases} \frac{2}{3}x = \frac{2}{3} - \frac{1}{6}y \\ y = 3x - 38 \end{cases}$

**33.** $\begin{cases} 0.6a = 3.2b + 4.48 \\ 2.9b = 0.3a + 4.91 \end{cases}$

**34.** $\begin{cases} p = 1.5b + 4 \\ 0.8p + 0.4b = 0 \end{cases}$

**CHALLENGE**

**35.** $\begin{cases} \frac{2}{3}x - \frac{3}{5}y = -\frac{17}{15} \\ \frac{8}{5}x - \frac{7}{6}y = -\frac{3}{10} \end{cases}$

**CONNECTION**

**36. GEOMETRY** The perimeter of a rectangle is 24, and its length is 3 times its width. Find the length and width of this rectangle.

**Use the equations $y = x - 2$ and $y = 2x$ for Exercises 37–39.**

**37.** Find the common solution by graphing.

**38.** Solve the system algebraically.

**39.** What method did you use in Exercise 38 and why?

**For Exercises 40–48 write a system of equations for each problem, and choose the best method to solve each system. Solve and check your answers.**

**APPLICATIONS**

**40.** **CONSUMER MATH** Mrs. Jones is celebrating Gauss's birthday by having a pizza party for her two algebra classes. She orders 3 pizzas and 3 bottles of soda for \$23.34 for her 2nd period class and 4 pizzas and 6 bottles of soda for \$32.70 for her 7th period class. How much does each pizza and each bottle of soda cost?

**41.** **INCOME** As a parking attendant, John earns a fixed salary for the first 15 hours he works each week and then additional pay for any time over 15 hours. During the first week, John worked 25 hours and earned \$240; the second week, he worked 22.5 hours and earned \$213.75. What is John's weekly salary and overtime pay per hour?

**42.** **INVESTMENTS** Mr. Moore sells his tractor for \$6000. He makes two investments, one at a bank paying 5% interest per year and the rest in stocks yielding 9% interest per year. If he earns a total of \$380 in interest in the first year, how much is invested at each rate?

**43.** **BUSINESS** At a local music store, single-recording tapes cost \$6.99 and concert tapes cost \$10.99. At the end of the day a total of 25 tapes were sold. If the sales for the day total \$230.75 for single-recording tapes and concert tapes, find the number of each type of tape sold.

**44.** **FINANCE** To rent an apartment, a one-time deposit is frequently required with the first month's rent. Roberto paid \$900 the first month and a total of \$6950 during the first year. Find the monthly rent and the amount of the deposit.

**45.** **RECREATION** The Shamrock Inn offers two holiday weekend specials. The first is 2 nights and 4 meals for \$205, while the other is 3 nights and 8 meals for \$342.50. At these rates, what is the room rate and the cost of a meal?

**46.** **CONSUMER MATH** Shopping at Super Sale Days, Martha buys her children 3 shirts and 2 pairs of pants for \$85.50. She returns during the sale and buys 4 more shirts and 3 more pairs of pants for \$123. What is the sale price of the shirts and pants?

**47.** **BUSINESS** At a local bakery, single-crust pies sell for \$5.99 and double-crust pies sell for \$9.99. The total number of pies sold on Monday was 25. If the sales total for Monday was \$189.75 for these two types of pies, find the number of each type sold.

**48.** **SALES** The owner of a health-food store sells a mixture of seeds and nuts. The seeds cost 20¢ per ounce, and the nuts cost 30¢ per ounce. Forty-two ounces of the mixture has a value of \$11.00. How many ounces of seeds and nuts did the owner use in the mix?

## Look Back

**Find the slope of each line.** ***(LESSON 5.4)***

**49.** $7x - 3y = 22$ **50.** $3x + 2y = 6$ **51.** $y = 2x$

**52. CULTURAL CONNECTION: AFRICA** The following problem is in the Rhind papyrus by Ahmes: A bag contains equal weights of gold, silver, and lead purchased for 84 *sha'ty*. What is the amount (in *deben* weight) of each metal if the price for 1 *deben* weight is 6 *sha'ty* for silver, 12 *sha'ty* for gold, and 5 *sha'ty* for lead? ***(LESSON 3.3)***

**Solve each equation for the variable.** ***(LESSONS 3.3 AND 3.4)***

**53.** $-5 = -x + 7$ **54.** $3x - 6 = 2x + 1$ **55.** $\frac{1}{2}x + 3 = 2$

**Draw a graph and give the slope for each linear equation.** ***(LESSON 5.5)***

**56.** $y = 3x$ **57.** $y = -3x + 7$ **58.** $3y + 2x = 7$ **59.** $7y = 2x$

**60.** $2x + 8y = 24$ **61.** $-7x = 3y + 9$ **62.** $-3y = 7x + 15$ **63.** $18x + 15y = 45$

**64.** Write the equation of the line that passes through the point $(-3, 8)$ and is parallel to $y = 0.8x - 7$. ***(LESSON 5.6)***

## Look Beyond

**65. TECHNOLOGY** Graph the lines represented by the equations below on the same coordinate plane and describe the graph. Use a graphics calculator if one is available.

**a.** $2x - 3y = 6$ **b.** $4x - 6y = 18$

PORTFOLIO ACTIVITY

Include this activity in your portfolio.

An entertainment promotor for a magic show must make $161,000 from ticket sales in order to cover the costs and make a reasonable profit. The auditorium has 3000 seats, with 1800 seats on the lower floor and 1200 balcony seats. Attendees will pay $20 more for the lower-floor seats than for the balcony seats.

- Write a system of equations with *f* for the floor-seat ticket price and *b* for the balcony-seat ticket price.
- Solve the system graphically.
- Solve the system algebraically by using either substitution or elimination.
- What should the entertainment promoter charge for each type of ticket?
- Suppose that the show will take place at a large arena with 6000 balcony seats and 7000 lower-floor seats. Ticket sales must total $500,000. You may choose the price difference between the tickets. Determine appropriate ticket prices.

**WORKING ON THE CHAPTER PROJECT**

You should now be able to complete Activity 1 of the Chapter Project on page 360.

# 7.4 Consistent and Inconsistent Systems

*According to the map, would it be possible to meet someone at the intersection of 2nd and 3rd Avenues? Is Fairview Parkway a different street from Route 243?*

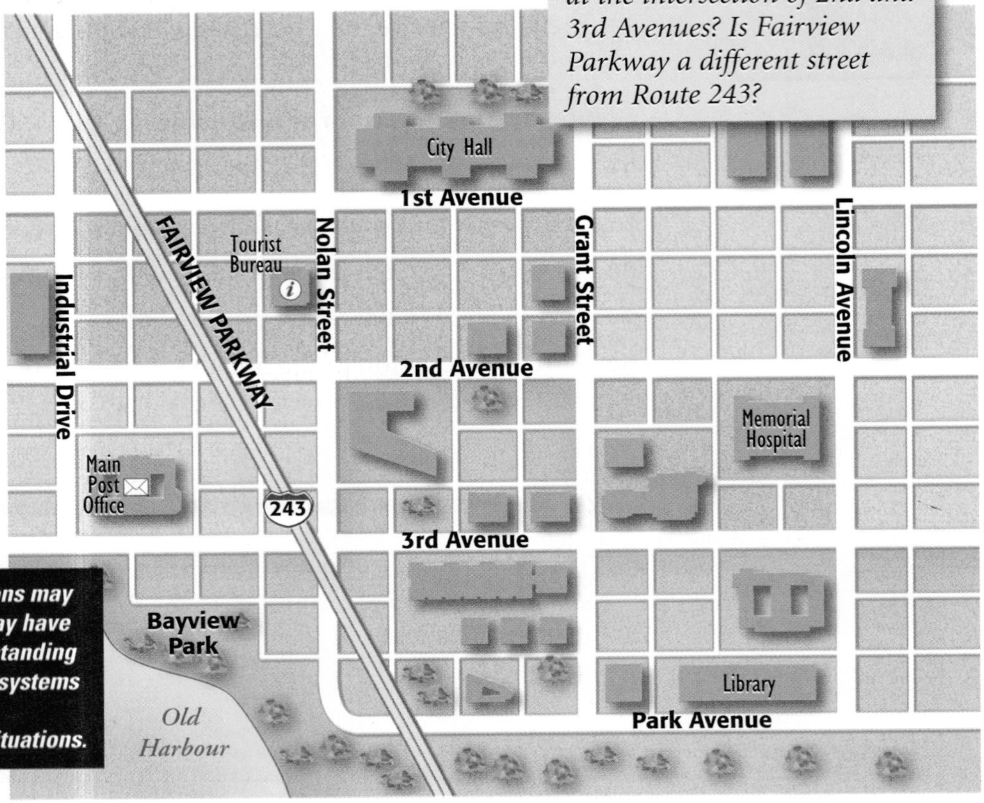


## Objectives

- Identify consistent and inconsistent systems of equations.
- Identify dependent and independent systems of equations.

**Why** *Some real-world situations may have no solutions and some may have many solutions. A clear understanding of consistent and inconsistent systems can help you model and better understand these real-world situations.*

The point of intersection for the graphs of the equations of a system is similar to the intersection of streets on the map above. However, some systems of equations are like the streets that do not intersect. These have no solution. Some systems of equations are like the street with two names. These have an infinite number of solutions.

## *Activity*
## Exploring No Solution and Many Solutions

TECHNOLOGY
GRAPHICS CALCULATOR

**You will need:** a graphics calculator or graph paper

1. Use a graphics calculator or graph paper to graph this system. $\begin{cases} y = -3x + 4 \\ y = -3x + 8 \end{cases}$
   **a.** Do the lines intersect?
   **b.** Is there a solution for this system? Why or why not?

CHECKPOINT ✔

2. Use a graphics calculator or graph paper to graph this system. $\begin{cases} y = 2x - 3 \\ 3y = 6x - 9 \end{cases}$
   **a.** Describe the graph of this system.
   **b.** Do the lines intersect?
   **c.** Is there a solution for this system? Why or why not?

CHECKPOINT ✔

## Inconsistent and Consistent Systems

A system of equations may have one unique solution, infinitely many solutions, or no solution at all. Systems that have one or many solutions are called **consistent**. Systems with no solution are called **inconsistent**.

| CONSISTENT SYSTEMS | | INCONSISTENT SYSTEMS |
|---|---|---|
| **One unique solution** | **Infinitely many solutions** | **No solution** |
| $\begin{cases} x + 2y = 3 \\ 2x - y = 1 \end{cases}$ 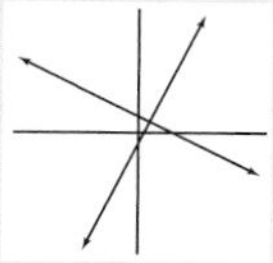 | $\begin{cases} x + y = 3 \\ 3x + 3y = 9 \end{cases}$ 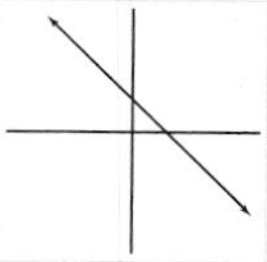 | $\begin{cases} -2x + y = -4 \\ -2x + y = 3 \end{cases}$ 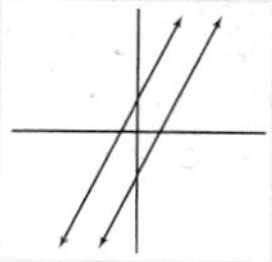 |
| The lines intersect. | The lines are the same. | The lines are parallel. |

**EXAMPLE 1** **Determine whether the system is consistent or inconsistent by solving it algebraically and graphically.**

$$\begin{cases} y = 4 - x \\ y = 3x - 6 \end{cases}$$

TECHNOLOGY
GRAPHICS CALCULATOR

**SOLUTION**

**Method 1**
Solve the system algebraically by substituting $3x - 6$ for $y$ in the first equation.

$$3x - 6 = 4 - x$$
$$4x = 10$$
$$x = 2.5$$

$$y = 4 - 2.5$$
$$y = 1.5$$

The solution is $(2.5, 1.5)$.
The system is consistent.

**Method 2**
Draw a graph.

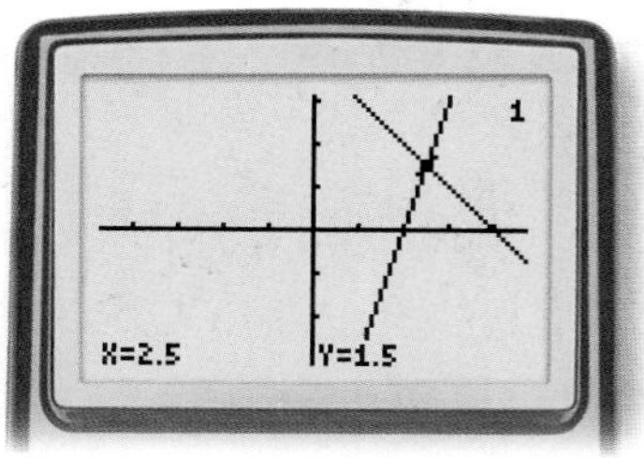


*The system is consistent.*

**EXAMPLE 2** **Determine whether the system is consistent or inconsistent by solving it algebraically and graphically.**

$$\begin{cases} 2x - y = -1 \\ 4x - 2y = 4 \end{cases}$$

**SOLUTION**

**Method 1**
Use elimination to solve the system algebraically. Multiply the first equation by $-2$.

$$\begin{cases} (-2)2x - (-2)y = (-2)(-1) \\ 4x - 2y = 4 \end{cases}$$
$$\Downarrow$$
$$-4x + 2y = 2$$
$$\underline{4x - 2y = 4}$$
$$0 \neq 6$$

There is no ordered pair that satisfies the original equations. The system is inconsistent.

**Method 2**
Draw a graph.

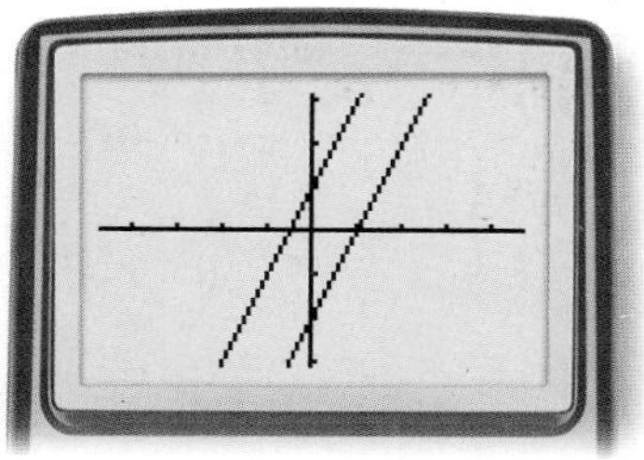

*The system is inconsistent.*

**TRY THIS** Determine whether each system is consistent or inconsistent by solving it algebraically and graphically.

**a.** $\begin{cases} y = 2x - 1 \\ y = 2x + 6 \end{cases}$ **b.** $\begin{cases} 2x - y = -1 \\ 4x + 2y = 7 \end{cases}$

---

### Definition of Consistent System

A system of equations is consistent if it has one or infinitely many solutions.

### Definition of Inconsistent System

A system of equations is inconsistent if it has no solution.

---

**EXAMPLE 3**

APPLICATION
**DEMOGRAPHICS**

**Based on the trend in the chart, will the average ages when first married for men and women ever be the same?**

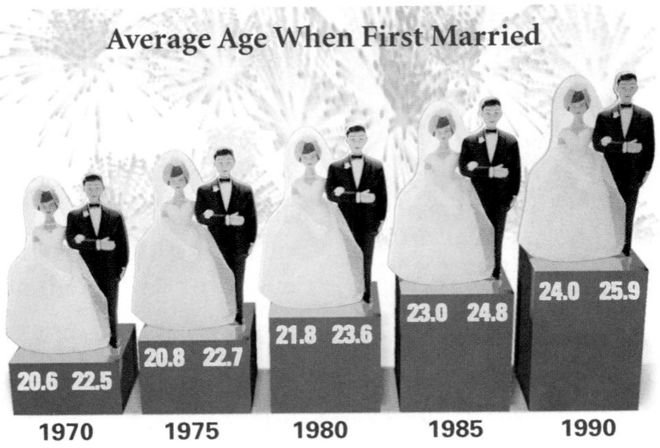

[*Source: Statistical Abstract of the United States, 1996*]

*The data shows that the average age when first married increased from 1970 to 1990 for both men and women.*

**SOLUTION**

The data in the chart can be compared by making two scatter plots on the same coordinate plane. The initial year, 1970, is assigned a value of 0.

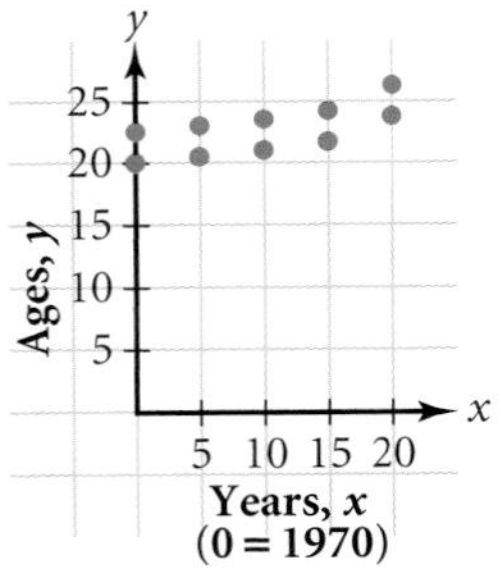


Use a graphics calculator to determine the equation for a line of best fit for each set of data. With coefficients rounded to the nearest hundredth, the equations form the following system:

$$\begin{cases} y = 0.18x + 20.24 & \text{Women's data} \\ y = 0.18x + 22.12 & \text{Men's data} \end{cases}$$

The slope is the same in both equations of the system. The lines are parallel. A graph of the lines of best fit confirms that the system is inconsistent.

*If the trend continues,* the average ages when first married for men and women will not coincide.

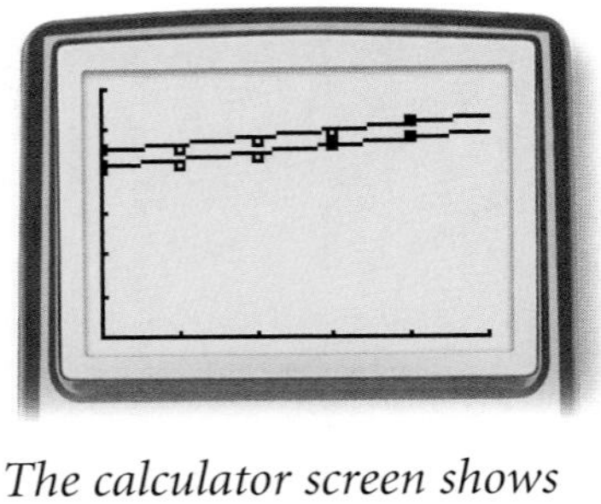

*The calculator screen shows the lines of best fit for the data.*

## Independent and Dependent Systems

Consistent systems of equations can be divided further into two categories—independent and dependent.

An **independent system** has only one solution. One unique ordered pair, $(x, y)$, satisfies both equations.

$$\begin{cases} x + 2y = 3 \\ 2x - y = 1 \end{cases}$$

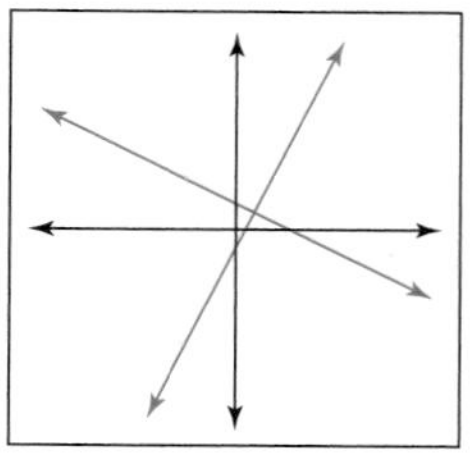

*The graph shows that the lines intersect at one unique point.*

A **dependent system** has infinitely many solutions. Every ordered pair that is a solution of the first equation is also a solution of the second equation.

$$\begin{cases} x - y = 2 \\ 2x - 2y = 4 \end{cases}$$

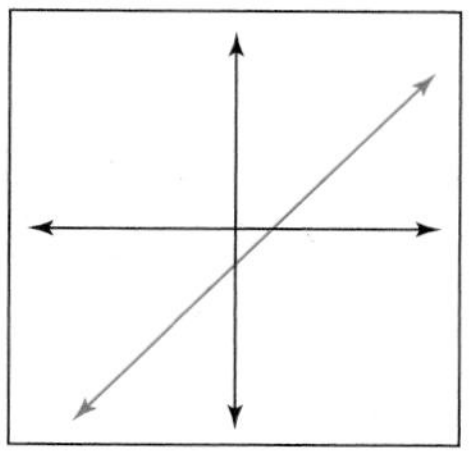

**EXAMPLE 4** **Determine whether the system is dependent or independent by solving it algebraically and graphically.**

$$\begin{cases} y = 2x - 1 \\ 3x - 1.5y = 1.5 \end{cases}$$

TECHNOLOGY
GRAPHICS CALCULATOR

**SOLUTION**

### Method 1

Solve the system algebraically by substituting $2x - 1$ for $y$ in the second equation.

$$\begin{aligned} 3x - 1.5(2x - 1) &= 1.5 \\ 3x - 3x + 1.5 &= 1.5 \\ 1.5 &= 1.5 \end{aligned}$$

Since the substitution resulted in the expression $1.5 = 1.5$, the equation is true for any value of $x$. This means that all of the ordered pairs on the line are solutions to both equations. Thus, the system is dependent.

### Method 2

Solve the system graphically by drawing a graph or by using a graphics calculator.

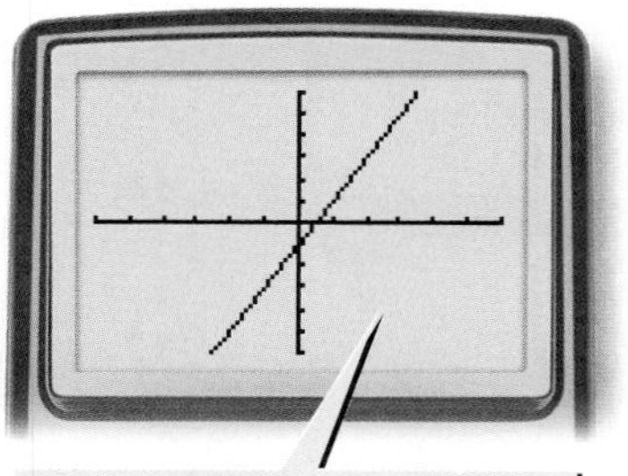

*The graph shows that the equations represent the same line.*

**CRITICAL THINKING** Use the equation $y = -3x + 4$ with different equations to create systems that are consistent and independent, consistent and dependent, and inconsistent.

# Exercises

## Communicate

**Explain how to identify each system of equations as inconsistent, dependent, or independent.**

1. $\begin{cases} y = 2x - 3 \\ 3y = 6x - 9 \end{cases}$

2. $\begin{cases} y = -3x + 2 \\ y = 2x + 2 \end{cases}$

3. $\begin{cases} x + y = 4 \\ x + y = 5 \end{cases}$

4. Explain how to write an equation that would form a dependent system with the equation $y = 2x + 3$.

5. Explain how to write an equation that would form an inconsistent system with the equation $y = -3x + 4$.

## Guided Skills Practice

**Determine whether each system is consistent or inconsistent by solving it graphically and algebraically.** ***(EXAMPLES 1, 2, AND 3)***

6. $\begin{cases} y = -4x + 6 \\ 2y = -8x - 8 \end{cases}$

7. $\begin{cases} c + d = 7 \\ 2c + 3d = 7 \end{cases}$

8. $\begin{cases} y = 0.34x + 0.45 \\ y = 0.34x + 4.5 \end{cases}$

9. Use the equation $y = 4x + 3$ with another equation to create a system that is dependent. ***(EXAMPLE 4)***

10. Use the equation $y = -2x + 7$ with another equation to create a system that is independent. ***(EXAMPLE 4)***

## Practice and Apply

**internet connect**
**Homework Help Online**
Go To: **go.hrw.com**
Keyword:
**MA1 Homework Help**
for Exercises 11-28

**Solve each system algebraically. Identify each system as consistent and dependent, consistent and independent, or inconsistent.**

11. $\begin{cases} x + y = 7 \\ x + y = -5 \end{cases}$

12. $\begin{cases} y = \frac{1}{2}x + 9 \\ 2y - x = 1 \end{cases}$

13. $\begin{cases} 3b = 2a - 24 \\ 4b = 3a - 3 \end{cases}$

14. $\begin{cases} y = -2x - 4 \\ 2x + y = 6 \end{cases}$

15. $\begin{cases} 2f + 3g = 11 \\ f - g = -7 \end{cases}$

16. $\begin{cases} 4x = y + 5 \\ 6x + 4y = -9 \end{cases}$

17. $\begin{cases} 4r + t = 8 \\ t = 4 - 2r \end{cases}$

18. $\begin{cases} y = \frac{3}{2}x - 4 \\ 2y - 8 = 3x \end{cases}$

19. $\begin{cases} 3p = 3m - 6 \\ p = m + 2 \end{cases}$

20. $\begin{cases} 2y + x = 8 \\ y = 2x + 4 \end{cases}$

21. $\begin{cases} k + 6h = 8 \\ k = -6h + 8 \end{cases}$

22. $\begin{cases} x - 5y = 10 \\ -5 = -x + 6 \end{cases}$

23. $\begin{cases} 8x + 3y = 7 \\ 3x + 2y = 7 \end{cases}$

24. $\begin{cases} 2x - \frac{3}{4}y = 17 \\ -3x - \frac{5}{8}y = -11 \end{cases}$

25. $\begin{cases} -6x + 8y = -16 \\ -2x - 3y = -28 \end{cases}$

26. $\begin{cases} 8x - 3y = -42 \\ \frac{2}{3}x + \frac{1}{3}y = 0 \end{cases}$

27. $\begin{cases} \frac{3}{5}x - \frac{8}{3}y = -105 \\ \frac{2}{3}x - \frac{4}{9}y = -26 \end{cases}$

28. $\begin{cases} \frac{5}{7}x - \frac{2}{3}y = 6 \\ \frac{3}{14}x - 2y = -9 \end{cases}$

29. The equations for lines $TU$ and $UV$ form a dependent system. If the slope of line $UV$ is $-3$ and the points $T(-3, p)$ and $U(5, 2-p)$ are given, find the value of $p$.

**CONNECTIONS**

30. **NUMBER THEORY** One number is 24 more than another number. If the sum of the numbers is 260, what are the numbers?

31. **GEOMETRY** The sum of the measures of two angles is 180°. What is the measure of each angle if one angle is 30° more than twice the other?

**COORDINATE GEOMETRY** Use the graph below to answer Exercises 32–37.

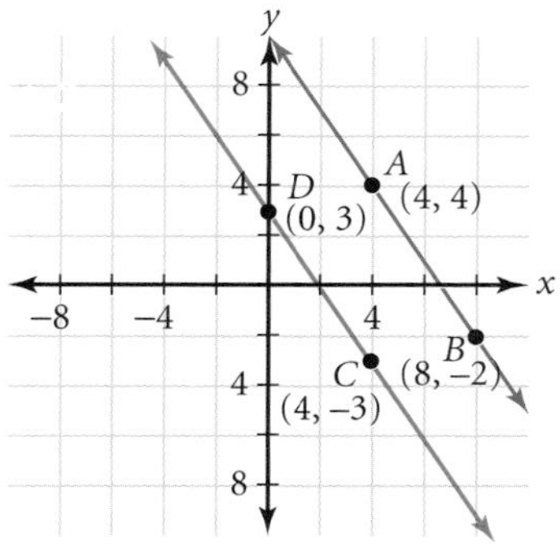


32. Write equations for lines $AB$ and $CD$.

33. What kind of system do lines $AB$ and $CD$ represent? Explain your answer.

34. Write the equation of a line that forms an independent system with line $AB$ and that contains point $A$. Graph your line.

35. Write the equation of a line that forms a dependent system with line $CD$ and that contains point $D$. Graph your line.

36. What is the point of intersection of $y = 6$ and line $AB$? of $y = 6$ and line $CD$?

37. Write the equation of a line that forms an inconsistent system with $y = 6$.

38. **GEOMETRY** Triangle $ABC$ below has a perimeter of 21, while rectangle $PQRS$ has a perimeter of 48. Write a system of equations for the perimeters of the figures, and find the dimensions of the rectangle.

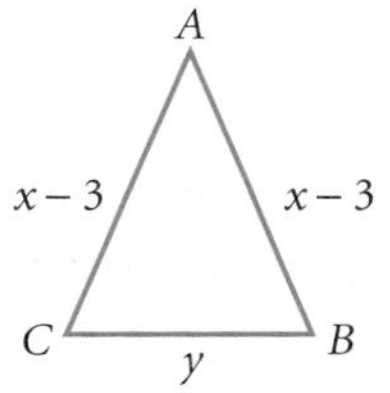


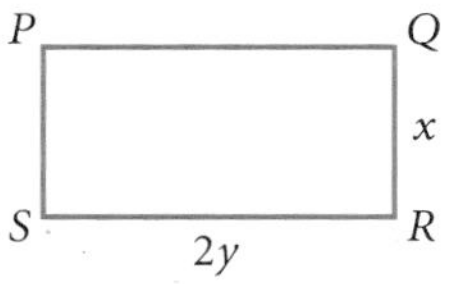

**APPLICATION**

**SALES** The chart shows the increase in total sales of two small produce companies from 1980 to 2000.

| | Vegiland Sales | Megaveggies Sales |
|---|---|---|
| 1980 | $40,070 | $43,750 |
| 1985 | $42,570 | $46,250 |
| 1990 | $45,070 | $48,750 |
| 1995 | $47,570 | $51,250 |
| 2000 | $50,070 | $53,750 |

39. Write a system of equations that can be used to predict the total sales, $y$, in a given year, $x$, for each company.

40. If the trend continues, will the sales of Vegiland ever equal the sales of Megaveggies? If so, when?

APPLICATION

**41.** **INCOME** Suppose that you are offered the choice of two different jobs. The first has a starting salary of \$2500 per month with a \$750-per-month expense account and an annual bonus of \$800. The other job pays \$3250 per month and has two semiannual bonuses of \$350 each. Explain which of the two jobs you would take and why.

## Look Back

**Find the value of each expression when $a = -1$, $b = -3$, and $c = 4$.** *(LESSON 2.4)*

**42.** $abc$ **43.** $-abc$ **44.** $-(a + b)$ **45.** $-a + bc$

**46.** $a^2 + b^2$ **47.** $a + b + c$ **48.** $-(a + b + c)$

**Solve for $c$.** *(LESSON 3.3 AND 3.6)*

**49.** $2c - 5 = 15$ **50.** $-3c + 4 = -14$ **51.** $4c + 7 = -5$ **52.** $ac + b = d$

**53.** $\frac{a+c}{d} = 3$ **54.** $ad - c = -b$ **55.** $-b(a - c) = d$ **56.** $c(b - d) = a$

**57.** Graph the equation $2x + 5y = 7$ and give the slope and $y$-intercept. *(LESSON 5.4)*

**58.** Write $3x + 2y = 6x + 9$ in standard form. *(LESSON 5.5)*

**Solve each system of equations.** *(LESSONS 7.2 AND 7.3)*

**59.** $\begin{cases} 2x + 3y = 6 \\ x + y = 4 \end{cases}$ **60.** $\begin{cases} -t + 3v = 9 \\ t - v = 12 \end{cases}$ **61.** $\begin{cases} -4p - 8q = 24 \\ p - 5q = 2 \end{cases}$

**62.** $\begin{cases} x + y = 3 \\ 2x = 7y + 7 \end{cases}$ **63.** $\begin{cases} y = 2x + 1 \\ x = 2y + 1 \end{cases}$ **64.** $\begin{cases} 2y + 3x = 7 \\ -3y + 2x = -3 \end{cases}$

## Look Beyond

**65.** To get to work, Charles takes Interstate 87 for 15 miles at 60 miles per hour and then takes Central Avenue for 12 miles at 40 mph. How many minutes does it take Charles to get to work?

7.5

# Systems of Inequalities

*Shakespeare's last play,* The Tempest, *is a favorite among costume and set designers because of the creative challenges and opportunities it presents. Costumes and sets are two of the major expenses in a production of the play.*

### Objectives

- Graph the solution to a linear inequality.
- Graph the solution to a system of linear inequalities.

**Why** ***Systems of inequalities are especially useful in discovering ranges of possibilities or, in some cases, for choosing the best option. For example, the producers of a stage play can use systems of inequalities to prevent cost from exceeding the budget.***

## Two-Variable Inequalities

**APPLICATION**
**FINANCE**

The amount of money spent by a theater company on costumes and sets for its production of *The Tempest* must not exceed its budget of \$5000. If $x$ represents the amount of money for costumes and $y$ the amount for sets, the situation can be represented by the following inequality:

$$x + y \leq 5000$$

costumes ($x$) sets ($y$) budget (5000)

Notice that the form of this inequality is like a *linear equation* with the equal sign replaced by an inequality symbol. Therefore, this inequality is a **linear inequality**.

The inequalities you studied in Chapter 6 were graphed on number lines because they had just one variable. Inequalities with two variables are graphed on a coordinate plane.

TECHNOLOGY
GRAPHICS CALCULATOR

*Activity*

## Modeling a Linear Inequality With a Graph

**You will need:** a graphics calculator or graph paper

Refer to the theater problem from the beginning of the lesson. An inequality that models the situation is $x + y \leq 5000$.

A graph of the inequality will show the possible solutions to the problem.

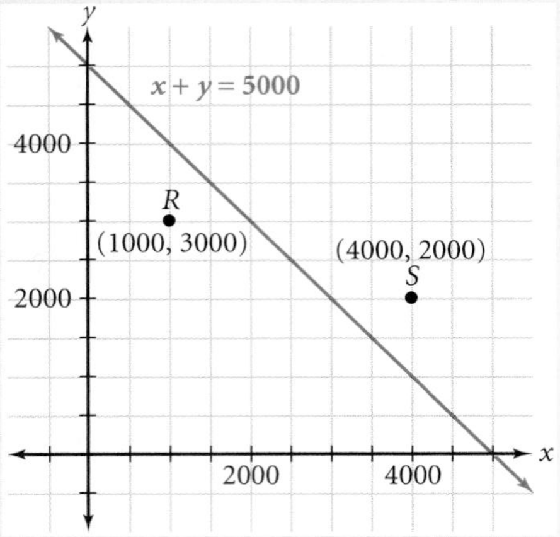


1. Replace the inequality sign with an equal sign. Then graph the resulting equation as shown at right. (Because the amount of money spent on costumes and sets is positive, draw the line only in the first quadrant.)

   The line that you graphed represents all combinations of costume and set expenses that add up to *exactly* \$5000.

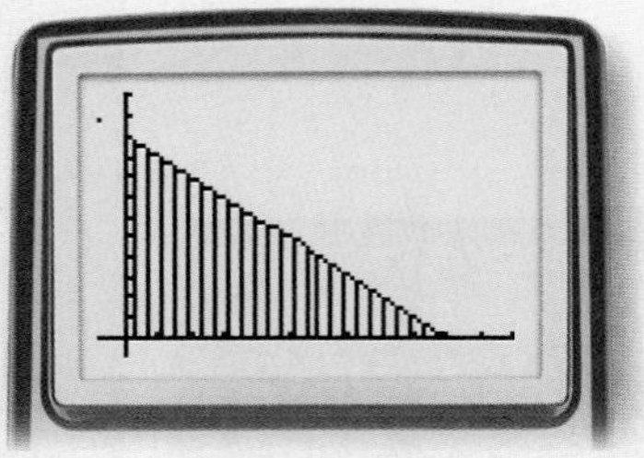

CHECKPOINT ✔

2. Look at points $R$ and $S$. Which one goes over budget? Which one goes under? What can you conjecture about other points that are above or below the line? Test your conjecture for a few randomly selected points.
3. Shade the part of the plane where the expenses are either on or under the budget of \$5000. (Because the amount of money spent on costumes and sets is positive, shade only in the first quadrant.) *The graph shows where the inequality is true.* The shaded part of your graph should include the straight line, which is known as a boundary line, because the expenses can equal \$5000.

## Graphing Linear Inequalities in Two Variables

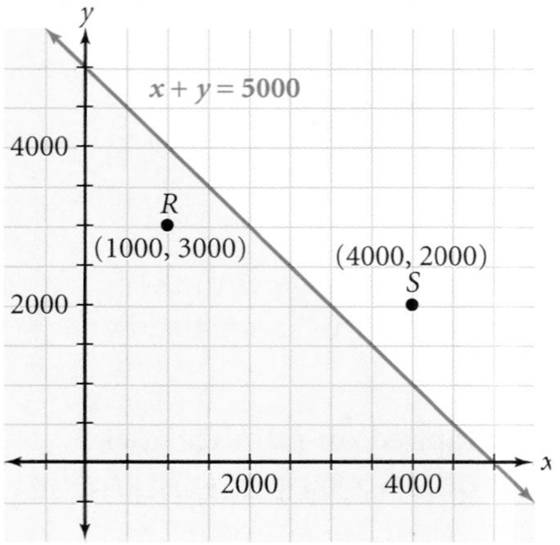


The graph at left shows the solution to the inequality $x + y \leq 5000$. The equation $x + y = 5000$ forms a **boundary line** that divides the coordinate plane into two half-planes. The boundary line is included in the solution because the inequality symbol, $\leq$, contains the equal sign. The shaded area and the boundary line contain all the ordered pairs that make the inequality true.

When the inequality symbol is $\leq$ or $\geq$, the points on the boundary line are included in the solution, and the line is solid.

When the inequality symbol is $<$ or $>$, the points on the boundary line are not included in the solution, and the line is dashed.

## EXAMPLE 1 Graph the inequality $x - 2y < 4$.

**SOLUTION**

Graph the boundary line by using the *equation* $x - 2y = 4$. Because the inequality symbol $<$ does not include the equal sign, use a dashed line for the boundary.

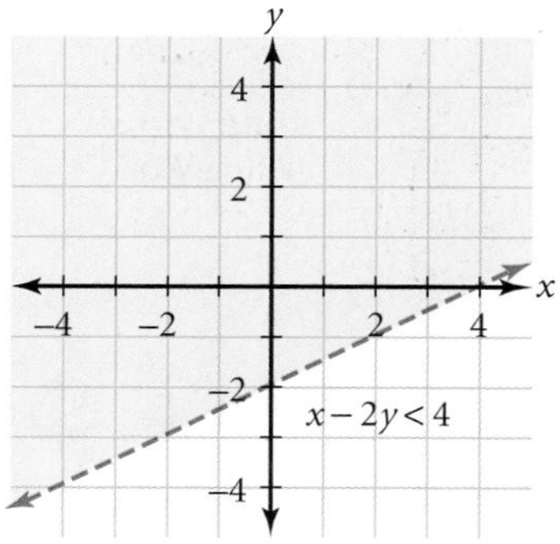


Solve the inequality in terms of positive $y$, with $y$ on the left side.

$$\begin{aligned} x - 2y &< 4 \\ -2y &< 4 - x \\ -y &< 2 - 0.5x \\ y &> -2 + 0.5x \end{aligned}$$

Since $y$ is *greater than* the expression on the right, shade the area of the graph *above* the boundary line.

Check the solution by selecting a point in the shaded region to see if it satisfies the inequality. For the point $(0, 0)$:

*You can choose any point in the region, but (0, 0) is an easy one to check.*

$$\begin{aligned} x - 2y &< 4 \\ 0 - 2(0) &\stackrel{?}{<} 4 \\ 0 &< 4 \quad \textit{True} \end{aligned}$$

The solution is correct.

## EXAMPLE 2 Graph the inequality $2x - y \geq -3$.

**SOLUTION**

Graph the boundary line by using the equation $2x - y = -3$. Because the inequality symbol $\geq$ includes the equal sign, use a solid line for the boundary.

Solve the inequality in terms of $y$, with $y$ on the left side.

$$\begin{aligned} 2x - y &\geq -3 \\ -y &\geq -3 - 2x \\ y &\leq 3 + 2x \end{aligned}$$

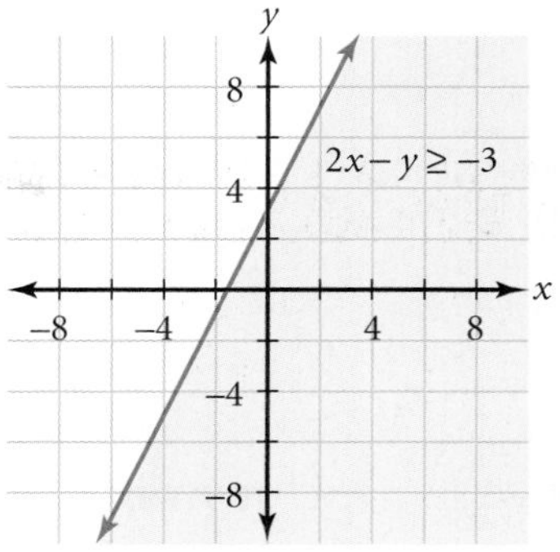


Since $y$ is *less than* the expression on the right, shade the area of the graph *below* the boundary line.

Check the solution by selecting a point in the shaded region to see if it satisfies the inequality. For the point $(0, 0)$:

$$\begin{aligned} 2x - y &\geq -3 \\ 2(0) - 0 &\stackrel{?}{\geq} -3 \\ 0 &\geq -3 \quad \textit{True} \end{aligned}$$

The solution is correct.

**TRY THIS** Graph each inequality. **a.** $3y < 2x + 7$ **b.** $y \geq 0.5x - 3$

## Systems of Linear Inequalities

A *system of linear inequalities* is like a system of linear equations, but with inequalities rather than equations. The solution of a system of linear inequalities is the *intersection* of the solutions of each inequality. Every point in the intersection region satisfies the system.

**EXAMPLE 3** **Solve by graphing.** $\begin{cases} y \leq -\frac{1}{2}x - 1 \\ y > x + 3 \end{cases}$

**SOLUTION**

Graph each inequality, and combine the graphs on one coordinate plane.

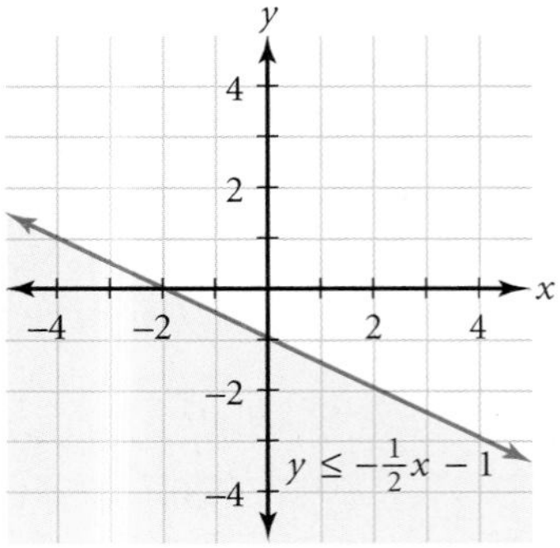


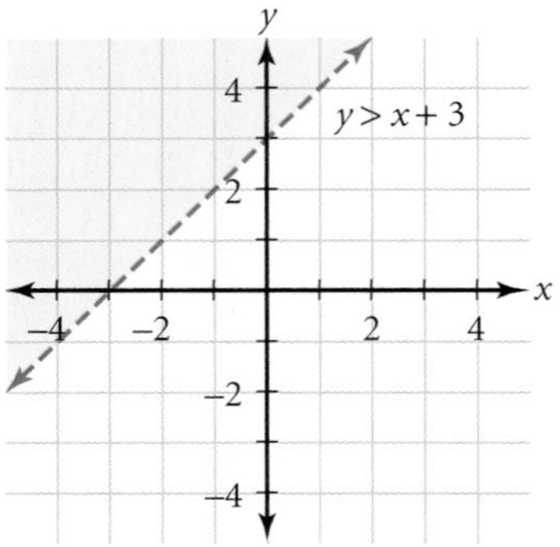


The solution of the system is the intersection of the two shaded regions (the darker region) in the combined graph at right, including the part of the solid boundary line that the intersection region touches.

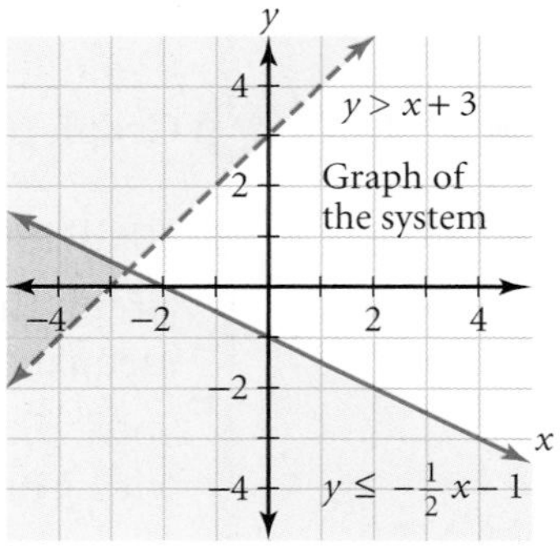


Try a point from the intersection region, such as (−4, 0), to test in both inequalities.

$y \leq -\frac{1}{2}x - 1$

$0 \overset{?}{\leq} -\frac{1}{2}(-4) - 1$

$0 \leq 1$ *True*

$y > x + 3$

$0 \overset{?}{>} -4 + 3$

$0 > -1$ *True*

**TRY THIS** Solve by graphing. $\begin{cases} y \leq -4x + 7 \\ y > -2x - 6 \end{cases}$

**EXAMPLE 4** During the summer, Kara wants to earn at least \$126 per week by working part-time. She can make \$6 an hour doing yard work and \$7 an hour working at the library. She can work no more than 30 hours each week. **Write a system of inequalities to represent Kara's situation, and use it to find all of the combinations of hours and jobs that she can work.**

PROBLEM SOLVING

APPLICATION
INCOME

**SOLUTION**

Define the variables.

Let $m$ represent the number of hours doing yard work.
Let $l$ represent the number of hours working at the library.

Write and graph inequalities that satisfy the conditions in the problem.

Kara can work no more than 30 hours per week.

$$m + l \leq 30$$

The shaded area shows the possible number of hours that Kara can work at each job.

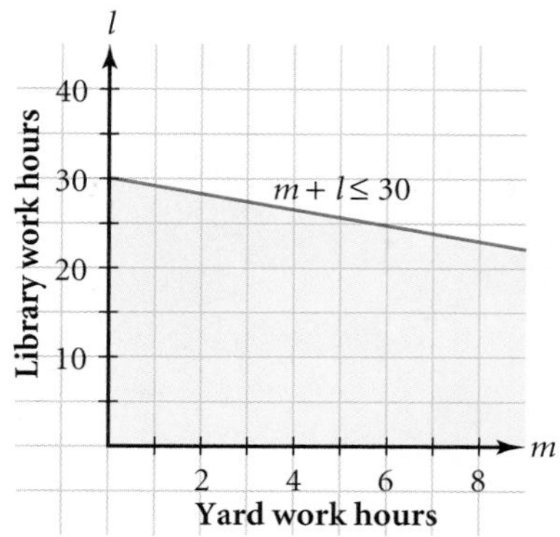


Kara wants to earn at least $126 per week.

$$6m + 7l \geq 126$$

The shaded area shows the possible number of hours that Kara can work to make at least $126.

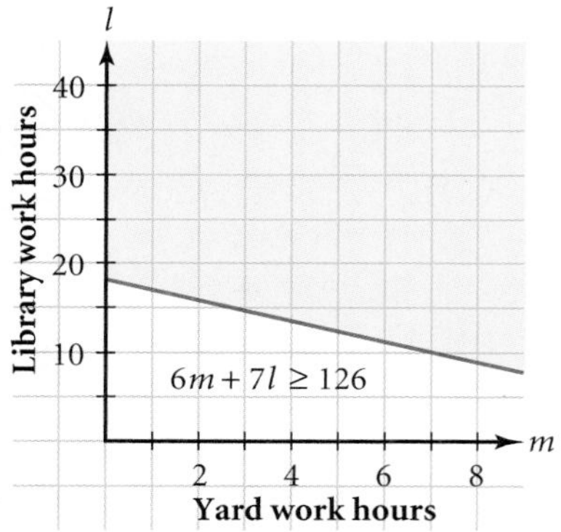


Combine the graphs of the inequalities on a single grid.

The intersection of the two shaded regions (the darker region) is the solution set of the system of inequalities.

You can check the solution by testing a point inside the intersection region. For example, the point (4, 20) is in the region.

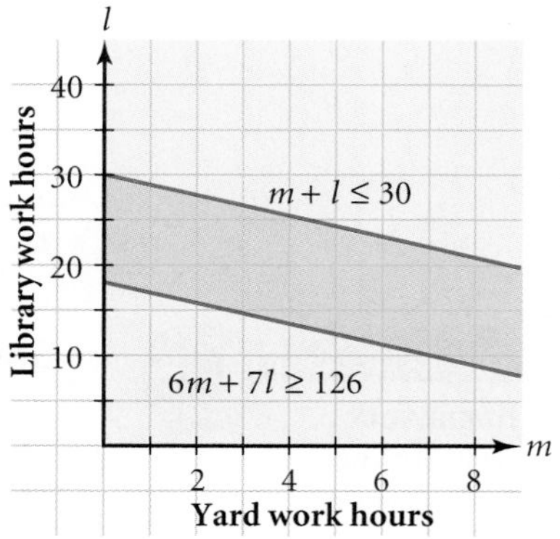


The number of hours worked is 4 + 20, or 24, which is within Kara's 30-hour maximum. The amount earned is 4(6) + 20(7), or $164, which meets her earnings requirement.

CRITICAL THINKING

Describe reasonable domains and ranges for the variables in the system from Example 4. Why should only points in the first quadrant be considered as a solution?

# Exercises

## Communicate

**1.** Explain how you would decide whether to draw a solid or a dashed line for the boundary for each of the following:
**a.** $y \geq x - 3$ **b.** $y < x - 3$

**2.** How do you choose the area to shade for the inequality $x + 2y < 2$? What is a good point to test in the solution region?

**Discuss how to graph the solution of the following systems of inequalities. Describe the process for determining the shaded region that satisfies both inequalities in each system.**

**3.** $\begin{cases} y < 1 \\ x < 1 \end{cases}$ **4.** $\begin{cases} 8x + 4y > 12 \\ y < 3 \end{cases}$ **5.** $\begin{cases} x \leq 3 \\ x - 2y \geq 2 \end{cases}$ **6.** $\begin{cases} y - 3 > x \\ y + 2x > 4 \end{cases}$

## Guided Skills Practice

**Graph each inequality.** ***(EXAMPLES 1 AND 2)***

**7.** $x - 3y < 12$ **8.** $4x - y \leq 7$ **9.** $-2y > 3x - 1$ **10.** $-3r - 5t \geq 10$

**Solve by graphing.** ***(EXAMPLE 3)***

**11.** $\begin{cases} 3x + y > 3 \\ 4x + 3y \leq 9 \end{cases}$ **12.** $\begin{cases} y > \frac{3}{4}x - 2 \\ -x + 6y \leq 12 \end{cases}$ **13.** $\begin{cases} -2c + 6 < d \\ c - d < 5 \end{cases}$ **14.** $\begin{cases} x > -2 \\ 5y - 3x \geq -20 \end{cases}$

**APPLICATION**

**15.** **GARDENING** Marta is planning to add at least 30 new plants to her garden. She wants to add perennials ($4.00 each) and annuals ($1.50 each) and spend no more than $100. Write a system of inequalities to represent the situation and solve by graphing. ***(EXAMPLE 4)***

## Practice and Apply

internet connect
**Homework Help Online**
Go To: **go.hrw.com**
Keyword:
**MA1 Homework Help**
for Exercises 16-31

**Graph each inequality.**

**16.** $y \geq \frac{1}{2}x + 2$ **17.** $-2x + 3y < -15$ **18.** $4y + x > -1$ **19.** $2x - 7y \leq -14$

**20.** $6y \leq -4x - 24$ **21.** $10x - 15y < 30$ **22.** $7x - y < 3$ **23.** $y < 2x$

**24.** $3x + 5y \leq 25$ **25.** $5x - 4y < -24$ **26.** $2m - 7n \geq 14$ **27.** $-3r - 5t < 10$

**28.** $k > \frac{5}{7}h - 9$ **29.** $8c - 2f > 10$ **30.** $4m - 3p < 9$ **31.** $j - k < -1$

**Solve by graphing.**

**32.** $\begin{cases} 2x - 3y > 6 \\ 5x + 4y < 12 \end{cases}$ **33.** $\begin{cases} x - 4y \leq 12 \\ 4y + x \leq 12 \end{cases}$ **34.** $\begin{cases} y < x - 5 \\ y \leq 3 \end{cases}$

**35.** $\begin{cases} 2x + y \leq 4 \\ 2y + x \geq 8 \end{cases}$ **36.** $\begin{cases} 5y < 2x - 5 \\ 4x + 3y \leq 9 \end{cases}$ **37.** $\begin{cases} 4y < 3x + 8 \\ y \leq 1 \end{cases}$

38. $\begin{cases} x \geq 4 \\ 0 < y \end{cases}$

39. $\begin{cases} x - 2y < 8 \\ x + y \geq 5 \end{cases}$

40. $\begin{cases} 3x - y > -2 \\ x - y > -1 \end{cases}$

41. $\begin{cases} 2y \leq -x + 6 \\ -3x + 4y < -8 \end{cases}$

42. $\begin{cases} -5a + b < -1 \\ -a + 5b > -5 \end{cases}$

43. $\begin{cases} 4w - z \geq 2 \\ 3w + z \geq 7 \end{cases}$

44. $\begin{cases} y < x \\ y > -x \end{cases}$

45. $\begin{cases} 7q - 5r < 30 \\ 5q + 7r \leq 42 \end{cases}$

46. $\begin{cases} 5c > 4b \\ c \leq 3 \end{cases}$

47. $\begin{cases} t < 5 \\ -3t - 7v > 14 \end{cases}$

48. $\begin{cases} z - x > 3 \\ z + x < 3 \end{cases}$

49. $\begin{cases} y - 2x \leq 2 \\ y + 2x \geq 2 \end{cases}$

50. $\begin{cases} 3m - 2n > 0 \\ 2m + 3n < -3 \end{cases}$

51. $\begin{cases} 8x + 2y \geq 1 \\ x - 3y < 4 \end{cases}$

52. $\begin{cases} 2l + 3m < 4 \\ -\frac{1}{2}l + \frac{1}{3}m > -5 \end{cases}$

53. $\begin{cases} x - 2y < 2 \\ 3x + \frac{1}{2}y \leq -2 \end{cases}$

54. $\begin{cases} w + v < 2 \\ v - w \geq -3 \end{cases}$

55. $\begin{cases} 8x + 5y \leq -2 \\ -5x \leq -3y + 2 \end{cases}$

**CONNECTION**

**56. GEOMETRY** Write the system of inequalities defined by each perimeter described below, and graph the common solution. Make sure that the solution represents real-world possibilities.

*Rectangle ABCD's perimeter is at most 30 centimeters, while rectangle PQRS's perimeter is at least 12 centimeters.*

**APPLICATIONS**

**57. SPORTS** In the Lincoln High School basketball game, Troy made only 2-point field goals and 1-point free throws. He made no more than 16 points in the game. Find the combinations of field goals and free throws Troy could have made.

**58. ACADEMICS** On most days Anna spends no more than 2 hours on her math and science homework. If math always takes about twice as long as science, what is the maximum time that Anna usually spends on science?

**CHALLENGE**

**59.** Write the system of inequalities that represents the shaded region. Use the points shown to find equations for the boundary lines.

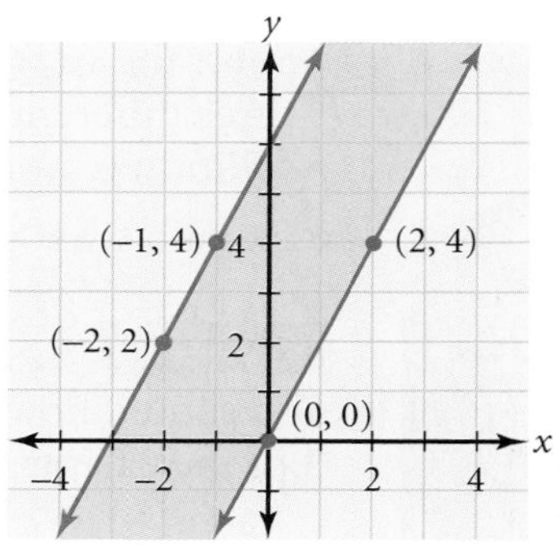

## Look Back

**60.** Find the first differences for the following sequence: 5, 17, 37, 65, 101, . . . ***(LESSON 1.1)***

**61.** Find the second differences for the following sequence: 5, 17, 37, 65, 101, . . . ***(LESSON 1.1)***

**62.** Simplify. $-8 - 2\left[3 + 6\left(5 - \frac{4}{2}\right) + 7\right]$ ***(LESSON 1.3)***

APPLICATION

**63.** **FUND-RAISING** Suppose that 20 businesses contribute equally at a charity benefit. If the total raised was $1200, how much did each business contribute? ***(LESSON 2.4)***

**Simplify.** ***(LESSON 2.6)***

**64.** $2x^2 + 4 + 3y - x^2 + 2$

**65.** $3a + a + z + 4a + 2b^2 - b$

**Use the slopes to determine whether each system represents lines that are parallel, perpendicular, or neither.** ***(LESSON 5.6)***

**66.** $\begin{cases} x + y = 7 \\ x + 3y = 9 \end{cases}$

**67.** $\begin{cases} -2x + y = -5 \\ -2x - y = 5 \end{cases}$

**68.** $\begin{cases} -x + 2y = 6 \\ -x + 2y = -3 \end{cases}$

**69.** $\begin{cases} x - 2y = 4 \\ 2x + y = 1 \end{cases}$

**70.** $\begin{cases} 3x = 2y \\ 3x - 2y = 0 \end{cases}$

**71.** $\begin{cases} x - y = 4 \\ y + x = 2 \end{cases}$

**72.** $\begin{cases} 3x = 4 \\ 3y = 2 \end{cases}$

**73.** $\begin{cases} 8x + 2y = 5 \\ x + y = -4 \end{cases}$

## Look Beyond

**For $f(x) = 2x + 4$, evaluate the following:**

**74.** $f(3)$

**75.** $f(-2)$

**76.** $f\left(-\frac{1}{2}\right)$

### PORTFOLIO ACTIVITY

Include the following activity in your portfolio:

The high-school yearbook staff is deciding how many color pages to include in the yearbook. In order to provide the appropriate visual appeal, the yearbook staff has decided that the number of black-and-white pages must be no more than twice the number of color pages. Due to cost constraints, there can be no more than 40 color pages in the yearbook.

- Use guess-and-check to determine three different combinations of pages that will work.
- Write a system of inequalities that models the situation.
- Solve the system graphically. Show that your three different combinations lie inside the solution region.
- Explain what the intersection region represents.

**WORKING ON THE CHAPTER PROJECT**

You should now be able to complete Activities 2 and 3 of the Chapter Project on page 360.

# 7.6 Classic Puzzles in Two Variables

A math phobic is a person who is afraid of math.

**Objective**

- Solve traditional math puzzles in two variables.

**Why** ***Traditional math puzzles may be a way of testing and sharpening your math skills. The concepts you learn from them may help you solve problems or make discoveries. Einstein used concepts from wind and current problems to investigate and make discoveries about the speed of light.***

**PROBLEM SOLVING**

Traditional math puzzles like the one the angel in the cartoon is presenting may indeed seem frightening. To someone who doesn't understand them, they may seem almost impossible. But there are specific ways to approach problems of this type, and a study of them will sharpen your algebra skills.

## Age Puzzles

The key to solving age puzzles is knowing how to represent a person's age in the future or the past.

If a person's age now is $x$, then 5 years from now the person's age will be $x + 5$. Five years ago, the person's age was $x - 5$.

Once you understand this part, you will need to set up a system of two equations in two unknowns. You already know how to do this.

**EXAMPLE 1** A father is 32 years older than his son. In 4 years the father will be 5 times as old as his son. **How old is each now?**

**SOLUTION**

Write two equations describing the age of the father and son at two different times. Let $f$ represent the father's present age. Let $s$ represent the son's present age. You are given the following information:

$$f = s + 32$$

In 4 years the father will be 5 times the son's age. Add 4 years to each age and express this information in an equation.

$$f + 4 = 5(s + 4)$$

Solving the system $\begin{cases} f = s + 32 \\ f + 4 = 5(s + 4) \end{cases}$ gives $f = 36$ and $s = 4$.

At the present time, the father is 36 years old, and his son is 4 years old.

**TRY THIS** June is 4 years older than her brother. Six years ago, she was twice as old as he was. How old is each now?

## Wind and Current Puzzles

Wind and current puzzles have a common idea. When a plane is traveling with the wind (a tailwind), the speed of the wind is added to the speed of the plane with no wind. When traveling against the wind (a headwind), the speed of the wind is subtracted from the plane's speed with no wind.

**EXAMPLE 2** A plane leaves New York City and heads for Chicago, which is 750 miles away. The plane, flying against the wind, takes 2.5 hours to reach Chicago. After refueling, the plane returns to New York City, traveling with the wind, in 2 hours. **Find the speed of the wind and the speed of the plane with no wind.**

**APPLICATION**
**AVIATION**

**SOLUTION**

Define the variables. Put the information in a table.

Let $x$ represent the speed of the plane with no wind.

Let $y$ represent the speed of the wind.

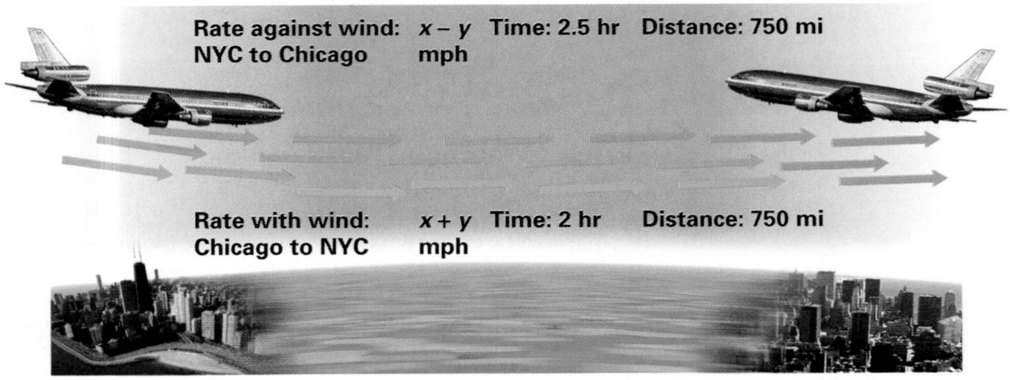

To write the system of equations, recall the relationship between rate, time, and distance.

$$\text{rate} \times \text{time} = \text{distance}$$

$$\begin{cases} (x-y)(2.5) = 750 \\ (x+y)(2) = 750 \end{cases} \Rightarrow \begin{cases} x-y = 300 \\ x+y = 375 \end{cases}$$

Solving the system gives $x = 337.5$ and $y = 37.5$.

The speed of the plane with no wind is 337.5 miles per hour. The speed of the wind is 37.5 miles per hour.

**TRY THIS** A riverboat traveling with the current takes 2.5 hours to make a trip of 45 miles. The return trip, against the current, takes 2.75 hours. Find the speed of the current and the speed of the boat in still water.

## Number-Digit Puzzles

In most number-digit problems the strategy is to write the value of a number in expanded form. You can write 52 as $5(10) + 2$. If you reverse the digits in 52, you get 25, which you can write as $2(10) + 5$.

**EXAMPLE 3** The sum of the digits of a two-digit number is 7. The original two-digit number is 3 less than 4 times the number with its digits reversed. **Find the original two-digit number.**

**SOLUTION**

Let $t$ represent the tens digit and $u$ represent the units digit of the original number.

Original number: $10t + u$

Number with digits reversed: $10u + t$

You are also given $t + u = 7$.

The original two-digit number is 3 less than 4 times the number with its digits reversed.

$$10t + u = 4(10u + t) - 3$$

Solve the system: $\begin{cases} t + u = 7 \\ 10t + u = 4(10u + t) - 3 \end{cases}$

Solving the system gives $t = 6$ and $u = 1$.

If $t$ is 6 and $u$ is 1, then the original number is $10(6) + 1$, or 61. The number 61 is a two-digit number that is 3 less than 4 times itself with its digits reversed. Since $4[10(1) + 6] - 3 = 61$, the solution is correct.

**TRY THIS** A two-digit number whose tens digit is 2 more than the units digit is 3 more than 6 times the sum of its digits. Find the original number.

## Coin Puzzles

Classic coin problems can usually be set up by using a system of equations—one involving the number of coins and the other involving the monetary value of the coins in cents.

**EXAMPLE 4** In a coin bank there are 250 dimes and quarters worth a total of $39.25. **Find how many dimes, *d*, and how many quarters, *q*, are in the bank.**

**SOLUTION**

Create a table to organize the information.

| | Quarters | Dimes | Total |
|---|---|---|---|
| **Number of coins** | $q$ | $d$ | 250 coins |
| **Value in cents** | $25q$ | $10d$ | 3925¢ |

From the table, write two equations based on the totals.

$$\begin{cases} q + d = 250 \\ 25q + 10d = 3925 \end{cases}$$

By solving the system, you find that $q = 95$ and $d = 155$.

The coin bank has 95 quarters and 155 dimes.

**TRY THIS** In the math department water-cooler fund there are 48 quarters and nickels in all, with a total value of $6.00. Find how many quarters and how many nickels are in the fund.

## Chemical-Solution Puzzles

Chemical-solution puzzles involve mixing solutions, typically acids, of different strengths. The strengths of the solutions are expressed as percents. The strategy is to keep track of the total amount of acid, in ounces or grams, in the original solutions and in the final solution.

**EXAMPLE 5** **A 1.5% acid solution is mixed with a 4% acid solution. How many ounces of each solution are needed to obtain 60 ounces of a 2.5% acid solution?**

**APPLICATION**
**CHEMISTRY**

**SOLUTION**

Draw a diagram and make a table to help solve this problem.

Define the variables.

Let $x$ represent the number of ounces of the 1.5% acid solution.

Let $y$ represent the number of ounces of the 4% acid solution.

Place the information in a table to help organize the facts.

| | First solution | Second solution | Third solution |
|---|---|---|---|
| **Amount of solution** | $x$ | $y$ | 60 oz |
| **Percent acid** | 1.5% | 4% | 2.5% |
| **Amount of acid** | $0.015x$ | $0.04y$ | $0.025(60)$ |

From the table, write two equations. $\begin{cases} x + y = 60 \\ 0.015x + 0.04y = 0.025(60) \end{cases}$

Solving the system gives the solution (36, 24).

To get the required 60 ounces of a 2.5% acid solution, mix 36 ounces of the 1.5% acid solution with 24 ounces of the 4% acid solution.

**TRY THIS** How many ounces of a 2% acid solution should be mixed with a 4% acid solution to produce 20 ounces of a 3.6% acid solution?

## Exercises

### Communicate

**1.** Discuss the problem-solving strategy you would use to solve the following problem, and explain how to set up the equations:

You are running a chemistry experiment that requires you mix a 25% alcohol solution with a 5% alcohol solution. This mixture produces 20 liters of a 13% alcohol solution. How many liters of the 25% alcohol solution must you use?

*Migratory birds sometimes fly several thousand miles to their winter nesting sites.*

**A bird can fly with the wind 3 times as fast as it can fly against the wind.**

**2.** How can you represent the rate of the bird flying in still air (no wind) and the rate of the wind?

**3.** How can you represent the rate of the bird flying with the wind and the rate against the wind?

**4.** Discuss how the rate of the bird flying with the wind compares with the rate of the bird flying in still air.

## Guided Skills Practice

**5.** Gabriel is 16 years older than Sam. In 4 years Gabriel will be twice as old as Sam. What are their ages now? ***(EXAMPLE 1)***

APPLICATIONS

**6.** **RECREATION** Chris can row a boat a distance of 20 miles in 2.5 hours when he is rowing with the current. It takes him 3 hours to row the same distance when he is rowing against the current. What is the rate of the current? ***(EXAMPLE 2)***

**7.** The sum of the digits of a two-digit number is 8. If 16 is added to the original number, the result is 3 times the original number with its digits reversed. Find the original number. ***(EXAMPLE 3)***

**8.** When Jim cleaned out the reflecting pool at the library, he found 20 nickels and quarters. The collection of nickels and quarters totaled $2.60. How many quarters did Jim find? ***(EXAMPLE 4)***

**9.** **CHEMISTRY** Janet is mixing a 15% glucose solution with a 35% glucose solution. This mixture produces 35 liters of a 19% glucose solution. How many liters of each solution is Janet using in the mixture? ***(EXAMPLE 5)***

## Practice and Apply

APPLICATION

**10.** **RECREATION** Catherine is camping along a river. It takes her 1.5 hours to paddle her canoe 6 miles upstream from her campsite. Catherine turns her canoe around and returns 6 miles downstream to her campsite in exactly 1 hour. What is the rate of the river's current and the rate of Catherine's paddling in still water?

**internet connect**
**Homework Help Online**
Go To: **go.hrw.com**
Keyword:
**MA1 Homework Help** for Exercises 11 and 16

**11.** The coin box of a vending machine contains 6 times as many quarters as dimes. If the total amount of money in quarters and dimes is $28.80, how many quarters and how many dimes are in the box?

**12.** Find the two-digit number whose tens digit is 4 less than its units digit if the original number is 2 more than 3 times the sum of the digits.

**13.** In 15 years, Maya will be twice as old as David is now. In 15 years, David will be as old as Maya will be 10 years from now. How old are they now?

**14.** The sum of Anjie's age and Carol's age is 66. Carol's age is 2 more than 3 times Anjie's age. How old are Carol and Anjie?

**15.** Wymon is 20 years older than Sabrina. In 8 years, Wymon will be twice as old as Sabrina. How old is each now?

16. A coin bank contains $17.00 in pennies and nickels. If there are 1140 coins in the bank, how many of the coins are nickels?

CONNECTION

17. **NUMBER THEORY** The sum of the digits of a positive two-digit number is 13. If twice the first digit is 1 less than the second digit, what is the two-digit number?

APPLICATIONS

18. **CHEMISTRY** How many ounces of a 20% salt solution should be mixed with a 10% salt solution to produce 45 ounces of a 14% salt solution?

19. **CHEMISTRY** A 4% salt solution is mixed with a 16% salt solution. How many milliliters of each solution are needed to obtain 600 milliliters of a 10% solution?

20. **AVIATION** With a tailwind, a jet flew 2000 miles in 4 hours, but the return trip against the same wind required 5 hours. Find the jet's speed and the wind speed.

21. **BUSINESS** A grocery store sells a mixture of peanuts and raisins for $1.75 per pound. If peanuts cost $1.25 per pound and raisins cost $2.75 per pound, what amount of raisins and peanuts go into 1 pound of the mixture?

## Look Back

**Evaluate each expression for $x = 2$, $y = 1$, and $z = 1$.** ***(LESSON 1.3)***

**22.** $2x + 4y - 3z$ **23.** $\frac{2x - z}{5}$ **24.** $\frac{x + y}{z}$ **25.** $3xyz$

**26.** Simplify. $(29 + 1) \div 3 \cdot 2 - 4$ ***(LESSON 1.3)***

**Simplify.** ***(LESSON 2.7)***

**27.** $-[2(a + b)]$ **28.** $-[-11(x - y)]$ **29.** $2(x - y) + 3[-2(x + y)]$

**Find the slope of each line.** ***(LESSON 5.5)***

**30.** $3x + 4y = 12$ **31.** $5y = 10x - 20$

## Look Beyond

**Find each square root.**

**32.** $\sqrt{\frac{25}{625}}$ **33.** $\sqrt{a^2}$

CHAPTER SEVEN PROJECT

# Minimum Cost Maximum Profit

The manager of a camping supply store wants to make a trail mix of nuts and raisins to provide all the required nutrients at the lowest cost. The chart shows the nutritional information and cost per ounce.

| | Calories/oz | Fat/oz | Cost/oz |
|---|---|---|---|
| Nuts | 150 | 13 g | \$0.40 |
| Raisins | 90 | 0 g | \$0.10 |

Backpackers usually have a weight limit for their pack, so each package of trail mix will contain no more than 18 ounces. The snack should supply at least 1800 Calories, with no more than 117 grams of fat. How many ounces each of raisins and nuts should each package of trail mix contain to minimize the costs?

This optimization problem can be solved by using a method called linear programming. The activities presented here lead to the solution produced by this method. Let $n$ represent the number of ounces of nuts. Let $r$ represent the number of ounces of raisins.

## Activity 1

1. Explain what the two equations in the following system represent:

$$\begin{cases} n + r = 18 \\ 150n + 90r = 1800 \end{cases}$$

2. Solve the system by substitution and by graphing.
3. Why can you represent the total amount of fat by the expression $13n$?
4. Explain what the two equations in each system below represent.

$$\begin{cases} 13n = 117 \\ 150n + 90r = 1800 \end{cases} \qquad \begin{cases} n + r = 18 \\ 13n = 117 \end{cases}$$

5. Solve each system by substitution and by graphing. Graph both systems on the same coordinate plane.

## Activity 2

1. Explain what each inequality in the following system represents:

$$\begin{cases} n + r \leq 18 \\ 150n + 90r \geq 1800 \\ 13n \leq 117 \end{cases}$$

2. Graph each inequality on the coordinate plane you used in Activity 1.
3. What does the common area (intersection region) represent in the problem?
4. Pick four different points in the intersection region and show that each one satisfies all of the conditions.

## Activity 3

1. Nuts cost \$0.40 per ounce and raisins cost \$0.10 per ounce. Complete the following exercises:
   a. Write an expression for the total cost of the nuts.
   b. Write an expression for the total cost of the raisins.
   c. Write an equation for the combined cost of nuts and raisins. This is known as the *optimization equation.*
2. Use graphing technology to graph each of the three *constraint inequalities* from Step 1 of Activity 2. Shade the region that represents the solution.
3. Explain what the vertices of the triangle represent in the problem.

   $A(3, 15)$ $\qquad$ $B(9, 9)$ $\qquad$ $C(9, 5)$

4. Substitute the coordinates of each vertex of the shaded region into the optimization equation that you wrote in part **c** of Step 1. Find the solution to the optimization equation. The equation with the lowest cost is the minimum.
5. How many ounces of each ingredient should be used per package to keep the costs at a minimum?

*The shaded region, $\Delta ABC$, represents the solution set. This region, determined by the linear inequalities containing the possible solutions, is called the feasibility region.*

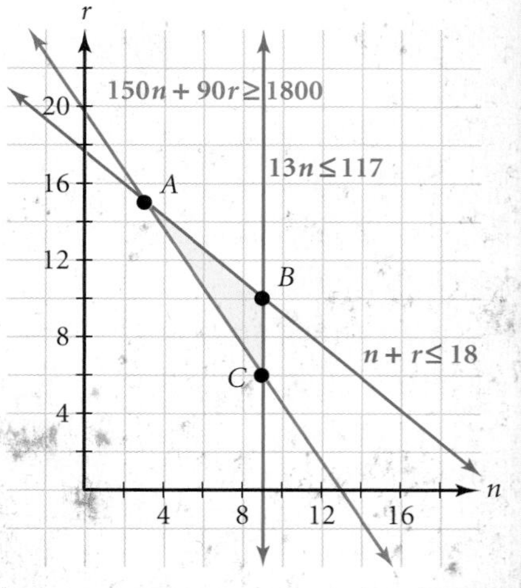

# 7 Chapter Review and Assessment

## VOCABULARY



## Key Skills & Exercises

### LESSON 7.1

#### Key Skills

**Graph a system of equations and estimate a solution from a graph by inspection.**

Solve by graphing. $\begin{cases} x + 3y = 12 \\ x - y = 4 \end{cases}$

Write the equations in slope-intercept form.

$x + 3y = 12$ $\qquad$ $x - y = 4$

$y = -\frac{x}{3} + 4$ $\qquad$ $y = x - 4$

Graph the lines on graph paper or on a graphics calculator. The solution is the point of intersection, (6, 2). Check algebraically.

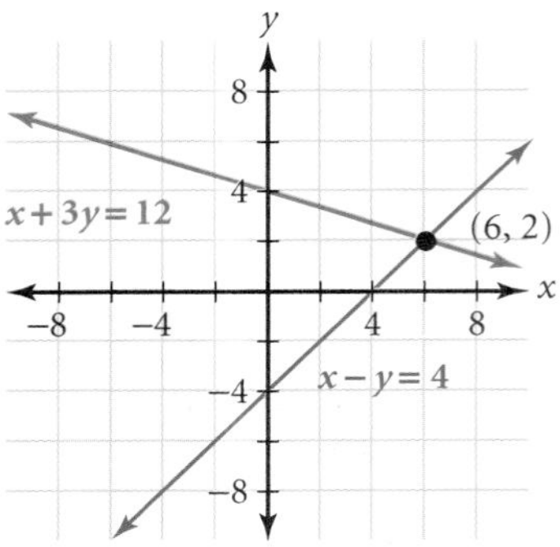


#### Exercises

**Solve each system by graphing.**

1. $\begin{cases} x - y = 6 \\ x + 2y = -9 \end{cases}$

2. $\begin{cases} y = -x \\ y = 2x \end{cases}$

3. $\begin{cases} x + 6y = 3 \\ 3x + y = -8 \end{cases}$

4. $\begin{cases} 3x + y = 6 \\ y + 2 = x \end{cases}$

5. $\begin{cases} 8x + 6y = 2 \\ x + 2y = -3 \end{cases}$

6. $\begin{cases} -x + 2y = 1 \\ x - 2y = 3 \end{cases}$

7. $\begin{cases} y = 3x + 7 \\ y = 2x - 6 \end{cases}$

8. $\begin{cases} y = -3x + 1 \\ y = -2x \end{cases}$

### LESSON 7.2

#### Key Skills

**Find an exact solution to a system of equations by using the substitution method.**

Solve by using the substitution method. $\begin{cases} x + 4y = 1 \\ 2x + 7y = 6 \end{cases}$

Solve $x + 4y = 1$ for $x$ to get $x = 1 - 4y$. Then substitute the expression $1 - 4y$ for $x$ in the equation $2x + 7y = 6$. Solve for $y$.

$$\begin{aligned} 2x + 7y &= 6 \\ 2(1 - 4y) + 7y &= 6 \\ 2 - 8y + 7y &= 6 \\ 2 - y &= 6 \\ -y &= 4 \\ y &= -4 \end{aligned}$$

#### Exercises

**Solve each system of linear equations by using substitution.**

9. $\begin{cases} 2x + y = 1 \\ x + y = 2 \end{cases}$

10. $\begin{cases} x - y = 6 \\ 2x - 4y = 28 \end{cases}$

11. $\begin{cases} x + 2y = 1 \\ 2x - 8y = -1 \end{cases}$

12. $\begin{cases} 4x = 3y + 44 \\ x + y = -3 \end{cases}$

13. $\begin{cases} y = \frac{1}{2}x + 2 \\ 3x + y = 7 \end{cases}$

14. $\begin{cases} \frac{3}{5}x - 2y = 5 \\ x = \frac{4}{5}y + 2 \end{cases}$

15. $\begin{cases} 4m + 5n = 6 \\ -3m - 2n = -4 \end{cases}$

16. $\begin{cases} 8a - \frac{1}{2}b = \frac{3}{4} \\ b = \frac{1}{2}a + 7 \end{cases}$

Now substitute –4 for $y$ in one of the original equations. Solve for $x$.

$$x + 4(-4) = 1$$
$$x - 16 = 1$$
$$x = 17$$

The solution is (17, –4).

LESSON 7.3

## Key Skills

**Find an exact solution to a system of equations by using the elimination method.**

Solve by using elimination. $\begin{cases} x + y = -3 \\ 2x - 3y = -11 \end{cases}$

Use the Multiplication Property of Equality to get opposite coefficients for one of the variables in the equations.

$$\begin{cases} 3(x + y) = 3(-3) \\ 2x - 3y = -11 \end{cases} \Rightarrow \begin{cases} 3x + 3y = -9 \\ 2x - 3y = -11 \end{cases}$$

Add the expressions on corresponding sides of the equations, and solve for the remaining variable.

$$3x + 3y = -9$$
$$\underline{2x - 3y = -11}$$
$$5x = -20$$
$$x = -4$$

Substitute –4 for $x$ in $x + y = -3$. Solve for $y$.

$$-4 + y = -3$$
$$y = 1$$

The solution is (–4, 1).

## Exercises

**Solve each system of linear equations by using elimination.**

**17.** $\begin{cases} 4x + 5y = 3 \\ 2x + 5y = -11 \end{cases}$

**18.** $\begin{cases} 2x + 3y = -6 \\ -5x - 9y = 14 \end{cases}$

**19.** $\begin{cases} 0.5x + y = 0 \\ 0.9x - 0.2y = -2 \end{cases}$

**20.** $\begin{cases} -2x + 4y = 12 \\ 3x - 2y = -10 \end{cases}$

**21.** $\begin{cases} 0.7m + 3n = 2.5 \\ 0.7m + 2n = 3 \end{cases}$

**22.** $\begin{cases} 5x + 7y = 4 \\ 10x - 2y = -2 \end{cases}$

**23.** $\begin{cases} 4x - 2y = 3 \\ x + 3y = 1 \end{cases}$

**24.** $\begin{cases} -7x - 3y = 3 \\ -5x + 6y = 2 \end{cases}$

LESSON 7.4

## Key Skills

**Identify consistent and inconsistent systems of equations. Identify consistent systems as dependent or independent.**

**a.** $\begin{cases} y = 2x + 6 \\ y = 2x - 5 \end{cases}$

The two equations have the same slope. This is an *inconsistent* system.

**b.** $\begin{cases} 5x + y = 3 \\ 10x + 2y = 6 \end{cases}$

The two equations are both the same equation: $y = -5x + 3$. This is a *consistent, dependent* system.

**c.** $\begin{cases} y = \frac{1}{2}x + 2 \\ y = 3x + 7 \end{cases}$

These two equations do not have the same slope. This is a *consistent, independent* system.

## Exercises

**Identify each system of equations as independent, dependent, or inconsistent.**

**25.** $\begin{cases} 3x - 2y = 7 \\ 4y = -14 + 6x \end{cases}$

**26.** $\begin{cases} 4y = 2x + 20 \\ 3x - y = -20 \end{cases}$

**27.** $\begin{cases} 2x + 5y = 7 \\ 3y = 2x + 17 \end{cases}$

**28.** $\begin{cases} x + y = 7 \\ 28 - 2y = 2x \end{cases}$

**29.** $\begin{cases} 3x + 6y = 2 \\ 6x + 6y = 4 \end{cases}$

**30.** $\begin{cases} 2x = y - 3 \\ y = -2x + y \end{cases}$

**LESSON 7.5**

### Key Skills

**Graph the solution to a system of linear inequalities.**

Graph the solution set for this system:

$$\begin{cases} y < -x - 1 \\ y \le x + 1 \end{cases}$$

Graph $y < -x - 1$ with a dashed boundary line and graph $y \le x + 1$ with a solid boundary line. In both inequalities, $y$ is less than the expression on the right, so shade the area of the graph below each boundary line. The solution is the region of the plane where the shaded regions overlap.

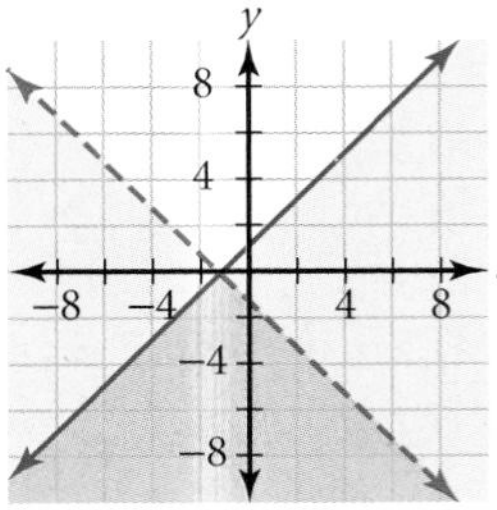


### Exercises

**Solve each system of linear inequalities by graphing.**

**31.** $\begin{cases} y \ge 4 \\ y \le x + 1 \end{cases}$

**32.** $\begin{cases} 4x - y > 1 \\ 2x + y > -2 \end{cases}$

**33.** $\begin{cases} y - 1 \le x \\ 4y \ge 2 \end{cases}$

**34.** $\begin{cases} 2x - y < 3 \\ 2y - 2 \le 6x \end{cases}$

**35.** $\begin{cases} x \le 1 \\ y \ge 4 \end{cases}$

**36.** $\begin{cases} -x < y + 7 \\ y \ge x - 7 \end{cases}$

**LESSON 7.6**

### Key Skills

**Solve traditional math puzzles.**

Mr. Carver invested \$5000, part at 5% interest and the rest at 3.9% interest. If he earned \$233.50 in interest the first year, how much did he invest at each rate?

Make a table to organize the information.

| **Amount** | $x$ | $y$ | \$5000 |
|---|---|---|---|
| **Interest** | $0.05x$ | $0.039y$ | \$233.50 |

From the table, write two equations. $\begin{cases} x + y = 5000 \\ 0.05x + 0.039y = 233.50 \end{cases}$

The solution to the system is (3500, 1500).
Mr. Carver invested \$3500 at 5% and \$1500 at 3.9%.

### Exercises

**Use a system of equations to solve each problem.**

**37.** Samantha has a total of 43 quarters and nickels in her purse. If her change is worth \$7.75, how many nickels and how many quarters does Samantha have?

**38.** A chemist wants to combine a solution that is 30% glycerin with pure (100%) glycerin. He wants to produce 42 milliliters of a solution that is 50% glycerin. How much of the 30% solution and how much of the pure glycerin should the chemist use?

## Applications

**39.** **CHEMISTRY** How many ounces of a 25% acid solution should be mixed with a 50% acid solution to produce 100 ounces of a 40% acid solution?

**40.** **GEOMETRY** The sum of the measures of angles $x$, $y$, and $z$ of triangle $XYZ$ is 180°. Angle $x$ has a measure of 60°. The measure of angle $y$ is 15° more than twice the measure of the angle $z$. What are the measures of angles $y$ and $z$?

# Chapter Test

**Solve each system by graphing. Round approximate solutions to the nearest tenth.**

1. $x + y = 6$
   $3x - y = 2$
2. $y = 2x - 4$
   $y = -\frac{1}{2}x + 1$
3. $x + 4y = 5$
   $-2x + 3y = 1$
4. $y = x + 3$
   $y = -4x$
5. MANUFACTURING It takes 3 hours to build an arm chair and 2 hours for a side chair. Finishing an arm chair requires 2 hours, while a side chair requires 1 hour. There are 18 hours available for building and 10 hours available for finishing. Set up and solve a system of equations to determine how many of each chair can be made.

**Solve each system of linear equations by using substitution.**

6. $y = 5x - 7$
   $y = -3x + 1$
7. $y = 2x + 3$
   $3x - y = -1$
8. $2x + 4y = 0$
   $3x + y = 5$
9. $x - 3y = -1$
   $2x - 4y = 2$
10. $\frac{1}{2}x + 3y = 1$
    $2x + 7y = -1$
11. $\frac{1}{4}x + \frac{1}{2}y = 3$
    $\frac{1}{2}x - \frac{1}{4}y = 1$
12. BUSINESS Hats and gloves sell for \$10 and \$12 respectively. If a store sells 5 fewer pairs of gloves than three times as many hats for \$308, write a system of equations and solve to determine the number of each sold.

**Solve each system of linear equations by elimination.**

13. $5x - 4y = 2$
    $-3x + 4y = 2$
14. $2x + y = -1$
    $-x + 2y = 13$
15. $2.3x + 4y = -8.5$
    $2.3x = 21.5 + 2y$
16. $5x + 7y = 4$
    $-x - 2y = -2$
17. NUMBER THEORY The sum of two numbers is 21. The first number is 3 more than twice the second. Set up a system of equations and solve it by elimination to find the numbers.

**Solve each system algebraically. Identify each system of equations as consistent and dependent, consistent and independent, or inconsistant.**

18. $y = 3x + 4$
    $3x - y = 7$
19. $2x + 8y = 6$
    $x = 3 - 4y$
20. $y = \frac{1}{2}x + 4$
    $y = -2x - \frac{1}{4}$
21. $y = 3 - 2x$
    $6x + 3y = 9$

**Solve by graphing.**

22. $y \geq -2$
23. $y < \frac{1}{2}x + 2$

**Solve each system of linear inequalities by graphing.**

24. $y \leq 2x + 3$
    $6x + 3y > 12$
25. $y < -\frac{1}{4}x + 2$
    $x \geq -1$
26. Xavier and Yolanda earned at least 100 points in a contest at school. Yolanda earned at most 20 points more than five times Xavier's points. Set up a system of inequalities for these statements, and graph the possible solutions.

**Use a system to solve each problem.**

27. COIN PUZZLE Martin has a total of 30 dimes and nickels in his pocket worth \$2.35. How many of each coin does he have?
28. BUSINESS Flavored coffee sells for \$8 per pound, and gourmet tea sells for \$12 per pound. How much of each should you buy to get 10 pounds for \$100?
29. CHEMISTRY How much pure water must be mixed with 2 liters of a 40% solution of antifreeze to get a 25% antifreeze solution?
30. TRAVEL A ship cruising on a river can travel the 135-mile distance between two cities in 15 hours when cruising against the current. When cruising with the current, the trip takes 5 fewer hours. Write and solve a system of equations to find the speed of the current and of the ship in still water.

# College Entrance Exam Practice

internet connect
Standardized Test Prep Online
Go To: go.hrw.com
Keyword: MM1 Test Prep.

**QUANTITATIVE COMPARISON** For Questions 1–5, write
A if the quantity in Column A is greater than the quantity in Column B;
B if the quantity in Column B is greater than the quantity in Column A;
C if the two quantities are equal; or
D if the relationship cannot be determined from the information given.

| | Column A | | Column B | Answers |
|---|---|---|---|---|
| **1.** | $x$ | $\begin{cases} 3x + 2y = 6 \\ x + y = 0 \end{cases}$ | $y$ | (A) (B) (C) (D) [Lesson 7.2] |
| **2.** | 0.37 | | $\frac{37}{10}$ | (A) (B) (C) (D) [Lesson 4.2] |
| **3.** | The eighth term of the sequence 2, 5, 8, 11, 14, . . . | | The eighth term of the sequence 32, 30, 28, 26, 24, . . . | (A) (B) (C) (D) [Lesson 1.1] |
| **4.** | The solution to $x + 3 \geq 4$ | | The solution to $x - 7 < 10$ | (A) (B) (C) (D) [Lesson 6.1] |
| **5.** | The slope of the line passing through the points (1, 2) and (0, 3) | | The slope of the line $4x + 4y = 12$ | (A) (B) (C) (D) [Lesson 5.5] |

**6.** Which expression is the solution to the equation $-4g = -64$? ***(LESSON 3.2)***

**a.** $\frac{-64}{4}$ **b.** $(-64)(-4)$
**c.** $-16$ **d.** 16

**7.** What is the reciprocal of $\frac{1}{6}$? ***(LESSON 2.4)***

**a.** 1.6 **b.** $-1.6$
**c.** 6 **d.** $-6$

**8.** What are the first three terms of a sequence if the fourth and fifth terms are 27 and 39 and the second differences are a constant 2? ***(LESSON 1.1)***

**a.** 21, 23, 25 **b.** 3, 9, 17
**c.** 6, 8, 10 **d.** 13, 15, 17

**9.** If a basketball league has 8 teams and each team plays each of the other teams once, how many games will be played? ***(LESSON 1.1)***

**a.** 28 **b.** 36
**c.** 72 **d.** 56

**10.** What is $|8.6|$? ***(LESSON 2.1)***

**a.** 8 **b.** $-8$
**c.** 8.6 **d.** $-8.6$

**11.** Which point is on the line $3x - 8y = 1$? ***(LESSON 5.1)***

**a.** $M(7, -4)$ **b.** $N(-2, -3)$
**c.** $O(3, 1)$ **d.** $P(4, 2)$

**12.** Which number line shows the solution set to the inequality $x - 6 > -4$? ***(LESSON 6.1)***

**a.** 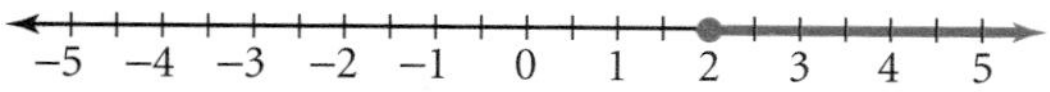


**b.** 

**c.** 

**d.** 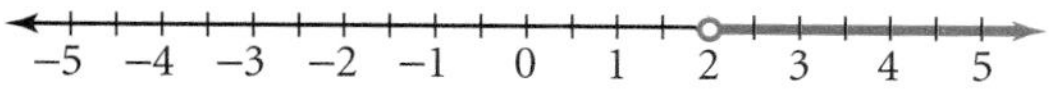


**13.** What is the solution to $\frac{3}{5}p = -3$? ***(LESSON 3.2)***

**a.** $-\frac{9}{5}$ **b.** $-\frac{5}{9}$

**c.** $-5$ **d.** $-\frac{1}{5}$

**14.** What is 65% of 120? ***(LESSON 4.2)***

**a.** 78 **b.** 185

**c.** 7800 **d.** 780

**15.** Which of the following lines is vertical? ***(LESSON 5.2)***

**a.** $y = -5$ **b.** $x = 3$

**c.** $y = -3x + 1$ **d.** $y = 2x - 2$

**Evaluate.** ***(LESSON 6.4)***

**16.** $|-2.2|$ **17.** ABS(7) **18.** ABS(–9.6)

**Find each sum.** ***(LESSON 2.2)***

**19.** $-25 + (-4)$ **20.** $-36 + 6 + (-6)$

**21.** $452 + (-452)$

**Solve each equation.** ***(LESSON 3.3)***

**22.** $3t + \frac{1}{6} = \frac{2}{3}$ **23.** $2a - 5 = 11$

**24.** $\frac{r}{4} + 6 = -4$

**25.** What is the $y$-intercept of the line that is parallel to the line $2x + 3y = 4$ and contains the point $(3, -1)$? ***(LESSON 5.6)***

**26.** Solve $4 - 3t \geq 19$. Show the solution set on a number line. ***(LESSON 6.2)***

**27.** Simplify: $\frac{15 - 36x}{3}$. ***(LESSON 2.7)***

**28.** Write an equation for the line passing through the points $(7, -7)$ and $(-2, 4)$. ***(LESSON 5.4)***

**29.** Mark has 3 nickels, 5 dimes, and 7 quarters in his pocket. He chooses one coin. What is the probability that the coin is a dime? ***(LESSON 4.3)***

**30.** The perimeter of a rectangle is 52 and its width is 6 less than its length. Find the length and width. ***(LESSON 7.2)***

**Solve each system.** ***(LESSONS 7.1, 7.2, AND 7.3)***

**31.** $\begin{cases} y = 7x \\ 2x - y = -10 \end{cases}$ **32.** $\begin{cases} 5x + 2y = 8 \\ x + y = 1 \end{cases}$

**FREE-RESPONSE GRID**

**The following questions may be answered by using a free-response grid such as that commonly used by standardized test services:**

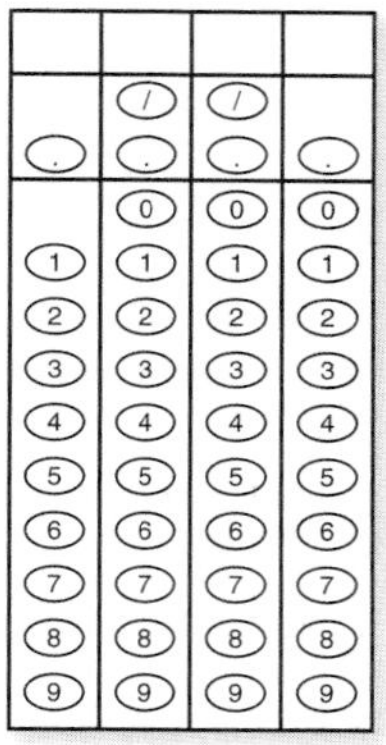

**33.** The sum of the digits of a two-digit number is 7. The tens digit is 1 less than the units digit. Find the number. ***(LESSON 7.6)***

**34.** What is the solution to the proportion $\frac{c}{16} = \frac{3}{8}$? ***(LESSON 4.1)***

**35.** A suit is marked down from an original price of \$75 to \$63.75. By what percent of the original price has the suit been marked down? ***(LESSON 4.2)***

**36.** Simplify: $4^2 \div 8 + 5(8 - 2) \cdot 2$. ***(LESSON 1.3)***

**37.** How much would each of 16 people need to contribute toward a gift that costs \$136? ***(LESSON 2.4)***

**38.** Solve $7(p + 4) = 49$. ***(LESSON 3.5)***

**39.** Find the slope of the line that passes through the origin and $(4, 2)$. ***(LESSON 5.2)***

**40.** 16 is 20% of what number? ***(LESSON 4.2)***

# 8

# Exponents and Exponential Functions

EXPONENTIAL FUNCTIONS MODEL A VARIETY OF important real-world activities. The growth of living organisms, population growth and decline, radioactive dating of archaeological artifacts, and the spread of certain diseases are all examples of exponential growth or decay.

## Lessons



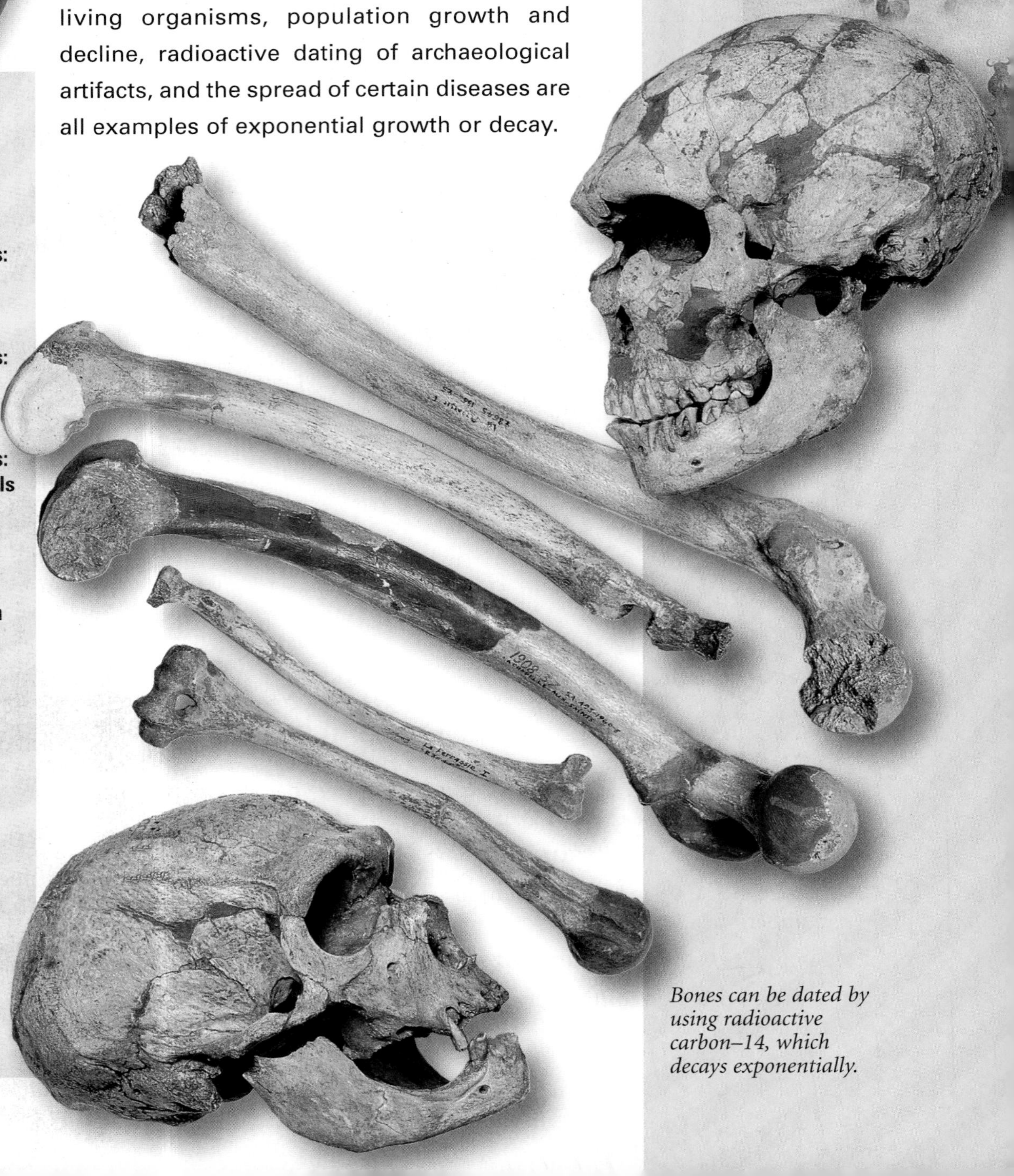

*Bones can be dated by using radioactive carbon–14, which decays exponentially.*

The explosive, exponential growth of certain bacteria can cause the rapid spread of a disease.

## About the Chapter Project

You have already seen how algebra makes it possible to use *functions* to model data. In the Chapter Project, *The Wolves' Den*, on page 416, you will compare linear and exponential models of data and decide which works better for the population that you study. After completing the Chapter Project, you will be able to do the following:

- Determine whether trends in given data sets are linear or exponential.

## About the Portfolio Activities

Throughout the chapter, you will be given opportunities to complete Portfolio Activities that are designed to support your work on the Chapter Project.

- Comparing linear and exponential growth is the topic of the Portfolio Activity on page 376.
- Population growth is the topic of the Portfolio Activity on page 395.
- Compound interest is the topic of the Portfolio Activity on page 415.

8.1

# Laws of Exponents: Multiplying Monomials

**Why** ***Real-world calculations frequently involve multiplying numbers by themselves. The area of a square, for example, requires that you multiply the length of a side by itself.***

## Objectives

- Define exponents and powers.
- Find products of powers.
- Simplify products of monomials.

The expression $2^{1000}$ represents 2 to the thousandth power. It is the product of a thousand 2s multiplied together. The number 2 is called the **base**, and 1000 is the **exponent.** Recall that a base and exponent together are called a power.

An exponent tells how many times a number is used as a factor.

### Exponents

For all real numbers $x$ and all positive integers $m$,

$$x^m = \underbrace{x \cdot x \cdot x \cdot \dots \cdot x}_{m \text{ factors}},$$

where $x$ is the base and $m$ is the exponent.

Notice that when $m = 1$, $x^m = x^1 = x$.

**EXAMPLE 1** **Evaluate.**

**a.** $2^8$ **b.** $5^3$ **c.** $3^4$ **d.** $6^1$

**SOLUTION**

**a.** $2^8 = 2 \cdot 2 \cdot 2 \cdot 2 \cdot 2 \cdot 2 \cdot 2 \cdot 2 = 256$
**b.** $5^3 = 5 \cdot 5 \cdot 5 = 125$
**c.** $3^4 = 3 \cdot 3 \cdot 3 \cdot 3 = 81$
**d.** $6^1 = 6$

**TRY THIS** Evaluate.

**a.** $2^6$ **b.** $4^3$ **c.** $3^3$ **d.** $8^1$

## Product of Powers

There is a simple and convenient way to multiply powers with the same base. In the activity that follows, you will discover this multiplication rule yourself.

*Activity*

### Multiplying Powers

**1.** Examine the example below.

In decimal notation: $1000 \cdot 10 = 10{,}000$

In factored form: $(10 \cdot 10 \cdot 10) \cdot (10) = 10 \cdot 10 \cdot 10 \cdot 10$

In exponential notation: $10^3 \cdot 10^1 = 10^4$

product form — simplified form

Complete the missing numbers as shown above.

**a.** $1000 \cdot 100 =$ ?

$10^3 \cdot 10^2 =$ ?

**b.** ? • ? = ?

$10^3 \cdot 10^3 =$ ?

**2.** Complete the table below for $10^4 \cdot 10^2$. Write the missing product in simplified form.

| | |
|---|---|
| **Decimal notation** | $10{,}000 \cdot 100 = 1{,}000{,}000$ |
| **Exponential notation** | $10^4 \cdot 10^2 =$ ? |

**3.** Write each term as a product of factors. Then use the definitions of exponents and powers to write each product below as a single power.

**a.** $2^3 \cdot 2^2$ **b.** $5^1 \cdot 5^3$ **c.** $10^2 \cdot 10^4$ **d.** $3^2 \cdot 3^2$

**CHECKPOINT ✔** **4.** For any positive number $x$ and any positive integers $m$ and $n$, how would you simplify $x^m x^n$? Fill in the box: $x^m x^n = x^{\square}$.

In the previous activity you may have discovered a property that lets you simplify the product of two powers with the same base. The property tells you to keep the base the same and add the exponents. For example:

$$x^3 \cdot x^5 = x^{3+5} = x^8$$

**Product-of-Powers Property**

For all real numbers $x$ and all integers $m$ and $n$,

$$x^m \cdot x^n = x^{m+n}.$$

**EXAMPLE 2** **Simplify.**

**a.** $2^3 \cdot 2^4$ **b.** $8 \cdot 8^3$ **c.** $y^2 \cdot y^5$ **d.** $5^m \cdot 5^p$

**SOLUTION**

**a.** $2^3 \cdot 2^4 = 2^{3+4} = 2^7 = 128$

**b.** $8 \cdot 8^3 = 8^{1+3} = 8^4 = 4096$

**c.** $y^2 \cdot y^5 = y^{2+5} = y^7$

**d.** $5^m \cdot 5^p = 5^{m+p}$

**TRY THIS** Simplify.

**a.** $3^5 \cdot 3^1$ **b.** $8^2 \cdot 8^2$ **c.** $z^6 \cdot z^3$ **d.** $a^z \cdot a^3$

**EXAMPLE 3**

**APPLICATION**
**BIOLOGY**

**Suppose that a colony of bacteria doubles in size every hour. If the colony contains 1000 bacteria at noon, how many bacteria will the colony contain at 3 P.M. and at 5 P.M. of the same day?**

**SOLUTION**

Between noon and 3 P.M. there are 3 hours so at 3 P.M. there will be $1000 \cdot 2^3$, or 8000, bacteria (where 2 stands for doubling and 3 stands for the number of hours). At 5 P.M., two hours later, the bacteria will double two more times. There will be $(1000 \cdot 2^3) \cdot 2^2 = 1000 \cdot 2^{3+2} = 1000 \cdot 2^5$, or 32,000, bacteria in the colony.

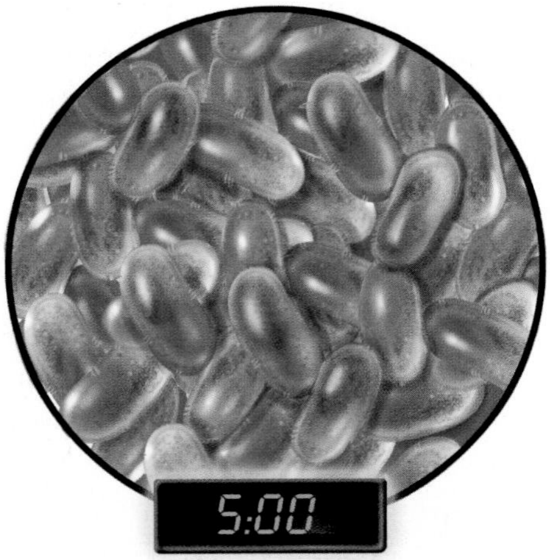


**CRITICAL THINKING** Describe two different ways of solving the following equations for $x$:

**a.** $2^x \cdot 2^3 = 2^7$ **b.** $2^x \cdot 2^4 = 2^6$

**TECHNOLOGY**
**SCIENTIFIC CALCULATOR**

You can use a graphics calculator to evaluate large numerical products involving exponents. Check the calculator you use to find out how to perform these computations. Most calculators use the $y^x$ or ^ key to indicate an exponent. To evaluate the product $8^3 \cdot 8^5$, you might use the following keystrokes:

8 [^] 3 [×] 8 [^] 5 [ENTER]

or 8 [$y^x$] 3 [×] 8 [$y^x$] 5 [=]

## Multiplying Monomials

The Product-of-Powers Property can be used to find the product of more complex expressions such as $5a^2b$ and $-2ab^3$. Expressions like $5a^2b$ and $-2ab^3$ are called *monomials*.

### Definition of Monomial

A **monomial** is an algebraic expression that is either a constant, a variable, or a product of a constant and one or more variables. The constant is called the **coefficient**.

To simplify a product of monomials, use the following steps:

1. Remove the parentheses and use the Commutative and Associative Properties of Multiplication to rearrange the terms. Group the constants together, and then group like variables together.
2. Simplify by using the Product-of-Powers Property where appropriate.

**EXAMPLE 4** **Simplify.**

**a.** $(5t)(-30t^2)$

**b.** $(-4a^2b)(-ac^2)(3b^2c^2)$

**SOLUTION**

**a.** $(5t)(-30t^2) = 5 \cdot (-30) \cdot t \cdot t^2 = -150t^3$

**b.** $(-4a^2b)(-ac^2)(3b^2c^2) = (-4) \cdot (-1) \cdot 3 \cdot a^2 \cdot a \cdot b \cdot b^2 \cdot c^2 \cdot c^2 = -12a^3b^3c^4$

**TRY THIS** **Simplify.**

**a.** $(3m^2)(60mp^2)$

**b.** $(8xz)(-10y)(-2yz^2)$

**EXAMPLE 5**

CONNECTION
GEOMETRY

The volume, $V$, of a right rectangular prism can be found by using the formula $V = lwh$, where $l$ is the length, $w$ is the width, and $h$ is the height. Suppose that a prism has a length of $2xy$, a width of $3xy$, and a height of $6xyz$. **Find the volume.**

**SOLUTION**

$$\begin{aligned} V &= lwh \\ &= 2xy \cdot 3xy \cdot 6xyz \\ &= 2 \cdot 3 \cdot 6 \cdot x \cdot x \cdot x \cdot y \cdot y \cdot y \cdot z \\ &= 36x^3y^3z \end{aligned}$$

**CRITICAL THINKING** Explain how to use the Product-of-Powers Property to simplify each of the following products:

**a.** $x^{3c} \cdot x^{4c}$

**b.** $(x + c)^d \cdot (x + c)^{2d}$

# Exercises

## Communicate

1. Use $10^{12}$ to explain the relationship between the exponent in a power of 10 and the number of zeros in the decimal notation.
2. How can you tell how many times the base is used as a factor when $3^6$ is rewritten in factored form?
3. Explain why you add the exponents when multiplying $7^3 \cdot 7^5$.
4. Explain how you could verify that the Product-of-Powers Property is true for $2^3 \cdot 2^4$.
5. The product $x \cdot x$ is equal to $x^2$. Explain how this illustrates the Product-of-Powers Property.

internet connect
**Activities Online**
Go To: **go.hrw.com**
Keyword:
**MA1 Number Sieves**

## Guided Skills Practice

**Evaluate.** ***(EXAMPLE 1)***

6. $5^4$
7. $2^3$
8. $3^4$
9. $11^1$

**Simplify each product.** ***(EXAMPLE 2)***

10. $2^5 \cdot 2^4$
11. $f^5 \cdot f^9$
12. $t^4 \cdot t^4$
13. $x^t \cdot x^r$

APPLICATION

14. **BIOLOGY** Suppose that a bacterial colony doubles in size every half-hour. If the colony contains 2000 bacteria at 10 A.M., how many bacteria will the colony contain at 1 P.M. and at 2 P.M.? ***(EXAMPLE 3)***

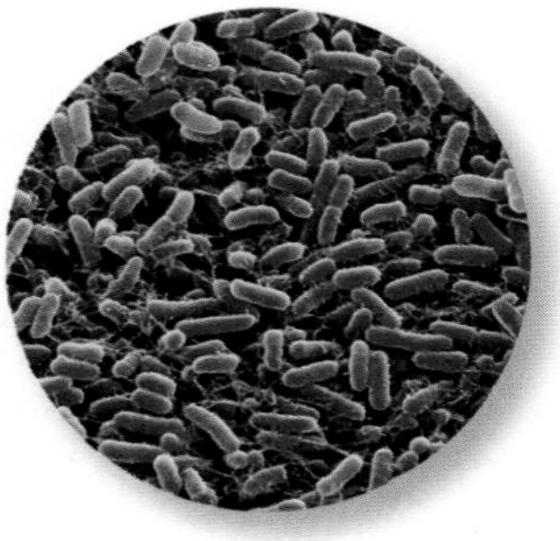

*Exponents can be used to model bacterial growth rates.*

**Simplify each product.** ***(EXAMPLE 4)***

15. $(8p^3)(40m^7p^6)$
16. $(-4x^3z^2)(-6y^5)(-y^3z^7)$

CONNECTION

17. **GEOMETRY** The volume, $V$, of a right rectangular prism can be found by using the formula $V = lwh$, where $l$ is the length, $w$ is the width, and $h$ is the height. Suppose that a prism has a length of $3rt$, a width of $5rt$, and a height of $2rty$. Find the volume. ***(EXAMPLE 5)***

## Practice and Apply

**Find the value of each expression.**

18. $4^5$
19. $2^6$
20. $3^6$
21. $5^3$
22. $10^4$
23. $10^7$
24. $100^3$
25. $6^4$
26. $8^3$
27. $3^7$
28. $10^8$
29. $9^4$
30. $2^8$
31. $2^4$
32. $2^5$
33. $2^2$

internet connect

**Homework Help Online**
Go To: **go.hrw.com**
Keyword:
**MA1 Homework Help** for Exercises 34-43

**Simplify each product. Leave the product in exponent form.**

**34.** $10^3 \cdot 10^4$ **35.** $10^3 \cdot 10^5$ **36.** $2^3 \cdot 2^4$ **37.** $3^4 \cdot 3^2$

**38.** $10^6 \cdot 10^2$ **39.** $10^4 \cdot 10^8$ **40.** $4^3 \cdot 4^5$ **41.** $6^2 \cdot 6^4$

**42.** Use your answers for Exercises 32 and 33 to evaluate $2^5 \cdot 2^2$.

**43.** Express the result of Exercise 42 as a single power of 2.

**Simplify each product.**

**44.** $(6x^2)(4x^2y^3)$ **45.** $(3x^3z^2)(-6y^5)$

**46.** $(5p^3)(-m^8p^2)$ **47.** $(-7m^5n^4)(8n^3)$

**48.** $(10g^3j^8v^6)(11gj^8)$ **49.** $(-13b^7)(2b^2f^4)$

**50.** $(4f^9h^3)(-5f^6)(-3h^2)$ **51.** $(3x^ay^bz^c)(-y^fz^g)$

Some of the following expressions can be written as a whole number power of 10; others cannot. If possible, rewrite the given expression as a whole number power of 10. If not possible, write the expression in decimal notation. Then simplify.

**52.** $10^4 \cdot 10^2$ **53.** $10^4 + 10^2$ **54.** $10^4 - 10^2$ **55.** $10^{2+3}$

**TECHNOLOGY** For exponents, $2^3$ is entered as 2 [^] 3 on some calculators and as 2 [$x^y$] 3 on others. Evaluate $2^3$. Familiarize yourself with the way the calculator you are using handles exponents. Then use the calculator to evaluate the following:

**56.** $2^{20}$ **57.** $3^{10}$ **58.** $0.5^4$ **59.** $0.8^5$

CONNECTION

**60. GEOMETRY** The area, $A$, of a trapezoid can be found by using the formula $A = \frac{1}{2}(b_1 + b_2)h$, where $b_1$ is one base, $b_2$ is the other base, and $h$ is the height. Suppose that a trapezoid has a base of $5x$, another base of $9x$, and a height of $8xy$. Find the area.

APPLICATION

**BIOLOGY** Food chains explain why even small organisms affect much larger ones. The following is an approximation of how a food chain works:

- A whale might eat 10 seals per day.
- A seal might eat 100 pounds of fish per day.
- Each pound of fish might need 10 pounds of zooplankton per day.
- Each pound of zooplankton might need 10 pounds of algae per day.

**Express the following as a power of 10:**

**61.** the number of seals needed to sustain a whale

**62.** the number of pounds of fish needed to sustain a whale

**63.** the number of pounds of zooplankton needed to sustain a whale

**64.** the number of pounds of algae needed to sustain a whale

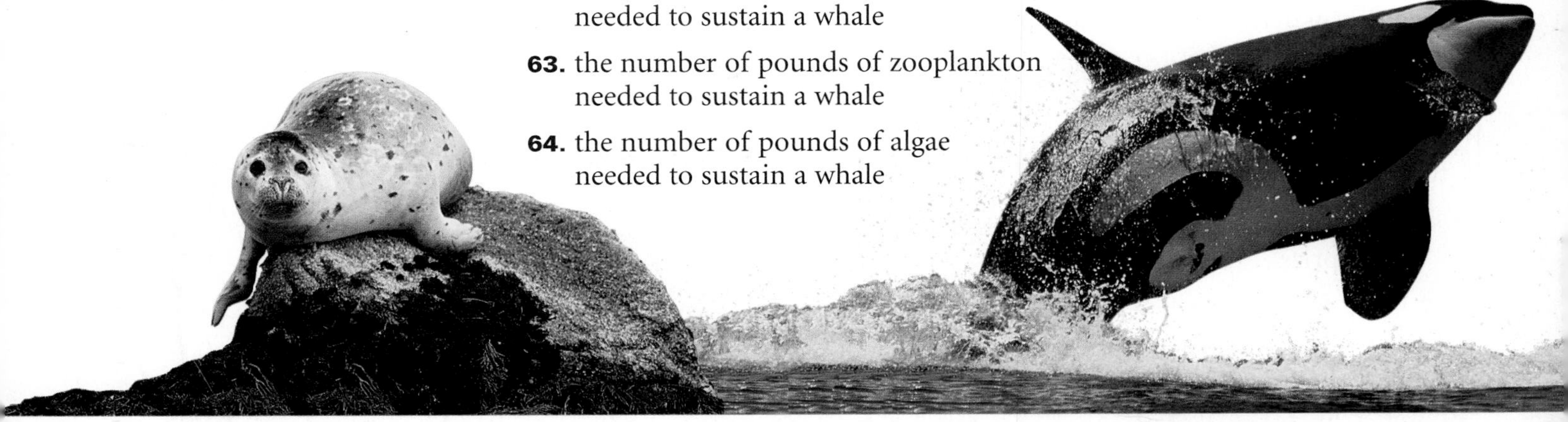

## Look Back

**Find each answer.** *(LESSON 4.2)*

**65.** 25 is 40% of what number?

**66.** 16 is 20% of what number?

**67.** What is 30% of 180?

**68.** What is 35% of 160?

**Graph each function by making a table and plotting points.** *(LESSON 6.4)*

**69.** $y = |x + 6|$

**70.** $y = |x - 5|$

**71.** $y = |3x|$

**Solve each system of equations.** *(LESSON 7.3)*

**72.** $\begin{cases} y = 2x + 11 \\ 3x = 29 - 5y \end{cases}$

**73.** $\begin{cases} 2x - 3y = 5 \\ -6x + 7y = -9 \end{cases}$

## Look Beyond

internet connect
**Portfolio Extension**
Go To: **go.hrw.com**
Keyword:
**MA1 Doubling**

**Use a calculator to evaluate the following:**

**74.** $4^{0.5}$

**75.** $9^{0.5}$

**76.** $81^{0.5}$

**77.** $100^{0.5}$

**78.** What is the value of $b^{0.5}$, where $b$ is any positive number?

## Portfolio Activity

**Prize A!**

**Start with $100. Repeatedly add $100 to the previous day's total.**

**Prize B!**

**Start with $0.01. Repeatedly multiply the previous day's total by 2.**

Include this activity in your portfolio.

After playing a certain game, the winner has 10 seconds to decide between the two prizes described above. For both prizes, the winner gets the amount of money calculated for the 20th day.

**1.** Make a table for the first four days of each prize.

**2.** After four days, which prize would you pick?

**3.** Do you think prize B will ever exceed prize A?

**4.** Continue the table for each prize through 20 days.

**5.** Did prize B ever exceed prize A?

**6.** Extend the table to 30 days. How much is each prize worth?

**WORKING ON THE CHAPTER PROJECT**

You should now be able to complete Activity 1 of the Chapter Project on page 416.

# 8.2 Laws of Exponents: Powers and Products

**Objectives**

- Find the power of a power.
- Find the power of a product.

**Why** *Variables with exponents are commonly found in real-world problems such as those involving area and volume.*

The volume of a sphere such as this hot-air balloon can be found with the formula $V = \frac{4}{3}\pi r^3$. If you substitute an expression for the value of the radius, $r$, the whole expression that is substituted must be cubed.

In Lesson 8.1 you learned about products of powers—that is, multiplying powers. Using the Product-of-Powers Property (page 371), you can simplify expressions by adding exponents to multiply powers.

$$a^2 \cdot a^3 = a^{2+3} = a^5$$

$$x^5 \cdot x^6 = x^{5+6} = x^{11}$$

In algebra you will sometimes need to raise *a power to a power.* For example, you might need to raise $a^2$ to the third power. This would be written as $(a^2)^3$. Do you think you should add the exponents in this case? You will examine this question in the following activity.

*Activity*

## Raising a Power to a Power

1. Study the following:
The expression $(a^2)^3$ means $a^2$ raised to the third power. By the definition of an exponent, $a^2$ is used as a factor 3 times.

$$\begin{aligned} a^2 \cdot a^2 \cdot a^2 &= (a \cdot a) \cdot (a \cdot a) \cdot (a \cdot a) \\ &= a \cdot a \cdot a \cdot a \cdot a \cdot a \\ &= a^6 \end{aligned}$$

As you can see, the result is *not* $a^5$, so the exponents are not added. Does the correct answer suggest a rule for raising a power to a power? Make a guess and keep it in mind as you do Steps 2 and 3.

2. How would you simplify the expression $(x^5)^2$? Following the pattern in Step 1, expand this expression and show how to simplify it. Does your result agree with the rule you guessed in Step 1?
3. Write down a power of a power of your own choosing. Then repeat the process in Step 2.

CHECKPOINT ✔

4. Write a rule for raising a power to a power. In your own words, explain why you think it is true.

In the activity, you may have discovered the following rule: To raise a power to a power, multiply the exponents.

**Power-of-a-Power Property**

For all real numbers $x$ and all integers $m$ and $n$,

$$(x^m)^n = x^{mn}.$$

**EXAMPLE 1** **Simplify and find the value of each expression when possible.**

**a.** $(2^3)^4$ **b.** $(10^3)^2$ **c.** $(p^2)^5$ **d.** $(x^m)^2$

**SOLUTION**

**a.** $(2^3)^4 = 2^{3 \cdot 4} = 2^{12} = 4096$

**b.** $(10^3)^2 = 10^{3 \cdot 2} = 10^6 = 1,000,000$

**c.** $(p^2)^5 = p^{2 \cdot 5} = p^{10}$

**d.** $(x^m)^2 = x^{m \cdot 2} = x^{2m}$

**TRY THIS** Simplify and find the value of each expression when possible.

**a.** $(2^6)^2$ **b.** $(10^4)^3$ **c.** $(y^3)^5$ **d.** $(m^3)^x$

## Raising a Monomial to a Power

Sometimes an exponential expression has a monomial for a base, as in the examples below.

$$(xy)^2 \qquad (xy^2)^2 \qquad (\pi r^2)^4$$

You can simplify expressions like these by using the definition of exponents.

**EXAMPLE 2** **Simplify $(xy^2)^3$ by using the definition of exponents.**

**SOLUTION**

The exponent, 3, outside the parentheses shows that the monomial factor, $(xy^2)$, is used as a factor 3 times. Remove the parentheses, and simplify by regrouping and multiplying.

$$\begin{aligned}(xy^2)^3 &= xy^2 \cdot xy^2 \cdot xy^2 \\ &= x \cdot x \cdot x \cdot y^2 \cdot y^2 \cdot y^2 \\ &= x^3y^2\end{aligned}$$

**TRY THIS** Simplify each expression by using the definition of exponents.

**a.** $(xy)^2$ **b.** $(xy^2)^2$ **c.** $(\pi r^2)^4$

**Power-of-a-Product Property**

For all real numbers $x$ and $y$, and all integers $n$,

$$(xy)^n = x^ny^n.$$

The Power-of-a-Product Property can be extended to include any number of factors inside the parentheses. For example:

$$(xyz)^n = x^ny^nz^n$$

**EXAMPLE 3** **Verify the Power-of-a-Product Property by finding the value of each expression in two different ways.**

**a.** $(2 \cdot 3)^3$ **b.** $(2 \cdot 5)^5$

**SOLUTION**

**a.** $(2 \cdot 3)^3 = 2^3 \cdot 3^3 = 8 \cdot 27 = 216$ or $(2 \cdot 3)^3 = 6^3 = 216$

**b.** $(2 \cdot 5)^5 = 2^5 \cdot 5^5 = 32 \cdot 3125 = 100{,}000$ or $(2 \cdot 5)^5 = 10^5 = 100{,}000$

**TRY THIS** Verify the Power-of-a-Product Property by finding the value of each expression in two different ways.

**a.** $(3 \cdot 3)^2$ **b.** $(2 \cdot 5)^6$

**EXAMPLE 4** **Simplify.**

**a.** $(x^2y)^3$ **b.** $(ab^2c^n)^5$

**SOLUTION**

**a.** $(x^2y)^3 = x^{2\cdot 3}y^3 = x^6y^3$

**b.** $(ab^2c^n)^5 = a^5b^{2\cdot 5}c^{n\cdot 5} = a^5b^{10}c^{5n}$

**TRY THIS** Simplify.

**a.** $(xy^3)^2$ **b.** $(x^2y^2z^2)^3$ **c.** $(abcd)^n$

**EXAMPLE 5**

CONNECTION
GEOMETRY

The volume, $V$, of a sphere is given by the formula $V = \frac{4}{3}\pi r^3$. Suppose that the radius, $r$, of a sphere is equal to $6s^2$, where $s$ is the radius of a smaller sphere. **Find the volume of the larger sphere in terms of the radius of the smaller sphere.**

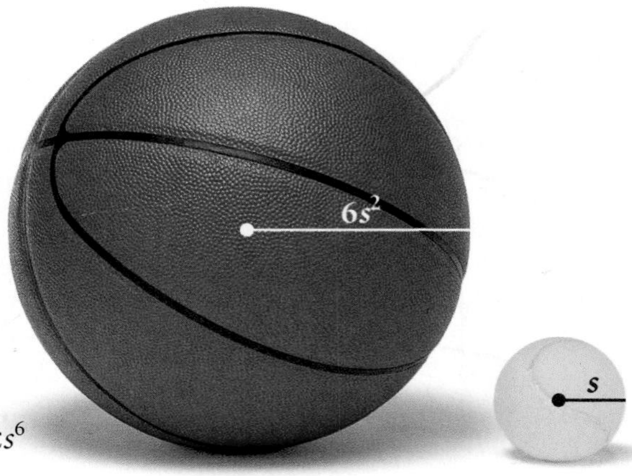


**SOLUTION**

$V = \frac{4}{3}\pi(6s^2)^3 = \frac{4}{3}\pi(216s^6) = 288\pi s^6$

## Powers of –1

Recall that $(-1) \cdot (-1) = 1$; that is, $(-1)^2 = 1$. However, $(-1) \cdot (-1) \cdot (-1) = -1$. Thus, $(-1)^3 = -1$. This pattern suggests the following rule:

> **Powers of –1**
>
> Even powers of $-1$ are equal to 1.
> Odd powers of $-1$ are equal to $-1$.

**EXAMPLE 6** **Simplify.**

**a.** $(-t)^5$ **b.** $-t^4$ **c.** $(-5x)^3$

**SOLUTION**

**a.** $(-t)^5 = (-1 \cdot t)^5 = (-1)^5 \cdot (t^5) = -1 \cdot t^5 = -t^5$

**b.** $-t^4$ is already in simplest form. The exponent applies only to $t$.

**c.** $(-5x)^3 = (-5)^3x^3 = -125x^3$

**TRY THIS** Simplify.

**a.** $(-y)^4$ **b.** $-y^{100}$ **c.** $(-3y)^4$

# Exercises

## Communicate

1. Explain the difference between $a^2a^3$ and $(a^2)^3$.
2. Explain why $(x^a)^b = (x^b)^a$ is true.
3. Explain why you multiply the exponents when you raise a power to a power.
4. Explain why $(-1)^2 > (-1)^3$ is true.
5. Explain why $(a^s)^t = a^{st}$ and $(xy)^n = x^ny^n$ are true.

## Guided Skills Practice

**Simplify and find the value of each expression when possible.** *(EXAMPLE 1)*

**6.** $(3^5)^2$ **7.** $(10^3)^2$ **8.** $(b^4)^9$ **9.** $(z^7)^y$

**Simplify each expression by using the definition of exponents.** *(EXAMPLE 2)*

**10.** $(ab)^8$ **11.** $(x^2y^4)^3$ **12.** $(c^2h^7)^4$ **13.** $(m^2n^5)^3$

**Verify the Power-of-a-Product Property by finding the value of each expression in two different ways.** *(EXAMPLE 3)*

**14.** $(5 \cdot 5)^3$ **15.** $(3 \cdot 10)^7$ **16.** $(4 \cdot 3)^2$ **17.** $(5 \cdot 11)^2$

**Simplify by using the Power-of-a-Product Property.** *(EXAMPLE 4)*

**18.** $(x^2y^3)^5$ **19.** $(a^4b^4c^4)^{11}$ **20.** $(qrs)^t$ **21.** $(x^2y^5)^3$

**Simplify.** *(EXAMPLE 5)*

**22.** $(-6)^3$ **23.** $(-b)^{10}$ **24.** $(-5k)^3$ **25.** $(-1)^{1001}$

## Practice and Apply

internet connect
**Homework Help Online**
Go To: **go.hrw.com**
Keyword:
**MA1 Homework Help** for Exercises 26-53

**Simplify each of the following:**

**26.** $(3x)^2$ **27.** $(6^2)^3$ **28.** $(-x^2)^3$ **29.** $(10^2)^3$

**30.** $(2x^4)^3$ **31.** $(-3b^2)^5$ **32.** $(-2r^3)^2$ **33.** $(-3x^6)^5$

**34.** $(-10m^4)^3$ **35.** $(5j^2k^3)^4$ **36.** $(-6xy^4z^5)^2$ **37.** $2(3a^2)^3$

**38.** $10(-5b^5)^2$ **39.** $(8n^2p)^3$ **40.** $(7j^2)^3$ **41.** $(-ab^5)(a^3)^2$

**42.** $(-v^4w^3)^2(-v^3)^4$ **43.** $(5f^3t^7)^3(d^4f^2)^5$ **44.** $(x^2)(y^2)^2$ **45.** $(3x^2)^2(12y^3)^3$

**46.** $(y^2)^4(x^3y^5)^6$ **47.** $(-4b^7e^3)^2(-h^3)^3$ **48.** $(-x^3)(-y^2)^3$ **49.** $(xy)^2(x^2y^2)^2$

CHALLENGE

**50.** $\left(\frac{1}{3x^2y}\right)^3$ **51.** $\left(\frac{3x^2}{2y^2}\right)^5$ **52.** $\left(\frac{8x^2}{2x^2}\right)^2$ **53.** $\left(\frac{-1}{5x^3y^5}\right)^{11}$

**Evaluate each monomial for $a = 10$.**

**54.** $2a^3$ **55.** $(2a)^3$ **56.** $-2a^2$ **57.** $(-2a)^2$

**Evaluate each of the following:**

**58.** $(-1)^1$ **59.** $(-1)^2$ **60.** $(-1)^3$

**61.** $(-1)^4$ **62.** $(-1)^5$ **63.** $(-1)^6$

CONNECTION

**64.** **GEOMETRY** Find the volume of the cube if each edge, $e$, is doubled.

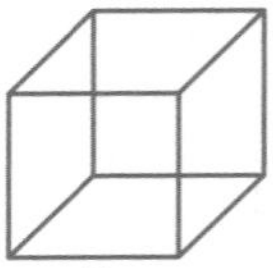


APPLICATION

**65.** **RECREATION** A summer camp needs to build a field for a new sport invented by one of its employees. The field needs to be rectangular and to have a length three times as long and a width twice as the long as the side of an existing square field. Use $s$ to represent the length of the sides of the square field. Find the area of the new field.

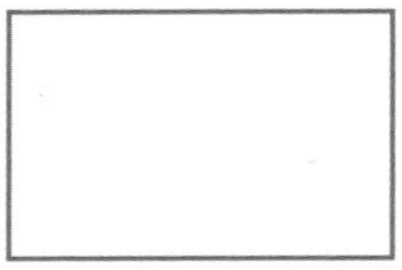

## Look Back

**66.** Distance, $d$, varies directly as time, $t$. If a car travels 195 miles in 3 hours, determine the constant, $k$, and write an equation for the direct variation. How far will the car travel in 4 hours? ***(LESSON 5.3)***

CONNECTION

**67.** **COORDINATE GEOMETRY** A line passes through the points $P(14, -3)$ and $Q(-5, 27)$. Write an equation for this line and an equation for a line perpendicular to this line. ***(LESSON 5.6)***

**Solve for *x*.** ***(LESSON 6.5)***

**68.** $|x - 3| \leq 7$ **69.** $|2x + 4| > 8$

## Look Beyond

**70.** Show that $1^2 + 8^2 = 4^2 + 7^2$. A consequence of this is $14^2 + 87^2 = 41^2 + 78^2$. How were the numbers 14, 87, 41, and 78 chosen?

**71.** Show that $14^2 + 87^2 = 41^2 + 78^2$ is true. Is $17^2 + 84^2 = 71^2 + 48^2$ true? How were the numbers 17, 84, 71, and 48 chosen?

8.3

# Laws of Exponents: Dividing Monomials

**Objectives**

- Simplify quotients of powers.
- Simplify powers of fractions.

***Why*** *The study of ratios, which is important in sciences such as biology and physics, requires an understanding of quotients, which may involve exponents.*

**APPLICATION**
**BIOLOGY**

A plant that has a large volume in comparison to its surface area will hold heat and moisture well. Do you think the ratio between volume and surface area is greater for small or large plants?

## Quotients of Powers

In a quotient of powers, one power is divided by another. You can simplify a quotient of powers by using the definition of powers and exponents. You can begin by expressing the powers in terms of their factors.

$$\frac{a^6}{a^4} = \frac{a \cdot a \cdot a \cdot a \cdot a \cdot a}{a \cdot a \cdot a \cdot a} = a \cdot a = a^2$$

In the activity that follows, you will use discover a rule for simplifying quotients of powers.

*Activity*

## Discovering a Rule for Quotients of Powers

1. Use the table to write the quotient $\frac{10^6}{10^2}$ as a power of 10.

| Form | Numerator | Denominator | Quotient |
|---|---|---|---|
| Decimal | 1,000,000 | 100 | 10,000 |
| Exponential | $10^6$ | $10^2$ | ? |

2. Make a similar table for each of the following:
   **a.** $\frac{10^5}{10^2}$ **b.** $\frac{2^6}{2^2}$ **c.** $\frac{3^4}{3^3}$
3. Guess the exponential notation for $\frac{10^5}{10^1}$. Make a table to check your guess.
4. For any positive number $a$ and any positive integers $m$ and $n$, where $m > n$, what is the simplified form for $\frac{a^m}{a^n}$?

**CHECKPOINT ✔** 5. Explain how to simplify $\frac{2^5}{2^3}$.

### Quotient-of-Powers Property

For all nonzero real numbers $x$ and all integers $m$ and $n$,

$$\frac{x^m}{x^n} = x^{m-n}.$$

**EXAMPLE 1** **Simplify.**

**a.** $\frac{2^6}{2^4}$ **b.** $\frac{10^9}{10^2}$

**SOLUTION**

**a.** $\frac{2^6}{2^4} = 2^{6-4} = 2^2 = 4$ **b.** $\frac{10^9}{10^2} = 10^{9-2} = 10^7 = 10{,}000{,}000$

**TRY THIS** Simplify.

**a.** $\frac{10^{12}}{10^4}$ **b.** $\frac{2^{10}}{2^5}$ **c.** $\frac{3^{10}}{3^7}$

**EXAMPLE 2** **Simplify. Assume that the conditions of the property are satisfied.**

**a.** $\frac{x^a}{x}$ **b.** $\frac{x^{a+b}}{x^c}$ **c.** $\frac{x^{m+1}}{x}$

**SOLUTION**

**a.** $\frac{x^a}{x} = \frac{x^a}{x^1} = x^{a-1}$ **b.** $\frac{x^{a+b}}{x^c} = x^{a+b-c}$ **c.** $\frac{x^{m+1}}{x} = \frac{x^{m+1}}{x^1} = x^{m+1-1} = x^m$

**TRY THIS** Simplify.

**a.** $\frac{m^n}{m^p}$ **b.** $\frac{y^{10}}{y}$ **c.** $\frac{10^{y+1}}{10}$

## Quotients of Monomials

To find quotients of monomials, treat the coefficients separately, and use the Quotient-of-Powers Property to simplify powers with the same base.

**EXAMPLE 3** **Simplify. Assume that the conditions of the property are satisfied.**

**a.** $\dfrac{-4x^2y^5}{2xy^3}$ **b.** $\dfrac{c^4b}{c^2a}$

**SOLUTION**

Simplify any numerical coefficients. Then subtract the exponents of the powers in the denominator from the exponents of the powers with the same base in the numerator.

**a.** $\dfrac{-4x^2y^5}{2xy^3} = \dfrac{-4}{2} \cdot x^{2-1} \cdot y^{5-3} = -2xy^2$

**b.** $\dfrac{c^4b}{c^2a} = c^{4-2} \cdot \dfrac{b}{a} = \dfrac{c^2b}{a}$

**TRY THIS** Simplify. Assume that the conditions of the property are satisfied.

**a.** $\dfrac{6ab^2}{-2ab}$ **b.** $\dfrac{(5y^2)(-81t^4)}{45t^3}$

**EXAMPLE 4**

**CONNECTION**
**GEOMETRY**

The volume, $V$, of a sphere can be found with the formula $V = \frac{4}{3}\pi r^3$. The surface area, $S$, of a sphere can be found with the formula $S = 4\pi r^2$. **Find the ratio of the volume of a sphere to its surface area. What happens to this ratio as the radius of the sphere increases?**

**SOLUTION**

Find the ratio by dividing the expression for the volume, $V$, by the expression for the surface area, $S$.

$$\frac{V}{S} = \frac{\frac{4}{3}\pi r^3}{4\pi r^2} = \frac{\frac{4}{3}}{4} \cdot \pi^{1-1} \cdot r^{3-2} = \frac{1}{3}r$$

By substituting different values for $r$, you will see that the value of the ratio increases as the value of $r$ increases.

**CRITICAL THINKING** Evaporation and heat loss take place at the surface of an object. Do you think that a large or a small object of the same shape will lose less heat and moisture? Explain how this relates to the ratio of volume to surface area.

*The anole at left has a smaller ratio of volume to surface area than the Komodo dragon above.*

## Powers of Fractions

In order to multiply a fraction by a fraction, recall that you must find the product of the numerators and the product of the denominators.

$$\frac{3}{4} \times \frac{7}{8} = \frac{3 \times 7}{4 \times 8} = \frac{21}{32}$$

This procedure leads to the rule for finding a power of a fraction.

$$\left(\frac{a}{b}\right)^n = \underbrace{\frac{a}{b} \cdot \frac{a}{b} \cdot \frac{a}{b} \cdot \dots \cdot \frac{a}{b}}_{n \text{ factors}} = \frac{a \cdot a \cdot a \cdot \dots \cdot a}{b \cdot b \cdot b \cdot \dots \cdot b} = \frac{a^n}{b^n}$$

**Power-of-a-Fraction Property**

For all real numbers $a$ and $b$ where $b \neq 0$, and all integers $n$

$$\left(\frac{a}{b}\right)^n = \frac{a^n}{b^n}.$$

**EXAMPLE 5** **Simplify each expression.**

**a.** $\left(\frac{3}{4}\right)^2$ **b.** $\left(\frac{10}{5}\right)^4$ **c.** $\left(\frac{x^n}{y^n}\right)^2$

**SOLUTION**

**a.** $\left(\frac{3}{4}\right)^2 = \frac{3^2}{4^2} = \frac{9}{16}$

**b.** $\left(\frac{10}{5}\right)^4 = \frac{10^4}{5^4} = \frac{10{,}000}{625} = 16$ or $\left(\frac{10}{5}\right)^4 = 2^4 = 16$

**c.** $\left(\frac{x^n}{y^n}\right)^2 = \frac{(x^n)^2}{(y^n)^2} = \frac{x^{2n}}{y^{2n}}$

**TRY THIS** Simplify each expression.

**a.** $\left(\frac{2}{3}\right)^4$ **b.** $\left(\frac{6}{2}\right)^3$ **c.** $\left(\frac{x^3}{2^y}\right)^z$

# Exercises

## Communicate

1. Explain how to simplify $\frac{2^5}{2^4}$.
2. Explain why you cannot use the same procedure to simplify $\frac{2^5}{2^4}$ and $\frac{5^2}{4^2}$.
3. Use $2^3 \cdot 2^2$ and $\frac{2^3}{2^2}$ to compare the Product-of-Powers Property with the Quotient-of-Powers Property.
4. Explain why $\left(\frac{f}{g}\right)^6 = \frac{f^6}{g^6}$ is true.

## Guided Skills Practice

**Use the Quotient-of-Powers Property to simplify each quotient. Then find the value of the result.** ***(EXAMPLE 1)***

5. $\frac{10^9}{10^3}$
6. $\frac{3^{12}}{3^9}$
7. $\frac{7^{10}}{7^7}$

**Use the Quotient-of-Powers Property to simplify each quotient. Assume that the conditions of the property are satisfied.** ***(EXAMPLE 2)***

8. $\frac{a^c}{a^d}$
9. $\frac{d^{12}}{d}$
10. $\frac{m^{y+1}}{m}$

**Simplify. Assume that the conditions of the Quotient-of-Powers Property are satisfied.** ***(EXAMPLES 3 AND 4)***

11. $\frac{-8x^3z^7}{2x^2yz}$
12. $\frac{h^3k^7}{h^2k^4}$
13. $\frac{p^2r^4s^5}{r^3s^2}$

**Simplify each expression.** ***(EXAMPLE 5)***

14. $\left(\frac{5}{8}\right)^2$
15. $\left(\frac{3}{4}\right)^4$
16. $\left(\frac{a^3}{4^t}\right)^n$

## Practice and Apply

**Use the Quotient-of-Powers Property to simplify each quotient. Then find the value of the result.**

17. $\frac{10^{11}}{10^6}$
18. $\frac{4^{11}}{4^7}$
19. $\frac{6^{12}}{6^4}$
20. $\frac{4^3}{4^2}$
21. $\frac{5^2}{5}$
22. $\frac{10^5}{10^2}$
23. $\frac{3^5}{3^3}$
24. $\frac{11^{42}}{11^{39}}$
25. $\frac{10^{15}}{10^5}$
26. $\frac{7^6}{7^2}$
27. $\frac{8^3}{8}$
28. $\frac{2^{16}}{2^{14}}$

internet connect

**Homework Help Online**
Go To: **go.hrw.com**
Keyword:
**MA1 Homework Help** for Exercises 29-64

**Simplify each expression. Assume that the conditions of the Quotient-of-Powers Property are met.**

**29.** $\left(\frac{a}{b}\right)^3$ **30.** $\left(\frac{10^x}{y^3}\right)^2$ **31.** $\left(\frac{5x}{y^6}\right)^3$ **32.** $\left(\frac{16x^2}{4y^5}\right)^3$

**33.** $\left(\frac{a^4}{b^2}\right)^3$ **34.** $\left(\frac{10g^7}{h^3}\right)^6$ **35.** $\left(\frac{9d^4}{e^6}\right)^2$ **36.** $\left(\frac{8w^7}{16}\right)^3$

**37.** $\left[\frac{c^5p^4}{(cp)^2}\right]^3$ **38.** $\left(\frac{6x^5}{24x^4}\right)^2$ **39.** $\left(\frac{2w^5}{3f^4}\right)^5$ **40.** $\left(\frac{4y^7}{40y^2}\right)^2$

**41.** $\left(\frac{7p^q}{f}\right)^d$ **42.** $\left(\frac{ac^5}{b^3}\right)^z$ **43.** $\left(\frac{wx^t}{w}\right)^t$ **44.** $\left(\frac{c^2a^3}{b^x}\right)^{xy}$

**Find each quotient. Assume that the conditions of the Quotient-of-Powers Property are met.**

**45.** $\frac{8r^5}{4r^3}$ **46.** $\frac{70x^4}{7x^3}$ **47.** $\frac{-2a^5}{4a^2}$ **48.** $\frac{-3z^5}{27z^3}$

**49.** $\frac{-p^6}{10p^3}$ **50.** $\frac{-2a^2b^7}{-5ab}$ **51.** $\frac{-x^2y^5}{-xy^2}$ **52.** $\frac{49a^7b^2}{7a^5b}$

**53.** $\frac{48a^4b^2}{-1.2ac^5}$ **54.** $\frac{0.08r^{12}}{0.004r^3}$ **55.** $\frac{4.38u^2v^{10}w^5}{0.1w^2v^3}$ **56.** $\frac{x^4y^3z^2}{x^3y^2z}$

**57.** $\frac{-r^4}{-5r^3}$ **58.** $\frac{-24a^5b^{14}}{-36a^3b}$ **59.** $\frac{-c^5d^7}{-c^4y^7}$ **60.** $\frac{-x^3y^7}{-x^3y^4}$

**61.** $\frac{5.6d^{15}e^{13}}{-7(de)^{12}}$ **62.** $\frac{0.27p^4}{0.09p^3}$ **63.** $\frac{156q^5(r^2s)^7}{12q^2s^3}$ **64.** $\frac{128a^4(bc^2)^3}{32abc}$

**TECHNOLOGY** Calculators can be used to evaluate expressions involving monomials.

**Store 10 for A and 2 for B in the memory of a calculator. Use the calculator to evaluate each of the following. Round each answer to the nearest hundredth if necessary.**

**65.** $\frac{A^2}{B^3}$ **66.** $\frac{1}{A^2 + B^3}$ **67.** $\frac{A^2}{B^3 + B}$ **68.** $\frac{3}{A^2B - B^3}$

**69.** $\frac{AB^2}{(AB)^2}$ **70.** $\frac{A^2 + B^2}{A^2B^2}$ **71.** $\frac{-A^2}{(-A)^2}$ **72.** $\frac{B^A}{A^B}$

**Use a calculator to evaluate each quotient for $A = 9.8$ and $B = 2.1$. Round each answer to the nearest hundredth if necessary.**

**73.** $\frac{A^2}{B^3}$ **74.** $\frac{1}{A^2 + B^3}$ **75.** $\frac{A^2}{B^3 + B}$ **76.** $\frac{3}{A^2B - B^3}$

**77.** $\frac{AB^2}{(AB)^2}$ **78.** $\frac{A^2 + B^2}{A^2B^2}$ **79.** $\frac{-A^2}{(-A)^2}$ **80.** $\frac{B^A}{A^B}$

CONNECTION

**81. GEOMETRY** Find the ratio of the volume of a sphere to the area of a circle with the same radius as the sphere.
Area of circle: $\pi r^2$
Volume of sphere: $\frac{4}{3}\pi r^3$

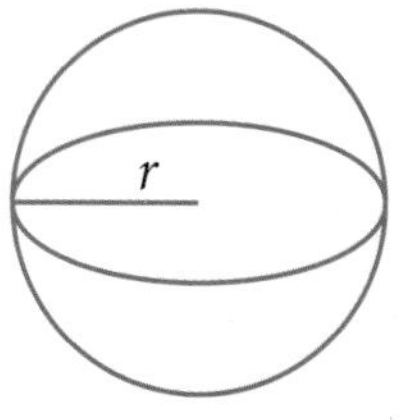

**APPLICATIONS**

**82. ECONOMICS** The 1996 population of the United States was 265 million people. The *Washington Post* reports that the national debt for 1996 was 3.7 trillion dollars. To find the amount of debt per person, divide the debt by the population. Simplify the following expression:

$$\frac{3.7 \times 10^{12}}{2.65 \times 10^{8}}$$

**CHALLENGE**

**83. PACKAGING** A box manufacturer has the option of producing hat boxes as right rectangular prisms or as cylinders. If the hats are $y$ inches high and $x^2$ inches wide at the widest part of the brim, write a ratio of the lateral surface area of the smallest possible prism to the lateral surface area of the smallest possible cylinder. Simplify the ratio as much as possible.

## Look Back

**Solve each equation for *n*.** ***(LESSONS 3.4 AND 3.5)***

**84.** $(4n - 20) - n = n + 12$

**85.** $9(n + k) = 5n + 17k + 32$

**Solve each system.** ***(LESSONS 7.2 AND 7.3)***

**86.** $\begin{cases} 5x - 3y = 32 \\ 2x = 12.8 \end{cases}$

**87.** $\begin{cases} 7x - 3y = 2 \\ 2x - y = -5 \end{cases}$

**88.** A canoer takes 2.5 hours to travel 7.5 miles upstream. The return trip takes 2 hours. Find the speed of the current and the speed of the canoe in still water. ***(LESSON 7.6)***

## Look Beyond

**89.** Find the ratio of the area of the square to the area of the circle.

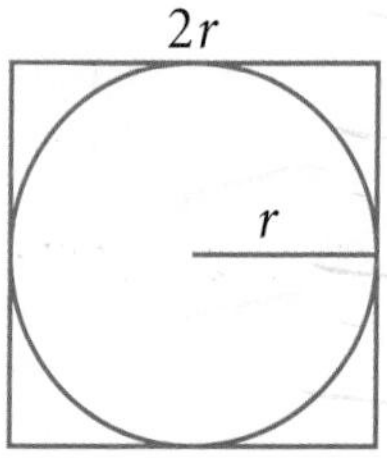


**90.** Notice that $r^2 \div r^2 = 1$, where $r \neq 0$. Does this suggest a definition of a zero exponent? Explain.

8.4

# Negative and Zero Exponents

**Objectives**

- Understand the concepts of negative and zero exponents.
- Simplify expressions containing negative and zero exponents.

*The approximate mass of a proton, such as those in the carbon atoms of coal, is $1.660 \times 10^{-24}$ grams.*

**Why** ***Negative exponents allow you to apply the ideas that you have learned to more real-world situations. They are also useful for representing small quantities.***

The expression $10^1$ is defined to be 10. How would you define $10^{-1}$?

When the Quotient-of-Powers Property is applied to an expression such as $\frac{10^4}{10^3}$, the result is $\frac{10^4}{10^3} = 10^{4-3} = 10^1$.

If the Quotient-of-Powers Property is applied when the exponent in the numerator is less than the exponent in the denominator, the result is a negative exponent. For example, $\frac{10^3}{10^4} = 10^{3-4} = 10^{-1}$.

If you write the powers as factors, the result is $\frac{1}{10}$.

The result above indicates that $10^{-1}$ can be defined as $\frac{1}{10}$.

**Definition of Negative Exponent**

For all nonzero real numbers $x$ and all integers $n$,

$$x^{-n} = \frac{1}{x^n}.$$

You can use this definition to simplify expressions with negative exponents. All the properties of exponents presented in Lessons 8.1 through 8.3 also apply to negative and zero exponents.

**EXAMPLE 1** **Evaluate each expression.**

**a.** $2^{-3} \cdot 2^2$ **b.** $\frac{10^3}{10^{-1}}$ **c.** $\frac{7^5}{7^7}$ **d.** $\frac{4}{3^{-2}}$

**SOLUTION**

**a.** Use the Product-of-Powers Property. Add the exponents and simplify.

$$2^{-3} \cdot 2^2 = 2^{-3+2} = 2^{-1} = \frac{1}{2^1} = \frac{1}{2}$$

**b.** Use the Quotient-of-Powers Property. Subtract the exponents and simplify.

$$\frac{10^3}{10^{-1}} = 10^{3-(-1)} = 10^4 = 10{,}000$$

**c.** Use the Quotient-of-Powers Property. Subtract the exponents and simplify.

$$\frac{7^5}{7^7} = 7^{5-7} = 7^{-2} = \frac{1}{7^2} = \frac{1}{49}$$

**d.** Use the Definition of Negative Exponent and simplify.

$$\frac{4}{3^{-2}} = 4 \cdot 3^2 = 4 \cdot 9 = 36$$

**TRY THIS** Evaluate each expression.

**a.** $\frac{10^5}{10^7}$ **b.** $3^{-5} \cdot 3^8$

**CRITICAL THINKING** Consider $a^n$ and $a^{-n}$. How are these two expressions related?

## *Activity*
## Defining $x^0$

1. Evaluate each expression.

   **a.** $\frac{10^5}{10^5}$ **b.** $\frac{2^3}{2^3}$ **c.** $\frac{5^4}{5^4}$

2. Use the Quotient-of-Powers Property to simplify each expression in Step 1.

**CHECKPOINT ✔** 3. Based on Steps 1 and 2, what are the numerical values of $10^0$, $2^0$, and $5^0$?

4. Why would you define $x^0$ to be 1 for any nonzero number $x$?

**Zero as an Exponent**

For any nonzero number $x$, $x^0 = 1$.

**EXAMPLE 2** **Simplify and write each expression with positive exponents only.**

**a.** $c^{-4} \cdot c^4$ **b.** $\frac{d^4f^3}{d^6f^3}$

**SOLUTION**

**a.** $c^{-4} \cdot c^4 = c^{-4+4} = c^0 = 1$

**b.** $\frac{d^4f^3}{d^6f^3} = (d^{4-6})(f^{3-3}) = d^{-2}f^0 = d^{-2} = \frac{1}{d^2}$

**TRY THIS** Simplify and write each expression with positive exponents only.

a. $\dfrac{b^6 \cdot b^{-2}}{b^4}$  b. $\dfrac{x^2y^3}{x^3y^2}$

Expressions containing variables with negative and zero exponents can be simplified by using the properties of exponents.

**EXAMPLE 3** **Simplify and write each expression with positive exponents only.**

a. $-3y^{-2}$  b. $\dfrac{4c^4d^5}{12c^4d^8}$  c. $\dfrac{m^2}{n^{-3}}$

**SOLUTION**

a. $-3y^{-2} = (-3)(y^{-2}) = \dfrac{-3}{y^2}$

b. $\dfrac{4c^4d^5}{12c^4d^8} = \dfrac{1}{3}(c^{4-4})(d^{5-8}) = \dfrac{1}{3}c^0d^{-3} = \dfrac{1}{3d^3}$

c. $\dfrac{m^2}{n^{-3}} = m^2n^{-(-3)} = m^2n^3$

**TRY THIS** Simplify and write each expression with positive exponents only.

a. $\dfrac{5a^2b^3}{25a^5b^2}$  b. $-4x^{-3}$

**CRITICAL THINKING** If $a$ and $b$ are nonzero numbers and $n$ is any integer, show that $\left(\frac{a}{b}\right)^{-n} = \left(\frac{b}{a}\right)^n$ is true.

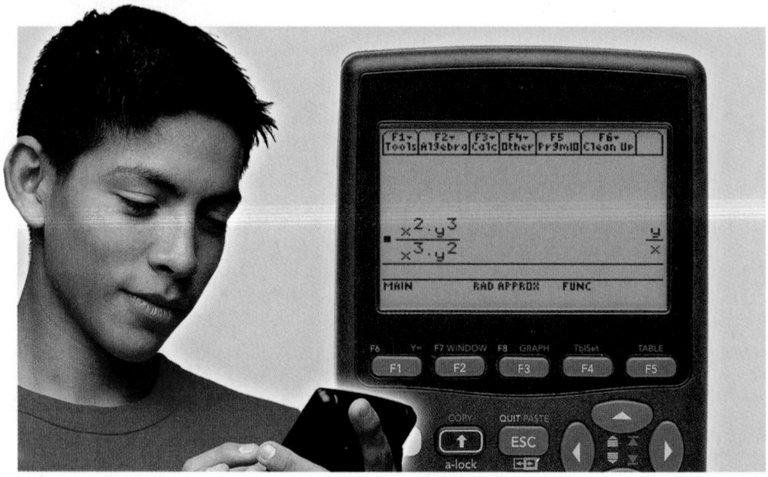


*Some calculators can perform algebraic manipulations such as simplifying rational expressions.*

**SUMMARY OF THE PROPERTIES OF EXPONENTS**

**Let $x$ and $y$ be any number and let $m$ and $n$ be integers.**

| Product of Powers | Quotient of Powers | Power of a Power | Power of a Product | Power of a Fraction |
|---|---|---|---|---|
| $x^m \cdot x^n = x^{m+n}$ | $\dfrac{x^m}{x^n} = x^{m-n}$ | $(x^m)^n = x^{mn}$ | $(xy)^m = x^m y^m$ | $\left(\dfrac{x}{y}\right)^n = \dfrac{x^n}{y^n}$ |

# Exercises

## Communicate

**1.** Explain how to show that $5^{-2} = \frac{1}{5^2}$ makes sense.

**2.** Why can $5^2 \cdot 4^{-3}$ not be simplified by using the properties of exponents?

**3.** Explain why $(3a)^{-2}$ does not equal $\frac{3}{a^2}$.

**4.** What is the meaning of a negative exponent?

**5.** What is the meaning of a zero exponent?

## Guided Skills Practice

**Evaluate each expression.** *(EXAMPLE 1)*

**6.** $7^{-5} \cdot 7^2$ **7.** $\frac{10^7}{10^{11}}$ **8.** $\frac{6^8}{6^2}$ **9.** $\frac{4^5}{4^7}$

**Simplify and write each expression with positive exponents only.** *(EXAMPLE 2)*

**10.** $\frac{a^5 \cdot a^{-5}}{a^8}$ **11.** $\frac{x^6y^7}{x^6y^6}$ **12.** $\frac{r^{-5}s^2}{r^{-3}s^2}$ **13.** $\frac{a^{-3}b^3}{a^3b^5}$

**Simplify and write each expression with positive exponents only.** *(EXAMPLE 3)*

**14.** $-5y^{-6}$ **15.** $\frac{3c^6d^3}{36c^4d^8}$ **16.** $\frac{j^3}{h^{-4}}$ **17.** $\frac{64a^7b^5}{25ab^{10}}$

## Practice and Apply

internet connect
**Homework Help Online**
Go To: **go.hrw.com**
Keyword:
**MA1 Homework Help**
for Exercises 18-33

**Evaluate each expression.**

**18.** $6^4 \cdot 6^{-2}$ **19.** $\frac{11^8}{11^{13}}$ **20.** $\frac{9^{12}}{9^3}$ **21.** $\frac{7^5}{7^{12}}$

**22.** $8^{-6} \cdot 8^3$ **23.** $\frac{13^8}{13^{15}}$ **24.** $\frac{5^3}{5^{10}}$ **25.** $3^4 \cdot 3^{-6}$

**Simplify and write each expression with positive exponents only.**

**26.** $\frac{f^7 \cdot f^{-7}}{f^2}$ **27.** $\frac{a^5b^8}{a^5b^3}$ **28.** $\frac{s^{-7}t^2}{s^{-3}t^3}$ **29.** $\frac{3x^4y^8}{27x^6y^4}$

**30.** $-6p^{-7}$ **31.** $\frac{4x^8y^4}{24x^3y^{11}}$ **32.** $\frac{r^5}{s^{-12}}$ **33.** $\frac{8a^4b^7c^{-4}}{3a^6b^{-6}c^{-4}}$

**Evaluate each expression.**

**34.** $-3^2$ **35.** $(-3)^2$ **36.** $3^{-2}$ **37.** $-(3^{-2})$

**38.** Copy the table and fill in the remaining entries by continuing the pattern.

| Decimal notation | 10,000 | 1000 | 100 | 10 | ? | ? | ? |
|---|---|---|---|---|---|---|---|
| Exponential notation | $10^4$ | $10^3$ | $10^2$ | $10^1$ | ? | ? | ? |

**39.** The corresponding numbers 10,000 and $10^4$ in the table on the previous page are equal. Are the other corresponding numbers that you wrote equal?

**Write each of the following without negative or zero exponents:**

**40.** $2^{-3}$ **41.** $10^{-5}$ **42.** $a^3b^{-2}$

**43.** $c^{-4}d^3$ **44.** $v^0w^2y^{-1}$ **45.** $(a^2b^{-7})^0$

**46.** $r^6r^{-2}$ **47.** $-t^{-1}t^{-2}$ **48.** $\frac{m^2}{m^{-3}}$

**49.** $\frac{2a^{-5}}{a^{-6}}$ **50.** $\frac{(2a^3)(10a^5)}{4a^{-1}}$ **51.** $\frac{b^{-2}b^4}{b^{-3}b^4}$

**TECHNOLOGY** Tell whether each expression is easier to evaluate on a calculator or mentally. Explain why you think so. Then evaluate each expression.

**52.** $\frac{2.56^7}{2.56^6}$ **53.** $\frac{2.56^6}{2.56^7}$

**54.** $0^7$ **55.** $7^0$

**56.** $(2.992 \times 9.554)^0$ **57.** $(19.43 \times 0)^{18}$

**58. TECHNOLOGY** What happens when you try to compute $2^{2^{36}}$ on a calculator?

**59. TECHNOLOGY** What happens when you try to compute $0^0$ on a calculator?

CONNECTION

CHALLENGE

**PROBABILITY** The probability, $p$, of correctly guessing a multiple-choice question with four answer choices is $\frac{1}{4}$, and the probability, $q$, of guessing incorrectly is $\frac{3}{4}$. Expressions for finding the probability of getting a certain number of answers correct on a test with 5 multiple-choice questions are shown in the chart below.

| No. correct | Probability | Value |
|---|---|---|
| 0 | $p^0q^5$ | $\left(\frac{1}{4}\right)^0\left(\frac{3}{4}\right)^5 = 0.237$ |
| 1 | $5p^1q^4$ | ? |
| 2 | $10p^2q^3$ | ? |
| 3 | $10p^3q^2$ | ? |
| 4 | $5p^4q^1$ | ? |
| 5 | $p^5q^0$ | ? |

**60.** Find the missing values in this chart, to the nearest thousandth.

**61.** Which number of correct answers is most likely?

**62.** Which number of correct answers is least likely?

**63.** Find the sum of all six probability values in the table. The sum of all probabilities should equal to 1. If the sum is not 1, explain the discrepancy.

## Look Back

**64.** What are the next three terms of the sequence 8, 11, 16, 23, 32, . . . ? ***(LESSON 1.1)***

**Solve each equation for $y$. How are the equations related?** ***(LESSON 3.6)***

**65.** $5(3y) = 3y - 24$

**66.** $aby = by - c$

**67.** What is the slope of a vertical line? How is it different from the slope of a horizontal line? ***(LESSON 5.2)***

**68.** Draw a number line showing the values of $x$ that solve the inequality $3x - 6 > -4x + 12$. Describe the solution set of this inequality in words. ***(LESSON 6.2)***

**Let $A = 3 \cdot 3 \cdot 3 \cdot 3 \cdot 3 \cdot x \cdot x \cdot x \cdot x \cdot x \cdot x \cdot y \cdot y \cdot y \cdot y$ and $B = 2 \cdot 2 \cdot 2 \cdot 2 \cdot 3 \cdot 3 \cdot 3 \cdot x \cdot x \cdot x \cdot x \cdot y$.**

**69.** Write $A$ and $B$ in exponential form. ***(LESSON 8.1)***

**70.** Write the product $A \cdot B$ in the original form and in exponential form. ***(LESSON 8.1)***

**internet connect** go.hrw.com

**Portfolio Extension**

Go To: **go.hrw.com**
Keyword:
**MA1 Centenarians**

## Look Beyond

*Pierre de Fermat*
*1601-1665*

**CULTURAL CONNECTION: EUROPE** Pierre de Fermat (1601–1665) was a French mathematician who helped develop number theory and probability. He studied numbers of the form $2^{2^n} + 1$, which are now called Fermat numbers.

**71. NUMBER THEORY** Fermat conjectured that each of the numbers of this form is prime. Though an unusually gifted mathematician, he was wrong in thinking that the sixth number, $2^{2^5} + 1$, is prime; in fact, it has a factor of 641. Use a calculator to find the other factor. Later research has suggested that *no* Fermat number with $n > 5$ is prime.

**PORTFOLIO ACTIVITY**

Include this activity in your portfolio.

In 1995 it was reported that there were 60,000 centenarians (people 100 or more years old) in the United States. The researcher predicted a 7% annual growth rate for this age group and claimed that the number of centenarians would reach 232,000 by the year 2015.

**1.** Verify the prediction numerically.

**2.** Write a function to model the problem. Use your calculator to determine when the number of centenarians will reach 1,000,000.

**WORKING ON THE CHAPTER PROJECT**

You should now be able to complete Activities 2 and 3 of the Chapter Project on page 416.

# EYEWITNESS MATH

## ALL MIXED UP?

You are playing a card game and it's your turn to deal. How many times do you shuffle the deck? Is that enough to be sure the cards are thoroughly mixed? Is there such a thing as shuffling too much? Read the article below to learn how two mathematicians, one of them a magician as well, found some answers that surprised shufflers around the world.

### In Shuffling Cards, 7 Is Winning Number

It takes just seven ordinary, imperfect shuffles to mix a deck of cards thoroughly, researchers have found. Fewer are not enough and more do not significantly improve the mixing.

The mathematical proof, discovered after studies of results from elaborate computer calculations and careful observation of card games, confirms the intuition of many gamblers, bridge enthusiasts and casual players that most shuffling is inadequate.

The finding has implications for everyone who plays cards and everyone, from casino operators to magicians, who has a stake in knowing whether a shuffle is random . . .

No one expected that the shuffling problem would have a simple answer, said Dr. Dave Bayer, a mathematician and computer scientist at Columbia who is co-author of the recent discovery. Other problems in statistics, like analyzing speech patterns to identify speakers, might be amenable to similar approaches, he said . . .

Dr. Persi Diaconis, a mathematician and statistician at Harvard University who is another author of the discovery, said the methods used are already helping mathematicians analyze problems in abstract mathematics that have nothing to do with shuffling or with any known real-world phenomena.

Dr. Diaconis, who is also a magician, has invented several card tricks and has been carefully watching casino dealers and casual card players shuffle for the past 20 years. The researchers studied the dovetail or riffle shuffle, in which the deck of cards is cut in half and the two halves are riffled together. They said this was the most commonly used method of shuffling cards. But, Dr. Diaconis said it produces a card order that is "far from random."...

Dr. Diaconis began working with Dr. Jim Reeds at Bell Laboratories and showed that a deck is perfectly mixed if it is shuffled between 5 and 20 times.

Next, Dr. Diaconis worked with Dr. Aldous and showed that it takes 5 to 12 shuffles to perfectly mix a deck. But, said Dr. Diaconis, "nobody in practice shuffles 12 times," adding, "We needed some new ideas."

In the meantime, he also worked on "perfect shuffles," those that exactly interlace the cards. Almost no one, except a magician can do perfect shuffles every time. But Dr. Diaconis showed several years ago that if a person actually does perfect shuffles, the cards would never be thoroughly mixed. He derived a mathematical proof showing that if a deck is perfectly shuffled eight times, the cards will be in the same order as they were before the shuffling.

To find out how many ordinary shuffles were necessary to mix a deck, Dr. Diaconis and Dr. Bayer watched players shuffle. He also watched Las Vegas dealers to see how perfectly they would interlace the cards they shuffled . . .

The researchers did extensive simulations of shuffling on a computer. To get the proof, the researchers looked at a lot of shuffles, guessed that the answer was seven, and finally proved it by finding an abstract way to describe what happens when cards are shuffled.

"When you take an honest description of something realistic and try to write it out in mathematics, usually it's a mess," Dr. Diaconis said. "We were lucky that the formula fit the real problem. That is just miraculous, somehow."

*By Gina Kolata*

# Getting Lost in the Shuffle

Even after a deck of cards is cut and shuffled, fragments of the original arrangement remain. In this example, the hearts are cut into two sequences: ace through 6, and 7 through king. Then the sequences are shuffled.

The cards are no longer ordered, but the sequence ace through 6 remains, with cards from the other fragment interspersed, also in sequence. The larger number of cards in play, the more shuffles are required to mix them.

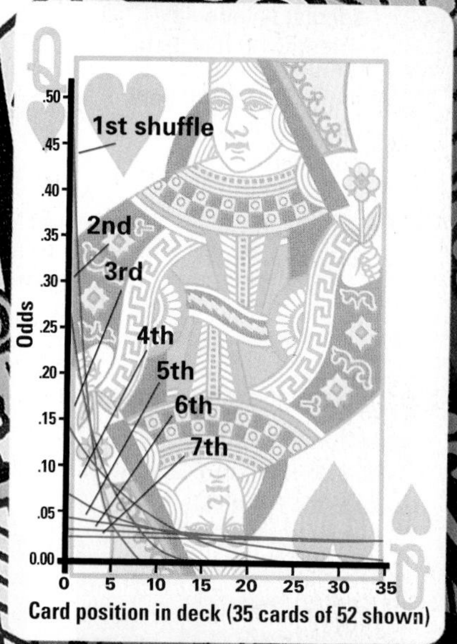


Curved lines show the odds that the first card in a deck will occupy any other position in the deck after shuffling from one to seven times. After one shuffle, for example, the first card is very likely to be one of the first few cards in the deck and very unlikely to be even five or six cards back. After four shuffles, it is still far more likely to be at the beginning of the deck than at the end. Only after seven shuffles does the card have about the same odds of being in any given position.

## Cooperative Learning

1. Based on the graph at right, after 4 shuffles, how many times more likely is it for the original 1st card to be in the 1st position than in the 35th position? Explain.

2. **a.** What is the difference between a perfect shuffle and an ordinary shuffle?
   **b.** Is the shuffle in the diagram above perfect? Explain.

3. Use a diagram or model to find out how many perfect shuffles it takes for a deck to return to its original order if the deck consists of
   **a.** $2^2$ cards. **b.** $2^3$ cards. **c.** $2^4$ cards.

4. **a.** How many perfect shuffles do you think it would take for a deck of 32 cards to return to its original order? Why?
   **b.** Use a diagram or model to check your prediction. What did you find?

5. Would a machine that always makes perfect shuffles be useful when you play a card game? Why or why not?

# 8.5 Scientific Notation

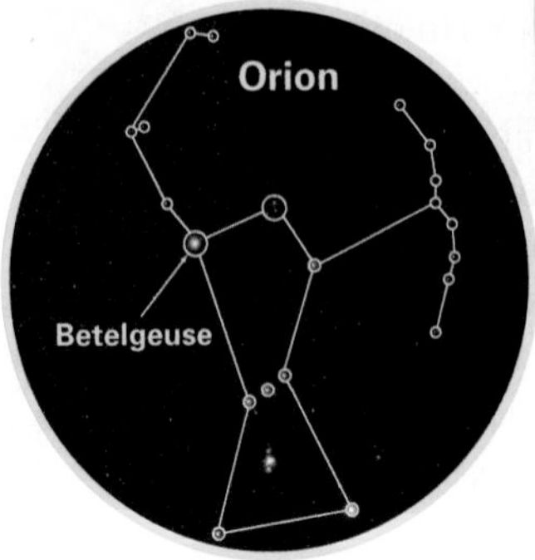


**Objectives**

- Recognize the need for special notation in scientific calculations.
- Perform computations involving scientific notation.

**Why** *The study of the universe involves measurements of very large and very small quantities. Scientific notation provides an easier way to do calculations with very large and very small numbers both in scientific fields and in other fields.*

**APPLICATION**
**ASTRONOMY**

Betelgeuse is one of the stars in the constellation Orion. The approximate distance between Earth and Betelgeuse is 6,000,000,000,000,000 miles. This number, with all its zeros, is hard to read and to use in calculations. A special way to express large and small numbers in a compact form is **scientific notation**. In scientific notation, the distance between Earth and Betelgeuse is written as $6 \times 10^{15}$ miles.

A number written in scientific notation is a product with two factors—a number from 1 to 10, including 1 but not including 10, and a power of 10. The first factor includes only significant digits (0s are included only if they lie between two other digits) and is often rounded to simplify the notation. Therefore, all numbers written in scientific notation are considered approximations.

Your knowledge of the powers of 10 will help you to write numbers in scientific notation.

## Activity
## Exploring Powers of 10

1. Complete the table for powers of 10.

| Value | 1000 | 100 | 10 | 1 | $\frac{1}{10}$ | $\frac{1}{100}$ | $\frac{1}{1000}$ |
|---|---|---|---|---|---|---|---|
| Power | $10^{?}$ | $10^{?}$ | $10^{?}$ | $10^{?}$ | $10^{?}$ | $10^{-2}$ | $10^{?}$ |

2. Copy and complete the table below.

| Decimal | Expanded form | Scientific notation |
|---|---|---|
| 1200 | $1.2 \times 1000$ | $1.2 \times 10^{?}$ |
| 120 | $1.2 \times 100$ | $1.2 \times 10^{?}$ |
| 12 | $1.2 \times 10$ | $1.2 \times 10^{?}$ |
| 1.2 | $1.2 \times 1$ | $1.2 \times 10^{?}$ |
| 0.12 | $1.2 \times \frac{1}{10}$ | $1.2 \times 10^{?}$ |
| 0.012 | $1.2 \times \frac{1}{100}$ | $1.2 \times 10^{?}$ |
| 0.0012 | $1.2 \times \frac{1}{1000}$ | $1.2 \times 10^{?}$ |

**CHECKPOINT ✔** 3. Explain how to write each of the following in scientific notation:
**a.** 0.000012 **b.** 120,000

**EXAMPLE 1** The distance between Earth and Proxima Centauri, the nearest star other than the sun, is about 24,000,000,000,000 miles—read as 24 trillion miles. **Write this number in scientific notation.**

**APPLICATION**
**ASTRONOMY**

**SOLUTION**

Place the decimal point between the 2 and the 4 (the first factor must be between 1 and 10), and count the number of digits it moved.

$$2\underset{\frown\frown\frown\frown\frown\frown\frown\frown\frown\frown\frown\frown\frown}{4,000,000,000,000.}$$

Because the decimal point was moved 13 places to the left, use $10^{13}$ as the other factor to maintain equality with the original number, 24 trillion.

$$24{,}000{,}000{,}000{,}000 = 2.4 \times 10^{13}$$

**TRY THIS** Write 875,000 in scientific notation.

**EXAMPLE 2** A typical X ray has a wavelength of about 0.000000000125 meter. **Write this number in scientific notation.**

APPLICATION
PHYSICS

SOLUTION

Since the first factor must be a number between 1 and 10, place the decimal after the 1. Because the decimal point moves 10 places to the right, use $10^{-10}$ as the other factor to maintain equality with the original number, 0.000000000125.

$$0.000000000125 = 1.25 \times 10^{-10}$$

**TRY THIS** Write 0.0000000402 in scientific notation.

**EXAMPLE 3** **Write each number in decimal notation.**

**a.** $3.67 \times 10^4$ **b.** $4.103 \times 10^{-5}$

SOLUTION

**a.** $10^4$ indicates that the decimal point should be moved 4 places to the right because 3.67 is being multiplied by 10,000.

$$3.6700 \qquad 3.67 \times 10^4 = 36700$$

**b.** $10^{-5}$ indicates that the decimal point should be moved 5 places to the left because 4.103 is being multiplied by 0.00001.

$$000004.103 \qquad 4.103 \times 10^{-5} = 0.00004103$$

**TRY THIS** Write each number in decimal notation.

**a.** $3.677 \times 10^{11}$ **b.** $4.36 \times 10^{-7}$

Answers to multiplication and division problems involving scientific notations should always be rounded to the least number of significant digits in the original problem.

**EXAMPLE 4** **Express the product and quotient in scientific notation.**

**a.** $(6 \times 10^{15}) \div 1000$ **b.** $(3 \times 10^3)(4 \times 10^{-5})$ **c.** $\frac{2.5 \times 10^6}{5 \times 10^2}$

SOLUTION

**a.** $(6 \times 10^{15}) \div 1000 = \frac{6 \times 10^{15}}{10^3} = 6 \times 10^{15-3} = 6 \times 10^{12}$

**b.**
$$\begin{aligned}(3 \times 10^3)(4 \times 10^{-5}) &= (3 \times 4)(10^3 \times 10^{-5}) \\ &= 12 \times 10^{-2} \\ &= (1.2 \times 10^1) \times 10^{-2} \\ &= 1.2 \times (10^1 \times 10^{-2}) \\ &= 1.2 \times 10^{-1}\end{aligned}$$

**c.**
$$\begin{aligned}\frac{2.5 \times 10^6}{5 \times 10^2} &= \left(\frac{2.5}{5}\right)\left(\frac{10^6}{10^2}\right) \\ &= 0.5 \times 10^4 \\ &= (5 \times 10^{-1}) \times 10^4 \\ &= 5 \times (10^{-1} \times 10^4) \\ &= 5 \times 10^3\end{aligned}$$

**TRY THIS** Express the product and quotient in scientific notation.

**a.** $3 \times 10^{28} \div 10^4$ **b.** $(2.5 \times 10^5)(1.56 \times 10^{-3})$ **c.** $\frac{3.6 \times 10^5}{6 \times 10^3}$

**EXAMPLE 5**

**TECHNOLOGY**
**SCIENTIFIC CALCULATOR**

The speed of light is about $2.98 \times 10^5$ kilometers per second. **If the sun is about $1.49 \times 10^8$ kilometers from Earth, how long does it take sunlight to reach Earth?**

**SOLUTION**

Use the [EE] key.

1.49 [EE] 8 [÷] 2.98 [EE] 5 [=] 500

Or use the [EXP] key.

1.49 [EXP] 8 [÷] 2.98 [EE] 5 [=] 500

Sunlight takes about 500 seconds to reach Earth.

Notice that on the calculator screen above, "E8" represents "$10^8$" and "E5" represents "$10^5$". This is just an easy way to display a number in scientific notation on a calculator or computer. The two forms are interchangeable.

# Exercises

## Communicate

1. Why is it helpful to be able to write a number such as 23,000,000,000 in scientific notation?
2. Explain what moving the decimal point 8 places to the left does to the value of a number. What power of 10 must you multiply by in order to compensate for this movement of the decimal point?
3. Explain what moving the decimal point 12 places to the right does to the value of a number. What power of 10 must you multiply by in order to compensate for this movement of the decimal point?

## Guided Skills Practice

**Write each number in scientific notation.** ***(EXAMPLE 1)***

**4.** 463,000,000 **5.** 25,000

**Write each number in scientific notation.** ***(EXAMPLE 2)***

**6.** 0.000597 **7.** 0.000000165 **8.** 0.00000457

**Write each number in decimal notation.** ***(EXAMPLE 3)***

**9.** $3.677 \times 10^8$ **10.** $5.28 \times 10^{-5}$

**Express each result in scientific notation.** ***(EXAMPLE 4)***

**11.** $6 \times 10^{14} \div 10^2$ **12.** $(3.5 \times 10^3)(2.67 \times 10^{-5})$ **13.** $\frac{2.4 \times 10^4}{6 \times 10^1}$

**14.** The speed of light is about $2.98 \times 10^5$ kilometers per second. If a certain star is about $2.36 \times 10^{14}$ kilometers from Earth, how long does it take the star's light to reach Earth? ***(EXAMPLE 5)***

## Practice and Apply

**internet connect**
**Homework Help Online**
Go To: **go.hrw.com**
Keyword:
**MA1 Homework Help** for Exercises 15-23, 32-50

**Write each number in scientific notation.**

**15.** 2,000,000 **16.** 8,000,000,000,000 **17.** 340,000

**18.** 58,000 **19.** 0.00008 **20.** 0.0000005

**21.** 0.000234 **22.** 0.000000082 **23.** 1,350,000,000

**Write each number in decimal notation.**

**24.** $3 \times 10^4$ **25.** $4 \times 10^8$ **26.** $6.7 \times 10^{10}$ **27.** $9.01 \times 10^5$

**28.** $4 \times 10^{-7}$ **29.** $5 \times 10^{-9}$ **30.** $8.8 \times 10^{-12}$ **31.** $7.2 \times 10^{-10}$

**Perform the following computations. Write your answers in scientific notation.**

**32.** $(2 \times 10^4)(3 \times 10^5)$ **33.** $(6 \times 10^8)(1 \times 10^6)$ **34.** $(8 \times 10^6)(2 \times 10^{10})$

**35.** $(9 \times 10^6)(7 \times 10^6)$ **36.** $(9 \times 10^6) + (7 \times 10^6)$ **37.** $(9 \times 10^6) - (7 \times 10^6)$

**38.** $(8.2 \times 10^6)(3.1 \times 10^6)$ **39.** $(1.9 \times 10^8)(2 \times 10^{10})$ **40.** $\frac{(2 \times 10^6)(9 \times 10^4)}{6 \times 10^4}$

**41.** $(2 \times 10^{10})(6 \times 10^7)(4 \times 10^{-8})$ **42.** $\frac{6 \times 10^5}{2 \times 10^2}$

**43.** $\frac{9 \times 10^8}{2 \times 10^4}$ **44.** $\frac{3 \times 10^4}{6 \times 10^7}$ **45.** $\frac{16 \times 10^5}{2 \times 10^{12}}$

**APPLICATIONS**

**SOCIAL STUDIES** The following are some large numbers that appeared in studies of different aspects of American culture. Write each number in scientific notation.

**46.** 125,000: the number of passengers bumped on United States airlines each year due to overbooking

**47.** $5 trillion: the total value of outstanding stock in the stock market

**48.** 5.4 million: the number of American businesses owned by women

**49.** 3.7 million: the number of square feet in the Pentagon, the world's largest office building

**ECONOMICS** The national debt for the United States in 1992 was listed as $4064.6 billion.

**50.** Write 4064.6 billion in scientific notation.

**51.** This number can be expressed as about $4 \times 10^{?}$.

**52.** Assuming that the population of the United States was about 250 million in 1992, how much was the debt per person?

**53.** Write your age at your last birthday in seconds. Use scientific notation.

**54. ASTRONOMY** The distance from Earth to the moon is about 248,000 miles. Write the distance in miles from Earth to the moon in scientific notation.

**APPLICATIONS**

**ASTRONOMY** The distance that light travels in one year, $5.87 \times 10^{12}$ miles, is called a light-year.

**55.** Write this number in decimal notation.

**56.** If the distance from Betelgeuse to Earth is $6 \times 10^{15}$ miles, how many years does it take light to travel from that star to Earth?

**57.** If the distance from Proxima Centauri to Earth is $2.14 \times 10^{13}$ miles, how many years does it take light to travel from Proxima Centauri to Earth?

**58. ASTRONOMY** According to Nobel Prize–winning physicist Leon Lederman, the universe is $10^{18}$ seconds old. How many years is this? Write your answer in words.

**internet connect**

**Activities Online**

Go To: **go.hrw.com**
Keyword:
**MA1 Scientific Notation**

**TECHNOLOGY** Write the following numbers in decimal notation:

**59.** 2.3 E 04 **60.** 5.6 E 03 **61.** 7.22 E –03

**62.** 1.01 E –04 **63.** –2.8 E 02 **64.** –9.303 E –04

**65.** What is the largest number that can be shown on your calculator without using scientific notation?

**66.** What is the largest number that can be shown on your calculator by using scientific notation?

**67.** What is the smallest positive number that can be shown on your calculator without using scientific notation?

**68.** What is the smallest positive number that can be shown on your calculator by using scientific notation?

## Look Back

**69.** Find the value of the expression. $3 + (-4) - (-9) - [-8 + 2(3 - 5)]$ ***(LESSONS 2.2 AND 2.3)***

**70.** Find the product. $(5)(-3)[-2(4)](-12)$ ***(LESSON 2.4)***

**71.** Simplify. $(6x + 4y - 8) - 5(-7x + y)$ ***(LESSON 2.6)***

**CONNECTION**

**72. COORDINATE GEOMETRY** Given points $A(3, 7)$ and $B(-3, 12)$, find the slope of the line joining the points. ***(LESSON 5.2)***

**73.** Graph the system of inequalities below and describe the solution region. ***(LESSON 7.5)***

$$\begin{cases} x - y > 5 \\ -3x + y \le 9 \end{cases}$$

## Look Beyond

**TECHNOLOGY** Use the constant feature of your calculator to write the given terms of the following sequences in decimal notation:

**74.** $0.9^1, 0.9^2, 0.9^3, 0.9^4, 0.9^5, \ldots$ **75.** $1.1^1, 1.1^2, 1.1^3, 1.1^4, 1.1^5, \ldots$

8.6

# Exponential Functions

**Why** *Exponential functions model problems of growth and decline in populations, financial trends, and scientific experiments.*

**Objectives**

- Understand exponential functions and how they are used.
- Recognize differences between graphs of exponential functions with different bases.

You have studied linear functions in which values in the range change by a fixed *amount*. The graphs of these functions form straight lines. **Exponential functions** apply to situations in which values in the range change instead by a fixed *rate*. The graphs of exponential functions do not form straight lines.

For example, $y = 2^x$ is an exponential function. The base, 2, is the number that is repeatedly multiplied. The exponent, $x$, represents the number of times that the base, 2, occurs in the multiplications. The function is graphed at right.

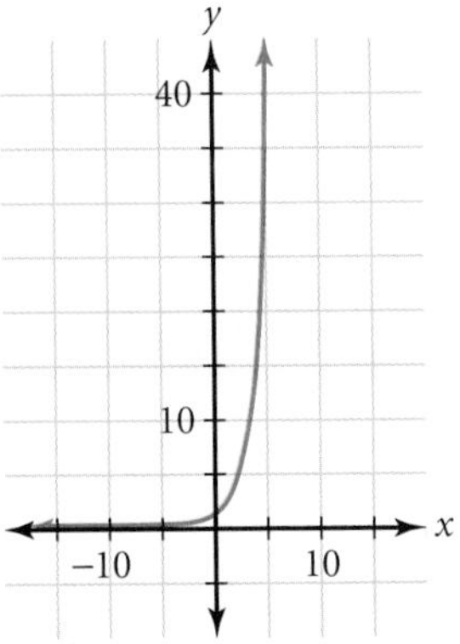


**APPLICATION**
**DEMOGRAPHICS**

According to the headline above, how many people are added to the population in one year? How many are added in one minute?

$$\begin{aligned} 1.52\% \times 5.85 \text{ billion} &= 0.0152 \times 5{,}850{,}000{,}000 \\ &= 88{,}920{,}000 \text{ people per year} \\ &\approx 243{,}616 \text{ people per day} \\ &\approx 10{,}151 \text{ people per hour} \\ &\approx 169 \text{ people per minute} \end{aligned}$$

In the time it took you to read this page, the population of the world probably increased by several dozen people.

**CRITICAL THINKING** How was the increase for one day obtained from the increase for one year?

*Activity*

## Population Growth

1. Complete the table below or build a spreadsheet to estimate the increase in the world's population from 1997 to 2007. Start with the 1997 population of 5.85 billion. Assuming that the population increases by 1.52% each year, the new population each year will be 101.52% of the previous year's population, so use 1.0152 as a multiplier.

| Year | Process | Exponential notation | Population (in billions) |
|---|---|---|---|
| 1997 | 5.85 | $5.85(1.0152)^0$ | 5.85 |
| 1998 | 5.85(1.0152) | $5.85(1.0152)^1$ | 5.94 |
| 1999 | 5.85(1.0152)(1.0152) | $5.85(1.0152)^2$ | 6.03 |
| 2000 | 5.85(1.0152)(1.0152)(1.0152) | $5.85(1.0152)^3$ | 6.12 |
| ... | ... | ... | ... |

2. At this rate, when will the population reach 20 billion?

CHECKPOINT ✔

3. Examine the pattern in the table in Step 1. What is the formula for finding the population if you know the original amount and rate of growth?

### General Growth Formula

Let $A$ be the amount after $t$ years at a yearly growth rate of $r$, where $r$ is expressed as a decimal. If the original amount is $P$, then the formula is as follows:

$$A = P(1 + r)^t$$

This formula works for population, compound interest, and many other situations in which growth takes place exponentially.

**EXAMPLE 1**

APPLICATION
DEMOGRAPHICS

TECHNOLOGY
SCIENTIFIC CALCULATOR

In 1992, India had an estimated population of about 886 million and was growing at a yearly rate of 1.9%. **At this rate, by how many people will the population increase in 10 years?**

**SOLUTION**

Use the general growth formula. If you have a calculator, use the $y^x$ or ^ key. You may also use a constant multiplier of 1.019 in your computation.

$$\begin{aligned} A &= P(1 + r)^t \\ &= 886(1.019)^{10} \\ &\approx 1069 \end{aligned}$$

The increase is 1069 – 886 = 183, or about 183 million people.

**TRY THIS** Suppose that the population of a country was 200,000,000 at the last census. If the population grows at an annual rate of 1.5%, by how many people will the population increase in the next 6 years?

**EXAMPLE 2** **Make a table of values and graph the function $y = 2^x$.**

**SOLUTION**

| $x$ | −3 | −2 | −1 | 0 | 1 | 2 | 3 |
|---|---|---|---|---|---|---|---|
| $2^x$ | 0.125 | 0.25 | 0.5 | 1 | 2 | 4 | 8 |

The graph of the function $y = 2^x$ shows the characteristics of an exponential function.

The graph remains above the $x$-axis for all values of $x$. When $x$ is 0, the function equals 1. After crossing the $y$-axis, the function increases rapidly.

Notice that, as with linear functions, each value in the domain, $x$, has a unique value in the range, $y$.

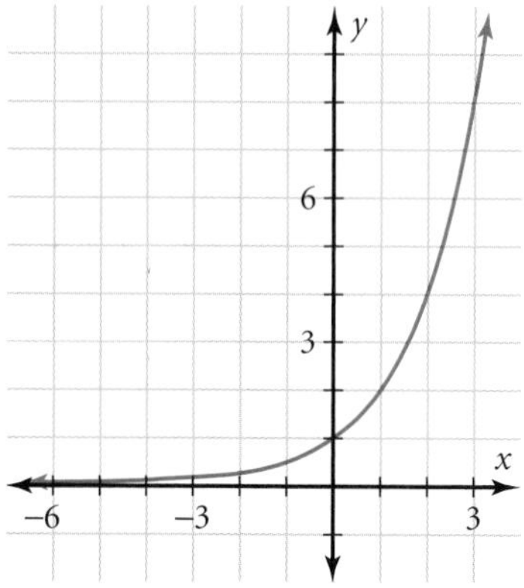


**EXAMPLE 3** **Graph $y = 3^x$, $y = 4^x$, and $y = \left(\frac{1}{2}\right)^x$ on the same axes. Explain how the graphs are similar and how they are different.**

**SOLUTION**

You can graph each function on a graphics calculator, or you can make a table of values and graph the functions on paper. To simplify your calculations, use integer values for $x$.

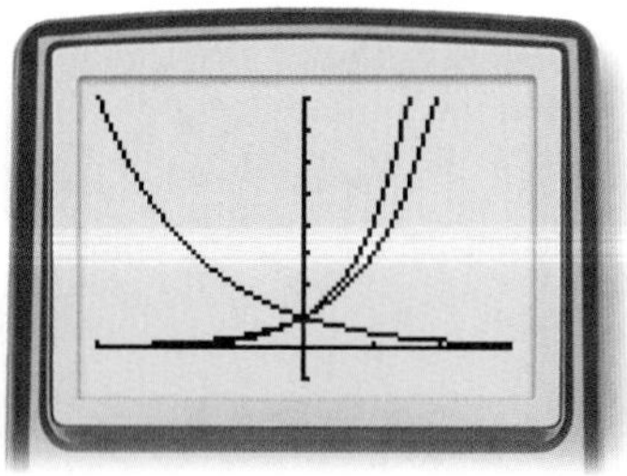

| $x$ | −3 | −2 | −1 | 0 | 1 | 2 | 3 |
|---|---|---|---|---|---|---|---|
| $3^x$ | 0.04 | 0.11 | 0.33 | 1 | 3 | 9 | 27 |
| $4^x$ | 0.0156 | 0.0625 | 0.25 | 1 | 4 | 16 | 64 |
| $\left(\frac{1}{2}\right)^x$ | 8 | 4 | 2 | 1 | $\frac{1}{2}$ | $\frac{1}{4}$ | $\frac{1}{8}$ |

The graphs of the first two functions are similar in that each rises from left to right. The third function does the opposite. Each graph crosses the $y$-axis at $y = 1$. Each gets closer and closer to the $x$-axis but never crosses it.

The first two functions are different in that the graph of $y = 4^x$ rises faster than the graph of $y = 3^x$ for values of $x$ greater than 0.

**CRITICAL THINKING** Consider the graphs of $y = 3^x$ and $y = 4^x$. For values of $x$ less than 0, which graph rises faster? Explain your reasoning.

# Exercises

## Communicate

**1.** Describe how an exponential function changes as the base increases.

**2.** Explain how population and money earning interest can both be represented by the formula $A = 1000(1.08)^t$.

**3.** Explain how you would determine whether the graph of $y = x^4$ represents an exponential function.

**4.** Explain how you would determine whether the graph of $y = 4^x$ represents an exponential function.

## Guided Skills Practice

**5.** **DEMOGRAPHICS** In 1995, the population of Zimbabwe was about 11 million and was growing at a rate of 1.9% per year. At this rate, by how many people will this population increase by 2005? ***(EXAMPLE 1)***

**6.** Make a table of values for $y = 5^x$ and $y = \left(\frac{1}{5}\right)^x$ and graph the functions on the same axes. Explain how the graphs are similar and how they are different. ***(EXAMPLES 2 AND 3)***

## Practice and Apply

**Graph each of the following:**

**7.** $y = 2^x$ **8.** $y = 4^x$ **9.** $y = 7^x$ **10.** $y = 12^x$

**11.** $y = 8^x$ **12.** $y = 10^x$ **13.** $y = 1^x$ **14.** $y = 6^x$

**15.** $y = 0.3^x$ **16.** $y = \left(\frac{1}{2}\right)^x$ **17.** $y = \left(\frac{1}{10}\right)^x$ **18.** $y = \left(\frac{1}{5}\right)^x$

**19.** $y = 0.4^x$ **20.** $y = \left(\frac{1}{4}\right)^x$ **21.** $y = \left(\frac{2}{3}\right)^x$ **22.** $y = 0.8^x$

**23.** $y = \left(\frac{3}{4}\right)^x$ **24.** $y = \left(\frac{4}{3}\right)^x$ **25.** $y = \left(\frac{2}{5}\right)^x$ **26.** $y = \left(\frac{5}{2}\right)^x$

**27.** $y = \left(\frac{1}{3}\right)^x$ **28.** $y = 3^x$ **29.** $y = \left(\frac{3}{7}\right)^x$ **30.** $y = \left(\frac{7}{3}\right)^x$

**APPLICATION**

**internet connect**

**Homework Help Online**
Go To: **go.hrw.com**
Keyword:
**MA1 Homework Help**
for Exercises 31-40

**DEMOGRAPHICS** The population of Spain was about 39 million in 1992 and was growing at a rate of about 0.3% per year.

**31.** What multiplier is used to find the new population each year?

**32.** Use this information to estimate the population in 1993.

**33.** Estimate the population in the year 2000.

**APPLICATION**

**DEMOGRAPHICS** The population of Italy was about 58 million in 1995 and was growing at a rate of about 0.1% per year. The population of France was about 57.8 million in 1995 and was growing at a rate of about 0.4% per year. Use this information to estimate the following populations:

**34.** Italy in 1996

**35.** Italy in 1998

**36.** Italy in 2000

**37.** France in 1996

**38.** France in 1998

**39.** France in 2000

**40.** In what year did the population of France surpass that of Italy?

## Look Back

**41.** Find the slope of the line containing the points (–1, 4) and (2, –3). *(LESSON 5.2)*

**APPLICATION**

**42.** **PHYSICS** A spring with a constant, *k*, of 17 is stretched a distance of 0.6 meters. Find the amount of force used to stretch the spring. *(LESSON 5.3)*

**internet connect**
**Activities Online**
Go To: **go.hrw.com**
Keyword:
**MA1 Snowflake**

**Graph each equation by using the slope and *y*-intercept.** *(LESSON 5.4)*

**43.** $y = -3x - 2$

**44.** $y = \frac{2}{5}x - 4$

**45.** $y = -\frac{1}{5}x + 7$

**Solve each system of equations.** *(LESSON 7.3)*

**46.** $\begin{cases} 2t + 3w = -4 \\ -2t + w = -1 \end{cases}$

**47.** $\begin{cases} 2x + 2y = 12 \\ -15x + 3y = 9 \end{cases}$

**48.** $\begin{cases} 5k + 4m = 10 \\ 5k - 2m = 10 \end{cases}$

**Simplify by using the order of operations and properties of exponents.** *(LESSONS 8.1 AND 8.2)*

**49.** $(x^4 \cdot x^2)^2 + (y^3 \cdot y^2)^3$

**50.** $(a^2b^4)^2(a^2b)^3$

## Look Beyond

*Maria Gaetana Agnesi*
*1718-1799*

**CULTURAL CONNECTION: EUROPE** Maria Gaetana Agnesi was born in Italy in 1718. She is considered to be the first woman in modern times to achieve a reputation for mathematical works. In her best known work, written in 1748, she discusses a curve that is now known as the Agnesi curve. This curve is represented by the function $f(x) = \frac{a^3}{x^2 + a^2}$. Let the domain of $f$ be {–3, –2, –1, 0, 1, 2, 3}.

**51.** Graph the equation for $a = 1$.

**52.** Graph the equation for $a = 5$.

**53.** How do these two graphs compare?

8.7

# Applications of Exponential Functions

**Objective**

- Use exponential functions to model applications that include growth and decay in different contexts.

**Why** *How much money should be invested to reach a target amount? How old is a wooden bowl found at an archaelogical site? These are two of the many questions that can be answered with the use of exponential functions.*

*Artifacts such as these can be dated by using the carbon-14 method.*

| Years | 0 | 5700 | 11,400 | 17,100 | 22,800 | 28,500 | 34,200 |
|---|---|---|---|---|---|---|---|
| Fraction of carbon-14 remaining | $1 = \left(\frac{1}{2}\right)^0$ | $\frac{1}{2} = \left(\frac{1}{2}\right)^1$ | $\frac{1}{4} = \left(\frac{1}{2}\right)^2$ | $\frac{1}{8} = \left(\frac{1}{2}\right)^3$ | $\frac{1}{16} = \left(\frac{1}{2}\right)^4$ | $\frac{1}{32} = \left(\frac{1}{2}\right)^5$ | $\frac{1}{64} = \left(\frac{1}{2}\right)^6$ |

**APPLICATION**
**ARCHAEOLOGY**

Exponential functions can be used to determine the age of fossils and archaeological finds. This method is based on the fact that the amount of radioactive carbon in an organic object decreases at a fixed rate over time.

*Carbon-14 dating* is a reliable method of determining the age of organic objects up to 40,000 years old. Carbon-14 is a radioactive form of carbon that has a half-life of 5700 years. This means that half of the carbon-14 is converted to nonradioactive carbon every 5700 years. For example, it can be concluded that an organic object with half as much carbon-14 as its living counterpart died 5700 years ago. Every 5700 years, half of the remaining amount of carbon-14 decays.

**EXAMPLE 1** **Use the table on the previous page to estimate the age of an object that has 10% of its original carbon-14 remaining.**

**SOLUTION**

Since 10%, or $\frac{1}{10}$, is between $\frac{1}{8}$ and $\frac{1}{16}$, the age is between 17,100 and 22,800 years. A reasonable estimate is about 19,000 years old.

**TRY THIS** Use the table to estimate the age of an object that has 22% of its original carbon-14 remaining.

Financial calculations are another common use of exponential functions.

**EXAMPLE 2** Suppose that you wish to have $1000 for a down payment on a car and that you want to buy the car when you graduate in 3 years. The current interest rate for a savings account is 5% compounded annually. **How much do you need to deposit in a savings account now in order to have enough to buy the car in 3 years?**

APPLICATION
PERSONAL FINANCE

**SOLUTION**

PROBLEM SOLVING

**Use a formula.** The general growth formula, $A = P(1 + r)^t$, can be used to find the value of an investment earning compound interest. Let $A$ represent the amount of money after $t$ years. Let $r$ represent the annual interest rate (expressed as a decimal). Substitute the values into the formula, and solve for $P$, the original amount.

$$A = P(1 + r)^t$$
$$1000 = P(1.05)^3$$
$$1000 = P(1.157625)$$
$$\frac{1000}{1.157625} = P$$
$$863.84 \approx P$$

You will need to deposit about $863.84.

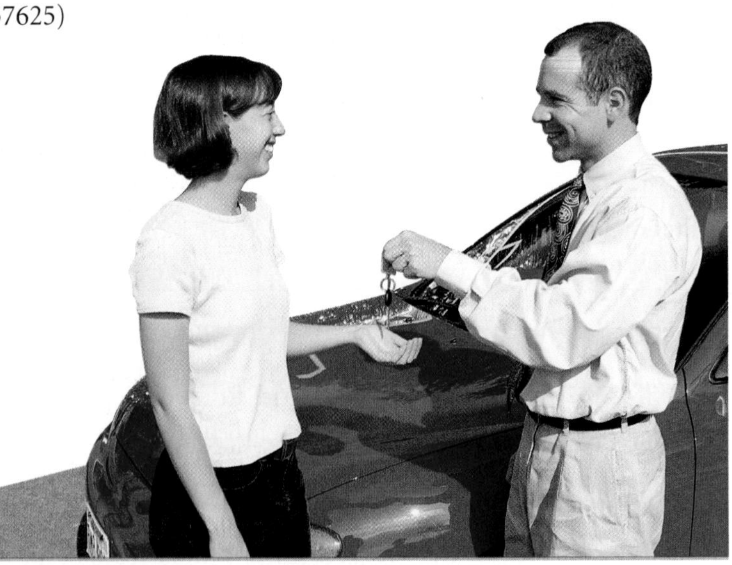

**TRY THIS** Suppose that Maria's parents want to have $3000 for tuition when she goes to college in 4 years. The current interest rate for a savings account is 4% compounded annually. How much do Maria's parents need to deposit in a savings account now in order to have $3000 in 4 years?

**EXAMPLE 3** The value of a sculpture has been growing at a rate of 5% per year for 4 years. **If the sculpture is worth $14,586.08 now, what was its value 4 years ago?**

**APPLICATION**
**INVESTMENTS**

**SOLUTION**

Use the general growth formula. Since you are trying to find the value 4 years ago, substitute −4 for *t*.

$$
\begin{aligned}
A &= P(1+r)^t \\
&= 14{,}586.08(1.05)^{-4} \\
&\approx 14{,}586.08(0.8227) \\
&\approx 12{,}000
\end{aligned}
$$

The value 4 years ago was about $12,000.

**TRY THIS** An art museum recently sold a painting for $4,500,000. The museum directors claim that its value has been growing at a rate of 7% per year for the last 50 years. What was the painting worth 50 years ago?

**EXAMPLE 4** Ginny has determined that she answers an individual step correctly in an algebra problem 90% of the time. She does 100 problems, each with 2 steps.

**a. In how many problems can she expect to answer both steps correctly?**
**b. Calculate how many she can expect to answer correctly if each problem has 3 steps.**

**SOLUTION**

**a.** In 100 problems she can expect to get the first step correct 90 times. Of these 90 times, she should get the second step correct 90% of the time. Because 90% of 90 is 81, she should get both steps correct 81 times out of 100. Notice that the answer is equal to $100(0.90)^2$.

**b.** From the solution in part **a**, you know that in 100 problems she should have the first 2 steps correct 81 times. She should also get the third step correct 90% of these times, or about 73 times. Notice that the answer is equal to $100(0.90)^3$.

**TRY THIS** Gavin gets a given step correct in an algebra problem 92% of the time. He does 100 problems, each with 3 steps. Calculate the number of problems in which he should expect to get all 3 steps correct.

In this lesson you have seen several examples of *exponential growth* and *exponential decay*. The general formula for exponential change gives the amount, $A$, when $P$ is the initial amount, $r$ is the rate of change expressed in decimal form, and $t$ is the time expressed in years:

$$A = P(1+r)^t$$

For **exponential growth**, $r$ is positive. For **exponential decay**, $r$ is negative.

**EXAMPLE 5** The population of a city in the United States was 152,494 in 2000 and is decreasing at a rate of about 1% per year. When will the city's population fall below 140,000? **Solve the problem by using the following methods:**

**a.** **a graph**
**b.** **a table**

**APPLICATION**
**DEMOGRAPHS**

## SOLUTION

Each year the population will be 100% – 1%, or 99%, of what it was the previous year, so the multiplier is 0.99. The population $x$ years after 2000 is given by $y = 152{,}494(0.99)^x$.

**PROBLEM SOLVING**

**TECHNOLOGY**
**GRAPHICS CALCULATOR**

**a.** **Draw a graph.** Graph **Y1**= $152{,}494(0.99)^x$ and **Y2**= 140,000, and find the point of intersection. The value of $y$ is 140,000 when $x$ is approximately 8.5. Therefore, the city's population will fall below 140,000 in about $8\frac{1}{2}$ years after 2000.

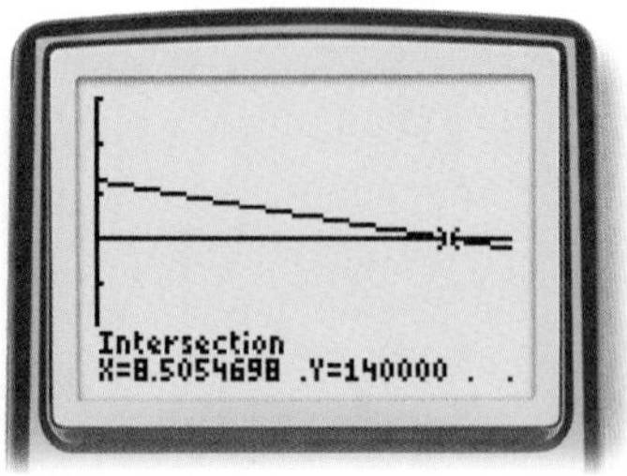


**b.** **Make a table.** The population decreases by a factor of 0.99 each year. Use the constant feature on a calculator to generate a table of values for the population. Continue the table until you reach the year in which the value falls below 140,000.

| No. of Years | Population |
|---|---|
| $x$ | $y = 152{,}494(0.99)^x$ |
| 0 | 152,494 |
| 1 | 150,969 |
| 2 | 149,459 |
| 3 | 147,965 |
| 4 | 146,485 |
| 5 | 145,020 |
| 6 | 143,570 |
| 7 | 142,134 |
| 8 | 140,713 |
| 9 | 139,306 |

According to the table, the population falls below 140,000 between 8 and 9 years after 2000.

**TRY THIS** The population of a small city in the United States was 16,225 in 2000 and was increasing at a rate of about 2% per year. When will the city's population be above 17,000? Solve the problem by using the following methods:

**a.** a graph
**b.** a table

## Exercises

### Communicate

1. Describe three real-world situations that can be modeled by exponential functions.
2. Explain the formula for exponential growth, $A = P(1 + r)^t$.
3. Explain the formula for exponential decay, $A = P(1 - r)^t$.
4. Explain why $r$ is subtracted from 1 in the formula for exponential decay.
5. Explain what is meant by the term *half-life* in relation to a radioactive substance.

### Guided Skills Practice

**APPLICATIONS**

*Henry Ford employed about 10 people in 1903, when the first Model A was made in Detroit.*

6. **ARCHAEOLOGY** Use the table on page 409 to estimate the age of an object that has 25% of its original carbon-14 remaining. ***(EXAMPLE 1)***
7. Suppose that David wants to have \$2500 saved at the end of 6 years. The current interest rate for a savings account is 3.5% compounded annually. How much will David need to invest in a savings account now in order to have \$2500 in 6 years? ***(EXAMPLE 2)***
8. **INVESTMENTS** Mike sold an antique car for \$32,000. The car's value has been growing at a rate of 7% per year for the last 25 years. What was the car valued at 25 years ago? ***(EXAMPLE 3)***
9. Sally gets a given step in an algebra problem correct 85% of the time. She does 100 problems, each with 5 steps. Calculate the number of problems in which she should expect to get all 5 steps correct. ***(EXAMPLE 4)***
10. **DEMOGRAPHICS** The population of a city in the United States was 160,525 in 1990 and was decreasing at a rate of about 1.5% per year. When will the city's population be below 155,000? ***(EXAMPLE 5)***

### Practice and Apply

**Each digit of a repeating decimal such as $0.33333\overline{3} = \frac{1}{3}$ can be described in terms of an exponential function. Evaluate each expression.**

**11.** $3\left(\frac{1}{10}\right)^1$ **12.** $3\left(\frac{1}{10}\right)^2$ **13.** $3\left(\frac{1}{10}\right)^3$ **14.** $3\left(\frac{1}{10}\right)^4$

**Using the table on page 409, estimate the age of an object that has the following amounts of its original carbon-14 remaining.**

**15.** 5% **16.** 28% **17.** 50% **18.** 100%

CONNECTION

19. **PROBABILITY** If the value of $x$ in $2^x$ was chosen from the list 1, 2, 3, 4, and 5 and each value came up twice in 10 trials, find the probability that $2^x$ is greater than 10.

20. **PROBABILITY** Repeat Exercise 19 if the value of $x$ was chosen from the list 1, 2, 3, . . ., 100 and each value came up 3 times in 300 trials.

APPLICATIONS

**ARCHAEOLOGY** Use the table on page 409 to estimate the age of an object that has

21. 25% of its original carbon-14 remaining.

22. 1% of its original carbon-14 remaining.

internet connect

**Homework Help Online**
Go To: **go.hrw.com**
Keyword:
**MA1 Homework Help** for Exercises 23, 32-35

23. **TECHNOLOGY** The population of a metropolitan area was 361,280 in 1998 and was decreasing at a rate of about 0.1% per year. Draw a graph that shows the population over the next 15 years.

**DEMOGRAPHICS** Use the graph from Exercise 23 to predict the population of the city in the following years:

24. 2000
25. 2003
26. 2005
27. 2010

**INVESTMENTS** An investment has been growing in value at a rate of 8% per year and now has a value of \$8200. Find the value of the investment

28. in 5 years.
29. in 10 years.
30. 5 years ago.
31. 10 years ago.

**INVESTMENTS** An investment is losing value at a rate of 2% per year and now has a value of \$94,000. Find the value of the investment

32. in 5 years.
33. in 10 years.
34. 5 years ago.
35. 10 years ago.

36. **ARCHAEOLOGY** Use the table on page 409 to estimate the age of an object that has 58% of its original carbon-14 remaining.

37. **INVESTMENTS** Suppose that Carl wants to have \$1800 saved at the end of 3 years. The current interest rate for a savings account is 6.5% compounded annually. How much will Carl need to invest in a savings account now in order to have \$1800 in 3 years?

*Paint used in cave paintings can be dated by using the carbon-14 method.*

38. **INVESTMENTS** Marla invested \$3000 in a new company. The value of her investment has been growing at a rate of 7.5% per year for the last 5 years. What is the value of her investment now? If the value of the investment continues to grow at this rate, what will be the value of her investment in 5 years?

39. **DEMOGRAPHICS** The population of a city is 100,000 and is increasing at a rate of about 2.5% per year. When will the city's population be above 125,000?

## Look Back

**40.** Suppose that the hourly wage at a job is \$18. Let $x$ represent the number of hours, and let $y$ represent the total wages earned. Write an equation of direct variation and graph it. *(LESSON 5.3)*

**Solve each system by using any method.** *(LESSONS 7.1, 7.2, AND 7.3)*

**41.** $\begin{cases} 4x + 2y = 15 \\ 6x - y = 3 \end{cases}$  **42.** $\begin{cases} 12x + 3y = 36 \\ 2x - 6y = -20 \end{cases}$  **43.** $\begin{cases} 1.6x + 0.7y = 2.8 \\ 8x - 5.6y = 1.7 \end{cases}$

**Use graphing to solve each system of inequalities.** *(LESSON 7.5)*

**44.** $\begin{cases} 2x + 8y < 10 \\ 4x - 3y \geq -4 \end{cases}$  **45.** $\begin{cases} 2b + 6d > -4 \\ b - 5d \geq 6 \end{cases}$  **46.** $\begin{cases} 3x - 2y \geq 5 \\ 3x + 2y \leq -4 \end{cases}$

**Graph each function.** *(LESSON 8.6)*

**47.** $y = 0.5^x$  **48.** $y = \left(\frac{1}{3}\right)^x$  **49.** $y = \left(\frac{1}{5}\right)^x$

## Look Beyond

**internet connect**
**Portfolio Extension**
Go To: **go.hrw.com**
Keyword:
**MA1 Compound Interest**

**50. TECHNOLOGY** A number that is often used to describe the growth of invested money is $e$, which is about 2.718. Find the $e^x$ key on your calculator. Write the first eight digits of $e$. (Use the $e^x$ key and use 1 for $x$.) Then draw the graph of $y = e^x$.

**PORTFOLIO ACTIVITY**

Include this activity in your portfolio.

The table below shows the growth of \$5000 earning a fixed rate of interest compounded annually.

1. Find the rate by looking at successive quotients.
2. Write a function for the relationship.
3. Calculate the amount through year 10.
4. When will the amount meet or exceed \$10,000? \$15,000?

| Year | Amount | Quotients |
|---|---|---|
| 0 | 5000 | |
| 1 | 5375 | $\frac{5375}{5000} = ?$ |
| 2 | 5778.13 | ? |
| 3 | 6211.48 | ? |
| ... | ... | ... |
| $x$ | $y$ | ? |

**WORKING ON THE CHAPTER PROJECT**

You should now be able to complete Activity 4 of the Chapter Project on page 416.

# THE WOLVES' DEN

An endangered species is one with so few individuals left that the species may become extinct. Wolves are among the species protected by the Endangered Species Act. Because the number of wolves in the United States was reduced and even eliminated in some areas, a program of re-introduction was started to rebuild the population under controlled circumstances. Re-introduction takes place when animals are set free to start a new population in areas where the species used to live. There are many factors that have to be considered when attempting such a large project with such far-reaching consequences. If the population grows too quickly, the resources available for the animals will be exhausted. However, if the population does not grow as quickly as expected, researchers must figure out what factors are causing the problem. The U.S. Fish and Wildlife Service is responsible for tracking and managing the re-introduction of wolves in various areas around the country.

## Activity 1

The table below shows the population of wolves living in a hypothetical wildlife preserve. In this simulation, 10 wolves were re-introduced to the area in 1989.

| Year | 1989 | 1990 | 1991 | 1992 | 1993 | 1994 | 1995 | 1996 | 1997 | 1998 |
|---|---|---|---|---|---|---|---|---|---|---|
| **Population** | 10 | 12 | 16 | 23 | 27 | 34 | 43 | 55 | 75 | 93 |

Predict the population of wolves in the wildlife preserve for the current year by looking at the table. Then make a scatter plot on your graphics calculator. Does the relationship among the data look linear? Write an equation for a line of best fit. Then use it to check your prediction.

You can use the idea of percent or growth rate to model population data with exponential functions. You can find the growth rate for each year by computing the ratio $\frac{\text{new population}}{\text{previous population}}$. The multiplier from 1989 to 1990 is $\frac{12}{10} = 1.2$, or 120%. This indicates a 20% percent increase over the first year.

## Activity 2

Use the model $y = 10(1.2)^x$ to build a table and graph and see if a 20% growth rate models the data well. Try other growth rates to refine your model. Observe the changes in the population for different rates. Make a prediction about the population over the next 10 years.

## Activity 3

Find data on the actual wolf population in a particular area or on another population that interests you. You can use your county, state, or city population or a wildlife population. Use a graphics calculator to find a reasonable line that fits the data and to make predictions about the population over next 10 years. Next, determine a reasonable rate of growth or decay. Use a graphics calculator to find a reasonable exponential function that fits the data (you may need to try several growth rates) and to make predictions about the population over next 10 years.

## Activity 4

Compare the two models from Activity 3 and explain which one you think is a better predictor.

# 8 Chapter Review and Assessment

## VOCABULARY



## Key Skills & Exercises

### LESSON 8.1

#### Key Skills

**Simplify the product of monomials containing exponents.**

Simplify $(4a^2b)(-3a^3b^2)$.

Multiply the constants. Rearrange the terms to group the like variables together.

$(4a^2b)(-3a^3b^2) = -12(a^2)(a^3)(b)(b^2)$

Simplify by using the Product-of-Powers Property. Add the exponents of like variables.

$-12(a^2)(a^3)(b)(b^2) = -12a^5b^3$

#### Exercises

**Simplify each product.**

1. $(5a^2)(4a^3)$
2. $(3m^2)(-2m^2n)$
3. $(7s^4t^3)(4t)$
4. $(-p^2)(6pq^2)$
5. $(4m^3)(3mn^2)$
6. $(2a^2bc^2)(5b^2c)$
7. $(-4xy)(-3x^2z)(2y^2z^2)$
8. $(2x^ay^b)(-3x^my)$

### LESSON 8.2

#### Key Skills

**Use the properties of exponents to simplify an expression.**

Simplify $(x^4)(x^3y^2)^2$.

Use the Power-of-a-Power Property and the Power-of-a-Product Property to simplify $(x^3y^2)^2$.

$$(x^4)(x^3y^2)^2 = (x^4)(x^6y^4)$$

Use the Product-of-Powers Property to simplify $(x^4)(x^6y^4)$.

$$(x^4)(x^6y^4) = x^{10}y^4$$

#### Exercises

**Simplify each expression.**

9. $(c^4)^2$
10. $(s^3)^a$
11. $(4x^2y)^3$
12. $2(r^2)^5$
13. $(-p^2)^3(p)$
14. $(-xy^2)^6$
15. $(-3mn^2)(2m^3)^3$
16. $(5b^3c^2)^4(2d^4)$

## LESSON 8.3

### Key Skills

**Use the properties of exponents to simplify expressions containing fractions.**

Simplify. **a.** $\frac{6x^2y^5}{2xy^2}$ **b.** $\left(\frac{4c^5}{5d^3}\right)^2$

**a.** Simplify by using the Quotient-of-Powers Property. First simplify any numerical coefficients. Then subtract the exponents in the denominator from the exponents in the numerator.

$$\frac{6x^2y^5}{2xy^2} = 3xy^3$$

**b.** Simplify by using the Power-of-a-Fraction Property.

$$\left(\frac{4c^5}{5d^3}\right)^2 = \frac{(4c^5)^2}{(5d^3)^2} = \frac{16c^{10}}{25d^6}$$

### Exercises

**Simplify each expression.**

**17.** $\left(\frac{x}{y}\right)^4$

**18.** $\left(\frac{2a^2}{b}\right)^3$

**19.** $\left(\frac{4w^2}{6w^4}\right)^3$

**20.** $\left(\frac{12fg^2}{6g}\right)^2$

**21.** $\left(\frac{y^4}{5xy^2}\right)^3$

**22.** $\left(\frac{7p^2q}{21q}\right)^2$

**23.** $\left[\frac{-15(ab)^2}{5a^2b}\right]^5$

**24.** $\left(\frac{st^2}{s^4}\right)^r$

## LESSON 8.4

### Key Skills

**Simplify expressions containing negative and zero exponents.**

Simplify $\frac{-8a^3c^5}{2a^4b^3c^2}$ and write the expression with positive exponents only.

Use the properties of exponents and the definition of negative exponents.

$$\frac{-8a^3c^5}{2a^4b^3c^2} = -4(a^{3-4})(b^{-3})(c^{5-2}) = -4a^{-1}b^{-3}c^3 = \frac{-4c^3}{ab^3}$$

### Exercises

**Simplify each expression. Write the expression with positive exponents only.**

**25.** $3^{-2}$

**26.** $a^2b^{-3}$

**27.** $a^0b^{-2}c^3$

**28.** $5x^{-3}y^2$

**29.** $\frac{12p^{-3}}{4p}$

**30.** $\left(\frac{4a^{-2}}{3b^{-3}}\right)^3$

**31.** $\frac{(15t^{-4})(t^3)}{-3t^{-2}}$

**32.** $(8b^2)(2b^{-8})(2b^6)$

## LESSON 8.5

### Key Skills

**Perform computations involving scientific notation.**

Write the product $(3 \times 10^3)(4 \times 10^5)$ in scientific notation.

Rearrange the factors: $(3 \times 4)(10^3 \times 10^5)$

Multiply: $12 \times 10^8$

Rewrite the first factor as a decimal between 1 and 10: $1.2 \times 10^1$

Complete the simplification: $12 \times 10^8 = (1.2 \times 10^1)(10^8) = 1.2 \times 10^9$

### Exercises

**Write each number in scientific notation.**

**33.** 5,900,000

**34.** 0.0000075

**Perform each computation. Write the answer in scientific notation.**

**35.** $(3 \times 10^2)(5 \times 10^5)$

**36.** $(2.1 \times 10^5)(3 \times 10^{-3})$

**37.** $(8 \times 10^2) + (2 \times 10^2)$

**38.** $(9 \times 10^5) - (3 \times 10^5)$

**39.** $\frac{9 \times 10^7}{3 \times 10^4}$

**40.** $\frac{8 \times 10^4}{2 \times 10^{-2}}$

LESSON 8.6

**Key Skills**

**Understand and graph exponential functions.**

The current population of a city is 120,000. It is estimated that the population of the city will grow 1.5% annually. What will the city's population be in 10 years?

Use the general growth formula, $A = P(1 + r)^t$, where $P$ is the original amount and $A$ is the amount after $t$ years at a yearly growth rate of $r$, expressed as a decimal.

$A = 120{,}000(1.015)^{10} = 139{,}264.899$

In 10 years, the population of the city will be about 139,265.

Graph the exponential function $y = 4^x$.

You can make a table or use a graphics calculator. Notice that the graph remains above the $x$-axis for all values of $x$.

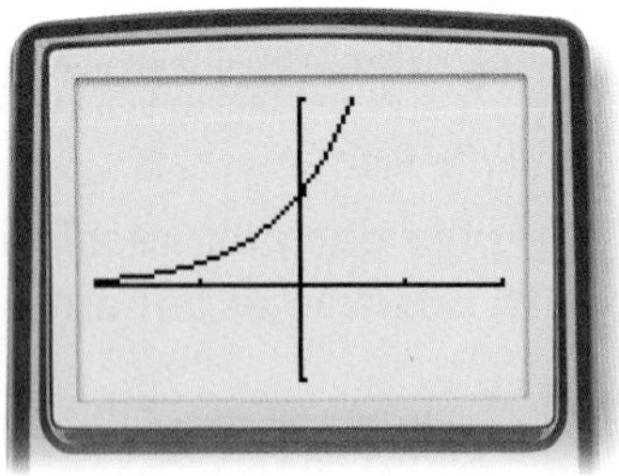

**Exercises**

**The population of New Zealand was about 3.3 million in 1991 and was growing at the rate of about 0.8% per year. Estimate the population**

**41.** in the year 2000.

**42.** in the year 2003.

**43.** in the year 2010.

**44.** in the year 2020.

**Graph each exponential function.**

**45.** $y = 2^x$

**46.** $y = 2^x + 3$

**47.** $y = 2^x + 7$

**48.** $y = 2^x - 13$

**49.** $y = 10^x$

LESSON 8.7

**Key Skills**

**Use exponential functions to model applications.**

An investment has been losing money at a rate of 3% per year and now has a value of $6500. What was the value of the investment 5 years ago?

Since you need to find the value 5 years ago, use the general growth formula with $t = -5$.

$A = P(1 + r)^t = 6500(1 - 0.03)^{-5} \approx \$7569.28$

**Exercises**

**An investment is growing at a rate of 5.5% per year and now has a value of $6300. Find the value of the investment**

**50.** in 5 years.

**51.** 5 years ago.

**52.** in 10 years.

**53.** 10 years ago.

## Applications

**54.** **ASTRONOMY** The sun has a diameter of about 864,000 miles. Write this number in scientific notation.

**55.** **FINANCE** The value of a famous painting is increasing at the rate of 6% per year and has a value of $800,000 in the year 2000. Find the value of the painting in the year 2005.

# 8 Chapter Test

**Find the value of each expression.**

1. $3^4$
2. $7^3$
3. $10^5$

**Simplify each product.**

4. $(3p^2)(2p^3)$
5. $(-2r^3)(7r^2s)$
6. $(b^2c^4)(3b^4c^2)$
7. $(4xy^2)(2x^3y)(x^2y^3)$

**Simplify each expression.**

8. $(g^2)^3$
9. $(2x^2)^3$
10. $(a^5)^b$
11. $(r^2s)^3(rs^2)$
12. $(-3pq^4)^2(2p^2q)$
13. $(5m^2n^3)^4(2m^3n^2)$
14. GEOMETRY The volume of a rectangular box is given by the formula $V = lwh$, where $l =$ length, $w =$ width, and $h =$ height. What happens to the volume if the length and height are doubled and the width is tripled?

**Simplify each expression.**

15. $\left(\frac{p}{q}\right)^3$
16. $\left(\frac{5a^3}{3b^6}\right)^3$
17. $\left(\frac{-3x^4yz^2}{9x^2z}\right)$
18. $\left(\frac{3a^4b}{a^2}\right)^c$
19. GEOMETRY The area of a circle is given by the formula $A = \pi r^2$, and the circumference of the circle is given by $C = 2\pi r$. Find the ratio of the area to the circumference. What happens to this ratio as the radius is quadrupled?

**Evaluate each expression.**

20. $5^4 \cdot 5^{-3}$
21. $\frac{6^5}{6^2}$
22. $\frac{9^2}{9^5}$

**Simplify and write each expression with positive exponents only.**

23. $5^{-3}$
24. $12x^{-5}$
25. $\frac{25h^{-2}}{5h^2}$
26. $\frac{(24a^{-2})(3a^5)}{9a^{-7}}$
27. $(5t^{-6})(2t^4)(3t^{-3})$

**Write each number in scientific notation.**

28. 520,000,000
29. 0.0000037

**Perform each computation. Write the answer in scientific notation.**

30. $(9.8 \times 10^3)(2.5 \times 10^7)$
31. $\frac{7.29 \times 10^{12}}{2.7 \times 10^{-3}}$
32. $(2.4 \times 10^2) + (5.7 \times 10^2)$

**Graph each exponential function.**

33. $y = 3^x - 4$
34. $y = \left(\frac{1}{3}\right)^x + 1$

DEMOGRAPHICS The population of Petosky, MI, was 6100 in 1990 and growing at the rate of 3.2% per year. Estimate the population in each year?

35. 1995
36. 2007
37. BANKING An investment is growing at a rate of 6.2% per year and now has a value of \$2400. Find its value in 3 years.
38. BANKING How much must be invested in a savings account paying an interest rate of 5.4% compounded annually to have \$2750 in 5 years?
39. DEMOGRAPHICS The population of a metropolitan area in 1997 was 503,247 and growing at a rate of 7.1% per year. When will the population be 750,000?

# College Entrance Exam Practice



**QUANTITATIVE COMPARISON** For Questions 1–5 write
A if the quantity in Column A is greater than the quantity in Column B;
B if the quantity in Column B is greater than the quantity in Column A;
C if the two quantities are equal; or
D if the relationship cannot be determined from the information given.

| | Column A | Column B | Answers |
|---|---|---|---|
| **1.** | $6 + 5^2 \cdot 2$ | $(3 + 7)^2 \div 2$ | Ⓐ Ⓑ Ⓒ Ⓓ [Lesson 1.3] |
| **2.** | The solution to $\frac{x}{5} = 1.2$ | The solution to $\frac{x}{10} = 2.4$ | Ⓐ Ⓑ Ⓒ Ⓓ [Lesson 3.2] |
| **3.** | $(2^4)^3$ | $2^3 \cdot 2^4$ | Ⓐ Ⓑ Ⓒ Ⓓ [Lesson 8.2] |
| **4.** | The slope of the line passing through (1, 3) and (4, 5) | The slope of the line passing through (–1, 3) and (4, 5) | Ⓐ Ⓑ Ⓒ Ⓓ [Lesson 5.2] |
| **5.** | 110% of 85 | 85% of 110 | Ⓐ Ⓑ Ⓒ Ⓓ [Lesson 4.2] |

**6.** Let $x = 3$ and $y = -4$. Simplify $x - 8 - y$. ***(LESSON 2.3)***
**a.** 15 **b.** –9 **c.** 1 **d.** –1

**7.** Solve the system. ***(LESSON 7.2)***
$$\begin{cases} 2x - 4 = -12 \\ y = x + 3 \end{cases}$$
**a.** (3, 0) **b.** (0, 3) **c.** (–3, 0) **d.** (–4, –1)

**8.** Solve the equation $5x + 5 = 10$. ***(LESSON 3.3)***
**a.** –1 **b.** 1 **c.** 3 **d.** –3

**9.** Which equation models the following situation: If each CD costs $12, how many CDs can you buy for $45? ***(LESSON 1.2)***
**a.** $x = 12 + 45$ **b.** $x = \frac{45}{12}$
**c.** $x = (45)(12)$ **d.** $x = 45 - 12$

**10.** Solve the system for $x$. ***(LESSON 7.2)***
$$\begin{cases} y = 1 - 4x \\ 3y = 2x + 2 \end{cases}$$
**a.** 14 **b.** $-\frac{1}{14}$ **c.** –14 **d.** $\frac{1}{14}$

**11.** Find the median of the data set: 45, 36, 17, 24, and 33. ***(LESSON 4.4)***
**a.** 33 **b.** 31 **c.** 28 **d.** 36

**12.** Find the slope of the line passing through $A(2, 9)$ and $B(4, -3)$. ***(LESSON 5.2)***
**a.** 6 **b.** $-\frac{1}{6}$ **c.** $\frac{1}{6}$ **d.** –6

**13.** Evaluate $y = 3^x$ for $x = 3$. ***(LESSON 8.6)***
**a.** 9 **b.** 3 **c.** 27 **d.** 1

**14.** Evaluate $3x^2 + 2y$ for $x = 3$ and $y = 2$. ***(LESSON 1.3)***
**a.** 24 **b.** 77 **c.** 31 **d.** 14

15. Solve $6x - 3 \le 12$. ***(LESSON 6.2)***
**a.** $x \le 1.5$ **b.** $x \le 2.5$ **c.** $x \le -2.5$ **d.** $x < 2.5$

16. Simplify: $\frac{15x - 6}{3}$. ***(LESSON 2.7)***

17. Solve $x - \frac{2}{3} = 1\frac{1}{3}$. ***(LESSON 3.1)***

18. Find the range for the data set: 45, 36, 17, 24, and 33. ***(LESSON 4.4)***

19. In the equation $y = 5x + 3$, which variable is the independent variable? ***(LESSON 1.5)***

20. Find the mean of the data set: 28, 36, 34, 29, and 31. ***(LESSON 4.4)***

21. Solve the system. $\begin{cases} x + y = 5 \\ y + 1 = x \end{cases}$ ***(LESSON 7.3)***

22. Find the next two terms in the sequence: 5, 6, 8, 11, . . . ***(LESSON 1.1)***

23. Write the equation of the line passing through $(0, 3)$ and $(-1, 0)$. ***(LESSON 5.5)***

24. Simplify $\frac{5x^2 - 10}{-5}$. ***(LESSON 2.7)***

25. Simplify $(3x + 4) - (8x - 12)$. ***(LESSON 2.6)***

26. Find the perimeter of a rectangle if $l = 3a + 2b$ and $w = 4a - 3b$. ***(LESSON 2.6)***

27. Solve for $x$ and graph the solution on a number line: $x - 7 \ge -4$. ***(LESSON 6.1)***

28. Solve for $a$ and graph the solution on a number line: $3a - 5 \le 16$. ***(LESSON 6.2)***

29. Simplify $\frac{16x + 4}{-2}$. ***(LESSON 2.7)***

30. Solve $|x - 4| = 7$. ***(LESSON 6.5)***

31. Solve $|a + 4| \le 12$. ***(LESSON 6.5)***

32. Write an equation for the line passing through the points $(3, 5)$ and $(5, 13)$. ***(LESSON 5.5)***

33. Write the equation for the line that passes through the point $(5, 3)$ and has a slope of $-2$. ***(LESSON 5.5)***

34. A new car being sold at $30,000 is advertised as "25% off." Find the original price of the car. ***(LESSON 4.2)***

35. Solve the system of equations by elimination, and check your solution. ***(LESSON 7.3)***

$$\begin{cases} 4x - 8y = 12 \\ 6x + 10y = -10 \end{cases}$$

36. Solve the system by graphing. ***(LESSON 7.5)***

$$\begin{cases} 2x - 4y \le 4 \\ -3x - 6y > 6 \end{cases}$$

37. The sum of the digits of a positive two-digit number is 10. If twice the first digit is 2 more than the second digit, what is the two-digit number? ***(LESSON 7.6)***

38. Simplify: $\frac{24a^7b^3}{48a^9}$. ***(LESSON 8.3)***

**FREE-RESPONSE GRID**

**The following questions may be answered by using a free-response grid such as that commonly used by standardized test services.**

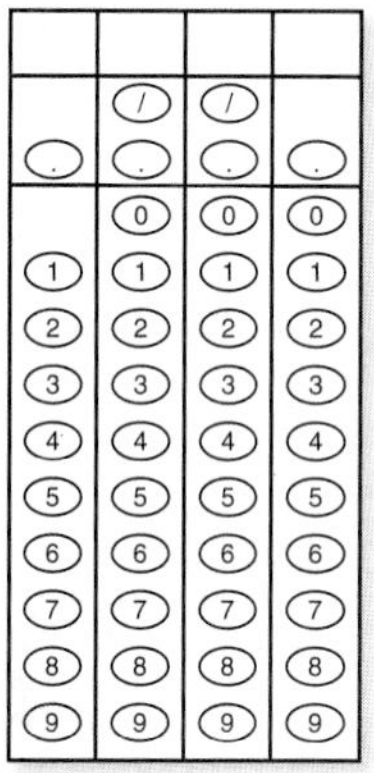

39. Evaluate $\frac{x + y}{2} + \frac{x - y}{2}$ for $x = 8$ and $y = 2$. ***(LESSON 1.3)***

40. Evaluate $3x^2 + 4y^2$ for $x = 0.5$ and $y = 1$. ***(LESSON 1.3)***

41. Solve $\frac{18}{a} = \frac{21}{14}$. ***(LESSON 4.1)***

42. Simplify $\frac{2^3 \cdot 2^5}{2^4}$. ***(LESSON 8.3)***

43. A weight of 6 pounds stretches a spring a distance of 12 inches. Find the constant, $k$, for the spring. ***(LESSON 5.3)***

44. Find the sum of the integers from 1 to 55. ***(LESSON 1.1)***

45. Solve $4a + 6 = -2a + 12$ for $a$. ***(LESSON 3.4)***

46. Solve for $x$: $\frac{x}{12} = \frac{9}{36}$. ***(LESSON 4.1)***

47. Find 30% of 180. ***(LESSON 4.2)***

9

# Polynomials and Factoring

POLYNOMIALS ARE USEFUL FOR REPRESENTING area and volume functions, problems involving gravity, consumer problems, and many other types of real-world situations.

## Lessons

*The volume and surface area of a hot-air balloon can be modeled using polynomial functions.*

CHAPTER PROJECT
PORTFOLIO ACTIVITIES

## About the Chapter Project

You have already seen how algebra makes it possible to use functions in many different types of situations. In the Chapter Project, *Powers, Pascal, and Probability,* on page 470, you will explore connections between Pascal's triangle and polynomial functions that can be used to simplify calculations. After completing the Chapter Project, you will be able to do the following:

- Apply Pascal's triangle to a probability problem.

## About the Portfolio Activities

Throughout the chapter, you will be given opportunities to complete Portfolio Activities that are designed to support your work on the Chapter Project.

- Forming Pascal's triangle is the topic of the Portfolio Activity on page 437.
- Modeling triangular numbers with polynomial functions is the topic of the Portfolio Activity on page 447.
- Modeling number sequences with polynomial functions is the topic of the Portfolio Activity on page 457.
- Modeling rectangular numbers with polynomial functions is the topic of the Portfolio Activity on page 470.

9.1

# Adding and Subtracting Polynomials

*The volume of the box on the left is $(x + 2)(x + 2)(x + 2)$, or $x^3 + 6x^2 + 12x + 8$.*

**Objective**

- Add and subtract polynomials.

**Why** *The arc of a football, the distance traveled by an accelerating car, and volume of an object can be modeled by polynomials.*

APPLICATION
PACKAGING

A **polynomial** is a monomial, or a sum or difference of monomials. The expression for the volume of the box above, $x^3 + 6x^2 + 12x + 8$, is an example of a polynomial in one variable, $x$.

The **degree** of a polynomial in one variable is determined by the exponent with the greatest value within the polynomial. The degree of $9 - 4x^2$ is 2. The degree of the polynomial above is 3.

The terms of a polynomial may appear in any order. However, in **standard form**, the terms of a polynomial are ordered from left to right in descending order, which means from the greatest exponent to the least.

EXAMPLE 1 **Write $9 + x - 4x^2$ in standard form.**

**SOLUTION**

Reorder the terms of the polynomial in descending order according to the exponents of the variables.

$9 + x - 4x^2$
$9 + x + (-4x^2)$
$-4x^2 + x + 9$

Some polynomial expressions have special names that are determined either by their degree or by the number of terms, as illustrated in the table.

| Polynomial | # of terms | Name by # of terms | Degree | Name by degree |
|---|---|---|---|---|
| 12 | 1 | monomial | 0 | constant |
| $8x$ | 1 | monomial | 1 | linear |
| $4x^2 + 3$ | 2 | binomial | 2 | quadratic |
| $5x^3 + x^2$ | 2 | binomial | 3 | cubic |
| $3x^2 - 4x + 6$ | 3 | trinomial | 2 | quadratic |
| $3x^4 - 4x^3 + 6x^2 - 7$ | 4 | polynomial | 4 | quartic |

*Activity*

## Polynomial Addition and Subtraction

**You will need:** algebra tiles

1. Use algebra tiles to model this sum of binomials: $(2x + 2) + (3x - 3)$.

$(2x + 2) \quad + \quad (3x - 3)$

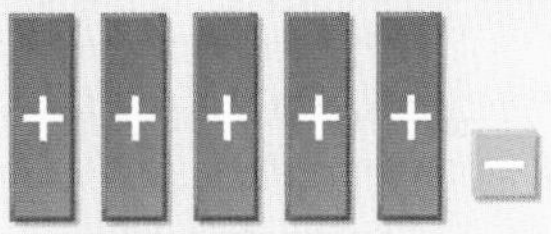

$(5x - 1)$

2. Check your operations with polynomials by substituting several values for $x$ to verify the following:

$$(2x + 2) + (3x - 3) = 5x - 1$$

3. Use algebra tiles to simplify the polynomial expression $(2x + 2) - (3x - 1)$, and complete the following:

$$(2x + 2) - (3x - 1) = ?$$

Check your operations with polynomials by substituting –3, 4, and 10 for $x$.

CHECKPOINT ✔ 4. Explain how to simplify the polynomial addition expression $(2x^2 + x) + (5x^2 - x)$.

Explain how to simplify the polynomial subtraction expression $(2x^2 + x) - (5x^2 - x)$.

## Adding Polynomials

Polynomials can be added in vertical or horizontal form. In vertical form, align the like terms and add.

$$\begin{array}{r} x^2 + \phantom{3}x + 1 \\ +\ 2x^2 + 3x + 2 \\ \hline 3x^2 + 4x + 3 \end{array}$$

In horizontal form, regroup like terms and add.

$$(x^2 + x + 1) + (2x^2 + 3x + 2)$$
$$(x^2 + 2x^2) + (x + 3x) + (1 + 2)$$
$$3x^2 + 4x + 3$$

**EXAMPLE 2** **Find the sum $(2x^2 - 3x + 5) + (4x^2 + 7x - 2)$ by using**

**a.** the vertical form.
**b.** the horizontal form.

**a.** Align like terms and add.

$$\begin{array}{r} 2x^2 - 3x + 5 \\ +\ 4x^2 + 7x - 2 \\ \hline 6x^2 + 4x + 3 \end{array}$$

**b.** Regroup like terms and add.

$$(2x^2 - 3x + 5) + (4x^2 + 7x - 2)$$
$$(2x^2 + 4x^2) + (-3x + 7x) + (5 - 2)$$
$$6x^2 + 4x + 3$$

**TRY THIS** Find the sum $(3x^2 + 5x) + (4 - 6x - 2x^2)$ by using the vertical form and the horizontal form.

**EXAMPLE 3** **Write a polynomial expression for the perimeter of each polygon.**

CONNECTION
GEOMETRY

**a.**

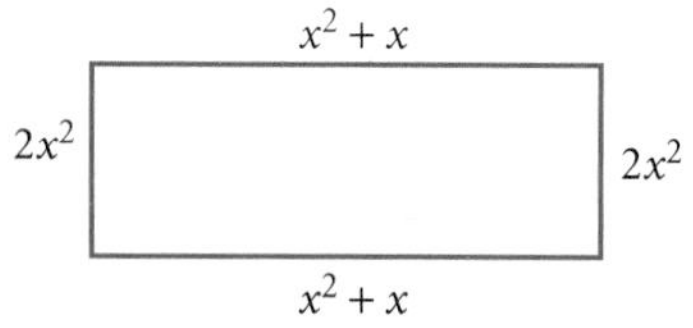


**b.**

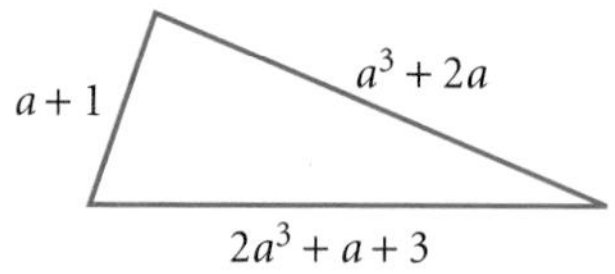


**SOLUTION**

**Vertical form**

**a.**

$$\begin{array}{l} x^2 + x \\ 2x^2 \\ 2x^2 \\ +\ x^2 + x \\ \hline 6x^2 + 2x \end{array}$$

**b.**

$$\begin{array}{r} a^3 + 2a \phantom{+ 3} \\ a + 1 \\ +\ 2a^3 + \phantom{2}a + 3 \\ \hline 3a^3 + 4a + 4 \end{array}$$

**Horizontal form**

**a.** $(x^2 + x) + 2x^2 + 2x^2 + (x^2 + x)$
$x^2 + x^2 + 2x^2 + 2x^2 + x + x$
$6x^2 + 2x$

**b.** $(a^3 + 2a) + (a + 1) + (2a^3 + a + 3)$
$a^3 + 2a^3 + 2a + a + a + 1 + 3$
$3a^3 + 4a + 4$

**TRY THIS** Write a polynomial expression for the perimeter of the polygon shown at right.

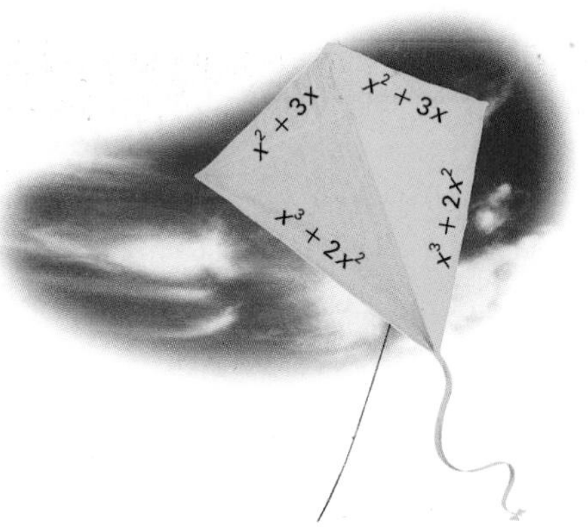


## Subtracting Polynomials

Recall that subtraction can be modeled by adding the opposite.

**EXAMPLE 4** **Find the difference.**

**a.** $(3x^2 - 2x + 8) - (x^2 - 4)$ **b.** $(4x^2 + 3x + 2) - (2x^2 - 3x + 7)$

**SOLUTION**

**a.** The binomial $x^2 - 4$ can be written as $x^2 + 0x - 4$, where $0x$ is a "placeholder" for the missing $x$-term. Remember to change signs when subtracting.

$$\begin{array}{r} 3x^2 - 2x + 8 \\ \underline{-(x^2 + 0x - 4)} \end{array} \quad \Rightarrow \quad \begin{array}{r} 3x^2 - 2x + 8 \\ \underline{-\ x^2 - 0x + 4} \\ 2x^2 - 2x + 12 \end{array}$$

**b.**
$$\begin{array}{r} 4x^2 + 3x + 2 \\ \underline{-(2x^2 - 3x + 7)} \end{array} \quad \Rightarrow \quad \begin{array}{r} 4x^2 + 3x + 2 \\ \underline{-2x^2 + 3x - 7} \\ 2x^2 + 6x - 5 \end{array}$$

**TRY THIS** Subtract $x^2 + 2x - 4$ from $3x^2 - 2x + 3$.

# *Exercises*

## *Communicate*

1. Explain how to use algebra tiles to represent $5x^2 - 2x + 3$.
2. How do you determine the degree of a polynomial?
3. Classify the polynomial $3x^3 + 4x^2 - 7$ in two different ways.
4. Explain how to find $(3b^3 - 2b + 1) - (b^3 + b - 3)$.
5. Use $5x^2 - 2x + 3x^4 - 6$ to explain how to write a polynomial in standard form.

## Guided Skills Practice

**Write each polynomial in standard form.** *(EXAMPLE 1)*

**6.** $x + 3x^3 - 2$ **7.** $15 + 2x - 3x^2$ **8.** $3x^4 + 23 - 2x + 2x^2$

**Find each sum by using the horizontal form and the vertical form.** *(EXAMPLE 2)*

**9.** $(3x^3 + 2x - 5) + (x^3 - 4)$ **10.** $(15x^2 + 5x) + (3x^2 - 3)$

**11.** $(12x^3 + 7x + 2) + (4x^2 + 3x - 6)$ **12.** $(10x^4 - 3x^2 + 2) + (3x^3 + 2x^2 - 13)$

**Write a polynomial expression for the perimeter of each polygon shown below.** *(EXAMPLE 3)*

**13.**

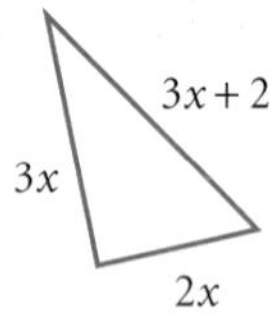


**14.**

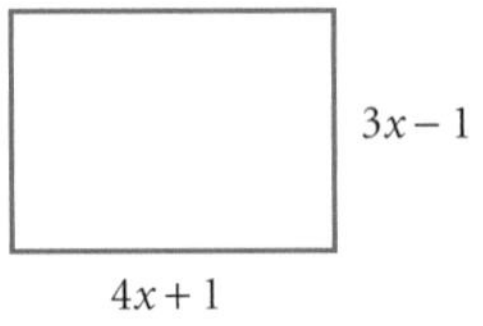


**Find each difference by using the horizontal form and the vertical form.** *(EXAMPLE 4)*

**15.** $(12x^2 + 5x + 11) - (10x^2 + 3x + 2)$ **16.** $(3x^4 + 2x^2) - (2x^4 + 3)$

## Practice and Apply

**17.** Draw an algebra-tile model of $x^2 + 4x + 4$.

**18.** Draw an algebra-tile model of $(3x^2 + 6) + (2x^2 - 1)$.

**Write each polynomial in standard form.**

**19.** $6 + c + c^3$ **20.** $5x^3 - 1 + 5x^4 + 5x^2$ **21.** $10 + p^7$

**Name each polynomial by the number of terms and by degree.**

**22.** $4r + 1$ **23.** $x^3 + x^4 + x - 1$ **24.** $y + y^3$

**25.** $3x + 1$ **26.** $8x^2 - 1$ **27.** $8x^2 - 2x + 3$

**Give an example of each type of polynomial.**

**28.** quadratic trinomial **29.** linear binomial **30.** cubic monomial

**Find the sum or difference by using the vertical form.**

**31.** $(3x^2 + 4x^4 - x + 1) + (3x^4 + x^2 - 6)$ **32.** $(2y^3 + y^2 + 1) + (3y^3 - y^2 + 2)$

**33.** $(4r^4 + r^3 - 6) + (r^3 + r^2)$ **34.** $(2c - 3) + (c^2 + c + 4)$

**35.** $(x^3 + x^2 + 7) - (x^2 + x)$ **36.** $(4y^2 - y + 6) - (3y^2 - 4)$

**37.** $(5c^3 + 10c + 5) - (4c^3 - c^2 - 1)$ **38.** $(x^3 - x + 4) - (8x^3)$

internet connect
go.hrw.com

**Homework Help Online**
Go To: **go.hrw.com**
Keyword:
**MA1 Homework Help**
for Exercises 39-49

**Find the sum or difference by using the horizontal form.**

**39.** $(y^3 - 4) + (y^2 - 2)$ **40.** $(x^3 + 2x - 1) + (3x^2 + 4)$

**41.** $(3s^2 + 7s - 6) + (s^3 + s^2 - s - 1)$ **42.** $(w^3 + w - 2) + (4w^3 - 7w + 2)$

**43.** $(y^2 + 3y + 2) - (3y - 2)$ **44.** $(3x^2 - 2x + 10) - (2x^2 + 4x - 6)$

**45.** $(3x^2 - 5x + 3) - (2x^2 - x - 4)$ **46.** $(2x^2 + 5x) - (x^2 - 3)$

**Simplify. Express all answers in standard form.**

**47.** $(1 - 4x - x^4) + (x - 3x^2 + 9)$

**48.** $(5 - 3x - 1.4x^2) - (13.7x - 62 + 5.6x^2)$

**49.** $(4x^5 - 3x^3 - 3x - 5) - (x - 2x^2 + 3x + 5)$

**CONNECTION**

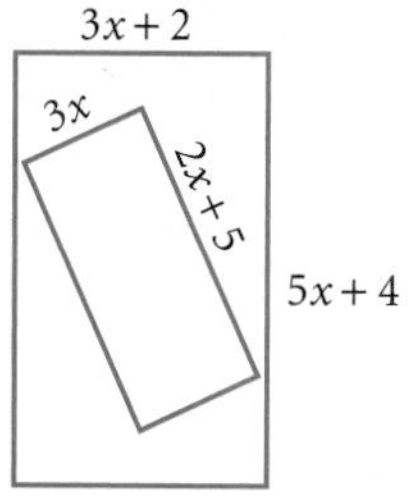


**GEOMETRY Exercises 50–52 refer to the dimensions of the figure at left.**

**50.** Find the polynomial expression that represents the perimeter of the large rectangle.

**51.** Find the polynomial expression that represents the perimeter of the small rectangle.

**52.** Find the difference between the perimeters of the two rectangles.

**APPLICATION**

**INTERIOR DECORATING** Tina is estimating the amount of wallpaper border trim that she will need for a rectangular room. The trim pieces that she plans to use are precut $x$ feet long. She estimates that one length of the room will take 3 pieces plus an extra 4 feet of trim and that one width of the room will take 2 pieces plus an extra 3 feet of trim.

**53.** Write a polynomial expression for the length of the walls to be trimmed and one for the width.

**54.** What is the total amount of border trim, in terms of $x$, needed for all four walls?

## Look Back

**Multiply.** ***(LESSON 3.5)***

**55.** $6(x^2 - 4x)$

**56.** $8(m^2 + 6m)$

**57.** Solve $\frac{6x - 12x + 18}{3} = 1$ for $x$. ***(LESSON 3.4)***

**58.** Solve $-2(a + 3) = 5 - 6(2a - 7)$ for $a$. ***(LESSON 3.5)***

**59.** Solve this system of equations. $\begin{cases} 6x = 4 - 2y \\ 12x - 4y = 16 \end{cases}$ ***(LESSON 7.3)***

**internet connect**

**Activities Online**

Go To: **go.hrw.com**
Keyword:
**MA1 Poly Ops**

**Write each number in scientific notation.** ***(LESSON 8.5)***

**60.** 7,100,000

**61.** 8,900,000,000

## Look Beyond

**62.** Use algebra tiles to model a square whose area can be represented by $x^2 + 6x + 9$. What is the length of a side of the square?

# 9.2 Modeling Polynomial Multiplication

### Objectives

- Use algebra tiles to model the products of binomials.
- Find products of binomials by using the Distributive Property.

**Why** ***Algebra tiles provide a physical representation of polynomial multiplication. This provides another way of thinking about the Distributive Property.***

Algebra tiles can be used as a model for multiplication. The model is related to the fact that the area of a rectangle is equal to the product of the length and the width.

In the product $(x+1)(2x+2)$, two binomials are multiplied. You can model this product by using algebra tiles, which will show you how to find the product.

*Activity*

## Exploring Multiplication of Polynomials

**You will need:** algebra tiles (or use paper-and-pencil sketches)

**Part I: Finding Products**

1. The model below shows how $x+1$ can be multiplied by 2. (See page 97.) The result is $2x+2$. This is an example of which property?

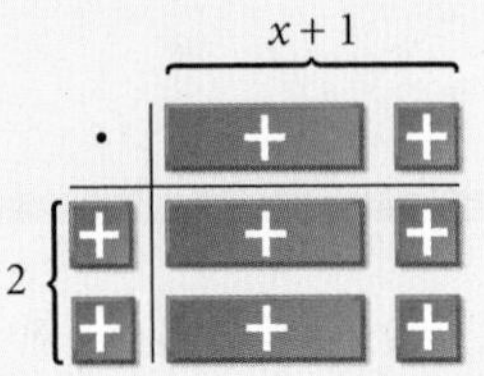


$2(x+1)$
$2(x)+2(1)$
$2x+2$

2. What two factors are multiplied in the model at right? What is the product shown by the tiles?

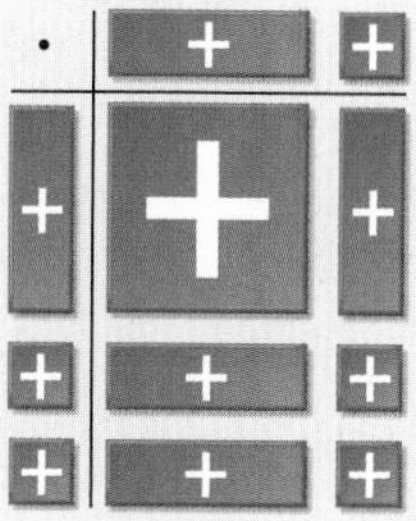

3. Check the product in Step 2 by substituting 10 for $x$.

CHECKPOINT ✔ 4. Use tiles to find the product $(2x+2)(x+4)$, and check your answer by substituting a value for $x$.

**Part II: Introducing Negative Numbers**

1. The tiles shown at right model the product of two binomials. Remove as many neutral pairs from the product as you can. What is the simplified product?

   $(x+2)(x-3) = \underline{\quad ? \quad}$

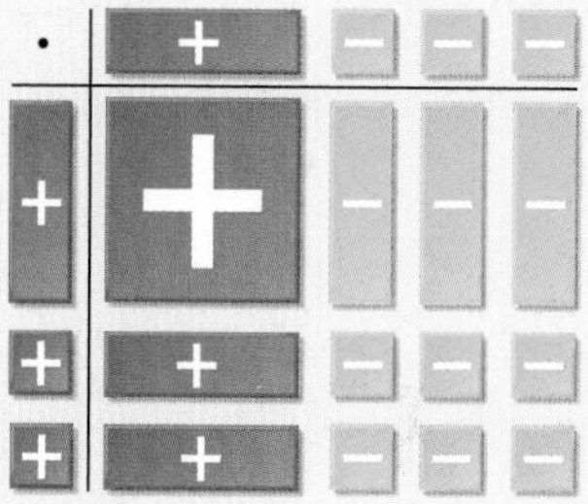

2. Check the product by substituting a value, such as 10, for $x$.

CHECKPOINT ✔ 3. Use tiles to find the product $(x+3)(2x-4)$, and check your answer by substituting a value for $x$.

**Part III: Special Products**

1. Use tiles to find the following products:

   $(x+2)(x-2)$ $\qquad$ $(x+3)(x-3)$

   What do these two products have in common? Use your observations to find the product $(x+5)(x-5)$ without using tiles.

2. Use tiles to find the following products:

   $(x+2)(x+2)$ $\qquad$ $(x+3)(x+3)$

   What do these two products have in common? Use your observations to find the product $(x+5)(x+5)$ without using tiles.

CHECKPOINT ✔ 3. Use tiles to find the following products:

   $(x-2)(x-2)$ $\qquad$ $(x-3)(x-3)$

   What do these two products have in common? Use your observations to find the product $(x-5)(x-5)$ without using tiles.

**CRITICAL THINKING**

Given the algebra tiles that model a second-degree polynomial, will you always be able to form a rectangle with the tiles? Explain.

**EXAMPLE 1** **Use algebra tiles to find each product. Check your results by substituting a value for $x$ in each expression.**

**a.** $(x + 2)(x + 4)$ **b.** $(x - 2)(x + 4)$

**SOLUTION**

**a.**

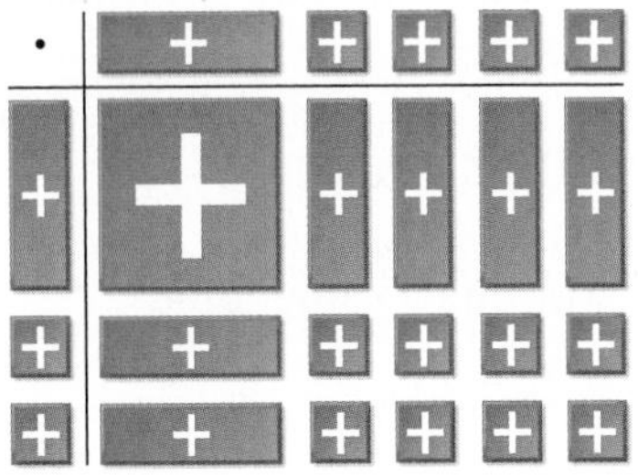

$(x + 2)(x + 4)$

$x^2 + 6x + 8$

Let $x = 5$, for example.

| | |
|---|---|
| $(5 + 2)(5 + 4)$ | $5^2 + 6(5) + 8$ |
| $7 \cdot 9$ | $25 + 30 + 8$ |
| $63$ | $63$ |

**b.**

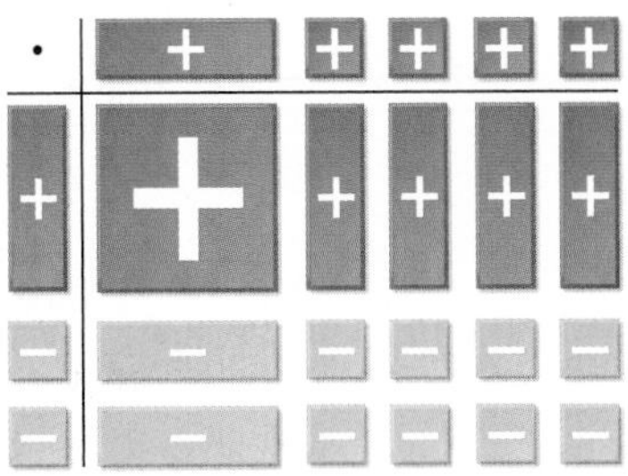

$(x - 2)(x + 4)$

$x^2 + 2x - 8$

Let $x = 3$, for example.

| | |
|---|---|
| $(3 - 2)(3 + 4)$ | $3^2 + 2(3) - 8$ |
| $1 \cdot 7$ | $9 + 6 - 8$ |
| $7$ | $7$ |

The Distributive Property can be used to multiply a monomial and a binomial. It can also be used to multiply two binomials.

## Using the Distributive Property

The Distributive Property can be used to find the product of $(x + 3)(x + 2)$.

| | |
|---|---|
| $(x + 3)(x + 2)$ | *Given* |
| $(x + 3)x + (x + 3)2$ | *Distribute the first binomial to each term of the second binomial.* |
| $x^2 + 3x + 2x + 6$ | *Simplify by using the Distributive Property.* |
| $x^2 + 5x + 6$ | *Add like terms.* |

**EXAMPLE 2** **Use the Distributive Property to find each product.**

**a.** $3x(2x - 5)$ **b.** $(x + 4)(x - 5)$

**SOLUTION**

**a.** $3x(2x - 5)$

$3x \cdot 2x - 3x \cdot 5$

$6x^2 - 15x$

**b.** $(x + 4)(x - 5)$

$(x + 4)x + (x + 4)(-5)$

$x^2 + 4x - 5x - 20$

$x^2 - x - 20$

**TRY THIS** Use the Distributive Property to find each product.

**a.** $5x(2x + 1)$ **b.** $(x - 1)(x + 9)$

# Exercises

## Communicate

**1.** Explain how you can use algebra tiles to simplify each product below.
**a.** $(x+2)(x+3)$ **b.** $(x-4)(x-4)$ **c.** $(x+6)(x+6)$

**2.** Give an example of a product represented by an algebra-tile model in which you would remove neutral pairs in order to simplify it.

**3.** Show how to use algebra tiles to model each product.
**a.** $(x+1)(x+2)$ **b.** $(x-1)(x-2)$

**4.** Show how to use the Distributive Property to model each product.
**a.** $(x+1)(x+2)$ **b.** $(x-1)(x-2)$

## Guided Skills Practice

**5.** Use algebra tiles to find each product. Check your results by substituting a value for $x$ in each expression. ***(EXAMPLE 1)***
**a.** $(x+1)(2x+3)$ **b.** $(2x+1)(3x-3)$ **c.** $(x+2)(x-4)$

**6.** Use the Distributive Property to find the product of $(x+5)(x-3)$.
***(EXAMPLE 2)***

13-21 Foil

## Practice and Apply

**Write a product of binomial factors for each model. Use the model to find the product. Check your results by substituting a value for *x* into each expression.**

**7.**

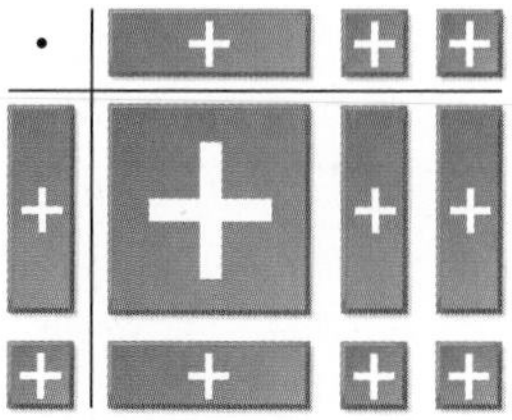

**8.**

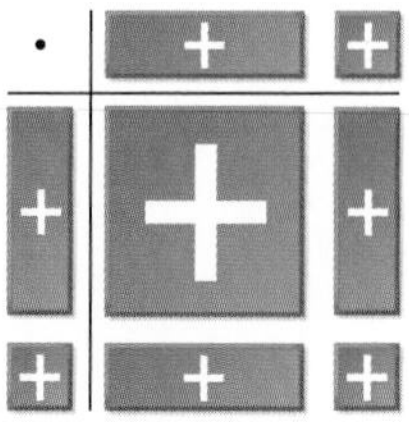

**9.**

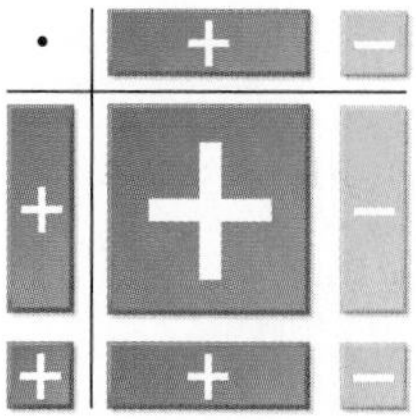

**10.**

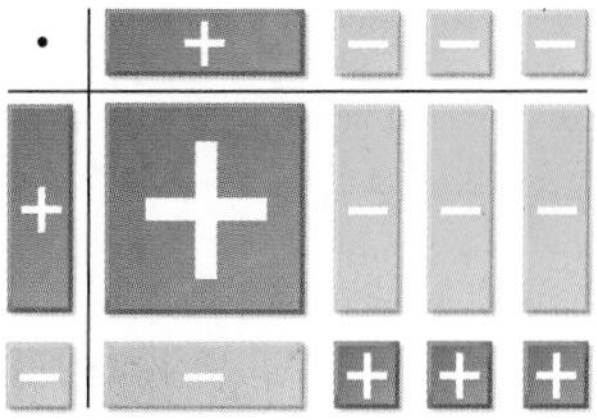

**11.**

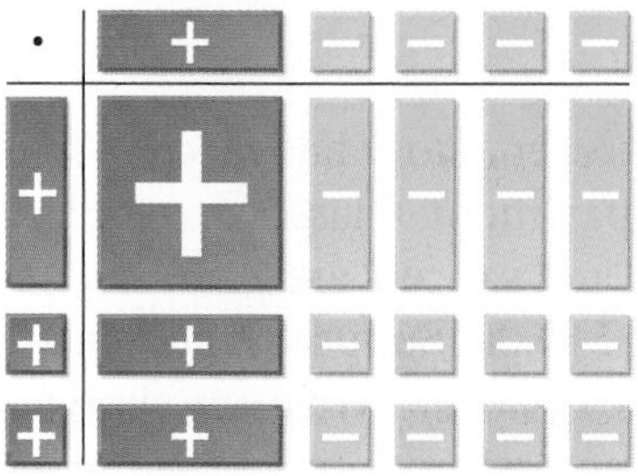

**12.**

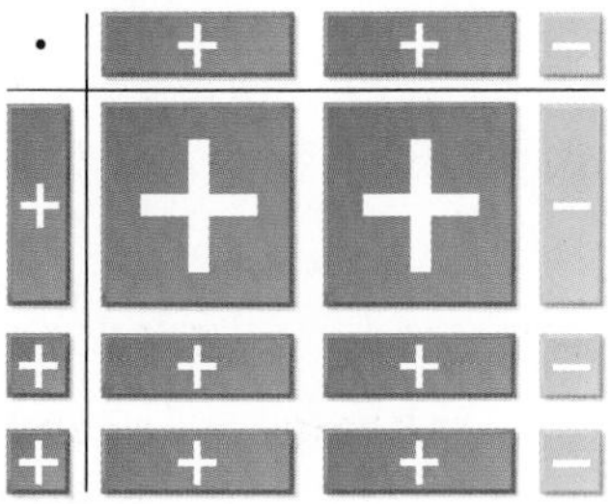

**Model each product with tiles, and give the simplified product. Check by substituting a value for *x*.**

**13.** $(x+5)(x+2)$ **14.** $(x+2)(x-1)$ **15.** $(2x+1)(x+1)$

**16.** $(x+5)(x-2)$ **17.** $(x+2)(x+1)$ **18.** $(3x+5)(x-7)$

**19.** $(x-2)(4x-3)$ **20.** $(x+5)(x-7)$ **21.** $(x-5)(2x+12)$

**Use the Distributive Property to find each product.**

**22.** $4(x+2)$ **23.** $6(2x+7)$ **24.** $5(x+10)$ **25.** $4(x-4)$

**26.** $3(x+8)$ **27.** $-2(x-3)$ **28.** $8(3x^2-4x)$ **29.** $-3x(x^2+4x)$

**30.** $x(1-x)$ **31.** $-x(2-x)$ **32.** $x(-x+5)$ **33.** $4x(3-x)$

**34.** $(x+2)(x+5)$ **35.** $(a-3)(a+4)$ **36.** $(b-1)(b-3)$

**37.** $(y+3)(y-2)$ **38.** $(c+5)(c+5)$ **39.** $(d+5)(d-5)$

**40.** $(2y)(y+3)$ **41.** $(w^2)(w+1)$ **42.** $(3b)(b+3)$

**Determine whether each statement is true by substituting first 10 and then another value for *x*.**

**43.** $x(2x) \stackrel{?}{=} 2x^2$ **44.** $x(2x+1) \stackrel{?}{=} 2x^2+2$

**45.** $x(2x-1) \stackrel{?}{=} 2x^2-x$ **46.** $(x+3)(x+1) \stackrel{?}{=} 2x^2+3x+3$

CHALLENGE

**47.** Explain why the statement $(x-10)(x+3) = (x-10)(x+4)$ is true for $x = 10$ but is false for $x \neq 10$.

CONNECTION

**GEOMETRY** A landscaper is designing Mary's backyard. Mary wants a rectangular flower bed and lawn in her backyard, as shown below.

**48.** Write an expression for the area of the flower bed.

**49.** Write an expression for the area of the lawn.

**50.** Write an expression for the area of the backyard.

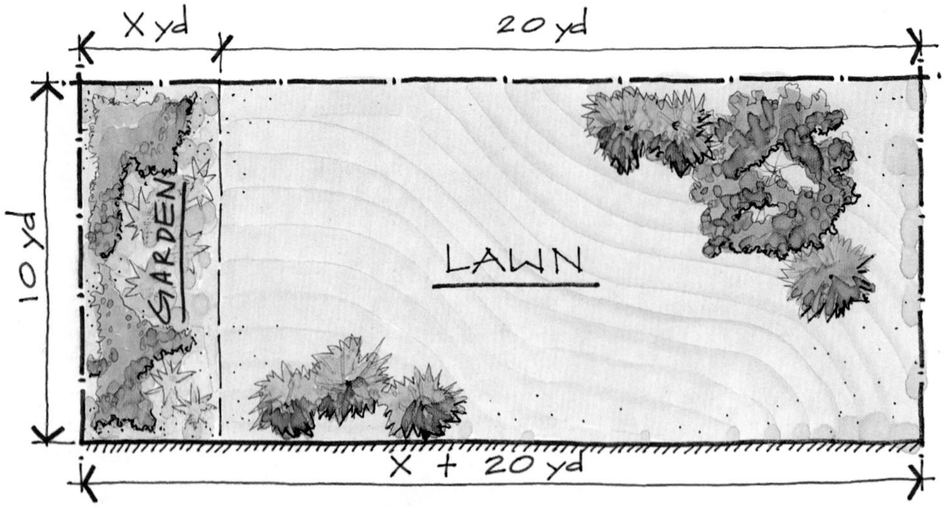


APPLICATION

**51. DISCOUNTS** A grocery store has pieces of dinnerware for sale in the housewares department. Glasses are \$1.50 each, plates are \$2.50 each, cups are \$1.25 each, and saucers are \$1.00 each. A 25% discount is given for purchases over \$10.00. Suppose that a customer buys *x* glasses and *y* plates and receives the discount. Write an expression for the amount of the discount that the customer receives.

## Look Back

**Solve each equation or inequality.** ***(LESSON 6.5)***

**52.** $|x - 2| = 5$ **53.** $|2x - 1| = -1$ **54.** $|x| \leq 5$

**55.** Solve this system of linear equations by graphing. ***(LESSON 7.1)*** $\begin{cases} 6y = 8x - 2 \\ 2x + 3y = 2 \end{cases}$

**56.** Solve this system of linear equations by substitution. ***(LESSON 7.2)*** $\begin{cases} x + y = 7 \\ 4x + 2y = 10 \end{cases}$

**57.** Solve this system of linear equations by elimination. ***(LESSON 7.3)*** $\begin{cases} -4x + 3y = -1 \\ -8x - 6y = -10 \end{cases}$

**Find each sum or difference.** ***(LESSON 9.1)***

**58.** $(x^2 + 2x) + (x + 1)$ **59.** $(x^2 + 2x) - (x + 1)$ **60.** $(2x^2 + x) - (2x^2 - x)$

## Look Beyond

**The zeros of the polynomial function $y = x^2 - 4$ are 2 and –2 because these values result in a value of 0 for $y$ when they are substituted for $x$. Find the zeros of the following functions:**

**internet connect**
**Portfolio Extension**
Go To: **go.hrw.com**
Keyword:
**MA1 Pascal**

**61.** $y = x^2 - 9$ **62.** $y = x^2 - 16$ **63.** $y = x^2 - 25$

**64. TECHNOLOGY** Graph the functions in Exercises 61–63. How can you identify the zeros on the graphs? Explain your reasoning.

**PORTFOLIO ACTIVITY**

Include this activity in your portfolio.

The first four rows of Pascal's triangle are shown below.

| | | |
|---|---|---|
| **Row 0** | **1** | **Sum = 1** |
| **Row 1** | **1 1** | **Sum = 2** |
| **Row 2** | **1 2 1** | **Sum = ?** |
| **Row 3** | **1 3 3 1** | **Sum = ?** |

**1.** Complete rows 4, 5, and 6.

**2.** Compute the sum for each row.

**3.** What should be the sums for rows 7, 8, and 9?

**WORKING ON THE CHAPTER PROJECT**

You should now be able to complete Activity 1 of the Chapter Project on page 470.

*Blaise Pascal 1623–1662*

9.3

# Multiplying Binomials

## Objectives

- Find products of binomials by using the FOIL method.
- Mentally simplify special products of binomials.

**Why** ***The Distributive Property can be used to develop a method for multiplying two binomials quickly.***

Mathematical software can find products like the one shown on the computer screen. To understand what such a product means and how the computer calculates it, you should first learn to work with products without the aid of a computer or calculator.

## The FOIL Method

A popular mnemonic (memory) device for multiplying two binomials mentally, which you may already know, is called the **FOIL method**.

- Multiply the **F**irst terms.
- Multiply the **O**utside terms.
- Multiply the **I**nside terms. Add the outside and inside products.
- Multiply the **L**ast terms.

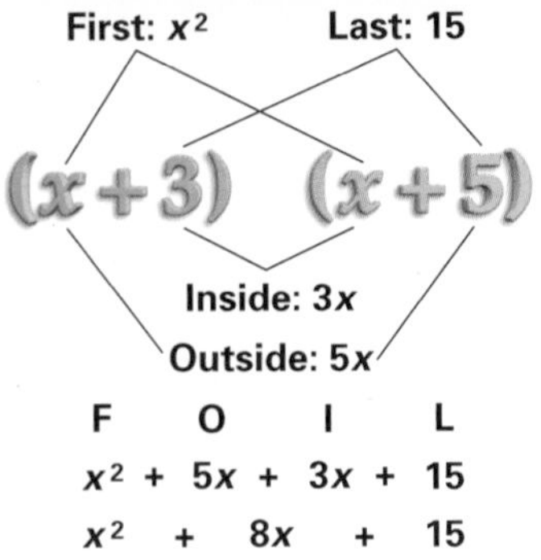


**EXAMPLE 1** **Use the FOIL method to find each product.**

**a.** $(3x+2)(5x+1)$ **b.** $(x+9)(2x-4)$ **c.** $(y+\frac{1}{4})(y-\frac{3}{4})$

**SOLUTION**

**a.** $(3x+2)(5x+1)$
$15x^2+3x+10x+2$
$15x^2+13x+2$

**b.** $(x+9)(2x-4)$
$2x^2-4x+18x-36$
$2x^2+14x-36$

**c.** $\left(y+\frac{1}{4}\right)\left(y-\frac{3}{4}\right)$
$y^2-\frac{3}{4}y+\frac{1}{4}y-\frac{3}{16}$
$y^2-\frac{1}{2}y-\frac{3}{16}$

**TRY THIS** Use the FOIL method to find each product.

**a.** $(x+4)(2x-5)$ **b.** $\left(2y-\frac{2}{3}\right)\left(3y+\frac{1}{2}\right)$

The FOIL method can be used to find products of special binomials.

| | | |
|---|---|---|
| $(a-b)(a+b)$ | $(a+b)(a+b)$ | $(a-b)(a-b)$ |
| $a^2-ab+ab-b^2$ | $a^2+ab+ab+b^2$ | $a^2-ab-ab+b^2$ |
| $a^2-b^2$ | $a^2+2ab+b^2$ | $a^2-2ab+b^2$ |

**Special Products**

For any real numbers $a$ and $b$ and for any expressions involving real numbers and variables:

$$(a+b)(a-b)=a^2-b^2$$
$$(a+b)(a+b)=(a+b)^2=a^2+2ab+b^2$$
$$(a-b)(a-b)=(a-b)^2=a^2-2ab+b^2$$

**EXAMPLE 2** **Find the following products using the rules for special products:**

**a.** $(x-6)(x+6)$ **b.** $(2x+6)(2x+6)$ **c.** $(2x-3)(2x-3)$

**SOLUTION**

**a.** Use the rule $(a+b)(a-b)=a^2-b^2$, where $a=x$ and $b=6$.

$(x-6)(x+6)$
$x^2-6^2$
$x^2-36$

**b.** Use the rule $(a+b)^2=a^2+2ab+b^2$, where $a=2x$ and $b=6$.

$(2x+6)^2$
$(2x)^2+2(2x)6+6^2$
$4x^2+24x+36$

**c.** Use the rule $(a-b)^2=a^2-2ab+b^2$, where $a=2x$ and $b=3$.

$(2x-3)^2$
$(2x)^2-2(2x)(3)+3^2$
$4x^2-12x+9$

**TRY THIS** Find each product using the rules for special products.

**a.** $(5x-2)(5x+2)$ **b.** $(2x-4)^2$

**EXAMPLE 3** Cruz's Frame Shop makes a mat by cutting out the inside of a rectangular mat board. **Use the measurements from the diagram at right to find the length and width of the original mat if the area of the mat board is 148 square inches.**

*Creating a mat to frame a picture requires precise cutting.*

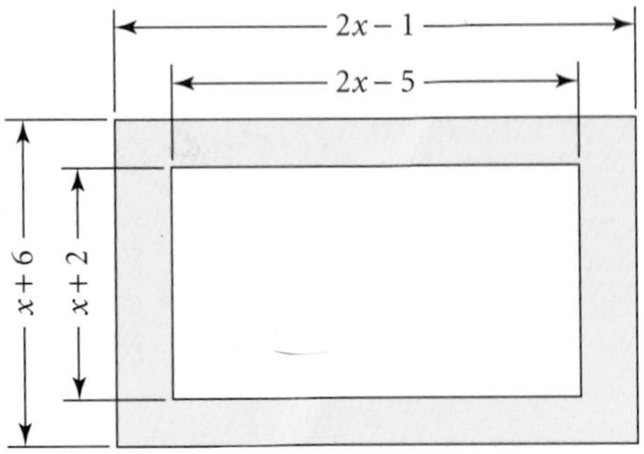


**SOLUTION**

Find the areas of the outer and inner rectangles formed by the mat.

| Area of the outer rectangle | Area of the inner rectangle |
|---|---|
| $(2x-1)(x+6)$ | $(x+2)(2x-5)$ |
| $2x^2+12x-x-6$ | $2x^2-5x+4x-10$ |
| $2x^2+11x-6$ | $2x^2-x-10$ |

Substitute each area into the formula below and then solve for $x$.

| Area of outer rectangle | – | Area of inner rectangle | = | Area of the mat |
|---|---|---|---|---|
| $2x^2+11x-6$ | – | $(2x^2-x-10)$ | = | 148 |
| $2x^2+11x-6$ | – | $2x^2+x+10$ | = | 148 |
| | | $12x+4$ | = | 148 |
| | | $12x$ | = | 144 |
| | | $x$ | = | 12 |

The length of the original rectangle was $2x-1$. Substitute 12 for $x$.

$$2x-1=2(12)-1=23$$

The width was $x+6$. Substitute 12 for $x$.

$$x+6=12+6=18$$

The original mat measured 23 inches by 18 inches.

**TRY THIS** The dimensions of a rug without its border are $x+3$ and $3x-3$. The dimensions of the rug with its border are $x+7$ and $3x+1$. If the area of the border is 160 square feet, find the length and width of the rug with its border.

# Exercises

## Communicate

1. Describe how to find $(2x + 3)(x - 4)$ by using the FOIL method.
2. Explain how you can multiply $(x + 7)(x - 7)$ mentally.
3. Explain how the FOIL method and Special Products are different.
4. Use the FOIL method to explain why $(x + 4)^2$ does not equal $x^2 + 16$.

## Guided Skills Practice

5. Use the FOIL method to find each product. ***(EXAMPLE 1)***
   **a.** $(x + 3)(x + 6)$ **b.** $(a - 4)(a - 9)$ **c.** $(m - 3)(m + 9)$
6. Find each product. ***(EXAMPLE 2)***
   **a.** $(x + 5)(x - 5)$ **b.** $(3x + 2)(3x + 2)$ **c.** $(2x - 8)(2x - 8)$

**CONNECTION**

7. **GEOMETRY** The area of the shaded part of the figure is 156. Find the length and width of the outer rectangle. ***(EXAMPLE 3)***

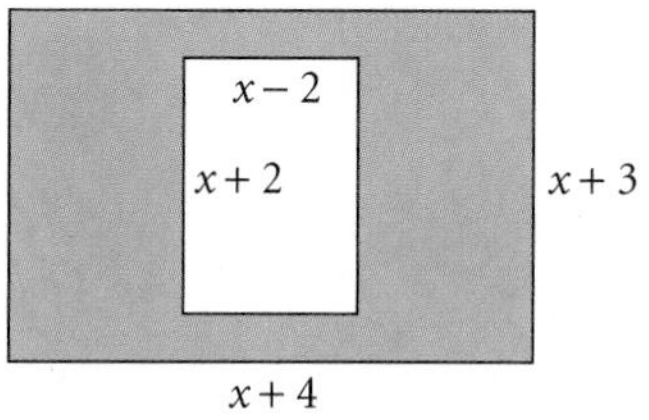


## Practice and Apply

**Use the FOIL method to find each product.**

8. $(3y - 2)(y - 1)$
9. $(5p + 3)(p + 1)$
10. $(2q - 1)(2q + 1)$
11. $(2x + 5)(2x - 3)$
12. $(4m + 1)(5m - 3)$
13. $(2w - 9)(3w - 8)$
14. $(3x - 5)(3x - 5)$
15. $(7s + 2)(2s - 3)$
16. $\left(y - \frac{1}{2}\right)\left(y + \frac{1}{2}\right)$
17. $\left(y - \frac{1}{3}\right)\left(y + \frac{5}{9}\right)$
18. $(c^2 + 1)(c^2 + 2)$
19. $(2a^2 + 3)(2a^2 + 3)$
20. $(a + c)(a + 2c)$
21. $(p + q)(p + q)$
22. $(a^2 + b)(a^2 - b)$
23. $(x^2 + y)(x + y)$
24. $(c + d)(2c + d)$
25. $(1.2m + 5)(0.8m - 4)$

**internet connect**

**Homework Help Online**
Go To: **go.hrw.com**
Keyword:
**MA1 Homework Help**
for Exercises 8-25; 26-34

**Find each product by using the rules for special products.**

26. $(2x + 2)(2x - 2)$
27. $(5x - 1)(5x + 1)$
28. $(x + 4)(x - 4)$
29. $(x + 1)^2$
30. $(x + 6)(x - 6)$
31. $(3x - 4)^2$
32. $(x + 8)^2$
33. $(x - 7)^2$
34. $(5x + 9)^2$

CONNECTIONS

**35.** **GEOMETRY** Find the area in terms of $y$ of a square rug whose sides are $y + 6$ units long.

**36.** **GEOMETRY** Which has the larger area, a square with sides that are $x + 1$ units long or a rectangle with a length of $x + 2$ units and a width of $x$ units? What is the difference in areas?

**37.** **GEOMETRY** Which has a greater area, a square with sides that are $x - 1$ units long or a rectangle with a length of $x$ units and a width of $x - 2$ units? What is the difference in areas?

APPLICATION

**38.** **CONSTRUCTION** An architect is designing a rectangular pavilion that has a length of $x + 8$ units and a width of $x - 4$ units. Draw a diagram and label the dimensions. Write an expression representing the area covered by the pavilion.

## Look Back

**Find the slope of each line.** ***(LESSON 5.4)***

**39.** $4x + 3y = 12$ **40.** $y = 4x$ **41.** $3y - 7x = 10$

**State the slope of each line, and indicate whether each pair of lines is parallel, perpendicular, or neither.** ***(LESSON 5.6)***

**42.** $2y = x + 6$; $2y = x - 3$ **43.** $x - 2y = 4$; $2x + y = 1$ **44.** $x = 7$; $y = 7$

**45.** Write the equation of the line through $(-2, -1)$ and parallel to $y = -2x - 4$. ***(LESSON 5.6)***

**Solve each system of equations by graphing. Check your solutions by using substitution.** ***(LESSON. 7.1)***

**46.** $\begin{cases} 2x + y = 9 \\ x - 2y = 2 \end{cases}$ **47.** $\begin{cases} x + y = 4 \\ x + y = 10 \end{cases}$ **48.** $\begin{cases} 3x + 2y = 7 \\ x + 5y = -10 \end{cases}$

**49.** Find the two-digit number whose tens digit is 2 less than its units digit if the original number is 4 times the sum of the digits. ***(LESSON 7.6)***

**Simplify each expression.** ***(LESSON 8.3)***

**50.** $\left(\frac{x^4}{y^2}\right)^3$ **51.** $\left(\frac{c^3b^4}{b^2}\right)^2$ **52.** $\left[\frac{x^3y^5}{(x^2)^3}\right]^4$

## Look Beyond

**At what point does the graph of each function cross the *x*-axis?**

**53.** $y = (x + 3)(x - 2)$ **54.** $y = (x + 2)^2$ **55.** $y = x^2 - 4$

9.4

# Polynomial Functions

**Objectives**

- Define polynomial functions.
- Solve problems involving polynomial functions.

**Why** *Surface area and volume formulas are two examples of polynomial functions that can be used in real-world situations such as manufacturing.*

*Shari, the production manager at Power Wash Inc., wants to change the size of the Super Suds box.*

The Super Suds box currently has a base with a length of 20 centimeters and a width of 10 centimeters. The height of the current box is 30 centimeters.

**EXAMPLE 1** **Find the volume of the current Super Suds box.**

**APPLICATION**
**PACKAGING**

**SOLUTION**

First write an equation that expresses the volume of the Super Suds box.

You can use the formula $V = Bh$, where $B$ is the area of the base and $h$ is the height, to find the volume, $V$, of the Super Suds box. You know that $B = lw$, so $V = lwh$.

You can substitute the current dimensions of the box into this formula to find the volume.

$$V = lwh$$
$$V = 20 \cdot 10 \cdot 30$$
$$V = 6000$$

The volume of the current Super Suds box is 6000 cubic centimeters.

**TRY THIS** Suppose that a box has a length of 25 centimeters, a width of 15 centimeters, and a height of 25 centimeters. Find the volume of the box.

You have already worked with several types of functions, including linear functions. Linear functions are a specific type of **polynomial function**, which can be used to help solve a wide variety of problems.

**Polynomial Function**

A *polynomial function* is a function that consists of a monomial, or the sum or difference of two or more monomials.

**APPLICATION**
**PACKAGING**

Consider the following problem: Power Wash Incorporated manufactures a variety of laundry detergents. The production manager notices that the production costs of Super Suds laundry detergent are rising. Instead of raising prices, the production manager considers lowering the production costs by changing the size of the packaging.

## *Activity*
## Exploring Volume and Surface Area

**You will need:** graphics calculator (optional)

1. Find the surface area of the original Super Suds box, using the dimensions given on the previous page.

2. If the production manager wants to reduce the volume of the original box by 10%, what would the new volume be? If the production manager wants to reduce the surface area of the original box by 10%, what would the new surface area be?

3. The production manager decides to reduce the size of the rectangular base, keeping the same height and keeping the length twice the width. Explain why you can use the following functions, where $x$ is the width of the base, to represent the volume and surface area of the Super Suds box.

$$V = 2x \cdot x \cdot 30 = 60x^2$$

$$S = 2(2x \cdot x) + 2(2x \cdot 30) + 2(x \cdot 30) = 4x^2 + 120x + 60x = 4x^2 + 180x$$

CHECKPOINT ✔ 4. Examine the table at right, where **X** is width, **Y1** is volume, and **Y2** is surface area. Does changing the width of the base have a similar effect on both the volume and the surface area? What width gives the original volume of the box? the original surface area?

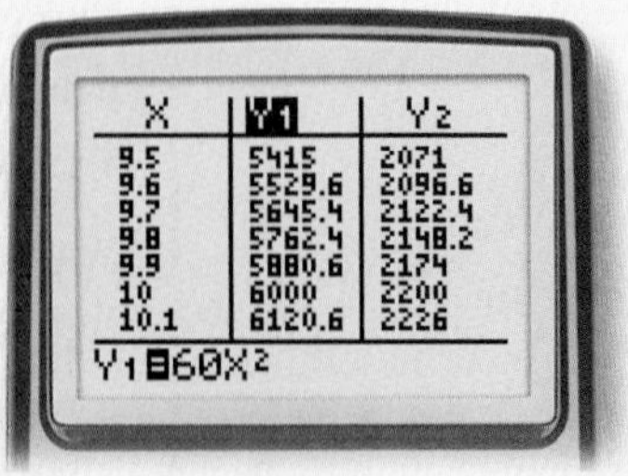

| X | Y1 | Y2 |
|---|---|---|
| 9.5 | 5415 | 2071 |
| 9.6 | 5529.6 | 2096.6 |
| 9.7 | 5645.4 | 2122.4 |
| 9.8 | 5762.4 | 2148.2 |
| 9.9 | 5880.6 | 2174 |
| 10 | 6000 | 2200 |
| 10.1 | 6120.6 | 2226 |



5. Suppose that the production manager decides to reduce the width of the base to 9.3 centimeters. What would be the volume and surface area of the new Super Suds box? What percent reduction in volume is this? What percent reduction in surface area is this?

**EXAMPLE 2**

**APPLICATION**
**PACKAGING**

The manufacturer of a powdered drink mix wants to increase the volume of the container by 20% by increasing the height. The current container is a cylinder with a radius of 6 centimeters and a height of 10 centimeters. **Find the volume of the original container, the new volume, and the height of the new container.**

**SOLUTION**

The formula for volume of a cylinder is $V = Bh$, or $V = \pi r^2 h$. First find the volume of the original container.

$$\begin{aligned} V &= \pi r^2 h \\ &= \pi \cdot 6^2 \cdot 10 \\ &= 360\pi \\ &\approx 1131 \end{aligned}$$

The volume of the original container is approximately 1131 cubic centimeters.

Find the new volume after a 20% increase.

$$\begin{aligned} \text{New volume} &= 1.20 \cdot 1131 \\ &\approx 1357.2 \end{aligned}$$

The volume after a 20% increase would be approximately 1357.2 cubic centimeters.

You can use a graphics calculator to make a table for the function $V = 36\pi h$, where $h$ is the height of the container. Substitute $x$ for $h$ and enter $y = 36\pi x$ into your graphics calculator.

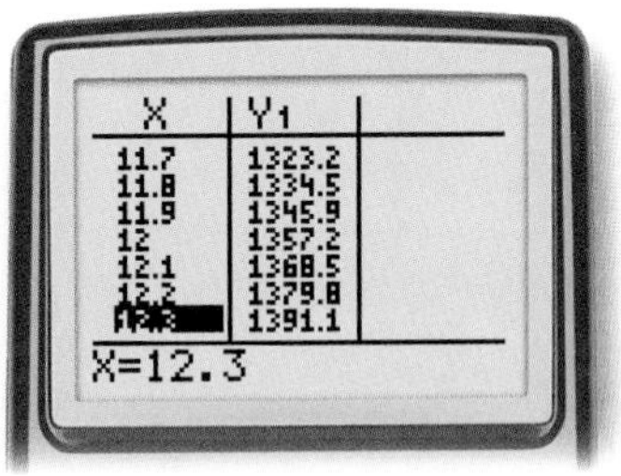


The height that increases the volume of the container by approximately 20% is 12 centimeters.

**TRY THIS**

Suppose that the container in Example 2 currently has a radius of 8 centimeters and a height of 10 centimeters and that the manufacturer wants to increase the volume by 25%. Find the volume of the original container, the new volume, and the height of the new container.

An **identity** is an equation that is true for all values of the variable or variables.

**EXAMPLE 3**

**TECHNOLOGY**
**GRAPHICS CALCULATOR**

**The equation $(x+2)^3 = x^3 + 6x^2 + 12x + 8$ is an identity. Show that each side of the equation has the same value by showing that the functions $y = (x+2)^3$ and $y = x^3 + 6x^2 + 12x + 8$ are equivalent for integer values of $x$ from −3 to 3.**

**SOLUTION**

Build a table of values for the functions. Enter **Y1 = (X+2)^3** and **Y2 = X^3 + 6X^2 + 12X + 8** into a graphics calculator, and compare the values in the table.

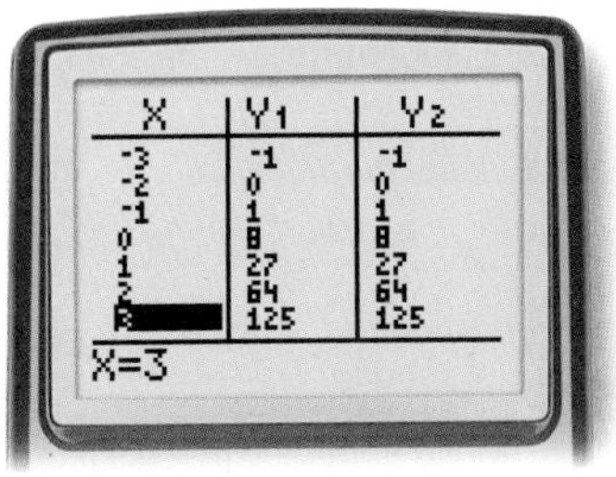


Since the columns for **Y1** and **Y2** have the same values, the functions are equivalent for integer values from −3 to 3.

# Exercises

## Communicate

**internet connect**
**Activities Online**
Go To: **go.hrw.com**
Keyword:
**MA1 Regression**

1. Define a polynomial function and give two examples.
2. Explain how to write the functions for the volume and surface area of a cube with an edge length of $x$.
3. Explain how to use a graphics calculator table to show that $x^2 - 4 = (x + 2)(x - 2)$ is true for integer values of $x$ from $-3$ to 3.

## Guided Skills Practice

4. Suppose that the base of a box has a length of 20 centimeters and a width of 15 centimeters and that the box is 25 centimeters high. Find the volume of the box. ***(EXAMPLE 1)***

APPLICATION

5. **PACKAGING** The manufacturer of a powdered drink mix wants to increase the volume of the container by 20% by increasing the height. The current container is a cylinder with a radius of 10 centimeters and a height of 20 centimeters. Find the volume of the original container, the new volume, and the height of the new container. ***(EXAMPLE 2)***
6. **TECHNOLOGY** The equation $(x + 3)^3 = x^3 + 9x^2 + 27x + 27$ is an identity. Use a graphics calculator to show that the functions $y = x^3 + 9x^2 + 27x + 27$ and $y = (x + 3)^3$ are equivalent for integer values of $x$ from $-3$ to 3. ***(EXAMPLE 3)***

## Practice and Apply

**internet connect**
**Homework Help Online**
Go To: **go.hrw.com**
Keyword:
**MA1 Homework Help** for Exercises 7-18

**The following equations are identities. Create a table to show that each side of the equation has the same value when *x* is replaced by integers from –3 to 3.**

7. $x^2 + 5x + 6 = (x + 2)(x + 3)$
8. $(x - 4)(x - 2) = x^2 - 6x + 8$
9. $x^2 - 5x + 6 = (x - 2)(x - 3)$
10. $(x + 7)(x - 1) = x^2 + 6x - 7$
11. $x^2 - 9 = (x - 3)(x + 3)$
12. $(x - 6)(x + 4) = x^2 - 2x - 24$
13. $x^3 - 8 = (x - 2)(x^2 + 2x + 4)$
14. $(x + 5)(x + 3) = x^2 + 8x + 15$
15. $x^3 + 12x^2 + 48x + 64 = (x + 4)^3$
16. $(x - 6)(x - 7) = x^2 - 13x + 42$
17. $x^2 - 6x + 9 = (x - 3)^2$
18. $(x + 5)^2 = x^2 + 10x + 25$

CONNECTION

**GEOMETRY** Write the indicated equation for each geometric solid.

19. the volume of a cube with an edge of $2x$ centimeters
20. the surface area of a cube with an edge of $x$ centimeters long
21. the surface area of a rectangular solid with a base length of 7 inches, a base width of 3 inches, and a height of $x$ inches
22. the volume of a cylinder with a height of 5 meters and a radius of $x$ meters

**APPLICATION**

**SPORTS** Sports balls come in different sizes. Complete the table for each radius given. Round your answers to the nearest tenth.

| | Radius $r$ | Circumference $C = 2\pi r$ | Surface area $S = 4\pi r^2$ | Volume $V = \frac{4\pi r^3}{3}$ |
|---|---|---|---|---|
| **23.** | 18 inches | ? | ? | ? |
| **24.** | 8 inches | ? | ? | ? |
| **25.** | 5 inches | ? | ? | ? |
| **26.** | 2 inches | ? | ? | ? |
| **27.** | 1 inch | ? | ? | ? |
| **28.** | 0.5 inch | ? | ? | ? |

## Look Back

**Simplify.** ***(LESSONS 2.2, 2.3, AND 2.4)***

**29.** $-2(-3) + 4(-8)$ **30.** $-11(7) - (-12)(-3)$ **31.** $22(-2) + (-9)(-7)$

**Write an equation for the line containing each pair of points.** ***(LESSON 5.5)***

**32.** $(2, 3), (-2, 3)$ **33.** $(0, -1), (4, 3)$ **34.** $(-1, 1), (-2, 2)$

internet connect

**Portfolio Extension**

Go To: **go.hrw.com**
Keyword:
**MA1 Figurate**

## Look Beyond

**TECHNOLOGY** Use a graphics calculator to graph each set of polynomial functions on the same coordinate plane. Describe the similarities and differences.

**35.** $y = x,\ y = x^3,\ y = x^5$ **36.** $y = x^2,\ y = x^4,\ y = x^6$

**37.** $y = x^2 - 2,\ y = x^3 - 2x,\ y = x^4 - 2x^2$

**PORTFOLIO ACTIVITY**

Include this activity in your portfolio.

The sequence of triangular numbers, 1, 3, 6, 10, . . ., can be represented by $\frac{n(n+1)}{2}$, where $n$ is the term number.

**1.** Draw the next two models in the sequence below. Use the models below to show that the triangular numbers can also be obtained from the polynomial function $y = \frac{x^2 + x}{2}$, where $x$ is the term number.

| **Term number** | 1 | 2 | 3 | 4 | 5 |
|---|---|---|---|---|---|
| | 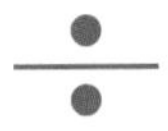 | 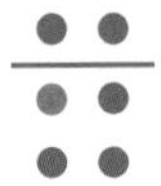 | 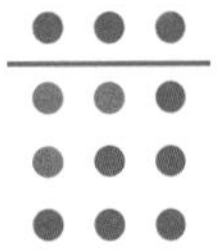 | ? | ? |

**2.** Verify the identity $\frac{x^2 + x}{2} = \frac{x(x+1)}{2}$ by building a table of values for the two functions $y = \frac{x^2 + x}{2}$ and $y = \frac{x(x+1)}{2}$.

9.5

# Common Factors

## Objectives

- Factor a polynomial by using the greatest common factor.
- Factor a polynomial by using a binomial factor.

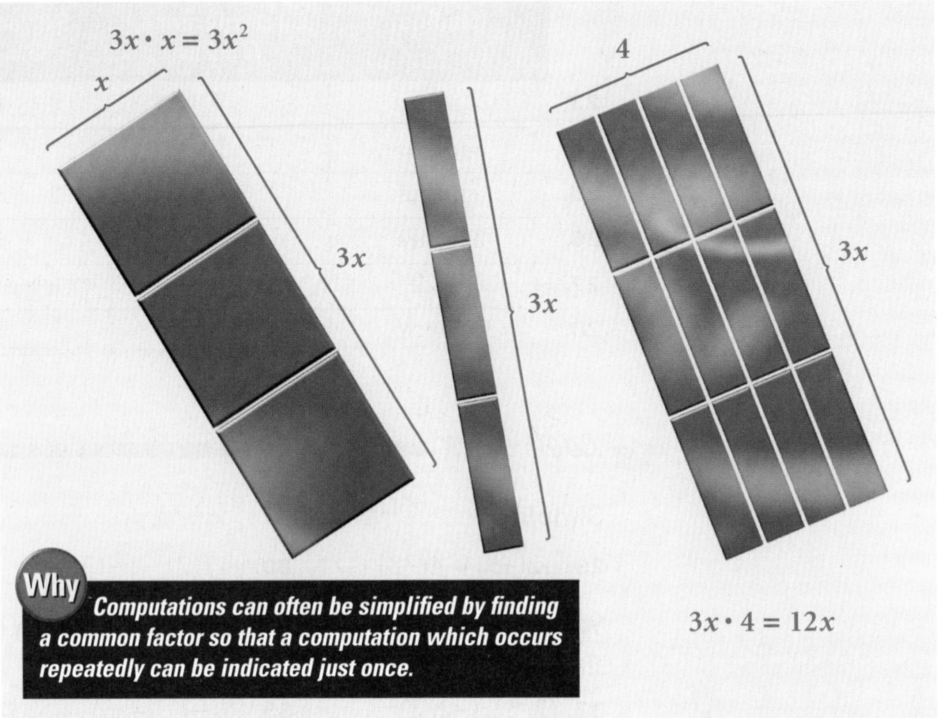


**Why** *Computations can often be simplified by finding a common factor so that a computation which occurs repeatedly can be indicated just once.*

The division of expressions you learned in Chapter 2 will help you in this lesson on common factors.

$$\left(20x^4 - 12x^3 + 8x^2\right) \div \left(4x^2\right)$$

$$\frac{20x^4 - 12x^3 + 8x^2}{4x^2}$$

$$\frac{20x^4}{4x^2} - \frac{12x^3}{4x^2} + \frac{8x^2}{4x^2}$$

$$5x^2 - 3x + 2$$

The division shown above indicates that $20x^4 - 12x^3 + 8x^2$ is the same as $4x^2\left(5x^2 - 3x + 2\right)$ where $4x^2$ and $5x^2 - 3x + 2$ are *factors* of $20x^4 - 12x^3 + 8x^2$.

When two numbers are multiplied, they form a product. Each number is called a **factor** of that product. When you are asked to factor a polynomial such as $3x^2 + 12x$, the first step is to examine the terms for the *greatest common factor*, or GCF. When you write $3x^2 + 12x$ as $(3x)x + (3x)4$, you can see that the two terms have $3x$ as the greatest common factor. You can then write the following:

$$3x^2 + 12x$$
$$3x(x + 4)$$

Some polynomials cannot be factored using the methods described in this lesson. Additional methods for factoring polynomials will be covered in later lessons.

**EXAMPLE 1** **Factor each polynomial.**

**a.** $5am - 5an$ **b.** $5x^3 - 3y^2$ **c.** $2c^4 - 4c^3 + 6c^2$

**SOLUTION**

Find the greatest common factor, or GCF, of the terms.

**a.** The GCF is $5a$.

$5am - 5an$

$5a(m - n)$

**b.** The GCF is 1.

$5x^3 - 3y^2$ cannot be factored using the method described above.

**c.** The GCF is $2c^2$.

$2c^4 - 4c^3 + 6c^2$

$2c^2(c^2 - 2c + 3)$

To check each factorization, multiply the two factors to find the original expression.

**TRY THIS** Factor each polynomial.

**a.** $6ab + 3a$ **b.** $5x^3 + 10x^2 - 20x$ **c.** $2x^2 - 3y^2$

A polynomial may have a common factor that is a binomial.

**EXAMPLE 2** **Factor $r(t + 1) + s(t + 1)$.**

**SOLUTION**

Notice that $r(t + 1) + s(t + 1)$ contains the **common binomial factor** $t + 1$. Use this common factor to write the following:

$$r(t + 1) + s(t + 1)$$
$$(t + 1)(r + s)$$

**TRY THIS** Factor $a(x - 3) + 6(x - 3)$.

Sometimes, polynomials can be factored by *grouping* terms.

**EXAMPLE 3** **Factor by grouping.**

**a.** $x^2 + x + 2x + 2$ **b.** $y^2 - y - 2y + 2$

**SOLUTION**

**a.** There are several ways to factor this polynomial by grouping. One way is to group the terms that have a common number or variable as a factor. Treat $x^2 + x$ as one expression, and treat $2x + 2$ as another expression.

$x^2 + x + 2x + 2$

$(x^2 + x) + (2x + 2)$ *Group terms.*

$x(x + 1) + 2(x + 1)$ *Factor the GCF from each group.*

$(x + 1)(x + 2)$ *$(x + 1)$ is a common factor.*

**b.** $y^2 - y - 2y + 2$

$(y^2 - y) - (2y - 2)$ *Note the sign change.*

$y(y - 1) - 2(y - 1)$

$(y - 1)(y - 2)$

To check, multiply the factors.

**TRY THIS** Factor by grouping.

**a.** $ax + bx + ay + by$ **b.** $ax + bx - ay - by$

# Exercises

## Communicate

1. Describe what a factor of a polynomial is.
2. What property of mathematics is used when you factor by using the GCF?
3. Explain how to find the GCF.
4. Once you have found the GCF of the following numbers and expressions, how do you find the remaining factor?
   a. 60 and 150  b. $x^3y^5$ and $x^5y^2$  c. $25(x+y)$ and $39(x+y)$
5. Explain how to group terms to factor the polynomial $5y^2 + 5y + 2y + 2$.

## Guided Skills Practice

**Factor each polynomial.** *(EXAMPLE 1)*

6. $10bc + 5b$
7. $3x^2 - 4y$
8. $4x^2y + 12xy^2 - 20xy$

**Factor.** *(EXAMPLE 2)*

9. $x(y+1) - 2(y+1)$
10. $m(n^2+3) + 4(n^2+3)$

**Factor by grouping.** *(EXAMPLE 3)*

11. $m^2 + m - 4m - 4$
12. $x^2 - 2ax + 3ax - 6a^2$

## Practice and Apply

internet connect
Homework Help Online
Go To: go.hrw.com
Keyword:
MA1 Homework Help for Exercises 13-42

**Factor each polynomial.**

13. $4x - 16$
14. $5r^2 + 10r$
15. $2n^2 + 4$
16. $5m^2 - 35$
17. $3p^2 + 12p$
18. $4f^2 + 4f$
19. $2x^2 - 4$
20. $5n^2 - 10$
21. $3x^2 + 6x$
22. $x^9 - x^2$
23. $k^5 + k^2$
24. $4a^8 - 20a^6 + 8a^4$
25. $4x^2 + 2x - 6$
26. $7x^2 - 28x - 14$
27. $27y^3 + 18y^2 - 81y$
28. $4n^3 - 16n^2 + 8n$
29. $100 + 25s^5 - 50s$
30. $2p^4r - 8p^3r^2 + 16p^2r^3$
31. $3m^3 - 9m^2 + 3m$
32. $90 + 15a^5 - 45a$
33. $2x^3y - 8x^2y^2 + 17xy^3$
34. $4x^4 - 24x^2$
35. $7y^3 - 21y^2 + 14y$
36. $42r - 14r^3$
37. $3x^2 + 21x^4$
38. $8a^4 + 4a^3 - 12a^2$
39. $2t^3 - 130t^5$

CHALLENGE

40. $x^{n+3} + x^n$
41. $9w^{2n} + 21w^{2n+1}$
42. $x^{3n+21} + 2x^{2n+14}$

**Write each polynomial as the product of two binomials.**

**43.** $x(x+1)+2(x+1)$ **44.** $5(y+3)-x(y+3)$ **45.** $a(x+y)+b(x+y)$

**46.** $(4+p)3q-4(4+p)$ **47.** $x(x-1)+2(x-1)$ **48.** $r(x-4)+t(x-4)$

**49.** $5a(a-3)+4(a-3)$ **50.** $2w(w+4)-3(w+4)$ **51.** $2(x-2)+x(2-x)$

**52.** $8(y-1)-x(y-1)$ **53.** $2r(r-s)^2-3(r-s)^2$ **54.** $3(m+7)-n(m+7)$

**Factor by grouping.**

**55.** $2x+2y+ax+ay$ **56.** $12ab-15a-8b+10$

**57.** $3(x+y)+12(x+y)$ **58.** $x^2+3x+4x+12$

CONNECTION

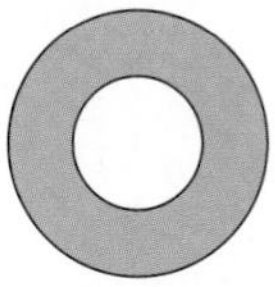

*The shaded area between two concentric circles is called an* ***annulus.***

**GEOMETRY** Donuts and bagels are in the shape of an annulus. The formula for the area of the annulus is $A=\pi R^2-\pi r^2$, where $R$ is the radius of the larger circle and $r$ is the radius of the smaller circle.

**59.** Write the formula for the area of an annulus in factored form.

**60.** Use the factored form to find the area of an annulus formed by concentric circles with radii of 8 and 5.

## Look Back

**61.** Evaluate $8-5(2-6)-3^2$. ***(LESSON 1.3)***

**62.** Solve the formula $A=\frac{1}{2}h(b_1+b_2)$ for $h$. ***(LESSON 3.6)***

**63.** Solve for $x$: $3(x-5)-2(x+8)=7x+3$. ***(LESSON 3.5)***

**64.** Graph $x-4y=8$. ***(LESSON 5.1)***

**65.** Solve. $\begin{cases} 4x=11+15y \\ 6x+5y=0 \end{cases}$ ***(LESSON 7.3)***

**66.** Is the sequence 3, 6, 12, 24, 48, . . . representative of a linear relationship or an exponential relationship? Why? ***(LESSON 8.6)***

**67.** Suppose that you want to save $6000 to buy a car when you graduate in 3 years. The current interest rate on a certificate of deposit (CD) is 8% compounded annually. How much will you need to invest in a CD now to buy the car in 3 years? ***(LESSON 8.7)***

## Look Beyond

APPLICATION

**68. PHYSICS** A certain projectile is fired vertically into the air at an initial velocity of 320 feet per second. Its motion can be modeled by the equation $h=-16t^2+320t$, where $h$ is the height at time $t$. Graph the equation, and find $t$ when $h=1200$.

9.6

# Factoring Special Polynomials

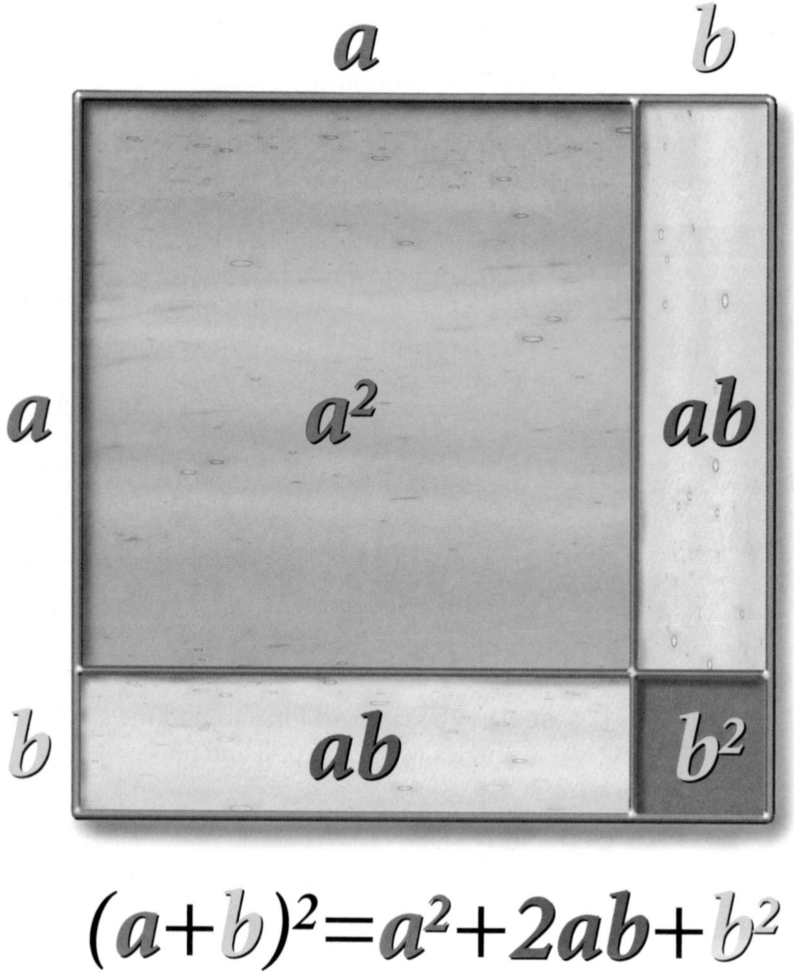


**Objectives**

- Factor perfect-square trinomials.
- Factor the difference of two squares.

**Why** *Perfect-square trinomials and the difference of two squares are polynomials that can be easily factored by applying simple rules.*

## Perfect-Square Trinomials

CONNECTION
GEOMETRY

You can see from the figure that the area of the large square is both $(a + b)^2$ and $a^2 + 2ab + b^2$. This figure illustrates why the product $a^2 + 2ab + b^2$ is called a **perfect-square trinomial.**

To decide whether a trinomial is a perfect-square trinomial, you will need to know something about squares and square roots.

Recall that $3^2 = 9$ and $(4x)^2 = 16x^2$. Terms 9 and $16x^2$ are called *perfect squares* and 3 and $4x$ are called *square roots* of 9 and $16x^2$, respectively.

To see if a trinomial such as $4x^2 + 12x + 9$ is a perfect square, use this test.

- Is the first term, $4x^2$, a perfect square? Yes: $(2x)^2$
- Is the last term, 9, a perfect square? Yes: $3^2$
- Is the middle term, $12x$, twice the product of the square roots of the first and last terms? Yes: $2(2x)(3) = 12x$

Thus, $4x^2 + 12x + 9$ is a perfect-square trinomial.

**EXAMPLE 1** **State whether each expression is a perfect-square trinomial. Explain.**

**a.** $x^2 + 8x + 16$ **b.** $3x^2 + 16x + 16$ **c.** $x^2 + 3x + 9$ **d.** $x^2 + 10x - 25$

**SOLUTION**

**a.** Yes; the first and last terms are perfect squares and the middle term is twice the product of the square roots of the first and last terms.

**b.** No; the first term, $3x^2$, is not a perfect square.

**c.** No; the middle term is not twice the product of the square roots of the first and last terms.

**d.** No; the last term must always be a *positive* perfect square.

**Rule for Factoring a Perfect-Square Trinomial**

For all numbers $a$ and $b$:

$$a^2 + 2ab + b^2 = (a + b)(a + b) = (a + b)^2, \text{ and}$$
$$a^2 - 2ab + b^2 = (a - b)(a - b) = (a - b)^2$$

**EXAMPLE 2** **Factor each expression.** **a.** $x^2 - 10x + 25$ **b.** $49y^4 + 14y^2 + 1$

**SOLUTION**

**a.** Since the expression is a perfect-square trinomial, apply the rule. Using the square roots of the first and last terms, $x$ and 5, write $x^2 - 10x + 25 = (x - 5)^2$. (Notice the minus sign.)

**b.** Since the expression is a perfect-square trinomial, apply the rule. Using the square roots of the first and last terms, $7y^2$ and 1, write $49y^4 + 14y^2 + 1 = (7y^2 + 1)^2$.

**TRY THIS** Factor each expression. **a.** $9s^2 + 24s + 16$ **b.** $64a^4 - 16a^2b + b^2$

## The Difference of Two Squares

When you multiply $(a + b)(a - b)$, the product is $a^2 - b^2$. This product is called the **difference of two squares**.

**EXAMPLE 3** **State whether each expression is a difference of two squares. Explain.**

**a.** $4a^2 - 25$ **b.** $9x^2 - 15$ **c.** $16r^2 + 49$

**SOLUTION**

**a.** Yes; the expression involves subtraction, $4a^2$ is the square of $2a$, and 25 is the square of 5.

**b.** No; 15 is not a perfect square.

**c.** No; it is the sum of two squares.

## Activity
## Patterns in Differences of Two Squares

1. Multiply the following pairs of binomials. Describe the pattern that you observe.
   - **a.** $(x+2)(x-2)$
   - **b.** $(x+3)(x-3)$
   - **c.** $(x+4)(x-4)$
   - **d.** $(x+5)(x-5)$
2. Use the pattern from Step 1 to factor the following differences of two squares:
   - **a.** $x^2-1$
   - **b.** $x^2-36$
   - **c.** $x^2-64$
   - **d.** $x^2-100$

CHECKPOINT ✔ 3. Based on the patterns from Steps 1 and 2, write a rule for factoring binomials of the form $a^2-b^2$.

There is a useful rule for factoring differences of two squares.

**Rule for Factoring a Difference of Two Squares**

For all numbers $a$ and $b$: $a^2-b^2=(a+b)(a-b)$

**EXAMPLE 4** **Factor each expression.**

**a.** $x^2-4$ **b.** $36a^2-49b^2$

**SOLUTION**

**a.** Apply the rule for factoring a difference of two squares.

$$x^2-4=(x+2)(x-2)$$

**b.** Apply the rule for factoring a difference of two squares.

$$\begin{aligned}36a^2-49b^2 &= (6a)^2-(7b)^2\\ &= (6a+7b)(6a-7b)\end{aligned}$$

**TRY THIS** Factor $25w^2-81$.

**CRITICAL THINKING** Explain why $a^8-b^8$ has more than two factors. What are all of the factors?

**EXAMPLE 5** **Find 17 • 13 by using the difference of two squares.**

**SOLUTION**

Think of 17 • 13 as $(15+2)(15-2)$.

The product is $15^2-2^2=225-4=221$.

**TRY THIS** Find 34 • 26 by using the difference of two squares.

# Exercises

## Communicate

**1.** Describe the process for finding the factors of the perfect-square trinomial $x^2 + 20x + 100$.

**2.** Explain how to factor $4x^2 - 12x + 9$.

**3.** Explain how to factor this difference of two squares: $p^2 - 121$.

## Guided Skills Practice

**State whether each expression is a perfect-square trinomial. Explain.** *(EXAMPLE 1)*

**4.** $x^2 + 16x + 64$ **5.** $4b^2 - 32bc + 64c^2$ **6.** $3x^2 + 2x - 7$

**Factor.** *(EXAMPLE 2)*

**7.** $x^2 + 10x + 25$ **8.** $36a^2 - 60a + 25$ **9.** $16x^2 + 8xy + y^2$

**State whether each expression is a difference of two squares. Explain.** *(EXAMPLE 3)*

**10.** $9q^2 - 16r^2$ **11.** $4k^4 - 25l^2$ **12.** $25x^2 - 5y^2$

**Factor.** *(EXAMPLE 4)*

**13.** $a^2 - b^2$ **14.** $36x^2 - 49y^2$ **15.** $4m^2 - 144n^2$

**Find each product by using the difference of two squares.** *(EXAMPLE 5)*

**16.** $47 \cdot 53$ **17.** $97 \cdot 103$

## Practice and Apply

**Factor each polynomial completely.**

**18.** $y^2 + 8y + 16$ **19.** $x^2 + 4x + 4$ **20.** $r^2 - 18r + 81$

**21.** $x^2 - 4$ **22.** $36d^2 + 12d + 1$ **23.** $4t^2 - 1$

**24.** $81 - 4m^2$ **25.** $25x^2 - 9$ **26.** $y^2 - 100$

**27.** $4x^2 - 20x + 25$ **28.** $100 - 36q^2$ **29.** $16c^2 - 25$

**30.** $p^2 - q^2$ **31.** $9c^2 - 4d^2$ **32.** $16x^2 + 72xy + 81y^2$

**33.** $9a^2 - 12a + 4$ **34.** $49x^2 - 42xy + 9y^2$ **35.** $a^2x^2 + 2axb + b^2$

**36.** $4m^2 + 4mn + n^2$ **37.** $81a^4 - 9b^2$ **38.** $x^4 - y^4$

CHALLENGE

**39.** $x^2(25 - x^2) - 4(25 - x^2)$ **40.** $(x - 1)x^2 - 2x(x - 1) + (x - 1)$

**41.** $(3x + 5)(x^2 - 3) - (3x + 5)$ **42.** $(x^2 - y^2)(x^2 + 2xy) + (x^2 - y^2)(y^2)$

internet connect
**Homework Help Online**
Go To: **go.hrw.com**
Keyword:
**MA1 Homework Help** for Exercises 43-48

**Find the missing term in each perfect-square trinomial.**

**43.** $x^2 - 14x + ?$
**44.** $16y^2 + ? + 9$
**45.** $25a^2 + 60a + ?$
**46.** $9x^2 + ? + 25$
**47.** $x^2 - 12x + ?$
**48.** $? - 36y + 81$

**49. CULTURAL CONNECTION: AFRICA** Abu Kamil, known as the "Egyptian calculator," used geometric models to solve problems around the year 900 C.E. You can make replicas of his models with algebra tiles or paper rectangles.

Use a sheet of paper to create a model of $a^2 - b^2$ as shown. Remove the $b^2$ section from the corner. Then cut along the dotted lines, and move the pieces to form a rectangle. Explain how this represents the factorization of $a^2 - b^2$.

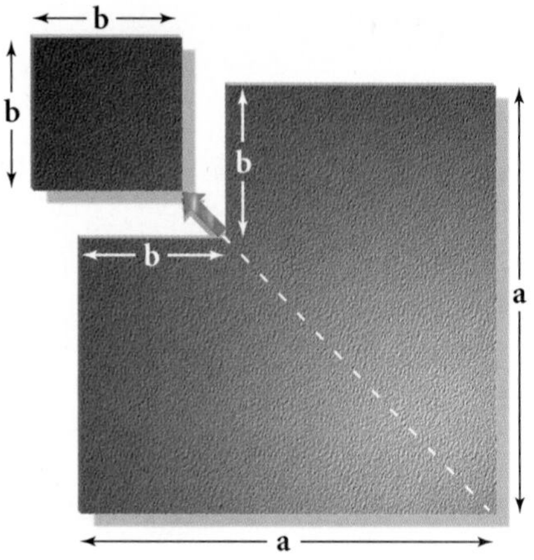


**CONNECTIONS**

**GEOMETRY** The area of a square is represented by $n^2 - 12n + 36$.

**50.** Find the length of each side of the square.

**51.** Find the perimeter of the square.

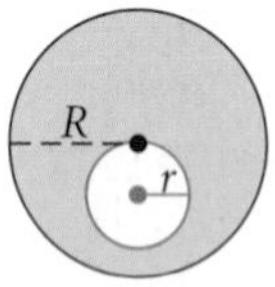


**52. GEOMETRY** A small circle with a radius of $r$ units is drawn within a large circle with a radius of $R$ units, as shown at left. What is the area of the large circle? What is the area of the shaded part? Factor the expression for the shaded area. (Hint: First factor out the common monomial factor, $\pi$. Then factor the binomial.)

**APPLICATION**

**CONSTRUCTION** A construction firm is designing a fountain to be built in several locations in varying sizes. The diagram below shows the dimensions of the proposed fountain.

**53.** What is the area of the top of one of the bases, including the top surface of the water?

**54.** What is the area of one of the square water surfaces in the center?

**55.** What is the area of the top of one of the stone surfaces surrounding the water?

**56.** Factor the polynomial that represents the area of the top of one of the stone surfaces.

**57.** Draw a rectangle by using the two factors you found in Exercise 56 for the dimensions.

**CHALLENGE**

**58.** Show that the area of this rectangle equals the area of the stone surface.

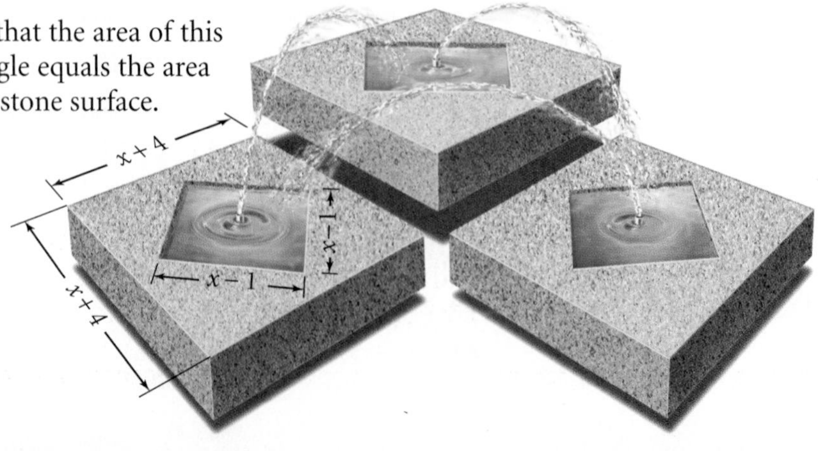

## Look Back

CONNECTION

**59.** STATISTICS Mark has scores of 78, 83, and 92 on his first three tests. What must his score be on the next test in order to have an average of 87? *(LESSON 4.4)*

**60.** Find the slope of a line that is parallel to the line represented by the equation $3y = 5 - 4x$. *(LESSON 5.6)*

**61.** Find the slope of a line that is perpendicular to the line you found in Exercise 60. *(LESSON 5.6)*

**62.** Graph the solution of $|x| < 4$ on a number line. *(LESSON 6.5)*

**63.** Graph this system of inequalities. $\begin{cases} y < x - 5 \\ y \geq 3 \end{cases}$ *(LESSON 7.5)*

## Look Beyond

**64.** Given the trinomial $x^2 + x - 42$, what are possible pairs of factors for $-42$? Which factors result in the sum of 1? What does the sign before the last term of the trinomial determine?

PORTFOLIO ACTIVITY

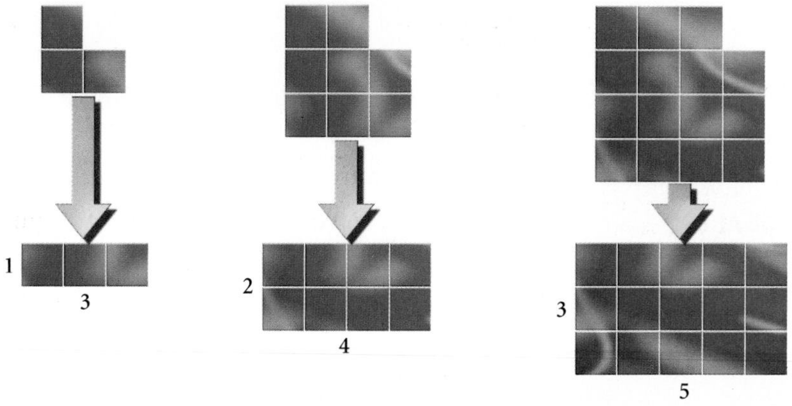


3 8 15

Include this activity in your portfolio.

1. Find the next three terms in the sequence 0, 3, 8, 15, . . . Compare this sequence with the sequence of square numbers, 1, 4, 9, 16, . . . Write a polynomial function to describe each sequence.
2. Draw the next two models in the sequence shown above. Label the dimensions.
3. Use the model to explain why $x^2 - 1 = (x + 1)(x - 1)$.
4. Show that each side of the identity $x^2 - 1 = (x + 1)(x - 1)$ has the same value by building a table for the two functions, $y = x^2 - 1$ and $y = (x + 1)(x - 1)$, using integer values from $-3$ to 3 for $x$.

**WORKING ON THE CHAPTER PROJECT**

You should now be able to complete Activities 2 and 3 of the Chapter Project on page 470.

# 9.7 Factoring Quadratic Trinomials

**Why** *If you examine certain products carefully, you will discover patterns that enable you to recognize the factors. This will simplify your work with real-world problems involving trinomials.*

**Objectives**

- Factor quadratic trinomials by using guess-and-check.
- Factor quadratic trinomials by using a grouping process.

**APPLICATION**
**FINE ARTS**

If a rectangular area is covered by 30 uniform squares, what could the length and width of the area be? To answer this question, you would need to find pairs of numbers that are factors of 30.

## Factoring by Guess-and-Check

Suppose that the area of a rectangle is represented by the quadratic trinomial $x^2 + 6x + 8$. How can you find expressions for its length and width? One way to find the length and width is to find the factors.

Because factoring is related to multiplication, a quadratic trinomial can be factored by working backward with the FOIL method, using guess-and-check.

**EXAMPLE 1** **Factor each trinomial.**

**a.** $x^2 + 6x + 8$ **b.** $x^2 - 6x + 8$

**SOLUTION**

PROBLEM SOLVING

**Work backward.** Because the constant term of the trinomial is positive, the constant terms in the binomial factors must be both positive or both negative.

$$(x + \_\_)(x + \_\_) \text{ or } (x - \_\_)(x - \_\_)$$

**a.** Because the coefficient of the middle term is positive, the constant terms in the binomial factors must both be positive. Find a pair of numbers that have a product of 8 and a sum of 6.

Try 2 and 4. *Guess*

$(x + 2)(x + 4)$

$x^2 + 4x + 2x + 8$ *Check*

$x^2 + 6x + 8$ *Correct*

Therefore, the factors are $(x + 2)$ and $(x + 4)$.

**b.** Because the coefficient of the middle term is negative, the constant terms in both binomial factors must be negative. Find a pair of numbers that have a product of 8 and a sum of –6.

Try –2 and –4. *Guess*

$(x - 2)(x - 4)$

$x^2 - 2x - 4x + 8$ *Check*

$x^2 - 6x + 8$ *Correct*

Therefore, the factors are $(x - 2)$ and $(x - 4)$.

**TRY THIS** Factor each trinominal.

**a.** $x^2 + 14x + 40$ **b.** $x^2 - 10x + 16$

**CRITICAL THINKING** If the constant term of a trinomial is positive, explain why the constant terms in the binomial factors must be both positive or both negative.

**EXAMPLE 2** **Factor each trinomial.**

**a.** $x^2 + 2x - 8$ **b.** $x^2 - 2x - 8$

**SOLUTION**

**Work backward.** Because the constant term of the trinomial is negative, one factor will contain $(x - \_\_\_)$, and the other will contain $(x + \_\_\_)$.

**a.** Because the coefficient of the middle term is positive, the positive constant term in one factor must be larger than the absolute value of the negative constant term in the other factor. Find a pair of numbers that have a product of –8 and a sum of 2.

Try –2 and 4. *Guess*

$(x - 2)(x + 4)$

$x^2 + 4x - 2x + 8$ *Check*

$x^2 + 2x - 8$ *Correct*

Thus, the factors are $(x - 2)$ and $(x + 4)$.

**b.** Because the coefficient of the middle term is negative, the positive constant term of one factor must be smaller than the absolute value of the negative constant term of the other factor. Find a pair of numbers that have a product of –8 and a sum of –2.

Try 2 and –4. *Guess*

$(x + 2)(x - 4)$

$x^2 - 4x + 2x - 8$ *Check*

$x^2 - 2x - 8$ *Correct*

Thus, the factors are $(x + 2)$ and $(x - 4)$.

**TRY THIS** Factor each trinomial.

**a.** $x^2 + 9x + 18$ **b.** $x^2 - 10x + 21$ **c.** $x^2 + 4x - 21$ **d.** $x^2 - x - 20$

**CRITICAL THINKING** For the trinomial in Example 2a, explain why the positive factor must be larger than the absolute value of the negative factor.

## Factoring by Grouping

In Lesson 9.5 polynomials were factored by a method called grouping. This method can be used to factor quadratic trinomials by first writing the trinomial as a polynomial.

In the polynomials $x^2 + 9x + 14$ and $x^2 + 7x + 2x + 14$ notice that $+7x + 2x$ in the second polynomial is equivalent to $+ 9x$, which is the middle term of the first trinomial. The two polynomials are equivalent. Note also that $+7x$ and $+2x$ are factors of the product of the first term, $x^2$, and the last term, $+14$, of the first trinomial.

$$(+7x) + (+2x) = +9x$$

$$(+7x)(+2x) = +14x^2$$

You can use this information to help you factor quadratic trinomials.

**EXAMPLE 3** **Factor $x^2 + 9x + 14$ by grouping.**

**SOLUTION**

Find two terms that are factors of the product of the first and the third terms and whose sum is the middle term. Use $+2x$ and $+7x$ because $2x \cdot 7x = 14x^2$ and $2x + 7x = 9x$.

$x^2 + 9x + 14$

$x^2 + 7x + 2x + 14$ *Replace +9x with +7x + 2x.*

$(x^2 + 7x) + (2x + 14)$ *Group terms.*

$x(x + 7) + 2(x + 7)$ *Factor the GCF from each group.*

$(x + 7)(x + 2)$ *(x + 7) is a common factor.*

Check your work by FOIL.

$$(x + 7)(x + 2)$$
$$x^2 + 2x + 7x + 14$$
$$x^2 + 9x + 14$$

**TRY THIS** Factor each polynomial by grouping.

**a.** $x^2 + 7x + 12$ **b.** $x^2 + 7x - 18$

To factor a quadratic trinomial such as $9x^2 - 15x + 4$ use the grouping method described above.

The next example is a trinomial that may look like a perfect-square trinomial but does not satisfy all the conditions.

**EXAMPLE 4** **Factor $9x^2 - 15x + 4$ by grouping.**

**SOLUTION**

The term $-15x$ will need to be replaced by two terms whose sum is $-15x$ and whose product is $(9x^2)(+4)$ or $+36x^2$. The two terms will be $-12x$ and $-3x$ because $(-12x) + (-3x) = -15x$ and $(-12x)(-3x) = +36x^2$.

| | |
|---|---|
| $9x^2 - 15x + 4$ | |
| $9x^2 - 12x - 3x + 4$ | *Replace $-15x$ with $-12x - 3x$.* |
| $(9x^2 - 12x) - (3x - 4)$ | *Use parentheses to group the terms. (Note the sign in the second parenthesis.)* |
| $3x(3x - 4) - 1(3x - 4)$ | *Factor the GCF from each group.* |
| $(3x - 4)(3x - 1)$ | *$(3x - 4)$ is a common factor.* |

Check your work by FOIL.

$$(3x - 4)(3x - 1)$$
$$9x^2 - 3x - 12x + 4$$
$$9x^2 - 15x + 4$$

**TRY THIS** Factor each trinomial by grouping.

**a.** $3x^2 - 10x + 3$ **b.** $6x^2 + 7x - 3$

**CRITICAL THINKING** Can perfect-square trinomials be factored by the grouping method? Can the difference of two squares be factored this way?

Sometimes the terms of a trinomial have a common monomial factor. To write the trinomial as a product, it may be convenient to factor out the common monomial first.

**EXAMPLE 5** **Write $2x^3 + 16x^2 + 24x$ as a product.**

**SOLUTION**

$$2x^3 + 16x^2 + 24x = 2x(x^2 + 8x + 12)$$
$$= 2x(x + 2)(x + 6)$$

**TRY THIS** Write $2x^3 + 9x^2 + 9x$ as a product.

| SUMMARY OF FACTORING METHODS | | |
|---|---|---|
| **Method** | **Example polynomial** | **Factored form** |
| Use the GCF | $4x^2 + 12x$ | $4x(x + 3)$ |
| Factor by grouping | $x^2 + x + 4x + 4$ | $(x + 4)(x + 1)$ |
| Perfect-square trinomial pattern | $x^2 + 8x + 16$<br>$x^2 - 8x + 16$ | $(x + 4)(x + 4)$<br>$(x - 4)(x - 4)$ |
| Difference of squares pattern | $x^2 - 25$ | $(x + 5)(x - 5)$ |
| Working backward (guess-and-check)<br>Grouping | $x^2 + 3x + 2$<br>$4x^2 + 12x - 7$ | $(x + 2)(x + 1)$<br>$(2x - 1)(2x + 7)$ |

# Exercises

## Communicate

**1.** Write each trinomial in factored form. Tell why the signs in the binomial factors are the same or opposite.

**a.** $x^2 + x - 6$ **b.** $x^2 - 7x + 10$ **c.** $x^2 + 2x - 15$

**2.** Explain how to use the guess-and-check method to factor $x^2 - 5x - 24$.

**3.** If the third term of a trinomial is 36, write the possible factor pairs.

**4.** Explain how to use grouping to factor each polynomial.

**a.** $x^2 - 5x + 4$ **b.** $3x^2 - 5x - 12$

## Guided Skills Practice

**Factor each trinomial by guess-and-check.**

**5. a.** $x^2 + 7x + 12$ **b.** $x^2 - 7x + 12$ ***(EXAMPLE 1)***

**6. a.** $x^2 + x - 12$ **b.** $x^2 - x - 12$ ***(EXAMPLE 2)***

**Factor each trinomial by grouping.**

**7.** $x^2 - 13x + 40$ ***(EXAMPLE 3)***

**8.** $4x^2 - 12x - 7$ ***(EXAMPLE 4)***

**9.** Write $3x^3 + 18x^2 + 27x$ as a product. ***(EXAMPLE 5)***

## Practice and Apply

**Factor each of the following trinomials:**

**10.** $x^2 + 5x + 6$ **11.** $x^2 - x - 12$ **12.** $x^2 - 6x + 5$ **13.** $x^2 - 6x + 9$

internet connect

**Homework Help Online**
Go To: **go.hrw.com**
Keyword:
**MA1 Homework Help**
for Exercises 14-37

**Factor each trinomial. If a trinomial cannot be factored, write *prime*.**

**14.** $a^2 - 2a - 35$ **15.** $p^2 + 4p - 12$ **16.** $y^2 - 5y + 6$

**17.** $b^2 - 5b - 24$ **18.** $n^2 - 11n + 18$ **19.** $z^2 + z - 20$

**20.** $3x^2 + 5x + 12$ **21.** $2a^2 - 2a + 7$ **22.** $3x^2 - 17x - 28$

**23.** $2s^2 - 27s + 63$ **24.** $5x^2 + 31x - 28$ **25.** $4m^2 - 3m + 17$

**Factor each polynomial. In some cases you will need to factor out a constant monomial term.**

**26.** $4x^3y - 20x^2y + 16xy$ **27.** $12y^3 - 30y^2 + 12y$

**28.** $x^2 - 18x + 81$ **29.** $(a + 3)(a^2 + 5a) - 6(a + 3)$

**30.** $5x^3 - 50x^2 + 45x$ **31.** $x^3 + 2x^2 - 36x - 72$

**32.** $-x^4 + 2x^2 + 8$ **33.** $64p^4 - 16$

**34.** $4xz^2 - 2xz - 72x$ **35.** $x^2 - 2x + 1$

**36.** $125x^2y - 5x^4$ **37.** $2ax + ay + 2bx + by$

CHALLENGE

**38.** Find all possible values of $b$ such that $x^2 + bx + 12$ can be factored.

**39.** Find all possible values of $b$ such that $x^2 + bx - 12$ can be factored.

CONNECTION

**40.** **GEOMETRY** A rectangle has an area of $49x^2 - 64$. The base of the rectangle is $7x + 8$. What is its height? Could the rectangle be a square?

## Look Back

**Solve.** ***(LESSONS 3.1 AND 3.2)***

**41.** $\frac{3}{4}a = 163$ **42.** $\frac{z}{-8} = \frac{11}{12}$ **43.** $w - \frac{7}{9} = 93$

APPLICATION

**44.** **SALES TAX** A freshman class is donating a tree to a local children's hospital. If sales tax is 5.5%, what is the total price of a tree marked at $139? ***(LESSON 4.2)***

**45.** A line with a slope of $-2$ passes through point $A(4, -1)$. Write the equation of the line that passes through this point and that is perpendicular to the original line. ***(LESSON 5.6)***

**46.** Represent the inequality $-4 < x < 5$ on a number line. ***(LESSON 6.3)***

## Look Beyond

**CULTURAL CONNECTION** Americas de Padilla, a Guatemalan mathematician, wrote this problem 250 years ago.

**47.** Find two numbers, given that the second is 3 times the first. If you multiply the first number by the second number and the second number by 4, the sum of the products is 420.

9.8

# Solving Equations by Factoring

**Objective**

- Solve equations by factoring.

**Why** *Factoring can be used to help solve equations that model real-world problems.*

*The height of a kicked soccer ball can be modeled approximately by a polynomial equation.*

**APPLICATION**
**PHYSICS**

The height of a soccer ball kicked with an initial upward velocity of 48 feet per second can be approximated by the function $y = -16x^2 + 48x$, where $y$ is the height in feet and $x$ is the time in seconds. You can use factoring to find the values of $x$ that make $y$ have a value of 0.

## The Zeros of a Function

A function such as $y = (x - 2)(x - 4)$ is a *polynomial function* because the product on the right simplifies to a polynomial, $x^2 - 6x + 8$. (You can verify this by using algebra tiles.) What happens when you substitute the numbers 2 and 4 for $x$ in the function?

$$y = (x - 2)(x - 4) = (2 - 2)(2 - 4) = (0)(-2) = 0 \qquad \textit{Substitute 2 for x.}$$
$$y = (x - 2)(x - 4) = (4 - 2)(4 - 4) = (2)(0) = 0 \qquad \textit{Substitute 4 for x.}$$

The numbers 2 and 4 are called the *zeros* of the function $y = x^2 - 6x + 8$ because the value of the function is 0 when these numbers are substituted for the independent variable, $x$.

When a polynomial function is written as a product of binomials, you can easily find the zeros of the polynomial function by using the Zero Product Property.

## Zero Product Property

If $a$ and $b$ are real numbers such that $ab = 0$, then $a = 0$ or $b = 0$.

**EXAMPLE 1** **Find the zeros of the polynomial function $y = (x + 2)(x - 3)$.**

**SOLUTION**

The function will equal 0 when either factor equals 0. Set each factor equal to 0 and solve for $x$.

| **First factor** | **Second factor** |
|---|---|
| $(x + 2) = 0$ | $(x - 3) = 0$ |
| $x = -2$ | $x = 3$ |

The zeros of the polynomial function are −2 and 3.

Check by substituting these values into the original polynomial function.

$y = (x + 2)(x - 3) = (-2 + 2)(-2 - 3) = (0)(-5) = 0$ *Substitute −2 for x.*

$y = (x + 2)(x - 3) = (3 + 2)(3 - 3) = (5)(0) = 0$ *Substitute 3 for x.*

Since the polynomial function equals 0 in each case, the zeros are correct.

**TECHNOLOGY**
**GRAPHICS CALCULATOR**

Use the table feature of your graphics calculator to build a table of values for the function $y = (x + 5)(x + 2)$.

Look for $x$-values that give a $y$-value of 0. You can see that the zeros of the function $y = (x + 5)(x + 2)$ are −5 and −2.

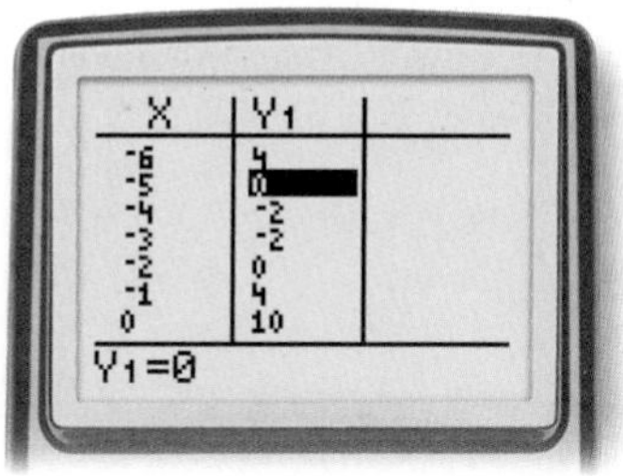


**EXAMPLE 2** **Find the zeros of the polynomial function $y = x^2 - 11x + 28$.**

**SOLUTION**

Write the function in factored form.

$y = x^2 - 11x + 28$
$y = (x - 4)(x - 7)$

The function will equal 0 when either factor equals 0. Set each factor equal to 0 and solve for $x$.

| **First factor** | **Second factor** |
|---|---|
| $(x - 4) = 0$ | $(x - 7) = 0$ |
| $x = 4$ | $x = 7$ |

The zeros of the polynomial function are 4 and 7.

Check by substitution these values into the original polynomial function.

$y = x^2 - 11x + 28 = (4)^2 - 11(4) + 28 = 16 - 44 + 28 = 0$ *Substitute 4 for x.*

$y = x^2 - 11x + 28 = (7)^2 - 11(7) + 28 = 49 - 77 + 28 = 0$ *Substitute 7 for x.*

Since the polynomial function equals 0 in each case, the zeros are correct.

## Using Factoring to Solve Equations

The solution to an equation like $-16x^2 + 96x = 0$ can be found by writing the equation in factored form and considering the values that make either factor equal to 0. The Zero Product Property allows you to solve many polynomial equations by factoring.

**EXAMPLE 3** **Solve by factoring: $-16x^2 + 96x = 0$.**

TECHNOLOGY
GRAPHICS CALCULATOR

**SOLUTION**

Write the equation in factored form.

$-16x^2 + 96x = 0$
$-16x(x - 6) = 0$

The equation is true when $-16x = 0$ or $x - 6 = 0$. Solve each factor for $x$.

$-16x = 0 \qquad x - 6 = 0$
$x = 0 \qquad x = 6$
$x = 0$ or $x = 6$ The solutions are 0 and 6.

You can check your solutions by using the table and graph features on some graphics calculators.

*The table and the graph show that $y = 0$ when $x = 0$ or $x = 6$.*

**TRY THIS** Solve by factoring: $-10x^2 + 100x = 0$.

**EXAMPLE 4** **Solve by factoring.**

**a.** $x^2 + 5x - 24 = 0$ **b.** $4x^2 + x - 5 = 0$

**SOLUTION**

**a.** Write the equation in factored form.

$x^2 + 5x - 24 = 0$
$(x + 8)(x - 3) = 0$

The equation is true when $x + 8 = 0$ or $x - 3 = 0$.

Solve each factor for $x$.

$x + 8 = 0 \qquad x - 3 = 0$
$x = -8 \qquad x = 3$
$x = -8$ or $x = 3$ The solutions are $-8$ and 3.

**b.** Write the equation in factored form.

$$4x^2 + x - 5 = 0$$
$$(4x + 5)(x - 1) = 0$$

The equation is true when $4x + 5 = 0$ or $x - 1 = 0$.

Solve each factor for $x$.

$$4x + 5 = 0 \qquad x - 1 = 0$$
$$4x = -5 \qquad x = 1$$
$$x = -\frac{5}{4}$$

$x = -\frac{5}{4}$ or $x = 1$ The solutions are $-\frac{5}{4}$ and 1.

**TRY THIS** Solve by factoring.

**a.** $x^2 + 6x - 16 = 0$

**b.** $2x^2 + 5x - 3 = 0$

**TECHNOLOGY GRAPHICS CALCULATOR**

In Example 3, part **b**, you found that the solutions to the equation $4x^2 + x - 5 = 0$ are $-\frac{5}{4}$ and 1. Thus, $-\frac{5}{4}$ and 1 are called the zeros of the polynomial function $y = 4x^2 + x - 5$. You can find the zeros of a polynomial function by using special menu commands on many graphics calculators.

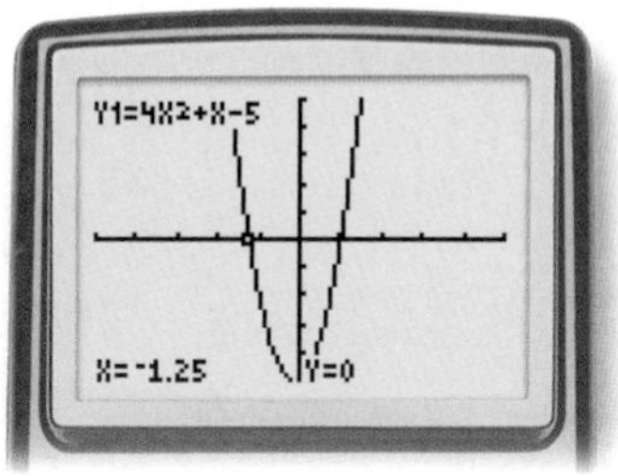


## Exercises

### Communicate

**internet connect**
**Activities Online**
Go To: **go.hrw.com**
Keyword:
**MA1 Zeros**

**1.** Explain what it means to solve an equation by factoring. Give two examples.

**2.** What are the zeros of a polynomial function?

**3.** Describe a polynomial function whose zeros are 2 and 3.

**4.** Is it possible for a polynomial function to have no real zeros? Explain.

**5.** The height of a projectile over time can be modeled by a polynomial function. Explain what the zeros of this function would represent.

## Guided Skills Practice

**Identify the zeros of each function.** *(EXAMPLE 1 AND 2)*

**6.** $y = (x + 5)(x + 2)$ **7.** $y = (x + 2)(x - 10)$ **8.** $y = x^2 + x - 42$

**Solve by factoring.** *(EXAMPLE 3)*

**9.** $-25x^2 + 5x = 0$ **10.** $12x^2 - 3x = 0$ **11.** $-15x^2 + 30x = 0$

**Solve by factoring.** *(EXAMPLE 4)*

**12.** $x^2 + 7x + 10 = 0$ **13.** $x^2 - 7x + 10 = 0$ **14.** $2x^2 + x - 10 = 0$

## Practice and Apply

internet connect
Homework Help Online
Go To: go.hrw.com
Keyword:
MA1 Homework Help for Exercises 15-35

**Identify the zeros of each function.**

**15.** $y = (x + 5)(x - 2)$ **16.** $y = (x + 15)(x - 7)$ **17.** $y = (x + 6)(x + 6)$

**18.** $y = (x + 4)(x - 3)$ **19.** $y = x^2 + 7x - 30$ **20.** $y = 4x^2 - 4$

**Solve by factoring.**

**21.** $x^2 - 6x + 9 = 0$ **22.** $x^2 - 6x - 7 = 0$ **23.** $x^2 - 10x + 9 = 0$

**24.** $x^2 - 9x + 18 = 0$ **25.** $x^2 + 3x - 18 = 0$ **26.** $2x^2 - 3x + 1 = 0$

**27.** $2x^2 - x - 1 = 0$ **28.** $x^2 - 5x - 36 = 0$ **29.** $x^2 - 15x + 36 = 0$

**30.** $x^2 - 16x = 36$ **31.** $x^2 - 9x = 36$ **32.** $x^2 + x = 12$

**33.** $x^2 - 7x = -12$ **34.** $x^2 + 7x = -12$ **35.** $x^2 - x = 12$

APPLICATION

**CONSTRUCTION** Alice has 250 feet of fencing. She decides to use it to build a pen. Use this information in Exercises 36–41.

**36.** Copy and complete the table of values to show the length, width, and area of each possible pen.

| Fixed perimeter | Length | Width | Area |
|---|---|---|---|
| 250 | 25 | ? | ? |
| 250 | 50 | ? | ? |
| 250 | 100 | ? | ? |

**37.** Which pen listed in the table has the largest area?

**38.** **TECHNOLOGY** Use your graphics calculator and the equation **Y1=X(125–X)** to build a table of values, where **Y1** is the area and **X** is the length. Explain why this function can be used to build the table above.

**39.** Find the zeros of the function $y = x(125 - x)$. What do the zeros mean in this problem?

**40.** Graph the function in Exercise 39. Find the dimensions that maximize the area of the pen. (What are the coordinates of the highest point on the graph?)

CHALLENGE

**41.** Solve the equation $2500 = x(125 - x)$ by factoring. What are the solutions of the equation? What does the equation $2500 = x(125 - x)$ mean in this problem?

## Look Back

**Solve each equation for *r*.** ***(LESSON 3.6)***

**42.** $C = 2\pi r$ **43.** $m = 4rt$ **44.** $rs = 4t$

**Solve each equation or inequality for *x*.** ***(LESSONS 3.6 AND 6.2)***

**45.** $2.5x - 2 = 1$ **46.** $5x = 2x + 16$ **47.** $x - 1 \leq 2x + 5$

**PROBABILITY** Two coins were tossed 50 times. Two heads appeared 23 times. Two tails appeared 21 times. Find the experimental probability of each outcome below. ***(LESSON 4.3)***

**48.** two heads **49.** two tails **50.** one head and one tail

## Look Beyond

**internet connect**
**Portfolio Extension**
Go To: **go.hrw.com**
Keyword:
**MA1 Rectangular**

**Solve each equation by graphing.**

**51.** $x^2 + 5x - 6 = 0$ **52.** $x^2 + x - 1 = 0$ **53.** $2x^2 - 2x + 3 = 0$

### PORTFOLIO ACTIVITY

Include this activity in your portfolio.

The first three numbers in the sequence of rectangular numbers are modeled at right.

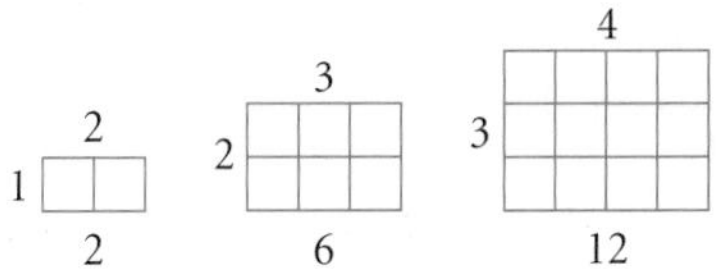


1. Draw the next two rectangular numbers by using the pattern shown.
2. Use the pattern to find the first 10 rectangular numbers.
3. Use the dimensions of each rectangle to write a function for the sequence.
4. Draw the next two rectangular numbers by using the model at right. Use this model to write a polynomial function for the sequence.

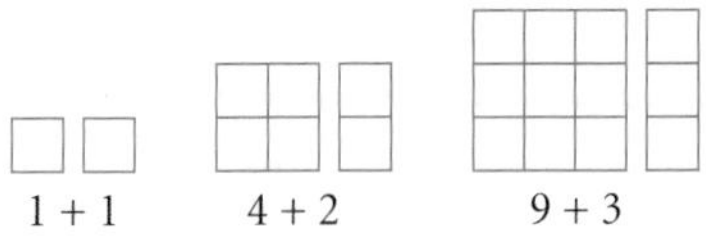


5. Write an identity by setting the two functions from Steps 3 and 4 equal to each other. Use the table feature on a graphics calculator to verify the identity.
6. What is the relationship between the rectangular numbers and the triangular numbers? (Hint: Recall the Portfolio Activity on page 447.)

**WORKING ON THE CHAPTER PROJECT**

You should now be able to complete Activity 4 of the Chapter Project on page 470.

CHAPTER PROJECT NINE

# POWERS, PASCAL, AND PROBABILITY

**CULTURAL CONNECTION: EUROPE AND ASIA** If a great mathematical discovery were made today, the discoverer would be discussed on the evening news. Throughout history, it has been much more difficult to share information about important mathematical ideas. Pascal's triangle, although known to the ancient Chinese for centuries, was discovered independently by French mathematician Blaise Pascal.

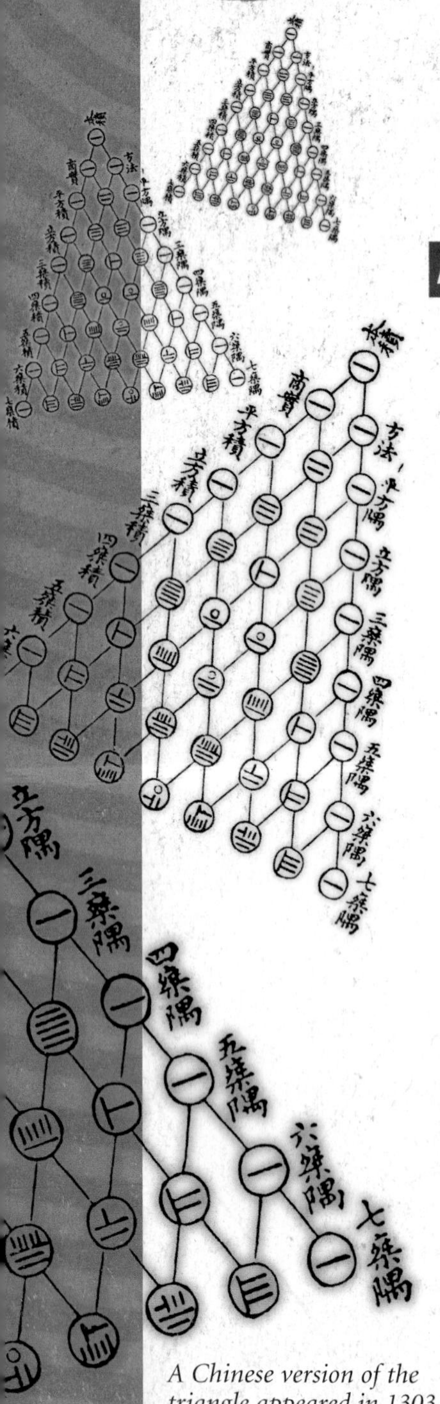

*A Chinese version of the triangle appeared in 1303.*

## Activity 1

Look at the different powers of the binomial $(a + b)$, and compare the expanded form with Pascal's triangle.

| Power | Expanded Form | Pascal's Triangle |
|---|---|---|
| $(a+b)^0$ | $1$ | 1 |
| $(a+b)^1 =$ | $1a + 1b$ | 1 1 |
| $(a+b)^2 =$ | $1a^2 + 2ab + 1b^2$ | 1 2 1 |
| $(a+b)^3 =$ | $1a^3 + 3a^2b + 3ab^2 + 1b^3$ | 1 3 3 1 |
| $(a+b)^4 =$ | $1a^4 + 4a^3b + 6a^2b^2 + 4ab^3 + 1b^4$ | 1 4 6 4 1 |
| $(a+b)^5 =$ | $1a^5 + 5a^4b + 10a^3b^2 + 10a^2b^3 + 5ab^4 + 1b^5$ | 1 5 10 10 5 1 |

1. Start with the expanded form of $(a + b)^2$, and multiply it by $a + b$. What is the result?
2. Do the same with the expanded forms of $(a + b)^3$ and $(a + b)^4$. What is the result?
3. How does the number of terms in each row of the expanded form compare with the exponent in the power of the binomial?
4. How many terms would you expect to be in the expansion of $(a + b)^7$?
5. Recalling from the Portfolio Activity on page 437 that new elements in the triangle are created by adding pairs of elements from a previous row, extend the triangle through row 10.
6. Write the coefficients of each term in the expansion of $(a + b)^7$.
7. Look at the terms in the expanded forms. Describe the pattern in the exponents as you read them from left to right in each row. What happens to the exponents of $a$? What happens to the exponents of $b$?
8. Write the complete algebraic expansion of $(a + b)^7$ without multiplying.

## Activity 2

1. Copy the table.

| | 3 heads | 2 heads, 1 tails | 1 heads, 2 tails | 3 tails |
|---|---|---|---|---|
| Tally marks | ? | ? | ? | ? |
| Totals | ? | ? | ? | ? |
| Totals ÷ 4 | ? | ? | ? | ? |

2. Toss three coins a total of 32 times. Record the results in the table.
3. Divide each total by 4, and round the answer to the nearest whole number. What are the results for each column?

## Activity 3

A tree diagram is a method for considering all possible outcomes of an experiment.

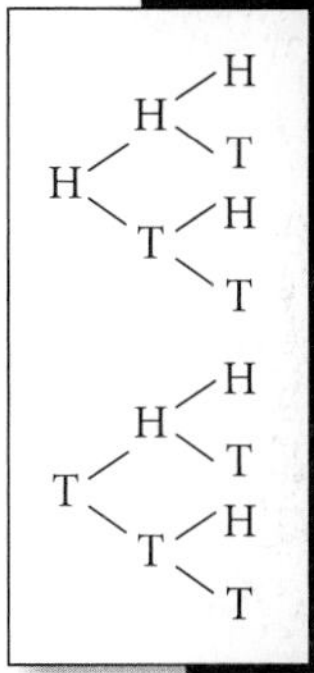


1. Look at the tree diagram shown at right for the coin tosses from Activity 2.
2. How many ways are there to toss 3 heads?
3. How many ways are there to toss 3 tails?
4. How many ways are there to toss 2 heads and 1 tails?
5. How many ways are there to toss 1 heads and 2 tails?
6. Assume that each outcome in the diagram occurred once in 8 trials. What is the probability of each outcome?

## Activity 4

1. Write the expanded form of $(h + t)^3$.
2. The probability of getting heads when tossing a coin is $\frac{1}{2}$. The probability of getting tails when tossing a coin is $\frac{1}{2}$. Let $h$ and $t$ each equal $\frac{1}{2}$, and find the value of *each term* in the expanded form of $(h + t)^3$.
3. How are the values of each term related to
   a. the probabilities you calculated in Activity 3?
   b. the results you obtained in Activity 2?
4. If you tossed 5 coins at a time, what would be the probability of getting 5 heads? 4 heads and 1 tails?

# 9 Chapter Review and Assessment

## VOCABULARY



## Key Skills & Exercises

### LESSON 9.1

#### Key Skills

**Use the horizontal form and the vertical form to add and subtract polynomials.**

Add $4x^2 + 6x$ and $2x^2 - 8x - 3$ vertically. Align the like terms and add.

$$\begin{array}{r} 4x^2 + 6x\phantom{{}-3} \\ \underline{+\,2x^2 - 8x - 3} \\ 6x^2 - 2x - 3 \end{array}$$

Subtract $6b^2 + 4b - 8$ from $10b^2 - 13$ horizontally. Remember that you can subtract by adding the opposite.

$(10b^2 - 13) - (6b^2 + 4b - 8)$

$10b^2 - 13 - 6b^2 - 4b + 8$

$4b^2 - 4b - 5$

#### Exercises

**Add or subtract.**

1. $(3x^2 - 4x + 2) + (2x^2 + 3x - 2)$
2. $(c^3 + 4c^2 + 6) + (c^2 + 3c - 5)$
3. $(8d^2 - d) - (2d^2 + 4d - 5)$
4. $(w^3 - 3w + 9) - (8w^3)$
5. $(10m^2 - m + 4) - (2m^2 + m)$
6. $(7c + 3) + (3c^2 - 7c - 2)$
7. $(8x^2 + x) - (2x^2 - 3x)$
8. $(5x^3 + 2x^2 - x) + (5x^3 + 3x^2 - 2)$
9. $(7t^5 + 2t^3 - t^2) - (3t^5 - 4t^4 + 3t^2)$

### LESSON 9.2

#### Key Skills

**Use the Distributive Property to find the product of a monomial and a binomial and the product of two binomials.**

Find the product: $8(x + 3)$.

$8(x + 3) = (8 \cdot x) + (8 \cdot 3) = 8x + 24$

Use the Distributive Property to find $(x + 6)(x + 2)$.

$$\begin{aligned}(x + 6)(x + 2) &= (x + 6)x + (x + 6)2 \\ &= x^2 + 6x + 2x + 12 \\ &= x^2 + 8x + 12\end{aligned}$$

#### Exercises

**Find each product.**

10. $5(x - 5)$
11. $4y(y + 2)$
12. $-x(2x - 3)$

**Use the Distributive Property to find each product.**

13. $(y + 9)(y - 2)$
14. $(2p - 9)(p + 5)$

## LESSON 9.3

### Key Skills

**Multiply two binomials by using the FOIL method.**

Use the FOIL method to find $(2x+3)(x-1)$.

$(2x+3)(x-1) = 2x^2 - 2x + 3x - 3 = 2x^2 + x - 3$

**Use the rules for special products to find the product of two binomials.**

Find the product: $(2a+3)^2$.

Use the special product rule

$(2a+3)^2 = (2a)^2 + 2(2a)(3) + (3)^2 = 4a^2 + 12a + 9$

### Exercises

**Find each product.**

**15.** $(x+3)(x-4)$

**16.** $(5d-8)(d-1)$

**17.** $(4w+3z)(w+z)$

**18.** $(3m+5)(m+5)$

**19.** $(x+7)(x-7)$

**20.** $(x-4)(x+4)$

**21.** $(x+1)(x+1)$

**22.** $(5x-2)^2$

**23.** $(3x-8)(3x+8)$

## LESSON 9.4

### Key Skills

**Use polynomial functions to model the volume and surface area of solid figures.**

The volume of a cylinder that is 10 inches high is $V = 10B$, where $B$ is the area of the circular base. The area of a circle with a radius of $r$ is $A = \pi r^2$, so the volume of the cylinder is modeled by the function $V = 10\pi r^2$.

**Use an identity to express a polynomial as the product of other polynomials.**

The equation $(x-1)^2 = x^2 - 2x + 1$ is an identity (true for all values of $x$). For example, if $x = -3$, then the left side, $(-3-1)^2$, is $(-4)^2$, or 16. The right side, $(-3^2) - 2(-3) + 1$, is $9 + 6 + 1$, or 16 also.

### Exercises

**24.** The formula for the volume of a cone is $V = \frac{1}{3}Bh$, where $B$ is the area of the circular base. Write a polynomial function that represents the volume of a cone with a height of 9 feet and a base that has a radius of $r$.

**25.** The equation $(x+2)(x-2) = x^2 - 4$ is an identity. Show that each side of the equation has the same value when $x$ is replaced by each of the following numbers: $-3, -2, -1, 0, 1, 2,$ and 3.

## LESSON 9.5

### Key Skills

**Factor a polynomial by using the GCF or by grouping.**

Factor $8m^6 + 4m^4 - 2m^2$.
The GCF is $2m^2$. Factor $2m^2$ from each term.

$8m^6 + 4m^4 - 2m^2 = 2m^2(4m^4 + 2m^2 - 1)$

Factor $a^2 + 3a - 4a + 4$ by grouping.
Group the terms with common factors and use the Distributive Property.

$$\begin{aligned} a^2 + 3a + 4a + 4 &= (a^2 + 3a) + (4a + 12) \\ &= a(a+3) + 4(a+3) \\ &= (a+4)(a+3) \end{aligned}$$

### Exercises

**Factor each polynomial by using the GCF.**

**26.** $x^3 + 3x^2$

**27.** $b^4 + 15b^3 + 5b$

**28.** $24m^5 + 16m^4 - 8m^3$

**29.** $2a^4y - 4a^3y^2 + 8a^2y$

**Factor each polynomial by grouping.**

**30.** $a(x+1) + b(x+1)$

**31.** $x^2 + 2x + 4x + 8$

**32.** $s^2 - s + 4s - 4$

**33.** $2m^2 + 3m - 10m - 15$

LESSON 9.6

**Key Skills**

**Use rules to factor special polynomials.**

Factor. **a.** $x^2 - 6x + 9$ **b.** $x^2 - 9$

**a.** $x^2 - 6x + 9$ is a perfect-square trinomial.

$$x^2 - 6x + 9 = (x - 3)^2$$

**b.** $x^2 - 9$ is a difference of two squares.

$$x^2 - 9 = (x + 3)(x - 3)$$

**Exercises**

**Factor each polynomial completely.**

**34.** $x^2 - 25$

**35.** $4x^2 + 4x + 1$

**36.** $9x^2 - 6x + 1$

**37.** $3s^2 - 6s + 3$

LESSON 9.7

**Key Skills**

**Factor a quadratic trinomial by working backward.**

Factor $x^2 + 7x + 12$ by using guess-and-check.

Since the constant term and the middle term are positive, the constant terms in both factors must be positive, and the binomial factors can be written in the form $(x + \_\_\_)(x + \_\_\_)$. The positive constant terms must have a product of 12 and a sum of 7. Thus, $x^2 + 7x + 12 = (x + 3)(x + 4)$.

**Exercises**

**Factor each trinomial. If a trinomial cannot be factored, write *not factorable*.**

**38.** $c^2 + 8c + 15$

**39.** $m^2 - 2m - 8$

**40.** $t^2 + 8t + 12$

**41.** $r^2 + r - 56$

**42.** $a^2 - 15x + 48$

**43.** $b^2 + 11b + 28$

LESSON 9.8

**Key Skills**

**Solve an equation by factoring.**

Solve $x^2 + x - 12 = 0$ by factoring.

Write the equation in factored form.
$(x + 4)(x - 3) = 0$

Set each factor equal to 0 and solve for $x$.

$x + 4 = 0$ $\quad$ $x - 3 = 0$
$x = -4$ $\quad$ $x = 3$

The solutions are −4 and 3.

**Exercises**

**Solve by factoring.**

**44.** $x^2 - 8x + 12 = 0$

**45.** $x^2 + 2x - 15 = 0$

**46.** $x^2 - 5x - 24 = 0$

**47.** $x^2 + 7x + 12 = 0$

**48.** $x^2 - 14x + 40 = 0$

## Applications

**49.** **GEOMETRY** The length of a rectangle is represented by $x + 1$. Its width is represented by $x - 3$. If the area of the rectangle is 117 square centimeters, what are the dimensions of the rectangle?

**50.** **LANDSCAPING** A rectangular garden is designed so that its length is twice its width. The number of feet in the perimeter of the garden is equal to the number of square feet in its area. What are the dimensions of the garden?

# Chapter Test

**Add or subtract.**

1. $(2a^2 - 5) + (4a + 7)$
2. $(12x^2 - 5x + 3) - (6x^2 + 2x - 4)$
3. $(z - 5) + (2z^2 - 3z - 4)$
4. $(4v^5 - 8v^3 + 5v) - (3v^5 + 2v^4 + 7v^2 - 5v)$

**Find each product.**

5. $3p(p - 4)$
6. $-r(5r + 9)$
7. $(m + 4)(m - 4)$
8. $(2b - 5)(2b + 5)$
9. $(3x - 5)^2$
10. $(2w + 3)(2w + 3)$
11. GEOMETRY A square with each side measuring 10 inches has a red piece $x$ inches long removed and a blue piece $x$ inches long added as shown in the figure below. Find the area of the white and blue rectangle.

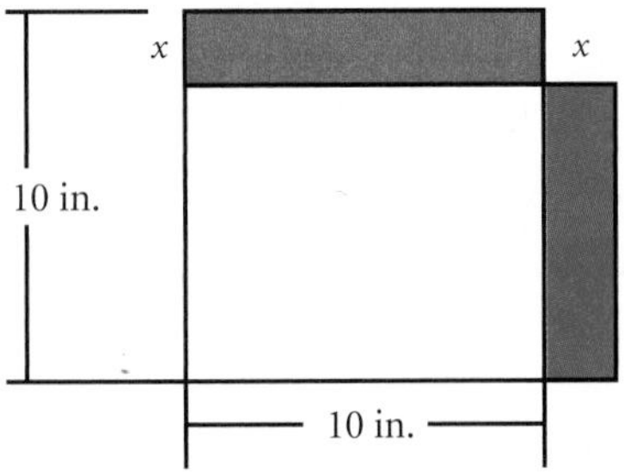


**Find each product using either the Distributive Property or the FOIL method.**

12. $(t + 4)(t + 5)$
13. $(3u - 7)(2u + 4)$
14. $(2y + 3)(2y - 3)$
15. $(5c - 1)(c + 4)$
16. $(k + 1)(5k - 5)$
17. $(3n - 2)(2n - 3)$
18. GEOMETRY Write a polynomial function that represents the volume of a rectangular box if the length of one side is $x$ inches, the width is 3 inches less, and the height is 2 inches more. What is the volume of the box when the length is 12 inches?
19. The following equation is an identity:

$$(x + 5)^3 = x^3 + 15x^2 + 75x + 125$$

Show that each side of the equation has the same value when $x$ is replaced by the integers from $-3$ to 3.

**Factor each polynomial.**

20. $3x + 18$
21. $9p^3 + 4p^2$
22. $3c^4 + 15c^3 - 12c^2$
23. $7m^2n - 21mn + 49mn^2$
24. $3p(k - 3) - 2q(k - 3)$

**Factor each polynomial completely.**

25. $y^2 - 81$
26. $4t^2 + 8t + 4$
27. $27b^2 - 18b + 3$
28. $72 - 8z^2$

**Find the missing term in each perfect-square trinomial.**

29. $4x^2 + ? + 25$
30. $16y^2 - 40y + ?$
31. $? + 8z + 1$
32. $9w^2 - ? + 16$

**Factor each trinomial. If a trinomial cannot be factored, write *not factorable*.**

33. $a^2 + 12a + 27$
34. $b^2 - 5b - 14$
35. $c^2 + 4c - 4$
36. $r^2 + 4r - 12$
37. $s^2 - 15s + 36$
38. $t^3 - t^2 - 2t$

**Solve by factoring.**

39. $x^2 + 3x + 2 = 0$
40. $x^2 - x - 12 = 0$
41. $x^2 - 8x + 15 = 0$
42. $2x^2 + 3x + 1 = 0$
43. The product of a certain number plus 3 and that same number minus 5 is 9. Set up an equation and solve the equation to find the number.
44. LANDSCAPING A design calls for a rectangular garden with an area of 180 square feet. If the length is to be 3 feet more than the width, set up an equation and solve that equation to find the dimensions of the garden.

# College Entrance Exam Practice

internet connect
**Standardized Test Prep Online**
Go To: **go.hrw.com**
Keyword: **MM1 Test Prep.**

**QUANTITATIVE COMPARISON** For Questions 1–5, write
A if the quantity in Column A is greater than the quantity in Column B;
B if the quantity in Column B is greater than the quantity in Column A;
C if the two quantities are equal; or
D if the relationship cannot be determined from the information given.

| | Column A | Column B | Answers |
|---|---|---|---|
| **1.** | reciprocal of $-2$ | reciprocal of $-3$ | Ⓐ Ⓑ Ⓒ Ⓓ [Lesson 2.4] |
| **2.** | $-18 + 4$ | $-10 + (-4)$ | Ⓐ Ⓑ Ⓒ Ⓓ [Lesson 2.2] |
| **3.** | $\lvert -5.2 \rvert$ | $\lvert 4.9 \rvert$ | Ⓐ Ⓑ Ⓒ Ⓓ [Lesson 2.1] |
| **4.** | 45% of 16 | 60% of 12 | Ⓐ Ⓑ Ⓒ Ⓓ [Lesson 4.2] |
| **5.** | $2 \cdot 3^2$ | $3 \cdot 2^2$ | Ⓐ Ⓑ Ⓒ Ⓓ [Lesson 8.1] |

**6.** What are the next three terms in the sequence 6, 12, 24, 48, . . . ? ***(LESSON 1.1)***
- **a.** 72, 96, 120
- **b.** 96, 192, 384
- **c.** 86, 162, 240
- **d.** 50, 52, 54

**7.** Solve: $16y = -120$. ***(LESSON 3.2)***
- **a.** 1920
- **b.** $-1920$
- **c.** 7.5
- **d.** $-7.5$

**8.** Find the equation of the line passing through $(-1, 2)$ and parallel to $y = 2x - 1$. ***(LESSON 5.6)***
- **a.** $y = -\frac{x}{2} + 4$
- **b.** $y = \frac{x}{2} + 2$
- **c.** $y = 2x + 4$
- **d.** $y = -x - 1$

**9.** Solve. $\begin{cases} 2x + 3y = 9 \\ x - 4y = -23 \end{cases}$ ***(LESSON 7.3)***
- **a.** $(3, 5)$
- **b.** $(-3, 5)$
- **c.** $(3, -5)$
- **d.** $(-3, -5)$

**10.** A television is marked at \$198. What is the total cost of the television if the sales tax is 7.5%? ***(LESSON 4.2)***
- **a.** \$212.85
- **b.** \$14.85
- **c.** \$183.15
- **d.** \$205.50

**11.** Simplify: $\frac{4x^3y^4}{2x^5y^2}$. ***(LESSON 8.3)***
- **a.** $2x^8y^6$
- **b.** $2x^{-2}y^{-2}$
- **c.** $2x^{-2}y^2$
- **d.** $2x^2y^2$

**12.** Simplify: $(2x - 1)(x + 3)$. ***(LESSON 9.3)***
- **a.** $2x^2 - 3$
- **b.** $2x^2 + 5x + 3$
- **c.** $2x^2 + 5x - 3$
- **d.** $2x^2 - 5x + 3$

**13.** Solve: $2(x + 2) = 12$. ***(LESSON 3.5)***
- **a.** 4
- **b.** 22
- **c.** $-4$
- **d.** 26

14. Solve: $|2x + 3| = 7$. ***(LESSON 6.5)***

**a.** 5, −3 **b.** 4, −4
**c.** 2, −5 **d.** 3, −6

15. Which number is not a solution of $|2x + 1| < 5$? ***(LESSON 6.5)***

**a.** −2 **b.** 0
**c.** 1 **d.** −3

16. Do the points $A(-2, 2)$, $B(1, 4)$, and $C(-4, 0)$ lie on a straight line? ***(LESSON 5.5)***

17. What is the slope of the line represented by $x = -4$? ***(LESSON 5.4)***

18. Solve: $4 - 3t \geq 19$. ***(LESSON 6.2)***

19. Find the opposite of $-(a - b)$. ***(LESSON 2.1)***

20. A jacket has been marked down from an original price of \$85 to \$59.50. By what percent of the original price has the jacket been marked down? ***(LESSON 4.2)***

21. The perimeter of a rectangle is 52 inches. Its width is 6 inches less than its length. Find the dimensions of the rectangle. ***(LESSON 3.3)***

22. Write 0.0000025 in scientific notation. ***(LESSON 8.5)***

23. Factor: $5xy^2 + 10x^2y^2 - 5xy$. ***(LESSON 9.7)***

24. What is the missing number? ***(LESSON 1.1)***

| | | | | |
|---|---|---|---|---|
| 0 | 1 | 2 | 3 | 4 |
| 1 | 2 | 5 | 10 | ? |

25. Simplify: $(3x^2 + 2x + 1) - (x^2 - 5x + 3)$. ***(LESSON 9.1)***

26. Simplify: $\frac{8x^2y^3}{2xy}$. ***(LESSON 8.3)***

27. Solve the system. $\begin{cases} 2x + y = 24 \\ y + 3 = x \end{cases}$ ***(LESSON 7.2)***

28. Evaluate. $\left[\left(\frac{x^3y^4}{y^2}\right)^3\left(\frac{xy^2}{x^2}\right)^4\right]^2$ ***(LESSON 8.3)***

29. Write an equation for the line passing through the points $(7, -8)$ and $(-4, -4)$. ***(LESSON 5.5)***

FREE-RESPONSE GRID

**The following questions may be answered by using a free-response grid such as that commonly used by standardized test services.**

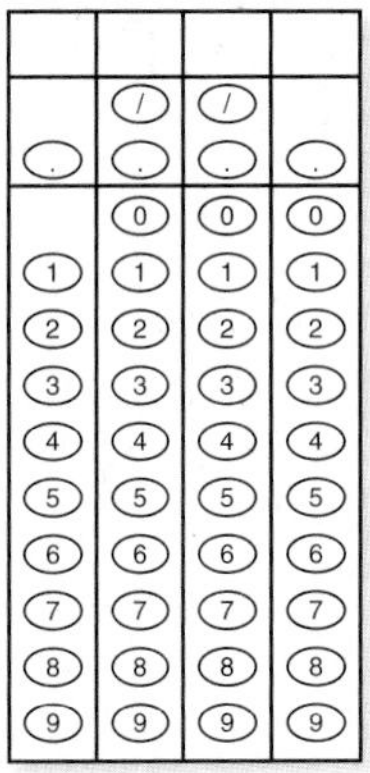

30. A city is growing at a rate of 7% per year. What multiplier is used to find the new population each year? ***(LESSON 8.6)***

31. Simplify: $4^2 \div 8 + 5(8 - 2) \cdot 2$. ***(LESSON 1.3)***

32. Find the slope of the line that passes through the origin and $(4, 2)$. ***(LESSON 5.4)***

33. Solve: $7(p + 4) = 49$. ***(LESSON 3.5)***

34. Simplify: $54 - 68 + |80|$. ***(LESSON 2.1)***

35. Evaluate $\frac{x + y}{3} + \frac{x - y}{2}$ for $x = 14$ and $y = 10$. ***(LESSON 1.3)***

36. Solve the system for $x$. $\begin{cases} y = 3 + 2x \\ 2y = 5x + 4 \end{cases}$ ***(LESSON 7.2)***

37. Find the slope of the line passing through the points $(-1, 3)$ and $(4, 5)$. ***(LESSON 5.5)***

38. Solve for $x$: $\frac{x}{12} = 13$. ***(LESSON 3.2)***

39. Find the next term in the sequence: 3, 5, 8, 13, 21, 33 . . . ***(LESSON 1.1)***

40. Evaluate $x^2 + 3y$ for $x = 4$ and $y = 0.5$. ***(LESSON 1.3)***

41. Solve for $d$: $\frac{15}{d} = \frac{5}{2}$. ***(LESSON 4.1)***

42. What percent of 80 is 20? ***(LESSON 4.2)***

43. Give the slope of a line that is perpendicular to $2x - 3y = 15$. ***(LESSON 5.6)***

44. Evaluate: $2^3 - 3^4$. ***(LESSON 1.3)***

45. Solve: $3(p - 5) = 30$. ***(LESSON 3.5)***

10

# Quadratic Functions

YOU HAVE SEEN THAT THE HEIGHT OF A PROJECTILE can be modeled by a quadratic function. Often the height versus time of the flight of a projectile will resemble a parabola, the graph of a quadratic function. Stopping distances for various vehicles can also be modeled by quadratic functions. In this chapter you will learn different methods for analyzing quadratic functions and quadratic equations.

## Lessons

*The height of a projectile, such as a baseball, can be modeled by using a quadratic function.*

*Quadratic functions can sometimes be used to model depreciation rates.*

*A quadratic function can be used to model braking distance.*

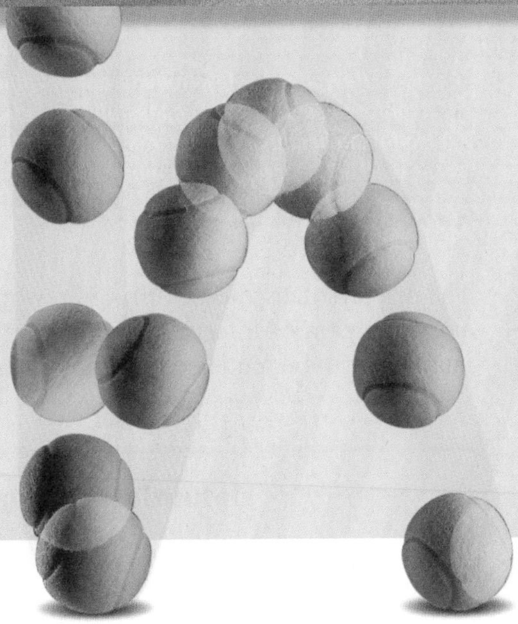

## About the Chapter Project

You have seen many circumstances in which quadratic functions can be used to model real-world situations. In the Chapter Project, *What's the Difference*, on page 516, you will explore connections between quadratic functions and the differences of a sequence. After completing the Chapter Project you will be able to do the following:

- Use the method of finite differences to determine a function rule.

## About the Portfolio Activities

Throughout the chapter, you will be given opportunities to complete Portfolio Activities that are designed to support your work on the Chapter Project.

- Using a motion detector to study a falling object is the topic of the Portfolio Activity on page 491.
- The general function for a falling object is the topic of the Portfolio Activity on page 497.
- Relating anxiety level and performance ratings is the topic of the Portfolio Activity on page 515.

10.1

# Graphing Parabolas

**Why** *Fireworks displays require careful planning in order to be safe and successful. Parabolas can model the height of a fireworks shell over time as well as other types of motion.*

## Objectives

- Discover how adding a constant to the parent function $y = x^2$ affects the graph of the function.
- Use the zeros of a quadratic function to find the vertex of the graph of the function.

APPLICATION
PHYSICS

The displays that you see during fireworks celebrations are typically launched at carefully timed intervals from mortars. The height of a shell after it is launched can be modeled by a quadratic function.

**Definition of Quadratic Function**

A quadratic function is a function of the form $y = ax^2 + bx + c$ where $a$, $b$, and $c$ are real numbers and $a \neq 0$.

Because it is the simplest quadratic function, the function $y = x^2$ is the parent function of the class of quadratic functions.

The graph of the parent function $y = x^2$ is shown at right. The shape is called a **parabola**. The lowest point on the parabola, which is the **minimum value** of the function, is the point $(0, 0)$. This point is the **vertex** of the parabola. The function $y = -x^2$ has a *maximum value* and is discussed on page 481.

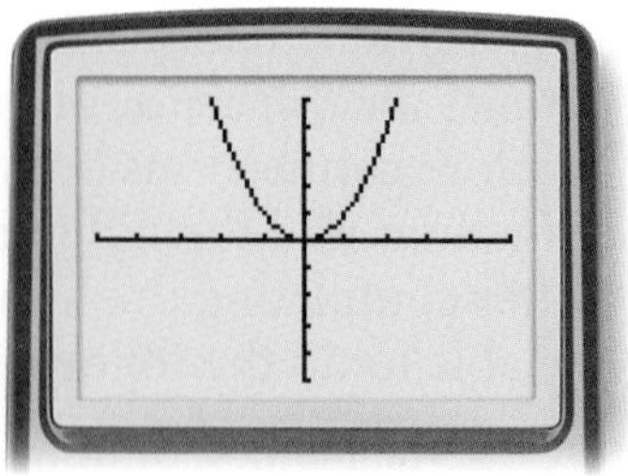

You can study the effect of adding a constant to the quadratic parent function in the following activity.

## *Activity*
## Transformations of $y = x^2$

**You will need:** pencil and graph paper, or a graphics calculator

1. Build a table of values and make a graph for each of these functions:

$$y = x^2 \qquad y = x^2 + 1 \qquad y = x^2 - 1$$

2. What effect does the addition or subtraction in Step 1 have on the graph of the parent function?
3. Build a table of values and make a graph for each of these functions:

$$y = x^2 \qquad y = (x + 2)^2 \qquad y = (x - 2)^2$$

CHECKPOINT ✔ 4. What effect does the addition or subtraction in Step 3 have on the graph of the parent function?

5. To observe the effect of combining Steps 1 and 3, build a table of values and make a graph for each of these functions:

$$y = x^2 \qquad y = (x - 2)^2 + 1 \qquad y = (x - 2)^2 - 1 \qquad y = (x + 2)^2 + 1$$

CHECKPOINT ✔ 6. What effect do addition and subtraction in Step 5 have on the graph of the parent function?

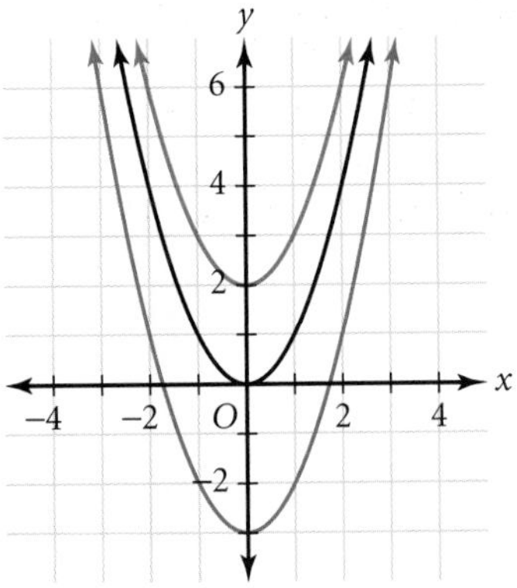


The functions $y = x^2 + 2$ and $y = x^2 - 3$ are **vertical translations** of the parent function $y = x^2$. The parabola is moved *up* or *down.*

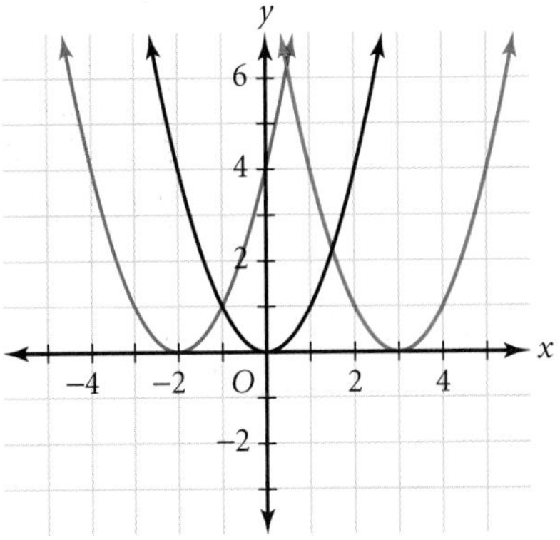


The functions $y = (x + 2)^2$ and $y = (x - 3)^2$ are **horizontal translations** of the parent function. The parabola is moved *left* or *right.*

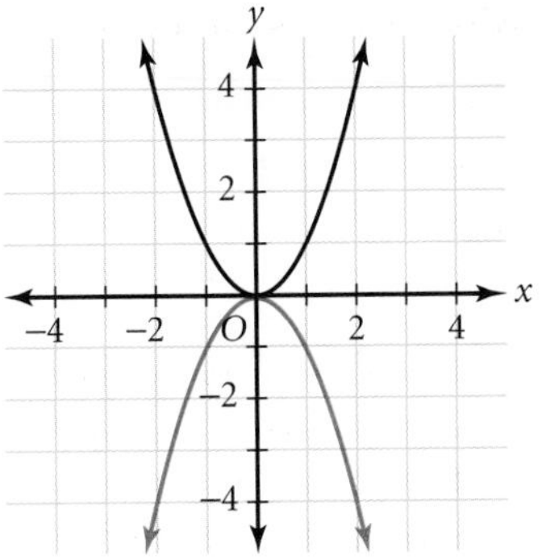


The function $y = -x^2$ is a **reflection**, or mirror image, of the parent function. The vertex, (0, 0), now corresponds to the **maximum value** of the function $y = -x^2$.

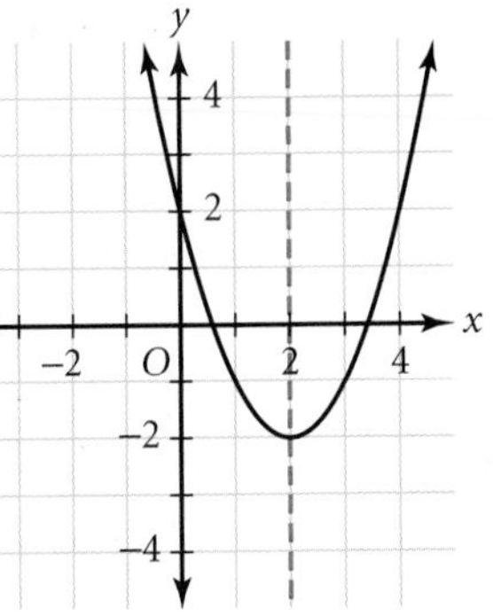


In a parabola, a vertical line drawn through the vertex is called the **axis of symmetry**. The half of the parabola on one side of the axis is a reflection of the half on the other side.

## The Vertex Form

### The Vertex Form of a Quadratic Function

The **vertex form** of the quadratic function is $y = a(x - h)^2 + k$, where $(h, k)$ is the vertex. The line $x = h$ is the axis of symmetry. When $a$ is positive, the vertex is the lowest point on the parabola. When $a$ is negative, the vertex is the highest point on the parabola.

**EXAMPLE 1** **Identify the vertex and the axis of symmetry for the graph of each function. Check by graphing.**

**a.** $y = 2(x - 3)^2 - 8$  **b.** $y = -(x + 3)^2 + 4$

**SOLUTION**

**a.** Vertex: $(3, -8)$
Axis of symmetry: $x = 3$

Since $a$ is positive, the vertex is the lowest point on the parabola.

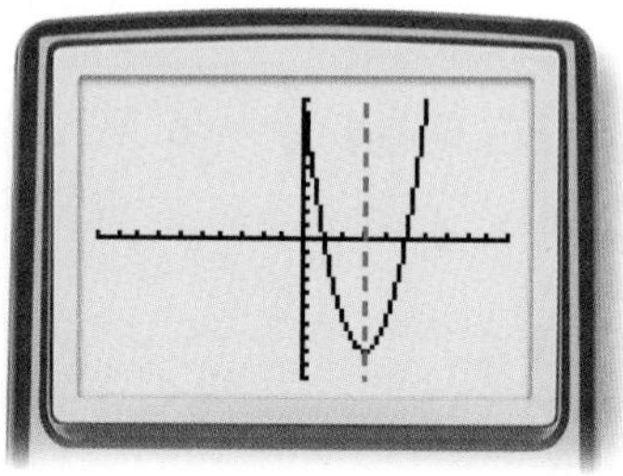

**b.** Vertex: $(-3, 4)$
Axis of symmetry: $x = -3$

Since $a$ is negative, the vertex is the highest point on the parabola.

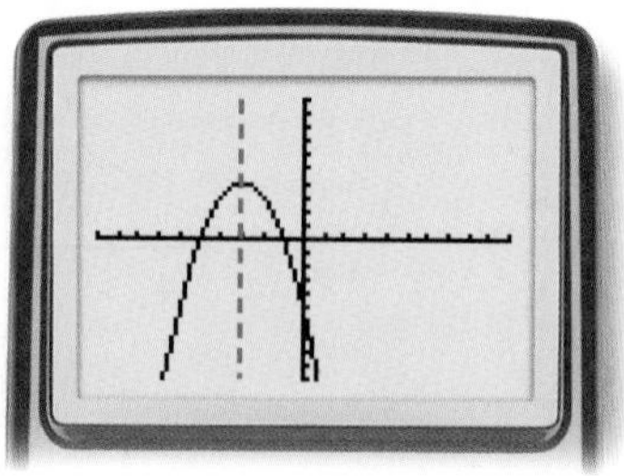

**TRY THIS** Identify the vertex and the axis of symmetry for each quadratic function.

**a.** $y = (x - 1)^2 - 2$  **b.** $y = -(x + 3)^2 + 1$

## Using Zeros to Find the Vertex

In Chapter 9 you learned how to find the zeros of quadratic functions by factoring. For example, the zeros of the quadratic function $y = x^2 - 8x + 7$ are 1 and 7.

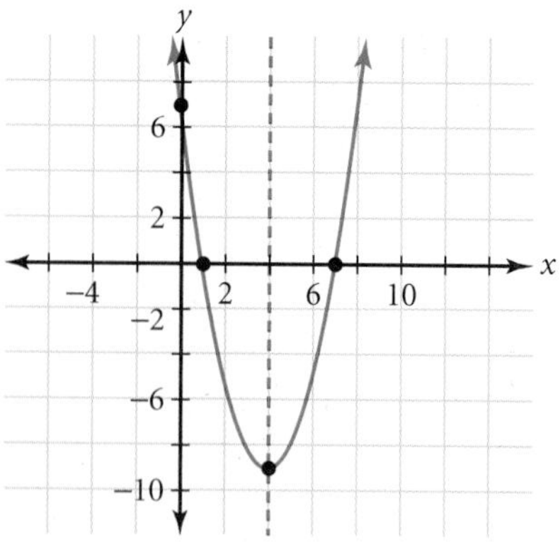


A parabola is symmetric. One half of the parabola is the mirror image of the other half. The axis of symmetry passes through a point midway between the zeros.

The zeros can be used to find the vertex of the parabola.

**EXAMPLE 2** **Use the zeros to find the vertex of the graph of $y = x^2 - 6x + 5$. Then write the vertex form of the equation.**

**SOLUTION**

Find the zeros by setting the function equal to 0. Start by factoring.

$$x^2 - 6x + 5 = 0$$
$$(x - 1)(x - 5) = 0$$

Set each binomial of the factored form equal to 0 and solve for $x$.

$$x - 1 = 0 \quad \text{or} \quad x - 5 = 0$$
$$x = 1 \qquad\qquad x = 5$$

The zeros of the function are 1 and 5.

Because the parabola is symmetrical, the vertex is halfway between the zeros. To find the $x$-coordinate of the vertex, find the midpoint, or average, of the $x$-coordinates of the zeros.

$$x = \frac{5 + 1}{2} = 3$$

Substitute 3 for $x$ in the function to find the $y$-coordinate of the vertex.

$$\begin{aligned} y &= x^2 - 6x + 5 \\ &= 3^2 - 6(3) + 5 \\ &= -4 \end{aligned}$$

Thus, the vertex is (3, –4). Write the equation in vertex form.

$$y = (x - 3)^2 - 4$$

**TRY THIS** Use the zeros to find the vertex and the vertex form of the quadratic function $y = x^2 - 7x + 10$.

**EXAMPLE 3** A fireworks shell is fired from a mortar. Its height is modeled by the function $y = -16(x - 7)^2 + 784$, where $x$ is the time in seconds and $y$ is the height in feet. **If the shell falls to the ground without exploding, how many seconds will it take for the shell to hit the ground?**

**APPLICATION**
**PHYSICS**

**SOLUTION**

In the form given, the vertex, the axis of symmetry, and the maximum value can be read directly from the function.

function: $y = -16(x - 7)^2 + 784$
vertex: (7, 784)
axis of symmetry: $x = 7$
maximum value: 784

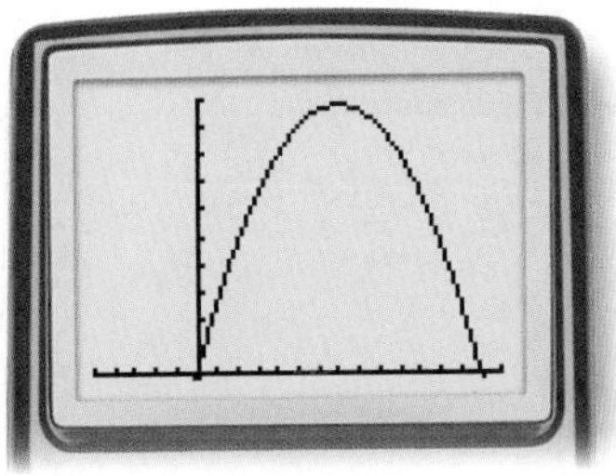

Sketch a graph of the function by using this information. The zeros are 0 and 14. Thus, the shell will hit the ground after 14 seconds of flight.

# Exercises

## Communicate

**internet connect**
**Activities Online**
Go To: **go.hrw.com**
Keyword:
**MA1 Parabolas**

1. Discuss how the graph of the quadratic function $y = x^2 - 8$ differs from the graph of the parent function $y = x^2$.
2. Discuss how the graph of the quadratic function $y = (x - 8)^2$ differs from the graph of the parent function $y = x^2$.
3. Explain how to find the vertex of $y = 2(x - 3)^2 - 8$.
4. Explain how to find the axis of symmetry of $y = 2(x - 3)^2 - 8$.

## Guided Skills Practice

**Identify the vertex and the axis of symmetry for each quadratic function.** *(EXAMPLE 1)*

5. $y = (x + 6)^2 - 4$
6. $y = (x - 3)^2 - 8$
7. $y = -4(x + 3)^2 + 6$

**Use the zeros to find the vertex and the vertex form of each quadratic function.** *(EXAMPLE 2)*

8. $y = x^2 - 12x + 20$
9. $y = x^2 - 3x - 10$
10. $y = x^2 + 11x + 18$

## Practice and Apply

**Compare the graphs of the following functions to the graph of $y = x^2$. Describe the vertical and horizontal translations of each vertex.**

11. $y = (x - 2)^2 + 3$
12. $y = 3(x - 5)^2 - 2$
13. $y = -(x - 2)^2 + 1$
14. $y = -7(x + 6)^2 - 2$
15. $y = -7(x - 3)^2 - 2$
16. $y = 5(x + 4)^2 - 7$
17. $y = \frac{1}{3}(x + 4)^2 - 5$
18. $y = \frac{1}{4}(x + 7)^2 + 14$
19. $y = 5(x + 2)^2 - 8$

**internet connect**
**Homework Help Online**
Go To: **go.hrw.com**
Keyword:
**MA1 Homework Help**
for Exercises 20-31

**Find the vertex and axis of symmetry for the graph of each function, and then sketch the graph.**

20. $y = -2(x + 4)^2 - 3$
21. $y = \frac{1}{2}(x - 2)^2 + 3$
22. $y = (x - 3)^2 - 7$
23. $y = -(x + 2)^2 - 4$
24. $y = -4(x - 3)^2 - 8$
25. $y = \frac{1}{2}(x - 4)^2 + 5$
26. $y = 3(x + 6)^2 - 3$
27. $y = (x + 5)^2 - 7$
28. $y = (x - 2)^2 + 9$
29. $y = -3(x - 5)^2 + 2$
30. $y = -(x + 3)^2 - 2$
31. $y = 2(x + 5)^2 + 7$

**Use factoring to find the zeros of each function.**

32. $y = x^2 + 8x - 9$
33. $y = x^2 - 20x + 100$
34. $y = x^2 - x - 72$
35. $y = x^2 + 6x - 7$
36. $y = x^2 + 4x - 5$
37. $y = x^2 + 2x - 24$
38. $y = x^2 + 18x + 81$
39. $y = x^2 + 2x - 63$
40. $y = x^2 - 5x + 6$

CONNECTION

**41.** **NUMBER THEORY** Use a quadratic equation to find two numbers whose difference is 5 and whose product is 24.

APPLICATION

*The longbow was used by soldiers in medieval times.*

**PHYSICS** The graph at right represents the relationship between the time, in seconds, when a projectile is propelled vertically into the air and its height.

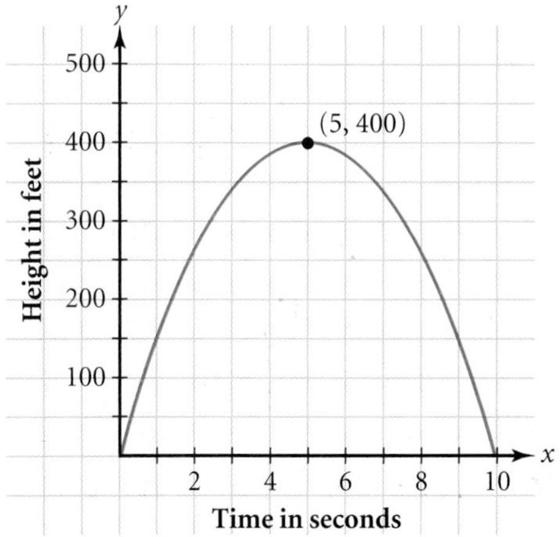


**42.** What is the maximum height reached by the projectile?

**43.** How long does it take for the projectile to reach its maximum height?

**44.** How long does it take for the projectile to return to the ground?

**45.** What is the axis of symmetry of the graph?

**46.** **PHYSICS** An arrow is shot upward at an initial velocity of 40 meters per second. Use the function $h = 40t - 5t^2$ to find the height in meters after 5 seconds. Round your answers to the nearest tenth.

## Look Back

**The expression $2(x - 3)^2 + 1$ involves multiplication, an operation inside parentheses, an exponent, and addition.** ***(LESSON 1.3)***

**47.** Which of these operations should be done first?

**48.** Which should be done second?

**49.** Which should be done third?

**Solve.** ***(LESSON 7.2)***

**50.** $\begin{cases} x = y \\ 2x + 3y = 2 \end{cases}$

**51.** $\begin{cases} y = 4 \\ 2x = 3y \end{cases}$

**52.** Find the two-digit number whose tens digit is 4 less than its units digit if the original number is 2 more than 3 times the sum of the digits. ***(LESSON 7.6)***

**Factor.** ***(LESSON 9.6)***

**53.** $x^2 + 10x + 25$

**54.** $x^2 - 14x + 49$

## Look Beyond

APPLICATION

**55.** **ECOLOGY** A student group hopes to collect 100,000 cans to recycle. Find how much space the cans will occupy if each can has a circumference of 21 centimeters and a height of 12.5 centimeters. Express your answer in cubic centimeters and in cubic meters ($1{,}000{,}000 \text{ cm}^3 = 1 \text{ m}^3$).

10.2

# Solving Equations by Using Square Roots

### Objectives

- Solve equations of the form $ax^2 = k$.
- Solve equations of the form $x^2 = k$ where $x$ is replaced by an algebraic expression.

**Why** *The solutions to quadratic equations can give you answers to problems in the real world, such as the height of a falling object.*

**APPLICATION**
**PHYSICS**

Galileo Galilei, born in Pisa, Italy, in 1564, discovered that objects fall with the same acceleration, regardless of the mass of the object. In a famous experiment, he hypothesized that if a cannon ball and a small stone were dropped from the Leaning Tower of Pisa, both would hit the ground at the same time.

Suppose that an object is dropped from the top of the Leaning Tower of Pisa, 185 feet above the ground. The object's height in feet, $h$, after $t$ seconds is given by the following function:

$$h = -16t^2 + 185$$

How long will it take for the object to hit the ground? You will investigate this question in the following activity.

## Activity
## Tables for Falling Objects

1. Build a table of values for the function $h = -16t^2 + 185$.

| **Time, $t$** | 0 | 1 | 2 | 3 | 4 | 5 |
|---|---|---|---|---|---|---|
| **Height, $h$** | ? | ? | ? | ? | ? | ? |

2. The value of $h$ is 0 when the object hits the ground. Use your table to find the approximate value of $t$ when $h = 0$.
3. Refine your estimate from Step 2 by building a new table with increments of 0.1 for $t$ near the value that you found in Step 2.

CHECKPOINT ✓
4. Refine your estimate further by building a table with increments of 0.01 for $t$ near the value that you found in Step 3.

## Using Square Roots

The equation $x^2 = 9$ can be stated as the question, "What number multiplied by itself is 9?" There are two such numbers, 3 and −3. Each of these is called a **square root** of 9.

The *positive* square root of a number is called the **principal square root.** The principal square root of a number is indicated by the **radical sign,** $\sqrt{\phantom{x}}$. In general, if $x^2 = k$ and $k > 0$, then $x = \sqrt{k}$ or $x = -\sqrt{k}$. You can write this as $\pm\sqrt{k}$, which is read as "plus or minus the square root of $k$."

Square roots are the solutions to equations of the form $x^2 = k$, as you will see in the following example.

**EXAMPLE 1** **Solve each equation.**

**a.** $x^2 = 25$ **b.** $x^2 = 10$ **c.** $4x^2 = 9$

TECHNOLOGY
SCIENTIFIC CALCULATOR

**SOLUTION**

**a.** The solution to $x^2 = 25$ is $x = \sqrt{25}$ or $x = -\sqrt{25}$. Thus, $x = 5$ or $x = -5$.

**b.** The solution to $x^2 = 10$ is $x = \sqrt{10}$ or $x = -\sqrt{10}$. There is no exact rational number answer, but an approximation can be found by using the $\sqrt{\phantom{x}}$ key on your calculator.

[ $\sqrt{\phantom{x}}$ ] 10 [ = ] 3.16227766

Thus, $x \approx 3.16$ or $x \approx -3.16$.

**c.** The solution to $4x^2 = 9$, or $x^2 = \frac{9}{4}$, is $x = \sqrt{\frac{9}{4}}$ or $x = -\sqrt{\frac{9}{4}}$.

Thus, $x = \frac{3}{2}$ or $x = -\frac{3}{2}$.

**TRY THIS** Solve each equation.

**a.** $x^2 = 100$ **b.** $4x^2 = 25$ **c.** $49x^2 = 16$

The results of Example 1 lead to a generalization for finding the solution to a quadratic equation of the form $ax^2 = k$.

**Solving $ax^2 = k$, Where $\frac{k}{a}$ Is Nonnegative**

An equation of the form $ax^2 = k$, where $\frac{k}{a} \geq 0$ and $a \neq 0$, has the following solutions:

$$x = \sqrt{\frac{k}{a}} \text{ or } x = -\sqrt{\frac{k}{a}}$$

**EXAMPLE 2** **Find the exact time that it takes for an object to fall from the top of the Leaning Tower of Pisa.**

APPLICATION

PHYSICS

**SOLUTION**

Use the equation obtained by setting the function from the beginning of the lesson equal to 0.

$$-16t^2 + 185 = 0$$
$$-16t^2 = -185$$
$$t^2 = \frac{-185}{-16} = \frac{185}{16}$$
$$t = \pm\sqrt{\frac{185}{16}} = \pm\frac{\sqrt{185}}{\sqrt{16}} = \pm\frac{\sqrt{185}}{4}$$
$$t = \frac{\sqrt{185}}{4} \text{ or } t = -\frac{\sqrt{185}}{4}$$

The *exact* time is $\frac{\sqrt{185}}{4}$ seconds.
This value can be approximated to the nearest hundredth as 3.40 seconds.

Although $-\frac{\sqrt{185}}{4}$ is a solution to the equation, it is not reasonable for this particular problem because time is not measured in negative units.

In the next example, an *expression* takes the place of $x$ in an equation of the form $ax^2 = k$.

**EXAMPLE 3** **Solve the equation $(x - 2)^2 - 9 = 0$.**

**SOLUTION**

$$(x - 2)^2 - 9 = 0$$
$$(x - 2)^2 = 9$$
$$x - 2 = \pm\sqrt{9}$$
$$x = 2 \pm 3$$
$$x = 2 + 3 \text{ or } x = 2 - 3$$
$$x = 5 \qquad x = -1$$

Check each solution in the original equation.

**TRY THIS** Solve each equation.

**a.** $25(t - \frac{2}{5})^2 - 144 = 0$ **b.** $(x - 4)^2 - 81 = 0$

**CRITICAL THINKING** In Example 3, the expression $(x - 2)$ takes the place of $x$ in an equation of the form $ax^2 = k$. What numbers represent $a$ and $k$?

**CONNECTION**
**COORDINATE GEOMETRY**

For the function $y = (x - 2)^2 - 9$, the vertex of the graph is $(2, -9)$ and the axis of symmetry is $x = 2$. Because the coefficient of the quadratic term is positive, the parabola opens upward and has a minimum value. For $y = 0$, the solutions are 5 and −1 (see Example 3). In other words, the graph of the function crosses the $x$-axis at $(5, 0)$ and $(-1, 0)$.

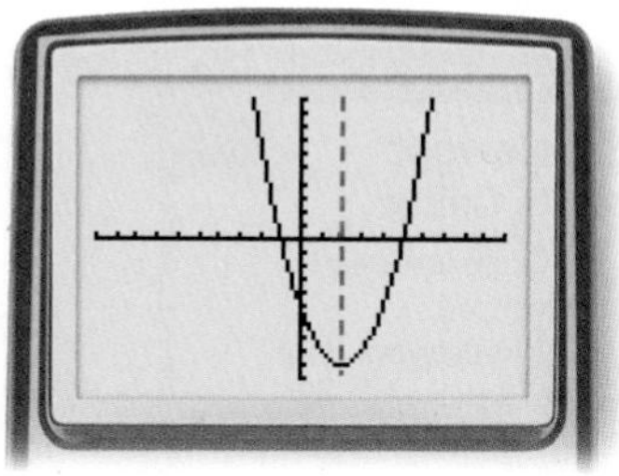

# Exercises

## Communicate

**1.** The function $h = -16t^2 + 300$ represents the height in feet, $h$, of a falling body in terms of time, $t$, in seconds. Explain how to make a table of values to find the point(s) where the function crosses the horizontal axis.

**2.** Explain how to find the square root of 100. Explain why you use the $\pm$ sign.

**3.** Explain how to solve the following equations:
**a.** $x^2 = 64$ **b.** $x^2 = 8$ **c.** $x^2 = \frac{16}{100}$

**4.** Discuss how to solve each equation by using square roots.
**a.** $(x + 3)^2 - 25 = 0$ **b.** $(x - 8)^2 = 2$

**5.** Explain how to make a rough sketch of the graph of $y = (x - 3)^2 - 4$ by using the vertex, axis of symmetry, and zeros.

## Guided Skills Practice

**Solve each equation.** ***(EXAMPLE 1)***

**6.** $x^2 = 64$ **7.** $25x^2 = 16$ **8.** $x^2 = 17$

**APPLICATION**

**9.** **PHYSICS** Find the time, $t$, in seconds it takes for an object to fall 600 feet by using the function $h = -16t^2 + 600$. ***(EXAMPLE 2)***

**10.** **PHYSICS** Find the time, $t$, in seconds it takes for an object to fall 800 feet by using the function $h = -16t^2 + 800$. ***(EXAMPLE 2)***

**11.** Solve the equation $(x + 2)^2 - 144 = 0$. ***(EXAMPLE 3)***

**12.** Solve the equation $(x - 3)^2 - 400 = 0$. ***(EXAMPLE 3)***

## Practice and Apply

**Find each positive square root. Round answers to the nearest hundredth when necessary.**

**13.** $\sqrt{121}$ **14.** $\sqrt{144}$ **15.** $\sqrt{625}$ **16.** $\sqrt{36}$

**17.** $\sqrt{44}$ **18.** $\sqrt{90}$ **19.** $\sqrt{88}$ **20.** $\sqrt{19}$

internet connect
**Homework Help Online**
Go To: **go.hrw.com**
Keyword:
**MA1 Homework Help**
for Exercises 21-48

**Solve each equation. Round answers to the nearest hundredth when necessary.**

**21.** $x^2 = 25$ **22.** $x^2 = 169$ **23.** $x^2 = 81$ **24.** $x^2 = 625$

**25.** $x^2 = 12$ **26.** $x^2 = 24$ **27.** $x^2 = 18$ **28.** $x^2 = 54$

**29.** $81x^2 = 25$ **30.** $121x^2 = 49$ **31.** $49x^2 = 36$ **32.** $100x^2 = 4$

**33.** $x^2 = 32$ **34.** $3x^2 = 135$ **35.** $2x^2 = 56$ **36.** $x^2 = 63$

**37.** $(x+4)^2 - 25 = 0$ **38.** $(x-5)^2 - 9 = 0$ **39.** $(x+1)^2 - 1 = 0$

**40.** $(x-2)^2 - 6 = 0$ **41.** $(x+7)^2 - 5 = 0$ **42.** $6(x-3)^2 - 12 = 0$

**43.** $(x+3)^2 = 36$ **44.** $(x-2)^2 = 144$ **45.** $(x-8)^2 = 81$

**46.** $(x-1)^2 = 11$ **47.** $(x+5)^2 = 10$ **48.** $7(x+6)^2 = 105$

**Find the vertex, axis of symmetry, and zeros of each function. Then sketch the graph of each.**

**49.** $y = (x-4)^2 - 3$ **50.** $y = (x+2)^2 - 1$ **51.** $y = (x-4)^2 - 2$

**Refer to the function $y = (x+4)^2 - 4$ to complete Exercises 52–55.**

**52.** Find the vertex.

**53.** Find the axis of symmetry.

**54.** Find the zeros.

**55.** Sketch the graph.

CHALLENGE

**56.** Explain why $x^2 + 100 = 0$ has no real solutions.

CONNECTION

**57.** **GEOMETRY** Use the formula $S = 4\pi r^2$ to find the radius of a sphere with a surface area of 90 square meters. (Use 3.14 for $\pi$.)

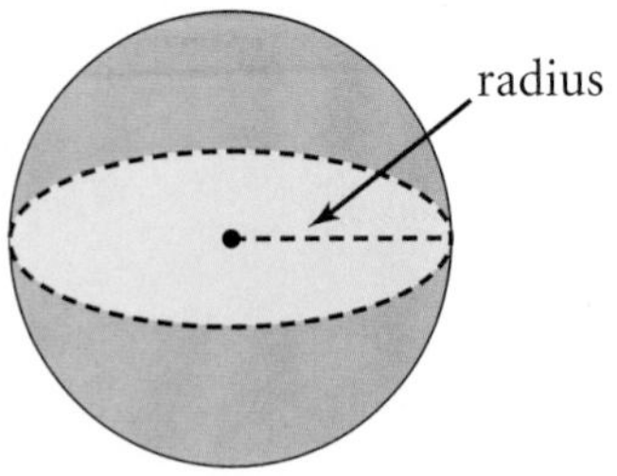


APPLICATION

**PHYSICS** An object is dropped from a point 700 feet above the ground. Its height in feet, $h$, after $t$ seconds is given by the function $h = -16t^2 + 700$. Find both approximate and exact answers for Exercises 58 and 59.

**58.** How long will it take for the object to reach the ground?

**59.** How long will it take for the object to reach a height of 100 feet?

## Look Back

**Find each slope.** ***(LESSON 5.4)***

**60.** $y = 2x + 3$ **61.** $2x + 3y = 4$ **62.** $x = y$

**63.** $4x + 7 = 5y$ **64.** $y - 2x = 3$ **65.** $y = 10$

**66.** Write an equation in slope-intercept form for a line that contains $(2, -4)$ and is perpendicular to the line $2x - y = 5$. ***(LESSON 5.6)***

**Simplify each expression.** ***(LESSONS 8.2 AND 8.3)***

**67.** $\frac{-6x^2y^2}{2x}$ **68.** $\frac{b^3c^4}{bc}$ **69.** $(p^2)^4(2a^2b^3)^2$

**Add or subtract each polynomial.** ***(LESSON 9.1)***

**70.** $(3x^2 - 2x + 1) + (2x^2 + 4x + 6)$ **71.** $(x^2 + 2x - 1) - (2x^2 - 5x + 7)$

## Look Beyond

**72.** Evaluate $\sqrt{5^2 - 4 \cdot 1 \cdot 2}$. **73.** Evaluate $\sqrt{6^2 - 4(1)(9)}$.

### PORTFOLIO ACTIVITY

Include this activity in your portfolio.

Use a motion detector to investigate a falling object.

1. Drop an object from 10 feet. Use the motion detector to find the object's height above the ground over time.
2. Use the time and height data to compute second differences in the height values. What are the second differences?
3. Make a scatter plot of the data. Write an equation of the form $y = -16x^2 + h$, where $h$ is the initial height, to fit the data.

# 10.3 Completing the Square

## Objectives

- Form a perfect-square trinomial from a given quadratic binomial.
- Write a given quadratic function in vertex form.

**Why** ***If you write the quadratic function $y = x^2 + bx + c$ in the form $y = (x - h)^2 + k$, you can find the vertex and the axis of symmetry of the parabola. A technique called completing the square will enable you to do this. This technique simplifies graphing and is helpful for complicated real-world problems.***

The algebra tiles shown above represent the factorization of a perfect-square trinomial. Do you see why?

You can represent $x^2 + 4x + 4$ with tiles, as shown in the illustration above. Notice that the tiles form a complete square. Recall from Lesson 9.6 that an expression such as $x^2 + 4x + 4$ is called a *perfect-square trinomial* because it can be written as $(x + 2)^2$, the *square of a binomial.*

These tiles represent the binomial $x^2 + 6x$. Can you form a square by rearranging them and adding a certain number of 1-tiles? How many 1-tiles do you need to add?

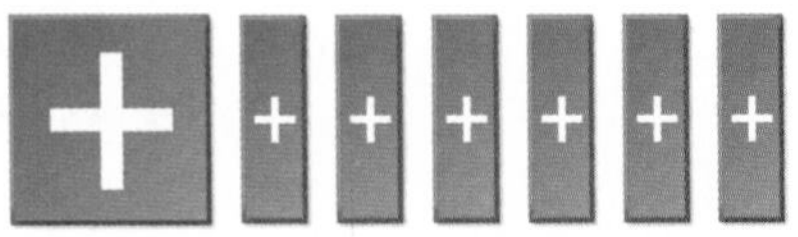

## Activity
## Completing the Square With Tiles

**You will need:** algebra tiles

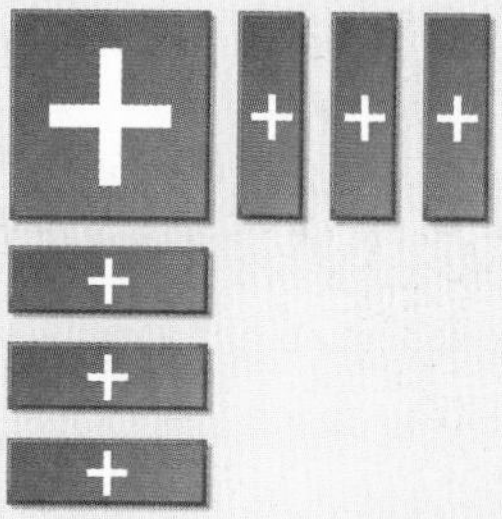

1. Arrange tiles as shown at right to represent $x^2 + 6x$.
2. Fill in the empty space in your new arrangement by adding 1-tiles so that a complete square is formed. How many 1-tiles did you add?
3. Your completed square represents a perfect-square trinomial. What is the trinomial? Rewrite the trinomial as the square of a binomial.
4. Repeat Steps 1–3 for the tile arrangement at right.

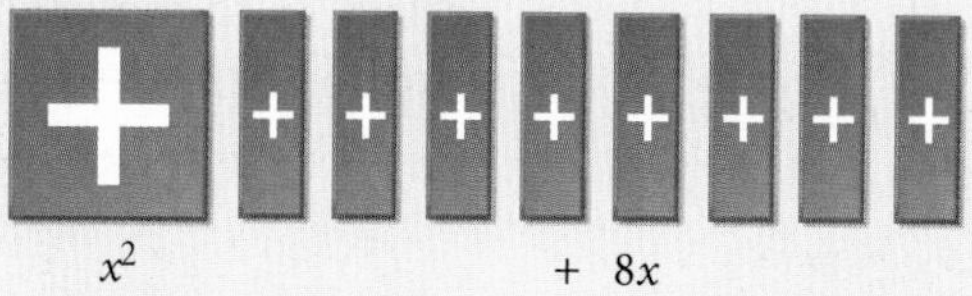


CHECKPOINT ✔ 5. Try to think of a rule for the number of 1-tiles you added in each case. Write a formula or a procedure for the rule.

In the activity you may have discovered the following rule:

### Completing the Square

To make a perfect-square trinomial from a binomial of the form $x^2 + bx$, add the square of half of the coefficient of the $x$-term, $\left(\frac{b}{2}\right)^2$.

$$x^2 + bx + \left(\frac{b}{2}\right)^2$$

In addition to being faster than using algebra tiles, the rule for completing the square can be used for cases in which algebra tiles cannot be used, such as with large numbers or fractions.

**EXAMPLE 1** **Complete the square by finding the value that makes a perfect-square trinomial when added to the binomial $x^2 + 5x$.**

**SOLUTION**

The coefficient of the $x$-term is 5. Take half of this number and square it.

$$\left(\frac{5}{2}\right)^2 = \frac{25}{4} = 6\frac{1}{4}$$

Add $6\frac{1}{4}$ to the original expression, $x^2 + 5x$, to make a perfect-square trinomial, $x^2 + 5x + 6\frac{1}{4}$. This can also be written as $\left(x + \frac{5}{2}\right)^2$.

**TRY THIS** Find the value that makes a perfect-square trinomial when added to the binomial $x^2 + 20x$.

The rule for completing the square also works when the coefficient of the $x$-term is negative, as the next example illustrates.

**EXAMPLE 2** **Complete the square for $x^2 - 10x$.**

**SOLUTION**

The coefficient of the $x$-term is $-10$. Take one-half this number and square it.

$$\left(\frac{b}{2}\right)^2 = \left(\frac{-10}{2}\right)^2 = (-5)^2 = 25$$

Add 25 to the original expression, $x^2 - 10x$, to make a perfect-square trinomial, $x^2 - 10x + 25$, or $(x - 5)^2$.

The diagram below represents the right side of the function $y = x^2 + 6x + 5$. It is easy to see that this is not a perfect-square trinomial because the tiles do not form a square.

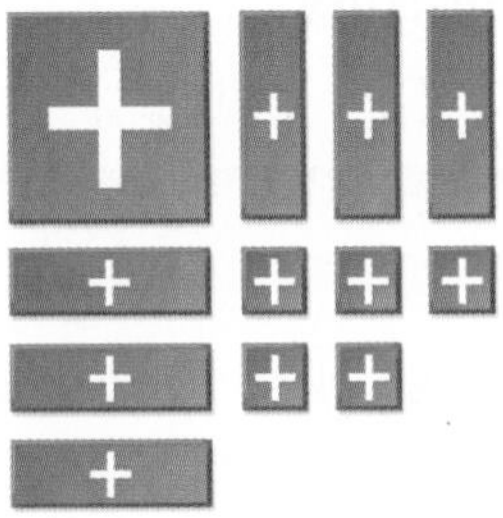

$y = x^2 + 6x + 5$

By adding neutral pairs of 1-tiles, you can create an arrangement that represents the same quadratic function—but includes a perfect square. The new arrangement suggests a way to rewrite the original function.

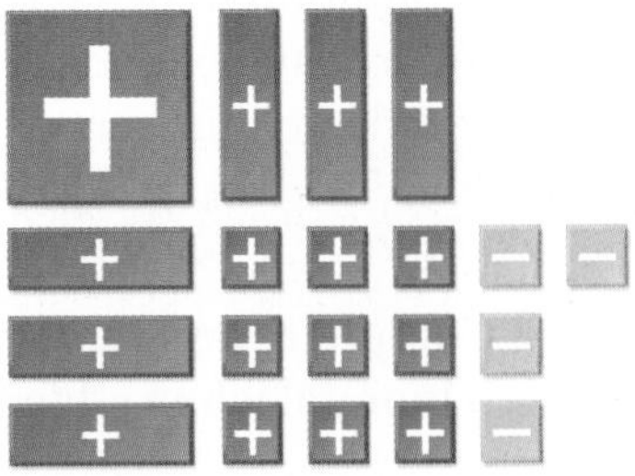

$y = x^2 + 6x + 5 + 4 - 4$
$y = x^2 + 6x + 9 - 4$
$y = (x + 3)^2 - 4$

The vertex $(h, k)$ of the function is $(-3, -4)$. (Recall Lesson 10.1.)

The algebra-tile model above suggests the following procedure for converting a quadratic function in which $a = 1$ to vertex form.

### Converting to Vertex Form

| | |
|---|---|
| **1.** Start with the quadratic function. | $y = x^2 + 6x + 5$ |
| **2.** Group the $x^2$-terms and $x$-terms. | $y = (x^2 + 6x) + 5$ |
| **3.** From the coefficient of the $x$-term, find the number you need to add to make a perfect-square trinomial. | $\frac{1}{2} \cdot 6 = 3;\ 3^2 = 9$ |
| **4.** Complete the square by adding the number to the expression inside the parentheses. Then subtract the number from the constant term outside the parentheses. | $y = (x^2 + 6x + 9) + 5 - 9$ |
| **5.** Simplify. | $y = (x^2 + 6x + 9) - 4$ |
| **6.** Write in the form $y = a(x - h)^2 + k$. | $y = (x + 3)^2 - 4$ |

**EXAMPLE 3**

**a.** **Rewrite $y = x^2 - 8x + 7$ in the form $y = a(x - h)^2 + k$.**
**b.** **Find the vertex of the parabola represented by this function.**

**SOLUTION**

**a.** Follow the steps for converting to vertex form.

1. $y = x^2 - 8x + 7$
2. $y = (x^2 - 8x) + 7$
3. $\frac{1}{2}(-8) = -4;\ (-4)^2 = 16$
4. $y = (x^2 - 8x + 16) + 7 - 16$
5. $y = (x^2 - 8x + 16) - 9$
6. $y = (x - 4)^2 - 9$

**b.** The vertex of the parabola is (4, –9).

## Exercises

### Communicate

1. Describe how to use algebra tiles to complete the square for $x^2 + 4x$.
2. Describe how to use algebra to complete the square for $x^2 - 7x$.
3. How do you rewrite $y = x^2 + 10x + 25 - 25$ in the form $y = a(x - h)^2 + k$?
4. Describe how to find the vertex of $y = x^2 + 2x$.
5. Explain how to rewrite $y = x^2 - 10x + 11$ in the form $y = a(x - h)^2 + k$. How can you find the vertex from this form of the equation?

## Guided Skills Practice

**6.** Find the value that makes a perfect-square trinomial when added to the binomial $x^2 + 19x$. ***(EXAMPLE 1)***

**7.** Find the value that makes a perfect-square trinomial when added to the binomial $x^2 - 3x$. ***(EXAMPLE 2)***

**8.** Rewrite $y = x^2 + 6x + 18$ in the form $y = a(x - h)^2 + k$. ***(EXAMPLE 3)***

## Practice and Apply

**Use algebra tiles to complete the square for each binomial. Draw a sketch of your algebra-tile model. Write each perfect-square trinomial in the form $x^2 + bx + c$.**

**9.** $x^2 + 14x$ **10.** $x^2 - 14x$ **11.** $x^2 + 8x$ **12.** $x^2 - 8x$

**13.** $x^2 + 4x$ **14.** $x^2 - 10x$ **15.** $x^2 + 12x$ **16.** $x^2 + 10x$

**Rewrite each function in the form $y = a(x - h)^2 + k$.**

**17.** $y = x^2 + 8x + 16 - 16$ **18.** $y = x^2 - 4x + 4 - 4$

**19.** $y = x^2 - 10x + 25 - 25$ **20.** $y = x^2 + 14x + 49 - 49$

**21.** $y = x^2 - 16x + 64 - 64$ **22.** $y = x^2 + 20x + 100 - 100$

internet connect — go.hrw.com
**Homework Help Online**
Go To: **go.hrw.com**
Keyword:
**MA1 Homework Help**
for Exercises 23-42

**Complete the square by using Example 2 as a model.**

**23.** $x^2 + 6x$ **24.** $x^2 - 2x$ **25.** $x^2 + 12x$ **26.** $x^2 - 12x$

**27.** $x^2 + 7x$ **28.** $x^2 - 10x$ **29.** $x^2 + 15x$ **30.** $x^2 - 5x$

**31.** $x^2 + 16x$ **32.** $x^2 + 20x$ **33.** $x^2 - 9x$ **34.** $x^2 + 40x$

**35.** $x^2 - 36x$ **36.** $x^2 + 44x$ **37.** $x^2 - 17x$ **38.** $x^2 + 30x$

**39.** $x^2 - 13x$ **40.** $x^2 + 13x$ **41.** $x^2 + 22x$ **42.** $x^2 - 44x$

**Refer to the function $y = x^2 + 10x$ to complete Exercises 43–45.**

**43.** Complete the square and rewrite the function in the form $y = a(x - h)^2 + k$.

**44.** Find the vertex of the function.

**45.** Find the maximum or minimum value of the function.

**Find the vertex of the parabolas represented by the following functions:**

**46.** $y = x^2 + 3$ **47.** $y = x^2 - 4$ **48.** $y = -x^2 + 2$

**49.** $y = x^2 - 1$ **50.** $y = -x^2 + 6$ **51.** $y = x^2 - 5$

**Rewrite each function in the form $y = a(x - h)^2 + k$ by using Example 3 as a model. Find the vertex of each.**

**52.** $y = x^2 - 4x + 7$ **53.** $y = x^2 + 6x + 5$ **54.** $y = x^2 - 4x + 6$

**55.** $y = x^2 - 12x + 35$ **56.** $y = 6x + 15 + x^2$ **57.** $y = -4x - 1 + x^2$

**58.** $y = x^2 + 4x - 1$ **59.** $y = x^2 - 2x - 3$ **60.** $y = x^2 + 10x - 12$

**61.** $y = x^2 - x + 2$ **62.** $y = -5 - 6x + x^2$ **63.** $y = x^2 + \frac{1}{3}x - 3$

**CHALLENGE**

**64.** $3y = 3x^2 - 24x + 18$

**APPLICATION**

**PHYSICS** For a certain projectile, the relationship between time in seconds, $t$, and height in feet, $h$, is given by the function $h = -16t^2 + 192t$.

**65.** Graph the function.

**66.** Use the zeros to find the vertex of the parabola represented by the function.

**67.** What is the maximum height reached by the projectile?

**68.** How long does it take for the projectile to reach its maximum height?

*Some volcanoes eject incandescent lava as high as 470 meters.*

## Look Back

**69.** Graph each line on the same coordinate plane. ***(LESSON 5.4)***

**a.** $y = -\frac{1}{3}x + 5$ **b.** $4x - 2y = 2$ **c.** $x = 2y$

**Solve each system algebraically. Check by graphing.**
***(LESSONS 7.2 AND 7.3)***

**70.** $\begin{cases} x - 3y = 3 \\ 2x - y = -4 \end{cases}$ **71.** $\begin{cases} x - 2y = 0 \\ x + y = 3 \end{cases}$

**Write each expression without negative exponents.** ***(LESSON 8.4)***

**72.** $a^2b^{-3}$ **73.** $\frac{m^5}{m^{-2}}$ **74.** $\frac{2n^{-6}}{n^{-4}}$

## Look Beyond

**75.** The height in feet, $h$, of a baseball $t$ seconds after it is hit is given by the function $h = -16t^2 + 100t$. Use the following methods to find $t$ when the ball is at a height of 100 feet:

**a.** graphing **b.** algebra

**Complete the square for each of the following:**

**76.** $2x^2 + 8x + 1$ **77.** $3x^2 + 5x - 4$

**internet connect**
**Portfolio Extension**
Go To: **go.hrw.com**
Keyword:
**MA1 Gravity**

**PORTFOLIO ACTIVITY**

Include this activity in your portfolio.

1. Build a table for the falling-object function $y = -16x^2 + 1000$ for $x$-values from 1 to 6.
2. Take successive differences of the $y$-values until they are constant. What is the constant?
3. The general form of the falling object function is $y = -32\left(\frac{1}{2}\right)x^2 - h$. The acceleration due to gravity is −32 feet per second squared. Use your data from the Portfolio Activity on page 491 to find the force of gravity from your experimental data.

10.4

# Solving Equations of the Form $x^2 + bx + c = 0$

**Objective**

- Solve quadratic equations by completing the square or by factoring.

**Why** *You have solved equations of the form $x^2 = k$. Equations of the form $x^2 + bx + c = 0$ can also be solved by completing the square. This type of equation can represent many different real-world situations.*

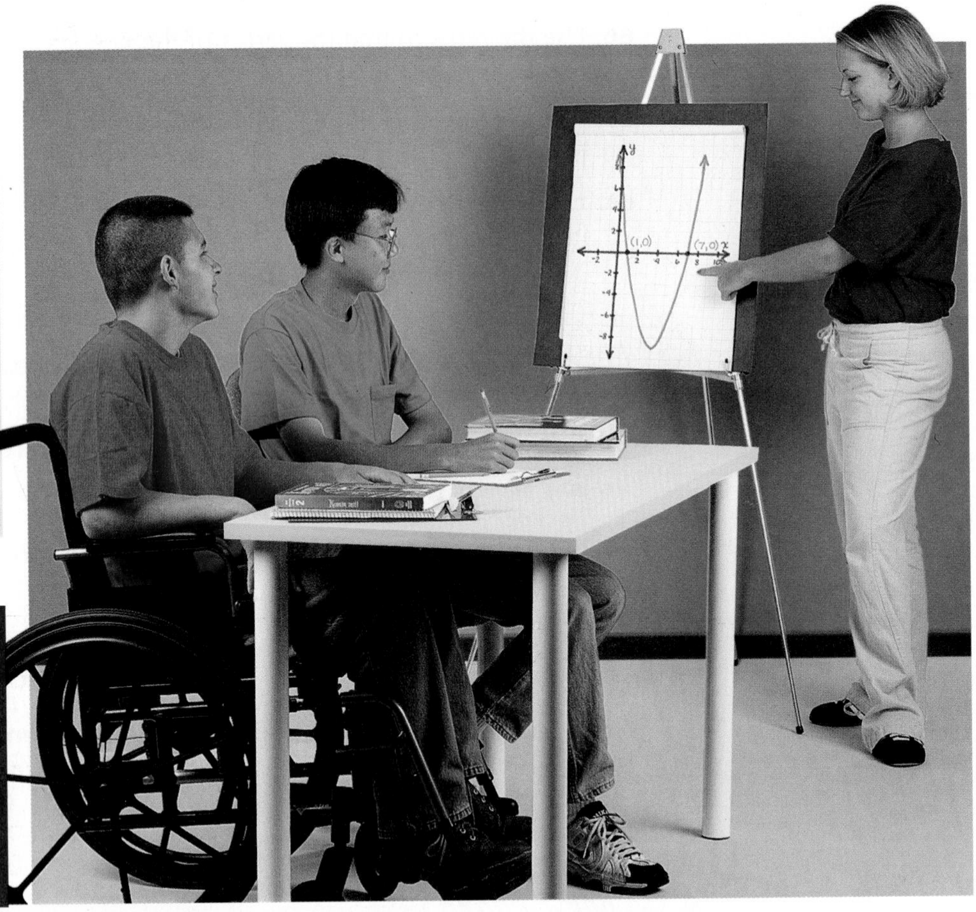

In Lesson 10.3 you found the vertex of $y = x^2 - 8x + 7$ by completing the square. When this function is rewritten in the form $y = (x - 4)^2 - 9$, the vertex, $(4, -9)$, can be found by inspection. To find the zeros of this function, substitute 0 for $y$, and rewrite the equation in the form $n^2 = k$, where $n$ is an expression involving $x$.

$$(x - 4)^2 - 9 = 0$$
$$(x - 4)^2 = 9$$
$$x - 4 = \pm 3$$

$$x - 4 = 3 \quad \text{or} \quad x - 4 = -3$$
$$x = 7 \qquad\qquad x = 1$$

The solutions are 7 and 1. These two solutions indicate the points where the graph of $y = x^2 - 8x + 7$ intersects the $x$-axis. Substitute 7 and 1 for $x$ in the original function to check.

$$y = 7^2 - 8(7) + 7 = 49 - 56 + 7 = 0$$
$$y = 1^2 - 8(1) + 7 = 1 - 8 + 7 = 0$$

The zeros of the function are $(7, 0)$ and $(1, 0)$. These are the points where the parabola intersects the $x$-axis.

**CRITICAL THINKING** Why are the zeros of the function $y = x^2 - 8x + 6$ and the solutions to the equation $x^2 - 8x + 6 = 0$ the same?

**EXAMPLE 1** **Find the zeros of $y = x^2 - 6x + 7$.**

**SOLUTION**

Replace $y$ with 0, complete the square, and solve.

$$(x^2 - 6x) + 7 = 0$$
$$(x^2 - 6x + 9) + 7 - 9 = 0$$
$$(x^2 - 6x + 9) - 2 = 0$$
$$(x - 3)^2 - 2 = 0$$
$$(x - 3)^2 = 2$$
$$x - 3 = \pm\sqrt{2}$$

$$x - 3 = \sqrt{2} \quad \text{or} \quad x - 3 = -\sqrt{2}$$
$$x = 3 + \sqrt{2} \qquad\qquad x = 3 - \sqrt{2}$$

The zeros are $3 + \sqrt{2}$ and $3 - \sqrt{2}$.

**TRY THIS** Find the zeros of $y = x^2 - 2x - 2$.

When a quadratic can be factored, you can quickly solve the quadratic equation. For example, $x^2 - 6x + 8$ can be factored as $(x - 4)(x - 2)$, so $x^2 - 6x + 8 = 0$ is the same as $(x - 4)(x - 2) = 0$.

The Zero Product Property (Lesson 9.8) can be used to solve $(x - 4)(x - 2) = 0$.

$$x - 4 = 0 \quad \text{or} \quad x - 2 = 0$$
$$x = 4 \qquad\qquad x = 2$$

The solutions are 4 and 2.

**EXAMPLE 2** **Solve $x^2 + x = 2$ by**
**a. completing the square.** **b. factoring.**

**SOLUTION**

**a.**

$$x^2 + x = 2$$
$$x^2 + x + \left(\frac{1}{2}\right)^2 = 2 + \left(\frac{1}{2}\right)^2$$
$$\left(x + \frac{1}{2}\right)^2 = \frac{9}{4}$$
$$x + \frac{1}{2} = \pm\frac{3}{2}$$
$$x + \frac{1}{2} = \frac{3}{2} \quad \text{or} \quad x + \frac{1}{2} = -\frac{3}{2}$$
$$x = 1 \qquad\qquad x = -2$$

**b.**

$$x^2 + x = 2$$
$$x^2 + x - 2 = 0$$
$$(x + 2)(x - 1) = 0$$
$$x + 2 = 0 \quad \text{or} \quad x - 1 = 0$$
$$x = -2 \qquad\qquad x = 1$$

The solutions are 1 and −2.

**CRITICAL THINKING** Sometimes both completing the square and factoring could be used to solve a quadratic equation of the form $x^2 + bx + c = 0$. How would you decide which method to use?

**EXAMPLE 3** **Solve $x^2 - 10x + 23 = 0$ by completing the square. Round your answer to the nearest hundredth.**

**SOLUTION**

$$x^2 - 10x + 23 = 0$$
$$x^2 - 10x = -23$$
$$x^2 - 10x + 25 = -23 + 25$$
$$(x - 5)^2 = 2$$
$$x - 5 = \pm\sqrt{2}$$

| | | |
|---|---|---|
| $x - 5 = \sqrt{2}$ | or | $x - 5 = -\sqrt{2}$ |
| $x = 5 + \sqrt{2}$ | | $x = 5 - \sqrt{2}$ |
| $x \approx 5 + 1.41$ | | $x \approx 5 - 1.41$ |
| $x \approx 6.41$ | | $x \approx 3.59$ |

The *exact* solutions are $5 + \sqrt{2}$ and $5 - \sqrt{2}$.

The *approximate* solutions are 6.41 and 3.59.

**TRY THIS** Solve $x^2 + 6x + 4 = 0$ by completing the square. Round your answer to the nearest hundredth.

**EXAMPLE 4** **Let $y = x^2 + 3x - 15$. Find the value(s) of $x$ when $y$ is $-5$.**

**SOLUTION**

**Method A**

Substitute $-5$ for $y$ in the function.

| | |
|---|---|
| $x^2 + 3x - 15 = -5$ | |
| $x^2 + 3x - 10 = 0$ | *Addition Property of Equality* |
| $(x - 2)(x + 5) = 0$ | *Factor.* |
| $x - 2 = 0$ or $x + 5 = 0$ | *Zero Product Property* |
| $x = 2$ $\quad$ $x = -5$ | |

The values of $x$ when $y$ is $-5$ are 2 and $-5$.

**Method B**

Graph $y = x^2 + 3x - 15$ and $y = -5$ on the same set of axes. Use a graphics calculator if you have one. Find the point(s) of intersection. You can see that $y = -5$ when $x = 2$ or $x = -5$.

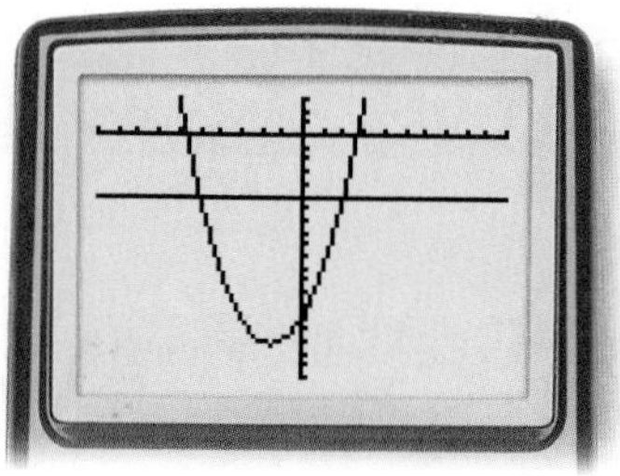

**TRY THIS** Let $y = x^2 - 8x + 22$. Find the value(s) of $x$ when $y$ is 10.

## EXAMPLE 5

**Find the points where the graph of $y = x^2 - 10x + 21$ intersects the graph of $y = x - 3$.**

TECHNOLOGY
GRAPHICS CALCULATOR

**SOLUTION**

**Method A**

Graph the functions, and find the points of intersection.

The graphs intersect at (8, 5) and (3, 0).

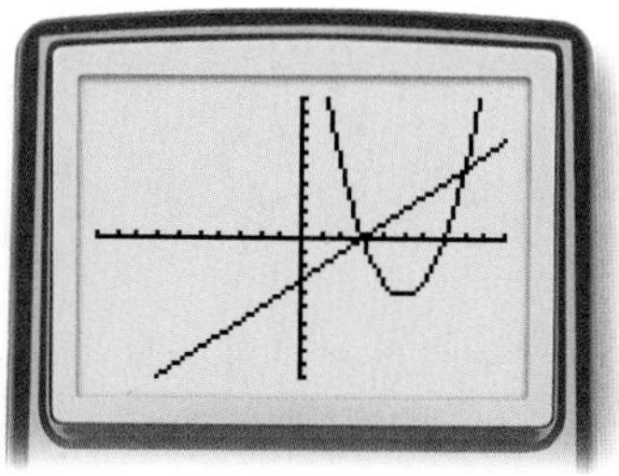

**Method B**

Use substitution to solve the following system: $\begin{cases} y = x^2 - 10x + 21 \\ y = x - 3 \end{cases}$

Substitute $x - 3$ from the second equation for $y$ in the first equation and simplify.

$$x^2 - 10x + 21 = x - 3$$
$$x^2 - 10x + 21 - x + 3 = 0$$
$$x^2 - 11x + 24 = 0$$

Solve the new quadratic equation by factoring.

$$(x - 8)(x - 3) = 0$$

$$x - 8 = 0 \quad \text{or} \quad x - 3 = 0$$
$$x = 8 \qquad\qquad x = 3$$

Substitute 8 and 3 for $x$ in the equation $y = x - 3$.

$$\begin{aligned} y &= x - 3 \\ y &= 8 - 3 \\ y &= 5 \end{aligned} \qquad\qquad \begin{aligned} y &= x - 3 \\ y &= 3 - 3 \\ y &= 0 \end{aligned}$$

The graphs intersect at (8, 5) and (3, 0).

**TRY THIS** Find the points where the graph of $y = -2x + 4$ intersects the graph of $y = x^2 - 4x + 1$.

# Exercises

## Communicate

1. How are the $x$-intercepts of the graph of $y = x^2 - 12x + 36$ related to the zeros of the function?
2. Explain how to find the zeros of $y = x^2 - 6x + 8$ by completing the square.
3. Describe how to solve $x^2 - 2x = 15$ by completing the square.
4. Graph $y = x^2 - 7x + 12$. Explain how to find the value of $x$ when $y$ is 2.
5. Describe two methods of finding the points where the graph of $y = x - 1$ intersects the graph of $y = x^2 - 3x + 2$.

## Guided Skills Practice

**Find the zeros of each function.** ***(EXAMPLE 1)***

6. $y = x^2 - 7x - 8$
7. $y = x^2 + 4x + 2$
8. $y = x^2 - 8x - 8$
9. Solve $x^2 + 12x = -27$ by completing the square and by factoring. ***(EXAMPLE 2)***
10. Solve $x^2 - 4x - 3 = 0$ by completing the square. Round your answer to the nearest hundredth. ***(EXAMPLE 3)***

**For $y = x^2 - 8x + 15$, find the value(s) of $x$ for each value of $y$ below.** ***(EXAMPLE 4)***

11. 3
12. −1
13. 8

**Find the points where the graphs of the two functions intersect.** ***(EXAMPLE 5)***

14. $y = x^2 - 9x + 25$ and $y = 3x - 11$
15. $y = x^2 + 4x - 12$ and $y = 9x + 2$

## Practice and Apply

**Find the zeros of each function. Graph to check.**

16. $y = x^2 - 2x - 8$
17. $y = x^2 + 6x + 5$
18. $y = x^2 - 4x + 4$

**Solve each equation by factoring.**

19. $x^2 - 2x - 3 = 0$
20. $x^2 + 4x - 5 = 0$
21. $x^2 + 7x + 12 = 0$
22. $x^2 - 10x + 24 = 0$
23. $x^2 - 3x = 10$
24. $x^2 - 8x = -15$
25. $x^2 - 6x + 9 = 0$
26. $x^2 + 10x = -25$
27. $x^2 - 2x + 1 = 0$

**Solve by completing the square. Round your answers to the nearest hundredth when necessary.**

28. $x^2 - 2x - 15 = 0$
29. $x^2 + 4x - 5 = 0$
30. $x^2 - x - 20 = 0$
31. $x^2 + 2x - 6 = 0$
32. $x^2 - 4x = 12$
33. $x^2 + x - 6 = 0$
34. $x^2 + 2x = 5$
35. $x^2 + 4x = 1$
36. $x^2 + 8x + 13 = 0$

internet connect
**Homework Help Online**
Go To: **go.hrw.com**
Keyword:
**MA1 Homework Help** for Exercises 23, 24, 26

**Solve each equation by factoring or by completing the square.**

**37.** $b^2 + 10b = 0$ **38.** $r^2 - 10r + 24 = 0$ **39.** $x^2 - x - 3 = 0$

**40.** $x^2 + 4x - 12 = 0$ **41.** $s^2 + 3s - 5 = 0$ **42.** $t^2 - 3t - 1 = 0$

**43.** $p^2 - 5p - 3 = 0$ **44.** $x^2 - x = 0$ **45.** $q^2 + 4q - 17 = 0$

**Use substitution to find the points where the graphs of each system intersect. Graph to check.**

**46.** $\begin{cases} y = 9 \\ y = x^2 \end{cases}$ **47.** $\begin{cases} y = 4 \\ y = x^2 - 2x + 1 \end{cases}$ **48.** $\begin{cases} y = x - 1 \\ y = x^2 - 3x + 3 \end{cases}$

**49.** $\begin{cases} y = x + 3 \\ y = x^2 - 4x + 3 \end{cases}$

**50.** Find two consecutive even integers whose product is 224.

CONNECTION

**51. GEOMETRY** The length of a rectangular flag is 4 yards longer than the width. The area of the flag is 140 square yards. Find the length and the width.

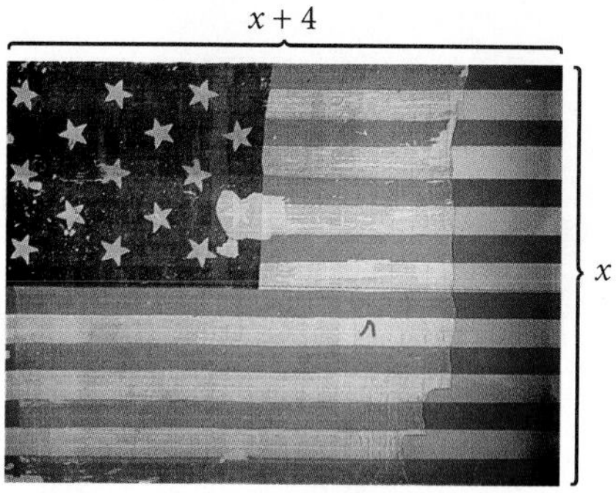


*The flag that flew over Fort McHenry inspired Francis Scott Key to write "The Star Spangled Banner".*

APPLICATION

**52. PHYSICS** A boat fires an emergency flare that travels upward at an initial velocity of 25 meters per second. To find the height in meters, $h$, of the flare at $t$ seconds, use the function $h = -5t^2 + 25t$. After how many seconds will the flare be at a height of 10 meters? Round your answer(s) to the nearest tenth.

**53. PHOTOGRAPHY** The perimeter of a rectangular photograph is 80 centimeters. Find the dimensions of the photograph if the length is 4 centimeters more than the width.

CHALLENGE

**54. MARCHING BAND** A marching band uses a rectangular marching formation with 8 columns and 10 rows. An equal number of columns and rows were added to the band's marching formation when 40 new members joined at the start of the school year. How many rows and columns were added?

## Look Back

**Solve each system of equations.** *(LESSONS 7.2 AND 7.3)*

**55.** $\begin{cases} x - 2y = 1 \\ 4x + 2y = -1 \end{cases}$ **56.** $\begin{cases} 5x - 2y = 11 \\ 3x + 5y = 19 \end{cases}$

**internet connect**

**Activities Online**

Go To: **go.hrw.com**
Keyword:
**MA1 Motion Equations**

**Write each number in standard notation.** *(LESSON 8.5)*

**57.** $4 \times 10^4$ **58.** $6.5 \times 10^7$ **59.** $9.6 \times 10^{-5}$

**Write each polynomial in factored form.** *(LESSON 9.5)*

**60.** $2b^2 + 6b - 36$ **61.** $6w(w^2 - w)$ **62.** $8p^4 - 16p$

## Look Beyond

**63.** Substitute values for $a$ and $b$ to determine whether the following statements are true:

**a.** $\sqrt{a + b} = \sqrt{a} + \sqrt{b}$ **b.** $\sqrt{a \cdot b} = \sqrt{a} \cdot \sqrt{b}$

# RESCUE AT 2000 FEET

After jumping out of a plane, the force of gravity causes the diver to accelerate downward.

As the sky diver's speed increases, the force of air resistance (or *drag*) plays a larger role.

Eventually the two forces balance, and the sky diver falls at a steady rate, called the *terminal velocity.*

## A MIRACULOUS SKY RESCUE

Robertson explains how he saved a life at 2,000 ft.

The jump began as a routine skydiving exercise, part of a convention of 420 parachutists sponsored by Skydive, Arizona, but it quickly turned into a test of nerve, instinct, and courage . . . Moments after he went out the open hatch of a four-engine DC-4 airplane near Coolidge, Arizona, sky diver Gregory Robertson, 35, could see that Debbie Williams, 31, a fellow parachutist with a modest 50 jumps to her credit, was in big trouble. Instead of "floating" in the proper stretched-out position parallel to the earth, Williams was tumbling like a rag doll. In attempting to join three other divers in a hand-holding ring formation, she had slammed into the backpack of another chutist and was knocked unconscious.

From his instructor's position above the other divers, Robinson reacted with instincts that had been honed by 1700 jumps during time away from his job as an AT&T engineer in Phoenix. He straightened into a vertical dart, arms pinned to his body, ankles crossed, head aimed at the ground in what chutists call a "no-lift" dive, and plummeted toward Williams . . . *At 3500 feet, about 10 seconds before impact, Robertson caught up with Williams,* almost hitting her but slowing his own descent by assuming the open-body, froglike position. He angled the unconscious sky diver so her chute could open readily and *at 2000 feet, with some 6 seconds left, yanked the ripcord* on her emergency chute, then pulled his own ripcord. The two sky divers floated to the ground. Williams, a fifth-grade teacher from Post, Texas, landed on her back, suffering a skull fracture and a perforated kidney—but alive. In the history of recreational skydiving, there has never been such a daring rescue in anyone's recollection.

*Time, May 4, 1987*

Does the article give you a complete picture of the rescue? How can you tell if the numbers in the article are accurate?

You can use what you know about distance, time, and speed to check the facts in the article and to get a fuller sense of what took place during this amazing feat. First, study these diagrams to get an idea of some of the forces that affect sky divers in free fall.

## Cooperative Learning

1. Answer parts **a**–**c** to determine whether the underlined parts of the article make sense. Use the fact that 88 feet per second (ft/s) equals 60 miles per hour (mph) for conversions.
   - **a.** Based on the data in the article, how long did it take Robertson to pull Williams' ripcord after he caught up to her?
   - **b.** During that time how far did the two divers fall, according to the article?
   - **c.** Given that the maximum speed of a person in free fall is about 190–200 mph, is the data in the article reliable? Explain.

2. After Robertson caught up with Williams, the two divers were probably falling at a speed between 125 and 150 mph. Suppose that they fell at 135 mph. You can use that value to find more plausible time estimates for the rescue.
   - **a.** About how many seconds from impact would the two sky divers be at 3500 feet? at 2000 feet?
   - **b.** Suppose that Robertson did catch up to Williams at 3500 feet and did pull her ripcord at 2000 feet. About how much time would have elapsed between those two events if the divers were falling at 135 mph?

3. While Robertson was catching up to Williams, he changed his body position to speed up and then slow down. Such motion is difficult to calculate exactly, but you can make estimates for this part of the rescue by using a single average speed for Robertson. Assume that Robertson was at 8500 feet and Williams was at 8300 feet when Robertson started trying to catch up.
   - **a.** Write an equation for $t$ in terms of $v_r$ and $v_w$, where $v_r$ is Robertson's average speed in ft/s while catching up to Williams, $v_w$ is Williams' speed in ft/s, and $t$ is the time in seconds from the time when Robertson starts to go after Williams to the time when he reaches her. (Hint: Compare the distances the two divers fell during the time in question.)
   - **b.** Use your equation to find $t$ if $v_r = 206$ (about 140 mph) and $v_w = 198$ (about 135 mph).
   - **c.** Based on your results from part **b**, at what altitude would Robertson catch up to Williams?

4. **Project** Imagine that you are a journalist writing about the rescue. Use your data from Steps 2 and 3 to write a paragraph that will give your readers a reasonably accurate sense of the event.

10.5

# The Quadratic Formula

## Objectives

- Use the quadratic formula to find solutions to quadratic equations.
- Use the quadratic formula to find the zeros of a quadratic function.
- Evaluate the discriminant to determine how many real roots a quadratic equation has and whether it can be factored.

**Why** *The height of an object thrown into the air, such as a javelin, can be modeled by a quadratic function. The zeros of the quadratic function tell when the object is at ground level.*

## Deriving the Quadratic Formula

**CULTURAL CONNECTION: ASIA** Since early Babylonian times, scholars have known how to solve the standard quadratic equation $ax^2 + bx + c = 0$ by using a formula. To see how the quadratic formula is developed, follow the steps as the equations $2x^2 + 12x + 4 = 0$ and $ax^2 + bx + c = 0$ are solved below.

**1.** Divide each side of each equation by the coefficient of $x^2$.

$$2x^2 + 12x + 4 = 0 \qquad ax^2 + bx + c = 0$$

$$x^2 + 6x + 2 = 0 \qquad x^2 + \frac{b}{a}x + \frac{c}{a} = 0$$

**2.** Subtract the constant term from each side of each equation.

$$x^2 + 6x = -2 \qquad x^2 + \frac{b}{a}x = -\frac{c}{a}$$

**3.** Complete the square. Then simplify.

$$x^2 + 6x + 9 = -2 + 9 \qquad x^2 + \frac{b}{a}x + \left(\frac{b}{2a}\right)^2 = -\frac{c}{a} + \left(\frac{b}{2a}\right)^2$$

$$(x + 3)^2 = 7 \qquad \left(x + \frac{b}{2a}\right)^2 = -\frac{c}{a} + \frac{b^2}{4a^2}$$

$$\left(x + \frac{b}{2a}\right)^2 = \frac{b^2 - 4ac}{4a^2}$$

**4.** Take the square root of each side. Then simplify.

$$x + 3 = \pm\sqrt{7}$$
$$x = -3 = \pm\sqrt{7}$$

$$x + \frac{b}{2a} = \pm\sqrt{\frac{b^2 - 4ac}{4a^2}}$$
$$x = \frac{-b \pm \sqrt{b^2 - 4ac}}{2a}$$

The equation above and to the right is known as the **quadratic formula.**

**The Quadratic Formula**

For $ax^2 + bx + c = 0$, where $a \neq 0$:

$$x = \frac{-b \pm \sqrt{b^2 - 4ac}}{2a}$$

**EXAMPLE 1** **Use the quadratic formula to solve $x^2 - 10 + 3x = 0$.**

**SOLUTION**

Rewrite $x^2 - 10 + 3x = 0$ in standard form, $x^2 + 3x - 10 = 0$. Identify the coefficients and the constant: $a = 1$, $b = 3$, and $c = -10$. Substitute these values into the quadratic formula. Then simplify.

$$x = \frac{-(3) \pm \sqrt{(3)^2 - 4(1)(-10)}}{2(1)}$$
$$= \frac{-(3) \pm \sqrt{49}}{2}$$
$$x = \frac{-3 + 7}{2} \quad \text{or} \quad x = \frac{-3 - 7}{2}$$
$$x = \frac{4}{2} = 2 \qquad x = \frac{-10}{2} = -5$$

The solutions are 2 and −5. Check by substitution.

**TRY THIS** Use the quadratic formula to solve $x^2 - 12 + 4x = 0$.

You now have five different methods for solving a quadratic equation.

1. taking square roots
2. factoring
3. completing the square
4. using the quadratic formula
5. graphing

**EXAMPLE 2** **Find the zeros of $y = 3x^2 + 2x - 4$.**

**SOLUTION**

Solve $0 = 3x^2 + 2x - 4$. Use the quadratic formula, with $a = 3$, $b = 2$, and $c = -4$.

$$x = \frac{-2 \pm \sqrt{4 - (-48)}}{6}$$
$$= \frac{-2 \pm \sqrt{52}}{6}$$

A calculator shows that $x \approx 0.87$ or $x \approx -1.54$. Check by substitution.

## The Discriminant

The discriminant of a quadratic equation allows you to determine how many real-number solutions the quadratic equation has.

**The Discriminant**

Given the quadratic equation $ax^2 + bx + c = 0$, where $a \neq 0$, the **discriminant** is defined as $b^2 - 4ac$.

- If the value of the discriminant is less than 0, the quadratic equation has no real solutions.
- If the value of the discriminant is 0, the quadratic equation has exactly one real solution.
- If the value of the discriminant is greater than 0, the quadratic equation has two real solutions.

**EXAMPLE 3** **Find the number of real solutions to each quadratic equation.**

**a.** $3x^2 - 2x + 1 = 0$ **b.** $4x = 4x^2 + 1$

**SOLUTION**

**a.** For $a = 3$, $b = -2$, and $c = 1$, the discriminant $b^2 - 4ac$ has a value of $-8$. Therefore, $3x^2 - 2x + 1 = 0$ has no real solutions.

**b.** Write the equation in standard form, $4x^2 - 4x + 1 = 0$. For $a = 4$, $b = -4$, and $c = 1$, the discriminant $b^2 - 4ac$ has a value of 0. Therefore, $4x = 4x^2 + 1$ has one real solution.

If the discriminant is a perfect square, the quadratic equation can be factored. You can use the quadratic formula to find the factors.

**CRITICAL THINKING** Examine the quadratic formula and explain why a quadratic equation can be factored if the discriminant is a perfect square.

**EXAMPLE 4** **Determine whether $6x^2 + 23x + 20$ can be factored and, if possible, determine the factored form.**

**SOLUTION**

For $a = 6$, $b = 23$, and $c = 20$, the discriminant $b^2 - 4ac$ has a value of 49, which is a perfect square.

Use the quadratic formula to find the factors.

$$x = \frac{-23 \pm \sqrt{49}}{12} = \frac{-23 \pm 7}{12}$$

Simplify. $$x = -\frac{30}{12} = -\frac{5}{2} \quad \text{or} \quad x = -\frac{16}{12} = -\frac{4}{3}$$

Work backward to find the factors.

| | | |
|---|---|---|
| $x = -\frac{5}{2}$ | $x = -\frac{4}{3}$ | *Given* |
| $2x = -5$ | $3x = -4$ | *Mult. Prop. of Equality* |
| $2x + 5 = 0$ | $3x + 4 = 0$ | *Add. Prop. of Equality* |

The factored form of $6x^2 + 23x + 20$ is $(2x + 5)(3x + 4)$.

**CULTURAL CONNECTION: EUROPE** In the sixteenth century, Italian mathematician Girolamo Cardano developed a way to work with square roots of negative numbers. Cardano defined $\sqrt{-1} \cdot \sqrt{-1} = -1$. Seventeenth century French mathematician Rene Descartes defined the **imaginary unit, *i*,** such that $i = \sqrt{-1}$ and $i^2 = -1$. **Imaginary numbers** have the form $i\sqrt{a}$, where $a$ is any positive real number.

The imaginary numbers and the real numbers together form the set of **complex numbers.** These numbers can be expressed in the form $a + bi$, where $a$ and $b$ are real numbers. Quadratic equations with no real-number solutions often have solutions that are complex numbers. You will study complex numbers in later mathematics courses.

# Exercises

## Communicate

internet connect
**Activities Online**
Go To: **go.hrw.com**
Keyword:
**MA1 Quadratic Formula**

1. Discuss how to identify the coefficients and constant ($a$, $b$, and $c$) in the equation $2n^2 + 3n = 7$ in order to use the quadratic formula.
2. Explain how to find the discriminant for $2t^2 + 3t - 7 = 0$.
3. Discuss how to use the discriminant to determine the number of real solutions to each equation.

   **a.** $2x^2 - x - 2 = 0$ **b.** $x^2 - 2x = -7$

## Guided Skills Practice

**Use the quadratic formula to solve each equation.** ***(EXAMPLE 1)***

**4.** $x^2 - 7x - 8 = 0$ **5.** $2x^2 + 4x + 2 = 0$ **6.** $4x^2 + 4x - 5 = 0$

**Find the zeros of each function.** ***(EXAMPLE 2)***

**7.** $y = 2x^2 - 8x - 8$ **8.** $y = 3x^2 - 2x + 5$ **9.** $y = x^2 + 7x - 3$

**Find the number of real solutions for each equation.** ***(EXAMPLE 3)***

**10.** $3x^2 + 6x + 3 = 0$ **11.** $x^2 - 2x + 3 = 0$ **12.** $-2x^2 + 7x = 5$

**Factor each expression.** ***(EXAMPLE 4)***

**13.** $6x^2 - 8x - 14$ **14.** $-15x^2 - 21x - 6$ **15.** $14x^2 + 37x - 42$

## Practice and Apply

**Find the value of the discriminant and determine the number of real solutions for each equation.**

**16.** $x^2 - x - 3 = 0$ **17.** $x^2 + 2x - 8 = 0$ **18.** $x^2 + 8x + 13 = 0$

**19.** $z^2 + 4z - 21 = 0$ **20.** $2y^2 - 8y = 8$ **21.** $8y^2 - 2 = 0$

internet connect
**Homework Help Online**
Go To: **go.hrw.com**
Keyword:
**MA1 Homework Help** for Exercises 22-30

**Use the quadratic formula to solve each equation. Round your answers to the nearest hundredth when necessary. Check by substitution.**

**22.** $a^2 - 4a - 21 = 0$ **23.** $t^2 + 6t - 16 = 0$ **24.** $m^2 + 4m - 5 = 0$

**25.** $w^2 - 4w = 0$ **26.** $x^2 + 9x = 0$ **27.** $x^2 - 9 = 0$

**28.** $x^2 + 2x = 0$ **29.** $3m^2 = 2m + 1$ **30.** $-x^2 + 6x - 9 = 0$

**Factor each expression by using the quadratic formula.**

**31.** $2x^2 - 14x + 20$ **32.** $2x^2 + 12x + 16$ **33.** $4x^2 + 8x - 12$

**34.** $2x^2 - 18x + 16$ **35.** $2x^2 - 3x + 1$ **36.** $3x^2 - 3x - 36$

**Choose any method to solve each quadratic equation. Round your answers to the nearest hundredth when necessary.**

**37.** $x^2 + 4x - 12 = 0$ **38.** $x^2 - 13 = 0$ **39.** $2x^2 + 5x - 3 = 0$

**40.** $x^2 + 7x + 10 = 0$ **41.** $2a^2 - 5a - 12 = 0$ **42.** $3x^2 + 10x - 5 = 0$

CONNECTION

**43.** **GEOMETRY** The seats in a theater are arranged in parallel rows that form a rectangular region. The number of seats in each row of the theater is 16 fewer than the number of rows. How many seats are in each row of a 1161-seat theater?

APPLICATION

**44.** **ACCOUNTING** To approximate the profit per day for her business, Mrs. Howe uses the formula $p = -x^2 + 50x - 350$. The profit, $p$, depends on the number of cases, $x$, of decorator napkins that are sold. How many cases of napkins must she sell to break even ($p = 0$)? to make the maximum profit? Find the maximum profit.

## Look Back

**Write each number in scientific notation.** ***(LESSON 8.5)***

**45.** 32,000,000 **46.** 67,000 **47.** 0.00654 **48.** 0.00000091

**Graph.** ***(LESSON 8.6)***

**49.** $y = 2^x$ **50.** $y = 0.3^x$ **51.** $y = \left(\frac{2}{3}\right)^x$

**Multiply.** ***(LESSON 9.3)***

**52.** $(x - 5)(x + 3)$ **53.** $(2x + 4)(2x - 2)$

## Look Beyond

**54.** From what you know about graphing quadratic equations and linear inequalities, describe what you think the graph of $y < x^2 + 3x + 2$ looks like.

# 10.6 Graphing Quadratic Inequalities

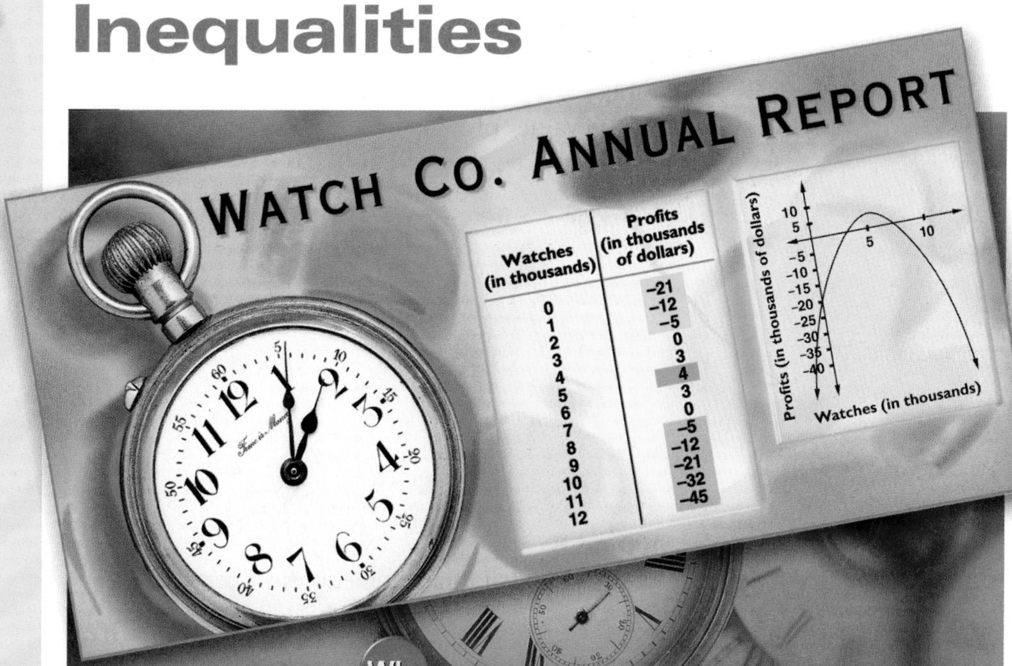


**Why** ***Not all problems involving quadratics are modeled by equations. Some, such as varying profit levels, are modeled by inequalities. Solving a quadratic inequality includes techniques that you already know plus one important step.***

### Objective

- Solve and graph quadratic inequalities and test solution regions.

**APPLICATION**
**BUSINESS**

To prepare the annual report, the president of a watch company gathers data about the number of watches produced per week and the corresponding profit. He uses the model $p = -x^2 + 10x - 21$, in which $p$ represents profit in thousands of dollars and $x$ represents the number of watches in thousands produced per week.

## *Activity*
## Production and Profit

**CONNECTION**
**MAXIMUM/MINIMUM**

Use the table of values and the graph of the profit function above to answer each question.

1. At what levels of production is the profit equal to 0?
2. At what levels of production is the profit greater than 0?
3. At what levels of production is the profit less than 0?
4. How many watches should be produced to maximize the profit?
5. According to the profit function, what is the maximum profit that the company can make? **CHECKPOINT ✔**

In the activity above, you solved a quadratic equation and two **quadratic inequalities.**

| $-x^2 + 10x - 21 = 0$ | $-x^2 + 10x - 21 > 0$ | $-x^2 + 10x - 21 < 0$ |
|---|---|---|
| Quadratic equation | Quadratic inequality | Quadratic inequality |

## EXAMPLE 1 Solve $x^2 - 2x - 15 > 0$.

**SOLUTION**

**Method A**

Graph $y = x^2 - 2x - 15$, and find the values of $x$ for which values of $y$ are greater than 0.

$x < -3$ or $x > 5$

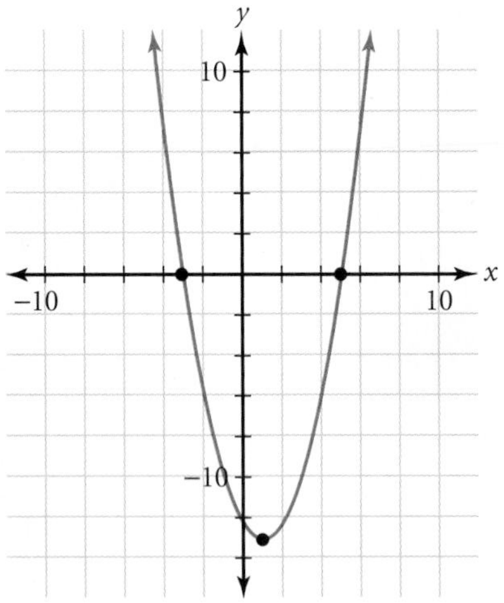


**Method B**

Solve the related quadratic equation, $x^2 - 2x - 15 = 0$.

$$x^2 - 2x - 15 = 0$$
$$(x-5)(x+3) = 0$$
$$x - 5 = 0 \quad \text{or} \quad x + 3 = 0$$
$$x = 5 \qquad\qquad x = -3$$

Use the $x$-axis as a number line, and plot the solutions. Since the inequality symbol is >, use open circles at −3 and 5.

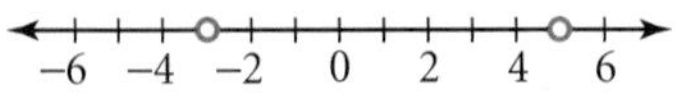


The solutions divide the $x$-axis into three intervals, to the left, between, and to the right of the two points. Test each interval. Choose a number from each interval, and substitute it for $x$ in the original inequality.

| Left of −3 | Between −3 and 5 | Right of 5 |
|---|---|---|
| $x^2 - 2x - 15 > 0$ | $x^2 - 2x - 15 > 0$ | $x^2 - 2x - 15 > 0$ |
| $(-5)^2 - 2(-5) - 15 \overset{?}{>} 0$ | $(0)^2 - 2(0) - 15 \overset{?}{>} 0$ | $(6)^2 - 2(6) - 15 \overset{?}{>} 0$ |
| $25 + 10 - 15 \overset{?}{>} 0$ | $0 - 0 - 15 \overset{?}{>} 0$ | $36 - 12 - 15 \overset{?}{>} 0$ |
| $20 > 0$ *True* | $-15 > 0$ *False* | $9 > 0$ *True* |
| **Solution interval** | **No** | **Solution interval** |

Graph the solutions.

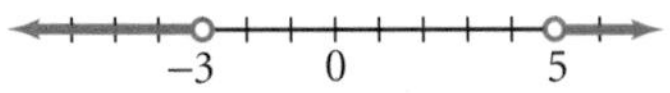


$x < -3$ or $x > 5$

**TRY THIS** Solve $x^2 + x - 12 \geq 0$.

**CRITICAL THINKING** Explain how you can determine which intervals to shade without testing numbers from all of the intervals.

## EXAMPLE 2 Graph $y > -x^2 + 1$.

**SOLUTION**

Graph $y = -x^2 + 1$. Draw the parabola with a dashed curve to show that points on the curve $y = -x^2 + 1$ are *not* included in the solution.

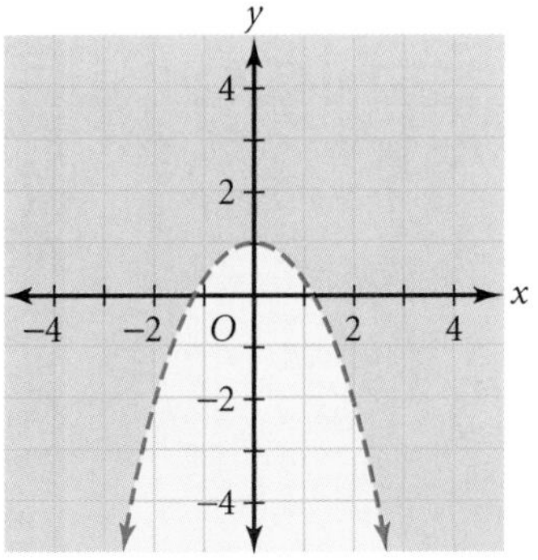


To graph $y > -x^2 + 1$, shade the points outside the parabola.

As a check, test points inside and outside the parabola.

Test $(0, 0)$, inside.

$$y > -x^2 + 1$$
$$0 \overset{?}{>} 0 + 1$$
$$0 > 1 \quad \textit{False}$$

Test $(0, 3)$, outside.

$$y > -x^2 + 1$$
$$3 \overset{?}{>} 0 + 1$$
$$3 > 1 \quad \textit{True}$$

The graph of $y > -x^2 + 1$ is the set of points outside the parabola.

**TRY THIS** Graph $y \geq x^2 - 7$.

## EXAMPLE 3

**APPLICATION**
**OPTICS**

*A flashlight has a parabolic reflector to direct the beam of light.*

A cross section of the parabolic light reflector shown at right is described by the equation $y = \frac{1}{14}x^2 + \frac{7}{2}$. The bulb inside the reflector is located at the point $(0, 7)$ Light from the bulb bounces off the reflector in parallel rays. This allows a flashlight to direct the light from the bulb in a narrow beam. **Determine which inequality below indicates the region in which the bulb is located.**

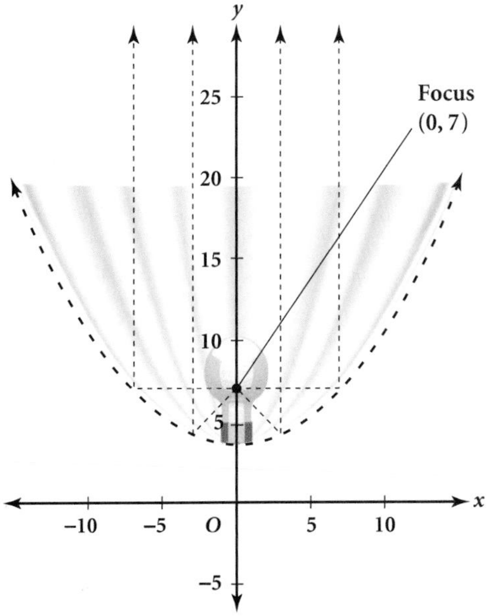


**a.** $y > \frac{1}{14}x^2 + \frac{7}{2}$

**b.** $y < \frac{1}{14}x^2 + \frac{7}{2}$

To determine which inequality indicates the region in which the bulb is located, test the focus point for part **a** and also for part **b.**

**SOLUTION**

**a.** $y > \frac{1}{14}x^2 + \frac{7}{2}$

Test $(0, 7)$.

$$7 \overset{?}{>} \frac{1}{14}(0) + \frac{7}{2}$$

$$7 > \frac{7}{2} \quad \textit{True}$$

**b.** $y < \frac{1}{14}x^2 + \frac{7}{2}$

Test $(0, 7)$.

$$7 \overset{?}{<} \frac{1}{14}(0) + \frac{7}{2}$$

$$7 < \frac{7}{2} \quad \textit{False}$$

The solution to the inequality $y > \frac{1}{14}x^2 + \frac{7}{2}$ includes the point where the bulb is located.

# Exercises

## Communicate

1. Explain how to solve $x^2 - 9x + 18 > 0$ by using the Zero Product Property.
2. Describe how to use a number line to graph and check the solution to the inequality from Exercise 1.
3. Explain how to determine whether the boundary line for the graph of each inequality is solid or dashed.
   **a.** $y < x^2 - 3$ **b.** $y \le x^2 - 3$ **c.** $y \ge x^2 - 3$
4. Describe how to graph each inequality. Explain two ways to decide whether to shade inside or outside the parabola for each solution.
   **a.** $y < x^2 - 3$ **b.** $y \le x^2 - 3$ **c.** $y \ge x^2 - 3$

## Guided Skills Practice

**Solve each inequality.** *(EXAMPLE 1)*

**5.** $x^2 + 6x + 9 > 0$ **6.** $x^2 - 4 < 0$ **7.** $x^2 + 7x + 12 > 0$

**Graph each inequality. Shade the solution region.** *(EXAMPLES 2 AND 3)*

**8.** $y > -x^2 + 4$ **9.** $y < x^2 - 4x + 4$ **10.** $y < x^2 + 3x + 2$

## Practice and Apply

**Solve each quadratic inequality by using the Zero Product Property. Graph each solution on a number line.**

**11.** $x^2 - 6x + 8 > 0$ **12.** $x^2 + 3x + 2 < 0$ **13.** $x^2 - 9x + 14 \le 0$

**14.** $x^2 + 2x - 15 > 0$ **15.** $x^2 - 7x + 6 < 0$ **16.** $x^2 - 8x - 20 < 0$

**17.** $x^2 - 8x > -15$ **18.** $x^2 - 3x \le 18$ **19.** $x^2 - 11x \ge -30$

**20.** $x^2 - 4x - 12 \ge 0$ **21.** $x^2 + 5x > 14$ **22.** $x^2 + 10x + 24 \ge 0$

internet connect
**Homework Help Online**
Go To: **go.hrw.com**
Keyword:
**MA1 Homework Help** for Exercises 23-34

**Graph each quadratic inequality. Shade the solution region.**

**23.** $y \ge x^2$ **24.** $y \ge x^2 + 4$ **25.** $y < -x^2$

**26.** $y > x^2$ **27.** $y > x^2 - 2x$ **28.** $y \ge 1 - x^2$

**29.** $y > -2x^2$ **30.** $y \le x^2 - 3x$ **31.** $y \le x^2 + 2x - 8$

**32.** $y \ge x^2 - 2x + 1$ **33.** $y \le x^2 - 4x + 4$ **34.** $y \le x^2 - 5x - 6$

CHALLENGE

**35.** Graph the solution region for $\begin{cases} y \ge x^2 - 9 \\ y \le -x^2 + 9 \end{cases}$.

**APPLICATION**

**PHYSICS** A projectile is fired vertically into the air. Its motion is described by $h = -16t^2 + 320t$, where $h$ is the projectile's height (in feet) after $t$ seconds. Use a quadratic inequality to find the answer to each question.

**36.** During what time interval(s) will the height of the projectile be below 1024 feet?

**37.** During what time interval(s) will the height of the projectile be above 1024 feet?

**38.** **TECHNOLOGY** Use graphing technology to graph $y = -16x^2 + 320x$. Use this graph to check your answers to Exercises 36 and 37.

## Look Back

**Find each product.** ***(LESSON 8.2)***

**39.** $(-ab^2c)(a^2bc^3)$

**40.** $(3p^2q^3r^4)(-2pqr)$

**Write each number in scientific notation.** ***(LESSON 8.5)***

**41.** 0.825

**42.** 0.000001

**43.** 0.0000074

**Factor each polynomial.** ***(LESSON 9.5)***

**44.** $12y^2 - 2y$

**45.** $2ax + 6x + ab + 3b$

**46.** $4x^2 - 24x + 32$

## Look Beyond

**47.** Three identical rectangles are stacked to form a square. If the perimeter of each rectangle is 24 inches, what is the area of the square?

**PORTFOLIO ACTIVITY**

Include this activity in your portfolio.

A scientist measures a rat's performance rating on a task after measuring various anxiety levels in the rat. The data is listed in the chart below.

| Anxiety level | 2 | 4 | 6 | 8 | 10 | 12 | 14 |
|---|---|---|---|---|---|---|---|
| Performance rating | 31.5 | 36 | 38.5 | 39 | 37.5 | 34 | 28.5 |

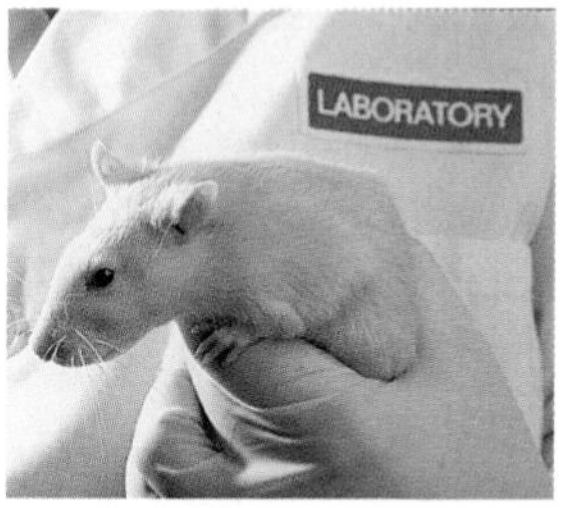


**1.** Make a scatter plot of the data. Use $x$ for anxiety level and $y$ for performance rating.

**2.** Use differences to determine the anxiety level when the performance rating is 0.

**3.** On a graphics calculator, use quadratic regression to find the equation that best fits the data.

**4.** Find the anxiety level that maximizes the performance rating.

**5.** Use the equation to find anxiety levels for performance ratings of at least 35.

**WORKING ON THE CHAPTER PROJECT**

You should now be able to complete the Chapter Project on page 516.

CHAPTER PROJECT TEN

# What's the Difference?

There are many sequences that are generated by a function which you can identify.

Consider the square numbers. If $n$ represents the dots on a side, the sequence shows the total number of dots in a square pattern.

| $n$ | 1 | 2 | 3 | 4 | 5 | 6 | 7 | ... | $n$ |
|---|---|---|---|---|---|---|---|---|---|
| $A(n)$ | 1 | 4 | 9 | 16 | 25 | 36 | 49 | ... | $n^2$ |

A function that can be used to produce this sequence is $A(n) = n^2$. The variable $n$ is usually used in place of $x$ when working with sequences. The variable $n$ represents the number of the term in the sequence. The function rule instructs you to square the number of the term in order to produce the value of the term.

How do you determine the function rule? One method is called **finite differences**. If a sequence eventually produces a constant difference, this method can provide a way to find the function rule for that sequence.

Examine another sequence: 0, 5, 12, 21, 32, 45, 60. You want to find a function rule that can be used to generate any term of this sequence. Begin by finding differences until they are constant.

| $n$ | 1 | 2 | 3 | 4 | |
|---|---|---|---|---|---|
| $P(n)$ | 0 | 5 | 12 | 21 | |
| | 5 | 7 | 9 | | ← First differences |
| | | 2 | 2 | | ← Second differences constant |

If the sequence produces constant second differences, you can expect a quadratic function. Write the quadratic function in the form of a polynomial function, $P(n) = an^2 + bn + c$. Evaluate the function for values of $n$ from 1 to at least 3. Then make a table of the terms and differences, using the expressions you got by evaluating the general quadratic polynomial.

$P(n) = an^2 + bn + c$
$P(1) = a(1)^2 + b(1) + c = a + b + c$
$P(2) = a(2)^2 + b(2) + c = 4a + 2b + c$
$P(3) = a(3)^2 + b(3) + c = 9a + 3b + c$
$P(4) = a(4)^2 + b(4) + c = 16a + 4b + c$

*These function rules correspond to the values of the terms in the actual sequence.*

| $n$ | 1 | 2 | 3 | 4 |
|---|---|---|---|---|
| $P(n)$ | $a+b+c$ | $4a+2b+c$ | $9a+3b+c$ | $16a+4b+c$ |

$3a+b$ $\quad$ $5a+b$ $\quad$ $7a+b$ ← First differences

$2a$ $\quad$ $2a$ ← Second differences constant

Set the corresponding terms and expressions in each layout equal to each other and solve for $a$, $b$, and $c$.

| Term 1 | First difference | Second difference |
|---|---|---|
| $a+b+c=0$ | $3a+b=5$ | $2a=2$ |

Use substitution to find the values of $a$, $b$, and $c$.

Use $2a = 2$ to solve for $a$. $\quad a = 1$

Use $3a + b = 5$ and the value for $a$ to solve for $b$. $\quad 3(1) + b = 5$, $b = 2$

Use $a + b + c = 0$ and the values for $a$ and $b$ to solve for $c$. $\quad 1+2+c=0$, $c = -3$

Finally, replace $a$, $b$, and $c$ in the quadratic polynomial, $P(n) = an^2 + bn + c$, with the values you found, $a = 1$, $b = 2$, and $c = -3$.

$$P(n) = n^2 + 2n - 3$$

Check your rule by substituting values for $n$ to see if it generates the appropriate sequence.

$$0, 5, 12, 21, 32, 45, 60, \ldots$$

## Activity 1 Find a Quadratic Function

Form groups of 2 to 5 students, and choose a leader. The leader begins the activity by choosing a quadratic function with integer coefficients and generating a sequence of five numbers. The leader then shows only the sequence to the other students in the group. The group then tries to find the quadratic function that generated the sequence.

## Activity 2 Find the Quadratic Function for Hexagonal Numbers

The design of the nested hexagons in the building shown at right can be represented by the sequence of hexagonal numbers. Let $n$ represent the number of dots per side. The terms of the sequence represent the total number of dots in each figure. Write the first four terms of the sequence of hexagonal numbers. Then use finite differences to identify the quadratic function that generates the sequence.

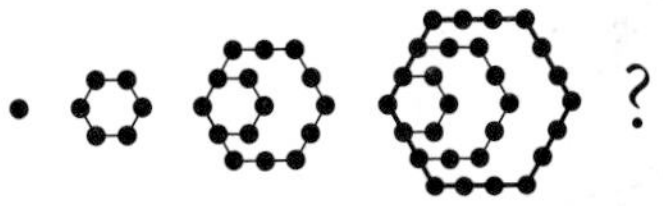

# 10 Chapter Review and Assessment

## VOCABULARY



## Key Skills & Exercises

### LESSON 10.1

#### Key Skills

**Find the vertex, axis of symmetry, and zeros of a quadratic function.**

Find the vertex, axis of symmetry, and zeros of $y = (x - 1)^2 - 4$ and sketch the graph.

The function $y = (x - 1)^2 - 4$ is in vertex form, $y = a(x - h)^2 + k$, where $(h, k)$ is the vertex, $x = h$ is the line of symmetry, and the zeros are the solution to $(x - 1)^2 - 4 = 0$.

Thus, the vertex is $(1, -4)$. The axis of symmetry is $x = 1$.

$y = (x - 1)^2 - 4$
$= x^2 - 2x - 3$
$= (x - 3)(x + 1)$

The zeros are 3 and −1.

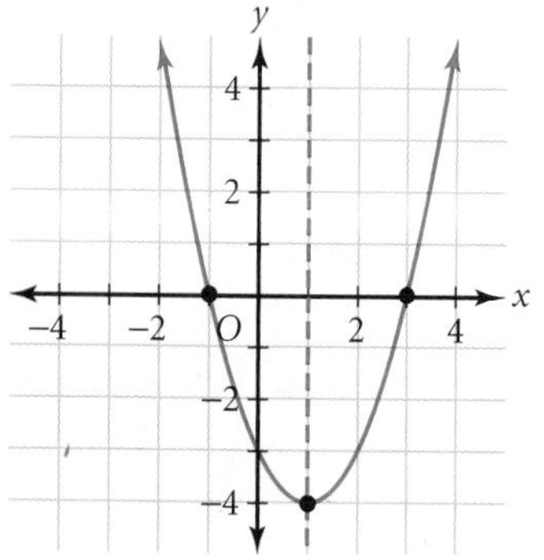


#### Exercises

**Find the vertex and axis of symmetry for the graph of each function. Sketch each graph.**

1. $y = (x + 1)^2 - 4$
2. $y = 2(x - 3)^2 - 2$

**Factor each quadratic function to find the zeros.**

3. $y = x^2 - 4x + 4$
4. $y = x^2 - 3x - 10$
5. $y = x^2 + 8x + 16$
6. $y = x^2 - 25$
7. $y = 2x^2 - 2x - 24$
8. $y = 2x^2 - 6x$

### LESSON 10.2

#### Key Skills

**Solve equations of the form $ax^2 = k$.**

Solve the equation $(x - 1)^2 - 16 = 0$.

Since $(x - 1)^2 - 16 = 0$, $(x - 1)^2 = 16$. The solutions for an equation of the form $ax^2 = k$ are $x = \sqrt{\frac{k}{a}}$ or $x = -\sqrt{\frac{k}{a}}$. For $(x - 1)^2 = 16$, $x - 1 = 4$ or $x - 1 = -4$. Thus, $x = 5$ or $x = -3$. The solutions are 5 and −3.

#### Exercises

**Solve each equation. Round answers to the nearest hundredth when necessary.**

9. $x^2 = 25$
10. $x^2 = \frac{9}{144}$
11. $x^2 - 4 = 0$
12. $x^2 = 12$
13. $(x + 1)^2 - 4 = 0$
14. $(x + 7)^2 - 81 = 0$
15. $(x - 3)^2 - 3 = 0$
16. $(x + 5)^2 - 16 = 0$

## LESSON 10.3

### Key Skills

**Find the vertex of a parabola by completing the square.**

Write $y = x^2 + 4x + 3$ in vertex form by completing the square. Find the vertex.

To complete the square, add and subtract the number that makes $x^2 + 4x + 3$ a perfect-square trinomial. To find this number take half the coefficient of $x$ and square it. Write the function in vertex form, $y = a(x - h)^2 - k$.

$$\begin{aligned} x^2 + 4x + 3 &= (x^2 + 4x + 4) + 3 - 4 \\ &= (x + 2)^2 - 1 \end{aligned}$$

The vertex is $(-2, -1)$.

### Exercises

**Find the number that completes the square.**

**17.** $x^2 + 8x$

**18.** $x^2 - 3x$

**Rewrite each function in the form $y = a(x - h)^2 - k$. Find each vertex.**

**19.** $y = x^2 - 8x + 18$

**20.** $y = x^2 - 14x + 50$

**21.** $y = x^2 + 12x + 16$

**22.** $y = x^2 - 2x + 5$

**23.** $y = x^2 - 5x + 2$

## LESSON 10.4

### Key Skills

**Solve a quadratic equation by completing the square.**

Solve $x^2 - 4x + 3 = 0$ by completing the square. Find the solutions of the equation.

$$x^2 - 4x + 3 = 0$$
$$x^2 - 4x = -3$$
$$x^2 - 4x + 4 = -3 + 4$$
$$(x - 2)^2 = 1$$
$$x - 2 = \pm 1$$
$$x - 2 = 1 \quad \text{or} \quad x - 2 = -1$$
$$x = 3 \qquad\qquad x = 1$$

The solutions to $x^2 - 4x + 3 = 0$ are 3 and 1.

### Exercises

**Find the solutions of each quadratic equation by factoring or completing the square.**

**24.** $x^2 + 2x = 24$

**25.** $x^2 + 5x + 6 = 0$

**26.** $x^2 - 8x + 16 = 0$

**27.** $x^2 - 2x - 35 = 0$

**28.** $x^2 + 6x - 12 = 0$

## LESSON 10.5

### Key Skills

**Use the quadratic formula to solve quadratic equations and functions.**

Solve $x^2 + 3x - 14 = 0$. What does the discriminant tell you about the solutions to the quadratic equation?

In the quadratic equation above, $a = 1$, $b = 3$, and $c = -14$. Substitute these values in the quadratic formula, $x = \frac{-b \pm \sqrt{b^2 - 4ac}}{2a}$, and simplify.

$$\begin{aligned} x &= \frac{-3 \pm \sqrt{9 - (4 \cdot 1)(-14)}}{2 \cdot 1} \\ &= \frac{-3 \pm \sqrt{65}}{2} \end{aligned}$$

Since the discriminant, $b^2 - 4ac = 65$, is positive, the equation has two real solutions, which are approximately 2.53 and −5.53.

### Exercises

**Find the value of the discriminant and determine the number of real solutions for each equation.**

**29.** $x^2 + 3x - 6 = 0$

**30.** $5x^2 - 6x + 2 = 0$

**Use the quadratic formula to solve each quadratic equation. Round your solutions to the nearest hundredth when necessary.**

**31.** $x^2 - 2x - 9 = 0$

**32.** $4x^2 - 5x - 4 = 0$

**33.** $4x^2 - 4x + 1 = 0$

**34.** $6x^2 - 7x - 3 = 0$

**35.** $8x^2 + 2x - 1 = 0$

LESSON 10.6

## Key Skills

### Solve and graph quadratic inequalities.

Solve and graph $x^2 + 3x - 4 > 0$.

To solve $x^2 + 3x - 4 > 0$, find the solutions for $x^2 + 3x - 4 = 0$. Since $x^2 + 3x - 4 = (x+4)(x-1)$, the solutions for the equation are $-4$ and 1. Since the numbers less than $-4$ or greater than 1 produce a true inequality, the solutions for $x^2 + 3x - 4 > 0$ are $x < -4$ or $x > 1$.

Graph $y \leq x^2 + 2$.

Graph $y \leq x^2 + 2$. Draw a solid parabola to show that the solutions include the graph of $y = x^2 + 2$. Since the inequality is $\leq$, shade the points below the parabola.

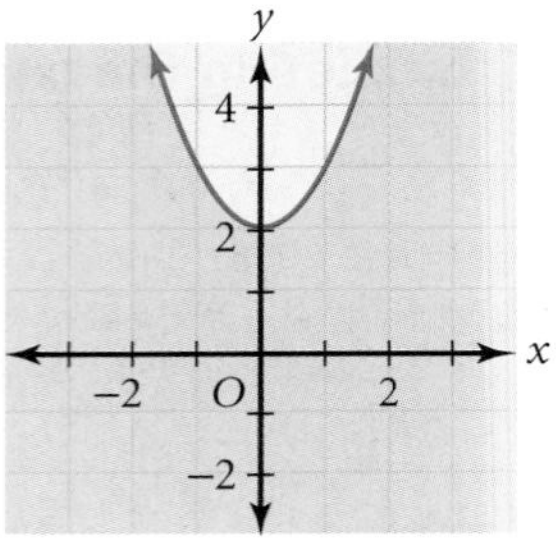


## Exercises

### Solve each quadratic inequality and graph the solution on a number line.

**36.** $x^2 + 6x + 8 \geq 0$

**37.** $x^2 - 3x - 10 < 0$

**38.** $x^2 + 9x + 18 > 0$

**39.** $x^2 + 2x - 15 \leq 0$

**40.** $x^2 - 3x - 28 \geq 0$

### Graph each inequality.

**41.** $y \leq x^2$

**42.** $y > x^2 - 4x$

**43.** $y \geq x^2 - x - 10$

**44.** $y > x^2 + 5x + 6$

**45.** $y < x^2 + 4x - 12$

# Applications

**SPORTS** The graph at right represents a relationship between the number of yards a football is thrown and its height above the ground.

**46.** What is the maximum height reached by the football?

**47.** How far does the football travel before it reaches its maximum height?

**48.** If a receiver catches the ball 30 yards from where the ball was thrown, what is the height of the ball when it is caught?

**49.** **GEOMETRY** A piece of wrapping paper has an area of 480 square inches. A square piece of paper is cut from the wrapping paper. If the remaining piece of paper has an area of 455 square inches, what is the length of each side of the square piece?

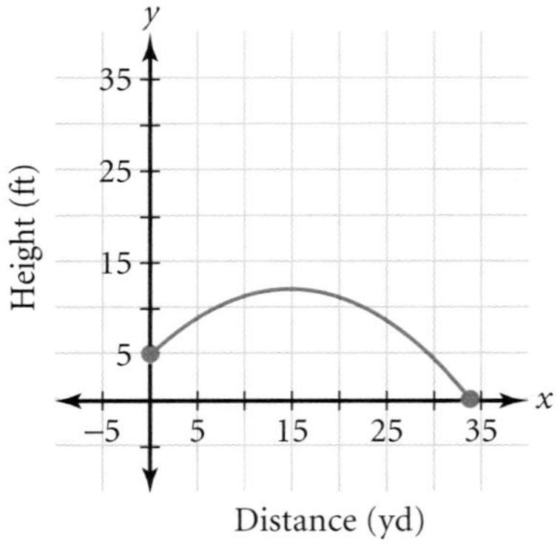

# 10 Chapter Test

**Find the vertex and axis of symmetry for the graph of each function. Sketch each graph.**

1. $y = (x - 4)^2 - 2$
2. $y = (x + 3)^2 + 1$

**Factor each quadratic function to find the zeros.**

3. $y = x^2 + 3x + 2$
4. $y = x^2 + 3x - 10$

**Describe the vertical and horizontal translations applied to the graph of the parent function $y = x^2$.**

5. $y = (x - 4)^2 - 10$
6. $y = -3(x + 6)^2 + 2$
7. NUMBER THEORY Use a quadratic equation to find two numbers whose sum is 14 and whose product is 48.

**Solve each equation. Round answers to the nearest hundredth when necessary.**

8. $x^2 = 49$
9. $81x^2 = 16$
10. $x^2 - 36 = 0$
11. $x^2 - 50 = 0$
12. $(x - 4)^2 - 25 = 0$
13. $(x + 3)^2 - 9 = 0$

PHYSICS An object at a point 480 feet above the water is dropped from a large suspension bridge. Its height after $t$ seconds is given by $h = -16t^2 + 480$. Round the answer to the nearest hundredth.

14. How long will it take the object to hit the water?

**Find the number that completes the square.**

15. $x^2 + 6x + \underline{?}$
16. $x^2 - 5x + \underline{?}$

**Rewrite the following functions in the form $y = a(x - h)^2 + k$. Find the vertex.**

17. $y = x^2 + 4x - 9$
18. $y = x^2 - 6x + 16$
19. $y = x^2 - 8x - 1$
20. $y = x^2 + 5x + 2$

**Find the solutions of each quadratic equation by factoring or completing the square.**

21. $x^2 - 4x = 5$
22. $x^2 + 6x = 16$
23. $x^2 - 4x + 3 = 0$
24. $x^2 - 7x + 12 = 0$
25. $x^2 + x - 2 = 0$
26. $x^2 + 2x - 15 = 0$

**Find the value of the discriminant and determine the number of real solutions for each equation.**

27. $x^2 - 5x + 4 = 0$
28. $x^2 + 7x - 18 = 0$

**Use the quadratic formula to solve each quadratic equation. Round your answer to the nearest hundredth when necessary.**

29. $x^2 + x - 1 = 0$
30. $4x^2 + 4x + 1 = 0$
31. $6x^2 - 5x + 1 = 0$
32. $4x^2 + 7x - 2 = 0$
33. GRAPHIC ARTS A print is 2 inches longer than it is wide. The matting that surrounds it is 3 inches on the top and sides and 4 inches on the bottom. If the area of the print and mat is 306 square inches, find the dimensions of the print itself.

**Solve each quadratic inequality, and graph the solution on a number line.**

34. $x^2 + 3x - 18 > 0$
35. $x^2 - x + 6 \leq 0$
36. $x^2 - 5x - 14 \geq 0$
37. $x^2 + 11x + 24 > 0$

**Graph each inequality.**

38. $y < x^2 - 2$
39. $y \geq x^2 + 2x$
40. LANDSCAPING A rectangular garden, 15 feet by 18 feet, has a border on one length and on one width of equal size. How wide must the border be to have a combined area, garden and border, that is at least 378 square feet?

# College Entrance Exam Practice

internet connect
Standardized Test Prep Online
Go To: go.hrw.com
Keyword: MM1 Test Prep.

**QUANTITATIVE COMPARISON** For questions 1–5 write
A if the quantity in Column A is greater than the quantity in Column B;
B if the quantity in Column B is greater than the quantity in Column A;
C if the quantities are equal; or
D if the relationship cannot be determined from the information given.

| | Column A | Column B | Answers |
|---|---|---|---|
| **1.** | $3c(c+1)$ for $c=-2$ | $3c^2+3c$ for $c=-2$ | Ⓐ Ⓑ Ⓒ Ⓓ [Lesson 9.2] |
| **2.** | slope of $2x+y=9$ | slope of $5x+y=3$ | Ⓐ Ⓑ Ⓒ Ⓓ [Lesson 5.4] |
| **3.** | perimeter of a square | area of a square | Ⓐ Ⓑ Ⓒ Ⓓ [Lesson 3.6] |
| **4.** | $(5+5)^2$ | $(5^2+5^2)$ | Ⓐ Ⓑ Ⓒ Ⓓ [Lesson 8.2] |
| **5.** | $\frac{-6}{3}$ | $\frac{-6}{-3}$ | Ⓐ Ⓑ Ⓒ Ⓓ [Lesson 2.4] |

**6.** Find the vertex for $y = 3(x-5)^2 + 8$. ***(LESSON 10.1)***

**a.** $(-5, 8)$ **b.** $(5, -8)$
**c.** $(5, 8)$ **d.** $(-5, 8)$

**7.** A shirt is marked down to \$24. The rate of the discount is 20%. What was the original price of the shirt? ***(LESSON 4.2)***

**a.** \$30 **b.** \$19.20
**c.** \$4.80 **d.** \$48

**8.** Simplify: $8 + 6 \cdot 4 \div 2 + 2$. ***(LESSON 1.3)***

**a.** 192 **b.** 9
**c.** 22 **d.** 14

**9.** Subtract: $5x^3 + x - 1 - (x^2 + x + 3)$. ***(LESSON 9.1)***

**a.** $5x^3 - x^2 - 4$ **b.** $4x^3 - x^2 - 4$
**c.** $5x^3 + 2x - 4$ **d.** $5x - 3$

**10.** A coin is tossed two times, resulting in one heads, and one tails. What is the experimental probability of one tails? ***(LESSON 4.3)***

**a.** $\frac{1}{4}$ **b.** $\frac{3}{4}$
**c.** $\frac{1}{2}$ **d.** 1

**11.** Find the slope of the line containing $A(3, 2)$ and $B(3, -4)$. ***(LESSON 5.2)***

**a.** 0 **b.** 1
**c.** $-6$ **d.** undefined

**12.** Evaluate $x - (y - z)$ for $x = 2$, $y = -3$, and $z = 4$. ***(LESSON 2.3)***

**a.** 9 **b.** $-5$
**c.** 1 **d.** 3

**13.** Solve: $(x+3)^2 - 16 = 0$. ***(LESSON 10.2)***

**a.** $-4$ and 4 **b.** $-7$ and 1
**c.** 7 and 1 **d.** $-1$ and 7

**14.** Simplify: $4^{-3} \cdot 4^3$. ***(LESSON 8.4)***

**a.** 4 **b.** 0
**c.** 12 **d.** 1

15. Solve. $\begin{cases} x + y = 5 \\ 2x - 3y = 0 \end{cases}$ ***(LESSON 7.2)***

**a.** $(2, 3)$ **b.** $(3, 2)$
**c.** $(-3, -2)$ **d.** $(-2, -3)$

16. Factor: $4x^2 - 4x + 1$. ***(LESSON 9.7)***

17. Solve: $36 = t - 24$. ***(LESSON 3.1)***

18. Solve. $\begin{cases} y = 2x + 3 \\ y = -3x + 2 \end{cases}$ ***(LESSON 7.3)***

19. Solve. $\begin{cases} 2x - 2y = -2 \\ 3x + 3y = 9 \end{cases}$ ***(LESSON 7.3)***

20. Solve: $|x - 8| = 12$. ***(LESSON 6.5)***

21. Write $3y = 5 - 2x$ in standard form. ***(LESSON 5.5)***

22. Solve: $\frac{n}{6} = \frac{15}{18}$. ***(LESSON 4.1)***

23. Solve: $2x + 9 \leq 23$. ***(LESSON 6.2)***

24. A city has a population of 3 million people. It is estimated that the population will increase at a yearly rate of 1.5%. At this rate, find the estimated population of the city in 5 years. ***(LESSON 8.7)***

25. The third and fourth terms of a sequence are 15 and 23. If the second differences are a constant 2, what are the first seven terms of the sequence? ***(LESSON 1.1)***

26. Simplify: $\frac{-100a^8b^{10}}{-20a^5b^9}$. ***(LESSON 8.3)***

27. Write an equation for a line that passes through the point $(-1, 0)$ and is perpendicular to a line that has a slope of $-2$. ***(LESSON 5.6)***

28. If tickets to a concert cost $24 plus a $7 handling charge per order, regardless of how many tickets are ordered, how many tickets could you order for $199? ***(LESSON 3.3)***

29. Factor $4x^2 - 25$. ***(LESSON 9.6)***

**FREE-RESPONSE GRID**

**The following questions may be answered by using a free-response grid such as that commonly used by standardized test services.**

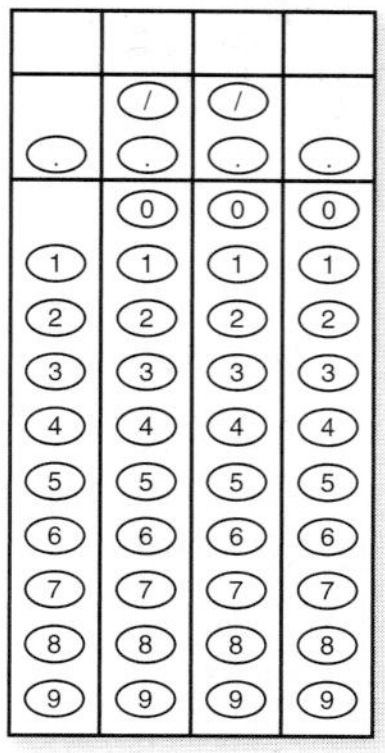

30. Todd scored 86 on his first test and 100 on his second test. What score does Todd need on his third test to have a 92 average for the three tests? ***(LESSON 4.4)***

31. The height of an object is expressed by the function $y = -16(x - 5)^2 + 345$, where $x$ is the time in seconds and $y$ is the height in feet. Find the maximum height of the object. ***(LESSON 10.1)***

32. Add: $-7.5 + (-1.8)$. ***(LESSON 2.2)***

33. The cost of 3 lunches and 4 dinners is $60. The cost of 5 lunches and 2 dinners is $44. Find the cost of one dinner. ***(LESSON 7.3)***

34. If $y$ varies directly as $x$ and $y = 35$ when $x = 7$, find $x$ when $y = 55$. ***(LESSON 5.3)***

35. Find the mean of 48, 62, 38, and 52. ***(LESSON 4.4)***

36. Simplify: $(3x^2y^3)^0$. ***(LESSON 8.4)***

37. Find the constant difference for the sequence 1, 2, 6, 13, 23, ... ***(LESSON 1.1)***

38. Solve: $\frac{2x}{3} = 12$. ***(LESSON 3.3)***

39. Find the value of the discriminant for $x^2 - 6x + 9 = 0$. ***(LESSON 10.5)***

40. How many real solutions are there for the equation in Exercise 39?

11

# Rational Functions

## Lessons

11.1 • Inverse Variation

11.2 • Rational Expressions and Functions

11.3 • Simplifying Rational Expressions

11.4 • Operations With Rational Expressions

11.5 • Solving Rational Equations

11.6 • Proof in Algebra

**Project**
Designing a Park

HAVE YOU EVER NOTICED THAT IT TAKES LONGER TO row upstream than downstream? Have you ever used a lever to lift a heavy object? Do you know what a gear ratio is?

The speed of a current affects the time spent rowing. The distance from the balance point affects the mechanical advantage of a lever. The gear ratio affects how fast you travel when riding a bicycle.

Rational expressions and functions can be used to model these types of relationships. A rational expression is a fraction that includes two algebraic expressions.

*You can make a bicycle easier or more difficult to pedal by adjusting the gear ratios.*

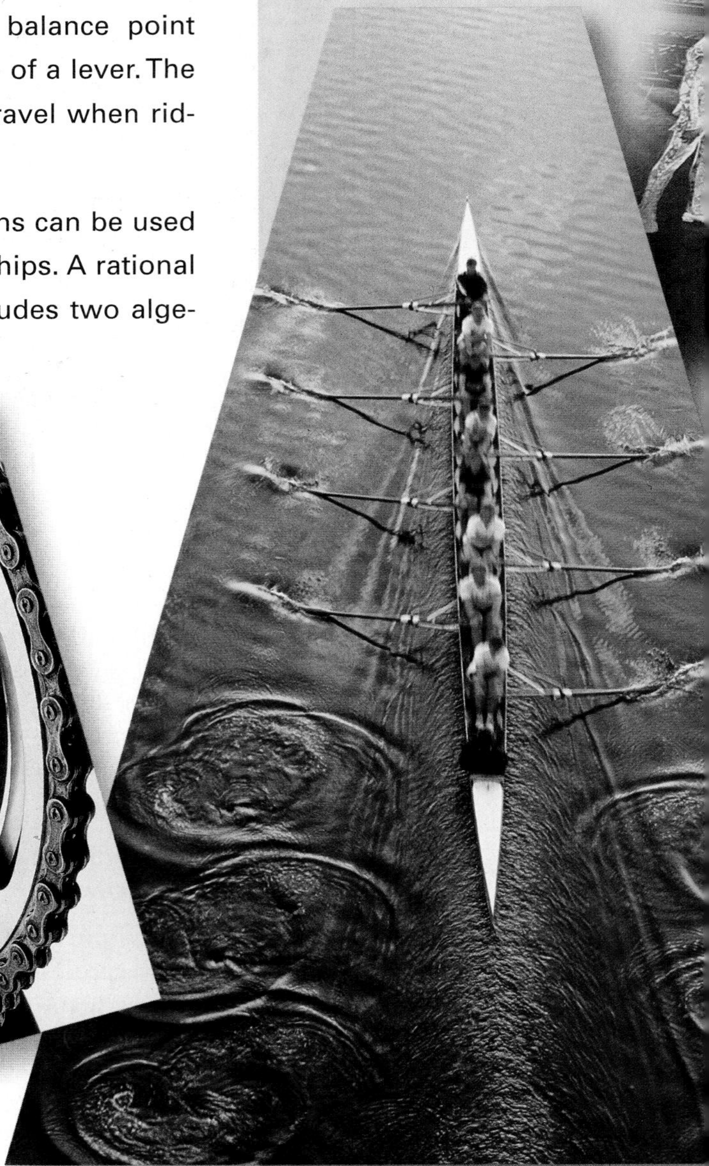

*Rowing with oars is an example of using levers to propel an object.*

*The ratio between the distances from the ends of the board to the balance point affects the height of the acrobats.*

*The various pulleys in a block and tackle utilize ratios so that heavy objects can be lifted with less force.*

CHAPTER PROJECT
PORTFOLIO ACTIVITIES

## About the Chapter Project

Many different types of situations can be modeled by rational expressions and functions. In the Chapter Project, *Designing a Park*, on page 566 you will use rational expressions and functions to calculate costs and make design decisions. After completing the Chapter Project, you will be able to do the following:

- Use equations containing rational expressions to solve a real-world problem.

## About the Portfolio Activities

Throughout the chapter, you will be given opportunities to complete Portfolio Activities that are designed to support your work on the Chapter Project.

- Balancing a lever is the topic of the Portfolio Activity on page 543.
- The relationship between gears on a bicycle is the topic of the Portfolio Activity on page 557.

# Inverse Variation

**Objective**

- Define and use two different forms of inverse variation to study real-world situations.

**Why** ***Recall from Chapter 5 that direct variation describes a linear relationship between two variables. Inverse variation describes a nonlinear relationship between two variables. This type of relationship models many real-world situations, such as comparing the time required to complete a project with the number of people working.***

Environmental projects such as tree planting are often made possible by the work of volunteer organizations. As the number of volunteers increases, the time it takes to plant a given number of trees decreases.

In *inverse variation*, as one quantity increases, another quantity decreases. In the activity that follows, you will explore an example of this type of variation.

*Activity*

## Exploring Inverse Variation

A volunteer organization has 5000 trees to plant in a particular area. It is estimated that a team of workers can plant 10 trees per hour.

1. How long would it take 1 team to plant all of the trees?
2. How long would it take 50 teams to plant all of the trees?
3. How long would it take 100 teams to plant all of the trees?

CHECKPOINT ✔

4. Write a function to represent the time in hours, $t$, that it takes $x$ teams to plant 5000 trees.

## Inverse Variation

In an inverse variation, $y$ is said to vary inversely as $x$. This is written as $y = \frac{k}{x}$, or $xy = k$, where $k$ is the constant of variation, $k \neq 0$, and $x \neq 0$.

**EXAMPLE 1** **If $y$ varies inversely as $x$ and $y = 5$ when $x = 15$, write an equation that shows how $y$ is related to $x$.**

**SOLUTION**

Use the first form of an inverse variation and the information given to find the constant of variation, $k$.

$$y = \frac{k}{x} \qquad \textit{Equation for inverse variation}$$

$$5 = \frac{k}{15} \qquad \textit{Substitute the given information.}$$

$$k = 75$$

Thus, the equation is $y = \frac{75}{x}$.

**TRY THIS** If $y$ varies inversely as $x$ and $y = 4$ when $x = 12$, write an equation that shows how $y$ is related to $x$.

**CULTURAL CONNECTION: AFRICA** One of the earliest references to inverse variation comes from an ancient Egyptian papyrus that describes how bread was used as a medium of exchange in some Egyptian cities. The value of bread was based on the number of *hekats* of wheat used to make the bread. A rating system was created to assign values to different types of bread. As you can see from the table below, a *lower* rating number meant a *higher* value. This is an example of inverse variation. The table shows the value of 1000 loaves of bread with various ratings.

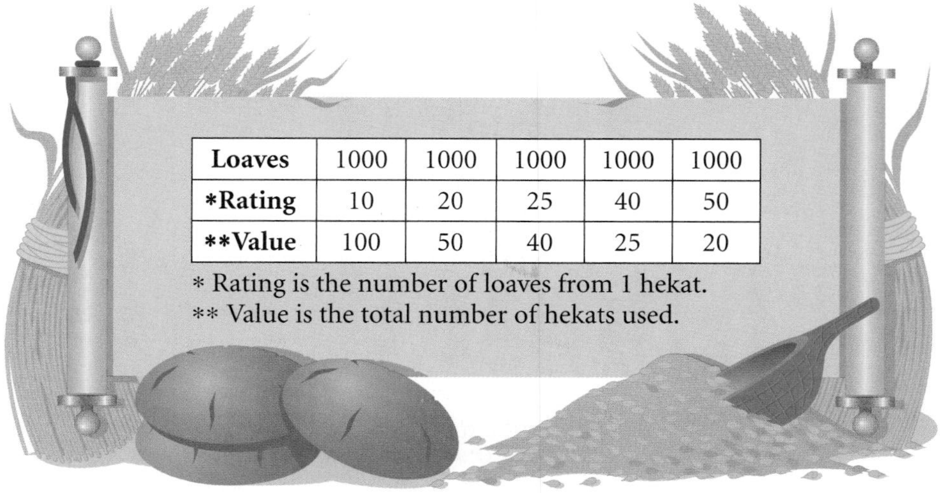

| **Loaves** | 1000 | 1000 | 1000 | 1000 | 1000 |
|---|---|---|---|---|---|
| ***Rating** | 10 | 20 | 25 | 40 | 50 |
| ****Value** | 100 | 50 | 40 | 25 | 20 |

* Rating is the number of loaves from 1 hekat.
** Value is the total number of hekats used.

Notice that the product of the rating and the value is always 1000. Thus, the relationship between the quantities in the table is an example of inverse variation of the form $xy = k$, or in this case, rating • value = 1000.

**EXAMPLE 2**

**APPLICATION**
**PHYSICS**

The physics student at right is investigating the properties of levers. On one side of the lever, a 30-gram mass is fixed 10 centimeters from the fulcrum. On the other side of the lever, distances from the fulcrum are marked in 5-centimeter intervals along the lever arm. The student is using the guess-and-check method to find the mass required at each distance to balance the 30-gram mass. The results of the investigation are shown in the table below.

| | | | | | |
|---|---|---|---|---|---|
| **Distance of mass from fulcrum** | 5 | 10 | 15 | 20 | 25 |
| **Mass required to balance 30 grams** | 60 | 30 | 20 | 15 | 12 |

**a. Examine the products of the corresponding distances and masses. Does there seem to be an inverse variation between these variables? Explain.**

**b. Write an equation that relates the variables. If an inverse variation exists, find the constant of variation, *k*.**

**c. Graph your equation.**

**SOLUTION**

**a.** The table seems to illustrate an inverse variation because the product of distance and mass is always 300.

**b.** The inverse variation is represented by the following:

$$\text{distance from fulcrum} \cdot \text{mass} = k$$

For 20 centimeters and 15 grams:

$$20 \cdot 15 = 300$$

Thus, the constant of variation, $k$, is 300.

An equation that represents this inverse variation is $d \cdot m = 300$.

**c.** If $d \cdot m = 300$, then $m = \frac{300}{d}$.
Use the information in the table to plot points contained in the graph. Sketch the graph using the points as a guide.

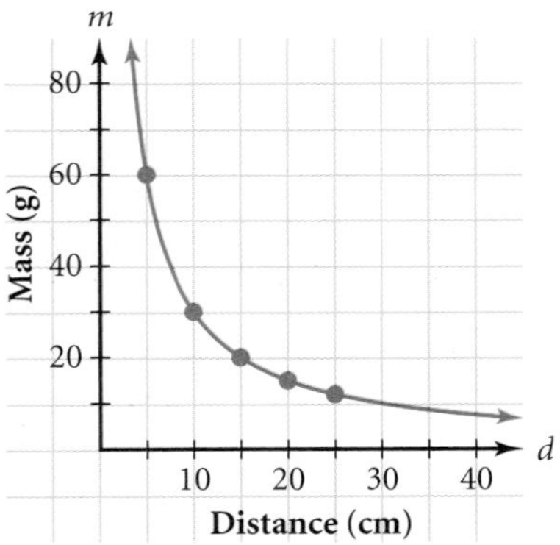


**TRY THIS**

Using the information in the table below, write an equation that shows the relationship between the variables $p$ and $q$.

| $p$ | 3 | 4 | 5 | 6 |
|---|---|---|---|---|
| $q$ | 20 | 15 | 12 | 10 |

**EXAMPLE 3**

**APPLICATION**
**INVESTMENTS**

The amount of time, $t$, in years that it takes to double a given amount of money with interest compounded annually can be approximated by the function $t = \frac{72}{r}$, where $r$ is the annual interest rate. In this application, percent is not converted to decimal form. For example, if the annual interest rate is 7%, then $r = 7$. In business, this function is usually referred to as the **rule of 72.**

**a. About how long does it take to double your money at 6% interest compounded annually?**

**b. Graph the function for the rule of 72.**

**SOLUTION**

**a.** To use the rule of 72, substitute 6 for $r$: $t = \frac{72}{6} = 12$. It takes about 12 years to double your money at 6% interest.

**b.** Choose values for $r$. Make a table and sketch the graph from the points.

**Rule of 72**

| $r$ | $t$ |
|---|---|
| 4 | 18 |
| 6 | 12 |
| 8 | 9 |
| 12 | 6 |

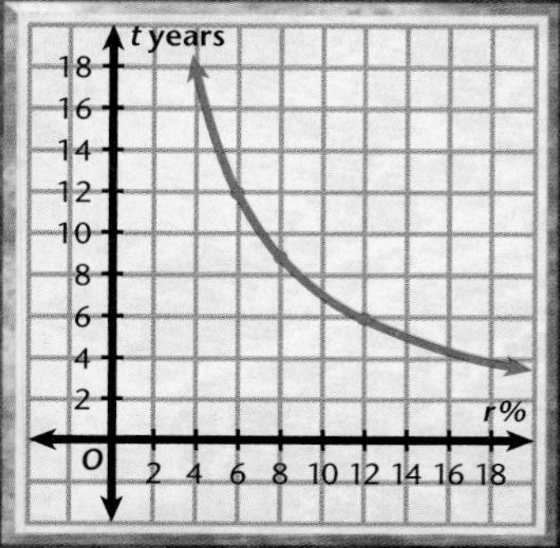


**TRY THIS** Use the rule of 72 to estimate how long it takes to double your money at 10% interest compounded annually.

# Exercises

## Communicate

**1.** Write two equivalent equations that express an inverse variation when $y$ varies inversely as $x$.

**2.** Explain what an inverse variation between rate and time means. Give a real-world example of an inverse variation involving these variables.

**3.** If $y$ is 3 when $x$ is 8 and $y$ varies inversely as $x$, explain how to find $y$ when $x$ is 2.

**4.** Explain how to use the rule of 72 to estimate how long it would take to double your money at 4% interest compounded annually.

## Guided Skills Practice

**If $y$ varies inversely as $x$, write an equation that shows how $y$ is related to $x$ for each situation below.** *(EXAMPLE 1)*

**5.** $y$ is 12 when $x$ is 60. **6.** $y$ is 3 when $x$ is 4. **7.** $y$ is 8 when $x$ is 4.

**Use the information in each table of values to write an equation that shows the relationship between the variables. Tell whether the relationship is an inverse variation.** *(EXAMPLE 2)*

**8.**

| | | | |
|---|---|---|---|
| $p$ | 2 | 3 | 4 |
| $q$ | 12 | 8 | 6 |

**9.**

| | | | | |
|---|---|---|---|---|
| $p$ | 4 | 15 | 20 | 60 |
| $q$ | 30 | 8 | 6 | 2 |

**10.**

| | | | | |
|---|---|---|---|---|
| $p$ | 2 | 3 | 4 | 5 |
| $q$ | 10 | 9 | 8 | 7 |

**11.**

| | | | | |
|---|---|---|---|---|
| $p$ | 5 | 15 | 25 | 75 |
| $q$ | 45 | 15 | 9 | 3 |

**Use the rule of 72 to estimate the amount of time that it takes to double a given amount of money at each interest rate, compounded annually.** *(EXAMPLE 3)*

**12.** 8% **13.** 4% **14.** 12%

## Practice and Apply

**Determine whether each equation represents an inverse variation.**

**15.** $rt = 400$ **16.** $y = \frac{-28}{x}$ **17.** $y = 10 - x$ **18.** $\frac{n}{5} = \frac{3}{m}$

**19.** $\frac{x}{y} = \frac{1}{2}$ **20.** $x = 10y$ **21.** $a = \frac{42}{b}$ **22.** $r = t$

**internet connect**
**Homework Help Online**
Go To: **go.hrw.com**
Keyword:
**MA1 Homework Help** for Exercises 23-28

**For Exercises 23–28, $y$ varies inversely as $x$.**

**23.** If $y$ is 8 when $x$ is 6, find $x$ when $y$ is 12.

**24.** If $y$ is 9 when $x$ is 12, find $y$ when $x$ is 36.

**25.** If $y$ is 3 when $x$ is 32, find $x$ when $y$ is 4.

**26.** If $y$ is 3 when $x$ is –8, find $x$ when $y$ is –4.

**27.** If $y$ is $\frac{3}{5}$ when $x$ is –60, find $y$ when $x$ is 2.

**28.** If $y$ is $\frac{3}{4}$ when $x$ is 12, find $y$ when $x$ is 27.

**CONNECTIONS**

**29.** **GEOMETRY** If the area of a triangle is constant, the length of the base varies inversely as the height. When the base is 22 centimeters, the height is 36 centimeters. What should the length of the base be when the height is 24 centimeters in order to maintain constant area?

**30.** **GEOMETRY** The area of a rectangle is 36 square centimeters. What is the length of a rectangle with the same area and a width of 3 centimeters? Given that the area remains constant, explain why the formula for the area of a rectangle is an inverse variation.

## APPLICATIONS

internet connect
**Activities Online**
Go To: **go.hrw.com**
Keyword:
**MA1 Gear Ratios**

**31.** **MECHANICS** The speed at which a gear revolves varies inversely as the number of teeth on the gear. If a gear with 16 teeth revolves at a speed of 500 revolutions per minute, at what speed should a gear with 20 teeth revolve?

**32.** **INVESTMENTS** Harold invests \$500 of his 4-H prize money at 5% interest compounded annually. Use the rule of 72 to determine how long will it take him to double his money.

**TRAVEL** Time varies inversely as the average rate of speed over a given distance.

**33.** Mike travels 6 hours at an average rate of 80 kilometers per hour. Find the time required to travel the same distance if Mike's average rate were 90 kilometers per hour.

**34.** A jet takes about 2.7 hours to fly from New York to Paris, averaging 2200 kilometers per hour. How long would it take a plane traveling at an average speed of 1760 kilometers per hour to make the same trip?

**35.** **PHYSICS** The frequency of a radio wave varies inversely as its wavelength. If a 200-meter wave has a frequency of 3000 kilocycles, what is the wavelength of a wave that has a frequency of 2000 kilocycles?

**36.** **MUSIC** A harp string vibrates to produce sound. The number of vibrations varies inversely as the length of the string. If a string that is 28 centimeters long vibrates 518 times per second, how long is a string that vibrates 370 times per second?

## *Look Back*

**37.** The first three terms of a sequence are 1, 5, and 12. The second difference is a constant 3. What are the next three terms of the sequence? ***(LESSON 1.1)***

**38.** A picture of an award-winning cat is surrounded by a frame with a width of $3x$ and a length of $4x$. What is the perimeter of the frame in terms of $x$? ***(LESSON 1.2)***

**39.** The total cost for an order of computer discs, including the \$1.95 handling charge, is \$57.60. The discs are sold in boxes of 10 and are on sale for \$7.95 per box. How many discs are in the order? ***(LESSON 3.3)***

**40.** What are the coordinates of the vertex of $y = -x^2 - 4x$? ***(LESSON 10.1)***

## *Look Beyond*

**41.** The intensity, $I$, of light varies inversely as the square of the distance, $d^2$, from the light source. If $I$ is 30 units when $d$ is 6 meters, find $I$ when $d$ is 3 meters.

11.2

# Rational Expressions and Functions

**Objective**

- Define and illustrate the use of rational expressions and functions.

**Why** *Rational functions provide convenient ways of expressing and solving some interesting and complex problems, such as calculating the average cost of pottery lessons.*

*A small pottery studio gives pottery lessons. The charge includes a \$55 initial fee plus \$4.50 for each lesson.*

A **rational expression** is an expression that can be written as a quotient of two polynomials.

### Rational Expressions

If $P$ and $Q$ are polynomials and $Q \neq 0$, then an expression of the form $\frac{P}{Q}$ is a rational expression.

You can create a rational expression to represent the average cost of a pottery lesson.

If a student takes $n$ pottery lessons, the total cost of the lessons will be $55 + 4.50n$. The average cost of a lesson can be found by dividing the total cost by the number of lessons taken. The result is the following rational expression:

$$\frac{55 + 4.50n}{n}$$

Notice that both the numerator and the denominator are first degree polynomials in the variable $n$.

**CRITICAL THINKING** Why do you think such expressions are called *rational*?

## Rational Functions

A function of the form $y = \frac{P}{Q}$, where $\frac{P}{Q}$ is a rational expression, is a **rational function**. The average cost per pottery lesson is as follows:

$$A = \frac{55 + 4.50n}{n}$$

In the following activity you will explore the domain and range of the rational function for the average cost of a pottery lesson shown above.

*Activity*

### Graphing a Rational Function

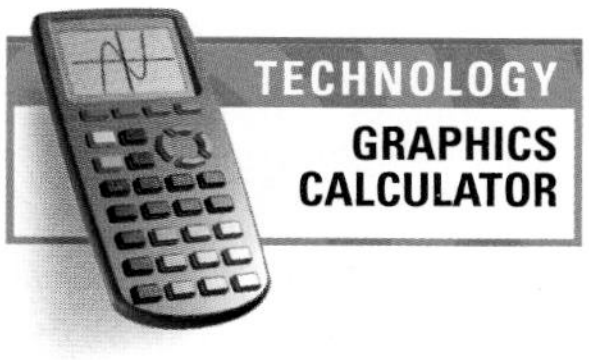


**You will need:** graph paper or a graphics calculator

1. Copy the table below and use your calculator to fill in the missing values.

| **Number of lessons, *n*** | **Average cost of a lesson,** $A = \frac{55 + 4.50n}{n}$ | **Cost** |
|---|---|---|
| 1 | $A = \frac{55 + 4.50(1)}{1}$ | 59.50 |
| 2 | ? | ? |
| 3 | ? | ? |
| 4 | ? | ? |
| 5 | ? | ? |
| 10 | ? | ? |
| 15 | ? | ? |
| 20 | ? | ? |
| 30 | ? | ? |
| 40 | ? | ? |

2. Does the function seem to be defined for all positive values of $n$? What happens when $n = 0$? Explain.

3. Negative and non-integer values of $n$ would have no meaning for the pottery-lesson example, but is the function defined for negative and non-integer values of $n$? Complete the table for two negative and two non-integer values of $n$.

4. Graph the function with your graphics calculator, or sketch the graph from the data in your table. (If you use your table data, you may want to complete the table for additional values of $n$.)

**CHECKPOINT ✓**

5. Describe the shape of your graph near the vertical line $n = 0$. As $n$ gets larger, what happens to $A$?

You may have seen in the activity that a situation may restrict the values that make sense for the domain of a particular function. In general, however, the domain of a rational function consists of all real-numbers that give a nonzero denominator. At each value that makes the denominator zero, the function is undefined, and there is either a vertical asymptote or hole. A **vertical asymptote** is a vertical line that the graph approaches but never crosses. A **hole** is an omitted point in the graph.

**EXAMPLE 1** **What is the domain of each of the following rational functions?**

**a.** $y = \frac{1}{x}$ **b.** $y = \frac{x^2 - 4}{x - 2}$ **c.** $n = \frac{m + 2}{m^2 - 5m + 6}$

**SOLUTION**

**a.** $y = \frac{1}{x}$ is undefined at $x = 0$. Thus, the domain of the function includes all real numbers except 0.

**b.** $y = \frac{x^2 - 4}{x - 2}$ is undefined when $x - 2 = 0$. Thus, the domain of the function includes all real numbers except 2.

**c.** $n = \frac{m + 2}{m^2 - 5m + 6}$ is undefined when $m^2 - 5m + 6 = 0$.
Since $m^2 - 5m + 6$ can be factored to $(m - 2)(m - 3)$, the function is undefined when $m = 2$ or $m = 3$. Thus, the domain of the function is all real numbers except 2 and 3.

**TRY THIS** What is the domain of each of the following rational functions?

**a.** $y = \frac{1}{x + 1}$ **b.** $y = \frac{x^2 + 0.5}{x^2 + 3x + 2}$

## Non-trivial Rational Functions

The function $y = \frac{1}{x}$, graphed at right, is the simplest **non-trivial** rational function. A function like $y = \frac{x}{2}$, whose denominator is a constant, is a **trivial** rational function because it can be written as $y = \frac{1}{2}x$, which is a polynomial.

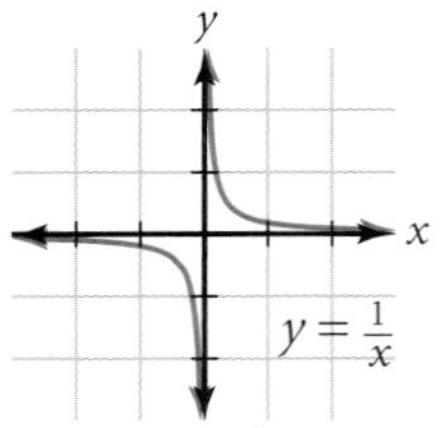


**CRITICAL THINKING** Name several examples of trivial and non-trivial functions. What are some differences that distinguish them?

**EXAMPLE 2** **a. Rewrite the function for the average price of a pottery lesson (page 533), $A = \frac{55 + 4.50n}{n}$, in simplest terms and graph.**

**b. Use the graph of the function from part a to determine how many lessons a person has taken if the average price of the lessons was \$5.60.**

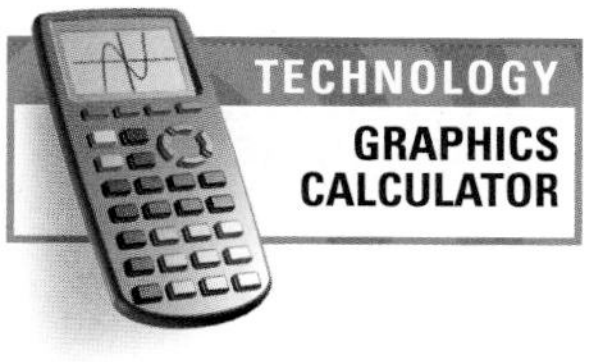


**SOLUTION**

**a.** Rewrite $A$ by applying the definition for adding fractions and simplifying.

$$A = \frac{55 + 4.50n}{n} = \frac{55}{n} + \frac{4.50n}{n} = \frac{55}{n} + 4.5$$

In order to use your graphics calculator to graph the function, you may need to let $A = y$ and $n = x$.

**b.** Use the graph of the function to find the $x$-value of the function for which $y = 5.60$.

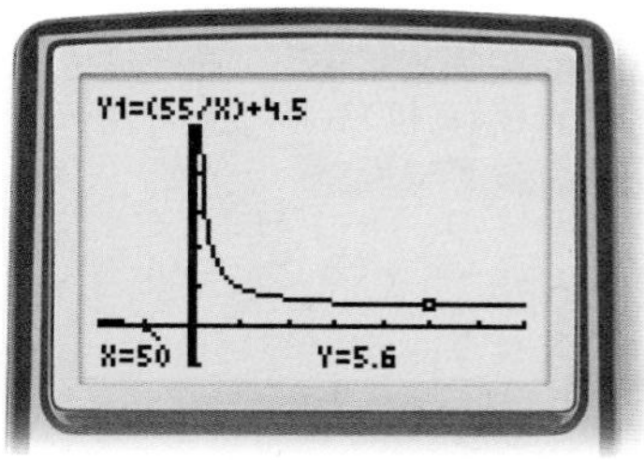


The $x$-value of 50 represents 50 lessons. This is the number of lessons for which the average cost is $5.60.

**TRY THIS**

**a.** Rewrite the function $y = \frac{83 - 44.3x}{x}$ in simplest terms and graph.

**b.** Use the graph of the function from part **a** to approximate the value of $x$ for which the function equals $-40$.

# Exercises

## Communicate

**1.** Define the terms *rational expression* and *rational function.* Give two examples of each.

**2.** Describe how to determine if a rational expression is undefined for any values of $x$.

**3.** Discuss how to express the function $y = \frac{12x - 7}{x}$ in simplest form.

## Guided Skills Practice

**Identify the domain of each function.** ***(EXAMPLE 1)***

**4.** $y = \frac{2}{x - 3}$　　**5.** $y = \frac{x}{x^2 - 5x}$　　**6.** $y = \frac{x^2 - 16}{x + 4}$

**Rewrite each function in simplest terms and graph.** ***(EXAMPLE 2)***

**7.** $y = \frac{2 - 3x}{x}$　　**8.** $y = \frac{5x + 7x^2}{x^2}$　　**9.** $y = \frac{3}{x} - 5$

## Practice and Apply

internet connect
**Homework Help Online**
Go To: go.hrw.com
Keyword:
**MA1 Homework Help**
for Exercises 10-17

**For what values of the variable is each rational expression undefined?**

**10.** $\frac{3x+9}{x}$ **11.** $\frac{6}{y-2}$ **12.** $\frac{2m-5}{6m^2-3m}$ **13.** $\frac{x^2+2x-3}{x^2+4x-5}$

**14.** $\frac{x^2-6x+8}{x^2-3x+2}$ **15.** $\frac{3x^2+2x}{x^2}$ **16.** $\frac{4z-5}{3z+4}$ **17.** $\frac{2l^2+4l}{2l+5}$

**Evaluate each rational function or expression for $x = 1$ and $x = -2$. Write *undefined* if appropriate.**

**18.** $y = \frac{1}{x-5}$ **19.** $y = \frac{2}{x}$ **20.** $y = \frac{-1}{x} + 3$

**21.** $y = \frac{1}{2x+3}$ **22.** $y = \frac{4}{x+2} - 5$ **23.** $y = \frac{-2}{x-1} + 4$

**24.** $y = \frac{3}{x+1} + 2$ **25.** $y = \frac{-6}{2x+5} - 3$ **26.** $y = \frac{2}{3x-4} - 7$

**27.** $\frac{5x-1}{x}$ **28.** $\frac{4x}{x-1}$ **29.** $\frac{x^2-1}{x^2-4}$

**30.** $\frac{x^2+2x}{x^2+x+2}$ **31.** $\frac{x^2-3}{x^2+4}$ **32.** $\frac{x^2-3x+4}{x^2+4x+4}$

**33.** $\frac{x^2-5}{x^2-4}$ **34.** $\frac{3x(x+2)}{x^2-1}$ **35.** $\frac{4x^2+2x-1}{x^2-2x+1}$

**Graph each rational function. List any values of $x$ for which the function is undefined.**

**36.** $y = \frac{1}{x}$ **37.** $y = \frac{1}{x-2}$ **38.** $y = \frac{1}{2(x-2)}$

**39.** $y = \frac{3}{2(x-2)}$ **40.** $y = \frac{-3}{2(x-2)}$ **41.** $y = \frac{-3}{2(x-2)} + 1$

**42.** $y = \frac{3}{x-1}$ **43.** $y = \frac{2}{x-3} + 4$ **44.** $y = \frac{-2}{x-4} + 3$

**Rewrite each function in simplest terms.**

**45.** $y = \frac{20+30x}{5x}$ **46.** $y = \frac{3-x}{3x}$ **47.** $y = \frac{x-10x^2}{2x^2}$

**48.** $y = \frac{x^{100}-x^{101}}{x^{101}}$ **49.** $y = \frac{9}{x} - 2$ **50.** $y = \frac{1-9x}{2x}$

**CONNECTION**

**51. COORDINATE GEOMETRY** Graph the trivial function $y = \frac{x+2}{4}$ and the non-trivial function $y = \frac{4}{x+2}$ on the same coordinate grid. List any values for which either function is undefined. List the coordinates of any points of intersection.

**APPLICATION**

**CONSUMER MATH** Ms. Kratochvil is financing her new car with a loan of \$23,000. At a simple interest rate of 6.5%, the total amount to be repaid is \$24,495. Her monthly payment, $P$, in dollars can be modeled by the rational function $P = \frac{24{,}495}{m}$, where $m$ is the term of the loan in months.

**52.** Use graphing technology to graph the rational function $P$.

**53.** If Ms. Kratochvil can spend no more than \$250 per month on her car payment, what is the shortest possible term for her loan?

## Look Back

**54.** Suppose that your class ordered 500 mugs with all the names of the class members. In the first week, 356 mugs were sold. What percent of the mugs was sold the first week? ***(LESSON 4.2)***

**55.** Write an equation for the line that contains the points (2, 1) and (−3, 5). ***(LESSON 5.4)***

**Rewrite each quadratic function in the form $y = (x - h)^2 + k$.**
***(LESSON 10.3)***

**56.** $y = x^2 - 4x + 6$

**57.** $y = x^2 + 14x + 39$

**58.** $y = x^2 - 10x + 32$

**59.** $y = x^2 + 4x + 5$

**60.** $y = x^2 - 2x - 3$

**61.** $y = x^2 + 6x - 8$

**62.** Solve $x^2 - 20 = -x$. ***(LESSON 10.4)***

## Look Beyond

*Benjamin Banneker, 1731–1806*

**63. CULTURAL CONNECTION: AMERICAS** Benjamin Banneker, an early African American astronomer, is famous for the almanacs he created from 1791 to 1797. Banneker also enjoyed solving mathematical problems such as the one above, which appeared in his journal. The ladders shown in the drawing are each 60 feet long. The height of one windowsill is 37 feet and the height of the other is 23 feet. If the two ladders meet at the bottom as shown, what is the width to the nearest hundredth of a foot of the street that runs between the two buildings? Use the Pythagorean Theorem (page 592) to solve Banneker's problem.

CONNECTION

**64. TRANSFORMATIONS** Graph the functions $y = x$, $y = x - 2$, and $y = 2x$ on the same coordinate plane. Describe the similarities and differences between the functions.

# 11.3 Simplifying Rational Expressions

**Why** *Most people do not look for the hardest way to do a job. Instead, they look for ways to simplify it. When working with rational expressions, the techniques of simplification will make your work much easier.*

## Objectives

- Factor the numerator and denominator to simplify rational expressions.
- State the restrictions on the variable of a simplified rational expression.
- Extend simplification techniques to other algebraic fractions.

Which rational expression is easier to evaluate for $x = 5$?

$$\frac{15x^4 - 120x^3 + 260x^2 - 120x + 245}{3x^5 - 24x^4 + 52x^3 - 24x^2 + 49x} \text{ or } \frac{5}{x}$$

Although the two rational expressions are equivalent, the second expression is much easier to evaluate. You can often simplify a rational function, making it easier to evaluate and graph.

## *Activity* Exploring Rational Functions

TECHNOLOGY GRAPHICS CALCULATOR

**You will need:** a graphics calculator or graph paper

Consider the functions $y = \frac{3x+6}{3x}$ and $y = \frac{x+2}{x}$.

1. Evaluate both functions for $x = -2$, $x = 5$, and $x = 10$.
2. What happens if you try to evaluate both functions for $x = 0$? CHECKPOINT ✔
3. Graph both functions.
4. Make a hypothesis about the relationship between $y = \frac{3x+6}{3x}$ and $y = \frac{x+2}{x}$. CHECKPOINT ✔

Recall the steps you would follow to rewrite $\frac{18}{24}$ in simplest form.

**1.** First write the numerator and denominator as products of prime factors.

$$\frac{18}{24} = \frac{2 \cdot 3 \cdot 3}{2 \cdot 2 \cdot 2 \cdot 3}$$

**2.** Separate out the fractions equal to one, and simplify.

$$\frac{2}{2} \cdot \frac{3}{3} \cdot \frac{3}{2 \cdot 2} = \frac{3}{4}$$

The same procedure is used to rewrite a rational expression in simplest form. A rational expression is in simplest form when the numerator and denominator have no common factors other than 1 or −1.

**EXAMPLE 1** **Simplify $\frac{5a+10}{5a}$. State any restrictions on the variable.**

**SOLUTION**

$$\frac{5a+10}{5a} = \frac{5(a+2)}{5a}$$ *Factor where possible.*

$$= \frac{5}{5} \cdot \frac{a+2}{a}$$ *Rewrite to show fractions equal to 1.*

$$= \frac{a+2}{a}, \text{ where } a \neq 0$$ *Simplify.*

**TRY THIS** Simplify $\frac{10-5x}{5x}$. State any restrictions on the variable.

### Steps to Simplify Rational Expressions

**1.** Factor the numerator and denominator.

**2.** Express each common-factor pair in the numerator and denominator as 1.

**3.** Simplify and state any restrictions on the variable.

Be sure that only common factors are eliminated when you simplify. Notice the common factors in the examples below.

| Expression | Common factor | Simplified form |
|---|---|---|
| $\frac{x(x+3)}{2x}$ | $x$ | $\frac{x+3}{2}, x \neq 0$ |
| $\frac{(x+4)}{x}$ | none | cannot be simplified further, $x \neq 0$ |
| $\frac{2x(x+1)}{4x^2}$ | $2x$ | $\frac{x+1}{2x}, x \neq 0$ |
| $\frac{2x^2+3}{x}$ | none | cannot be simplified further, $x \neq 0$ |

When you simplify an expression, the simplified result must carry the *same restrictions on the variable* as the original expression.

**EXAMPLE 2** **Simplify $\frac{3x-6}{x^2+x-6}$. State any restrictions on the variable.**

**SOLUTION**

$$\frac{3x-6}{x^2+x-6} = \frac{3(x-2)}{(x+3)(x-2)} \quad \textit{Factor the numerator and denominator.}$$

$$= \frac{x-2}{x-2} \cdot \frac{3}{x+3} \quad \textit{Rewrite to show fractions equal to 1.}$$

$$= \frac{3}{x+3} \quad \textit{Simplify.}$$

Since the denominator of the original expression, $x^2+x-6$, equals 0 when $x=2$ or $x=-3$, the restricted values of the variable are 2 and $-3$.

**TRY THIS** Simplify $\frac{x+2}{x^2+5x+6}$. State any restrictions on the variable.

**CRITICAL THINKING** The simplified expression in Example 2 above would have just one restriction, $x \neq -3$. Why must the restriction $x \neq 2$ be added?

**EXAMPLE 3** **Simplify $\frac{x^2+x-20}{16-x^2}$. State any restrictions on the variable.**

**SOLUTION**

$$\frac{x^2+x-20}{16-x^2} = \frac{x^2+x-20}{(-1)(x^2-16)} \quad \textit{Factor } -1 \textit{ from the denominator.}$$

$$= \frac{(x+5)(x-4)}{(-1)(x+4)(x-4)} \quad \textit{Factor numerator and denominator.}$$

$$= \frac{x+5}{(-1)(x+4)} \quad \textit{Simplify.}$$

$$= \frac{x+5}{-x-4}, \text{ where } x \neq 4 \text{ and } x \neq -4$$

**TRY THIS** Simplify $\frac{x^2+x-30}{36-x^2}$. State any restrictions on the variable.

**CRITICAL THINKING** How do the graphs of $y = \frac{x^2-36}{x+6}$ and $y = x-6$ differ?

## Other Algebraic Fractions

In the examples you have studied so far in this lesson, both the numerator and denominator have been polynomials with just a single variable. However, the techniques you have learned can be extended to other algebraic fractions.

In the example that follows, both the numerator and the denominator contain two variables.

**EXAMPLE 4**

CONNECTION
GEOMETRY

**Find the ratio of the volume of a right circular cylinder, with a radius of *r* and a height of *h*, to its surface area. Simplify if possible.**

SOLUTION

PROBLEM SOLVING

**Use a formula.** The formula for the volume of a right circular cylinder is $V = \pi r^2 h$. The formula for the surface area of a right circular cylinder is $S = 2\pi r^2 + 2\pi rh$.

$$\frac{\text{volume}}{\text{surface area}} = \frac{\pi r^2 h}{2\pi r^2 + 2\pi rh}$$
$$= \frac{\pi r^2 h}{2\pi r(r+h)}$$
$$= \frac{\pi r}{\pi r} \cdot \frac{rh}{2(r+h)}$$
$$= \frac{rh}{2(r+h)}, \text{ where } r \neq -h \text{ and } r \neq 0$$

Therefore, the ratio $\frac{\text{volume}}{\text{surface area}}$ is $\frac{rh}{2(r+h)}$.

**TRY THIS** Find the ratio of the surface area of a cube, $6s^2$, to its volume, $s^3$. Simplify if possible.

**CRITICAL THINKING** In the final statement of the solution in Example 4, no restrictions on the variable were mentioned. Explain.

# Exercises

## Communicate

1. When is a rational expression undefined?
2. Write a rational expression that is undefined for $x = 7$.
3. Discuss how to find the restrictions on a variable in a rational expression.
4. Explain what is meant by a common factor.
5. Describe the process that is used to write $\frac{x+1}{x^2+2x+1}$ in simplest form.

## Guided Skills Practice

**Simplify each expression and state any restrictions on the variables.**
***(EXAMPLES 1, 2, 3, AND 4)***

**6.** $\frac{6b-18}{3}$ **7.** $\frac{4x+16}{8x}$ **8.** $\frac{30n^2-18n}{12n}$ **9.** $\frac{24}{16x+32}$

**10.** $\frac{14ab}{49ab+98b}$ **11.** $\frac{2x-10}{x-5}$ **12.** $\frac{a-6}{a^2-6a}$ **13.** $\frac{x^2-y^2}{x^2+2xy+y^2}$

**14.** $\frac{x^2-2x+1}{2x-2}$ **15.** $\frac{x+4}{x^2+x-12}$ **16.** $\frac{y^2-49}{y^2+4y-21}$ **17.** $\frac{m^2+4m}{m^2+m-12}$

## Practice and Apply

**For what value(s) of the variable is each rational expression undefined?**

**18.** $\frac{10x}{x-3}$ **19.** $\frac{10}{5-y}$ **20.** $\frac{r-6}{r}$ **21.** $\frac{7p}{p-4}$

**22.** $\frac{k-3}{3-k}$ **23.** $\frac{(a+3)(a+4)}{(a-3)(a+4)}$ **24.** $\frac{c-4}{2c-10}$ **25.** $\frac{3}{y(y^2-5y+6)}$

**Name the common factors of the numerator and denominator.**

**26.** $\frac{9}{12}$ **27.** $\frac{3(x+4)}{6x}$ **28.** $\frac{x-y}{(x+y)(x-y)}$ **29.** $\frac{r+3}{r^2+5r+6}$

**internet connect**
**Homework Help Online**
Go To: **go.hrw.com**
Keyword:
**MA1 Homework Help**
for Exercises 30-45

**Simplify each expression and state any restrictions on the variables.**

**30.** $\frac{16(x+1)}{30(x+2)}$ **31.** $\frac{3(a+b)}{6(a-b)}$ **32.** $\frac{4(c+2)}{10(2+c)}$ **33.** $\frac{3(x+y)(x-y)}{6(x+y)}$

**34.** $\frac{6m+9}{6}$ **35.** $\frac{7t+21}{t+3}$ **36.** $\frac{12+8x}{4x}$ **37.** $\frac{3d^2+d}{3d+1}$

**38.** $\frac{b+2}{b^2-4}$ **39.** $\frac{x-2}{x^2+2x-8}$ **40.** $\frac{-(a+1)}{a^2+8a+7}$ **41.** $\frac{4-k}{k^2-k-12}$

**42.** $\frac{c^2-9}{3c+9}$ **43.** $\frac{3n-12}{n^2-7n+12}$ **44.** $\frac{y^2+2y-3}{y^2+7y+12}$ **45.** $\frac{a^2-b^2}{(a+b)^2}$

CONNECTION

**46. GEOMETRY** The area of a certain rectangle is represented by $x^2+7x+12$ and its length is represented by $x+4$. Find the width in terms of $x$.

APPLICATION

**47. BIOLOGY** Two equal populations of animals, one a predator and the other its prey, were released into the wild. Five years later a survey showed that the predator population was 3 times the original predator population, while the prey population was 8 times the original prey population. Then 12 more predators and 32 more prey were released into the area. What is the ratio of predator to prey after the release? Write your answer in simplest form.

## Look Back

APPLICATION

**48.** **DISCOUNTS** Chin-Lyn buys a new sweater at a 20% discount and pays $23.60, before tax. What was the original price of the sweater? *(LESSON 4.2)*

**49.** Write an equation for the line that contains the points $\left(\frac{2}{3}, -2\right)$ and $\left(-\frac{5}{6}, 3\frac{1}{3}\right)$. *(LESSON 5.4)*

**50.** Multiply: $(x+2)(x-2)$. *(LESSON 9.3)*

**51.** Write the equation of the parabola whose vertex is (2, 3) and whose graph opens downward. *(LESSON 10.1)*

**52.** Which of the following parabolas open downward? *(LESSON 10.1)*

**a.** $y = 3x^2 - 8$ **b.** $y = 9 - x^2$ **c.** $y = -x^2 + 4$ **d.** $y = \frac{3}{4}x^2 - 6$

## Look Beyond

**53.** Eight squares are the same size, but each is a different color. The squares are stacked on top of each other as shown at right. Identify the order of the squares numerically from top to bottom. The top square is number 1.

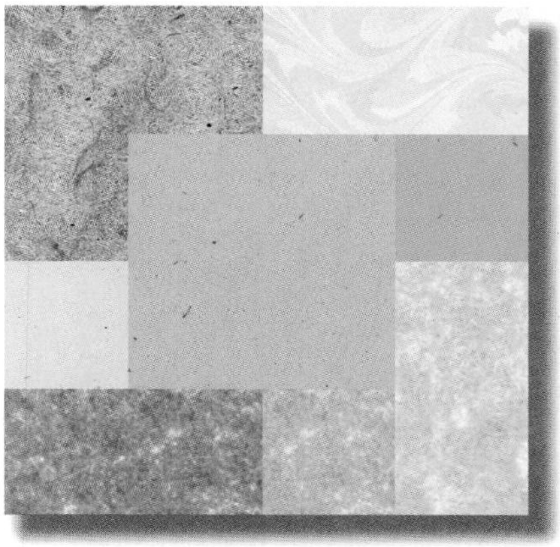

internet connect

**Portfolio Extension**

Go To: **go.hrw.com**
Keyword:
**MA1 Lever**

PORTFOLIO ACTIVITY

Include this activity in your portfolio.

Levers can be found in many real-world settings, from playground equipment to scissors in the office and pliers in the garage. When used correctly, levers are useful because the amount of effort required to apply a large force decreases as the lever arm length increases. To see how this works, perform the experiment described in Example 2 on page 528.

1. Use three different fixed masses. Place the balancing masses at five different points along the lever arm for each fixed mass.
2. For each fixed mass, record your data in a table similar to the one shown in the example.
3. Graph your data from each table on separate coordinate planes, and write a short report that describes your results. Be sure to include a function for each fixed mass that describes the relationship between the balanced mass and the distance from the fulcrum.

# HOW WORRIED SHOULD YOU BE?

NOW, THE TESTING QUESTION

*U.S. NEWS & WORLD REPORT, SEPT. 21, 1992*

Concerns about iron's role in heart disease, if ultimately confirmed, surely will touch off a debate over iron tests. Who should be tested? And should testing be a matter of public-health policy?...

The virtue of such broad screening would be to help ferret out many of the estimated 32 million Americans who carry a faulty gene that prompts their body to harbor too much iron. There is no test for the gene itself, but anyone with a high blood level of iron is a suspect. Siblings and children would be candidates for testing as well, since the condition is inherited . . .

Two widely used blood tests for iron generally cost from $25 to $75 each. The first measures ferritin, a key iron-storing protein . . .

The results should always be discussed with a doctor, since some physicians take recent studies seriously enough to worry about even moderately elevated ferritin levels. Moreover, the test might bear repeating. Erroneously high readings are relatively common.

Although people submit to medical tests because they want to find out if something is wrong, the results may cause them to worry. Read the following excerpt about an imaginary case and see how you would react.

Assume that there is a test for cancer which is 98 percent accurate; i.e., if someone has cancer, the test will be positive 98 percent of the time, and if one doesn't have it, the test will be negative 98 percent of the time. Assume further that .5 percent—one out of two hundred people—actually have cancer. Now imagine that you've taken the test and that your doctor somberly informs you that you've tested positive. The question is: How depressed should you be?

To find out whether your reaction would be justified, you need to look at the numbers more closely.

## Cooperative Learning

**1.** Use the data in the previous excerpt to complete steps a–g. Explain each answer.

**a.** If 10,000 people are tested, how many would you expect to actually have cancer?

**b.** How many of those people that actually have cancer would you expect to test positive? (Such cases are called *true positives.*)

**c.** If 10,000 people are tested, how many would you expect not to have cancer?

**d.** How many of those people that do not have cancer would you expect to test positive? (Such cases are called false positives.)

**e.** Add your results from parts **b** and **d** to find the total number of people expected to test positive.

**f.** What is the probability, *P*, that if you tested positive, you actually have cancer?

**g.** How does your answer from part **f** compare with your response to the question, How worried should you be?

$$\text{Probability} = \frac{\text{favorable outcomes}}{\text{total number of outcomes}}$$

$$\text{Thus, } P = \frac{\text{number of true positives}}{\text{total number expected to test positive}}$$

**2.** You used fictional data in Step 1. To explore what happens with different sets of data, you can find a general formula for *P*, the probability that if you tested positive, you actually have cancer. Use the following variables:

$N$ = the number of people tested
$a$ = the accuracy of the test (If the test is 98% accurate, then $a = 0.98$.)
$r$ = the portion of people tested that actually have what's being tested for (In Step 1, $r = 0.005$.)

**a.** Write a formula for the number of true positives. (Hint: Look back at parts **a** and **b** of Step 1.)

**b.** Write a formula for the number of false positives. (Hint: Look back at parts **c** and **d** of Step 1.)

**c.** Write a formula for *P*. (Hint: Look back at parts **e** and **f** of Step 1.)

**3.** Look at your formula from part c of Step 2.

**a.** Find $P$ when $r = 1$. Explain why that result makes sense.

**b.** For the situation in Activity 1, suppose that $r = 0.5$ (instead of $r = 0.005$). Would the results in part **f** of Step 1 have been as surprising? Explain.

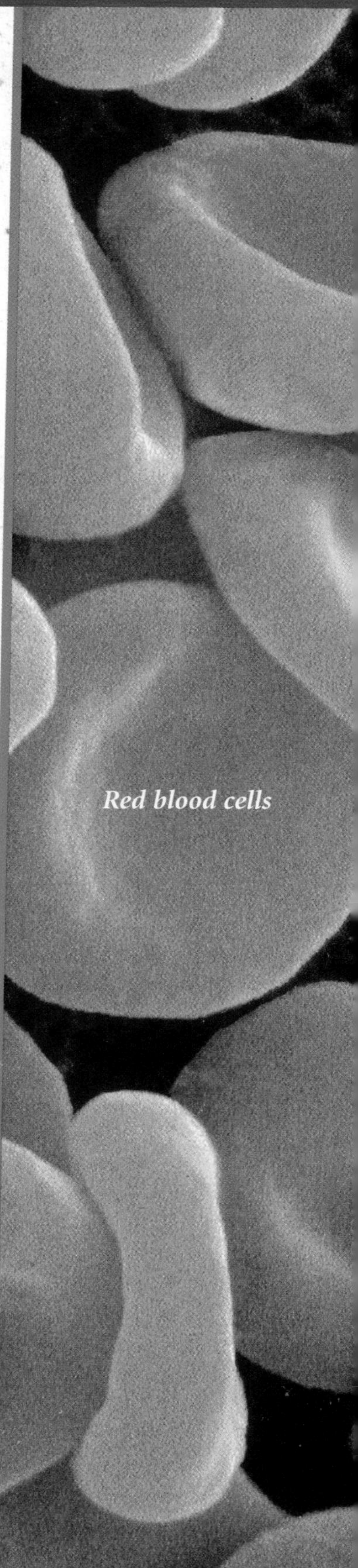

*Red blood cells*

11.4

# Operations With Rational Expressions

## Objectives

- Perform operations with rational expressions in a real-world example.
- Add, subtract, multiply, and divide rational expressions.

**Why** *Calculating travel times with and against a current is a problem that can be solved by using rational expressions. In solving real-world problems involving rational expressions, you will often need to perform arithmetic operations such as addition, subtraction, multiplication, and division.*

APPLICATION
**RECREATION**

At a rowing club's annual picnic, the "Current Challenge" is a popular event. The challenge requires participants to paddle a kayak up and back down a 1-mile portion of the river. The above kayaker's average rate in still water is 4 times the speed of the river's current. You can use this information to study travel times for different river trips. As you will see, a single rational expression can be used to represent a kayaker's round-trip time.

You can write an equation for the total time it takes to paddle upstream and downstream.

$$\text{total time} = \text{time } \underbrace{\text{upstream}}_{\text{against current}} + \text{ time } \underbrace{\text{downstream}}_{\text{with current}}$$

You can rewrite this equation for the total time in terms of the distance and rate.

Since distance = rate • time, time = $\frac{\text{distance}}{\text{rate}}$. Thus, you can rewrite the equation for total time as follows:

$$\text{total time} = \frac{\text{distance}}{\text{rate against current}} + \frac{\text{distance}}{\text{rate with current}}$$

Let $x$ represent the rate of the current. The rate of the kayaker in the illustration on the previous page is $4x$. The distance each way is 1 mile. Substituting this information into the above equation gives the following equation:

$$\text{total time} = \frac{1 \text{ mile}}{4x \text{ mph} - x \text{ mph}} + \frac{1 \text{ mile}}{4x \text{ mph} + x \text{ mph}}$$

By combining like terms in the denominator of each rational expression, you obtain the following equation:

$$t = \frac{1}{3x} + \frac{1}{5x}, \text{ where } x \neq 0$$

To find a single rational expression for $\frac{1}{3x} + \frac{1}{5x}$, add them in the same way that you would add fractions with unlike denominators. Compare the methods for adding fractions and rational expressions.

| Fractions | Rational expressions | Simplifying fractions and expressions |
|---|---|---|
| $\frac{1}{30} + \frac{1}{50}$ | $\frac{1}{3x} + \frac{1}{5x}$ | Given |
| $\frac{5}{5} \cdot \frac{1}{30} + \frac{3}{3} \cdot \frac{1}{50}$ | $\frac{5}{5} \cdot \frac{1}{3x} + \frac{3}{3} \cdot \frac{1}{5x}$ | Find a common denominator. |
| $\frac{5}{150} + \frac{3}{150}$ | $\frac{5}{15x} + \frac{3}{15x}$ | Change to common denominators. |
| $\frac{8}{150}$ | $\frac{8}{15x}$ | Simplify. |

Thus, the total time for the kayaker in the illustration is $\frac{8}{15x}$, where $x \neq 0$.

**CRITICAL THINKING**

If you knew the time it took the kayaker to complete the trip, how could you find the rate of the current?

Multiplying two rational expressions is relatively simple. Just multiply the numerators and multiply the denominators.

**EXAMPLE 1** **Multiply: $\frac{x-2}{x+3} \cdot \frac{x+3}{x-5}$. State any restrictions on the variable.**

**SOLUTION**

$\frac{x-2}{x+3} \cdot \frac{x+3}{x-5} = \frac{(x-2)(x+3)}{(x+3)(x-5)}$ *Multiply the numerators and the denominators.*

$= \frac{x-2}{x-5} \cdot \frac{x+3}{x+3}$ *Separate fractions that equal 1.*

$= \frac{x-2}{x-5}$, where $x \neq 5$ and $x \neq -3$ *Simplify.*

The restrictions on the variable must include the restrictions on the original expressions.

**EXAMPLE 2** **Multiply: $\frac{x+2}{6x^2} \cdot \frac{x-3}{x+2} \cdot \frac{3x}{5}$. State any restrictions on the variable.**

**SOLUTION**

$$\frac{x+2}{6x^2} \cdot \frac{x-3}{x+2} \cdot \frac{3x}{5} = \frac{(x+2)(x-3)(3x)}{(6x^2)(x+2)(5)}$$ *Multiply the numerators and the denominators.*

$$= \frac{x-3}{(2x)(5)} \cdot \frac{x+2}{x+2} \cdot \frac{3x}{3x}$$ *Separate fractions that equal 1.*

$$= \frac{x-3}{10x}, \text{ where } x \neq 0 \text{ and } x \neq -2$$ *Simplify.*

The restrictions on the variable must include the restrictions on the original expressions.

**TRY THIS** Multiply: $\frac{7}{x} \cdot \frac{x^2}{14} \cdot \frac{5}{x}$. State any restrictions on the variable.

Recall that division by a fraction is equivalent to multiplication by the reciprocal of the fraction. The same is true for rational expressions.

**EXAMPLE 3** **Divide: $\frac{x-2}{x} \div \frac{x-2}{x-3}$. State any restrictions on the variable.**

**SOLUTION**

Invert the divisor and multiply:

$$\frac{x-2}{x} \div \frac{x-2}{x-3} = \frac{x-2}{x} \cdot \frac{x-3}{x-2}$$
$$= \frac{x-3}{x}$$

The restrictions are $x \neq 0$, $x \neq 3$, and $x \neq 2$. The restrictions must include $x \neq 2$ because the divisor, $\frac{x-2}{x-3}$, is zero when $x = 2$.

**TRY THIS** Divide the following expressions, and state any restrictions on the variable.

**a.** $\frac{x+5}{x-2} \div \frac{x-3}{x-2}$ **b.** $\frac{x-2}{x+7} \div (x-2)$

To add or subtract rational expressions, you will need to find a common denominator. To do this, you will need to use multiplication.

**EXAMPLE 4** **Add: $\frac{x}{x+2} + \frac{x}{x-1}$. State any restrictions on the variable.**

**SOLUTION**

Multiply each expression by the appropriate equivalent of 1. The resulting expressions must have the same denominator.

$$\frac{x}{x+2} + \frac{x}{x-1} = \frac{x-1}{x-1} \cdot \frac{x}{x+2} + \frac{x+2}{x+2} \cdot \frac{x}{x-1}$$

Once the expressions have the same denominator, add the numerators.

$$\frac{(x-1)x}{(x-1)(x+2)} + \frac{(x+2)x}{(x-1)(x+2)} = \frac{(x^2-x)+(x^2+2x)}{(x-1)(x+2)}$$
$$= \frac{2x^2+x}{x^2+x-2}, \text{ where } x \neq 1 \text{ and } x \neq -2$$

**TRY THIS** Add: $\frac{2x}{x+3} + \frac{x-1}{x-4}$. State any restrictions on the variable.

**CRITICAL THINKING** How do you choose the appropriate equivalents for 1 in an addition problem like the one in Example 4?

**EXAMPLE 5** **Subtract: $\frac{3a}{a^2-4} - \frac{2}{a+2}$. State any restrictions on the variable.**

**SOLUTION**

Since $a^2 - 4 = (a+2)(a-2)$, multiply the second expression by $\frac{a-2}{a-2}$. Then subtract.

$$\frac{3a}{a^2-4} - \frac{2}{a+2} = \frac{3a}{(a+2)(a-2)} - \frac{2(a-2)}{(a+2)(a-2)}$$

$$= \frac{3a-2a+4}{(a+2)(a-2)}$$

$$= \frac{a+4}{a^2-4}, \text{ where } a \neq 2 \text{ and } a \neq -2$$

**TRY THIS** Add or subtract the expressions below.

**a.** $\frac{x}{x^2-1} + \frac{2}{x+1}$ **b.** $\frac{4x-1}{8x} - \frac{2}{x}$

When completing operations with rational expressions that involve several steps, it is important to work carefully in order to avoid errors. You can check your work by substituting a number in the place of the variable in the initial and final expressions. The values of the two expressions should equal one another.

# Exercises

## Communicate

**Refer to the expressions $\frac{x}{x+1}$ and $\frac{3}{x}$ for Exercises 1–5.**

1. Explain how to find a common denominator for these two rational expressions.
2. Describe how to find the sum of these rational expressions.
3. Describe how to find the difference of the first expression subtracted from the second expression.
4. Describe how to find the product of these rational expressions.
5. Describe how to find the quotient of the first expression divided by the second expression.

## Guided Skills Practice

**Multiply, and state the restrictions on the variable.**
***(EXAMPLES 1 AND 2)***

**6.** $\frac{x+3}{x-5} \cdot \frac{x-5}{x-4}$

**7.** $\frac{k-4}{k+7} \cdot \frac{k+7}{k+8}$

**8.** $\frac{3k-6}{2k+4} \cdot \frac{2k-8}{k-2}$

**Divide, and state the restrictions on the variable.** ***(EXAMPLE 3)***

**9.** $\frac{x+4}{3(x-2)} \div \frac{x+1}{4(x-2)}$

**10.** $\frac{k+1}{3(k+6)} \div \frac{2k+1}{2(k+6)}$

**11.** $\frac{y+3}{2y-18} \div \frac{y}{y-9}$

**Simplify, and state the restrictions on the variable.**
***(EXAMPLES 4 AND 5)***

**12.** $\frac{x}{x^2-1} + \frac{2}{x+1}$

**13.** $\frac{x}{x^2-9} + \frac{4}{x-3}$

**14.** $\frac{k}{k^2-4k+4} + \frac{5}{k-2}$

**15.** $\frac{z}{z^2+6z+9} + \frac{3}{z+3}$

**16.** $\frac{y}{y^2-8y+16} + \frac{3}{y-4}$

**17.** $\frac{x+4}{x-1} - \frac{x-3}{x-1}$

**18.** $\frac{x}{x+3} - \frac{x}{x-2}$

**19.** $\frac{x}{x+5} - \frac{x}{x-4}$

**20.** $\frac{a}{a-1} - \frac{a}{a+1}$

## Practice and Apply

internet connect
**Homework Help Online**
Go To: **go.hrw.com**
Keyword:
**MA1 Homework Help** for Exercises 21-50

**Perform the indicated operations. Simplify, and state the restrictions on the variable.**

**21.** $\frac{5}{3x} + \frac{2}{3x}$

**22.** $\frac{8}{x+1} - \frac{5}{x+1}$

**23.** $\frac{2}{y+4} \div \frac{5}{y+4}$

**24.** $\frac{5x+4}{x-2} - \frac{7+3x}{x-2}$

**25.** $\frac{2}{a} + \frac{3}{a+1}$

**26.** $\frac{7}{3t} - \frac{8}{2t}$

**27.** $\frac{5}{2r} + \frac{3}{r}$

**28.** $\frac{5}{p} \div \frac{3}{p^2}$

**29.** $\frac{-2}{x+1} + \frac{3}{2(x+1)}$

**30.** $x - \frac{x-4}{x+4}$

**31.** $\frac{-3-d}{d-1} + 2$

**32.** $\frac{1}{b} + \frac{3}{b^2}$

**33.** $\frac{5}{3+n} \cdot \frac{3+n}{2+n}$

**34.** $\frac{x+2}{x(x+1)} \cdot \frac{x^2}{(x+2)(x+3)}$

**35.** $\frac{q^2-1}{q^2} \cdot \frac{q}{q+1}$

**36.** $\frac{1}{x+1} \cdot \frac{2}{x}$

**37.** $\frac{3}{x-1} \cdot \frac{5}{x}$

**38.** $\frac{y}{y-4} - \frac{1}{y}$

**39.** $\frac{-2}{x+1} + \frac{3}{x}$

**40.** $\frac{a-2}{a+1} + \frac{5}{a+3}$

**41.** $\frac{m+5}{m+2} \cdot \frac{m-3}{m-1}$

**42.** $\frac{2r}{(r+5)(r+1)} - \frac{r}{r+5}$

**43.** $\frac{3}{b^2+b-6} - \frac{5}{b-2}$

**44.** $\frac{4}{y-2} + \frac{5y}{y^2-4y+4}$

**45.** $\frac{2}{y-5} - \frac{5y-3}{y^2+y-30}$

**46.** $\frac{x}{x-1} - \frac{1}{x-1}$

**47.** $\frac{2}{x^2-9} - \frac{1}{2x+6}$

**48.** $\frac{13}{3x+12} + \frac{2x}{x+4}$

**49.** $\frac{x+5}{3x-21} \div \frac{x}{x-7}$

**50.** $\frac{4p}{p^2-2p-3} \cdot \frac{p-3}{p+1}$

CONNECTION

**51. COORDINATE GEOMETRY** On the same coordinate plane, sketch the graphs of $y = x$ and $y = (x-4)(x+4)$. Make a table of values for each function. Then sketch the graph of $y = x + (x-4)(x+4)$ on the same coordinate plane by using the values in your tables and looking at the two graphs that you drew. You may want to use a graphics calculator to check your work.

**APPLICATION**

**ECOLOGY** Students are creating a plan for an exhibit of a bay ecosystem. They determine that the water in the bay exhibit should be 4% salt by mass concentration. The water in the tank for a sea exhibit is 6% salt by mass concentration. The students' plan calls for them to dilute 750 kilograms of water taken from the sea tank by adding enough pure water to get the 4% salt concentration needed for the bay tank. The concentration can be determined by using the following equation:

$$\text{concentration} = \frac{\text{salt concentration of sea water} \cdot \text{mass of sea water}}{\text{mass of sea water} + \text{mass of pure water}}$$

(Note: Because pure water is 0% salt, it is not necessary to account for its concentration in the numerator of the expression.)

**52.** Write a rational function to represent the resulting concentration when $w$ kilograms of pure water are added.

**53.** Graph the function from Exercise 52.

**54.** How much pure water needs to be added to 750 kilograms of sea water in order to produce a salt concentration of 4.0%?

## Look Back

**55.** Identify the next three terms of the sequence below. ***(LESSON 1.1)***

$$-1, 6, 15, 26, \_\_\_, \_\_\_, \_\_\_$$

**56.** Solve $0.3(x-90) = 0.7(2x-70)$. ***(LESSON 3.5)***

**57.** Graph the solution to $\begin{cases} y < -2x+6 \\ y \leq 3x-4 \end{cases}$. ***(LESSON 7.5)***

**58.** Express 25,000,000,000 in scientific notation. ***(LESSON 8.5)***

**internet connect**

**Activities Online**

Go To: **go.hrw.com**
Keyword:
**MA1 Coaster Math**

## Look Beyond

A proportion is the equality of two ratios, such as $\frac{5}{10} = \frac{2}{4}$. Each of these ratios equals $\frac{1}{2}$. This simplified fraction is called the constant of the proportion. Find the value of the variable in each proportion below, and state the constant of the proportion.

**59.** $\frac{26}{39} = \frac{x}{9}$

**60.** $\frac{14}{x} = \frac{49}{28}$

**61.** $\frac{15}{40} = \frac{27}{x}$

11.5

# Solving Rational Equations

### Objectives

- Solve rational equations by using the common-denominator method.
- Solve rational equations by graphing.

**Why** *Many quantities, such as time, work, and average cost, can be modeled by rational equations.*

A cyclist left home and rode his bike 7 miles before his chain broke; then he walked home. During his return trip, the cyclist estimated that he bikes 5 times faster than he walks. If the entire trip took him 2 hours and 24 minutes, what are his walking and riding speeds? You can approach the problem by guessing and checking.

## *Activity* Guessing and Checking to Solve a Problem

1. It took the cyclist 2 hours and 24 minutes to bike 7 miles and then walk home. Convert the total time to a decimal.
2. Copy and complete the table below for a walking speed of 2 mph.

| Walking speed (mph) | Walking time: $\frac{7 \text{ miles}}{\text{walking speed}}$ | Riding speed: 5 • walking speed | Riding time: $\frac{7 \text{ miles}}{\text{riding speed}}$ | Total time (hours) |
|---|---|---|---|---|
| 1.0 | $7 \div 1 = 7$ | $5 \cdot 1 = 5$ | $7 \div 5 = 1.4$ | 8.4 |
| 2.0 | ? | ? | ? | ? |

**CHECKPOINT ✓** 3. Use the table and the guess-and-check method to find the walking speed that results in a total time of 2 hours and 24 minutes. What is the riding speed?

You can also determine the cyclist's walking and riding speeds by solving an equation. The time for each part of the trip can be represented by the expression $\frac{\text{distance}}{\text{rate}}$. The total time can be represented by the equation $\frac{\text{riding distance}}{\text{riding speed}} + \frac{\text{walking distance}}{\text{walking speed}} = \text{total time}$. If $x$ is the cyclist's walking speed, then $5x$ is his riding speed. You can rewrite the equation as follows:

$$\frac{7 \text{ miles}}{5x \text{ mph}} + \frac{7 \text{ miles}}{x \text{ mph}} = 2 \text{ hours and } 24 \text{ minutes, or } 2.4 \text{ hours}$$

Therefore, the equation that models the problem is $\frac{7}{5x} + \frac{7}{x} = 2.4$, where $x \neq 0$. The equation can be solved by two methods.

### Common-Denominator Method

Multiply all of the terms in the original equation by the lowest common denominator to clear the equation of fractions. Using the lowest common denominator saves extra steps when solving the equation.

$$\frac{7}{5x} + \frac{7}{x} = 2.4$$

$$5x\left(\frac{7}{5x} + \frac{7}{x}\right) = 5x(2.4)$$ *Multiply both sides by 5x, the least common denominator.*

$$7 + 35 = 12x$$

$$12x = 42$$

$$x = 3.5$$

Thus, the cyclist walks at a rate of 3.5 mph. His biking speed is 5 times his walking speed, or 17.5 mph.

### Graphing Method

Graph $y = \frac{7}{5x} + \frac{7}{x}$ and $y = 2.4$. You will find that the intersection is (3.5, 2.4).

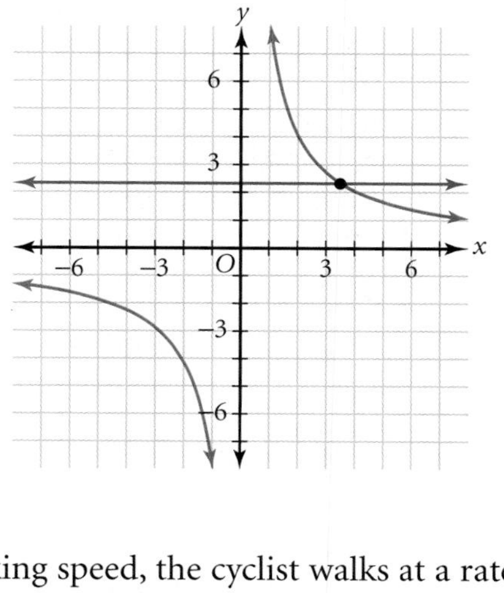


Since $x$ represents walking speed, the cyclist walks at a rate of 3.5 mph. As in the common-denominator method, his biking speed is 5 times his walking speed, or 17.5 miles per hour.

**CRITICAL THINKING** What is the real-world domain for the equation that describes the cyclist's total time? For what value(s) of the variable are the rational expressions in the equation undefined?

**EXAMPLE 1** **Solve $\frac{6}{x-1} + 2 = \frac{12}{x^2-1}$ by using the common-denominator method and the graphing method.**

**SOLUTION**

**Common-Denominator Method**

*Factor the denominator of the second expression. Mutiply both sides by the least common denominator.*

$$\frac{6}{x-1} + 2 = \frac{12}{(x+1)(x-1)}$$

$$(x+1)(x-1)\left(\frac{6}{x-1} + 2\right) = (x+1)(x-1)\left[\frac{12}{(x+1)(x-1)}\right]$$

$$6(x+1) + 2(x+1)(x-1) = 12$$

$$6x + 6 + 2x^2 - 2 = 12$$

$$2x^2 + 6x + 4 = 12$$

$$2x^2 + 6x - 8 = 0$$

$$2(x+4)(x-1) = 0$$

$$x + 4 = 0 \quad \text{or} \quad x - 1 = 0$$

$$x = -4 \qquad\qquad x = 1$$

*Zero Product Property*

It is necessary to always check the original equation for restrictions on the variable. In this case, $x \neq -1$ and $x \neq 1$. Since 1 is excluded, $x = -4$ is the only solution.

TECHNOLOGY
GRAPHICS CALCULATOR

**Graphing Method**

Graph $y = \frac{6}{x-1} + 2$ and $y = \frac{12}{x^2-1}$. Be sure that you enter the equations correctly; for example, the first equation should be entered as $6 \div (x-1) + 2$. The two equations have a common asymptote at $x = 1$ and intersect at $x = -4$.

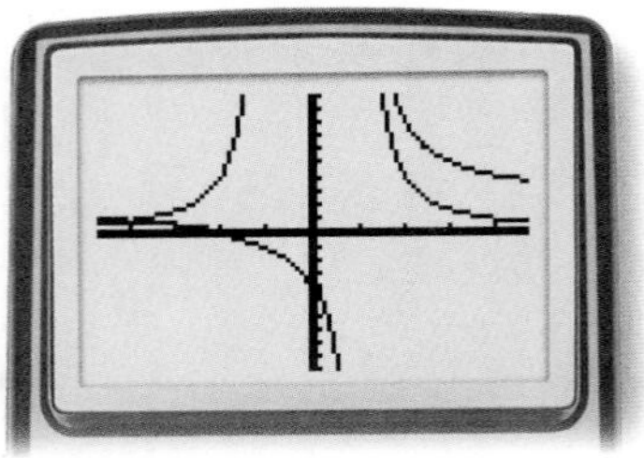

The solution is $x = -4$.

**TRY THIS** Solve $\frac{7}{x-2} + 2 = \frac{18}{(x-2)(x+3)}$ by using the common-denominator method and the graphing method. (Hint: Be careful to use parentheses properly when entering the expressions into your graphics calculator.)

**EXAMPLE 2** Max and his younger brother Carl are earning extra money by delivering papers. Max can deliver all of the papers in 80 minutes when he works alone. Carl can deliver them in 2 hours when he works alone. **If Max and Carl work together, how many minutes will it take them to finish this job?**

**SOLUTION**

To solve the problem, write an expression for each brother's rate of work. Then, by multiplying by the time in minutes, *t*, that each brother works, you obtain the *fractional part* of the job that each brother completes. (Notice that when they work together, each brother works the same length of time although they work at different rates.) The finished job is equal to the sum of all the fractional parts, which represents 1 complete job.

PROBLEM SOLVING

**Make a table.**

| Fractional part of the job done by Max | + | Fractional part of the job done by Carl | = | The complete job, or 1 whole unit |
|---|---|---|---|---|
| $\left(\text{rate of Max's work}\right) \cdot \left(\text{time worked}\right)$ | + | $\left(\text{rate of Carl's work}\right) \cdot \left(\text{time worked}\right)$ | = | 1 |
| $\frac{1}{80} \cdot t$ | + | $\frac{1}{120} \cdot t$ | = | 1 |
| $\frac{t}{80}$ | + | $\frac{t}{120}$ | = | 1 |

Solve $\frac{t}{80} + \frac{t}{120} = 1$.

Multiply each term by the lowest common denominator.

$$240 \cdot \frac{t}{80} + 240 \cdot \frac{t}{120} = 240 \cdot 1$$
$$3t + 2t = 240$$
$$5t = 240$$
$$t = 48$$

Working together, it will take Max and Carl 48 minutes to deliver all of the papers.

**TRY THIS** Suppose that Max can deliver all of the papers in 2 hours and Carl can deliver all of the papers in 3 hours. If Max and Carl work together, how many minutes will it take them to deliver all of the papers?

# Exercises

## Communicate

internet connect
**Activities Online**
Go To: **go.hrw.com**
Keyword:
**MA1 Rational Functions**

**1.** Explain the significance of restricted values of the variable in a rational equation.

**2.** How is the real-world domain of a rational equation determined in applications?

**In Exercises 3–5, refer to $\frac{3}{c} - \frac{2}{c-1} = \frac{6}{c}$.**

**3.** Describe how to identify the lowest common denominator.

**4.** Explain how to solve the rational equation by graphing.

**5.** How would you determine the values of the variable for which the equation is undefined?

## Guided Skills Practice

**Solve each equation by using the common-denominator method and the graphing method.** *(EXAMPLE 1)*

**6.** $\frac{10}{x+2} - 1 = \frac{9}{x^2-4}$ **7.** $\frac{7}{x+1} - 1 = \frac{-20}{x^2-1}$ **8.** $\frac{3}{x-2} - 3 = \frac{24}{x^2-4}$

**Solve each problem below by using rational equations.** *(EXAMPLE 2)*

**9.** Dan paints a house in 6 days. It takes Joan 9 days to paint the same house. How long will it take them to paint the house if they work together?

**10.** Sue and Ed can clean their house in 7 hours. It takes Sue 15 hours to do the job alone. How long would it take Ed to do the job alone?

## Practice and Apply

**internet connect**
**Homework Help Online**
Go To: **go.hrw.com**
Keyword:
**MA1 Homework Help**
for Exercises 11-19

**Solve each rational equation by using the lowest common denominator.**

**11.** $\frac{x-1}{x} + \frac{7}{3x} = \frac{9}{4x}$ **12.** $\frac{10}{2x} + \frac{4}{x-5} = 4$ **13.** $\frac{x-3}{x-4} = \frac{x-5}{x+4}$

**14.** $\frac{x+3}{x} + \frac{6}{5x} = \frac{7}{2x}$ **15.** $\frac{x}{x-3} = 5 + \frac{x}{x-3}$ **16.** $\frac{x+3}{x^2-9} - \frac{6}{x-3} - 5$

**17.** $\frac{4}{y-2} + \frac{5}{y+1} = \frac{1}{y+1}$ **18.** $\frac{10}{x+3} - \frac{3}{5} = \frac{10x+1}{3x+9}$ **19.** $\frac{3}{x-2} - \frac{6}{x^2-2x} = 1$

**Solve each rational equation by graphing.**

**20.** $\frac{2}{x} + \frac{1}{3} = \frac{4}{x}$ **21.** $\frac{1}{2} + \frac{1}{x} = \frac{1}{2x}$ **22.** $\frac{20-x}{x} = x$

**23.** $\frac{12x-7}{x} = \frac{17}{x}$ **24.** $\frac{3}{7x+x^2} = \frac{9}{x^2+7x}$ **25.** $\frac{x-3}{x-4} = \frac{x-5}{x-4}$

**26.** $\frac{3}{x-2} - \frac{6}{x^2-2x} = 1$ **27.** $\frac{1}{x-2} + \frac{16}{x^2+x-6} = -3$ **28.** $\frac{1}{x-2} + 3 = \frac{-16}{x^2+x-6}$

**29.** The sum of two numbers is 56. If the larger number is divided by the smaller, the quotient is 1 with a remainder of 16. Find the two numbers.

CONNECTION

**30.** **COORDINATE GEOMETRY** Find any points of intersection for the graphs of the function $y = \frac{1}{x} + 3$ and the function $y = \frac{x-4}{x+2}$.

APPLICATION

**RECREATION** Marisa pedals her paddleboat downstream at a still-water speed of 20 meters per minute for 500 meters before turning around and pedaling upstream to where she started. If Marisa maintains the same still-water speed, she will pedal only 300 meters upstream in the same time it took her to pedal 500 meters downstream.

**31.** Write rational expressions that represent the time it takes Marisa to pedal downstream and upstream in the current.

**32.** What is the speed of the current?

APPLICATION

**SMALL BUSINESS** At a car body shop, Kayla needs 5 hours working alone to paint a car. It takes Emily 7 hours to paint the same car.

**33.** Write rational expressions that represent the fractional parts of the work that Kayla and Emily complete when painting the car together.

**34.** Write and solve a rational equation to find the time it would take them to paint the car if they work together.

## Look Back

**35.** Solve: $x - 4 = -\frac{3}{4}(x + 2)$. *(LESSON 3.5)*

**36.** Solve: $|2x - 7| \leq 7$. *(LESSON 6.5)*

**37.** Solve: $\begin{cases} x + 5y = 9 \\ 3x - 2y = 10 \end{cases}$. *(LESSON 7.3)*

**38.** Multiply: $(3x - 8)(5x + 2)$. *(LESSON 9.3)*

**39.** Identify the vertex of $y = -2(x + 1)^2 + 7$. *(LESSON 10.1)*

**internet connect**
**Portfolio Extension**
Go To: **go.hrw.com**
Keyword:
**MA1 Gears**

## Look Beyond

**40.** As a way to vary his exercise, Antonio climbs 6 steps in 3 seconds and then carefully descends 4 steps in 2 seconds. Continuing this sequence of climbing and descending, how long will it take him to reach the 100th step?

## PORTFOLIO ACTIVITY

Include this activity in your portfolio.

The relationship between the number of teeth on the pedal gear and the number of teeth on the wheel gear of a bicycle affects the speed and effort of pedaling. For example, if the wheel gear has 16 teeth and the pedal gear has 54 teeth, then the gear ratio is $\frac{16}{54} \approx 0.296$. The reciprocal is $\frac{54}{16} = 3.375$. The reciprocal 3.375 represents the number of turns of the wheel for every full turn of the pedal.

**1.** Assume for this activity that the circumference of a bicycle tire is 2.19 meters. How far will a rider travel for each full turn of the pedal if the gear ratio is 0.25?

**2.** The wheel gear could have 13, 14, 15, 16, 18, 21, or 24 teeth and the pedal gear could have 49 or 54 teeth. Write a formula for the possible gear ratios, *r*, based on the number of teeth on the wheel gear, *w*, and the number of teeth on the pedal gear, *p*. Find the ratios and reciprocals for each possible combination.

**3.** Suppose that a bicycle rider makes 60 full turns of the pedal per minute. Choose two possible gear combinations and calculate the rate in miles per hour for each. (Note that 1 kilometer is approximately equal to 0.62 miles.)

**WORKING ON THE CHAPTER PROJECT**

You should now be able to complete the Chapter Project on page 566.

11.6

# Proof in Algebra

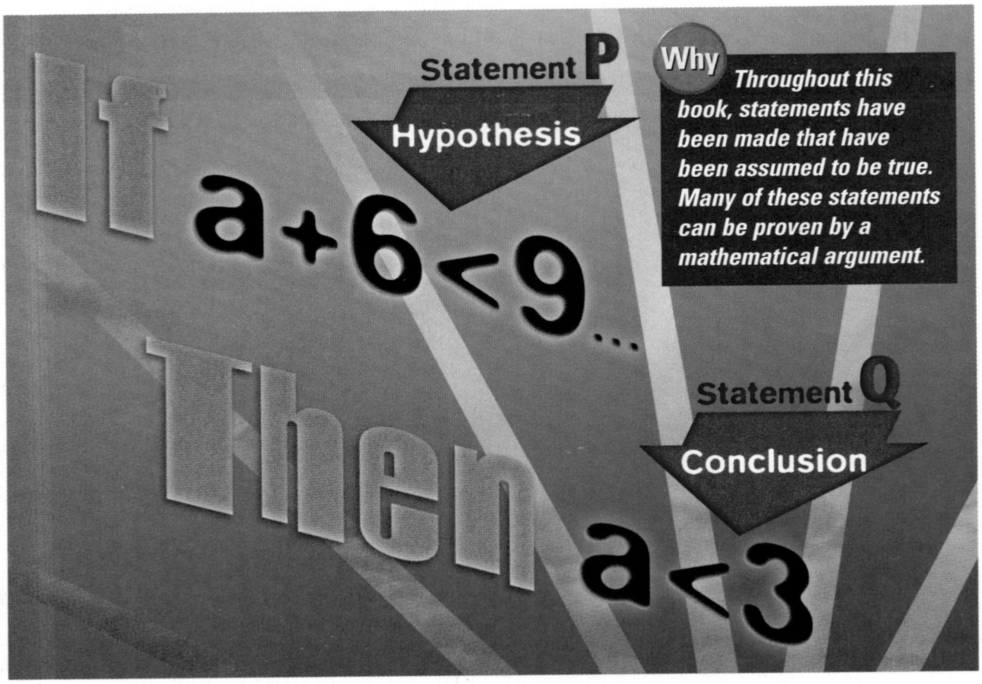


**Objectives**

- Define the parts of a conditional statement.
- Define *converse.*
- Define *proof.*
- Prove theorems stated in conditional form.

## Theorems as "If-*p*-then-*q*" Statements

Mathematical statements can often be written as "if-*p*-then-*q*" statements, also called **conditional statements**. Once such statements are proven, they are called **theorems**.

In a conditional statement, the *p* portion (following the word *if*) is called the **hypothesis**. The *q* portion (following the word *then*) is called the **conclusion**. A **proof** is a series of statements in a logical sequence which demonstrate that the conclusion is true *whenever the hypothesis is true.*

In a proof, the hypothesis is assumed to be true. Each step in a proof is based on established *definitions, axioms* or *postulates* (basic statements that are accepted as true without proof), or previously proven theorems. Study the following logical argument, which is actually a proof of a very simple conditional statement:

Prove: If $a + 6 < 9$, then $a < 3$.

| Statement | Reason |
|---|---|
| $a + 6 < 9$ | Hypothesis or given |
| $a + 6 - 6 < 9 - 6$ | Subtraction Property of Inequality |
| $a < 3$ | Simplified form |

Therefore, if $a + 6 < 9$, then $a < 3$. Notice the importance of the hypothesis. The conclusion, $a < 3$, taken by itself cannot be proven mathematically.

If you interchange the hypothesis and the conclusion, the new statement is called the **converse** of the original statement. "If *q*, then *p*" is the converse of "If *p*, then *q*." What is the converse of the conditional statement "If $a + 6 < 9$, then $a < 3$?" Is the converse true?

Mathematical theorems often start out as **conjectures**, or statements that seem to be true but have not yet been proven.

There are two types of reasoning that are used in mathematics: inductive and deductive. Both are necessary parts of mathematical thinking.

**Inductive reasoning** is the process of reasoning that a general principle is true because the special cases that you have seen are true. For example, you might conclude that all of the food from a particular restaurant is bad because you did not like the meal that you ate there. Inductive reasoning is not logically valid and does not constitute mathematical proof except in the special method called "mathematical induction." However, inductive reasoning is often used to formulate conjectures.

**Deductive reasoning** is the process of concluding that a special case of a proven general principle is true. For example, you might conclude that the sum of the interior angles of a particular triangle is 180 degrees because you know that the sum of the angles in any triangle is always 180 degrees. Deductive reasoning is logically valid and is the fundamental method by which mathematical facts are shown to be true.

## *Activity*
## Inductive Reasoning

1. Consider the equation $\frac{a}{b} = \frac{c}{d}$. Find nonzero values for *a*, *b*, *c*, and *d* that make the equation true. Substitute the same values for *a*, *b*, *c*, and *d* in each equation below.

**a.** $\frac{b}{a} = \frac{d}{c}$ **b.** $\frac{a+b}{c} = \frac{c+b}{d}$ **c.** $\frac{a+b}{b} = \frac{c+d}{d}$

**d.** $\frac{a}{c} = \frac{b}{d}$ **e.** $\frac{a-b}{b} = \frac{c-b}{d}$ **f.** $\frac{a}{b} = \frac{a+c}{b+d}$

2. Which of the equations in Step 1 are true for your values of *a*, *b*, *c*, and *d*?
3. Write if-then statements relating the original equation, $\frac{a}{b} = \frac{c}{d}$, with each equation in Step 1.

CHECKPOINT ✓ 4. Are your conditional statements conjectures or theorems? Explain.

In the activity, you may have made the following conjecture:

$$\text{If } \frac{a}{b} = \frac{c}{d}, \text{ then } \frac{a+b}{b} = \frac{c+d}{d}.$$

This conjecture can be proven by adding 1 to each side of the equation in the hypothesis.

| Statement | Reason |
|---|---|
| $\frac{a}{b} = \frac{c}{d}$ | Hypothesis or given |
| $\frac{a}{b} + 1 = \frac{c}{d} + 1$ | Addition Property of Equality |
| $\frac{a}{b} + \frac{b}{b} = \frac{c}{d} + \frac{d}{d}$ | Substitution: $\frac{b}{b} = 1$ and $\frac{d}{d} = 1$ |
| $\frac{a+b}{b} = \frac{c+d}{d}$ | Definition of addition |

**EXAMPLE 1** **Prove the following conjecture:**

**If $\frac{a}{b} = \frac{c}{d}$, then $ad = bc$.**

SOLUTION

| Statement | Reason |
| --- | --- |
| $\frac{a}{b} = \frac{c}{d}$ | Hypothesis or given |
| $bd\left(\frac{a}{b}\right) = bd\left(\frac{c}{d}\right)$ | Multiplication Property of Equality |
| $da = bc$ | Simplified form |
| $ad = bc$ | Commutative Property of Multiplication |

Therefore, if $\frac{a}{b} = \frac{c}{d}$, then $ad = bc$.

The definitions of even and odd numbers can also be used in proofs.

**Definitions of Even and Odd Numbers**

An even number is a number of the form $2n$, where $n$ is an integer.

An odd number is a number of the form $2n + 1$, where $n$ is an integer.

**EXAMPLE 2** **Prove the following conjecture:**

**If two numbers are both odd, then their sum is even.**

SOLUTION

| Statement | Reason |
| --- | --- |
| $a$ and $b$ are two odd numbers. | Hypothesis or given |
| $a = 2k + 1$ and $b = 2p + 1$, where $k$ and $p$ are integers | Definition of odd numbers |
| $a + b = (2k + 1) + (2p + 1)$ | Addition of real numbers |
| $= 2k + 2p + 2$ | Simplified form |
| $= 2(k + p + 1)$ | Distributive Property |

Since $k + p + 1$ is an integer, $2(k + p + 1)$ is in the form $2n$, where $n$ is an integer. Thus, $2(k + p + 1)$ is an even number.

Therefore, if two numbers are both odd, then their sum is even.

**EXAMPLE 3** **Prove the following conjecture:**

**If $k$ is an integer, then $2k + 2$ is an even number.**

**SOLUTION**

| Statement | Reason |
|---|---|
| $k$ is an integer. | Hypothesis or given |
| $k + 1$ is an integer. | Definition of integer |
| $2(k + 1)$ is even. | Definition of even numbers *Think of k + 1 as n.* |
| $2k + 2$ is even. | Distributive Property |

Therefore, if $k$ is an integer, then $2k + 2$ is an even number.

In addition to the sum shown in Example 2 on the previous page, it is easy to prove that the following are true:

odd + even = even + odd = odd

even + even = even

odd • even = even • odd = even

odd • odd = odd

even • even = even

Using these facts, you can see that a quadratic trinomial with all odd coefficients, such as $3x^2 + 25x + 15$, cannot be factored into linear terms with integer coefficients.

Suppose that a trinomial like $3x^2 + 25x + 15$ has the factors $(ax + b)$ and $(cx + d)$, where $a$, $b$, $c$, and $d$ are integers. Then the products $ac$ and $bd$, in this case 3 and 15, must both be odd. Therefore, $a$, $b$, $c$, and $d$ must all be odd.

The middle term, in this case 25, is the sum of $ad$ and $bc$, both of which must be odd since all four factors ($a$, $b$, $c$, and $d$) are odd. But the sum of two odd numbers will always be even and will never result in an odd middle coefficient. Thus, it is impossible to find integer values of $a$, $b$, $c$, and $d$ that will work.

These facts about sums and products can also be used to solve simpler problems.

**EXAMPLE 4** **Find four even integers whose sum is 101, or explain why it is impossible.**

**SOLUTION**

The sum of any two even numbers will always be an even number, so adding four even integers will also result in an even integer. Thus, it is not possible to add four even integers and get an odd number such as 101 as the sum.

**TRY THIS** Find three odd integers whose product is 100, or explain why it is impossible.

Most mathematical statements can be proven true or shown to be false with a **counterexample**. There are some cases in which the statement is sometimes true and sometimes false, often depending on the value of a variable within the statement. It is important for you to be able to recognize these kinds of statements because statements that are only sometimes true cannot be used as reasons in proofs. However, they can often be rewritten with further specifications that will make them either always true or never true.

**EXAMPLE 5** **Tell whether each statement is always true, sometimes true, or never true.**

**a.** $a^2b^2 = (ab)^2$ **b.** $-b^2 = (-b)^2$ **c.** $|a + 2| = |a| + 2$

**SOLUTION**

**a.** This statement is always true based on the laws of exponents that you studied in Chapter 8.

**b.** This statement is sometimes true, depending on the value of $b$. If $b$ is 0, then the equation is true. The equation is false for all other values of $b$.

**c.** This statement is sometimes true, depending on the value of $a$. If $a$ is a positive number or 0, then the equation is true. But if $a$ is negative, the equation is false. For example, $|3 + 2| = |3| + 2$, but $|-3 + 2| \neq |-3| + 2$.

**TRY THIS** Tell whether each statement is always true, sometimes true, or never true.

**a.** $a - b = -a + (-b)$ **b.** $ab = |a| \cdot |b|$ **c.** If $|x| < a$, then $-a < x < a$.

You have already studied many properties of real numbers, including the Associative, Commutative, and Distributive Properties. Another property that is valuable for studying sets of numbers is the **Closure Property**.

A set of numbers is *closed* under addition if the sum of any two numbers in the set is also in the set. Since the sum of any two integers is also an integer, the set of integers is closed under addition. The Closure Property applies in the same way to any set of numbers and any mathematical operation.

### Closure Property

A set of numbers is said to be closed, or to have closure, under a given operation if the result of the operation on any two numbers in the set is also in the set.

Only one case in which the result is not in the set, a counterexample, is required to show that a set is not closed. For example, the set $\{0, 1, 2\}$ is not closed under addition because $1 + 2 = 3$ and 3 is not in the set.

**CRITICAL THINKING** Is it possible to prove that a set is closed by using several examples? Why or why not?

**EXAMPLE 6** **Determine whether each set is closed under the given operation.**

**a.** whole numbers; multiplication
**b.** $\{-1, 0, 1\}$; multiplication
**c.** whole numbers; subtraction
**d.** integers; subtraction

**SOLUTION**

**a.** Since the product of two whole numbers is always a whole number, the set of whole numbers is closed under multiplication.

**b.** Since the product of any two numbers in the set is also in the set, $\{-1, 0, 1\}$ is closed under multiplication.

**c.** Since $4 - 8 = -4$, the set of whole numbers is not closed under subtraction.

**d.** Since the difference of any two integers is always an integer, the set of integers is closed under subtraction.

**TRY THIS** Determine whether the set of integers is closed under division.

The Closure Property is used in the study of *fields.* The set of real numbers is probably the best-known field. The set of real numbers has the properties shown below for addition and multiplication. Any set that has the properties shown in the table below for any two operations is a **field.**

| Property | Addition | Multiplication |
|---|---|---|
| Associative | $(a+b)+c=a+(b+c)$ | $(a \cdot b) \cdot c = a \cdot (b \cdot c)$ |
| Closure | $a+b$ is in the set. | $a \cdot b$ is in the set. |
| Commutative | $a+b=b+a$ | $a \cdot b = b \cdot a$ |
| Distributive | $a \cdot (b+c) = (a \cdot b) + (a \cdot c)$ | |
| Identity | A unique identity element, 0, exists such that $a+0=0+a=a$. | A unique identity element, 1, exists such that $a \cdot 1 = 1 \cdot a = a$. |
| Inverse | For any $a$, there is an inverse, $-a$, such that $a+(-a)=-a+a=0$. | For any $a \neq 0$, there is an inverse, $\frac{1}{a}$, such that $a \cdot \frac{1}{a} = \frac{1}{a} \cdot a = 1$. |

**CRITICAL THINKING** Is the set $\{-1, 0, 1\}$ a field? Explain.

# Exercises

## Communicate

**1.** Name three types of reasons that can be used in a proof.

**2.** Identify the part of an if-then statement that is assumed to be true. What part has to be proven?

**3.** Explain what is meant by the converse of a conditional statement.

## Guided Skills Practice

**4.** Prove: If $\frac{2}{3} = \frac{x}{9}$, then $x = 6$. ***(EXAMPLE 1)***

**5.** Prove the following conjecture: If an even number is added to an odd number, then the sum is an odd number. ***(EXAMPLE 2)***

**6.** Prove the following conjecture: If $k$ is an integer, then $4k + 4$ is an even integer. ***(EXAMPLE 3)***

**7.** Find four odd integers whose product is 100, or explain why it is impossible. ***(EXAMPLE 4)***

**8.** Is the statement $ab = |a| \cdot |b|$ always true, sometimes true, or never true? ***(EXAMPLE 5)***

**9.** Determine whether the set of integers is closed under multiplication. ***(EXAMPLE 6)***

## Practice and Apply

**State a reason for each step in the proof of the following conjecture: If $3m - 4 = 0$, then $m = \frac{4}{3}$.**

| | Statement | Reason |
|---|---|---|
| | $3m - 4 = 0$ | Hypothesis or given |
| **10.** | $3m - 4 + 4 = 0 + 4$ | __________ |
| **11.** | $3m + 0 = 0 + 4$ | __________ |
| **12.** | $3m = 4$ | __________ |
| **13.** | $\frac{3m}{3} = \frac{4}{3}$ | __________ |
| **14.** | $1 \cdot m = \frac{4}{3}$ | __________ |
| **15.** | $m = \frac{4}{3}$ | __________ |

**State a reason for each step in the proof of the following conjecture: If $x = (a + b) + (-a)$, then $x = b$.**

| | Statement | Reason |
|---|---|---|
| **16.** | $x = (a + b) + (-a)$ | __________ |
| **17.** | $= (b + a) + (-a)$ | __________ |
| **18.** | $= b + [a + (-a)]$ | __________ |
| **19.** | $= b + 0$ | __________ |
| **20.** | $= b$ | __________ |

**For Exercises 21–26, write a proof for each conjecture. Give a reason for each step. Let all variables represent real numbers.**

**21.** If $y = 5x - 3x$, then $y = 2x$.

**22.** If $n = (ax + b) + ay$, then $n = a(x + y) + b$.

**23.** If $a = b$, then $a + c = b + c$.

**24.** If $a$ is an even number, then its square, $a^2$, is an even number.

25. The expressions $a - b$ and $b - a$ are opposites. (Hint: Prove that $(a - b) + (b - a) = 0$ is true. Then refer to the definition of opposites.)

26. If $x = \frac{a}{b} + \frac{c}{d}$, then $x = \frac{da + bc}{bd}$.

27. Find three odd integers whose sum is 50, or explain why it is impossible.

28. Find five even integers whose sum is 105, or explain why it is impossible.

**internet connect**
**Homework Help Online**
Go To: **go.hrw.com**
Keyword:
**MA1 Homework Help** for Exercise 33

**For Exercises 29–32, tell whether each statement is always true, sometimes true, or never true.**

29. $(-x)^n = x^n$ 30. $(a)(-a) = -1$ 31. $|x^2| = x^2$ 32. $a \cdot \frac{1}{a} = -1$

33. Determine whether the set $\{-2, -1, 0, 1, 2\}$ is closed under subtraction.

APPLICATION

34. **CONTESTS** With each purchase, a fast-food restaurant gives out a card with a number on it. Anyone with cards whose numbers add up to 100 wins a major prize. A group of students collects cards with the following numbers: 3, 3, 3, 6, 12, 15, 24, 42, 54, 60, 72, 84. Use the methods learned in this lesson to prove that it is impossible to form a sum of 100 from the numbers on these cards.

## Look Back

35. Solve $x = 16x + 45$. *(LESSON 3.3)*

36. **STATISTICS** What is the average length in centimeters of the hand spans (from the tip of the thumb to the tip of the little finger) in this sample: 16.2, 18.5, 23.0, 21.2, 22.9, 21.1, 20.6, 19.2? *(LESSON 4.4)*

37. Write the equation of the line that is perpendicular to $2x - 3y = 9$ and contains the point $(-6, 12)$. *(LESSON 5.6)*

38. Solve: $x^2 - 8 = 2x$. *(LESSON 10.4)*

## Look Beyond

39. What is wrong with the following proof, which gives the conclusion that $1 = 0$?

$$\text{Let } a = b$$
$$a - b = 0$$
$$\frac{a-b}{a-b} = \frac{0}{a-b}$$
$$1 = 0$$

CHAPTER PROJECT ELEVEN

# designing a PARK

Suppose that your city has set aside a square piece of land with an area of 1 square mile on which to build a park. You are on the committee to design the park. Your committee has decided that the basic design will include a road, three parking lots, and a jogging trail. The road will run diagonally across the park. The parking lots will be at each end of this road and in the center of the park. The jogging trail will be constructed around the perimeter of the park.

**Activity 1**

## The Basics

1. **a.** Find the length of the road. If the road is 50 feet wide, what is its area? Assume that the road forms a rectangle.
   **b.** Suppose that materials for the road cost \$100 per square foot of area plus an additional \$20 per foot of length. If the city requires an increase of 5 feet in the planned width of the road, calculate the new area and the total cost of the widened road.
   **c.** Using the cost information from part **b.**, write a rational function to determine the total cost per square foot based on the length and width of the road.
2. **a.** Find the length of the jogging trail. Suppose that the trail is 10 feet wide. Find the area of the jogging trail.
   **b.** Suppose that the jogging trail costs \$20 per square foot of area plus \$2 per foot for the outer perimeter. After an initial meeting, the committee wants to increase the width of the trail by 2 feet in order to accommodate more people. Calculate the new area and the total cost for the widened trail.
   **c.** Using the cost information from part **b.**, write a rational function to determine the total cost per square foot based on the length and width of the trail.

## Activity 2

### Adding On

The first facility that the committee wants to build on the land is a recreation building. Based on successful parks in other areas, the committee determines that the recreation building needs to have an area of 14,400 square feet. In order to simplify the design and keep costs as low as possible, the building will be rectangular in shape and have the smallest possible perimeter.

1. Make a table of possible dimensions for the building. Use widths of 20, 40, 60, . . . , 200. Which dimensions have the smallest perimeter?
2. Write and graph a rational function for the perimeter, $p$, in terms of the width, $x$.
3. Which dimensions are possible if you want to keep the perimeter under 525 feet?

## Activity 3

### The Final Design

Suppose that after presenting the basic design to the city and getting feedback, the committee decides to add a pool, a baseball stadium, and six playground areas to the park. The pool will be Olympic size; the baseball stadium will seat 1000 people; and each playground area will be 80 yards by 120 yards.

1. Each of the three parking lots has dimensions of $\frac{1}{8}$ mile by $\frac{1}{8}$ mile. Find the area of the land that is not used by the parking lots, the jogging trail, the road, or the recreation facility.
2. Determine the total area of the six playgrounds. Do some research to estimate the areas of the pool and baseball, and determine how much land will remain undeveloped after all of the features have been added.
3. Create a complete scale drawing or model of the park that includes all of the features discussed by the committee.

# 11 Chapter Review and Assessment

## VOCABULARY

| | | | | | |
|---|---|---|---|---|---|
| Closure Property | 562 | field | 563 | rational expression | 532 |
| conclusion | 558 | hole | 534 | rational function | 533 |
| conditional statement | 558 | hypothesis | 558 | rule of 72 | 529 |
| conjecture | 559 | inductive reasoning | 559 | theorem | 558 |
| converse | 558 | inverse variation | 526 | trivial rational function | 534 |
| counterexample | 561 | non-trivial rational function | 534 | vertical asymptote | 534 |
| deductive reasoning | 559 | proof | 558 | | |

## Key Skills & Exercises

### LESSON 11.1

#### Key Skills

**Identify and use inverse variation.**

If $y$ varies inversely as $x$ and $y$ is 10 when $x$ is 3, the constant of variation is found by solving the equation $10 = \frac{k}{3}$. Thus, $k = 30$.

To find $x$ when $y = 5$, use the constant of variation.

$$5 = \frac{30}{x} \qquad 5x = 30 \qquad x = 6$$

#### Exercises

**1.** If $x$ varies inversely as $y$ and $x = 75$ when $y = 20$, find $y$ when $x = 25$.

**2.** If $p$ varies inversely as $q$ and $q = 36$ when $p = 15$, find $p$ when $q = 24$.

**3.** If $m$ varies inversely as $n$ and $m = 0.5$ when $n = 5$, find $m$ when $n = 50$.

**4.** If $y$ varies inversely with $x$ and if $y$ is $\frac{1}{2}$ when $x$ is 20, find $x$ when $y$ is 5.

### LESSON 11.2

#### Key Skills

**Evaluate rational functions. List the values of the variable for which functions are undefined.**

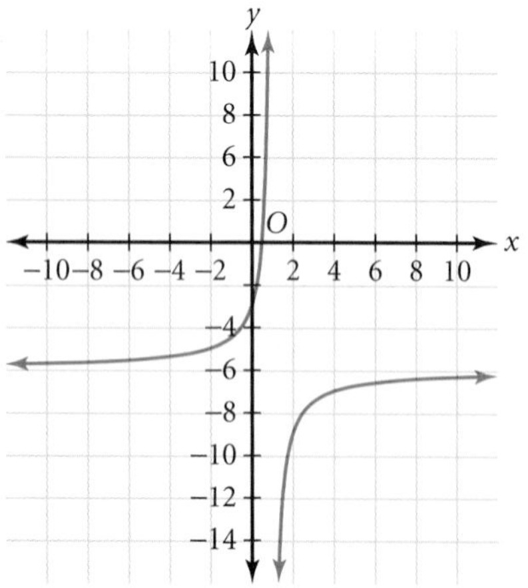


The function $y = \frac{-3}{x-1} - 6$ is graphed above. The graph has a vertical asymptote at $x = 1$. The function is undefined when $x - 1 = 0$, or $x = 1$.

#### Exercises

**Evaluate each rational expression for $x = -1$ and $x = 2$. Write "undefined" if appropriate.**

**5.** $\frac{4}{x}$

**6.** $\frac{3}{x+1}$

**7.** $\frac{-12}{2x-4}$

**8.** $\frac{-12}{x^2 - x - 2}$

**Graph each of the following functions. List any values of the variable for which the function is undefined.**

**9.** $y = \frac{-2}{x} + 4$

**10.** $y = \frac{-3}{x+4} - 1$

**11.** $y = \frac{2}{x-2} + 5$

## LESSON 11.3

### Key Skills

**Simplify rational expressions.**

To simplify $\frac{12(x+y)}{3x^2-3y^2}$, factor the numerator and denominator, remove any common factors, and list any restrictions for the original expression.

$$\frac{12(x+y)}{3x^2-3y^2} = \frac{(3)(4)(x+y)}{3(x+y)(x-y)} = \frac{4}{x-y},\ x \neq \pm y$$

### Exercises

**Simplify. List any restrictions.**

12. $\frac{2x(x+3)}{8x^2}$
13. $\frac{8(x+1)}{10(x+1)^2}$
14. $\frac{2x^2+6x}{6x-30}$
15. $\frac{2x^2-x-1}{1-x^2}$

## LESSON 11.4

### Key Skills

**Add, subtract, multiply, and divide rational expressions.**

To add $\frac{1}{4x}+\frac{3}{5x^2}$, rewrite the expression with common denominators and add. List any restrictions for the original expressions.

$$\frac{1}{4x}+\frac{3}{5x^2} = \frac{1(5x)}{4x(5x)}+\frac{3(4)}{5x^2(4)}$$
$$= \frac{(5x)}{20x^2}+\frac{12}{20x^2} = \frac{5x+12}{20x^2},\ x \neq 0$$

To multiply $\frac{x+y}{5x} \cdot \frac{3x}{x^2-y^2}$, multiply the numerators and multiply the denominators. Simplify, and list any restrictions for the original expressions.

$$\frac{x+y}{5x} \cdot \frac{3x}{x^2-y^2} = \frac{(x+y)(3x)}{5x(x^2-y^2)}$$
$$= \frac{(x+y)(3x)}{5x(x+y)(x-y)}$$
$$= \frac{3}{5(x-y)}\ x \neq 0, \text{ and } x \neq \pm y$$

To divide $\frac{x^2-4}{x-5} \div \frac{x+2}{x-5}$, invert the divisor and multiply. List any restrictions for the original expressions. Also, restrict any values for the variable that make the divisor zero.

$$\frac{x^2-4}{x-5} \div \frac{x+2}{x-5} = \frac{x^2-4}{x-5} \cdot \frac{x-5}{x+2}$$
$$= \frac{(x+2)(x-2)(x-5)}{(x-5)(x+2)}$$
$$= x-2,\ x \neq 5 \text{ and } x \neq -2$$

### Exercises

**Perform the indicated operations and simplify. List any restrictions.**

16. $\frac{8mt}{3s} \cdot \frac{s^2}{16t}$
17. $\frac{2}{5x}+\frac{5}{10y}$
18. $\frac{16}{y^2-16}-\frac{2}{y+4}$
19. $\frac{t^2-9}{6} \div \frac{3-t}{9}$
20. $\frac{3}{4x}+\frac{5}{2}$
21. $\frac{a^2-2a-15}{6} \cdot \frac{4}{3a+9}$
22. $\frac{5x}{6}-\frac{3}{2x}$
23. $\frac{5r^2q}{6q} \div \frac{3r}{q^2r}$
24. $\frac{1}{2}+\frac{10}{5x}$
25. $\frac{x^2+x-12}{x-3} \cdot \frac{x^2-25}{x^2-x-20}$

## LESSON 11.5

### Key Skills

**Solve rational equations by the common-denominator method.**

To solve $\frac{2}{x-1} + \frac{3}{x+1} = 1$, multiply both sides of the equation by the common denominator, $x^2 - 1$.

$$(x^2 - 1)\left(\frac{2}{x-1} + \frac{3}{x+1}\right) = (x^2 - 1)1$$
$$2(x+1) + 3(x-1) = x^2 - 1$$
$$2x + 2 + 3x - 3 = x^2 - 1$$
$$x^2 - 5x = 0$$
$$x(x-5) = 0$$
$$x = 0 \text{ or } x = 5$$

### Exercises

**Solve each equation by using the common-denominator method.**

**26.** $\frac{5}{2y} - \frac{3}{10} = \frac{1}{y}$

**27.** $\frac{z+2}{2z} + \frac{z+3}{z} = 5$

**28.** $\frac{6-x}{x} = \frac{4}{x+1}$

**29.** $\frac{5}{w+6} - \frac{2}{w} = \frac{9w+6}{w^2+6w}$

## LESSON 11.6

### Key Skills

**Prove algebraic statements.**

To prove that if $x = a + 2(b + a)$, then $x = 3a + 2b$, list each step and give the reason next to the step.

| | Statement | Reason |
|---|---|---|
| **1.** | $x = a + 2(b + a)$ | Hypothesis or given |
| **2.** | $= a + 2b + 2a$ | Distributive Property |
| **3.** | $= a + 2a + 2b$ | Commutative Property |
| **4.** | $= 3a + 2b$ | Simplified form |

### Exercises

**Prove each of the following:**

**30.** If $2a + 3 = 5$, then $a = 1$.

**31.** If $x = (a - b) - (b - a)$, then $x = 2a - 2b$.

**32.** If $3x - 9 = 6$, then $x = 5$.

**33.** If $n = (2x + y) - (x - 3y)$, then $n = x + 4y$.

**34.** If $-6a - 5 = -95$, then $a = 15$.

## Applications

**35.** **FITNESS** Millie joins a fitness club. There is a \$300 fee to join and a monthly charge of \$20. Write a rational expression to represent Millie's average cost per month for belonging to the fitness club.

**36.** **GEOMETRY** The area $\left(\frac{1}{2} \cdot \text{base} \cdot \text{height}\right)$ of a certain triangle is given by $x^2 - 2x - 15$ and its height is given by $x - 5$. Write a simplified expression for the length of its base.

**37.** **TRAVEL** A trip of 189 miles can take 1 hour less if Gerald increases the average speed of his car by 12 miles per hour. What was the original average speed of the car?

**38.** **GEOMETRY** The area (length • width) of certain rectangle is represented by $6x^2 + x - 35$. Its length is represented by $2x + 5$. Find the width of the rectangle.

# 11 Chapter Test

1. If $x$ varies inversely as $y$ and $x = 30$ when $y = 12$, find $y$ when $x = 15$.
2. If $x$ varies inversely as $y$ and $x = 12$ when $y = 25$, find $y$ when $x = 60$.
3. If $x$ varies inversely as $y$ and $x = 10$ when $y = 20$, find $y$ when $x = 25$.
4. If $x$ varies inversely as $y$ and $x = 48$ when $y = 4$, find $y$ when $x = 3$.
5. TRAVEL Juan drives his car at an average rate of 60 miles per hour for 8 hours. How long would it take for him to cover the same distance driving a truck at 45 miles per hour?

**Evaluate each rational function or expression for $x = -1$ and 3. Write *undefined* if appropriate.**

6. $y = \frac{3x}{x-1}$
7. $y = \frac{2x+1}{x-3}$
8. $y = \frac{x+1}{5x}$
9. $\frac{x-1}{x^2-4x+3}$

**Graph each of the following functions. List any values of the variable for which the function is undefined.**

10. $y = \frac{3}{x+2}$
11. $y = \frac{5}{x-1} + 2$

**Simplify. List any restrictions.**

12. $\frac{-3x(x-1)}{9x^2}$
13. $\frac{x(x+2)}{x^2+4x+4}$
14. $\frac{2x(x-4)}{4x^2-64}$
15. $\frac{m^2-n^2}{3(m+n)^2}$
16. NUMBER THEORY The product of two integers is given by $x^2 - 3x - 10$. If one integer is $x + 2$, find the other integer in terms of $x$.

**Perform the indicated operations and simplify. List any restrictions.**

17. $\frac{3}{r} + \frac{4}{r-1}$
18. $\frac{q+1}{q-3} - q$
19. $\frac{2p}{p-4} \cdot \frac{3p-12}{8p^2}$
20. $\frac{a^2-b^2}{(a-b)^2} \div \frac{(a+b)^2}{a^2-ab}$
21. $\frac{3t}{t+2} + \frac{6}{t+2}$
22. $\frac{-9z}{3z+12} \div \frac{3}{z+4}$
23. $\frac{3}{x-4} - \frac{1}{x^2-16}$
24. $\frac{6w+12}{w^2+5w+6} \cdot \frac{w+3}{3}$

**Solve each rational equation by using the lowest common denominator.**

25. $\frac{2}{x+4} - \frac{1}{2} = -\frac{1}{3}$
26. $\frac{x+1}{x+3} = \frac{x+4}{x+2}$
27. $\frac{a+2}{a} - \frac{3}{2a} = \frac{1}{3a}$
28. $\frac{h}{h+4} - 3 = \frac{2}{h+4}$
29. NUMBER THEORY The difference of two integers is 23. If the larger integer is divided by the smaller, the quotient is 2 with a remainder of 1. Find the two integers.

**For Exercises 30–35, state a reason for each step in the proof of this conjecture:**
**If $3k + 5 = 17$, then $k = 4$.**

| Statements | Reasons |
|---|---|
| $3k + 5 = 17$ | 30. ? |
| $3k + 5 - 5 = 17 - 5$ | 31. ? |
| $3k + 0 = 12$ | 32. ? |
| $3k = 12$ | 33. ? |
| $\frac{3k}{3} = \frac{12}{3}$ | 34. ? |
| $k = 4$ | 35. ? |

36. Determine if the set of integers is closed under division.

**Tell whether each statement is always true, sometimes true, or never true.**

37. $(2n)^2 = 2^2n^2$
38. $x^2 + 1^2 = (x+1)^2$
39. $|-3n| = -3n$
40. $|a - b| = |b - a|$

# College Entrance Exam Practice



**QUANTITATIVE COMPARISON** For Questions 1–5, write
A if the quantity in Column A is greater than the quantity in Column B;
B if the quantity in Column B is greater than the quantity in Column A;
C if the two quantities are equal; or
D if the relationship cannot be determined from the information given.

| | Column A | Column B | Answers |
|---|---|---|---|
| **1.** | $(-6)(9)$ | $(6)(-9)$ | Ⓐ Ⓑ Ⓒ Ⓓ [Lesson 2.4] |
| **2.** | $x$ Given: $\begin{cases} x+y=3 \\ x-y=1 \end{cases}$ | $y$ | Ⓐ Ⓑ Ⓒ Ⓓ [Lesson 7.3] |
| **3.** | slope of a line that rises left to right | slope of a line that is horizontal | Ⓐ Ⓑ Ⓒ Ⓓ [Lesson 5.2] |
| **4.** | 70% of $x$ | 90% of $x$ | Ⓐ Ⓑ Ⓒ Ⓓ [Lesson 4.2] |
| **5.** | $\frac{5 \times 10^{-2}}{7 \times 10^{-5}}$ | $\frac{5 \times 10^{2}}{7 \times 10^{5}}$ | Ⓐ Ⓑ Ⓒ Ⓓ [Lesson 8.5] |

**6.** Solve. $\begin{cases} 2x+y=1 \\ x-y=-7 \end{cases}$ ***(LESSON 7.3)***

**a.** $(5, -2)$ **b.** $(-2, 5)$
**c.** $(2, -5)$ **d.** $(5, 2)$

**7.** What is the factored form of $4x^2 - 25$? ***(LESSON 9.6)***

**a.** $(2x-5)^2$ **b.** $(2x+5)(2x-5)$
**c.** $(2x+5)^2$ **d.** cannot be factored

**8.** Evaluate $x-(z-y)$ for $x=-4$, $y=2$, and $z=-3$. ***(LESSON 2.3)***

**a.** $-9$ **b.** $-5$
**c.** $1$ **d.** $-1$

**9.** Which equation represents the axis of symmetry for the quadratic function $y=(x-2)^2-3$? ***(LESSON 10.1)***

**a.** $x=3$ **b.** $x=-3$
**c.** $x=2$ **d.** $x=-2$

**10.** The length of each side of a square is 3 cm. What is the square's perimeter? ***(LESSON 2.4)***

**a.** 15 cm **b.** 12 cm
**c.** 3 cm **d.** 9 cm

**11.** Solve: $\frac{n}{12} = \frac{12}{18}$. ***(LESSON 4.1)***

**a.** $-2$ **b.** $8$
**c.** $1$ **d.** $-1$

**12.** Simplify: $\frac{x^2-1}{x-1}$. ***(LESSON 11.3)***

**a.** $x^2+1$ **b.** $x+1$
**c.** $x-1$ **d.** $x^2-1$

**13.** Solve: $-3x-5>-11$. ***(LESSON 6.2)***

**a.** $x>2$ **b.** $x>-2$
**c.** $x<6$ **d.** $x<2$

14. Simplify: $(5x^2y)(2xy^{-1})^2$. ***(LESSON 8.4)***

**a.** $10x^4y^{-1}$ **b.** $20x^4y$
**c.** $20x^4y^2$ **d.** $20x^4y^{-1}$

15. Evaluate $3(x+1)^2+1$ for $x=2$. ***(LESSON 1.3)***

16. Which number must be added to $x^2+4x$ to complete the square? ***(LESSON 10.3)***

17. Write an equation for the line passing through the origin and $(3, -4)$. ***(LESSON 5.5)***

18. Tickets to a concert cost \$24 per person, plus a \$7 handling charge per order. How many tickets can be ordered for \$199? ***(LESSON 3.3)***

19. Simplify: $\frac{-100a^8b^{10}}{-20a^5b^9}$. ***(LESSON 8.3)***

20. Vernon scored 86 on his first two tests, 90 on his third test, and 78 on his fourth test. Find the median test score. ***(LESSON 4.4)***

21. Write an equation for the line with a slope of $-2$ passing through $(-1, 0)$. ***(LESSON 5.5)***

22. Factor: $3a(2c+d)^2-5(2c+d)^2$. ***(LESSON 9.5)***

23. Solve: $\frac{2}{x+1}=\frac{x}{6}$. ***(LESSON 11.5)***

24. Solve. $\begin{cases} 3x+y=-2 \\ 2x+3y=8 \end{cases}$ ***(LESSON 7.3)***

25. Simplify: $(3x^2y^3)^0$. ***(LESSON 8.4)***

26. Solve: $|x-5|=17$. ***(LESSON 6.5)***

27. Solve: $\frac{7}{9}=\frac{x}{27}$. ***(LESSON 3.2)***

28. Find 60% of 240. ***(LESSON 4.2)***

29. 16 is 20% of what number? ***(LESSON 4.2)***

30. Solve: $(x+3)^2=25$. ***(LESSON 10.2)***

31. If you invest \$2000 in a CD that earns 5% interest, compounded annually, how much will the CD be worth in 5 years? ***(LESSON 8.7)***

32. Write an equation in slope-intercept form for the line that contains the point $(3, -7)$ and is perpendicular to $5x+2y=10$. ***(LESSON 5.6)***

33. Solve: $x^2-3x-18=0$. ***(LESSON 9.8)***

34. Solve. $\begin{cases} 3x-4y=-18 \\ 2x+6y=14 \end{cases}$ ***(LESSON 7.3)***

35. Solve. $\begin{cases} x=y \\ 2x+3y=7 \end{cases}$ ***(LESSON 7.2)***

### FREE-RESPONSE GRID

**The following questions may be answered by using a free-response grid such as that commonly used by standardized test services.**

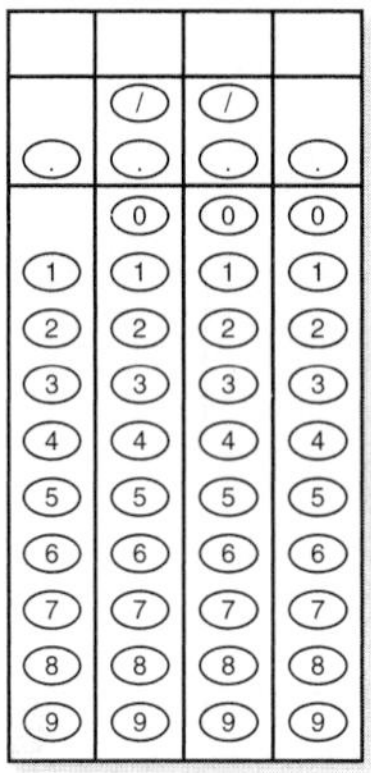

36. Evaluate $2x^2+7x-20$ for $x=5$. ***(LESSON 1.3)***

37. Solve: $\frac{2x}{3}=12$. ***(LESSON 3.2)***

38. What is the slope of the line whose equation is $y=2x+5$? ***(LESSON 5.4)***

39. Tim must score a total of at least 270 points on his tests to earn an A for the semester. On the first test, he scored an 85 and on the second test he scored an 89. What is the lowest score that Tim can receive on the third and final test and still earn an A? ***(LESSON 6.1)***

40. Find the slope of a line that passes through the points $(4, 3)$ and $(-2, -3)$. ***(LESSON 5.2)***

41. Solve: $3x^2-6x=-3$. ***(LESSON 9.8)***

42. Solve: $\frac{x}{7}=\frac{12}{42}$. ***(LESSON 3.2)***

43. Find the slope of the line represented by $2x+3y=5$. ***(LESSON 5.5)***

44. Solve: $\frac{5}{11}=\frac{15}{w}$. ***(LESSON 4.1)***

45. Find 15% of 260. ***(LESSON 4.2)***

# 12 Radicals, Functions & Coordinate Geometry

MONUMENTS SUCH AS THOSE PICTURED HERE are great accomplishments that can be even more fully appreciated with an understanding of the mathematical properties demonstrated by their construction. In order to understand the relationships between the various distances and angles used to construct such monuments, it is necessary to understand radicals, or square roots.

Radicals play an important role in algebra and geometry because they are essential to both the quadratic formula and the distance formula. In this chapter, you will learn techniques for solving equations that contain radicals and study several geometric properties and trigonometric functions that relate to radicals. You will have opportunities to apply these techniques to real-world situations ranging from construction to aviation.

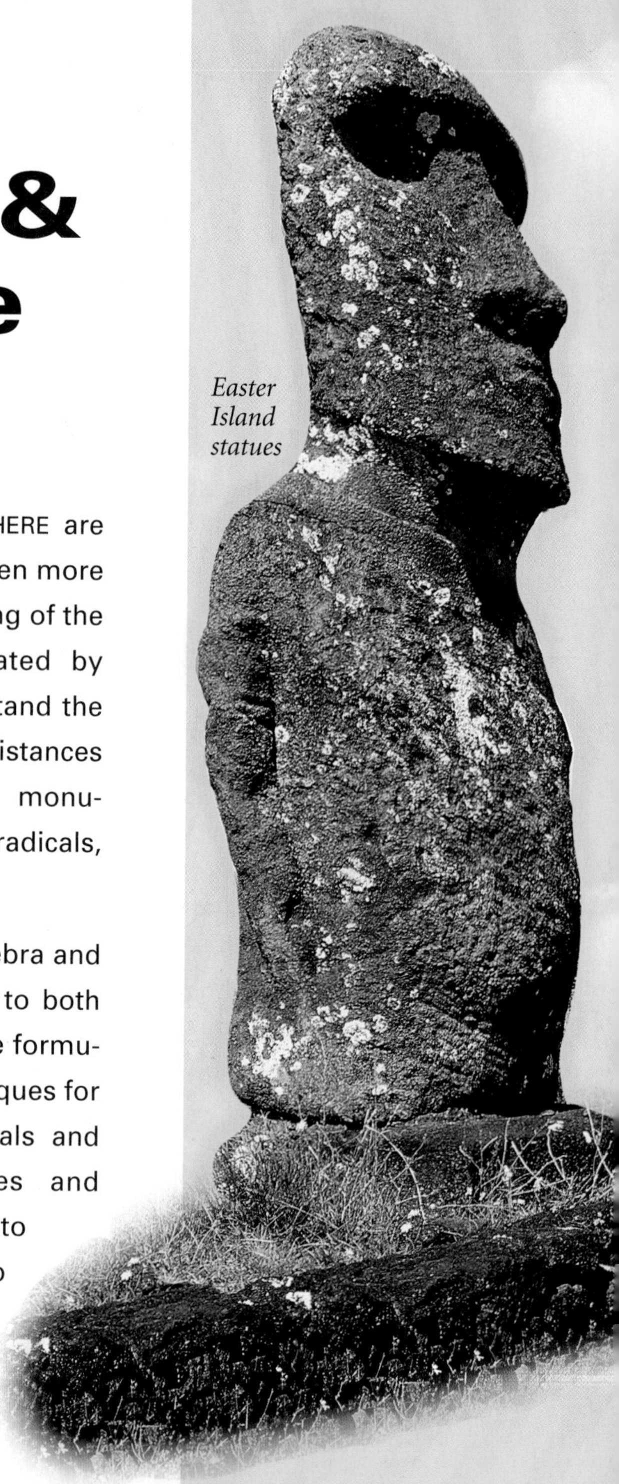

*Easter Island statues*

## Lessons



*Taj Mahal, India*

The Great Wall of China

Stonehenge, England

CHAPTER PROJECT

PORTFOLIO ACTIVITIES

## About the Chapter Project

Solving equations with radicals is an important skill that allows you to solve complex problems involving distance and various geometric and trigonometric properties. In the Chapter Project, *Working the Angles*, on page 636, you will learn how to build a hypsometer, use it to measure angles of elevation and depression, and calculate heights and distances from your observations.

After completing the Chapter Project, you will be able to do the following:

- Apply the inverse tangent function to a real-world situation.

## About the Portfolio Activities

Throughout the chapter, you will be given opportunities to complete Portfolio Activities that are designed to support your work on the Chapter Project.

- Examining and graphing the action of pendulums is the topic of the Portfolio Activity on page 590.
- The "Wheel of Theodorus" is the topic of the Portfolio Activity on page 605.
- Golden rectangles are the topic of the Portfolio Activity on page 627.

12.1

# Operations With Radicals

**Why** *Radicals often appear when computing statistics or evaluating formulas in physics, but they can also be found in everyday activities like tiling.*

### Objectives

- Identify or estimate square roots.
- Define and write square roots in simplest radical form.
- Perform mathematical operations with radicals.

## Estimating Square Roots

The blue square has an area of 12. How can the length of a side of the blue square be determined from the area?

Recall that the area of a square is found by squaring the length of a side.

$$Area = s^2$$

Thus, the square root of the area is the length of a side.

To estimate $\sqrt{12}$, find the two nearest perfect-square numbers, one that is less than 12 and another that is greater than 12—the numbers 9 and 16.

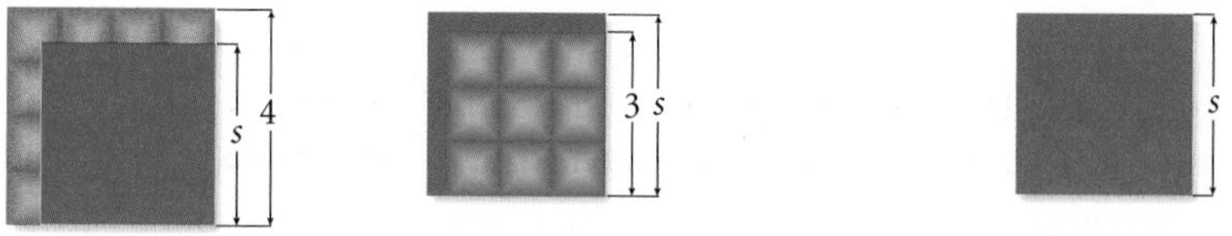


The area of the square, 12, is a little closer to 9 than it is to 16, so a good estimate for $\sqrt{12}$ is somewhere between 3.4 and 3.5. Multiply to refine the estimate to the nearest hundredth.

$$3.45^2 = 11.9025$$
$$3.46^2 = 11.9716$$
$$3.47^2 = 12.0409$$
$$3.48^2 = 12.1104$$

Thus, a reasonable estimate for $\sqrt{12}$ is 3.46. Recall from Lesson 10.2 that the principal square root, indicated by the radical sign, is always positive.

### Definition of Square Root

If $a$ is a number greater than or equal to zero, $\sqrt{a}$ represents the principal, or positive, square root of $a$ and $-\sqrt{a}$ represent the negative square root of $a$. The square roots of $a$ have the following property:

$$\sqrt{a} \cdot \sqrt{a} = a \qquad (-\sqrt{a})(-\sqrt{a}) = a$$

You were introduced to the radical sign, or just *radical*, in Lesson 10.2. The number or expression under the radical is the **radicand**. You also learned in 10.2 that the solutions to $x^2 = a$ are $x = \sqrt{a}$ or $x = -\sqrt{a}$. This is often abbreviated as $x = \pm\sqrt{a}$, "$x$ equals plus or minus the square root of $a$."

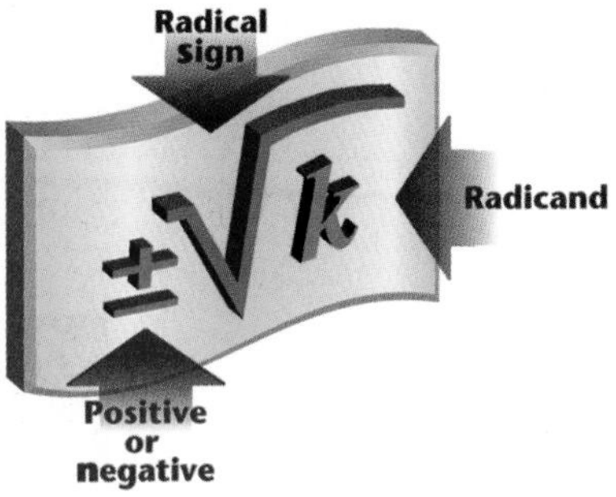


**EXAMPLE 1** **Evaluate each square root.**

**a.** $\sqrt{25}$ **b.** $-\sqrt{144}$ **c.** $\sqrt{7}$ **d.** $\pm\sqrt{9}$

**SOLUTION**

**a.** $\sqrt{25} = 5$ because $5 \cdot 5 = 25$ and $\sqrt{25}$ means the principal, or positive, square root.

**b.** $-\sqrt{144} = -12$ because $-(\sqrt{144}) = -(12) = -12$.

**c.** $\sqrt{7}$ is irrational because 7 is not a perfect square; 4 and 9 are perfect squares, so $\sqrt{7}$ must be between 2 and 3. Multiplication will confirm that 2.65 is a good estimate to the nearest hundredth.

**d.** $\pm\sqrt{9} = \pm 3$ because both the positive and negative roots are indicated.

**TRY THIS** Evaluate each square root.

**a.** $\sqrt{196}$ **b.** $-\sqrt{225}$

## Simplifying Radicals and Radical Expressions

In order to add or subtract expressions that contain radicals, radicands must be equal. Terms that contain $\sqrt{5}$, for example, can be added by applying the Distributive Property.

$$2\sqrt{5} + 4\sqrt{5} = (2 + 4)\sqrt{5} = 6\sqrt{5}$$

**EXAMPLE 2** **Simplify.**

**a.** $5\sqrt{6} - 2\sqrt{6}$ **b.** $5 + 6\sqrt{7} - 2\sqrt{7} - 3$

**c.** $8\sqrt{3} + 6\sqrt{2} - \sqrt{3} + 2\sqrt{2}$ **d.** $a\sqrt{x} + b\sqrt{x}$

**SOLUTION**

**a.** Use the Distributive Property to combine like terms. In this case, $\sqrt{6}$ is the common factor. Then simplify.

$$5\sqrt{6} - 2\sqrt{6} = (5 - 2)\sqrt{6} = 3\sqrt{6}$$

**b.** Rearrange in order to combine like terms.

$$5 + 6\sqrt{7} - 2\sqrt{7} - 3 = (5 - 3) + (6 - 2)\sqrt{7} = 2 + 4\sqrt{7}$$

**c.** Rearrange in order to combine like terms.

$$8\sqrt{3} + 6\sqrt{2} - \sqrt{3} + 2\sqrt{2} = (8 - 1)\sqrt{3} + (6 + 2)\sqrt{2} = 7\sqrt{3} + 8\sqrt{2}$$

**d.** Variables are treated in a similar way.

$$a\sqrt{x} + b\sqrt{x} = (a + b)\sqrt{x}$$

**TRY THIS** Simplify. **a.** $4\sqrt{6} - \sqrt{6}$ **b.** $a + 2\sqrt{b} - 2a - 3\sqrt{b}$

*Activity*

## Operations and Radical Expressions

You can use your understanding of radicals to perform mathematical operations.

1. Compute $\sqrt{9 \cdot 16}$ in two different ways:

   **a.** $\sqrt{9 \cdot 16} = \sqrt{144} = ?$ **b.** $\sqrt{9 \cdot 16} = \sqrt{9} \cdot \sqrt{16} = ?$

2. Give three examples illustrating that $\sqrt{ab} = \sqrt{a}\sqrt{b}$ is true.

**CHECKPOINT ✔**

3. Compute $\sqrt{9 + 16}$ and $\sqrt{9} + \sqrt{16}$.
4. Give three examples illustrating that $\sqrt{a + b} = \sqrt{a} + \sqrt{b}$ is *not* always true.

### Multiplication Property of Square Roots

For all numbers $a$ and $b$, where $a \geq 0$ and $b \geq 0$:

$$\sqrt{ab} = \sqrt{a}\sqrt{b}$$

Radical expressions that are simplified are easier to manipulate algebraically. A square-root expression is in **simplest radical form** when all the following conditions are met:

1. No factor of the radicand is a perfect square other than 1.
2. The radicand contains no fractions.
3. No radical appears in the denominator of a fraction.

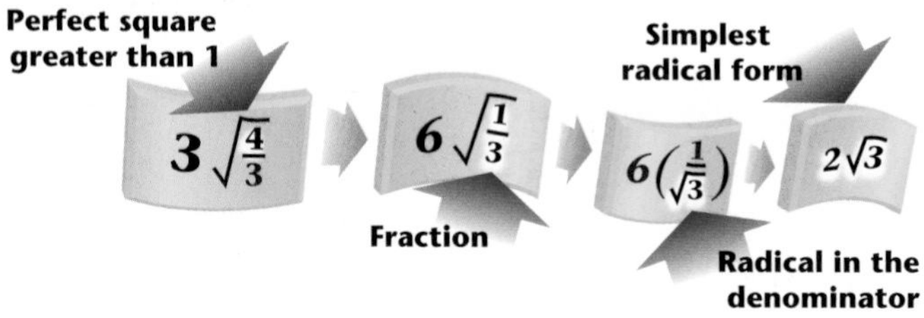

**EXAMPLE 3** **Write in simplest radical form.**

**a.** $\sqrt{12}$ **b.** $\sqrt{400}$ **c.** $\sqrt{x^2}$ **d.** $\sqrt{a^5b^{10}}$

**SOLUTION**

Look for perfect-square factors, and apply the Multiplication Property of Square Roots. Then find the square roots of the perfect squares. Leave any factor that is not a perfect square in radical form.

**a.** $\sqrt{12} = \sqrt{4 \cdot 3} = \sqrt{4} \cdot \sqrt{3} = 2\sqrt{3}$

**b.** $\sqrt{400} = \sqrt{4 \cdot 100} = \sqrt{4} \cdot \sqrt{100} = 2 \cdot 10 = 20$

Because the radical sign designates the principle square root, the value of $\sqrt{x^2}$ must be positive. Use the absolute value sign to indicate this when the exponent of a variable in the radical is **even** and the simplified exponent outside of the radical is **odd**.

**c.** $\sqrt{x^2} = |x^1| = |x|$

**d.** $\sqrt{a^5b^{10}} = \sqrt{(a^2)^2 \cdot a \cdot (b^5)^2} = a^2|b^5|\sqrt{a}$

**TRY THIS** Simplify $\sqrt{72m^2n^5}$.

**CRITICAL THINKING** Why is the absolute-value sign not necessary for the following?

**a.** $\sqrt{x^3} = x\sqrt{x}$ **b.** $\sqrt{x^4} = x^2$

**EXAMPLE 4** **Simplify.**

**a.** $(5\sqrt{3})^2$ **b.** $\sqrt{3}\sqrt{6}$ **c.** $\sqrt{2}(6 + \sqrt{12})$ **d.** $(3 - \sqrt{2})(4 + \sqrt{2})$

**SOLUTION**

**a.** Recall from the properties of exponents that the second power means the product of two identical factors. Rearrange the factors and multiply.

$(5\sqrt{3})^2 = (5\sqrt{3})(5\sqrt{3}) = (5 \cdot 5)(\sqrt{3}\sqrt{3}) = 25 \cdot 3 = 75$

**b.** The Multiplication Property of Square Roots allows you to multiply separate radicals. You can then factor the new product in a different way and simplify.

$\sqrt{3}\sqrt{6} = \sqrt{3 \cdot 6} = \sqrt{18} = \sqrt{9 \cdot 2} = \sqrt{9}\sqrt{2} = 3\sqrt{2}$

**c.** Use the Distributive Property to multiply, factor, and simplify the result. Notice that the radical is usually written last in each term.

$\sqrt{2}(6 + \sqrt{12}) = \sqrt{2} \cdot 6 + \sqrt{2} \cdot \sqrt{12} = 6\sqrt{2} + \sqrt{2 \cdot 12} = 6\sqrt{2} + \sqrt{24}$

$6\sqrt{2} + \sqrt{24} = 6\sqrt{2} + \sqrt{4 \cdot 6} = 6\sqrt{2} + 2\sqrt{6}$

**d.** To multiply differences and sums, multiply the two binomials by using the FOIL method.

$(3 - \sqrt{2})(4 + \sqrt{2}) = 12 + 3\sqrt{2} - 4\sqrt{2} - 2 = 10 - \sqrt{2}$

**TRY THIS** Simplify. **a.** $(2\sqrt{7})^2$ **b.** $\sqrt{2}(4 - \sqrt{8})$

## Division Property of Square Roots

For all numbers $a \geq 0$ and $b > 0$:

$$\sqrt{\frac{a}{b}} = \frac{\sqrt{a}}{\sqrt{b}}$$

**EXAMPLE 5** **Simplify.**

**a.** $\sqrt{\frac{16}{25}}$ **b.** $\sqrt{\frac{7}{16}}$ **c.** $\sqrt{\frac{a^2b^3}{c^2}}$ **d.** $\sqrt{\frac{9}{5}}$

**SOLUTION**

Rewrite each square root by using the Division Property of Square Roots. Then simplify the numerator and denominator separately.

**a.** $\sqrt{\frac{16}{25}} = \frac{\sqrt{16}}{\sqrt{25}} = \frac{4}{5}$

**b.** $\sqrt{\frac{7}{16}} = \frac{\sqrt{7}}{\sqrt{16}} = \frac{\sqrt{7}}{4}$

**c.** $\sqrt{\frac{a^2b^3}{c^2}} = \frac{\sqrt{a^2b^2b}}{\sqrt{c^2}} = \frac{|a|b\sqrt{b}}{|c|}$

**d.** $\sqrt{\frac{9}{5}} = \frac{\sqrt{9}}{\sqrt{5}} = \frac{3}{\sqrt{5}}$

A radical remains in the denominator of the answer for part **d.** Change this expression to simplest radical form by multiplying by $\frac{\sqrt{5}}{\sqrt{5}}$, which is equivalent to multiplying by 1:

$$\frac{3}{\sqrt{5}} \cdot \frac{\sqrt{5}}{\sqrt{5}} = \frac{3\sqrt{5}}{\sqrt{5}\sqrt{5}} = \frac{3\sqrt{5}}{5}$$

Removing a radical expression from the denominator is known as **rationalizing the denominator.**

**TRY THIS** Simplify. **a.** $\sqrt{\frac{4}{25}}$ **b.** $\sqrt{\frac{4}{3}}$ **c.** $\sqrt{\frac{ab^2}{c}}$

# Exercises

## Communicate

**1.** Describe how you can use graph paper to determine the square root of 16, a perfect square.

**2.** Describe how you can use graph paper to estimate the square root of 19, a number that is not a perfect square.

**3.** Explain how to estimate $\sqrt{7}$ without a calculator or graph paper.

**4.** How is factoring used to simplify expressions that contain radicals such as $5\sqrt{90x^3y^4}$ ?

**5.** What is *simplest radical form*?

## Guided Skills Practice

**Evaluate each square root.** *(EXAMPLE 1)*

**6.** $\sqrt{36}$ **7.** $-\sqrt{64}$ **8.** $\pm\sqrt{81}$ **9.** $-\sqrt{121}$

**Simplify.** *(EXAMPLE 2)*

**10.** $8\sqrt{3} - 6\sqrt{3}$ **11.** $9 + 3\sqrt{7} - 5\sqrt{7} + 4$

**Express in simplest radical form.** *(EXAMPLE 3)*

**12.** $\sqrt{32}$ **13.** $\sqrt{x^2y^7}$ **14.** $\sqrt{27x^6}$ **15.** $\sqrt{19a^7b^3}$

**Simplify.** *(EXAMPLE 4)*

**16.** $(7\sqrt{11})^2$ **17.** $\sqrt{2}\sqrt{10}$ **18.** $(2 - \sqrt{3})(5 + \sqrt{3})$

**Simplify.** *(EXAMPLE 5)*

**19.** $\sqrt{\frac{9}{4}}$ **20.** $\sqrt{\frac{6}{49}}$ **21.** $\sqrt{\frac{225}{18}}$ **22.** $\sqrt{\frac{x^7y^{14}}{z^3}}$

## Practice and Apply

**Find each square root. If the square root is irrational, approximate the value to the nearest hundredth.**

**23.** $\sqrt{225}$ **24.** $-\sqrt{169}$ **25.** $\pm\sqrt{11}$ **26.** $\sqrt{\frac{4}{9}}$ **27.** $-\sqrt{40}$

**28.** $-\sqrt{27}$ **29.** $\sqrt{1000}$ **30.** $\sqrt{10{,}000}$ **31.** $-\sqrt{0.04}$ **32.** $\sqrt{0.059}$

**Simplify each radical by factoring.**

**33.** $\sqrt{49}$ **34.** $\sqrt{196}$ **35.** $\sqrt{576}$ **36.** $\sqrt{3600}$ **37.** $\sqrt{192}$

**38.** $\sqrt{75}$ **39.** $\sqrt{98}$ **40.** $\sqrt{1620}$ **41.** $\sqrt{264}$ **42.** $\sqrt{648}$

**Decide whether each statement is true or false. Assume that $a > 0$ and $b > 0$.**

**43.** $\sqrt{a+b} = \sqrt{a} + \sqrt{b}$ **44.** $\sqrt{ab} = \sqrt{a}\sqrt{b}$ **45.** $\sqrt{\frac{a}{b}} = \frac{\sqrt{a}}{\sqrt{b}}, b \neq 0$

**internet connect**

**Homework Help Online**

Go To: **go.hrw.com**
Keyword:
**MA1 Homework Help** for Exercises 46-57

**Express in simplest radical form.**

**46.** $\sqrt{3}\sqrt{12}$ **47.** $\sqrt{8}\sqrt{18}$ **48.** $\sqrt{48}\sqrt{3}$ **49.** $\sqrt{54}\sqrt{6}$

**50.** $\sqrt{\frac{64}{16}}$ **51.** $\sqrt{\frac{96}{2}}$ **52.** $\frac{\sqrt{50}}{\sqrt{8}}$ **53.** $\frac{\sqrt{150}}{\sqrt{6}}$

**54.** $\sqrt{5}\sqrt{15}$ **55.** $\sqrt{98}\sqrt{14}$ **56.** $\sqrt{\frac{56}{8}}$ **57.** $\frac{\sqrt{96}}{\sqrt{8}}$

**Simplify each of the following. Assume that all denominators are nonzero.**

**58.** $\sqrt{a^4b^6}$ **59.** $\sqrt{x^8y^9}$ **60.** $\sqrt{\frac{p^9}{q^{10}}}$ **61.** $\sqrt{\frac{x^3}{y^6}}$

**If possible, perform the indicated operations, and simplify your answer.**

**62.** $3\sqrt{5} + 4\sqrt{5}$ **63.** $4\sqrt{5} + 2\sqrt{5} - 5\sqrt{5}$ **64.** $\sqrt{6} + 2\sqrt{3} - \sqrt{6}$

**65.** $\left(4 + \sqrt{3}\right) + \left(1 - \sqrt{2}\right)$ **66.** $\frac{6 + \sqrt{18}}{3}$ **67.** $\frac{\sqrt{15} + \sqrt{10}}{\sqrt{5}}$

**Simplify.**

**68.** $(3\sqrt{5})^2$ **69.** $(4\sqrt{25})^2$ **70.** $\sqrt{12}\sqrt{6}$

**71.** $\sqrt{72}\sqrt{32}$ **72.** $3(\sqrt{5}+9)$ **73.** $\sqrt{5}(6-\sqrt{15})$

**74.** $\sqrt{6}(6+\sqrt{18})$ **75.** $(\sqrt{5}-2)(\sqrt{5}+2)$ **76.** $(\sqrt{3}-4)(\sqrt{3}+2)$

CHALLENGE

**77.** $\sqrt{3}(\sqrt{3}+2)^2$ **78.** $\sqrt{12}(\sqrt{3}+8)^2$ **79.** $\sqrt{5}(\sqrt{5}-4)^2$

CONNECTION

**GEOMETRY** Determine the side length of a square with each given area.

**80.** 250 square meters **81.** 144 square inches **82.** 28 square miles

APPLICATIONS

**83. LANDSCAPING** The Smiths have a yard in the shape of a square. If the area of the yard is 676 square feet, what is the length of each side?

## Look Back

**Simplify.** ***(LESSONS 8.2 AND 8.3)***

**84.** $(-a^2b^2)^3(a^4b)^2$ **85.** $\frac{x^5y^7}{x^2y^3}$ **86.** $\left(\frac{20x^3}{-4x^2}\right)^3$

**Find each product.** ***(LESSON 9.3)***

**87.** $(2x-4)(2x-4)$ **88.** $(3a+5)(2a-6)$ **89.** $(6b+1)(3b-1)$

**internet connect**
**Activities Online**
Go To: **go.hrw.com**
Keyword:
**MA1 Continued Fractions**

## Look Beyond

**Roots and Fractional Exponents**

The radical sign can be used to indicate other roots of numbers. For example:

$\sqrt[3]{27}$ means "the principal third (or cubed) root of 27"

$\sqrt[4]{16}$ means "the principal fourth root of 16"

$\sqrt[3]{27}=3$, since $3^3=27$ $\sqrt[4]{16}=2$, since $2^4=16$

**Evaluate.**

**90.** $\sqrt[3]{125}$ **91.** $\sqrt[5]{32}$ **92.** $\sqrt[4]{256}$ **93.** $\sqrt[5]{100000}$

Fractional exponents can also be used to indicate roots. **For any real numbers $a \geq 0$ and integer $n \geq 2$, $a^{\frac{1}{n}}=\sqrt[n]{a}$.** For example:

$\sqrt{5}=5^{\frac{1}{2}}$ $\sqrt[4]{7}=7^{\frac{1}{4}}$ $\sqrt[3]{a^2}=a^{\frac{2}{3}}$

The laws of exponents can be used to simplify expressions with fractional exponents. For example:

$\left(x^{\frac{1}{2}}\right)^5\left(x^{\frac{3}{2}}\right)=\left(x^{\frac{5}{2}}\right)\left(x^{\frac{3}{2}}\right)=x^{\frac{8}{2}}=x^4$ $\left(a^{\frac{1}{3}}\right)\left(a^{\frac{1}{3}}\right)\left(a^{\frac{1}{3}}\right)=\left(a^{\frac{1}{3}}\right)^3=a^1=a$

**Simplify.**

**94.** $\left(x^{\frac{1}{3}}\right)^4(x^5)^{\frac{1}{3}}$ **95.** $(xy)^{\frac{1}{2}}\left(x^{\frac{1}{3}}\right)^6\left(y^{\frac{1}{2}}\right)^2$ **96.** $\left(x^3y^{\frac{3}{2}}\right)^6(xy)^{\frac{1}{2}}$

# 12.2 Square-Root Functions and Radical Equations

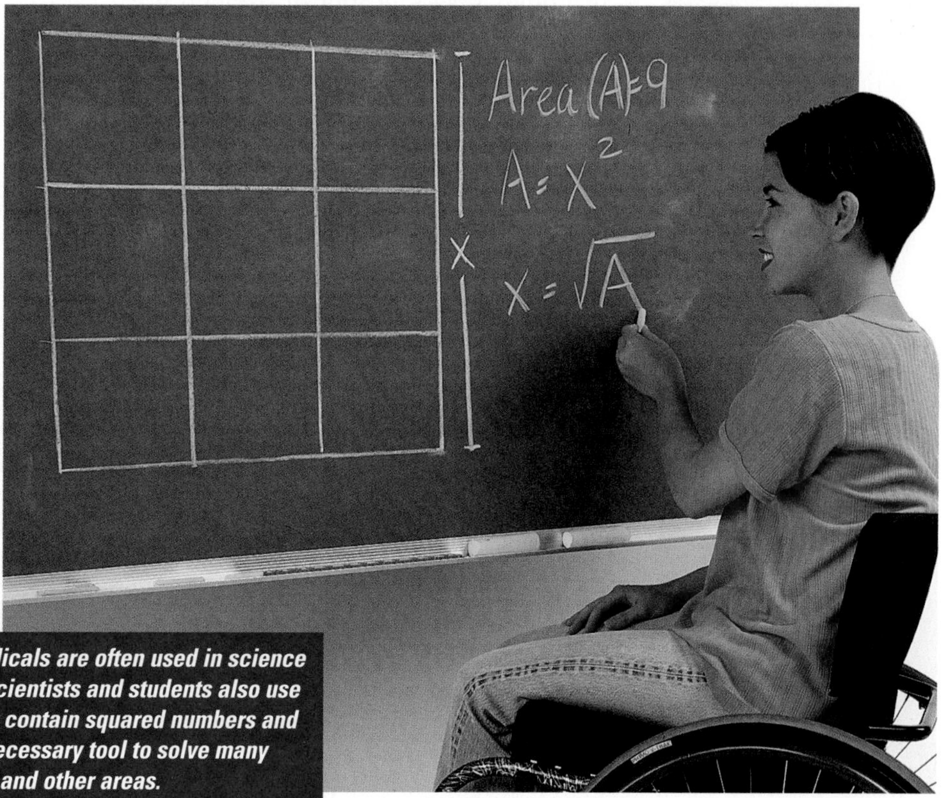


**Objectives**

- Solve equations containing radicals.
- Solve equations by using radicals.

**Why** *Equations that contain radicals are often used in science to model natural phenomena. Scientists and students also use radicals to solve equations that contain squared numbers and variables. Radicals provide a necessary tool to solve many problems that occur in science and other areas.*

## The Square-Root Function

The function $y = \sqrt{x}$ is the simplest **square-root function**. Because $\sqrt{x}$ means the positive, or principal, square root of $x$, the range of $y = \sqrt{x}$ is limited to 0 and all positive real numbers. The domain is also limited to 0 and all positive real numbers because square roots of negative numbers are not real numbers.

The function $y = \sqrt{x}$ is graphed below. The graph of the function lies entirely in the first quadrant because the domain and range are both limited to 0 and all the positive real numbers.

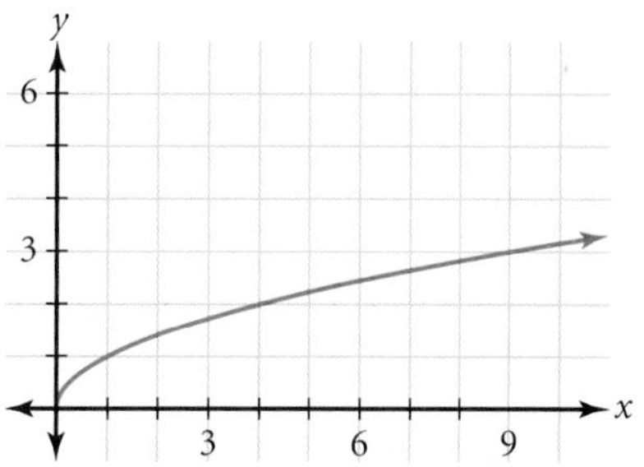


**CRITICAL THINKING** Describe similarities and differences between the graph of a square-root function and the graph of a quadratic function. Explain how the two functions are related.

# Solving Radical Equations

The equation relating the length of a pendulum to the number of swings per minute is one example of a radical equation.

**APPLICATION**
**PHYSICS**

In a physics class, groups of students are given six strings of different lengths, each tied to a washer. The students time one complete swing of each pendulum and then put all of their data together, make a table, and draw a graph of the data. The point (0, 0) represents the time for one complete swing of a pendulum with a length of 0.

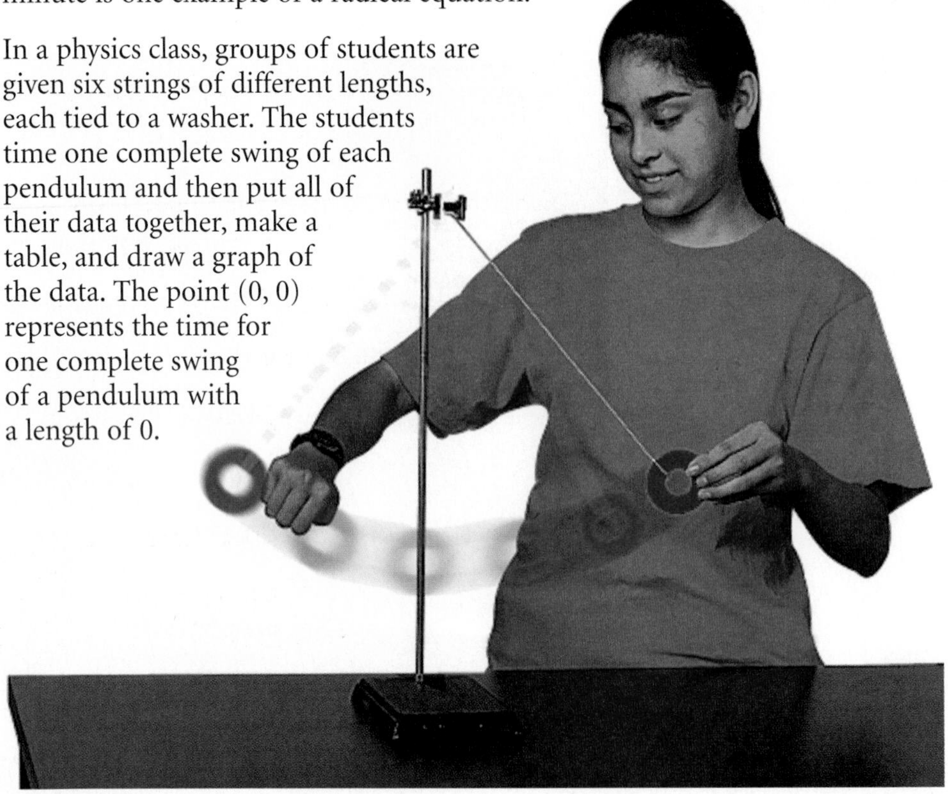

Let $l$ represent the length of the pendulum in centimeters and $t$ represent the time of one complete swing in seconds. Examine the data in the table and on the graph.

| $l$ | $t$ |
|---|---|
| 0 | 0 |
| 6 | 0.5 |
| 25 | 1.0 |
| 58 | 1.5 |
| 100 | 2.0 |
| 155 | 2.5 |

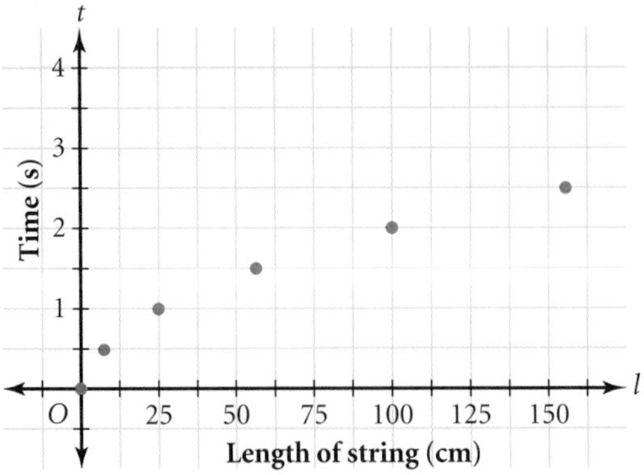


What conclusions can be drawn from the information in the graph? How can you use the graph to predict the time that it will take for a string of any length to make one complete swing?

Notice that the curve created by connecting the points resembles a transformation of the parent function $y = \sqrt{x}$.

The equation that models the motion of a pendulum is $t = 2\pi\sqrt{\frac{l}{g}}$. The variable $t$ is the time in seconds, $l$ is the length in centimeters, and $g$ is the acceleration due to gravity (980 centimeters per second squared). This formula can be used to calculate the time that a pendulum of a given length should take to make one complete swing.

**EXAMPLE 1**

**APPLICATION**
**PHYSICS**

**a. Determine the time in seconds that it takes a 75-centimeter pendulum to make one complete swing.**
**b. Determine the length in centimeters of a pendulum that takes 3 seconds to make one complete swing.**

**SOLUTION**

**a.** Substitute the values into the pendulum formula.

$$t = 2\pi\sqrt{\frac{75}{980}}$$

Use a calculator to help in the calculation.

$$t \approx 1.738$$

One swing of a 75-centimeter pendulum takes about 1.74 seconds.

**b.** Substitute the known information into the formula, and solve for *l*.

$3 = 2\pi\sqrt{\frac{l}{980}}$ *Given*

$3\sqrt{980} = 2\pi\sqrt{l}$ *Multiply both sides by $\sqrt{980}$.*

$\frac{3\sqrt{980}}{2\pi} = \sqrt{l}$ *Divide both sides by $2\pi$.*

$\left(\frac{3\sqrt{980}}{2\pi}\right)^2 = l$ *Use the definition of square root.*

$223.413 \approx l$ *Use a calculator.*

The length of the pendulum is about 223.41 centimeters.

*A Foucault pendulum*

**TRY THIS**

**a.** Find the time in seconds that it takes a 50-centimeter pendulum to make one complete swing.
**b.** Find the length in centimeters of a pendulum that takes 5 seconds to make one complete swing.

**EXAMPLE 2** **Solve $\sqrt{x+2} = 3$.**

**SOLUTION**

The square-root and squaring operations are inverses of each other and can be used to solve equations.

$\sqrt{x+2} = 3$

$\left(\sqrt{x+2}\right)^2 = 3^2$ *Square both sides of the equation.*

$x + 2 = 9$

$x = 7$

Check by substituting. Because $\sqrt{7+2} = \sqrt{9} = 3$, the solution is 7.

**TRY THIS** Solve $\sqrt{2x-3} = 4$

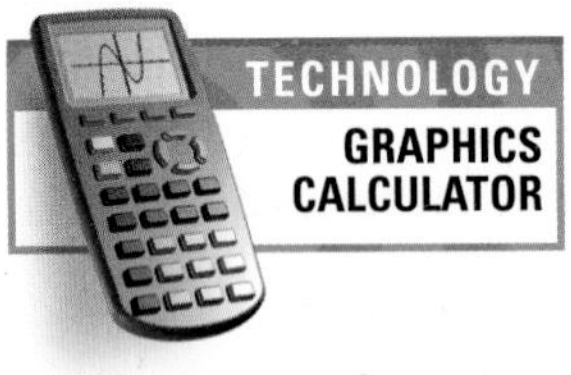


You can also solve the equation with a graphics calculator. Set an appropriate window. Enter the left side of the equation in the graphics calculator as **Y1** = $\sqrt{x+2}$. Then enter the right side of the equation as **Y2** = 3.

The calculator will graph both equations in the same window. You can then use the trace feature to find the point(s) of intersection. The point of intersection is (7, 3), so $x = 7$.

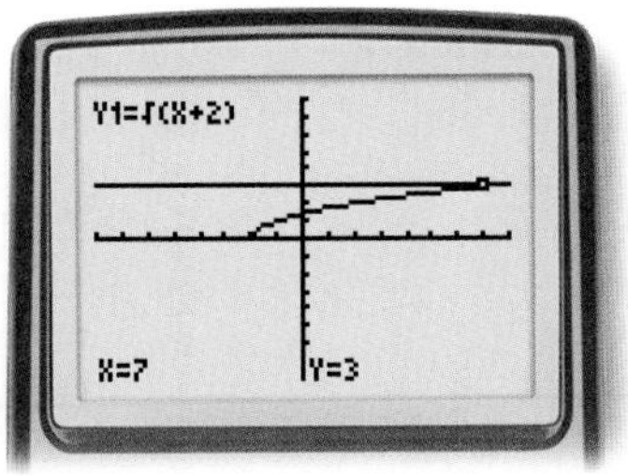


## EXAMPLE 3 Solve $\sqrt{x+6} = x$.

**SOLUTION**

Notice that the variable appears on both sides of the equation.

$$\sqrt{x+6} = x$$
$$\left(\sqrt{x+6}\right)^2 = x^2 \quad \textit{Square both sides of the equation.}$$
$$x+6 = x^2$$
$$0 = x^2 - x - 6 \quad \textit{Form a quadratic equation equal to 0.}$$
$$0 = (x-3)(x+2) \quad \textit{Factor.}$$
$$x-3=0 \quad \text{or} \quad x+2=0 \quad \textit{Set each factor equal to zero.}$$

The possible solutions are 3 and –2. You must check each possible solution by substituting it into the original equation to determine whether it is true or an *extraneous* solution.

**CHECK**

For $x = 3$, $\sqrt{3+6} = \sqrt{9} = 3$, so 3 is a solution.

For $x = -2$, $\sqrt{-2+6} = \sqrt{4} \neq -2$. Because $\sqrt{4}$ is defined as the positive square root of 4, –2 does not satisfy the original equation. It is an **extraneous solution.**

**TRY THIS** Solve $\sqrt{3x+4} = x$ for $x$, and check. Verify by graphing.

You can solve the equation $x^2 = 225$ by using the definition on page 577. Recall that the square root may have two solutions because the value of $x$ can be positive or negative.

$$x^2 = 225$$
$$x = \pm\sqrt{225}$$
$$x = \pm 15$$

## EXAMPLE 4 Solve each equation for $x$.

**a.** $x^2 = 150$ **b.** $x^2 = 4^2 + 3^2$ **c.** $x^2 = y^2 + z^2$

**SOLUTION**

**a.**
$$x^2 = 150$$
$$x = \pm\sqrt{150} = \pm\sqrt{25 \cdot 6} = \pm 5\sqrt{6}$$
$$x = 5\sqrt{6} \quad \text{or} \quad x = -5\sqrt{6}$$

**b.**

$$x^2 = 4^2 + 3^2$$
$$x = \pm\sqrt{4^2 + 3^2} = \pm\sqrt{25} = \pm 5$$
$$x = 5 \quad \text{or} \quad x = -5$$

**c.** For an equation with variables, follow the same procedure that you use with numbers. Remember to include everything on the right side of the equation under one radical. Consider both positive and negative values as possible solutions.

$$x^2 = y^2 + z^2$$
$$x = \pm\sqrt{y^2 + z^2}$$
$$x = \sqrt{y^2 + z^2} \quad \text{or} \quad x = -\sqrt{y^2 + z^2}$$

**TRY THIS** Solve the following equations for $x$:

**a.** $x^2 = 200$ **b.** $x^2 = 5^2 + 12^2$ **c.** $x^2 = y^2 + z^3$

## EXAMPLE 5

The area of a circular flower garden is 23 square yards. **What is the radius of the garden?**

### SOLUTION

The equation for the area of a circle is $A = \pi r^2$. Substitute 23 for $A$, and solve for $r$.

Since $\pi r^2 = 23$ and $r^2 = \frac{23}{\pi}$,

$$r = \pm\sqrt{\frac{23}{\pi}} \approx \pm\sqrt{\frac{23}{3.14}} \approx \pm\sqrt{7.325} \approx \pm 2.706.$$

Only the positive square root is meaningful in this case. The radius is approximately 2.7 yards.

## EXAMPLE 6

**Solve the equation $x^2 - 8x + 16 = 36$ by using radicals.**

### SOLUTION

The trinomial $x^2 - 8x + 16$ is a perfect square. When it is factored, it can be written as $(x - 4)^2$. Write the equation in this form, and solve.

$$(x - 4)^2 = 36$$
$$x - 4 = \pm\sqrt{36}$$
$$x - 4 = \pm 6$$
$$x = 10 \quad \text{or} \quad x = -2$$

Since $x$ can be any number, you must consider both positive and negative values.

Check by substitution to confirm the values are actual solutions to the original equation.

**TRY THIS** Solve the equation $x^2 + 10x + 25 = 49$ by using radicals.

# Exercises

## Communicate

1. State the domain of the function $y = \sqrt{x}$. Explain the reason for the restriction on the domain values.
2. Discuss which values of $a$ make $\sqrt{a^2} = a$ true.
3. Explain why it is helpful to know the general form of an equation like the one that models the motion of a pendulum.
4. Explain why it is important to check all solutions found for an equation such as $\sqrt{6x+7} = x$.

## Guided Skills Practice

*The regular rhythm of a pendulum helps some clocks to keep time.*

**Find the time in seconds it takes a pendulum with the given length to make one complete swing. Round your answers to the nearest hundredth.** ***(EXAMPLE 1)***

5. 60 cm
6. 150 cm
7. 100 cm
8. 45 cm

**Find the length in centimeters of a pendulum that takes the given amount of time to make one complete swing. Round your answers to the nearest hundredth.** ***(EXAMPLE 1)***

9. 2 seconds
10. 2.5 seconds
11. 3.5 seconds
12. 4 seconds

**Solve each equation.** ***(EXAMPLE 2)***

13. $\sqrt{x-2} = 3$
14. $\sqrt{x+3} = 1$
15. $\sqrt{x-1} = 2$
16. $\sqrt{x+5} = 3$

**Solve each equation, if possible. If there is no solution, write *no solution*.** ***(EXAMPLE 3)***

17. $\sqrt{x-1} = x$
18. $\sqrt{4x-4} = x$
19. $\sqrt{3x-2} = x$
20. $\sqrt{2x+24} = x$

**Solve each equation for *x*. Give answers in simplified form.** ***(EXAMPLE 4)***

21. $x^2 = 800$
22. $x^2 = 250$
23. $x^2 = 8^2 + 6^2$
24. $x^2 = v^2 - t^2$
25. Find the radius of a circular flower garden that has an area of 18 square yards. ***(EXAMPLE 5)***

**Solve each equation by using radicals.** ***(EXAMPLE 6)***

26. $x^2 - 8x + 16 = 25$
27. $x^2 + 8x + 16 = 4$
28. $x^2 + 4x + 4 = 16$

## Practice and Apply

**internet connect**

**Homework Help Online**
Go To: **go.hrw.com**
Keyword:
**MA1 Homework Help** for Exercises 29-43

**Solve each equation algebraically. Be sure to check your solution(s).**

**29.** $\sqrt{x-5}=2$

**30.** $\sqrt{x+7}=5$

**31.** $\sqrt{2x}=6$

**32.** $\sqrt{10-x}=3$

**33.** $\sqrt{2x+9}=7$

**34.** $\sqrt{2x-1}=4$

**35.** $\sqrt{x+2}=x$

**36.** $\sqrt{6-x}=x$

**37.** $\sqrt{5x-6}=x$

**38.** $\sqrt{x-1}=x-7$

**39.** $\sqrt{x+3}=x+1$

**40.** $\sqrt{2x+6}=x-1$

**41.** $\sqrt{x^2+3x-6}=x$

**42.** $\sqrt{x-1}=x-1$

**43.** $\sqrt{x^2+5x+11}=x+3$

**Solve each equation and write the solution(s) in simplest form.**

**44.** $x^2=90$

**45.** $4x^2=7$

**46.** $3x^2-27=0$

**47.** $2x^2=48$

**48.** $x^2-8x+16=0$

**49.** $x^2-12x+36=0$

**Solve each equation, if possible. If not possible, explain why.**

**50.** $x^2=9$

**51.** $\sqrt{x}=9$

**52.** $|x|=9$

**53.** $x^2=-9$

**Graph each side of the equation on the same coordinate plane. Solve the equations that have solutions by finding the points of intersection of the graphs.**

**54.** $\sqrt{x+12}=x$

**55.** $\sqrt{x-2}=x$

**56.** $\sqrt{x}=\frac{1}{3}x+\frac{2}{3}$

**57.** Find three ordered pairs for each function. Plot the ordered pairs for each pair of functions on the same coordinate axes. Sketch the graphs, and state the domain and range of each pair of graphs.

**a.** $y=\sqrt{4-x}$ and $y=-\sqrt{4-x}$ **b.** $y=\sqrt{-x}+3$ and $y=-\sqrt{-x}+3$

**PHYSICS** The motion of a pendulum can be modeled by $t=2\pi\sqrt{\frac{l}{32}}$, where $l$ is the length of the pendulum in feet and $t$ is the number of seconds required for one complete swing. ($g$ is 32 feet per second squared.)

**58.** The mechanism of a grandfather clock is based on the motion of a pendulum. About how long should a pendulum be so that the time required for one complete swing is 1 second?

**59.** If the time required for one complete swing is doubled, by what value is the length multiplied?

**60.** The time required for one complete swing of a pendulum is multiplied by $c$. By what value is the corresponding length multiplied? Generalize.

**61.** **PHYSICS** The formula for kinetic energy is $E=\frac{1}{2}mv^2$, where E is the kinetic energy in joules, $m$ is the mass of the object in kilograms, and $v$ is the velocity of the object in meters per second.

**a.** Solve the formula for $v$. Write in simplest radical form.

**b.** If a baseball with a mass of 0.14 kilograms has 50 joules of kinetic energy, what is its velocity?

## Look Back

**62.** Solve $-2x + 3 < 11$. ***(LESSON 6.2)***

**63.** Factor $4x^4 - 16y^4$ completely. ***(LESSON 9.6)***

**64.** What values of the discriminant indicate that a quadratic equation has no real solution? Give an example. ***(LESSON 10.5)***

**Solve and graph each inequality on a number line.** ***(LESSON 10.6)***

**65.** $x^2 - x - 12 > 0$ **66.** $y^2 + 5y + 6 < 0$ **67.** $a^2 - 6a - 16 < 0$

## Look Beyond

internet connect
**Activities Online**
Go To: **go.hrw.com**
Keyword:
**MA1 Planets**

**68.** What is the midpoint of the segment connecting the points (3, 2) and (9, 8)? (Hint: It may be helpful to accurately plot the points and connect them with a line.)

**Graph each pair of functions on the same set of axes. Describe the relationship between the two graphs.**

**69.** $y = x + 3$
$y = (x - 2) + 3$

**70.** $y = x^2 + 3$
$y = (x - 2)^2 + 3$

**71.** $y = |x| + 3$
$y = |x - 2| + 3$

## PORTFOLIO ACTIVITY

Include this activity in your portfolio.

Ideally, a graph of the length of a pendulum versus the time for one complete swing of the pendulum will be a transformation of the graph of the square-root parent function, $y = \sqrt{x}$, where $x$ is the length of the string and $y$ is the time for one complete swing of the pendulum.

1. Tie a piece of string to an object, such as a washer, and tie the other end to a fixed point where the washer can swing freely. Measure and record the length of the string (in centimeters) and time (in seconds) for ten complete swings of the pendulum. Divide the total time by 10 to get the average time of one swing.
2. Repeat Step 1 for six different string lengths.
3. Using the horizontal axis for the length of the string, $l$, and the vertical axis for the time for one complete swing, $t$, plot your data.
4. Does your graph appear to be a transformation of a square-root function?
5. The function that models the motion of a pendulum is $t = 2\pi\sqrt{\frac{l}{980}}$. Does this function model your data? If not, what are some possible reasons for the discrepancy?

12.3

# The Pythagorean Theorem

## Objectives

- Find a side length of a right triangle given the lengths of its other two sides.
- Apply the Pythagorean Theorem to real-world problems.

**Why** ***Just as the Parthenon is a familiar and significant landmark, the Pythagorean Theorem is one of the most familiar and significant theorems in mathematics. To apply the theorem, you will need to use your knowledge of squares and square roots.***

*The Babylonian tablet lists numbers that show that the Babylonians knew about the relationship among the sides of right triangles many years before Pythagoras.*

Recall from your earlier mathematics classes that a *right triangle* is one that has one 90° (or *right*) angle. The famous relationship among the sides of right triangles that you will study in this lesson has long been associated with the ancient Greeks. It is now known, however, that knowledge of this relationship goes back much further than the Greeks.

The conventional labeling of the sides and angles of a right triangle is shown at right.

The angles are labeled with capital letters. The right angle is commonly labeled $C$.

*1. The side opposite the right angle is the hypotenuse.*

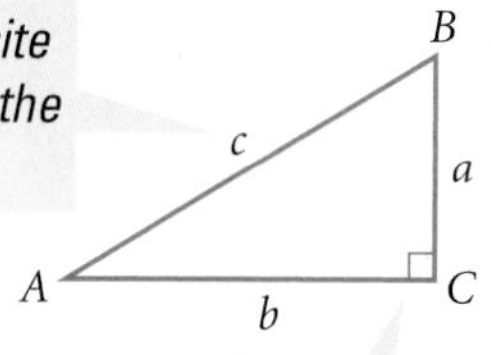


*2. The 90° angle is indicated by a square sign.*

The side opposite an angle is labeled in lowercase with the same letter as the angle or in segment notation. In the diagram, side $c$, $\overline{AB}$, is the **hypotenuse**. Sides $a$, $\overline{BC}$, and $b$, $\overline{AC}$, are called **legs**.

The ancient Babylonians, Chinese, and Hindus all knew of the right-triangle relationship that is known as the Pythagorean Theorem. European culture gave this name to the relationship in honor of Pythagoras, who was a famous Greek philosopher and mathematician of the sixth century B.C.E.

## The Pythagorean Theorem

If given a right triangle with legs of length $a$ and $b$ and hypotenuse of length $c$, then $a^2 + b^2 = c^2$.

*Activity*

## Exploring Right Triangles

**You will need:** large-grid graph paper or calculator

1. On a piece of graph paper, draw right triangles with the side lengths given in the table. (Make the legs, $a$ and $b$, of the triangle horizontal and vertical. You can use a strip of graph paper as a ruler to measure the hypotenuse, $c$.)

| Side $a$ | Side $b$ | Side $c$ | $a^2$ | $b^2$ | $c^2$ | $a^2+b^2$ |
|---|---|---|---|---|---|---|
| 3 | 4 | 5 | ? | ? | ? | ? |
| 5 | 12 | 13 | ? | ? | ? | ? |
| 16 | 30 | 34 | ? | ? | ? | ? |
| 13 | 84 | 85 | ? | ? | ? | ? |

CHECKPOINT ✓

2. Compare the entries under $c^2$ with those under $a^2 + b^2$. Explain how the relationships between them support the theorem given on the previous page.
3. Do you think that the theorem works for triangles other than right triangles? Test this idea by drawing and measuring some nonright triangles.
4. The converse of a theorem stated in if-then form is a new statement in which the phrases following the words *if* and *then* are interchanged. Write the converse of the Pythagorean theorem, which should begin with "If, in a triangle with sides $a$, $b$, and $c$, $c^2 = a^2 + b^2$, then . . . " Based on Step 3, do you think the converse of the Pythagorean Theorem is true?

CONNECTION
GEOMETRY

Over the years, illustrations like the one at right have been used to show that the sum of the squares of the legs of a right triangle is equal to the square of the hypotenuse. Notice that the combined areas of the red and blue squares are equal to the area of the purple square.

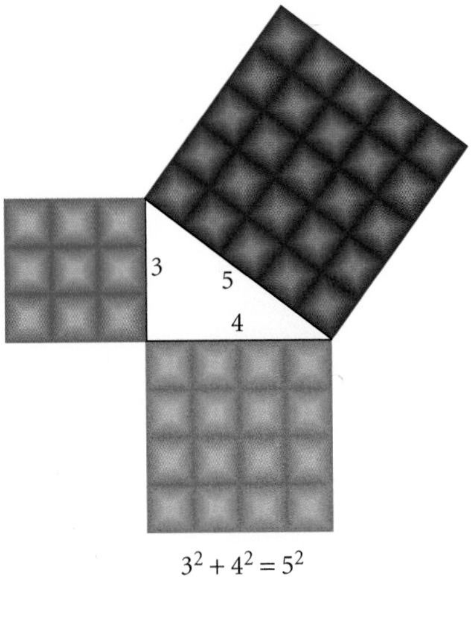

$3^2 + 4^2 = 5^2$

**EXAMPLE 1** **Find the length of sides $x$, $y$, and $z$.**

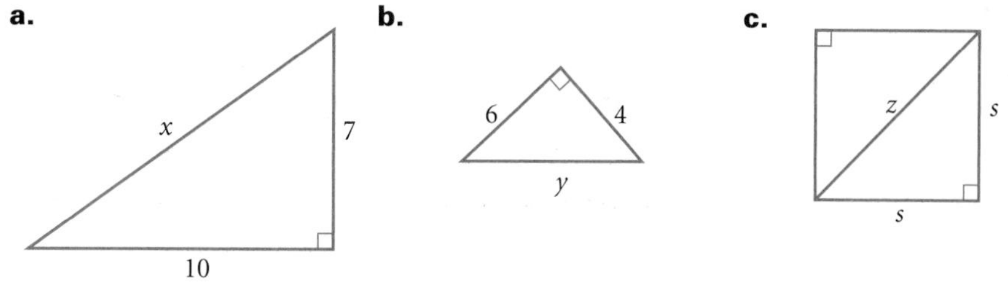


**SOLUTION**

**a.** Since $x$ is the side opposite the right angle, $x$ is the hypotenuse.

$$x^2 = 7^2 + 10^2$$
$$x^2 = 149$$
$$x = \sqrt{149} \approx 12.21$$

Since $x$ represents the length of a side, it must be positive.

**b.** The lengths of the legs are 4 and 6.

$$y^2 = 4^2 + 6^2$$
$$y^2 = 52$$
$$y = \sqrt{52} \approx 7.21$$

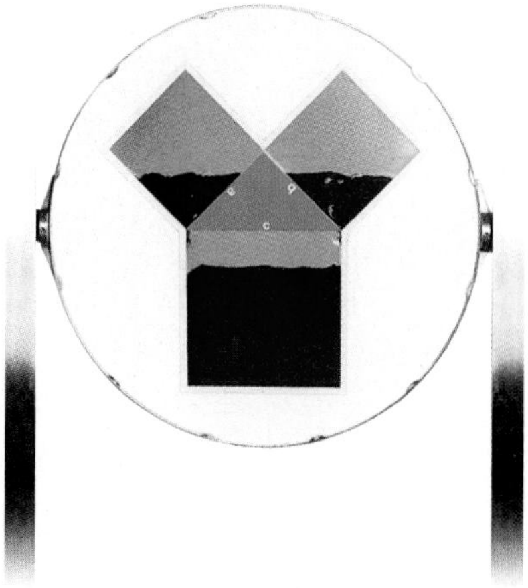

*Colored water is contained in the squares attached to the sides of a right triangle. As the device rotates, the water flows from the two squares opposite the hypotenuse, filling the square on the hypotenuse.*

**c.** The diagonal of the square is the hypotenuse of a right triangle. Each leg has length $s$.

$$z^2 = s^2 + s^2$$
$$z^2 = 2s^2$$
$$z = \sqrt{2s^2}$$
$$z = s\sqrt{2}$$

Thus, the formula for the diagonal, $d$, of a square with side length $s$ is $d = s\sqrt{2}$.

**TRY THIS** Find the length of sides $x$, $y$, and $z$.

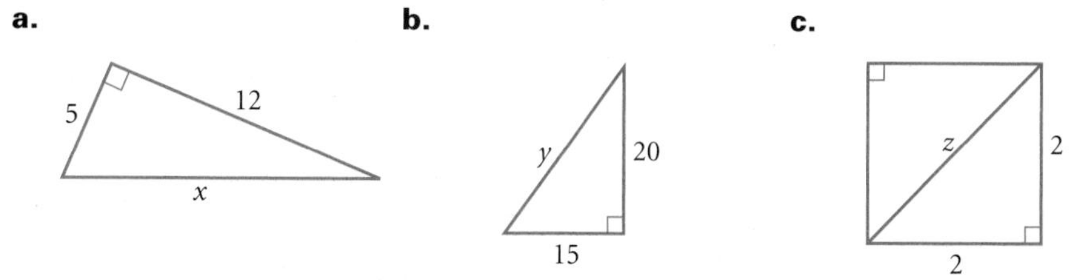


**CRITICAL THINKING** You have seen that the Pythagorean Theorem can be used to find the length of the hypotenuse of a right triangle. Could it also be used to find the length of a leg? What information would be necessary?

**EXAMPLE 2**

**APPLICATION**
**RECREATION**

Your boat is traveling due north at 20 miles per hour. A friend's boat left at the same time from the same location and headed due west at 15 miles per hour. After an hour you get a call from your friend who tells you that he has just stopped because of engine trouble. **How far must you travel to reach your friend?**

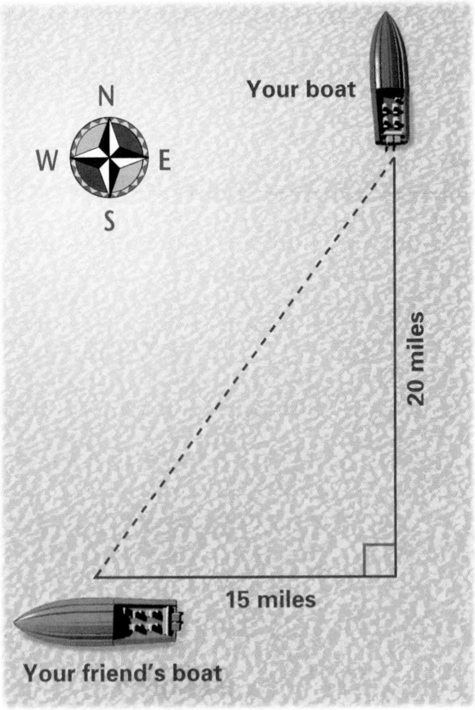


**SOLUTION**

The legs of the triangle in the diagram are 15 miles and 20 miles long. Apply the Pythagorean Theorem to the triangle.

$$c^2 = a^2 + b^2$$
$$c^2 = 15^2 + 20^2$$
$$c^2 = 625$$
$$c = \sqrt{625} = 25$$

Since the answer represents a distance, only the positive root is used. You will need to travel 25 miles.

**TRY THIS** Find the hypotenuse of a right triangle whose legs are 36 and 48 units long.

**CRITICAL THINKING** What do you notice about multiples of the side lengths of a 3-4-5 right triangle (lengths of 6, 8, and 10, for example)?

**EXAMPLE 3**

The height of a pyramid with a square base can be found from the measurements of the base and the slant height. **Find the height of a pyramid whose square base measures 40 meters on each side and whose slant height is 52 meters.**

**CONNECTION**
**GEOMETRY**

**SOLUTION**

In the diagram below, notice that the measure of leg $\overline{AC}$ of the triangle is equal to half the length of a side of the base.

Apply the Pythagorean Theorem to $\triangle ABC$ with $AB = 52$ and $AC = \frac{1}{2} \cdot 40 = 20$.

$$52^2 = BC^2 + 20^2$$
$$BC^2 = 52^2 - 20^2$$
$$BC^2 = 2304$$
$$BC = \sqrt{2304} = 48$$

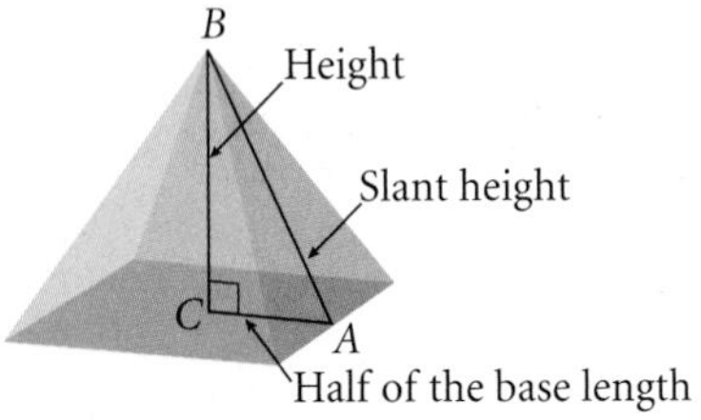


The height of the pyramid is 48 meters.

**TRY THIS** Find the height of a pyramid whose square base measures 36 meters on each side and whose slant height is 60 meters.

# Exercises

*Pythagoras*
*580-500 B.C.E.*

## Communicate

1. Identify the parts of a right triangle.
2. State the Pythagorean Theorem in your own words.
3. Explain how to find the hypotenuse of a right triangle if you know the lengths of the legs.
4. Describe how to rewrite the basic equation in the Pythagorean Theorem to determine the length of one of the legs.
5. Explain how the Pythagorean Theorem can be used to find the length of the diagonal of a square when given the length of its sides.

## Guided Skills Practice

**Find the length of sides *x*, *y*, and *z* in simplified form.** ***(EXAMPLE 1)***

**6.**

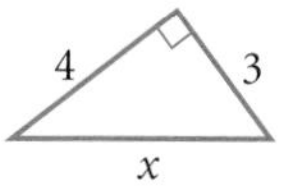


**7.**

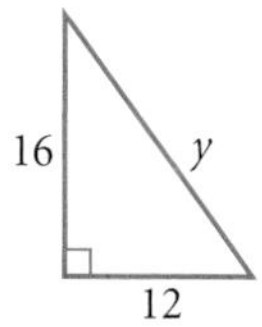


**8.**

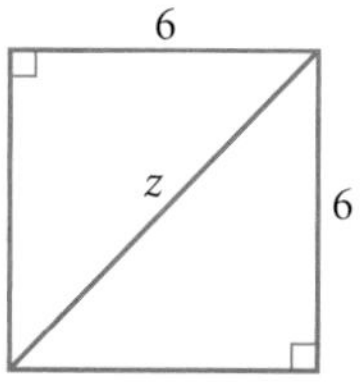


9. Calculate the length of the hypotenuse of a right triangle in simplified form whose legs are 6 and 10 units long. ***(EXAMPLE 2)***
10. Determine the height of a pyramid whose square base measures 30 meters on each side and whose slant height is 25 meters. ***(EXAMPLE 3)***

## Practice and Apply

**Identify the hypotenuse and legs in each right triangle, and label them appropriately.**

**11.**

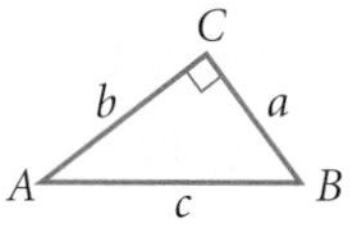


**12.**

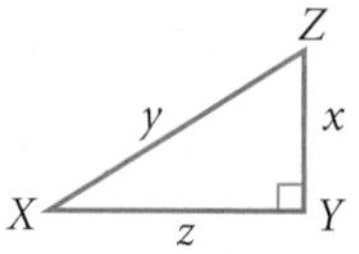


**13.**

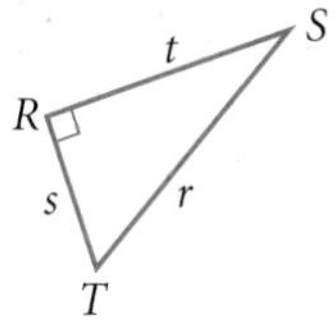



**internet connect**
**Homework Help Online**
Go To: **go.hrw.com**
Keyword:
**MA1 Homework Help**
for Exercises 14-21


**For Exercises 14–21, the tables give the lengths of various sides of right triangles. Copy and complete the tables. Use a calculator and round each answer to the nearest tenth.**

| | Leg | Leg | Hypotenuse |
|---|---|---|---|
| **14.** | 24 | 45 | ? |
| **16.** | ? | 8 | 17 |
| **18.** | 12 | ? | 25 |
| **20.** | ? | 40 | 50 |

| | Leg | Leg | Hypotenuse |
|---|---|---|---|
| **15.** | 10 | ? | 26 |
| **17.** | 5 | 9 | ? |
| **19.** | ? | 6 | 25 |
| **21.** | 0.75 | ? | 1.25 |

**Decide whether each set of numbers can represent the side lengths of a right triangle.**

**22.** 3, 9, 7 **23.** 10, 6, 8 **24.** 5, 9, 11

**25.** $\sqrt{5}, \sqrt{6}, \sqrt{11}$ **26.** $4, 4\sqrt{3}, 8$ **27.** $\sqrt{3}, \sqrt{4}, \sqrt{5}$

**CONNECTION**

**28. GEOMETRY** Recall from your previous math classes that the *altitude* of a triangle forms a 90° angle with the base. The altitude of an *equilateral* triangle divides the triangle into 2 right triangles of the same size and shape. In the triangle at right, *a* is the altitude.

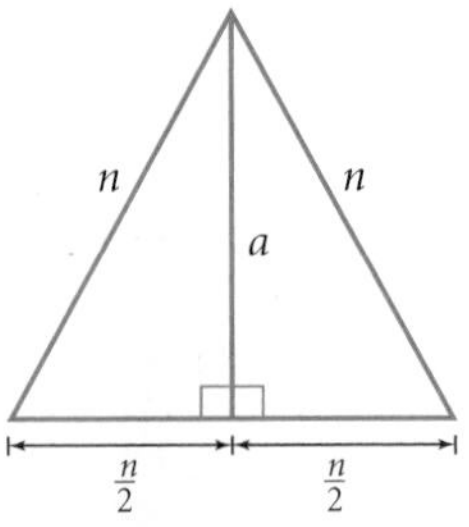


**a.** Find the length of the altitude, *a*, in terms of the length of the sides, *n*.

**b.** Copy and complete the table below. Simplify each answer.

| Equilateral Triangles | | | | | | | | |
|---|---|---|---|---|---|---|---|---|
| Side | 4 | 6 | 8 | 10 | 12 | 20 | $n$ | ? |
| Half of a side | 2 | ? | ? | ? | ? | ? | ? | ? |
| Altitude | ? | ? | ? | ? | ? | ? | ? | $17\sqrt{3}$ |

**APPLICATIONS**

**LANDSCAPING** A garden in the shape of a right triangle is represented by the figure at right.

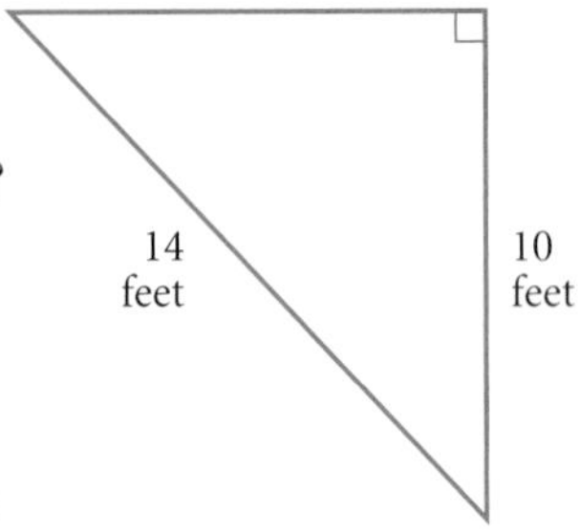


**29.** What is the length of the third side of the garden?

**30.** How many feet of fencing must be bought to enclose the garden? (Assume that fencing is sold by the foot.)

**31.** If the cost of fencing is $4.98 per foot, how much will it cost to enclose the garden?

**32. SPORTS** A baseball diamond is a square with sides of 90 feet. To the nearest foot, how long is a throw from third base to first base?

## Look Back

APPLICATION

**INCOME** A student is paid a weekly salary of \$20 plus \$2.50 for every lamp that he sells. Use this information in Exercises 33–36. *(LESSON 1.2)*

**33.** Copy and complete the table.

| Number of lamps sold | 0 | 10 | 20 | 30 | 40 | 50 |
|---|---|---|---|---|---|---|
| Week's pay | **a.** ? | **b.** ? | **c.** ? | **d.** ? | **e.** ? | **f.** ? |

**34.** Write an equation that describes $w$, one week's pay, in terms of $l$, the number of lamps sold.

**35.** If the student sells 29 lamps in a given week, what will his week's pay be?

**36.** If the student needs to earn \$149 in a given week, how many lamps must he sell?

**Decide whether the given pairs of lines are parallel, perpendicular, or neither.** *(LESSON 5.6)*

**37.** $y = \frac{1}{2}x + 3$
$y = 4x + 3$

**38.** $y = 3x - 4$
$y = -\frac{1}{3}x + 2$

**39.** $-2x + y = 8$
$-6x + 3y = 15$

## Look Beyond

CONNECTION

**GEOMETRY** A **Pythagorean triple** is a set of three positive integers that satisfy the Pythagorean Theorem. For example, {3, 4, 5} is a Pythagorean triple. The three expressions listed below can be used to create Pythagorean triples in a three-step process.

If $p$ and $q$ are integers and $p > q > 0$, then the following expressions give the elements of a Pythagorean triple:

$$\{p^2 - q^2,\ 2pq,\ p^2 + q^2\}$$

For $p = 2$ and $q = 1$:

$$p^2 - q^2 = 2^2 - 1^2 = 3$$
$$2pq = 2 \cdot 2 \cdot 1 = 4$$
$$p^2 + q^2 = 2^2 + 1^2 = 5$$

Thus, {3, 4, 5} is a Pythagorean triple.

**40.** Using the expressions above, make a list of at least 6 different Pythagorean triples.

**41.** Explain why the three expressions give infinitely many Pythagorean triples.

# 12.4 The Distance Formula

## Objectives

- Use the distance formula to find the distance between two points in a coordinate plane.
- Determine whether a triangle is a right triangle.
- Apply the midpoint formula.

**Why** ***In an emergency situation, rescue personnel must be able to quickly determine how long it will take to reach an accident scene. The ability to calculate the distance between two points is of critical importance not only to pilots, but also to sailors, surveyors, architects, astronomers, and drafting professionals.***

**APPLICATION**
**EMERGENCY SERVICES**

A medical center learns of an auto accident located 7 miles east and 11 miles north of the center. If the medical center's helicopter is now located 1 mile east and 3 miles north of the center, what is the shortest distance the helicopter can travel to get to the scene of the accident?

The *distance* between two points, $A$ and $B$, is the *length* of $\overline{AB}$, which is the segment with $A$ and $B$ as endpoints.

The shortest distance between the helicopter and the accident is $HA$, which is the length of the hypotenuse of right triangle $HVA$. $\triangle HVA$ can be visualized by picturing $\overline{HV}$ as an east-west segment and $\overline{VA}$ as a north-south segment.

CONNECTION
GEOMETRY

You can find the length of $\overline{HV}$ by subtracting the first coordinates of V and H.

$$\text{length of } \overline{HV} = 7 - 1 = 6$$

You can find the length of $\overline{AV}$ by subtracting the second coordinates of *A* and *V*.

$$\text{length of } \overline{AV} = 11 - 3 = 8$$

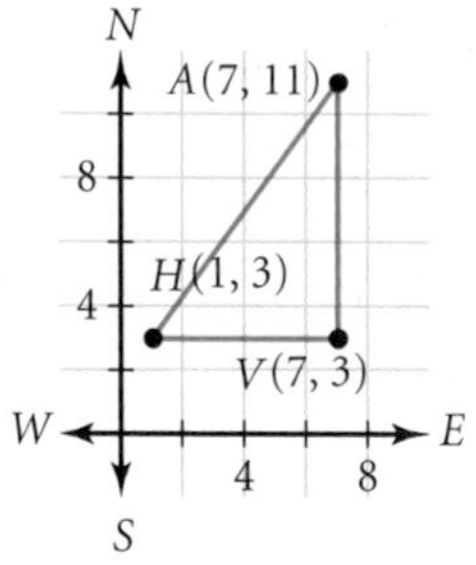


Now use the Pythagorean Theorem.

$$c^2 = a^2 + b^2$$
$$(\text{length of } \overline{AH})^2 = 6^2 + 8^2$$
$$(\text{length of } \overline{AH})^2 = 100$$
$$\text{length of } \overline{AH} = \sqrt{100} = 10$$

The shortest distance that the medical center's helicopter can travel is 10 miles.

**Distance Formula**

The distance, *d*, between two points $A(x_1, y_1)$ and $B(x_2, y_2)$ is given by

$$d = \sqrt{(x_2 - x_1)^2 + (y_2 - y_1)^2}.$$

EXAMPLE 1 **Find the distance between each pair of points. Round answers to the nearest hundredth.**

**a.** (2, –3) and (8, 2)

**b.** (–4, –3) and (2, 0)

**SOLUTION**

Use the distance formula.

**a.**
$$d = \sqrt{(x_2 - x_1)^2 + (y_2 - y_1)^2}$$
$$d = \sqrt{(8 - 2)^2 + [2 - (-3)]^2}$$
$$d = \sqrt{6^2 + 5^2}$$
$$d = \sqrt{36 + 25}$$
$$d = \sqrt{61}$$
$$d \approx 7.81$$

**b.**
$$d = \sqrt{(x_2 - x_1)^2 + (y_2 - y_1)^2}$$
$$d = \sqrt{[2 - (-4)]^2 + [0 - (-3)]^2}$$
$$d = \sqrt{6^2 + 3^2}$$
$$d = \sqrt{36 + 9}$$
$$d = \sqrt{45}$$
$$d \approx 6.71$$

**TRY THIS** Find the distance between (4, 5) and (–5, 7).

## The Converse of the Pythagorean Theorem

Recall from Lesson 11.6 that when the *if* and *then* portions of a theorem are interchanged, the new statement is called the *converse* of the given theorem.

**Converse of the Pythagorean Theorem**

If $a$, $b$, and $c$ are the lengths of the sides of a triangle and $a^2 + b^2 = c^2$, where $c > a$ and $c > b$, then the triangle is a right triangle.

**CRITICAL THINKING** Is the converse of an if-then statement always true? If not, give an example.

**EXAMPLE 2** **Given vertices $P(1, 2)$, $Q(3, -1)$, and $R(-5, -2)$, determine whether triangle $PQR$ is a right triangle.**

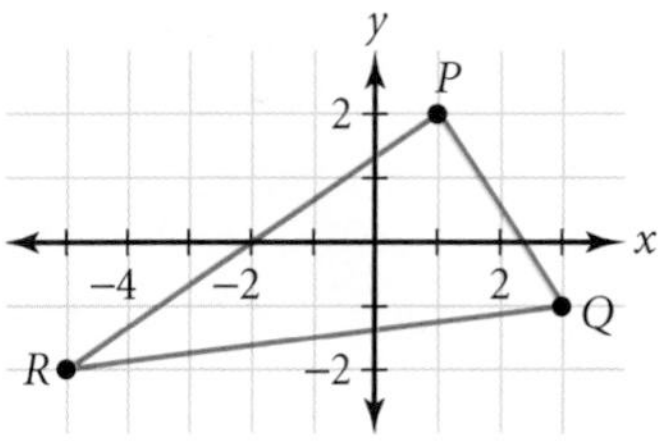


**SOLUTION**

Draw triangle $PQR$ on graph paper.

Use the distance formula to find the lengths of the sides.

$$PR = \sqrt{(-5-1)^2 + (-2-2)^2} = \sqrt{(-6)^2 + (-4)^2} = \sqrt{52}$$

$$PQ = \sqrt{(3-1)^2 + (-1-2)^2} = \sqrt{2^2 + (-3)^2} = \sqrt{13}$$

$$RQ = \sqrt{[3-(-5)]^2 + [-1-(-2)]^2} = \sqrt{8^2 + 1^2} = \sqrt{65}$$

Now use the converse of the Pythagorean Theorem to test whether the three lengths can be the sides of a right triangle.

Is $(\sqrt{52})^2 + (\sqrt{13})^2 = (\sqrt{65})^2$ true?

Since $52 + 13 = 65$, the triangle is a right triangle.

**TRY THIS** Given vertices $P(-1, -2)$, $Q(5, 2)$, and $R(-3, 1)$, determine whether triangle $PQR$ is a right triangle.

**CRITICAL THINKING** How can you use the slopes of $PR$ and $PQ$ to show that triangle $PQR$ in Example 2 is a right triangle?

## The Midpoint Formula

*Activity*

### Deriving the Midpoint Formula

**You will need:** graph paper

1. Plot the following pairs of points on grid paper. Guess the coordinates of the midpoint, $M$, of $\overline{PQ}$ for each pair.
   **a.** $P(1, 6)$, $Q(7, 10)$ **b.** $P(-2, 3)$, $Q(6, 7)$ **c.** $P(-4, -1)$, $Q(8, 7)$
2. Use the distance formula to verify that your proposed midpoints are the midpoints of the given segments.
3. What rule can be applied to the $x$-coordinates of the endpoints of a segment to obtain the $x$-coordinate of the midpoint of the segment?
4. What rule can be applied to the $y$-coordinates of the endpoints of a segment to obtain the $y$-coordinate of the midpoint of the segment?

CHECKPOINT ✔

5. Apply your rules from Steps 3 and 4 to calculate the coordinates of the midpoint of $\overline{AB}$, for points $A(12, 63)$ and $B(43, 20)$.
6. Use the distance formula to verify your answer.

**Midpoint Formula**

Given $\overline{PQ}$, where $P(x_1, y_1)$ and $Q(x_2, y_2)$, the coordinates of the midpoint, $M$, of $\overline{PQ}$, are $\left(\frac{x_1 + x_2}{2}, \frac{y_1 + y_2}{2}\right)$.

**EXAMPLE 3** The streets of a certain city are laid out like a coordinate plane, with City Hall at the origin. Jacques lives 3 blocks east and 2 blocks north of City Hall. His friend Alise lives 11 blocks east and 8 blocks north of City Hall. They want to meet at the point midway between their homes. **At what point should they meet?**

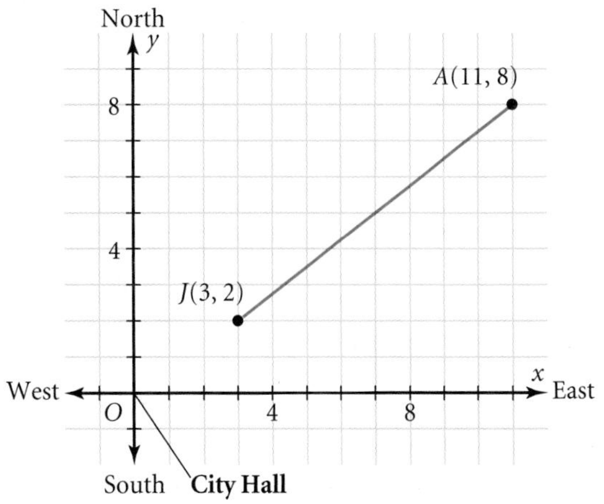

**SOLUTION**

Plot $J(3, 2)$ and $A(11, 8)$ on a coordinate plane, as shown in the figure. To find the point midway between $J$ and $A$, use the midpoint formula, which gives the "average" coordinates, written as $(\overline{x}, \overline{y})$.

$$\overline{x} = \frac{x_1 + x_2}{2} = \frac{3 + 11}{2} = \frac{14}{2} = 7$$

$$\overline{y} = \frac{y_1 + y_2}{2} = \frac{2 + 8}{2} = \frac{10}{2} = 5$$

They should meet at $M(7, 5)$, the midpoint of $\overline{JA}$.

**EXAMPLE 4** The center of a circle is $M(3, 4)$. **If one endpoint of a diameter is $A(-4, 6)$, what is the other endpoint, $B(x_2, y_2)$?**

CONNECTION
GEOMETRY

**SOLUTION**

Substitute the coordinates of $M$ and $A$ into the midpoint formula, and solve.

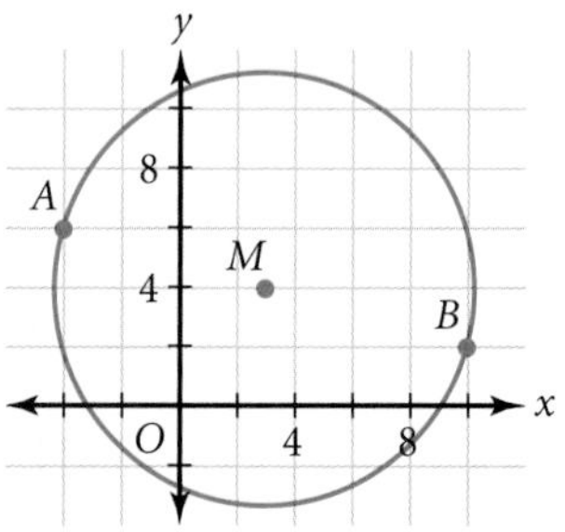


$$\frac{x_1 + x_2}{2} = \overline{x} \qquad \frac{y_1 + y_2}{2} = \overline{y}$$

$$\frac{-4 + x_2}{2} = 3 \qquad \frac{6 + y_2}{2} = 4$$

$$-4 + x_2 = 6 \qquad 6 + y_2 = 8$$

$$x_2 = 10 \qquad y_2 = 2$$

The other endpoint is $B(10, 2)$.

**TRY THIS** If the center of a circle is $M(-1, -2)$ and one endpoint of a diameter is $A(4, 3)$, what is the other endpoint, $B(x_2, y_2)$?

# Exercises

## Communicate

1. Explain, in your own words, the difference between points and coordinates.
2. Discuss the relationship between the Pythagorean Theorem and the distance formula.
3. Explain how to find the distance between the points (2, 5) and (5, 7) in a coordinate plane.
4. Describe how to find the midpoint of the segment connecting the points from Exercise 3.
5. Explain what the converse of an if-then statement is.

## Guided Skills Practice

**Find the distance between each pair of points.** *(EXAMPLE 1)*

**6.** $A(2, 3), B(5, 7)$ **7.** $C(-5, 6), D(7, 1)$ **8.** $E(-3, -2), F(5, -8)$

**Determine whether each set of points represents the vertices of a right triangle.** *(EXAMPLE 2)*

**9.** $A(3, 6), B(1, 2), C(-3, 4)$ **10.** $D(2, 3), E(-2, 9), F(-5, 7)$

**11.** $G(4, -2), H(-1, 0), I(6, -2)$ **12.** $J(6, 4), K(3, 0), L(2, 7)$.

**What is the midpoint of a segment with the following endpoints?** *(EXAMPLE 3)*

**13.** $A(5, -1), B(-3, 7)$ **14.** $J(-3, -2), K(4, -6)$

**15.** $M(-18, 13), N(44, -13)$ **16.** $O(17, -13), P(14, 12)$

**Given the coordinates of the midpoint, *M*, and one endpoint, *E*, of a segment, find the other endpoint, $F(x_2, y_2)$.** *(EXAMPLE 4)*

**17.** $M(4, -3), E(7, -7)$ **18.** $M(-6, -2), E(4, 8)$

**19.** $M\left(\frac{3}{2}, \frac{7}{2}\right), E(4, 2)$ **20.** $M\left(\frac{9}{2}, \frac{1}{2}\right), E(-4, 3)$

## Practice and Apply

**Find the distance between each pair of points. Round answers to the nearest hundredth.**

**21.** $A(4, 7), B(1, 3)$ **22.** $P(5, 6), Q(17, 11)$ **23.** $R(-5, -2), S(-9, 3)$

**24.** $T(5, -2), U(-6, 3)$ **25.** $P(-3, 8), Q(4, 8)$ **26.** $J(5, -3), K(5, 6)$

**27.** $F(7, 2), G(-1, -1)$ **28.** $M(3, -2), N(5, 4)$ **29.** $S(5, 6), T(4, -3)$

**30.** $K(4, 2), L(7, 1)$ **31.** $C(-1, -5), D(3, -3)$ **32.** $Y(-4, 2), Z(10, 6)$

**33.** $P(15, 2), Q(-1, 2)$ **34.** $T(5, 7), U(5, -10)$ **35.** $V(4, 8), W(-7, 12)$

**For Exercises 36–41, plot $\triangle PQR$ on graph paper. Decide which of the following terms apply to $\triangle PQR$: scalene (no sides equal), isosceles (two sides equal), equilateral (three sides equal), or right.**

**36.** $P(-1, 3), Q(4, 6), R(4, 0)$ **37.** $P(-2, 2), Q(2, 5), R(8, -3)$

**38.** $P(1, -1), Q(-3, 2), R(-3, -4)$ **39.** $P(5, 2), Q(1, 1), R(6, -2)$

**40.** $P(-4, -2), Q(0, -2), R(-2, 0)$ **41.** $P(1, -4), Q(2, 1), R(8, 0)$

internet connect
Homework Help Online
Go To: go.hrw.com
Keyword:
MA1 Homework Help
for Exercises 42-49

**Plot each segment in a coordinate plane, and identify the coordinates of the midpoint of the segment.**

**42.** $\overline{AB}$, where $A(2, 5)$ and $B(9, 5)$ **43.** $\overline{EF}$, where $E(-4, -1)$ and $F(-2, -7)$

**44.** $\overline{GH}$, where $G(-5, 6)$ and $H(-1, -2)$ **45.** $\overline{IJ}$, where $I(4, 3)$ and $J(9, -4)$

**In Exercises 46–47, the locations of two points are given relative to the origin, (0, 0). Calculate the coordinates of the midpoint of the segment connecting each point and the origin.**

**46.** *A* is 2 units east and 3 units north of the origin.

**47.** *C* is 2 units west and 5 units south of origin.

48. Use the midpoint formula to determine the coordinates of the midpoint of $\overline{PQ}$, where $P(10, 7)$ and $Q(2, 3)$.

49. If the coordinates of $P$ are $(a, b)$ and the coordinates of $Q$ are $(c, d)$, what are the coordinates of the midpoint of $\overline{PQ}$?

**The midpoint of $\overline{PQ}$ is $M$. Calculate the missing coordinates.**

50. $P(2, -5)$, $Q(6, 7)$, $M(?, ?)$

51. $P(4, 8)$, $Q(?, ?)$, $M(10, 7)$

52. $P(?, ?)$, $Q(6, -2)$, $M(9, 4)$

53. $P(3, ?)$, $Q(?, 5)$, $M(2, 8)$

CONNECTION

54. **COORDINATE GEOMETRY** Draw $\overline{RS}$, where $R(2, 1)$ and $S(6, 3)$, on a coordinate plane.
   - **a.** Determine the coordinates of $M$, the midpoint of $\overline{RS}$, and plot $M$.
   - **b.** Identify the slope of $\overline{RS}$.
   - **c.** What is the slope of any line perpendicular to $\overline{RS}$?
   - **d.** Determine an equation for $k$, the line that is perpendicular to $\overline{RS}$ and contains point $M$. Draw line $k$ on your graph. Line $k$ is called the *perpendicular bisector* of $\overline{RS}$ because it is perpendicular to $\overline{RS}$ and contains the midpoint of $\overline{RS}$.
   - **e.** Choose any two points on $k$ other than $M$. Name them $A$ and $B$. Compute the distance from $A$ to $R$ and from $A$ to $S$. Compute the distance from $B$ to $R$ and from $B$ to $S$.

CHALLENGE

55. Use right triangles and the Pythagorean Theorem to show that $(4, 7)$ is the midpoint of the segment connecting the points $(1, 3)$ and $(7, 11)$.

CONNECTION

**COORDINATE GEOMETRY** For Exercises 56–58, state the coordinates of $Q$, and calculate the length of the hypotenuse of each right triangle.

56.
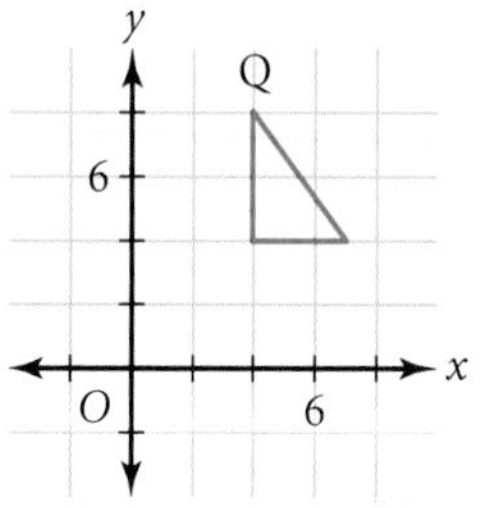


57.
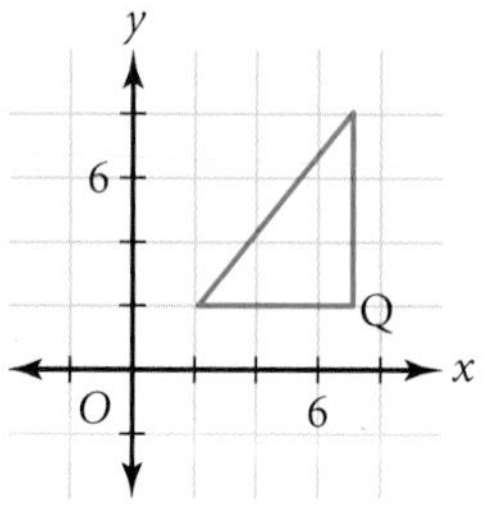


58.
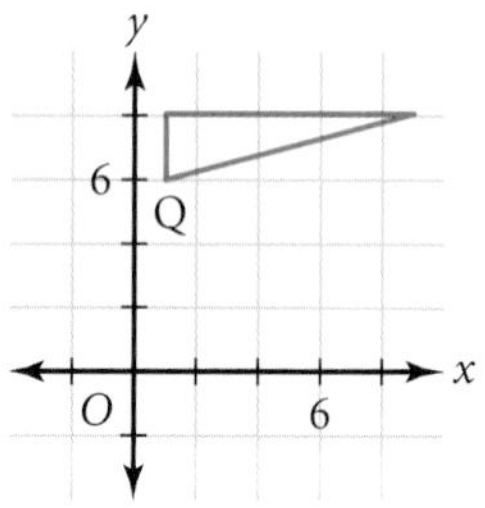


APPLICATION

59. **AVIATION** A rescue helicopter located 3 miles west and 2 miles north of a Coast Guard communications center must respond to a ship located 5 miles east and 7 miles south of the center. What distance must the helicopter travel to reach the ship?

## Look Back

**Factor each trinomial.** ***(LESSON 9.7)***

**60.** $y^2 + 35y + 300$

**61.** $x^2 + 30x + 216$

**Find the coordinates of the vertex of each parabola by inspection.** ***(LESSON 10.1)***

**62.** $y = (x - 2)^2 + 3$

**63.** $y = (x + 3)^2 - 1$

**64.** $y = -2(x - 1)^2 - 4$

**65.** $y = -(x + 4)^2 + 1$

**66.** Use algebra to find the $x$-intercept(s) of the graph of $y = x^2 - 6x - 10$. Give the exact solution(s). ***(LESSON 10.5)***

**Use properties of radicals to simplify each expression. Use a calculator to check your answers.** ***(LESSON 12.1)***

**67.** $\sqrt{3}\left(\sqrt{12} - \sqrt{75}\right)$

**68.** $\dfrac{\sqrt{75}}{\sqrt{3}}$

**69.** $\dfrac{\sqrt{36} + \sqrt{81}}{\sqrt{9}}$

## Look Beyond

**internet connect**
**Portfolio Extension**
Go To: **go.hrw.com**
Keyword:
**MA1 Theodorus**

**TECHNOLOGY** Use a scientific calculator to evaluate each expression.

**70.** $9^{\frac{1}{2}}$

**71.** $16^{\frac{1}{2}}$

**72.** $39^{\frac{1}{2}}$

**73.** $2^{\frac{1}{2}}$

## PORTFOLIO ACTIVITY

Include this activity in your portfolio.

**CULTURAL CONNECTION: AFRICA** The figure at left is known as the "Wheel of Theodorus" and is named after a Greek-speaking mathematician who lived in Alexandria, Egypt, in the mid-fifth century B.C.E.

Notice that the hypotenuse of each triangle acts as one of the legs of the next tringle in the spiral.

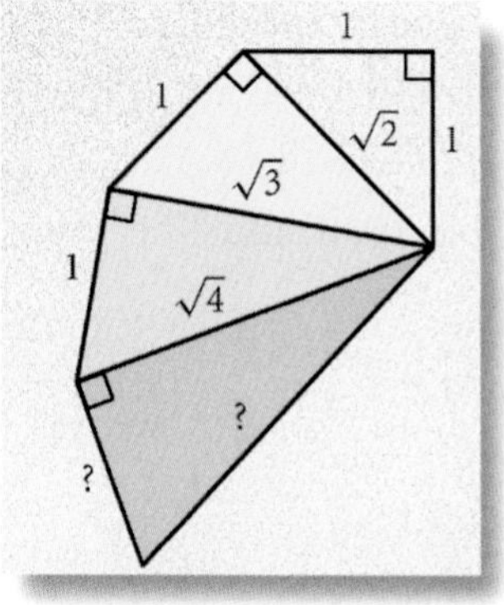


1. Continue the spiral by building each new right triangle on the hypotenuse of the previous one. Check the side lengths of each triangle with the Pythagorean Theorem. Add 4 new triangles.
2. Make a table relating the number of a given triangle in the spiral to the length of its hypotenuse.
3. Predict the length of the hypotenuse of the 10th triangle in the spiral.
4. Describe in terms of $n$ the length of the hypotenuse of the $n$th right triangle in the spiral.

**WORKING ON THE CHAPTER PROJECT**

You should now be able to complete Activities 1 and 2 of the Chapter Project beginning on page 636.

12.5

# Geometric Properties

**Why** *Studying a monument like Stonehenge can be difficult because of its size and the restrictions protecting the site. By representing objects such as Stonehenge as geometric figures on a coordinate plane, algebra can be used to study their basic geometric properties.*

### Objectives

- Define and use the equation of a circle.
- Use the coordinate plane to investigate the diagonals of a rectangle and the midsegment of a triangle.
- Use given information about a focus and directrix to write an equation describing a parabola.

## The Equation of a Circle

CONNECTION
GEOMETRY

A **circle** is the set of all points in a plane that are the same distance from a given point called the **center** of the circle. The distance from the center of the circle to any point on the circle is the **radius** of the circle.

You can represent the outer circle of stones at Stonehenge by using an equation for a circle with its center at (0, 0) and a radius of 160. You will need to use the distance formula (Lesson 12.4).

If a point $(x, y)$ is on the circle, then its distance, $d$, from (0, 0) must be 160.

$$d = \sqrt{(x-0)^2 + (y-0)^2} = 160$$
$$\sqrt{x^2 + y^2} = 160$$
$$x^2 + y^2 = 160^2$$
$$x^2 + y^2 = 25{,}600$$

The equation $x^2 + y^2 = 25{,}600$ represents all possible points on the circle. If you replace the radius, 160, in the Stonehenge example with the variable $r$, the resulting equation is the *general equation for a circle centered at the origin,* (0, 0):

$$x^2 + y^2 = r^2$$

**EXAMPLE 1** **Use the distance formula to find the equation of a circle with a radius of 3 and its center at (1, 2).**

**SOLUTION**

If a point $(x, y)$ is on the circle, then its distance, $d$, from the center, $(1, 2)$, must be 3.

$$\sqrt{(x-1)^2+(y-2)^2}=3$$
$$(x-1)^2+(y-2)^2=9$$

The method in the example above leads to the following equation:

**The Equation of a Circle**

The equation of a circle with a radius of $r$ and its center at $(h, k)$ is

$$(x-h)^2+(y-k)^2=r^2.$$

**TRY THIS** Find the equation of a circle with the given radius, $r$, and the given center, $C$.

**a.** $r = 1$, $C(0, 0)$ **b.** $r = 5$, $C(-2, 5)$

*Activity*

## The Diagonals of a Rectangle

**You will need:** graph paper

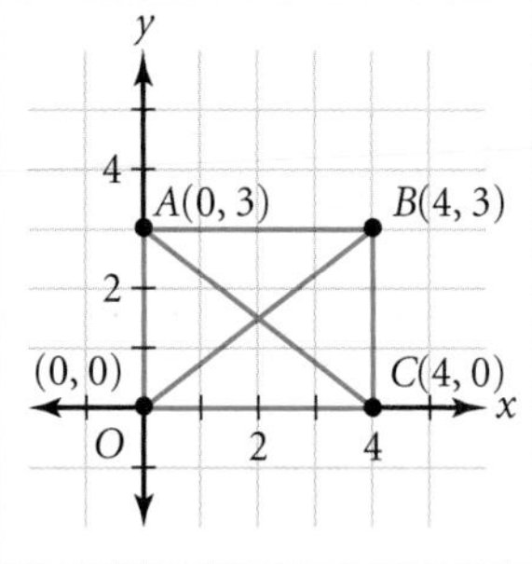


1. Use the distance formula to find the length of each diagonal of rectangle $ABCO$.
2. Draw two other rectangles and find the lengths of their diagonals. In each case, place one vertex at the origin and two of the sides along the axes.
3. Make a conjecture about the diagonals of a rectangle.

**CHECKPOINT ✔**

4. A rectangle is shown at right. Notice that the length of the rectangle is $a$ and the width is $b$. Write an expression for the length of the diagonals of a rectangle with length $a$ and width $b$.
5. Your result from Step 4 will be true for any rectangle with length $a$ and width $b$. Explain how this result proves your conjecture from Step 3 about the diagonals of a rectangle that is positioned as shown.

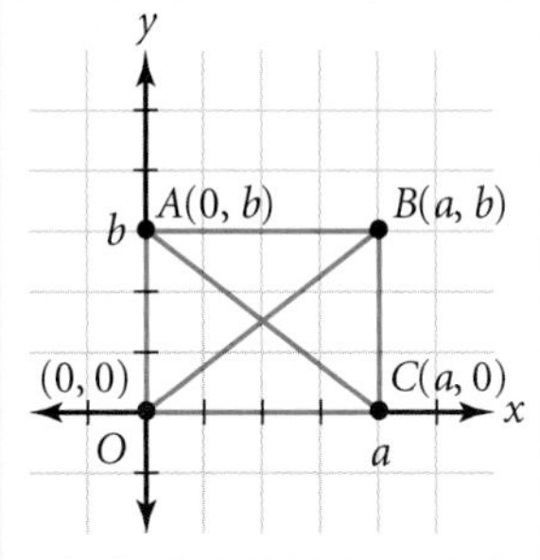


In the exercise set at the end of this lesson, you will be asked to discover and demonstrate other properties of rectangles.

## Segments and Midpoints

CONNECTION
GEOMETRY

You can use the midpoint formula to test an important geometry theorem, the Triangle Midsegment Theorem.

**Triangle Midsegment Theorem**

The segment joining the midpoints of two sides of a triangle is parallel to the third side of the triangle and is half the length of the third side.

**EXAMPLE 2** **Test the Triangle Midsegment Theorem for the triangle shown at right.**

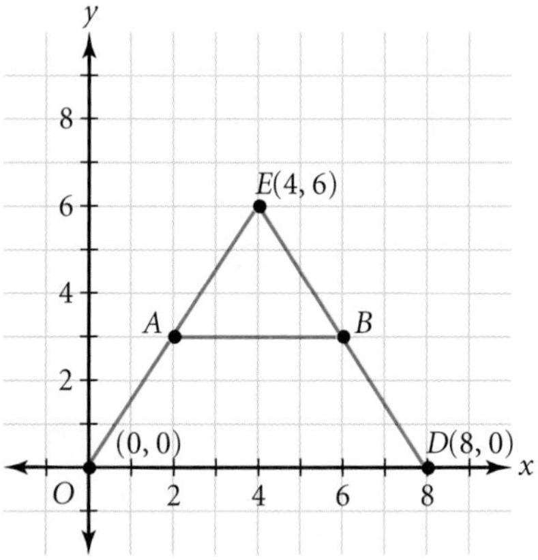


**SOLUTION**

1. Begin by finding the midpoints, $A$ and $B$, of sides $\overline{OE}$ and $\overline{ED}$ of the triangle.

**Midpoint *A***

$x$-coordinate: $x = \frac{0+4}{2} = 2$

$y$-coordinate: $y = \frac{0+6}{2} = 3$

Thus, the midpoint is (2, 3).

**Midpoint *B***

$x$-coordinate: $x = \frac{4+8}{2} = 6$

$y$-coordinate: $y = \frac{6+0}{2} = 3$

Thus, the midpoint is (6, 3).

2. Determine the slope of the **midsegment**, which connects the midpoints of two sides of the triangle, and the slope of the third side.

**Slope of midsegment**

$m = \frac{3-3}{6-2} = \frac{0}{4} = 0$

**Slope of third side**

$m = \frac{0-0}{8-0} = \frac{0}{8} = 0$

The slope is 0 in each case, so the lines are parallel, as stated in the theorem.

3. Calculate the length of the midsegment and of the third side of the triangle.

**Length of midsegment**

$d = 6 - 2 = 4$

**Length of third side**

$d = 8 - 0 = 8$

The length of the midsegment is one-half of the length of the third side, as stated in the theorem.

## The Focus and Directrix of a Parabola

One of the interesting properties of a parabola is that it consists of the set of all points equidistant from a given point, known as the **focus**, and a given line, known as the **directrix**.

**EXAMPLE 3**

**APPLICATION**
**ENGINEERING**

Engineers plan to locate a maintenance facility at a point that is equidistant from a TV tower and a road. The TV tower is 2 miles from the road. The diagram below shows several possible locations for the facility. **Write an equation that describes all possible locations of the maintenance facility.**

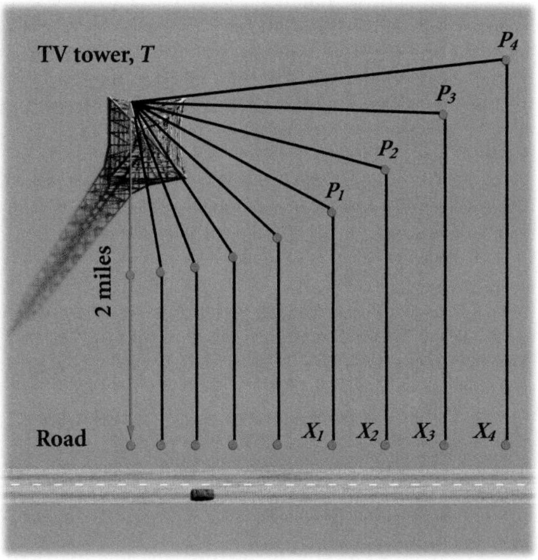


**SOLUTION**

PROBLEM SOLVING

**Make a diagram.** Use a coordinate grid to model the problem. Let $T(0, 2)$ be the TV tower. Let $R(x, 0)$ be a point on the $x$-axis, which represents the road. Finally, let $P(x, y)$ be any point equidistant from points $T$ and $R$.

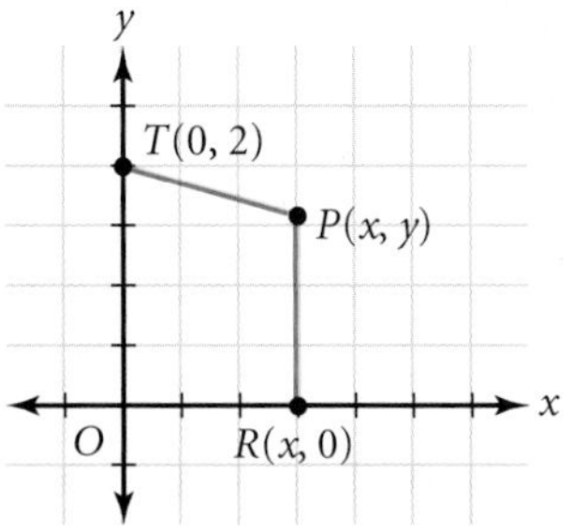


$$PR = PT$$
$$\sqrt{(x-x)^2 + (y-0)^2} = \sqrt{(x-0)^2 + (y-2)^2}$$
$$\sqrt{y^2} = \sqrt{x^2 + (y-2)^2}$$
$$y^2 = x^2 + (y-2)^2$$
$$y^2 = x^2 + y^2 - 4y + 4$$
$$4y = x^2 + 4$$
$$y = \frac{1}{4}x^2 + 1$$

The equation that describes all possible locations of the maintenance facility is $y = \frac{1}{4}x^2 + 1$. All possible locations of the facility, $P(x, y)$, lie on a parabola, with the TV tower representing the focus and the road representing the directrix.

**TRY THIS** Write an equation that describes all possible locations for another facility that are equidistant from a road and a radio tower that is 3 miles from the road.

# Exercises

## Communicate

**1.** Explain how the distance formula can be used to find the set of all points equidistant from a given point.

**2.** How can algebra help you to solve geometry problems? Include examples in your answer.

**3.** State the Triangle Midsegment Theorem in your own words. Why is it convenient to locate one vertex of a triangle at the origin when proving the theorem?

**4.** Define the focus and directrix of a parabola in your own words.

## Guided Skills Practice

**5.** Find the equation of a circle with the given radius, $r$, and the given center, $C$. ***(EXAMPLE 1)***

**a.** $r = 3$, $C(0, 0)$ **b.** $r = 4$, $C(-3, 4)$

**6.** Test the Triangle Midsegment Theorem for the triangle shown at right. ***(EXAMPLE 2)***

**7.** Write an equation for all points that are equidistant from a given point, $(2, 3)$, and a given line, the $x$-axis. ***(EXAMPLE 3)***

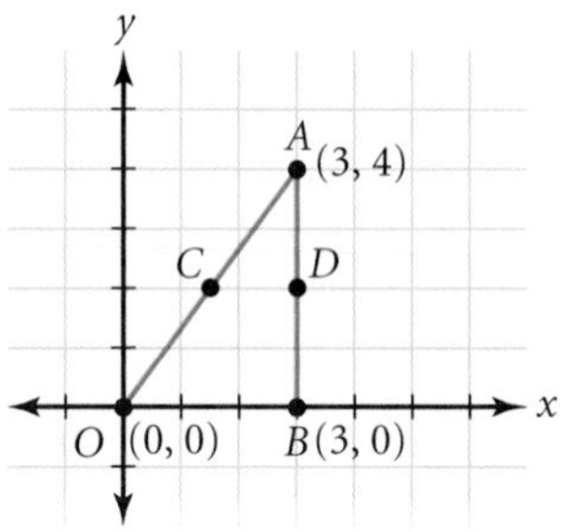


## Practice and Apply

internet connect
Homework Help Online
Go To: go.hrw.com
Keyword: MA1 Homework Help for Exercises 8-15

**Write the equation of a circle with its center at the origin and the given radius.**

**8.** radius = 6 **9.** radius = 1.2

**10.** radius = $\sqrt{3}$ **11.** radius = $5\sqrt{3}$

**Find the equation of a circle with the given radius, *r*, and the given center, *C*.**

**12.** $r = 3$, $C(1, 1)$ **13.** $r = 4$, $C(-1, 4)$

**14.** $r = 3$, $C(4, -2)$ **15.** $r = 4$, $C(-3, -6)$

**From each equation of a circle, give the center and the radius.**

**16.** $(x - 1)^2 + (y - 3)^2 = 9$ **17.** $(x + 1)^2 + (y + 3)^2 = 9$

**18.** $(x - 5)^2 + (y + 3)^2 = 625$ **19.** $(x + 6)^2 + (y + 4)^2 = 100$

**20.** $(x - 1.5)^2 + (y - 3.5)^2 = \frac{625}{36}$ **21.** $(x + 1)^2 + (y + 4)^2 = \frac{4}{25}$

**Write an equation for each circle. *C* represents each center.**

**22.**

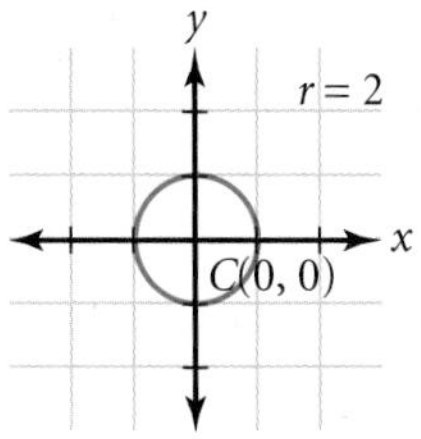


**23.**

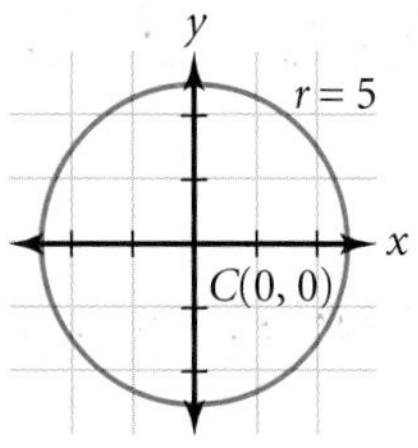


**24.**

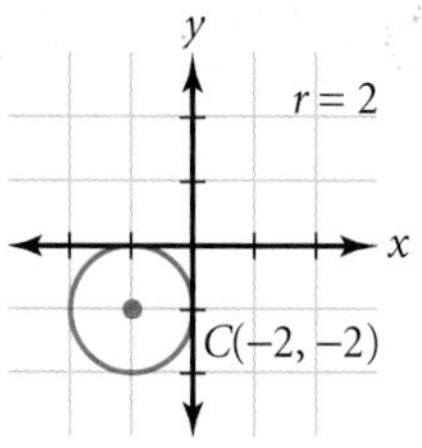


**25.**

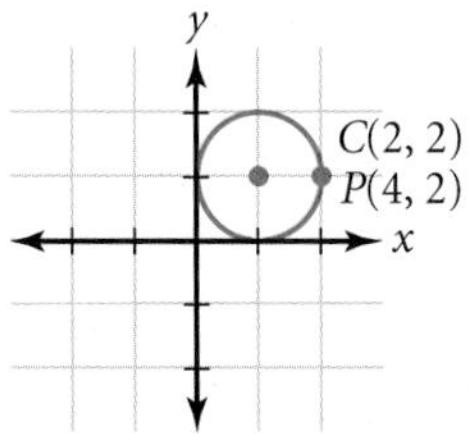


**26.**

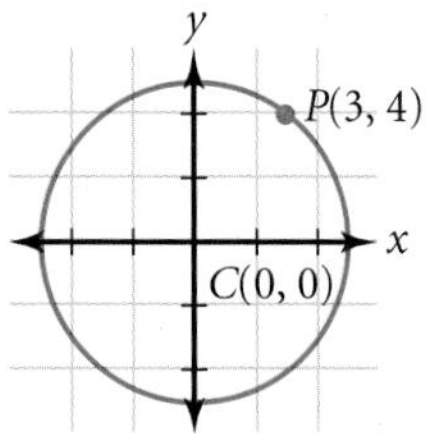


**27.**

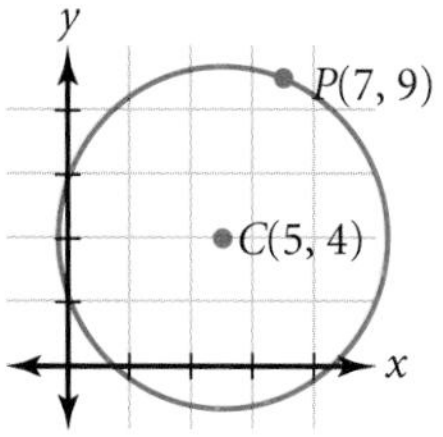


**28. a.** Copy and complete the table of values below.

| Values that satisfy the equation $x^2 + y^2 = 100$ | | | | | | | |
|---|---|---|---|---|---|---|---|
| $x$ | 0 | 6 | −6 | 8 | −8 | 10 | −10 |
| $y$ | ±10 | ±8 | ? | ? | ? | ? | ? |

**b.** Plot the values on a coordinate plane.
**c.** What shape is the graph?
**d.** What are the $x$-intercepts of the graph?
**e.** What are the $y$-intercepts of the graph?

**Plot triangle *PQR* on graph paper, where *P*(2, 1), *Q*(4, 7), and *R*(12, 3). Refer to your drawing for Exercises 29–35.**

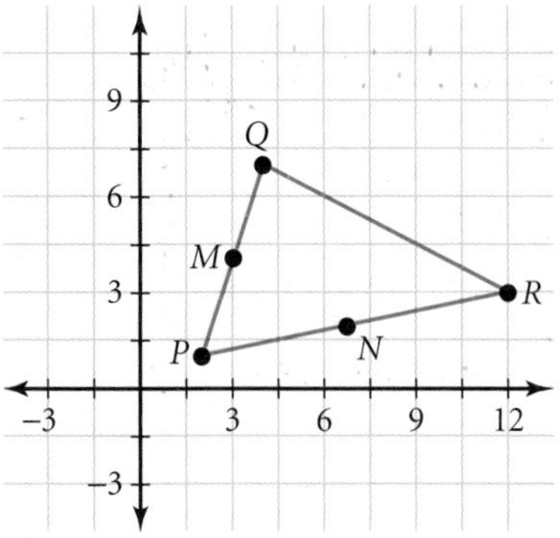


**29.** Identify the coordinates of $M$, the midpoint of $\overline{PQ}$, and plot $M$.

**30.** Find the coordinates of $N$, the midpoint of $\overline{PR}$, and plot $N$.

**31.** Draw $\overline{MN}$, and give the length of $\overline{MN}$ in simplest terms.

**32.** Find the length of $\overline{QR}$.

**33.** Compare the lengths of $\overline{MN}$ and $\overline{QR}$.

**34.** Find the slopes of $\overline{MN}$ and $\overline{QR}$.

**35.** What do the slopes tell you about $\overline{MN}$ and $\overline{QR}$?

**CHALLENGES**

**36.** Use the distance formula to write the equation for all points $P(x, y)$ such that the sum of the distances from $P$ to $A(-8, 0)$ and from $P$ to $B(8, 0)$ is 20.

**37.** Write the equation for all points $P(x, y)$ such that the sum of the distances from $P$ to $A(0, -4)$ and from $P$ to $B(0, 4)$ is 10.

**For Exercises 38–41, refer to the graph below.**

**38.** Find the midpoint of each diagonal. Are the midpoints the same or not?

**39.** In this lesson you may have discovered that the diagonals of a rectangle have equal lengths. Make a new conjecture about the midpoints of the diagonals based on your answer to Exercise 38.

**40.** Find the slope of each diagonal. What is the relationship between the slopes? Are the diagonals perpendicular?

**41.** Construct a graph of a square (a rectangle with all sides of equal length). What conjecture can you make about the diagonals of a square?

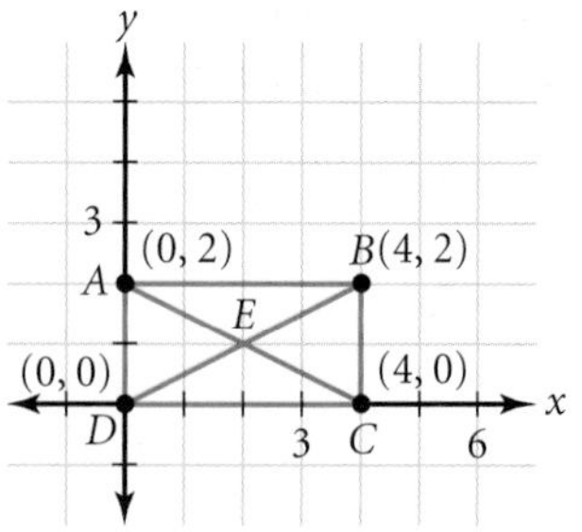


## Look Back

**State whether each expression correctly represents the distance between 2 and 5 on a number line.** ***(LESSON 2.1)***

**42.** $2 + 5$ **43.** $2 - 5$ **44.** $5 + 2$ **45.** $5 - 2$

**46.** $|2 + 5|$ **47.** $|2 - 5|$ **48.** $|5 + 2|$ **49.** $|5 - 2|$

**Write each number in scientific notation.** ***(LESSON 8.5)***

**50.** 2,300,000 **51.** 0.00000125

**Simplify.** ***(LESSON 9.1)***

**52.** $(x^2 + 3x + 5) + (7x^2 - 5x - 10)$ **53.** $(8b^2 - 15) - (2b^2 + b + 1)$

**Find the value of the discriminant for each quadratic equation.** ***(LESSON 10.5)***

**54.** $3x^2 - 6x + 3 = 0$ **55.** $2x^2 + 3x - 2 = 0$

**Graph each quadratic inequality.** ***(LESSON 10.6)***

**56.** $y < x^2 - 4$ **57.** $y > x^2 - 6x + 5$

## Look Beyond

**58.** Explain why the figure at right is a square for any positive value of $a$. Prove your conjecture from Exercise 41 about the diagonals of a square.

**59.** Graph a rectangle with length $a$ and width $b$. Prove that the diagonals bisect each other.

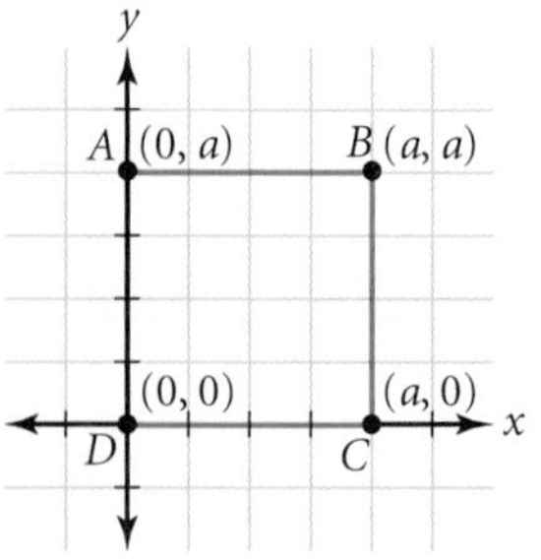

12.6

# The Tangent Function

## Objectives

- Identify and use the tangent ratio in a right triangle.
- Find unknown side and angle measures in right triangles.

**Why** ***A pilot must be able to determine not only the distance to the destination, but also the exact direction. By using the Pythagorean Theorem and the tangent function, you can determine both side lengths and angle measures in right triangles.***

**APPLICATION**
**AVIATION**

The direction in which a plane is traveling is referred to by pilots as the *heading* of the plane. The heading is an angle from 0 to 360 degrees, measured clockwise from the magnetic north.

A pilot at point $A$ is 300 miles due south of a town at point $C$. A town at point $B$ is 200 miles due east of point $C$. Assuming that there is no wind, in what direction must the pilot head to reach point $B$ by a straight path?

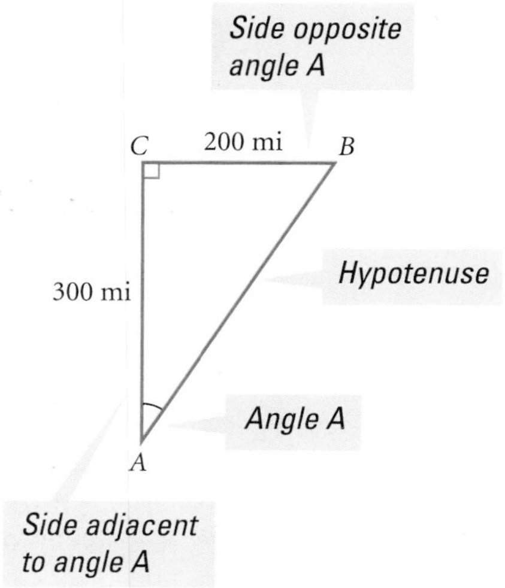


To answer the question, you will need to find the measure of angle $A$. Examine the ratio given with respect to angle $A$. On a right triangle this ratio is called the **tangent** of angle $A$.

$$\text{tangent } A = \tan A = \frac{BC}{AC} = \frac{200}{300} = \frac{2}{3}$$

### The Tangent Ratio

On a right triangle:

$$\text{tangent of angle } A = \tan A = \frac{\text{length of the leg opposite angle } A}{\text{length of the leg adjacent to angle } A}$$

$$\tan A = \frac{3}{4}$$

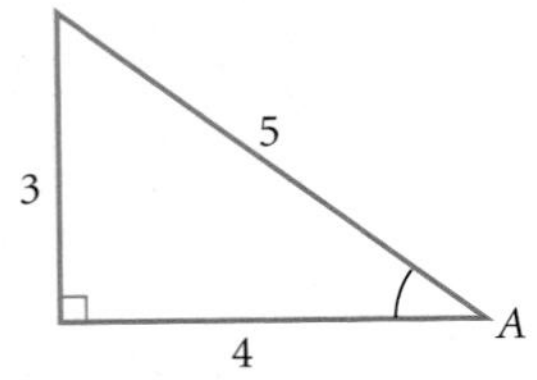


The tangent ratio is one of the fundamental ratios of the branch of mathematics known as **trigonometry** (from Greek words meaning "triangle measure").

Before returning to the plane problem and calculating the proper heading, consider the following question: Do you think the tangent of an angle depends on the size of the triangle it is a part of? In the activity that follows, you will investigate this question.

*Activity*

## The Tangent Ratio in Similar Triangles

**You will need:** graph paper, a centimeter ruler, and a calculator

$$\tan A = \frac{opp}{adj}$$

1. On a piece of graph paper, draw triangle $ABC$ as shown at right. Identify side $BC$ as the side *opposite* angle $A$, side $AC$ as the side *adjacent* to angle $A$, and side $AB$ as the *hypotenuse.*

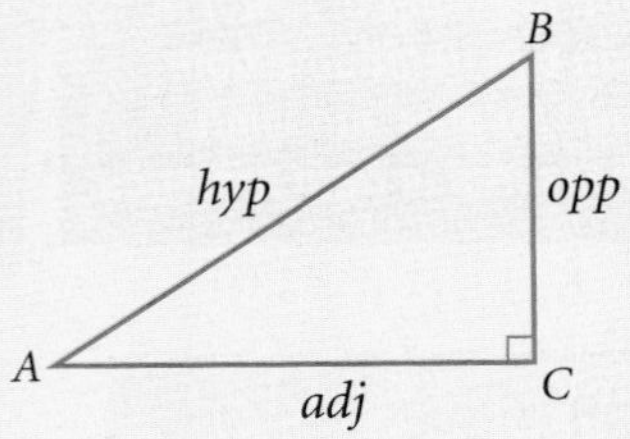


2. Draw segments $B'C'$ and $B''C''$ perpendicular to side $AC$. Each new segment is part of a smaller triangle that contains angle $A$. What leg is opposite angle $A$ in triangle $AB'C'$? What leg is adjacent to angle $A$ in triangle $AB'C'$?

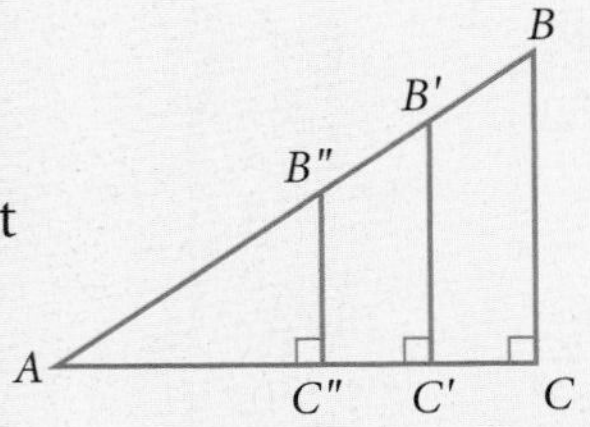


3. For each of the triangles containing angle $A$, measure the legs opposite and adjacent to angle $A$ in millimeters. Use your calculator to find the tangent ratio for angle $A$ in each triangle, rounding your answers to the nearest tenth. Keep a record of your results.

CHECKPOINT ✔ 4. Based on your approximate measurements and rounded results, does the tangent ratio for an angle depend on the size of the triangle in which the angle appears? Explain.

As you may have conjectured in the activity, the tangent ratio of an angle is independent of the size of the triangle in which the angle appears. In fact, the tangent ratio is a function of the angle measure. For any angle, there is exactly one value for the tangent of the angle. The domain of the tangent function is the set of angle measures, and the range is the set of numbers known as tangents.

The value of the tangent function increases as the angle measure increases from 0° to 90°, as the illustrations below indicate. (The side adjacent to angle $X$ is 10 units in all of the triangles.)

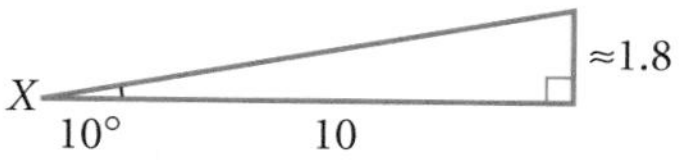


$\tan 10° \approx 0.18$

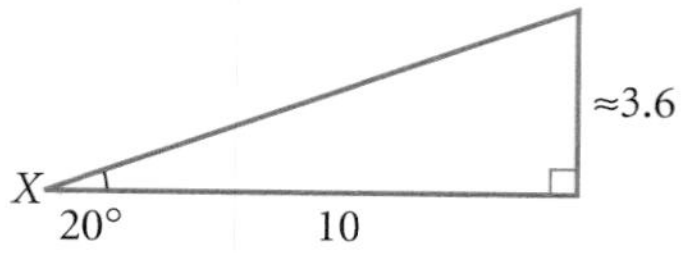


$\tan 20° \approx 0.36$

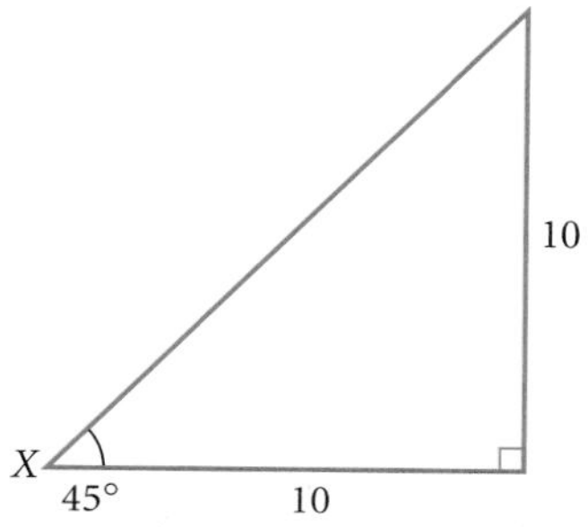


$\tan 45° = 1.00$ (exact)

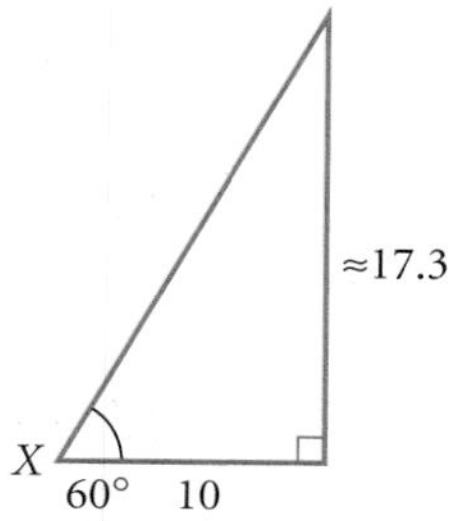


$\tan 60° \approx 1.73$

You can use a calculator to find the tangent of an angle by using the TAN key. Because you are using degrees to measure the angles, be sure the calculator mode is set for degrees. For example, to find tan 45°, press TAN 45 ENTER (or = ). The calculator will display the value 1.

In the problem of finding the angle for the plane's heading, the tangent of the angle is known. The problem is to find an angle between 0° and 90° whose tangent is $\frac{2}{3}$. The tangent function constantly increases over the interval from 0° to 90° so you can be sure that there is just one angle in the interval whose tangent has this value. "The angle whose tangent is $\frac{2}{3}$" is written as $\tan^{-1}\left(\frac{2}{3}\right)$, which is read as "the inverse tangent of $\frac{2}{3}$." You can use your calculator to find the angle by using the TAN⁻¹ key or INV TAN keys. The plane's direction, or heading, should be approximately 33.7°.

*A precision flying team breaks from a formation.*

**EXAMPLE 1**

CONNECTION
TRIGONOMETRY

**Find the tangent of the acute angles in each triangle. Then find the measures of the angles by using the TAN⁻¹ key on your calculator. Round your answers for the angle measures to the nearest degree.**

**a.**

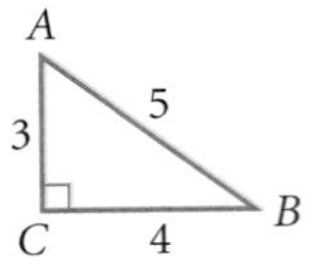


**b.**

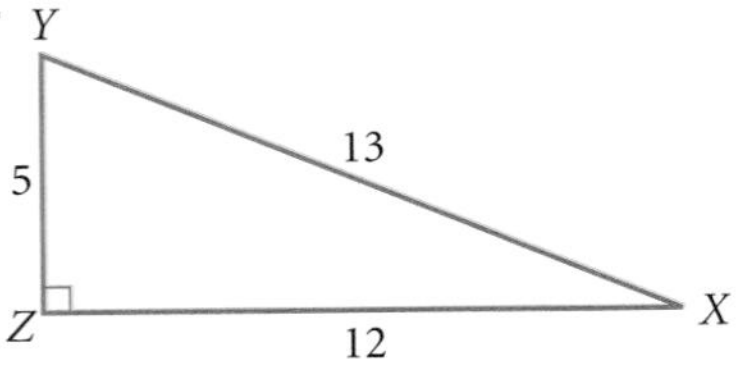


**SOLUTION**

TECHNOLOGY
SCIENTIFIC CALCULATOR

**a.** $\tan A = \dfrac{\text{opposite}}{\text{adjacent}} = \dfrac{4}{3}$

$A = \tan^{-1}\left(\dfrac{4}{3}\right) \approx 53°$

$\tan B = \dfrac{\text{opposite}}{\text{adjacent}} = \dfrac{3}{4}$

$B = \tan^{-1}\left(\dfrac{3}{4}\right) \approx 37°$

**b.** $\tan Y = \dfrac{\text{opposite}}{\text{adjacent}} = \dfrac{12}{5}$

$Y = \tan^{-1}\left(\dfrac{12}{5}\right) \approx 67°$

$\tan X = \dfrac{\text{opposite}}{\text{adjacent}} = \dfrac{5}{12}$

$X = \tan^{-1}\left(\dfrac{5}{12}\right) \approx 23°$

**TRY THIS**

Find the tangent of the acute angles in each triangle below. Then find the measures of the angles by using the TAN⁻¹ key or INV TAN keys on your calculator. Round your answers for the angle measures to the nearest degree.

**a.**

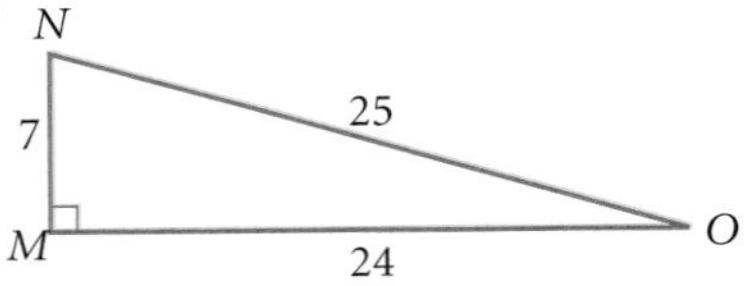


**b.**

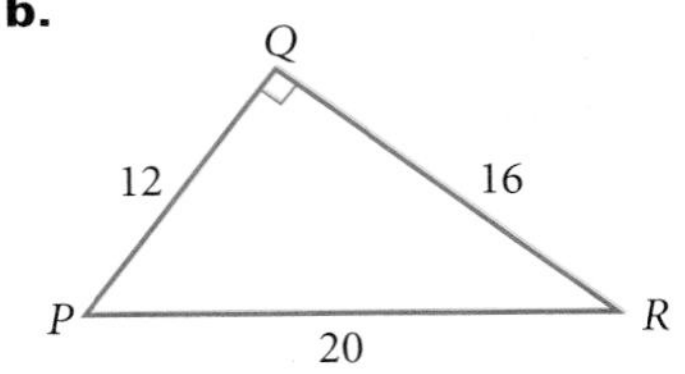


**EXAMPLE 2**

APPLICATION
AVIATION

PROBLEM SOLVING

TECHNOLOGY
SCIENTIFIC CALCULATOR

A plane climbs steadily to 30,000 feet. **If this altitude is reached when the airplane has covered a horizontal distance of 35 miles, what is its angle of ascent (angle of elevation)?**

**SOLUTION**

**Draw a diagram** to model the problem. Since 1 mile = 5280 feet, convert 35 miles to feet by multiplying by 5280.

$\tan A = \dfrac{30{,}000}{184{,}800}$

$A = \tan^{-1}\left(\dfrac{30{,}000}{184{,}800}\right) \approx 9°$

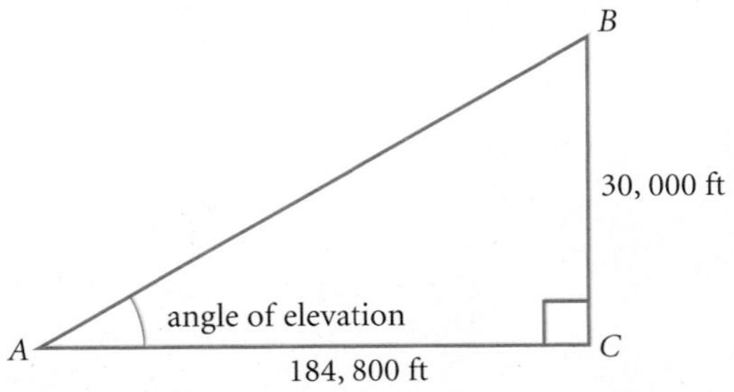


The airplane's angle of ascent is about 9°.

**TRY THIS** A 1500-foot building casts a 2000-foot shadow. What angle does the ray of sunlight shown form with the ground?

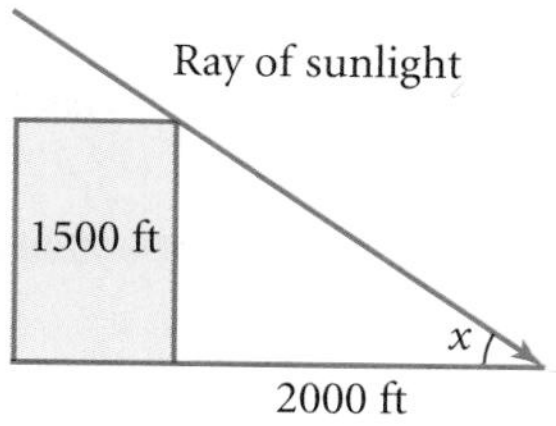


**EXAMPLE 3**

**APPLICATION**
**RECREATION**

A group of skateboarders wants to build a ramp with an angle of 12°. **What should be the rise for every 10 meters of run in order to achieve a 12° angle?**

**SOLUTION**

Draw a diagram, and write the tangent ratio for the information that you have.

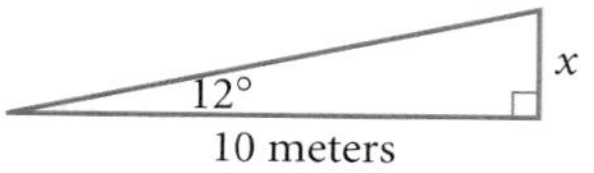


**TECHNOLOGY**
**SCIENTIFIC CALCULATOR**

Solve for $x$ to find the rise.

$$\tan 12° = \frac{x}{10}$$
$$x = 10(\tan 12°)$$
$$x \approx 2.126$$

The rise should be about 2.1 meters for every 10 meters of run.

**TRY THIS** A park ranger on a 200-foot tower spots a wolf at an angle of depression of 4°. How far is the wolf from the base of the tower?

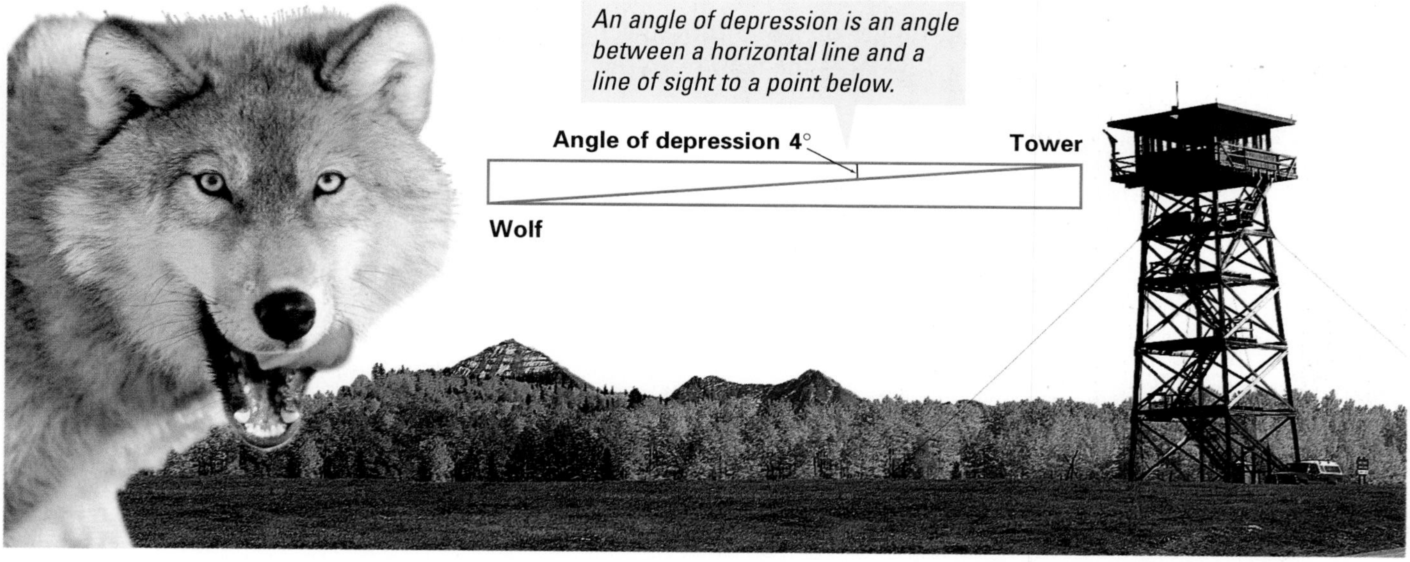

# Exercises

## Communicate

**1.** Explain how the tangent function is related to the slope of a line.

**2.** How can you find the length, $a$, of the side opposite angle $A$ in a right triangle when you know the measure of angle $A$ and the length, $b$, of the other leg?

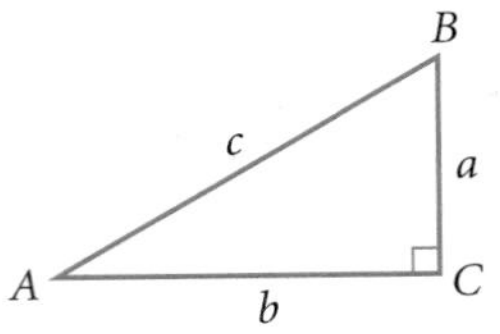


**3.** How can you find the measure of an acute angle in a right triangle when you know the lengths of the legs opposite and adjacent to that angle?

## Guided Skills Practice

CONNECTION

**4.** **TRIGONOMETRY** Find the tangent of each acute angle in the triangles below. Then find the measure of the angles by using the TAN⁻¹ key of your calculator. Round answers to the nearest degree. ***(EXAMPLE 1)***

**a.**

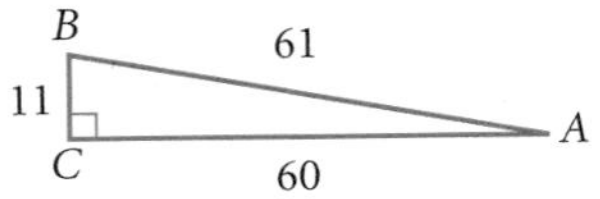


**b.**

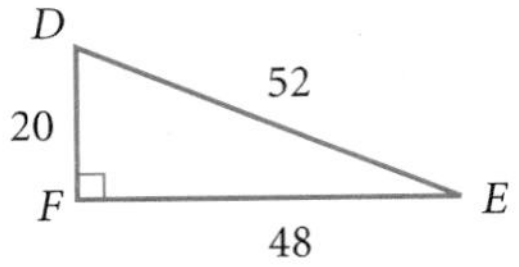


**5.** The sun casts a 400-foot shadow from a 1000-foot building. What angle does the ray of sunlight shown form with the ground? ***(EXAMPLE 2)***

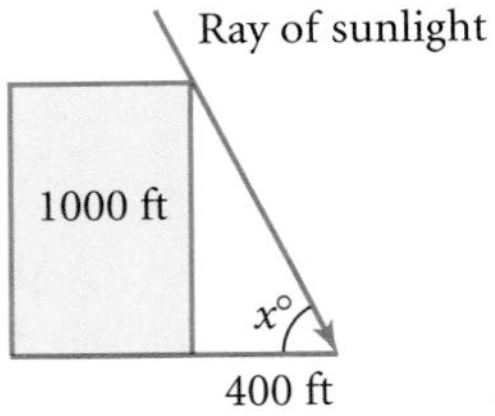


**6.** A park ranger on a 300-foot tower spots a wolf at an angle of depression of 6°. How far is the wolf from the base of the tower? ***(EXAMPLE 3)***

## Practice and Apply

**Use a calculator to find the tangent of each angle to the nearest ten-thousandth.**

**7.** 45° **8.** 30° **9.** 60° **10.** 0°

**11.** 38° **12.** 57° **13.** 89° **14.** 89.9°

**Find the measure of the angle whose tangent is given. Give each angle measure to the nearest whole number of degrees.**

**15.** 1.7321 **16.** 1.1918 **17.** 0.2679 **18.** 3.7321

**CONNECTIONS**

**internet connect**
**Homework Help Online**
Go To: **go.hrw.com**
Keyword:
**MA1 Homework Help**
for Exercises 19-24

**GEOMETRY** Use the diagram at right for Exercises 19–24. Round your answers to the nearest hundredth.

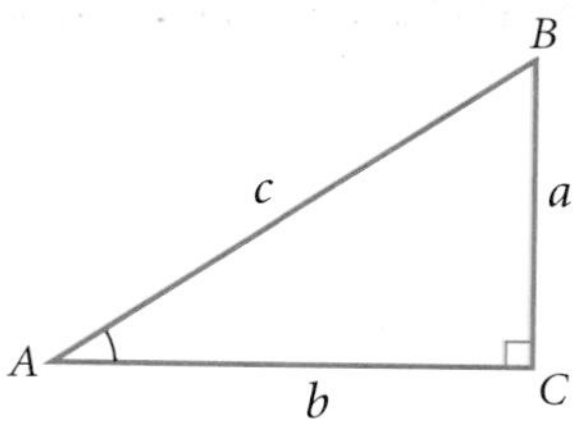


**19.** If $m\angle A = 30°$ and $b = 12$ meters, find $a$.

**20.** If $m\angle B = 60°$ and $a = 10$ feet, find $b$.

**21.** If $a$ is 3 centimeters and $\tan A$ is 0.75, what is $b$?

**22.** If $b$ is 23 feet and $\tan A$ is 1.0, what is $a$?

**23.** If $a$ is 6 inches and $m\angle A \approx 31°$, what is $b$?

**24.** If $m\angle A \approx 35°$ and $b$ is 20 millimeters, what is $a$?

**APPLICATIONS**

**25. CONSTRUCTION** A home buyer wants the roof of his house to have an angle of 20°, as shown. How high does the roof rise above the horizontal line shown if the width of the roof is 28 meters?

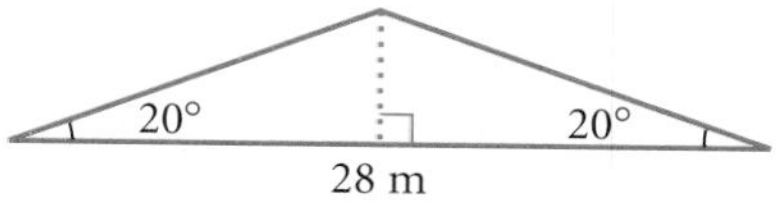


**26. ECOLOGY** A park ranger on a 200-foot tower spots the wildlife listed below. Determine the distances from the tower for the given angles of depression.

| Animal | Angle of depression | Distance |
|---|---|---|
| bear | 3° | ? |
| raccoon | 15° | ? |
| fox | 20° | ? |
| moose | 2° | ? |

**APPLICATION**

**27. ASTRONOMY** A science class conducts an experiment to determine the angle of elevation of the sun at different times during the day by measuring the shadow cast by a 28-foot goalpost. Copy the table shown below. Compute the angle of elevation from ground level for each shadow length. (The angle of elevation is the angle between a horizontal line and a line of sight to a point *above* the horizontal line.)

| Shadow length in feet | 80 | 48 | 13 | 8 | 2 | 4 |
|---|---|---|---|---|---|---|
| Angle of elevation | ? | ? | ? | ? | ? | ? |

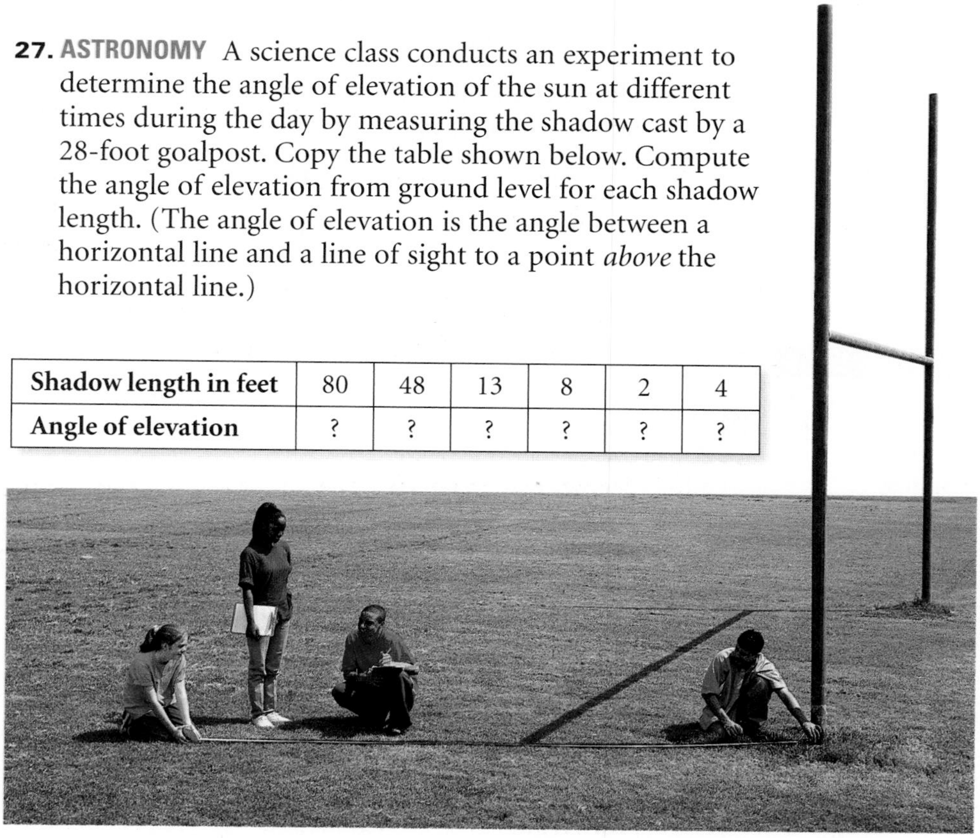

APPLICATIONS

**28.** **HOBBIES** A model airplane and a car begin traveling from the same point. The airplane stays directly above the car. After 15 seconds, the car has traveled one-tenth of a mile, and the angle of elevation of the airplane is 30°. What is the altitude of the model airplane?

**29.** **CIVIL ENGINEERING** If a road has a 7% grade, it rises 7 feet per 100 feet of run. Thus, its slope is $\frac{7}{100}$. What is the angle of the road?

**30.** **CIVIL ENGINEERING** A water tower is 600 feet from an observer. A line of sight to the top of the water tank forms an angle of 28° with the ground. A line of sight to the bottom of the tank forms an angle of 26°. What is the distance from the bottom of the water tank to the top of the water tank?

## Look Back

**31.** Solve $4x - 5 = 7(x - 3) + 2$ for $x$. ***(LESSON 3.5)***

**32.** State whether the quadratic equation $2x^2 + x - 15 = 0$ has any real solutions. If it does, use the quadratic formula to determine the solution(s). ***(LESSON 10.5)***

**33.** Solve $x^2 - 5x + 6 \geq 0$. ***(LESSON 10.6)***

**34.** Graph the solution to $x^2 - 16 < 0$. ***(LESSON 10.6)***

**35.** Points $P(4, 9)$ and $Q(10, 5)$ are plotted on a coordinate plane. Find the coordinates of $M$, the midpoint of $\overline{PQ}$. ***(LESSON 12.4)***

**Using the information given in Exercise 35, find the length of each segment in simplified form.** ***(LESSON 12.4)***

**36.** $\overline{PQ}$ **37.** $\overline{PM}$ **38.** $\overline{MQ}$

## Look Beyond

CONNECTION

internet connect
**Activities Online**
Go To: **go.hrw.com**
Keyword:
**MA1 Right Triangles**

**TRIGONOMETRY** Like the tangent (tan) ratio, the *sine* (sin) and *cosine* (cos) ratios are basic trigonometric functions. They provide additional tools for solving problems by using indirect measure.

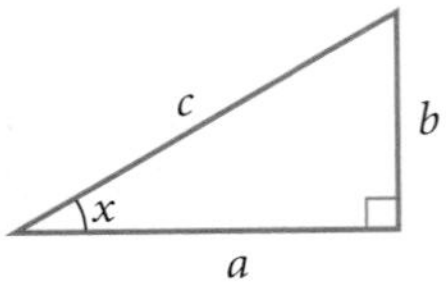


$\tan x = \frac{b}{a}$ $\sin x = \frac{b}{c}$ $\cos x = \frac{a}{c}$

**39.** If the tangent of angle $x$ is $\frac{3}{4}$, what are the sine and the cosine of $x$? (Hint: Use the Pythagorean Theorem to find $c$.)

**40.** Show why $(\sin x)^2 + (\cos x)^2 = 1$ is true for any right triangle.

12.7

# The Sine and Cosine Functions

**Objectives**

- Define the sine and cosine ratios in a right triangle.
- Find unknown side and angle measures in right triangles.

**Why** *Directly measuring the steepness of something like a water slide or ramp can be difficult. The sine and cosine functions, like the tangent function, can be used in indirect measurement.*

**APPLICATION**
**RECREATION**

As you go down the water slide, it feels almost like you are dropping straight down. How steep is the slide? You can use trigonometry to find an answer from the information in the diagram.

Recall the tangent ratio from Lesson 12.6.

$$\tan A = \frac{\text{length of the leg opposite } \angle A}{\text{length of the leg adjacent to } \angle A}$$

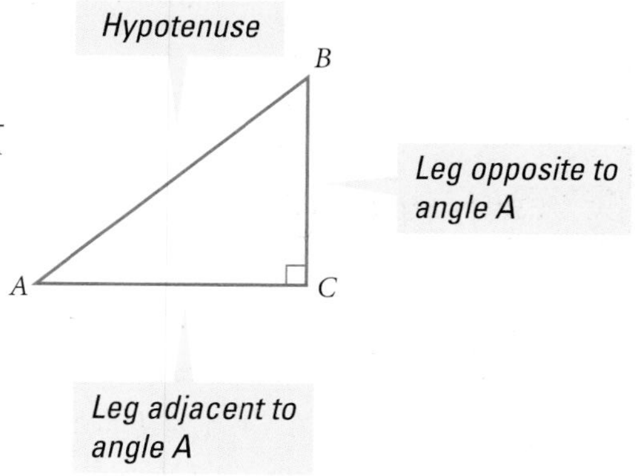


Two other important ratios, the *sine* and the *cosine*, are also based on the sides of a right triangle.

### Sine and Cosine

$$\text{sine of } \angle A = \sin A = \frac{\text{length of the leg opposite } \angle A}{\text{length of hypotenuse}} = \frac{opp}{hyp}$$

$$\text{cosine of } \angle A = \cos A = \frac{\text{length of the leg adjacent to } \angle A}{\text{length of hypotenuse}} = \frac{adj}{hyp}$$

*Activity*

## Exploring Sines and Cosines

**You will need:** graph paper and a scientific calculator

1. In the illustrations below, the hypotenuse is the same in each triangle. What happens to the value of sin *A* as the measure of angle *A* increases? Does it increase or decrease? Write a conjecture about the sine of 0° and the sine of 90°.

$$\sin A = \frac{opp}{hyp}$$

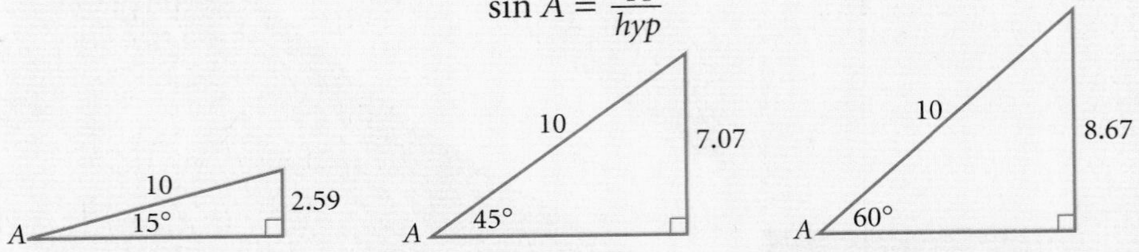


2. In the illustrations below, the hypotenuse is the same in each triangle. What happens to the value of cos *A* as the measure of angle *A* increases? Does it increase or decrease? Write a conjecture about the cosine of 0° and the cosine of 90°.

$$\cos A = \frac{adj}{hyp}$$

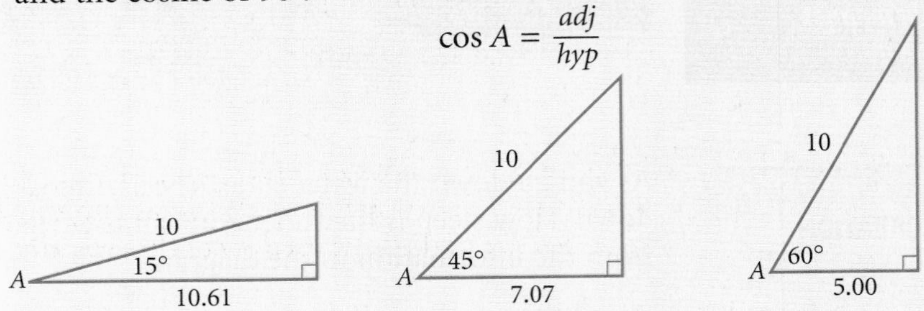


**CHECKPOINT ✓**

3. Write a statement contrasting the behavior of the sine and cosine ratios as the measure of an angle increases from 0° to 90°.

**TECHNOLOGY**
**SCIENTIFIC CALCULATOR**

You can use your calculator to find the sine or cosine of an angle. Because the angle measure is given in degrees, your calculator must be in degree mode. To determine sin 10°, use the SIN key.

SIN 10 ENTER $\approx 0.174$

**TRY THIS** Use the COS and SIN keys on your calculator to find cos 65° and sin 25°.

**EXAMPLE 1**

**APPLICATION**
**RECREATION**

A parasail is towed behind a boat. The rope that is attached to the parasail is 35 feet long. The parasail is at an angle of 30° with the surface of the water. **Approximately how high above the water is the parasailer?** (The rope is connected to the boat at a point 1 foot above the surface of the water.)

**SOLUTION**

**TECHNOLOGY**
**SCIENTIFIC CALCULATOR**

Use the sine ratio to find $b$, the length of the side opposite the 30° angle.

$$\sin 30° = \frac{\text{opposite}}{\text{hypotenuse}}$$

$$\sin 30° = \frac{b}{35}$$

$$b = 35(\sin 30°)$$

$$b = 17.5 \quad \text{Use a calculator.}$$

Since the rope is connected to the boat 1 foot above the water, add 1 to 17.5. Thus, the parasail is 18.5 feet above the surface of the water.

**TRY THIS** Find the length of side $a$ in the diagram at right.

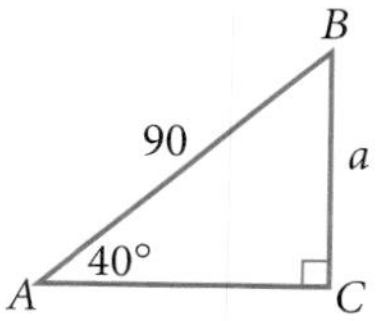


**EXAMPLE 2** Suppose that the horizontal distance of a proposed water slide is 115 feet and that the slide is 200 feet long. **If the maximum allowable angle of the slide is 35°, does the slide described meet this requirement?**

**SOLUTION**

$$\cos B = \frac{\text{adjacent}}{\text{hypotenuse}} = \frac{115}{200}$$

$$B = \cos^{-1}\left(\frac{115}{200}\right) \quad \text{(Use the } \boxed{\cos^{-1}} \text{ key.)}$$

$$B \approx 55°$$

The ramp described would be too steep.

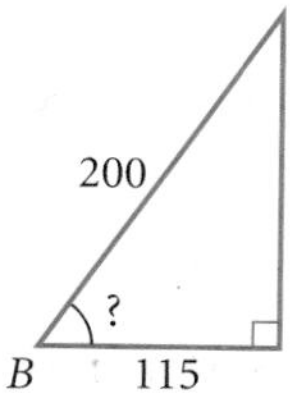


**TRY THIS** Find the measure of angle $M$ in the diagram at right.

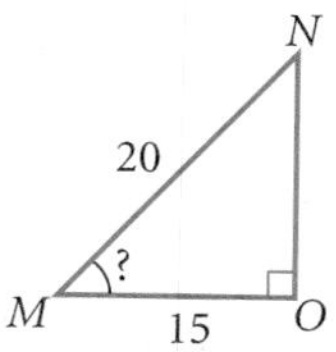

# Exercises

## Communicate

1. Explain how to find the sine of an angle in a right triangle.
2. Explain how to find the cosine of an angle in a right triangle.
3. Explain how to find the length of a ramp that makes a 4° angle up to a 3-inch curb.
4. What are the maximum and minimum values of the sine and cosine ratios in a right triangle?

## Guided Skills Practice

CONNECTION

5. **TRIGONOMETRY** Find the length of side $s$ in the diagram at right. ***(EXAMPLE 1)***

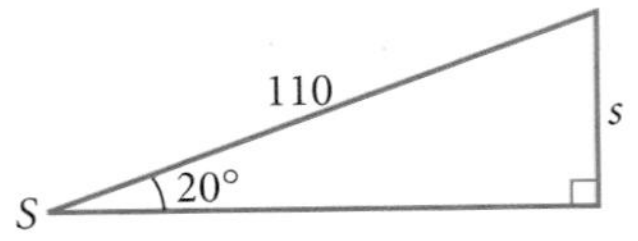


6. Find the measure of angle $X$ in the diagram at right. ***(EXAMPLE 2)***

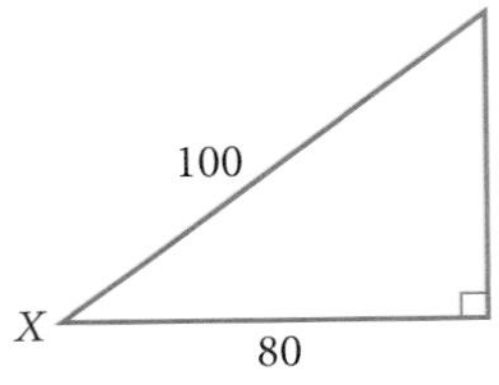

7. **RECREATION** Suppose that the horizontal distance of a proposed rollercoaster drop is 125 feet and that the track is 200 feet long. If the maximum allowable angle of the ride is 45°, does the ride described meet this requirement? ***(EXAMPLE 2)***

## Practice and Apply

**Approximate each sine and cosine. Give your answers to the nearest ten-thousandth.**

8. sin 24°
9. cos 32°
10. cos 88°
11. sin 18°
12. sin 66.8°
13. cos 84.2°
14. cos 27.7°
15. sin 48.6°

**Find the acute angle measure for each approximate sine value below. Give your answers to the nearest tenth of a degree.**

| | | | |
|---|---|---|---|
| **16.** 0.56 | **17.** 0.892 | **18.** 0.129 | **19.** 0.759 |
| **20.** 0.5 | **21.** 0.707 | **22.** 0.563 | **23.** 0.445 |
| **24.** 0.26 | **25.** 0.972 | **26.** 0.0 | **27.** 1.0 |

**Find the acute angle measure for each approximate cosine value below. Give your answers to the nearest tenth of a degree.**

| | | | |
|---|---|---|---|
| **28.** 0.126 | **29.** 0.5 | **30.** 0.886 | **31.** 0.707 |
| **32.** 0.99 | **33.** 0.81 | **34.** 0.54 | **35.** 0.78 |
| **36.** 0.612 | **37.** 0.643 | **38.** 0.0 | **39.** 1.0 |

**internet connect** go.hrw.com

**Homework Help Online**
Go To: **go.hrw.com**
Keyword:
**MA1 Homework Help**
for Exercises 40-45

**For each triangle, state whether the sine or cosine function should be used to find the value of *x*. Then find *x* to the nearest tenth.**

**40.**

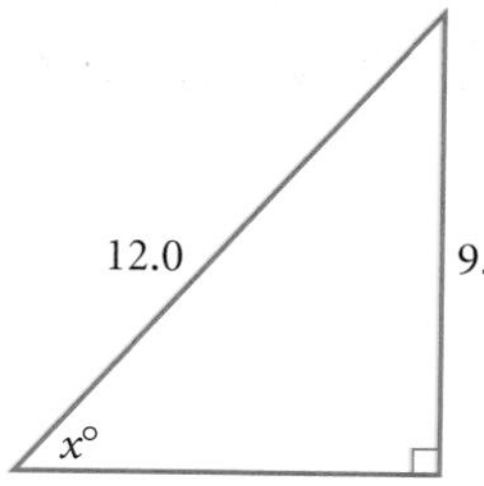


**41.**

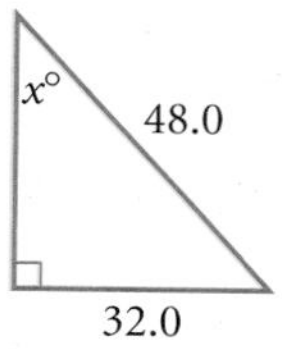


**42.**

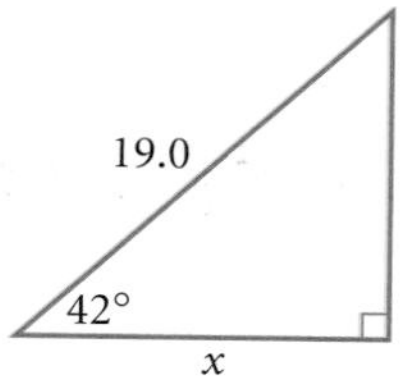


**43.**

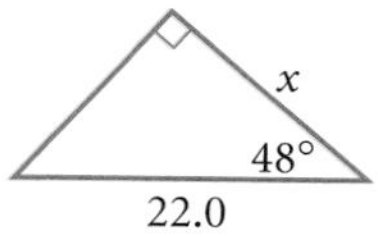


**44.**

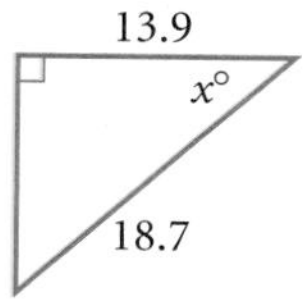


**45.**

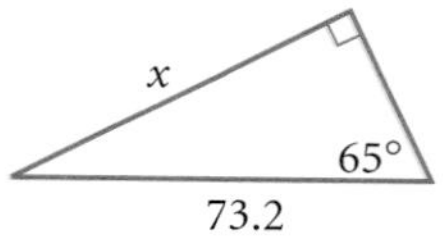


**For Exercises 46–55, use right triangle *ABC* at right. Find the indicated side length or angle measure to the nearest tenth.**

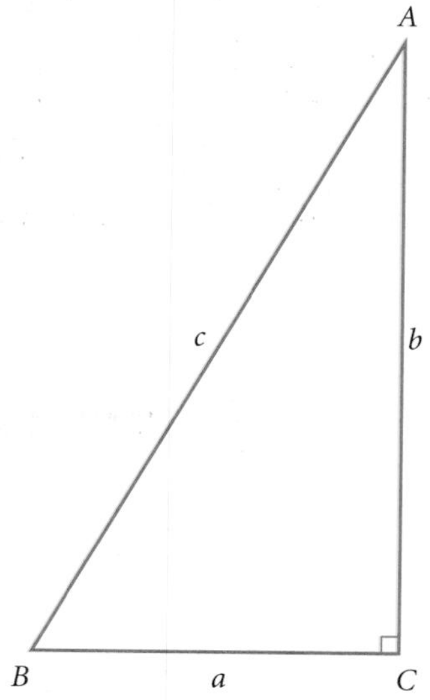


**46.** $b = 7, c = 10, m\angle B = \underline{\ ?\ }$

**47.** $b = 6, c = 13, m\angle A = \underline{\ ?\ }$

**48.** $m\angle B = 35°, c = 65, b = \underline{\ ?\ }$

**49.** $m\angle B = 51°, c = 22, a = \underline{\ ?\ }$

**50.** $b = 154, a = 25, m\angle B = \underline{\ ?\ }$

**51.** $a = 27, c = 36, m\angle B = \underline{\ ?\ }$

**52.** $m\angle B = 41°, c = 6, b = \underline{\ ?\ }$

**53.** $m\angle B = 30°, b = 11, c = \underline{\ ?\ }$

**54.** $a = 7, c = 14, m\angle B = \underline{\ ?\ }$

**55.** $m\angle B = 45°, c = 4, b = \underline{\ ?\ }$

**CONNECTION**

**56. GEOMETRY** If $m\angle B = 60°$ and $c = 100$ meters, find the area of triangle *ABC*.

**APPLICATIONS**

**CONSTRUCTION** A wheelchair ramp is to have an angle of 4.5° with the ground. The deck at the top of the ramp is 18 inches above ground level. Use this information for Exercises 57 and 58.

**57.** How long should the ramp be?

**58.** How far from the deck should the ramp begin?

**59.** **CONSTRUCTION** A 12-foot ladder is leaning against a building. The ladder makes a 75° angle with the ground. How far is the base of the ladder from the base of the building?

**60.** **CIVIL ENGINEERING** A guy wire attached to the top of a tower makes an angle of 25° with the tower. If the wire is anchored to the ground 30 feet from the tower, how long is the wire?

**CHALLENGE**

**61.** **AVIATION** An airplane traveling at 500 mph is descending from 30,000 feet to 20,000 feet at an angle of depression of 3°. How long, in seconds, does the descent take?

**internet connect**
**Activities Online**
Go To: **go.hrw.com**
Keyword:
**MA1 Trig Functions**

## Look Back

**Solve each equation. Round your answers to the nearest hundredth.** *(LESSON 10.2)*

**62.** $x^2 = 36$

**63.** $x^2 = 144$

**64.** $(x + 4)^2 - 36 = 0$

**65.** $(x - 1)^2 = 11$

**Complete the square.** *(LESSON 10.3)*

**66.** $x^2 + 10x$

**67.** $x^2 - 6x$

**68.** $x^2 + 9x$

**69.** $x^2 - 7x$

**Use the quadratic formula to solve each equation. Round your answers to the nearest hundredth.** *(LESSON 10.5)*

**70.** $t^2 + 6t - 22 = 0$

**71.** $h^2 + 6h + 5 = 0$

**Find *x* for each triangle. Round your answers to the nearest tenth.** *(LESSON 12.6)*

**72.**

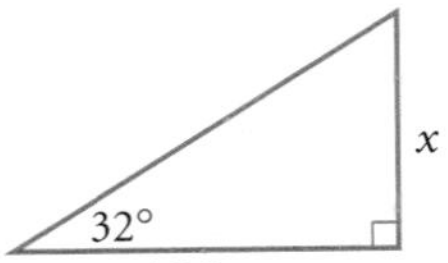


**73.**

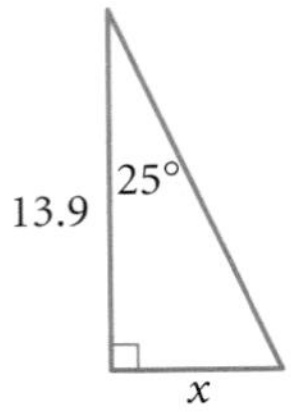


**74.**

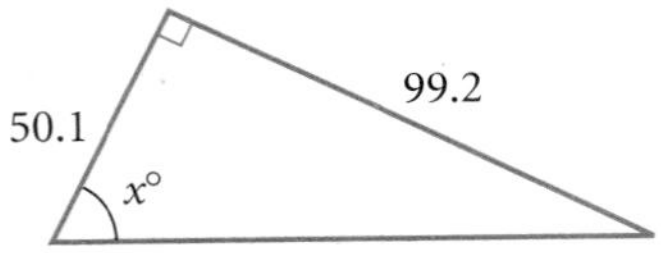


**75.**

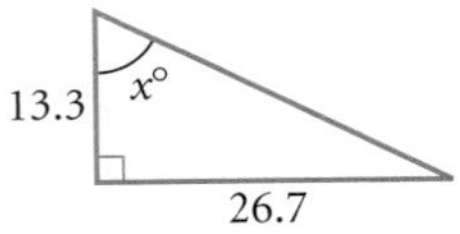

## Look Beyond

**CONNECTIONS**

**TRIGONOMETRY** Use trigonometry to find the area of each figure.

**76.**

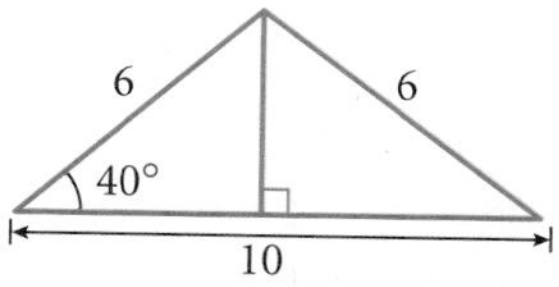


**77.**

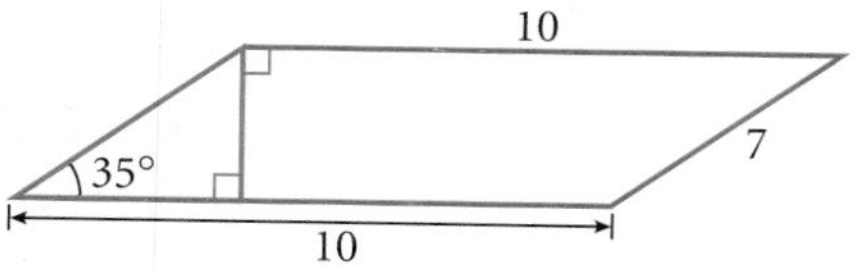


**78.** Use a calculator to complete the table, and sketch a graph of $y = \sin x$.

| $x$ | 0° | 30° | 60° | 90° | 120° | 150° | 180° | 210° | 240° | 270° | 300° | 330° |
|---|---|---|---|---|---|---|---|---|---|---|---|---|
| **sin $x$** | ? | ? | ? | ? | ? | ? | ? | ? | ? | ? | ? | ? |

**79.** Use a calculator to complete the table, and sketch a graph of $y = \cos x$.

| $x$ | 0° | 30° | 60° | 90° | 120° | 150° | 180° | 210° | 240° | 270° | 300° | 330° |
|---|---|---|---|---|---|---|---|---|---|---|---|---|
| **cos $x$** | ? | ? | ? | ? | ? | ? | ? | ? | ? | ? | ? | ? |

**80.** Compare the graphs you sketched in Exercises 78 and 79. How are they different? How are they similar?

**internet connect**
**Portfolio Extension**
Go To: **go.hrw.com**
Keyword:
**MA1 Golden**

**PORTFOLIO ACTIVITY**

Include this activity in your portfolio.

The golden ratio was used in ancient civilizations to form "golden" rectangles. Golden rectangles are pleasing to the eye. A golden rectangle has the property that the ratio of the width to the length is equal to the ratio of the length to the length plus the width:

*This television screen has dimensions of a golden rectangle.*

$$\frac{W}{L} = \frac{L}{L + W}$$

You can compute the golden ratio by considering a rectangle with a width of 1.

1. Find the golden ratio by substituting 1 for $W$ and solving for $L$, using the quadratic formula.
$$\frac{1}{L} = \frac{L}{L + 1}$$
2. Explain why a $3 \times 5$ card is close to being a golden rectangle.
3. Find at least three other real-world rectangles that are close to being golden rectangles.

**WORKING ON THE CHAPTER PROJECT**

You should now be able to complete Activities 3 and 4 of the Chapter Project beginning on page 636.

12.8

# Introduction to Matrices

**Why** *Matrices are an easy way of combining information from different tables.*

*Costs associated with car ownership can be studied by using matrices.*

### Objectives

- Determine the dimensions and addresses of a matrix.
- Determine whether two matrices are equal.
- Add, subtract, and multiply matrices.
- Determine the multiplicative identity of a matrix.

## Organizing Data by Using Matrices

**APPLICATION**
**TRANSPORTATION**

The table shows the costs associated with owning and operating a car for the given years.

| Item | Unit | 1980 | 1993 |
|---|---|---|---|
| Cost per mile | Cents/mile | 27.95 | 45.14 |
| Gas and oil | Cents/mile | 7.62 | 6.00 |
| Maintenance | Cents/mile | 5.86 | 2.40 |
| Tires | Cents/mile | 1.12 | 0.90 |
| Additional costs (insurance, license and registration, depreciation, and finance charges) | Dollars/year | 2033 | 4486 |

*[Source: Statistical Abstract of the United States, 1996]*

Data that are organized in a table are easy to read and manipulate. In mathematics, data can often be arranged in a special table of rows and columns enclosed in a set of brackets, [ ]. This arrangement is called a **matrix**.

A matrix is usually named with a capital letter. Matrix $C$ contains the data from the table above.

$$C = \begin{bmatrix} 27.95 & 45.14 \\ 7.62 & 6.00 \\ 5.86 & 2.40 \\ 1.12 & 0.90 \\ 2033 & 4486 \end{bmatrix}$$

Matrix $C$ has 5 rows and 2 columns. The number of rows by the number of columns gives the row-by-column dimensions of the matrix. The dimensions of $C$ are 5 by 2. This can be written as $5 \times 2$ or $C_{5\times2}$.

The information that appears in each position of the matrix is called an **entry**. Each entry in a matrix can be located by its **matrix address.** The entry at the address $c_{32}$, 2.4, is the entry at row 3 and column 2 of matrix $C$.

**Addresses for matrix $C$**

| | Column 1 | Column 2 |
|---|---|---|
| Row 1 | $c_{11}$ | $c_{12}$ |
| Row 2 | $c_{21}$ | $c_{22}$ |
| Row 3 | $c_{31}$ | $c_{32}$ |
| Row 4 | $c_{41}$ | $c_{42}$ |
| Row 5 | $c_{51}$ | $c_{52}$ |

The dimensions are $5 \times 2$.

A matrix is called a **square matrix** if its row-by-column dimensions are equal.

## Matrix Equality

Two matrices are equal when their dimensions are the same and their corresponding entries are equal.

$$U = \begin{bmatrix} 1 & 2 & 3 \\ 4 & 5 & 6 \end{bmatrix} \quad V = \begin{bmatrix} 1 & 2 & 3 \\ 4 & 5 & 6 \end{bmatrix} \quad W = \begin{bmatrix} 1 & 4 \\ 2 & 5 \\ 3 & 6 \end{bmatrix} \quad X = \begin{bmatrix} 1 & 3 & 5 \\ 2 & 4 & 6 \end{bmatrix}$$

Matrix $U$ and matrix $V$ have the same dimensions and their corresponding entries are equal, so $U = V$. Matrix $V$ and matrix $W$ do not have the same dimensions, so $V \neq W$. Matrix $U$ and matrix $X$ have the same dimensions, but corresponding entries are not equal, so $U \neq X$.

**EXAMPLE 1** **Are matrices $S$ and $T$ equal? Explain.**

$$S = \begin{bmatrix} 2^2 & (12-3) \\ \sqrt{49} & 11 \\ -3(9) & \frac{24}{3} \end{bmatrix} \qquad T = \begin{bmatrix} 2(2) & \sqrt{81} \\ 7 & \frac{44}{4} \\ -(30-3) & 2^3 \end{bmatrix}$$

**SOLUTION**

Think of the characteristics that make two matrices equal.

The two matrices must have the same dimensions. In this example, both matrices have dimensions of 3 by 2.

The entries in both matrices simplify to $\begin{bmatrix} 4 & 9 \\ 7 & 11 \\ -27 & 8 \end{bmatrix}$, so the corresponding entries in each matrix are equal. Thus, matrices S and T are equal.

## Adding and Subtracting Matrices

A matrix is an effective tool for storing data in a way that makes sense visually. But what if the data changes? You may need to add or subtract matrices.

To add or subtract matrices, both matrices *must have the same dimensions.* Then add or subtract corresponding entries to create a new matrix.

**EXAMPLE 2** **Add the following matrices:**

$$A = \begin{bmatrix} 3 \\ 7 \\ 9 \end{bmatrix} \qquad B = \begin{bmatrix} 11 \\ 15 \\ 16 \end{bmatrix}$$

**SOLUTION**

$$A + B = \begin{bmatrix} 3 \\ 7 \\ 9 \end{bmatrix} + \begin{bmatrix} 11 \\ 15 \\ 16 \end{bmatrix} = \begin{bmatrix} 3+11 \\ 7+15 \\ 9+16 \end{bmatrix} = \begin{bmatrix} 14 \\ 22 \\ 25 \end{bmatrix}$$

**TRY THIS** Add the following matrices: $A = \begin{bmatrix} 1.5 \\ 4.35 \end{bmatrix}$ $B = \begin{bmatrix} 7.32 \\ 9.7 \end{bmatrix}$

### Addition of Matrices

If $A$ and $B$ are matrices with the same dimensions, then the matrix $(A + B)$ has the same dimensions with elements that are the sum of the corresponding elements of $A$ and $B$.

$$\begin{bmatrix} a & b & c \\ d & e & f \end{bmatrix} + \begin{bmatrix} g & h & i \\ j & k & l \end{bmatrix} = \begin{bmatrix} a+g & b+h & c+i \\ d+j & e+k & f+l \end{bmatrix}$$

**EXAMPLE 3** **Let $A = \begin{bmatrix} 1 & 5 & 7 \\ 9 & 8 & 2 \end{bmatrix}$ and $B = \begin{bmatrix} 4 & 3 & 6 \\ 2 & 1 & 5 \end{bmatrix}$. Perform the following operations:**

**a.** $A + B$ **b.** $B - A$

**SOLUTION**

**a.** $A + B = \begin{bmatrix} 1 & 5 & 7 \\ 9 & 8 & 2 \end{bmatrix} + \begin{bmatrix} 4 & 3 & 6 \\ 2 & 1 & 5 \end{bmatrix} = \begin{bmatrix} 1+4 & 5+3 & 7+6 \\ 9+2 & 8+1 & 2+5 \end{bmatrix} = \begin{bmatrix} 5 & 8 & 13 \\ 11 & 9 & 7 \end{bmatrix}$

**b.** $B - A = \begin{bmatrix} 4 & 3 & 6 \\ 2 & 1 & 5 \end{bmatrix} - \begin{bmatrix} 1 & 5 & 7 \\ 9 & 8 & 2 \end{bmatrix} = \begin{bmatrix} 4-1 & 3-5 & 6-7 \\ 2-9 & 1-8 & 5-2 \end{bmatrix} = \begin{bmatrix} 3 & -2 & -1 \\ -7 & -7 & 3 \end{bmatrix}$

## Scalar Multiplication of a Matrix

To multiply a matrix by a number, simply multiply each entry in the matrix by the number and write the product in the appropriate address.

$$3\begin{bmatrix} 16 & 18 \\ 14 & 22 \end{bmatrix} = \begin{bmatrix} (3)(16) & (3)(18) \\ (3)(14) & (3)(22) \end{bmatrix} = \begin{bmatrix} 48 & 54 \\ 42 & 66 \end{bmatrix}$$

This type of multiplication is called **scalar multiplication**. The **scalar** is the number multiplied by each entry in the matrix. In this case, the scalar is 3.

**EXAMPLE 4** **Perform the scalar multiplication.**

$$5\begin{bmatrix} 3 & -2 \\ 4 & 5 \end{bmatrix}$$

**SOLUTION**

Multiply each entry by 5.

$$5\begin{bmatrix} 3 & -2 \\ 4 & 5 \end{bmatrix} = \begin{bmatrix} (5)(3) & (5)(-2) \\ (5)(4) & (5)(5) \end{bmatrix} = \begin{bmatrix} 15 & -10 \\ 20 & 25 \end{bmatrix}$$

## Multiplying Matrices

Multiplying two matrices creates a new matrix called the **product matrix**. In order for two matrices to be compatible for multiplication, the number of columns in the matrix on the left must match the number of rows in the matrix on the right.

To multiply, first multiply each entry in row 1 of the left matrix by the corresponding entry in column 1 of the right matrix. Then add the products and place the sum at the address $p_{11}$ in a product matrix, $P$.

Repeat the procedure for the second row and second column, placing the sum in $p_{21}$ of $P$, and so on.

$$A = \begin{bmatrix} a & d \\ b & e \\ c & f \end{bmatrix} \quad B = \begin{bmatrix} p & r \\ q & s \end{bmatrix} \quad A \times B = \begin{bmatrix} ap + dq & ar + ds \\ bp + eq & br + es \\ cp + fq & cr + fs \end{bmatrix}$$

As you proceed, the address of the sum will reflect the row number of the values used from $A$ and the column number of the values used from $B$.

$$P = \begin{bmatrix} p_{11} & p_{12} \\ p_{21} & p_{22} \\ p_{31} & p_{32} \end{bmatrix}$$

**CHECKPOINT ✔** What is the size of the product matrix when $A$ is a $3 \times 5$ matrix and B is a $5 \times 1$ matrix?

**EXAMPLE 5** **Find the product. If multiplication is not possible, explain why.**

$$\begin{bmatrix} 4 & -8.6 \\ -7 & 6 \end{bmatrix}\begin{bmatrix} -1 & -2 \\ 3 & 4.2 \end{bmatrix}$$

**SOLUTION**

The matrices are compatible for multiplication. The number of columns in the matrix on the left matches the number of rows in the matrix on the right.

$$\begin{bmatrix} 4 & -8.6 \\ -7 & 6 \end{bmatrix}\begin{bmatrix} -1 & -2 \\ 3 & 4.2 \end{bmatrix} = \begin{bmatrix} (4)(-1)+(-8.6)(3) & (4)(-2)+(-8.6)(4.2) \\ (-7)(-1)+(6)(3) & (-7)(-2)+(6)(4.2) \end{bmatrix}$$

$$= \begin{bmatrix} -29.8 & -44.12 \\ 25 & 39.2 \end{bmatrix}$$

**TRY THIS** Find the products $AB$ and $BA$, if multiplication is possible.

$$A = \begin{bmatrix} 3 & 7 \end{bmatrix} \quad B = \begin{bmatrix} 2 \end{bmatrix}$$

One special matrix is the **identity matrix** for multiplication. The identity matrix is always a square matrix. The number of rows equals the number of columns. Each entry on the **main diagonal** is 1; all other entries are 0.

$$\text{identity matrix} = \begin{bmatrix} 1 & 0 \\ 0 & 1 \end{bmatrix}$$

Any matrix multiplied by its identity matrix leaves that matrix unchanged.

**EXAMPLE 6** **What is the identity matrix for $B = \begin{bmatrix} 3 & 2 & 1 \\ 2 & 3 & 1 \\ 4 & 4 & 5 \end{bmatrix}$?**

**SOLUTION**

Since $B$ has dimensions of $3 \times 3$, the identity matrix for multiplication must be a $3 \times 3$ matrix.

The identity matrix for $B = \begin{bmatrix} 3 & 2 & 1 \\ 2 & 3 & 1 \\ 4 & 4 & 5 \end{bmatrix}$ is $I = \begin{bmatrix} 1 & 0 & 0 \\ 0 & 1 & 0 \\ 0 & 0 & 1 \end{bmatrix}$.

**TRY THIS** What is the identity matrix for $D = \begin{bmatrix} 3 & 2 & 1 & 0 \\ 5 & 1 & 2 & 6 \\ 4 & 3 & 1 & 7 \\ 4 & 6 & 5 & 1 \end{bmatrix}$?

# Exercises

## Communicate



**1.** Create a $3 \times 5$ matrix, $A$, and enter each address. Describe how the addresses are notated.

**2.** Is matrix addition commutative? Explain.

**3.** Is matrix multiplication commutative? Explain.

## Guided Skills Practice

**4.** Determine which two of the matrices below are equivalent. *(EXAMPLE 1)*

$$H = \begin{bmatrix} 3 & -4 & 4(5) \\ 1 & 0 & 2 \end{bmatrix} \quad I = \begin{bmatrix} 3 & -4 & 20 \\ 1 & 0 & 12 \end{bmatrix} \quad J = \begin{bmatrix} (2+1) & -4 & 2(10) \\ 1 & 0 & 4-2 \end{bmatrix}$$

**Add or subtract as indicated.** *(EXAMPLES 2 AND 3)*

**5.** $\begin{bmatrix} 7 & 9 \\ -4 & 5 \end{bmatrix} + \begin{bmatrix} 18 & 4 \\ -3 & -16 \end{bmatrix}$

**6.** $\begin{bmatrix} -5 & 6 \\ 12 & 8 \end{bmatrix} - \begin{bmatrix} 4 & -19 \\ 17 & 7 \end{bmatrix}$

**7.** $\begin{bmatrix} 4 & -6 & 12 \\ -3 & 9 & 8 \\ 2 & 16 & -3 \end{bmatrix} - \begin{bmatrix} 1 & -2 & 3 \\ 27 & 8 & 7 \\ 5 & -3 & 14 \end{bmatrix}$

**Perform each scalar multiplication.** *(EXAMPLE 4)*

**8.** $5\begin{bmatrix} 6 & -4 \\ -3 & 5 \end{bmatrix}$

**9.** $-2\begin{bmatrix} 3 & 9 \\ -6 & -4 \end{bmatrix}$

**10.** $3\begin{bmatrix} 2 & -5 & 1 \\ 12 & -3 & -8 \\ 14 & 10 & 4 \end{bmatrix}$

**Find each product matrix, if it exists.** *(EXAMPLE 5)*

**11.** $\begin{bmatrix} -6 & 4 \\ -23 & 3 \end{bmatrix}\begin{bmatrix} 9 & -4 \\ 6 & 2 \end{bmatrix}$

**12.** $\begin{bmatrix} 13 & 12 \\ -4 & 3 \end{bmatrix}\begin{bmatrix} 3 & 9 \\ -6 & -4 \end{bmatrix}$

**13.** $\begin{bmatrix} 3 & 8 & 7 \\ -3 & 14 & -6 \\ 8 & 3 & -7 \end{bmatrix}\begin{bmatrix} 2 & 5 \\ -7 & 9 \end{bmatrix}$

**14.** $\begin{bmatrix} 4 \\ 2 \\ 9 \end{bmatrix}\begin{bmatrix} 7 \\ 7 \\ 4 \end{bmatrix}$

**Find the identity matrix for each matrix below.** *(EXAMPLE 6)*

**15.** $\begin{bmatrix} 18 & 4 \\ -5 & -6 \end{bmatrix}$

**16.** $\begin{bmatrix} 3 & 2 & 1 \\ 2 & 3 & 1 \\ 4 & 4 & 5 \end{bmatrix}$

## Practice and Apply

**Perform the indicated matrix operations. If a solution is not possible, explain why.**

**17.** $\begin{bmatrix} -3 & 4 \\ -4 & 0 \\ 2 & -5 \end{bmatrix} + \begin{bmatrix} 0 & -1 \\ -6 & 9 \\ -7 & 5 \end{bmatrix}$

**18.** $\begin{bmatrix} 0.4 & -1.5 & 0.9 \\ 2.6 & 6.9 & 3.7 \end{bmatrix} + \begin{bmatrix} -4.7 & 2.6 & 6.9 \\ -7.3 & 9.8 & -5.5 \end{bmatrix}$

**19.** $\begin{bmatrix} -25 & 32 & 14 \\ 36 & -42 & -45 \\ -71 & 65 & 29 \end{bmatrix} - \begin{bmatrix} 16 & -34 & -55 \\ 21 & 11 & 22 \\ -43 & -67 & -44 \end{bmatrix} + \begin{bmatrix} 57 & 79 & 64 \\ -38 & -22 & -48 \\ -56 & 88 & 26 \end{bmatrix}$

**20.** $\begin{bmatrix} -1.4 & 4.3 & -9.6 \\ 15.8 & 1 & -3.5 \end{bmatrix} - \begin{bmatrix} 6.2 & 3.2 \\ -9.6 & 7.1 \\ 2.6 & -8.5 \end{bmatrix}$

**21.** $\begin{bmatrix} 1 \\ 2 \\ 3 \end{bmatrix} - \begin{bmatrix} 3 \\ 2 \\ 1 \end{bmatrix}$

**22.** $\begin{bmatrix} 4 & 2 \\ 6 & -7 \\ 3 & 9 \end{bmatrix} + \begin{bmatrix} -5 & 7 \\ -3 & 9 \\ 3 & -6 \end{bmatrix} - \begin{bmatrix} -11 & 3 \\ 8 & -15 \\ -7 & -2 \end{bmatrix}$

**23.** $\begin{bmatrix} -2 & 3 \\ 3 & -2 \end{bmatrix} - \begin{bmatrix} 4 & 6 \\ 4 & 6 \end{bmatrix}$

**24.** $\begin{bmatrix} -3 \\ -2 \\ -5 \end{bmatrix} - \begin{bmatrix} 1 & 2 & 3 \end{bmatrix}$

**25.** $\begin{bmatrix} 3 & 2 \\ 6 & 4 \\ 9 & 6 \end{bmatrix} - \begin{bmatrix} -5 & 3 \\ 2 & -2 \\ -1 & 7 \end{bmatrix}$

**26.** $5\begin{bmatrix} 3 & -2 \\ 7 & 8 \end{bmatrix}$

**27.** $-3\begin{bmatrix} 5 & 17 & -3 \\ 12 & -8 & 2 \\ -7 & 14 & 12 \end{bmatrix}$

**28.** $-7\begin{bmatrix} 5 & -3 \\ 4 & -12 \\ 8 & 6 \end{bmatrix}$

**29.** $\begin{bmatrix} 4 & -6 \\ 12 & 8 \end{bmatrix}\begin{bmatrix} 3 & -8 \\ 4 & 9 \end{bmatrix}$

**30.** $\begin{bmatrix} 3 & -7 & 18 \end{bmatrix}\begin{bmatrix} 4 \\ 13 \\ 9 \end{bmatrix}$

**31.** $\begin{bmatrix} 3 & -8 \\ 1 & -3 \\ 4 & 16 \end{bmatrix}\begin{bmatrix} 8 & -6 & 9 \\ 12 & 4 & -2 \end{bmatrix}$

**32.** $\begin{bmatrix} 3 & 12 \\ -4 & -7 \end{bmatrix}\begin{bmatrix} -4 & 7 \\ 9 & 12 \end{bmatrix}$

**33.** $\begin{bmatrix} 4 & 12 & 9 \\ -3 & -4 & 6 \\ 3 & 2 & -4 \end{bmatrix}\begin{bmatrix} 3 \\ -3 \\ 14 \end{bmatrix}$

**34.** $\begin{bmatrix} 3 & -12 \\ 2 & 4 \end{bmatrix}\begin{bmatrix} 6 \\ 2 \end{bmatrix}$

**Matrices *M* and *N* are equal. Find the values of *a*, *b*, *c*, *d*, *g*, and *k*.**

**35.** $M = \begin{bmatrix} 2(a+4) & 77 & \frac{1}{3}c \\ -5d-1 & 30 & -\frac{1}{2}k \end{bmatrix}$ $\quad N = \begin{bmatrix} -12 & 11b & 5 \\ -(3-d) & 0.4g & \frac{3}{4}k-3 \end{bmatrix}$

**CONNECTION**

**STATISTICS** In the Woodlake public school system, there are two junior high schools, Glenn and Kelly. The enrollment data for the electives music (Mu), art (Ar), technology (Te), and health (He) appear in the matrices below.

**Glenn Middle School**

| | Mu | Ar | Te | He |
|---|---|---|---|---|
| Boys | 447 | 199 | 514 | 389 |
| Girls | 498 | 352 | 432 | 399 |

**Kelly Middle School**

| | Mu | Ar | Te | He |
|---|---|---|---|---|
| Boys | 387 | 276 | 489 | 367 |
| Girls | 505 | 392 | 387 | 437 |

**36.** How many girls are enrolled in art? How many boys are enrolled in health?

**37.** Create a matrix to show the total enrollment in these electives.

**38.** How many students are enrolled in technology this year?

APPLICATION

**SPORTS** Suppose that you keep all the statistics for the girls' basketball team in a matrix, a portion of which is shown below. You record 3-point baskets, 2-point baskets, and 1-point free throws. The matrices show the statistics for the regular season and the playoff series. Use these matrices to complete Exercises 39–41.

**Baskets Scored in Regular Season**

| | 3 pts | 2 pts | 1 pt |
|---|---|---|---|
| Alesia | 3 | 15 | 12 |
| Jennel | 1 | 20 | 16 |
| Betsy | 3 | 17 | 13 |
| Katie | 4 | 6 | 9 |
| Nancy | 0 | 14 | 18 |

**Baskets Scored in Playoff Series**

| | 3 pts | 2 pts | 1 pt |
|---|---|---|---|
| Alesia | 2 | 10 | 3 |
| Jennel | 1 | 2 | 4 |
| Betsy | 2 | 3 | 2 |
| Katie | 1 | 6 | 3 |
| Nancy | 0 | 3 | 4 |

**39.** Create matrices to show the total points scored by each player in the regular season and during the playoff series.

**40.** Add the two matrices from Exercise 39 to create a new matrix.

**41.** How many total points were scored by the team during the regular season and the playoff series combined?

## Look Back

**Find each sum or difference.** ***(LESSONS 2.2 AND 2.3)***

**42.** $-8 + 6$ **43.** $-7 - (-3)$ **44.** $8 + (-4)$ **45.** $-5 + (-6)$

**Match the name of the property with the equation that best illustrates it.** ***(LESSON 2.5)***

**46.** $7 + (11 + 4) = (7 + 11) + 4$
**47.** $4 + 9 = 9 + 4$
**48.** $12 + 18 = 6(2 + 3)$

**a.** Commutative Property for Addition
**b.** Associative Property for Addition
**c.** Distributive Property

**Solve each equation.** ***(LESSON 3.1)***

**49.** $x + 1.4 = -5.6$ **50.** $6 = c - (-6)$ **51.** $m + (-47) = 31$

**Find the slope of the line that contains each pair of points.** ***(LESSON 5.2)***

**52.** M(2, 5), N(−6, 7) **53.** M(7, 4), N(−12, 3)

**Solve each equation by factoring.** ***(LESSON 9.8)***

**54.** $x^2 + x - 12 = 0$ **55.** $x^2 - 2x - 35 = 0$ **56.** $x^2 - 49 = 0$

**Use the quadratic formula to find exact answers.** ***(LESSON 10.5)***

**57.** $2x^2 + 12x + 14 = 0$ **58.** $3x^2 - 3x - 36 = 0$

## Look Beyond

CONNECTION

**59.** **PROBABILITY** If you flipped a coin 100 times, how many times would you expect it to land heads up?

**60.** If you rolled a 6-sided number cube 100 times, how many times would you expect a 3 or a 5 to appear?

CHAPTER PROJECT TWELVE

# Working the Angles

## Activity 1

**Building a Hypsometer**

**You will need:** a protractor, a straw, tape, string, a metal washer (or other weight), and a tape measure (or other measuring device)

**a.** Attach a straw to the edge of a protractor.

**b.** Suspend a string tied to a washer from the hole in the protractor. This will give you the angle formed when you sight objects through the straw.

**c.** Find three objects with known heights, and use the hypsometer to measure the angle from your position to the top of each object. Record your measurements in a table, including your horizontal distance from the object when you measured the angle.

**d.** Compare the ratio of height to distance with the tangent of the angle created by the line of sight to each object.

| Object | Height | Distance | Ratio (height/distance) | Angle | Tangent |
|---|---|---|---|---|---|
| 1 | | | | | |
| 2 | | | | | |
| 3 | | | | | |

## Activity 2

### Measuring Up

**a.** Use your hypsometer to measure the *angle of elevation* of three objects with unknown heights.

**b.** Measure the horizontal distance between the hypsometer and the objects.

**c.** Record your measurements in a table. Use the $\tan^{-1}$ function on your calculator to compute the height.

| Object | Distance | Angle | Height |
|---|---|---|---|
| 1 | ? | ? | ? |
| 2 | ? | ? | ? |
| 3 | ? | ? | ? |

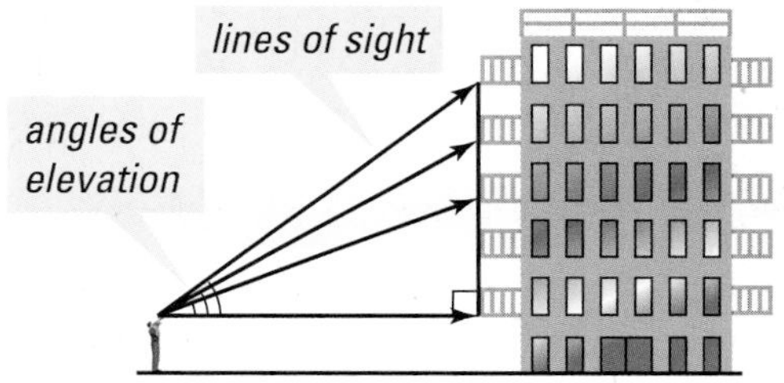


## Activity 3

### Measuring Down

**a.** Use your hypsometer to measure the angle between the building and your line of sight from three unknown heights to a specific object on the ground. Call these angles *A, A', and A''*. (The unknown heights can be different floors of a building.)

**b.** Measure the horizontal distance on the ground between the object and a point beneath the unknown heights.

**c.** Use the $\tan^{-1}$ function on your calculator and the formula $\tan A = \frac{d}{h}$, where $d$ is distance and $h$ is height, to compute the heights.

| Object | Distance | Angle | Height |
|---|---|---|---|
| 1 | ? | ? | ? |
| 2 | ? | ? | ? |
| 3 | ? | ? | ? |

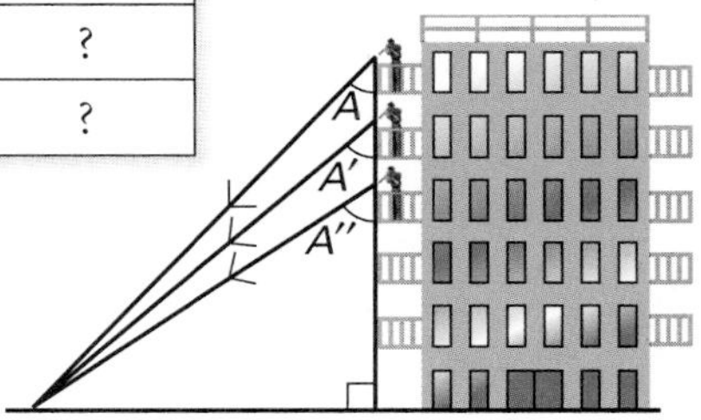

# 12 Chapter Review and Assessment

## VOCABULARY



## Key Skills & Exercises

### LESSON 12.1

#### Key Skills

**Simplify radical expressions, and estimate square roots.**

To express $\sqrt{12} + 2\sqrt{3}$ in simplest radical form, first use the Multiplication Property of Square Roots.

$$\begin{aligned}\sqrt{12} + 5\sqrt{3} &= \sqrt{4 \cdot 3} + 5\sqrt{3} \\ &= \sqrt{4} \cdot \sqrt{3} + 5\sqrt{3} \\ &= 2\sqrt{3} + 5\sqrt{3} = 7\sqrt{3} \\ &\approx 12.12\end{aligned}$$

#### Exercises

**Find the value of each square root. If the square root is irrational, find the value to the nearest hundredth.**

**1.** $\sqrt{20}$

**2.** $\sqrt{\frac{9}{16}}$

**Simplify each radical expression.**

**3.** $\sqrt{a^2b^7}$

**4.** $\sqrt{2} + 3\sqrt{7} - 3\sqrt{2}$

**5.** $(2\sqrt{3})^2$

**6.** $\sqrt{3}(2 - \sqrt{12})$

**7.** $(\sqrt{5} - 6)(\sqrt{5} + 6)$

**8.** $(\sqrt{17} - 8)(\sqrt{17} + 4)$

### LESSON 12.2

#### Key Skills

**Solve radical equations, and use radicals to solve equations.**

To solve $\sqrt{x+2} = 5$, square both sides of the equation. $(\sqrt{x+2})^2 = 5^2$

$$x + 2 = 25,\ x = 23$$

To solve $2x^2 = 36$, use the definition of square root. $2x^2 = 36,\ x^2 = 18$

$$x = \pm\sqrt{18} = \pm\sqrt{9 \cdot 2} = \pm 3\sqrt{2}$$

#### Exercises

**Solve each equation. Give exact answers. Remember to check your solutions.**

**9.** $\sqrt{x-7} = 2$

**10.** $\sqrt{3x+4} = 1$

**11.** $\sqrt{x^2 + 6x - 1} = x + 4$

**12.** $\sqrt{x^2 - 2x + 1} = x - 5$

**13.** $x^2 = 40$

**14.** $2x^2 - 32 = 0$

**15.** $5x^2 + 14 = 139$

**16.** $x = \sqrt{x + 12}$

## LESSON 12.3

### Key Skills

**Use the Pythagorean Theorem to find the length of a missing side of a right triangle.**

The Pythagorean Theorem, $a^2 + b^2 = c^2$, can be used to find the length of the hypotenuse of the right triangle.

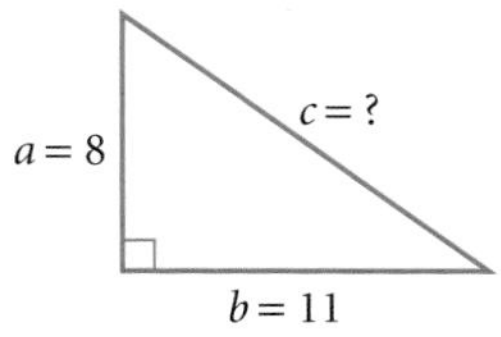


$$a^2 + b^2 = c^2$$
$$8^2 + 11^2 = c^2$$
$$64 + 121 = c^2$$
$$185 = c^2$$
$$c = \pm\sqrt{185} \approx \pm 13.60$$

Since the length of a segment must be positive, the length of the hypotenuse is approximately 13.6 meters.

### Exercises

**Use the Pythagorean Theorem to find the length of the missing side of each right triangle.**

**17.**

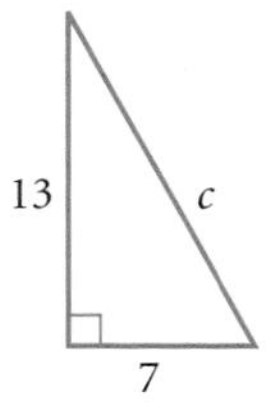


**18.**

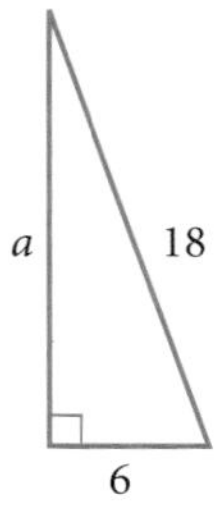


**19.** Find the hypotenuse of a right triangle whose legs have lengths of 12 and 5.

**20.** Can a triangle with side lengths of 2, $\sqrt{3}$, and $\sqrt{5}$ form a right triangle? Explain.

**21.** Can a triangle with side lengths of $\sqrt{5}$, 2, and 3 form a right triangle? Explain.

## LESSON 12.4

### Key Skills

**Use the distance formula to find the distance between two points, and use the midpoint formula to find the midpoint of a segment.**

To find the distance between (2, 3) and (−4, −1), use the distance formula.

$$d = \sqrt{(x_2 - x_1)^2 + (y_2 - y_1)^2}$$
$$= \sqrt{(-4 - 2)^2 + (-1 - 3)^2}$$
$$= \sqrt{36 + 16} = \sqrt{52} = 2\sqrt{13} \approx 7.21$$

To find the midpoint of a segment whose endpoints are (−2, 6) and (3, 4), use the midpoint formula, $\left(\frac{x_1 + x_2}{2}, \frac{y_1 + y_2}{2}\right)$

$$\left(\frac{x_1 + x_2}{2}, \frac{y_1 + y_2}{2}\right) \rightarrow \left(\frac{-2 + 3}{2}, \frac{6 + 4}{2}\right) \rightarrow \left(\frac{1}{2}, 5\right)$$

Notice that the midpoint formula simply finds the average of the $x$-coordinates and the average of the $y$-coordinates.

### Exercises

**Find the distance between the two points.**

**22.** $A(0, 3)$ and $B(2, 8)$

**23.** $X(-1, 5)$ and $Y(3, -8)$

**24.** $M(5, 2)$ and $N(3, 6)$

**25.** S(6, 4) and T(9, 0)

**Find the midpoint of each segment with the given endpoints.**

**26.** $M(5, 2)$ and $N(3, 6)$

**27.** $P(-3, 4)$ and $Q(-7, -1)$

**28.** $R(8, -2)$ and $S(-1, 4)$

**29.** $M(-4, 3)$ and $N(5, -2)$

**30.** $R(5, -7)$ and $S(-8, -12)$

**31.** $Q(3, 7)$ and $R(-8, -8)$

**32.** $A(7, -4)$ and $B(8, 4)$

**33.** $A(a, b)$ and $B(3a, 2b)$

## LESSON 12.5

### Key Skills

**Find the equation of a circle whose center and radius are given.**

To find the equation of a circle with its center at (2, −3) and a radius of 5, use the equation of a circle, $(x-h)^2+(y-k)^2=r^2$, where $(h, k)$ is the center and $r$ is the radius. Thus, the equation is $(x-2)^2+(y+3)^2=25$.

### Exercises

**Find the equation of each circle with the given center and radius.**

**34.** center at (0, 0) and radius of 2

**35.** center at (−2, 5) and radius of 5

**36.**

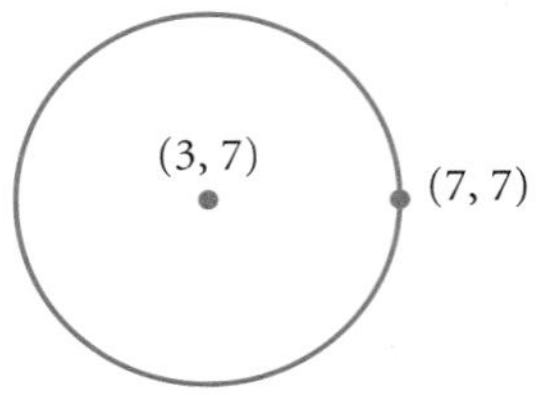


**37.**

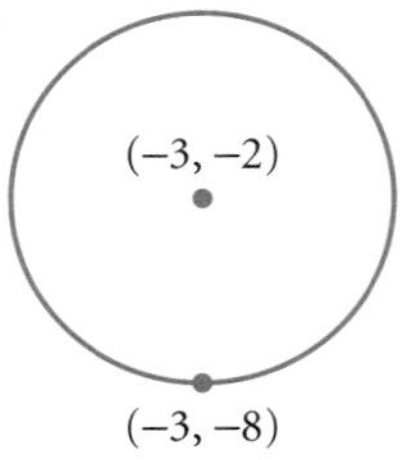


**From each equation of a circle, give the center and radius.**

**38.** $(x-2)^2+(y-3)^2=36$

**39.** $(x+0.25)^2+y^2=6.25$

## LESSON 12.6

### Key Skills

**Find the tangent of an acute angle of a right triangle.**

The tangent of an acute angle in a right triangle is the ratio of the length of its opposite side to the length of its adjacent side.

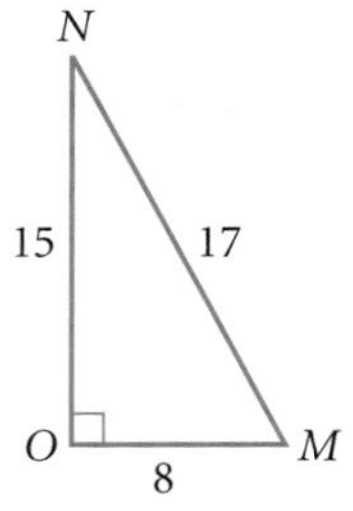


The tangent of angle $M$ is $\frac{opp}{adj}=\frac{15}{8}$. The tangent of angle $N$ is $\frac{opp}{adj}=\frac{8}{15}$.

### Exercises

**Find the indicated tangent ratio for the right triangles below.**

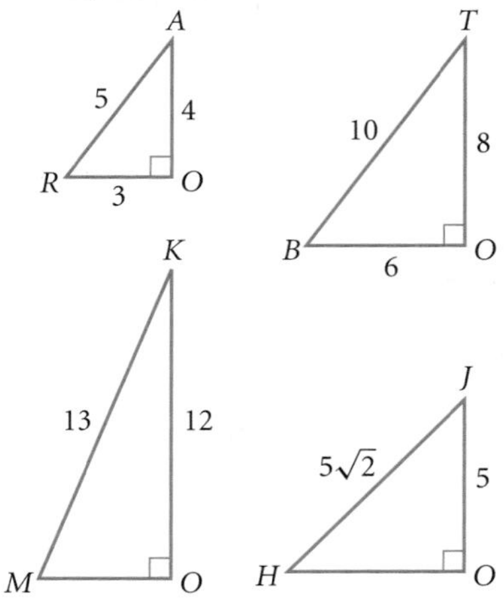


**40.** tan $M$

**41.** tan $K$

**42.** tan $A$

**43.** tan $B$

**44.** tan $R$

**45.** tan $T$

**46.** tan $H$

**47.** tan $J$

## LESSON 12.7

### Key Skills

**Find the sine and cosine of an acute angle of a right triangle.**

The sine of an angle in a right triangle is the ratio of the length of the side opposite the angle to the length of the hypotenuse of the triangle.

The cosine of an angle in a right triangle is the ratio of the length of the side adjacent to the angle to the length of the hypotenuse of the triangle.

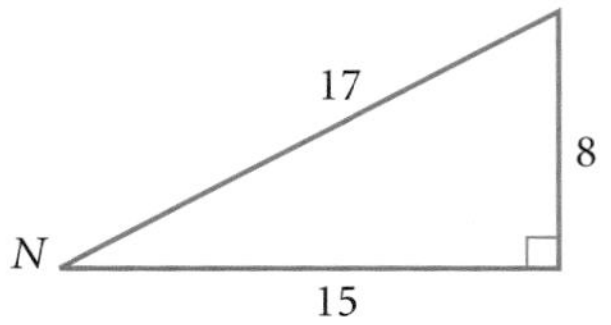


The sine of angle $N$ is $\sin N = \frac{opp}{hyp} = \frac{8}{17}$.

The cosine of angle $N$ is $\cos N = \frac{adj}{hyp} = \frac{15}{17}$.

If you know the sine or cosine of an angle, the degree measure can be found by using the $\sin^{-1}$ or $\cos^{-1}$ function on your calculator.

The measure of angle $N$ can be found by using either $\sin^{-1}$ or $\cos^{-1}$.

$$\sin^{-1}(N) = \sin^{-1}\left(\frac{8}{17}\right) = 28.07°$$

$$\cos^{-1}(N) = \cos^{-1} = \cos^{-1}\left(\frac{15}{17}\right) = 28.07°$$

### Exercises

**Find the indicated sine or cosine to the nearest hundredth.**

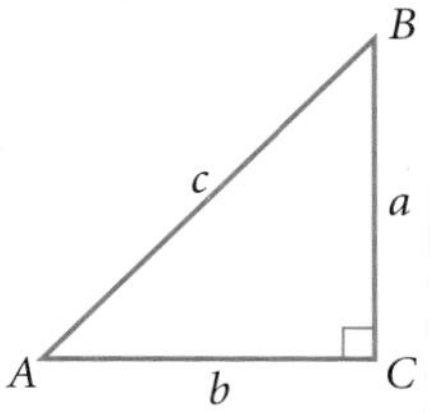


**48.** $a = 5,\ c = 9,\ \cos\angle B = ?$

**49.** $b = 6,\ c = 12,\ \cos\angle A = ?$

**50.** $a = 1,\ c = 2,\ \sin\angle A = ?$

**51.** $b = 8,\ c = 15,\ \sin\angle B = ?$

**52.** $a = 5,\ c = 14,\ \cos\angle B = ?$

**53.** $a = 6,\ c = 10,\ \cos\angle B = ?$

**54.** $b = 3,\ c = 18,\ \sin\angle B = ?$

**55.** $a = 8,\ c = 15,\ \sin\angle A = ?$

**Find the degree measure of each angle to the nearest hundredth.**

**56.** $b = 7,\ c = 24,\ m\angle A = ?$

**57.** $a = 14,\ c = 27,\ m\angle B = ?$

**58.** $a = 3,\ c = 9,\ m\angle A = ?$

**59.** $a = 9,\ c = 20,\ m\angle A = ?$

## LESSON 12.8

### Key Skills

**Add and subtract matrices.**

In order to add or subtract matrices, the matrices must have the same dimensions. Then the corresponding entries from each matrix are added or subtracted.

$$\begin{bmatrix} 5 & 7 \\ -3 & -9 \end{bmatrix} - \begin{bmatrix} 9 & 3 \\ -4 & -8 \end{bmatrix} = \begin{bmatrix} 5-9 & 7-3 \\ -3-(-4) & -9-(-8) \end{bmatrix}$$

$$= \begin{bmatrix} -4 & 4 \\ 1 & -1 \end{bmatrix}$$

### Exercises

**Perform the indicated matrix operations.**

**60.** $\begin{bmatrix} -2 & 8 \\ -12 & 4 \end{bmatrix} + \begin{bmatrix} 4 & 6 \\ -1 & -7 \end{bmatrix}$

**61.** $\begin{bmatrix} 2 & 14 \\ 3 & -10 \end{bmatrix} + \begin{bmatrix} -4 & 1 \\ -3 & 8 \end{bmatrix}$

**62.** $\begin{bmatrix} 5 & -1 \\ 3 & 8 \end{bmatrix} - \begin{bmatrix} 2 & 4 \\ 7 & -1 \end{bmatrix}$

**63.** $\begin{bmatrix} 11 & 14 \\ 2 & 16 \end{bmatrix} - \begin{bmatrix} 4 & 3 \\ -2 & -14 \end{bmatrix}$

**64.** $\begin{bmatrix} 253 & 786 \\ 212 & 1452 \end{bmatrix} - \begin{bmatrix} 221 & 753 \\ 819 & 1221 \end{bmatrix}$

**Multiply a matrix by a scalar and by another matrix.**

To multiply a matrix by a scalar, multiply each element in the matrix by the scalar.

$$-7\begin{bmatrix} 4 & -8 \\ -4 & 7 \end{bmatrix} = \begin{bmatrix} -7(4) & -7(-8) \\ -7(-4) & -7(7) \end{bmatrix}$$
$$= \begin{bmatrix} -28 & 56 \\ 28 & 49 \end{bmatrix}$$

To multiply a matrix by another matrix, the number of columns in the matrix on the left must match the number of rows in the matrix on the right. To multiply, find the product of each entry in a row of the matrix on the left with the corresponding column entry in the matrix on the right. Then add the products.

$$\begin{bmatrix} -6 & 4 \\ -3 & 8 \end{bmatrix}\begin{bmatrix} 8 & -2 \\ 4 & -1 \end{bmatrix}$$
$$= \begin{bmatrix} -6(8) + 4(4) & -6(-2) + 4(-1) \\ -3(8) + 8(4) & (-3)(-2) + 8(-1) \end{bmatrix}$$
$$= \begin{bmatrix} -32 & 8 \\ 8 & -2 \end{bmatrix}$$

**Perform the indicated matrix operations.**

**65.** $4\begin{bmatrix} -3 & 12 \\ 24 & -8 \end{bmatrix}$

**66.** $-12\begin{bmatrix} 4 & -12 \\ -14 & -11 \end{bmatrix}$

**67.** $23\begin{bmatrix} 14 & 2 \\ 13 & -6 \end{bmatrix}$

**68.** $\begin{bmatrix} 4 & 3 \\ 8 & 7 \end{bmatrix}\begin{bmatrix} 6 & 3 \\ 8 & 9 \end{bmatrix}$

**69.** $\begin{bmatrix} -3 & 8 \\ -14 & 7 \end{bmatrix}\begin{bmatrix} -2 & -8 \\ 3 & -6 \end{bmatrix}$

**70.** $\begin{bmatrix} -6 & -2 \\ 4 & 8 \end{bmatrix}\begin{bmatrix} -3 & -8 \\ 9 & 10 \end{bmatrix}$

**71.** $\begin{bmatrix} 1 & 0 \\ 0 & 1 \end{bmatrix}\begin{bmatrix} 8 & 9 \\ 12 & 356 \end{bmatrix}$

**72.** $\begin{bmatrix} 1 & 0 \\ 0 & 1 \end{bmatrix}\begin{bmatrix} 856 & 231 \\ -876 & 329 \end{bmatrix}$

**73.** $\begin{bmatrix} 845 & -765 \\ -156 & -943 \end{bmatrix}\begin{bmatrix} 1 & 0 \\ 0 & 1 \end{bmatrix}$

## Applications

**74.** **CONSTRUCTION** A ladder is leaning against a house. It forms an angle of 55° with the house. If the ladder touches the house at a point 12 feet from the ground, how far is the bottom of the ladder from the bottom of the house?

**75.** **SPORTS** At the end of a ski lift, riders are 120 feet above their starting point. The ski lift travels a horizontal distance of 500 feet.
- **a.** How long is the cable that carries the ski lift?
- **b.** What angle does the cable make with the ground at the bottom of the lift?

**76.** **TRANSPORTATION** Suppose you are driving up a hill whose total height is 2000 feet. Your odometer shows that you drove a total of 6 miles to reach the top. What horizontal distance have you traveled? (Hint: 1 mile = 5280 feet)

# 12 Chapter Test

**Find each square root. If the square root is irrational, approximate the value to the nearest hundredth.**

**1.** $\sqrt{400}$ **2.** $\sqrt{\frac{49}{225}}$ **3.** $\sqrt{125}$

**Simplify each radical expression.**

**4.** $\sqrt{25}\sqrt{18}$ **5.** $\sqrt{5} + 2\sqrt{20}$

**6.** $\left(\sqrt{2} - 1\right)^2$ **7.** $\sqrt{r^3 s^6}$

**Solve each equation, and check your solutions. Give exact answers.**

**8.** $a^2 = 45$ **9.** $\sqrt{m + 4} = -3$

**10.** $\sqrt{z^2 + 4z - 18} = z$

**11.** $x + 2 = \sqrt{x^2 + 2x + 4}$

**12.** PHYSICS An object falling from rest covers distance of $d = 9800t^2$mm in $t$ seconds. Solve this equation for $t$ and determine how many seconds it takes for an object to fall 22,050 mm.

**Use the Pythagorean Theorem to find the length of the missing side of each right triangle.**

**13.**

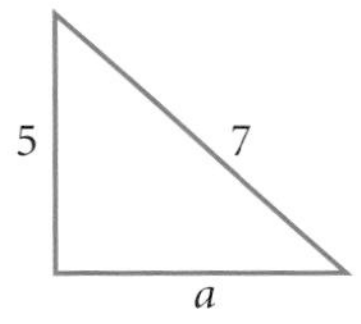


**14.**

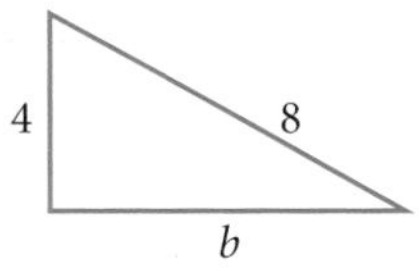


**15.** GEOMETRY An isosceles right triangle has two equal legs. If an isosceles right triangle has a hypotenuse of 12, what are the lengths of the two legs?

**Find the distance between the two points.**

**16.** $R(5, 7)$ and $S(2, 9)$

**17.** $P(-3, 2)$ and $Q(4, -5)$

**Find the midpoint of each segment with the given endpoints.**

**18.** $G(8, 2)$ and $H(2, 3)$

**19.** $M(-13, -5)$ and $N(-3, -4)$

**Find the equation of the circle with the given center and radius.**

**20.** Center at (3, 4) and radius of 8.

**21.**

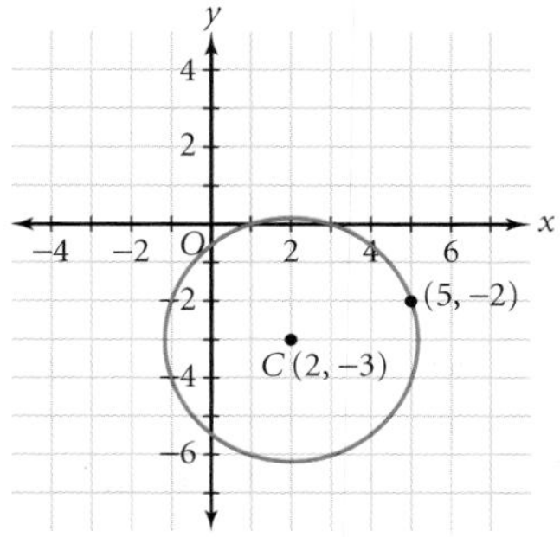


**22.** Find the center and radius of the circle with equation $(x + 4)^2 + (y - 1)^2 = 5$

**Use the right triangles for Exercises 23–30.**

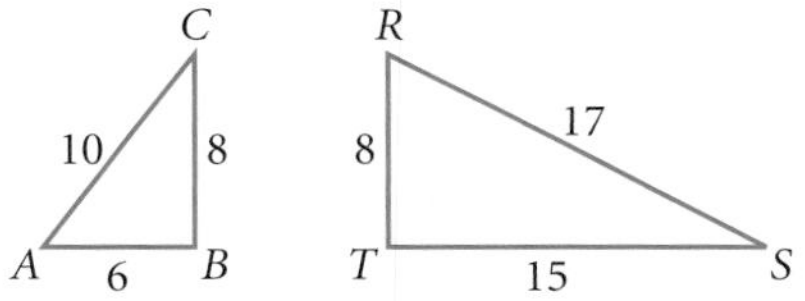


**Find the indicated tangent ratio.**

**23.** tan $A$ **24.** tan $R$

**25.** tan $S$ **26.** tan $C$

**Find the indicated sine or cosine to the nearest hundredth.**

**27.** sin $A$ **28.** cos $R$

**Find the degree measure of each angle to the nearest hundredth.**

**29.** $m\angle A$ **30.** $m\angle R$

**31.** CIVIL ENGINEERING A guy wire attached to the top of an antenna makes an angle of 52° with the antenna. If the wire is anchored to the ground 40 feet from the antenna, how long is the wire?

**Perform the indicated matrix operations.**

**32.** $\begin{bmatrix} 2 & -3 \\ 7 & 1 \end{bmatrix} - \begin{bmatrix} -2 & 5 \\ 7 & 4 \end{bmatrix}$ **33.** $\begin{bmatrix} 1 & -3 \\ 5 & 2 \end{bmatrix}\begin{bmatrix} -2 & 1 \\ 0 & 4 \end{bmatrix}$

**34.** $\begin{bmatrix} 1 & 4 \\ -2 & 5 \end{bmatrix} + \begin{bmatrix} 3 & -8 \\ 2 & -4 \end{bmatrix}$ **35.** $15\begin{bmatrix} 5 & -7 \\ -3 & 4 \end{bmatrix}$

# College Entrance Exam Practice

internet connect
Standardized Test Prep Online
Go To: go.hrw.com
Keyword: MM1 Test Prep.

**QUANTITATIVE COMPARISON** For Questions 1–5 write:
A if the quantity in Column A is greater than the quantity in Column B;
B if the quantity in Column B is greater than the quantity in Column A;
C if the quantities are equal; or
D if the relationship cannot be determined from the information given.

| | Column A | Column B | Answers |
|---|---|---|---|
| 1. | The mean of 16, 26, 21, 23, 19 | The median of 16, 26, 21, 23, 19 | Ⓐ Ⓑ Ⓒ Ⓓ [Lesson 4.4] |
| 2. | The solution to $\frac{6}{n} = \frac{9}{15}$ | The solution to $\frac{n}{15} = \frac{9}{25}$ | Ⓐ Ⓑ Ⓒ Ⓓ [Lesson 4.1] |
| 3. | The solution to $5(x + 4) = 25$ | The solution to $4(x - 3) = 16$ | Ⓐ Ⓑ Ⓒ Ⓓ [Lesson 3.5] |
| 4. | $(2 + \sqrt{3})(2 - \sqrt{3})$ | 8 | Ⓐ Ⓑ Ⓒ Ⓓ [Lesson 12.1] |
| 5. | $\frac{8b - 2}{-6}$, where $b = 4$ | $3(b - 6)$, where $b = 4$ | Ⓐ Ⓑ Ⓒ Ⓓ [Lesson 2.4] |

**6.** If $y$ varies directly as $x$ and $y$ is 3 when $x$ is 8, what is the constant of variation? ***(LESSON 5.3)***

**a.** $\frac{8}{3}$ **b.** 3
**c.** 8 **d.** $\frac{3}{8}$

**7.** Let $f(x) = 2^x$. Find $f(x)$ when $x = 3$. ***(LESSON 8.6)***

**a.** 6 **b.** $\frac{1}{6}$
**c.** 4 **d.** 8

**8.** Simplify: $(x - y) - (y + x)$. ***(LESSON 2.3)***

**a.** $2x + 2y$ **b.** $2x - 2y$
**c.** $-2y$ **d.** $2y - 2x$

**9.** Solve. $\begin{cases} y = x + 3 \\ 2x - 3y = -12 \end{cases}$ ***(LESSON 7.2)***

**a.** $(-3, -6)$ **b.** $(3, -6)$
**c.** $(3, 6)$ **d.** $(-3, 6)$

**10.** Solve: $|x + 5| < 5$. ***(LESSON 6.5)***

**a.** $-10 < x < 0$ **b.** $-5 < x < 5$
**c.** $-10 < x < 5$ **d.** $-10 < x < 10$

**11.** Find 30% of 50. ***(LESSON 4.2)***

**a.** 1.5 **b.** 150
**c.** 1500 **d.** 15

**12.** Find the slope of the line connecting the points $M(3, -5)$ and $N(-6, 14)$. ***(LESSON 5.2)***

**a.** $\frac{7}{3}$ **b.** $\frac{19}{9}$
**c.** $-\frac{19}{9}$ **d.** $\frac{9}{19}$

**13.** Find the equation of the line passing through the points $(3, 4)$ and $(-1, -2)$. ***(LESSON 5.5)***

**a.** $y = 3x - 1$ **b.** $y = \frac{2}{3}x - 2$
**c.** $y = \frac{3}{2}x - 1$ **d.** $y = \frac{3}{2}x - \frac{1}{2}$

**14.** Multiply: $(2x+4)(2x-4)$. ***(LESSON 9.3)***

**a.** $4x^2 - 16$ **b.** $x^2 - 4$
**c.** $2x^2 - 16x - 16$ **d.** $2x^2 + 16x - 16$

**15.** What is the slope of a line perpendicular to $2x + 4y = 5$? ***(LESSON 5.6)***

**a.** 2 **b.** $\frac{1}{2}$
**c.** $-2$ **d.** $-\frac{1}{2}$

**16.** Find the distance between $A(0, 1)$ and $B(4, 5)$. ***(LESSON 12.4)***

**a.** 16 **b.** $8\sqrt{2}$
**c.** 2 **d.** $4\sqrt{2}$

**17.** Solve: $5x + 8 \le 3x$. ***(LESSON 6.2)***

**a.** $x \le -4$ **b.** $x \le 4$
**c.** $x \ge 4$ **d.** $x \ge -4$

**18.** Factor: $2x^2 + 12x + 18$. ***(LESSON 9.6)***

**a.** $(2x+3)^2$ **b.** $(2x+3)(2x-3)$
**c.** $(x+3)^2$ **d.** $2(x+3)^2$

**19.** Let $P = \begin{bmatrix} 3 & 0 \\ -5 & 1 \end{bmatrix}$ and $Q = \begin{bmatrix} 5 & -1 \\ -9 & -1 \end{bmatrix}$. Find $P - Q$. ***(LESSON 12.8)***

**20.** Solve: $9x^2 = 1$. ***(LESSON 10.2)***

**21.** List the domain of the following relation: $\{(2, -3), (2, 3), (4, 5), (4, -5)\}$. ***(LESSON 5.1)***

**22.** Solve: $\frac{1.2}{7.5} = \frac{n}{2.5}$. ***(LESSON 4.1)***

**23.** Find the midpoint of $\overline{MN}$, where $M = (3, -4)$ and $N = (7, 8)$. ***(LESSON 12.4)***

**24.** What is the $y$-intercept of the graph of $3x - 2y = 6$? ***(LESSON 5.5)***

**25.** Factor: $6x^2 + 9x + 3$. ***(LESSON 9.6)***

**26.** Mary has a solution that is 60% alcohol and another that is 20% alcohol. How much of each should she use to make 100 milliliters of a solution that is 52% alcohol? ***(LESSON 7.6)***

**27.** Simplify: $8 - (3 - 5) - 4$. ***(LESSON 2.3)***

**28.** Solve: $x + \frac{1}{2} = \frac{x}{2} + 3$. ***(LESSON 3.4)***

### FREE-RESPONSE GRID

**The following questions may be answered by using a free-response grid such as that commonly used by standardized test services.**

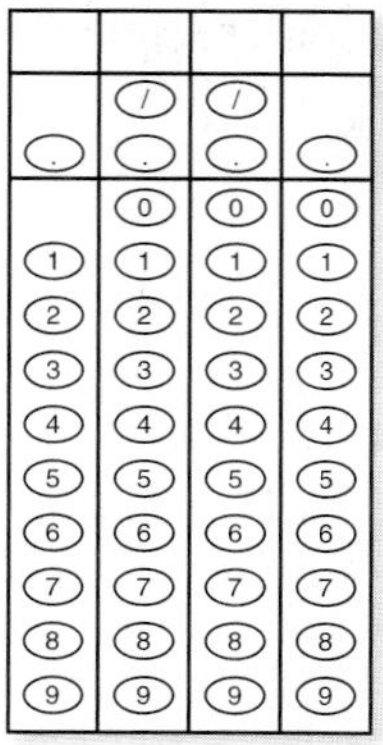

**29.** Let $g(x) = x^2 - 5x + 6$. Evaluate $g(3)$. ***(LESSON 14.1)***

**30.** Find the slope of $5x - 2y = 8$. ***(LESSON 5.5)***

**31.** One month, Latoya spent \$300 more than one-third of her usual budget. If Latoya spent \$390 that month, how much is Latoya's usual budget? ***(LESSON 3.3)***

**32.** Simplify: $5 - |-3| - 4$. ***(LESSON 2.2)***

**33.** Simplify: $\left(\frac{2}{3}\right)^{-3}$. ***(LESSON 8.4)***

**34.** Name the next term in the sequence 8, 13, 21, 34, 55, … ***(LESSON 1.1)***

**35.** Solve $3x - 5 = 16$ for $x$. ***(LESSON 3.3)***

**36.** What number is 45% of 60? ***(LESSON 4.2)***

**37.** Find the slope of the line containing the points $(3, 4)$ and $(6, 10)$. ***(LESSON 5.2)***

**38.** Find the first differences of the sequence 2, 8, 14, 20, 26, … ***(LESSON 1.1)***

**39.** Name the constant differences of the sequence 20, 21, 26, 35, 48, … ***(LESSON 1.1)***

**40.** 9 is what percent of 60? ***(LESSON 4.2)***

**41.** Evaluate: $\sqrt{121}$. ***(LESSON 12.1)***

**42.** What is the length of the hypotenuse of a right triangle with legs of 5 and 12 units? ***(LESSON 12.3)***

**43.** Find the distance between the points $M(5, -6)$ and $N(-7, -1)$. ***(LESSON 12.4)***

13

# Probability

PROBABILITY TOUCHES YOUR LIFE IN MANY WAYS. Games, insurance, matters of health and diet, and countless other situations involve an element of chance. Statistics makes extensive use of probability to study, predict, and draw conclusions from data. Sometimes probabilities are found experimentally by looking at data. Other times, probabilities are found theoretically by considering all the possibilities. In this chapter, you can learn about both kinds of probability and explore examples of probability in action.

## Lessons

*The pictures on these pages illustrate rare events such as the birth of triplets, coins coming up all heads, and misprinted postage stamps.*

CHAPTER PORTFOLIO ACTIVITIES PROJECT

## About the Chapter Project

The study of probability provides the power to predict the outcome of chance events over a series of trials. In the Chapter Project, *Happy Birthday!*, on page 682, you will perform an experiment that will allow you to predict the chances that two people in a group will have the same birthday or birth month. After completing the Chapter Project, you will be able to do the following:

- Perform a probability experiment.

## About the Portfolio Activity

Throughout the chapter, you will be given opportunities to complete Portfolio Activities that are designed to support your work on the Chapter Project.

- Theoretical probability is the topic of the Portfolio Activity on page 653.
- Experimental probability is the topic of the Portfolio Activity on page 666.
- The results of a three-game series between two evenly matched teams is the topic of the Portfolio Activity on page 673.
- A comparison of the probabilities of independent and dependent events is the topic of the Portfolio Activity on page 681.

13.1

# Theoretical Probability

*The coin toss before a football game is considered a fair way to determine initial possession because heads or tails are equally likely.*

**Objectives**

- List or describe the sample space of an experiment.
- Find the theoretical probability of a favorable outcome.

**Why** *Theoretical probability is a mathematical model that is helpful in making predictions. Exact models can be found without actually doing experiments.*

**APPLICATION**
**SPORTS**

The result of a coin toss is used to determine which team first gets the ball in a football game. Whether the coin lands heads or tails is an interesting question of probablility.

Consider the following probability experiment:

*Toss a coin one time and see whether it lands heads or tails.*

For this experiment, which is very simple, there are two possible *outcomes*—heads or tails. The outcomes are said to be **equally likely outcomes** because a tossed coin is just as likely to be heads as tails.

In the long run, a tossed coin tends to come up heads half of the time and tails half of the time (recall Lesson 4.3). Thus, the probability of getting heads is said to be $\frac{1}{2}$, and the probability of getting tails is said to be $\frac{1}{2}$. In probability notation:

$$P(\text{heads}) = \frac{1}{2} \quad \text{and} \quad P(\text{tails}) = \frac{1}{2}$$

The sum of the probabilities is 1.

Similarly, for a single roll of a number cube there are six possible outcomes— 1, 2, 3, 4, 5, and 6—and each of them is equally likely.

$P(1) = \frac{1}{6}$ $\quad P(2) = \frac{1}{6}$ $\quad P(3) = \frac{1}{6}$

$P(4) = \frac{1}{6}$ $\quad P(5) = \frac{1}{6}$ $\quad P(6) = \frac{1}{6}$

The sum of the probabilities is again 1.

The set of all possible outcomes of a probability experiment is called the **sample space** of the experiment. For the experiment of tossing a coin once, a sample space is {heads, tails} or {H, T}. For rolling a number cube once, a sample space is {1, 2, 3, 4, 5, 6}.

**EXAMPLE 1** **Find the sample space for the experiment of tossing a coin twice.**

**SOLUTION**

There are four different possible results from tossing a coin twice.

| First Toss | Second Toss |
|---|---|
| H | H |
| H | T |
| T | H |
| T | T |

You can write these in set notation as {HH, HT, TH, TT}.

**TRY THIS** Find the sample space for the experiment of tossing a coin three times.

In a probability experiment, to say that an outcome is "favorable" does not mean that it is good or bad. A **favorable outcome** is the outcome you are looking for when you do the experiment.

**EXAMPLE 2** A coin is tossed twice. You are interested in the probability that both tosses will be heads or both will be tails. **What are the favorable outcomes in the sample space for the experiment? How many favorable outcomes are there?**

**SOLUTION**

The sample space for the experiment is {HH, HT, TH, TT}. (See Example 1.) The outcomes in which the two tosses are both heads or both tails (the favorable outcomes) are shown in red. There are two favorable outcomes.

**TRY THIS** A coin is tossed three times. You are interested in the probability that two or more of the tosses will be heads. What is the sample space for the experiment? What are the favorable outcomes? How many favorable outcomes are there?

To find the theoretical probability of something happening, you first need to define the sample space and the favorable outcomes.

**Theoretical Probability**

Let $n$ be the number of elements in a sample space in which each element is equally likely to occur. Let $E$ be an event for which there are $f$ favorable outcomes in the sample space.

$$P(E) = \frac{f}{n}$$

**EXAMPLE 3** **A number cube is rolled twice. What is the probability that the sum of the rolls will be 9?**

**SOLUTION**

The sample space for the experiment is {(1, 1), (1, 2), (1, 3), (1, 4), (1, 5), (1, 6), (2, 1), (2, 2), (2, 3), (2, 4), (2, 5), (2, 6), (3, 1), (3, 2), (3, 3), (3, 4), (3, 5), (3, 6), (4, 1), (4, 2), (4, 3), (4, 4), (4, 5), (4, 6), (5, 1), (5, 2), (5, 3), (5, 4), (5, 5), (5, 6), (6, 1), (6, 2), (6, 3), (6, 4), (6, 5), (6, 6)}

Out of 36 possibilities, there are 4 favorable outcomes. The probability is $P(\text{sum of } 9) = \frac{4}{36} = \frac{1}{9}$.

**TRY THIS** A coin is tossed three times. What is the probability that only one or only two of the tosses will be heads?

**EXAMPLE 4** On the first day of class, a teacher puts a sticker under 3 of the 20 chairs in her classroom. A second teacher puts a sticker under 4 of the 28 chairs in her classroom. Students who sit in the chairs with stickers will receive a $5 gift certificate to the school cafeteria. **In which classroom is the probability of receiving a gift certificate greater?**

**SOLUTION**

The sample space for the first teacher consists of 20 chairs, and there are 3 favorable outcomes. The sample space for the second teacher consists of 28 chairs, and there are 4 favorable outcomes.

$P(\text{receiving certificate in classroom 1}) = \frac{3}{20}$, or 15%

$P(\text{receiving certificate in classroom 2}) = \frac{4}{28}$, or about 14%

Thus, a student in the first classroom has a slightly greater probability of receiving a gift certificate than a student in the second classroom.

**TRY THIS** Which has the greater probability: 19 chances out of 91 or 17 chances out of 71?

# Exercises

## Communicate

1. Based on his record so far this season the probability that Nat will make a successful free throw is 40%. Is this a theoretical probability or an experimental probability (recall Lesson 4.3)? Explain.
2. In a two-player game, 4 coins are tossed. The first player wins if the coins come up all heads or all tails. Otherwise the second player wins. Does each player have an equally likely chance of winning? Explain.
3. When finding the theoretical probability of rolling a certain number on a number cube, why must each possibility be equally likely?

## Guided Skills Practice

**Find the sample space for each experiment.** ***(EXAMPLE 1)***

4. rolling a number cube twice
5. tossing a coin four times

**Find the number of *favorable* outcomes in the sample space for each experiment.** ***(EXAMPLE 2)***

6. A number cube is rolled twice. The sum of the rolls is 7.
7. A coin is tossed three times. All three tosses are heads.

**Find the probability of each outcome.** ***(EXAMPLE 3)***

8. the probability that two rolls of a number cube will have a sum of 7
9. the probability that three tosses of a coin will be three heads

**Which has the greater probability?** ***(EXAMPLE 4)***

10. 12 chances out of 53 or 18 chances out of 74
11. 34 chances out of 71 or 53 chances out of 117

## Practice and Apply

**An integer between 1 and 25, inclusive, is drawn at random. Find each probability.**

12. The integer is even.
13. The integer is odd.
14. The integer is a multiple of 3.
15. The integer is a multiple of 5.

internet connect

**Homework Help Online**
Go To: **go.hrw.com**
Keyword:
**MA1 Homework Help** for Exercises 16-19

**Find the probability of each outcome.**

**16.** the probability that two rolls of a number cube will have a sum of 3

**17.** the probability that two tosses of a coin will be one heads and one tails

**18.** the probability that two rolls of a number cube will have a sum of 11

**19.** the probability that three tosses of a coin will be at least two heads

**Suppose that you select a letter of the English alphabet at random. Find each probability.**

**20.** The letter is *r*.

**21.** The letter is a vowel.

**22.** The letter is a consonant.

**23.** The letter is in the word *mathematics*.

CHALLENGE

**24.** John is trying to solve the equation $2x - 1 = 15$ on an algebra quiz. He doesn't remember how to do it, but he reasons as follows: The answer, $x$, is multiplied by 2 and comes out to be around 15. He assumes that the answer must be a positive one-digit number, and he makes a guess. Find the probability that he guesses the correct solution.

APPLICATIONS

**25.** **GEOGRAPHY** Find the probability that the name of a state chosen at random from the United States begins with the letter *M*.

**26.** **ASTRONOMY** Find the probability that a planet chosen at random in our solar system is closer to the Sun than Earth is. (Hint: Of the 9 planets, only Mercury and Venus are closer to the Sun than Earth is.)

**27.** **LANGUAGE ARTS** List all possible arrangements of the letters in the word *tap*. Find the probability that an arrangement chosen at random is a word.

**CULTURAL CONNECTION: AMERICAS** The Blackfoot people of Montana played a game of chance in which they threw 4 decorated game pieces made from buffalo ribs. Three were blank on one side, but the fourth, called the chief, was engraved on both sides like a coin with heads and tails. The highest score was 6 points, for a throw of 3 blanks and the chief landing heads up. For 3 blanks and the chief reversed, the score was 3 points.

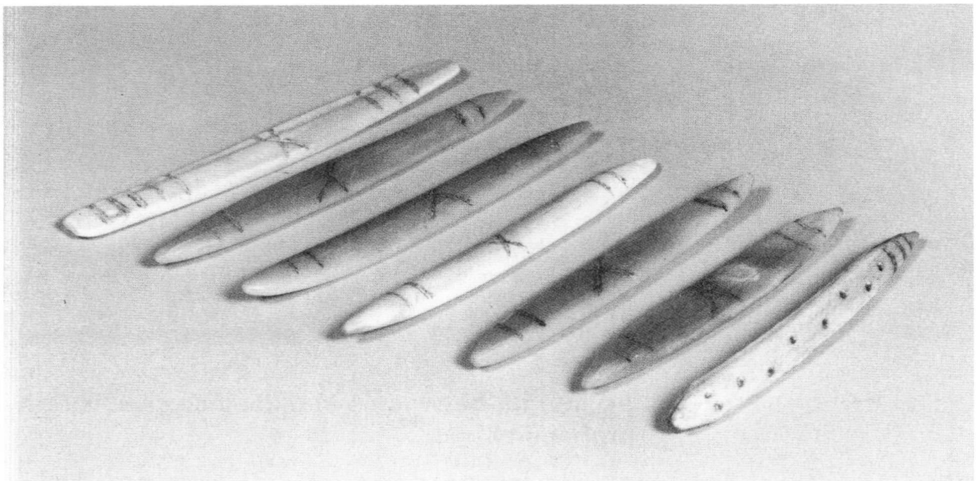

**28.** What is the probability of throwing 3 blanks and the chief landing heads up?

**29.** What is the probability of throwing 3 blanks and the chief reversed?

## Look Back

**30.** Solve the equation $y = mx + b$ for $m$. *(LESSON 3.6)*

**31.** An item that costs \$53 totals \$56.18 with sales tax. Use the proportion method to find the rate of the sales tax. *(LESSON 4.2)*

**Karl's typing speeds on 5 tests were 19, 21, 26, 21, and 27 words per minute.**

**32.** What is the mode of Karl's typing speeds? *(LESSON 4.4)*

**33.** What is the median of Karl's typing speeds? *(LESSON 4.4)*

**34.** Given $-6 \le n - 4 < 10$, graph the solution to the inequality on a number line for values of $n$ that are integers. *(LESSON 6.3)*

**35.** Draw the graph of the function $y = \frac{1}{x}$ for the interval $\frac{1}{4} < x \le 4$. What is the range of the function? *(LESSON 11.2)*

## Look Beyond

CONNECTION

**36.** GEOMETRY A dart is thrown at a circular target. Assuming that it is equally likely to land anywhere on the target, what is the probability that it hits the bull's-eye, the red region in the center? (The measurements indicate the diameter of each circular region.)

internet connect
**Portfolio Extension**
Go To: **go.hrw.com**
Keyword:
**MA1 Blaise**

PORTFOLIO ACTIVITY

Include this activity in your portfolio.

A game is played by rolling two number cubes. Player A wins if the sum is 2, 3, 4, 5, 10, 11, or 12. Player B wins if the sum is 6, 7, 8, or 9.

| Sum | 2 | 3 | 4 | 5 | 6 | 7 | 8 | 9 | 10 | 11 | 12 |
|---|---|---|---|---|---|---|---|---|---|---|---|
| # of ways to Roll Sum | 1 | 2 | ? | ? | ? | 6 | 5 | ? | ? | ? | 1 |

1. Complete the table above.
2. How many favorable sums are possible for player A?
3. How many favorable sums are possible for player B?
4. Find the theoretical probability of player A winning the game.
5. Find the theoretical probability of player B winning the game.
6. Which player has the better chance of winning? Explain.

13.2

# Counting the Elements of Sets

**Objectives**

- Find the union and intersection of sets.
- Count the elements of sets.
- Apply the Addition of Probabilities Principle.

**Why** *In many situations, you need to count the number of favorable outcomes that occur in subsets of favorable outcomes.*

*Maria keeps a database of addresses of her friends and relatives. The user of a database can list records and fields in various ways. Maria will need these capabilities in order to write thank-you notes for her Quinceañera gifts.*

**APPLICATION**
**INTERPERSONAL SKILLS**

A database user, such as Maria in the photo, may want to create a number of different lists. For example, Maria can print out a list of all the people in her database who are her relatives.

| | A | B | C | D | E | F | G |
|---|---|---|---|---|---|---|---|
| 1 | Last name | First name | Street | City | State | Zip | Relative? |
| 2 | Sanchez | Richardo | 1823 Putnam | Santa Fe | NM | 87010 | Yes |
| 3 | Gutierrez | Alfred | 3201 Swan's Crossing | Victoria | TX | 77901 | Yes |
| 4 | Lopez | Pauline | 1573 Piedmont | Austin | TX | 78746 | Yes |
| 5 | Ortiz | Mary | 809 Millwood | Pensacola | FL | 32514 | Yes |
| 6 | Hays | Laura | 2337 Sneed | Edinburg | TX | 78539 | Yes |
| 7 | Trimble | Alicia | 408 Elm Creek | Berea | OH | 44017 | Yes |

She can also print out a list of people in her database who live in her home state, Texas.

| | A | B | C | D | E | F | G |
|---|---|---|---|---|---|---|---|
| 1 | Last name | First name | Street | City | State | Zip | Relative? |
| 2 | Gutierrez | Alfred | 3201 Swan's Crossing | Victoria | TX | 77901 | Yes |
| 3 | Lopez | Pauline | 1573 Piedmont | Austin | TX | 78746 | Yes |
| 4 | Hays | Laura | 2337 Sneed | Edinburg | TX | 78539 | Yes |
| 5 | Vasquez | David | 784 Fairfield Dr | San Antonio | TX | 78034 | No |
| 6 | Gomez | George | 2305 Main | San Antonio | Tx | 78042 | No |

If she wishes, she can also print out a list of people in her database who are her relatives AND who live in Texas.

| | A | B | C | D | E | F | G |
|---|---|---|---|---|---|---|---|
| 1 | Last name | First name | Street | City | State | Zip | Relative? |
| 2 | Gutierrez | Alfred | 3201 Swan's Crossing | Victoria | TX | 77901 | Yes |
| 3 | Lopez | Pauline | 1573 Piedmont | Austin | TX | 78746 | Yes |
| 4 | Hays | Laura | 2337 Sneed | Edinburg | TX | 78539 | Yes |

In computer software and in logic, the word AND is a **logical operator**. To understand the logical operator AND, Venn diagrams are useful.

## The "AND" Relationship in Venn Diagrams

In a **Venn diagram**, circles are used to represent sets. When circles overlap, the overlapping region is known as the *intersection* of the sets.

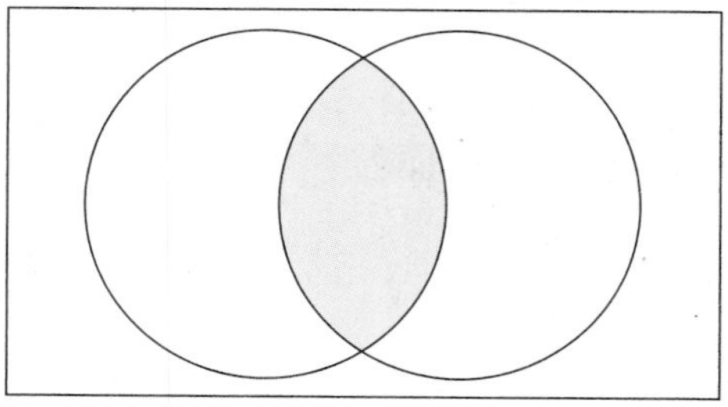

*The intersection of the sets represented by the circles is the shaded region.*

The **intersection** of two sets consists of the elements that are common to both sets. The logical relationship AND represents the intersection of two sets.

The intersection of sets is shown by the symbol $\cap$. The intersection of sets $A$ and $B$ is written as $A \cap B$.

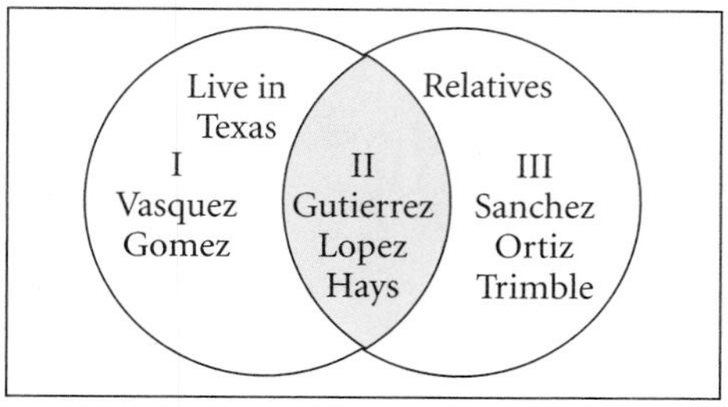


Compare this Venn diagram with Maria's three separate printouts on the previous page.

## The "OR" Relationship in Venn Diagrams

Maria can print out a list of people in her database who live in Texas OR who are relatives, as shown.

| | A | B | C | D | E | F | G |
|---|---|---|---|---|---|---|---|
| 1 | Last name | First name | Street | City | State | Zip | Relative? |
| 2 | Sanchez | Richardo | 1823 Putnam | Santa Fe | NM | 87010 | Yes |
| 3 | Gutierrez | Alfred | 3201 Swan's Crossing | Victoria | TX | 77901 | Yes |
| 4 | Lopez | Pauline | 1573 Piedmont | Austin | TX | 78746 | Yes |
| 5 | Ortiz | Mary | 809 Millwood | Pensacola | FL | 32514 | Yes |
| 6 | Hays | Laura | 2337 Sneed | Edinburg | TX | 78539 | Yes |
| 7 | Vasquez | David | 784 Fairfield Dr | San Antonio | TX | 78034 | No |
| 8 | Trimble | Alicia | 408 Elm Creek | Berea | OH | 44017 | Yes |
| 9 | Gomez | George | 2305 Main | San Antonio | Tx | 78042 | No |

The logical relationship OR represents the *union* of sets.

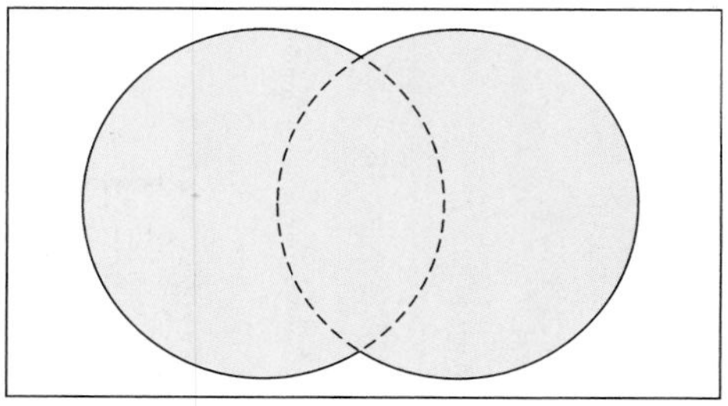

*Union of sets with some elements in common*

The **union** of two sets consists of all elements of both sets. In a Venn diagram of the union of two sets, the whole circles are shaded.

The union of two sets is shown by the symbol $\cup$. The union of sets $A$ and $B$ is written as $A \cup B$.

| | A | B | C | D | E | F | G |
|---|---|---|---|---|---|---|---|
| 1 | Last name | First name | Street | City | State | Zip | Relative? |
| 2 | Sanchez | Richardo | 1823 Putnam | Santa Fe | NM | 87010 | Yes |
| 3 | Gutierrez | Alfred | 3201 Swan's Crossing | Victoria | TX | 77341 | Yes |
| 4 | Lopez | Pauline | 1573 Piedmont | Austin | TX | 78746 | Yes |
| 5 | Ortiz | Mary | 809 Millwood | Pensacola | FL | 32514 | Yes |
| 6 | Hays | Laura | 2337 Sneed | Edinburg | TX | 78539 | Yes |
| 7 | Shwartz | Ricky | 489 King Blvd. | Baton Rouge | LA | 70422 | No |
| 8 | Velez | Henry | 6097 Nash | Oklahoma City | OK | 73114 | No |
| 9 | Vasquez | David | 784 Fairfield | San Antonio | TX | 78034 | No |
| 10 | Trimble | Alicia | 408 Elm Creek | Berea | OH | 44017 | Yes |
| 11 | Gomez | George | 2305 Main | San Antonio | Tx | 78042 | No |

*Use the database printout to complete the activity below.*

## Activity
## Venn Diagrams for AND and OR

Make two large copies of the Venn diagram at right. The circles in one of your diagrams should be labeled Texas and Ohio. The other diagram should have labels for Ohio and Relative. Fill in each diagram with last names from Maria's database placed in the appropriate places. Then answer the following questions:

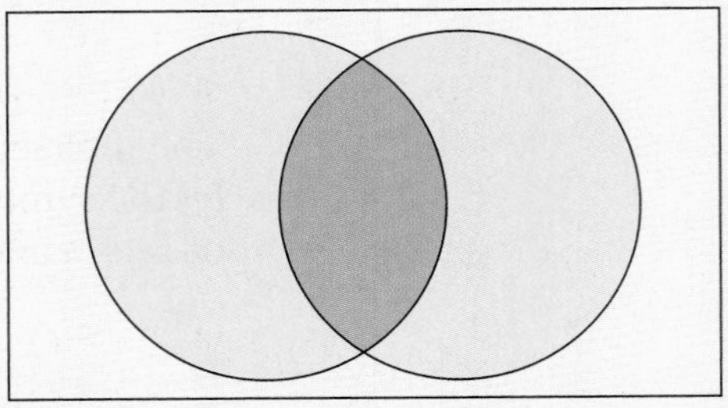

1. How many people in the database live in Ohio? in Texas?
2. How many people in the database live in Ohio AND Texas? in Ohio OR Texas?
3. How many people in the database are relatives of Maria?
4. How many people in the database live in Ohio OR are relatives of Maria? live in Ohio AND are relatives of Maria?
5. Draw new diagrams and shade the portions that represent people in the database who
   - **a.** live in Ohio OR Texas.
   - **b.** live in Ohio AND Texas.
   - **c.** live in Ohio AND are relatives of Maria.
   - **d.** live in Ohio OR are relatives of Maria.

**CHECKPOINT ✔**

6. Write a rule for finding the number of items in one set OR the other. (Hint: Use the number of people who live in Ohio, the number who are relatives, and the number who are relatives OR who live in Ohio in your answer.)

You may have noticed that it is important to not count the elements in the intersection of the circles twice.

### Counting Elements of Sets

Suppose that there are $m$ elements in set $M$ and $n$ elements in set $N$; then the total number of elements in the two sets is $m + n - t$, where $t$ is the number of elements in the intersection of $M$ and $N$.

**EXAMPLE 1**

**APPLICATION**
**GAMES**

In a card game called Crazy Eights, if the queen of hearts is showing, you can play only a queen, a heart, or an eight. **How many different cards can be played in this situation, assuming that all cards except the queen of hearts are available?**

**SOLUTION**

Do not count the queen of hearts (it has already been played). There are 3 other queens, 12 other hearts, and 4 eights—with the eight of hearts being counted twice. Thus, there are $3 + 12 + 4 - 1$, or 18, cards that can be played.

**TRY THIS** How many cards in a deck of playing cards are spades OR aces?

**EXAMPLE 2**

**If you draw a card at random from a complete deck of 52 playing cards, what is the probability that you will draw a king? What is the probability that you will draw a queen? What is the probability that you will draw a king OR a queen?**

**SOLUTION**

There are 4 kings and 4 queens in the deck.

$$P(\text{king}) = \frac{4}{52} = \frac{1}{13} \quad \text{and} \quad P(\text{queen}) = \frac{4}{52} = \frac{1}{13}$$

Since there are 8 cards in the set of favorable outcomes for drawing a king or a queen, the probability is as follows:

$$P(\text{king OR queen}) = \frac{8}{52} = \frac{2}{13}$$

**TRY THIS** If you draw a card at random from a complete deck of 52 playing cards, what is the probability that you will draw a spade OR a club?

Example 2 suggests that you can add the probabilities of two outcomes to find the probability that one OR the other will occur.

$$P(\text{king OR queen}) = P(\text{king}) + P(\text{queen}) = \frac{4}{52} + \frac{4}{52} = \frac{8}{52}, \text{ or } \frac{2}{13}$$

When the two outcome sets from a sample space do not overlap, they are said to be **disjoint**.

### Addition of Probabilities Principle

If $M$ and $N$ are *intersecting* sets of outcomes in the same sample space, then

$$P(M \text{ OR } N) = P(M) + P(N) - P(M \cap N).$$

If $M$ and $N$ are *disjoint* sets of outcomes in the same sample space, then

$$P(M \text{ OR } N) = P(M) + P(N).$$

**EXAMPLE 3** **What is the probability of drawing a face card OR a spade from a deck of playing cards?** (Note: The face cards are the 4 kings, 4 queens, and 4 jacks. There are 13 spades.)

**SOLUTION**

There are 12 face cards and 13 spades. Three of the 13 spades are also face cards, so there are 3 cards in the intersection of the two sets. Let $F$ be the set of face cards and $S$ the set of spades.

$$P(F \text{ OR } S) = P(F) + P(S) - P(F \cap S) = \frac{12}{52} + \frac{13}{52} - \frac{3}{52} = \frac{22}{52} = \frac{11}{26} \approx 0.42$$

**TRY THIS** What is the probability of drawing an odd-numbered spade OR an ace from a deck of playing cards? (Note: The numbered cards are ace through 10 in each of the four suits; an ace counts as a one.)

**SUMMARY OF RELATIONSHIPS BETWEEN TWO SETS**

| Intersection | Union | Disjoint |
| --- | --- | --- |

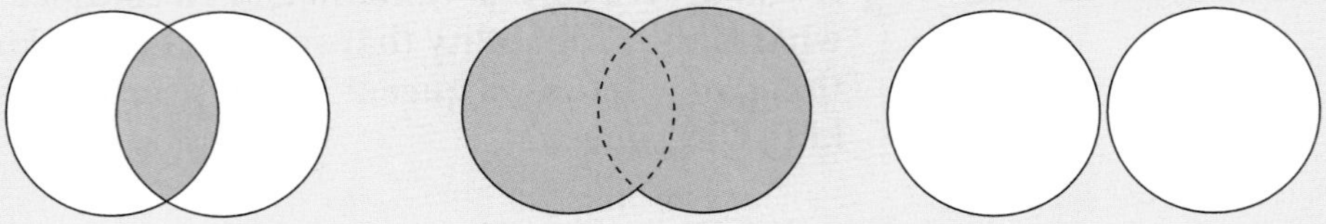

# Exercises

## Communicate

**internet connect**
**Activities Online**
Go To: **go.hrw.com**
Keyword:
**MA1 Fair Polyhedra**

1. Explain some ways that a database can be used to organize sets of data.
2. In the database at the beginning of this lesson, name a person other than Richardo Sanchez who is in the intersection of the set of people who are relatives and the set of people who do not live in Texas.
3. In the Crazy Eights situation (Example 1), explain how the eight of hearts is counted twice.

**CONNECTION**

**STATISTICS** A survey of student participation in school music programs had the results shown in the table below. Explain how to find each of the following:

4. How many students are girls?
5. How many students participated?
6. How many students are girls AND participated?
7. How many students are girls OR participated?

| | Participated | Did not participate |
| --- | --- | --- |
| **Boys** | 43 | 59 |
| **Girls** | 49 | 57 |

## Guided Skills Practice

**In a deck of 52 playing cards, how many are the type listed below?** ***(EXAMPLE 1)***

**8.** a queen OR a heart **9.** a heart OR red **10.** a seven OR an eight

**If you draw a card at random from a complete deck of 52 playing cards, what is the probability that you will draw the cards below?** ***(EXAMPLE 2)***

**11.** a two OR a club

**12.** a spade OR an even-numbered card

**13.** a jack OR an eight

**If you draw a card at random from a complete deck of 52 playing cards, what is the probability that you will draw the cards below?** ***(EXAMPLE 3)***

**14.** an even numbered card OR a red card

**15.** a face card OR a black card (Note: There are 12 face cards: 4 kings, 4 queens, and 4 jacks.)

**16.** a spade OR a nine

## Practice and Apply

**List the integers from 1 to 10 inclusive that are**

**17.** even. **18.** multiples of 3.

**19.** even AND multiples of 3. **20.** even OR multiples of 3.

CONNECTION

internet connect

**Homework Help Online**
Go To: **go.hrw.com**
Keyword:
**MA1 Homework Help** for Exercises 21-34

**STATISTICS** Mark took a class survey to get student opinions about a new rule concerning students driving to school. Use the table to at right answer the questions below.

| | Favor rule | Oppose rule | Total |
|---|---|---|---|
| **Boys** | 4 | 9 | 13 |
| **Girls** | 7 | 10 | 17 |
| **Total** | 11 | 19 | 30 |

**21.** How many of those surveyed are boys?

**22.** How many of those surveyed are girls AND oppose the rule?

**23.** How many of those surveyed are girls OR oppose the rule?

**24.** How many of those surveyed are boys AND favor the rule?

**25.** How many of those surveyed are boys OR favor the rule?

**List the integers from 1 to 20 inclusive that are multiples of the following numbers:**

**26.** 5 **27.** 3 **28.** 5 AND 3 **29.** 5 OR 3

**30.** People waiting to hear a concert are given numbers. At one point, the people with tickets numbered from 100 to 150 inclusive are allowed to enter. How many people are allowed to enter?

CONNECTIONS

**STATISTICS** In a class of 28 students, 17 have brown eyes, and 13 have black hair. This includes 10 students who have both.

**31.** Show this information in a Venn diagram.

**32.** How many students have either brown eyes OR black hair?

**STATISTICS** A survey of student participation in team sports had the results shown in the table below.

| | Participated | Did not participate |
|---|---|---|
| **Boys** | 124 | 239 |
| **Girls** | 111 | 250 |

**33.** How many students are girls AND participated?

**34.** How many students are girls OR participated?

CHALLENGE

**35.** If it takes 15 minutes to saw a log into 3 pieces, how long does it take to saw it into 4 pieces? (Hint: The answer is not 20.)

APPLICATION

**GAMES** Tell how many of the following are in an ordinary deck of 52 cards:

**36.** aces **37.** red cards **38.** cards that are red OR aces

## Look Back

**Simplify.** ***(LESSON 1.3)***

**39.** $-[-3 + (-7)] + 23 - 2(7 - 5)$ **40.** $4(a + 2b) - 5(b - 3a)$

CONNECTION

**41.** **GEOMETRY** The volume of a cone is $V = \frac{1}{3}\pi r^2 h$. Solve the formula for $h$. ***(LESSON 3.6)***

**Solve.**

**42.** $3(x + 5) - 23 = 2x - 47$ ***(LESSON 3.5)***

**43.** $\frac{x}{3} = -3x + 5$ ***(LESSON 3.4)***

APPLICATION

**44.** **INVESTMENTS** Chou deposits $150 in a bank that pays 5% interest per year, compounded annually. How much money will she have in this account after 5 years? ***(LESSON 8.7)***

## Look Beyond

**45.** Two coins are tossed. Find the probability that neither one will be heads.

# 13.3 The Fundamental Counting Principle

*Abi is scheduling her classes for next year. After she registers for her required courses, she may choose one elective from cluster 1 and one from cluster 2.*

## Objectives

- Use tree diagrams to count the number of choices that can be made from sets.
- Use the Fundamental Counting Principle to count the number of choices that can be made from sets.

**APPLICATION**
**SCHEDULING**

In certain situations you need to count the number of combined choices that are possible when one choice comes from one set of things and one choice comes from another set of things.

**EXAMPLE 1** **Use a tree diagram to find the number of elective pairs (from the illustration above) that are possible for Abi.**

**SOLUTION**

1. Make branches for band, orchestra, and chorus. A dot indicates a decision point.

   To each branch, add home economics and woodworking.

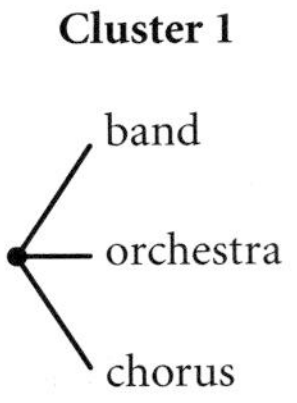


2. Count the total number of choices by following all of the possible pathways, starting with the first decision point. There are 6 possible sets of choices.

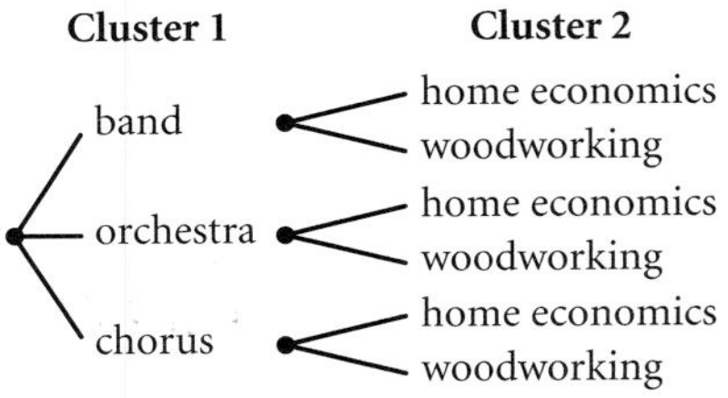

**TRY THIS** A diner at a cafe may select 1 of 6 main dishes and 1 of 2 desserts. Use a tree diagram to find how many different meals are possible.

You can use multiplication to find the number of sets of choices that are possible as illustrated by the following activity.

## *Activity* Sets of Choices

APPLICATION
CLOTHING

Michael has 3 dress shirts (blue, grey, and white) and 2 ties (black and blue). In the steps that follow, you will find how many different shirt-and-tie combinations he can make from the 3 shirts and 2 ties.

1. Suppose that Michael chooses the blue shirt. How many choices are possible for the tie?
2. How many choices for a shirt-and-tie combination are available to Michael?
3. How might your answer to Step 2 involve multiplication? (CHECKPOINT ✔)
4. Repeat this activity for 5 shirts and 3 ties.

In the activity, you may have discovered the following principle:

**Fundamental Counting Principle**

If there are $m$ ways to make a first choice and $n$ ways to make a second choice, then there are $m \cdot n$ ways to make a first choice and a second choice.

**EXAMPLE 2** **If you choose 1 boy out of 12 boys and 1 girl out of 13 girls, how many ways can this be done?**

**SOLUTION**

Since there are 12 boys, there are 12 ways to make the first choice. Thus, $m = 12$. Since there are 13 girls, there are 13 ways to make the second choice. Thus, $n = 13$. In all there are $m \cdot n = 12 \cdot 13 = 156$ ways to make a first and a second choice.

**TRY THIS** A car buyer can choose one of three different models of a new car. Five different colors are available for each. How many ways can the buyer make the choices? Relate your answer to the Fundamental Counting Principle.

The Fundamental Counting Principle can be extended to more than two choices, as the next example illustrates.

**EXAMPLE 3**

**APPLICATION**
**INVENTORY**

A store stocks 20 styles of shoes. Each style has 4 sizes and 3 colors. **How many pairs of shoes must the store carry in order to have one pair of each choice of size, style, and color?**

**SOLUTION**

For each of the 20 styles there are 4 sizes. This means there are 20 • 4, or 80, choices of style and size.

For each of these 80 choices, there are 3 colors. Thus, there are 80 • 3, or 240, choices in all.

You can also perform all of the multiplications at once:

$$20 \cdot 4 \cdot 3 = 240$$

**TRY THIS**

Tammy is selecting colored pens. The first characteristic is the point, which may be either felt tip or ball-point. The second characteristic is the color—red, blue, green, or black. The third characteristic is the brand—Able or Blakely. How many possible selections of a pen can Tammy make?

**EXAMPLE 4**

A deck of playing cards has 26 red cards and 26 black cards. There are 4 aces —2 red and 2 black. Two decks of cards are shuffled separately. A card is drawn from each deck. **How many ways are there of drawing a black card from the first deck AND an ace from the second deck?**

**SOLUTION**

There are 26 ways of drawing a black card from the first deck, and 4 ways of drawing an ace from the second deck. By the Multiplication Counting Principle, there are 26 • 4, or 104, ways to draw a black card from one deck AND an ace from a second deck.

**TRY THIS** A card is drawn from a deck of playing cards. Then it is replaced. A second card is drawn. How many ways are there of drawing a red card followed by a face card? (The face cards are the 4 kings, 4 queens, and 4 jacks. For each kind of face card, 2 are red, and 2 are black.)

**CRITICAL THINKING** In Example 4, two different decks were used. Would the result be the same or different if both cards were drawn from one deck? Would the result be the same or different if one deck were used and the first card chosen was *replaced* before drawing the second card?

## Exercises

### Communicate

1. Explain the difference between these two sentences:
   - I am going to the mall AND I am going to study.
   - I am going to the mall OR I am going to study.
2. Suppose that you are playing a game in which the first step is to flip a coin. If the coin lands heads up, you will then roll a number cube. If the coin lands tails up, you will choose a card from a stack of 10. Explain how to find the number of branches on a tree diagram representing this situation.
3. Two cards are drawn from an ordinary deck of 52 cards. The first card is replaced before the second card is drawn. Explain how to find the number of possible ways to draw a 3 followed by a red card.

internet connect

**Activities Online**

Go To: **go.hrw.com**

Keyword: **MA1 Bells**

### Guided Skills Practice

**Use a tree diagram to find the combined number of choices that are possible in each situation.** *(EXAMPLE 1)*

4. choosing a food item and a drink from a menu with 6 food and 4 drink choices
5. flipping a coin and then rolling a number cube

**Use the Fundamental Counting Principle to find the number of choices that are possible for each situation.** *(EXAMPLE 2)*

6. A pasta dish is offered with 2 choices of pasta and 4 choices of sauces.
7. A businessman has 6 ties that will go with 3 shirts.

**Use the Fundamental Counting Principle to find the number of possibilities for each situation.** *(EXAMPLE 3)*

8. A car has the following options: 2 different types of engines, 2 body styles, 6 colors, 2 seat fabrics, and 3 interior colors. How many choices are there?
9. A store stocks 10 styles of shorts. Each style has 5 sizes and 6 colors. How many pairs of shorts must the store carry to have at least one pair of each size, style, and color?

**Use the Fundamental Counting Principle to answer each question. Two decks of playing cards are shuffled separately. A card is drawn from each deck.** ***(EXAMPLE 4)***

**10.** How many ways are there of drawing a king from the first deck AND a spade from the second deck?

**11.** How many ways are there of drawing a red card from the first deck AND a face card from the second deck?

## *Practice and Apply*

internet connect
**Homework Help Online**
Go To: **go.hrw.com**
Keyword:
**MA1 Homework Help** for Exercises 12-19

**12.** In how many ways can Leanna choose 1 CD AND 1 tape from a collection of 4 CDs and 5 tapes?

**13.** Juliana is trying to choose 1 CD AND 1 tape from a collection of 2 CDs and 3 tapes that she likes. In how many ways can she do this?

**14.** A menu contains 4 appetizers, 3 salads, 5 main courses, and 6 desserts. How many ways are there to choose one of each?

CONNECTION

**15.** **GEOMETRY** How many ways are there to name the triangle shown at right?

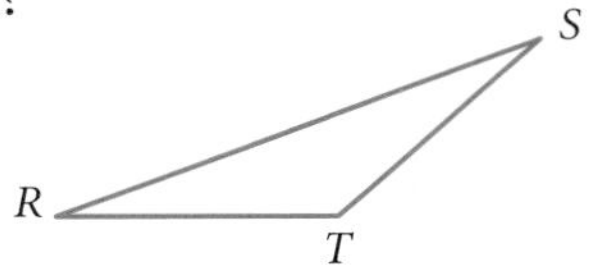


APPLICATIONS

**16.** **CONSUMER ECONOMICS** The Rahalls are ordering a new car. They can choose from 9 exterior colors. For each of these, there are 4 interior colors. How many versions are possible?

**17.** **GAMES** At the start of a game of chess, white moves first and has 20 possible moves. For each of these possibilities, black has 20 possible countermoves. How many possibilities are there for the first round?

**18.** **TELECOMMUNICATIONS** How many different 7-digit phone numbers are possible if the first 3 digits are 555? Assume that 555-0000 is allowed.

**19.** **TELECOMMUNICATIONS** How many different 7-digit phone numbers are possible? Assume that phone numbers cannot start with 0.

**LANGUAGE ARTS** A word that is obtained by rearranging the letters of another word is called an *anagram*. For example, *art* is an anagram of *rat*.

**20.** Find another anagram of *rat*.

**21.** How many ways are there to arrange the letters of the word *rat*, including non-words, using each letter exactly once?

**22.** Write three anagrams of *stake*.

**23.** How many ways are there to arrange the six letters of the word *foster*?

**24.** Find two anagrams of *foster*.

## Look Back

**25.** Hexagonal numbers form a sequence: 1, 6, 15, 28, 45, 66, 91, . . . Find the next two hexagonal numbers. ***(LESSON 1.1)***

**Solve.** ***(LESSON 3.3)***

**26.** $\frac{3}{2}x = 6$ **27.** $\frac{x}{-4} = -\frac{5}{8}$ **28.** $\frac{x}{5} = 10$

**CONNECTION**

**29.** **COORDINATE GEOMETRY** Points $P(7, -4)$ and $Q(-2, 5)$ are on a line. What is the slope of that line? ***(LESSON 5.2)***

**30.** If two lines are parallel, how are their slopes related? ***(LESSON 5.6)***

**APPLICATION**

**31.** **CHEMISTRY** How many milliliters of a 9% acid solution should be mixed with 450 milliliters of a 1.6% acid solution to produce a 5% acid solution? ***(LESSON 7.6)***

## Look Beyond

**APPLICATION**

**ACADEMICS** Find the probability that Abi from page 661 chooses the following:

**32.** band from cluster 1 **33.** woodworking from cluster 2

**34.** band OR woodworking **35.** band AND woodworking

**internet connect**

**Portfolio Extension**

Go To: **go.hrw.com**
Keyword:
**MA1 Probability**

## PORTFOLIO ACTIVITY

Include this activity in your portfolio.

Design and conduct an experiment with two number cubes to find the probability of getting the following sums as outcomes:

**1.** 2 OR 12
**2.** 1
**3.** 7 OR 11
**4.** 6 OR 8

Perform at least 100 trials, and list each outcome as either favorable (the sum indicated) or unfavorable (not the sum indicated).

Calculate the theoretical probabilities of the four outcomes listed at left and compare with the results of your experiment.

**WORKING ON THE CHAPTER PROJECT**

You should now be able to complete Activity 1 of the Chapter Project on page 682.

13.4

# Independent Events

## Objective

- Find the probability of independent events.

**APPLICATION**
**SCHEDULING**

In Lesson 13.3, Abi selects one elective from cluster 1 (band, orchestra, or chorus) and one elective from cluster 2 (home economics or woodworking). If all of her choices are equally likely, how can you find the probability that she chooses band AND woodworking?

## Modeling Independent Events

You can model the sample space of Abi's choices geometrically. Draw two vertical lines to divide the model into three equal regions, one for each elective in cluster 1.

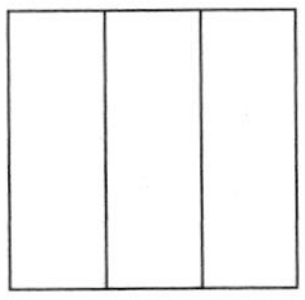

Then draw a horizontal line to divide the model into two equal regions, one for each elective in cluster 2.

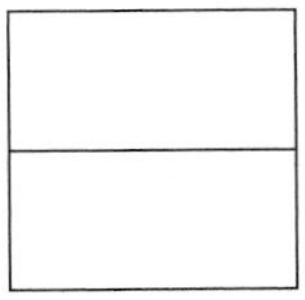

Thus, as you saw in Lesson 13.3, there are 6 possibilities. That is, the sample space for the Abi's decision has 6 elements, one of which is band AND woodworking. The probability of her selecting band and woodworking is as follows:

$$P(\text{band AND woodworking}) = \frac{1}{6}$$

Now consider each of the choices separately. The probability of choosing woodworking is $\frac{1}{2}$.

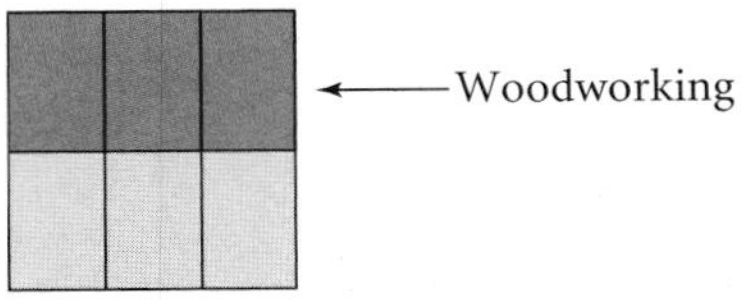

The probability of choosing band is $\frac{1}{3}$.

Notice that $\frac{1}{2} \times \frac{1}{3} = \frac{1}{6}$.
This suggests that you can multiply the probabilities of the individual events to find the probability of both of them occurring. This is in fact true whenever two events, such as Abi's choices, are *independent.*

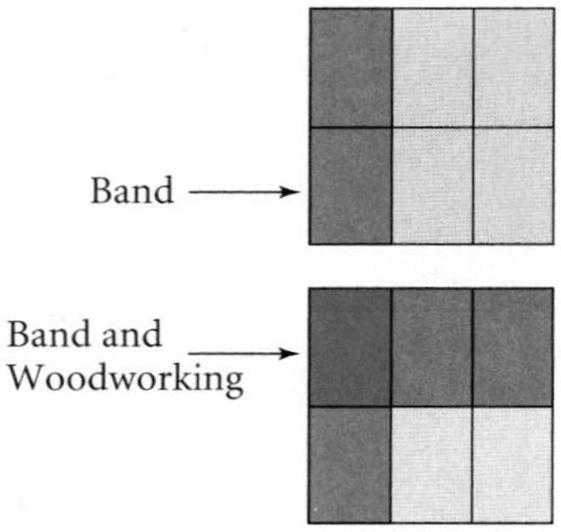


**Independent Events**

Two events are independent if the occurrence of one does not affect the probability of the occurrence of the other.

Abi's choices are independent because her choice from cluster 1 does not affect her choice from cluster 2.

**Probabilities of Independent Events**

If $A$ and $B$ are independent events, then $P(A \text{ AND } B) = P(A) \cdot P(B)$.

*Activity*

## Using a Model

1. Extend the model of Abi's choices in the situation above by adding dance to cluster 1 and auto shop to cluster 2.
2. Assuming that Abi's choices are all equally likely, how many equally likely possibilities are there?
3. What is the probability that she will choose band? woodworking? band AND woodworking?
4. In Step 3, should you add or multiply the first two probabilities to find the third probability?
5. What is the probability that she will choose chorus? dance? chorus OR dance?
6. In Step 5, should you add or multiply the first two probabilities to find the third probability?

CHECKPOINT ✔

7. When is multiplication appropriate in the questions above? When is addition appropriate in the questions above? (Relate your answer to the use of the logical relationships OR and AND.)

## EXAMPLE 1

Todd rolls a red and a green number cube. **What is the probability that the red cube shows an odd number AND the green cube shows a number greater than or equal to 5?**

**SOLUTION**

Because the result on one cube does not affect the result on the other, the events are independent. Thus, the probabilities can be multiplied.

Three of the six numbers on a cube are odd.

$$P(\text{rolling an odd number on the red cube}) = \frac{3}{6} = \frac{1}{2}$$

Two of the six numbers on a cube are greater than or equal to 5.

$$P(\text{rolling a 5 or greater on the green cube}) = \frac{2}{6} = \frac{1}{3}$$

To find the probability of both happening, multiply the individual probabilities. $\frac{1}{2} \times \frac{1}{3} = \frac{1}{6}$

**TRY THIS** A red and a green number cube are rolled. Find the probability that the red cube shows a number that is greater or equal to 5 AND the green cube shows a number less than or equal to 2.

## EXAMPLE 2

**APPLICATION**
**SCHEDULING**

John and Debbie have trouble being punctual. John is late 40% of the time, and Debbie is late 20% of the time. **Assuming that they are just as likely to be late on any particular occasion, what is the probability that both will be on time for a lunch date?**

**SOLUTION**

On graph paper, divide a 10-by-10 square into two parts with a vertical line, as shown in the first figure. This shows a 60% probability that John is on time. Now divide the square into two parts with a horizontal line, as shown in the second figure. This shows an 80% probability that Debbie is on time.

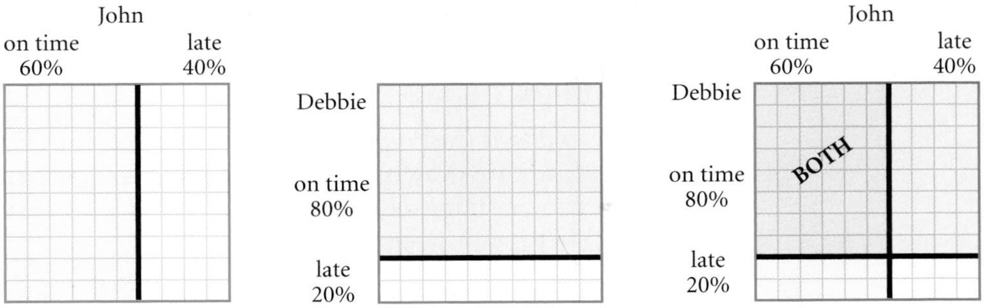


When the squares for John and Debbie are combined, the lines divide the 100-block square into four regions. The region marked BOTH represents the probability that both are on time. It contains $8 \cdot 6$, or 48, out of $10 \cdot 10$, or 100 squares. The probability that both John and Debbie are on time is $\frac{48}{100}$, or 48%. Notice that $\frac{48}{100}$ is the product of $\frac{6}{10}$ and $\frac{8}{10}$.

**TRY THIS** Find the probability that Mark and Tracy will both be on time if Mark is on time 40% of the time and Tracy is on time 70% of the time.

**EXAMPLE 3** Sarah draws a card, replaces it, and draws another card from a regular 52-card deck. **Find the probability that the 2 cards she draws are black. In a regular deck, half of the cards are black.**

**SOLUTION**

There are 52 ways of drawing the first card. After Sarah replaces the first card, there are 52 ways of drawing the second card. Since she draws a first card and a second card, use the Fundamental Counting Principle. There are $52 \cdot 52$, or 2704, equally likely ways of drawing the two cards when the first card is replaced.

A successful event occurs if Sarah draws two black cards. There are 26 ways to draw the first black card. After the first black card is replaced, there are 26 ways to draw the second black card. Apply the Fundamental Counting Principle. There are $26 \cdot 26$, or 676, ways to draw two black cards.

The probability of drawing two black cards is $\frac{676}{2704}$, or 25%.

**TRY THIS** Thelma draws a card from a regular 52-card deck. After replacing it, she draws another card from the same deck. Find the probability that the two cards she draws are aces.

## Finding Probabilities With the Complement

Finding the probability of certain events can often require several calculations if the probability is found directly. In these cases, it is often easier to count the ways that something does *not* happen.

**EXAMPLE 4** Two cards are drawn with replacement from an ordinary deck of 52 cards. **Find the probability that at least one of them is red.**

**SOLUTION**

The event *at least one card is red* includes every event that can happen when two cards are drawn *except* the event of getting two black cards. Mathematically this can be stated as $P(\text{at least one red}) = 1 - P(\text{two black cards})$.

In the previous example, we found that the probability of drawing two black cards is $\frac{676}{2704}$. Therefore, $P(\text{at least one red}) = 1 - \frac{676}{2704} = \frac{2028}{2704}$, or 75%.

**TRY THIS** If two number cubes are rolled, what is the probability that the sum is greater than or equal to 3? Solve the problem two different ways.

When you find the number of ways that an event can occur by considering the number of ways that it does *not* occur, you are using the **complement** of the set of ways that it can occur.

# Exercises

## Communicate

1. Explain what it means for two events to be independent.
2. Name two ways to represent the probabilities of independent events with a diagram.
3. Explain why you use the Fundamental Counting Principle when determining the probabilities of independent events.
4. Explain how to find a probability by using the complement.

## Guided Skills Practice

**A red number cube and a green number cube are rolled together. Find each probability.** *(EXAMPLE 1)*

5. Both numbers are even.
6. Both numbers are greater than 3.
7. The green cube is even AND the red cube is less than 4.

**Use area models to find each probability below.** *(EXAMPLE 2)*

8. Melissa is interviewing for a job in the suburb of another city and decides that the best way to get there is to fly to the city and take a train to the suburb. If the plane is on time 90% of the time and the train is on time 70% of the time, what is the probability that both will be on time?
9. When commuting by train, a passenger notices that his first train is on time 90% of the time and his connecting train is on time 80% of the time. What is the probability that both will be on time?

**internet connect**
**Homework Help Online**
Go To: **go.hrw.com**
Keyword:
**MA1 Homework Help**
for Exercises 13-15, 26, 30

**Suppose that you draw two cards with replacement from a regular deck of playing cards. Find each probability.** *(EXAMPLE 3)*

10. Both cards are kings.
11. Both cards are spades.
12. Both cards are red aces.

**Two cards are drawn with replacement from an ordinary deck of 52 cards. Find each probability.** *(EXAMPLE 4)*

13. At least one card is not a king.
14. At least one card is not a spade.
15. At least one card is not a red ace.

## Practice and Apply

**A coin is tossed twice. Tell whether the events in each pair below are dependent or independent.**

**16.** The first toss is heads, and the second toss is heads.

**17.** The first toss is heads, and the second toss is tails.

**18.** The first toss is tails, and the second toss is heads.

**Use the grid below to find the following probabilities:**

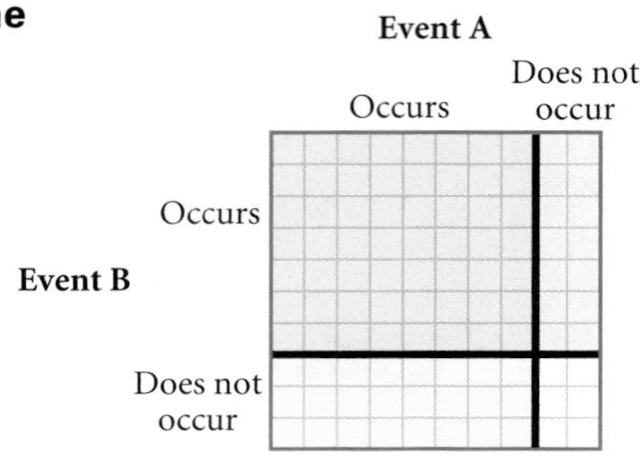


**19.** Event A occurs.

**20.** Event B occurs.

**21.** Event A occurs AND event B occurs.

CHALLENGE

**22.** Event A occurs OR event B occurs.

APPLICATIONS

**GAMES** One number is selected from the list {1, 3, 5, 7}. Another is selected from the list {5, 6, 7}. Find each probability.

**23.** The two numbers selected are even.

**24.** The two numbers selected are the same.

**25.** The sum of the two numbers selected is even.

**26.** The number selected from the second list is greater than the number selected from the first list.

**GAMES** Five chips numbered 1 through 5 are placed in a bag. A chip is drawn and replaced. Then a second chip is drawn. Find the following probabilities:

**27.** Both chips are even.

**28.** Both chips are odd.

**29.** The first chip is even AND the second chip is odd.

**30.** One chip is even AND the other is odd.

**RECREATION** John and Jim are each hiking from Wapitu Falls to Songbird Lake on the trails shown on the map. Assume that each trail is equally likely to be chosen.

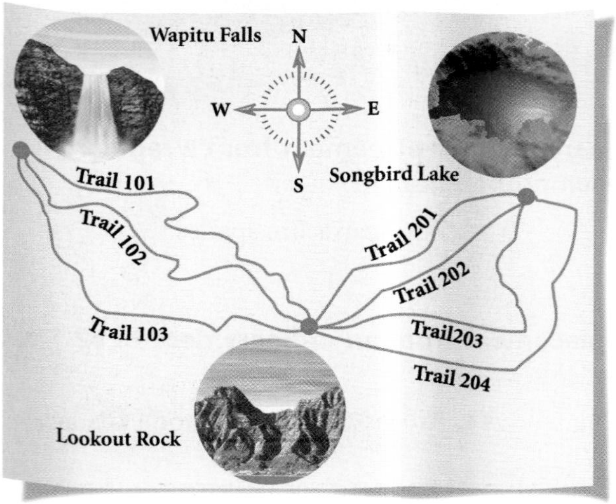


**31.** Find the probability that John takes trail 101 from Wapitu Falls to Lookout Rock.

**32.** Find the probability that John takes trail 201 from Lookout Rock to Songbird Lake.

**33.** Find the probability that John takes trail 101 from Wapitu Falls to Lookout Rock AND takes trail 201 from Lookout Rock to Songbird Lake.

**34.** Find the probability that John takes trail 101 from Wapitu Falls to Lookout Rock OR he takes trail 201 from Lookout Rock to Songbird Lake (or both).

## Look Back

**35.** Find the product: $-6(2)\left(\frac{5}{-3}\right)$. *(LESSON 2.4)*

**Find the slope of each line.** *(LESSON 5.4)*

**36.** $y = 4x$ **37.** $y = 4x - 3$ **38.** $y = -4x$ **39.** $y = \frac{x}{4}$

**40.** Tim leaves home on his bike and rides at a rate of 12 miles per hour. Two hours after he leaves, Tim's mother remembers that Tim has a dentist's appointment. She sets out after him in her car at 40 miles per hour. How long does it take her to catch up to him? *(LESSON 7.6)*

APPLICATIONS

**41.** **ACADEMICS** At one school, 48 students signed up to take Spanish, and 23 signed up to take French. If there were 12 students that signed up for both, how many students total signed up for these foreign languages? *(LESSON 13.2)*

**42.** **TRAVEL** There are 3 highways from Birchville to Pine Springs. From Pine Springs to Clearwater, there are 5 roads. How many possible ways are there to travel from Birchville to Clearwater? *(LESSON 13.3)*

## Look Beyond

APPLICATION

**43.** **CONTESTS** Two radio stations are each having a contest. At the first station, the prize is \$1000, and your probability of winning is $\frac{1}{100{,}000}$. At the second station, the prize is \$25, and your probability of winning is $\frac{1}{2000}$. Which contest provides a better return for the listener's time? Compare the probabilities of winning each contest. (Hint: What would you expect to happen if you played each contest 100,000 times?)

PORTFOLIO ACTIVITY

Include this activity in your portfolio.

Design and conduct an experiment for the following situation: Two evenly matched teams play a three-game series. Perform at least 100 trials and list the experimental probability of the outcomes below.

Team 1 wins the following number of games:
**a.** exactly 1 game **b.** exactly 2 games **c.** 3 games

Team 2 wins the following number of games:
**a.** exactly 1 game **b.** exactly 2 games **c.** 3 games

Calculate the theoretical probabilities of each outcome above and compare them with the results of your experiment.

# EYEWITNESS MATH

## HOT HANDS OR HOOP-LA?

### 'Hot Hands' Phenomenon: A Myth?

The gulf between science and sports may never loom wider than in the case of hot hands. Those who play, coach, or otherwise follow basketball believe almost universally that a player who has successfully made his last shot or last few shots—a player with hot hands—is more likely to make his next shot. An exhaustive statistical analysis led by a Stanford University psychologist examined thousands of shots in actual games and found otherwise: the probability of a successful shot does not depend at all on the shots that came before.

To the psychologist, Amos Tversky, the discrepancy between reality and belief highlights the extraordinary differences between events that are random and events that people perceive as random. When events come in clusters and streaks, people look for explanations; they refuse to believe they are random, even though clusters and streaks do occur in random data.

To test the theory, researchers got the records of every shot taken from the field by the Philadelphia 76ers over one and a half seasons. When they looked at every sequence of two shots by the same player—hit-hit, hit-miss, miss-hit, or miss-miss—they found that a hit followed by a miss was actually a tiny bit likelier than a hit followed by a hit.

They also looked at sequences of more than two shots. Again, the number of long successful streaks was no greater than would have been expected in a random set of data, with every event independent of its predecessor.

*The New York Times, April 19, 1988*

### Researchers' Survey of 100 Basketball Fans

Does a player have a better chance of making a shot after having just made his last two shots than he does after having just missed his last two shots?

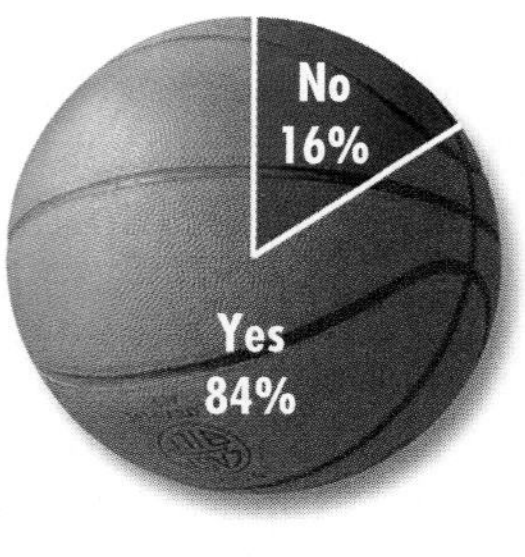


Is it important for players to pass the ball to someone who has just made several (2, 3, or 4) shots in a row?

**DISCUSS** Do you agree with the majority of basketball fans surveyed? Why or why not?

## Cooperative Learning

To form your own opinion about the hot-hands study, it helps to know how the researchers tabulated their data. For simplicity, you will use a portion of their data, which shows field-goal attempts by two players for several games. Both players usually make about half of their shots from the field.

| | |
|---|---|
| A | 11100010 01110111010010 000011101 0100000101<br>010101001110 1001111 01001010001 011110110 |
| B | 10101000100 11100000110011 0101010111001111<br>111001001101 0001101010 10010001111 |

**(1 = hit, 0 = miss, space = new game)**

1. First, look at what happens after a shot is made within each game.
   - **a.** Copy and complete Table 1 below, using the data for either player A or player B.
   - **b.** What percent of the time was a hit followed by another hit? What percent of the time was a hit followed by a miss?
   - **c.** What would you expect your results in part **b** to be if there were no such thing as *hot hands*, that is, if the shots were hit or missed just by chance? Explain. (Assume that the player makes 50% of his shots on average.)

**Table 1: After a hit**

| No. of times next shot is made | No. of times next shot is missed |
|---|---|
| | |

**Table 2: After a miss**

| No. of times next shot is missed | No. of times next shot is made |
|---|---|
| | |

2. Now look at what happens after a shot is missed within each game.
   - **a.** Copy and complete Table 2. Use the same player you used for Table 1.
   - **b.** What percent of the time was a miss followed by a hit? What percent of the time was a miss followed by another miss?
   - **c.** What would you expect your results in part **b** to be if the shots were hit or missed just by chance?
3. Do you think there is such a thing as hot hands in basketball? Why or why not? How does your belief affect the strategy you would use if you were coaching a basketball team?

# 13.5 Simulations

*Amy usually makes 50% of her field-goal attempts in games.*

**Objective**

- Design and perform simulations to find experimental probabilities.

**Why** *Experiments that use lists of random numbers can be used to simulate outcomes in order to find experimental probabilities. Simulations are used to study games, sports, weather, political elections, and many other situations.*

**APPLICATION**
**SPORTS**

In a certain game, Amy made four consecutive baskets. How often do you think Amy makes four baskets in a row in a game in which she takes 20 shots? If you had the actual shot-by-shot records of Amy's performance, you could see how often this happens. Even without the records, it is possible to find the **experimental probability** by using a *simulation.* A **simulation** is an experiment with mathematical characteristics that are similar to the actual event.

Suppose that Amy usually makes exactly half of the shots she takes in a game. If Amy takes 20 shots, a simulation can be designed to find the experimental probability that she will make four successful shots in a row.

**TECHNOLOGY**
**GRAPHICS CALCULATOR**

Outcomes are random if all possible outcomes are equally likely. A graphics calculator can be used to generate random numbers. (See the Calculator Keystroke Guide on page 782.)

1. Set up your calculator to randomly generate 0s and 1s.
2. Let 1 represent a successful shot, and let 0 represent a missed shot.
3. Each randomly generated number represents a shot. Generate 20 random numbers (these 20 numbers are considered one **trial**), and write them down in order.
4. Repeat Step 3 at least 10 times. A larger number of simulations should result in an experimental probability that is closer to the theoretical probability.
5. Count the number of trials in which a sequence of at least four consecutive 1s appears, and divide this result by the number of trials.

| Trial | Shot | | | | | | | | | |
|---|---|---|---|---|---|---|---|---|---|---|
| | 1 | 2 | 3 | 4 | 5 | 6 | 7 | 8 | 9 | 10 |
| 1 | 0 | 0 | 1 | 0 | 0 | 0 | 0 | 1 | 0 | 0 |
| 2 | 1 | 1 | 1 | 0 | 0 | 0 | 1 | 1 | 1 | 1 |
| 3 | 0 | 1 | 1 | 0 | 1 | 1 | 1 | 1 | 0 | 1 |
| 4 | 0 | 1 | 0 | 1 | 1 | 0 | 0 | 1 | 0 | 0 |
| 5 | 0 | 1 | 0 | 1 | 1 | 0 | 0 | 1 | 1 | 0 |
| 6 | 0 | 0 | 0 | 1 | 1 | 0 | 0 | 0 | 1 | 1 |
| 7 | 1 | 1 | 1 | 0 | 0 | 1 | 0 | 1 | 0 | 1 |
| 8 | 0 | 0 | 0 | 0 | 0 | 0 | 1 | 1 | 1 | 0 |
| 9 | 1 | 1 | 0 | 1 | 1 | 0 | 0 | 0 | 1 | 1 |
| 10 | 1 | 1 | 1 | 0 | 0 | 1 | 0 | 1 | 0 | 1 |

Of the 10 trials shown, there were at least four consecutive 1s recorded in 2 of the trials (which ones?). The experimental probability, in this case, that Amy will make four baskets in a row is $\frac{2}{10}$. Note that this is a relatively small number of trials, so this experimental probability may be significantly different from the theoretical probability.

When designing a simulation of an actual event to approximate an experimental probability, you must do the following:

1. Choose a random-number generator, such as a coin toss, number-cube roll, calculator, computer, or other method, that can be used to simulate the situation. Decide how to perform the experiment so that it simulates one trial.
2. Perform a large number of trials, and record the results.
3. Use the number of favorable outcomes, $f$, and the number of trials, $t$, to compute the experimental probability: $P = \frac{f}{t}$.

## *Activity*
## Basketball Simulation

APPLICATION
**SPORTS**

**You will need:** a coin

A coin can be used to simulate the basketball situation previously described. Let heads represent a success, and tails represent a miss.

1. Toss a coin 20 times, and record the sequence of heads and tails. This represents one game.
2. Does your sequence contain at least four consecutive heads?
3. Does your trial represent making four baskets in a row or not?
4. Conduct several trials, each representing one game with 20 attempted shots. Divide the total number of four consecutive heads by the total number of trials.

CHECKPOINT ✔

5. What is the experimental probability of four consecutive baskets?

**EXAMPLE 1**

**APPLICATION**
**METEOROLOGY**

When watching weather reports to plan outdoor activities for the weekend, the *chance of rain* is often mentioned. This is given as a percent. The given percent is simply an *experimental probability*. The weather conditions that exist have resulted, in the past, in rain the given percent of the time. Suppose that the weather report is for a 40% chance of rain on Saturday and a 40% chance of rain on Sunday. **Create a simulation and discuss the results. According to your results, on how many weekends did it rain at least one day?**

**SOLUTION**

1. Using a random-number generator such as a graphics calculator or slips of paper numbered from 1 to 10, generate pairs of numbers from 1 to 10 inclusive. Each pair of numbers that you generate will represent one trial—the first number represents Saturday and the second number represents Sunday. Because the numbers 1 through 4 are 40% of the numbers 1 through 10, let 1, 2, 3, and 4 represent rain. Let any other number represent no rain.

2. Perform a large number of trials, and record the results.

   For example, the following table shows a simulation of 20 trials:

| **Trial** | 1 | 2 | 3 | 4 | 5 | 6 | 7 | 8 | 9 | 10 | 11 | 12 | 13 | 14 | 15 | 16 | 17 | 18 | 19 | 20 |
|---|---|---|---|---|---|---|---|---|---|---|---|---|---|---|---|---|---|---|---|---|
| **1st number** | 3 | 5 | 10 | 6 | 3 | 9 | 8 | 9 | 4 | 8 | 6 | 2 | 4 | 8 | 10 | 7 | 1 | 7 | 7 | 5 |
| **2nd number** | 10 | 1 | 9 | 10 | 6 | 10 | 5 | 5 | 2 | 6 | 5 | 1 | 3 | 2 | 4 | 7 | 2 | 7 | 8 | 10 |

In Trial 1, the number 3 indicates rain on Saturday and the number 10 indicates no rain on Sunday.

In Trial 2, the number 5 indicates no rain on Saturday and the number 1 represents rain on Sunday.

In Trial 9, the 4 and the 2 represent rain on both Saturday and Sunday.

In Trial 11, the 6 and the 5 represent no rain on both Saturday and Sunday.

According to the results, rain occurred on at least one day of the weekend 9 out of 20 times, or 45% of the time.

**TRY THIS**

Suppose that the weather report gives a 60% chance of rain on Saturday and a 60% chance of rain on Sunday. Create a simulation and discuss the results. According to your results, what is the probability that it will rain on both days?

**CRITICAL THINKING**

Explain how the number of trials affects the accuracy of the probability of an event determined by experiment.

Computer spreadsheets are powerful tools for many applications. One useful feature of spreadsheets is their ability to use formulas to perform many calculations with a single set of instructions. To set up a random number generator for Example 1 in a spreadsheet, you would perform the following steps:

1. Select a cell and enter the formula Int(10*Rand())+1. This generates a random number from 1 to 10 inclusive.

2. Copy that cell, select a row or column of cells, and use the "Paste Special" feature from the Edit menu. You will select "Paste Formulas" from the options given.

3. The selected cells will then automatically fill with randomly generated numbers from 1 to 10 inclusive.

| | A | B | C | D | E | F |
|---|---|---|---|---|---|---|
| 1 | 2 | 2 | 4 | 8 | 2 | 3 |
| 2 | 3 | 1 | 6 | 7 | 7 | 4 |
| 3 | 3 | 1 | 5 | 7 | 7 | 5 |
| 4 | 4 | 1 | 9 | 5 | 5 | 8 |
| 5 | 7 | 2 | 10 | 10 | 7 | 8 |
| 6 | 4 | 2 | 4 | 4 | 3 | 5 |
| 7 | 2 | 9 | 7 | 10 | 2 | 5 |
| 8 | 4 | 1 | 6 | 4 | 10 | 3 |
| 9 | 5 | 1 | 6 | 8 | 6 | 7 |
| 10 | 9 | 3 | 5 | 1 | 7 | 4 |
| 11 | | | | | | |

# Exercises

internet connect

**Activities Online**

Go To: **go.hrw.com**
Keyword:
**MA1 Simulation**

## Communicate

1. What are the three steps for designing a simulation?

2. Assume that the chance of making a shot in a basketball game is 50%. Give three different ways to simulate making or missing a shot.

3. Suppose that you want to simulate a 30% chance of rain. Tell how to do this by using slips of paper numbered from 1 to 10 inclusive.

4. Name three things that could be simulated by tossing a coin one time.

5. Tell how you might simulate the situation of correctly guessing the answer to a multiple-choice question with 5 possible responses.

6. Suppose that 90% of the flights for an airline are on time. Explain how to design a simulation to find the experimental probability that three consecutive flights are on time.

## Guided Skills Practice

**Use coin tosses to simulate a 50% chance of rain for three days. Perform 20 trials and combine your results with those of your classmates to answer the questions below.** *(EXAMPLE 1)*

7. For the entire class, what was the percent of trials that had rain on all three days?

8. For the entire class, what was the percent of trials that had no rain?

## Practice and Apply

**APPLICATIONS**

**SPORTS** Refer to the data for the basketball simulation on page 677.

**9.** How many baskets did Amy make in the first trial?

**10.** In how many trials did Amy make baskets for more than 50% of her shots?

**11.** What formula is used to generate random numbers in this spreadsheet?

**12.** What does a value of 0 mean in this chart? What does a value of 1 mean in this chart?

**METEOROLOGY** Suppose that you want to simulate an 80% chance of rain on each of two consecutive days in order to find the probability of there being rain on at least one of the two days. Explain how you would perform the simulation by using the following methods:

**13.** generating random numbers from 1 to 10 inclusive

**14.** using slips of paper

**Describe one way to simulate the random selection of the following:**

**15.** a day of the week

**16.** a day of the year

**17.** 1 student out of 24 students

**18.** 1 of 8 team captains

internet connect
**Homework Help Online**
Go To: **go.hrw.com**
Keyword:
**MA1 Homework Help** for Exercises 19-22

**Design a simulation for each situation in Exercises 19–22, and give the results of your simulations.**

**19.** Pick two random numbers from 1 to 100 inclusive. Give the experimental probability that both numbers are less than or equal to 20.

**20.** A person has a 30% chance of correctly guessing a number. What is the experimental probability that a person will correctly guess the number on 2 out of 5 tries?

**21.** Design a simulation and then give the experimental probability that a family with 4 children has 2 boys and 2 girls, assuming that the births of boys and girls are equally likely.

**22.** Design a simulation and then give the experimental probability that a student guesses correctly on all 3 questions of a true-false quiz.

*"Guess your weight" is a game of chance that is popular at some carnivals.*

## Look Back

**23.** Write an expression that will generate the $n$th term of the following sequence: 4, 8, 12, 16, 20, 24, … ***(LESSON 1.1)***

**24.** Write an equation to represent the length, $l$, of a piece of fabric whose width, $w$, is 74 centimeters less than the length. ***(LESSON 1.2)***

**25.** Solve Exercise 24 for the width, $w$, if the length, $l$, is 86 centimeters. ***(LESSON 3.1)***

CONNECTION

**26.** **COORDINATE GEOMETRY** A line passes through the origin and the point $A(3, -2)$. Write an equation for the line that passes through point $A$ and is perpendicular to the original line. ***(LESSON 5.6)***

**27.** Find each product. Are the products equal? ***(LESSON 12.8)***

$$\begin{bmatrix} -1 & 3 \\ 5 & 2 \end{bmatrix}\begin{bmatrix} 2 & -4 \\ 1 & -3 \end{bmatrix} \qquad \begin{bmatrix} 2 & -4 \\ 1 & -3 \end{bmatrix}\begin{bmatrix} -1 & 3 \\ 5 & 2 \end{bmatrix}$$

## Look Beyond

**28.** Draw the graphs of $y = x^2$ and $y = 2^x$, and explain the differences between them.

PORTFOLIO ACTIVITY

Include this activity in your portfolio.

A bag contains 12 marbles: 6 red (R), 4 green (G), and 2 yellow (Y). Two marbles are randomly drawn. Design a simulation and find the experimental probabilities of the events below. Copy and complete the table.

| | First marble returned (independent events) | First marble *not* returned (dependent events) |
|---|---|---|
| $P$(R, then R) | ? | ? |
| $P$(R, then G) | ? | ? |
| $P$(R, then Y) | ? | ? |
| $P$(G, then R) | ? | ? |
| $P$(G, then G) | ? | ? |
| $P$(G, then Y) | ? | ? |
| $P$(Y, then R) | ? | ? |
| $P$(Y, then G) | ? | ? |
| $P$(Y, then Y) | ? | ? |
| SUM | ? | ? |

Write a short summary that describes your results. Describe and explain differences in the results when returning the marble and when not returning the marble.

**WORKING ON THE CHAPTER PROJECT**

You should now be able to complete Activities 2 and 3 of the Chapter Project on page 683.

CHAPTER PROJECT THIRTEEN

# Happy Birthday!

## Activity 1

### Matching Birth Months

How many of your friends or family were born in your birth month? on your birthday? In this experiment you will explore the probability of matching birth months and birthdays.

1. Do you think all birthdays are equally likely? Explain your reasoning.
2. Estimate the following probability: If you choose 4 people at random, at least 2 will have the same birth month.
3. Estimate the following probability: If you choose 6 people at random, at least 2 will have the same birth month.
4. Devise a method for randomly picking 6 students from your class and execute it. Check whether 2 of the students chosen have the same birth month.
5. Repeat this experiment 10 times. Compute the experimental probability 2 people having the same birth month in a group of 6 people.
6. How does the experimental probability compare with your prediction from Step 3?

If you assume that all birth months are equally likely, you can use a coin and number cube to conduct this experiment. The sample space is shown below.

| J | F | M | A | M | J | J | A | S | O | N | D |
|---|---|---|---|---|---|---|---|---|---|---|---|
| H1 | H2 | H3 | H4 | H5 | H6 | T1 | T2 | T3 | T4 | T5 | T6 |

Or use a calculator by numbering the months 1–12.

For the TI-83, the keystrokes for randomly generating the integers from 1 to 12 inclusive are given below.

MATH ► ► ► 5 1 , 1 2 ) ENTER

7. Use a coin and number cube, or a calculator, to repeat the experiment for 10 trials. Compute the experimental probability of 2 people having the same birth month in a group of 6 people. Combine the trials for the whole class, and record the experimental probability for the class data.
8. Andrea said, "Since there are 12 months and I am choosing 6 people at random there is a 50% probability that I will get at least 2 with the same birth month." Is she right?
9. What is the probability of 2 people having the same birth months in a group of 13 people chosen at random?

## Activity 2

### Matching Birthdays

1. What do you think is the probability 2 people having the same birthday in a group of 35 people? Check the birthdays of students in your class for matches.

   You can use a graphics calculator or spreadsheet software to simulate picking 35 people at random and looking for at least one birthday match. One trial is shown below. Working with a partner, one person can generate the data, and the other can record the birthdays generated by crossing out values on a chart. Two strikes through a single date will represent a match.

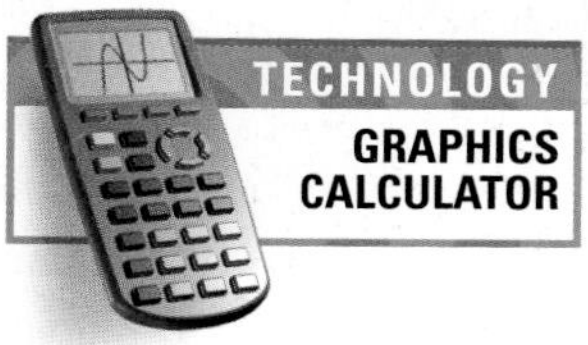

   For the TI-83, the keystrokes for randomly generating the integers from 1 to 365 are:

   MATH ► ► ► 5 1 , 3 6 5 ) ENTER.

2. Explain why the function INT(365 rand) + 1 is used to generate random birthdays.

3. Perform 10 trials of the experiment. Compute the experimental probability of a birthday match in a group of 35 people.

4. Collect the data from your class, and compute the experimental probability of a match. Does the experimental probability agree with your guess from Step 1?

## Activity 3

In Activity 2 you may have discovered that probabilities are often difficult to estimate. You will now explore the birthday problem in a different way.

1. Simulate the birthday problem for 35 people at least 100 times, either by performing all of the simulations yourself or by combining data from the entire class.

2. Find the number of matches for trials 1–5, 6–10, 11–15, and so on. Create a histogram to display the frequency distribution. It will show the frequency of the matches for each block of 5 numbers.

3. How many people do you think you need in order to have
   a. a 50% chance of a match?
   b. a 90% chance of a match?
   c. a 99% chance of a match?

# 13 Chapter Review and Assessment

## VOCABULARY



## Key Skills & Exercises

### LESSON 13.1

#### Key Skills

**Find the theoretical probability of a favorable outcome.**

Suppose that you roll a number cube once. The sample space is 1, 2, 3, 4, 5, and 6. The probability that you will roll a number greater than 4 is the ratio of the favorable outcomes ($f$), 5 and 6, to the number of elements in the sample space ($n$), 6.

$$P(E) = \frac{f}{n}$$

$$P(\text{greater than } 4) = \frac{2}{6} = \frac{1}{3}$$

#### Exercises

**A chip is drawn from a bag containing chips numbered with integers from −10 to 10 inclusive. Find the probability that the number drawn is**

**1.** negative.

**2.** even.

**3.** equal to 0.

**4.** less than or equal to −5.

**A number cube is rolled once. Find the probability that you roll**

**5.** an even number.

**6.** a prime number.

**7.** a number not evenly divisible by 3.

**8.** a number between 2 and 6 inclusive.

### LESSON 13.2

#### Key Skills

**Use the Addition of Probabilities Principle.**

To find the probability of drawing an ace or a ten in an ordinary deck of 52 playing cards, use the Addition of Probabilities Principle.

There are 4 aces in the deck. There are 4 tens in the deck.

$P(\text{ace}) = \frac{4}{52} = \frac{1}{13}$ $\qquad$ $P(\text{ten}) = \frac{4}{52} = \frac{1}{13}$

Use the Addition of Probabilities Principle.

$$P(\text{ace OR ten}) = P(\text{ace}) + P(\text{ten}) = \frac{4}{52} + \frac{4}{52} = \frac{8}{52}, \text{ or } \frac{2}{13}$$

#### Exercises

**9.** In a music class, 11 students play the clarinet and 15 play the flute, including 8 who play both. How many students are in the class?

**10.** In King High School, there are 245 freshmen in science classes and 238 in math classes. If 150 freshmen are in science AND math classes, how many are in math OR science?

**In an ordinary deck of cards,**

**11.** how many cards are black OR face cards?

**12.** how many cards are red OR numbered cards?

**If you toss a number cube, what is the probability that you will toss**

**13.** a 4 OR a number greater than 3?

**14.** an odd OR a prime number?

LESSON 13.3

**Key Skills**

**Use the Fundamental Counting Principle.**

To find the number of T-shirts stocked by a craft store that stocks 5 styles, 4 sizes, and 10 different colors, use the Fundamental Counting Principle, which states that if there are $m$ ways to make a first choice and $n$ ways to make a second choice, then there are $m \cdot n$ ways to make a first choice and a second choice. Thus, there are $5 \cdot 4 \cdot 10$, or 200, T-shirts stocked by the craft store.

**Exercises**

**15.** How many ways are there to arrange A, B, C, and D?

**16.** A dress shop stocks 6 styles of blouses. Each blouse comes in 7 sizes and 5 colors. How many different selections are possible?

**17.** If you use only the digits 1–9, how many 3-digit numbers can you write?

**18.** How many different results can you get from flipping a coin and then rolling a number cube?

**Two decks of ordinary playing cards are shuffled separately. A card is drawn from each deck. How many ways are there to draw**

**19.** a queen from the first deck AND a face card from the second deck?

**20.** a face card from the first deck AND a red card from the second deck?

LESSON 13.4

**Key Skills**

**Find the probability of two independent events.**

There are 2 red marbles and 4 blue marbles in a bag. You draw a marble, replace it, and then draw another marble. The probability of selecting a red marble and a blue marble is $\frac{2}{6} \cdot \frac{4}{6} = \frac{8}{36} = \frac{2}{9}$.

**Exercises**

**There are 10 cards numbered 1 through 10 in a box. One card is drawn and then replaced. Then another card is drawn. Find the probability that**

**21.** the first card is 4 and the second is 6.

**22.** both cards are less than 5.

**23.** both cards are multiples of 3.

**24.** both cards are the same.

**There are 5 blue marbles and 3 red marbles in a bag. One marble is drawn and replaced. A second marble is drawn. Find the probability that**

**25.** the first marble is red and the second is blue.

**26.** both marbles are red.

**27.** both marbles are blue.

LESSON 13.5

## Key Skills

**Use a simulation to find experimental probability.**

To simulate randomly choosing the correct response to a multiple-choice question with 5 possibilities, use a number cube to simulate the experiment. Let 1, 2, 3, 4, or 5 represent the possible responses, with 1 representing a correct response. Roll again if the result is a 6. One roll represents one trial. The following chart indicates a possible simulation:

| Trial | 1 | 2 | 3 | 4 | 5 | 6 | 7 | 8 | 9 | 10 |
|---|---|---|---|---|---|---|---|---|---|---|
| Roll | 1 | 4 | 2 | 1 | 3 | 3 | 1 | 4 | 3 | 5 |

## Exercises

**Suppose that you want to simulate answering all of the questions on a 10-question true-false quiz correctly. Design a simulation using**

**28.** random numbers.

**29.** coins.

## Applications

MARKETING The Hamburger Hut is giving away free food during their grand opening. They are distributing 1000 cards—100 are for a free hamburger, 150 are for free fries, and 750 are for a free soft drink.

If one card is drawn, what is the probability of winning

**30.** a hamburger?

**31.** a soft drink?

If two cards are drawn with replacement, what is the probability of winning

**32.** a hamburger OR soft drink?

**33.** a hamburger AND soft drink?

**34.** FINANCE Tim, Mary, and Paul have applied for a loan to buy a used car. The probability that Tim's application is approved is 75%. The probability that Mary's application is approved is 80%. The probability that Paul's application is approved is 62.5%. What is the probability that all three loans will be approved?

**35.** FINANCE In Exercise 34, what is the probability that only Mary's application will be approved?

# 13 Chapter Test

**A coin is tossed once, and a number cube is rolled once. Use this information for Exercises 1–4.**

1. Find the sample space for the experiment.
2. Find the probability that the coin will be heads.
3. Find the probability that the coin will be heads and the number cube an even number.
4. Find the probability that the number cube will be a prime number and the coin tails.
5. There are 31 days in January and 28 days in February. Find the probability of being born on a prime number date in each month. In which month is the probability of being born on a prime number date greater.

**Exercises 6 and 7 refer to a standard deck of 52 cards.**

6. How many cards are either sevens OR clubs?
7. How many cards are red cards OR twos?
8. RECREATION There are 28 students in a class. If 15 decide to go to a movie this weekend and 20 decide to go to the zoo, what is the smallest number that went to both the movie and the zoo?

**If you toss a number cube, what is the probability of each result?**

9. a 3 or an even number
10. a 1 or a prime number
11. DEMOGRAPHICS In a survey of a small city of 5000 households, 2500 have a VCR, 750 have a DVD player and 250 have both a VCR and a DVD player. If you randomly select one household, what is the probability that it owns either a VCR or a DVD player?
12. GOVERNMENT A license plate has 2 letters first, followed by 4 numbers. How many license plates are available with this combination?
13. LANGUAGE ARTS How many different ways could the letters in the state of Maine be rearranged?
14. CONSUMER SCIENCE When purchasing a new computer, you are given the choice of 4 styles of monitors, 3 keyboards, 7 storage devices, and 2 processors. How many different computers could be purchased?
15. Two number cubes are rolled. How many ways are there to get an even number on the first cube and a multiple of 3 on the second cube?

**Tell whether the events are independent or dependent.**

16. choosing one boy in the class and one girl in the class to attend a meeting
17. choosing three integers, without replacement, from the set of integers 1 to 48, inclusive

**A letter is selected from the name Wayne and a number is chosen between 1 and 10, inclusive. Find the probability that**

18. the letter selected is *a* and the number chosen is 7.
19. the letter selected is a vowel and the number chosen is a prime number.
20. the letter selected is a capital letter and the number chosen is a multiple of 5.

**There are 6 red marbles and 4 white marbles in a bag. One marble is drawn and replaced and then a second marble is drawn. Find each probability.**

21. Both marbles are red.
22. The first marble is red and the second is white.

**Suppose you want to simulate the results of a multiple-choice test where each question has 4 possible answers. Design a simulation using the given devices.**

23. coins
24. random numbers

# College Entrance Exam Practice

**internet connect**

**Standardized Test Prep Online**
Go To: **go.hrw.com**
Keyword: **MM1 Test Prep.**

**QUANTITATIVE COMPARISON** For questions 1–5 write
A if the quantity in Column A is greater than the quantity in Column B;
B if the quantity in Column B is greater than the quantity in Column A;
C if the quantities are equal; or
D if the relationship cannot be determined from the information given.

| | Column A | Column B | Answers |
|---|---|---|---|
| **1.** | The number of \$15 tickets you can buy for \$90 | The number of \$12 tickets you can buy for \$84 | (A) (B) (C) (D) [Lesson 3.2] |
| **2.** | $(-18)(-4)$ | $-18(-2)$ | (A) (B) (C) (D) [Lesson 2.4] |
| **3.** | $\frac{2\sqrt{2}}{3}$ | $\frac{2\sqrt{10}}{5}$ | (A) (B) (C) (D) [Lesson 12.1] |
| **4.** | $(9 \times 10^6)(7 \times 10^6)$ | $(1.9 \times 10^8)(3 \times 10^4)$ | (A) (B) (C) (D) [Lesson 8.5] |
| **5.** | The sine of angle $A$ | The cosine of angle $A$ | (A) (B) (C) (D) [Lesson 12.7] |

**6.** Find $|-3.6| - |4.5|$. ***(LESSON 2.1)***
**a.** 0.9 **b.** −0.9
**c.** 8.1 **d.** −8.1

**7.** Solve: $2x^2 = 288$. ***(LESSON 10.2)***
**a.** 12 **b.** 72
**c.** ±12 **d.** ±72

**8.** What is the slope of a line parallel to $2x + 3y = -5$? ***(LESSON 5.6)***
**a.** 2 **b.** $-\frac{2}{3}$
**c.** $\frac{3}{2}$ **d.** 3

**9.** What is the probability of choosing a blue chip from a bag containing 4 blue chips and 3 red chips? ***(LESSON 13.1)***
**a.** $\frac{3}{4}$ **b.** $\frac{4}{3}$
**c.** $\frac{4}{7}$ **d.** $\frac{3}{7}$

**10.** Simplify: $3\sqrt{2} - \sqrt{5} + 6\sqrt{2} + 9\sqrt{5}$. ***(LESSON 12.1)***
**a.** $9\sqrt{2} + 9$ **b.** $-3\sqrt{2} + 8\sqrt{5}$
**c.** $-3\sqrt{2} + 9$ **d.** $9\sqrt{2} + 8\sqrt{5}$

**11.** Find 65% of 20. ***(LESSON 4.2)***
**a.** 1300 **b.** 0.0325
**c.** 13 **d.** 3.25

**12.** Which spreadsheet command generates a random number from the list 3, 4, 5? ***(LESSON 13.5)***
**a.** INT(RAND()*5)+5
**b.** INT(RAND()*3)+5
**c.** INT(RAND()*5)+3
**d.** INT(RAND()*3)+3

13. Which function has a graph that is identical to that of $s = t^2$? ***(LESSON 10.1)***

**a.** $y = x^2$ **b.** $y = x$

**c.** $y = |x|$ **d.** $y = \frac{1}{x}$

14. Simplify: $\frac{x^2 + 5x + 6}{x + 3}$. ***(LESSON 11.3)***

**a.** $x + 2, x \neq -3$ **b.** $x + 2, x \neq 3$

**c.** $-2$ **d.** $-2, -3$

15. What is the simplified form of $(-2m^2n^4)^2$? ***(LESSON 8.2)***

**a.** $-4m^4n^6$ **b.** $4m^4n^6$

**c.** $-4m^4n^8$ **d.** $4m^4n^8$

16. There are 48 students in the choir and 50 students in the orchestra. If 15 students are in both the choir and orchestra, what is the total number of students in the choir and the orchestra? ***(LESSON 13.2)***

17. Write an inequality that describes the points graphed on the number line. ***(LESSON 6.1)***

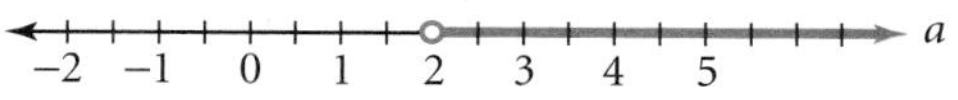


18. If $x$ varies inversely as $y$ and $x = 5$ when $y = 7$, find $x$ when $y = 10$. ***(LESSON 11.1)***

19. Write the common denominator for $\frac{x}{x + 3} + \frac{2}{x - 3}$. ***(LESSON 11.4)***

20. Find two numbers whose sum is 32 and whose product is 255. ***(LESSON 7.6)***

21. Factor: $4x^2 - 16$. ***(LESSON 9.6)***

22. Solve: $(x + 1)^2 - 4 = 0$. ***(LESSON 10.2)***

23. The first three terms of a sequence are 4, 7, and 12. The second differences are a constant 2. Find the next three terms. ***(LESSON 1.1)***

24. Simplify: $3w(4w + 5) - (w + 10)$. ***(LESSON 9.2)***

25. Write and solve an equation for this situation: A computer is on sale for $1800. The sale price is $\frac{2}{3}$ of the regular price. Find the regular price of the computer. ***(LESSON 3.2)***

26. Evaluate: $\frac{-15}{-3}$. ***(LESSON 2.4)***

27. Solve: $13 - 3x = -8$. ***(LESSON 3.3)***

28. Solve: $|5x + 14| < 11$. ***(LESSON 6.5)***

29. What percent of 160 is 60? ***(LESSON 4.2)***

30. What is the probability of getting an even number when rolling a number cube? ***(LESSON 4.3)***

31. What is the slope of the line containing the points $A(8, 4)$ and $B(-3, 10)$? ***(LESSON 5.2)***

32. A force of 5 Newtons stretches a spring 35 centimeters. Find the constant, $k$, for the spring. ***(LESSON 5.3)***

33. Solve the system of equations. $\begin{cases} 4p + 6q = 36 \\ 10p - 2q = 22 \end{cases}$ ***(LESSON 7.3)***

## FREE-RESPONSE GRID

**The following questions may be answered by using a free-response grid such as that commonly used by standardized test services.**

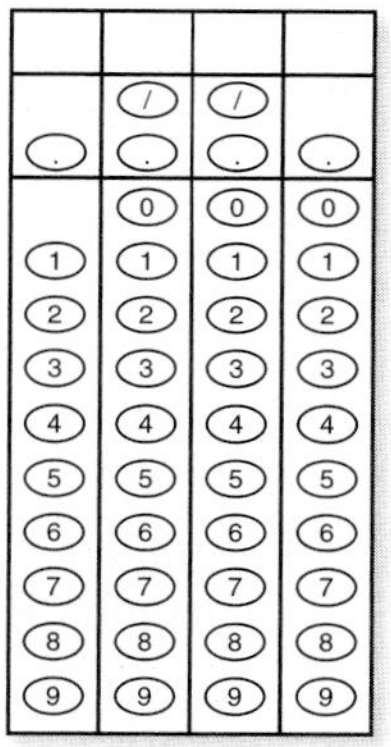

34. Evaluate $2x - (3y + 4)$ for $x = -3$ and $y = -2$. ***(LESSON 1.3)***

35. How many ways can you arrange 6 different letters? ***(LESSON 13.3)***

36. A large truck has a load capacity of 9 tons more than a smaller truck. The ratio of their load capacities is 5 to 2. Find the load capacity of the smaller truck. ***(LESSON 4.1)***

37. Evaluate $x^2 - 4x + 7$ for $x = 3$. ***(LESSON 10.1)***

38. Find the slope of the line containing the points $M(-1, -2)$ and $N(3, 4)$. ***(LESSON 5.2)***

# 14

# Functions and Transformations

ONCE A BASIC FUNCTION AND ITS GRAPH ARE understood, it is possible to create many adaptations that help fit the function to specific needs. Adaptations, known as transformations in mathematics, appear in such fields as science, business, advertising, and art. In mathematics, the use of transformations lets you explore a broad range of functions. Using transformations, you will be able to graph many interesting functions.

## Lessons



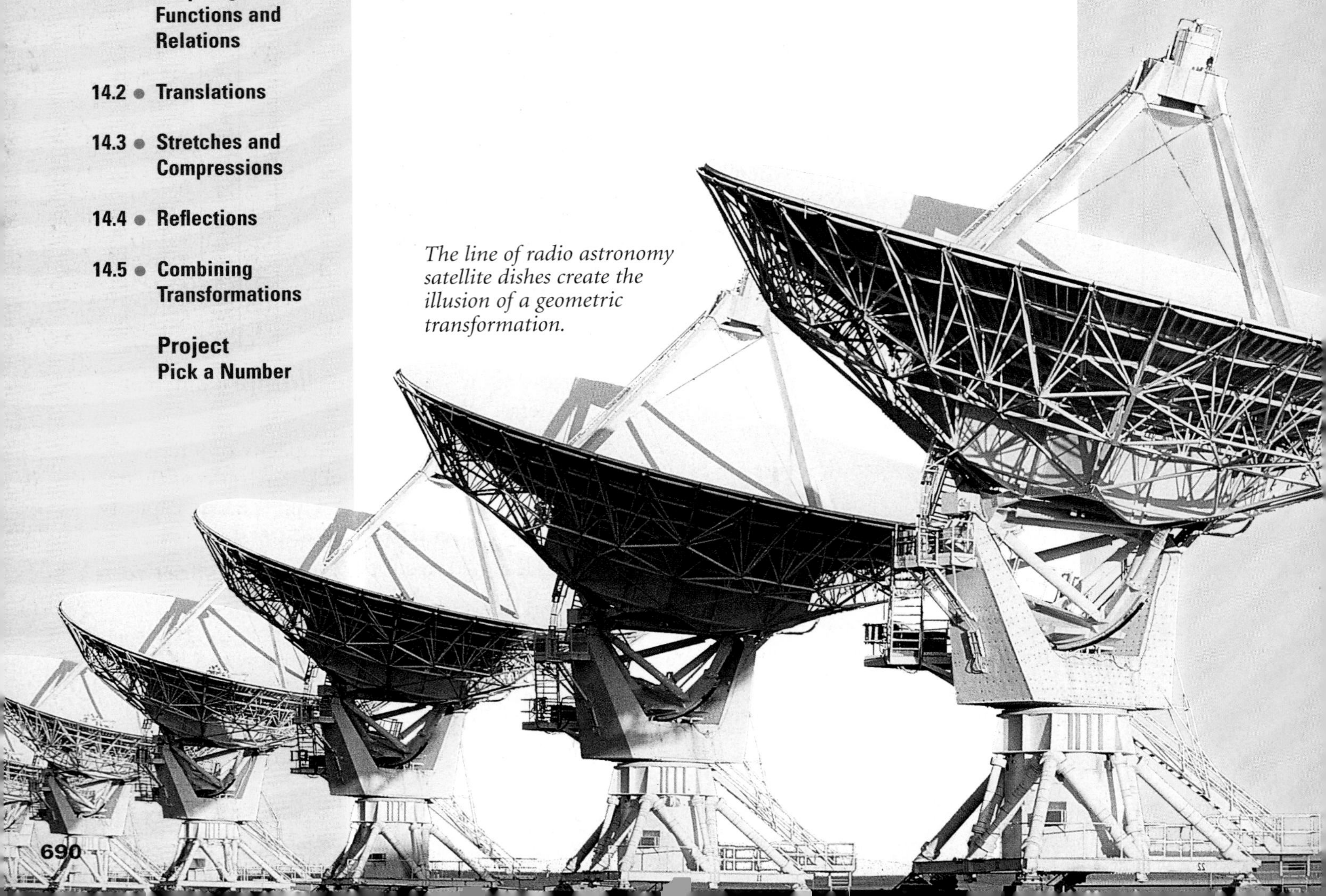

*The line of radio astronomy satellite dishes create the illusion of a geometric transformation.*

*Patterns in architecture, agriculture, and industry model geometric transformations.*

CHAPTER PORTFOLIO ACTIVITIES PROJECT

## About the Chapter Project

Algebra provides the power to simplify many seemingly complex situations through the use of functions. In the Chapter Project, *Pick a Number,* on page 728, you will investigate the relationship between number games and composite functions. After completing the Chapter Project you will be able to do the following:

- Define and apply the concept of composite functions.

## About the Portfolio Activities

Throughout the chapter, you will be given opportunities to complete Portfolio Activities that are designed to support your work on the Chapter Project.

- Recognizing transformed functions is the topic of the Portfolio Activity on page 719.
- Comparing flying time and mileage between cities is the topic of the Portfolio Activity on page 727.

14.1

# Graphing Functions and Relations

## Objectives

- Use models to understand functions and relations.
- Evaluate functions by using function rules.
- Identify the parent functions of some important families of functions.

| National Park | State | Camping | Lodge | Hiking | Fishing | Bike trails |
|---|---|---|---|---|---|---|
| Arches | Utah | • | | • | | |
| Bryce Canyon | Utah | • | • | • | | |
| Canyonlands | Utah | • | | • | | |
| Carlsbad Caverns | New Mexico | | | • | | |
| Capitol Reef | Utah | • | | • | | |
| Chaco Culture | New Mexico | • | | • | | • |
| Grand Canyon | Arizona | • | • | • | • | |
| Mesa Verde | Colorado | • | • | • | | |
| Petrified Forest | Arizona | | | • | | |
| Rocky Mountain | Colorado | • | | • | • | |
| Zion | Utah | • | • | • | | |

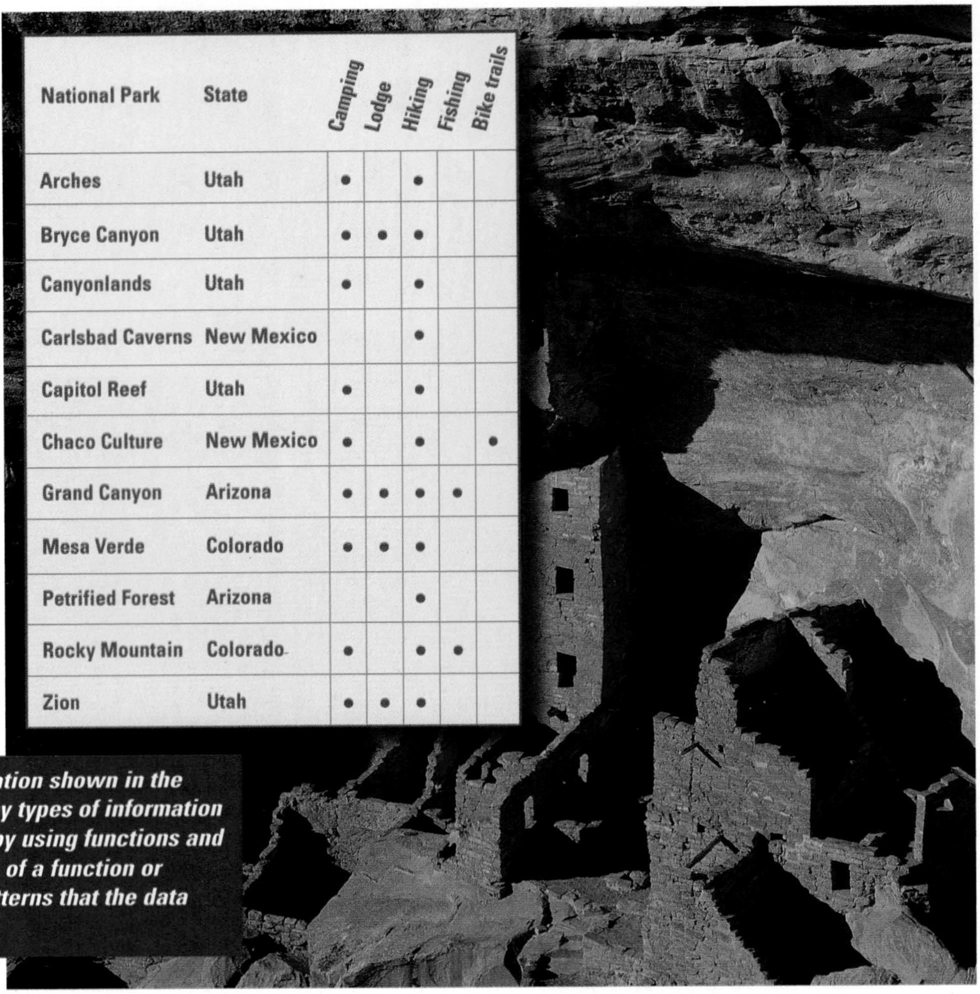

**Why** ***The national park information shown in the chart is an example of the many types of information that can be better understood by using functions and relations as models. The graph of a function or relation often shows visual patterns that the data alone would not reveal.***

*Mesa Verde National Park covers 52,122 acres in southwest Colorado. It contains some of the best-preserved cliff dwellings in the United States.*

Recall that a relation is a set of ordered pairs. The first elements of the ordered pairs form a set called the *domain*, and the second elements of the ordered pairs form a set called the *range*. A function is a special type of relation in which each element of the domain is paired with exactly one element of the range.

**EXAMPLE 1**

**APPLICATION**
**RECREATION**

Consider each relation below. **Draw a model of each and tell whether the relation is a function.**

**a.** {(Canyonlands, camping), (Capitol Reef, hiking), (Bryce Canyon, lodge), (Grand Canyon, fishing)}
**b.** {(Arches, hiking), (Chaco Culture, bike trails), (Zion, lodge), (Petrified Forest, hiking)}
**c.** {(Rocky Mountain, fishing), (Mesa Verde, lodge), (Carlsbad Caverns, hiking), (Mesa Verde, camping)}

## SOLUTION

Represent each relation by listing the elements of the domain and the range in two seperate columns. Connect the elements of the ordered pairs with lines.

**a.**

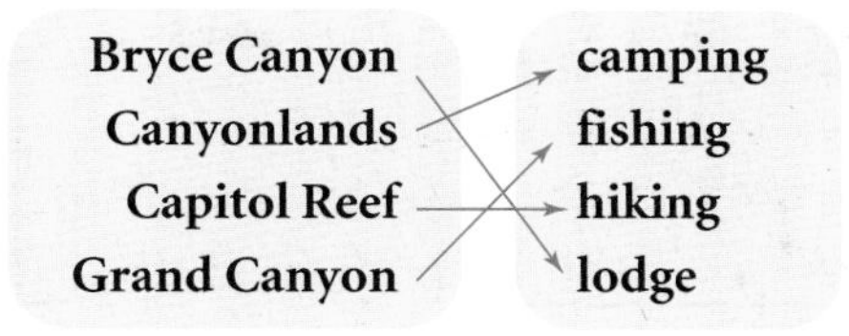


The relation in **a** is a function because each element in the first set is paired with exactly one element in the second set.

**b.**

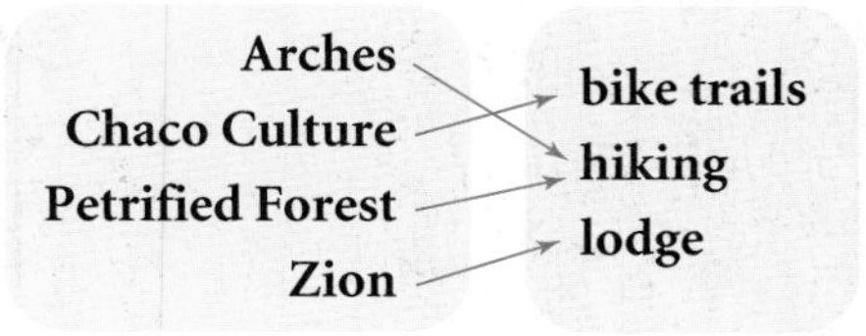


The relation in **b** is a function even though one element in the second set is paired with two different elements in the first set.

**c.**

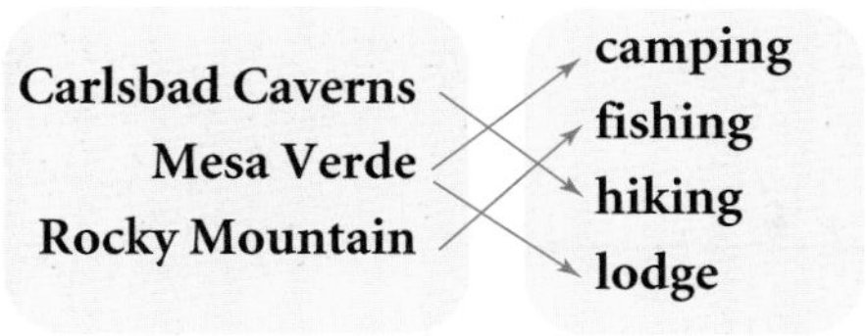


The relation in **c** is not a function because an element of the first set is paired with more than one element in the second set.

# Graphs of Functions and Relations

Relations can be graphed. Once a relation is graphed, the **vertical-line test** can be used to determine whether the relation is a function.

## *Activity* Vertical-Line Test

**You will need:** graph paper

1. Draw a model of each relation below.
   **a.** $\{(2, 3), (3, 5), (8, 6), (7, 9)\}$
   **b.** $\{(2, 3), (0, 4), (-9, 5), (8, 3)\}$
   **c.** $\{(2, 3), (2, 9), (3, 5), (8, 3)\}$
   **d.** $\{(2, 3), (-2, 3), (2, -3), (-2, -3)\}$
2. Determine whether or not each relation in Step 1 is a function.
3. Graph each relation from Step 1 on a separate coordinate grid.
4. For each relation, try to draw a vertical line that intersects the graph in more than one place.

**CHECKPOINT ✓**

5. What do you notice about the graphs of the relations that are functions compared with those of the other relations? Explain how the vertical-line test distinguishes functions from other relations.

As you may have discovered in the activity, if an element of the domain is paired with more than one element of the range, a vertical line can be drawn through both points on a graph. Therefore, a relation is a function if it is impossible to draw a vertical line that intersects the graph of the relation more than once.

**E X A M P L E 2** **Which of the following graphs represent functions?**

**a.**

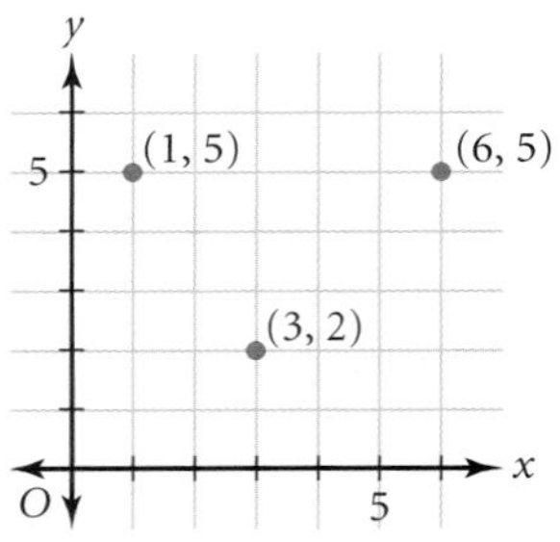


**b.**

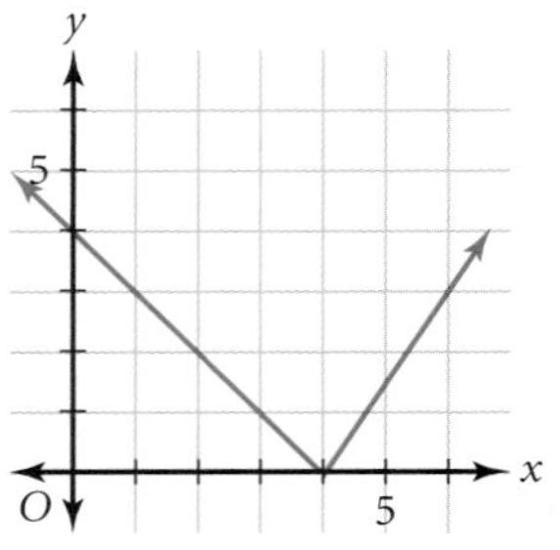


**c.**

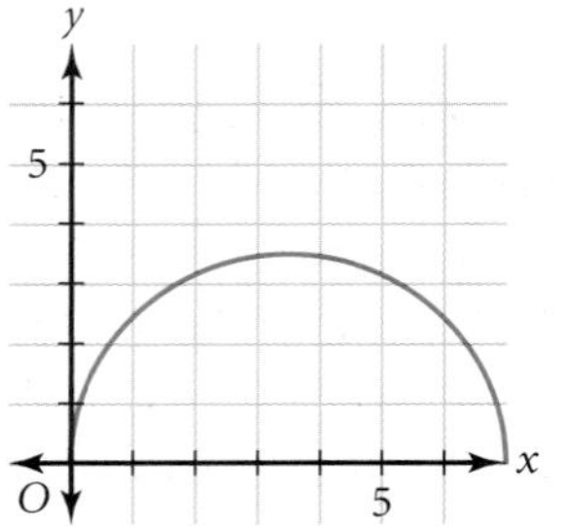


**d.**

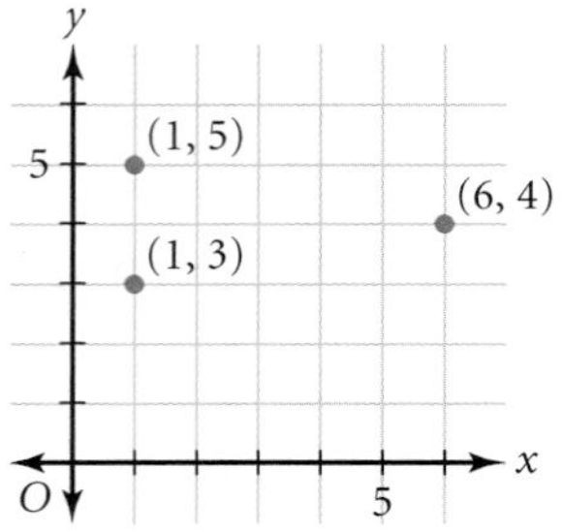


**e.**

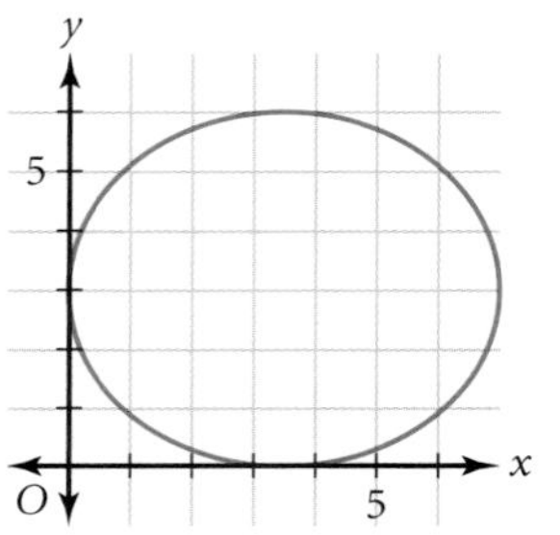


**f.**

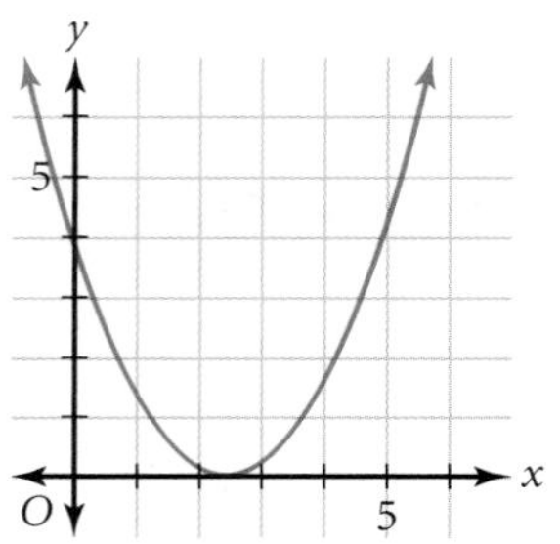


**SOLUTION**

It is impossible to draw a vertical line that will intersect graphs **a**, **b**, **c**, and **f** more than once, so these graphs represent functions. Neither graph **d** nor graph **e** represents a function because the vertical line $x = 1$ intersects graph **d** twice and the vertical line $x = 2$, for example, intersects graph **e** twice.

## Function Notation and Function Rules

The statement $y = x^2$ represents a quadratic function. Another way to express the same function is to use **function notation**. In function notation, the dependent variable, $y$, is written as $f(x)$—or as $g(x)$, $h(x)$, etc. Thus, $y = x^2$ would be written as $f(x) = x^2$, $g(x) = x^2$, $h(x) = x^2$, etc. Notice that the independent variable is placed inside the parentheses.

The statement $f(x) = x^2$ is read as "$f$ of $x$ equals $x$ squared." The expression on the right side of the statement is called the **function rule**. To evaluate the function for a given value of $x$, replace $x$ in the function rule with that value. For example, if $x = 3$, then $f(3) = 3^2 = 9$.

The elements of the domain of a function are sometimes called the *input* of the function rule. The elements of the range are sometimes called the *output* of the function rule.

input (an element from the domain) ➡ function rule ➡ output (an element from the range)

You can think of a function rule as a machine that accepts domain values and returns range values.

It is important to consider what domain and range values are appropriate for a particular situation. If no particular domain is given, the domain is understood to be all real numbers for which the function rule is defined.

**EXAMPLE 3** **If $h(x) = x^2 - 2x + 1$, what is $h(4)$?**

**SOLUTION**

The notation $h(4)$ indicates that you substitute 4 for $x$ in the function rule $x^2 - 2x + 1$. Next, simplify to find the value of the function when $x$ is 4.

$h(x) = x^2 - 2x + 1$
$h(4) = 4^2 - 2(4) + 1$
$h(4) = 9$

**TRY THIS** If $g(x) = 2x^2 - 3x - 2$, what is $g(3)$?

**EXAMPLE 4** Consider the function $f(x) = x^2$, where the domain of the function consists of the integers from −2 to 2 inclusive. **Make a list of the ordered pairs of the function and draw a model of the function.**

**SOLUTION**

Use the function rule $x^2$ to find the range values for each of the domain values. For example, for the domain value −2, the range value is $(-2)^2$, or 4. Thus, (−2, 4) is an ordered pair of the function. The complete list of ordered pairs of the function is as follows:

$\{(-2, 4), (-1, 1), (0, 0), (1, 1), (2, 4)\}$

Use the list of ordered pairs to draw the function model.

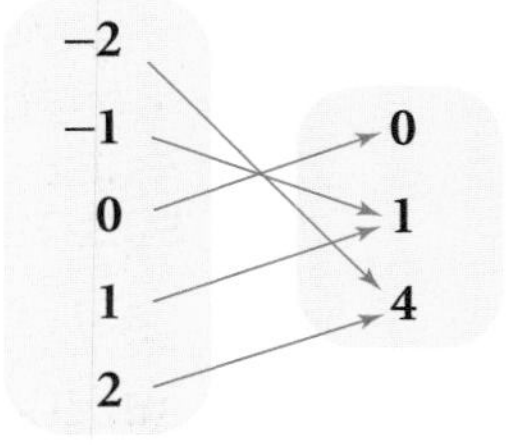

**TRY THIS** Consider the function $f(x) = x^2 + 2$, where the domain of the function consists of the integers from $-3$ to 3 inclusive. Make a list of the ordered pairs of the function and draw a model of the function.

## Parent Functions

In nature, a mother or a father may have many offspring. In mathematics, a **parent function** can be transformed into many—in fact, *infinitely* many—new functions. A *family* of functions is comprised of these new functions and the parent function. Every function can be classified as a member of a family. Certain functions are identified as parent functions of these families. For example, the function $f(x) = x^2$ or $y = x^2$ is the parent function of the family of quadratic functions. Other parent functions are shown in the table below.

*Characteristics of parents are passed from one generation to the next.*

| Parent Function | Family |
|---|---|
| $f(x) = x$ | linear |
| $f(x) = \|x\|$ | absolute value |
| $f(x) = a^x$ | exponential |
| $f(x) = x^2$ | quadratic |
| $f(x) = \frac{1}{x}$ | rational |
| $f(x) = \sqrt{x}$ | square root |

**EXAMPLE 5** **Name the parent function for each graph.**

**a.**
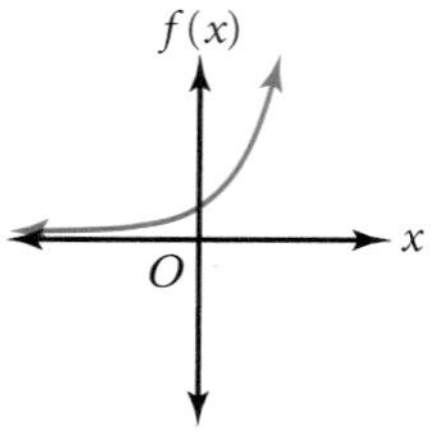


**b.**
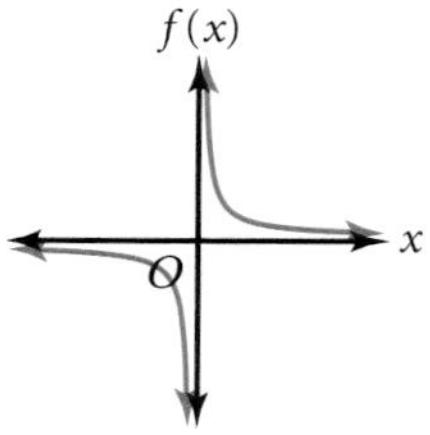


**c.**
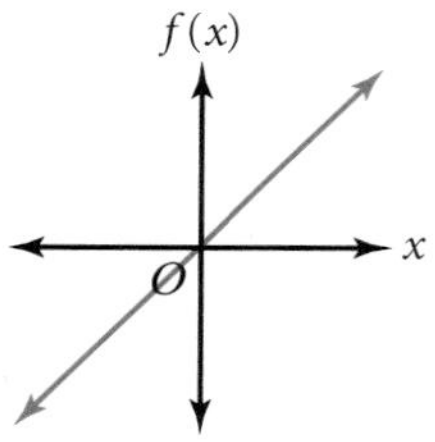


**d.**
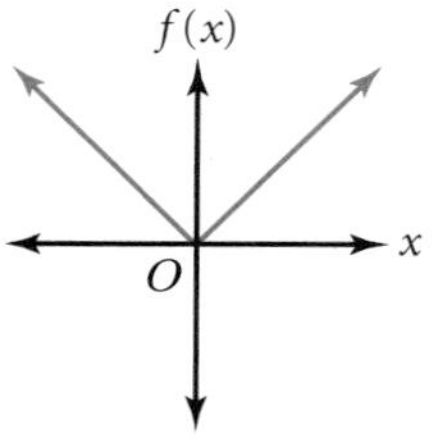


**e.**
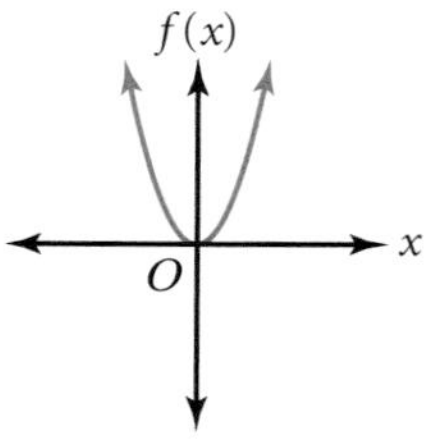


**f.**
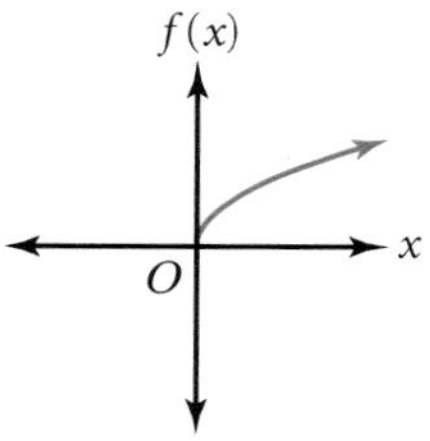


**SOLUTION**

**a.** $f(x) = a^x, (a > 1)$ **b.** $f(x) = \frac{1}{x}$ **c.** $f(x) = x$

**d.** $f(x) = |x|$ **e.** $f(x) = x^2$ **f.** $f(x) = \sqrt{x}$

# Exercises

## Communicate

**1.** Given a graph on a coordinate plane, explain how to use the vertical-line test. What is the purpose of the vertical-line test?

**2.** Explain how to find the range of the function $f(x) = 2x^2 - 5x$ if the domain is restricted to integers from $-2$ to 2 inclusive.

## Guided Skills Practice

**Draw a model for each relation and tell whether the relation is a function. Give a reason for each answer.** ***(EXAMPLE 1)***

**3.** $\{(-7, 5), (3, 2), (7, -1), (8, 8)\}$

**4.** $\{(2, 3), (-6, 5), (8, 4), (-2, 3)\}$

**5.** $\{(-1, 4), (-3, 6), (-1, -2), (4, 4)\}$

**6.** $\{(10, -1), (-6, -6), (3, 1), (10, -5)\}$

**Which of the following graphs represent functions? Give a reason for each answer.** ***(EXAMPLE 2)***

**7.**

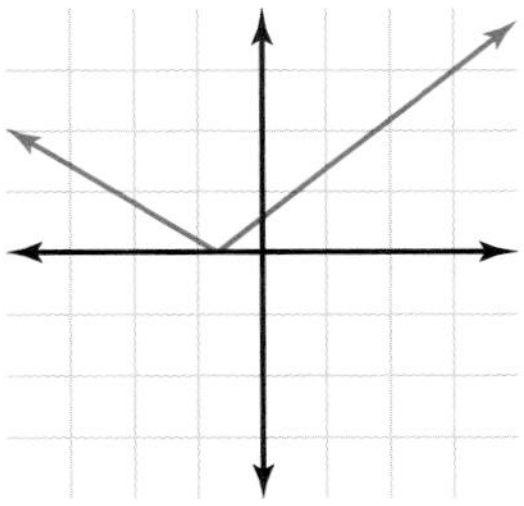

**8.**

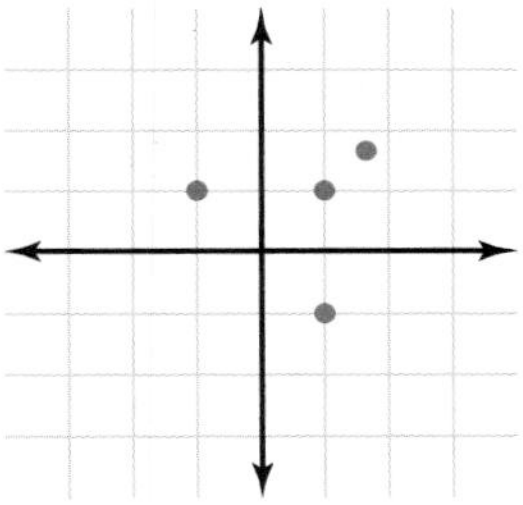

**Find $f(2)$, $f(-1)$, and $f(4)$ for each function.** ***(EXAMPLE 3)***

**9.** $f(x) = 3x^2$

**10.** $f(x) = |2x - 5|$

**11.** Consider the function $f(x) = |x|$, where the domain of the function consists of the integers from $-3$ to 3 inclusive. Make a list of the ordered pairs of the function and draw a model of the function. ***(EXAMPLE 4)***

**Name the parent function for each graph.** ***(EXAMPLE 5)***

**12.**

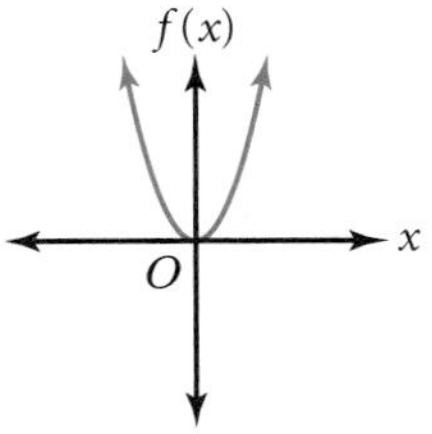


**13.**

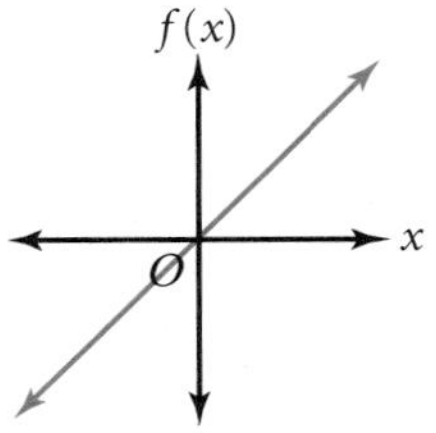


**14.**

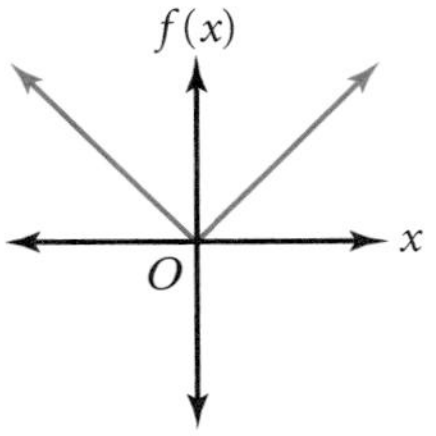


**15.**

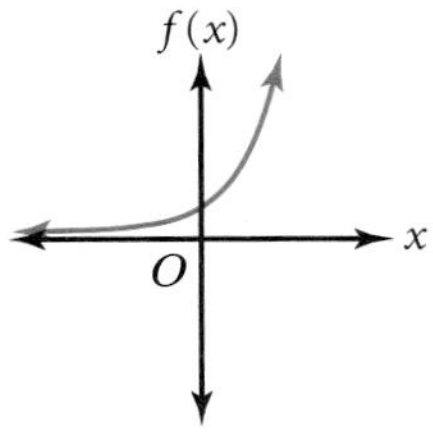


**16.**

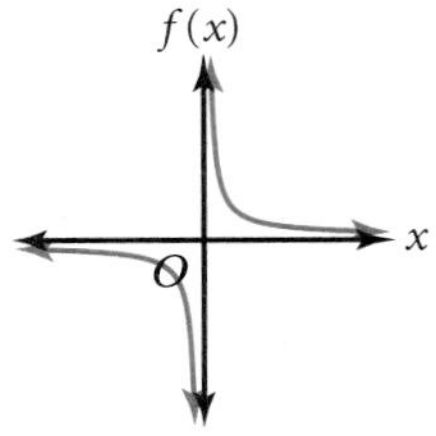


**17.**

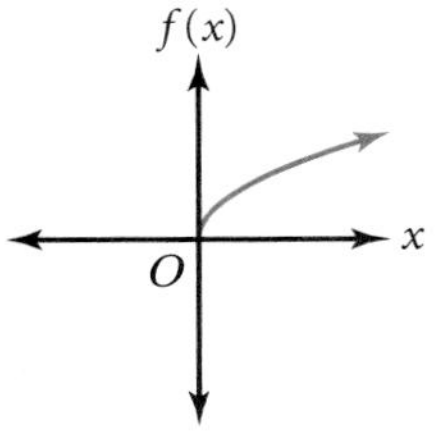

## Practice and Apply

internet connect
**Homework Help Online**
Go To: **go.hrw.com**
Keyword:
**MA1 Homework Help**
for Exercises 18-21

**Draw a model for each relation and tell whether the relation is a function. Give a reason for each answer.**

**18.** $\{(3, 4), (4, 4), (5, 4)\}$

**19.** $\{(1, 1)\}$

**20.** $\{(9, 5), (8, 3), (9, -5), (7, 6)\}$

**21.** $\{(5, 9), (4, 3), (6, -7), (-5, 9)\}$

**Which of the following graphs represent functions? Give a reason for each answer.**

**22.**

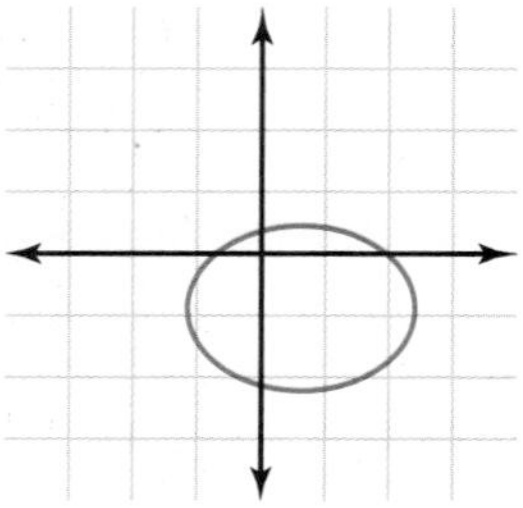

**23.**

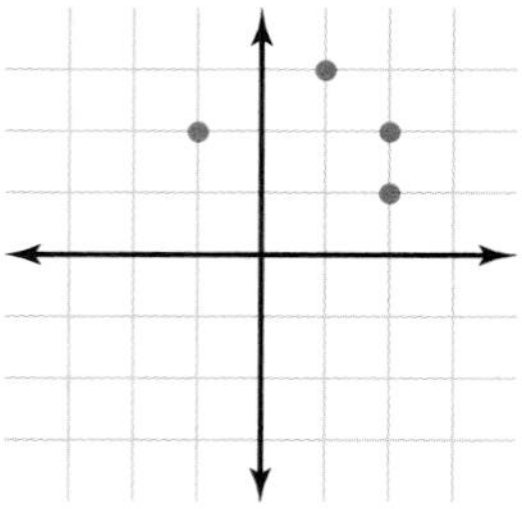

**Find the domain and range of each function.**

**24.** $\{(3, 4), (4, 3), (5, 2)\}$

**25.** $\{(0, 1), (1, 3), (2, 5), (3, 7)\}$

**26.** $\left\{\left(1, \frac{1}{2}\right), \left(2, \frac{1}{2}\right), \left(3, \frac{1}{3}\right), \left(4, \frac{1}{4}\right)\right\}$

**27.** $\{(0.1, 1), (0.2, 2), (0.3, 3)\}$

**28.** $\{(5, 6), (-2, 1), (4, -7)\}$

**29.** $\{(2.2, 3), (5, 9), (4.1, 5.3)\}$

**30.** $\{(17, 100), (23, -12), (14, 54)\}$

**31.** $\{(70, 13), (20, 53), (-11, 91)\}$

**32.** Given $y = \frac{1}{x}$, write the ordered pair $(x, y)$ for $x = -5$.

**For $h(x) = x + 3$, evaluate the following:**

**33.** $h(7)$ **34.** $h(4)$ **35.** $h(-8)$ **36.** $h(-3)$

**For $f(x) = 5x$, evaluate the following:**

**37.** $f(3)$ **38.** $f(0)$ **39.** $f(-2)$ **40.** $f(-6)$

**Evaluate each function for $x = 3$.**

**41.** $g(x) = x^2$ **42.** $f(x) = 2^x$ **43.** $h(x) = |x|$ **44.** $k(x) = \frac{1}{x}$

**For each function, describe the domain and range.**

**45.** $f(x) = x - 5$

**46.** $f(x) = x^2$

**47.** $f(x) = (x + 4)^2$

**48.** $f(x) = (x - 7)^2$

**49.** $f(x) = -3x^2$

**50.** $f(x) = -|x|$

**51.** $f(x) = |x + 2|$

**52.** $f(x) = |x - 5|$

**53.** $f(x) = -\frac{1}{x}$

**Given $f(x) = x^2 - 1$, write an ordered pair $(x, y)$ for each of the following elements of the domain.**

**54.** $x = 0$ **55.** $x = -9$ **56.** $x = 2$

**Identify the parent function for each of the following functions.**

**57.** $g(x) = -(3x - 2)^2 - 5$

**58.** $y = \sqrt{2x + 5}$

**59.** $h(x) = \frac{5}{6 + x}$

APPLICATIONS

internet connect
**Activities Online**
Go To: **go.hrw.com**
Keyword:
**MA1 Weather**

**In Exercises 60 and 61, use the formula $F = \frac{9}{5}C + 32$. ($F$ stands for Fahrenheit and $C$ stands for Celsius.)**

**60.** METEOROLOGY If you know the Celsius temperature, is it possible to find the Fahrenheit temperature? Is Fahrenheit temperature a function of Celsius temperature? Explain.

**61.** METEOROLOGY If you know the Fahrenheit temperature, is it possible to find the Celsius temperature? Is Celsius temperature a function of Fahrenheit temperature? Explain.

**62.** TRAVEL A family is driving between national parks as a part of their summer vacation. The points below represent the relationship between time in minutes, $t$, and speed in miles per hour, $s$, for one part of the drive. Graph the points and connect them to represent the relation between $t$ and $s$. Is this relation a function?
$\{(2, 25), (5, 0), (7, 30), (10, 30), (12, 0), (15, 25), (20, 55), (25, 55)\}$

**63.** TRAVEL Describe the relationship between speed and time that would be necessary in order for the graph from Exercise 62, with speed on the horizontal axis, to represent a function.

## Look Back

**64.** There are 5 friends who play tennis. How many games must occur in order for each player to play each of the others one time? ***(LESSON 1.1)***

**Solve each equation.** ***(LESSON 3.4)***

**65.** $3r + 4 = -2 + 6r$

**66.** $8x - 7 = 3x + 4$

**67.** Solve $A = \frac{h}{2}(B + b)$ for $B$. ***(LESSON 3.6)***

**68.** Solve. $\begin{cases} 2x + 3y = 7 \\ 3x - 2y = 5 \end{cases}$ ***(LESSON 7.3)***

internet connect
**Activities Online**
Go To: **go.hrw.com**
Keyword:
**MA1 Graphs**

## Look Beyond

**69.** Given the graph of the function $f(x) = 10^x$, what would you do to the function rule in order to move the graph up by 1 unit? (Hint: What would you do to the function rule to increase each resulting $y$-value by 1 unit?)

**70.** Given the graph of the function $f(x) = 10^x$, what would you do to the function rule in order to reflect the graph across the $x$-axis?

CONNECTIONS

**71.** COORDINATE GEOMETRY Plot several points and sketch the graph of each function. Name the parent function for each.

**a.** $f(x) = 4 + |x|$ **b.** $g(x) = -2x$ **c.** $p(s) = \frac{1}{s+2}$

**d.** $h(t) = 5t^2$ **e.** $z(u) = \frac{1}{u} + 3$ **f.** $f(r) = |r - 5|$

14.2

# Translations

**Objective**

- Describe how changes to the rule of a function correspond to the translation of its graph.

**Why** *Some designs in architecture involve repeated figures that appear to have been shifted vertically or horizontally. When the graph of a function looks like the graph of a parent function that has been shifted vertically or horizontally there is a simple relationship between the functions.*

"Tile facade for the hall of a school in the Hague by M.C. Escher. "

A **transformation** of a function is an alteration of the function rule that produces a new function. Changes to the rule of a function correspond to changes to its graph. A transformation that shifts the graph of a function horizontally or vertically is called a **translation**. In the Escher drawing above, point $A$ is translated to point $B$ by shifting point $A$ 2 units to the right.

TECHNOLOGY
GRAPHICS CALCULATOR

## Activity
## Translations

**You will need:** graph paper or a graphics calculator

You can determine the characteristics of translations by exploring the functions $f(x) = |x|$, $g(x) = |x| - 10$, and $h(x) = |x - 10|$.

1. Graph the functions $f(x) = |x|$ and $g(x) = |x| - 10$ on the same set of axes.
2. Compare the graph of $g(x) = |x| - 10$ with the graph of its parent function.
   a. In which direction was the parent function shifted?
   b. By how much?
3. Graph the functions $f(x) = |x|$ and $h(x) = |x - 10|$ on the same set of axes.

CHECKPOINT ✓

4. Compare the graph of $h(x) = |x - 10|$ with the graph of its parent function.
   a. In which direction was the parent function shifted?
   b. By how much?
5. Compare the graphs of $g$ and $h$.

## Vertical Translation

A transformation that shifts the graph of a function either up (in a positive direction) or down (in a negative direction) is called a **vertical translation.**

**EXAMPLE 1** Compare the graph of $h(x) = x^2 - 3$ with the graph of the parent function $f(x) = x^2$. **Describe how to obtain the graph of $h(x)$ from the graph of $f(x)$.**

**SOLUTION**

Graph the parent function $f(x) = x^2$.
Graph the function $h(x) = x^2 - 3$.

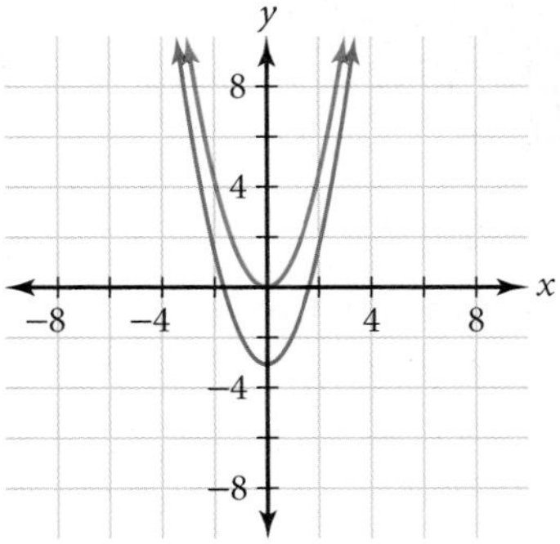


Notice that the graph of $h(x)$ is the same as the graph of $f(x)$ translated 3 units down.

To determine the answer algebraically, notice that $h(x) = x^2 - 3$ contains the parent function $f(x) = x^2$. Therefore, $h$ can be rewritten as $h(x) = f(x) - 3$. This means that each function value of $h$ is 3 units less than the corresponding function value of $f$. In other words, the graph of $h(x)$ is identical to the graph of the parent function translated vertically 3 units down.

**TRY THIS** Describe how to obtain the graph of $f(x) = |x| + 4$ from its parent function.

### Vertical Translation

The graph of $y = f(x) + k$ is identical to the graph of $y = f(x)$ translated vertically $|k|$ units.

- When $k$ is positive, the translation is in a positive direction, or up.
- When $k$ is negative, the translation is in a negative direction, or down.

The following translations show how this formula applies to the parent function $y = x^2$. The graph of a function is shifted vertically when a constant is added to or subtracted from the function.

The parent function of each function graphed at right is $y = x^2$. In each case, the parent function has been translated vertically.

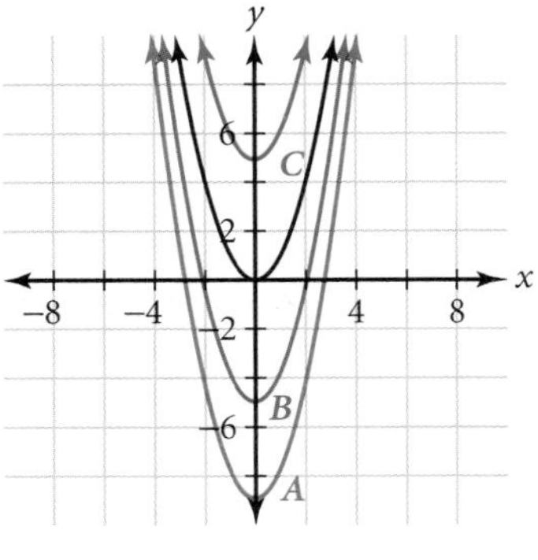


The graph of $A$ is identical to the graph of $y = x^2$ translated 9 units down ($k = -9$): $y = x^2 - 9$.

The graph of $B$ is identical to the graph of $y = x^2$ translated 5 units down ($k = -5$): $y = x^2 - 5$.

The graph of $C$ is identical to the graph of $y = x^2$ translated 5 units up ($k = 5$): $y = x^2 + 5$.

## Horizontal Translation

A transformation that shifts the graph of a function either to the right (in a positive direction) or to the left (in a negative direction) is referred to as a **horizontal translation**.

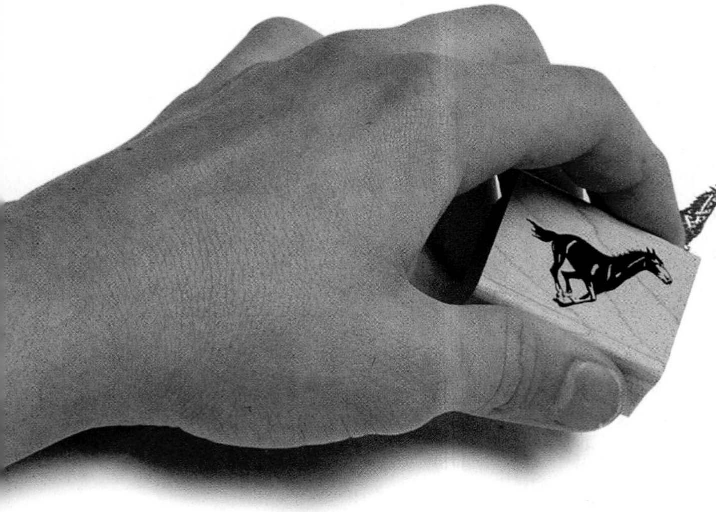

Horizontal translations are useful in creating simple patterns and pictures like the one shown above, but they are also used in much more complicated situations. For example, some animations used on the Internet show an object moving across the page. This is a form of horizontal translation.

**EXAMPLE 2** Compare the graph of $g(x) = (x-3)^2$ with the graph of its parent function, $f(x) = x^2$. **Describe how to obtain the graph of $g(x)$ from its parent function.**

**SOLUTION**

Graph the parent function $f(x) = x^2$. The function $g(x) = (x-3)^2$ can be interpreted as "Subtract 3 from *x*, and then square this quantity." The table shows some *y*-values for *f* and *g*. Sketch *g* on the same set of axes as its parent function.

| $x$ | $f(x) = x^2$ | $g(x) = (x-3)^2$ |
|---|---|---|
| −2 | 4 | 25 |
| −1 | 1 | 16 |
| 0 | 0 | 9 |
| 1 | 1 | 4 |
| 2 | 4 | 1 |
| 3 | 9 | 0 |
| 4 | 16 | 1 |

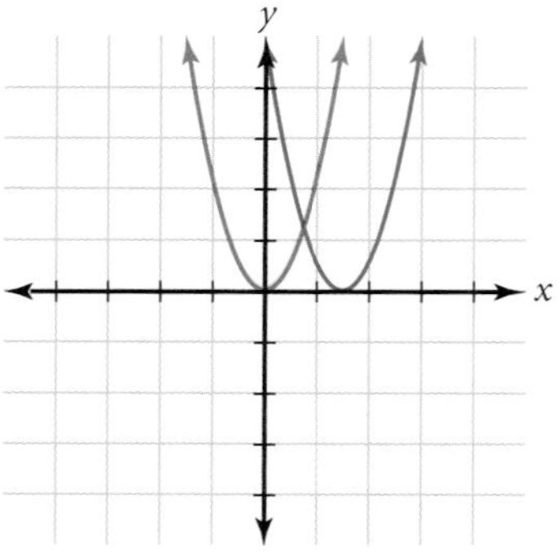


Notice that a translation does not change the shape of the graph. If you subtract 3 from *x* *before* you square the quantity, you shift the graph horizontally 3 units to the right.

To get the same output values, the input values for $g(x) = f(x-3)$ have to be 3 units greater than the input values for $f(x)$.

**TRY THIS** Compare the graph of $h(x) = |x - 6|$ with the graph of its parent function, $f(x) = |x|$. Describe how to obtain the graph of $h(x)$ from its parent function.

**Horizontal Translation**

The graph of $y = f(x - h)$ is translated horizontally $|h|$ units from the graph of $y = f(x)$.

- When $h$ is positive, the translation is in a positive direction, or to the right.
- When $h$ is negative, the translation is in a negative direction, or to the left.

The parent function of each function graphed below is $y = |x|$. In each case, the parent function has been translated horizontally.

To determine $h$, you can look at a distinct feature of the original function, such as a vertex or line of symmetry, and note how many units and in what direction this feature is translated to obtain the new graph.

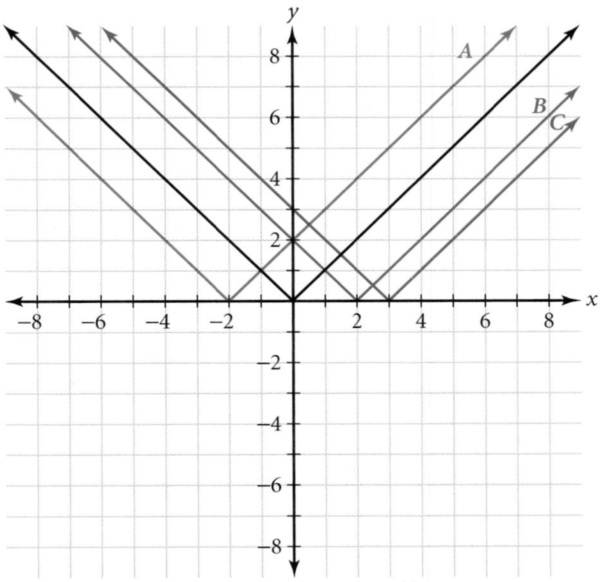


The graph of *A* is identical to the graph of $y = |x|$ translated 2 units to the left $(h = -2)$: $y = |x - (-2)|$ or $y = |x + 2|$.

The graph of *B* is identical to the graph of $y = |x|$ translated 2 units to the right $(h = 2)$: $y = |x - (+2)|$ or $y = |x - 2|$.

The graph of *C* is identical to the graph of $y = |x|$ translated 3 units to the right $(h = 3)$: $y = |x - (+3)|$ or $y = |x - 3|$.

**CRITICAL THINKING** If a reciprocal function is translated, what happens to the asymptotes? (Hint: Graph $y = \frac{1}{x}$, $y = \frac{1}{x} - 3$, and $y = \frac{1}{x - 3}$ on the same coordinate plane and observe the asymptotes.)

**EXAMPLE 3** **Which of the functions below represent translations of the parent graph?**

**a.** $y = \frac{1}{x} + 3$ **b.** $y = \frac{3}{4} + 10^x$ **c.** $y = \frac{|x|}{9}$ **d.** $y = (x - 10)^2$

**SOLUTION**

**a.** Because 3 is added to $\frac{1}{x}$, the graph of the parent function $y = \frac{1}{x}$ is translated 3 units up.

**b.** Because $\frac{3}{4}$ is added to $10^x$, the graph of the parent function $y = 10^x$ is translated $\frac{3}{4}$ of a unit up.

**c.** Because $|x|$ is *divided* by 9, this is a not a translation.

**d.** Because 10 is subtracted from $x$ *before* the quantity is squared, this is a horizontal translation. The graph of the parent function $y = x^2$ is translated 10 units to the right.

**EXAMPLE 4**

**APPLICATION**
**BOOK CLUB**

A book club for computer professionals charges a membership fee of \$20 and then \$12 for each book purchased. **Write and graph a function to represent cost of membership in the club, *c*, based on the number of books purchased, *b*. Write and graph a second function to represent the cost of membership if the club raises its membership fee to \$30. Describe the relationship between the two graphs.**

**SOLUTION**

The function $c = 20 + 12b$ represents the cost of membership in the club at the current fee. The graph of this function is linear and begins at (0, 20).

The function $c = 30 + 12b$ represents the cost of membership in the club at the new fee. The graph of this function is also linear and begins at (0, 30).

The second function represents a vertical translation of the original function by 10 units upward.

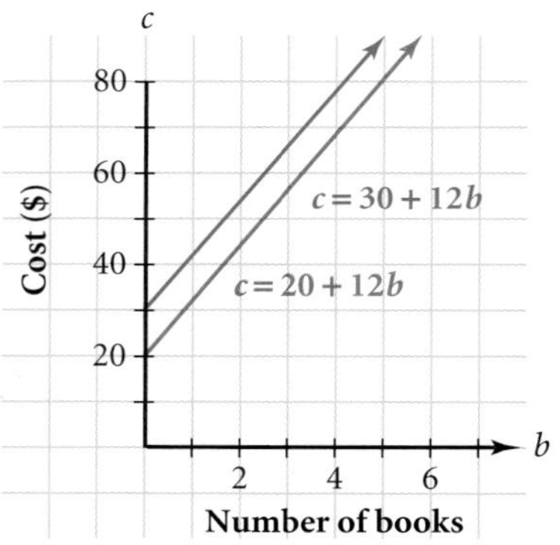


**Summary**

Translations of the graph $y = f(x)$ are represented as follows:

| | |
|---|---|
| Vertical shift $k$ units up | $y = f(x) + k$ |
| Vertical shift $k$ units down | $y = f(x) + (-k) = f(x) - k$ |
| Horizontal shift $h$ units right | $y = f(x - h)$ |
| Horizontal shift $h$ units left | $y = f(x - (-h)) = f(x + h)$ |

# Exercises

## Communicate

**1.** What is meant by a *translation* in this lesson?

**2.** The general form for a transformation of the quadratic parent function, $f(x) = x^2$, is $g(x) = a(x - h)^2 + k$. Which variable affects the distance and direction of a horizontal translation? Explain.

**3.** The function $g(x) = |x - 3| + 5$ is a transformation of the absolute-value parent function. Which value tells you the distance and direction of the vertical translation? Explain.

## Guided Skills Practice

**Describe how to obtain the graph of each function from the graph of its parent function.** ***(EXAMPLE 1)***

**4.** $y = x^2 - 2$ **5.** $y = x^2 + 3$ **6.** $y = 10^x + 1$ **7.** $y = \frac{1}{x} - 3$

**Is the graph of the given function a translation of the graph of the parent function? If so, describe how to obtain the graph of the function from the graph of its parent function.** ***(EXAMPLES 2 AND 3)***

**8.** $y = (x - 2)^2$ **9.** $y = 3x^2$ **10.** $y = 10^{x+1}$ **11.** $y = \frac{1}{x - 3}$

**12.** Refer to the book club problem in Example 4. Let the cost of membership, *c*, be represented on the horizontal axis and the number of books received for that cost, *b*, be represented on the vertical axis. ***(EXAMPLE 4)***

**a.** Using the same conditions as in Example 4, write and graph the two functions that give *b* in terms of *c*.

**b.** Describe the relationship between the two graphs.

## Practice and Apply

**For exercises 13–16:**

**a. Identify the parent function for each function.**

**b. Describe how to obtain the graph of each function from the graph of its parent function.**

**c. Graph and label the parent function and the transformed function.**

**13.** $y = (x - 6)^2 - 1$ **14.** $y = x^2 - 6$ **15.** $y = |x| - 1$ **16.** $y = |x - 1|$

**internet connect**

**Homework Help Online**

Go To: **go.hrw.com**

Keyword:

**MA1 Homework Help**

for Exercises 17-20

**The point (3, 9) is on the graph of the parent function $f(x) = x^2$. Use this fact to describe how $f(x)$ is translated to obtain each function below.**

**17.** $g(x) = x^2 + 5$ contains the point $(3, 14)$.

**18.** $g(x) = x^2 - 5$ contains the point $(3, 4)$.

**19.** $g(x) = (x + 5)^2$ contains the point $(-2, 9)$.

**20.** $g(x) = (x - 5)^2$ contains the point $(8, 9)$.

**The point (5, 8) is on the graph of a function. Give the coordinates of the corresponding point on the new graph when the following translations are applied:**

**21.** vertical translation of 6

**22.** vertical translation of $-2$

**23.** vertical translation of $-10$

**24.** vertical translation of 3

**25.** horizontal translation of 3

**26.** horizontal translation of $-1$

**27.** horizontal translation of $-12$

**28.** horizontal translation of 10

**Tell whether the graph of each function is a translation of its parent function.**

**29.** $y = \frac{2}{3}|x|$

**30.** $y = \frac{2}{3} + |x|$

APPLICATION

**31. ARCHITECTURE** A stadium in Seattle, Washington, was made with a convertible roof. The lower trusses of the roof are 22 feet below the upper trusses and are the same parabolic shape. Suppose that the function that represents the height in feet of the upper trusses is $y = -\left(\frac{x}{40}\right)^2 + 230$, where $x$ represents the horizontal distance in feet from the axis of symmetry of the parabola. Write the equation that represents the height in feet of the lower trusses, and describe the transformation that relates the two functions.

## Look Back

**32.** Solve $7x - (24 + 3x) = 0$ for $x$. *(LESSON 3.5)*

CONNECTION

**33. STATISTICS** On a certain street, the rents being asked for available 2-bedroom apartments are: \$475, \$490, \$530, \$545, \$550, and \$1025. *(LESSON 4.4)*

**a.** Find the mean, median and mode of this data.

**b.** Which is higher, the mean or the median? Why?

## Look Beyond

**34.** Graph $y = x^3$, $y = (-x)^3$, and $y = -x^3$ on the same coordinate plane.

**a.** Describe the two transformations of the parent function.

**b.** Compare the results of the transformations.

**c.** Write two other functions whose reflections across the $x$-axis and $y$-axis give graphs that coincide. What do these functions have in common?

14.3

# Stretches and Compressions

**Objective**

- Describe how changes to the rule of a function stretch or compress its graph.

**Why** *The graph of a function can be stretched or compressed without changing its basic shape. Once you understand how changes to the rule of a function correspond to these stretches and compressions, it will be easy to graph many functions.*

The height of a waterfall can be approximated if you know the time it takes a drop of water to fall from the top to the pool below. This relationship can be modeled by the function $h = 16t^2$, where $h$ is the height in feet and $t$ is the time in seconds. For example, if it takes a drop of water 3 seconds to fall to the pool below, the height of the waterfall is about $16(3^2)$, or 144, feet. You may have recognized that the parent function for $h = 16t^2$ is the quadratic parent function, $h = t^2$. You can obtain the function $h = 16t^2$ from its parent function by using a transformation that you will learn about in this lesson.

TECHNOLOGY
GRAPHICS CALCULATOR

*Activity*

## Vertically Stretching a Function

**You will need:** graph paper or a graphics calculator

1. Make a table of values for the function $h(t) = t^2$ using $t$-values 0, 1, 2, 3, 4, and 5. Graph the function using all real numbers greater than or equal to 0 for the domain.
2. Repeat Step 1 for the function $j(t) = 16t^2$. Extend your previous table and use the same coordinate grid.
3. Compare the range values of the two functions. What do you notice about the output value for each function for the same input value?
4. Complete Steps 1–3 for the functions $f(x) = |x|$ and $f(x) = 10|x|$ using integer values from $-3$ to 7 inclusive for $x$.

CHECKPOINT ✓

5. What is the effect on the graph of a function when the function is multiplyed by a number greater than 1? What happens to the range values of the function? Are the domain values affected?

## Vertical Stretch and Compression

**EXAMPLE 1** **Graph each pair of functions, and identify the transformation from *f* to *g*.**

TECHNOLOGY
GRAPHICS CALCULATOR

**a.** $f(x) = x^2 - 3$ and $g(x) = 2(x^2 - 3)$

**b.** $f(x) = x^2 - 3$ and $g(x) = \frac{1}{2}(x^2 - 3)$

**SOLUTION**

**a.**

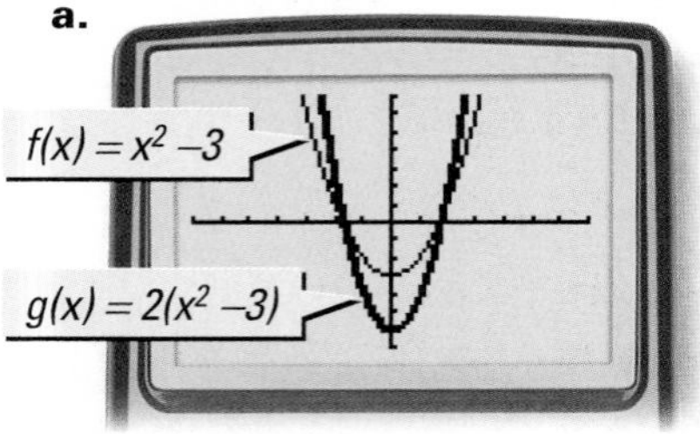


Notice that $g(x) = 2 \cdot f(x)$. The function $g(x) = 2(x^2 - 3)$ is a *vertical stretch* of *f* by a factor of 2. The point (3, 6), for example, is transformed to the point (3, 12).

**b.**

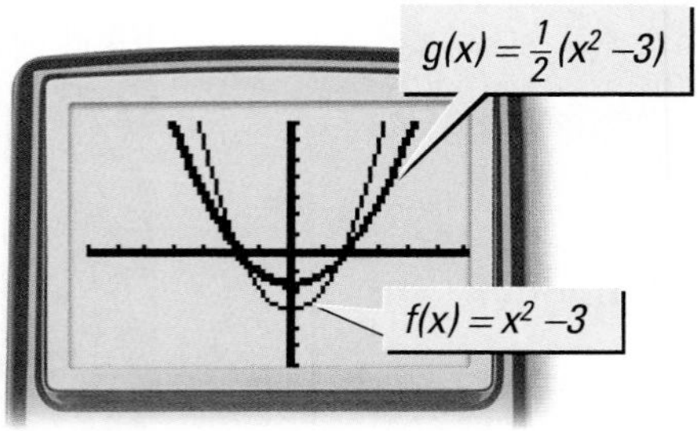


Notice that $g(x) = \frac{1}{2} \cdot f(x)$. The function $g(x) = \frac{1}{2}(x^2 - 3)$ is a *vertical compression* of *f* by a factor of $\frac{1}{2}$. The point (3, 6), for example, is transformed to the point (3, 3).

For a given value in the domain of a function, a vertical stretch or compression multiplies the corresponding value in the range of the function by the scale factor.

**TRY THIS** Graph each pair of functions, and identify the transformation from *f* to *g*.

**a.** $f(x) = \sqrt{x}$ and $g(x) = 3\sqrt{x}$ **b.** $f(x) = \sqrt{x}$ and $g(x) = \frac{1}{3}\sqrt{x}$

A vertical stretch or compression moves the graph away from or toward the *x*-axis, respectively. This can be generalized as follows:

### Vertical Stretch and Vertical Compression

If $y = f(x)$, then $y = af(x)$ gives a **vertical stretch** or **vertical compression** of the graph of *f*.

- If $a > 1$, the graph is stretched vertically by a factor of *a*.
- If $0 < a < 1$, the graph is compressed vertically by a factor of *a*.

Vertical Stretch

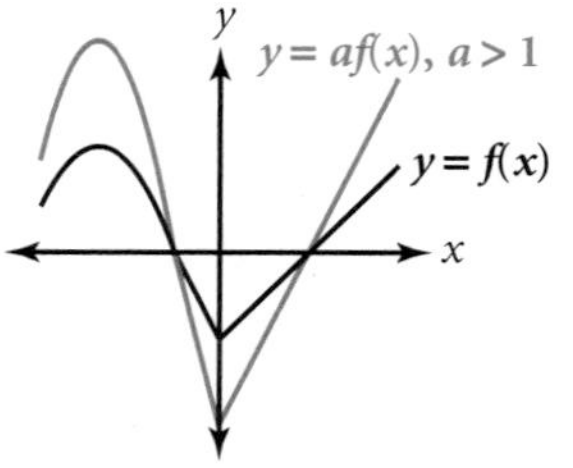


Vertical Compression

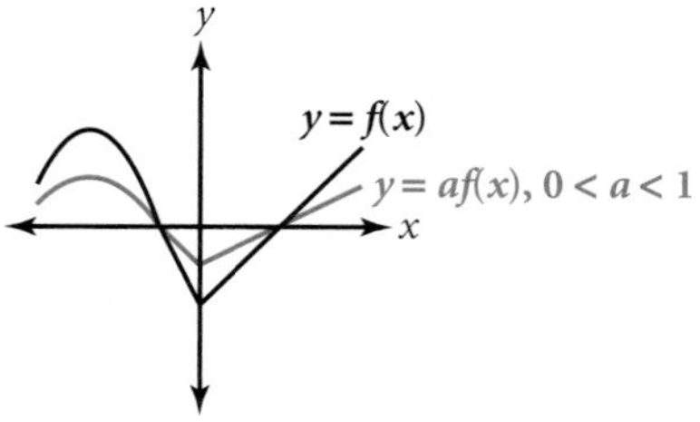


**CRITICAL THINKING** Let $0 < r < 1$ and $s > 1$. Compare the appearance of the graphs of $f(x) = |x|$, $h(x) = r|x|$, and $g(x) = s|x|$.

## EXAMPLE 2

**a.** **Use $y = x^2$ to sketch the graph of $y = \frac{x^2}{3}$.**

**b.** **If you know the value of $y$ for a given value of $x$ in the parent function $y = x^2$, describe how you would determine the corresponding $y$-value for the function $y = \frac{x^2}{3}$.**

### SOLUTION

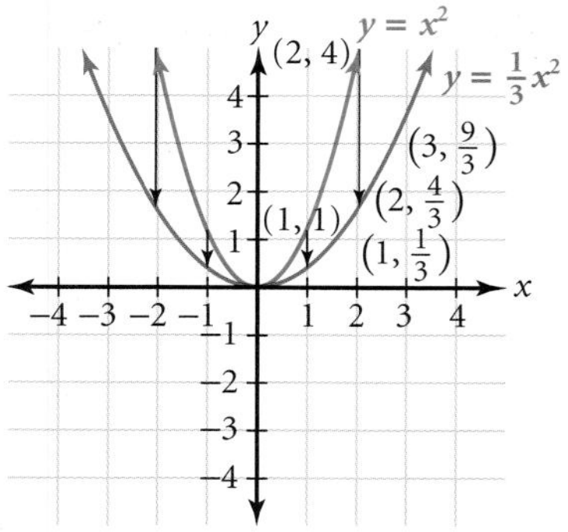


**a.** The function can be written as $y = \frac{1}{3}x^2$. Its parent function is the function, $y = x^2$. Each point of the parent graph is compressed by a factor of $\frac{1}{3}$.

**b.** To find the $y$-value of $y = \frac{x^2}{3}$ from a known value of $y$ in the parent function, divide the known value by 3. For example, if the $y$-value in the parent function is 9, then the corresponding $y$-value for $y = \frac{x^2}{3}$ will be $\frac{9}{3}$, or 3.

**TRY THIS** Use $y = |x|$ to sketch the graph of $y = \frac{1}{4}|x|$.

## EXAMPLE 3

**Which functions are stretches or compressions of their parent graph?**

**a.** $y = \frac{x^2}{3}$ **b.** $y = \frac{3}{4} + x^2$ **c.** $y = 5|x|$

**TECHNOLOGY**
**GRAPHICS CALCULATOR**

### SOLUTION

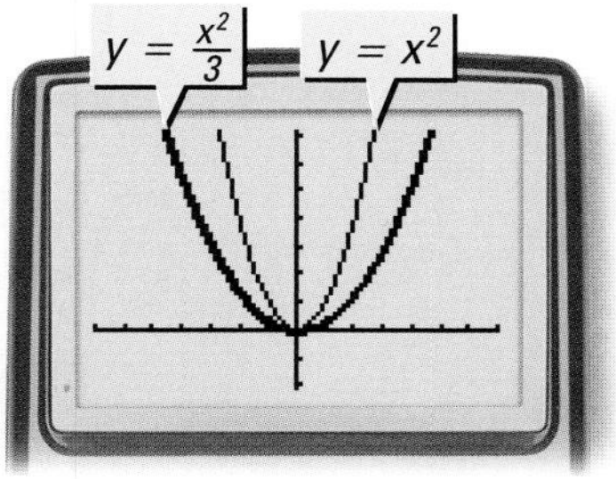


**a.** Because $x^2$ is multiplied by $\frac{1}{3}$, this is a vertical compression by a factor of $\frac{1}{3}$. The $y$-values of this function will be $\frac{1}{3}$ of the $y$-values of the parent function.

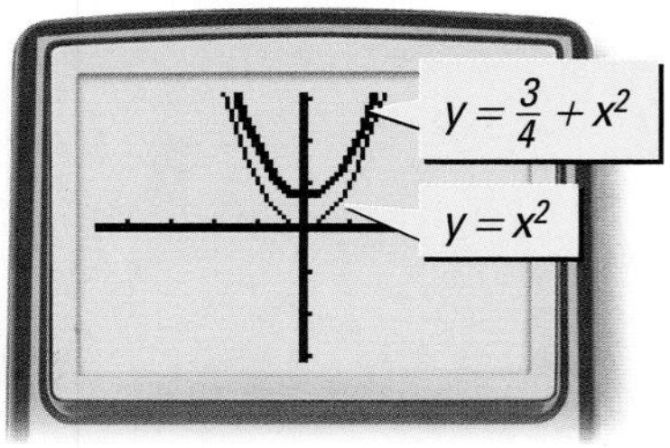


**b.** Because $\frac{3}{4}$ is *added* to $x^2$, this is not a stretch or compression. The graph of the function is identical to the graph of the parent function *translated* $\frac{3}{4}$ of a unit up.

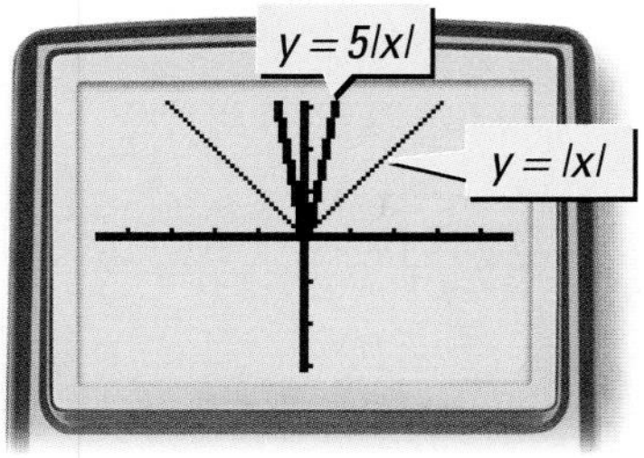


**c.** Because $|x|$ is multiplied by 5, each $y$-value of the parent function is stretched vertically by a factor of 5.

## Horizontal Stretch and Compression

**EXAMPLE 4** **Graph each pair of functions, and identify the transformation from $f$ to $g$.**

**a.** $f(x) = |x|$ and $g(x) = |\frac{1}{3}x|$ **b.** $f(x) = |x|$ and $g(x) = |2x|$

**SOLUTION**

**a.**

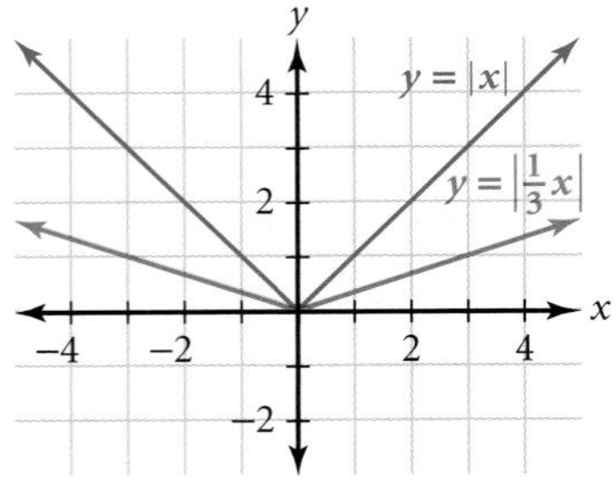


Notice that $g(x) = f\left(\frac{1}{3}x\right)$.

The function $g(x) = \left|\frac{1}{3}x\right|$ is a *horizontal stretch* of $f$ by a factor of 3 (the reciprocal of $\frac{1}{3}$). The point (1, 1), for example, is transformed to the point (3, 1).

**b.**

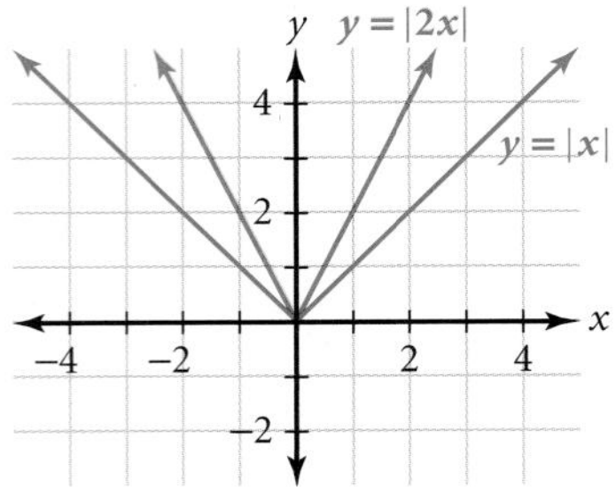


Notice that $g(x) = f(2x)$. The function $g(x) = |2x|$ is a *horizontal compression* of $f$ by a factor of $\frac{1}{2}$. The point (1, 1), for example, is transformed to the point $(\frac{1}{2}, 1)$.

**TRY THIS** Graph each pair of functions, and identify the transformation from $f$ to $g$.

**a.** $f(x) = 10^x$ and $g(x) = 10^{3x}$ **b.** $f(x) = 10^x$ and $g(x) = 10^{\frac{x}{3}}$

A horizontal stretch or compression moves the graph away from or toward the $y$-axis, respectively. This can be generalized as follows:

### Horizontal Stretch and Horizontal Compression

If $y = f(x)$, then $y = f(bx)$ gives a **horizontal stretch** or **horizontal compression** of the graph of $f$.

- If $b > 1$, the graph is compressed horizontally by a factor of $\frac{1}{b}$.
- If $0 < b < 1$, the graph is stretched horizontally by a factor of $\frac{1}{b}$.

**Horizontal Stretch**

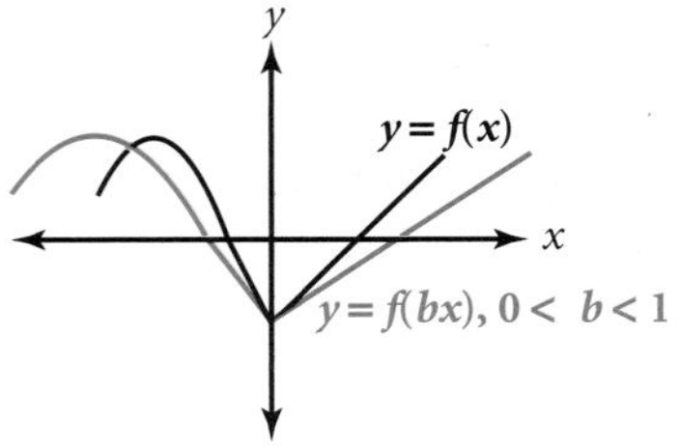


**Horizontal Compression**

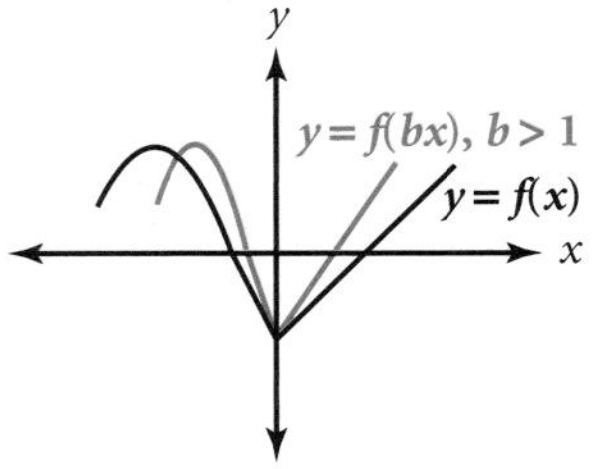

You may have noticed that with many graphs, vertical stretches and horizontal compressions look exactly the same. This is because many functions can be written in more than one way. For example, $y = 4x^2$ (indicating a vertical stretch of $y = x^2$) is equivalent to $y = (2x)^2$.

**GRAPHING A STRETCH OR COMPRESSION**

To graph a stretch or compression, you should follow these steps.

1. Identify and graph the parent function.
2. Determine if the transformation is vertical or horizontal.
3. Determine if the transformation is a stretch or compression.
4. Identify the factor by which the function is stretched or compressed.
5. Transform several points on the original function.
6. Plot the new points and use them as a guide to graph the transformed function.

# Exercises

## Communicate

1. Explain how the graph of $y = ax^2$ changes as $a$ increases from 1 to 4 and as $a$ decreases from 1 to 0.
2. Explain how the graph of $y = 10^{bx}$ changes as $b$ increases from 1 to 4 and as $b$ decreases from 1 to 0.
3. Given the graph of a function $f(x)$, explain how you could use transformations to graph the function $g(x) = \frac{1}{2}f(x)$.
4. Given the graph of a function $f(x)$, explain how you could use transformations to graph the function $h(x) = f(2x)$.

## Guided Skills Practice

**Use the appropriate parent function to sketch the graph of each function.** ***(EXAMPLES 1, 2, AND 4)***

**5.** $y = 3|x|$

**6.** $y = \frac{1}{4}x^2$

**7.** $y = \frac{\sqrt{x}}{3}$

**8.** $y = \left|\frac{1}{2}x\right|$

**9.** $y = (3x)^2$

**10.** $y = \frac{1}{2x}$

**Determine which functions are stretches or compressions of their parent graph.** ***(EXAMPLE 3)***

**11.** $y = \frac{1}{2}x^2$

**12.** $y = |x| + 3$

**13.** $y = (4x)^2$

**14.** $f(x) = \sqrt{x+7}$

**15.** $f(x) = \frac{1}{52 + x}$

**16.** $f(x) = 10^{2x}$

## Practice and Apply

**Graph each of the following:**

**17.** $y = 2|x|$ **18.** $y = 0.5x^2$ **19.** $y = \sqrt{\frac{1}{4}x}$

**Which functions are stretches or compressions of their parent function?**

internet connect
Homework Help Online
Go To: go.hrw.com
Keyword:
MA1 Homework Help
for Exercises 20-22

**20.** $y = 3x^2$ **21.** $y = |x| + 1$ **22.** $y = \frac{5}{x}$

**23. Consider the function $y = 5x^2$.**

**a.** Identify the parent function.
**b.** Sketch the graphs of both functions on the same coordinate plane.
**c.** Each point $(x, y)$ on the graph of the parent function is transformed to what point on the new graph?

**24. Consider the function $y = \frac{2}{x}$.**

**a.** Identify the parent function.
**b.** Sketch the graphs of both functions on the same coordinate plane.
**c.** Each point $(x, y)$ on the graph of the parent function is transformed to what point on the new graph?

**Tell whether each statement is true or false.**

**25.** The $y$-values of $y = 5|x|$ are 5 times the corresponding $y$-values of $y = |x|$.

**26.** The $x$-values of $y = (5x)^2$ must be $\frac{1}{5}$ as large as the $x$-values of $y = x^2$ in order to get the same $y$-values.

**Write two equations for each graph.**

**27.**

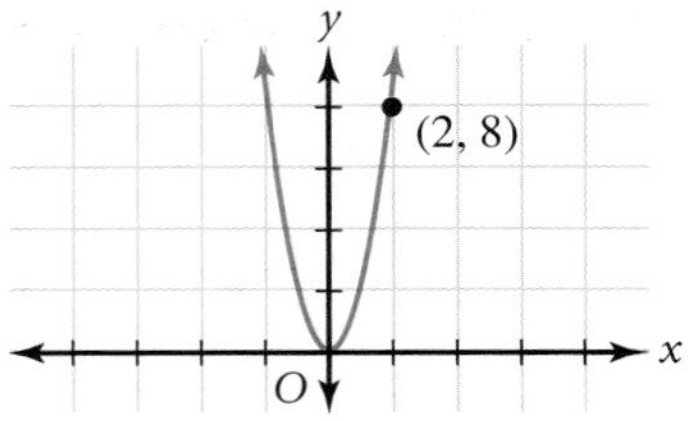


**28.**

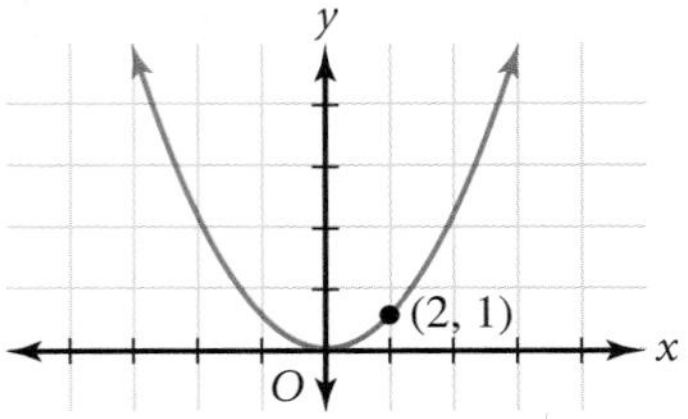


**For each function below, identify the parent function, draw both graphs, and tell how the graph of each function is obtained from its parent function.**

**29.** $y = |3x|$ **30.** $y = 3x^2$ **31.** $y = \left(\frac{x}{3}\right)^2$

CONNECTION

**32. STATISTICS** Draw a bar graph for the frequency distribution of Ms. Tienert's students' test scores, shown in the table below.

| Score | 50 | 55 | 60 | 65 | 70 | 75 | 80 | 85 | 90 | 95 | 100 |
|---|---|---|---|---|---|---|---|---|---|---|---|
| Frequency | 1 | 0 | 2 | 5 | 10 | 6 | 2 | 0 | 0 | 0 | 0 |

**33.** Because the students didn't have enough time and the scores were low, Ms. Tienert decided to add 20 points to every student's score. Draw a new bar graph that reflects this change, and explain how it compares with the original bar graph.

**34.** Ms. Tienert also considered multiplying each score by $\frac{5}{4}$. Draw a new bar graph that reflects this change, and explain how it compares with the original bar graph.

APPLICATIONS

**35. ART** Pictures can be "stretched" or enlarged by using a grid. One method to enlarge a picture is to place a grid over it and copy what is in each part of the grid onto a larger grid.

**a.** On a coordinate plane, draw a triangle with vertices at (1, 1), (3, 1), and (2, 3).

**b.** Create a coordinate plane with grid squares 3 times larger. Use this grid to make an enlargement of the triangle.

**c.** Use the language of transformations to describe this process for enlarging a picture.

*Artist Chuck Close uses a grid technique to enlarge portraits adding his unique artistic interpretation to each grid space.*

Photograph by Bill Jacobson, Courtesy of Pace Wildenstein

CHALLENGE

**36. INCOME** The workers in an electronics plant received a raise of 5%. Find the scale factor, or the factor by which their salaries were "stretched." (Hint: The scale factor is not 0.05.)

## Look Back

**37.** Solve the equation $5x - (7x + 4) = 3(x + 5) + 4(5 - 2x)$. ***(LESSON 3.5)***

**Solve each system of linear equations.** ***(LESSONS 7.2 AND 7.3)***

**38.** $\begin{cases} 3x + 7y = -6 \\ x - 2y = 11 \end{cases}$

**39.** $\begin{cases} 5y - 3x = -31 \\ 4y = 16 \end{cases}$

**40.** $\begin{cases} 1.25x + 2y = 5 \\ 3.75x + 6y = 15 \end{cases}$

**41.** Dean and Deanna are siblings. Deanna has twice as many brothers as she has sisters. Dean has the same number of brothers and sisters. How many girls and how many boys are there in the family? ***(LESSON 7.6)***

**42.** Consider the relation consisting of the ordered pairs (3, 7), (5, –1), (1, 0), (2, 5), (16, –6), (0, 15), (1, –1), and (–2, 5). Is this relation a function? Why or why not? ***(LESSON 14.1)***

## Look Beyond

**Match each equation with its graph.**

**43.** $y = -\sqrt{x}$

**44.** $y = -\sqrt{\frac{x}{4}}$

**45.** $y = -\frac{\sqrt{x}}{4}$

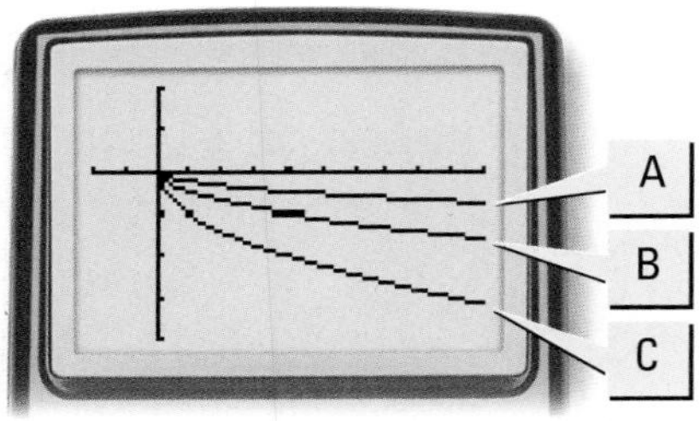

14.4

# Reflections

**Objective**

- Describe how a change to the rule of a function corresponds to a reflection of its graph.

**Why** ***You can graph many functions by understanding how an alteration to the rule of a function corresponds to the reflection of its graph in the x- or y-axis.***

A mirror reflects a figure, forming the illusion of its image behind the mirror. The graph of a function can be reflected across the $x$- and $y$-axes in a similar manner.

*Activity*

## Reflection Across the $x$-Axis

**You will need:** a reflection tool (optional)

1. Make a table of values for the function $f(x) = x^2$, using $x$-values $-2, -1, 0, 1,$ and $2$, and graph the function.
2. Reflect $f$ across the $x$-axis. That is, create a mirror image of $f$ as if a mirror were placed on the $x$-axis. Let $g$ represent the function for this reflected image.
3. Make a table of values for the reflected function, $g$. Use the same $x$-values that you used for $f$.
4. Compare the tables for the two graphs. The point $(2, 4)$ is on the parent graph. What is the function value on the reflected graph when $x$ is 2? If $(-3, 9)$ is a point on the original graph, what is the corresponding point on the reflected graph?

CHECKPOINT ✔

5. For the following values of $x$, give the corresponding function values on the reflected graph.
   **a.** 3 **b.** 0 **c.** $-3$ **d.** $a$
6. If $f(x) = x^2$ is reflected across the $x$-axis, what is the equation of the new function?

Example 1 illustrates a reflection across the $x$-axis and a reflection across the $y$-axis.

**EXAMPLE 1** **Graph each pair of functions, and identify the transformation from $f$ to $g$.**

**a.** $f(x) = x^2$ and $g(x) = -(x^2)$

**b.** $f(x) = 2x + 3$ and $g(x) = 2(-x) + 3$

**SOLUTION**

**a.**

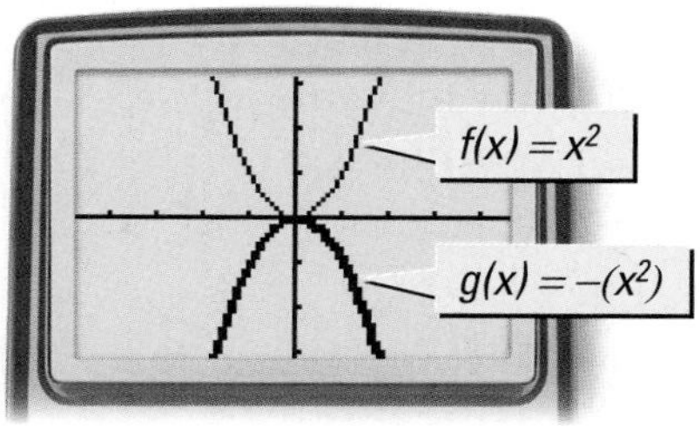


Notice that $g(x) = -f(x)$. The function $g(x) = -(x^2)$ is the reflection of $f$ across the $x$-axis. The point $(2, 4)$, for example, is transformed to the point $(2, -4)$.

**b.**

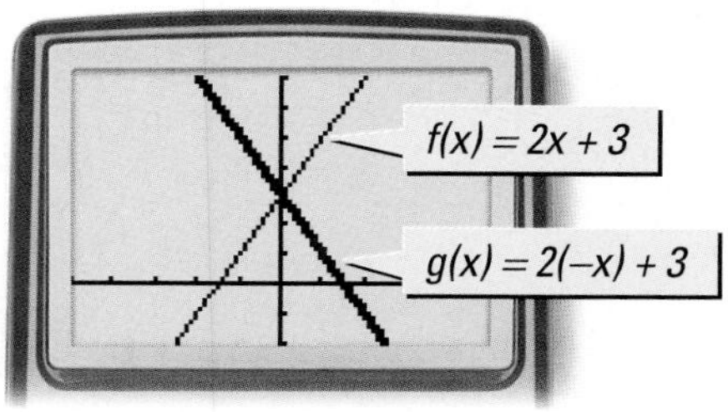


Notice that $g(x) = f(-x)$. The function $g(x) = 2(-x) + 3$ or $g(x) = -2x + 3$, is the reflection of $f$ across the $y$-axis. The point $(1, 5)$, for example, is transformed to the point $(-1, 5)$.

In general, a reflection across the $x$-axis transforms each point $(a, b)$ on the original graph to the point $(a, -b)$, and a reflection across the $y$-axis transforms each point $(a, b)$ to the point $(-a, b)$.

**TRY THIS** Graph each pair of functions, and identify the transformation from $f$ to $g$.

**a.** $f(x) = |x|$ and $g(x) = -|x|$

**b.** $f(x) = 2x - 1$ and $g(x) = 2(-x) - 1$

Reflections across the $x$- and $y$-axes can be generalized as follows:

### Reflection Across the *x*-Axis and *y*-Axis

If $y = f(x)$, then $y = -f(x)$ gives the reflection of the graph of $f$ across the $x$-axis.

If $y = f(x)$, then $y = f(-x)$ gives the reflection of the graph of $f$ across the $y$-axis.

**Reflection Across the $x$-Axis**

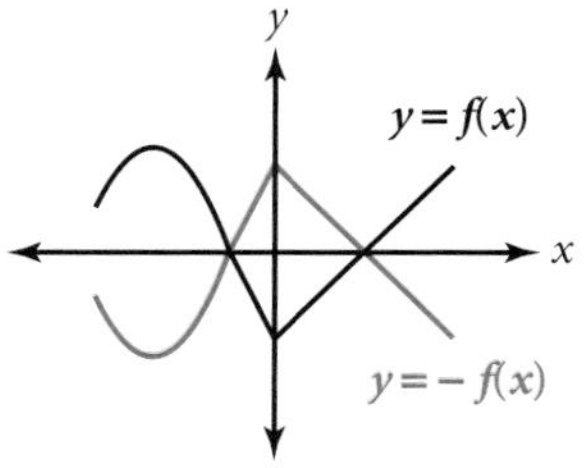


**Reflection Across the $y$-Axis**

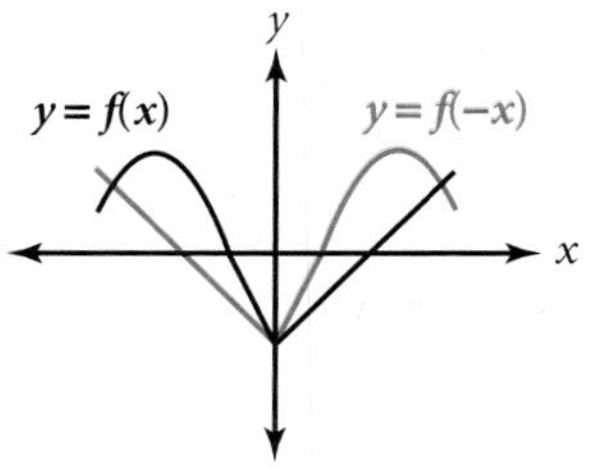


**CRITICAL THINKING** Let $f(x) = |x|$. What does $f(-x) = |-x| = |x|$ tell you about the graph of $f(-x)$? Justify your response.

**EXAMPLE 2** **Graph each function with its parent function on the same coordinate plane. Describe the transformation of the parent function.**

**a.** $h(x) = -|x|$ **b.** $y = -\frac{1}{x}$ **c.** $g(x) = \sqrt{-x}$ **d.** $y = (-x)^2$

**SOLUTION**

**a.** The parent function is $f(x) = |x|$. Since $h(x) = -f(x)$, $h$ is the reflection of $f$ across the $x$-axis. To obtain the graph of $h$, leave the $x$-coordinates of $f$ the same and replace the $y$-coordinates with their opposites.

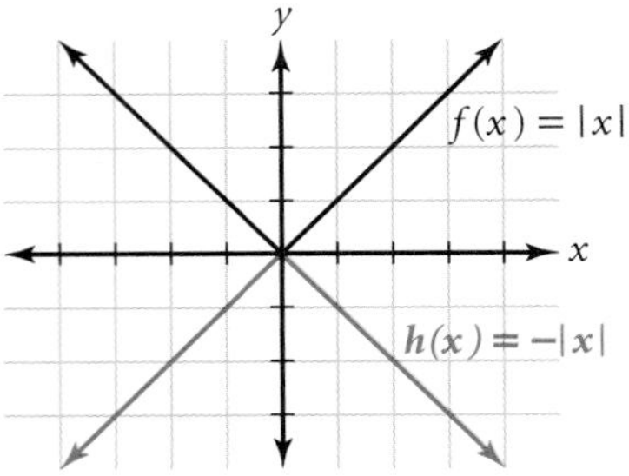


**b.** The parent function is $y = \frac{1}{x}$. For every point $\left(x, \frac{1}{x}\right)$ on the graph of the parent function, the point $\left(x, -\frac{1}{x}\right)$ is on the graph of $y = -\frac{1}{x}$. Thus, $y = -\frac{1}{x}$ gives the reflection of the graph of $y = \frac{1}{x}$ across the $x$-axis.

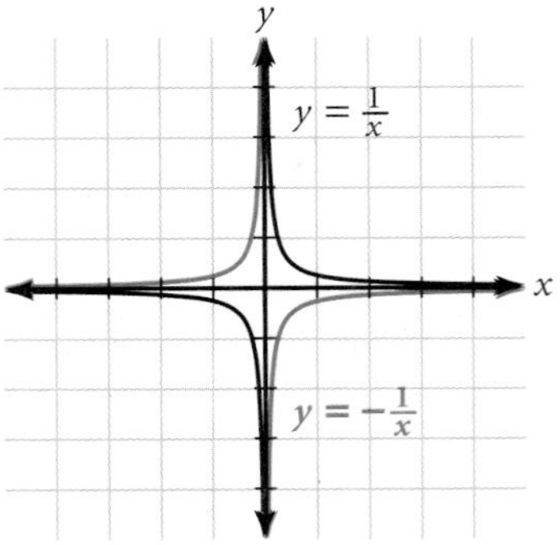


**c.** The parent function is $f(x) = \sqrt{x}$. Since $g(x) = f(-x)$, $g$ is the reflection of $f$ across the $y$-axis. To obtain the graph of $g$, replace the $x$-coordinates of $f$ with their opposites and leave the $y$-coordinates the same.

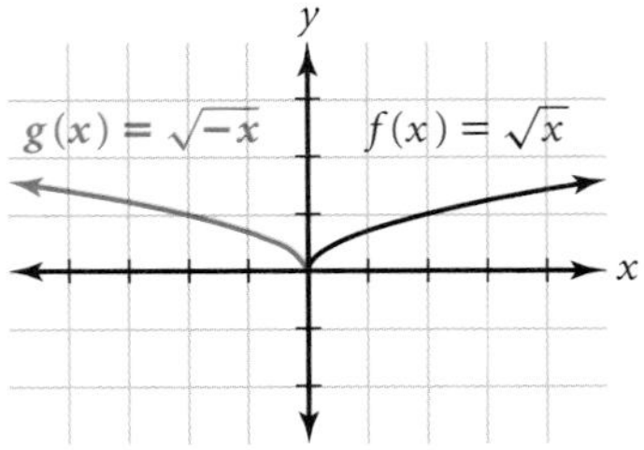


**d.** The parent function is $y = x^2$. For every point $(x, x^2)$ on the graph of the parent function, the point $\left(x, (-x)^2\right)$ is on the graph of $y = (-x)^2$. Thus, $y = (-x)^2$ gives the reflection of the graph of $y = x^2$ across the $y$-axis.

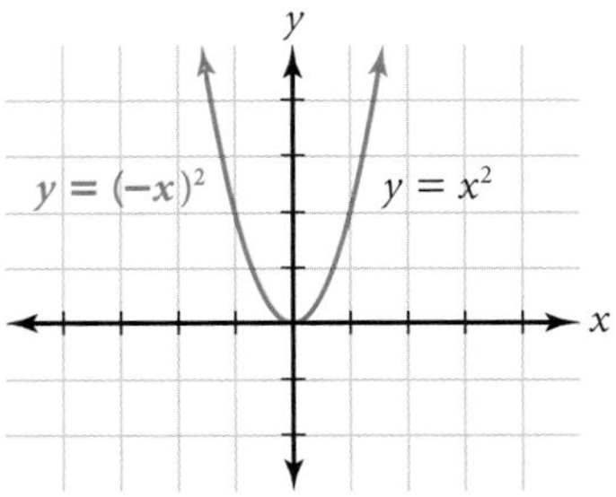


Since the graph of $y = x^2$ is symmetric with respect to the $y$-axis, the function is reflected onto itself. This can also be explained algebraically by the equation $(-x)^2 = x^2$.

**TRY THIS** Graph each function with its parent function on the same coordinate plane. Describe the transformation of the parent function.

**a.** $f(x) = -2^x$ **b.** $g(x) = 2^{-x}$

**CRITICAL THINKING** Does the line $y = -x$ reflect the line $y = x$ across the $x$-axis or across the $y$-axis? Explain.

**Summary**

To graph the reflection of a function, follow these steps.

1. Graph the original function.
2. Determine if the reflection is across the $x$- or $y$-axis.
3. Transform several points on the original graph.
   a. If the reflection is across the $x$-axis, leave the $x$-coordinates the same and replace the $y$-coordinates with their opposites.
   b. If the reflection is across the $y$-axis, replace the $x$-coordinates with their opposites and leave the $y$-coordinates the same.
4. Plot these new points and use them as a guide to graph the reflected function.

# Exercises

## Communicate

1. Explain the effect of the negative sign in the function $y = -ax^2$.
2. If $h(x) = 2\sqrt{x} + 1$ is reflected across the $y$-axis, how is its function rule changed to express this reflection?
3. Describe how to change the function $f(x) = 2x$ to reflect its graph:
   a. across the $x$-axis.
   b. across the $y$-axis.
4. Is the reflection of a mountain in a lake similar to a reflection across the $x$-axis? Explain why or why not.

*A lake creates a natural reflection of a mountain.*

## Guided Skills Practice

**Graph each pair of functions and identify the transformation from *f* to *g*.** ***(EXAMPLE 1)***

**5.** $f(x) = \sqrt{x+1}$
$g(x) = -\sqrt{x+1}$

**6.** $f(x) = 2x^2$
$g(x) = -2x^2$

**7.** $f(x) = \frac{1}{x} + 2$
$g(x) = \frac{1}{-x} + 2$

**8.** $f(x) = |x+1|$
$g(x) = |-x+1|$

**Graph each function with its parent function on the same coordinate plane. Describe the transformation of the parent function.** ***(EXAMPLE 2)***

**9.** $y = -\sqrt{x}$  **10.** $y = -3^x$  **11.** $y = |-x|$  **12.** $y = \frac{1}{-x}$

## Practice and Apply

> **internet connect**
> **Homework Help Online**
> Go To: **go.hrw.com**
> Keyword:
> **MA1 Homework Help** for Exercises 13-18

**Graph each pair of functions and identify the transformation from *f* to *g*.**

**13.** $f(x) = (x+1)^2$, $g(x) = (-x+1)^2$

**14.** $f(x) = \frac{1}{x-2}$, $g(x) = \frac{1}{-x-2}$

**15.** $f(x) = |x| - 3$, $g(x) = -|x| + 3$

**16.** $f(x) = -(x^2+2)$, $g(x) = (x^2+2)$

**17.** Write the functions for the graph of $f(x) = 5x + 2$ reflected across the $x$-axis and across the $y$-axis.

**18. CULTURAL CONNECTION: ASIA** On a piece of graph paper, construct the line segments shown in red. Point $E$ is at (0, 2), and point $D$ is at (0, 6). Segment $EA$ is twice the length of segment $OE$, and segment $DB$ is twice the length of segment $OD$. Segment $FC$ connects the midpoints of the segments $EA$ and $DB$. Reflect all of these segments across the $x$-axis, and then reflect both graphs across the $y$-axis. This basic pattern appears in a carving found in a tomb dating back to the Han dynasty in China (206 B.C.E. to 222 C.E.). Patterns constructed in this manner have *bilateral symmetry.*

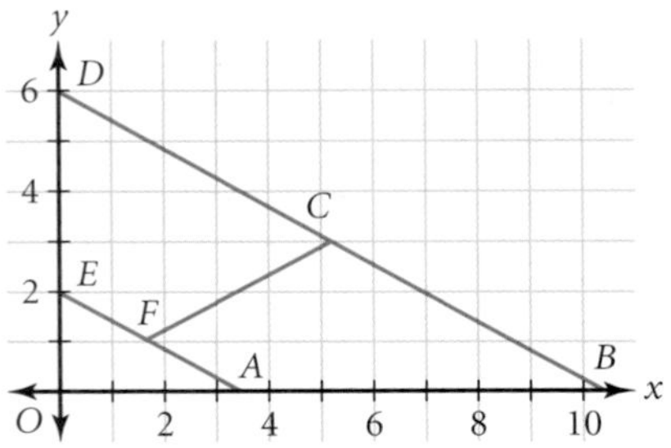


## Look Back

**APPLICATION**

**19. PHYSICS** Seven seconds after a flash of lightning, Mark heard thunder. If the speed of sound through air is approximately 330 meters per second, how far away did the lightning strike? ***(LESSON 2.4)***

**Solve each equation.** ***(LESSON 3.5)***

**20.** $6(3 - 2r) = -(5r + 3)$

**21.** $\frac{x}{2} + 5 = \frac{3x-2}{4}$

**22.** $3x = -9(x - 4)$

**CONNECTIONS**

**23.** **GEOMETRY** The length of a rectangular garden is 3 times its width. If the perimeter of the rectangular garden is 72 meters, what are the dimensions? *(LESSON 3.6)*

**24.** **STATISTICS** A factory worker measures 5 randomly selected glass rods in a 15-minute production run. What is the average length of the glass rods tested in this run based on the following measurements? *(LESSON 4.4)*

| Sample | 1 | 2 | 3 | 4 | 5 |
|---|---|---|---|---|---|
| Measurement (in cm) | 23.5 | 23.9 | 23.6 | 24.0 | 23.7 |

**25.** Graph the solution set $x \geq 3$ or $x < -1$ on a number line. *(LESSON 6.3)*

## Look Beyond

**Match each function with the appropriate graph.**

**26.** $y = \sqrt{x}$ **27.** $y = \sqrt{-x}$ **28.** $y = -\sqrt{x}$ **29.** $y = -\sqrt{-x}$

**a.**

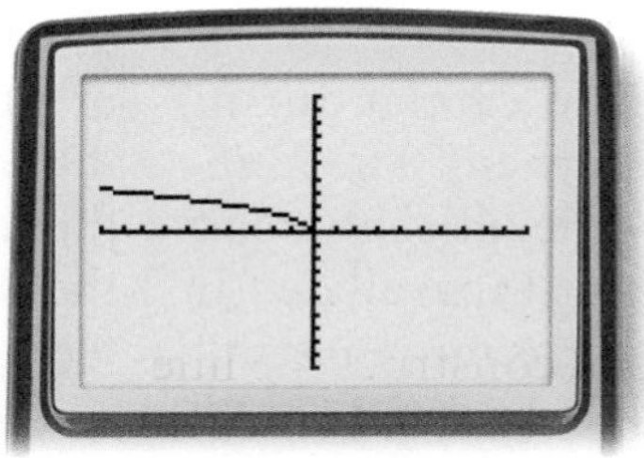

**b.**

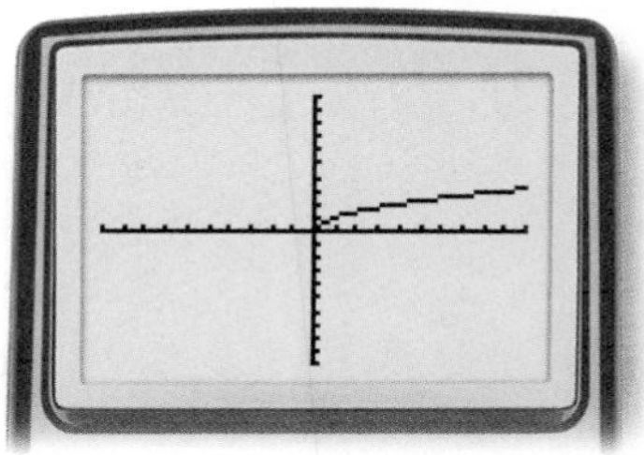

**c.**

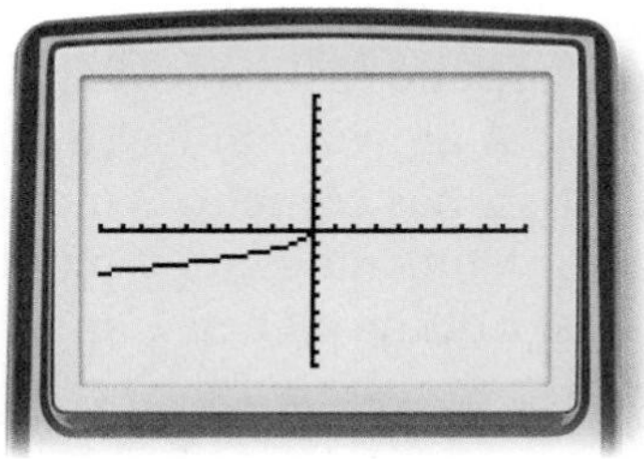

**d.**

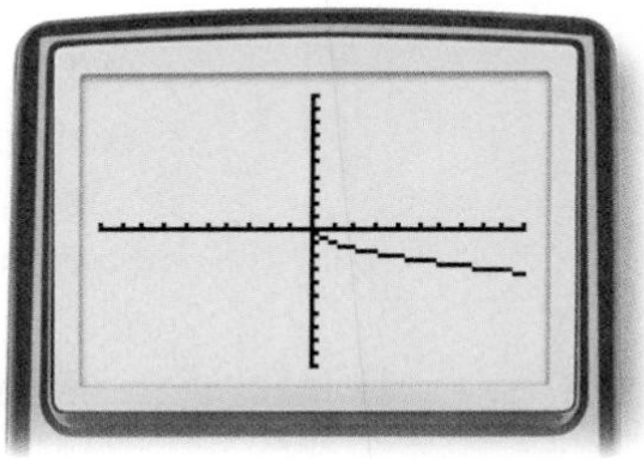

**internet connect**

**Activities Online**

Go To: **go.hrw.com**
Keyword:
**MA1 Friezes**

**PORTFOLIO ACTIVITY**

Include this activity in your portfolio.

Play a game of "What's My Function?" with a partner.

1. On a coordinate grid, give your partner a graph of a function that has been translated, stretched, or reflected and the graph of its parent function. You can use a graphics calculator in "Grid On" format for this activity.
2. Your partner must guess your function. Record a score of 1 for each correct response.
3. Take turns drawing the graph and guessing the function for 10 turns each.

**WORKING ON THE CHAPTER PROJECT**

You should now be able to complete Activity 1 of the Chapter Project on page 728.

# Is There ORDER in CHAOS?

## FINDING ORDER IN *DISORDER*

***The science of chaos reveals nature's secrets.***

Stock prices lurch and careen during weeks of financial mayhem. The number of measles cases inexplicably soars and crashes. A storm unparalleled in more than 40 years slams into England, killing at least 13 people. Such apparently random events have always seemed far beyond the understanding of even the most powerful computers and brilliant researchers. But are they? A growing and eclectic band of scientists has come to suspect that such chaotic happenings are governed by laws just as orbiting planets and falling apples are . . .

The premise of chaos theory is the oldest cliché in science: beneath disorder lurks order. "Chaologists" find that, although it may never be possible to precisely predict the weather or the stock market or even the path of a roulette ball, one can foresee the patterns of their behavior. These patterns are the order within the chaos . . .

Chaos's power to explain diverse phenomena has encouraged researchers to seek it everywhere. Cardiologists studying the human heart find that its normal rhythm is subtly chaotic, rather than regular like a metronome. Climatologists pondering the ice ages find that their timing follows no regular pattern.

*Newsweek, December 21, 1987*

*ORDER* Turn on a faucet just a little. The flow of water looks as smooth as glass tube. This is order.

*CHAOS* Turn the handle a little more. The stream changes. Its outer edges flicker in unpredictable turbulence. This is chaos.

Some things, like the water stream in the photos, are sometimes orderly and sometimes chaotic. You can use equations to model such dual behavior. To see how this works, explore a biological system that is simpler to model than a stream of water—an imaginary population of insects.

The insects hatch in the spring and die after laying eggs in the fall. The population each year depends on the population of the prior year. The more insects there are, the more eggs there are to hatch next spring. Other factors, such as limited food supply, keep the population from becoming infinitely large.

If you know the size of the insect population one year, you can find the size the next year by using the following equation:

$$P_2 = r \cdot P_1(1 - P_1)$$

$P_1$ is the size of the population in year 1, $P_2$ is the size of the population in year 2, and $r$ is a constant.

To find the population in year 3, use the equation $P_3 = r \cdot P_2(1 - P_2)$.

What equation would you use for year 4?

In these equations, all values of $P$ must stay in the range from 0 to 1. A value of 0 represents no insects, and a value of 1 represents the final population size.

Now you are ready to see how these equations can model behavior that is sometimes orderly and predictable and sometimes chaotic and unpredictable.

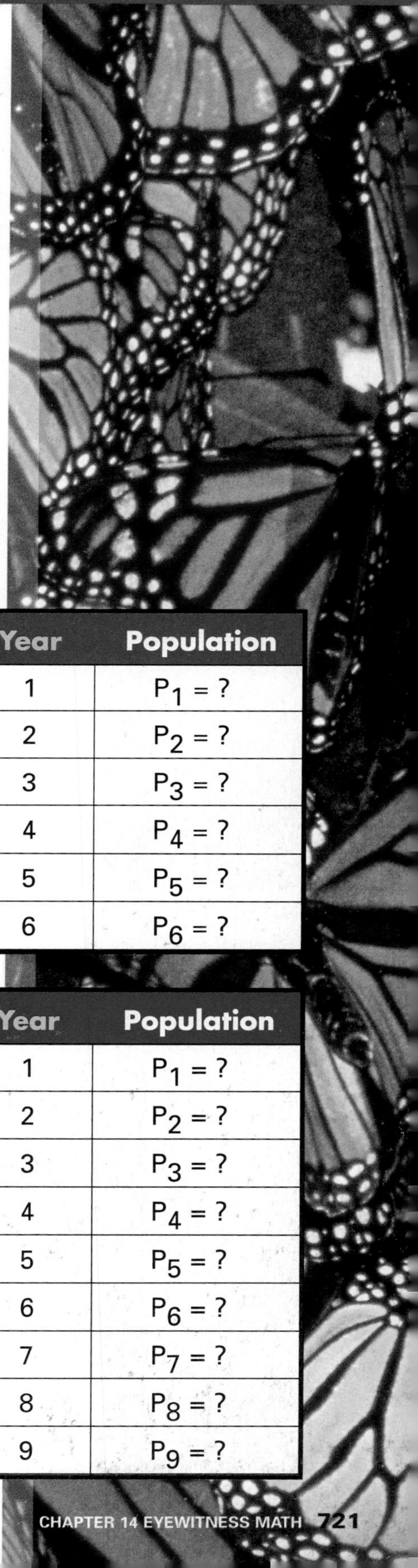

| Year | Population |
|---|---|
| 1 | $P_1 = ?$ |
| 2 | $P_2 = ?$ |
| 3 | $P_3 = ?$ |
| 4 | $P_4 = ?$ |
| 5 | $P_5 = ?$ |
| 6 | $P_6 = ?$ |

## Cooperative Learning

**1.** First see how the population equations behave for $r = 2$.

**a.** Copy and complete the chart at right. To start, choose any value between 0 and 1 for $P_1$. Round each answer to 3 decimal places.

**b.** In your model, what happens to the size of the insect population?

**c.** Compare your results with the results of groups that started with a different value for $P_1$.

**d.** Could you make a good prediction about the size of the population in year 6 even if your year 1 data were off? Explain.

**2.** Now see what happens for $r = 4$.

**a.** Copy and complete the chart. Remember to choose a value for $P_1$ that is between 0 and 1.

**b.** Do you see a pattern in your results? Explain.

**c.** Repeat part **a**, but this time start with a value for $P_1$ that is 0.001 less or 0.001 greater than the value you originally used in part **a**.

**d.** Compare your results from parts **a** and **c**.
For $r = 4$, could you make a good prediction about the size of the population in year 9 if your year 1 data were off a little? Explain.

**e.** Did other groups get similar results even though they started with different values for $P_1$?

| Year | Population |
|---|---|
| 1 | $P_1 = ?$ |
| 2 | $P_2 = ?$ |
| 3 | $P_3 = ?$ |
| 4 | $P_4 = ?$ |
| 5 | $P_5 = ?$ |
| 6 | $P_6 = ?$ |
| 7 | $P_7 = ?$ |
| 8 | $P_8 = ?$ |
| 9 | $P_9 = ?$ |

**3.** In chaos, a tiny difference at the beginning can make a big difference later on. For which value of $r$ did the equations model chaos? Explain.

14.5

# Combining Transformations

## Objectives

- Study a real-world application of transformed functions.
- Graph functions that involve more than one transformation.

**Why** ***Many everyday situations like biking can be represented by functions. Once a function is recognized as being related to a known parent function, the graph can be sketched by identifying and applying one or more transformations to the parent graph.***

**APPLICATION**
**RECREATION**

When planning their 100-mile bike ride, the two friends in the photograph above decide to allow 2 hours for rest stops. They want to know how fast they must ride to complete the trip in certain lengths of time.

Using $d = rt$, where $d$ is the distance in miles, $r$ is the rate in miles per hour, and $t$ is the time in hours, you can write $r = \frac{d}{t}$.

In this function, $t$ is the time actually spent riding. The total number of hours it takes to complete the trip, $h$, is equal to $t + 2$. Thus, $t = h - 2$, and the function that gives the rate of travel necessary to complete the trip in $h$ hours is $r = \frac{100}{h-2}$.

You can substitute different values of $h$ into the function to determine how fast the friends must ride.

Notice that the function is a transformation of the parent function $r = \frac{1}{h}$.

To sketch the graph of $r = \frac{100}{h-2}$ proceed as follows:

First graph the parent function, $r = \frac{1}{h}$.

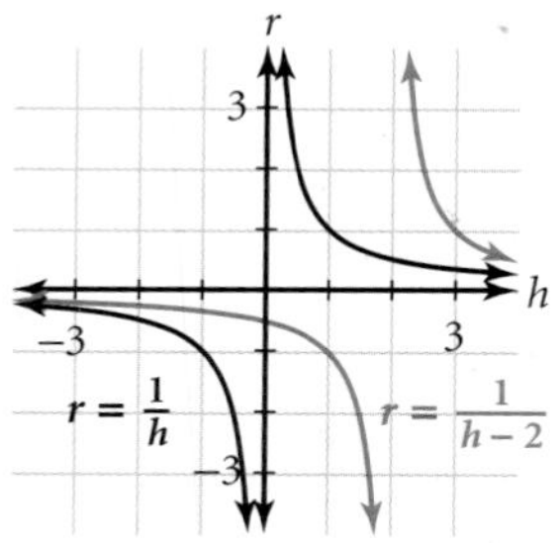


Then determine whether the graph of the parent function is horizontally translated, stretched, or compressed, or reflected in the $y$-axis. The parent graph is horizontally translated 2 units right. The resulting function is $r = \frac{1}{h-2}$.

Next determine whether the graph of $r = \frac{1}{h-2}$ is vertically stretched or compressed, reflected in the $x$-axis, or vertically translated. The graph is vertically stretched by a factor of 100. The resulting function is $r = \frac{100}{h-2}$.

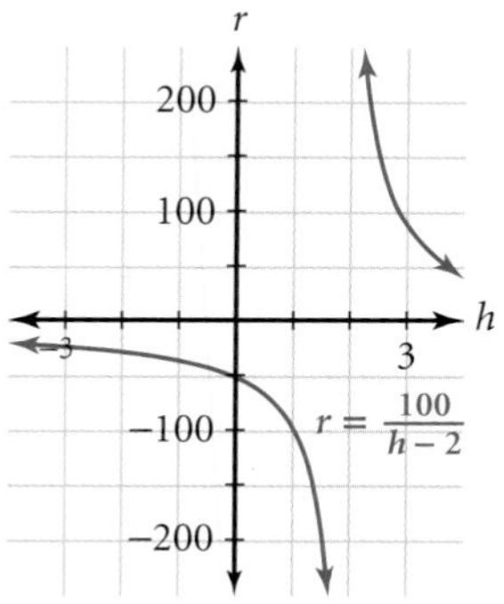


Because neither the number of hours nor the number of miles per hour can be negative, only the right-hand branch of the graph represents the values relevant to the problem.

**CRITICAL THINKING** What happens to the graph of $r = \frac{100}{h-2}$ for values of $h$ close to 2? What happens at $h = 2$? Explain these areas of the graph as they relate to the biking problem.

*Activity*

## Order of Vertical Transformations

**You will need:** graph paper

For steps 1–3 below, begin with the graph of the parent function $f(x) = |x|$ and combine the transformations in the order listed. Show each set of transformations on a separate coordinate plane.

1. Translate 3 units up, stretch vertically by a factor of 2, and reflect across the $x$-axis.
2. Stretch vertically by a factor of 2, translate 3 units up, and reflect across the $x$-axis.
3. Stretch vertically by a factor of 2, reflect across the $x$-axis, and translate 3 units up.
4. Graph the function $g(x) = -2|x| + 3$ by making a table of values and plotting several points. Compare the graphs from steps 1–3 with the graph of $g$. Which one matches the graph of $g$? Is order important when combining transformations?

**CHECKPOINT ✔**

5. When combining vertical transformations, are translations done before or after stretches and reflections?

You may have noticed in the activity that the order of vertical transformations follows the order of operations for real numbers. Multiplication (stretches, compressions, and reflections) is done before addition (translation). The order of horizontal transformations is the *opposite* of the order of operations for real numbers. Addition (translation) is done before multiplication (stretches, compressions, and reflections).

The following list will help you to proceed methodically when graphing transformed functions:

**GRAPHING TRANSFORMED FUNCTIONS**

1. Do horizontal translations first.
2. Do horizontal stretches or compressions second.
3. Do reflections across the $y$-axis third.
4. Do vertical stretches or compressions fourth.
5. Do reflections across the $x$-axis fifth.
6. Do vertical translations sixth.

**EXAMPLE 1** **Apply transformations to the graph of $y = x^2$ in order to graph $y = -2(-3x + 1)^2 - 1$.**

**SOLUTION**

Six transformations are combined to obtain the graph of $y = -2(-3x + 1)^2 - 1$ from its parent function, $y = x^2$.

First graph $y = x^2$. Then apply the transformations in the correct order.

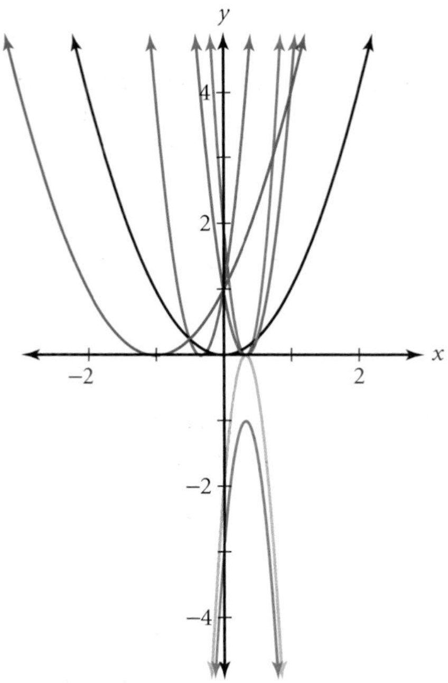


1. The graph of $y = x^2$ is translated 1 unit left to obtain the graph of $y = (x + 1)^2$.
2. The graph of $y = (x + 1)^2$ is compressed horizontally by $\frac{1}{3}$ to obtain the graph of $y = (3x + 1)^2$.
3. The graph of $y = (3x + 1)^2$ is reflected across the $y$-axis to obtain the graph of $y = (-3x + 1)^2$.
4. The graph of $y = (-3x + 1)^2$ is stretched vertically by 2 to obtain the graph of $y = 2(-3x + 1)^2$.
5. The graph of $y = 2(-3x + 1)^2$ is reflected across the $x$-axis to obtain the graph of $y = -2(-3x + 1)^2$.
6. The graph of $y = -2(-3x + 1)^2$ is translated 1 unit down to obtain the graph of $y = -2(-3x + 1)^2 - 1$.

**TRY THIS** Apply transformations to the graph of $y = 2^x$ in order to graph $y = -\frac{1}{3}(2^{-2x-1}) + 1$.

# Exercises

## Communicate

1. For the bike-ride problem at the beginning of the lesson, explain how to identify which number in the function rule represents a translation.
2. For the bike-ride problem, explain how to identify which number in the function rule represents a stretch.
3. Explain why:
   **a.** $y = x^2$ has the same graph as $y = (-x)^2$;
   **b.** $y = |x - 3|$ has the same graph as $y = |3 - x|$.
4. Explain how to graph $y = -a(-bx - h)^2 + k$ using transformations of the parent function $y = x^2$.

## Guided Skills Practice

internet connect

**Homework Help Online**
Go To: **go.hrw.com**
Keyword:
**MA1 Homework Help** for Exercises 11-19

**Sketch the graph of each function by applying transformations to its parent function.** *(EXAMPLE 1)*

5. $f(x) = 2(-x - 3)^2 + 1$
6. $f(x) = -2(-3x + 3)^2 + 1$
7. $f(t) = \frac{1}{2}\left|\frac{1}{2}t + 1\right| + 3$
8. $f(t) = -\frac{1}{2}\left|\frac{1}{4}t + 3\right| - 2$
9. $f(g) = -\frac{3}{2g + 1} + 2$
10. $f(x) = \frac{2}{2x - 4} - 6$

## Practice and Apply

**Sketch the graph of each function.**

11. $g(x) = 2(x + 5)^2$
12. $m(x) = -2x^2 + 5$
13. $t(x) = |x + 3| - 4$
14. $v(x) = 2|5x - 4|$
15. $f(x) = \frac{3}{x - 2}$
16. $g(x) = \frac{3}{-x} - 2$
17. $h(x) = 0.5(2^{2x}) - 1$
18. $z(x) = \frac{3}{4} + \frac{2}{x}$
19. $p(x) = \frac{3}{4} - \frac{2}{4 + x}$

**Name the point that corresponds to the point (4, 10) after applying the specified transformations to the graph of $y = \frac{5}{2}x$. (Hint: If you are not sure, draw a sketch.)**

20. a vertical stretch by a factor of 2, followed by a vertical translation of 3
21. a reflection across the $x$-axis followed by a vertical stretch by a factor of 2
22. a reflection across the $y$-axis followed by a vertical translation of $-3$
23. a horizontal stretch by a factor of 3, followed by a vertical translation of 5
24. a reflection across the $y$-axis, then a vertical compression by a factor of $\frac{1}{2}$, followed by a  translation of $-5$

25. Use the information in the table to determine the transformed function that fits the data. Give the parent function also.

| *x* | −3 | −2 | −1 | 0 | 1 | 2 | 3 | 4 | 5 | 6 | 7 |
|---|---|---|---|---|---|---|---|---|---|---|---|
| *y* | 21 | 16 | 11 | 6 | 1 | −4 | −9 | −14 | −19 | −24 | −29 |

CHALLENGE

26. **a.** Find a function rule for the graph shown at right.

**b.** Is there more than one possible function rule? Explain.

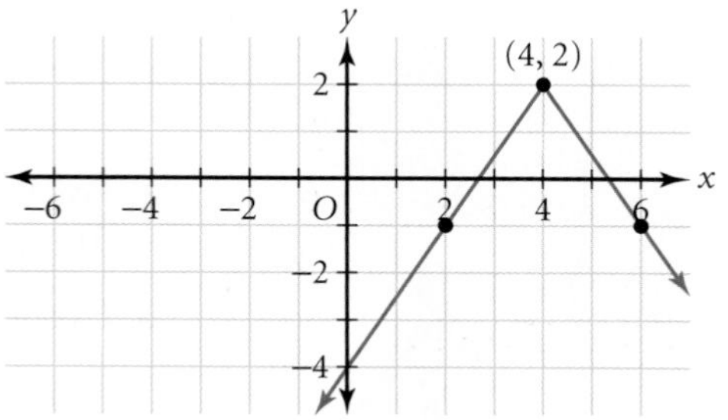


CONNECTION

27. **GEOMETRY** In the figure at right, a triangle is shown in the first quadrant. A reflection, followed by a translation of −18 units, is also shown. The coordinates of $A$ are $(a, b)$. Find the coordinates of $A'$ and $A''$. Do the other vertices change according to the same pattern?

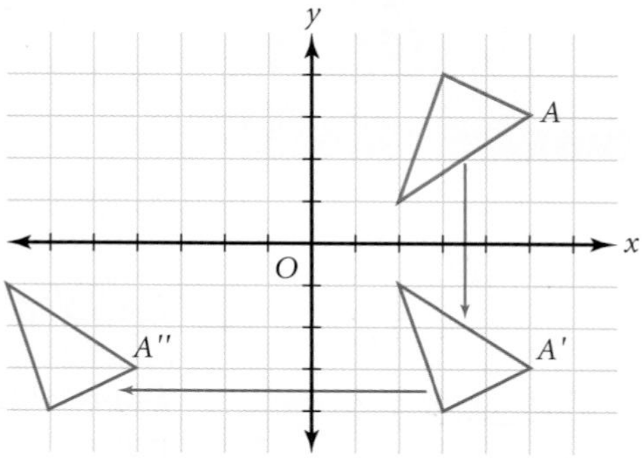


APPLICATION

28. **RECREATION** Two friends are planning a 200-mile bike ride, allowing 5 hours for rest stops.

**a.** Write the function that describes the rate, $r$, that they must travel in order to arrive in a certain number of hours, $h$.

**b.** Give the parent function.

**c.** Graph both functions. (Hint: Refer to the biking problem at beginning of the lesson.)

## Look Back

APPLICATION

29. **PACKAGING** A box requires 75 centimeters of tape to be sealed for delivery. How many boxes can be sealed with 32 meters of tape? *(LESSON 1.2)*

30. Graph the function $4x - 5y = 3$. Outline your procedure for graphing this function. *(LESSON 5.5)*

31. If the graph of a line contains the points $A(-3, 7)$ and $B(2, -9)$, what is an equation for the line that is perpendicular to the given line and passes through point $A$? *(LESSON 5.6)*

32. Solve $|x - 5| \leq 17$, and graph the solution on a number line. *(LESSON 6.5)*

**Solve the system of equations by using the methods below.** $\begin{cases} 3x - y = 5 \\ x - 2y = -10 \end{cases}$

33. graphing *(LESSON 7.1)*

34. elimination *(LESSON 7.3)*

## Look Beyond

**Reflect the graph of each function across the line $y = x$. (Hint: Interchange the $x$ and the $y$ in the original equation and solve for $y$.)**

**35.** $y = -2x + 1$ **36.** $y = \sqrt{x}$ **37.** $y = \frac{1}{x}$

**38.** $y = |x|$ **39.** $y = x^2$

internet connect

**Activities Online**

Go To: **go.hrw.com**
Keyword:
**MA1 Inverse**

PORTFOLIO ACTIVITY

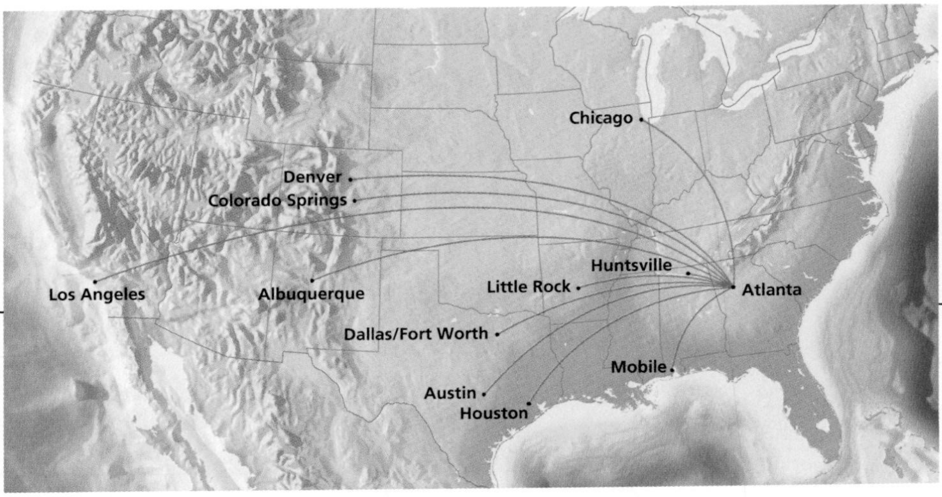


Include this activity in your portfolio.

The table below shows the flight time for one airline and the mileage from Atlanta to the cities shown.

1. Make a scatter plot and graph the line of best fit. Use $t$, time in minutes, as the independent variable.
2. **a.** Write the function for the line of best fit as a transformation of the parent function $y = x$.
   **b.** What is the slope given by the function, and what does the slope represent in the problem?
   **c.** What is the $y$-intercept given by the function, and what does the $y$-intercept represent in the problem?
   **d.** Where does the line of best fit meet the $x$-axis? What could account for this value in actual plane travel?

| City | Time (hours:minutes) | Distance (miles) |
|---|---|---|
| Albuquerque, NM | 3:20 | 1269 |
| Austin, TX | 2:23 | 812 |
| Chicago | 1:44 | 596 |
| Colorado Springs | 3:20 | 1185 |
| Dallas/Fort Worth | 2:10 | 731 |
| Denver | 3:32 | 1201 |
| Houston | 2:13 | 689 |
| Huntsville, AL | 0:48 | 151 |
| Little Rock, AR | 1:27 | 453 |
| Mobile, AL | 1:04 | 302 |
| Los Angeles | 5:08 | 1946 |

CHAPTER PROJECT FOURTEEN

# Pick a Number

**Ask a classmate to do the following:**

*Pick a number from 1 through 9. Triple the number and add 6. Then divide the result by 3.*

**Ask for the final number. If you subtract 2, you can tell your classmate the original number.**

**Try this activity with other numbers. What do you notice?**

Algebra can be used to help explain why this happens. Let $x$ be the number chosen.

| | Arithmetic | Algebra |
|---|---|---|
| **Pick a number.** | 5 | $x$ |
| **Triple the number and add 6.** | $15 + 6 = 21$ | $3x + 6$ |
| **Divide by 3.** | $21 \div 3 = 7$ | $x + 2$ |

The algebra shows you how to go directly to the final number. This is an example of finding a function of a function.

In terms of a function machine, the output from the first machine is used as the input for the second machine.

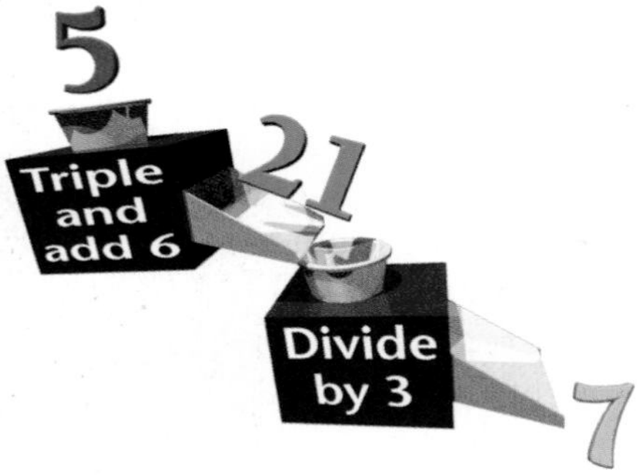


When you find a function of a function, you are forming a composite function. If $f$ is one function and $g$ is another function, you can write $g(f(x))$ to indicate the value of a **composite function** for any $x$-value.

If $f(x) = 3x + 6$ and $g(x) = \frac{x}{3}$, ***then*** $g(f(x)) = \frac{f(x)}{3} = \frac{3x+6}{3} = x + 2$.

The composition of two functions produces a new function with a new graph.

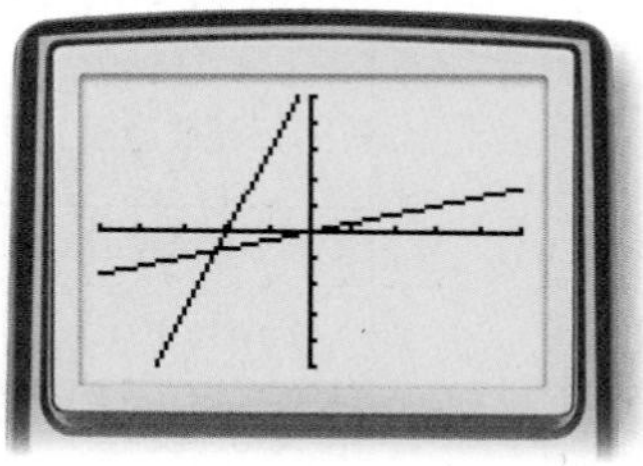

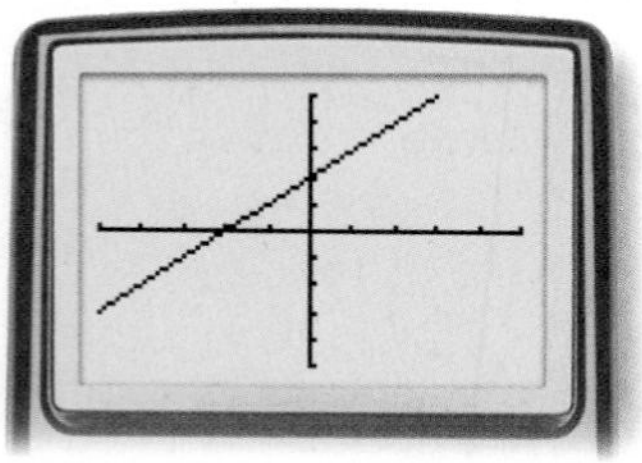

Composition of functions is not a commutative operation.

$g(f(x)) = x + 2$, but $f(g(x)) = 3(g(x)) + 6 = 3\left(\frac{x}{3}\right) + 6 = x + 6$

## Activity 1

Make up a two-step computation game like the one at the beginning of this project. Show how the domain and range are related through the steps. Then show the simplified composition of the functions that lets you perform the calculation easily. Try the game on your friends.

## Activity 2

Use what you have learned about composing functions to graph the composition of two functions. Begin by selecting a parent function from those you studied in Lesson 14.1. Replace the $x$ in the function rule with another function, and graph the composition.

**a.** Identify the domains and ranges of the original function and the composite function.

**b.** Make a table of the input values, the intermediate values for the input in the second step, and the final composite values.

**c.** Use a graphics calculator or graph paper to draw the graph of the new composite function, and compare it to the original functions.

**d.** Try reversing the order of the two functions you composed. Graph the new composite function and complete parts **a–c** above. What do you notice?

# 14 Chapter Review and Assessment

## VOCABULARY

| | | |
|---|---|---|
| function notation . . . . . . . . . . 694 | parent function . . . . . . . . . . . 696 | vertical compression . . . . . . 708 |
| function rule . . . . . . . . . . . . . 694 | reflection . . . . . . . . . . . . . . . . 715 | vertical stretch . . . . . . . . . . . 708 |
| horizontal compression . . . . 710 | scale factor . . . . . . . . . . . . . . 708 | vertical translation . . . . . . . . 701 |
| horizontal stretch . . . . . . . . . 710 | transformation . . . . . . . . . . . . 700 | vertical-line test . . . . . . . . . . 693 |
| horizontal translation . . . . . . 702 | translation . . . . . . . . . . . . . . . 700 | |

## Key Skills & Exercises

### LESSON 14.1

#### Key Skills

**Determine if a set of ordered pairs or a graph represents a function; identify the domain and range of a function; and evaluate a function.**

The set of ordered pairs $\{(0, 1), (1, 2), (2, 5), (3, 10)\}$ is a function because there is exactly one second element for each first element in the set. The domain of the function is 0, 1, 2, and 3. The range of the function is 1, 2, 5, and 10.

The graph below is a function because no vertical line will intersect the graph more than once.

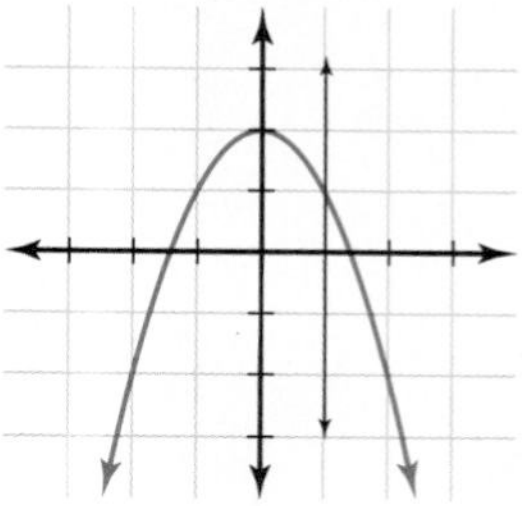

To evaluate the function $f(x) = x^2 + 1$ for $f(-3)$, substitute $-3$ for $x$, and simplify.

$f(x) = x^2 + 1$

$f(-3) = (-3)^2 + 1 = 10$

**Identify parent functions.**

These are the six basic parent functions:

**a.** $y = x^2$ **b.** $y = x$

**c.** $y = \frac{1}{x}$ **d.** $y = a^x$

**e.** $y = |x|$ **f.** $y = \sqrt{x}$

#### Exercises

**Determine whether each set of ordered pairs represents a function or a relation that is not a function.**

**1.** $\{(4, 1), (3, 2), (4, 3), (3, 4)\}$

**2.** $\{(1, 4), (2, 3), (3, 4), (4, 3)\}$

**Determine if each of the following are functions.**

**3.**

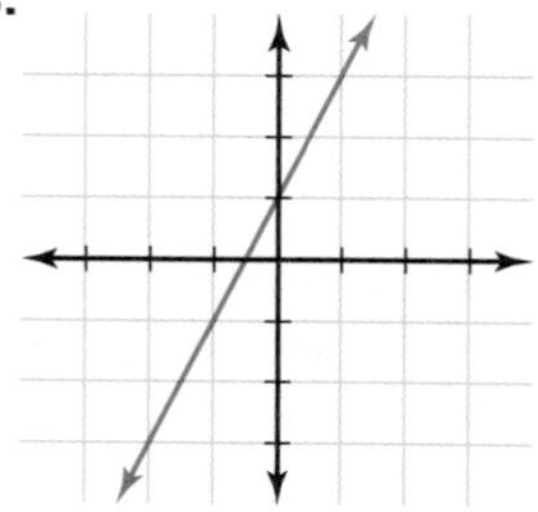

**4.**

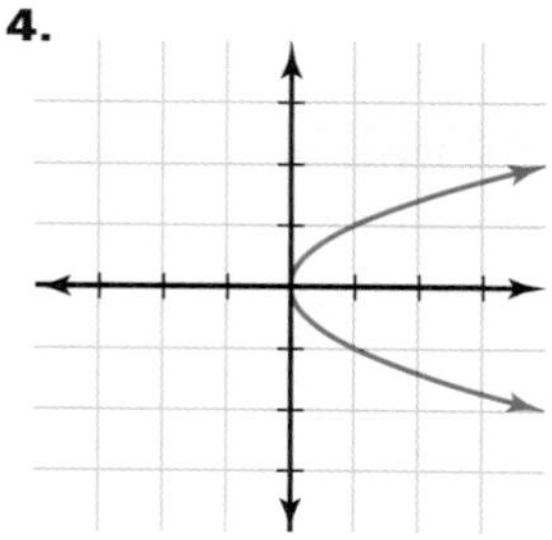

**Evaluate the following functions for $x = 4$:**

**5.** $f(x) = x - 8$

**6.** $g(x) = |x + 2|$

**7.** $h(x) = 3^x$

**8.** $j(x) = 4x^2$

**9.** $k(x) = -5|x - 3|$

**10.** $f(x) = \frac{2}{x}$

**Name the parent function for each function graphed below.**

**11.**

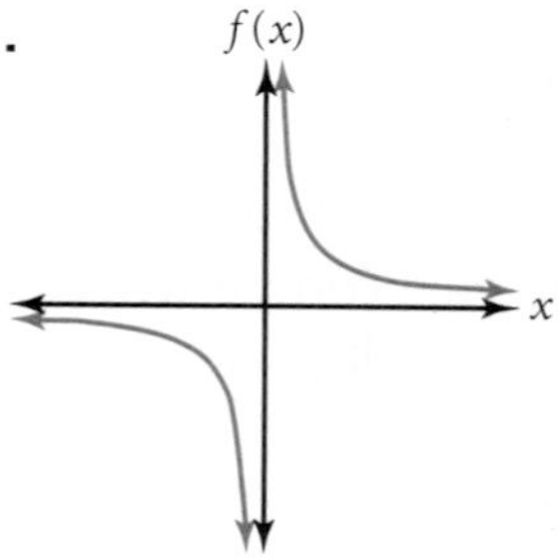


**12.**

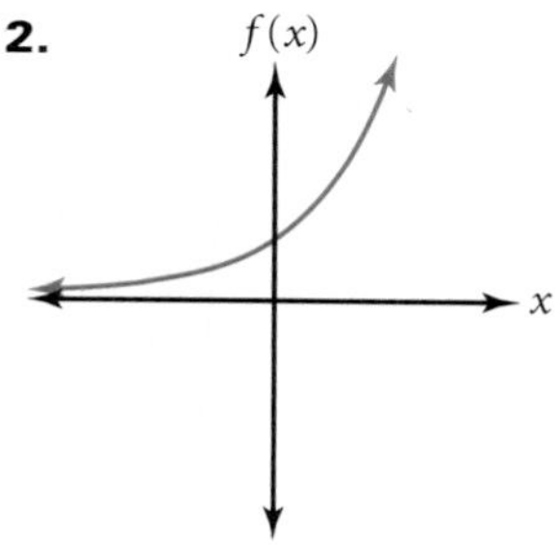

## LESSON 14.2

### Key Skills

**Describe the effect of a translation on the graph of a function.**

The parent function of $y = |x| + 2$ is $y = |x|$. The graph of $y = |x| + 2$ is shifted 2 units up from the graph of $y = |x|$. The graph of $y = |x + 1|$ is shifted 1 unit to the left of the graph of $y = |x|$. The graph of $y = |x - 1|$ is shifted 1 unit to the right of the graph of $y = |x|$.

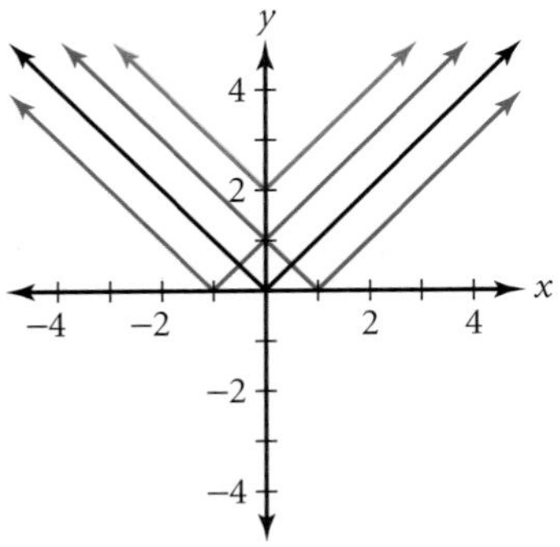


### Exercises

Identify the parent function. Tell what type of transformation is applied to the parent function to obtain each function, and then graph each function.

**13.** $y = x^2 + 1$

**14.** $y = (x + 1)^2$

**15.** $y = |x| - 4$

**16.** $y = \frac{1}{x - 5}$

**17.** $y = \frac{1}{x} + 3$

**18.** $y = (x - 2)^2$

**19.** $y = 2^{x + 3}$

## LESSON 14.3

### Key Skills

**Describe the effect of a vertical stretch or compression on the graph of a function.**

The parent function of $y = 3|x|$ is $y = |x|$. Each of the $y$-values of the transformed function is 3 times the corresponding $y$-value of the parent function.

| $x$ | $\|x\|$ | $3\|x\|$ |
|---|---|---|
| −1 | 1 | 3 |
| 0 | 0 | 0 |
| 2 | 2 | 6 |

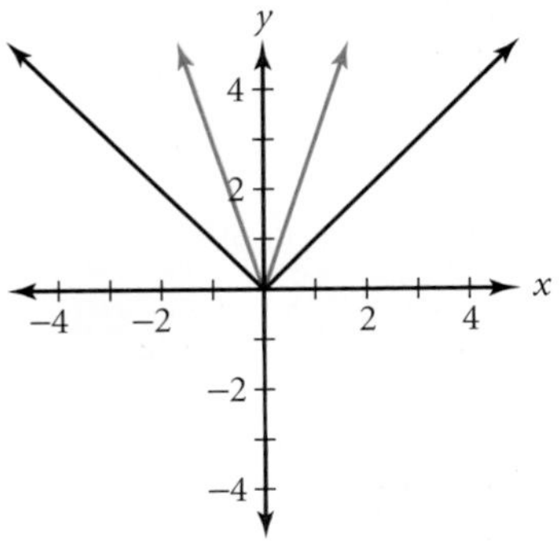


### Exercises

**Graph each of the following:**

**20.** $y = 4|x|$

**21.** $y = 5|x|$

**22.** $y = |5x|$

**23.** $y = 3x^2$

**24.** $y = 2x^2$

**25.** $y = \frac{1}{4x}$

**26.** $y = \left(\frac{x}{4}\right)^2$

**27.** $y = \sqrt{\frac{1}{2}x}$

## LESSON 14.4

### Key Skills

**Describe the effect of a reflection across the *x*- or *y*-axis on the graph of a function.**

The parent function of $y = -|x|$ is $y = |x|$. The values of $-|x|$ are opposite those of $|x|$. This is a reflection of the parent graph $y = |x|$ across the $x$-axis.

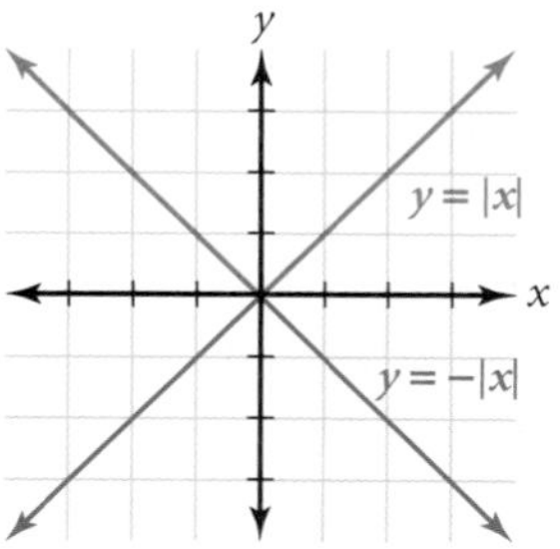


### Exercises

**Graph each function along with its parent function on the same coordinate plane.**

**28.** $y = -\frac{1}{x}$

**29.** $y = \sqrt{-x}$

**30.** $y = -\sqrt{-x}$

**31.** $y = 2^{-x}$

**32.** $y = -(2)^x$

**33.** $y = -x^2$

## LESSON 14.5

### Key Skills

**Apply more than one transformation to a parent function to sketch a graph.**

The parent function of $y = -2|x + 2|$ is $y = |x|$. The parent function has been reflected across the $x$-axis, stretched vertically by 2, and translated horizontally 2 units to the left.

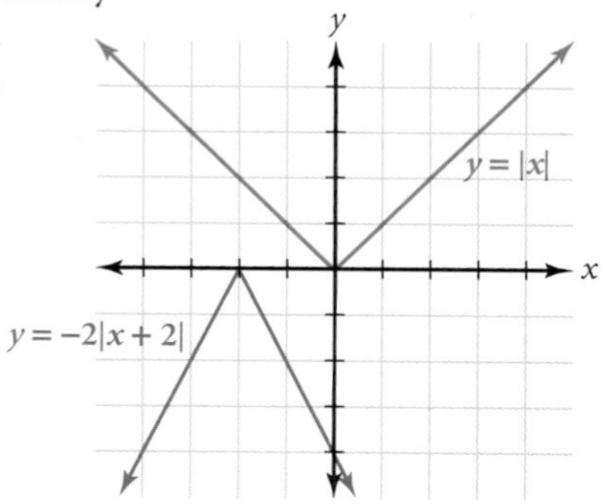


### Exercises

**Graph each function and list each transformation.**

**34.** $f(x) = 3x^2 + 1$

**35.** $g(x) = |x - 6| + 2$

**36.** $h(x) = -\frac{2}{x} + 1$

**37.** $f(x) = (x - 3)^2 + 4$

**38.** $g(x) = 2^{x+1} - 3$

**39.** $h(x) = |x + 4| - 2$

**40.** $f(x) = -6\left|\frac{1}{3}x - 1\right| + 2$

## Applications

**41. PHYSICS** Kelvins are often used in science to measure the temperature of substances. To change from degrees Celsius to kelvins, add 273.16 to degrees Celsius. What type of transformation relates degrees Celsius to kelvins?

**42. BUSINESS** Thomas decides to have a sale at his art store. He decides to reduce the price of everything in the store by 20%. Find the scale factor that shows the amount by which the prices of the items were "stretched."

# Chapter Test

**Determine whether each of the following is a function.**

1. $\{(5, 7), (6, 8), (7, 7), (8, 8), (9, 1)\}$
2. 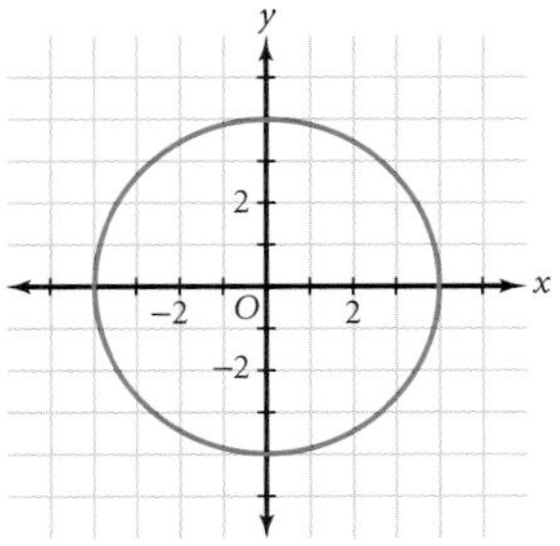

**Evaluate the following functions for $x = 3$.**

3. $f(x) = 2x - 5$
4. $g(x) = 4^x$
5. $h(x) = 2|3 - x| + 4$
6. $k(x) = \frac{x-1}{x+1}$

**Name the parent function for each function graphed below.**

7. 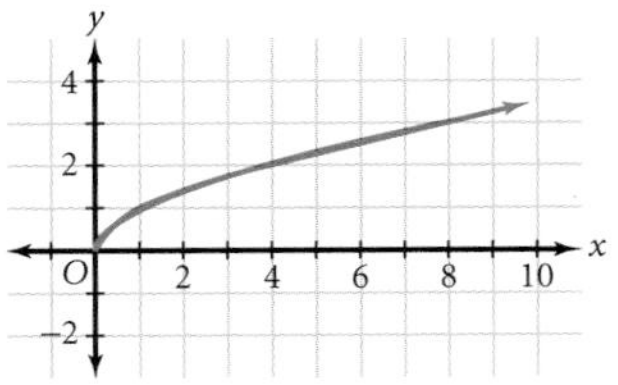

8. 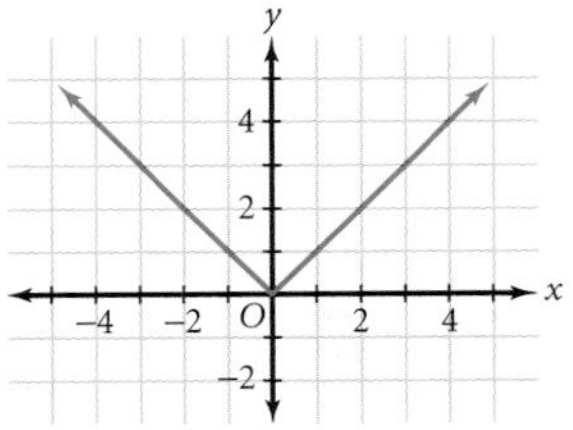

**Identify the parent function for each function. Describe how to obtain the graph of each function from the graph of its parent function.**

9. $y = \sqrt{x+2}$
10. $y = 3^x - 5$
11. $y = \frac{1}{x} - 1$
12. $y = |x + 4|$
13. $y = (x - 2)^2 - 2$
14. $y = \frac{1}{x+3}$

15. BUSINESS A bicycle messenger service charges $2 per half mile to deliver a package. The service recently added an additional $7 fee per trip to cover the cost of new bicycles. What type of transformation is this?

**Determine how the graph of each function is related to its parent function. Identify the transformation.**

16. $y = \frac{1}{2x}$
17. $y = \frac{1}{2}x^2$
18. $y = \sqrt{\frac{1}{4}x}$
19. $y = \frac{3}{x}$

20. CURRENCY To exchange U.S. dollars to Canadian dollars, it is necessary to multiply by an exchange factor as each U.S. dollar is equal to 1.45 Canadian dollars. What type of transformation is this?

**Give the equation of the resulting function.**

21. The function $f(x) = \sqrt{3x}$ is reflected across the $x$-axis.
22. The function $f(x) = 3^x$ is reflected across the $y$-axis.
23. The function $f(x) = |x - 4|$ is reflected across the $x$-axis.

**Graph each function and list each transformation from the parent function.**

24. $f(x) = \frac{1}{2}x^2 - 2$
25. $g(x) = 2|x + 1| - 2$
26. $h(x) = 2^{-(x-1)} - 4$
27. $f(x) = -\sqrt{x+3}$
28. Write the function that describes the new pricing plan in Exercise 15 above. Identify the parent function.
29. The point (2, 4) lies on the graph of $y = x^2$. Name the corresponding point after a vertical stretch by a factor of 3 and a right horizontal translation of 5 units.

# College Entrance Exam Practice

**internet connect**
**Standardized Test Prep Online**
Go To: **go.hrw.com**
Keyword: **MM1 Test Prep.**

**QUANTITATIVE COMPARISON** For Questions 1–4, write
A if the quantity in Column A is greater than the quantity in Column B;
B if the quantity in Column B is greater than the quantity in Column A;
C if the two quantities are equal; or
D if the relationship cannot be determined from the information given.

| | Column A | Column B | Answers |
|---|---|---|---|
| **1.** | $(x+y)^2$ | $x^2+y^2$ | Ⓐ Ⓑ Ⓒ Ⓓ [Lesson 9.4] |
| **2.** | $5 \times 10^{-3}$ | 0.005 | Ⓐ Ⓑ Ⓒ Ⓓ [Lesson 8.5] |
| **3.** | slope of $y = -2$ | slope of $y = x - 2$ | Ⓐ Ⓑ Ⓒ Ⓓ [Lesson 5.4] |
| **4.** | $7^0$ | $15(-9+9)$ | Ⓐ Ⓑ Ⓒ Ⓓ [Lesson 8.4] |
| **5.** | solution to $3(x+9) = 15$ | solution to $x^2 = 15$ | Ⓐ Ⓑ Ⓒ Ⓓ [Lesson 10.2] |

**6.** What is the reciprocal of 4? ***(LESSON 2.4)***
**a.** $-4$ **b.** $\frac{1}{4}$
**c.** 1 **d.** $-\frac{1}{4}$

**7.** Find the slope of $3y + 2x = 5$. ***(LESSON 5.4)***
**a.** $-\frac{3}{2}$ **b.** $-\frac{2}{3}$
**c.** $-\frac{5}{3}$ **d.** $\frac{5}{3}$

**8.** Simplify: $5r - (3r + 2)$. ***(LESSON 9.1)***
**a.** $8r + 2$ **b.** $2r + 2$
**c.** $2r - 2$ **d.** $8r - 2$

**9.** Solve: $6 - 4t > 18$. ***(LESSON 6.2)***
**a.** $t < -3$ **b.** $t > -3$
**c.** $t > -6$ **d.** $t < -6$

**10.** Simplify: $(5x + 2) - (x - 4)$. ***(LESSON 9.1)***
**a.** $5x - 4$ **b.** $6x + 6$
**c.** $4x + 6$ **d.** $4x - 6$

**11.** Which function is graphed below? ***(LESSON 10.1)***

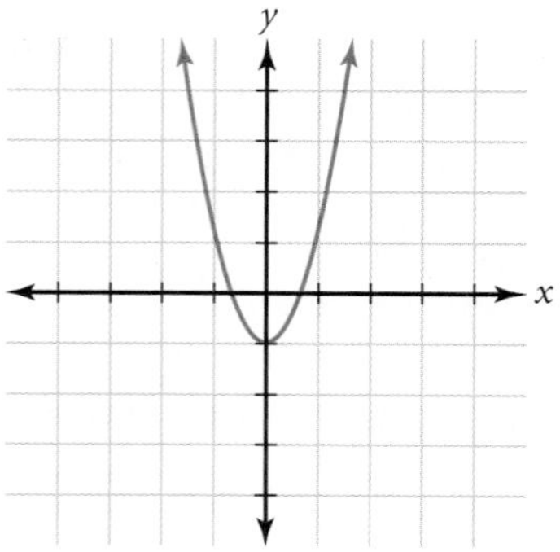


**a.** $y = 2x^2 - 1$ **b.** $y = -2x^2 - 1$
**c.** $y = \frac{1}{2}x^2 - 1$ **d.** $y = -\frac{1}{2}x^2 - 1$

**12.** Find the slope of a line containing the points $M(-1, 2)$ and $N(4, -3)$. ***(LESSON 5.2)***
**a.** $-2$ **b.** 2
**c.** 1 **d.** $-1$

13. Solve. $\begin{cases} 2x - y = -1 \\ 3x + 2y = -5 \end{cases}$ ***(LESSON 7.3)***

**a.** $(1, 1)$ **b.** $(1, -1)$
**c.** $(-1, 1)$ **d.** $(-1, -1)$

14. Which expression is *not* equivalent to $\frac{3x+9}{-3}$? ***(LESSON 2.7)***

**a.** $-x - 3$ **b.** $-(x + 3)$
**c.** $-x + 9$ **d.** $-3 - x$

15. Solve: $a + 19 = 25$. ***(LESSON 3.1)***

16. Simplify: $(2^3)^{-2}$. ***(LESSON 8.4)***

17. Cathi has \$17.40 in dimes and quarters. If she has a total of 81 coins, how many of each type does she have? ***(LESSON 7.6)***

18. Solve: $\frac{x}{-10} = \frac{-3}{5}$. ***(LESSON 4.1)***

19. The first three terms of a sequence are 2, 4, and 9. The second differences are a constant 3. What are the next three terms? ***(LESSON 1.1)***

20. During one year, a total of 10,924,000 individual tax returns were filed electronically. Write this number in scientific notation. ***(LESSON 8.5)***

21. Identify the parent function of $y = \frac{-2}{x}$. Draw the graph of $y = \frac{-2}{x}$. ***(LESSON 14.5)***

22. Solve: $3x + 9 = 6x - 12$. ***(LESSON 3.4)***

23. A store marked all of its items down by 10%. After the sale, all of the items were marked up by 10% from the sale price. Before the sale, a sweater cost \$50. What does the same sweater cost after the sale? ***(LESSON 4.2)***

24. Solve: $2x^2 + 3x - 2 = 0$. ***(LESSON 9.8)***

25. Solve by using substitution. ***(LESSON 7.2)***

$$\begin{cases} 2x + 3y = 16 \\ -3x + y = -2 \end{cases}$$

26. Show the solution by using a graph. ***(LESSON 7.5)***

$$\begin{cases} x + y \geq 3 \\ y \leq 3 \end{cases}$$

27. Factor: $x^2 + 4x - 32$. ***(LESSON 9.7)***

## FREE-RESPONSE GRID

**The following questions may be answered by using a free-response grid such as that commonly used by standardized test services.**

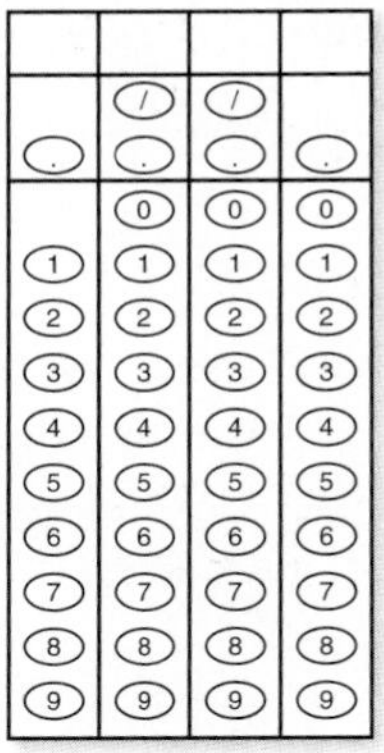

28. The perimeter of a rectangle is 30 centimeters and its length is twice its width. Find the length of the rectangle. ***(LESSON 7.3)***

29. Simplify: $3(4 + 5) \cdot (4 - 3)2 \div 5 \cdot 3$. ***(LESSON 1.3)***

30. Find the slope of the lines that are parallel to the line passing through $(-7, -5)$ and $(1, -3)$. ***(LESSON 5.6)***

31. Find the root(s) of $4x^2 - 4x + 1 = 0$. ***(LESSON 9.8)***

32. Find the sum: $1 + 2 + 3 + \cdots + 50$. ***(LESSON 1.1)***

33. How many \$22 concert tickets can you buy with \$125? ***(LESSON 1.2)***

34. A set of golf clubs is marked for sale at \$180, a reduction of 25%. What was the original price in dollars of the golf clubs? ***(LESSON 4.2)***

35. Find the slope of the line containing the points $(3, 4)$ and $(7, 7)$. ***(LESSON 5.2)***

36. Find the slope of a line containing the points $(-4, -2)$ and $(0, 1)$. ***(LESSON 5.2)***

37. Solve: $5(x + 2) = 3x + 13$. What does $x$ equal? ***(LESSON 3.5)***

38. Give the decimal form of $1.1 \times 10^{-2}$. ***(LESSON 8.5)***

39. Solve: $\frac{r}{10} = \frac{20}{25}$. ***(LESSON 4.1)***

Extra Practice
Tables
Calculator Keystroke Guide
Selected Answers
Index

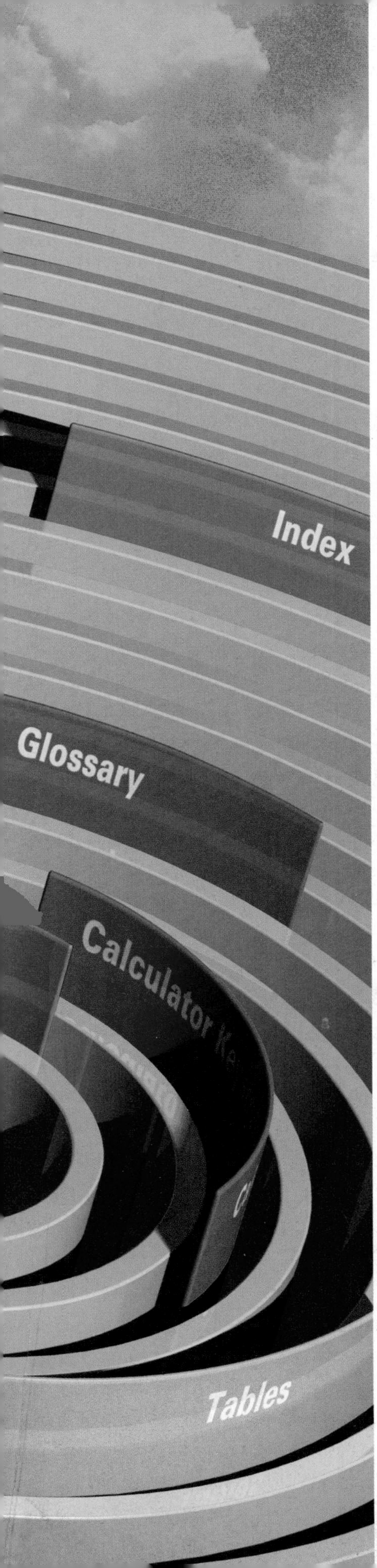


# Info Bank

# Extra Practice

## CHAPTER 1

### LESSON 1.1

**Find the next three terms in each sequence.**

**1.** 6, 9, 12, 15, 18, 21, . . .
**2.** 80, 75, 70, 65, 60, 55, . . .
**3.** 3, 8, 13, 18, 23, . . .
**4.** 3, 5, 7, 9, 11, 13, . . .
**5.** 18, 22, 26, 30, 34, 38, . . .
**6.** 100, 93, 86, 79, 72, 65, . . .
**7.** 34, 38, 42, 46, 50, 54, . . .
**8.** 28, 25, 22, 19, 16, 13, . . .
**9.** 7, 13, 19, 25, 31, 37, . . .
**10.** 1, 1, 2, 3, 5, 8, . . .
**11.** 11, 23, 36, 50, 65, 81, . . .
**12.** 8, 15, 22, 29, 36, 43, . . .
**13.** 3, 9, 15, 21, 27, . . .
**14.** 2, 9, 16, 23, 30, . . .
**15.** 0, 15, 30, 45, 60, . . .

**16.** There are 8 players in a chess tournament. Each player plays every other player. How many games of chess will be played?

**17.** There are 5 people in a room. If each person talks to every other person, how many conversations occur?

**18.** There are 16 people at a meeting. Each person shakes hands with every other person. How many handshakes are there?

**Find each sum. Think of a geometric dot pattern, but do not draw a sketch.**

**19.** $1 + 2 + . . . + 90$
**20.** $1 + 2 + . . . + 200$
**21.** $1 + 2 + . . . + 110$

**22.** The third and fourth terms of a sequence are 7 an 10. If the first differences are a constant 3, what are the first five terms of the sequence?

**23.** The third and fourth terms of a sequence are 11 and 18. If the second differences are a constant 2, what are the first five terms of the sequence?

**24.** The second differences of a sequence are a constant 4, the first of the first differences is 2, and the first term is 1, find the first 5 terms of the sequence.

### LESSON 1.2

**Identify each of the following as an expression or an equation.**

**1.** $5x$
**2.** $t = 4$
**3.** $3x - 3$
**4.** $2 = q$
**5.** $0.5p$
**6.** $y + 1$
**7.** $3w + 4$
**8.** $3m - 1 = 14$

**Make a table showing the variable and the value of the expressions below. Use 1, 2, 3, 4, and 5 as values for the variables.**

**9.** $3x$
**10.** $8y$
**11.** $2z + 4$
**12.** $3d - 1$
**13.** $9q + 3$
**14.** $15c$
**15.** $9v - 33$
**16.** $2 + 8z$

**Use guess and check to solve each equation.**

**17.** $4x + 1 = 9$
**18.** $15 = 5x + 5$
**19.** $24x - 8 = 112$
**20.** $10x - 3 = 67$
**21.** $12 = 2x - 4$
**22.** $3 + 19x = 98$
**23.** $7 + 6x = 55$
**24.** $9 - x = 3$
**25.** $115 - 3x = 100$

## LESSON 1.3

**Evaluate each expression by using the algebraic order of operations.**

**1.** $21 - 7 + 2 \cdot 5$
**2.** $(21 - 7) + 2 \cdot 5$
**3.** $(21 - 7 + 2) \cdot 5$
**4.** $32 + 2 \cdot 6 \div 3$
**5.** $(32 + 2) \cdot 6 \div 3$
**6.** $32 + 2 \cdot (6 \div 3)$
**7.** $24 + 2^2 \div 4$
**8.** $42 \cdot 20 + 6$
**9.** $5^2(30 + 12)$
**10.** $15 \div 3 \cdot 3 - 4$
**11.** $45(2) + 3(16)$
**12.** $6(4) - 4$
**13.** $(3 + 7) \div (3 + 2)$
**14.** $(4 + 5 \cdot 2^2) \div (2 \cdot 3)$
**15.** $(7 - 4 \cdot 2 + 2) \div 4$
**16.** $14 + 4 \div 2 - 5$
**17.** $6 \div 2 \cdot 5 + 2^2$
**18.** $4 + 2^2 - 10 + 3$

**Place grouping symbols according to the algebraic order of operations to make each equation true.**

**19.** $24 - 6 \cdot 2 + 2^2 = 40$
**20.** $16 \div 2 + 2 - 2^2 = 0$
**21.** $6 \cdot 8 + 1 \div 3 = 18$
**22.** $7 + 3 \cdot 8 - 5 = 30$
**23.** $15 + 5 \div 5 - 2 = 2$
**24.** $64 + 2^2 \cdot 2 \div 2 = 68$
**25.** $39 - 7 \div 4 + 3 = 11$
**26.** $10^2 \div 5 \cdot 2^2 + 15 = 20$

**Given that $a = 4$, $b = 5$, and $c = 6$, evaluate each expression.**

**27.** $a + b + c$
**28.** $a \cdot b - c$
**29.** $b - a \cdot c$
**30.** $a + b - c$
**31.** $a \cdot (b + c)$
**32.** $(a + b) \cdot c$
**33.** $a^2 - b + c$
**34.** $a^2 + c^2 - b$
**35.** $a^2 - c^2 - b^2$

## LESSON 1.4

**Graph each list of ordered pairs. State whether the points lie on a straight line.**

**1.** (1, 2), (2, 3), (3, 4)
**2.** (1, 6), (2, 8), (3, 6)
**3.** (2, 4), (5, 3), (11, 1)
**4.** (−1, 2), (3, 2), (−5, 2)
**5.** (−4, 3), (−2, 1), (0, 1)
**6.** (3, 0), (4, 2), (5, 4)
**7.** (3, 2), (0, 2), (0, 0)
**8.** (−1, −1), (−2, −2), (−3, −3)
**9.** (1, 4), (1, −3), (1, 0)

**Make a table of values using 1, 2, 3, 4, and 5 for *x*.**

**10.** $y = x + 2$
**11.** $y = x + 5$
**12.** $y = x - 6$
**13.** $y = x - 4$
**14.** $y = 3x$
**15.** $y = -4x$
**16.** $y = 3x + 1$
**17.** $y = 2x - 3$
**18.** $y = -x + 1$
**19.** $y = -4x - 1$
**20.** $y = 0.5x + 2$
**21.** $y = 2x - 0.5$
**22.** $y = 10x$
**23.** $y = -7x + 5$
**24.** $y = x + 11$
**25.** $y = 12x$

## LESSON 1.5

**Find the first differences and write an equation that describes the pattern shown in each table.**

**1.**

| $x$ | 5 | 6 | 7 | 8 | 9 |
|---|---|---|---|---|---|
| $y$ | 27 | 34 | 41 | 48 | 55 |

**2.**

| $x$ | 2 | 3 | 4 | 5 | 6 |
|---|---|---|---|---|---|
| $y$ | 1 | −1 | −3 | −5 | −7 |

**Make a table of values for each equation using 1, 2, 3, 4, and 5 as values for *x*.**

**3.** $y = 13 - 2x$ **4.** $y = 6x + 4$ **5.** $y = -4x + 2$ **6.** $y = -3x + 5$

**Pens cost 69 cents each. Write and simplify an expression for the cost of**

**7.** 0 pens. **8.** 5 pens. **9.** 10 pens. **10.** $p$ pens.

**Hamburgers cost $2.50 each. Write and simplify an expression for the cost of**

**11.** 5 hamburgers. **12.** 30 hamburgers. **13.** 8 hamburgers. **14.** $h$ hamburgers.

**15.** If tickets for a concert cost $12 each, how many tickets can you buy with $48?

**16.** If videos rent for $2.99 each, how many videos can you rent with $11.96?

## LESSON 1.6

**Describe the correlation as strong positive, strong negative, or little to none.**

**1.**

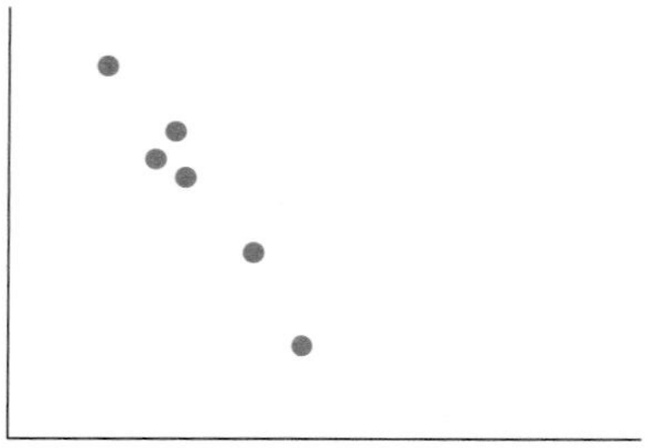

**2.**

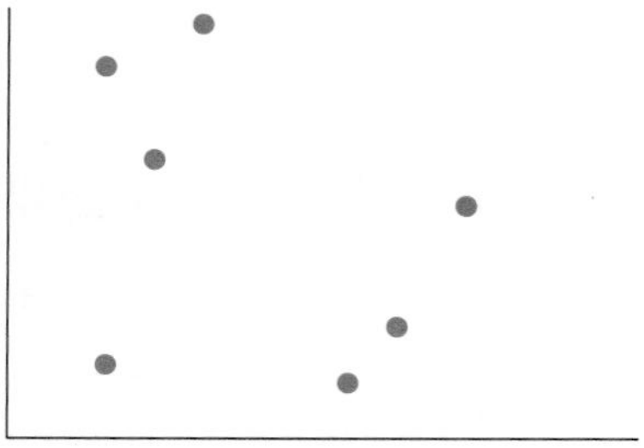

**3.**

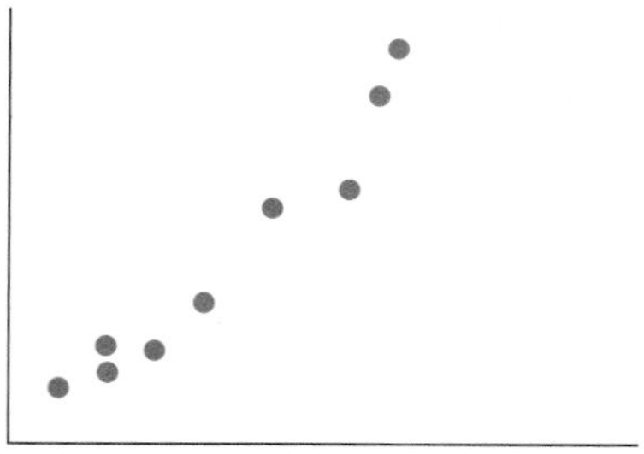

**4.**

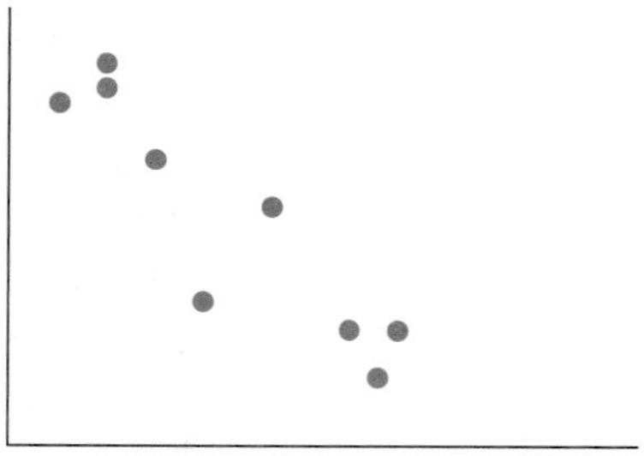

**Use the chart of winning times for the women's 200-meter run in the Olympics below for Exercises 5–7.**

| Year | 1964 | 1968 | 1972 | 1976 | 1980 | 1984 | 1988 | 1992 |
|---|---|---|---|---|---|---|---|---|
| Seconds | 23.00 | 22.50 | 22.40 | 22.37 | 22.03 | 21.81 | 21.34 | 21.81 |

**5.** Make a scatter plot of the data.

**6.** Estimate a line of best fit and draw it on your scatter plot.

**7.** Describe the correlation between the years and the winning times.

# CHAPTER 2

## LESSON 2.1

**Insert an ordering symbol to make each statement true.**

**1.** 0 __ 8 **2.** −3.2 __ −3 **3.** $\frac{2}{3}$ __ $\frac{3}{4}$ **4.** $-\frac{2}{3}$ __ $-\frac{3}{4}$

**Find the opposite of each number.**

**5.** 13 **6.** −13 **7.** −2.36
**8.** 743 **9.** 18 **10.** −12
**11.** $-4\frac{5}{6}$ **12.** 73

**Find the absolute value of each number.**

**13.** 15 **14.** −11 **15.** −3.47
**16.** 469 **17.** $-7\frac{2}{3}$ **18.** −16
**19.** −842 **20.** 89

## LESSON 2.2

**Find each sum.**

**1.** −4 + 2 **2.** 3 + (−9)
**3.** −8 + (−5) **4.** −8 + 102
**5.** 8 + (−10) **6.** −8 + (−10)
**7.** −5 + 4 + (−6) **8.** 15 + (−9) + (−12)
**9.** −14 + 3 + (−9) **10.** −15 + 36
**11.** 26 + (−40) **12.** 15 + (−34)
**13.** −42 + 16 **14.** −89 + (−23)
**15.** −25 + (−25) **16.** −25 + 25
**17.** −91 + 78 **18.** −32 + (−40)
**19.** −4 + (−35) + 26 **20.** 16 + (−39) + 21
**21.** −56 + (−56) + (−13) **22.** 17 + (−73) + 68
**23.** −34 + 43 + (−43) **24.** −8 + (−120) + 30
**25.** $-8 + |7|$ **26.** $|-8| + 7$
**27.** $|-9| + |-4|$ **28.** $|-13| + |12|$
**29.** $|-43| + |-18|$ **30.** $-5 + |-15| + |7|$
**31.** $-600 + |-25| + |-30|$ **32.** $78 + (-43) + |-5|$

Extra Practice

## LESSON 2.3

**Find each difference.**

1. $7 - 11$
2. $8 - (-7)$
3. $6 - 15$
4. $-5 - 9$
5. $16 - 20$
6. $-7 - (-12)$
7. $30 - (-30)$
8. $-17 - (-18)$
9. $-52 - 19$
10. $-24 - (-24)$
11. $16 - (-90)$
12. $-34 - 43$
13. $-52 - 17$
14. $72 - 75$
15. $-68 - (-10)$
16. $-5 - (-5) - 10$
17. $-3 - 5 - 16$
18. $12 - 18 - 7$
19. $-27 - (-15) - 2$
20. $-100 - 75 - (-75)$
21. $-36 - 21 - 16$
22. $20 - 60 - 32$
23. $42 - (-72) - 6$
24. $-83 - 14 - (-36)$
25. $32 - 41 - 51$
26. $12 - (-17) - 36$
27. $-45 - 61 - (-27)$
28. $-7 - 24 - 2$
29. $26 - 81 - 6$
30. $95 - 52 - (-6)$
31. $14 - 86 - 15$
32. $-9 - (-27) - 56$
33. $-38 - 74 - 109$

**Find the distance between each pair of points on a number line.**

34. 5, −2
35. −7, 13
36. 8, 92
37. −4, −7
38. −12, −32
39. −6, 14
40. 12, 43
41. 14, −14
42. 15, −12
43. −13, −19
44. −63, −162
45. 19, −44

## LESSON 2.4

**Evaluate.**

1. $(3)(-4)$
2. $(5)(-6)$
3. $(-4)(1)$
4. $(-3)(9)$
5. $(-2)(-4)$
6. $(7)(-6)$
7. $(16) \div (-4)$
8. $(-25) \div (-5)$
9. $\frac{-36}{9}$
10. $(-100) \div (-10)$
11. $26 + (-40)$
12. $\frac{36}{-3}$
13. $\frac{-56}{-2}$
14. $(-30) \div (3)$
15. $(-5)(14)$
16. $(-12)(-30)$
17. $(45)(-5)$
18. $(-24) \div (8)$
19. $(-35) \div (-7)$
20. $(-2)(-1)(6)$
21. $(-5)(-3)(7)$
22. $(-2)[(-3) + (-2)]$
23. $(-4)[(-1) - (-4)]$
24. $(-6)[7 + (-5)]$
25. $\frac{-36}{(-12) - (-3)}$
26. $(-12) \div [4 - 6]$
27. $(7)(-2) \div (-7)$

## LESSON 2.5

**Name the opposite of each expression.**

**1.** 16 **2.** −28 **3.** $5z$ **4.** −7

**5.** $-5r$ **6.** $3a + 2b$ **7.** $a - 7$ **8.** $-8w + 4$

**9.** $5x - 7y$ **10.** $-s - 6$ **11.** $8a - c$ **12.** $9 - 3x$

**13.** $-10m + 4n$ **14.** $-y - 6z$ **15.** $-15 - v$ **16.** $x + y + z$

**17.** $-2c + 4d - 5$ **18.** $t - w - 5$ **19.** $8a + 4b - c$ **20.** $5r - s + 3t$

**Use the Associative, Commutative, and Distributive Properties to find each sum or product. Show your work and explain each step.**

**21.** $(56 + 48) + 22$ **22.** $(25 \cdot 3) \cdot 4$ **23.** $5 \cdot (6 \cdot 12)$

**24.** $(137 + 149) + 113$ **25.** $5 \cdot (138 \cdot 2)$ **26.** $4 \cdot (76 \cdot 25)$

**27.** $(157 + 29) + 43$ **28.** $40 \cdot (36 \cdot 50)$ **29.** $(167 \cdot 345) \cdot 0$

**30.** $16 + (24 + 608)$ **31.** $6(5 + 20)$ **32.** $25(3 + 4)$

**Name the property illustrated. Be specific.**

**33.** $54 + 15 = 15 + 54$ **34.** $6 \cdot (3 \cdot 14) = (6 \cdot 3) \cdot 14$

**35.** $2(2.5 + 5.9) = 2 \cdot 2.5 + 2 \cdot 5.9$ **36.** $8 \cdot 40 - 8 \cdot 25 = 8(40 - 25)$

**37.** $(6 + 34) + 17 = 6 + (34 + 17)$ **38.** $18 \cdot 3 = 3 \cdot 18$

**39.** $7(5x) = (7 \cdot 5)x$ **40.** $7(27.2 - 6.5) = 7 \cdot 27.2 - 7 \cdot 6.5$

**41.** $(6 \cdot 9) \cdot 5 = 6 \cdot (9 \cdot 5)$ **42.** $7 + (18 + 26) = (7 + 18) + 26$

## LESSON 2.6

**Use the Distributive Property to show that each statement is true.**

**1.** $12x + 3x = 15x$ **2.** $-2x - 7x = -9x$ **3.** $-5x + 4x = -x$

**4.** $65x + 13x = 78x$ **5.** $7x - 12x = -5x$ **6.** $73x - 100x = -27x$

**Simplify the following expressions.**

**7.** $(6a + 1) + (2a - 3)$ **8.** $(4x - 4) + (3x - 6)$ **9.** $(2c - 5) + (3c + 2)$

**10.** $(4y + 2) + (3 - y)$ **11.** $(3z - 4) + (7 - 2z)$ **12.** $(2w + 1) + (4w - 9)$

**13.** $(1 + 4r) + (-5r - 3)$ **14.** $9r - 4r$ **15.** $6a - 5a$

**16.** $8c - (3c + 2)$ **17.** $9w - (5w + 8)$ **18.** $6z - (5 - 7z)$

**19.** $7 - (3y - 4y)$ **20.** $16a - (5a - 3)$ **21.** $(5r - s) - (3t + s)$

**22.** $(2.2a + 1.1b) + (3a + 0.9b)$ **23.** $(3.5x + 4.2y) + (0.6x + y)$

**24.** $(4a + 2) + (3a - 6)$ **25.** $(6w - 3) + (3w + 2)$

**26.** $(7 - 3w) + (6 - 8w)$ **27.** $(9q + 3) - (5 - 5q)$

**28.** $(10q + 8q) - (4m + 2m)$ **29.** $(3y + 1) - (4y - 1)$

**30.** $(3v - 7t) - (7v + 2t)$ **31.** $(2h + 7j) - (-4h - 5j)$

**32.** $(3d + 1) - (4d + 3) + (5d - 6)$ **33.** $(5m - 4) - (4m + 1) + (8 - m)$

**34.** $(3x - 5) + (2x + 3y - 5z)$ **35.** $(3a + b) + (2b + c) + (a + b + c)$

**36.** $(3 - 4t + 8r) - (7r + t) - (6 + t)$ **37.** $(5 + p) - (p + q) - (3p + 4q - 6)$

## LESSON 2.7

**Simplify the following expressions. Use the Distributive Property if needed.**

1. $3 \cdot 4m$
2. $-7a \cdot 5$
3. $2x \cdot 5x$
4. $-12y \cdot 4y$
5. $-44x \cdot 10x$
6. $-4(3b + 5)$
7. $7x(3 + 2x)$
8. $2.5c \cdot 5c$
9. $\frac{14x}{2}$
10. $\frac{-64x}{4}$
11. $\frac{-48x}{-3}$
12. $-2(11b - 3)$
13. $5n - (3 - 4n)$
14. $4f \cdot 2 + 3f \cdot (-2)$
15. $5(h - 2) - (3 - 2h)$
16. $(4r - 3) - 9(r + 3)$
17. $\frac{-18 - 36e}{-9}$
18. $8(2y + 3z) - 5(7y - 7z)$
19. $(-4t - 2e) + 7(-t - e)$
20. $9(2x + 3y) - 6(4x - 2y)$
21. $\frac{7 - 21g}{7}$
22. $-(y - x) - 4(5y - 2x)$
23. $\frac{4x + 20}{4}$
24. $-4(2q - 3p) - 5(3q - 2p)$
25. $\frac{-6 + 24u}{-3}$
26. $3(z - 4w) - (5z - 5w)$
27. $\frac{-15x + 10}{5}$
28. $4c(6a + 1)$
29. $8e(3x - 6)$
30. $\frac{6a(5b + 7)}{3a}$
31. $\frac{-5t(t + r)}{t}$
32. $7a(4x - 4)$
33. $15c(2c - 5)$
34. $10q(16x + 11)$
35. $\frac{2y^2(y + 1)}{2y}$
36. $(6c - 5) + 8(3c + 4)$

# CHAPTER 3

## LESSON 3.1

**Solve each equation. Check the solution.**

1. $x + 2 = 1$
2. $t + 6 = 2$
3. $w - 4 = -3$
4. $5 = x + 1$
5. $y + 3 = -3$
6. $y + 3 = -4$
7. $h - 5 = 6$
8. $x - 9 = -9$
9. $8 = 6 + z$
10. $-7 = 3 + g$
11. $t + 5 = 11$
12. $16 = f - 2$
13. $-8 = 9 + n$
14. $d - 12 = 7$
15. $6 = -9 + c$
16. $(-9) + r = 17$
17. $b + 10 = 2$
18. $8 = h - 3$
19. $x - 3 = 5$
20. $d - 8 = -10$
21. $-6 = 11 + j$
22. $h - 14 = -10$
23. $(-2) + w = -9$
24. $14 = d + (-2)$
25. $x + 9 = -1$
26. $t - 6 = 6$
27. $w - 4 = -2$
28. $h - 5 = 7$
29. $5 = y + 1$
30. $y + 3 = -8$
31. $x - 15 = 20$
32. $r + 23 = 10$
33. $n - 65 = 59$
34. $76 + k = 12$
35. $w - 52 = -45$
36. $y + 19 = -10$
37. $l + 48 = 50$
38. $b - 350 = -80$
39. $x + 28 = -20$
40. $450 = 100 + f$
41. $88 + u = 44$
42. $498 + x = 500$
43. $q - 90 = 19$
44. $10 + d = 72$
45. $s - 25 = 36$
46. $54 = 112 + d$
47. $p + 365 = 425$
48. $b + 350 = -80$

## LESSON 3.2

**Write the property needed to solve the equation. Then solve it.**

**1.** $\frac{x}{2} = -12$
**2.** $\frac{b}{-10} = -4$
**3.** $\frac{x}{2} = 2$
**4.** $5x = 4$
**5.** $-15y = 5$
**6.** $\frac{c}{7} = -70$
**7.** $42x = 6$
**8.** $5x = 3$
**9.** $-12y = 144$
**10.** $8 = -64z$
**11.** $\frac{a}{-5} = -3.5$
**12.** $7 = 4x$

**Solve each equation.**

**13.** $44x = 55$
**14.** $k - 1.5 = 7.5$
**15.** $\frac{r}{3} = -18$
**16.** $4t = -36$
**17.** $\frac{c}{5} = 12$
**18.** $7n = 15$
**19.** $\frac{x}{2} = 17$
**20.** $-3c = 18$
**21.** $45d = 0$
**22.** $9p = -72$
**23.** $\frac{h}{-1} = -99$
**24.** $3500t = 4000$
**25.** $\frac{m}{0.6} = 7$
**26.** $0.25 = 0.25f$
**27.** $-2y = -35.24$
**28.** $0.75v = 75$
**29.** $-50 = -2z$
**30.** $\frac{j}{-19} = 30$
**31.** $12m = 20$
**32.** $\frac{r}{32} = 1$
**33.** $2.2 = 1.1w$
**34.** $\frac{65}{x} = 13$
**35.** $-6v = -582$
**36.** $56x = -1$

## LESSON 3.3

**Solve each equation and check your solution.**

**1.** $7y - 1 = 27$
**2.** $4a - 2 = 30$
**3.** $8y - 5 = 19$
**4.** $6 + 4w = -26$
**5.** $3d + 2 = 23$
**6.** $4x + \frac{1}{2} = \frac{1}{4}$
**7.** $\frac{6c}{5} + 8 = -2$
**8.** $7 + 7a = 28$
**9.** $36 = 3m + 6$
**10.** $-2 - 14z = -30$
**11.** $8 - 2n = -4$
**12.** $-10 - 5x = 25$
**13.** $15 = 2p - 5$
**14.** $-5w + 3 = -72$
**15.** $50 = -4g + 2$
**16.** $58 = -8s + 18$
**17.** $-51 = 8x + 5$
**18.** $\frac{151 + r}{3} = 80$
**19.** $3j + \frac{2}{5} = \frac{1}{3}$
**20.** $\frac{3}{4} + 5h = \frac{1}{8}$
**21.** $2b - \frac{1}{9} = \frac{1}{3}$
**22.** $3k - \frac{1}{3} = \frac{4}{9}$
**23.** $\frac{350}{x + 10} = 5$
**24.** $6g - \frac{7}{10} = -\frac{1}{5}$

## LESSON 3.4

**Solve each equation and check your solution.**

1. $7x = 5x + 6$
2. $b + 3 = 9 - b$
3. $8m - 14 = 6m$
4. $8g + 1 = 4g + 9$
5. $4c - 5 = 2c - 9$
6. $2 - 4h = 3h - 33$
7. $24 - 6y = 2y - 40$
8. $15 + 2k = 5k + 6$
9. $6s - 5 = 19 - 2s$
10. $6y - 2 = 4y + 5$
11. $3r - 8 = 5r - 20$
12. $1 - 3x = 2x + 8$
13. $8 - 2b = 7b - 4$
14. $0.75n + 100 = 0.25n - 1$
15. $2.6x - 3.9 = 4.5 + 1.2x$
16. $8w + 20 = 12w - 9$
17. $\frac{3a}{5} = 2a + 1$
18. $4x + 9 = 2x + 6$
19. $2c + 5 = \frac{c}{5} - 4$
20. $t + \frac{1}{2} = \frac{t}{4} + 2$

## LESSON 3.5

**Solve each equation and check your solution.**

1. $2(g - 4) = 6$
2. $5(r + 3) = 35$
3. $4(z - 20) = 16$
4. $2(5 - x) = -25$
5. $2(3 - 4d) = -42$
6. $2(4t + 1) = 74$
7. $14 = 2(f + 3)$
8. $3(y - \frac{2}{3}) = \frac{1}{4}$
9. $3r + 2(r - 4) = 12$
10. $7k - 2(k + 6) = -2$
11. $3v + 5 + 2v = 5(2 - v)$
12. $3(2b + 6) = 5b + 20$
13. $3x - (5 - 2x) = x - 21$
14. $4(s + \frac{1}{2}) = 8(s + \frac{3}{4})$
15. $2(u - 3) = 3(u - 4)$
16. $5(x - 2) + 8x - 14 = 9x - 2$
17. $2(4m + 1) = 3(2m + 4)$
18. $14(n + 7) = 3(n - 4)$
19. $p - \frac{1}{4} = \frac{p}{3} + 7\frac{3}{4}$
20. $\frac{x}{3} + \frac{4}{5} = 2x - \frac{5}{6}$

## LESSON 3.6

**Solve each equation for the indicated variable.**

1. $P = s - c$, for $s$
2. $h = R - k$, for $k$
3. $a + b = c$, for $a$
4. $I = C + T$, for $C$
5. $P = 4s$, for $s$
6. $f = \frac{i}{12}$, for $i$
7. $d = rt$, for $t$
8. $a - b = -c$, for $a$
9. $s = \frac{m}{h}$, for $m$
10. $V = lwh$, for $h$
11. $I = prt$, for $r$
12. $A = \frac{bh}{2}$, for $b$
13. $x + 4y = r$, for $y$
14. $a = \frac{m}{g}$, for $m$
15. $P = 15 + \frac{1}{2}D$, for $D$
16. $P = a + b + c$, for $a$
17. $d + e + f = 180$, for $f$
18. $C = 2\pi r$, for $r$
19. $S = (n - 2)180$, for $n$
20. $3x + g + 2h = t - y$, for $h$

# CHAPTER 4

## LESSON 4.1

**Determine whether each proportion is true.**

1. $\frac{36}{7} = \frac{108}{21}$
2. $\frac{8}{15} = \frac{15}{20}$
3. $\frac{64}{72} = \frac{8}{9}$
4. $\frac{12}{7} = \frac{60}{35}$
5. $\frac{14}{18} = \frac{9}{20}$
6. $\frac{4}{5} = \frac{28}{35}$
7. $\frac{13}{26} = \frac{25}{50}$
8. $\frac{9}{30} = \frac{10}{40}$
9. $\frac{5}{6} = \frac{25}{36}$
10. $\frac{15}{16} = \frac{30}{32}$
11. $\frac{7}{9} = \frac{18}{14}$
12. $\frac{9}{15} = \frac{15}{25}$

**Solve each proportion. Round your answers to the nearest hundredth if necessary.**

13. $\frac{16}{18} = \frac{48}{x}$
14. $\frac{28}{15} = \frac{n}{20}$
15. $\frac{32}{27} = \frac{24}{r}$
16. $\frac{m}{42} = \frac{60}{84}$
17. $\frac{6}{9} = \frac{30}{a}$
18. $\frac{14}{8} = \frac{21}{f}$
19. $\frac{c}{11} = \frac{22}{10}$
20. $\frac{15}{21} = \frac{h}{6}$
21. $\frac{44}{7} = \frac{11}{t}$
22. $\frac{9.1}{3.5} = \frac{g}{5}$
23. $\frac{86}{4} = \frac{r}{3}$
24. $\frac{9}{25} = \frac{3}{w}$
25. $\frac{16}{b} = \frac{25}{36}$
26. $\frac{68}{12} = \frac{4}{r}$
27. $\frac{a}{7} = \frac{7}{15}$
28. $\frac{26}{13} = \frac{13}{b}$
29. $\frac{1.8}{0.9} = \frac{2}{x}$
30. $\frac{45}{y} = \frac{9}{5}$
31. $\frac{q}{4.2} = \frac{0.8}{0.6}$
32. $\frac{14}{b} = \frac{21}{3}$
33. $\frac{20.5}{x} = \frac{82}{46}$
34. $\frac{y}{21.4} = \frac{8}{53.5}$
35. $\frac{50.8}{124} = \frac{z}{10}$
36. $\frac{m}{75} = \frac{27}{40}$
37. $\frac{f}{6} = \frac{2}{2.4}$
38. $\frac{4.2}{5.6} = \frac{7}{x}$
39. $\frac{4}{7} = \frac{11}{b}$
40. $\frac{9}{100} = \frac{10}{m}$

## LESSON 4.2

**Write each percent as a decimal.**

**1.** 65%
**2.** 5.5%
**3.** 9%
**4.** 0.6%
**5.** 2.4%
**6.** 400%
**7.** 225%
**8.** 0.267%

**Write each decimal as a percent.**

**9.** 0.87
**10.** 0.45
**11.** 0.06
**12.** 1.28
**13.** 0.0256
**14.** 5.08
**15.** 0.073
**16.** 0.00098

**Write each percent as a fraction or mixed number in lowest terms.**

**17.** 25%
**18.** 45%
**19.** 6.5%
**20.** 0.02%
**21.** 350%
**22.** 9%
**23.** 44%
**24.** 16%

**Find each answer.**

**25.** What number is 30% of 60?
**26.** 130% of what number is 78?
**27.** What percent of 70 is 35?
**28.** 96 is 80% of what number?
**29.** What number is 20% of 75?
**30.** 45% of what number is 15?
**31.** What percent of 99 is 18?
**32.** 57 is 79% of what number?
**33.** What number is 22% of 1000?
**34.** 250% of what number is 80?
**35.** What percent of 18 is 24?
**36.** 29 is 2% of what number?
**37.** What number is 2% of 25?
**38.** 44% of what number is 12?
**39.** What percent of 50 is 30?
**40.** 65 is 37% of what number?
**41.** What number is 290% of 11?
**42.** 180% of what number is 115?
**43.** What percent of 20.8 is 1.5?
**44.** 7.4 is 30% of what number?

## LESSON 4.3

The results of spinning a spinner with regions I, II, III, and IV 100 times are shown in the table below.

| Region | I | II | III | IV |
|---|---|---|---|---|
| Frequency | 34 | 29 | 22 | 15 |

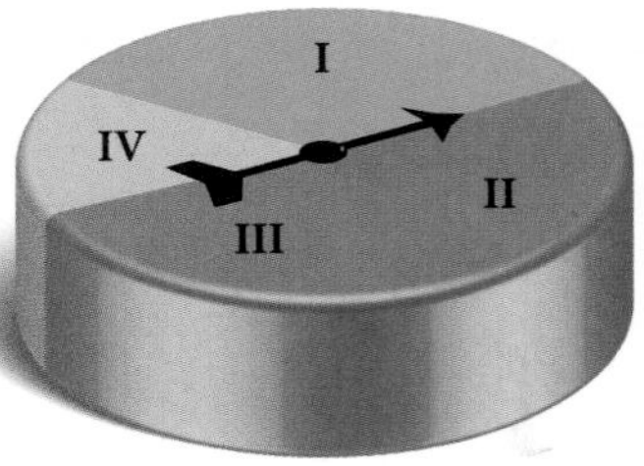


**1.** Is it equally likely for the spinner to land in any region? Explain.
**2.** Find the experimental probability of landing in each region.
**3.** Find the sum of the experimental probabilities of landing in Regions I and IV.
**4.** Find the sum of the experimental probabilities for landing in Regions II and III.
**5.** Are these sums you calculated in Exercises 3 and 4 almost the same? Why or why not?
**6.** Find the sum of the experimental probabilities for landing in Regions I and II.
**7.** Find the sum of the experimental probabilities for landing in Regions III and IV.
**8.** Are these sums you calculated in Exercises 6 and 7 almost the same? Why or why not?

## LESSON 4.4

**Find the mean, median, mode(s), and range for each set of data.**

1. 12, 12, 18, 14, 8, 9, 10, 20, 19
2. 8, 11, 15, 25, 35, 62, 20, 40
3. 4, 2, 5, 1, 1, 9, 8, 4, 6, 3, 2
4. 150, 320, 200, 41, 700, 210, 300
5. 130, 135, 132, 120, 145, 136
6. 30, 35, 200, 42, 95, 100, 300, 25
7. 60, 54, 45, 72, 83, 64, 51, 75, 77, 59, 48, 61
8. 55, 26, 34, 21, 37, 48, 34, 27, 19, 51, 27, 32

**The coaches at Midville Middle School kept track of the runs scored by each member of the baseball team. Use the data shown in the table below for Exercises 9–13.**

9. Make a frequency table for the data.
10. Find the median number of runs scored.
11. Find the mean of the data.
12. What is the mode(s)?
13. Which measure of central tendency would you use to give the best impression of the players? Explain.

**Runs Scored**

| | | | |
|---|---|---|---|
| 3 | 5 | 5 | 6 |
| 7 | 5 | 1 | 0 |
| 8 | 9 | 2 | 2 |
| 6 | 1 | 2 | 3 |
| 0 | 0 | 2 | 4 |
| 4 | 7 | 8 | 9 |

## LESSON 4.5

**The circle graph at right shows the percent of the population in Carsonville in certain age groups. In 1995, the total population was 150,600.**

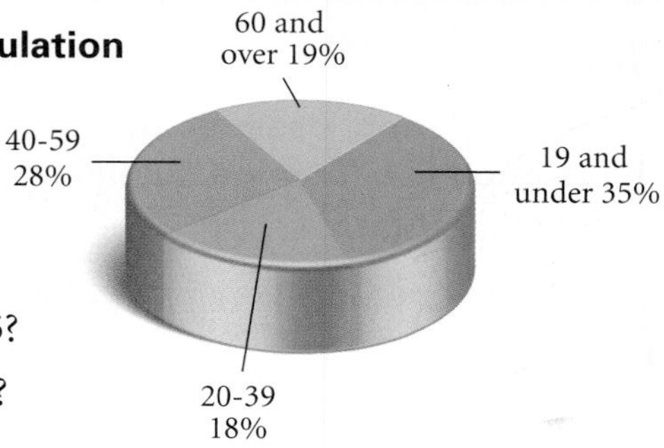


1. How many residents were 60 and over in 1995?
2. How many residents were 19 and under in 1995?
3. How many residents were 40 to 59 years of age in 1995?
4. How many residents were 20 to 59 years of age in 1995?

**The population of Carsonville is expected to grow to 190,000 by the year 2005, and the percent of the population in each age group is expected to remain the same.**

5. How many residents are expected to be 19 years of age or less in 2005?
6. How many residents are expected to be 60 years of age or more in 2005?
7. How many residents are expected to be between 20 and 39 years of age in 2005?
8. How many residents are expected to be between 40 and 59 years of age in 2005?

## LESSON 4.6

**Ms. Smith compiled the list of scores shown in the table below for a recent test given in her class.**

**1.** Make a stem-and-leaf plot for the data.

**2.** What is the median of the data?

**3.** What is the lower quartile for this data?

**4.** What is the upper quartile for this data?

**5.** Make a box-and-whisker plot for this data.

Test scores

| | | | |
|---|---|---|---|
| 80 | 90 | 92 | 68 |
| 75 | 60 | 68 | 93 |
| 95 | 69 | 75 | 76 |
| 80 | 80 | 83 | 91 |
| 67 | 87 | 77 | 97 |
| 95 | 58 | 73 | 59 |

The data shown below are the numbers of fish caught by a fishing crew on each day in April.

**6.** Make stem-and-leaf plot of the data.

**7.** Make a histogram of the data.

Number of fish caught

| | | | |
|---|---|---|---|
| 10 | 25 | 21 | 21 |
| 12 | 2 | 24 | 10 |
| 20 | 10 | 23 | 18 |
| 8 | 27 | 10 | 19 |
| 0 | 18 | 9 | 22 |
| 17 | 22 | 23 | 22 |

# CHAPTER 5

## LESSON 5.1

**For each relation, describe the domain and range. Determine whether the relation is a function. Explain your reasoning.**

**1.** $\{(1, 2), (2, 4), (3, 6), (4, 8)\}$

**2.** $\{(-1, 5), (3, 6), (2, 3), (-1, 8)\}$

**3.** $\{(200, 5), (150, 4), (100, 3), (-200, 5)\}$

**4.** $\{(1.5, 2), (3.2, 5), (1.2, 4), (2.3, 7.1)\}$

**5.** $\{(-5, 6), (-2, 7), (1, 8), (4, 9), (-5, 10)\}$

**6.** $\{(25, 1), (32, 8), (25, 2), (46, 5), (31, -2)\}$

**7.** $\{(2, 20), (3, 25), (4, 30), (5, 20)\}$

**8.** $\{(-2, 12), (5, 7.8), (-1, 12), (-2, 10), (6, 12)\}$

**Complete each ordered pair so that it is a solution to $3x - 2y = 16$.**

**9.** (1, ?)

**10.** (−2, ?)

**11.** (?, 1)

**12.** (?, 3)

**13.** (4, ?)

**14.** (?, −2)

**15.** (?, 4)

**16.** (2, ?)

**17.** (−12, ?)

**18.** (?, −10)

**19.** (7, ?)

**20.** (?, −8)

**21.** (10, ?)

**22.** (−6, ?)

**23.** (?, 5)

**24.** (?, −5)

## LESSON 5.2

**Find the slope for each rise and run.**

**1.** rise 4, run 2
**2.** rise 1, run 8
**3.** rise 4, run 1
**4.** rise –6, run 3
**5.** rise 0, run 10
**6.** rise –9, run –5
**7.** rise 4, run –5
**8.** rise 2, run 16
**9.** rise 18, run 8

**Find the slope of each line graphed below.**

**10.**

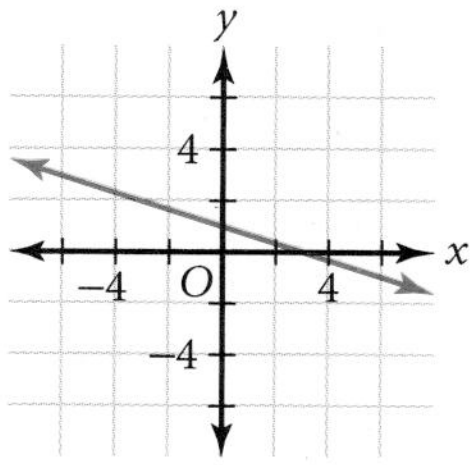


**11.**

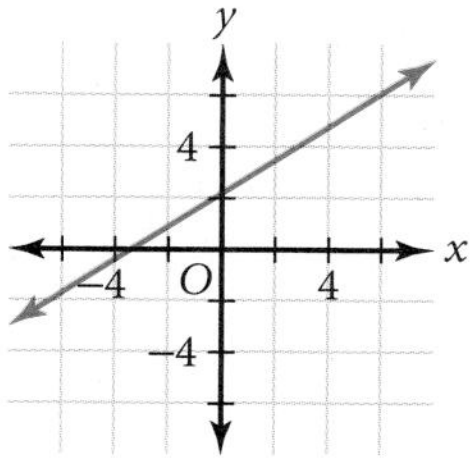


**12.**

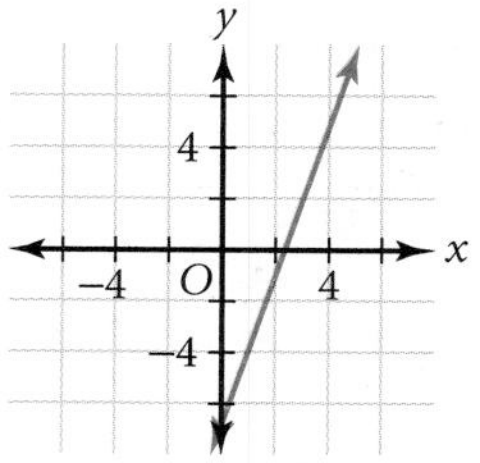


**Find the slope of the line that contains each pair of points.**

**13.** $M(5, 3)$, $N(3, 4)$
**14.** $M(1, 7)$, $N(4, 1)$
**15.** $M(-3, 2)$, $N(1, -4)$
**16.** $M(-5, 3)$, $N(0, 6)$
**17.** $M(1, -7)$, $N(4, -1)$
**18.** $M(-5, 9)$, $N(3, -3)$
**19.** $M(6, 7)$, $N(3, 4)$
**20.** $M(9, -4)$, $N(3, 2)$
**21.** $M(-3, -1)$, $N(6, -4)$

## LESSON 5.3

**The graph at the right shows cost of renting in-line skates from each of two different places.**

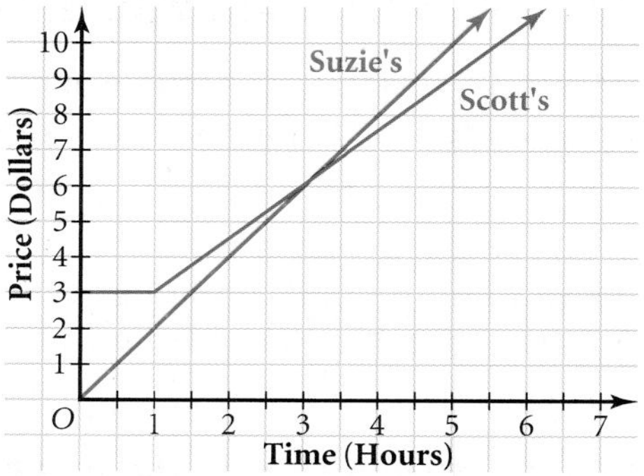


**1.** Suzie's Skate Land requires no deposit. Find the rate of change, or cost per hour.

**2.** Scott's Skates requires a non-refundable deposit, and the first hour is free. What is the amount of the deposit?

**3.** Maria wants to rent roller blades for 4 hours. At which shop will the rental be less expensive?

**4.** For what rental time will the skate rentals cost the same?

**In Exercises 5–19, *y* varies directly as *x*. Find the constant of variation, and write an equation for the direct variation.**

**5.** $y = 10$ when $x = 5$
**6.** $y = 1.5$ when $x = 25$
**7.** $y = 6$ when $x = 2$
**8.** $y = 11$ when $x = 8$
**9.** $y = 4$ when $x = 3$
**10.** $y = 1.7$ when $x = 2.1$
**11.** $y = 14$ when $x = 16$
**12.** $y = 2$ when $x = 12$
**13.** $y = 9$ when $x = 18$
**14.** $y = 23$ when $x = 42$
**15.** $y = 9$ when $x = 6$
**16.** $y = 3.6$ when $x = 24$
**17.** $y = 4.1$ when $x = 0.84$
**18.** $y = 22$ when $x = 16$
**19.** $y = 7$ when $x = 11$

## LESSON 5.4

**Identify the slope and the *x*- and *y*-intercepts.**

**1.** $y = 3x + 1$ **2.** $y = 2x - 1$ **3.** $y = 4x$ **4.** $y = -6x + 8$

**5.** $y = 5x + 4$ **6.** $y = x - 7$ **7.** $y = -11x + 10$ **8.** $y = 13x - 5$

**9.** $y = x + 9$ **10.** $y = 3x - 6$ **11.** $y = 7x - 12$ **12.** $y = x$

**Write an equation in slope-intercept form for the line that contains each pair of points.**

**13.** $M(3, 5), N(-3, 1)$ **14.** $M(1, 0), N(2, 1)$ **15.** $M(-1, 3), N(1, -1)$

**16.** $M(3, 1), N(-3, 3)$ **17.** $M(-2, 0), N(2, 4)$ **18.** $M(0, 1), N(1, 3)$

**19.** $M(2, 1), N(-2, -3)$ **20.** $M(-1, 4), N(1, -2)$ **21.** $M(4, 3), N(8, 4)$

**22.** $M(-5, 1), N(5, -3)$ **23.** $M(-1, -3), N(1, -2)$ **24.** $M(1, 2), N(2, 4)$

**Write an equation in slope-intercept form for each line graphed below.**

**25.**

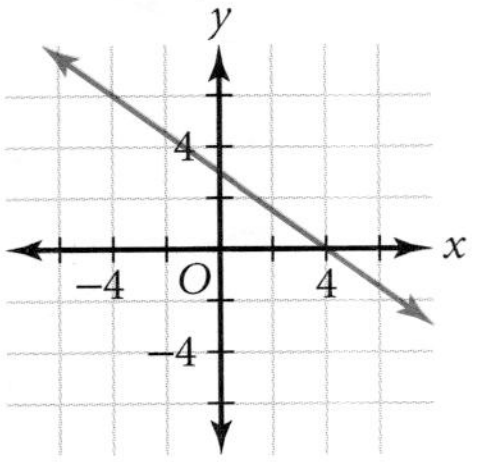


**26.**

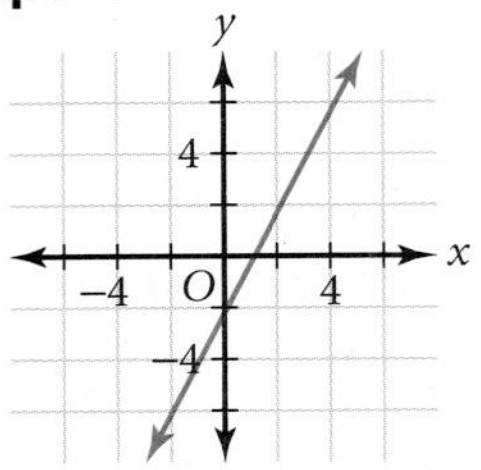


**27**

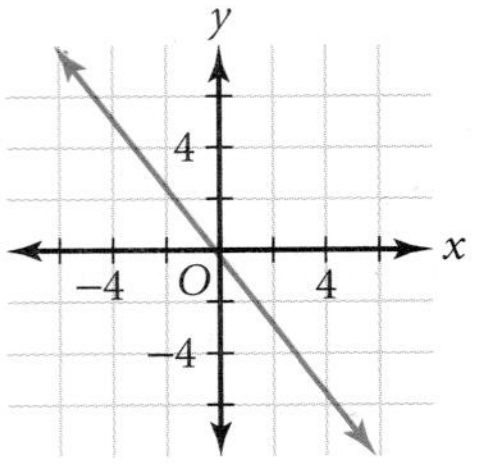


## LESSON 5.5

**Write each equation in standard form.**

**1.** $3x = 6y + 18$ **2.** $5y = -4x - 20$ **3.** $-2x + 5y = 1$

**4.** $y = -7x + 2$ **5.** $5x = y + 4$ **6.** $6 = x - y$

**7.** $12 = -3x + 7y$ **8.** $2x - 4 = 3y + 6$ **9.** $4x = 8y$

**10.** $21x = 7y + 13$ **11.** $3y - 2x = 1$ **12.** $15y + x + 17 = 0$

**13.** $\frac{1}{2}x + \frac{1}{4}y = 1$ **14.** $\frac{1}{3}y = 2x + \frac{2}{3}$ **15.** $y = 0.5x + 6$

**Write an equation in point-slope form for the line that has the given slope and that contains the given point.**

**16.** slope $-1$, $(2, 3)$ **17.** slope 2, $(4, 5)$ **18.** slope 0, $(4, 5)$

**Find the *x*- and *y*-intercepts for the graph of each equation.**

**19.** $x + y = 2$ **20.** $x - y = 4$ **21.** $x + 2y = 8$ **22.** $3x + y = 9$

**23.** $4x - 5y = 20$ **24.** $2x - 7y = 14$ **25.** $9x + y = 18$ **26.** $4x - 6y = 12$

**27.** $x = 3y$ **28.** $x = -5y + 1$ **29.** $x = 2y$ **30.** $9x + y = 4$

**Write an equation for the line that contains each pair of points.**

**31.** $(2, 1), (-2, -3)$ **32.** $(-1, 4), (1, -2)$ **33.** $(4, 3), (8, 4)$

**Write an equation in slope-intercept form for each line described below.**

**34.** containing $(4, 8)$ and $(-2, -1)$ **35.** containing $(2, 3)$ and with slope $-1$

**36.** containing $(4, 5)$ and with slope 2 **37.** containing $(-1, -5)$ and with slope 0

**38.** crossing the $x$-axis at $x = -5$ and the $y$-axis at $y = 0$

**39.** crossing the $x$-axis at $x = 2$ and the $y$-axis at $y = 2$

LESSON 5.6

**Identify the slope of a line that is parallel to each line.**

**1.** $y = 4x + 10$
**2.** $3x + y = 7$
**3.** $10 = -5x + 2y$
**4.** $4x - 3y = 12$
**5.** $y = x - 3$
**6.** $4x - y = 16$
**7.** $2x + 6y = 6$
**8.** $7x - y = 7$
**9.** $4x = 8y + 1$
**10.** $2x = 7y + 10$
**11.** $9y - x = 1$
**12.** $5y + 2x = 7$

**Write the slope of a line that is perpendicular to each line.**

**13.** $y = 4x + 10$
**14.** $y = 2x + 11$
**15.** $y = 5x + 6$
**16.** $2x + 3y = 2$
**17.** $x - 7y = 4$
**18.** $3x + 2y = 9$
**19.** $3x + 4y = -9$
**20.** $4x - 8y = 24$
**21.** $2x - y = 1$
**22.** $9x + 6y = 18$
**23.** $4x - 2y = 12$
**24.** $y = x$

**Tell whether each pair of lines are parallel, perpendicular, or neither.**

**25.** $y = 0.5x$
$y = -2x$

**26.** $y = 3x - 2$
$y = -3x + 4$

**27.** $x - 3y = 4$
$3x + y = 1$

**Write an equation in slope-intercept form for the line specified.**

| Contains | Is parallel to | Contains | Is perpendicular to |
|---|---|---|---|
| **28.** $(2, -4)$ | $3x - y = 5$ | **29.** $(2, -4)$ | $3x - y = 5$ |
| **30.** $(-1, 4)$ | $y = 2x + 5$ | **31.** $(-1, 4)$ | $y = 2x + 5$ |
| **32.** $(-3, -1)$ | $2y = x + 2$ | **33.** $(-3, -1)$ | $2y = x + 2$ |
| **34.** $(0, 5)$ | $y = -2x - 1$ | **35.** $(0, 5)$ | $y = -2x - 1$ |

# CHAPTER 6

LESSON 6.1

**Solve each inequality and graph the solution on a number line.**

**1.** $x - 1 < 2$
**2.** $y + 5 > 1$
**3.** $w - 9 \leq -1$
**4.** $p + 6 \geq -3$
**5.** $3 + m > 2$
**6.** $k - 3 \leq 4$

**Solve each inequality.**

**7.** $x + 9 \leq -4$
**8.** $11s - 7 < 3$
**9.** $b + \frac{1}{4} \geq 1$
**10.** $3p + 0.9 < 3$
**11.** $t - 6.25 > 2.5$
**12.** $r + 0.05 < 0.01$
**13.** $w - \frac{2}{3} > 2$
**14.** $h + 10 \leq 90$
**15.** $15 \leq d - 1.25$
**16.** $0.85 < -2.65 + n$
**17.** $\frac{5}{8} \leq \frac{1}{4} + t$
**18.** $-5.7 + v > 3.6$
**19.** $c - \frac{1}{4} \leq \frac{1}{3}$
**20.** $z + 5 \geq 15$
**21.** $4 \leq 17 + x$
**22.** $q + 0.5 > 9$
**23.** $7 \leq r - 4$
**24.** $d - 2\frac{1}{3} > -6$
**25.** $8 > m - 13$
**26.** $t + 1.2 < 32.5$
**27.** $10.5 + f \leq 1.5$
**28.** $h + 6.2 \leq 0.7$
**29.** $k + 11.4 < 0$
**30.** $4.2 - n < -7$

## LESSON 6.2

**Solve each inequality.**

**1.** $2x + 6 \geq 12$
**2.** $\frac{t}{2} \geq -36$
**3.** $5y - 8 \leq 17$
**4.** $9 - y > 14$
**5.** $17 - h > 5$
**6.** $\frac{4c}{2} \geq 7$
**7.** $\frac{c}{5} + 8 < -2$
**8.** $-3 - x < -9$
**9.** $2x - 15 \geq -14$
**10.** $-3p + 7 < 2$
**11.** $\frac{5}{m} \geq 2$
**12.** $8n < 7n - 4$
**13.** $2 < 3(2b - 4)$
**14.** $-w + 3 \leq -32$
**15.** $\frac{w}{-8} \geq 4$
**16.** $2n + 7 \geq 16 + n$
**17.** $\frac{9}{n} > -5$
**18.** $-3p > -21$
**19.** $6t - 12 < -9$
**20.** $12y + 2 \geq 18$
**21.** $7 - \frac{m}{6} \leq 1$
**22.** $\frac{r}{5} - 10 \geq 2$
**23.** $2z - 2 \leq 10$
**24.** $-8x < 20$

**Write an inequality to represent each statement.**

**25.** $W$ is less than $L$.
**26.** $x$ is greater than 5.
**27.** $v$ is less than or equal to 4.6.
**28.** $w$ is greater than 3.
**29.** $x$ is less than or equal to –2.
**30.** $a$ is greater than or equal to $b$.
**31.** $d$ is positive.
**32.** $r$ is not negative.

## LESSON 6.3

**Graph the solution set for each inequality.**

**1.** $-5 < x \leq 3$
**2.** $y > 2$ or $y < 0$
**3.** $4 \leq s \leq 11$
**4.** $d \geq -2$ or $d < -6$
**5.** $5 \leq r < 13$
**6.** $4\frac{1}{2} < f \leq 6$
**7.** $z > 0$ or $z < -2$
**8.** $-1.3 < g \leq 2.4$
**9.** $b \geq 5\frac{2}{3}$ or $b < 2$
**10.** $-2.7 < p < 1$
**11.** $4 \leq x < 9$
**12.** $t > 2.5$ or $t \leq 1.2$
**13.** $h > 5.3$ or $h < -2$
**14.** $5.8 < y < 9\frac{1}{2}$
**15.** $-3 \leq t < -1$
**16.** $x > 2$ or $x < 1$

**Solve and graph each inequality.**

**17.** $-15 < 3s \leq 12$
**18.** $2f - 1 > 3$ or $3f < 3$
**19.** $5 \leq 5t < 12$
**20.** $4s > 16$ or $-2s > 4$
**21.** $-11 < 2g - 5 \leq 15$
**22.** $3x > 12$ or $4x < 10$
**23.** $-12 \leq 3h - 15 < 0$
**24.** $11 < d + 3 < 14$
**25.** $3t + 2 < 17$ or $5t + 1 > 31$
**26.** $-2w + 1 > 15$ or $w + 2 \geq -3$
**27.** $-3 < 2t + 5 \leq 16$
**28.** $5 < t + 3$ or $3t < -6$
**29.** $-2 < 4b \leq 12.8$
**30.** $5\frac{1}{2} > 2f$ or $3f + 1 > 10$
**31.** $5 < d$ or $9 \geq d + 7$
**32.** $-12 < 4r + 8 \leq 12$

## LESSON 6.4

**Find the absolute value of each number.**

**1.** 16 **2.** $-24$ **3.** $-4.62$ **4.** $\frac{3}{4}$

**5.** $-1.03$ **6.** 0.25 **7.** $-1\frac{3}{10}$ **8.** 9.1

**9.** $-\frac{7}{8}$ **10.** $-144$ **11.** $-0.064$ **12.** $7\frac{2}{9}$

**Evaluate.**

**13.** $-8 + |7|$ **14.** $|-8| + 7$ **15.** $|-8| + |7|$

**16.** $|-9 - (-4)|$ **17.** $|-13 + 12|$ **18.** $|-43 + (-18)|$

**19.** $-5 + |-15 + 7|$ **20.** $-600 - |-25 + (-30)|$ **21.** $|78 + (-43) - 5|$

**Find the domain and range of each function.**

**22.** $y = |-5 + x|$ **23.** $y = |-5 - x|$ **24.** $y = |-5| - x$

**25.** $y = -|3 + x|$ **26.** $y = -3 + |x|$ **27.** $y = -5|x|$

**28.** $y = \frac{1}{4}|-x|$ **29.** $y = -5|x| + 6$ **30.** $y = -5|x + 6|$

## LESSON 6.5

**Solve each equation.**

**1.** $|x + 2| = 4$ **2.** $|x - 4| = 2$ **3.** $|x - 1| = 6$ **4.** $|x - 5| = 1$

**5.** $|x + 10| = 8$ **6.** $|7 + x| = 5$ **7.** $|10 - x| = 2$ **8.** $|-2 + x| = 2$

**9.** $|3x + 1| = 13$ **10.** $|2x - 2| = 4$ **11.** $|4x - 1| = 7$ **12.** $|3x + 6| = 15$

**13.** $|-3 - 4x| = 1$ **14.** $|3 - 7x| = 11$ **15.** $|6 + 5x| = 16$ **16.** $|-5 - 2x| = 13$

**17.** $|4x - 14| = 0$ **18.** $|9x - 9| = 9$ **19.** $|5 - 2x| = 4$ **20.** $|-5 - 6x| = 7$

**Solve each inequality.**

**21.** $|x + 2| > 1$ **22.** $|x - 1| < 2$ **23.** $|x - 7| \geq 4$ **24.** $|x + 5| \leq 4$

**25.** $|x - 3| < 5$ **26.** $|x - 3| \leq 9$ **27.** $|x - 2| \leq 0$ **28.** $|x + 6| > 2$

**29.** $|5 - 3x| \geq 9$ **30.** $|x| < 4$ **31.** $|x| \geq 4$ **32.** $|x| > 0$

**33.** $|x - 5| < 3$ **34.** $|x + 4| < 7$ **35.** $|x - 2| \geq -4$ **36.** $|2x - 8| > 6$

**37.** $|3x + 3| \leq 9$ **38.** $|4x - 8| \geq 20$ **39.** $|7x - 14| < 7$ **40.** $|x + 2| < 5$

Extra Practice

# CHAPTER 7

## LESSON 7.1

**Solve by graphing. If necessary, round answers to the nearest tenth. Check by substituting the approximate solutions into the original equations.**

**1.** $\begin{cases} 3x + 4y = 2 \\ x - y = 1 \end{cases}$

**2.** $\begin{cases} 2y = 5x - 1 \\ x = y + 2 \end{cases}$

**3.** $\begin{cases} -3y = 2x - 6 \\ x + 4y = 8 \end{cases}$

**4.** $\begin{cases} x = 300 - y \\ x - 200 = y \end{cases}$

**5.** $\begin{cases} 4x + 3y = 18 \\ 5x - 6y = 15 \end{cases}$

**6.** $\begin{cases} x = 5 \\ 3y = 2x - 1 \end{cases}$

**7.** $\begin{cases} y = 4 \\ 2x - y = 1 \end{cases}$

**8.** $\begin{cases} 4x - 2y = 8 \\ 3y = x + 9 \end{cases}$

**9.** $\begin{cases} y - 5x = 10 \\ x - 5y = 10 \end{cases}$

**10.** $\begin{cases} x = -4 \\ y = -2 \end{cases}$

**11.** $\begin{cases} 15x + y = 50 \\ y = 35x - 40 \end{cases}$

**12.** $\begin{cases} 7x - 5y = 1 \\ 3x + 3y = 2 \end{cases}$

**Determine whether the point (3, 8) is a solution to each system of equations.**

**13.** $\begin{cases} 2x + y = 14 \\ x + y = 11 \end{cases}$

**14.** $\begin{cases} y = -x - 5 \\ y = x + 5 \end{cases}$

**15.** $\begin{cases} 4x - y = -4 \\ 3x - 2y = 7 \end{cases}$

## LESSON 7.2

**Solve by substitution. Check your answers.**

**1.** $\begin{cases} 2x = y - 1 \\ y = 3x \end{cases}$

**2.** $\begin{cases} x = -y \\ 2x + y = -2 \end{cases}$

**3.** $\begin{cases} 2x + 2y = 4 \\ x = 10 - 3y \end{cases}$

**4.** $\begin{cases} y = 7 - x \\ 2x + 3y = -1 \end{cases}$

**5.** $\begin{cases} x + y = -1 \\ 2x + y = 3 \end{cases}$

**6.** $\begin{cases} 2x + 3y = -8 \\ y = 9 - x \end{cases}$

**7.** $\begin{cases} y + 3x = 1 \\ 2y + 5x = 5 \end{cases}$

**8.** $\begin{cases} 5 = 2y + x \\ x = 20y \end{cases}$

**9.** $\begin{cases} 4x + 4y = 1 \\ -x = 2y \end{cases}$

**10.** $\begin{cases} x + 4y = 5 \\ 8y - x = -8 \end{cases}$

**11.** $\begin{cases} 2y + x = 4 \\ 4y = x - 1 \end{cases}$

**12.** $\begin{cases} x = y \\ 3x + 6y = 6 \end{cases}$

**13.** $\begin{cases} x = -y \\ 16x + 4y = 9 \end{cases}$

**14.** $\begin{cases} 5x - y = 3 \\ 10x = y \end{cases}$

**15.** $\begin{cases} 5x + y = 3 \\ 2y = 5x \end{cases}$

**16.** $\begin{cases} x + y = 8 \\ 2x - y = 7 \end{cases}$

**17.** $\begin{cases} 2x + y = -5 \\ x + 2y = -2 \end{cases}$

**18.** $\begin{cases} 2x + 2y = 8 \\ y - x = 24 \end{cases}$

Extra Practice

## LESSON 7.3

**Solve each system of equations by elimination, and check your solution.**

1. $\begin{cases} x - y = 9 \\ x + y = 7 \end{cases}$

2. $\begin{cases} x + 3y = 7 \\ 2x - 3y = -4 \end{cases}$

3. $\begin{cases} 5x + 4y = 12 \\ 3x - 4y = 4 \end{cases}$

4. $\begin{cases} x + y = 1 \\ x - 2y = 2 \end{cases}$

5. $\begin{cases} 3x + 3y = 6 \\ 2x - y = 1 \end{cases}$

6. $\begin{cases} x + 8y = 3 \\ 4x - 2y = 7 \end{cases}$

7. $\begin{cases} 4x - y = 4 \\ x + 2y = 3 \end{cases}$

8. $\begin{cases} 3x - 5y = -13 \\ 4x + 3y = 2 \end{cases}$

9. $\begin{cases} 2x + 3y = 8 \\ 3x + 2y = 17 \end{cases}$

10. $\begin{cases} x - y = 1 \\ 3x - y = 3 \end{cases}$

11. $\begin{cases} y = 3x - 1 \\ 3x + 4y = 16 \end{cases}$

12. $\begin{cases} 4x + 2y = -8 \\ x = 2y - 7 \end{cases}$

13. $\begin{cases} x + y = 15 \\ \frac{1}{6}x = \frac{1}{9}y \end{cases}$

14. $\begin{cases} 3x + \frac{1}{3}y = 10 \\ 2x - 5 = \frac{1}{3}y \end{cases}$

15. $\begin{cases} 2x = y + 36 \\ 3x = \frac{1}{2}y + 26 \end{cases}$

16. $\begin{cases} y = -2x \\ 5x + 3x = 1 \end{cases}$

17. $\begin{cases} x + y = 6 \\ -2x + y = -3 \end{cases}$

18. $\begin{cases} 0.2x - 0.3y = 0 \\ 0.4x - 0.2y = 0.2 \end{cases}$

19. $\begin{cases} 3x + 7y = 5 \\ 3y = -7 - 2x \end{cases}$

20. $\begin{cases} 2x + y = 45 \\ 3x - y = 5 \end{cases}$

21. $\begin{cases} 4x + 2y = -8 \\ \frac{1}{2}x - y = -\frac{7}{2} \end{cases}$

## LESSON 7.4

**Solve each system using any method.**

1. $\begin{cases} x - 6y = 2 \\ 5x - y = 3 \end{cases}$

2. $\begin{cases} x - 5y = -42 \\ x - y = -5 \end{cases}$

3. $\begin{cases} 4x + 5y = 14 \\ -x - y = -10 \end{cases}$

4. $\begin{cases} 2y - 3 = x \\ 2x + 2y = 7 \end{cases}$

5. $\begin{cases} 3x = y - 4 \\ 2y - 6x = 12 \end{cases}$

6. $\begin{cases} 3x + y = 13 \\ 2x - y = 2 \end{cases}$

7. $\begin{cases} 2x - 3y = 5 \\ x - 5y = 0 \end{cases}$

8. $\begin{cases} y - 4x = 11 \\ 2y + x = 6 \end{cases}$

9. $\begin{cases} x + 2y = 3 \\ 5x - 3y = 2 \end{cases}$

10. $\begin{cases} y = -3x \\ x = 38 + 6y \end{cases}$

11. $\begin{cases} 2x + 4y = 8 \\ x + 2y = 4 \end{cases}$

12. $\begin{cases} 3x - 8y = 4 \\ 6x = 42 + 16y \end{cases}$

**Determine which of the following systems are dependent, independent, or inconsistent.**

13. $\begin{cases} x + 2y = 6 \\ x + 2y = 8 \end{cases}$

14. $\begin{cases} 2x + 3y = 4 \\ 8 - 6y = 4x \end{cases}$

15. $\begin{cases} y = 3x - 4 \\ 6x + 2y = -8 \end{cases}$

16. $\begin{cases} x + y = 3 \\ x + y = 4 \end{cases}$

17. $\begin{cases} 9x - 2 = 4y \\ x = 6 - y \end{cases}$

18. $\begin{cases} 3x + 6 = 7y \\ x = 11 - 2y \end{cases}$

19. $\begin{cases} 8x + 3y = 48 \\ 4x = 24 - 1.5y \end{cases}$

20. $\begin{cases} 4x - 3y = 12 \\ 12x + 9y = -36 \end{cases}$

21. $\begin{cases} 2x + y = 1 \\ x = 1 - 0.5y \end{cases}$

## LESSON 7.5

**Solve by graphing. Check your solution.**

**1.** $\begin{cases} x > 3 \\ y < 6 \end{cases}$

**2.** $\begin{cases} y > 2 \\ y > -x + 2 \end{cases}$

**3.** $\begin{cases} y \geq 2x + 1 \\ y \leq -x + 1 \end{cases}$

**4.** $\begin{cases} y \geq 3x \\ 3y \leq 5x \end{cases}$

**5.** $\begin{cases} y + 3 \geq x \\ x + y \geq -1 \end{cases}$

**6.** $\begin{cases} y > x + 1 \\ y < x + 3 \end{cases}$

**7.** $\begin{cases} y - x < 1 \\ y - x > 3 \end{cases}$

**8.** $\begin{cases} 2y + x < 4 \\ 3x - y > 6 \end{cases}$

**9.** $\begin{cases} y + 2 < x \\ 2y - 3 > 2x \end{cases}$

**Match the system of inequalities to the graph that represents the solution.**

**10.** $\begin{cases} y \leq 3x + 6 \\ y > -\frac{1}{2}x + 1 \end{cases}$

**11.** $\begin{cases} y > 3x + 6 \\ y \leq -\frac{1}{2}x + 1 \end{cases}$

**a.**

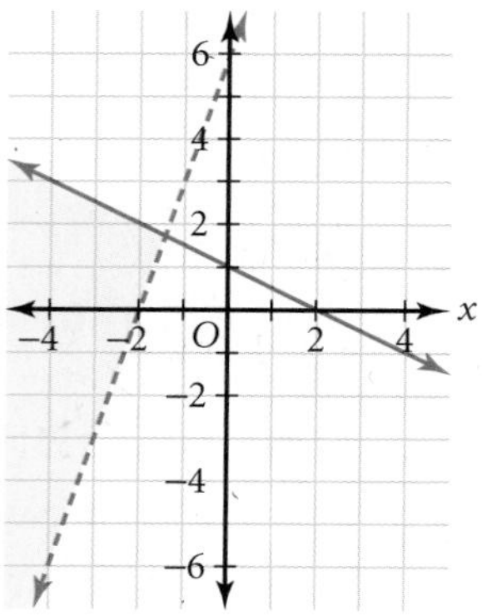


**b.**

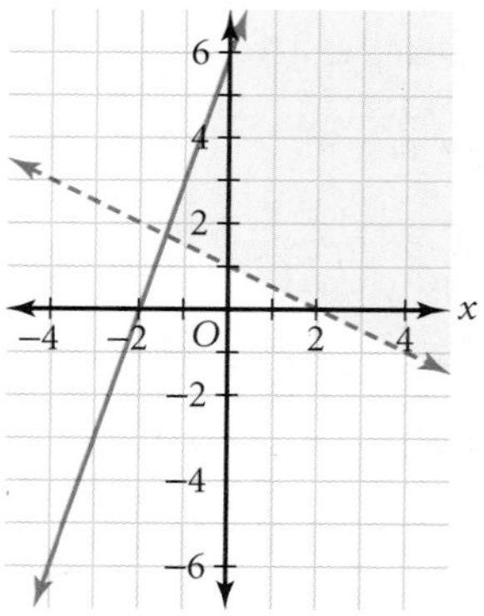


## LESSON 7.6

**Solve each problem.**

**1.** Davida has 83 coins in nickels and dimes. She has a total of \$6.95. How many of each coin does she have?

**2.** How many ounces of a 25% acid solution should be mixed with a 50% solution to produce 100 ounces of a 45% acid solution?

**3.** A plane leaves Chicago and heads for Columbus, which is 300 miles away. The plane, flying with the wind, takes 40 minutes. The plane returns to Chicago, traveling against the wind, in 45 minutes. Find the speed of the wind and the speed of the plane with no wind.

**4.** A two-digit number is equal to 7 times the units digit. If 18 is added to the number, its digits are reversed. Find the original two-digit number.

**5.** Tonia is 25 years younger than her father. The sum of their ages is 75. How old is each now?

**6.** How many ounces of a 50% saline solution should be mixed with a 10% solution to produce 100 ounces of a 30% saline solution?

**7.** Angel has 27 coins in quarters and nickels. She has a total of \$5.35. How many of each coin does she have?

# CHAPTER 8

## LESSON 8.1

**Find the value of each expression.**

**1.** $4^3$ **2.** $5^3$ **3.** $8^4$ **4.** $10^5$

**5.** $7^2$ **6.** $6^6$ **7.** $9^4$ **8.** $11^3$

**9.** $30^4$ **10.** $45^2$ **11.** $100^2$ **12.** $1^6$

**13.** $2^9$ **14.** $4^5$ **15.** $6^3$ **16.** $10^3$

**Simplify each product.**

**17.** $10^2 \cdot 10^3$ **18.** $20^2 \cdot 20^2$ **19.** $3^3 \cdot 3^4$

**20.** $4^4 \cdot 4^2$ **21.** $5^3 \cdot 5^4$ **22.** $6^2 \cdot 6^3$

**23.** $9^2 \cdot 9^2$ **24.** $12^2 \cdot 12^3$ **25.** $7^2 \cdot 7$

**26.** $(5a^2)(7ab^2)$ **27.** $(4d^3e^5)(8d^4e^6)$ **28.** $(-12s^6t)(5st^2)$

**29.** $(8q^3p^9)(-2q^5p^8)$ **30.** $(9a^3b^4)(5a^2b^2c^2)$ **31.** $(4w^4x^8y)(-4w^6xy^{10})$

**32.** $(6h^4g^9)(-12h^6g^8)$ **33.** $(8c^2d^5)(4c^2d^3e^4)$ **34.** $(-j^3k^7l^5)(-j^6k^2l^{11})$

## LESSON 8.2

**Simplify each expression.**

**1.** $(4x)^3$ **2.** $(9^2)^3$ **3.** $(-x^3)^2$ **4.** $(3^2)^4$

**5.** $(7x^2)^4$ **6.** $(-5g^2)^5$ **7.** $(-3r^2)^6$ **8.** $(-6a^8)^2$

**9.** $(-5wr2)^6$ **10.** $(30t^2x^2)^3$ **11.** $(-4z^4a^3)^5$ **12.** $(-20m^2n^2)^4$

**13.** $(2r^2s^2)^4$ **14.** $(x^1y^3)^3$ **15.** $(-3c^5d^4)^2$ **16.** $(r^3t^5)^5$

**17.** $(ab^3c^4)^7$ **18.** $(5g^7h^1k)^5$ **19.** $(-6y^3z^5)^3$ **20.** $(w^2z^3)^9$

**21.** $(-w^3z^2)^6$ **22.** $(4n^6n^2)^3$ **23.** $(j^4k^6)^5$ **24.** $(-7x^3y^3)^2$

**25.** $(-c^3v^5)(c^2)^6$ **26.** $(7g^2h^5)^3(8a^2)^2$

**27.** $(-a^3)^2(a^3b)^5$ **28.** $(-x^3y^5)^4(-xy)^5$

**29.** $(b^3c^6)^5(b^3c^7d)^3$ **30.** $(5j^3k^3)^4(j^2k^2)^6$

**31.** $(8np^2)^3(n^2p^5)^6$ **32.** $(-ab^5)^3(2a^3b^4)^5$

**33.** $(3a^2b)^2(5b^3)^4$

Extra Practice

## LESSON 8.3

**Simplify each expression. Assume that the conditions of the Quotient-of-Powers Property are met.**

**1.** $\frac{10^9}{10^4}$ **2.** $\frac{5^8}{5^3}$ **3.** $\frac{7^{13}}{7^{10}}$ **4.** $\frac{12^7}{12^5}$

**5.** $\frac{f^6}{f^3}$ **6.** $\frac{h^{6a}}{h^{2a}}$ **7.** $\frac{v^x}{v^y}$ **8.** $\frac{t^{x+3}}{t^x}$

**9.** $\frac{k^{x+7}}{k^{x+2}}$ **10.** $\frac{30k^7}{6k^4}$ **11.** $\frac{4^3h^3r^5}{4^2h^2r^2}$ **12.** $\frac{15w^3s^6}{5ws}$

**13.** $\left(\frac{s}{t}\right)^3$ **14.** $\left(\frac{12}{d}\right)^4$ **15.** $\left(\frac{11}{x^3}\right)^5$ **16.** $\left(\frac{2c}{s^8}\right)^7$

**17.** $\left(\frac{h^4}{g^6}\right)^3$ **18.** $\left(\frac{12c^6}{c^2d^5}\right)^2$ **19.** $\left(\frac{2z^5}{zy^3}\right)^8$ **20.** $\left(\frac{w^4s^7}{w^2s^3}\right)^{10}$

**21.** $\left(\frac{-4q^3r^5}{12q^5}\right)^3$ **22.** $\left(\frac{x^3z^2}{x^5z}\right)^3$ **23.** $\left(\frac{-j^4k^5l^3}{j^5k^2l^6}\right)^4$ **24.** $\left(\frac{4p^5f^3}{t^3h^7}\right)^4$

**25.** $\frac{0.45a^6y^8}{0.9a^2y^5}$ **26.** $\frac{16f^5g^6}{4f^3\left(g^2h\right)^3}$ **27.** $\frac{\left(a^3g^6y^9\right)^8}{a^2g^3y^9}$ **28.** $\frac{0.18s^5z^{10}}{0.03s^2z}$

## LESSON 8.4

**Evaluate each expression.**

**1.** $4^{-2}$ **2.** $-4^2$ **3.** $-4^2$ **4.** $-4^{-2}$

**5.** $(-4)^{-2}$ **6.** $2^{-4}$ **7.** $10^{-3}$ **8.** $(-10)^3$

**9.** $-6^2$ **10.** $(-6)^2$ **11.** $5^{-2}$ **12.** $-5^2$

**Write each expression without negative or zero exponents.**

**13.** $r^2s^{-3}$ **14.** $x^{-2}y^4$ **15.** $c^0d^4$ **16.** $r^{-3}t^5$

**17.** $ab^0c^{-4}$ **18.** $g^{-7}h^{-1}k$ **19.** $-y^3z^{-5}$ **20.** $\left(w^2z^{-3}\right)^0$

**21.** $-w^{-3}w^{-2}$ **22.** $n^6n^{-2}$ **23.** $j^4j^{-6}$ **24.** $-x^3x^{-3}$

**25.** $\frac{n^4}{n^{-2}}$ **26.** $\frac{t^{-5}}{t^2}$ **27.** $\frac{g^{-4}}{g^{-3}}$ **28.** $\frac{rs^{-6}}{r^{-3}}$

**29.** $\frac{4d^{-5}}{d^4}$ **30.** $\frac{-3a^{-3}}{2a^2b}$ **31.** $\frac{2r^4}{r^{-3}s}$ **32.** $\frac{10w^{-3}}{w^2v^{-2}}$

**33.** $\frac{\left(3a^2\right)\left(5a^3\right)}{a^{-1}}$ **34.** $\frac{\left(2n^4\right)\left(10n^2\right)}{5n^{-3}}$ **35.** $\frac{\left(-4b^{-2}\right)\left(-2b^4\right)}{2b}$ **36.** $\frac{\left(6d^{-5}\right)\left(-3d^{-4}\right)}{2d^{-3}}$

**37.** $\frac{c^{-3}c^2}{c^{-4}c^3}$ **38.** $\frac{a^{-3}a^{-6}}{b^{-5}b^{-2}}$ **39.** $\frac{-t^{-7}t^2}{r^{-5}r^2}$ **40.** $\frac{-w^2w^{-5}}{-2w^3}$

## LESSON 8.5

**Write each number in scientific notation.**

1. 4,000,000
2. 7,000,000,000
3. 2,500,000
4. 730,000
5. 0.000044
6. 0.0028
7. 0.00056
8. 0.00000021
9. 0.000345
10. 0.0056
11. 795,000
12. 8,430,000

**Write each number in decimal notation.**

13. $4 \times 10^3$
14. $8 \times 10^5$
15. $7 \times 10^7$
16. $8.2 \times 10^2$
17. $1.7 \times 10^{-3}$
18. $6.65 \times 10^{-9}$
19. $-9.7 \times 10^5$
20. $-8.033 \times 10^{-7}$

**Simplify. Write your answers in scientific notation.**

21. $(3 \times 10^3)(4 \times 10^6)$
22. $(7 \times 10^2)(1 \times 10^8)$
23. $(7 \times 10^5)(2 \times 10^{11})$
24. $(8 \times 10^8)(9 \times 10^3)$
25. $(6 \times 10^4) + (4 \times 10^4)$
26. $(8 \times 10^9) + (9 \times 10^9)$
27. $(7 \times 10^5) - (6 \times 10^5)$
28. $(8 \times 10^2) - (3 \times 10^2)$
29. $(2.8 \times 10^6)(1.4 \times 10^3)$
30. $(3.1 \times 10^4)(5.5 \times 10^4)$
31. $(5.8 \times 10^6)(6.2 \times 10^{10})$
32. $(3 \times 10^2)(4.9 \times 10^{12})$
33. $\dfrac{8 \times 10^4}{4 \times 10^2}$
34. $\dfrac{6 \times 10^8}{2 \times 10^3}$
35. $\dfrac{7 \times 10^{12}}{2 \times 10^5}$
36. $\dfrac{(3 \times 10^5)(6 \times 10^2)}{9 \times 10^4}$
37. $\dfrac{(5 \times 10^{10})(6 \times 10^2)}{5 \times 10^9}$
38. $\dfrac{(4 \times 10^3)(6 \times 10^8)}{8 \times 10^5}$

## LESSON 8.6

**Graph each of the following.**

1. $y = 3^x$
2. $y = 6^x$
3. $y = 8^x$
4. $y = 0.2^x$

**Graph each of the following. In each case describe the effect of the 2 or 4 on the graph of the parent function $y = 2^x$.**

5. $y = 2^x + 2$
6. $y = 2 \cdot 2^x$
7. $y = 2^{x+4}$
8. $y = 2^{4x}$

**Let $f(x) = 3^x$. Evaluate.**

9. $f(2)$
10. $f(0)$
11. $f(1)$
12. $f(3)$
13. $f(-2)$
14. $f(-3)$
15. $f(-1)$
16. $f(-4)$
17. $f(4)$

**The population of a city was about 300,000 in 1994 and was growing at a rate of about 0.4% per year.**

18. What multiplier should be used to find the new population each year?
19. Use the information from Exercise 18 to estimate the population of the city in the year 2000.
20. Use the information from Exercise 18 to estimate the population of the city in the year 2005.

## LESSON 8.7

**Use the method of carbon–14 dating to estimate the age of an object which has**

1. 30% of its original carbon–14 remaining.
2. 2% of its original carbon–14 remaining.
3. 50% of its original carbon–14 remaining.
4. 15% of its original carbon–14 remaining.
5. 80% of its original carbon–14 remaining.
6. 90% of its original carbon–14 remaining.

**An investment is growing at a rate of 7% per year and now has a value of $7600. Find the value of the investment**

7. in 2 years.
8. in 5 years.
9. in 8 years.
10. in 10 years
11. 2 years ago.
12. 3 years ago.
13. 5 years ago.
14. 8 years ago.

**An investment is losing money a rate of 1% per year and now has a value of $48,000. Find the value of the investment**

15. in 2 years.
16. in 5 years.
17. in 8 years.
18. in 10 years.
19. 2 years ago.
20. 3 years ago.
21. 5 years ago.
22. 8 years ago.

**The population of a city in the United States was 164,998 in 1990 and was decreasing at a rate of about 1.2% per year. In how many years will the city's population fall below**

23. 160,000?
24. 150,000?
25. 130,000?
26. 125,000?
27. 110,000?
28. 100,000?
29. 90,000?
30. 50,000?

# CHAPTER 9

## LESSON 9.1

**Rewrite each polynomial in standard form.**

1. $7 + x + x^2$
2. $9x^3 - 4x - 2 + x^2$
3. $8 - 3w^2 + w$
4. $y - 4y^4 + 3 - 2y^2$
5. $r - r^3$
6. $5w^3 - 7w^4 + 5w - 5$

**Identify the degree of each polynomial.**

7. $6t^2 - t + 1$
8. $3x - 2$
9. $m + m^3$
10. $p^5 - p^2 + p$

**Use vertical form to add or subtract.**

11. $(4x^4 + 3x^2 - 3) + (6x^4 + 2x^2 + 5)$
12. $(7w^3 - 4w + 3) + (5w^3 - 3w - 3)$
13. $(8z^4 - 4z^2 - z + 4) + (z^2 - z)$
14. $(x^2 + 3x + 4) + (3x + 7)$
15. $(2x^3 + x^2 - 1) - (x^3 - x^2)$
16. $(3w^3 - 5w + 4) - (6w + 2)$
17. $(9x^4 - 5x^2 + 7) - (8x^4 + 2)$
18. $(3z^5 + z^3 - 8) - (z^5 - z^3 + 8)$

**Use horizontal form to add or subtract.**

19. $(w^4 + 3w^2 - 5) + (w^2 - 5w)$
20. $(3z^3 - 4z + 5) + (8z^3 - 5z^2 - 1)$
21. $(9x + 3) + (4x^2 - 5x + 2)$
22. $y^3 - y^2 - 1 - (4y + 3)$
23. $4x^2 - 6x + 4 - (2x^2 + 4x - 6)$
24. $7x^2 + 4x - (x^2 - 7)$

**Simplify. Express all answers in standard form.**

25. $(2 - x - x^4) + (x + 4.5x^4 + 5)$
26. $6 + 4x - 3.6x^2 - (10.6x - 65 + 6.8x^2)$

## LESSON 9.2

**Use the Distributive Property to find each product.**

**1.** $5(x+4)$ **2.** $6(2x+6)$ **3.** $7(x+11)$ **4.** $4(x-7)$

**5.** $3(2x-6)$ **6.** $-4(x+2)$ **7.** $-3(x-2)$ **8.** $5(3x-4)$

**9.** $x(2-x)$ **10.** $x(3+3x)$ **11.** $-x(4+x)$ **12.** $-x(x-4)$

**13.** $5x(x+3)$ **14.** $2x(4-x)$ **15.** $x(-x+6)$ **16.** $-2x(x-4)$

**Use the Distributive Property to find each product.**

**17.** $(x+2)(x+3)$ **18.** $(z-5)(z-1)$ **19.** $(a+2)(a-4)$

**20.** $(b-2)(b-3)$ **21.** $(d+3)(d+4)$ **22.** $(y-2)(y+5)$

**23.** $(c+6)(c+6)$ **24.** $(w+4)(w-3)$ **25.** $(m-2)(m-4)$

**26.** $(a+5)(a-6)$ **27.** $(s+4)(s-3)$ **28.** $(x-6)(x+1)$

**29.** $(s+7)(s-2)$ **30.** $(b-9)(b+3)$ **31.** $(w+6)(w-1)$

**32.** $(q+5)(q-3)$ **33.** $(w+8)(w-8)$ **34.** $(r-9)(r-2)$

**35.** $(m+10)(m-11)$ **36.** $(n-7)(n+9)$ **37.** $(a-1)(a-10)$

## LESSON 9.3

**Use the FOIL method to find each product.**

**1.** $(w+10)(w+2)$ **2.** $(c-5)(c-5)$ **3.** $(y-3)(y-9)$

**4.** $(r-4)(r+7)$ **5.** $(w+8)(w-3)$ **6.** $(3z-1)(2z+1)$

**7.** $(4x+3)(x-2)$ **8.** $(2t-1)(2t+3)$ **9.** $(5x-2)(3x+2)$

**10.** $(8s-1)(s+4)$ **11.** $(4w-5)(3w-1)$ **12.** $(a+b)(a+b)$

**13.** $(r+s)(r-s)$ **14.** $(2x-z)(x-z)$ **15.** $(a+2b)(a-b)$

**16.** $(y+3x)(2y-x)$ **17.** $(p+q)^2$ **18.** $(c^2+d)^2$

**19.** $(m^2-n)(m^2+n)$ **20.** $(3c+4d)(2c-4d)$ **21.** $(x+3)(x+6)$

**Find each product by using the rules for special products.**

**22.** $(2x+6)(2x-6)$ **23.** $(x-11)(x+11)$ **24.** $(x-7)(x-7)$

**25.** $(2x-6)(2x-6)$ **26.** $(4x+5)(4x-5)$ **27.** $(x-6)(x+6)$

**28.** $(3x-4)(3x+4)$ **29.** $(2+x)(2-x)$ **30.** $(3+5x)(3+5x)$

**31.** $(8-x)(8-x)$ **32.** $(x-4)(x-4)$ **33.** $(7x+3)(7x+3)$

**34.** $(4-x)(4+x)$ **35.** $(-5x+6)(-5x+6)$ **36.** $(2x-9)(2x-9)$

**37.** $(x+4)(x+4)$

## LESSON 9.4

**Write surface area and volume functions for each geometric solid.**

1. A cube with edge length $x$.
2. A rectangular solid with a 5 inch by 6 inch base and height $x$.
3. A rectangular solid with a 4 cm by 9 cm base and height $x$.
4. A cylinder with a height of 6 inches and a radius $x$.
5. A rectangular solid with a base length of 9.5 centimeters and base width of 4.2 centimeters and a height $x$.
6. A cube with edge length $5x$.

**Verify whether each equation could be an identity or not by substituting −1, 0, and 1 for *x*.**

7. $x^2 + 4x + 3 \stackrel{?}{=} (x+1)\,(x+3)$
8. $x^2 - 7x + 10 \stackrel{?}{=} (x-5)(x-2)$
9. $x^2 - 7x + 12 \stackrel{?}{=} (x-4)(x-3)$
10. $x^2 + 7x + 6 \stackrel{?}{=} (x+6)\,(x+1)$
11. $(x+4)\,(x-3) \stackrel{?}{=} x^2 + x - 12$
12. $(x-4)(x-5) \stackrel{?}{=} x^2 - 9x + 20$
13. $x^2 - 16 \stackrel{?}{=} (x+4)\,(x-4)$
14. $(x-5)(x+5) \stackrel{?}{=} x^2 - 25$
15. $(x+4)(x-6) \stackrel{?}{=} x^2 - 2x - 24$
16. $x^2 + 9x + 14 \stackrel{?}{=} (x+7)(x+2)$
17. $x^2 - 100 \stackrel{?}{=} (x-10)\,(x-10)$
18. $(x-4)(x+3) \stackrel{?}{=} x^2 - x + 12$
19. $(x-2)^2 \stackrel{?}{=} x^2 - 4x + 4$
20. $x^2 + 4x + 4 \stackrel{?}{=} (x+2)^2$
21. $(x+2)^3 \stackrel{?}{=} x^3 + 4x^2 + 8x + 8$
22. $(x-4)^2 \stackrel{?}{=} x^2 - 8x - 16$

## LESSON 9.5

**State whether each polynomial can be factored using the methods studied in Lesson 9.5.**

1. $5x^2 - 15$
2. $t^2 + 6$
3. $w^2 - 4$
4. $4m^2 + 9$
5. $7t^2 - 8t$
6. $25c - 25d$
7. $16x^2 - 1$
8. $w^3 - w^2$
9. $3x - 2$
10. $8x^3 - 9$
11. $7x^2 + x$
12. $16c - 8$

**Factor each polynomial by using the GCF.**

13. $4x^2 + 16$
14. $7w^2 + 21$
15. $8a^2 - 4a$
16. $5n^3 - 10p^2 + 15$
17. $25d^2 - 75d^3$
18. $6x^2y - 14xy$
19. $3s^2t^3 + 15st^3$
20. $4z^3w - 16w^3 + 24w$
21. $25ab^4 + 20a^3b^2$

**Write each expression as the product of two binomials.**

22. $x(x-1) + 2(x-1)$
23. $7(y+4) - x(y+4)$
24. $b(c+d) + a(c+d)$
25. $4(w-3) - v(w-3)$
26. $x(y-z) + 4(y-z)$
27. $s(x+5) - t(x+5)$
28. $q(4-s) - 5(4-s)$
29. $3w(x+y) - 4(x+y)$
30. $mn(p-q) + ab(p-q)$

**Factor by grouping.**

31. $15a - 3ay - 4y + 20$
32. $xy + 5x + 3y + 15$
33. $mr + 3m + 2r + 6$
34. $ch + c - 4h - 4$
35. $zr + 4z + 3r + 12$
36. $mr + 3r - 2m - 6$
37. $cl + 2c + 7l + 14$
38. $kx - 3x + k - 3$
39. $rs - 6r - 5s + 30$
40. $kj + 3k - 4j - 12$
41. $am - 6m - 2a + 12$
42. $nx + 12n - x - 12$

## LESSON 9.6

**Find each product.**

1. $(x+2)^2$
2. $(2x+1)^2$
3. $(m-4)^2$
4. $(a-6)^2$
5. $(4y-5)^2$
6. $(9q+1)^2$
7. $(x+3y)(x-3y)$
8. $(7a-b)(7a+b)$
9. $(5a-3)(5a+3)$

**Find the missing terms in each perfect-square trinomial.**

10. $a^2 - 10a + \underline{?}$
11. $25z^2 + \underline{?} + 4$
12. $9b^2 - 12b + \underline{?}$
13. $\underline{?} - 4y + 1$
14. $9r^2 + 30r + \underline{?}$
15. $\underline{?} + 42x + 49$

**Factor each polynomial completely.**

16. $y^2 - 9$
17. $9t^2 - 1$
18. $z^2 - 144$
19. $x^2 - 4x + 4$
20. $y^2 + 16y + 64$
21. $25w^2 - 16$
22. $1 - 81q^2$
23. $x^2 - y^2$
24. $x^2 + 12x + 36$
25. $q^2 - 20q + 100$
26. $49m^2 - 14m + 1$
27. $25r^2 + 20r + 4$
28. $64t^2 + 16t + 1$
29. $25b^2 - 10b + 1$
30. $w^2 + 22w + 121$
31. $2x^2 - 8$
32. $16 - 49c^2$
33. $x^2 - 10x + 25$
34. $x^2 - 36y^2$
35. $x^2 + 2xy + y^2$
36. $49a^2 - 4b^2$

## LESSON 9.7

**Write each trinomial in factored form.**

1. $x^2 - 3x - 10$
2. $x^2 - 3x - 28$
3. $x^2 + 9x + 20$
4. $x^2 - 5x - 24$
5. $x^2 + 8x - 20$
6. $x^2 - 4x - 45$
7. $x^2 - 9x + 8$
8. $x^2 - 12x - 45$
9. $x^2 + 10x + 16$
10. $b^2 + 12b + 27$
11. $x^2 - 8x + 15$
12. $z^2 + 2z - 3$
13. $s^2 + 22s + 21$
14. $n^2 - n - 20$
15. $h^2 + 15h + 26$
16. $q^2 - 10q + 16$
17. $x^2 - 9x + 14$
18. $c^2 + 13c - 30$

**Write each polynomial in factored form. If it cannot be factored, write *prime*.**

19. $y^2 - 49$
20. $w^2 - 16w + 64$
21. $z^2 + 14z + 24$
22. $y^2 - 8y + 7$
23. $28n^2 + 18n$
24. $5a^2 + 20b^2$
25. $s^2 - 12a - 13$
26. $r^2 - r - 72$
27. $by + 4my + 3xb + 12mx$
28. $3x^2 - 12$
29. $t^2 - t - 65$
30. $m^2 + 10m - 39$
31. $y^2 + 7y + 12$
32. $a^2 - 4ac + ab - 4bc$
33. $4y^2 + 2y - 6$
34. $4t^2 - 6t - 40$
35. $6n^2 - 24n + 6$
36. $a^2 + 18a + 51$

## LESSON 9.8

**Identify the zeros for each function.**

1. $y = (x - 5)(x + 4)$
2. $y = (x + 6)(x - 1)$
3. $y = (x + 11)(x - 10)$
4. $y = (x - 8)(x + 3)$
5. $y = (x + 2)(x + 2)$
6. $y = (x + 12)(x - 12)$
7. $y = (x + 9)(x + 6)$
8. $y = (x - 16)(x + 15)$
9. $y = (x - 4)(x - 4)$
10. $y = (x - 7)(x + 1.6)$
11. $y = (x - 2.9)(x + 3.2)$
12. $y = (x - 8)(x + 8)$
13. $y = (x - 5)(x - 5)$
14. $y = (x - 3)(x + 24)$
15. $y = (x - 13)(x - 13)$
16. $y = x(x + 3)$
17. $y = x(x - 5)$
18. $y = (x + 16)(x + 16)$

**Solve by factoring.**

19. $x^2 + 6x + 5 = 0$
20. $x^2 + 4x - 21 = 0$
21. $x^2 - 4 = 0$
22. $x^2 - 13x + 42 = 0$
23. $x^2 + 6x + 9 = 0$
24. $x^2 - 36 = 0$
25. $x^2 - 64 = 0$
26. $x^2 - 121 = 0$
27. $x^2 + 23x + 132 = 0$
28. $x^2 + 10x = -9$
29. $x^2 + 10x + 16 = 0$
30. $x^2 + 10x + 21 = 0$
31. $x^2 + 10x + 25 = 0$
32. $x^2 + 16x + 64 = 0$
33. $x^2 + 20x = -100$
34. $x^2 + 13x + 30 = 0$
35. $x^2 - 2x = -1$
36. $x^2 - 8x + 16 = 0$
37. $x^2 + 13x + 22 = 0$
38. $x^2 + 4x - 45 = 0$
39. $x^2 - 9x + 18 = 0$
40. $x^2 - 3x = 18$
41. $x^2 + 3x = 18$
42. $x^2 - 11x = -10$

# CHAPTER 10

## LESSON 10.1

**Compare the graph of each function to the graph of the parent function, $y = x^2$. Describe any translations or reflections.**

1. $y = (x - 3)^2 + 1$
2. $y = (x + 2)^2 - 4$
3. $y = 2(x + 1)^2 - 2$
4. $y = (x + 3)^2$
5. $y = -(x - 4)^2$
6. $y = 3(x - 6)^2$
7. $y = 4(x - 1)^2 - 1$
8. $y = -2(x - 5)^2 + 2$
9. $y = -5(x + 2)^2 - 5$
10. $y = \frac{1}{2}(x - 1)^2 + 4$
11. $y = -\frac{1}{3}(x + 9)^2 - 3$
12. $y = \frac{2}{5}(x - 4)^2 - 5$

**Find the vertex and axis of symmetry for the graph of each function, and then sketch the graph.**

13. $y = -2(x + 1)^2 - 2$
14. $y = 3(x - 2)^2 + 3$
15. $y = (x - 3)^2 + 4$
16. $y = \frac{1}{2}(x + 2)^2 + 1$
17. $y = -3(x + 3)^2$
18. $y = -5(x + 1)^2 + 4$

**Find the zeros of each quadratic function.**

19. $y = x^2 + 7x + 12$
20. $y = x^2 + 4x + 4$
21. $y = x^2 - x - 30$
22. $y = x^2 + 9x + 8$
23. $y = x^2 + x - 20$
24. $y = x^2 + 2x - 15$
25. $y = x^2 - 2x + 1$
26. $y = x^2 + 7x - 8$
27. $y = x^2 + 10x + 9$

**First graph the function. Then find the zeros, the axis of symmetry, and the minimum value of the function.**

28. $y = x^2 + 10x + 25$
29. $y = x^2 + 4x + 3$
30. $y = x^2 - 2x - 8$
31. $y = x^2 + 8x + 15$
32. $y = x^2 - 6x - 7$
33. $y = x^2 - 2x - 24$

## LESSON 10.2

**Find each square root. Round your answers to the nearest hundredth when necessary.**

**1.** $\sqrt{100}$ **2.** $\sqrt{169}$ **3.** $\sqrt{484}$ **4.** $\sqrt{49}$

**5.** $\sqrt{50}$ **6.** $\sqrt{89}$ **7.** $\sqrt{28}$ **8.** $\sqrt{14}$

**9.** $\sqrt{99}$ **10.** $\sqrt{110}$ **11.** $\sqrt{200}$ **12.** $\sqrt{65}$

**13.** $\sqrt{95}$ **14.** $\sqrt{250}$ **15.** $\sqrt{135}$ **16.** $\sqrt{61}$

**Solve each equation. Round your answers to the nearest hundredth when necessary.**

**17.** $x^2 = 36$ **18.** $x^2 = 121$ **19.** $x^2 = 64$

**20.** $x^2 = 400$ **21.** $x^2 = \frac{16}{25}$ **22.** $x^2 = \frac{64}{121}$

**23.** $(x+4)^2 - 16 = 0$ **24.** $(x+2)^2 - 1 = 0$ **25.** $(x-3)^2 - 36 = 0$

**26.** $(x+3)^2 = 64$ **27.** $(x-1)^2 = 12$ **28.** $(x+5)^2 = 35$

**Find the vertex, axis of symmetry, and zeros of each function. Then sketch the graph.**

**29.** $f(x) = (x-9)^2 - 16$ **30.** $g(x) = (x+3)^2 - 4$ **31.** $h(x) = (x-4)^2 - 2$

## LESSON 10.3

**Complete the square.**

**1.** $x^2 + 12x$ **2.** $x^2 + 10x$ **3.** $x^2 + 4x$ **4.** $x^2 + 20x$

**5.** $x^2 - 30x$ **6.** $x^2 + 11x$ **7.** $x^2 + 30x$ **8.** $x^2 - 21x$

**Find the minimum or maximum value for each quadratic function.**

**9.** $f(x) = x^2 + 5$ **10.** $f(x) = x^2 - 9$ **11.** $f(x) = x^2 + 15$

**12.** $f(x) = x^2 + 1$ **13.** $f(x) = x^2 + 16$ **14.** $f(x) = x^2 - 2$

**Rewrite each function in the form $y = (x - h)^2 + k$. Find each vertex.**

**15.** $y = x^2 - 6x$ **16.** $y = x^2 - 12x$ **17.** $y = x^2 + 2x$

**18.** $y = x^2 + 6x + 5$ **19.** $y = x^2 - 2x - 3$ **20.** $y = x^2 - 8x + 12$

**21.** $y = x^2 - 10x + 21$ **22.** $y = x^2 - 12x + 32$ **23.** $y = x^2 + 5x + 4$

**24.** $x^2 + \frac{1}{2}x - 2$ **25.** $x^2 - \frac{1}{4}x + 6$ **26.** $x^2 - \frac{1}{3}x + 5$

## LESSON 10.4

**Find the zeros of each function. Graph to check.**

**1.** $y = x^2 - 2x - 3$ **2.** $y = x^2 + 5x + 4$ **3.** $y = x^2 - 8x + 12$
**4.** $y = x^2 - 2x - 35$ **5.** $y = x^2 - 6x + 8$ **6.** $y = x^2 - 6x + 5$

**Solve each equation by factoring.**

**7.** $x^2 - x - 2 = 0$ **8.** $x^2 + 6x + 5 = 0$ **9.** $x^2 - x - 6 = 0$
**10.** $x^2 + 3x - 10 = 0$ **11.** $x^2 + 3x - 28 = 0$ **12.** $x^2 + 7x - 30 = 0$

**Solve each equation by completing the square.**

**13.** $x^2 + 6x - 7 = 0$ **14.** $x^2 + 2x - 8 = 0$ **15.** $x^2 + 4x - 45 = 0$
**16.** $x^2 + 3x - 18 = 0$ **17.** $x^2 - x - 12 = 0$ **18.** $x^2 - 2x - 3 = 0$

**Solve each equation by factoring or completing the square.**

**19.** $x^2 - 12x + 20 = 0$ **20.** $x^2 - 3x - 4 = 0$ **21.** $x^2 + 7x = -12$
**22.** $x^2 + 2x - 15 = 0$ **23.** $x^2 - 4x = 0$ **24.** $x^2 + 5x - 24 = 0$
**25.** $x^2 - x = 2$ **26.** $x^2 - 9x + 20 = 0$ **27.** $x^2 + 5x = 6$

**Find the point(s) where the graphs of each system intersect. Graph to check.**

**28.** $\begin{cases} y = 4 \\ y = x^2 \end{cases}$ **29.** $\begin{cases} y = x - 2 \\ y = x^2 - 4x + 4 \end{cases}$ **30.** $\begin{cases} y = x + 4 \\ y = x^2 + 3x - 4 \end{cases}$

## LESSON 10.5

**Identify *a*, *b*, and *c* for each quadratic equation.**

**1.** $x^2 - 4x + 5 = 0$ **2.** $x^2 - 6x + 10 = 0$ **3.** $8x^2 - 5x + 2 = 0$
**4.** $3x^2 - 6 + 3x = 0$ **5.** $-x^2 + 7x - 6 = 0$ **6.** $-5x + x^2 - 9 = 0$

**Find the value of the discriminant and determine the number of real solutions for each equation.**

**7.** $x^2 + 3x - 4 = 0$ **8.** $x^2 + 4x + 1 = 0$ **9.** $x^2 + 6x - 2 = 0$
**10.** $x^2 - 6x + 1 = 0$ **11.** $x^2 - 5x + 9 = 0$ **12.** $x^2 + 6x - 10 = 0$
**13.** $4x^2 + 8x + 3 = 0$ **14.** $5x^2 - 125 = 0$ **15.** $x^2 + x - 12 = 0$

**Use the quadratic formula to solve each equation. Give answers to the nearest hundredth when necessary. Check by substitution.**

**16.** $x^2 + 7x + 6 = 0$ **17.** $y^2 + 8y + 15 = 0$ **18.** $x^2 + 4x + 3 = 0$
**19.** $w^2 - 6w = 0$ **20.** $x^2 - 16 = 0$ **21.** $y^2 + 10y = 0$
**22.** $-4x^2 + 16x + 13 = 0$ **23.** $8x^2 + 10x + 3 = 0$ **24.** $2x^2 + 7x - 15 = 0$

**Choose any method to solve the following quadratic equations. Round your answers to the nearest hundredth.**

**25.** $x^2 - 9x + 20 = 0$ **26.** $x^2 + 10x - 2 = 0$ **27.** $x^2 + 13x = -42$
**28.** $3x^2 - 5x - 2 = 0$ **29.** $2x^2 + x = 5$ **30.** $4x^2 - 7x - 2 = 0$
**31.** $5x^2 + 9x + 3 = 0$ **32.** $x^2 - 2x - 2 = 0$ **33.** $3x^2 - 5x + 1 = 0$
**34.** $x^2 - 4x - 2 = 0$ **35.** $x^2 - 5x - 7 = 0$ **36.** $3x^2 - 7x - 3 = 0$

## LESSON 10.6

**Solve each quadratic inequality using the Zero Product Property.**

**1.** $x^2 - x - 12 > 0$
**2.** $x^2 + 3x - 10 < 0$
**3.** $x^2 - 4 \geq 0$
**4.** $x^2 + 7x + 12 \leq 0$
**5.** $x^2 - 10x + 21 < 0$
**6.** $x^2 + 4x - 12 \leq 0$
**7.** $x^2 - 9 < 0$
**8.** $x^2 + 4x + 3 \geq 0$
**9.** $x^2 - 4x + 3 > 0$
**10.** $x^2 - 3x - 10 < 0$
**11.** $x^2 + 2x - 3 \leq 0$
**12.** $x^2 - 10x - 24 \geq 0$

**Graph each quadratic inequality. Shade the solution region.**

**13.** $y < x^2$
**14.** $y < x^2 + 2$
**15.** $y > -x^2$
**16.** $y \geq x^2$
**17.** $y < x^2 - x$
**18.** $y < 3 - x^2$
**19.** $y < -3x^2$
**20.** $y \leq x^2 - 9$
**21.** $y > x^2 + 3x + 4$
**22.** $y \leq x^2 + 2x + 1$
**23.** $y \geq x^2 - 6x + 9$
**24.** $y \leq x^2 + 5x$

# CHAPTER 11

## LESSON 11.1

**Determine whether each equation is an inverse variation equation.**

**1.** $xy = 200$
**2.** $x = \frac{25}{y}$
**3.** $x + y = 36$
**4.** $\frac{a}{3} = \frac{5}{b}$
**5.** $\frac{c}{d} = \frac{2}{3}$
**6.** $r = 50t$
**7.** $m = \frac{n}{75}$
**8.** $w = 7t$
**9.** $\frac{-12}{x} = y$
**10.** $p = \frac{1}{q}$
**11.** $bh = 14$
**12.** $\frac{1}{10}m = n$

**For Exercises 13–30, *y* varies inversely as *x*.**

**13.** If $y$ is 24 when $x$ is 8, find $y$ when $x$ is 4.
**14.** If $y$ is 3 when $x$ is 12, find $x$ when $y$ is 4.
**15.** If $y$ is 6 when $x$ is 2, find $x$ when $y$ is 4.
**16.** If $y$ is 12 when $x$ is 15, find $x$ when $y$ is 18.
**17.** If $y$ is 15 when $x$ is 21, find $x$ when $y$ is 27.
**18.** If $y$ is $-6$ when $x$ is $-2$, find $y$ when $x$ is 5.
**19.** If $y$ is $-8$ when $x$ is 2, find $x$ when $y$ is 7.
**20.** If $y$ is 99 when $x$ is 11, find $x$ when $y$ is 11.
**21.** If $y$ is 2 when $x$ is 5, find $y$ when $x$ is 20.
**22.** If $y$ is 6.9 when $x$ is 1.7, find $y$ when $x$ is 5.1.
**23.** If $y$ is $\frac{1}{3}$ when $x$ is 5, find $x$ when $y$ is $6\frac{2}{3}$.
**24.** If $y$ is 7 when $x$ is $\frac{2}{3}$, find $x$ when $y$ is $\frac{2}{3}$.
**25.** If $y$ is 5.6 when $x$ is 2.8, find $x$ when $y$ is 4.48.
**26.** If $y$ is 8.1 when $x$ is 2.7, find $y$ when $x$ is 3.6.
**27.** If $y$ is 7 when $x$ is 18, find $y$ when $x$ is 30.
**28.** If $y$ is 7 when $x$ is 11, find $y$ when $x$ is 54.
**29.** If $y$ is 3.6 when $x$ is 8.1, find $y$ when $x$ is 2.4.
**30.** If $y$ is $-5$ when $x$ is 7, find $y$ when $x$ is $-3$.

## LESSON 11.2

**For what value(s) of the variable is each rational expression undefined?**

**1.** $\frac{3c+7}{2c}$ **2.** $\frac{6x+2}{x-1}$ **3.** $\frac{b}{3+b}$ **4.** $\frac{12}{x-9}$

**5.** $\frac{4c}{2c+1}$ **6.** $\frac{2t^2}{3t+2}$ **7.** $\frac{5y-11}{4y-8}$ **8.** $\frac{2r+10}{r^2}$

**9.** $\frac{g+3}{2g^2}$ **10.** $\frac{3a+9}{a^2-4}$ **11.** $\frac{5z+7}{z^2+5z+6}$ **12.** $\frac{q}{2q^2-15q}$

**Evaluate each rational expression for $x = 1$ and $x = -2$. Write "undefined" if appropriate.**

**13.** $\frac{2x}{x+6}$ **14.** $\frac{3x+4}{2x+1}$ **15.** $\frac{4x+1}{2x+4}$ **16.** $\frac{7x+7}{x+1}$

**17.** $\frac{5x}{x^2-2x+1}$ **18.** $\frac{12x}{12x+3}$ **19.** $\frac{8x+9}{x^2-4}$ **20.** $\frac{2x+3}{x^2+x-2}$

**21.** $\frac{1}{x-3}$ **22.** $\frac{2}{x}+2$ **23.** $\frac{6}{x+3}-4$ **24.** $\frac{1}{x-5}+3$

**Graph each rational function. List any values for which the function is undefined.**

**25.** $h(x) = \frac{2}{x}+1$ **26.** $h(x) = \frac{1}{x-2}+3$ **27.** $h(x) = \frac{2}{x-3}-1$ **28.** $h(x) = \frac{1}{2x}-3$

## LESSON 11.3

**For what value(s) of the variable is each rational expression undefined?**

**1.** $\frac{5x}{x+15}$ **2.** $\frac{12}{7-y}$ **3.** $\frac{8-w}{w}$ **4.** $\frac{9q}{q-4}$

**5.** $\frac{t-6}{6-t}$ **6.** $\frac{(a+1)(a-3)}{(a-1)(a-3)}$ **7.** $\frac{s-5}{3s-6}$ **8.** $\frac{4}{d(d^2-2d-8)}$

**Name the common factors of the numerator and the denominator.**

**9.** $\frac{4(x-3)}{8x}$ **10.** $\frac{6w-12}{3}$ **11.** $\frac{x^2-9}{x+3}$ **12.** $\frac{r+4}{r^2+8r+16}$

**Simplify each expression and state any restrictions on the variable.**

**13.** $\frac{12(x+3)}{24(x-1)}$ **14.** $\frac{4(x-y)}{16(x+y)}$ **15.** $\frac{8(d+1)}{28(1+d)}$ **16.** $\frac{5(c^2-81)}{10(c-9)}$

**17.** $\frac{7w-14}{7}$ **18.** $\frac{6x-18}{x-3}$ **19.** $\frac{10a+15b}{5a}$ **20.** $\frac{4x^2-3x}{4x-3}$

**21.** $\frac{w^2-25}{w+5}$ **22.** $\frac{y+4}{y^2+3y-4}$ **23.** $\frac{g^2-9}{g^2+6g-27}$ **24.** $\frac{n^2-n-20}{n^2+9n+20}$

Extra Practice

## LESSON 11.4

**Perform the indicated operations. Simplify, and state the restrictions on the variable.**

1. $\frac{4}{3y} + \frac{5}{3y}$
2. $\frac{7}{s-2} - \frac{3}{s-2}$
3. $\frac{4m}{n+3} - \frac{9m}{n+3}$
4. $\frac{7w-3}{x-y} - \frac{4+5w}{x-y}$
5. $\frac{x}{x+1} + \frac{1}{x+1}$
6. $\frac{z}{2} + \frac{z-6}{2}$
7. $\frac{6}{ab} - \frac{c}{ab^2}$
8. $\frac{-3}{w-2} + \frac{4}{5(w-2)}$
9. $\frac{m+n}{m-2} + \frac{m-n}{2-m}$
10. $a - \frac{a-5}{a+5}$
11. $\frac{-4-g}{g-1} + 3$
12. $\frac{a}{b} - \frac{c}{d}$
13. $\frac{7-x}{4+y} \cdot \frac{4+y}{m+n}$
14. $\frac{a^2}{(a+4)(a+1)} \cdot \frac{a+1}{a(a+4)}$
15. $\frac{x}{x+3} \cdot \frac{x^2-9}{x^2}$
16. $\frac{4}{y+2} \cdot \frac{1}{y}$
17. $\frac{-6}{r-3} \cdot \frac{2}{r}$
18. $\frac{t^2-t}{3} \cdot \frac{6}{t-1}$
19. $\frac{5}{(x-1)(x+2)} + \frac{4}{x-1}$
20. $\frac{3}{c+5} - \frac{4}{(c+2)(c+5)}$
21. $\frac{v-2}{v+2} + \frac{3}{v-3}$
22. $\frac{1}{y-1} + \frac{y}{1-y}$
23. $\frac{4}{h+2} - \frac{3h-1}{h^2+6h+8}$
24. $\frac{2}{x^2-4} + \frac{3}{4x+8}$

## LESSON 11.5

**Solve each rational equation.**

1. $\frac{11}{2x} - \frac{2}{3x} = \frac{1}{6}$
2. $\frac{1}{x} + \frac{5x}{x+1} = 5$
3. $\frac{x-1}{x+3} = \frac{x+4}{3-x}$
4. $\frac{1}{4} + \frac{2}{x} = \frac{11}{12}$
5. $\frac{2w}{3} = 2 + \frac{w+3}{6}$
6. $a + \frac{a}{a-1} = \frac{4a-3}{a-1}$
7. $1 + \frac{3}{z-1} = \frac{4}{3}$
8. $\frac{1}{2x} = \frac{1}{x^2} - \frac{1}{9}$
9. $\frac{4y}{3y-2} + \frac{2y}{3y+2} = 2$
10. $\frac{5}{p-1} + \frac{p}{p+1} = 1$
11. $\frac{x+3}{x} + \frac{x-12}{x} = 5$
12. $\frac{w-1}{w+1} + 1 = \frac{2w}{w-1}$
13. $\frac{2q}{q-1} + \frac{q-5}{q^2-1} = 1$
14. $\frac{14}{y-6} - \frac{1}{2} = \frac{6}{y-8}$
15. $\frac{a-2}{a} = \frac{1}{a} + \frac{a-3}{a-6}$
16. $\frac{c-1}{3-c} = c - \frac{2}{c-3}$
17. $\frac{x-5}{x^2-1} + \frac{2x}{x-1} = 1$
18. $\frac{m}{3m+6} = \frac{2}{5} + \frac{m}{5m+10}$

**Solve each rational equation by graphing.**

19. $\frac{16}{x+8} = \frac{4}{9}$
20. $\frac{2}{x+1} = \frac{3}{x+2}$
21. $\frac{2x}{x-4} - \frac{3}{5} = 5$

## LESSON 11.6

**Give a reason for each step in the proof of the following statement:**
**If $2x + 6 = 0$, then $x = -3$**

| | Proof | Reasons |
|---|---|---|
| | $2x + 6 = 0$ | Hypothesis or given |
| **1.** | $2x + 6 - 6 = 0 - 6$ | ______________ |
| **2.** | $2x = -6$ | ______________ |
| **3.** | $\frac{2x}{2} = \frac{-6}{2}$ | ______________ |
| **4.** | $x = -3$ | ______________ |

**Prove each statement. Give a reason for each step.**

**5.** For all $x$, $3x + 4x = 7x$.

**6.** For all $y$, $7y - 6y = y$.

**7.** For all $m$, $(-4m)(-3m) = 12m^2$.

**8.** If $8x - 3x = 35$, then $x = 7$.

**9.** If $3c - 1 = 11$, then $c = 4$.

**10.** If $4(y + 2) = -24$, then $y = -8$.

**11.** If $\frac{3}{4}x + 5 = 2$, then $x = -4$.

**12.** If $\frac{3}{5}n - 6 = 0$, then $n = 10$.

**13.** If $\frac{d}{3} - 7 = -4$, then $d = 9$.

**14.** If $5 - \frac{q}{4} = 8$, then $q = -12$.

# CHAPTER 12

## LESSON 12.1

**Simplify each radical by factoring.**

**1.** $\sqrt{36}$ **2.** $\sqrt{225}$ **3.** $\sqrt{324}$ **4.** $\sqrt{2500}$

**5.** $\sqrt{108}$ **6.** $\sqrt{80}$ **7.** $\sqrt{192}$ **8.** $\sqrt{1440}$

**9.** $\sqrt{605}$ **10.** $\sqrt{784}$ **11.** $\sqrt{1944}$ **12.** $\sqrt{768}$

**13.** $\sqrt{490}$ **14.** $\sqrt{270}$ **15.** $\sqrt{133}$ **16.** $\sqrt{525}$

**17.** $\sqrt{81}$ **18.** $\sqrt{99}$ **19.** $\sqrt{67}$ **20.** $\sqrt{140}$

**21.** $\sqrt{1600}$ **22.** $\sqrt{825}$ **23.** $\sqrt{44}$ **24.** $\sqrt{121}$

**Simplify.**

**25.** $\sqrt{5}\sqrt{20}$ **26.** $\sqrt{2}\sqrt{18}$ **27.** $\sqrt{21}\sqrt{98}$ **28.** $\sqrt{45}\sqrt{5}$

**29.** $\sqrt{\frac{100}{25}}$ **30.** $\sqrt{\frac{72}{3}}$ **31.** $\frac{\sqrt{60}}{\sqrt{6}}$ **32.** $\frac{\sqrt{216}}{\sqrt{6}}$

**33.** $(4\sqrt{3})^2$ **34.** $(6\sqrt{25})^2$ **35.** $(8\sqrt{2})^2$ **36.** $(5\sqrt{15})^2$

**37.** $3(\sqrt{6} + 4)$ **38.** $3(\sqrt{12} - 8)$ **39.** $\sqrt{6}(3 - \sqrt{18})$ **40.** $\sqrt{3}(7 + \sqrt{15})$

**41.** $(\sqrt{6} - 4)(\sqrt{6} + 4)$ **42.** $(\sqrt{7} + 2)^2$ **43.** $(\sqrt{3} - 2)(\sqrt{3} + 5)$ **44.** $\sqrt{2}(\sqrt{5} - 1)^2$

## LESSON 12.2

**Solve each equation if possible. If not possible, explain why.**

**1.** $\sqrt{x-3}=1$

**2.** $\sqrt{x+5}=3$

**3.** $\sqrt{x-9}=8$

**4.** $\sqrt{3x}=6$

**5.** $\sqrt{5x}=10$

**6.** $\sqrt{x-4}=2$

**7.** $\sqrt{8-x}=2$

**8.** $\sqrt{17-x}=4$

**9.** $\sqrt{3x+1}=5$

**10.** $\sqrt{x+2}=x-3$

**11.** $\sqrt{1-2x}=x+1$

**12.** $\sqrt{x-2}=x-4$

**13.** $\sqrt{x+3}=x-3$

**14.** $\sqrt{3x-5}=x-5$

**15.** $\sqrt{3x-14}=6-x$

**16.** $\sqrt{x^2-3x+10}=x+1$

**17.** $\sqrt{x^2+x-8}=x-2$

**18.** $\sqrt{x^2+6x-4}=x-5$

**19.** $x^2=80$

**20.** $x^2=27$

**21.** $x^2=98$

**22.** $9x^2=10$

**23.** $16x^2=5$

**24.** $25x^2=1$

**25.** $3x^2=27$

**26.** $5x^2=75$

**27.** $6x^2=72$

**28.** $x^2-6x+9=0$

**29.** $x^2+4x+2=0$

**30.** $x^2-8x+16=0$

**31.** $x^2=16$

**32.** $\sqrt{x}=16$

**33.** $|x|=16$

**34.** $x^2=-16$

**35.** $\sqrt{x}=-16$

**36.** $|x|=-16$

## LESSON 12.3

**Copy and complete the table of lengths of sides of right triangles. Use a calculator and round your answers to the nearest tenth.**

| | Leg | Leg | Hypotenuse |
|---|---|---|---|
| **1.** | 2 | 6 | ? |
| **3.** | 6 | ? | 10 |
| **5.** | 7 | 15 | ? |
| **7.** | ? | 9 | 20 |
| **9.** | 15 | 18 | ? |
| **11.** | 10 | 24 | ? |
| **13.** | ? | 40 | 41 |

| | Leg | Leg | Hypotenuse |
|---|---|---|---|
| **2.** | 8 | ? | 17 |
| **4.** | ? | 12 | 13 |
| **6.** | 7 | 12 | ? |
| **8.** | 6 | 7 | ? |
| **10.** | ? | 12 | 14 |
| **12.** | 8 | ? | 30 |
| **14.** | 1.5 | 2 | ? |

**Solve for *x*.**

**15.**

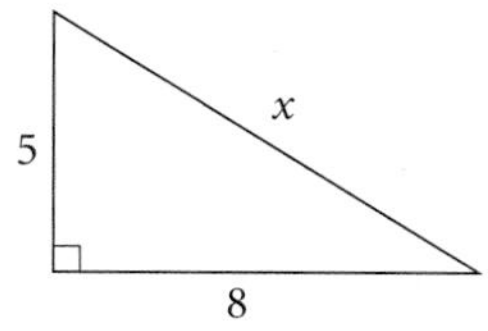


**16.**

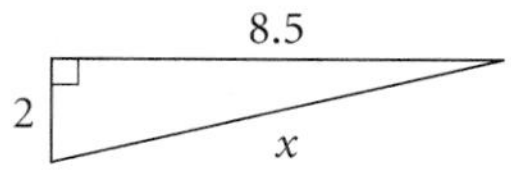


**17.**

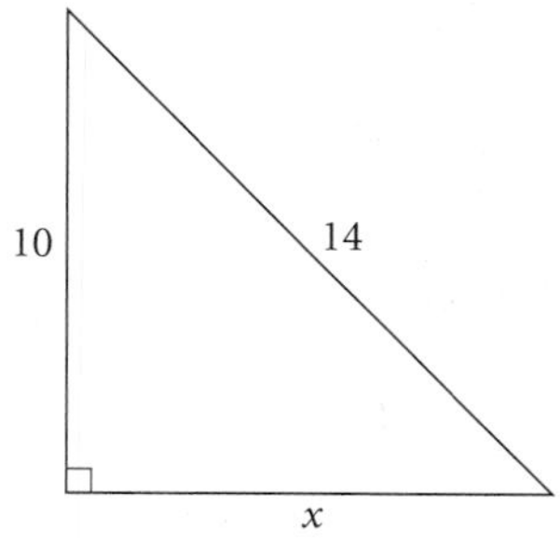

## LESSON 12.4

**For each figure, find the coordinates of *Q* and the lengths of the three sides of the trianale.**

**1.**

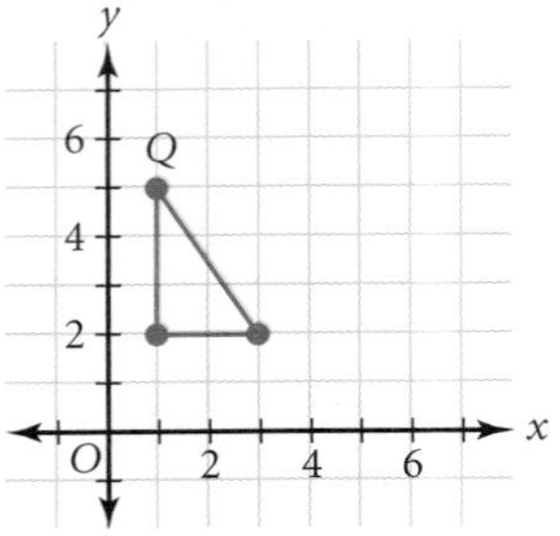


**2.**

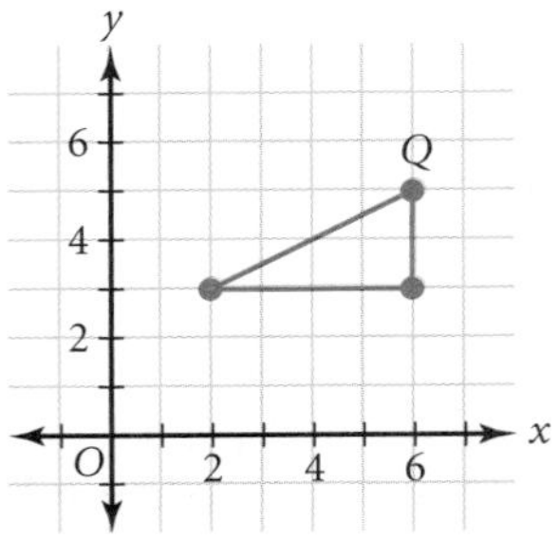


**3.**

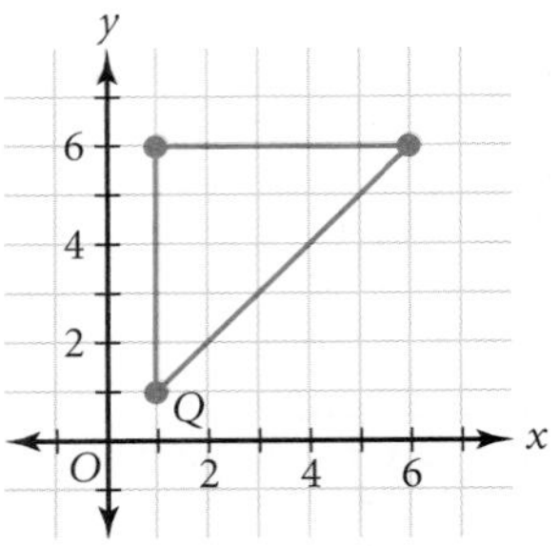


**Find the distance between each pair of points.**

**4.** $A(3, 6), B(1, 2)$ **5.** $M(3, 7), N(12, 19)$ **6.** $X(-5, -2), Y(-8, 3)$

**7.** $D(6, 7), E(-1, 1)$ **8.** $P(-5, 3), Q(5, 5)$ **9.** $R(3, 4), S(-4, -1)$

**Plot each segment in a coordinate plane, and find the coordinates of the midpoint of the segment.**

**10.** $\overline{AB}$, with $A(4, 6)$ and $B(8, 8)$ **11.** $\overline{MN}$, with $M(-4, 4)$ and $N(0, -8)$

**12.** $\overline{XY}$, with $X(-6, 5)$ and $Y(-2, 3)$ **13.** $\overline{CD}$, with $C(7, 8)$ and $D(4, 1)$

**The midpoint of segment *PQ* is *M*. Find the missing coordinates.**

**14.** $P(5, 8), Q(1, 2), M(?, ?)$ **15.** $P(6, 2), Q(?, ?), M(8, 6)$ **16.** $P(-4, 6), Q(8, 9), M(?, ?)$

## LESSON 12.5

**Find an equation of a circle with a center at the origin and having the given radius.**

**1.** radius = 3 **2.** radius = 5 **3.** radius = 10 **4.** radius = 16

**5.** radius = 1 **6.** radius = 7 **7.** radius = 11 **8.** radius = 4

**9.** radius = 20 **10.** radius = 25 **11.** radius = 32 **12.** radius = 60

**Use graph paper to graph each circle below. Find an equation of a circle with**

**13.** center (1, 2) and radius = 2 **14.** center (−1, 2) and radius = 2

**15.** center (1, −2) and radius = 2 **16.** center (−1, −2) and radius = 2

**17.** center (2, 2) and radius = 3 **18.** center (1, 3) and radius = 1

**19.** center (−3, 0) and radius = 1 **20.** center (−1, 1) and radius = 3

**Identify the center and radius for each circle.**

**21.** $(x - 1)^2 + (y - 2)^2 = 16$ **22.** $(x - 4)^2 + (y - 5)^2 = 1$

**23.** $(x + 3)^2 + (y - 4)^2 = 25$ **24.** $(x + 2)^2 + (y - 6)^2 = 9$

**25.** $(x + 7)^2 + (y - 1)^2 = 3$ **26.** $(x + 10)^2 + (y - 9)^2 = 7$

**27.** $(x - 3)^2 + (y - 9)^2 = 11$ **28.** $(x + 15)^2 + (y - 1)^2 = 49$

**29.** $x^2 + (y - 1)^2 = 13$ **30.** $(x + 17)^2 + (y - 9)^2 = 81$

**31.** $(x - 3)^2 + y^2 = 121$ **32.** $(x - 5)^2 + (y - 10)^2 = 100$

## LESSON 12.6

**Find the tangent of each acute angle of each triangle.**

**1.**

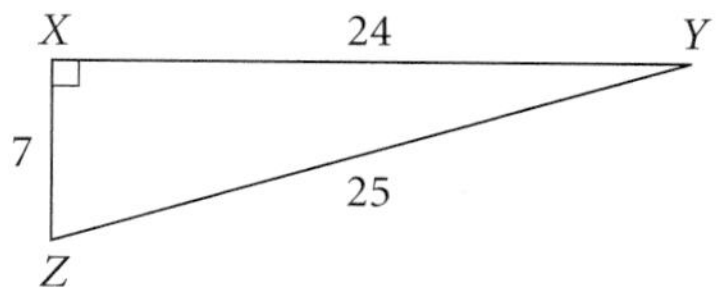


**2.**

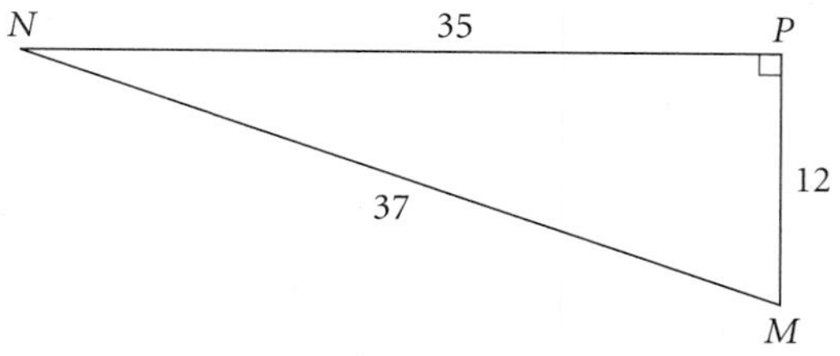


**Use a calculator to find the tangents of the following angles to the nearest ten-thousandth.**

**3.** 60° **4.** 35° **5.** 2° **6.** 52° **7.** 49.9° **8.** 40.5°

**Find the angle measure with the approximate tangent value. Give your answer to the nearest tenth of a degree.**

| | | | |
|---|---|---|---|
| **9.** 0.4663 | **10.** 5.1446 | **11.** 0.1228 | **12.** 0.2493 |
| **13.** 0.6745 | **14.** 14.3007 | **15.** 1.2800 | **16.** 0.9004 |

**Use the diagram at right for Exercises 17–21. Give your answer to the nearest tenth.**

**17.** $b = 12,\ m\angle A = 42°,\ a = ?$

**18.** $b = 35,\ m\angle B = 66°,\ a = ?$

**19.** $a = 9,\ m\angle B = 56°,\ b = ?$

**20.** $b = 6,\ a = 11,\ m\angle B = ?$

**21.** $b = 25,\ a = 18,\ m\angle A = ?$

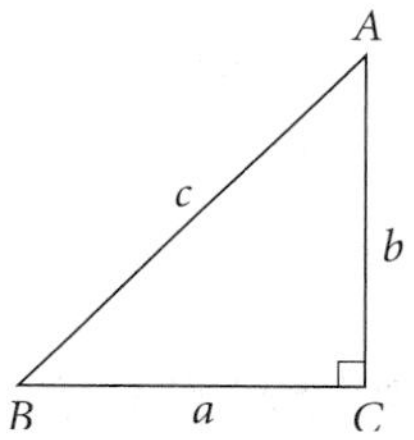


## LESSON 12.7

**Evaluate each expression to the nearest ten-thousandth.**

| | | | |
|---|---|---|---|
| **1.** sin 45° | **2.** cos 45° | **3.** sin 83° | **4.** cos 62° |
| **5.** sin 2° | **6.** cos 58.5° | **7.** cos 10.9° | **8.** sin 26.4° |
| **9.** cos 37° | **10.** sin 46.7° | **11.** cos 15° | **12.** sin 10° |

**Find the acute angle measure with the approximate sin value. Round your answers to the nearest tenth of a degree.**

**13.** 0.2924 **14.** 0.9781 **15.** 0.0523 **16.** 0.4778

**Find the acute angle measure with the approximate cosine value. Round your answers to the nearest tenth of a degree.**

**17.** 0.8090 **18.** 0.5211 **19.** 0.6901 **20.** 0.0322

**Find the indicated angle measure to the nearest tenth.**

**21.** $b = 6,\ c = 9,\ m\angle B = ?$

**22.** $b = 2,\ c = 7,\ m\angle A = ?$

**23.** $a = 9,\ c = 12,\ m\angle B = ?$

**24.** $a = 5,\ b = 12,\ m\angle B = ?$

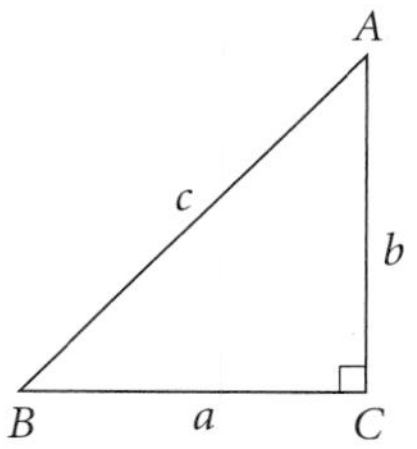

Extra Practice

## LESSON 12.8

**Perform the following matrix operations. If a solution is not possible, explain why.**

1. $\begin{bmatrix} -6 & 15 \\ 7 & -14 \end{bmatrix} - \begin{bmatrix} 0 & -12 \\ 8 & 6 \end{bmatrix}$

2. $\begin{bmatrix} 8 & 5 \\ 6 & -23 \end{bmatrix} + \begin{bmatrix} 14 & 0 \\ 1 & 2 \end{bmatrix} - \begin{bmatrix} 2 & 3 & -11 \\ -6 & 2 & 4 \end{bmatrix}$

3. $\begin{bmatrix} 5 & 7 & -9 \\ -8 & 23 & -4 \end{bmatrix} - 2\begin{bmatrix} -2 & 17 & 23 \\ -40 & 21 & -6 \end{bmatrix}$

4. $\begin{bmatrix} 2.1 & 3.5 \\ 7.8 & 1.9 \\ 0.6 & 3.3 \end{bmatrix} + \begin{bmatrix} 3.4 & 7.1 \\ 7.9 & 6.4 \\ 1.4 & 0.6 \end{bmatrix}$

5. $\begin{bmatrix} 4.2 & 2.4 \\ 1.9 & 2.5 \\ 3.7 & 5.4 \end{bmatrix} + \begin{bmatrix} 2.1 & 4.9 \\ 7.2 & 0.6 \\ 4.2 & 4.8 \end{bmatrix}$

6. $\begin{bmatrix} -\frac{1}{2} & \frac{2}{5} \\ \frac{1}{4} & \frac{3}{4} \end{bmatrix} + \begin{bmatrix} \frac{1}{2} & \frac{4}{5} \\ \frac{3}{4} & 2 \end{bmatrix}$

7. $\begin{bmatrix} 15 & 8.5 & 1 \\ 0 & 7 & 9.25 \\ 6 & 12 & 6 \end{bmatrix} + \begin{bmatrix} -2 & 5 & 6.5 \\ -1 & -2 & 0.75 \\ 15 & -3 & 2 \end{bmatrix}$

8. $\begin{bmatrix} 0 & 3.5 & -4 \\ 9 & -6 & 9 \\ 4 & 11 & -10 \end{bmatrix} + \begin{bmatrix} 3 & -5 & 2.4 \\ -11 & -1 & 6 \\ 4 & 0 & 0.5 \end{bmatrix}$

9. $\begin{bmatrix} 6 & 11 & 6 \\ 5 & 15 & 18 \\ 2 & 12 & 20 \end{bmatrix} \begin{bmatrix} 9 & 11 \\ 12 & 27 \end{bmatrix}$

10. $\begin{bmatrix} 2 & -1 \\ -3 & 7 \end{bmatrix} \begin{bmatrix} 8 & 6 \\ 0 & 8 \end{bmatrix}$

# CHAPTER 13

## LESSON 13.1

**An integer from 50 to 100 is drawn at random. Find the probability it is the following:**

1. The integer is even.
2. The integer is odd.
3. The integer is a multiple of 10.
4. The integer is a multiple of 9.
5. The integer is a multiple of 3.
6. The integer is prime.
7. The integer is a multiple of 10 AND a multiple of 9.
8. The integer is a multiple of 10 OR a multiple of 9.

**A letter of the alphabet is selected at random. Find the probability that it is the following:**

9. *q*
10. *a* or *z*
11. *r*, *s*, or *t*
12. The letter is one of the letters in the word *math*.
13. The letter is a letter from *a* to *m*, inclusive.
14. The letter is one of the letters of the word *games*.

## LESSON 13.2

**List the integers from 20 to 40, inclusive, which are the following:**

**1.** even AND odd.
**2.** even OR odd.
**3.** even AND multiples of 3.
**4.** even OR multiples of 3.
**5.** odd OR multiples of 3.
**6.** even AND multiples of 5.
**7.** even OR multiples of 5.
**8.** odd AND multiples of 5.
**9.** odd OR multiples of 5.

**A survey about favorite items on the lunch menu had the results shown in the table.**

| | Pizza | Salad | Tacos |
|---|---|---|---|
| **Boys** | 194 | 102 | 148 |
| **Girls** | 91 | 136 | 129 |

How many surveyed students

**10.** are girls?
**11.** like pizza the best?
**12.** like salad the best?
**13.** are girls AND like salad the best?
**14.** are girls OR like salad the best?
**15.** are girls AND like pizza the best?
**16.** are boys?
**17.** like tacos the best?
**18.** are boys AND like tacos the best?
**19.** are boys OR like tacos the best?

## LESSON 13.3

**A menu contains 3 appetizers, 4 salads, 6 main courses, and 5 desserts.**

**1.** How many ways are there to choose one of each?
**2.** How many combinations include a piece of apple pie, one of the 5 desserts?
**3.** How many combinations include a shrimp cocktail, one of the 3 appetizers?
**4.** How many combinations include a shrimp cocktail and a piece of apple pie?
**5.** How many combinations include a steak, one of the 6 main courses?

**A regular deck of 52 playing cards contains 26 red cards and 26 black cards. The deck contains 4 queens, 2 black and 2 red. One card is drawn. Find the number of ways to draw each.**

**6.** a black queen
**7.** a queen
**8.** a red card
**9.** a red card OR a queen

**Two cards are drawn form a regular deck of 52 playing cards. The first card is replaced before drawing the second card. Find the number of ways to draw each pair in order.**

**10.** a red card, then a queen
**11.** a queen, then a black card
**12.** a queen, then a red card
**13.** a black card, then a red card

**How many ways are there to arrange the letters in each word? (Each arrangement will not necessarily spell another word.)**

**14.** dog
**15.** mail
**16.** heart
**17.** computer

## LESSON 13.4

**Nine cards numbered 1 to 9 are in a box. A card is drawn and replaced. Then a second card is drawn. Find the probability that**

1. both are prime.
2. both are multiples of 3.
3. both are odd.
4. both are even.
5. one is even and the other is odd.
6. one is prime and the other is even.
7. the first one is even and the second is odd.
8. the first one is even and the second is a multiple of 3.

**A box contains 5 quarters, 8 dimes, 2 nickels, and 6 pennies. One coin is drawn and replaced. Then a second coin is drawn. Find the probability that**

9. both are quarters.
10. both are dimes.
11. both are nickels.
12. both are pennies.
13. one is a quarter and one is a penny.
14. at least one is a penny or a dime.
15. the first one is a nickel and the other is a quarter.

## LESSON 13.5

**Describe a way to simulate each random selection.**

1. a day of a thirty-day month
2. one student out of a class of 28 students
3. which of 6 athletes crosses the finish line first
4. which team of 10 teams wins a tournament
5. selecting one book from a collection of 15 books
6. choosing one number between the numbers 1 and 50, inclusive

# CHAPTER 14

## LESSON 14.1

**Which of the following relations are functions? Explain.**

**1.** $\{(2, 3), (3, 5), (2, 8)\}$ **2.** $\{(6, -2), (5, -2), (8, -2)\}$ **3.** $\{(0, 3), (2, 5), (-2, -5)\}$

**4.**

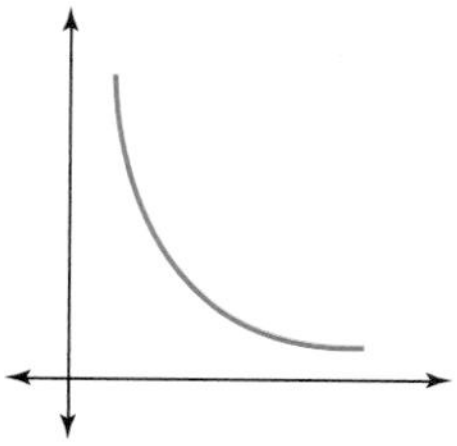

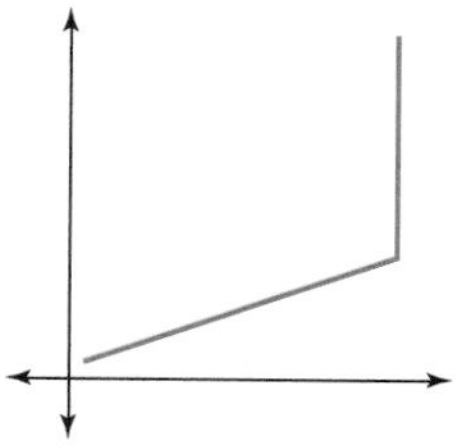

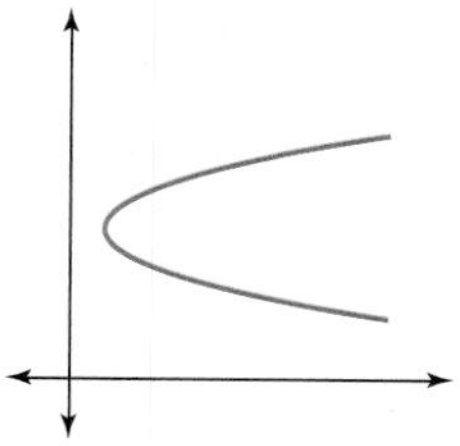

**If *f*(*x*) = 4*x*, evaluate each of the following:**

**7.** $f(3)$ **8.** $f(-1)$ **9.** $f(0)$ **10.** $f(6)$ **11.** $f(-12)$ **12.** $f(0.5)$

**Evaluate each fraction for *x* = 2.**

**13.** $h(x) = \frac{1}{x}$ **14.** $f(x) = x^2 - 5x$ **15.** $h(x) = -2|x|$

**16.** $h(x) = \frac{-4}{x}$ **17.** $k(x) = 3x^2 + 9$ **18.** $g(x) = -x^2 + x - 7$

**19.** $k(x) = -5x^2 - 3x + 1$ **20.** $h(x) = |x - 6|$ **21.** $f(x) = \frac{5}{x}$

## LESSON 14.2

**Graph each function and identify the parent function. Describe the transformation applied.**

**1.** $y = -2x^2$ **2.** $y = x^2 - 1$ **3.** $y = (x - 1)^2$ **4.** $y = 2(-x)^2$

**5.** $y = -3|x|$ **6.** $y = |x| + 2$ **7.** $y = |x + 2| + 2$ **8.** $y = |-3x|$

**9.** $y = -2|x|$ **10.** $y = |x| + 3$ **11.** $y = |x + 1|$ **12.** $y = 3x + 1$

**13.** $y = x + 6$ **14.** $y = x^2 - 3$ **15.** $y = (x - 2)^2$ **16.** $y = x - 4$

**Consider the parent function $f(x) = x^2$. Note that $f(2) = 4$, so the point (2, 4) is on the graph. Use this fact to describe how the parent function is translated.**

**17.** $f(x) = x^2 + 2$ contains the point (2, 6). **18.** $f(x) = x^2 - 2$ contains the point (2, 2).

**19.** $f(x) = (x + 2)^2$ contains the point (2, 16). **20.** $f(x) = (x - 2)^2$ contains the point (2, 0).

**The point (6, 9) is on the graph of a function. Give the coordinates of the corresponding point on the new graph after each of the following transformations are applied to the function.**

**21.** vertical translation by 5 **22.** vertical translation by –3

**23.** vertical translation by –9 **24.** vertical stretch by 4

**25.** horizontal translation by 2 **26.** horizontal translation by –1

**27.** horizontal translation by –10 **28.** vertical stretch by 5

## LESSON 14.3

**Given the function $y = 3x^2$,**

1. identify the parent function.
2. graph the function and the parent function on the same coordinate plane.
3. explain how the graph of the parent function was changed by the 3.

**Given the function $y = \frac{4}{x}$,**

4. identify the parent function.
5. graph the function and the parent function on the same coordinate plane.
6. explain how the graph of the parent function was changed by the 4.

**Given the function $y = 5|x|$,**

7. identify the parent function.
8. graph the function and the parent function on the same coordinate plane.
9. Explain how the graph of the parent function was changed by the 5.

**Determine whether each function has a graph that is a stretch of the parent function.**

10. $y = x^2 + 4$
11. $y = 3|x|$
12. $y = \frac{x}{4}$
13. $y = \frac{1}{x} + 5$

## LESSON 14.4

**Graph each function.**

1. $y = -2x^2$
2. $y = -|x|$
3. $y = \frac{-2}{x}$
4. $y = -3x$
5. $y = 4x^2$
6. $y = \frac{4}{x}$

**Determine whether each function is a vertical reflection of the parent function.**

7. $y = x^2$
8. $y = |x| - 1$
9. $y = \frac{-1}{x}$
10. $y = (-x)^2$
11. $y = -x$
12. $y = x^2 - 1$
13. $y = -x^2$
14. $y = |-x|$
15. $y = x - 1$

**Find $f(3)$, and use this information to decide whether a vertical reflection has been applied to $f(x) = x^2 - 6$.**

16. $f(x) = -(x^2 - 6)$
17. $f(x) = (-x)^2 - 6$

**Tell whether the graph of each function shows a stretch or a translation.**

18. $y = 5x$
19. $y = 5 + x$
20. $y = \frac{3}{5}x$
21. $y = \frac{3}{5} + x$

## LESSON 14.5

**Sketch the graph of each function.**

1. $k(x) = 3(x + 1)^2$
2. $n(x) = 3x^2 + 1$
3. $h(x) = |x + 2| - 1$
4. $q(x) = 5|x - 2|$
5. $f(x) = 2^x - 2$
6. $m(x) = \frac{2}{x} - 1$

**Name the point that corresponds to the point (4, 6) on the graph of $y = \frac{3}{2}x$ after each transformation is applied. HINT: If you are not sure, draw a sketch.**

7. a vertical stretch by 2, followed by a vertical translation by 2
8. a vertical reflection, followed by a vertical stretch by 3
9. a vertical reflection, followed by a vertical translation by –2
10. a vertical stretch by –2, followed by a horizontal translation by 4

**Name the point that corresponds to the point (4, 8) on the graph of $y = \frac{1}{2}x^2$ after each transformation is applied. HINT: If you are not sure, draw a sketch.**

11. a vertical stretch by 2, followed by a vertical translation by 2
12. a vertical reflection, followed by a vertical stretch by 3
13. a vertical reflection, followed by a vertical translation by –2
14. a vertical stretch by –2, followed by a horizontal translation by 4

# Calculator Keystroke Guide

## Chapter 1

*Essential keystroke sequences (using the TI-82, TI-83, or TI-83 Plus graphics calculator) are presented below for the most significant activities and examples found in this chapter that require or suggest the use of a graphics calculator.*

**internet** connect

For keystrokes of other graphing calculators models, visit the HRW web site at **go.hrw.com** and enter the keyword **MA1 CALC**.

### LESSON 1.2

**EXAMPLE 4**
*Page 14*

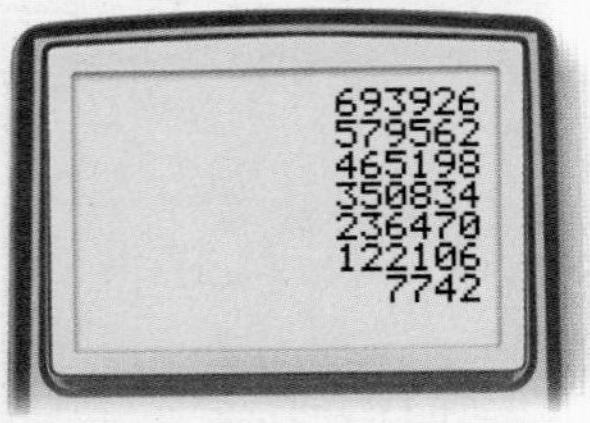


**Part a.** **Find the solution by using the constant feature of a calculator.**

**Enter 922,654 and subtract 114,364 repeatedly, counting the number of times you subtract:**

922654 ENTER − 114364 ENTER ENTER ENTER . . . (Press ENTER a total of 8 times.)

**Part b.** **Write an equation and solve it by using the table feature.**

**Enter the equation:**

Y= (Y1=) 922654 − 114364 X,T,θ,n

**Use a table:**

2nd TBLSET WINDOW (TblStart=) 1 ENTER (ΔTbl=)1 ENTER
⇑TI-82: (TblMin =)

(Indpnt:) Auto ▼ ENTER (Depend:) Auto ENTER 2nd TABLE GRAPH

### LESSON 1.3

***Activity***
*Page 19*

**For Step 3a, predict the value of 2 + 3 • 5.**

2 + 3 × 5 ENTER

Use a similar keystroke sequence for Steps 3**b**–3**h**.

### LESSON 1.3

**EXAMPLE 4**
*Page 20*

**Use a graphics or scientific calculator to evaluate**

$$\frac{57+95}{16}+\frac{220}{88+104}.$$

( 57 + 95 ) ÷ 16 + 220 ÷ ( 88 + 104 ) ENTER

*Activity*

**Page 31**

*The window dimension 4.7 produces a "friendly" window which is easy to use with the TRACE function.*

**Graph and trace the equation $d = 53t$ on a graphics calculator.**

Use friendly viewing window [0, 4.7] by [0, 250].

**Set the viewing window:**

WINDOW **(Xmin =) 0** ▼ **(Xmax=) 4.7** ▼ **(Xscl=) 1** ▼ **(Ymin=) 0** ▼ **(Ymax=) 250** ▼ **(Yscl=) 10**

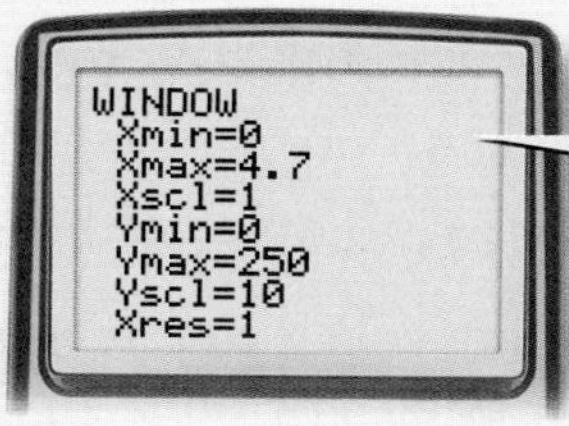


*The scales of the x-axis and y-axis should be chosen according to the difference between the minimum and maximum values. (Normally, **Xres=1** should not be changed.)*

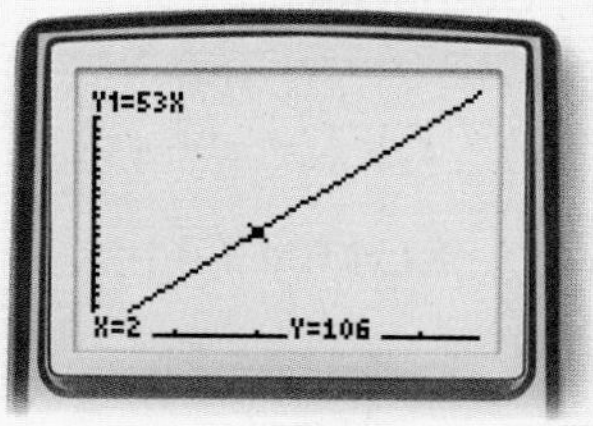


**Graph and trace the equation:**

Y= **(Y1=) 53** X,T,θ,n GRAPH TRACE (Use ◄ or ► to move along the graph.)

**E X A M P L E 2** **Graph the equation $d = 280t$.**

**Pages 32 and 33**

*The window dimension 9.4 also produces a friendly viewing window because it is a multiple of 4.7.*

Use friendly viewing window [0, 9.4] Xscl = 1 by [0, 3000] Yscl = 500.

**Graph the equation:**

Y= **(Y1=) 280** X,T,θ,n GRAPH TRACE

(Use ◄ or ► to move along the graph to the desired point.)

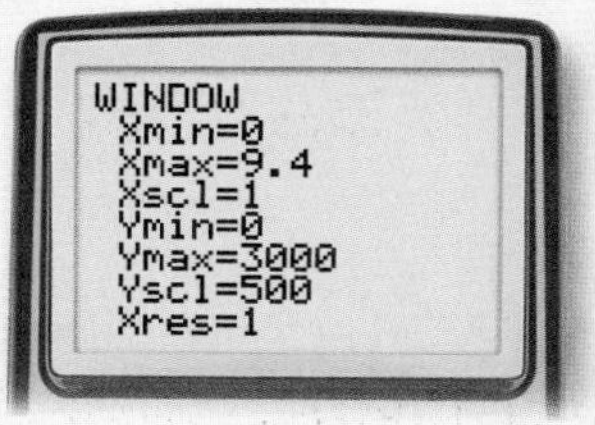


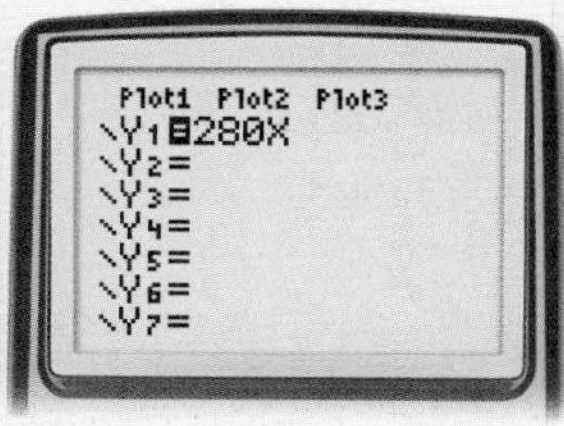


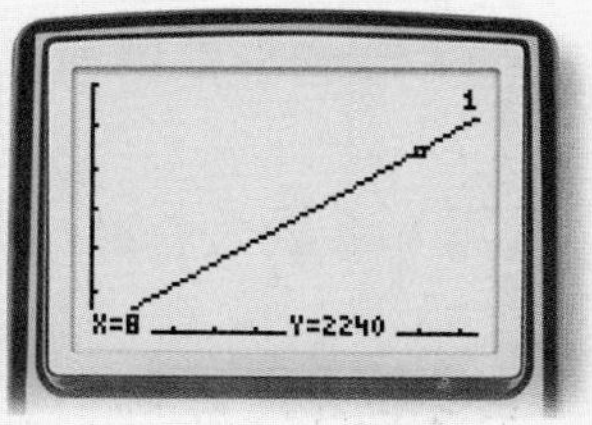


**E X A M P L E 3** **Graph the equation $r = 220 - a$ on your graphics calculator.**

**Page 33**

Use friendly viewing window [0, 94] Xscl = 10 by [0, 230] Yscl = 25.

Y= **(Y1=) 220** – X,T,θ,n ENTER GRAPH TRACE

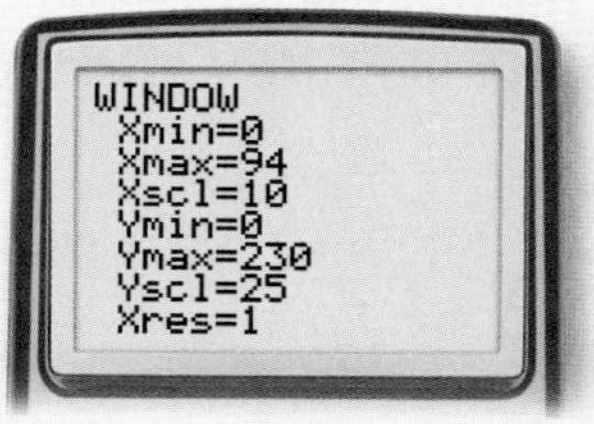


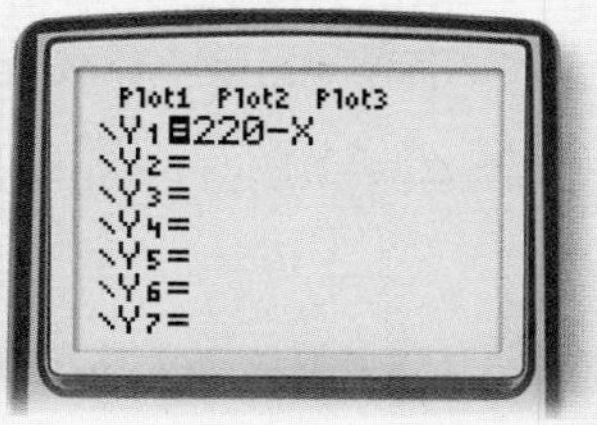


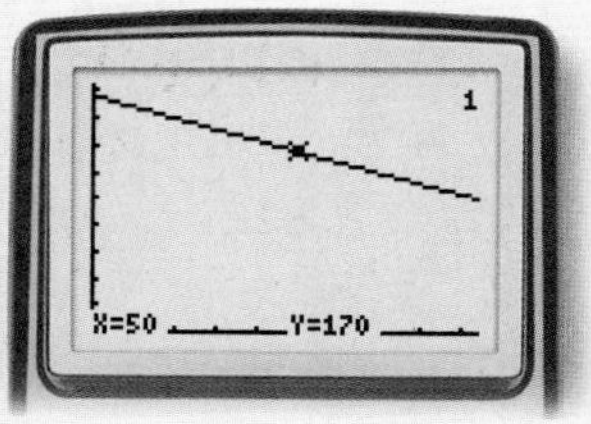

## LESSON 1.6

*Activity*

*Pages 38 and 39*

**For Step 1, use a graphics calculator to draw a scatter plot of the data.**

Use viewing window [0, 20] Xscl = 1 by [0, 250] Yscl = 10.

**Make a table of values:**
STAT EDIT 1:EDIT ENTER

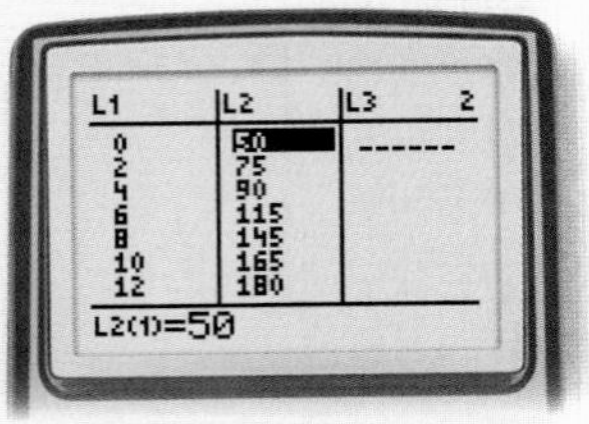


**Under the heading L1, enter the data values that represent the year:**
0 ENTER 2 ENTER 4 ENTER 6 ENTER 8 ENTER 10 ENTER 12 ENTER 14 ENTER 16 ENTER (Use ► and ▲ to move to the top of the **L2** list.)

**Under L2, enter the data values that represent population:**
50 ENTER 75 ENTER 90 ENTER 115 ENTER 145 ENTER 165 ENTER 180 ENTER 200 ENTER 230 ENTER 2nd Y= STAT PLOTS

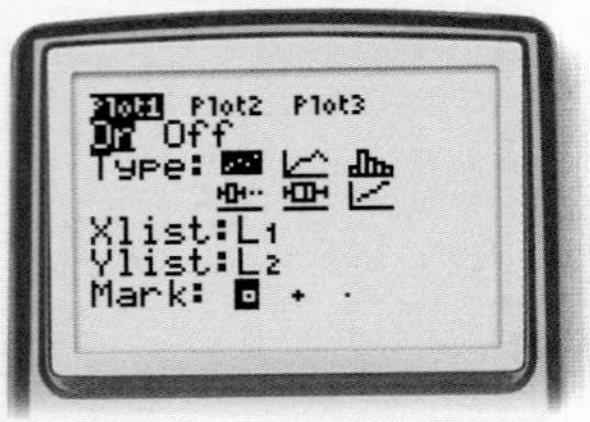


**1:Plot 1 ENTER On ENTER ▼ (TYPE:) ENTER ▼ (XLIST:) L1 ENTER (YLIST:) L2 ENTER Mark: ▪ ENTER GRAPH**

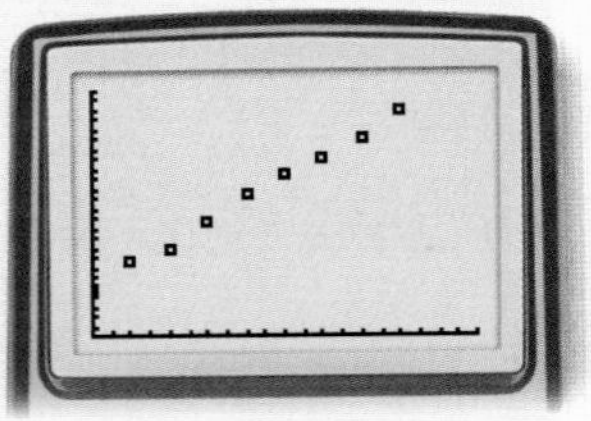

**For Step 3, graph the equation $y = 10x + 50$:**
Y= (Y1= ) 10 X,T,θ,n + 50 GRAPH

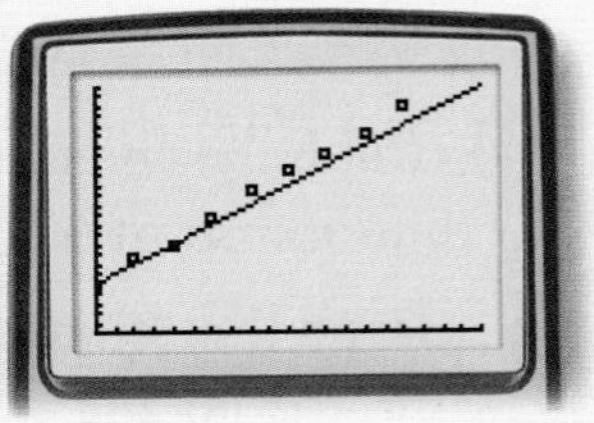

For Step 4, adjust the equation to fit the data by using a guess and check strategy.

# Chapter 3

*Essential keystroke sequences (using the TI-82, TI-83, or TI-83 Plus graphics calculator) are presented below for the most significant activities and examples found in this chapter that require or suggest the use of a graphics calculator.*

**internet** connect

For keystrokes of other graphing calculators models, visit the HRW web site at **go.hrw.com** and enter the keyword **MA1 CALC**.

## LESSON 3.1

**EXAMPLE** 5

*Page 118*

**Graph the equations $x + 7 = y$ and $y = 10$ on the same screen and find the $x$-value at the intersection point.**
Use friendly viewing window [0, 9.4] by [0, 15].

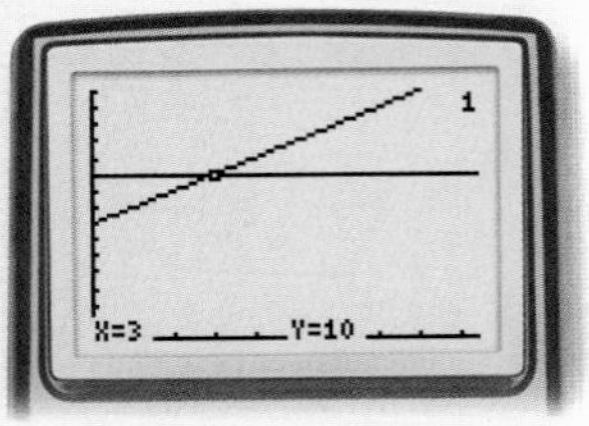


**Graph the equations:**
Y= (Y1=) X,T,θ,n + 7 ENTER (Y2=) 10 GRAPH

**Find the point of intersection using the TRACE feature:**
TRACE (Use ◄ or ► to move the cursor to $y = 10$.)

**Set up a table:**

2nd WINDOW (TBLSET) (TblStart=)1 ENTER ($\Delta$Tbl=)1 ENTER
⇑TI-82: (TblMin=)

(Indpnt:) **Auto** ENTER ▼ (Depend:) **Auto** ENTER

Use the TABLE feature to confirm that $x = 3$ when $y = 10$ by determining the $x$-value when both **Y1** and **Y2** equal 10.

**Locate values on the table:** 2nd GRAPH (TABLE)

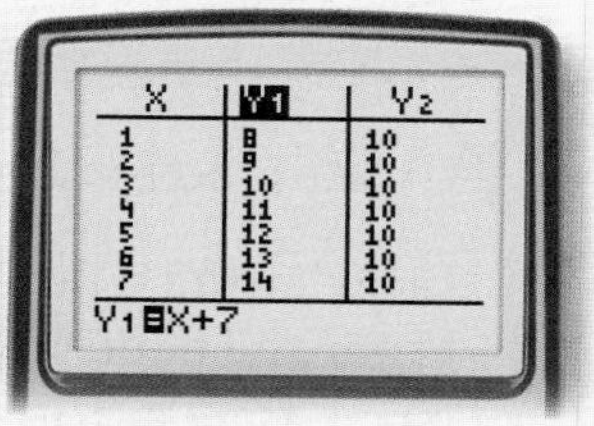


## LESSON 3.3

**TECHNOLOGY**
*Page 132*

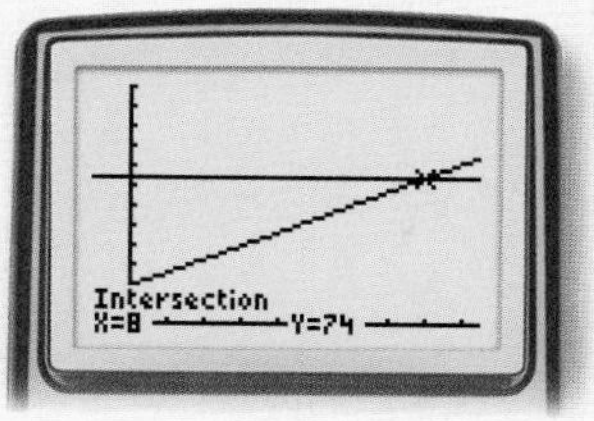


**Graph the equations $y = 18 + 7x$ and $y = 74$ on the same screen.**
Use window $[-1, 9.4]$ by $[-5, 120]$.

Y= (Y1=) 18 + 7 X,T,θ,n ENTER (Y2=) 74 GRAPH

**Find the point of intersection using the CALCULATE function:**

2nd TRACE (CALC) **5:intersect** ENTER (First curve?) ENTER
(Second curve?) ENTER (Guess?) ENTER

## LESSON 3.4

*Page 137*

**Solve the equation $5x - 6 = -6x + 16$.**
Use window $[-5, 10]$ by $[-5, 10]$.
Graph the equations $y = 5x - 6$ and $y = -6x + 16$ on the same calculator display and find their point of intersection.

**Graph the equations:**

Y= (Y1=) 5 X,T,θ,n − 6 ENTER (Y2=) (−) 6 X,T,θ,n + 16 GRAPH

**Find the point of intersection using the CALCULATE function:**

2nd TRACE (CALC) **5:intersect** ENTER (First curve?) ENTER

(Second curve?) ENTER (Guess?) ENTER

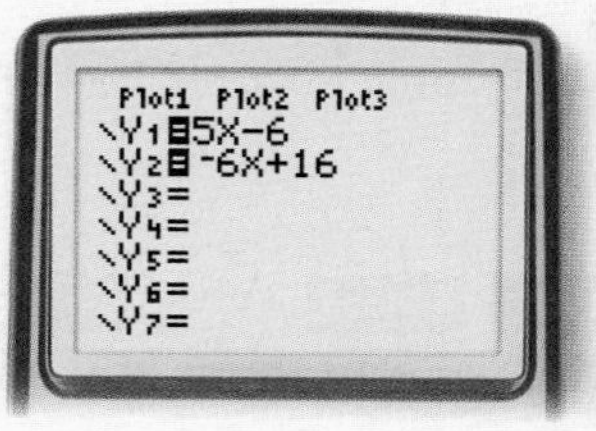


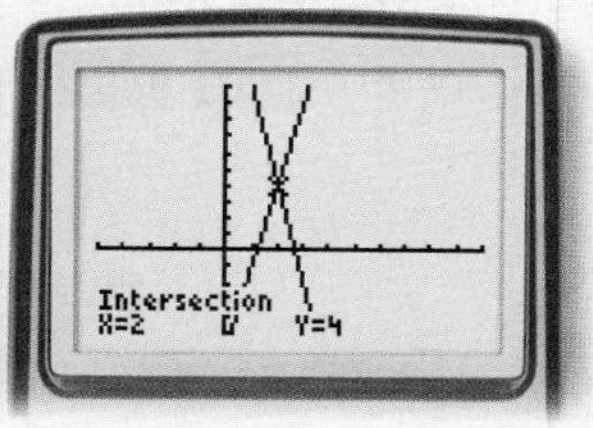

# Chapter 4

*Essential keystroke sequences (using the TI-82, TI-83, or TI-83 Plus graphics calculator) are presented below for the most significant activities and examples found in this chapter that require or suggest the use of a graphics calculator.*

**internet** connect

For keystrokes of other graphing calculators models, visit the HRW web site at **go.hrw.com** and enter the keyword **MA1 CALC**.

## LESSON 4.6

**EXAMPLE 2**

*Pages 203 and 204*

**Create a box-and-whisker plot from the data given in Example 2.**

Use viewing window [12, 16] Xscl = 0.2 by [0, 10] Yscl = 1.

**Enter the data:**

STAT EDIT 1:EDIT ENTER (LI) 12.7 ENTER 12.8 ENTER 12.8 ENTER 12.8 ENTER 12.9 ENTER 13.0 ENTER 13.1 ENTER 13.1 ENTER 1 ENTER (Continue using a similar keystroke sequence to enter the remaining data.)

**Create a box-and-whisker plot:**

2nd Y= STAT PLOTS 1:Plot 1 ENTER On ENTER ▼ (TYPE:) ENTER ▼ (XLIST:) L1 ENTER (Freq:) 1 ENTER GRAPH

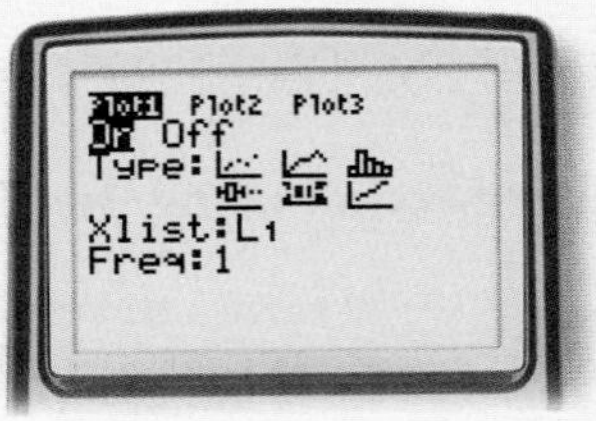


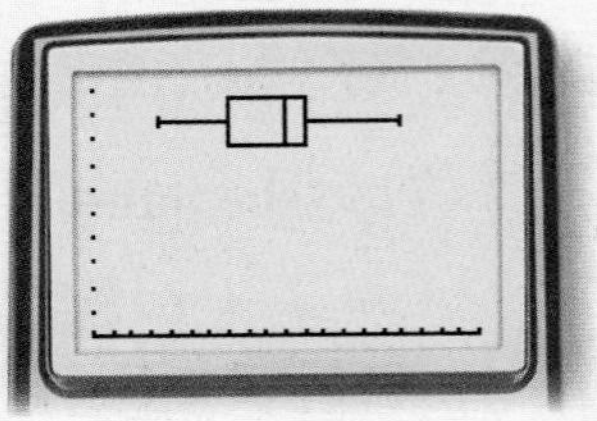

# Chapter 5

*Essential keystroke sequences (using the TI-82, TI-83, or TI-83 Plus graphics calculator) are presented below for the most significant activities and examples found in this chapter that require or suggest the use of a graphics calculator.*

**internet** connect

For keystrokes of other graphing calculators models, visit the HRW web site at **go.hrw.com** and enter the keyword **MA1 CALC**.

## LESSON 5.1

**EXAMPLE 3**

*Page 221*

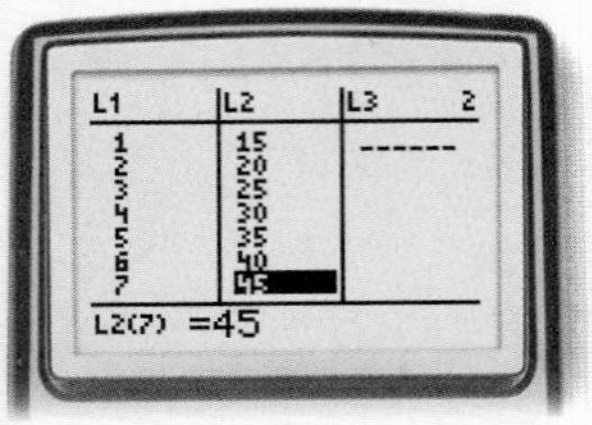


**Use your graphing calculator to list the ordered pairs and plot the corresponding points.**

Use viewing window [0, 12] Xscl = 1 by [0, 70] Yscl = 5.

**Enter the data:**

STAT EDIT 1:EDIT ENTER (LI) 1 ENTER 2 ENTER 3 ENTER and so on through L1 and L2.

**Create a graph:**

2nd Y= STAT PLOTS 1:Plot 1 ENTER On ENTER ▼ (TYPE:) ENTER ▼ (XLIST:) L1 ENTER (YLIST:) L2 ENTER (Mark:) ▫ ENTER GRAPH

## LESSON 5.2

### *Activity*

**For Step 1, graph the point A(3, 6).**
Use viewing window [–5, 10] by [–10, 10].

**Graph the point:**
STAT EDIT 1:EDIT ENTER (LI) 3 ENTER (L2) 6 ENTER 2nd Y=
STAT PLOTS 1:Plot 1 ENTER On ENTER ▼ (TYPE:) ⋰ ENTER ▼
(XLIST:) L1 ENTER (YLIST:) L2 ENTER (Mark:) ▫ ENTER GRAPH

For Step 2, find values for $m$ such that line $y = mx$ intersects with point $A$. For example, let $m = 4$: Y= (Y1=)4 X,T,θ,n GRAPH

For Step 3, use a similar keystroke sequence to graph the other given points.

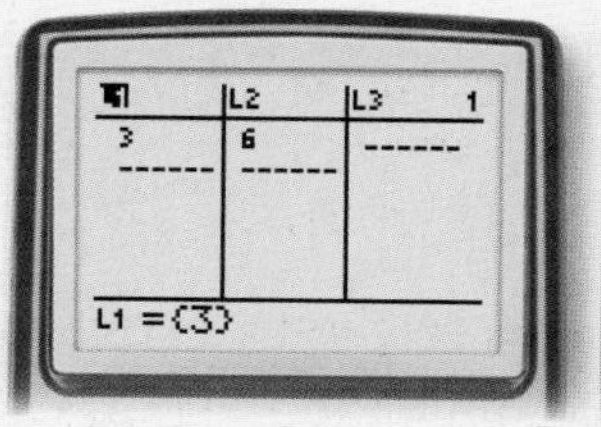


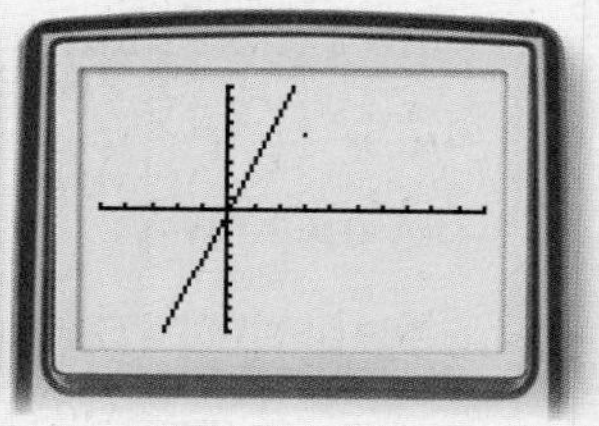

## LESSON 5.4

### *Activity*

*Page 246*

**For Step 1, graph the equations $y = 2x$, $y = 2x + 3$, $y = 2x + 5$, and $y = 2x - 3$ on the screen.**

**To format the *standard viewing window* [–10, 10] by [–10, 10]:** ZOOM ZOOM 6:ZStandard ENTER

**Enter the equations:**
Y= (Y1=) 2 X,T,θ,n ENTER (Y2=) 2 X,T,θ,n + 3 ENTER
(Y3=) 2 X,T,θ,n + 5 ENTER (Y4=) 2 X,T,θ,n – 3 ENTER GRAPH

For Step 4, use a similar keystroke sequence to graph $y = 2x - 4$.

## LESSON 5.5

**EXAMPLE 2**

*Pages 253 and 254*

**Enter the equation from Example 2 in a graphics calculator.**
(First rewrite the equation in slope-intercept form, $y = -2x + 300$.)

Use viewing window [–10, 200] by [–10, 350] with scales of 10.

**Graph the equation:**
Y= (Y1=) (–) 2 X,T,θ,n + 300 GRAPH

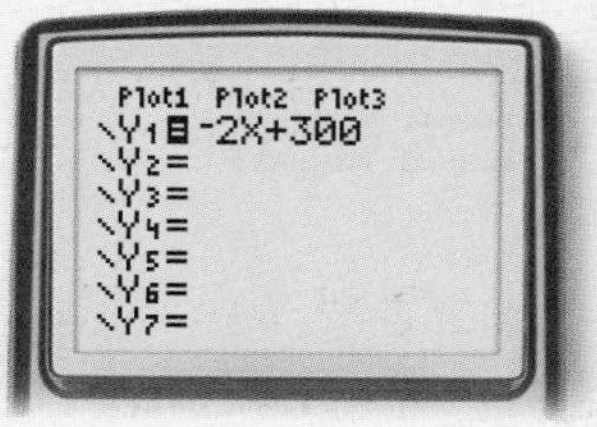


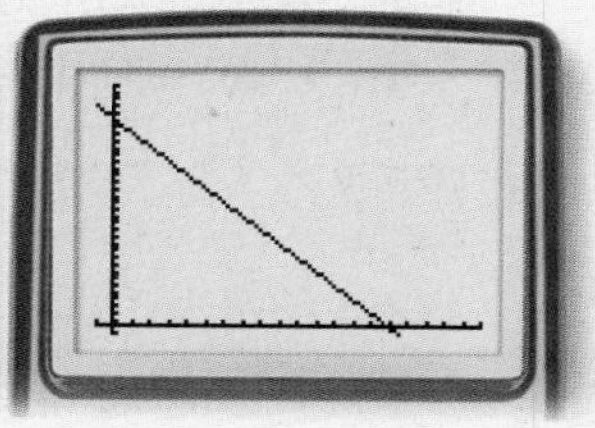

# Chapter 6

*Essential keystroke sequences (using the TI-82, TI-83, or TI-83 Plus graphics calculator) are presented below for the most significant activities and examples found in this chapter that require or suggest the use of a graphics calculator.*

**internet connect**

For keystrokes of other graphing calculators models, visit the HRW web site at **go.hrw.com** and enter the keyword **MA1 CALC**.

## LESSON 6.2

**EXAMPLE 4** *Page 285*

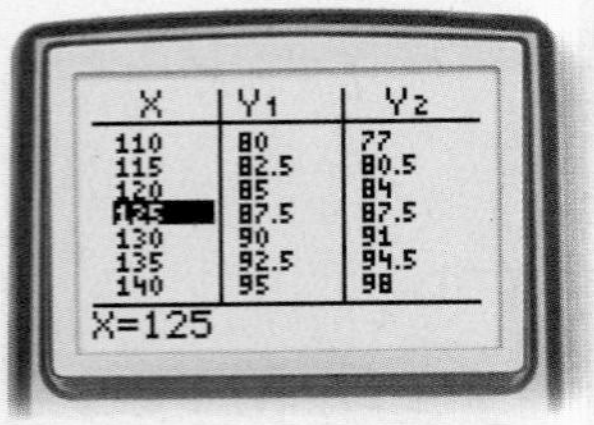


**Use the TABLE feature to detemine the more economical calling plan.**

**Enter the equations:**

[Y=] **(Y1=)** 25 [+] 0.5 [X,T,θ,n] [ENTER] **(Y2=)** 0.7 [X,T,θ,n]

**Set up a table:**

[2nd] [WINDOW] (TBLSET) **(TblStart=)** 100 [ENTER] **(ΔTbl=)** 5 [ENTER]

⇑TI-82: (TblMin =)

**(Indpnt:)** [Auto] [ENTER] [▼] **(Depend:)** [Auto] [ENTER]

**Locate values on the table:** [2nd] [GRAPH] (TABLE)

Use the [▼] key to locate the $x$-value when **Y1 = Y2**.

## LESSON 6.4

***Activity*** *Page 296*

**For Step 1, graph $y = |x|$ and $y = |x - 3|$ on the same screen.**

Use viewing window [−10, 10] by [−5, 5].

**Graph the equations:**

[Y=] **(Y1=)** [MATH] [►] [NUM] [1:abs (] [ENTER] [X,T,θ,n] [ ) ]

⇑TI-82: [2nd] [$x^{-1}$] (ABS) [ ( ]

**(Y2=)** [MATH] [►] [NUM] [1:abs (] [ENTER] [X,T,θ,n] [−] 3 [ ) ] [GRAPH]

⇑TI-82: [2nd] [$x^{-1}$] (ABS) [ ( ]

Use the same sequence of keystrokes to graph the equations in Steps 3 and 5.

## LESSON 6.5

***Activity*** *Page 304*

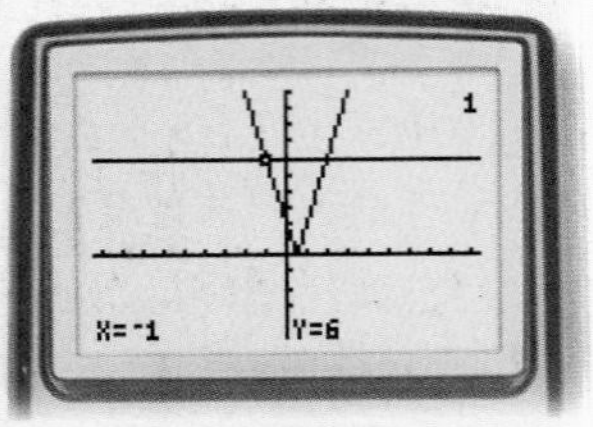


**For Step 1, graph the equations $y = |4x - 2|$ and $y = 6$ on the same screen.**

Use viewing window [−10, 10] by [−5, 10].

**Graph the equations:**

[Y=] **(Y1=)** [MATH] [►] [NUM] [1:abs (] [ENTER] 4 [X,T,θ,n] [−] 2 [ ) ] [ENTER]

⇑TI-82: [2nd] [$x^{-1}$] (ABS) [ ( ]

**(Y2=)** 6 [GRAPH]

Use a similar keystroke sequence for Steps **4a–4c**.

# Chapter 7

*Essential keystroke sequences (using the TI-82, TI-83, or TI-83 Plus graphics calculator) are presented below for the most significant activities and examples found in this chapter that require or suggest the use of a graphics calculator.*

**internet connect**

For keystrokes of other graphing calculators models, visit the HRW web site at **go.hrw.com** and enter the keyword **MA1 CALC**.

## LESSON 7.1

### EXAMPLE 1
*Page 319*

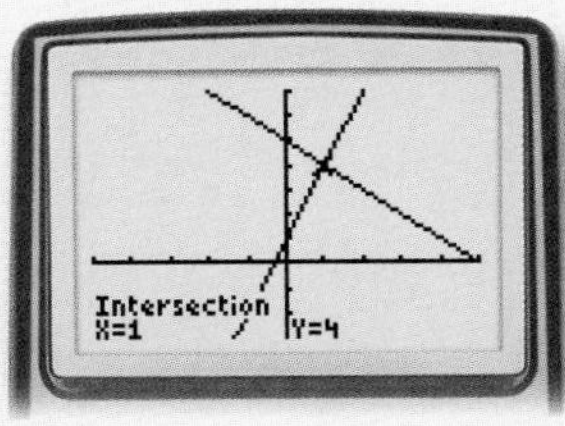


**Graph $y = 3x + 1$ and $y = -x + 5$ on the same screen.**
Use viewing window [–5, 5] by [–3, 7].

**Graph the equations:**

Y= **(Y1=)** 3 X,T,θ,n + 1 ENTER **(Y2=)** (–) X,T,θ,n + 5 GRAPH

**Find the point of intersection:**

2nd TRACE (CALC) 5:intersect ENTER **(First curve?)** ENTER **(Second curve?)** ENTER **(Guess?)** ENTER

### EXAMPLE 2
*Page 320*

**Graph $y = -3x + 4$ and $y = \frac{1}{2}x - 3$ on the same screen.**

Use viewing window [–3, 7] by [–7, 3].

Use a keystroke sequence similar to Example 1 in this lesson.

### *Activity*
*Page 321*

**For Step 3, graph the equations $4x + 3y = 6$ and $2y = x + 2$ on the same screen.**

Use the standard viewing window (shown in the activity in 5.4).

Enter the equations in slope–intercept form. Use a keystroke sequence similar to that in Example 1 of this lesson to graph the equations and to find the point of intersection.

## LESSON 7.4

### *Activity*
*Page 338*

**For Step 1, graph $y = -3x + 4$ and $y = -3x + 8$ on the same screen.**

Use viewing window [–4.7, 4.7] by [–3.1, 3.1].

**Graph the equations:**

Y= **(Y1=)** (–) 3 X,T,θ,n + 4 ENTER **(Y2=)** (–) 3 X,T,θ,n + 8 GRAPH

*To enter $3y = 6x - 9$ in slope-intercept form you need to divide both sides by 3.*

**For Step 2, graph $y = 2x - 3$ and $3y = 6x - 9$ on the same screen:**
[Y=] (Y1=) 2 [X,T,θ,n] [–] 3 [ENTER] (Y2=) 2 [X,T,θ,n] [–] 3 [GRAPH]

**EXAMPLE 1** *Page 339*

**Determine whether the system** $\begin{cases} y = 4 - x \\ y = 3x - 6 \end{cases}$ **is consistent or inconsistent.**
Use a keystroke sequence similar to that in the Activity in this lesson.

**EXAMPLE 3** *Page 340*

**Determine the line of best fit for the two sets of data in Example 3.**
Use viewing window [0, 25] by [0, 30] with scales of 5.

**Make a table of values:**
[STAT] [EDIT] [1:EDIT] [ENTER]

Under the heading **L1**, enter the data values that represent the year:
0 [ENTER] 5 [ENTER] 10 [ENTER] 15 [ENTER] 20 [ENTER]

Use [►] and [▲] to move to the top of the **L2** list.

Under **L2**, enter the data values that represent women's ages:
20.6 [ENTER] 20.8 [ENTER] 21.8 [ENTER] 23 [ENTER] 24 [ENTER]

**Under L3, enter the data values that represent men's ages:**
22.5 [ENTER] 22.7 [ENTER] 23.6 [ENTER] 24.8 [ENTER] 25.9 [ENTER]

**Make a scatterplot of the women's ages:**
[2nd] [Y=] [STAT PLOTS] [1:Plot 1] [ENTER] [ON] [ENTER] [▼] (TYPE:) [scatter] [ENTER] [▼] (XLIST:) [L1] [ENTER] (YLIST:) [L2] [ENTER] Mark: [▫] [ENTER] [GRAPH]
Make a scatterplot of the men's ages using the same keystrokes but substituting **2:PLOT2** for **1:PLOT1** and **YLIST: L3** for **YLIST: L2**. In order to differentiate between the plots, you may also select one of the other marks.

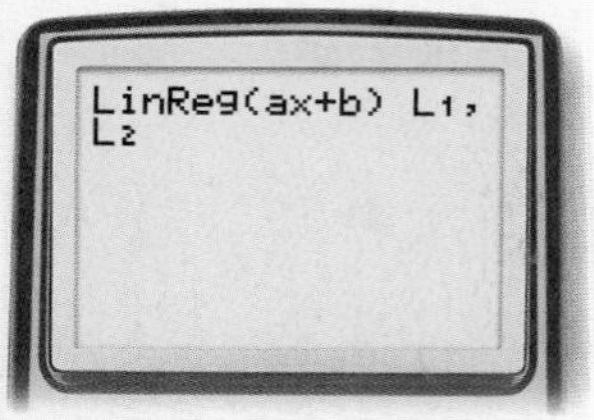

**Find the equation for the line of best fit for Plot 1 using the linear regression feature:**
[STAT] [CALC] [4:LinReg(ax+b)] [ENTER] (LinReg(ax+b)) [2nd] [1] (L1) [,] [2nd] [2] (L2) [ENTER]

Use the same keystroke sequence for Plot 2, entering [2nd] [3] (L3) in place of [2nd] [2] (L2).

**EXAMPLE 4** *Page 341*

**Determine whether the system** $\begin{cases} y = 2x - 1 \\ 3x - 1.5y = 1.5 \end{cases}$ **is dependent or independent.**

Use a keystroke sequence similar to that in the Activity in this lesson.

## LESSON 7.5

*Activity*

Page 346

**Graph the inequality $x + y \leq 5000$.**
Use viewing window [−1000, 6000] by [−1000, 6000] with scales of 1000.

For Step 1, replace the inequality sign with an equal sign in the equation $x + y \leq 5000$, then graph the resulting equation $y = -x + 5000$.

**Graph the equation:**

Y= **(Y1=)** (−) X,T,θ,n + **5000** GRAPH

**For Step 3, shade the values that satisfy the inequality:**
Go to the Y= menu and press ◄ two times or until the cursor is on the left of **Y1=**. Then press ENTER three times or until you see the shaded area flashing, then press GRAPH.

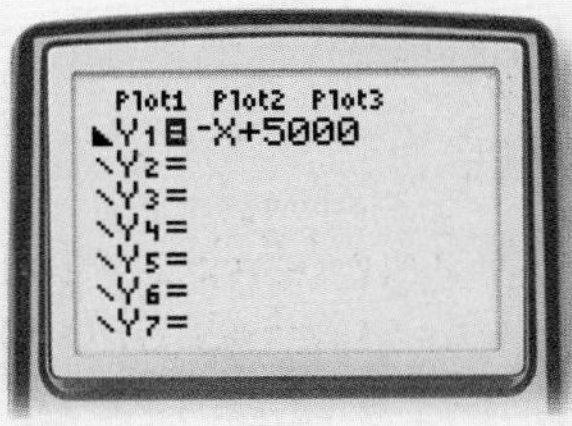


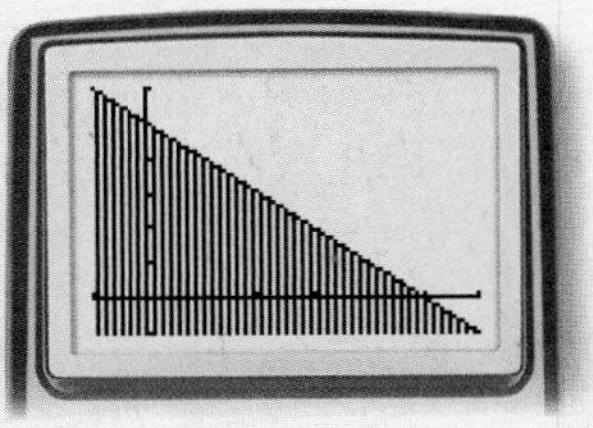

# Chapter 8

*Essential keystroke sequences (using the TI-82, TI-83, or TI-83 Plus graphics calculator) are presented below for the most significant activities and examples found in this chapter that require or suggest the use of a graphics calculator.*

**internet** connect

For keystrokes of other graphing calculators models, visit the HRW web site at **go.hrw.com** and enter the keyword **MA1 CALC**.

## LESSON 8.7

**EXAMPLE 5**

Page 412

**Solve the population problem in Example 5.**
Use viewing window [0, 10] Xscl = 1 by [110 000, 170 000] Yscl = 10000.

Graph $y = 152{,}494(0.99)^x$ and $y = 140{,}000$ on the same screen, and find their point of intersection.

**Graph the equations:**

Y= **(Y1=) 152494** ( **.99** ) ^ X,T,θ,n ENTER
**(Y2=) 140000** GRAPH

**Find the point of intersection:**

2nd CALC TRACE **5:intersect** ENTER **(First curve?)** ENTER

**(Second curve?)** ENTER **(Guess?)** ENTER

# Chapter 9

*Essential keystroke sequences (using the TI-82, TI-83, or TI-83 Plus graphics calculator) are presented below for the most significant activities and examples found in this chapter that require or suggest the use of a graphics calculator.*

**internet connect**

For keystrokes of other graphing calculators models, visit the HRW web site at **go.hrw.com** and enter the keyword **MA1 CALC**.

## LESSON 9.4

**EXAMPLE 2** *Page 445*

**Use a table to determine the value for $h$ in the equation $1357.2 = 36\pi h$.**

To enter the equation, substitute $x$ for $h$ and enter $y = 36\pi h$.

**Enter the equation:**

[Y=] **(Y1=)** 36 [2nd] [^] (π) [X,T,θ,n]

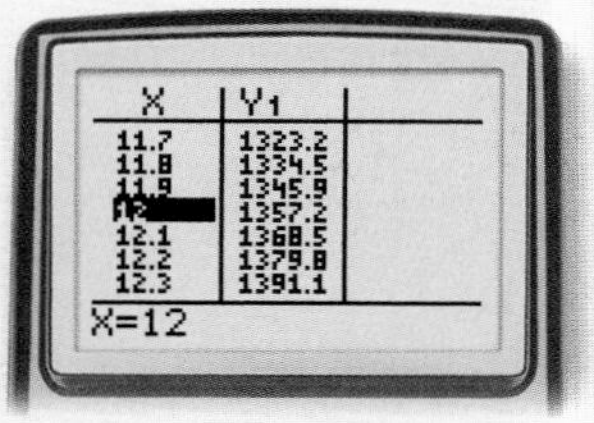


**Make a table:**

[2nd] [WINDOW] (TBLSET) **(TblStart=)** 11.7 [ENTER] **(ΔTbl=)** .1 [ENTER]
⇑TI-82: (TblMin=)

**(Indpnt:)** **Auto** [ENTER] [▼] **(Depend:)** **Auto** [ENTER] [2nd] [GRAPH] (TABLE)

Use [▼] to move the cursor and determine that when $y = 1357.2$, $x = 12$.

**EXAMPLE 3** *Page 445*

**Show that $(x + 2)^3 = x^3 + 6x^2 + 12x + 8$ is an identity.**

**Enter each side as a separate equation:**

[Y=] **(Y1=)** [(] [X,T,θ,n] [+] 2 [)] [^] 3 [ENTER]

**(Y2=)** [X,T,θ,n] [^] 3 [+] 6 [X,T,θ,n] [^] 2 [+] 12 [X,T,θ,n] [+] 8

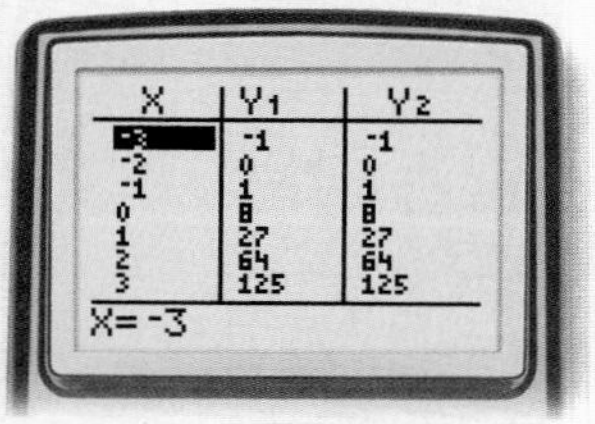


**Make a table of values:**

[2nd] [WINDOW] (TBLSET) **(TblStart=)** [(-)] 3 [ENTER] **(ΔTbl=)** 1 [ENTER]
⇑TI-82: (TblMin=)

**(Indpnt:)** **Auto** [ENTER] [▼] **(Depend:)** **Auto** [ENTER] [2nd] [GRAPH] (TABLE)

Use the table to confirm that Y1 = Y2 for integer values greater than –3.

## LESSON 9.8

*Activity*

**For Step 6, graph the equation $y = -16x^2 + 48x$.**

Use friendly viewing window [0, 4.7] Xscl = 1 by [0, 50] Yscl = 10.

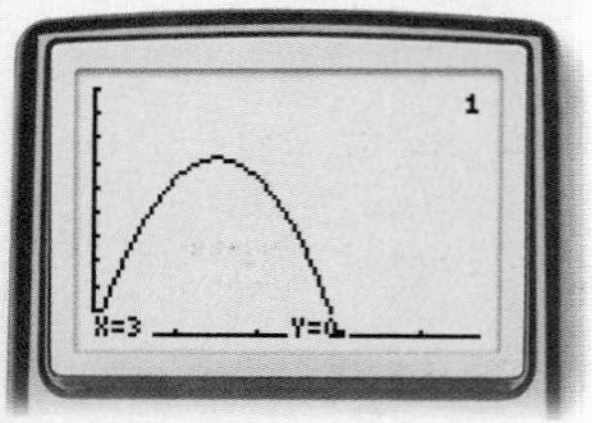


**Graph the equation:**

[Y=] **(Y1=)** [(-)] 16 [X,T,θ,n] [$x^2$] [+] 48 [X,T,θ,n] [GRAPH]

**Locate the $x$-value(s) for the point $y = 0$:** [TRACE].

Then use [►] until the cursor arrives at $y = 0$. The $x$-value will appear on the screen.

**TECHNOLOGY**
*Page 465*

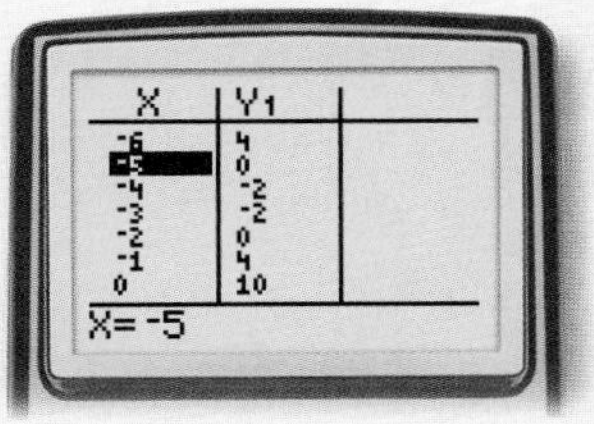


**Find the zeros of the polynomial function $y = (x + 5)(x + 2)$.**

**Enter the function:**

[Y=] **(Y1=)** [ ( ] [X,T,θ,n] [ + ] 5 [ ) ] [ ( ] [X,T,θ,n] [ + ] 2 [ ) ]

**Make a table of values:**

[2nd] [WINDOW] (TBLSET) **(TblStart=)** [(–)] 6 [ENTER] **(ΔTbl=)1** [ENTER]
⇑TI-82: (TblMin=)

**(Indpnt:)** **Auto** [ENTER] **(Depend:)** **Auto** [ENTER] [2nd] [GRAPH] (TABLE)

Use [▼] to locate $x$-values that give a $y$-value of 0.

**EXAMPLE 3**
*Page 466*

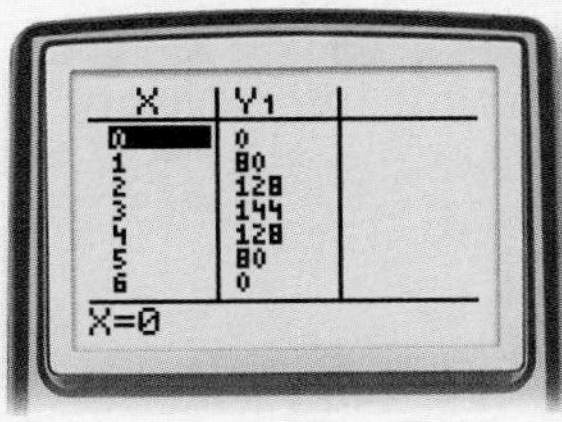


**Check your solutions to the equation $-16x^2 + 96x = 0$ by using the table and graph features.**
Use friendly viewing window [0, 9.4] by [0, 150].

**Graph the function:**

[Y=] **(Y1=)** [(–)] 16 [X,T,θ,n] [$x^2$] [ + ] 96 [X,T,θ,n] [GRAPH]

**Make a table of values:**

[2nd] [WINDOW] (TBLSET) **(TblStart=) 0** [ENTER] **(ΔTbl=) 1** [ENTER]
⇑TI-82: (TblMin=)

**(Indpnt:)** **Auto** [ENTER] [▼] **(Depend:)** **Auto** [ENTER] [2nd] [GRAPH] (TABLE)

Identify $x$-values that give a $y$-value of 0.

**TECHNOLOGY**
*Page 467*

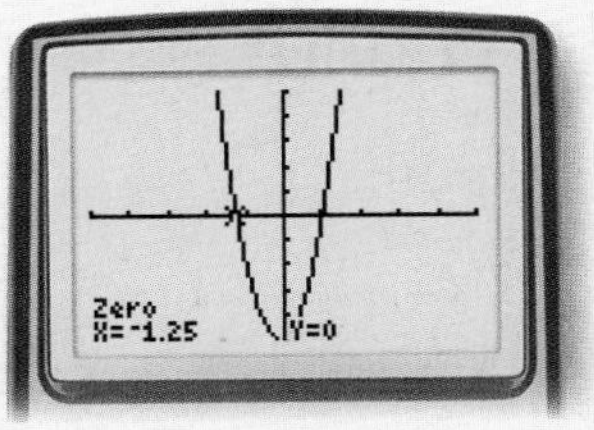


**Find the zeros of the polynomial function $y = 4x^2 + x - 5$.**
Use viewing window [–5, 5] by [–5, 5].

**Graph the function:**

[Y=] **(Y1=) 4** [X,T,θ,n] [$x^2$] [ + ] [X,T,θ,n] [ – ] 5 [GRAPH]

When you use the zero operation, you will have to provide left and right bounds. These bounds can be any points you choose on the indicated side of the $x$-intercept. Use the [◄] and [►] keys to move to the indicated side.

**Find the zeros:**

[2nd] [TRACE] (CALC) **2:zero** [ENTER] **(Left Bound?)** (move the cursor to any point left of one $x$-intercept.) [ENTER] **(Right Bound?)** (move the cursor to any point right of the same $x$-intercept.) [ENTER] **(Guess?)** [ENTER]
⇑TI-82: **2:root** ⇑TI-82: (Lower Bound?) ⇑TI-82: (Upper Bound?)

Use a similar keystroke sequence to find the other zero.

# Chapter 10

*Essential keystroke sequences (using the TI-82, TI-83, or TI-83 Plus graphics calculator) are presented below for the most significant activities and examples found in this chapter that require or suggest the use of a graphics calculator.*

**internet** connect

For keystrokes of other graphing calculators models, visit the HRW web site at **go.hrw.com** and enter the keyword **MA1 CALC**.

## LESSON 10.4

**EXAMPLE 4**
*Page 500–501*

**Solve $-5 = x^2 + 3x - 15$.**

Use a friendly viewing window [−9.4, 9.4] by [−20, 10].

**Graph the function:**

[Y=] **(Y1=)** [X,T,θ,n] [$x^2$] [+] 3 [X,T,θ,n] [−] 15 [ENTER]
**(Y2=)** [(−)] 5 [GRAPH]

**Find any points of intersection:**

[2nd] [TRACE] (CALC) [5:intersect] [ENTER] **(First curve?)** [ENTER]
**(Second curve?)** [ENTER] **(Guess?)** [ENTER]

Use similar keystrokes to identify the other point(s) of intersection. However, when **Guess?** is displayed, move the cursor closer to the desired intersection before you press [ENTER].

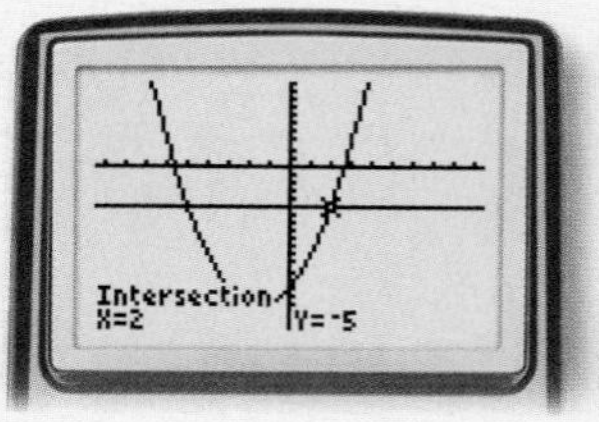


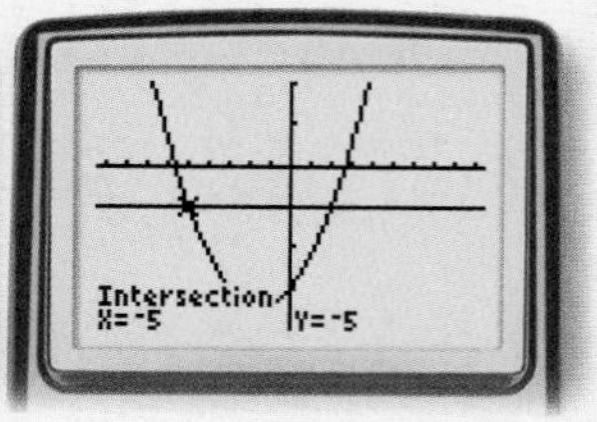


**EXAMPLE 5**
*Page 501*

**Find the point(s) where the graph of $y = x^2 - 10x + 21$ intersects the graph of $y = x - 3$.**

Use a friendly viewing window [−9.4, 9.4] by [−10, 10].

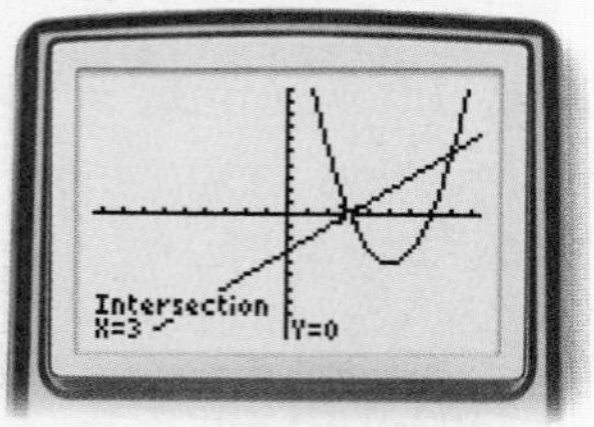


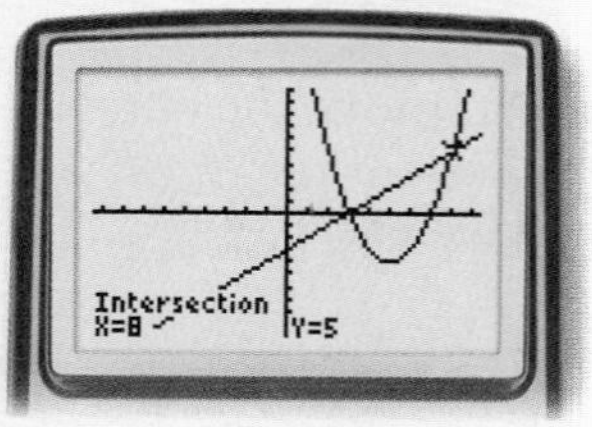


**Graph the functions:**

[Y=] **(Y1=)** [X,T,θ,n] [$x^2$] [−] 10 [X,T,θ,n] [+] 21 [ENTER]
**(Y2=)** [X,T,θ,n] [−] 3 [GRAPH]

**Find any points of intersection:**

[2nd] [TRACE] (CALC) [5:intersect] [ENTER] **(First curve?)** [ENTER]
**(Second curve?)** [ENTER] **(Guess?)** [ENTER]

Use similar keystrokes to find the other point of intersection.

# Chapter 11

*Essential keystroke sequences (using the TI-82, TI-83, or TI-83 Plus graphics calculator) are presented below for the most significant activities and examples found in this chapter that require or suggest the use of a graphics calculator.*

**internet connect**

For keystrokes of other graphing calculators models, visit the HRW web site at **go.hrw.com** and enter the keyword **MA1 CALC**.

## LESSON 11.2

*Activity*
Page 533

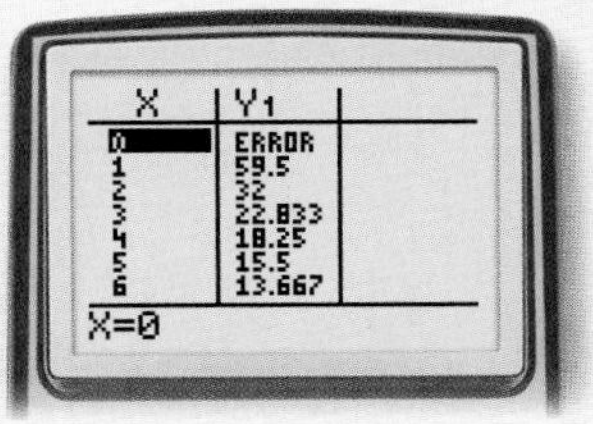


**For Step 4, graph the function $A(n) = \dfrac{55 + 4.50n}{n}$.**

Use a friendly viewing window [−47, 47] by [−31, 31]with scales of 10.

**Graph the function:**

Y= (Y1=) ( 55 + 4.5 X,T,θ,n ) ÷ X,T,θ,n GRAPH

**Make a table:** 2nd GRAPH (TABLE)

**EXAMPLE 2**
Pages 534 and 535

**Use the graph of $y = \dfrac{55 + 4.50n}{n}$ to find the $n$-value of the function for which $y = 5.60$.**

**Graph the function:**

Y= (Y1=) ( 55 + 4.5 X,T,θ,n ) ÷ X,T,θ,n GRAPH

Use the TRACE feature and the ◄ and ► keys to locate $y = 5.6$.

The $n$-value will be shown on the screen as $x = 50$.

## LESSON 11.3

*Activity*
Page 538

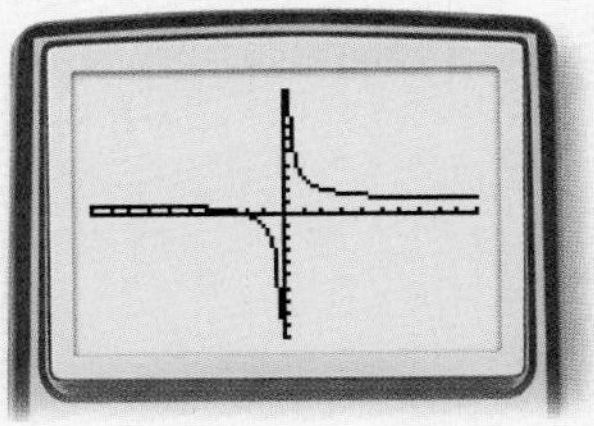

**For Step 3, graph $y = \dfrac{3x + 6}{3x}$ and $y = \dfrac{x + 2}{x}$.**

Use the standard viewing window.

**Graph the equations:**

Y= (Y1=) ( 3 X,T,θ,n + 6 ) ÷ ( 3 X,T,θ,n )
ENTER (Y2=) ( X,T,θ,n + 2 ) ÷ X,T,θ,n GRAPH

Go to TRACE and use ▼ and ▲ to switch from one graph to the other; observe that the functions are the same for all values of $x$.

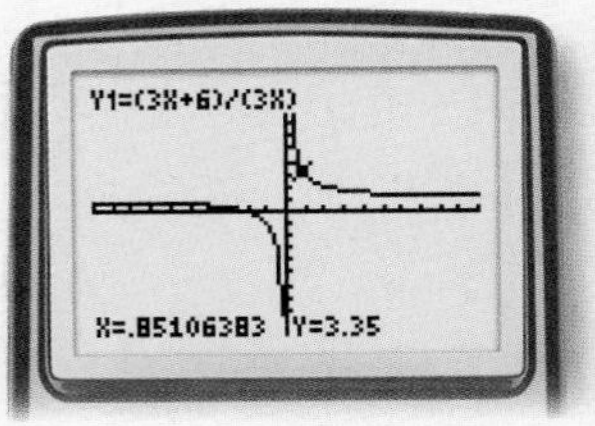


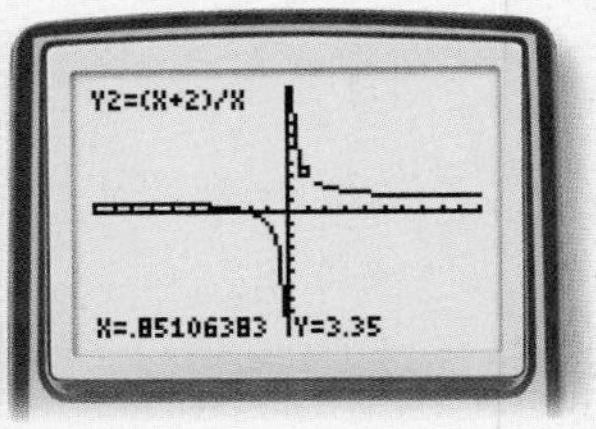

## LESSON 11.5

**EXAMPLE 1**
*Page 554*

**Use graphing to solve $\frac{6}{x-1} + 2 = \frac{12}{x^2 - 1}$.**

Use a friendly viewing window [−4.7, 4.7] by [−10, 10].

Graph each side of the equation separately.

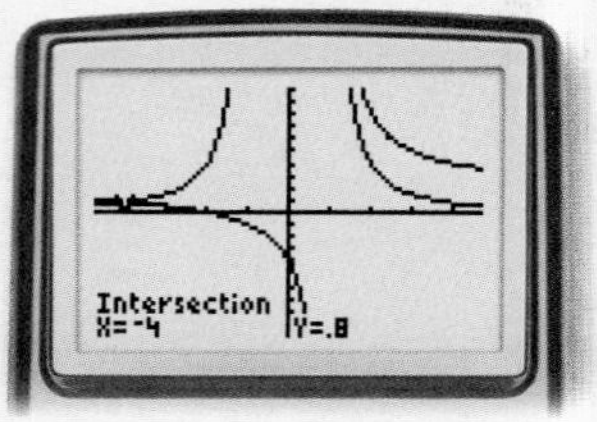


**Graph the equations:**

[Y=] **(Y1=)** 6 [÷] [(] [X,T,θ,n] [−] 1 [)] [+] 2 [ENTER]
**(Y2=)** 12 [÷] [(] [X,T,θ,n] [$x^2$] [−] 1 [)] [GRAPH]

**Look for any points of intersection:**

[2nd] [TRACE] (CALC) [5:intersect] [ENTER] **(First curve?)** [ENTER] **(Second curve?)** [ENTER] **(Guess?)** [ENTER]

# Chapter 12

*Essential keystroke sequences (using the TI-82, TI-83, or TI-83 Plus graphics calculator) are presented below for the most significant activities and examples found in this chapter that require or suggest the use of a graphics calculator.*

**internet** connect

For keystrokes of other graphing calculators models, visit the HRW web site at **go.hrw.com** and enter the keyword **MA1 CALC**.

## LESSON 12.2

**EXAMPLE 2**
*Pages 585 and 586*

**Solve $\sqrt{x+2} = 3$.**

Use the standard viewing window.

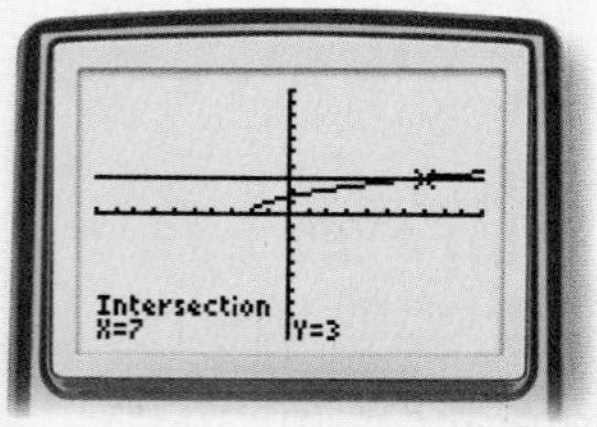


**Enter the left and right sides of the equation as separate functions:**

[Y=] **(Y1=)** [2nd] [$x^2$] ($\sqrt{\ }$) [X,T,θ,n] [+] 2 [)] [ENTER] **(Y2=)** 3 [GRAPH]

**Look for the intersection:**

[2nd] [TRACE] (CALC) [5:intersect] [ENTER] **(First curve?)** [ENTER]

**(Second curve?)** [ENTER] **(Guess?)** [ENTER]

# Chapter 13

*Essential keystroke sequences (using the TI-82, TI-83, or TI-83 Plus graphics calculator) are presented below for the most significant activities and examples found in this chapter that require or suggest the use of a graphics calculator.*

**internet connect**

For keystrokes of other graphing calculators models, visit the HRW web site at **go.hrw.com** and enter the keyword **MA1 CALC**.

## LESSON 13.5

**TECHNOLOGY**
*Page 676*

**Generate 0's and 1's with the random integer feature:**

MATH PRB 5:randInt( ENTER 0 , 1 ) ENTER

(Push ENTER once for every number you want to generate.)

# Chapter 14

*Essential keystroke sequences (using the TI-82, TI-83, or TI-83 Plus graphics calculator) are presented below for the most significant activities and examples found in this chapter that require or suggest the use of a graphics calculator.*

**internet connect**

For keystrokes of other graphing calculators models, visit the HRW web site at **go.hrw.com** and enter the keyword **MA1 CALC**.

## LESSON 14.2

*Activity*
*Page 700*

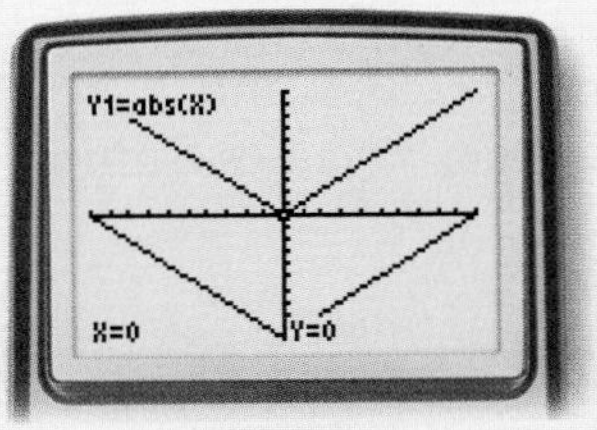


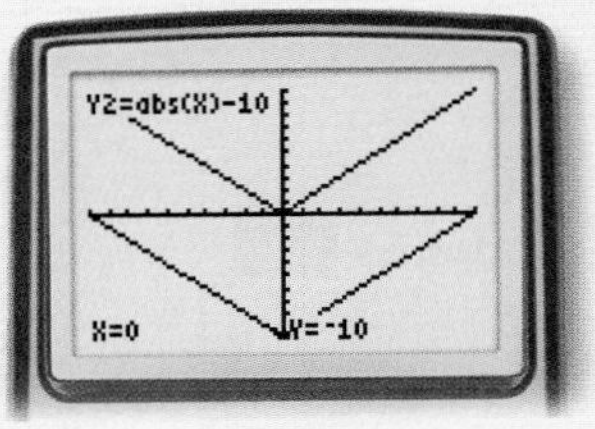


**For Step 1, graph the functions $f(x) = |x|$ and $g(x) = |x| - 10$ on the same screen.**

Use the standard viewing window.

**Graph the functions:**

Y= (Y1=) MATH ► NUM 1:abs( ENTER X,T,θ,n ) ENTER

↑TI-82: 2nd $x^{-1}$ ( (ABS)

(Y2=) MATH NUM 1:abs( ENTER X,T,θ,n ) – 10 GRAPH

↑TI-82: 2nd $x^{-1}$ 2nd (ABS)

**For Step 2, use the TRACE function and the ▼ ▲ buttons to determine the number of units shifted:**

**For Step 3, graph the functions $f(x) = |x|$ and $h(x) = |x - 10|$ on the same screen.**

Use viewing window [−10, 20] by [−10, 10].

[Y=] (Y1=) [MATH] [NUM] [1:abs (] [ENTER] [X,T,θ,n] [)] [ENTER]
TI-82: [2nd] [$x^{-1}$] (ABS) [2nd]

(Y2=) [MATH] [NUM] [1:abs (] [ENTER] [X,T,θ,n] [–] 10 [)] [GRAPH]
TI-82: [2nd] [$x^{-1}$] (ABS) [2nd]

**For Step 4, make a table:** [2nd] [GRAPH] (TABLE)

Locate the $x$-values where **Y1** = 0 and where **Y2** = 0. The difference between these $x$-values will give the number of units shifted.

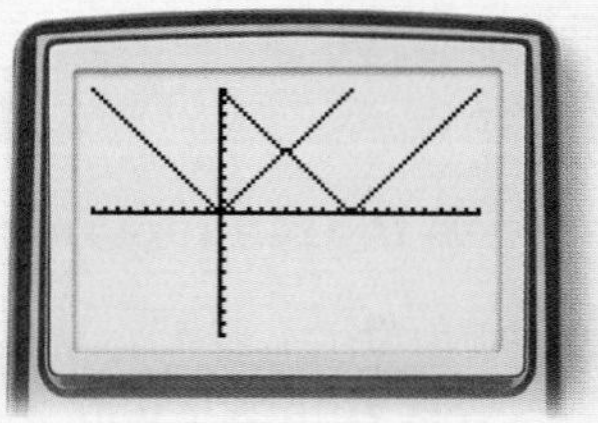

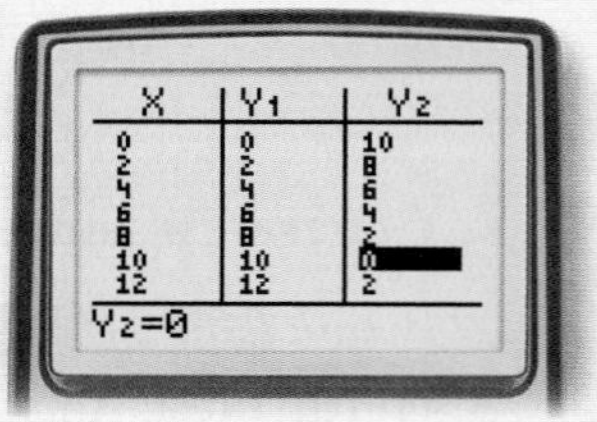


**LESSON 14.3**

*Activity*
**Page 707**

**For Step 1, make a table of values for the function $h(t) = t^2$.**

Use viewing window [0, 5] by [0, 25].

**Enter the function:**

[Y=] (Y1=) [X,T,θ,n] [$x^2$]

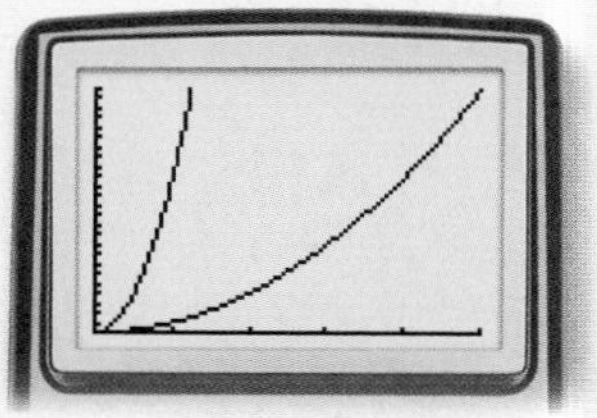

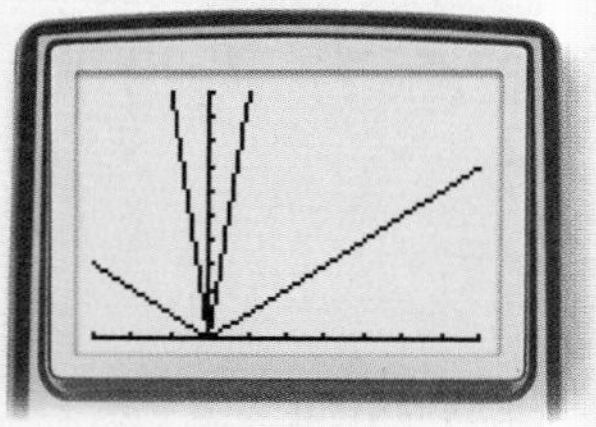

**Make a table:**

[2nd] [WINDOW] (TBLSET) (TblStart=) 0 [ENTER] (ΔTbl=) 1 [ENTER]
TI-82: (TblMin=)

(Indpnt:) [Auto] [ENTER] [▼] (Depend:) [Auto] [ENTER]

**View your table:** [2nd] [GRAPH] (TABLE) **View your graph:** [GRAPH]

For Steps 2 and 3, enter the function $j(t) = 16t^2$ as **Y2** $= 16x^2$ using a keystroke sequence similar to that used in Step 1. Enter [2nd] [GRAPH] (TABLE) to view the expanded table and then [GRAPH] to view the graphs of both functions on the same viewing screen.

For Step 4, complete Steps 1–3 for the functions $f(x) = |x|$ and $f(x) = 10|x|$ and then graph the functions for integer $x$-values from 3 to 7. Use viewing window [–3, 7] by [0, 10].

[Y=] (Y1=) [MATH] [►] [NUM] [1:abs] [ENTER] [X,T,θ,n] [)] [ENTER]
TI-82: [2nd] [$x^{-1}$] (ABS) [(]

(Y2=) 10 [MATH] [►] [NUM] [1:abs] [ENTER] [X,T,θ,n] [)]
TI-82: [2nd] [$x^{-1}$] (ABS) [(]

[2nd] [WINDOW] (TBLSET) (TblStart=) 0 [ENTER] (ΔTbl=) 1 [ENTER] (Indpnt:) [Auto]
TI-82: (TblMin=)

[ENTER] [▼] (Depend:) [Auto] [ENTER] [2nd] [GRAPH] (TABLE) [GRAPH]

# Tables

## Table of Squares, Cubes, and Roots

| No. | Squares | Cubes | Square Roots | Cube Roots |
|---|---|---|---|---|
| **1** | 1 | 1 | 1.000 | 1.000 |
| **2** | 4 | 8 | 1.414 | 1.260 |
| **3** | 9 | 27 | 1.732 | 1.442 |
| **4** | 16 | 64 | 2.000 | 1.587 |
| **5** | 25 | 125 | 2.236 | 1.710 |
| **6** | 36 | 216 | 2.449 | 1.817 |
| **7** | 49 | 343 | 2.646 | 1.913 |
| **8** | 64 | 512 | 2.828 | 2.000 |
| **9** | 81 | 729 | 3.000 | 2.080 |
| **10** | 100 | 1,000 | 3.162 | 2.154 |
| **11** | 121 | 1,331 | 3.317 | 2.224 |
| **12** | 144 | 1,728 | 3.464 | 2.289 |
| **13** | 169 | 2,197 | 3.606 | 2.351 |
| **14** | 196 | 2,744 | 3.742 | 2.410 |
| **15** | 225 | 3,375 | 3.873 | 2.466 |
| **16** | 256 | 4,096 | 4.000 | 2.520 |
| **17** | 289 | 4,913 | 4.123 | 2.571 |
| **18** | 324 | 5,832 | 4.243 | 2.621 |
| **19** | 361 | 6,859 | 4.359 | 2.668 |
| **20** | 400 | 8,000 | 4.472 | 2.714 |
| **21** | 441 | 9,261 | 4.583 | 2.759 |
| **22** | 484 | 10,648 | 4.690 | 2.802 |
| **23** | 529 | 12,167 | 4.796 | 2.844 |
| **24** | 576 | 13,824 | 4.899 | 2.884 |
| **25** | 625 | 15,625 | 5.000 | 2.924 |
| **26** | 676 | 17,576 | 5.099 | 2.962 |
| **27** | 729 | 19,683 | 5.196 | 3.000 |
| **28** | 784 | 21,952 | 5.292 | 3.037 |
| **29** | 841 | 24,389 | 5.385 | 3.072 |
| **30** | 900 | 27,000 | 5.477 | 3.107 |
| **31** | 961 | 29,791 | 5.568 | 3.141 |
| **32** | 1,024 | 32,768 | 5.657 | 3.175 |
| **33** | 1,089 | 35,937 | 5.745 | 3.208 |
| **34** | 1,156 | 39,304 | 5.831 | 3.240 |
| **35** | 1,225 | 42,875 | 5.916 | 3.271 |
| **36** | 1,296 | 46,656 | 6.000 | 3.302 |
| **37** | 1,369 | 50,653 | 6.083 | 3.332 |
| **38** | 1,444 | 54,872 | 6.164 | 3.362 |
| **39** | 1,521 | 59,319 | 6.245 | 3.391 |
| **40** | 1,600 | 64,000 | 6.325 | 3.420 |
| **41** | 1,681 | 68,921 | 6.403 | 3.448 |
| **42** | 1,764 | 74,088 | 6.481 | 3.476 |
| **43** | 1,849 | 79,507 | 6.557 | 3.503 |
| **44** | 1,936 | 85,184 | 6.633 | 3.530 |
| **45** | 2,025 | 91,125 | 6.708 | 3.557 |
| **46** | 2,116 | 97,336 | 6.782 | 3.583 |
| **47** | 2,209 | 103,823 | 6.856 | 3.609 |
| **48** | 2.304 | 110,592 | 6.928 | 3.634 |
| **49** | 2,401 | 117,649 | 7.000 | 3.659 |
| **50** | 2,500 | 125,000 | 7.071 | 3.684 |
| **51** | 2,601 | 132,651 | 7.141 | 3.708 |
| **52** | 2,704 | 140,608 | 7.211 | 3.733 |
| **53** | 2,809 | 148,877 | 7.280 | 3.756 |
| **54** | 2,916 | 157,464 | 7.348 | 3.780 |
| **55** | 3,025 | 166,375 | 7.416 | 3.803 |
| **56** | 3,136 | 175,616 | 7.483 | 3.826 |
| **57** | 3,249 | 185,193 | 7.550 | 3.849 |
| **58** | 3,364 | 195,112 | 7.616 | 3.871 |
| **59** | 3,481 | 205,379 | 7.681 | 3.893 |
| **60** | 3,600 | 216,000 | 7.746 | 3.915 |
| **61** | 3,721 | 226,981 | 7.810 | 3.936 |
| **62** | 3,844 | 238,328 | 7.874 | 3.958 |
| **63** | 3,969 | 250,047 | 7.937 | 3.979 |
| **64** | 4,096 | 262,144 | 8.000 | 4.000 |
| **65** | 4,225 | 274,625 | 8.062 | 4.021 |
| **66** | 4,356 | 287,496 | 8.124 | 4.041 |
| **67** | 4,489 | 300,763 | 8.185 | 4.062 |
| **68** | 4,624 | 314,432 | 8.246 | 4.082 |
| **69** | 4,761 | 328,509 | 8.307 | 4.102 |
| **70** | 4,900 | 343,000 | 8.367 | 4.121 |
| **71** | 5,041 | 357,911 | 8.426 | 4.141 |
| **72** | 5,184 | 373,248 | 8.485 | 4.160 |
| **73** | 5,329 | 389,017 | 8.544 | 4.179 |
| **74** | 5,476 | 405,224 | 8.602 | 4.198 |
| **75** | 5,625 | 421,875 | 8.660 | 4.217 |
| **76** | 5,776 | 438,976 | 8.718 | 4.236 |
| **77** | 5,929 | 456,533 | 8.775 | 4.254 |
| **78** | 6,084 | 474,552 | 8.832 | 4.273 |
| **79** | 6,241 | 493,039 | 8.888 | 4.291 |
| **80** | 6,400 | 512,000 | 8.944 | 4.309 |
| **81** | 6,561 | 531,441 | 9.000 | 4.327 |
| **82** | 6,724 | 551,368 | 9.055 | 4.344 |
| **83** | 6,889 | 571,787 | 9.110 | 4.362 |
| **84** | 7,056 | 592,704 | 9.165 | 4.380 |
| **85** | 7,225 | 614,125 | 9.220 | 4.397 |
| **86** | 7,396 | 636,056 | 9.274 | 4.414 |
| **87** | 7,569 | 658,503 | 9.327 | 4.431 |
| **88** | 7,744 | 681,472 | 9.381 | 4.448 |
| **89** | 7,921 | 704,969 | 9.434 | 4.465 |
| **90** | 8,100 | 729,000 | 9.487 | 4.481 |
| **91** | 8,281 | 753,571 | 9.539 | 4.498 |
| **92** | 8,464 | 778,688 | 9.592 | 4.514 |
| **93** | 8,649 | 804,357 | 9.644 | 4.531 |
| **94** | 8,836 | 830,584 | 9.695 | 4.547 |
| **95** | 9,025 | 857,375 | 9.747 | 4.563 |
| **96** | 9,216 | 884,736 | 9.798 | 4.579 |
| **97** | 9,409 | 912,673 | 9.849 | 4.595 |
| **98** | 9,604 | 941,192 | 9.899 | 4.610 |
| **99** | 9,801 | 970,299 | 9.950 | 4.626 |
| **100** | 10,000 | 1,000,000 | 10.000 | 4.642 |

# Table of Random Digits

| Line / Column | (1) | (2) | (3) | (4) | (5) | (6) | (7) | (8) | (9) | (10) | (11) | (12) | (13) | (14) |
|---|---|---|---|---|---|---|---|---|---|---|---|---|---|---|
| **1** | 10480 | 15011 | 01536 | 02011 | 81647 | 91646 | 69179 | 14194 | 62590 | 36207 | 20969 | 99570 | 91291 | 90700 |
| **2** | 22368 | 46573 | 25595 | 85393 | 30995 | 89198 | 27982 | 53402 | 93965 | 34095 | 52666 | 19174 | 39615 | 99505 |
| **3** | 24130 | 48360 | 22527 | 97265 | 76393 | 64809 | 15179 | 24830 | 49340 | 32081 | 30680 | 19655 | 63348 | 58629 |
| **4** | 42167 | 93093 | 06243 | 61680 | 07856 | 16376 | 39440 | 53537 | 71341 | 57004 | 00849 | 74917 | 97758 | 16379 |
| **5** | 31570 | 39975 | 81837 | 16656 | 06121 | 91782 | 60468 | 81305 | 49684 | 60672 | 14110 | 06927 | 01263 | 54613 |
| **6** | 77921 | 06907 | 11008 | 42751 | 27756 | 53498 | 18602 | 70659 | 90655 | 15053 | 21916 | 81825 | 44394 | 42880 |
| **7** | 99562 | 72905 | 56420 | 69994 | 98872 | 31016 | 71194 | 18738 | 44013 | 48840 | 63213 | 21069 | 10634 | 12952 |
| **8** | 96301 | 91977 | 05463 | 07972 | 18876 | 20922 | 94595 | 56869 | 69014 | 60045 | 18425 | 84903 | 42508 | 32307 |
| **9** | 89579 | 14342 | 63661 | 10281 | 17453 | 18103 | 57740 | 84378 | 25331 | 12566 | 58678 | 44947 | 05585 | 56941 |
| **10** | 85475 | 36857 | 53342 | 53988 | 53060 | 59533 | 38867 | 62300 | 08158 | 17983 | 16439 | 11458 | 18593 | 64952 |
| **11** | 28918 | 69578 | 88231 | 33276 | 70997 | 79936 | 56865 | 05859 | 90106 | 31595 | 01547 | 85590 | 91610 | 78188 |
| **12** | 63553 | 40961 | 48235 | 03427 | 49626 | 69445 | 18663 | 72695 | 52180 | 20847 | 12234 | 90511 | 33703 | 90322 |
| **13** | 09429 | 93969 | 52636 | 92737 | 88974 | 33488 | 36320 | 17617 | 30015 | 08272 | 84115 | 27156 | 30613 | 74952 |
| **14** | 10365 | 61129 | 87529 | 85689 | 48237 | 52267 | 67689 | 93394 | 01511 | 26358 | 85104 | 20285 | 29975 | 89868 |
| **15** | 07119 | 97336 | 71048 | 08178 | 77233 | 13916 | 47564 | 81056 | 97735 | 85977 | 29372 | 74461 | 28551 | 90707 |
| **16** | 51085 | 12765 | 51821 | 51259 | 77452 | 16308 | 60756 | 92144 | 49442 | 53900 | 70960 | 63990 | 75601 | 40719 |
| **17** | 02368 | 21382 | 52404 | 60268 | 89368 | 19885 | 55322 | 44819 | 01188 | 65225 | 64835 | 44919 | 05944 | 55157 |
| **18** | 01011 | 54092 | 33362 | 94904 | 31273 | 04146 | 18594 | 29852 | 71585 | 85030 | 51132 | 01915 | 92747 | 64951 |
| **19** | 52162 | 53916 | 46369 | 58586 | 23216 | 14513 | 83149 | 98736 | 23495 | 64350 | 94738 | 17752 | 35156 | 35749 |
| **20** | 07056 | 97628 | 33787 | 09998 | 42698 | 06691 | 76988 | 13602 | 51851 | 46104 | 88916 | 19509 | 25625 | 58104 |
| **21** | 48663 | 91245 | 85828 | 14346 | 09172 | 30168 | 90229 | 04734 | 59193 | 22178 | 30421 | 61666 | 99904 | 32812 |
| **22** | 54164 | 58492 | 22421 | 74103 | 47070 | 25306 | 76468 | 26384 | 58151 | 06646 | 21524 | 15227 | 96909 | 44592 |
| **23** | 32639 | 32363 | 05597 | 24200 | 13363 | 38005 | 94342 | 28728 | 35806 | 06912 | 17012 | 64161 | 18296 | 22851 |
| **24** | 29334 | 27001 | 87637 | 87308 | 58731 | 00256 | 45834 | 15398 | 46557 | 41135 | 10367 | 07684 | 36188 | 18510 |
| **25** | 02488 | 33062 | 28834 | 07351 | 19731 | 92420 | 60952 | 61280 | 50001 | 67658 | 32586 | 86679 | 50720 | 94953 |
| **26** | 81525 | 72295 | 04839 | 96423 | 24878 | 82651 | 66566 | 14778 | 76797 | 14780 | 13300 | 87074 | 79666 | 95725 |
| **27** | 29676 | 20591 | 68086 | 26432 | 46901 | 20849 | 89768 | 81536 | 86645 | 12659 | 92259 | 57102 | 80428 | 25280 |
| **28** | 00742 | 57392 | 39064 | 66432 | 84673 | 40027 | 32832 | 61362 | 98947 | 96067 | 64760 | 64584 | 96096 | 98253 |
| **29** | 05366 | 04213 | 25669 | 26422 | 44407 | 44048 | 37937 | 63904 | 45766 | 66134 | 75470 | 66520 | 34693 | 90449 |
| **30** | 91921 | 26418 | 64117 | 94305 | 26766 | 25940 | 39972 | 22209 | 71500 | 64568 | 91402 | 42416 | 07844 | 69618 |
| **31** | 00582 | 04711 | 87917 | 77341 | 42206 | 35126 | 74087 | 99547 | 81817 | 42607 | 43808 | 76655 | 62028 | 76630 |
| **32** | 00725 | 69884 | 62797 | 56170 | 86324 | 88072 | 76222 | 36086 | 84637 | 93161 | 76038 | 65855 | 77919 | 88006 |
| **33** | 69011 | 65795 | 95876 | 55293 | 18988 | 27354 | 26575 | 08625 | 40801 | 59920 | 29841 | 80150 | 12777 | 48501 |
| **34** | 25976 | 57948 | 29888 | 88604 | 67917 | 48708 | 18912 | 82271 | 65424 | 69774 | 33611 | 54262 | 85963 | 03547 |
| **35** | 09763 | 83473 | 73577 | 12908 | 30883 | 18317 | 28290 | 35797 | 05998 | 41688 | 34952 | 37888 | 38917 | 88050 |
| **36** | 91567 | 42595 | 27958 | 30134 | 04024 | 86385 | 29880 | 99730 | 55536 | 84855 | 29080 | 09250 | 79656 | 73211 |
| **37** | 17955 | 56349 | 90999 | 49127 | 20044 | 59931 | 06115 | 20542 | 18059 | 02008 | 73708 | 83517 | 36103 | 42791 |
| **38** | 46503 | 18584 | 18845 | 49618 | 02304 | 51038 | 20655 | 58727 | 28168 | 15475 | 56942 | 53389 | 20562 | 87338 |
| **39** | 92157 | 89634 | 94824 | 78171 | 84610 | 82834 | 09922 | 25417 | 44137 | 48413 | 25555 | 21246 | 35509 | 20468 |
| **40** | 14577 | 62765 | 35605 | 81263 | 39667 | 47358 | 56873 | 56307 | 61607 | 49518 | 89656 | 20103 | 77490 | 18062 |
| **41** | 98427 | 07523 | 33362 | 64270 | 01638 | 92477 | 66969 | 98420 | 04880 | 45585 | 46565 | 04102 | 46880 | 45709 |
| **42** | 34914 | 63976 | 88720 | 82765 | 34476 | 17032 | 87589 | 40836 | 32427 | 70002 | 70663 | 88863 | 77775 | 69348 |
| **43** | 70060 | 28277 | 39475 | 46473 | 23219 | 53416 | 94970 | 25832 | 69975 | 94884 | 19661 | 72828 | 00102 | 66794 |
| **44** | 53976 | 54914 | 06990 | 67245 | 68350 | 82948 | 11398 | 42878 | 80287 | 88267 | 47363 | 46634 | 06541 | 97809 |
| **45** | 76072 | 29515 | 40980 | 07391 | 58745 | 25774 | 22987 | 80059 | 39911 | 96189 | 41151 | 14222 | 60697 | 59583 |
| **46** | 90725 | 52210 | 83974 | 29992 | 65831 | 38857 | 50490 | 83765 | 55657 | 14361 | 31720 | 57375 | 56228 | 41546 |
| **47** | 64364 | 67412 | 33339 | 31926 | 14883 | 24413 | 59744 | 92351 | 97473 | 89286 | 35931 | 04110 | 23726 | 51900 |
| **48** | 08962 | 00358 | 31662 | 25388 | 61642 | 34072 | 81249 | 35648 | 56891 | 69352 | 48373 | 45578 | 78547 | 81788 |
| **49** | 95012 | 68379 | 93526 | 70765 | 10592 | 04542 | 76463 | 54328 | 02349 | 17247 | 28865 | 14777 | 62730 | 92277 |
| **50** | 15664 | 10493 | 20492 | 38391 | 91132 | 21999 | 59516 | 81652 | 27195 | 48223 | 46751 | 22923 | 32261 | 85653 |

# Glossary



## A

**absolute error** The absolute value of the difference between the actual measure and the specified measure. (300)

**absolute value** The absolute value of a real number $x$ is the distance from $x$ to 0 on a number line; the symbol $|x|$ means the absolute value of $x$. (57, 295)

**absolute-value equation** An equation that includes an absolute value; it will have 2 solutions. (300)

**absolute-value function** A function written in the form $y = |x|$ or $y = ABS(x)$. (295)

**absolute-value inequality** An inequality that includes an absolute value. (302)

**addition of matrices** If $A$ and $B$ are matrices with the same dimensions, then the matrix $(A + B)$ has the same dimensions; each of its entries is the sum of the corresponding entries of $A$ and $B$. (630)

**Addition of Probabilities Principle** If $M$ and $N$ are intersecting sets of outcomes in the same sample space, then $P(M \text{ OR } N) = P(M) + P(N) - P(M \cap N)$. If $M$ and $N$ are disjoint sets of outcomes in the same sample space, then $P(M \text{ OR } N) = P(M) + P(N)$.(657)

**Addition Property of Equality** For all real number $a$, $b$, and $c$, if $a = b$, then $a + c = b + c$. Adding equal amounts to each side of an equation results in an equivalent equation. (116)

**Addition Property of Inequality** Let $a$, $b$, and $c$ be real numbers. If $a < b$, then $a + c < b + c$. If $a > b$, then $a + c > b + c$. If equal amounts are added to the expressions on each side of an inequality, the resulting inequality is true. (277)

**Additive Inverse Property** For every real number $a$, there is exactly one real number $-a$. such that $a + (-a) = 0$ and $-a + a = 0$. (64)

**additive inverses** Two numbers $a$ and $-a$ whose sum is 0; see *opposites.* (64)

**address** The location of an entry in a matrix, given by the row and column in which the entry appears. (629)

**algebraic expression** Variables and numbers combined by operations. (11)

**algebraic order of operations** A logical system of rules for computation. (19)

**AND** A logical operator representing the intersection of two sets. (655)

**angle of depression** The angle formed by a horizontal line and a line of sight to a point below. (617)

**angle of elevation** The angle formed by a horizontal line and a line of sight to a point above. (616)

**approximate solution** A reasonable estimate for a point of intersection for a system of equations. (321)

**Associative Property of Addition** For all numbers $a$, $b$, and $c$, $(a + b) + c = a + (b + c)$. (81)

**Associative Property of Multiplication** For all numbers $a$, $b$, and $c$, $(a \cdot b) \cdot c = a \cdot (b \cdot c)$. (81)

**asymptote** A line that the graph of a function approaches but never touches or crosses. Asymptotes are found at each value for which the denominator of the function would be 0 (and thus where the function's value would be undefined). (534)

**axis of symmetry** A line drawn through the graph of a function that divides the graph into two halves that are reflections of each other. (481)

## B

**bar graph** A method of displaying data that uses rectangular bars or objects to represent the data. (193)

**base (of a percentage)** The number of which a percentage is calculated. (172)

**base (of a power)** The number that is raised to an exponent. In an expression of the form $x^a$, $x$ is the base. (370)

**boundary line** A line that divides a coordinate plane into two half-planes. (346)

**box-and-whisker-plot** A method of showing how data is distributed by using the median, the upper and lower quartiles, and the greatest and least values; also called a box plot. (203)

## C

**center** A point inside a circle that is the same distance from every point on the circle. (606)

**circle graph** A method of displaying data in which the data are represented by parts of a circle. (194)

**circle** The set of all points in a plane that are the same distance from a given point called the center of the circle. The equation for a circle with a radius of $r$ and its center at $(h, k)$ is $(x - h)^2 + (y - k)^2 = r^2$. (606)

**Closure Property** A set of numbers is said to be closed, or to have closure, under a given operation if the result of the operation on any two numbers in the set is also in the set. (70, 562)

**coefficient** A number that is multiplied by a variable. (89, 373)

**common binomial factor** A binomial factor that is common to all terms of a polynomial. (449)

**common monomial factor** A monomial factor that is common to all terms of a polynomial. (448)

**Commutative Property of Addition** For any numbers $a$ and $b$, $a + b = b + a$. (81)

**Commutative Property of Multiplication** For any numbers $a$ and $b$, $a \cdot b = b \cdot a$. (81)

**complement** The complement of an event includes all the outcomes in the sample space that are not that event. (670)

**completing the square** To make a perfect-square trinomial from a binomial of the form $x^2 + bx$, add the square of one-half the coefficient of the $x$-term, $\left(\frac{b}{2}\right)^2$. The completed perfect-square trinomial is $x^2 + bx + \left(\frac{b}{2}\right)^2$. (493)

**compound inequality** Two inequalities that are combined into one statement by the word *and* or *or*. (289)

**conclusion** In a conditional statement, the portion following the word *then*. In a proof of a conditional statement, the conclusion is the part that must be proven to be true. (558)

**conditional statement** A statement of the form "If $p$, then $q$." (558)

**conjecture** A statement that is believed to be true but is not yet proven. (5, 559)

**conjunction** A compound inequality whose solution region is an intersection. (290)

**consistent system** A system of equations that has one or infinitely many solutions. (339)

**constant** A term in an algebraic expression that does not contain variables. (89)

**constant difference** The difference that is always the same from one term to the next in a number sequence. (4)

**constant of variation** In a direct variation of the form $\frac{y}{x} = k$, $k$ is the constant of variation. (238)

**converse** A statement formed by interchanging the if and then portions of a conditional statement. "If $q$, then $p$" is the converse of "If $p$, then $q$." (558)

**Converse of the Pythagorean** Theorem If $a$, $b$, and $c$ are the lengths of the sides of a triangle and $a^2 + b^2 = c^2$, where $c > a$ and $c > b$, then the triangle is a right triangle. (600)

**coordinate plane** A plane that is divided into four regions by a horizontal and a vertical number line. The addresses, or coordinates, of points in the plane are given by ordered pairs. (24)

**coordinates** An ordered pair of numbers that locates a point in the coordinate plane in relation to the $x$- and $y$-axes. (24)

**correlation** A relationship between two variables in a scatter plot. (38)

**cosine** In a right triangle, the ratio of the length of the leg adjacent to an acute angle of the triangle to the length of the hypotenuse; cosine of angle $A = \cos A = \dfrac{\text{length of the leg adjacent to } \angle A}{\text{length of the hypotenuse}}$. (622)

**Counting Elements of Sets** Suppose that there are $m$ elements in set $M$ and $n$ elements in set $N$; then the total number of elements in the two sets is $m + n - t$, where $t$ is the number of elements in the intersection of $M$ and $N$. (656)

**cross products** In the proportion $\frac{a}{b} = \frac{c}{d}$, the product of the means, $bc$, and the product of the extremes, $ad$, are the cross products. (165)

**Cross Products Property** For any real numbers $a$, $b$, $c$, and $d$, where $b \neq 0$ and $d \neq 0$, if $\frac{a}{b} = \frac{c}{d}$, then $ad = bc$. (165)

## D

**deductive reasoning** The process of concluding that something must be true because it is a special case of a general principle that is known to be true. (559)

**degree of a polynomial** The degree of a polynomial in one variable is determined by the exponent with the greatest value within the polynomial. (426)

**dependent system** A consistent system of equations that has an infinitely many solutions. The graph of the equations in the system is the same line. (341)

**dependent variable** The variable in a function whose value depends on the value of the other variable. (32)

**difference of two squares** A polynomial of the form $a^2 - b^2$, which can be written as the product of two factors, $(a + b)(a - b)$. (453)

**difference** The result of subtraction. (62)

**direct variation** If $y$ varies directly as $x$, then $y = kx$, or $\frac{y}{x} = k$, where $k$ is the constant of variation. (238)

**directrix of a parabola** The fixed line from which a parabola is the same distance as it is from a fixed point, called the focus. (609)

**discriminant** For a quadratic equation of the form $ax^2 + bx + c = 0$, where $a \neq 0$, the discriminant is $b^2 - 4ac$. (508)

**disjoint** Two outcome sets of a sample space that do not intersect are said to be disjoint. (657)

**disjunction** A compound inequality whose solution region is outside an intersection. (290)

**distance formula** The distance, $d$, between two points $A(x_1,y_1)$ and $B(x_2,y_2)$ is given by $d = \sqrt{(x_2 - x_1)^2 + (y_2 - y_1)^2}$. (599)

**Distributive Property** For all numbers $a$, $b$, and $c$, $a(b + c) = ab + ac$ and $(b + c)a = ba + ca$, and $a(b - c) = ab - ac$ and $(b - c)a = ba - ca$. (82)

**dividing an expression** When an expression is divided by a number, each term in the expression is divided by that number. For all numbers $a$, $b$, and $c$, $\frac{a+b}{c} = \frac{a}{c} + \frac{b}{c}$ and $\frac{a-b}{c} = \frac{a}{c} - \frac{b}{c}$. (99)

**Division Property of Equality** For all real number $a$, $b$, and $c$, if $a = b$ and $c \neq 0$, then $\frac{a}{c} = \frac{b}{c}$. Dividing each side of an equation by equal amounts results in an equivalent equation. (123)

**Division Property of Inequality** Let $a$, $b$, and $c$ be nonzero real numbers. For $c > 0$, if $a > b$, then $\frac{a}{c} > \frac{b}{c}$, and if $a < b$, then $\frac{a}{c} > \frac{b}{c}$. For $c < 0$, if $a < b$, then $\frac{a}{c} > \frac{b}{c}$ and if $a > b$, then $\frac{a}{c} < \frac{b}{c}$. (283)

**Division Property of Square Roots** For all numbers $a \geq 0$ and $b > 0$, $\sqrt{\frac{a}{b}} = \frac{\sqrt{a}}{\sqrt{b}}$. (580)

**domain** The first coordinates in a set of ordered pairs of a relation or function. (219)

## E

**elimination method** A method used to solve a system of equations in which one variable is eliminated by adding opposites. (332)

**entry** The information that appears at each address of a matrix. (629)

**equally likely outcomes** Outcomes of an experiment that have the same probability of happening. (648)

**equation** Two equivalent algebraic expressions separated by an equal sign. (12)

**equation method** A method for finding an unknown part of a percent by setting up an equation. (172)

**equation of a circle** The equation of a circle with radius $r$ and its center at $(h, k)$ is $(x - h)^2 + (y - h)^2 = r^2$. (607)

**error** The difference between the actual measure and the specified measure. (300)

**even numbers** An even number is a number of the form $2n$, where $n$ is an integer. (560)

**experimental probability** The experimental probability, $P(E)$, of an event is approximated by the formula $P(E) = \frac{\text{number of favorable outcomes}}{\text{total number of trials)}}$. The larger the number of trials, the closer the approximation. (182, 676)

**exponent** In a power, the number that tells how many times the base is used as a factor. In an expression of the form $x^a$, $a$ is the exponent. (97, 379)

**exponential decay** A situation modeled by the general growth formula, $A = P(1 + r)^t$, in which $r$ is negative. (411)

**exponential function** A function of the form $y = 2^x$ in which a base number is raised to a variable exponent. Exponential functions model situations in which values in the range change by a fixed rate rather than a fixed amount. (404)

**exponential growth** A situation modeled by the general growth formula, $A = P(1 + r)^t$, in which $r$ is positive. (411)

**extraneous solution** A solution that is obtained but that does not satisfy the original equation. (586)

**extremes** In the proportion $\frac{a}{b} = \frac{c}{d}$, $a$ and $d$ are the extremes. (165)

## F

**factor** A number or polynomial that is multiplied by another number or polynomial to form a product. (448)

**favorable outcome** The outcome that is desired in a probability experiment. (649)

**field** Any set that has the Associative, Closure, Commutative, Distributive, Identity, and Inverse Properties for any two operations. (563)

**focus of a parabola** The fixed point from which a parabola is the same distance as it is from a fixed line, called the directrix. (609)

**FOIL method** A mnemonic (memory) device for a method of multiplying two binomials.
- Multiply the **F**irst terms.
- Multiply the **O**utside terms.
- Multiply the **I**nside terms. Add the inside and outside products.
- Multiply the **L**ast terms. (438)

**formula** A literal equation that states a rule for a relationship among quantities. (147)

**frequency table** A method of displaying data in which tally marks are used to show how often each element of the set occurs. (188)

**function** A pairing between two sets of numbers in which each element of the first set is paired with no more than one element of the second set. (219)

**function notation** A method of writing a function in which the dependent variable is written in the form $f(x)$. The independent variable, $x$, is placed inside the parentheses. (223, 694)

**function rule** The expression on the right side of the written form of a function. (694)

**Fundamental Counting Principle** If there are $m$ ways to make a first choice and $n$ ways to make a second choice, then there are $m \cdot n$ ways to make a first choice and a second choice. (662)

## G

**general growth formula** Let $P$ be the amount after $t$ years at a yearly growth rate of $r$, where $r$ is expressed as a decimal. If the original amount is $A$, then the general growth formula is $P = A(1 + r)^t$. (405)

**greatest common factor (GCF)** The greatest factor that is common to all terms of a polynomial. (448)

**greatest-integer function** A function written in the form $y = \text{INT}(x)$ in which the number $x$ is rounded down to the nearest integer. (309)

**grouping symbols** Symbols such as parentheses, ( ), brackets, [ ], braces, { }, and the fraction bar, —, that are used to group numbers and variables. (19)



Glossary

## H

**histogram** A bar graph that has no space between the bars and that shows the frequency of the data. (202)

**hole (in a graph)** An omitted point in a graph. (534)

**Hooke's law** A mathematical relationship between the amount of force, $F$, applied to a spring and the distance, $d$, that the spring stretches, $F = kd$, where $k$ is a constant. (239)

**horizontal line** The equation for a horizontal line is $y = b$, where $b$ is the $y$-intercept. The slope of a horizontal line is 0. (230, 248)

**horizontal stretch/compression** If $y = f(x)$, then $y = f(bx)$ gives a horizontal stretch or compression of the graph of $f$ by a factor of $\frac{1}{b}$. (710)

**horizontal translation** A transformation that shifts the graph of a function either to the right (in a positive direction) or to the left (in a negative direction). The graph of $y = f(x - h)$ is translated horizontally $h$ units from the graph of $y = f(x)$. (481, 703)

**hypotenuse** The side opposite the right angle in a right triangle. (591)

**hypothesis** In a conditional statement, the portion following the word *if.* In a proof of a conditional statement, the hypothesis is assumed to be true. (558)

## I

**identity** An equation that is true regardless of the values of the variable(s). (445)

**identity matrix for multiplication** A square matrix with 1s along the main diagonal and 0s elsewhere. The product of a given matrix and the identity matrix is the given matrix. (632)

**Identity Property for Addition** For all real numbers $a$, $a + 0 = a$ and $0 + a = a$; 0 is the *identity element* for addition. (64, 84, 563)

**Identity Property for Multiplication** For all real numbers $a$, $a \cdot 1 = a$ and $1 \cdot a = a$; 1 is the *identity element* for multiplication. (75, 84, 563)

**inconsistent system** A system of equations that has no solution. The graphs of the equations in the system are parallel lines. (339)

**independent events** Two events are independent if the occurrence of one does not affect the probability of the occurrence of the other. (668)

**independent system** A consistent system of equations that has only one solution. The graphs of the equations are lines that intersect at one point. (341)

**independent variable** The variable in a function whose value does not depend on the value of the other variable. (32)

**inductive reasoning** The process of reasoning that general principle is true because special cases are true. (559)

**inequality** A statement that two expressions are not equal. Inequalities contain one of the following signs: $<, >, \leq, \geq$, or $\neq$. (276)

**integers** The set of whole numbers and their opposites. (54)

**intersection (of sets)** The intersection of two sets consists of the elements that are common to both sets. The logical relationship AND represents the intersection of two sets. (659)

**intersection (of graphs)** The solution to a system of linear equalities or inequalities, consisting of the solutions common to each. (290)

**inverse variation** A relationship in which $y$ varies inversely as $x$; written as $y = \frac{k}{x}$, or $xy = k$, where $k$ is the constant of variation, $k \neq 0$, and $x \neq 0$. (526)

**irrational numbers** Numbers whose decimal part never terminates or repeats. Irrational numbers cannot be expressed in the form $\frac{a}{b}$, where $a$ and $b$ are integers and $b \neq 0$. (56)

## L

**leaves** In a stem-and-leaf plot, the leaves represent the second part each element and are listed to the right of the vertical line. (201)

**legs of a right triangle** The sides that form the right angle of a right triangle. (591)

**like terms** Terms that contain the same form of a variable. (89)

**line graph** A type of graph that uses line segments between known data points to show changes that have occurred over time. (193)

**line of best fit** A line that shows the overall trend of the data in a scatter plot. (39)

**line of reflection** The line across which a graph is reflected. (297)

**linear equation** An equation whose graph is a straight line. (32)

**linear inequality** An inequality whose form is like that of a linear equation with the equal sign replaced by an inequality symbol. (345)

**literal equation** An equation that involves two or more variables. (147)

**lower quartile** The median of the elements in the lower half of a data set. (204)

## M

**main diagonal (of a matrix)** The entries in a square matrix that form a diagonal from the upper left corner to the lower right corner. (632)

**matrix** An arrangement of data in rows and columns and enclosed by brackets [ ]. The plural of *matrix* is *matrices.* (628)

**matrix address** See **address**.

**matrix equality** Two matrices are equal when their dimensions are the same and their corresponding entries are equal. (629)

**maximum value** The $y$-value of the vertex of a parabola that opens down. (481)

**mean** The quotient that results when the sum of all the elements in a data set is divided by the total number of elements in the set. (194)

**means** In the proportion $\frac{a}{b} = \frac{c}{d}$, $b$ and $c$ are the means. (165)

**median** The middle number in a data set when the elements are placed in numerical order. If there is an even number of elements in the data set, the median is the average of the two middle numbers. (186)

**midpoint formula** Given $\overline{PQ}$, where $P(x_1, y_1)$ and $Q(x_2, y_2)$, the coordinates of the midpoint, $M$, of $\overline{PQ}$ are $M = \left(\frac{x_1 + x_2}{2}, \frac{y_1 + y_2}{2}\right)$. (601)

**midsegment** A line segment connecting the midpoints of 2 sides of a triangle. (608)

**minimum value** The $y$-value of the vertex of a parabola that opens up. (480)

**mode** The element, if any, that occurs most often in a data set. There may be no mode, one mode, or several modes for a set of data. (186)

**monomial** An algebraic expression that is either a constant, a variable, or a product of a constant and one or more variables. (373)

**Multiplication Property of Equality** For all real numbers $a$, $b$, and $c$, if $a = b$, then $ac = bc$. Multiplying each side of an equation by equal amounts results in an equivalent equation. (124)

**Multiplication Property of Inequality** Let $a$, $b$, and $c$ be nonzero real numbers. For $c > 0$, if $a > b$, then $ac > bc$, and if $a < b$, then $ac < bc$. For $c < 0$, if $a < b$, then $ac > bc$, and if $a > b$, then $ac < bc$. (283)

**Multiplication Property of Square Roots** For all numbers $a$ and $b$, where $a \geq 0$ and $b \geq 0$, $\sqrt{ab} = \sqrt{a}\sqrt{b}$. (678)

**Multiplicative Inverse Property** For every nonzero real number a, there is exactly one number $\frac{1}{a}$ such that $a \cdot \frac{1}{a} = 1$ and $\frac{1}{a} \cdot a = 1$. (75)

**multiplicative inverse** See *reciprocal.*

## N

**natural numbers** The numbers 1, 2, 3, 4, 5, . . . (54)

**negative correlation** A relationship between two variables in a scatter plot in which one variable increases while the other decreases. (38)

**negative exponent** If $x$ is any number except zero and $n$ is a positive integer, then $x^{-n} = \frac{1}{x^n}$. (390)

**negative numbers** Numbers that are less than zero. Negative numbers lie to the left of zero on a number line. (55)

**neutral pair** One positive and one negative algebra tile of the same value. (61)

**non-trivial rational function** A rational function whose denominator is not a constant. (534)

**number sequence** A list of numbers in a certain order. (4)

## O

**odd numbers** An odd number is a number of the form $2n + 1$, where $n$ is an integer. (560)

**opposites** Two numbers that lie on opposite sides of zero and are the same distance from zero on a number line. (57)

**OR** A logical operator which represents the union of two sets. (655)

**ordered pair** The $x$- and $y$-coordinates that give the location of a point in a coordinate plane, indicated by two numbers in parentheses, $(x, y)$. (25)

**origin** The point of intersection of the $x$- and $y$-axes in the coordinate plane. (24)

## P

**parabola** The shape of the graph of a quadratic function. A parabola is defined as the set of all points that are the same distance from a fixed point, called the focus, and a fixed line, called the directrix. (480)

**parallel lines** If two different lines have the same slope, the lines are parallel. If two nonvertical lines are parallel, they have the same slope. Two parallel, vertical lines have undefined slopes. (259)

**parent function** The most basic function of a family of functions, or the original function before a transformation is applied. (296, 696)

**percent** A ratio that compares a number with 100. (171)

**percent rate** The percent of the base calculated or indicated in a percent statement. (172)

**percentage** The amount obtained by multiplying a base by a given percent rate. (172)

**perfect-square trinomial** A trinomial of the form $a^2 + 2ab + b^2$ or $a^2 - 2ab + b^2$. The factored form of $a^2 + 2ab + b^2$ is $(a + b)(a + b) = (a + b)^2$, and the factored form of $a^2 - 2ab + b^2$ is $(a - b)(a - b) = (a - b)^2$. (452)

**perpendicular lines** If the slopes of two lines are $m$ and $-\frac{1}{m}$, the lines are perpendicular. (259)

**point-slope form** The point-slope form for the equation of a line is $y - y_1 = m(x - x_1)$, where the coordinates $x_1$ and $y_1$ are taken from a given point, $(x_1, y_1)$, and $m$ is the slope. (254)

**polynomial** The sum or difference of two or more monomials. (426)

**polynomial function** A function that consists of the sum or difference of two or more monomials. (443)

**positive correlation** A relationship between two variables in a scatter plot in which both variables increase. (38)

**positive numbers** Numbers that are greater than zero. Positive numbers lie to the right of zero on a number line. (55)

**power** An expression of the form $x^a$, where $x$ is the base and $a$ is the exponent. (97)

**Power-of-a-Fraction Property** If $n$ is an integer and $a$ and $b$ are numbers, where $b \neq 0$, then $\left(\frac{a}{b}\right)^n = \frac{a^n}{b^n}$. (386)

**Power-of-a-Power Property** If $x$ is any number and $m$ and $n$ are integers, then $(x^m)^n = x^{mn}$. (378)

**Power-of-a-Product Property** If $x$ and $y$ are any numbers and $n$ is an integer, then $(xy)^n = x^n y^n$. (379)

**Powers of −1** Even powers of −1 are equal to 1. Odd powers of −1 are equal to −1. (380)

**principal square root** The positive square root of a number; indicated by the radical sign. (487)

**Probabilities of Independent Events** If $A$ and $B$ are independent events, then $P(A \text{ AND } B) = P(A) \cdot P(B)$. (668)

**probability** The likelihood that an event will occur. (180)

**product matrix** The result of multiplying two matrices. If matrix $A$ has dimensions $m \times n$ and matrix $B$ has dimensions $n \times p$, then the product matrix $AB$ has dimensions $m \times p$. (631)

**Product-of-Powers Property** If $x$ is any number and $m$ and $n$ are integers, then $x^m \cdot x^n = x^{m+n}$. (371)

**proof** A series of statements in a logical sequence which demonstrates that the conclusion of a conditional statement is true whenever the hypothesis is true. (558)

**Properties of Zero** Let $a$ represent any number.
1. The product of any number and zero is zero. $a \cdot 0$ and $0 \cdot a = 0$
2. Zero divided by any nonzero number is zero. $\frac{0}{a} = 0$, where $a \neq 0$
3. Division by zero is undefined. That is, division by zero is not possible. (76)

**proportion** A statement that two ratios are equal, $\frac{a}{b} = \frac{c}{d}$. (164)

**proportion method** A method of solving percent problems by setting two proportions equal to each other. (172)

**Pythagorean Theorem** If given a right triangle with legs of length $a$ and $b$ and hypotenuse of length $c$, then $a^2 + b^2 = c^2$. (592)

**Pythagorean Triple** A set of three positive integers that satisfy the Pythagorean Theorem. (597)

## Q

**quadrant** One of the four regions into which a horizontal and vertical number line divide a plane. (24)

**quadratic formula** For $ax^2 + bx + c = 0$, where $a \neq 0$, $x = \frac{-b \pm \sqrt{b^2 - 4ac}}{2a}$. (507)

**quadratic function** A function of the form $y = ax^2 + bx + c$, where $a$, $b$, and $c$ are real numbers and $a \neq 0$. (480)

**quadratic inequality** An inequality whose form is like that of a quadratic equation with the equal sign replaced by an inequality symbol. (511)

**Quotient-of-Powers Property** If $x$ is any number except 0 and $m$ and $n$ are any positive integers, where $m > n$, then $\frac{x^m}{x^n} = x^{m-n}$. (384)

## R

**radical expression** An expression that contains a square root. (577)

**radical sign** The sign $\sqrt{\ }$ used to denote a square root. (487)

**radicand** The number under a radical sign. (577)

**radius** The distance from the center of a circle to any point on the circle. (608)

**range** The difference between the greatest and least values in a data set, or the set of second coordinates in the ordered pairs of a relation or function. (186, 219)

**rate of change** The rate of change of a linear function is equal to the slope of the graph of the function. (236)

**ratio** The comparison of two quantities by division. (164)

**rational expression** If $P$ and $Q$ are polynomials and $Q \neq 0$, then an expression of the form $\frac{P}{Q}$ is a rational expression. (532)

**rational function** A function of the form $y = \frac{P}{Q}$, or $f(x) = \frac{P(x)}{Q(x)}$ where is a rational expression. (533)

**rational numbers** A number that can be expressed in the form $\frac{a}{b}$, where $a$ and $b$ are integers and $b \neq 0$. (55)

**rationalizing the denominator** Removing a radical expression from the denominator of a fraction. (580)

**real numbers** The set of rational numbers together with the set of irrational numbers. (56)

**reciprocal** The number $\frac{1}{a}$ is called the reciprocal of $a$. (75)

**reflection** A transformation that creates a mirror image of a given function. (297, 481, 715)

**Reflexive Property of Equality** For all real numbers $a$, $a = a$ (a number is equal to itself). (84)

**relation** A pairing between two sets of numbers. (218)

**rise** The difference of the $y$-coordinates of two points on a line. (226)

**rule of 72** The time, $t$, required to double the value of an investment earning compound interest is give by the function $t = \frac{72}{r}$, where $r$ is the annual interest rate. (529)

**run** The difference of the $x$-coordinates of two points on a line. (226)

# S

**sample space** The set of all possible outcomes of a probability experiment. (673)

**scalar** A number by which each entry in a matrix is multiplied. (631)

**scalar multiplication** The multiplication of a matrix by a scalar. (631)

**scale factor** The number by which a parent function is multiplied to create a vertical stretch. (708)

**scatter plot** A display of ordered pairs of data graphed on a coordinate plane that is used to show how two variables relate to each other. (37)

**scientific notation** A number written in scientific notation is a product of two factors—a number from 1 to 10, including 1 but not including 10, and a power of 10. (398)

**simplest radical form** A radical expression is in simplest radical form if the expression under the radical sign contains no perfect squares greater that 1, contains no fractions, and is not in the denominator of a fraction. (578)

**simplified** An algebraic expression is said to be simplified when all of the like terms have been combined and all parentheses have been removed. (89)

**simulation** A probability experiment with mathematical characteristics that are similar to the actual event. (676)

**sine** In a right triangle, the ratio of the length of the leg opposite an acute angle of the triangle to the length of the hypotenuse; sine of angle $A = \sin\ A = \frac{\text{length of the leg opposite } \angle A}{\text{length of the hypotenuse}}$. (622)

**slope** A measure of the steepness of a line. Given two points with coordinates $(x_1, y_1)$ and $(x_2, y_2)$ on a line, the slope, $m$, of the line is given $m = \frac{\text{rise}}{\text{run}} = \frac{y_2 - y_1}{x_2 - x_1}$. (228)

**slope-intercept form** The slope-intercept form for a line with a slope of $m$ and a $y$-intercept of $b$ is $y = mx + b$. (245)

**solution (to an equation)** A set of values for the variables in an equation that make a true statement when substituted into the equation. (13)

**solution (to a system of equations)** An ordered pair of numbers that is the solution to each equation in the system. (319)

**special products of polynomials** For any real numbers $a$ and $b$ and for any expressions involving real numbers and variables: $(a + b)(a - b) = a^2 - b^2$; $(a + b)(a + b) = (a + b)^2 = a^2 + 2ab + b^2$; $(a - b)(a - b) = (a - b)^2 = a^2 - 2ab + b^2$. (439)

**square matrix** A matrix with equal row and column dimensions. (629)

**square root** If $a$ is a number greater than or equal to zero, $\sqrt{a}$ represents the positive, or principal, square root of $a$ and $-\sqrt{a}$ represents the negative square root of $a$; The square roots of a have the following property: $(-\sqrt{a})(-\sqrt{a}) = a$. (487, 577)

**square-root function** A function of the form $f(x) = \sqrt{x}$. (583)

**standard form (of a linear equation)** A linear equation in the form $Ax + By = C$ is in standard form when $A$, $B$, and $C$ are real numbers, $A$ and $B$ are not both zero, and $A$ is not negative. (252)

**standard form (of a polynomial)** A polynomial is in standard form when the terms of the polynomial are ordered from left to right in descending order, which is from the greatest exponent to the least. (426)

**stem-and-leaf plot** A method of displaying data by dividing each element into two parts, called stems and leaves. (201)

**stems** In a stem-and-leaf plot, the stems represent the first part each element and are listed to the left of the vertical line. (201)

**substitution method** A method used to solve a system of equations in which variables are replaced with known values or algebraic expressions. (327)

**Substitution Property of Equality** For all real numbers $a$ and $b$, if $a = b$, then a can be replaced by $b$ and $b$ can be replaced by $a$ in any expression. (84)

**subtraction** For all real numbers $a$ and $b$, $a - b = a + (-b)$. (68)

**Subtraction Property of Equality** For all real numbers $a$, $b$, and $c$, if $a = b$, then $a - c = b - c$. Subtracting equal amounts from each side of an equation results in an equivalent equation. (115)

**Subtraction Property of Inequality** Let $a$, $b$, and $c$ be real numbers. If $a < b$, then $a - c < b - c$. If $a > b$, then $a - c > b - c$. If equal amounts are subtracted from the expressions on each side of an inequality, the resulting inequality is true. (277)

**sum** The result of addition. (62)

**Symmetric Property of Equality** For all real numbers $a$ and $b$, if $a = b$ then $b = a$. (84)

**system of equations** Two or more equations in two or more variables. (319)

**system of linear inequalities** Two or more inequalities in two or more variables. (345-349)

# T

**tangent** In a right triangle, the ratio of the length of the leg opposite an acute angle of the triangle to the length of the leg adjacent to the acute angle; $A = \tan\ A = \frac{\text{length of the leg opposite } \angle A}{\text{length of the hypotenuse}}$. (613)

**term** Each number in a number sequence; a number, a variable, or a product or quotient of numbers and variables that is added or subtracted in an algebraic expression. (5, 89)

**theorem** A statement that can be proved to be true. (85, 558)

**theoretical probability** Let $n$ be the number of elements in a sample space in which each element is equally likely to occur. Let $E$ be an event for which there are $f$ favorable outcomes in the sample space. Then, the theoretical probability that the event will occur is $P(E) = \frac{f}{n}$. (650)

**transformation** A variation such as a stretch, reflection, or translation of a parent function. (296, 700)

**Transitive Property of Equality** For all real numbers $a$, $b$, and $c$, if $a = b$ and $b = c$, then $a = c$. (84)

**translation** A transformation that shifts the graph of a function horizontally or vertically. (296, 700)

**tree diagram** A diagram that is used to find all the possible choices in a situation in which one choice AND another choice need to be made. (661)

**trial** The simulation of one event in an experiment. (181, 676)

**Triangle Midsegment Theorem** The segment joining the midpoints of two sides of a triangle is parallel to the third side of the triangle and is half the length of the third side. (608)

**trigonometry** A branch of mathematics that combines arithmetic, algebra, and geometry and is named from the Greek words meaning "triangle measure."(614)

**trivial rational function** A rational function whose denominator is a constant. (534)

## U

**union** The union of two sets consists of all elements from both sets. The logical relationship OR represents the union of sets. (290, 655)

**upper quartile** The median of the elements in the upper half of a data set. (204)

## V

**variable** A letter that is used to represent numbers in an algebraic expression. (11)

**Venn diagram** A diagram used to illustrate the relationships among different sets of data represented by circles. (655)

**vertex** The point where a parabola changes direction. (480)

**vertex form of a quadratic function** The vertex form of a quadratic function is $y = a(x - h)^2 + k$, where $(h, k)$ is the vertex and the line $x = h$ is the axis of symmetry. (482)

**vertical asymptote** A vertical line that a graph approaches but never crosses. (534)

**vertical line** The equation for a vertical line is $x = a$, where $a$ is the $x$-intercept. The slope of a vertical line is undefined. (229-230, 248)

**vertical stretch/compression** A transformation in which, for a given value of the domain of a function, the corresponding range value is multiplied by a scale factor. (708)

**vertical translation** A transformation that shifts the graph of a function either up (in a positive direction) or down (in a negative direction). The graph of $y = f(x) + k$ is translated vertically $k$ units from the graph of $y = f(x)$. (481, 701)

**vertical-line test** A test used to determine whether a relation is a function. A relation is a function if any vertical line intersects the graph of the relation no more than once. (693)

## W

**whole numbers** The numbers 0, 1, 2, 3, 4, . . . (54)

## X

***x*-axis** The horizontal number line in a coordinate plane. (24)

***x*-coordinate** A number that indicates distance along the $x$-axis in a coordinate plane. (24)

***x*-intercept** The $x$-coordinate of the point where a line crosses the $x$-axis. (247)

## Y

***y*-axis** The vertical number line in a coordinate plane. (24)

***y*-coordinate** A number that indicates distance along the $y$-axis in a coordinate plane. (24)

***y*-intercept** The $y$-coordinate of the point where a line crosses the $y$-axis. (245)

## Z

**zero as an exponent** For any nonzero number $x$, $x^0 = 1$. (391)

**zero of a function** A number that gives a function value of 0 when substituted for the independent variable of a function. (465-467)

**Zero Product Property** If $a$ and $b$ are real numbers such that $ab = 0$, then $a = 0$ or $b = 0$. (465)

# Selected Answers



## Chapter 1

### LESSON 1.1

**TRY THIS** (p. 5, Ex. 1)
**a.** 16, 19, 22
**b.** 5, 0, –5

**TRY THIS** (p. 6, Ex. 2)
**a.** 42, 56, 72
**b.** 48, 50, 50

**TRY THIS** (p. 6, Ex. 3)
212°F; the boiling point of water

**TRY THIS** (p. 7)
Sample Answer: They are products of two consecutive integers.

**Exercises**
**4.** 11, 13, 15 **5.** 20, 22, 24 **6.** 42, 33, 24 **7.** 36, 49, 64 **8.** 65, 82, 101 **9.** 93, 123, 157 **11.** 113, 129, 145 **13.** 29, 26, 23 **15.** 65, 81, 100 **17.** 69, 97, 130 **19.** 18, 24.5, 32 **21.** 36, 29, 22, 15, 8 **23.** 137, 119, 101, 83, 65 **25. a.** The first difference is constant. The constant difference is 1. **b.** The second difference is constant. The constant difference is 2. **c.** The third difference is constant. The constant difference is 6. **d.** You need to compute four differences. **27.** 55 **29.** 820 **31.** 3240 **33. a.** The boiling point at Colorado Springs, Colorado, is approximately 201.2°. **b.** Continuing in the pattern, the boiling point is approximately 179.6°. **35.** There will be 56 cables. **37.** 8.62 **39.** $3\frac{1}{3}$ **41.** 0.4 **43.** $1\frac{3}{16}$

### LESSON 1.2

**TRY THIS** (p. 13)

| Hours | Expression | Cost |
|---|---|---|
| 1 | 2.5(1) + 2 | 4.50 |
| 2 | 2.5(2) + 2 | 7.00 |
| 3 | 2.5(3) + 2 | 9.50 |
| $h$ | $2.5(h) + 2$ | $2.5h + 2$ |

The equation is $c = 2.5h + 2$.

**TRY THIS** (p. 14)
25 students

**Exercises**
**5.** 12, 17, 22, 27 **6.** The equation representing the situation is $C = 2.5h + 7$ when $h = 1, 2, 3,$ and 4. The values are $9.50, $12, $14.50, and $17. **7.** Let $h$ equal the number of hours Bob can skate. $2.5h + 7 = 32$ $h = 10$ **8.** The 28th term of the sequence is zero. **9.** $d = 24t$ **11.** 5, 10, 15, 20, 25 **13.** 3, 6, 9, 12, 15 **15.** 1, 3, 5, 7, 9 **17.** 5, 14, 23, 32, 41 **19.** 5, 8, 11, 14, 17 **21.** 0, 3, 6, 9, 12 **23.** $x = 2$ **25.** $x = 8$ **27.** $x = 10$ **29.** $s = 17$ **31.** 0 pencils cost $0. **33.** 10 pencils cost $2.00. **35.** $r$ pens cost $0.30$r$. **37.** You can buy 15 tickets. **39.** $c = 16g$ **41.** Enter 15 and repeatedly add 10. The 20th term is 205. **43.** Enter 52 and repeatedly subtract 4. The 20th term is –24. **45.** First differences 6, 16, 27, 39 Second differences 10, 11, 12 **47.** The first five terms are 5, 12, 21, 32, and 45.

### LESSON 1.3

**TRY THIS** (p. 19, Ex. 1)
**a.** 23 **b.** 73 **c.** 39

**TRY THIS** (p. 19, Ex. 2)
$80 \div (6 + 2) \cdot 5 = 50$

**TRY THIS** (p. 20, Ex. 3)
600 feet

**TRY THIS** (p. 20, Ex. 4)
$(17 + 97) \div 2 - (42 - 6) \div 12 = 54$

**TRY THIS** (p. 20, Ex. 5)
289

**Exercises**
**6.** 8 **7.** 37 **8.** 2 **9.** $14 \div (5 + 2) + 8 = 10$ **10.** $(4 + 7) \cdot 3 + 1 = 34$ **11.** $(8 + 25) \cdot 2 \div 6 = 11$ **12.** 52 **13.** 4 **14.** 0 **15.** 2.4 **16.** 46 **17.** 1 **19.** 7068 **21.** 54 **23.** 3591 **25.** 19 **27.** 2 **29.** 6 **31.** 48 **33.** 7 **35.** 70 **37.** 35 **39.** 23 **41.** 51 **43.** $59 - 4 \cdot (6 - 4) = 51$ **45.** $(22 - 4) \cdot 3 = 54$ **47.** $108 - 17 \cdot (2 + 3) = 23$ **49.** 34 **51.** 32 **53.** 18 **55.** 20 **57.** $(97 - 17) \div (72 + 8)$. The value is 1. **59.** $(28 + 59) \div (97 - 17)$. The value after rounding is 1.088. **61.** $(43 \cdot 91) \div (8 \cdot 25)$. The value is 19.565. **63.** 44 **65.** $(100 - 99) \cdot [3 + 7 \cdot (2 + 1)] \div 24 = 1$ **67. a.** For house A, 423 $ft^2$ of carpet are needed. **b.** For house B, 531 $ft^2$ of carpet are needed. **c.** For house A, the cost of carpeting is $610.53. **d.** For house B the cost of carpeting is $766.41. **69. a.** $1 + 3 = 4, 3 + 6 = 9, 10 + 15 = 25$ **b.** Yes. The number 1, which is both square and triangular, cannot be represented as the sum of two triangular numbers (0 is not a triangular number). **71.** The next two terms are 34 and 29. **73.** Jean could ice skate for 13 hours and have $1 left. **75.** Length = 4, Width = 4

### LESSON 1.4

**TRY THIS** (p. 25, Ex. 1)
$G(7, 7), H(3, 2), J(10, 5)$

**TRY THIS** (p. 25, Ex. 2)

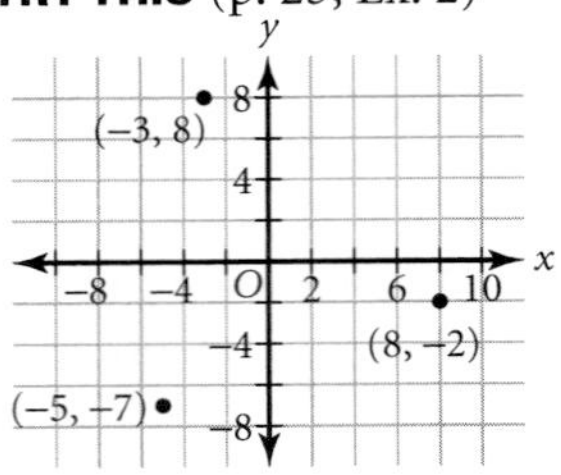


**TRY THIS** (p. 26)

| x | 1 | 2 | 3 | 4 | 5 |
|---|---|---|---|---|---|
| y | 3 | 6 | 9 | 12 | 15 |

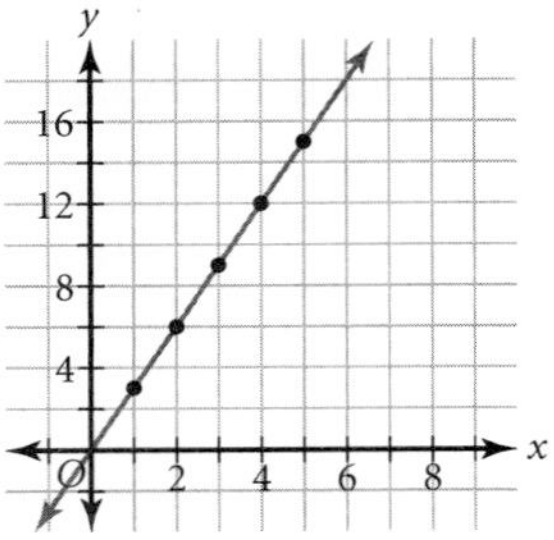

**Exercises**

**6.** II, (−3, 8) **7.** I, (4, 8) **8.** II, (−5, 2) **9.** I, (7, 9) **10.** III, (−8, 5) **11.** IV, (6, −8) **12.** III, (−4, −3) **13.** I, (5, 5) **14.** IV, (9, −3) **15.** IV, (2, −5) **16.** III, (−2, −3) **17.** II, (−8, 6)

**18.–20.**

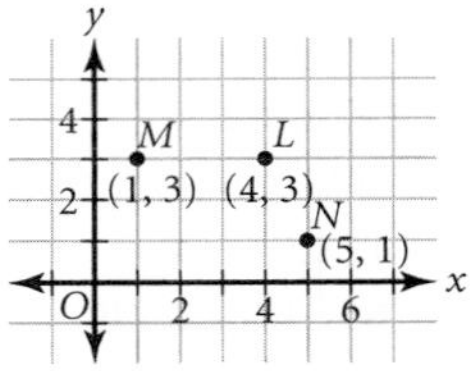


**21.** 3, 6, 9, 12, 15 **22.** 3, 5, 7, 9, 11 **23.** 1, 4, 7, 10, 13 **25.** No **27.** Yes **29.** Yes **31.** I, (10, 7) **33.** IV, (7, −8) **35.** I, (4, 10) **37.** II, (−2, 5) **39.** III, (−3, −5) **41.** II, (−6, 8) **43.** IV, (4, −3) **45.** −3, −2, −1, 0, 1 **47.** 7, 9, 11, 13, 15 **49.** 7, 12, 17, 22, 27 **51.** 2, 9, 16, 23, 30

**53.**

| x | 1 | 2 | 3 | 4 | 5 |
|---|---|---|---|---|---|
| y | 4 | 5 | 6 | 7 | 8 |

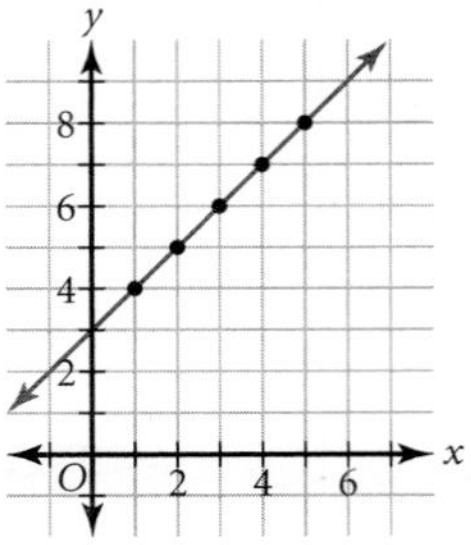

**55.**

| x | 1 | 3 |
|---|---|---|
| y | 8 | 10 |

**57.**

| x | 2 | 5 |
|---|---|---|
| y | 5 | 2 |

**59. a.** An order for 2 T-shirts costs $19. **b.** An order for 5 T-shirts costs $43. **c.** The points lie on a straight line. **d.** 9 T-shirts. **61.** 171 **63.** 49 pairs of numbers can be formed without repetition. **65.** 5050

## LESSON 1.5

**TRY THIS** (p. 31)

$c = 10n + 21$

**TRY THIS** (p. 33)

$d = 54t$

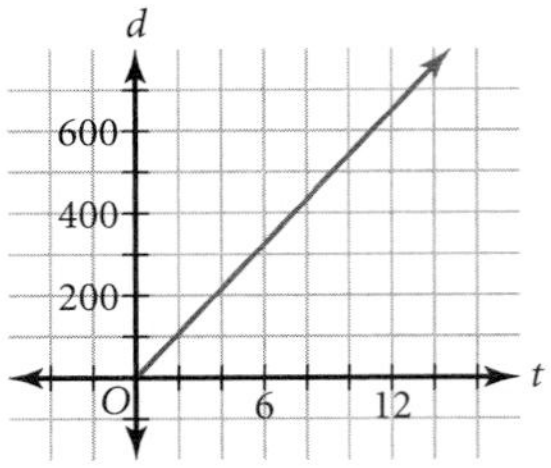

**Exercises**

**5.** The first differences are 12. Use $c$ for cost and $h$ for hours. The equation is $c = 12h + 25$.

**6.** $d = 47t$

| x | 0 | 1 | 2 | 3 |
|---|---|---|---|---|
| y | 0 | 47 | 94 | 141 |

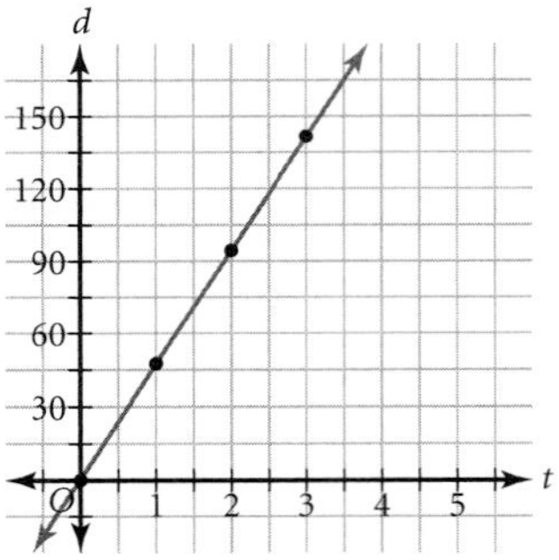

**7.** profit = 200 − 6$b$ **9.** 6, $y = 6x + 25$ **11.** 3, $y = 3x$ **13.** −45, $y = 500 - 45x$ **15.** 21, $y = 21x + 5$

**17.**

| x | 1 | 2 | 3 | 4 | 5 |
|---|---|---|---|---|---|
| y | 19 | 26 | 33 | 40 | 47 |

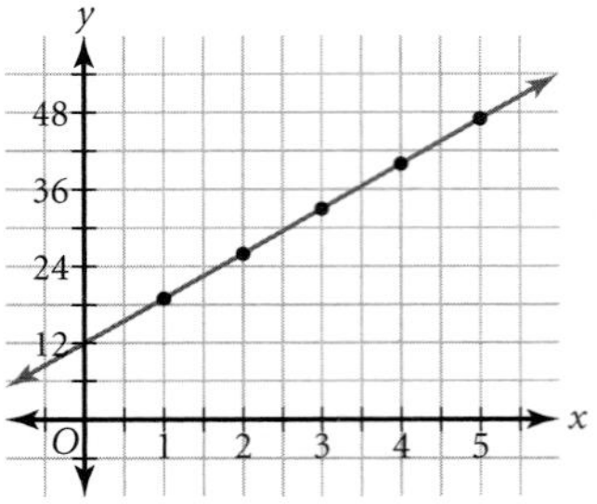

**19.**

| x | 1 | 2 | 3 | 4 | 5 |
|---|---|---|---|---|---|
| y | 30 | 60 | 90 | 120 | 150 |

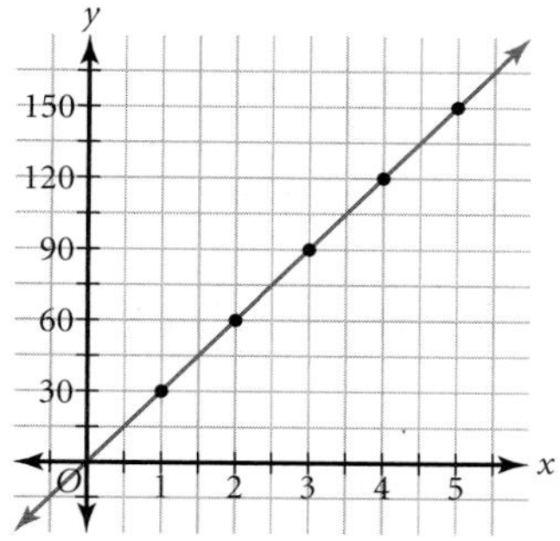


**21.**

| x | 1 | 2 | 3 | 4 | 5 |
|---|---|---|---|---|---|
| y | 298 | 296 | 294 | 292 | 290 |

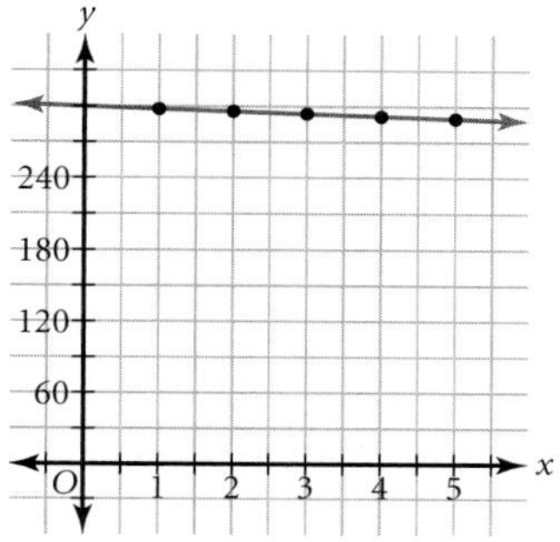


**23.**

| x | 1 | 2 | 3 | 4 | 5 |
|---|---|---|---|---|---|
| y | 22 | 29 | 36 | 43 | 50 |

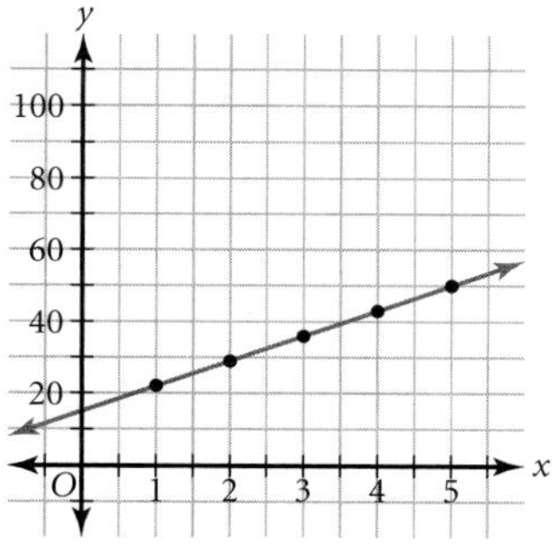


**25.**

| x | 1 | 2 | 3 | 4 | 5 |
|---|---|---|---|---|---|
| y | 110 | 100 | 90 | 80 | 70 |

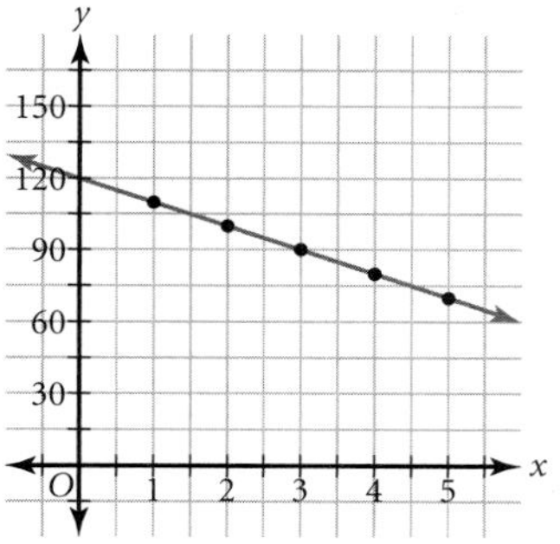


**27.**

| x | 1 | 2 | 3 | 4 | 5 |
|---|---|---|---|---|---|
| y | 110 | 220 | 330 | 440 | 550 |

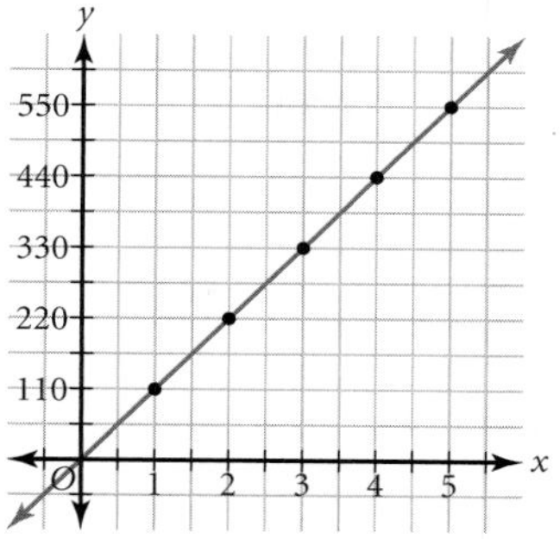


**29.**

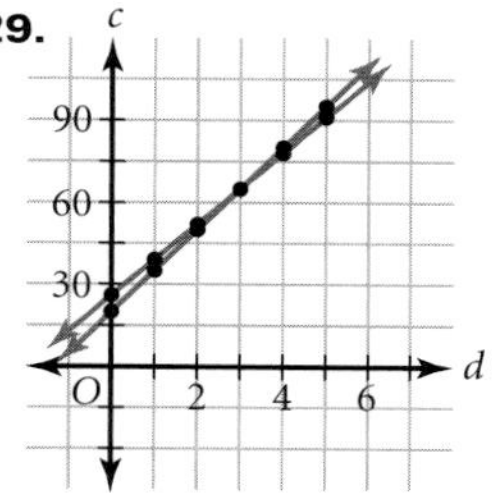


The lines intersect at $d = 3$, $c = 65$. This point represents when the costs are the same.

**31.** $d = 14h$

| *h*, hours | 0 | 1 | 2 | 3 | 4 | 5 |
|---|---|---|---|---|---|---|
| *d*, miles | 0 | 14 | 28 | 42 | 56 | 70 |

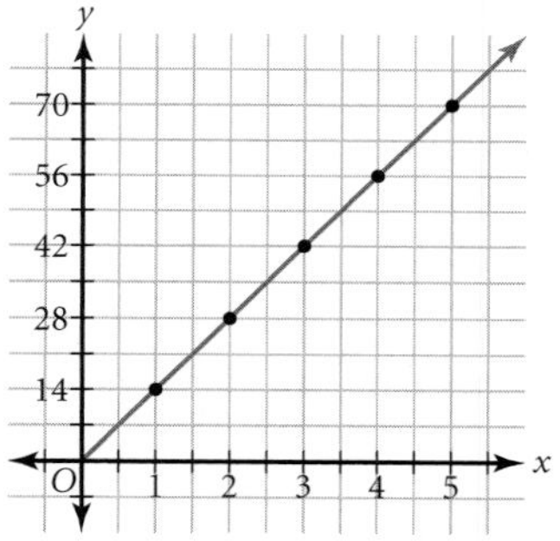


**33.**

| n | 1 | 2 | 3 | 4 | 5 |
|---|---|---|---|---|---|
| c | 37.25 | 48.50 | 59.75 | 71 | 82.25 |

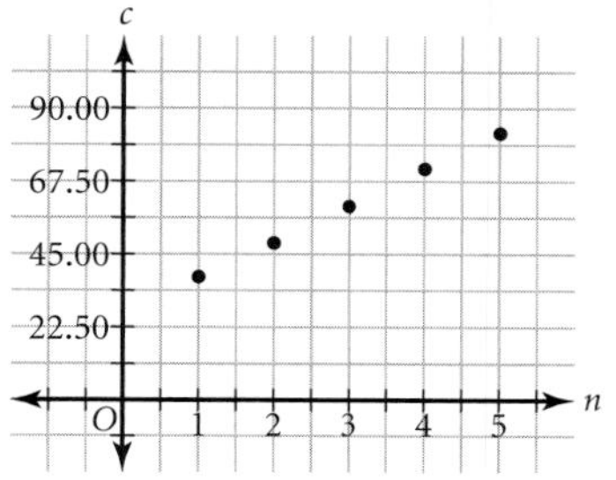


The equation is $c = 11.25n + 26$.

**35.** 184 beats per minute **37.** To raise \$32, you must sell 7 tickets. **39.** Exercises 36 and 37 deal with whole numbers of objects, but Exercise 38 allows us to divide \$32 evenly by 5. **41.** Three differences are needed to reach a constant difference. The constant is 2. **43.** 19 **45.** 6

## LESSON 1.6

**TRY THIS** (p. 38)

Answers may vary. Sample answer: The variables *average daily temperature* and *number of people swimming* would likely show a strong positive correlation. The variables *average speed* and *traveling time* would likely show a strong negative correlation.

**TRY THIS** (p. 40)

There may be 1300 tornadoes reported.

### Exercises

**5.**

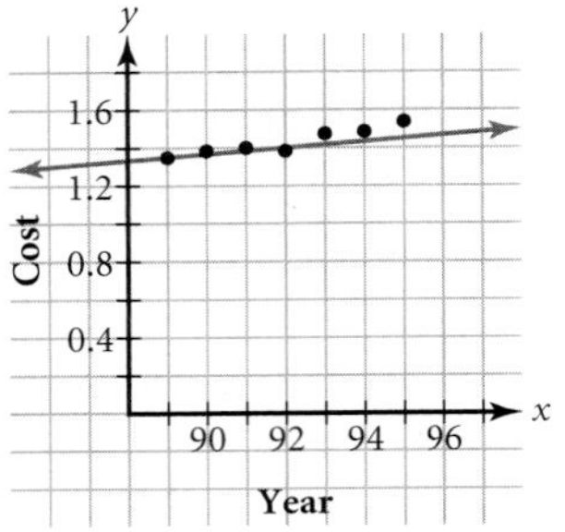


**6.** There is a strong positive correlation between the rising cost of milk and passing time. **7.** \$1.50 **9.** Strong positive correlation **11.** There is a strong negative correlation between the number of people working on farms as the years pass. **13.** There is little or no correlation.

**15.**

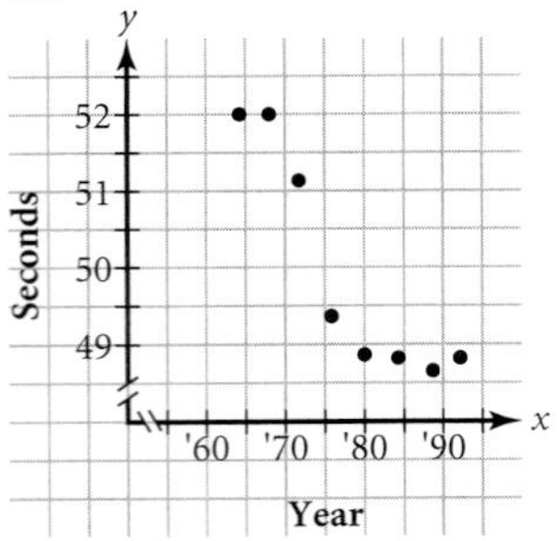


**17.**

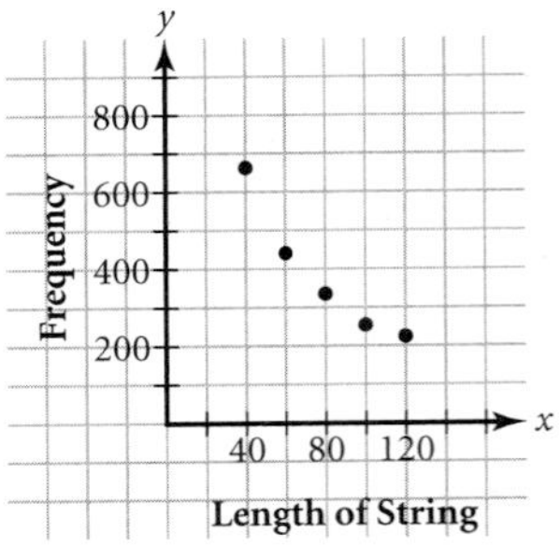


**19.** 12 **21.** 13 **23.** 15

## CHAPTER REVIEW AND ASSESSMENT

**1.** 41, 55 **3.** 30, 30 **5.** 52, 68 **7.** The first five terms are 6, 8, 13, 21, and 32. **9.** 7, 12, 17, 22 **11.** 6, 8, 10, 12 **13.** −1, 3, 7, 11 **15.** 5 **17.** 36 **19.** 126 **21.** Yes **23.** No

**25.**

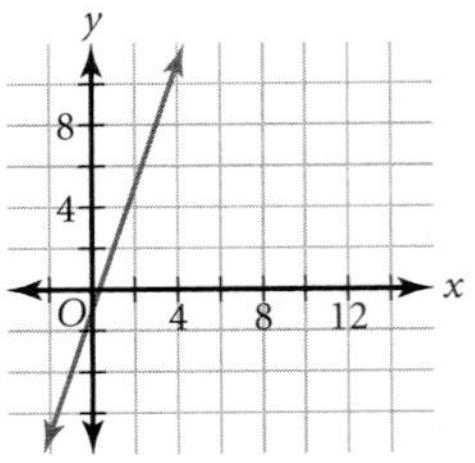


**27.**

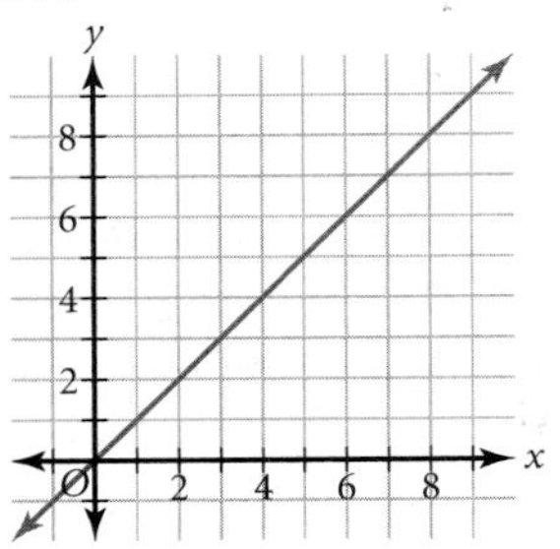


**29.** $d = 5h$

| *h* | *d* |
|---|---|
| 0 | 0 |
| 1 | 5 |
| 2 | 10 |
| 3 | 15 |

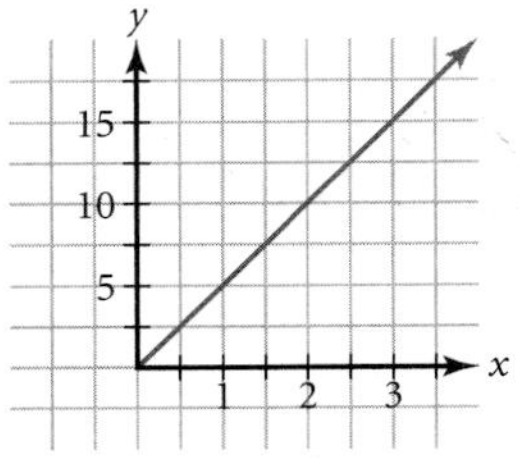


**31.** There is a strong positive correlation between the cost of food items and the passing of time.
**33.** There is a strong negative correlation between the temperature and increasing altitude. **35. a.** \$93 **b.** \$181 **c.** $c = 22t + 5$

# Chapter 2

## LESSON 2.1

**TRY THIS** (p. 56)

**a.** $-9 > -12$
**b.** $\frac{1}{2} < \frac{3}{5}$
**c.** $-6.3 < 5$
**d.** $\frac{4}{12} = \frac{7}{21}$

**TRY THIS** (p. 57)

**a.** $\frac{4}{7}$
**b.** 0
**c.** −3.5

**TRY THIS** (p. 58)

**a.** 13.2 **b.** 5

#### Exercises

**4.** terminating **5.** repeating **6.** repeating **7.** terminating **8.** > **9.** > **10.** < **11.** > **12.** −2 **13.** $\frac{4}{5}$ **14.** −0.5 **15.** 0 **16.** 15 **17.** $12\frac{1}{2}$ **18.** 8 **19.** 1 **21.** < **23.** < **25.** > **27.** < **29.** < **31.** > **33.** 17 **35.** −7.8 **37.** 1.2 **39.** $\frac{13}{20}$ **41.** 18 **43.** 10.3 **45.** 3.8 **47.** $\frac{3}{4}$ **49.** 10 **51.** −24 **53.** 3 **55.** 4 **57.** 12 **59.** 8 **61.** True **63.** False **65.** $-5\frac{1}{2}$; $-7-3+4\frac{1}{2}=-5\frac{1}{2}$ **67.** 141, 207 **69.**

| *m* | 25 | 50 | 75 | 100 | 125 |
|---|---|---|---|---|---|
| *c* | \$32.50 | \$40 | \$47.50 | \$55 | \$62.50 |

## LESSON 2.2

**TRY THIS** (p. 63, Ex. 1)
**a.** −15
**b.** −15
**c.** 3

**TRY THIS** (p. 63, Ex. 2)
**a.** −5.3
**b.** $5\frac{1}{4}$
**c.** −8.8
**d.** $3\frac{7}{15}$

#### Exercises

**5.** −18 **6.** 5 **7.** 13 **8.** −4.11 **9.** 6.73 **10.** $-4\frac{1}{4}$ **11.** −7 **13.** −4 **15.** 4 **17.** 3 **19.** −22 **21.** 8 **23.** 2.5 **25.** $-\frac{8}{17}$ **27.** −15 **29.** −388 **31.** −10 **33.** 7 **35.** 7.81 **37.** 6.78 **39.** −0.25 **41.** 3 **43.** −11.47 **45.** 3.6 **47.** $-\frac{11}{15}$ **49.** $-\frac{1}{6}$ **51.** $1\frac{17}{20}$ **53.** 0 **55.** 5 **57.** 1 **59.** −3 **61.** −8 **63.** 0 **65.** 1 **67. a.** −2 **b.** 5 **69. a.** 2 **b.** −2 **c.** 0 **71.** 28, 25 **73.** 15, 21 **75.** 32, 64 **77.** 23 **79.** 24 **81.** 9

**83.–85.**

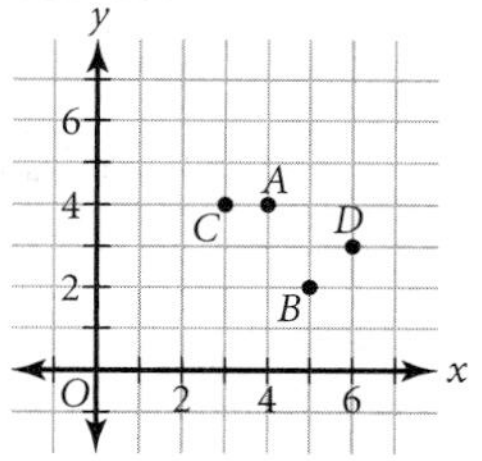


**87.** 17 **89.** $\frac{13}{20}$

## LESSON 2.3

**TRY THIS** (p. 68)

After we take away 10 positive tiles, there will be 6 negative tiles left. 4 − 10 = −6

**TRY THIS** (p. 69, Ex. 2)
**a.** $-5\frac{1}{2}$
**b.** −8
**c.** −3
**d.** 2.2

**TRY THIS** (p. 69, Ex. 3)
**a.** 15
**b.** $6\frac{1}{2}$

#### Exercises

**6.** −13 **7.** 2 **8.** −19 **9.** 14 **10.** −187 **11.** 10 **12.** 1 **13.** 26 **14.** 16 **15.** 14 **17.** −13 **19.** 14 **21.** 35 **23.** −71 **25.** 96.7 **27.** 110 **29.** 51 **31.** −52 **33.** −123 **35.** −185 **37.** 0 **39.** 97 **41.** 7.87 **43.** −0.5 **45.** $\frac{5}{8}$ **47.** $\frac{9}{8}$ **49.** −2 **51.** 10 **53.** 13 **55.** 2 **57.** 21 **59.** 86 **61.** 27 **63.** 63 **65. a.** 2, 5, 8 **b.** 0, −15, −30 **c.** 10, 15, 19 **67.** −4°F. **69.** 12 **71.** 102 **73.** 38 **75.** Yes **77.** Yes **79.** −17 **81.** 24 **83.** 1.4 **85.** −2.2

## LESSON 2.4

**TRY THIS** (p. 75, Ex. 1)
**a.** −48
**b.** 21.2
**c.** −18.2

**TRY THIS** (p. 75, Ex. 2)
**a.** −4
**b.** 4
**c.** −5

**TRY THIS** (p. 76)
**a.** 7
**b.** $-\frac{5}{27}$
**c.** $-\frac{5}{6}$

**TRY THIS** (p. 77)
**a.** 0
**b.** 0
**c.** 0
**d.** undefined

#### Exercises

**6.** 24 **7.** 30 **8.** −135 **9.** −51 **10.** −5 **11.** −7 **12.** 3 **13.** 1 **14.** 0 **15.** 0 **16.** undefined **17.** 0 **19.** −35 **21.** −132 **23.** 51 **25.** −102 **27.** −12 **29.** 64 **31.** −8 **33.** 18.72 **35.** −2.1276 **37.** $-1\frac{65}{96}$ **39.** $-11\frac{1}{9}$ **41.** $\frac{1}{17}$ **43.** $\frac{-9}{4}$ **45.** 22 **47.** 0.2 **49.** 180 **51.** 28 **53.** $-\frac{1}{2}$ **55.** −1081 **57.** 1 **59.** −144 **61.** False **63.** True **65. a.** \$100 **b.** \$50 **c.** 8 deposits **67.** 2 **69.** 4 **71.** 1 **73.** −74 **75.** −101 **77.** 10 **79.** −3

## LESSON 2.5

**TRY THIS** (p. 83)
\$38; \$38.50

**TRY THIS** (p. 84)
**a.** $-a+b$
**b.** $x-8$

**TRY THIS** (p. 85, Ex. 4)

| | |
|---|---|
| $a+(b+c)=(a+b)+c$ | Associative Property |
| $=(b+a)+c$ | Commutative Property |
| $a+(b+c)=(b+a)+c$ | Transitive Property |

**TRY THIS** (p. 85, Ex. 5)

| | |
|---|---|
| $a=b$ | Given |
| $ca=ca$ | Reflexive Property |
| $ca=cb$ | Substitution Property |

#### Exercises

**5.** 36, 64, Associative, 100,115 **6.** 5, Commutative, 20, Associative, 100, 72, 7200 **7.** \$186 **8.** $-x-y$ **9.** $a+b$ **10.** $-x+y$

**11.**

| | |
|---|---|
| $a+(b+c)=a+(c+b)$ | Commutative Property |
| $=(a+c)+b$ | Associative Property |

**13.** 7, 25, Distributive, 175, 100, 275 **15.** 0
**17.** 118,000; Commutative then Associative
**19.** $-2y+9$ **21.** $a-2b-8$ **23.** 10, 8, 400, 320, 720
**25.** 70, 500, 70, 40, 35,000, 2800, 37,800
**27.** $r(s+q)$ **29.** $p(4q+r)$ **31.** $5s(7t+4r)$
**33.** Distributive **35.** Associative **37.** Distributive

**39.** $(ab)c = a(bc)$ Associative Property
$= a(cb)$ Commutative Property

**41.** $a = b$ Given
$c + a = c + a$ Reflexive Property
$c + a = c + b$ Substitution Property of Equality

**43.** \$150 **45.** \$10,512

**47.**

| $x$ | 1 | 2 | 3 | 4 | 5 | 10 |
|---|---|---|---|---|---|---|
| $y$ | 3 | 6 | 9 | 12 | 15 | 30 |

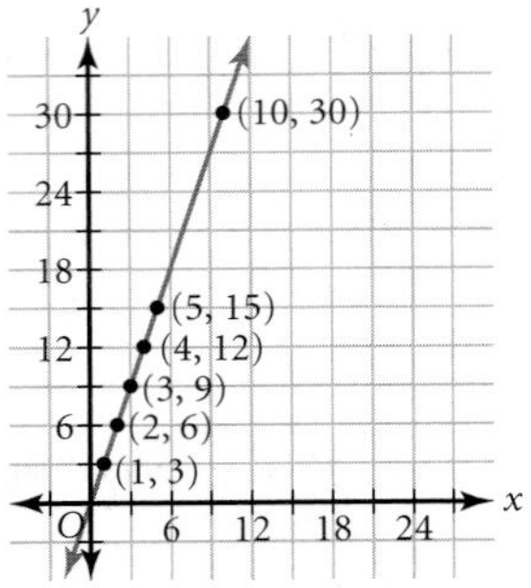


**49.**

| $h$ | 1 | 2 | 3 | 4 | 5 | 6 | 7 | 8 |
|---|---|---|---|---|---|---|---|---|
| $c$ | 145 | 190 | 235 | 280 | 325 | 370 | 415 | 460 |

$c = 45h + 100$

**51.** 25

## LESSON 2.6

**TRY THIS** (p. 90)
**a.** $7x + 2x = (7+2)x$
$= 9x$
**b.** $9x - 3x = (9-3)x$
$= 6x$

**TRY THIS** (p. 91)
**a.** $-x + 11$
**b.** $2m$
**c.** $6x - y$
**d.** $-2m + 2n$

### Exercises

**5.** $-7x + 10x = (-7+10)x = 3x$
**6.** $5h - 9h = (5-9)h = -4h$
**7.** $3d + (-2d) = (3-2)d = d$ **8.** $10x - 4$ **9.** $9c - 5$
**10.** $2f - 2g$ **11.** $16x$
**13.** $4m - 10m = (4-10)m = -6m$
**15.** $\frac{1}{3}p + \frac{2}{3}p = \left(\frac{1}{3} + \frac{2}{3}\right)p = p$
**17.** $11f - 15f = (11-15)f = -4f$
**19.** $-8x - 4x = (-8-4)x = -12x$ **21.** $6x$ **23.** $7c - 3$
**25.** $13r + 5s$ **27.** $7a + 4$ **29.** $3x + 14$ **31.** $20r - 13$
**33.** $2x - 2$ **35.** $2x - 1$ **37.** $v + w$ **39.** $17 - 5y$ **41.** $-2y$
**43.** $6.2m - 6.2n$ **45.** $4x + 14y$ **47.** $\frac{1}{6}x + 2$
**49.** $xy + xz = x(y+z)$ **51.** $6a + 10b - 2c$

**53.** $l = 12x + 3$ **55.** 320, 640, 1280
**57.** $16 \div (5+3) \div 2 = 1$ **59.** $-5$ **61.** 5 **63.** 125

## LESSON 2.7

**TRY THIS** (p. 98, Ex. 1)
**a.** $-15x^2 - 12x$
**b.** $-12x - 16$

**TRY THIS** (p. 98, Ex. 2)
$-10b - 9c$

**TRY THIS** (p. 99)

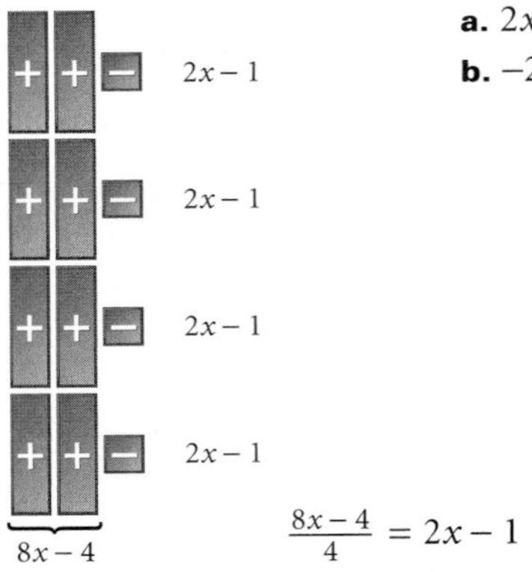


**TRY THIS** (p. 100)
**a.** $2x^3 - 3$
**b.** $-2x^2 + 3$

### Exercises

**6.** $12x - 6$ **7.** $-15x - 6$ **8.** $-32 + 8x$ **9.** $8x + 6y - 6$
**10.** $-7b + 13c - 4$

**11.**

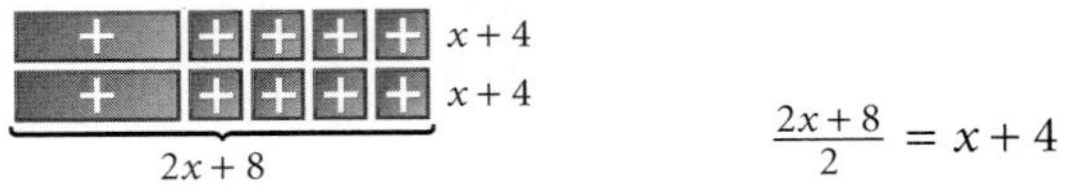


**12.**

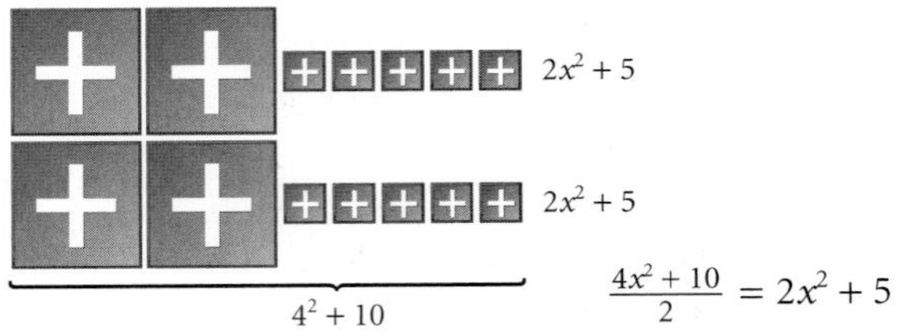


**13.**

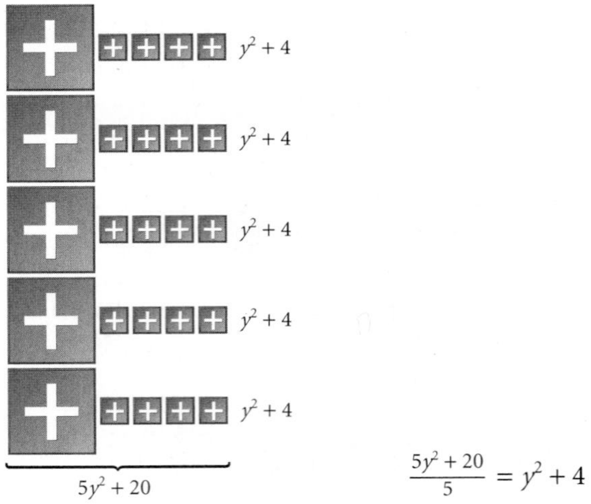


**14.** $3x - 5$ **15.** $-2x + 3$ **16.** $x^2 - 7$ **17.** $12x$ **19.** $12x^2$
**21.** $-33x$ **23.** $-3.6x^2$ **25.** $8x^2 - 3$ **27.** $3x^2 - 6$
**29.** $-16x$ **31.** $16x^2$ **33.** $-21x + 9$ **35.** $63x^2$

**37.** $13x^2 - 2$ **39.** $1 - 3y$ **41.** $10x - 0.3$ **43.** $4m + 5$ **45.** $y^2 + 1$ **47.** $-6x^2 - 4x + 1.6$ **49.** $\frac{4.95}{x} + 1$ **51. a.** $1 - 2x$ **b.** $1 - 2x$ **c.** $1 - 2x$ **d.** $1 - 12x$ (a), (b) and (c) are equivalent. **53.** 35 in$^2$ **55.** 96 in$^3$ **57.** $18xyz$ cubic units **59.** $12 + 2h$ **61. a.** \$90.00 **b.** \$230.00 **c.** \$545.00 **63.** 6.4 **65.** −9.4 **67.** −2.3 **69.** −29 **71.** 10 **73.** 6.3 **75.** −2.6 **77.** −6 **79.** $1\frac{5}{6}$ **81.** 1.3 **83.** $10m - 4$ **85.** $3x + 2y$

## CHAPTER REVIEW AND ASSESSMENT

**1.** < **3.** > **5.** 73 **7.** 15.2 **9.** $\frac{23}{54}$ **11.** −11 **13.** −45 **15.** −22 **17.** −10 **19.** 24 **21.** −41 **23.** −25 **25.** 60 **27.** −3 **29.** 16 **31.** 0 **33.** −25 **35.** 47 (Associative) **37.** 17.1 (Associative and Commutative)
**39.** 20 (Associative and Commutative)
**41.** 31.1 (Associative and Commutative)
**43.** 7, 23, Associative, 30, 78 **45.** $11a - 5$ **47.** $2n - 1$ **49.** $\frac{5}{6}x + 2$ **51.** $x + 3$ **53.** $27x$ **55.** $-14b + 4$ **57.** $10v - 1.2$ **59.** $2x + 2$ **61.** $22w$ **63.** $4x + 3y - 1$

# Chapter 3

## LESSON 3.1

**TRY THIS** (p. 115)
**a.** $a = -1$
**b.** $n = 11$
**c.** $h = -18.4$

**TRY** This (p. 116)
**a.** $a = 12$
**b.** $n = -5$
**c.** $h = 6.8$

### Exercises

**5.** $x = 24$ **6.** $x = -10.7$ **7.** $m = 2\frac{1}{8}$ **8.** $x + 137 = 547$; Jill needs \$410. **9.** $y = 4.8$ **10.** $m = 35$ **11.** $r = 5\frac{5}{6}$ **12.** $x = 17$ **13.** $p = -5$ **14.** $m = 11$ **15.** $m = -3$ **16.** $x = 8$ **17.** $x = -6$ **18.** The flood level is 719 feet. **19.** $a = 31$ **21.** $t = -18$ **23.** $r = -70$ **25.** $b = 167$ **27.** $y = -147$ **29.** $g = 744$ **31.** $x = -3$ **33.** $x = -10$ **35.** $g = 4.2$ **37.** $c = 1.4$ **39.** $k = -4.2$ **41.** $g = 1\frac{1}{6}$ **43.** $y = 15.7$ **45.** $x = -10.3$ **47.** $x = 12$ **49.** $p = -5$ **51.** $m = -1$ **53.** $a = 985$ **55.** $t = -\frac{4}{21}$ **57.** $k = -9.0$ **59.** $n = \frac{17}{18}$ **61.** $m = 4.7$ **63.** $x = \frac{1}{2}$ **65.** $92 + s = 180$ The measure of the angle is 88°. **67.** $x + 30 = 70$ The measure of angle $SRT$ is 40°. **69.** $m + 53 = 67$ Lisa will have \$14 left. **71.** $s + 23 = 49$ Sandy saved \$26. **73.** $m + 149 = 23{,}580$ The odometer reading was 23,431 mi. **75.** $d + 288 = 355$ Carl's birthday is 67 days away. **77.** 78 **79.** $t = 23 \cdot \frac{1}{2} = \frac{23}{2} = 11.5$; 11.5 seconds **81.** 6 poster boards **83.** $8 - 5y$

## LESSON 3.2

**TRY THIS** (p. 122)
$x = 12$

**TRY THIS** (p. 123, Ex. 2)
$b = -\frac{1}{5}$

**TRY THIS** (p. 123, Ex. 3)
\$0.40 per foot

**TRY THIS** (p. 124)
**a.** $t = -16$
**b.** $d = -16.2$
**c.** $m = 4.5$

**TRY THIS** (p. 125)
**a.** $m = -\frac{3}{32}$ **b.** $d = \frac{15}{4}$ **c.** $x = -\frac{4}{15}$

### Exercises

**6.** $x = -2$ **7.** $x = -20$ **8.** $x = \frac{1}{4}$ **9.** $m = -5$ **10.** $8a = 940$ The measure of each angle is 117.5°. **11.** $x = -6$ **12.** $x = -6$ **13.** $x = 8.4$ **14.** $m = 6.3$ **15.** $x = -12$ **16.** $x = \frac{1}{20}$ **17.** $x = 3$ **18.** $m = -\frac{9}{4}$ **19.** $\frac{c}{8} = 14$; $c = 112$ **21.** $m = 2$ **23.** $v = 1.25$ **25.** $b = -54$ **27.** $p = -8.1$ **29.** $x = -8$ **31.** $x = 807$ **33.** $x = 3$ **35.** $f = -5$ **37.** $x = 1$ **39.** $a = \frac{1}{2}$ **41.** $a = \frac{13}{2}$ **43.** $b = 15$ **45.** $x = -\frac{3}{14}$ **47.** $m = -8$ **49.** $x = -35$ **51.** $w = -50$ **53.** $1.15x = 5.75$; $x = 5$; Max can buy 5 rolls. **55.** $4.50x = 18$; $x = 4$; He can buy 4 tape measures. **57.** $4x = 2.52$; $x = 0.63$; The price of one battery is 63 cents. **59.** Maria should drive 40 mph. **61.** 0 **63.** $\frac{1}{6}$ **65.** 1 **67.** $-\frac{1}{2}$ **69.** $x = 2$ **71.** $x = -\frac{11}{15}$

## LESSON 3.3

**TRY THIS** (p. 130)
**a.** $y = 2$
**b.** $p = -22.5$

**TRY THIS** (p. 131)
**a.** $m = -12$
**b.** $x = -52$

### Exercises

**6.** $p = 10$ **7.** $x = -1$ **8.** $m = 22$ **9.** $x = 96$ **10.** $g = 60$ **11.** $m = 34.8$ **12.** $6x + 6(3.99) = 65.88$; $x = 6.99$; \$6.99 **13.** $x = 6$ **15.** $d = -18$ **17.** $m = 6$ **19.** $x = 9$ **21.** $d = 6$ **23.** $x = 2$ **25.** $w = -2$ **27.** $y = -8$ **29.** $m = 0.36$ **31.** $x = 55$ **33.** $m = -4$ **35.** $x = -5$ **37.** $x = 28.5$ **39.** $a = 51$ **41.** $p = -162$ **43.** $z = 10$ **45.** $d = -300$ **47.** $x = 2$ **49.** $x = 9$ **51.** $x = 4\frac{1}{6}$ $m = 34$ **53.** $3c + 35 = 347$ $c = 104$ **55.** $15.95d + 2.95 = 98.65$ $d = 6$ **57.** $1.5h + 4 = 10$ $h = 4$ The basket holds 4 hekats. **59.** 56 **61.** −30 **63.** −12.6 **65.** −22.5 **67.** −5 **69.** −8 **71.** −8 **73.** −5 **75.** $g = 2$ **77.** $h = 1$ **79.** $m = 3$ **81.** $p = 9.8$ **83.** $x = 12$ **85.** $y = 3$ **87.** $m = -60$ **89.** $p = -2$

## LESSON 3.4

**TRY THIS** (p. 136, Ex. 2)
**a.** $x = 10$
**b.** $m = -3$

**TRY THIS** (p. 136, Ex. 3)
**a.** $x = 7$
**b.** $m = \frac{7}{4}$

**TRY THIS** (p. 137)
$x = 10$

**TRY THIS** (p. 138)
$x = 40$

**Exercises**

**6.** $b = -6$ **7.** $h = -1$ **8.** $f = 6$ **9.** $x = 3$ **10.** $x = \frac{1}{15}$
**11.** $n = \frac{1}{16}$ **12.** $a = -\frac{1}{2}$
**13.** $\frac{84 + 92 + 71 + 94 + x}{5} = 85, x = 84$
**15.** $x = -3$ **17.** $x = -2$ **19.** $y = 3$ **21.** $m = -2$
**23.** $p = 3$ **25.** $h = 20$ **27.** $n = -11$ **29.** $a = 4$
**31.** $t = -\frac{8}{3}$ **33.** $a + 5 = 2a - 3$, area $= 8$
**35.** $\frac{27 + 18 + 27 + 32 + 21 + x}{6} = 25, x = 25$
**37.** $6t = t + 80; t = 16$ **39.** $x = 7$ **41.** $m = -15$
**43.** $m = 8$ **45.** $x = -10$ **47.** $x = 4.25$ **49.** $y = -1$

## LESSON 3.5

**TRY THIS** (p. 143, Ex. 2)
**a.** $t = 4$
**b.** $p = -1$
**c.** $m = 2$

**TRY THIS** (p. 143, Ex. 3)
**a.** All real numbers
**b.** No solution

**TRY THIS** (p. 144)
$y = -5$

**Exercises**

**5.** $s = 3$ **6.** $f = 3$ **7.** $r = 21$ **8.** All real numbers
**9.** No solution **10.** $y = 1$ **11.** $y = 4$ **13.** $r = 7$
**15.** No solution **17.** $y = -4$ **19.** $w = 5$ **21.** $y = 0.4$
**23.** $m = 12$ **25.** $m = 1$ **27.** $y = 4$ **29.** $x = -4$
**31.** $m = -8$ **33.** $x = 0.25$ **35.** $r = 3$ **37.** No solution
**39.** $y = -1$ **41.** All real numbers **43.** $w = -9$
**45.** $14 + 2(x + 6) = 4x;\ x = 13$
**47.** $x + 2x + (2x + 7) = 327, x = 64$, Jill 64 votes, Morgan 128 votes, Trisha 135 votes **49.** 21, 31, 43
**51.** $\frac{1}{16}, \frac{1}{32}, \frac{1}{64}$ **53.** 1.9 **55.** 0.9 **57.** −1 **59.** −3
**61.** 6.21 **63.** $\frac{1}{4}$ **65.** −1 **67.** −120 **69.** $x = 10$
**71.** $x = 6.1$ **73.** $r = -20$ **75.** $x = -17$ **77.** $x = 0.5$
**79.** $x = 11$ **81.** $m = -2$ **83.** $p = 0.4$

## LESSON 3.6

**TRY THIS** (p. 147)
**a.** 10°C
**b.** 100°C
**c.** 40°C

**TRY THIS** (p. 148)
**a.** $P = A - I$
**b.** $r = \frac{C}{2\pi}$

**TRY THIS** (p. 150)
**a.** $r = 5.41$
**b.** $r = 11.15$

**Exercises**

**5.** 15°C **6.** −10°C **7.** 65°C **8.** $A = I + P$ **9.** $m = \frac{y}{x}$
**10.** $b = \frac{d - a}{c}$ **11.** $w = 4, l = 12$ **12.** $b = 5$
**13.** $r \approx 4.78$ **15.** 30°C **17.** 51.7°C **19.** $T = M + R$
**21.** $b = c - a$ **23.** $a = -b - c$ **25.** $d = \frac{C}{\pi}$ **27.** $w = \frac{A}{l}$
**29.** $c = -\frac{4p}{9}$ **31.** $x = \frac{7 - r}{a}$ **33.** $x = \frac{5 - 4y}{3}$
**35.** $b = \frac{y - 3x}{3}$ **37.** $r = 0.80$ centimeters
**39.** $r = 2.39$ meters **41.** $r = 8.28$ feet
**43.** $r = 1.75$ inches **45.** $b_2 = 16$ **47.** $b_2 = 6$
**49.** $b_2 = 4$ **51.** $w = 20$ $l = 80$ **53.** 22 square feet
**55.** 511 **57.** 9 **59.** −11 **61.** $-8y + 4$ **63.** $x = 7$
**65.** $x = -11$

## CHAPTER REVIEW AND ASSESSMENT

**1.** $r = -10$ **3.** $t = -12$ **5.** $x = 43$ **7.** $h = \frac{5}{6}$ **9.** $m = 18$
**11.** $x = -5$ **13.** $h = -15$ **15.** $w = -0.0625$
**17.** $t = -120$ **19.** $w = -152$ **21.** $a = 8$ **23.** $y = 182.5$
**25.** $x = -33$ **27.** $x = \frac{2}{3}$ **29.** $x = 0$ **31.** $x = 5$
**33.** $m = -385$ **35.** $m = -5$ **37.** $m = -2$ **39.** $m = 6\frac{2}{3}$
**41.** $f = 3$ **43.** $m = 1$ **45.** $x = 3$ **47.** $m = -10$
**49.** $x = 2.5$ **51.** $x = 5$ **53.** $x = 9$ **55.** $x = -1$
**57.** $m = \frac{y - b}{x}$ **59.** $b = c - ax$ **61.** 15.92 meters
**63.** 2.39 feet **65.** 4.46 inches

# Chapter 4

## LESSON 4.1

**TRY THIS** (p. 166)
**a.** $x = 7$
**b.** $n = 10$

**TRY THIS** (p. 167)
$x = 21$

**Exercises**

**5.** True **6.** Not true **7.** True **8.** $x = 15$ **9.** $p = 18$
**10.** $m = 98$ **11.** $b = 18\frac{2}{3}$ **12.** 384 fish of species B are needed. **13.** True **15.** True **17.** True **19.** Not true
**21.** True **23.** Not true **25.** $n = 28$ **27.** $n = 24$
**29.** $x = 6$ **31.** $y = 2.75$ **33.** $m = 10$ **35.** $f = 22.4$
**37.** $d \approx 8.23$ **39.** $p = 4$ **41.** $x = 121$ **43.** $q = 2$
**45.** $p = 37$ **47.** 19.5 feet **49.** 6.4 pounds **51. a.** 12.5 pounds **b.** 16.25 pounds **53.** $15x$ **55.** $11x - 26$
**57.** $-12t^2$ **59.** $w = -156$ **61.** $x = 2.16$ **63.** $y = -31$

## LESSON 4.2

**TRY THIS** (p. 172)
**a.** $\frac{2}{5}$ or 0.4
**b.** $1\frac{3}{4}$ or 1.75
**c.** $\frac{1}{40}$ or 0.025

**TRY THIS** (p. 173)
**a.** 18.75%
**b.** 230
**c.** 112.5

**TRY THIS** (p. 174, Ex. 4)
6.5%

**TRY THIS** (p. 174, Ex. 5)
25%

**TRY THIS** (p. 175)
21%

**Exercises**

**5.** 0.34; $\frac{17}{50}$ **6.** 0.2; $\frac{1}{5}$ **7.** 1.3; $1\frac{3}{10}$ **8.** 0.075; $\frac{3}{40}$ **9.** 105
**10.** 7 **11.** 190.4 **12.** 34 **13.** 62.5 **14.** 10% **15.** 20%
**16.** 10% **17.** 20% **18.** 15% **19.** 40% **20.** 20%
**21.** $8\frac{1}{3}$% **22.** 45% **23.** 0.55 **25.** 0.08 **27.** 0.005
**29.** 1.86 **31.** 0.045 **33.** $\frac{47}{100}$ **35.** $\frac{7}{200}$ **37.** $1\frac{1}{4}$ **39.** $\frac{19}{50}$
**41.** $2\frac{1}{10}$ **43.** $x = 28$ **45.** $x = 150$ **47.** $x = 120$ **49.** 20

**51.** 2% **53.** 50 **55.** 72 **57.** 36 **59.** 75% **61.** 45 **63.** 15% **65.** 5% **67.** 5% **69.** 9% **71.** 200 points **73.** 25% **75.** $168 **77.** 4 **79.** 4 **81.** $13x - 2$ **83.** $-3c - 3$ **85.** $2x + 9x^2$ **87.** $x^2 - 7$

## LESSON 4.3

**TRY THIS** (p. 182, Ex. 1)

$\frac{29}{100}$ or 29%

**TRY THIS** (p. 182, Ex. 2)

$\frac{52}{200} = 0.26 = 26\%$

### Exercises

**6.** 0.53 or 53% **7.** 0.07 or 7% **8.** 0.29 or 29% **9.** 0.23 or 23% **10.** 0.255 or 25.5% **11.** 0.53 or 53% **13.** 0.76 or 76% **15.** 0.24 or 24% **17.** 0.03 or 3% **19.** 0.865 or 86.5% **21.** 0.335 or 33.5% **23.** 0.11 or 11% **25.** 0.25 or 25% **27.** 0.1 or 10% **29.** Answers may vary. Check students' responses. **31.** favorable, 69; unfavorable, 531 **33.** Answers will vary. Roll a pair of number cubes and record the results. Repeat for 5, 10, 15, 20, 25, and 30 trials. Divide the number of trials where doubles occurred by 5, 10, 15, 20, 25, 30 respectively. **35.** $x = 5$ **37.** $-29$ **39.** $\frac{1}{6}$ **41.** $t = -\frac{1}{10}$

## LESSON 4.4

**TRY THIS** (p. 188)

Mean is 0.3324, Median is 0.329, Mode is 0.328, Range is 0.019

### Exercises

**5.** 5; 5; 2 and 8; 6 **6.** 20; 15; none; 28 **7.** 7; 7.5; none; 7 **8.** 0.6; 0.65; none; 0.7 **9.** 5.175; 4.6; none; 7.3 **10.** 4.8; 4; 2 and 4; 11 **11.** 27 games **12.** ≈3.7 **13.** 3, 3, 8 **15.** 10 **17.** 87 **19.** 80 **21.** 110.4 **23.** 51 **25.** 10 **27.** 87 **29.** 84.5 **31.** 114 **33.** 52 **35.** 8 **37.** 20 **39.** 50 **41.** 25 **43.** 55 **45.** 19 **47.** 98 **49.** $4\frac{1}{3}$; 3.5; 8 **51.** 52; 50; 9 **53.** 94; 94; 13 **55.** 92 **57.** mode **59.** $x = 13$ **61.** 166.1 **63.** $a = 12$ **65.** $x = -368$ **67.** $b = 4$ **69.** $q = 16$

## LESSON 4.5

**TRY THIS** (p. 196, Ex. 1)

1997 MRI = $340, 1999 CBC = $160

**Don's CBC Stock**

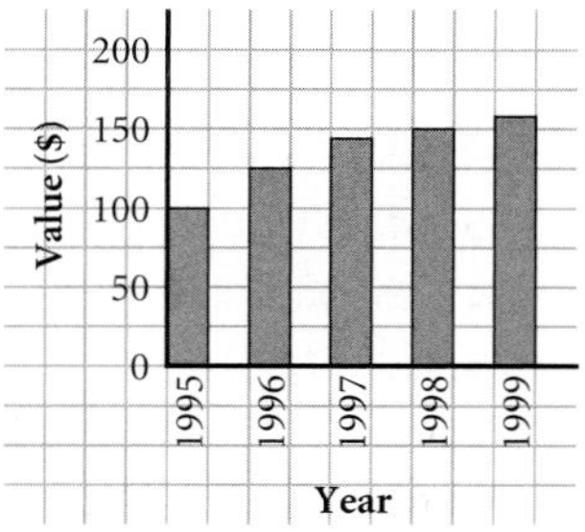


**TRY THIS** (p. 196, Ex. 2)

Investments: $2560; Clothing: $1920; Miscellaneous: $7040.

**TRY THIS** (p. 198)

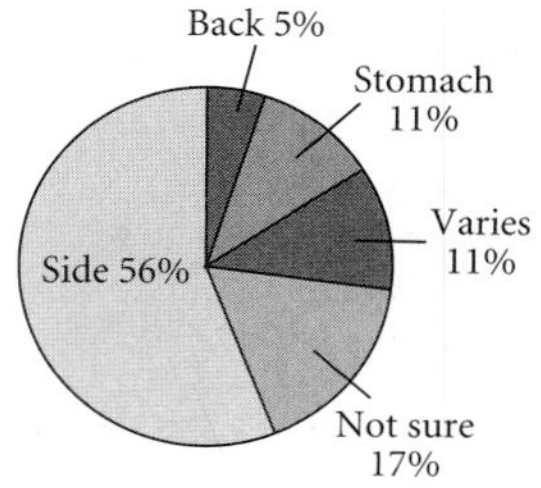


### Exercises

**5.** Smallest = under 5, Second smallest = 65 and over **6.** Northeast: 88,550,000, West: 60,720,000, South: 53,130,000, Midwest: 50,600,000 **7.** 92,750,000

**8.** **Movie Poll**

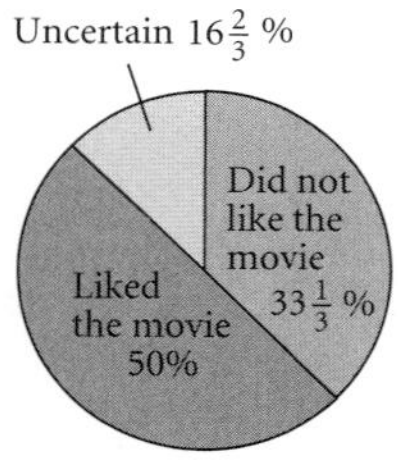


**9.** High Technology Products **11.** The scales of the two graphs are different. **13.** Moscow; January

**15.** **Canine Companions Program**

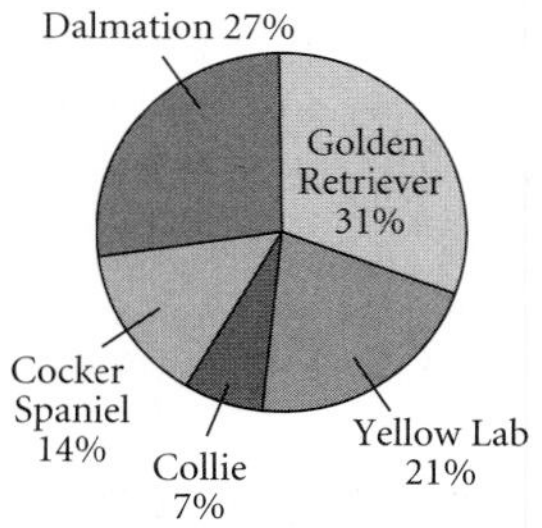


**17.** 248,820 **19.** Answers may vary. Sample answer: You could check the electrical wiring in your home. **21.** Other **23.** about 100 **25.** $8x - 5$ **27.** $t = 8$ **29.** $x = 4$ **31.** $m = 0$ **33.** 56 **35.** 9.25; 8; 4; 14

## LESSON 4.6

**TRY THIS** (p. 202)

| Stems | Leaves |
|---|---|
| 6 | 4 |
| 7 | 6, 9, 9 |
| 8 | 0, 0, 1, 6 |
| 9 | 1, 6, 7, 9 |

**TRY THIS** (p. 204)

**a.** 30 **b.** 80 **c.** 65 **d.** 45 **e.** 55 **f.** 20

## Exercises

**5.** Stems:ones, Leaves:tenths

| Stems | Leaves |
|---|---|
| 10 | 3, 7 |
| 11 | 5 |
| 12 | 1, 1, 2, 6 |
| 13 | 4, 7, 8, 8, 9 |
| 14 | 1, 3, 4 |
| 15 | 2, 2, 2, 5 |
| 16 | 6, 7 |
| 17 | 3 |

**6.**

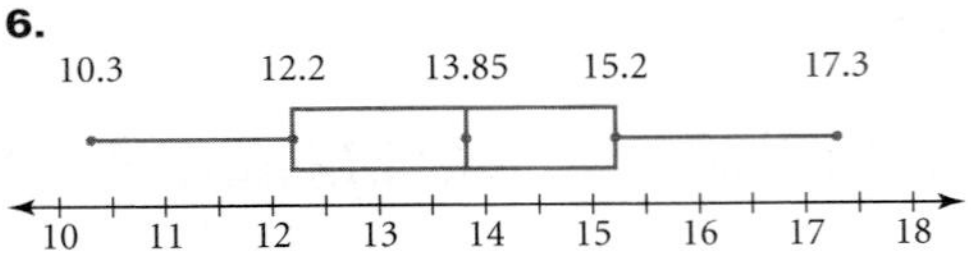


3.0

**7.** Stems:ones, Leaves:tenths

| Stems | Leaves |
|---|---|
| 12 | 2, 5 |
| 13 | 3 |
| 14 | 8 |
| 15 | 1, 7, 9 |
| 16 | 3, 4, 4, 6 |
| 17 | 2 |
| 18 | 7, 9 |
| 19 | 1 |
| 20 | 3 |

**9.** Stems:tens, Leaves:ones

| Stems | Leaves |
|---|---|
| 12 | 5, 7, 7, 8 |
| 13 | 3, 5, 6, 6, 8, 9 |
| 14 | 1, 2, 4, 7, 8 |
| 15 | 3, 3, 3, 4, 9 |
| 16 | 2 |
| 17 | 5, 6, 9 |
| 18 | 3, 4 |

**11.** Stems:ones, Leaves:tenths

| Stems | Leaves |
|---|---|
| 1 | 2 |
| 2 | 5, 5, 7 |
| 3 | 4 |
| 4 | 6, 8 |
| 5 | 2, 6 |
| 6 | 3, 5, 7, 8 |
| 7 | 8 |
| 8 | 4, 4 |
| 9 | 1, 2, 7 |

**13.** Stems:tens, Leaves:ones

| Stems | Leaves |
|---|---|
| 0 | 4, 5, 5, 5, 6, 7, 7, 8, 9 |
| 1 | 0, 2, 2, 6, 7, 8, 8, 9, 9 |
| 2 | 3, 6, 9 |

**15.** Stems:ones, Leaves:tenths

| Stems | Leaves |
|---|---|
| 20 | 8 |
| 21 | 4, 6 |
| 23 | 1, 6, 9 |
| 24 | 7 |
| 25 | 8 |
| 26 | 1, 4 |
| 27 | 6 |
| 29 | 4, 4, 5, 6 |

**17.**

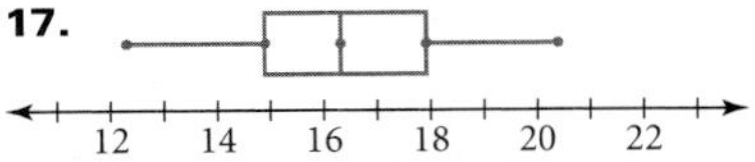


3

**19.**

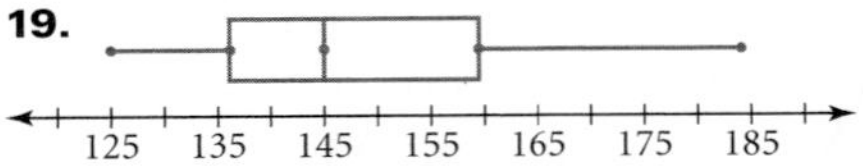


23

**21.**

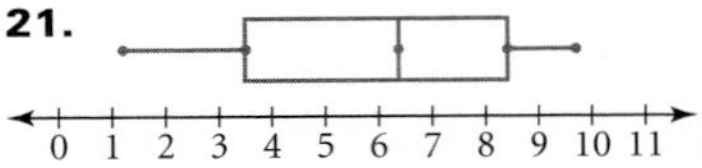


5.0

**23.**

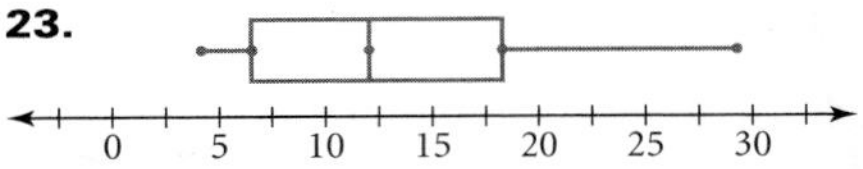


12.0

**25.**

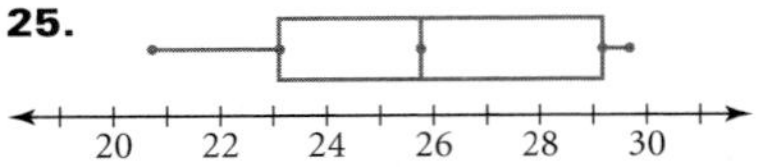


6.3

**27.** The new mean would be slightly lower since the new data is less than the original mean. The new mode would be 59 and 60. The median and range would stay the same. **29.** 44.8%

**31.**

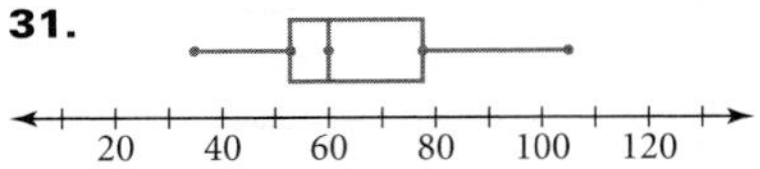


24.5

**33.** 5.9 **35.** 17.2 and 19.3

**37.**

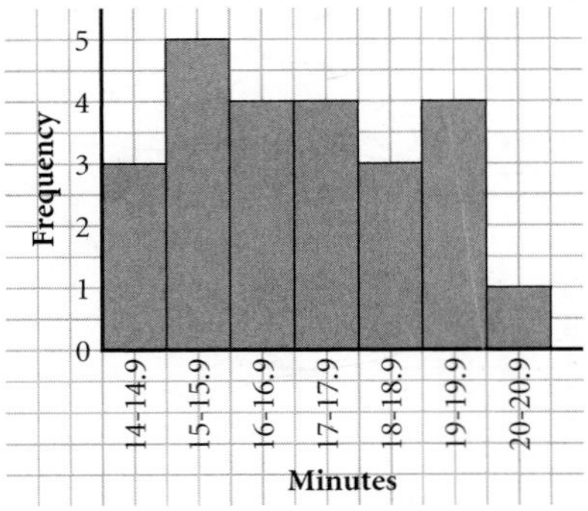

**39.** Stems:tens, Leaves:ones

| Stems | Leaves |
|---|---|
| 24 | 9 |
| 27 | 5 |
| 34 | 3, 5 |
| 38 | 3 |
| 39 | 9 |
| 47 | 8 |
| 48 | 4 |
| 49 | 1, 7, 7, 9 |
| 51 | 2 |
| 54 | 9 |
| 56 | 7 |
| 57 | 4, 6 |
| 58 | 6, 8, 9 |

**41.** 497 **43.** Yes; 497 **45.** 10 **47.** $\frac{1023}{50}$ **49.** $7y - 4$ **51.** $y = -\frac{1}{2}$ **53.** $\frac{19}{50}$

## CHAPTER REVIEW AND ASSESSMENT

**1.** $n = 8$ **3.** $x = 32$ **5.** $x = 1$ **7.** $x = 3.75$ **9.** 33 **11.** 80 **13.** 45% **15.** 25% **17.** 62 **19.** 1305 **21.** ≈ 44.7% **23.** ≈ 7.3% **25.** 8; 8; 8; 5 **27.** 97; 90.5; 55; 151 **29.** 0.3752; 0.376; none; 0.043 **31.** $18.\overline{3}$; 12; 8; 39

**33.** Car-Expense Budget

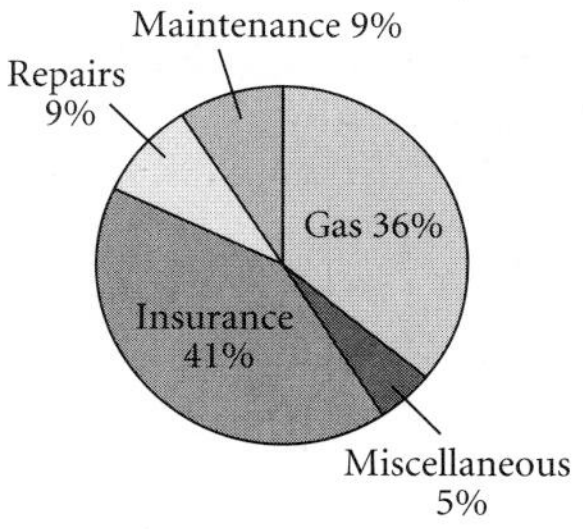


**35.** 46 **37.** 38

**39.** Stems: tens
Leaves: ones

| Stems | Leaves |
|---|---|
| 5 | 0, 8 |
| 6 | 0, 5, 8 |
| 7 | 4, 5, 6, 8, 9 |
| 8 | 0, 1, 4, 5, 5, 8 |
| 9 | 2, 2, 3, 8, 9 |

**41.**

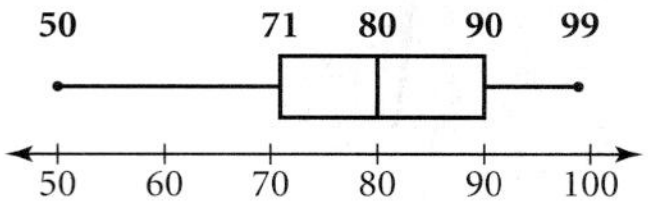


**43.** McGroty Family Budget

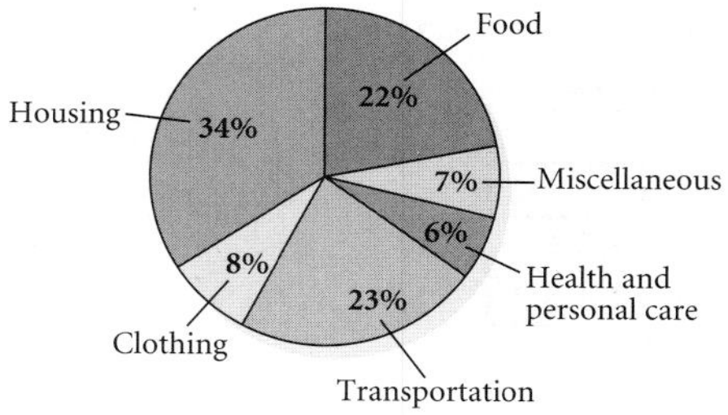


# Chapter 5

## LESSON 5.1

**TRY THIS** (p. 220, Ex. 1)

Answers may vary. Sample: A set of ordered pairs that is a function is {(0, 1), (1, 2), (2, 3), (3, 4)}. A set of ordered pairs that is not a function is {(1, 5), (2, 6), (1, 7), (3, 8)}.

**TRY THIS** (p. 220, Ex. 2)

**a.** (3, 6) **b.** (4, 3)
**c.** (0, 15) **d.** (5, 0)

**TRY THIS** (p. 222, Ex. 3)

$y = 3.5x + 15${(1, 18.50), (2, 22), (3, 25.50), (4, 29), (5, 32.50), (6, 36), (7, 39.50), (8, 43), (9, 46.50), (10, 50)} The domain is {1, 2, 3, 4, 5, 6, 7, 8, 9, 10}. The range is {18.50, 22, 25.50, 29, 32.50, 36, 39.50, 43, 46.50, 50}.

**TRY THIS** (p. 222, Ex. 4)

The dependent variable is distance. The independent variable is time. The domain and range are both the set of real numbers greater than or equal to 0.
$y = 50x$

### Exercises

**5.** The domain is {3, 4, 9}. The range is {–5, –2, 5}. This relation is a function. **6.** The domain is {1, 2, 3, 4}. The range is {1}. This relation is a function. **7.** (5,1) **8.** (3,9) **9.** (0,21) **10.** $\left(5\frac{1}{4}, 0\right)$ **11.** $y = 0.5x + 2$ The domain is {1, 2, 3, 4, 5, 6} and the range is {2.5, 3, 3.5, 4, 4.5, 5}.

**12.** $y = 2x$ **13.** Domain: {–2, 5.7, 9.2} Range: {–3.4, 3.4, 4} This relation is a function, since no element of the domain is repeated.
**15.** Domain: {–3, –1, 0, 1, 3,} Range: {–2, –1, 4} This relation is a function, since no element of the domain is repeated. **17.** Domain: {3, 4, 5, 9, 12} Range: {3, 4, 5, 9, 12} This relation is a function, since no element of the domain is repeated.
**19.** (10,6) **21.** (7,0) **23.** (–5, –24) **25.** $\left(8\frac{1}{2}, 3\right)$ **27.** (5,–4) **29.** (12,10)

**31.**

| $x$ | 0 | 1 | 2 | 3 | 4 |
|---|---|---|---|---|---|
| $y$ | 0 | 5 | 10 | 15 | 20 |

The equation is $y = 5x$. The domain and range are both the set of real numbers greater than or equal to zero.

**33.** $y = 8x + 35$ **35.** 3, 5, 7, 9, 11, 21

**37.** Graphs may vary.

**39.** Answers may vary. Approximate sample: $y = 1.2x + 12$ **41.** 104 **43.** 30 **45.** 49 square centimeters

## LESSON 5.2

**TRY THIS** (p. 227)
slope $= \frac{12}{24} = \frac{1}{2}$

**TRY THIS** (p. 229, Ex. 3)
slope $= \frac{15 - (-3)}{-2 - 2} = \frac{18}{-4} = -\frac{9}{2}$

**TRY THIS** (p. 229, Ex. 4)

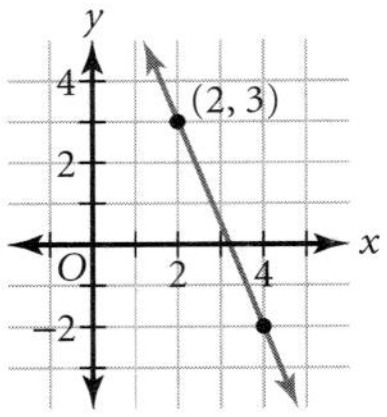


**TRY THIS** (p. 230)
**a.** Vertical line. Slope is undefined.
**b.** Horizontal line. Slope is zero.

### Exercises

**6.** $-\frac{6}{7}$ **7.** $\frac{1}{2}$ **8.** $-\frac{10}{9}$

**9.**

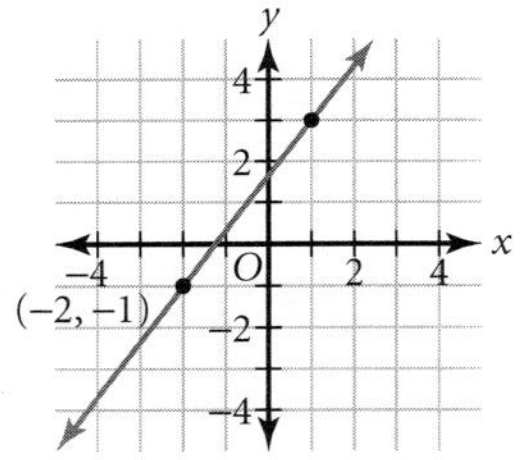


**10a.** Horizontal; 0 **b.** Vertical, undefined **11.** 1 **13.** 0 **15.** −2 **17.** undefined **19.** 2 **21.** $\frac{3}{10}$ **23.** −2 **25.** $-\frac{5}{2}$ **27.** $\frac{4}{3}$ **29.** $-\frac{5}{21}$ **31.** undefined **33.** 0 **35.** $\frac{7}{2}$

**37.**

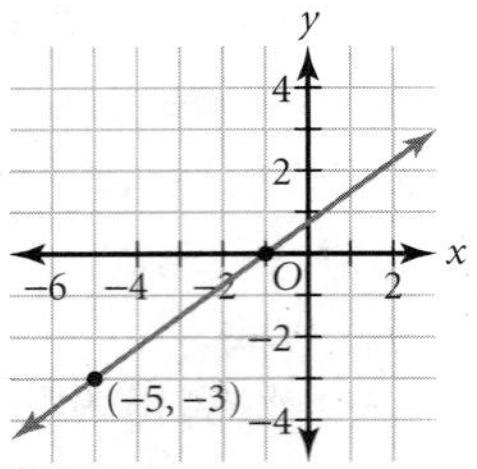


**39.**

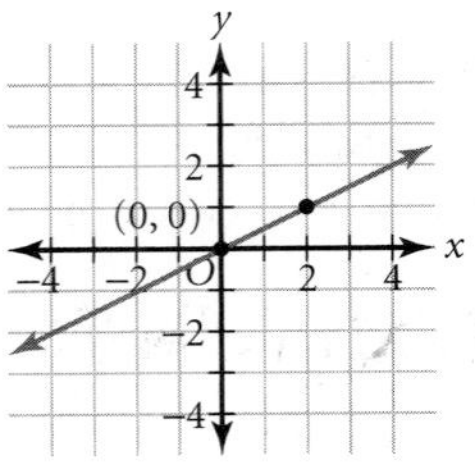


**41.**

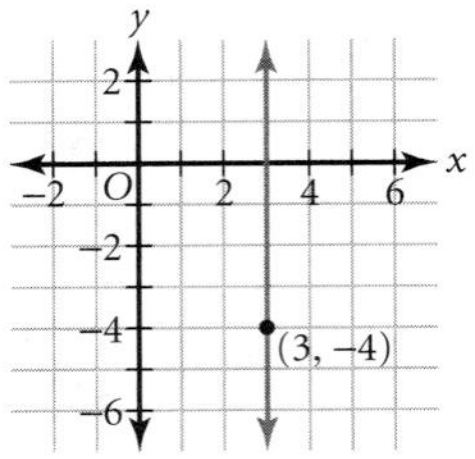


**43.**

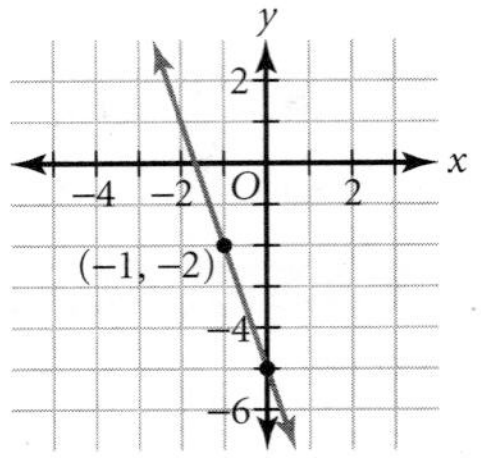


**45.** $\frac{d - b}{c - a}$ **47.** The change in temperature is −5°F per minute. **49.** $\frac{27}{25} = 1.08$ **51.** 7 **53.** −29 **55.** $12.25 per person **57.** $x = 8$ **59.** $x = \frac{10}{7}$ **61.** $x = \frac{a + 15}{2}$

## LESSON 5.3

**TRY THIS** (p. 237)
$$\frac{\text{change in height}}{\text{change in length}} = \frac{4}{24} = \frac{1}{6} \approx 0.17$$
The ramp rises about 0.17 inches for every foot.

**TRY THIS** (p. 238, Ex. 4)
$k = \frac{10}{2}$, $y = 5x$

**TRY THIS** (p. 238, Ex. 3)
$\frac{35}{7} = \frac{y}{84}$, $y = 420$

**TRY THIS** (p. 239)
$F = kd = 28(0.25) = 7$ Newtons

### Exercises

**5.** The cost per video is $3. **6.** $10 per hour **7.** $15 per hour **8.** $k = \frac{1}{4}$; $y = \frac{1}{4}x$ **9.** $x = 25$ **10.** $k = 32$ **11.** $2.50 per video **13.** $k = 7$; $y = 7x$ **15.** $k = 0.625$; $y = 0.625x$ **17.** $k = 11$; $y = 11x$ **19.** $k = 0.3$; $y = 0.3x$ **21.** $k = 7$; $y = 7x$ **23.** $x = 3$ **25.** $x = 1$ **27.** $p = 4l$ **29.** $y = 8x$ **31.** $y = 4x$ **33.** 2 miles per hour **35.** 8 hours

**37.**

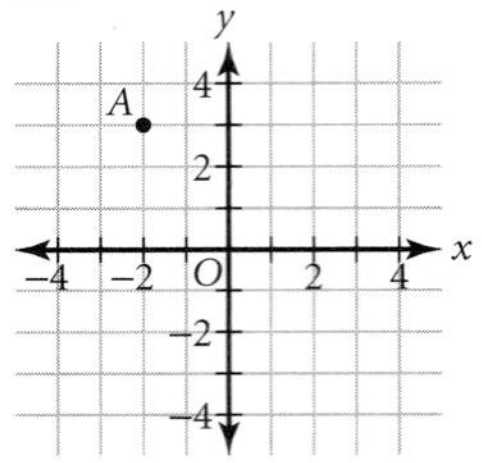


**39.**

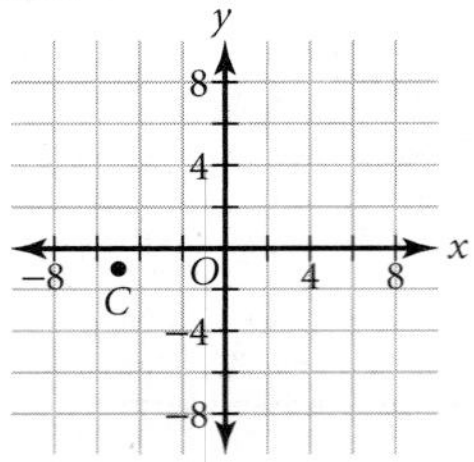

**41.** −2.5, −2, −1.5, −1, −0.5
**43.** $-7x + 11x = (-7 + 11)x = 4x$
**45.** $3d + 5d = (3 + 5)d = 8d$ **47.** $17c + 1$
**49.** $2h - 5$ **51.** $-5f + 48$ **53.** $-4b + 13f - 14g$

## LESSON 5.4

**TRY THIS** (p. 246)

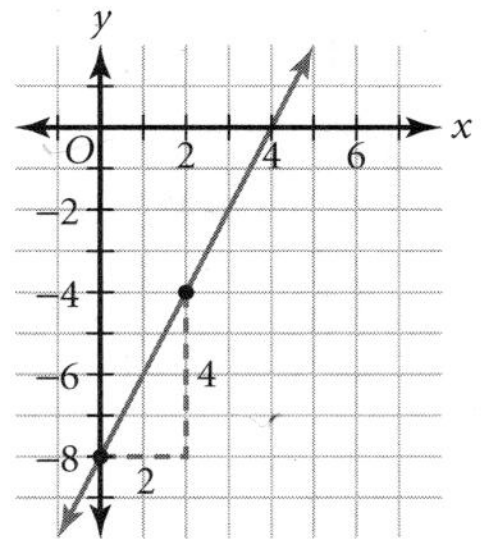

**TRY THIS** (p. 247, Ex. 2)
$y = 2x - 3$

**TRY THIS** (p. 247, Ex. 3)

**a.** The $y$-intercept is 8. The $x$-intercept is 2.

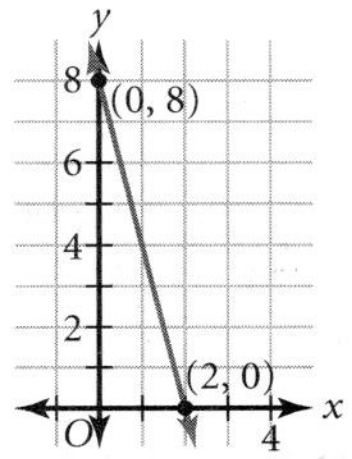


**b.** The $y$-intercept is −15, the $x$-intercept is 5.

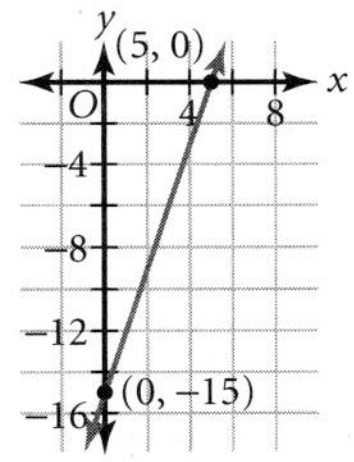


**TRY THIS** (p. 248)

**a.** $y = -3$
**b.** $x = 3$

**Exercises**

**6.**

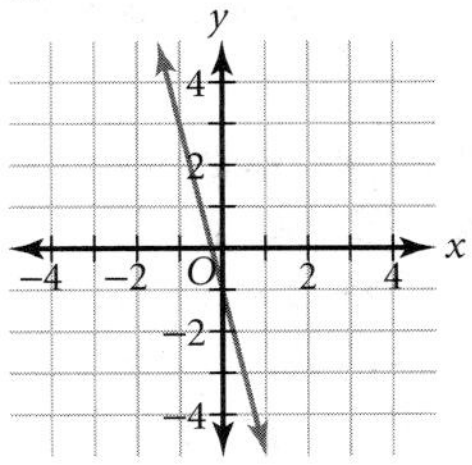

**7.**

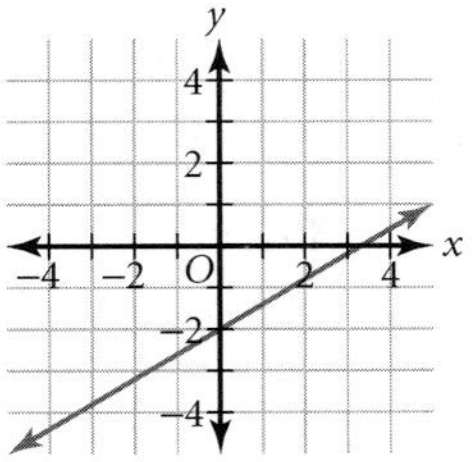

**8.** $y = -\frac{7}{2}x + 19$
**9.** The $x$-intercept is 7. The $y$-intercept is 7.

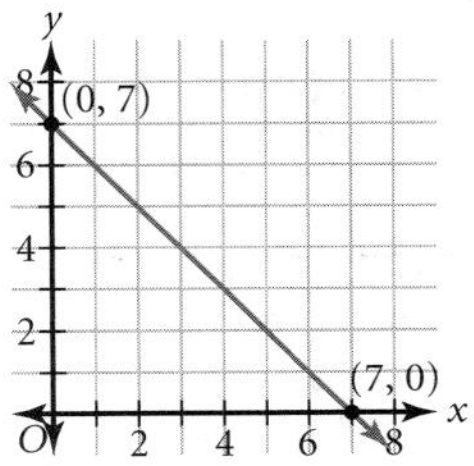


**10.** $y = 0$ **11.** $x = 0$
**13.** $\frac{1}{8}, -1$

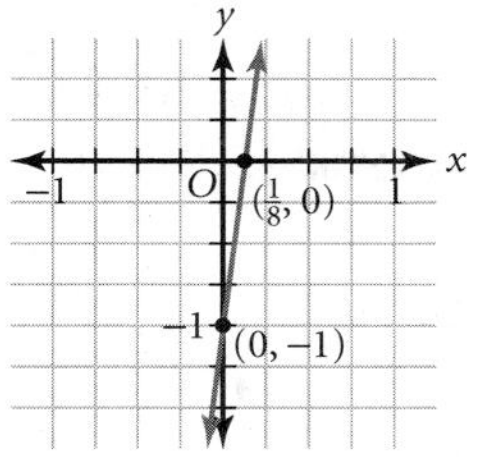


**15.** $\frac{13}{2}$, 13

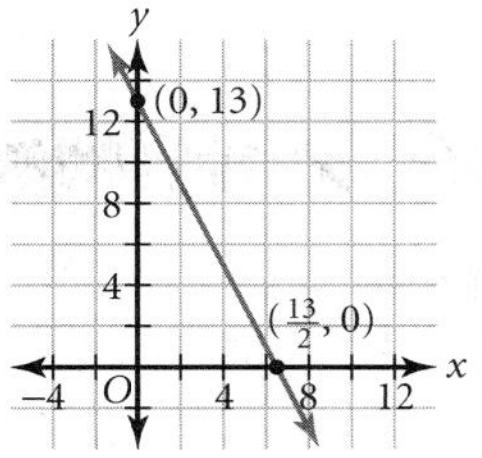


**17.** $-\frac{9}{5}, -9$

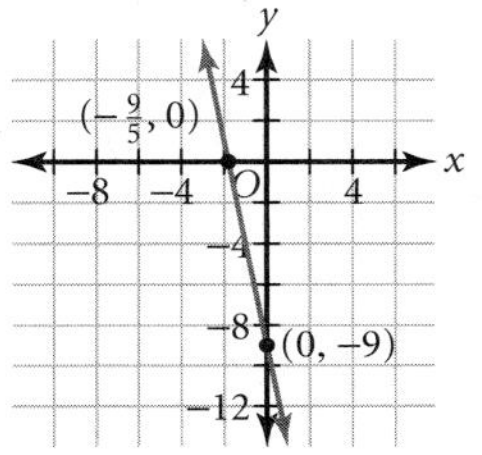


**19.**

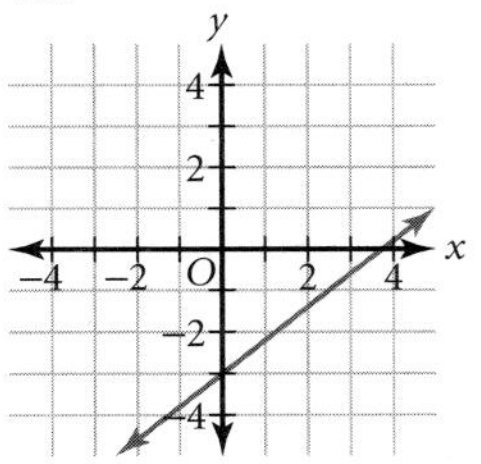

**21.**

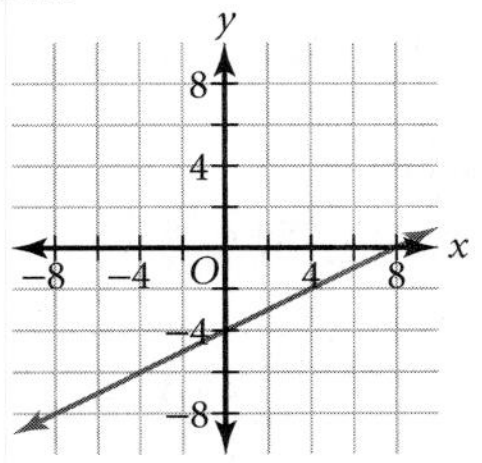

**23.**

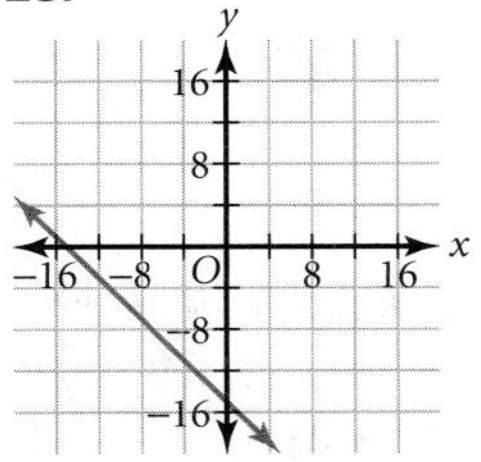

**25.** $y = \frac{3}{2}x - \frac{1}{2}$ **27.** $y = \frac{4}{11}x - \frac{6}{11}$ **29.** $y = -\frac{4}{11}x - \frac{49}{11}$
**31.** $y = x$ **33.** $y = -x$ **35.** $y = \frac{1}{2}x - 2$ **37.** $y = \frac{3}{2}x + 4$
**39.** $x = 3$ **41.** $y = -5$ **43.** $y = -2x$ **45.** $y = 5x - 1$
**47.** $y = \frac{4}{9}x + \frac{10}{3}$ **49.** $y = \frac{4}{9}x + \frac{16}{3}$ **51.** Slope = −5 $y$-intercept = 3 **53.** Slope is undefined, No $y$-intercept, Vertical line through (7, 0) **55.** Slope = 2, $y$-intercept = −8 **57.** Slope = $\frac{4}{5}$, $y$-intercept = 9 **59.** Slope = $\frac{1}{8}$, $y$-intercept = 3 **61.** Slope = $\frac{6}{2} = 3$, $y$-intercept = −1 **63.** e **65.** a **67.** $L = -0.5d + 34$; 16 days **69.** \$10 **71.** \$12.50 **73.** $d = \frac{7 - 3a}{9}$

## LESSON 5.5

**TRY THIS** (p. 253)
**a.** $4x + y = -12$
**b.** $3x - 2y = -5$
**c.** $5.7x - 4.3y = 5.3$

**TRY THIS** (p. 254)
$y - 2 = -4(x - 4)$

**TRY THIS** (p. 255, Ex. 4)
**a.** $y = 5x + 3$
**b.** $5x - y = -3$

**TRY THIS** (p. 255, Ex. 5)
$y - 65 = 3(x - 5)$ or
$y - 71 = 3(x - 7)$;
$y = 3x + 50$

**Exercises**

**5.** $3x - y = -7$ **6.** $3x - 2y = 4$ **7.** $3x + 7y = -17$
**8.** $y - 4 = 2(x - 3)$ **9.** $y - 4 = -2(x - (-3))$
**10.** $y - (-4) = \frac{1}{3}(x - 3)$ **11.** $y = 8x + 18$, $8x - y = -18$ **12.** $y = 3x + 5, 3x - y = -5$
**13.** $y = -40x + 30, 40x + y = 30$
**14.** $y - (-2) = -1(x - 5)$; $y = -x + 3$
**15.** $y - 3 = -1(x - (-3))$; $y = -x$ **16.** $y - 2 = \frac{2}{3}(x - 3)$; $y = \frac{2}{3}x$ **17.** $4x + 3y = 24$ **19.** $6x + 4y = -12$
**21.** $6x - 2y = 14$ **23.** $4x + 14y = -10$
**25.** $7x + 2y = -14$ **27.** $x - y = 6$
**29.** $y - 6 = 3(x - 2)$; $y = 3x$
**31.** $y - (-2) = \frac{1}{5}(x - (-4))$; $y = \frac{1}{5}x - \frac{6}{5}$
**33.** $x = 4$; $y = -6$

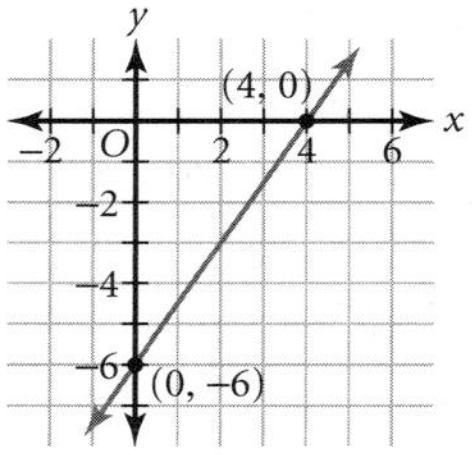


**35.** $x = 5$; $y = -4$

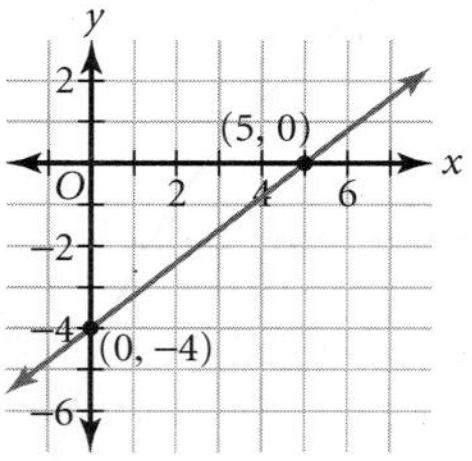


**37.** $x = 2$; $y = 18$

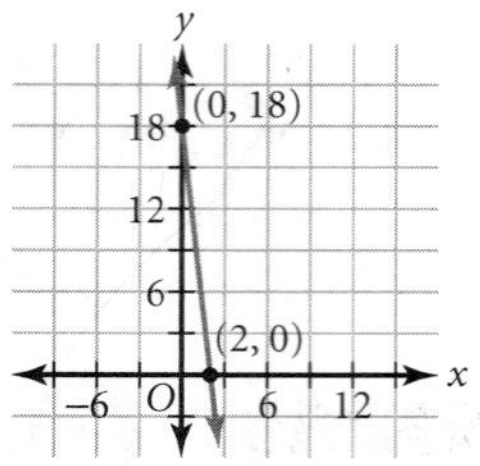


**39.** $y = 2x + 3$ **41.** Slope $= -3$

| | Slope-intercept form | Standard form |
|---|---|---|
| **43.** | None, because the slope is undefined | $1x + 0y = 1$ |
| **45.** | $y = -x + 5$ | $1x + 1y = 5$ |
| **47.** | $y = \frac{1}{4}x + 0$ | $1x - 4y = 0$ |

**49.** A scatter plot with a strong negative correlation indicates that as the $x$-value increases, the $y$-value decreases. **51.** 2 **53.** $-4$ **55.** $x = 140$ **57.** $x = 216$
**59.** 4; $y = 4x$ **61.** $\frac{7}{5}$; $y = \frac{7}{5}x$

## LESSON 5.6

**TRY THIS** (p. 259)
$y = 0.5x - 2$

**TRY THIS** (p. 260, Ex. 2)
$y = -\frac{1}{4}x + 6$

**TRY THIS** (p. 260, Ex. 3)
$y - (-2) = -\frac{1}{2}(x - 3)$

**Exercises**

**5.** $y = 2x + 5$ **6.** $y = -3x + 5$ **7.** $y = \frac{1}{4}x + 5$
**8.** $y = 6x + 5$ **9.** $y = -\frac{1}{3}x + 4$ **10.** $y = \frac{1}{3}x + 4$
**11.** $y = -5x + 4$ **12.** $y = 6x + 4$ **13.** $y - 5 = \frac{3}{2}(x - 4)$
**14.** $y - 5 = -3(x - 4)$ **15.** $y - 5 = 4(x - 4)$ **17.** $-3$
**19.** $\frac{4}{3}$ **21.** 3 **23.** $-\frac{3}{2}$ **25.** $-\frac{4}{3}$ **27.** 16 **29.** 2 **31.** 4
**33.** $\frac{1}{3}$ **35.** $-1$ **37.** $-2$ **39.** $-\frac{1}{5}$ **41.** $y - 3 = \frac{3}{7}(x - 2)$
**43.** $y - 3 = \frac{2}{3}(x - 2)$ **45.** $y - 3 = -\frac{2}{3}(x - 2)$
**47.** $y = 3x + 13$ **49.** $y = 3x - 10$ **51.** $y = \frac{2}{5}x - \frac{31}{5}$
**53.** $x = 2$ **55.** $y = -\frac{1}{2}x + \frac{7}{2}$ **57.** Answers may vary. Sample: $y = -\frac{1}{5}x$ **59.** Undefined **61.** 0
**63.** Answers may vary. Sample: $y = x$, $y = x + 3$, $y = -x$, $y = -x + 3$ **65.** $y = \frac{5}{9}x - \frac{160}{9}$, $y = \frac{5}{9}x + \frac{2300}{9}$
**67.** These lines are less steep. **69.** $-6$ **71.** $-4$ **73.** $-20$
**75.** $-2$ **77.** $5x^2 + 7y$ **79.** $5x^2 + 10xy + 2x$
**81.** $15x^2 + 8x + 8$ **83.** $13x^2 + 5xy + 2x$

## CHAPTER REVIEW AND ASSESSMENT

**1.** Not a function **3.** Not a function **5.** The domain is {2, 3, 6, 8}. The range is {5, 12}. **7.** The domain is {1, 2, 4, 8}. The range is {5, 7, 9}. **9.** $\frac{1}{2}$ **11.** $\frac{2}{3}$ **13.** $\frac{1}{5}$
**15.** 0 **17.** $-\frac{1}{2}$ **19.** $\frac{3}{2}$ **21.** 2 **23.** $\frac{1}{2}$ **25.** 25 **27.** 50
**29.** $k = \frac{2}{5}$, $y = \frac{2}{5}x$ **31.** $k = -\frac{1}{3}$, $y = -\frac{1}{3}x$ **33.** $k = \frac{1}{4}$, $y = \frac{1}{4}x$ **35.** $k = -\frac{1}{3}$, $y = -\frac{1}{3}x$ **37.** $y = 2x + 1$
**39.** $y = 8x - 4$ **41.** $y = -1$ **43.** $y = -3x + 6$
**45.** $y = 5x - 2$

**47.**

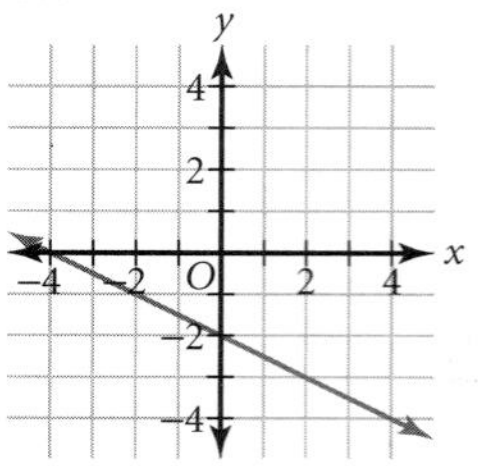


**49.**

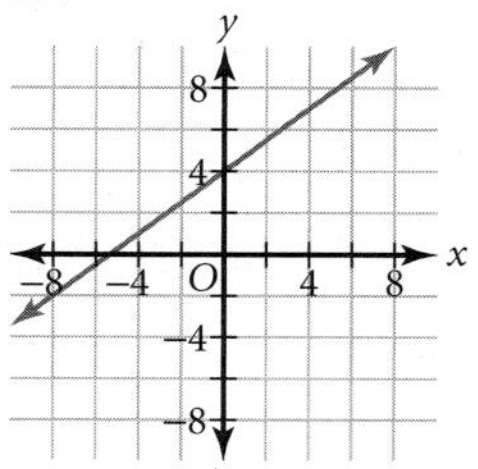

**51.**

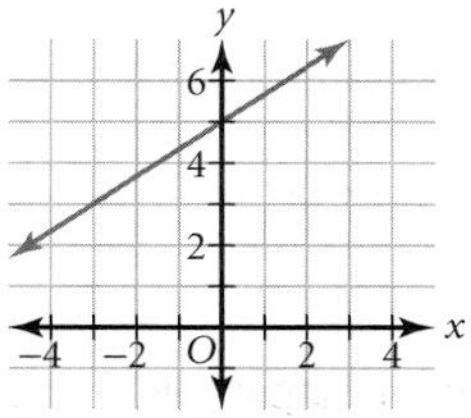

**53.**

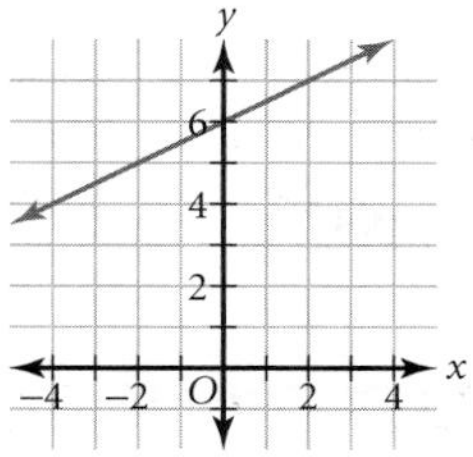

**55.**

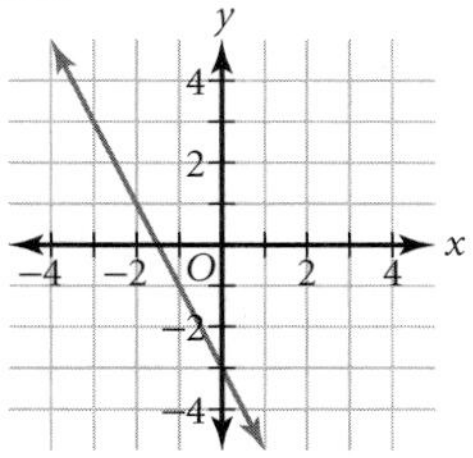

**57.** $4x - y = 17$ **59.** $2x - y = -17$ **61.** $y - (-1) = 2(x - 0)$ **63.** $y - 5 = 4(x - 3)$ **65.** $y - (-3) = 3(x - 0)$ **67.** $y - 0 = -2(x - 3)$ or $y - 6 = -2(x - 0)$ **69.** Slope of parallel line: $\frac{2}{3}$, Slope of perpendicular line: $-\frac{3}{2}$ **71.** $y - 5 = -2(x - 1)$ **73.** 0.00097 or 5 feet per mile **75.** 255 miles **77.** 2 degrees per hour **79.** \$12.50 per hour

# Chapter 6

## LESSON 6.1

**TRY THIS** (p. 277)

$d + 2.50 \le 3.75$

**TRY THIS** (p. 278, Ex. 2)

$c < -18$

**TRY THIS** (p. 278, Ex. 3)

$t < -10$;

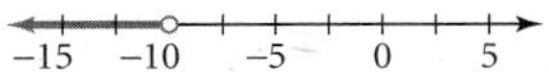

**TRY THIS** (p. 279)

**a.** $x > 2$

**b.** $x \le -3$

### Exercises

**5.** $18 + s \le 27$ **6.** $b \le -12$ **7.** $x > 14$ **8.** $m \ge 6$ **9.** $t > -5$ **10.** $x \ge 7$ **11.** $y < -5$ **12.** $x \ge 20$ **13.** $x < -9$ **14.** $x < 4$

**15.** $x > -9$ 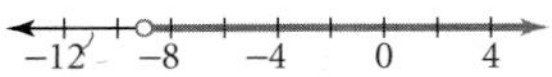

**17.** $x > 25$ 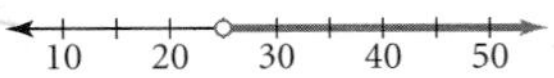

**19.** $y \le 39$ 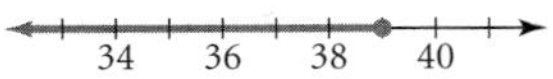

**21.** $m \le 49$ 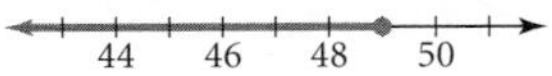

**23.** $x > 0.56$ 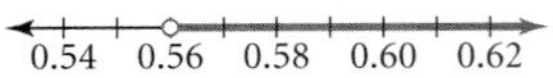

**25.** $x \ge -\frac{1}{4}$ 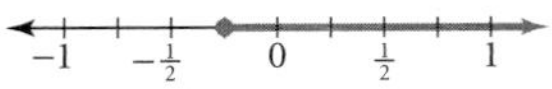

**27.** $v \le 20$ 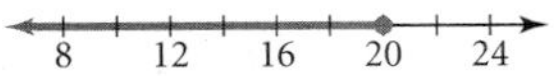

**29.** $t \le 126$ 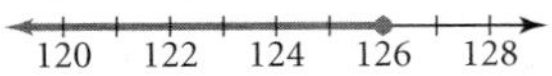

**31.** $v \ge 1.9$ 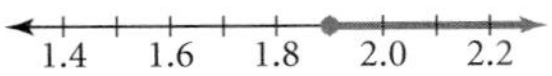

**33.** $x \le -\frac{1}{9}$ 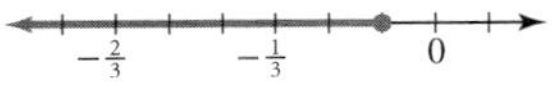

**35.** $c \ge \frac{81}{91}$ 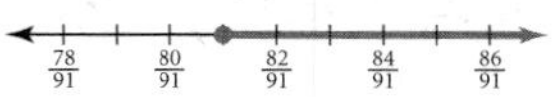

**37.** $x \le 3.5$ **39.** any number less than $-2$ **41.** $450 \ge x + 74$; $x \le 376$ **43.** $t \ge 40$ **45.** $t > 34$ **47.** $70 \le t + 5 \le 80$ (or $65 \le t \le 75$) **49.** negative **51.** 9 **53.** $x = 3$ **55.** $x = -7$

## LESSON 6.2

**TRY THIS** (p. 283)

**a.** $y > 2$ 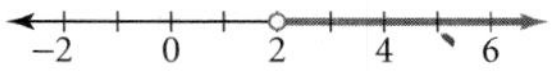

**b.** $x > 6$ 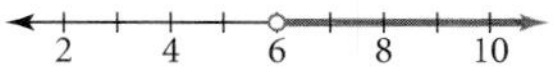

**c.** $m \le -7$ 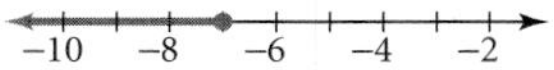

**d.** $x \ge -42$ 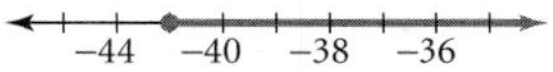

**TRY THIS** (p. 284, Ex. 2)

**a.** $x > -12$

**b.** $x < -6$

**c.** $x \ge -2$

**d.** $x \ge 972$

**TRY THIS** (p. 284, Ex. 3)

**a.** $m \ge \frac{11}{3}$

**b.** $p \ge \frac{7}{4}$

**TRY THIS** (p. 285)

Option 2 pays more if Jennifer works more than 150 hours/month.

### Exercises

**8.** $x \le 6$ 

**9.** $x < 16$ 

**10.** $x \ge -20$ 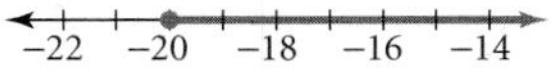

**11.** $d \ge -63$ 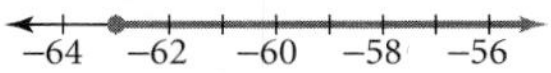

**12.** $b < -9$ **13.** $x > 4$ **14.** $x \le -36$ **15.** $x > 21$ **16.** $x \ge -20$ **17.** $y < -3$ **18.** Plan B costs less for fewer than 80 minutes/month time used.

**19.** $h > 16$ 

**21.** $y \le \frac{22}{5}$ 

**23.** $y > 102$ 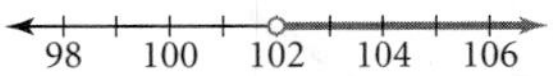

**25.** $x < 8$ 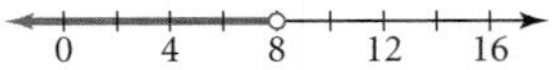

**27.** $y \le -16$ 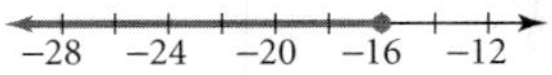

**29.** $m \le -40$ 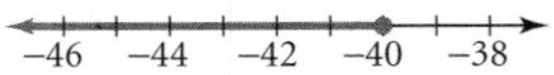 **31.** $x \ge \frac{3}{2}$

**33.** $x > \frac{25}{3}$ **35.** $h < -1$ **37.** $b \ge -\frac{14}{5}$ **39.** $y \ge \frac{21}{4}$ **41.** $m > 8$ **43.** $q > -4.375$ **45.** $f > -20$ **47.** $x \le 60$ **49.** $x \le -4$ **51.** $c > 5.8$ **53.** $p \le 8$ **55.** $p > 13$ **57.** $m \le \frac{17}{10}$ **59.** $y \ge 7$ **61.** $q < 5.\overline{81}$ **63.** $p \le 4$ **65.** $d > \frac{28}{3}$ **67. a.** $l \ge w + 5$ **b.** $l \ge 25$ **69.** 25 gallons **71. a.** $550 + 6.5g \le 1200$ **b.** $g \le 100$; 100 guests **73.** $-511$ **75.** 111 **77.** 19 **79.** 16 **81.** $-64$ **83.** $-2500$ **85.** $11x + 10y$ **87.** $x = 24$ **89.** $m = 11$ **91.** $x = -3$

## LESSON 6.3

**TRY THIS** (p. 291, Ex. 3)

**a.** $q < 5$ or $q > 16$

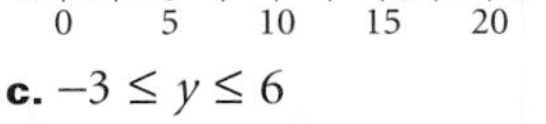

**b.** $x < -5$ and $x > -2$

no solution

**c.** $-3 \le y \le 6$

**TRY THIS** (p. 291, Ex. 4)

$-5 < x \leq 3$

**TRY THIS** (p. 292)

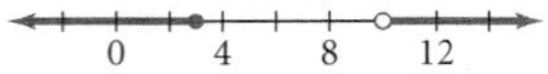


$x \leq 3$ or $x > 10$

### Exercises

**5.** $-7 < x < 5$

**6.** $-2 \leq y \leq 4$

**7.** $q < 5$ or $q \geq 7$

**8.** $45 < z \leq 47$

**9.** $y < 4$ or $v \geq 12$

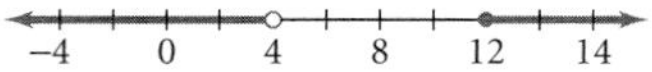


**10.** $-5 \leq x \leq 5$

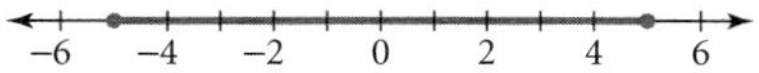


**11.** $-3 \leq x \leq 2$ **12.** $-2 \leq x \leq 4$ **13.** $3 < x \leq 4$ **14.** $x \leq 3$ or $x > 4$ **15.** $x > 3$ or $x < -1$ **16.** $x > \frac{7}{2}$ or $x < \frac{7}{3}$ **17.** $x \leq 5$ or $x > 0$, All real numbers **18.** $x > 1$ or $x < 2$, All real numbers

**19.** $2 < x < 5$

**21.** $-1 \leq x < 7$

**23.** $-4 < n \leq 6$

**25.** $-6 < v \leq 3$

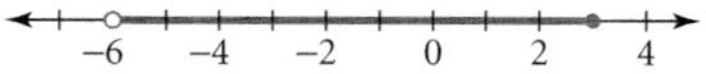


**27.** $3.5 < t \leq 7.5$

**29.** $t < 6$ or $t \geq 8.4$

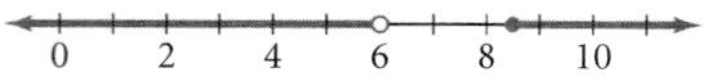


**31.** $-9 \leq x < 3$ **33.** All real numbers. **35.** $1 < x < 7$ **37.** $-\frac{5}{2} < x < \frac{7}{2}$ **39.** $78 \leq t \leq 98$ **41.** 21 **43.** 12,006

## LESSON 6.4

**TRY THIS** (p. 295)

**a.** 9 **b.** 10

**TRY THIS** (p. 296)

**a.** domain: all real numbers
range: all non-negative real numbers

**b.** domain: all real numbers
range: all non-negative real numbers

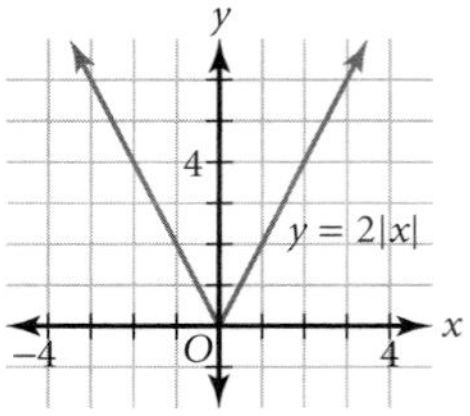


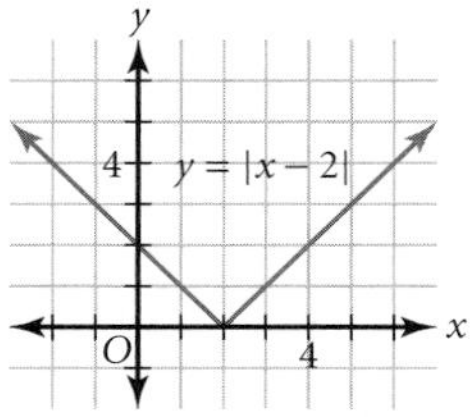


### Exercises

**6.** 7 **7.** 1 **8.** 6 **9.** 0 **10. a.** domain: all real numbers, range: all non-negative numbers **b.** domain: all real numbers, range: all real numbers ≥ 1
**11. a.** translation by 1 unit to the left **b.** translation by 1 unit up **c.** reflection across the $x$-axis, then translation 1 unit up **13.** 33 **15.** 8.67 **17.** 2.5 **19.** 79.2 **21.** 3 **23.** 10 **25.** 26 **27.** 0 **29.** 7 **31.** 22 **33.** 5 **35.** 15 **37.** domain: all real numbers, range: all non-negative real numbers **39.** domain: all real numbers, range: all real numbers equal to or greater than −4 **41.** domain: all real numbers, range: all non-positive real numbers **43.** domain: all real numbers, range: all real numbers equal to or less than −4 **45.** domain: all real numbers, range: all non-negative real numbers **47.** domain: all real numbers, range: all non-negative real numbers **49.** horizontal translation 5 units right **51.** vertical translation 4 units down **53.** horizontal translation 5 units right, then reflection **55.** reflection then vertical translation 4 units down **57.** 8 **59.** 2
**61. a.** $y = |x - (-5)| - 2 = |x + 5| - 2$ **b.** $y = -(|x| + 5)$ **c.** $y = -|x| + 2$ **63.** 160 miles; 80 miles south of Dallas **65.** 22, 26, 30 **67.** 1 **69.** 0 **71.** −5 **73.** −6
**75.** Graphs a and c have the same slope, 1, but different $y$-intercepts, 2 and 0. Graph b has a different slope from the others, $\frac{2}{3}$, and a different $y$-intercept, −1. The $x$-intercepts are all different: $-2, \frac{3}{2}$, and 0, respectively. **77.** $y = 2x$ **79.** $y = x - \frac{17}{2}$ **81.** $x \geq 5$ **83.** $x > 7$

## LESSON 6.5

**TRY THIS** (p. 301, Ex. 1)
$x = 14.5$ or $x = 5.5$

**TRY THIS** (p. 301, Ex. 2)
$x = 6$ or $x = -2$

**TRY THIS** (p. 303, Ex. 4)
$-10 \leq x \leq -2$

**TRY THIS** (p. 303, Ex. 5)

**a.** $x > -4$ or $x < -10$

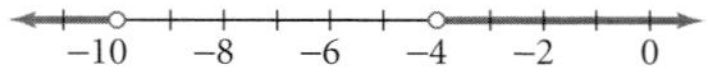


**b.** $x \geq 7$ or $x \leq 3$

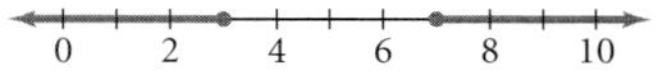


### Exercises

**5.** $x = 5$ or $x = 6$ **6.** $x = 7$ or $x = -3$ **7.** $a = 10.3$ or $a = -3.7$ **8.** $x = 2.4$ or $x = 3.4$ **9.** $z = -7$ or $z = \frac{11}{3}$ **10.** $v = 1$ or $v = -7$ **11.** $|x - 10| \leq 4$; $6 \leq x \leq 14$ **12.** $2.4 \leq x \leq 3.4$ **13.** $-\frac{26}{3} \leq x \leq 7$ **14.** $-7 \leq x \leq 4.25$ **15.** $x < 7$ or $x > 13$ **16.** $x \geq 2$ or $x \leq -3$ **17.** $1 < x < 3$ **19.** $x = 7$ or $x = -5$ **21.** $x = 13$ or $x = 3$ **23.** $x = \frac{3}{2}$ or $x = -\frac{11}{2}$

**25.** no solution

**27.** $-4 < x < 10$ 

**29.** $x < -12$ or $x > 4$ 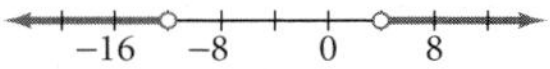


**31.** no solution

**33.** $x < -4$ or $x > 8$ 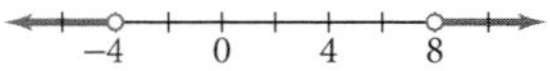


**35.** $14 \leq x \leq 18$ 

**37.** $x > -2$ or $x < -6$ 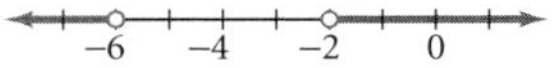


**39.** $d > -\frac{2}{3}$ or $d < -4$ **41.** $\frac{2}{7} \leq y \leq \frac{6}{7}$ **43.** $x \geq \frac{13}{8}$ or $x \leq -\frac{1}{8}$ **45.** $-\frac{57}{8} < x < \frac{39}{8}$ **47.** $\frac{2}{5} < x < \frac{14}{5}$ (or $0.4 < x < 2.8$) **49.** $y > 2.34$ or $y < 0.66$

**51.** $x > 0.404$ or $x < -0.404$ **53.** $x > 7300$ or $x < 3100$

**55. a.**

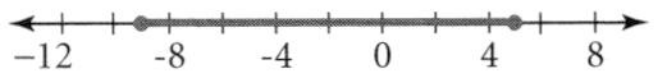


**b.** $|x - (-2)| = 7$; $x = 5$; $x = -9$

**57.** $-1 < x < 7$

**59.** 2, 0, 5, 6, 3

**61. a.** boundary values: 165, 171; acceptable weights: all weights between and including 165 and 171 **b.** $|x - 168| = 3$; $x = 171$ or $x = 165$ **c.** $|x - 168| \leq 3$; $165 \leq x \leq 171$ **63. a.** upper: 55%, lower: 49% **b.** $|b - 52| = 3$ **c.** Given the error margin of ±3%, candidate A could be preferred by as few as 49% to 50% of the voters. This range does not represent a majority. **65.** 2 **67.** $-40c - 15$ **69.** $20k - 32$ **71.** $x = 2$ **73.** $y = 14$ **75.** $p = 2$ **77.** $y > \frac{5}{2}$

## CHAPTER REVIEW AND ASSESSMENT

**1.** $x > 5$ **3.** $g < 5$ **5.** $y < 3.00$ **7.** $x \geq 1$ **9.** $x \leq 3.5$ **11.** $m > -3$ **13.** $x < 1$ **15.** $r > 12$ **17.** $b < -28$ **19.** $m > \frac{30}{11}$ **21.** $x \leq -\frac{5}{6}$ **23.** $k < 2$ **25.** $k \geq 10$ **27.** $1 < x \leq 4$ **29.** $-2 < y \leq -1$ **31.** $-5 \leq x \leq 7$ **33.** $x < 6$ or $x \geq 9$ **35.** All real numbers **37.** 2 **39.** 4 **41.** 17 **43.** 14 **45.** domain: all real numbers, range: $y \geq 0$ **47.** domain: all real numbers, range: $y \geq 0$ **49.** domain: all real numbers, range: $y \geq -2$ **51.** domain: all real numbers, range: $y \leq 0$ **53.** domain: all real numbers, range: $y \geq 3$ **55.** $n = 12$ or $n = -4$ **57.** $d = 10$ or $d = -5$ **59.** $x = \frac{2}{3}$ or $x = -2$ **61.** $2 \leq k \leq 4$ **63.** $0 \leq m \leq 4$ **65.** no solution **67.** $r \leq -5$ or $r \geq 1$ **69.** $m \leq -4$ or $m \geq 1$ **71.** all real numbers **73.** $x - 5.45 < 5$, $x < 10.45$ **75.** $x + 90 \geq 280$, $x \geq 190$

# Chapter 7

## LESSON 7.1

**TRY THIS** (p. 320, Ex. 1)

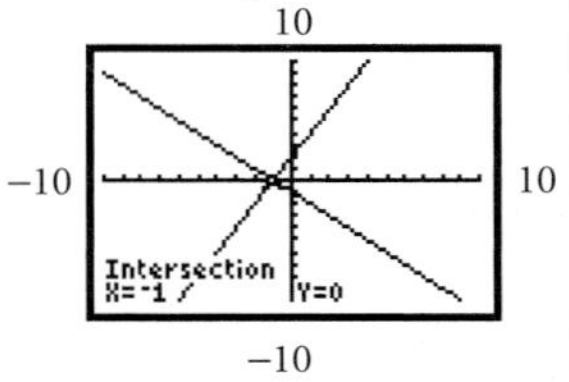


The solution is (−1, 0)

**TRY THIS** (p. 320, Ex. 2)

**a.** $y = -2x + 2$
$y = -x + 1$

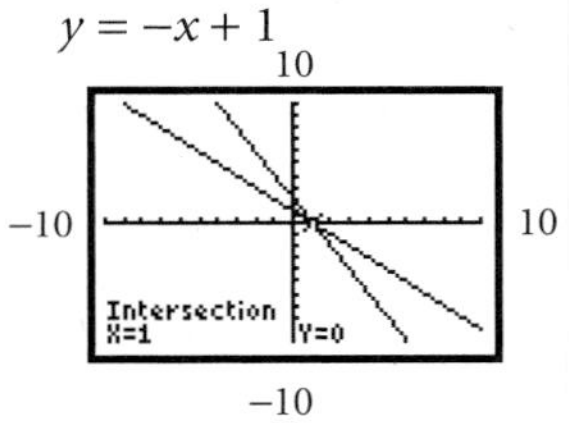


The solution is (1, 0).

**b.**

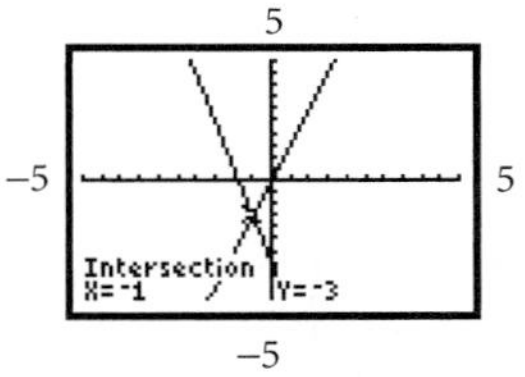


The solution is (−1, −3).

**TRY THIS** (p. 322)

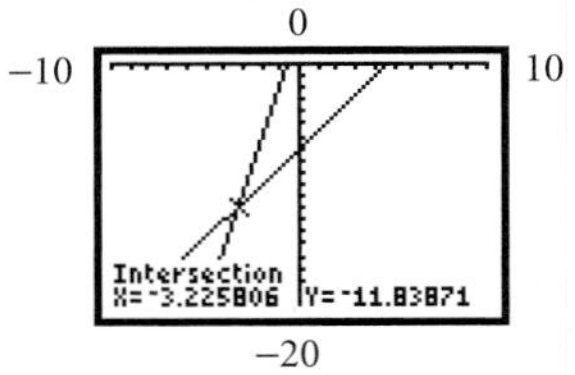


An approximate solution is (−3.2, −11.8).

### Exercises

**4.**

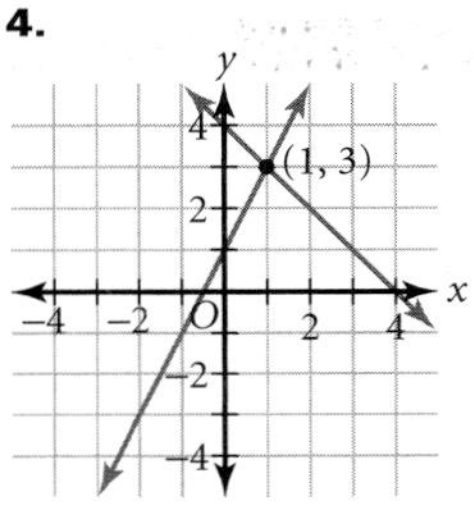


(1, 3)

**5.**

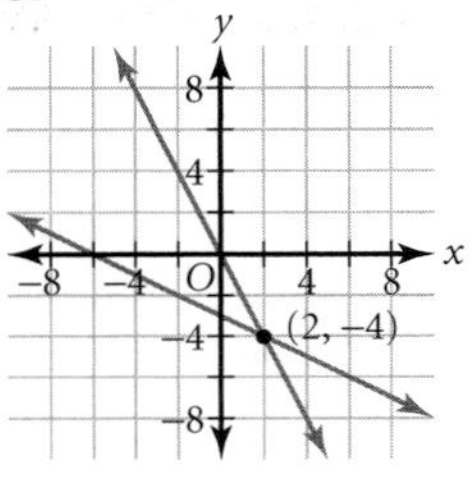


(2, −4)

**6.**

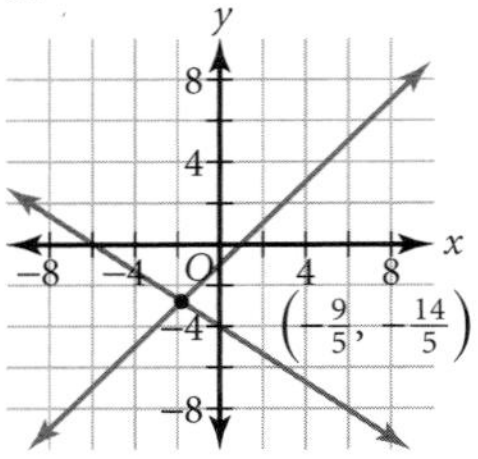


$\left(-\frac{9}{5}, -\frac{14}{5}\right)$

**7.**

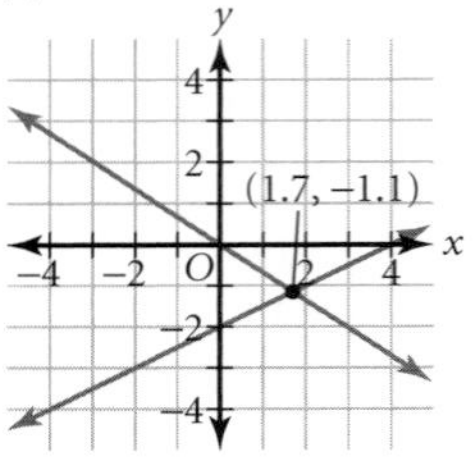


(1.7, –1.1)

**8.**

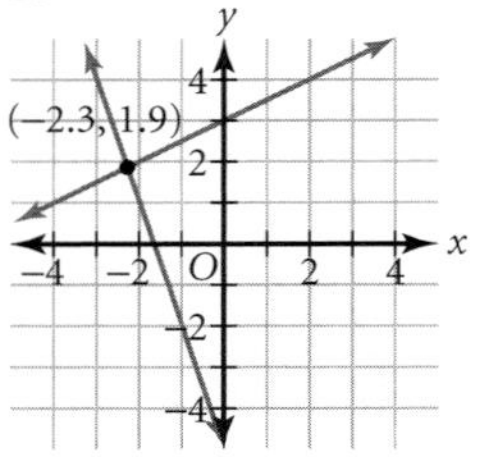


(–2.3, 1.9)

**9.**

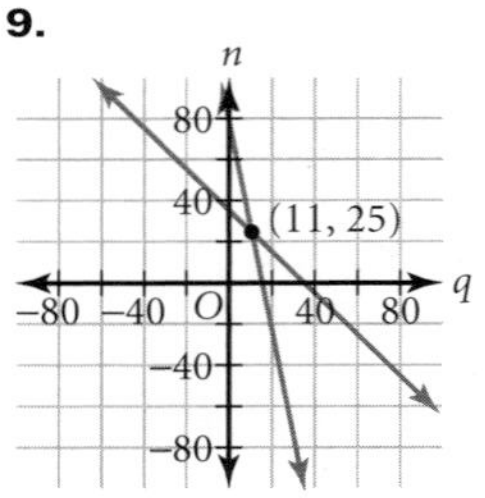


$n = 25, q = 11$

**11.**

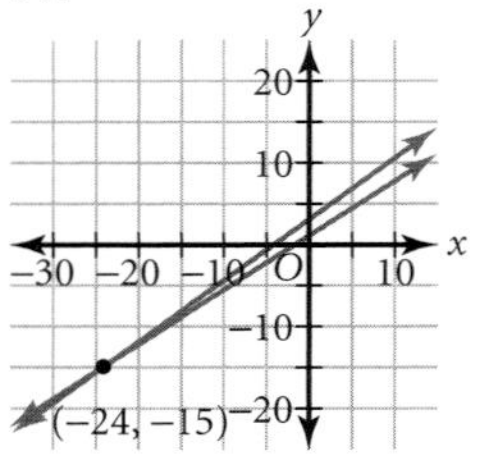


(–24, –15)

**13.**

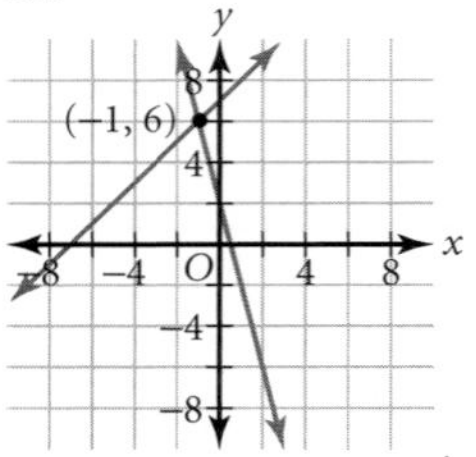


(–1, 6)

**15.**

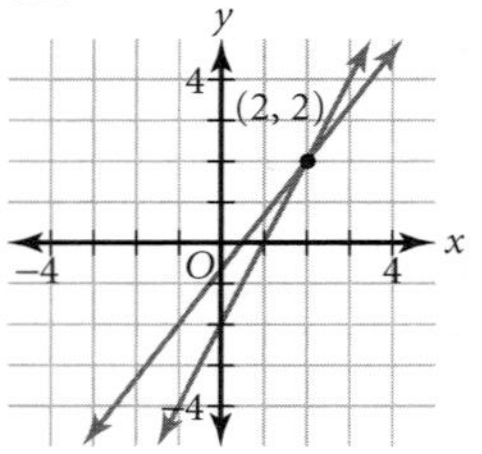


(2, 2)

**17.**

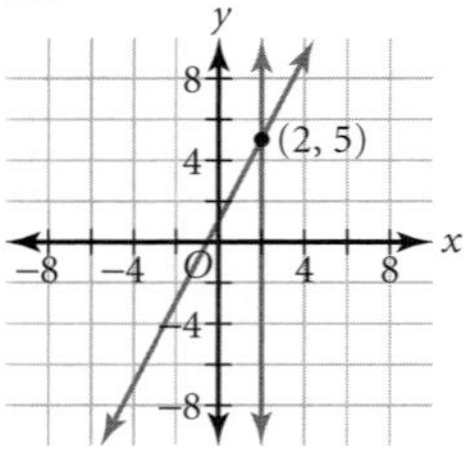


(2, 5)

**19.** (–0.5, 2.3) **21.** (2, 2) **23.** (–2, 0) **25.** (4, –1) **27.** (29, 21.1) **29.** No **31.** No **33.** $\left(4\frac{2}{3}, -3\frac{1}{3}\right)$ **35.** $y = -6x + 23$

**37.**

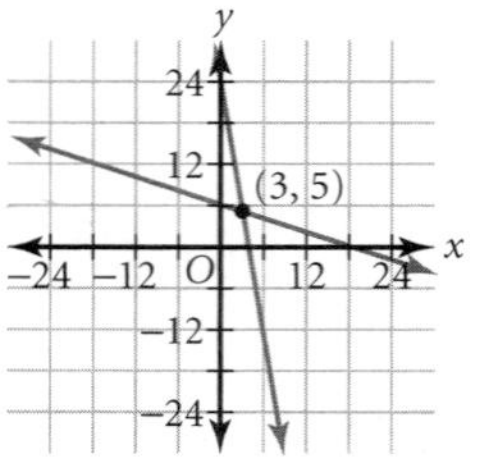


**39.**

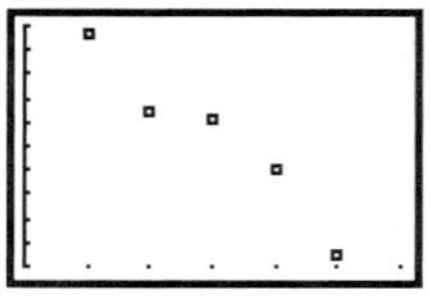

**41.** education: $y = -210x + 7667.4$, engineering: $y = 301x + 2623.4$ **43.** $A = -1800r + 28{,}000$, $A = 2300r + 1500$, about 6.5 minutes **44.** $21 \div (3 - 4) + 0 + 6 \cdot (-3) = -39$ **45.** little or no correlation **47.** 3 **49.** –6 **51.** $-\frac{28}{3}$ or $-9\frac{1}{3}$ **53.** $-2x$ **55.** $x = 3$

## LESSON 7.2

**TRY THIS** (p. 327)

**a.** $\left(-\frac{11}{2}, 5\right)$

**b.** (55, 98)

**TRY THIS** (p. 328)

$\left(\frac{41}{20}, \frac{13}{20}\right)$

### Exercises

**4.** $\left(5, \frac{13}{3}\right)$ **5.** (–18, –28) **6.** $\left(\frac{1}{4}, \frac{1}{4}\right)$ **7.** $\left(\frac{1}{3}, \frac{1}{3}\right)$ **8.** 12 and 15 **9.** $\left(\frac{1}{6}, \frac{1}{12}\right)$ **11.** (3, –4) **13.** (–43, –6) **15.** (0, 3) **17.** (0.2, 2.2) **19.** (6, –1) **21.** (4, 1) **23.** (2, –3) **25.** (–5, 7) **27.** (–3, 4) **29.** $W = 34\frac{2}{3}$, $L = 69\frac{1}{3}$ **31.** 8 and 4 **33. a.** $6a + 3.50c = 935$ **b.** $c + a = 210$ **c.** 80 adults, 130 children **35.** Tom, Sam, Rob, Mark, Joe **37.** $x = \frac{1}{5}$ **39.** $x = 45$ **41.** perpendicular **43.** perpendicular

## LESSON 7.3

**TRY THIS** (p. 334)

**a.** (3, –1) **b.** (4, 3)

### Exercises

**10.** (1.5, 0.25) **11.** (2.5, 1) **12.** (0, –2) **13.** (–4, 3) **15.** (3, 4) **17.** $\left(2, -\frac{1}{3}\right)$ **19.** $\left(\frac{1}{2}, \frac{4}{3}\right)$ **21.** (–2, 8) **23.** (–4, –7) **25.** (3, 4) **27.** (–2, 7) **29.** $\left(0, -\frac{7}{3}\right)$ **31.** (–9, –2) **33.** (36.8, 5.5) **35.** $\left(\frac{257}{41}, \frac{363}{41}\right)$ **37.** (–2, –4) **39.** Answers may vary.

Sample answer: Substitution, since both equations are solved for *y*. **41.** John's weekly salary is \$135, plus \$10.50 per hour for overtime. **43.** 11 single tapes, 14 concert tapes **45.** Rooms cost \$67.50 per night, and meals cost \$17.50 each. **47.** 15 pies at \$5.99, 10 pies at \$9.99 **49.** $\frac{7}{3}$ **51.** 2 **53.** $x = 12$ **55.** $x = -2$ **57.** slope = –3 **59.** slope = $\frac{2}{7}$ **61.** slope = $-\frac{7}{3}$ **63.** slope = $-\frac{6}{5}$

## LESSON 7.4

**TRY THIS** (p. 340)

**a.** *Method 1:* Substitute $y = 2x + 6$ for $y$ in the first equation. The system is inconsistent.

*Method 2:*

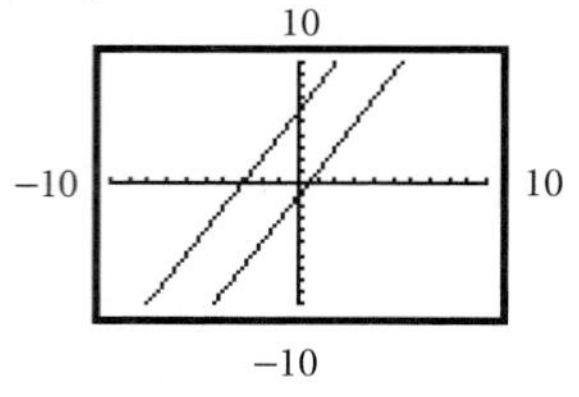


The system is inconsistent.

**b.** *Method 1:* Use elimination to solve the system. The solution is $x = \frac{5}{8}$ and $y = \frac{9}{4}$. The system is consistent.

*Method 2:*

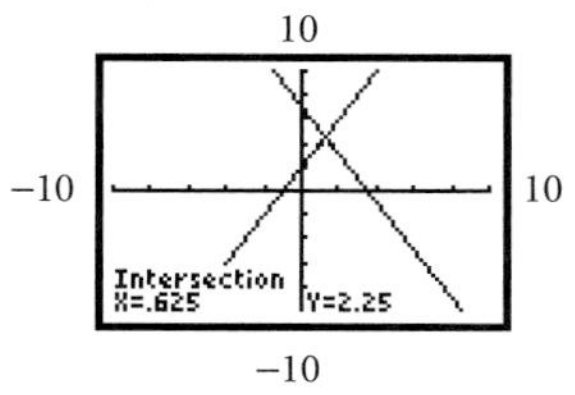


The solution is $x = 0.625$ $\left(\text{or } \frac{5}{8}\right)$ and $y = 2.25$ $\left(\text{or } \frac{9}{4}\right)$. The system is consistent.

### Exercises

**6.** inconsistent **7.** consistent **8.** inconsistent

**9.** Answers may vary. Sample answer: $\begin{cases} y = 4x + 3 \\ 2y = 8x + 6 \end{cases}$

**10.** Answers may vary. Sample answer: $\begin{cases} y = -2x + 7 \\ y = \frac{1}{2}x + 7 \end{cases}$

**11.** no solution, inconsistent **13.** $(-87, -66)$, consistent and independent **15.** solution $(-2, 5)$, consistent and independent **17.** solution $(2, 0)$, consistent and independent **19.** no solution, inconsistent **21.** infinitely many solutions, consistent and dependent **23.** $(-1, 5)$, consistent and independent **25.** $(8, 4)$, consistent and independent **27.** $(-15, 36)$, consistent and independent **29.** $p = 13$ **31.** 50° and 130° **33.** The lines are parallel so they form an inconsistent system. **35.** Answers may vary. Sample answer: $2y = -3x + 6$ **37.** Answers may vary. Sample answer: $y = 8$ **39.** Vegiland: $y = 500x - 949$, 930; Megaveggies: $y = 500x - 946$, 250 **41.** Choose Job 1. It pays $100 more per year. **43.** −12 **45.** −11 **47.** 0 **49.** 10 **51.** −3 **53.** $c = 3d - a$ **55.** $c = \frac{d + ab}{b}$ **57.** slope $= -\frac{2}{5}$, $y$-intercept $= \frac{7}{5}$ **59.** $(6, -2)$ **61.** $\left(-\frac{26}{7}, -\frac{8}{7}\right)$ **63.** $(-1, -1)$

## LESSON 7.5

**TRY THIS** (p. 347)

**a.** $y < \frac{2}{3}x + \frac{7}{3}$

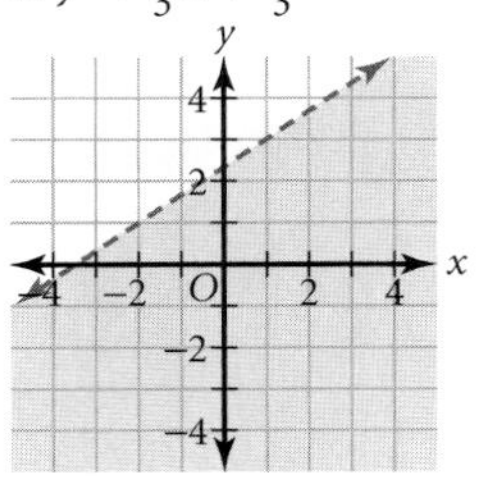

**b.** $y \geq 0.5x - 3$

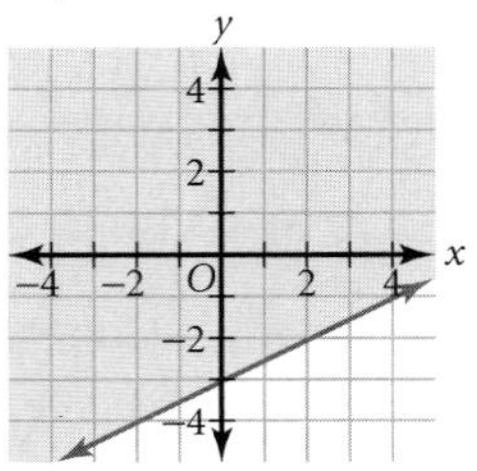

**TRY THIS** (p. 348)

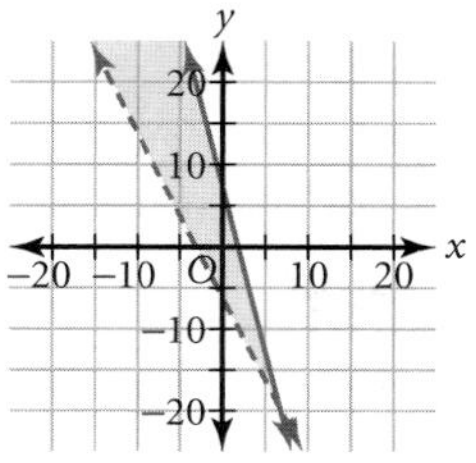

### Exercises

**7.**

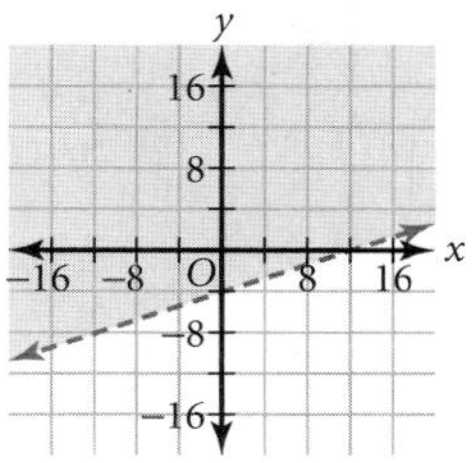

**8.**

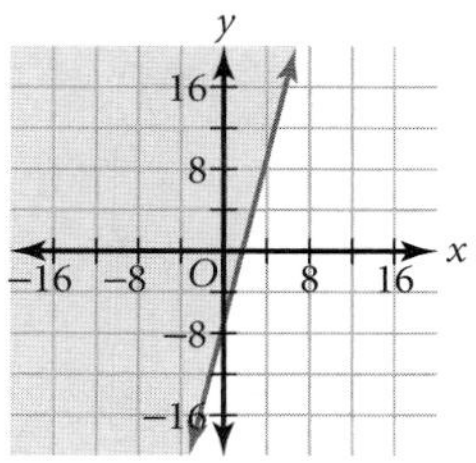

**9.**

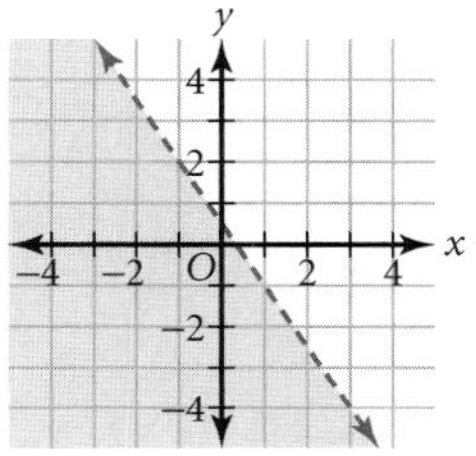

**10.**

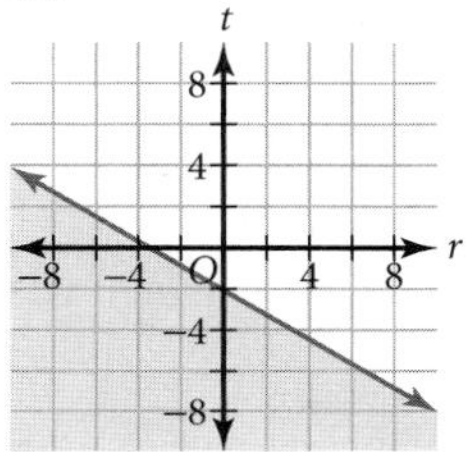

**11.**

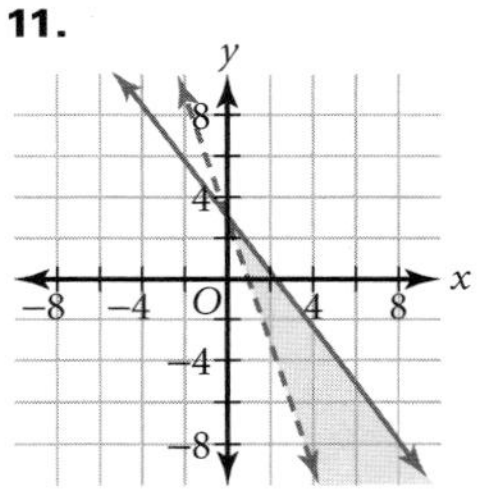

**12.**

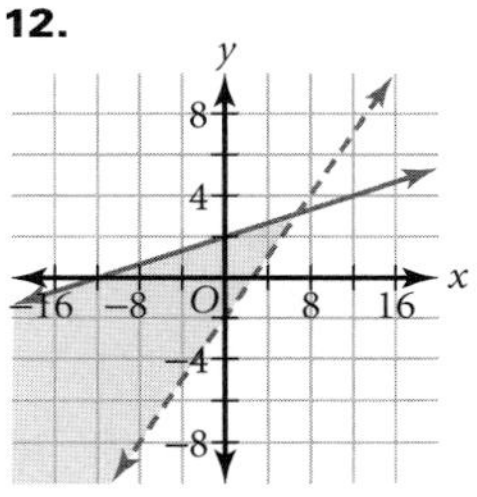

**13.**

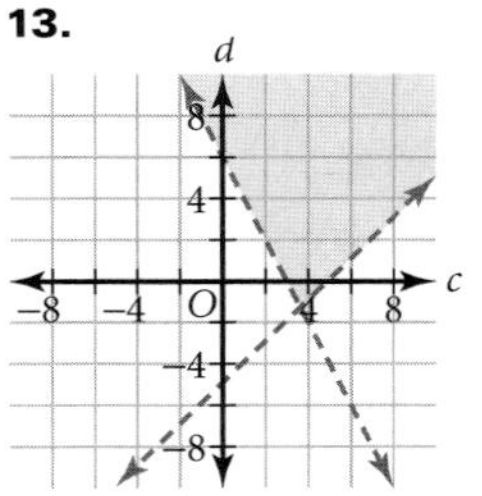

**14.**

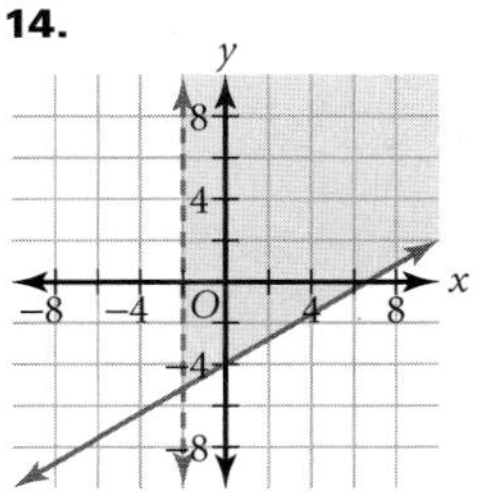

**15.**

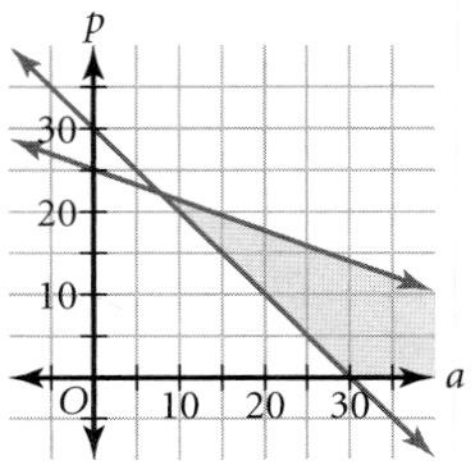

$$\begin{cases} 4p - 1.5a \leq 100 \\ p + a \geq 30 \\ p \geq 0 \\ a \geq 0 \end{cases}$$

The shaded region contains all the possible combinations.

**17.**

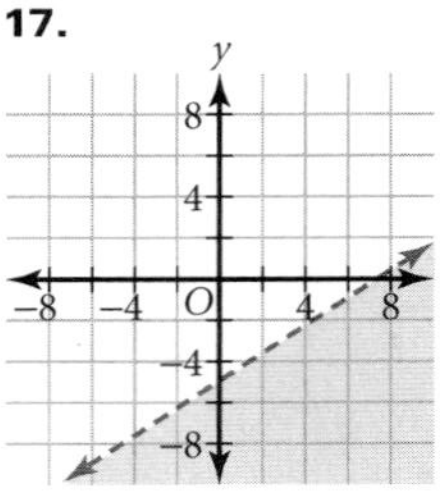


**19.**

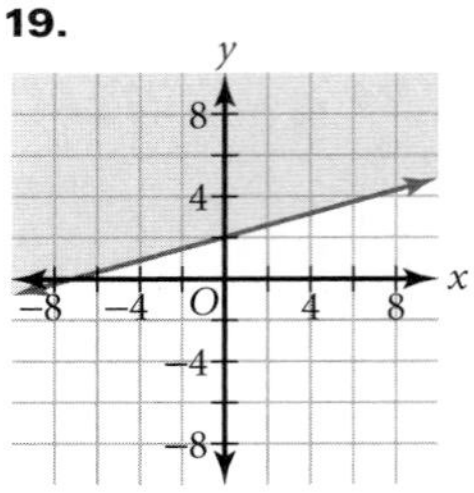


**21.**

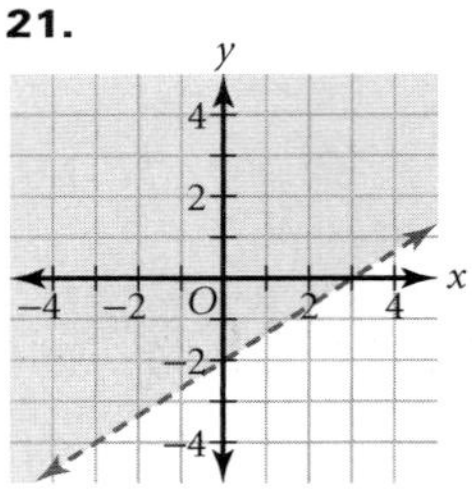


**23.**

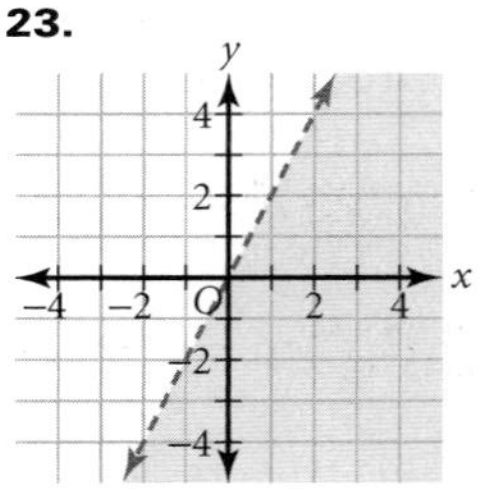


**25.**

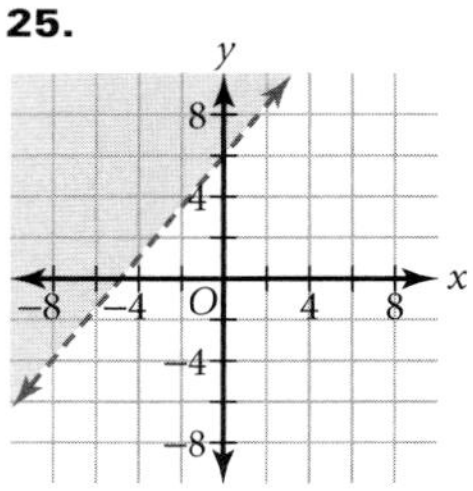


**27.**

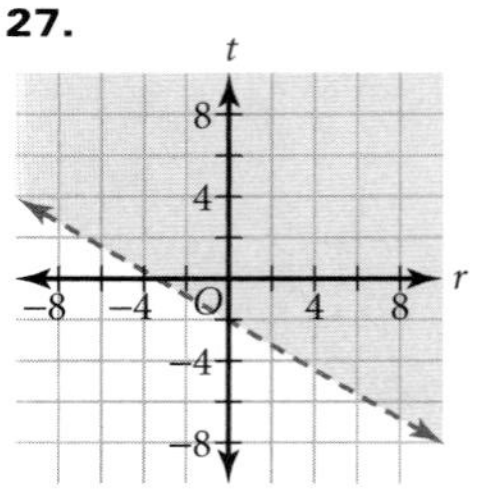


**29.**

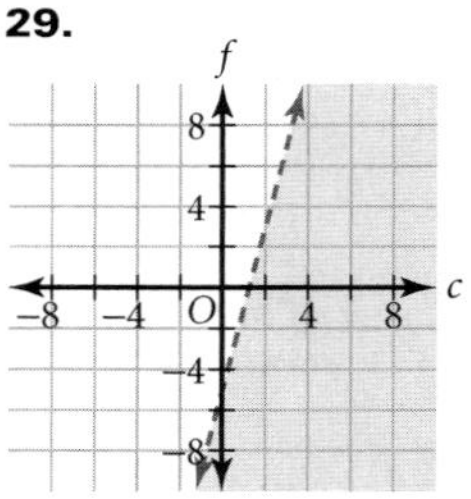


**31.**

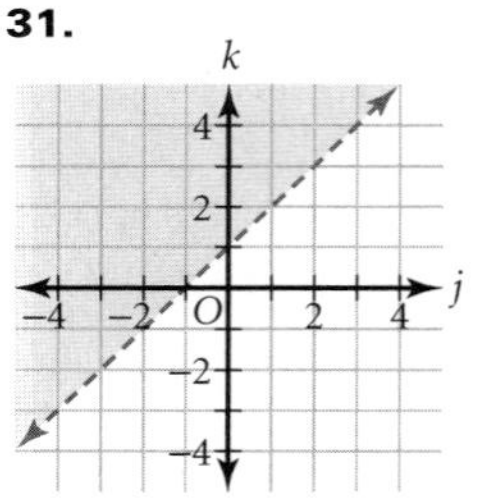


**33.**

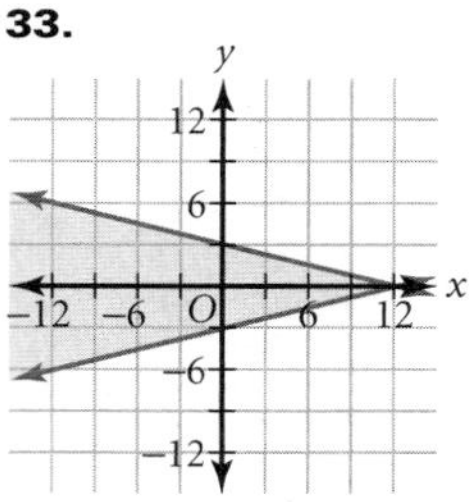


**35.**

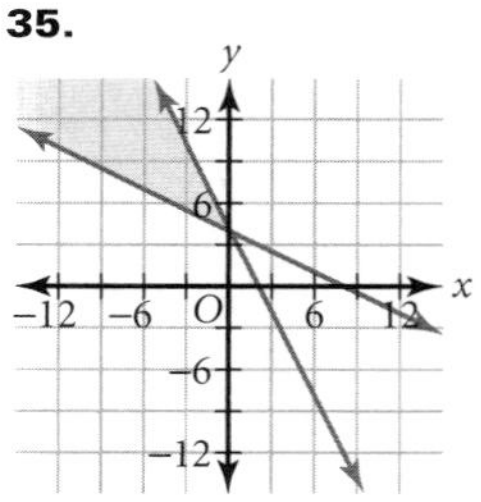


**37.**

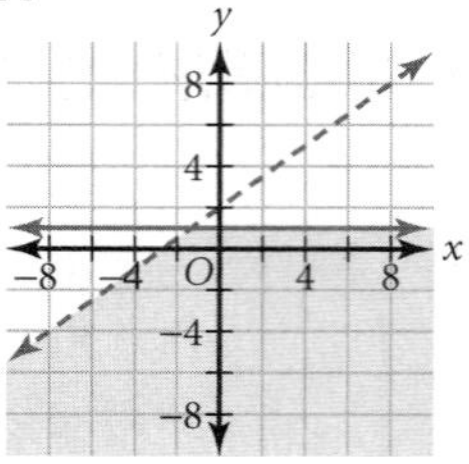


**39.**

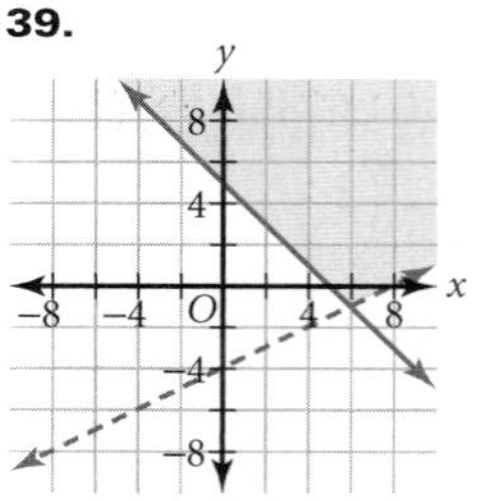


**41.**

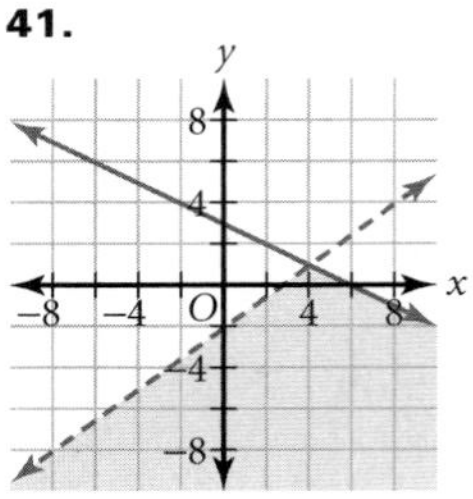


**43.**

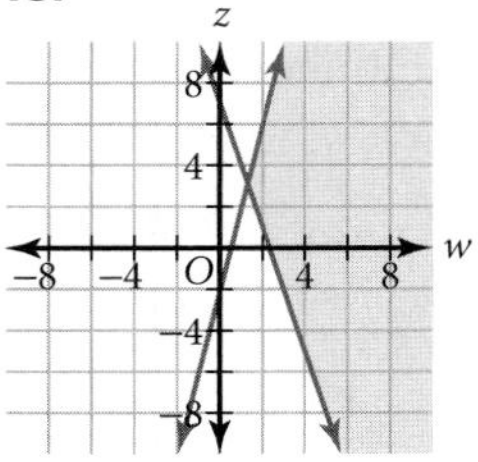


**45.**

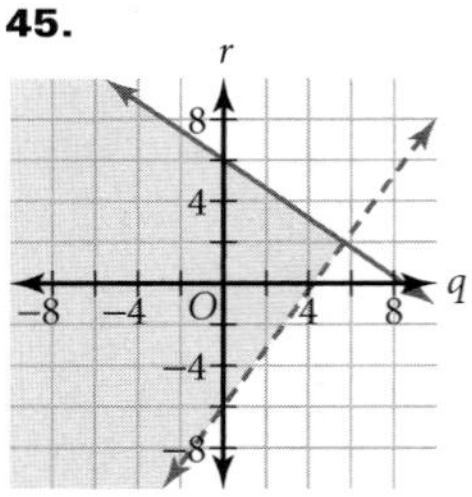


**47.**

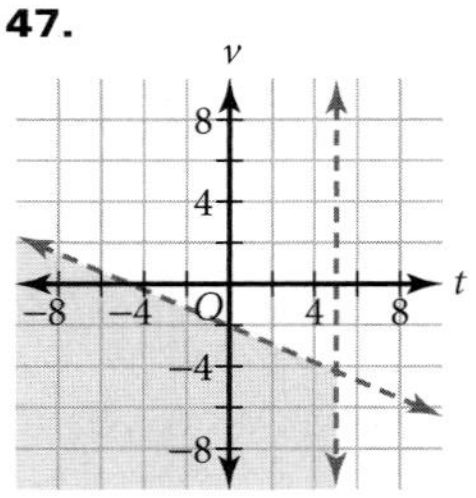


**49.**

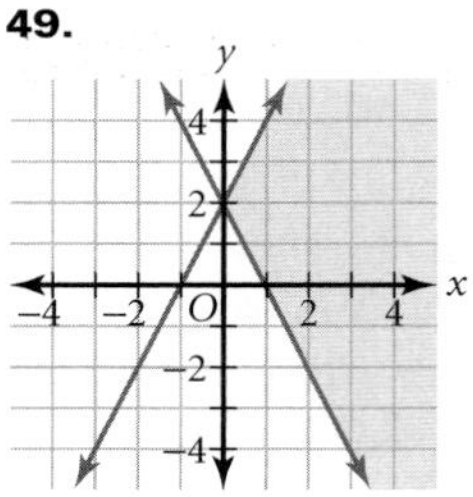


**51.**

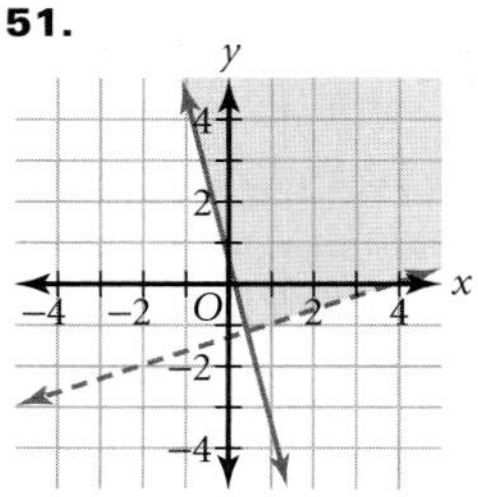


**53.**

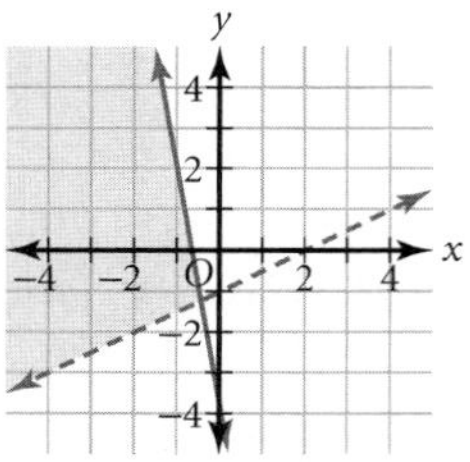


**55.**

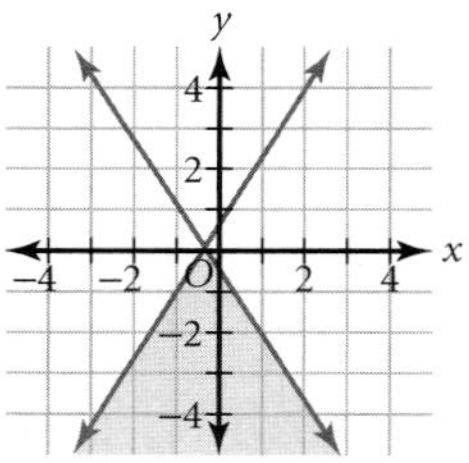


**57.** $2x + y \leq 16$

| $x$ | 0 | 1 | 2 | 3 | 4 | 5 | 6 | 7 | 8 |
|---|---|---|---|---|---|---|---|---|---|
| $y$ | 16 | 14 | 12 | 10 | 8 | 6 | 4 | 2 | 0 |

**59.** $y \geq 2x$ and $y \leq 2x + 6$ **61.** The second differences are a constant 8. **63.** \$60 **65.** $8a + 2b^2 - b + z$ **67.** neither **69.** perpendicular **71.** perpendicular **73.** neither

## LESSON 7.6

**TRY THIS** (p. 354)

June is 14 years old and her brother is 10 years old.

**TRY THIS** (p. 355, Ex. 2)

The speed of the current is $\frac{9}{11}$ miles per hour. The speed of the boat in still water is $17\frac{2}{11}$ miles per hour.

**TRY THIS** (p. 355, Ex. 3)

The original number is 75.

**TRY THIS** (p. 356)

There are 18 quarters and 30 nickels.

**TRY THIS** (p. 357)

Mix 4 ounces of the 2% acid solution with 16 ounces of the 4% acid solution.

**Exercises**

**5.** Gabriel is 28, Sam is 12 **6.** $\frac{2}{3}$ mph **7.** 62 **8.** 8 quarters **9.** 28 liters of the 15% glucose solution and 7 liters of the 35% glucose solution **11.** 18 dimes, 108 quarters **13.** Maya is 25, David is 20 **15.** Wymon is 32, Sabrina is 12 **17.** 49 **19.** 300 ml of the 4% salt solution, and 300 ml of the 16% salt solution **21.** $\frac{2}{3}$ pound of peanuts and $\frac{1}{3}$ pound of raisins **23.** $\frac{3}{5}$ **25.** 6 **27.** $-2a - 2b$ **29.** $-4x - 8y$ **31.** 2

## CHAPTER REVIEW AND ASSESSMENT

**1.**

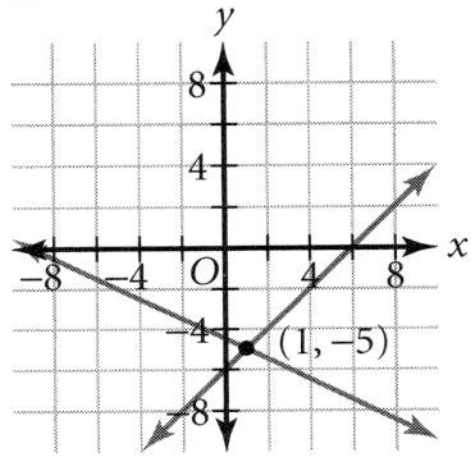


(1, −5)

**3.**

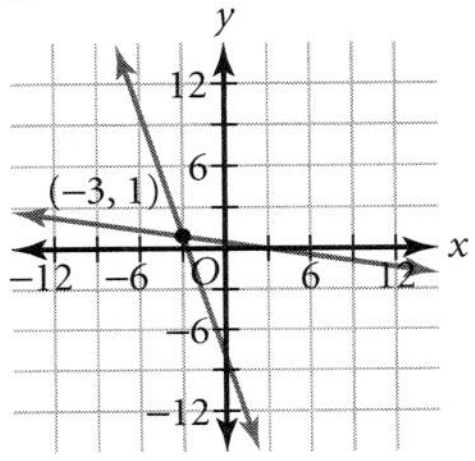


(−3, 1)

**5.**

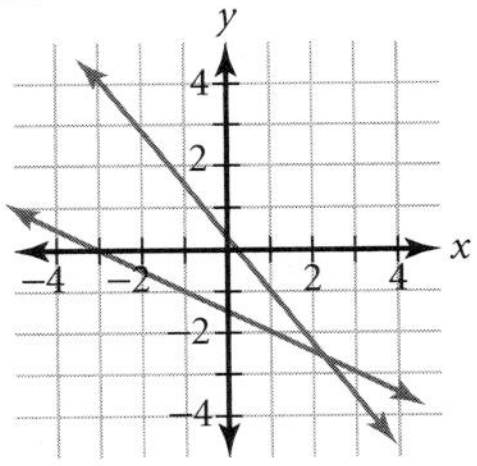

$\left(2\frac{1}{5}, -2\frac{3}{5}\right)$

**7.**

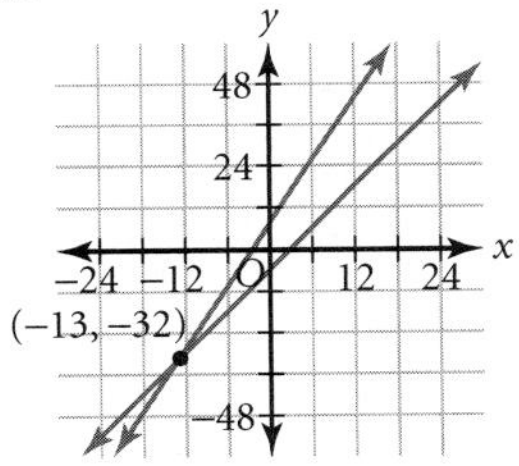


(−13, −32)

**9.** (−1, 3) **11.** $\left(\frac{1}{2}, \frac{1}{4}\right)$ **13.** $\left(\frac{10}{7}, \frac{19}{7}\right)$ **15.** $\left(\frac{8}{7}, \frac{2}{7}\right)$ **17.** (7, −5) **19.** (−2, 1) **21.** $\left(\frac{40}{7}, -\frac{1}{2}\right)$ **23.** $\left(\frac{11}{14}, \frac{1}{14}\right)$ **25.** dependent **27.** independent **29.** independent

**31.**

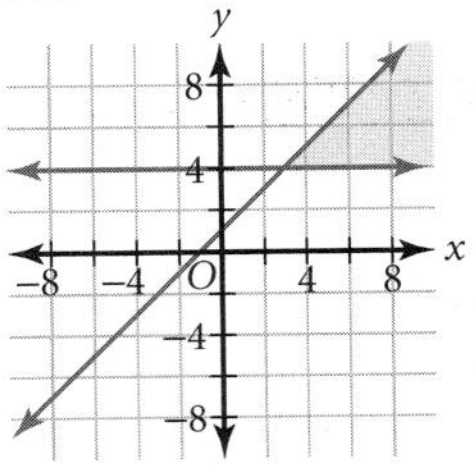

**33.**

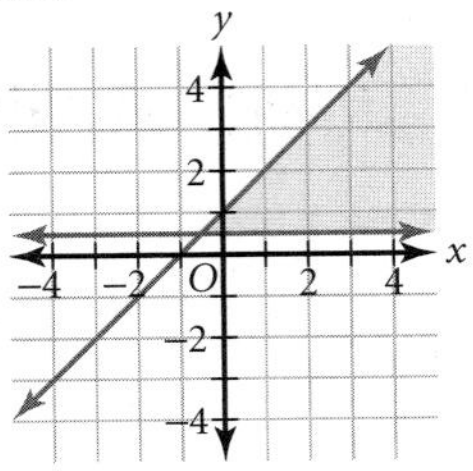

**35.**

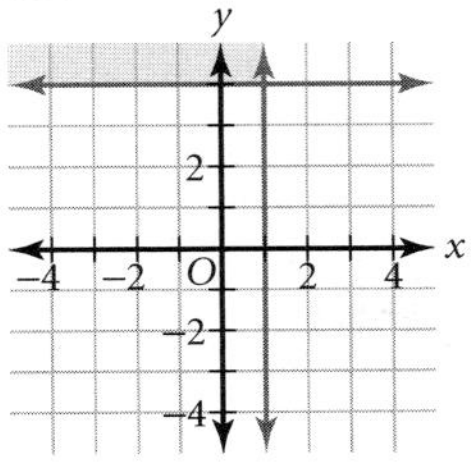

**37.** 15 nickels and 28 quarters **39.** 40 ounces of the 25% acid solution

# Chapter 8

## LESSON 8.1

**TRY THIS** (p. 371)

**a.** 64

**b.** 64

**c.** 27

**d.** 8

**TRY THIS** (p. 372)

**a.** $3^6 = 729$

**b.** $8^4 = 4096$

**c.** $z^9$

**d.** $a^{z+3}$

**TRY THIS** (p. 373)

**a.** $180m^3p^2$

**b.** $160xy^2z^3$

**Exercises**

**6.** 625 **7.** 8 **8.** 81 **9.** 11 **10.** 512 **11.** $f^{14}$ **12.** $t^8$ **13.** $x^{t+r}$ **14.** 128,000; 512,000 **15.** $320m^7p^9$ **16.** $-24x^3y^8z^9$ **17.** $30r^3t^3y$ **19.** 64 **21.** 125 **23.** 10,000,000 **25.** 1296 **27.** 2187 **29.** 6561 **31.** 16 **33.** 4 **35.** $10^8$ **37.** $3^6$ **39.** $10^{12}$ **41.** $6^6$ **43.** $2^7$ **45.** $-18x^3y^5z^2$ **47.** $-56m^5n^7$ **49.** $-26b^9f^4$ **51.** $-3x^ay^{b+f}z^{c+g}$ **53.** 10,100 **55.** $10^5$ **57.** 59,049 **59.** 0.32768 **61.** $10^1$ **63.** $10^4$ pounds **65.** 62.5 **67.** 54

**69.**

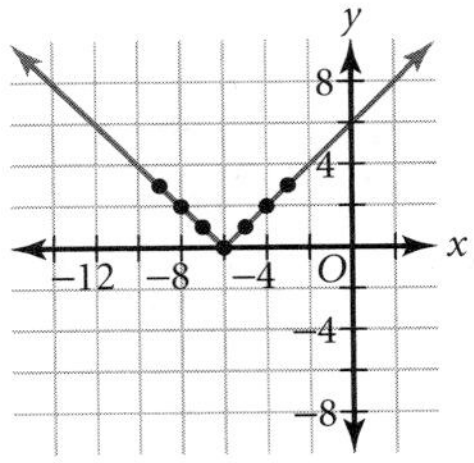

**71.**

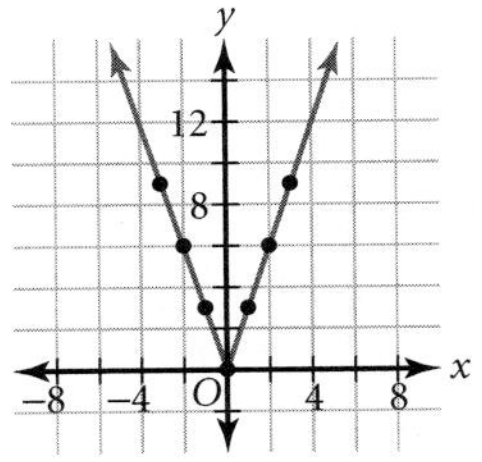

**73.** (−2, −3)

## LESSON 8.2

**TRY THIS** (p. 378)

**a.** $2^{12} = 4096$ **b.** $10^{12} = 1{,}000{,}000{,}000{,}000$ **c.** $y^{15}$ **d.** $m^{3x}$

**TRY THIS** (p. 379, Ex. 2)

**a.** $x^2y^2$ **b.** $x^2y^4$ **c.** $\pi^4r^8$

**TRY THIS** (p. 379, Ex. 3)

**a.** $(3 \cdot 3)^2 = 3^2 \cdot 3^2 = 9 \cdot 9 = 81$ or $(3 \cdot 3)^2 = 9^2 = 81$

**b.** $(2 \cdot 5)^6 = 2^6 \cdot 5^6 = 64 \cdot 15{,}625 = 1{,}000{,}000$ or $(2 \cdot 5)^6 = 10^6 = 1{,}000{,}000$

**TRY THIS** (p. 380, Ex. 4)
**a.** $x^2y^6$ **b.** $x^6y^6z^6$ **c.** $a^nb^nc^nd^n$

**TRY THIS** (p. 380, Ex. 6)
**a.** $y^4$ **b.** $-y^{100}$ **c.** $81y^4$

### Exercises

**6.** 59,049 **7.** 1,000,000 **8.** $b^{36}$ **9.** $z^{7y}$ **10.** $a^8b^8$ **11.** $x^6y^{12}$ **12.** $c^8h^{28}$ **13.** $m^6n^{15}$ **14.** 15,625 **15.** 21,870,000,000 **16.** 144 **17.** 3025 **18.** $x^{10}y^{15}$ **19.** $a^{44}b^{44}c^{44}$ **20.** $q^tr^ts^t$ **21.** $x^6y^{15}$ **22.** $-216$ **23.** $b^{10}$ **24.** $-125k^3$ **25.** $-1$ **27.** 46,656 **29.** 1,000,000 **31.** $-243b^{10}$ **33.** $-243x^{30}$ **35.** $625f^8k^{12}$ **37.** $54a^6$ **39.** $512n^6p^3$ **41.** $-a^7b^5$ **43.** $125d^{20}f^{19}t^{21}$ **45.** $15{,}552x^4y^9$ **47.** $-16b^{14}e^6h^9$ **49.** $x^6y^6$ **51.** $\frac{243x^{10}}{32y^{10}}$ **53.** $\frac{-1}{48{,}828{,}125x^{33}y^{55}}$ **55.** 8000 **57.** 400 **59.** 1 **61.** 1 **63.** 1 **65.** $6s^2$ **67.** $y-(-3) = -\frac{30}{19}(x-14)$ or $y-27 = -\frac{30}{19}(x+5)$; $y-(-3) = \frac{19}{30}(x-14)$ or $y-27 = \frac{19}{30}(x+5)$ **69.** $x < -6$ or $x > 2$

## LESSON 8.3

**TRY THIS** (p. 384, Ex. 1)
**a.** $10^8 = 100{,}000{,}000$
**b.** $2^5 = 32$
**c.** $3^3 = 27$

**TRY THIS** (p. 384, Ex. 2)
**a.** $m^{n-p}$
**b.** $y^9$
**c.** $10^y$

**TRY THIS** (p. 385)
**a.** $-3b$
**b.** $-9ty^2$

**TRY THIS** (p. 386)
**a.** $\frac{16}{81}$
**b.** 27
**c.** $\frac{x^{3z}}{2^{yz}}$

### Exercises

**5.** 1,000,000 **6.** 27 **7.** 343 **8.** $a^{c-d}$ **9.** $d^{11}$ **10.** $m^y$ **11.** $\frac{-4xz^6}{y}$ **12.** $hk^3$ **13.** $p^2rs^3$ **14.** $\frac{25}{64}$ **15.** $\frac{81}{256}$ **16.** $\frac{a^{3n}}{4^{tn}}$ **17.** 100,000 **19.** 1,679,616 **21.** 5 **23.** 9 **25.** 10,000,000,000 **27.** 64 **29.** $\frac{a^3}{b^3}$ **31.** $\frac{125x^3}{y^{18}}$ **33.** $\frac{a^{12}}{b^6}$ **35.** $\frac{81d^8}{e^{12}}$ **37.** $c^9p^6$ **39.** $\frac{32w^{25}}{243f^{20}}$ **41.** $\frac{7^dp^{pd}}{f^d}$ **43.** $x^{t^2}$ **45.** $2r^2$ **47.** $-\frac{a^3}{2}$ **49.** $-\frac{p^3}{10}$ **51.** $xy^3$ **53.** $-\frac{40a^3b^2}{c^5}$ **55.** $43.8u^2v^7w^3$ **57.** $\frac{r}{5}$ **59.** $\frac{cd^7}{y^7}$ **61.** $-0.8d^3e$ **63.** $13q^3r^{14}s^4$ **65.** 12.50 **67.** 10 **69.** 0.10 **71.** $-1.00$ **73.** 10.37 **75.** 8.45 **77.** 0.10 **79.** $-1.00$ **81.** $\frac{4}{3}r$ **83.** lateral surface area ratio is $\frac{4}{\pi}$ **85.** $n = 2k+8$ **87.** (17, 39)

## LESSON 8.4

**TRY THIS** (p. 391)
**a.** $10^{-2} = \frac{1}{100}$ **b.** $3^3 = 27$

**TRY THIS** (p. 392, Ex. 2)
**a.** $b^0 = 1$ **b.** $\frac{y}{x}$

**TRY THIS** (p. 392, Ex. 3)
**a.** $\frac{b}{5a^3}$ **b.** $-\frac{4}{x^3}$

### Exercises

**6.** $\frac{1}{343}$ **7.** $\frac{1}{10{,}000}$ **8.** 46,656 **9.** $\frac{1}{16}$ **10.** $\frac{1}{a^8}$ **11.** $y$ **12.** $\frac{1}{r^2}$ **13.** $\frac{1}{a^6b^2}$ **14.** $-\frac{5}{y^6}$ **15.** $\frac{c^2}{12d^5}$ **16.** $f^3h^4$ **17.** $\frac{64a^6}{25b^5}$ **19.** $\frac{1}{161{,}051}$ **21.** $\frac{1}{823{,}543}$ **23.** $\frac{1}{62{,}748{,}517}$ **25.** $\frac{1}{9}$ **27.** $b^5$ **29.** $\frac{y^4}{9x^2}$ **31.** $\frac{x^5}{6y^7}$ **33.** $\frac{8b^{13}}{3a^2}$ **35.** 9 **37.** $-\frac{1}{9}$ **39.** yes **41.** $\frac{1}{10^5}$ **43.** $\frac{d^3}{c^4}$ **45.** 1 **47.** $-\frac{1}{t^3}$ **49.** $2a$ **51.** $b$ **53.** $\frac{1}{2.56}$ **55.** 1 **57.** 0 **59.** You get an error message because $0^0$ is undefined. **61.** 1 **63.** 1.001 is the sum. Rounding errors explain the discrepancy. **65.** $y = -2$ **67.** The slope of a vertical line is undefined. The slope of a horizontal line is 0.
**69.** $A = 3^5x^6y^4$
$B = 2^43^3x^4y$

## LESSON 8.5

**TRY THIS** (p. 399)
$8.75 \times 10^5$

**TRY THIS** (p. 400, Ex. 2)
$4.02 \times 10^{-8}$

**TRY THIS** (p. 400, Ex. 3)
**a.** 367,700,000,000
**b.** 0.000000436

**TRY THIS** (p. 400, Ex. 4)
**a.** $3 \times 10^{24}$
**b.** $(2.5)(1.56)(10^{5+(-3)}) = 3.9 \times 10^2$
**c.** $\left(\frac{3.6}{6}\right)(10^{5-3}) = 0.6 \times 10^2$
$= 6 \times 10^{-1} \times 10^2$
$= 6 \times 10^1$

### Exercises

**4.** $4.63 \times 10^8$ **5.** $2.5 \times 10^4$ **6.** $5.97 \times 10^{-4}$ **7.** $1.65 \times 10^{-7}$ **8.** $4.57 \times 10^{-6}$ **9.** 367,700,000 **10.** 0.0000528 **11.** $6 \times 10^{12}$ **12.** $9.345 \times 10^{-2}$ **13.** $4 \times 10^2$ **14.** $7.92 \times 10^8$ sec or 25.1 years **15.** $2 \times 10^6$ **17.** $3.4 \times 10^5$ **19.** $8 \times 10^{-5}$ **21.** $2.34 \times 10^{-4}$ **23.** $1.35 \times 10^9$ **25.** 400,000,000 **27.** 901,000 **29.** 0.000000005 **31.** 0.00000000072 **33.** $6 \times 10^{14}$ **35.** $6.3 \times 10^{13}$ **37.** $2 \times 10^6$ **39.** $3.8 \times 10^{18}$ **41.** $4.8 \times 10^{10}$ **43.** $4.5 \times 10^4$ **45.** $8 \times 10^{-7}$ **47.** $5 \times 10^{12}$ **49.** $3.7 \times 10^6$ **51.** $4 \times 10^{12}$ **53.** Answers may vary. Sample answer: $4.41504 \times 10^8$ seconds **55.** 5,870,000,000,000 **57.** 3.6 years **59.** 23,000 **61.** 0.00722 **63.** $-280$ **65.** Answers may vary. Sample answer: 9,999,999,999 **67.** Answers may vary. Sample answer: $-9{,}999{,}999{,}999$ **69.** 20 **71.** $41x - y - 8$

**73.**

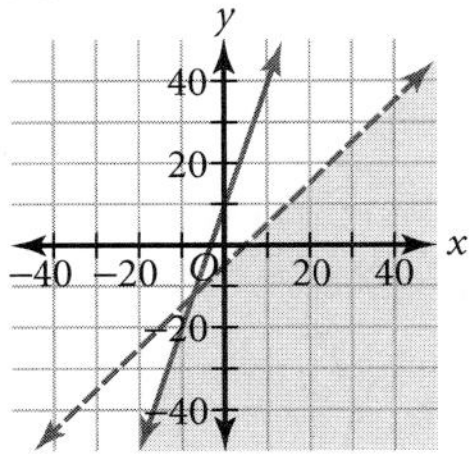

## LESSON 8.6

**TRY THIS** (p. 406)

Approximately 18,690,000

**Exercises**

**5.** 2,278,057 people

**6.** $y = 5^x$ $\qquad$ $y = \left(\frac{1}{5}\right)^x$

The graphs are mirror images of each other. One increases and the other decreases as $x$ gets larger. They both pass through the point (0, 1).

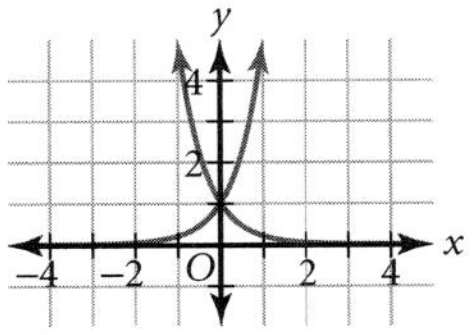

**7.**

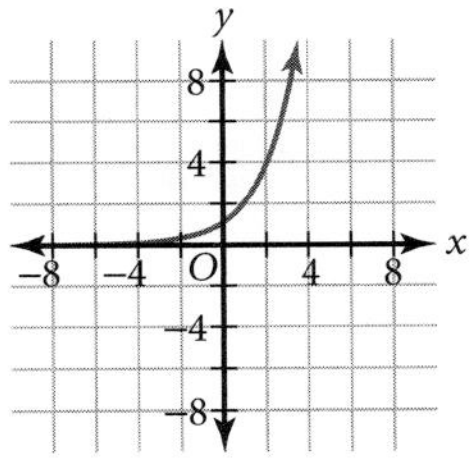

**9.**

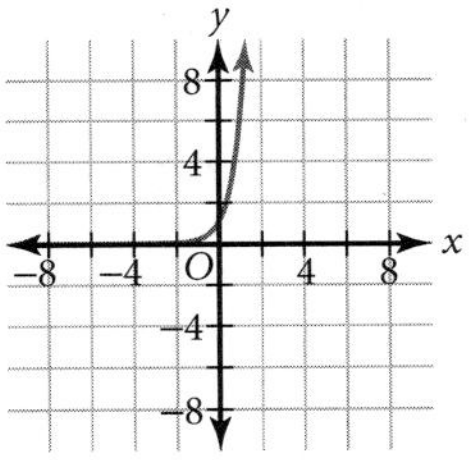

**11.**

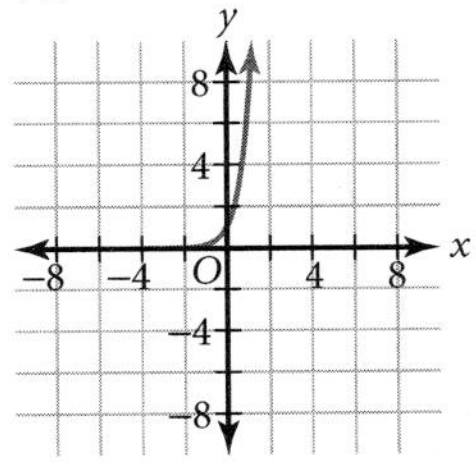

**13.**

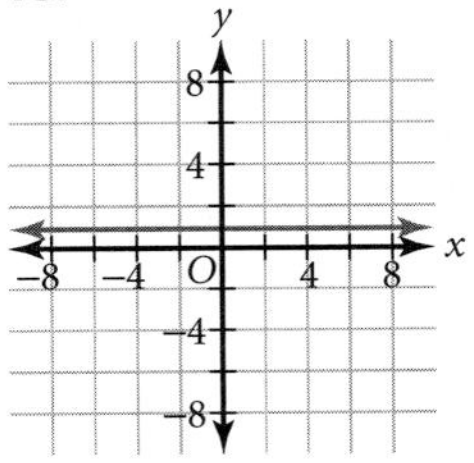

**15.**

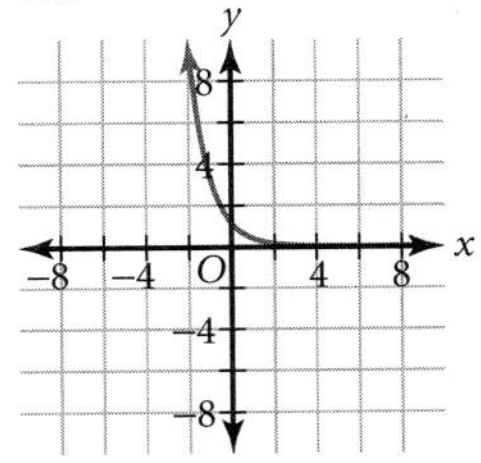

**17.**

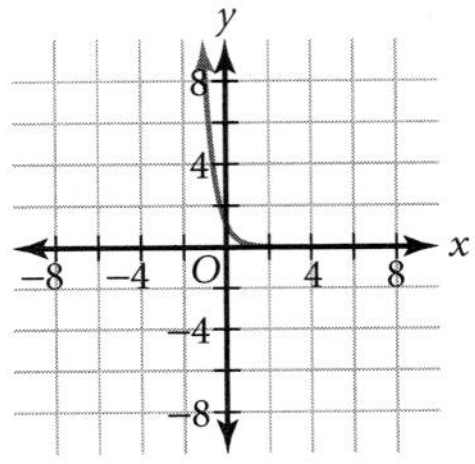

**19.**

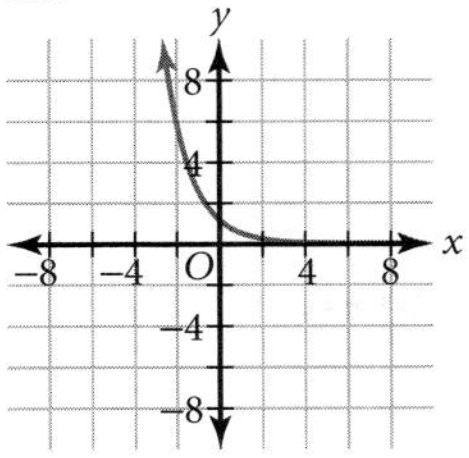

**21.**

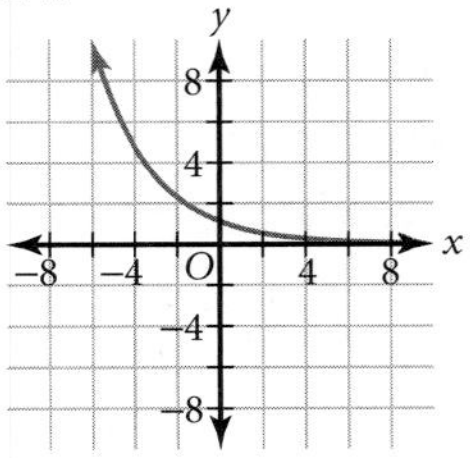

**23.**

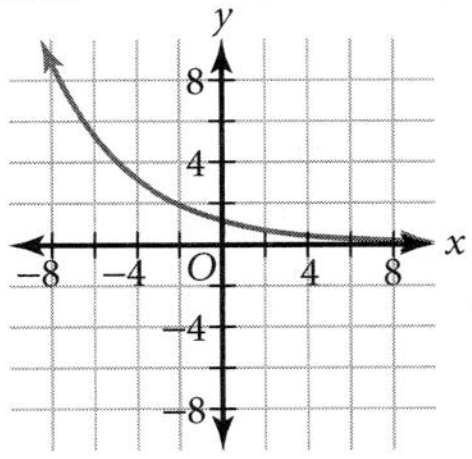

**25.**

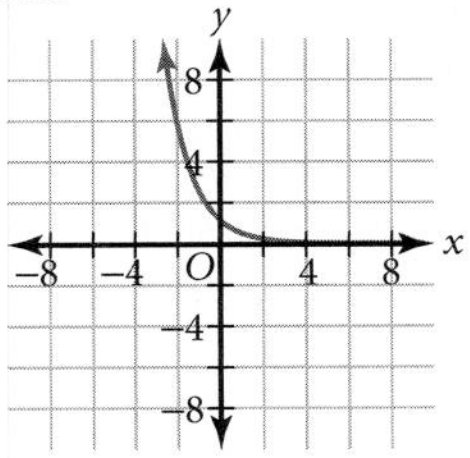

**27.**

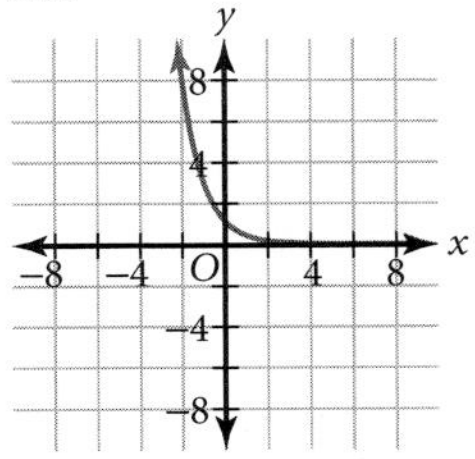

**29.**

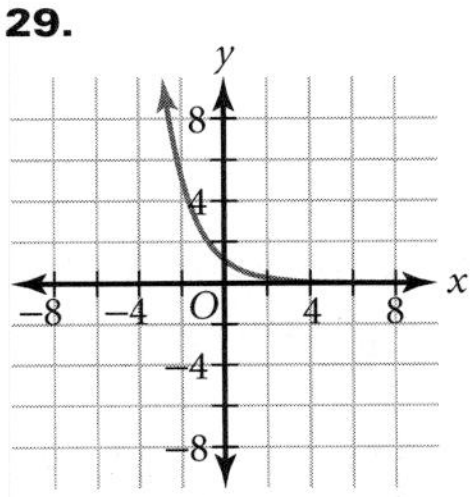

**31.** 1.003 **32.** ≈ 39,117,000 **33.** ≈ 39,945,887 **34.** ≈ 58,058,000 **35.** ≈ 58,174,174 **36.** ≈ 58,290,581 **37.** ≈ 58,031,200 **38.** ≈ 58,496,378 **39.** ≈ 58,965,285 **40.** 1997 **41.** $-\frac{7}{3}$

**43.** Slope = −3, $y$-intercept = −2

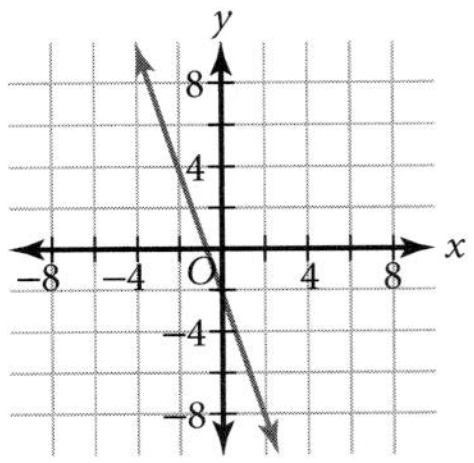

**45.** Slope = $-\frac{1}{5}$, $y$-intercept = 7

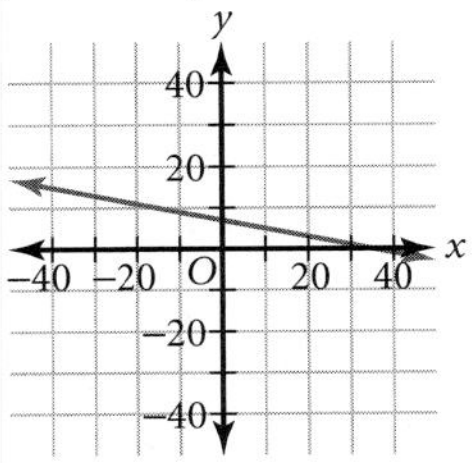

**47.** $x = \frac{1}{2}, y = \frac{11}{2}$ **49.** $x^{12} + y^{15}$

## LESSON 8.7

**TRY THIS** (p. 410, Ex. 1)

22% is slightly less than $\frac{1}{4}$, so the object is a little more than 11,400 years old. A reasonable estimate is about 12,000 years.

**TRY THIS** (p. 410, Ex. 2)

$$3000 = P(1 + 0.04)^4$$
$$P \approx \$2565$$

**TRY THIS** (p. 411, Ex. 3)

$$A = 4{,}500{,}000(1 + 0.07)^{-50}$$
$$A \approx \$152{,}765$$

**TRY THIS** (p. 411, Ex. 4)

$100(0.92)^3 \approx 78$ problems

**TRY THIS** (p. 412)

**a.** $y = 16{,}225(1.02)^x$

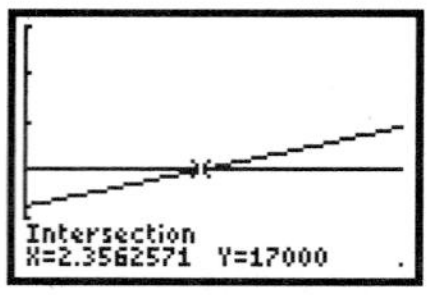


In about 2.4 years, or during the year 2003.

**b.**

| Year | Population |
|---|---|
| $x$ | $y = 16{,}225(1.02)^x$ |
| 0 | 16,225 |
| 1 | 16,550 |
| 2 | 16,880 |
| 3 | 17,218 |

**Exercises**

**6.** 11,400 years old **7.** \$2033.75 **8.** \$5895.97 **9.** 44 problems **10.** 1993 **11.** 0.3 **13.** 0.003 **15.** about 24,000 years **17.** about 5700 years **19.** $\frac{2}{5}$
**21.** 11,400 years old

**23.**

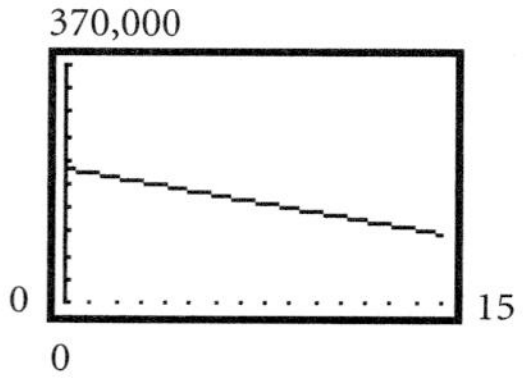


**25.** About 359,477 **27.** About 356,968 **29.** 17,703.18 **31.** 3,798.19 **33.** ≈ \$76,804.84 **35.** ≈ \$115,044.83 **37.** \$1490.13 **39.** 10 years from now **41.** $x = \frac{21}{16}$, $y = \frac{39}{8}$ **43.** $x \approx 1.159, y \approx 1.352$

**45.**

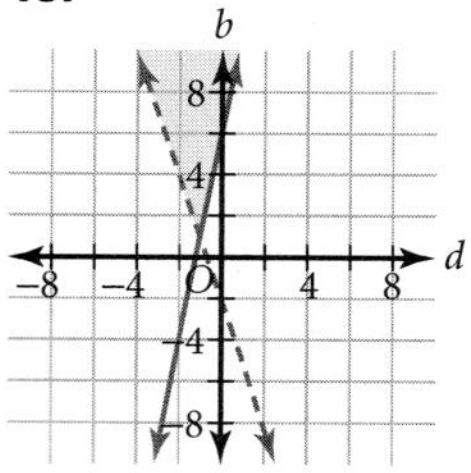

**47.**

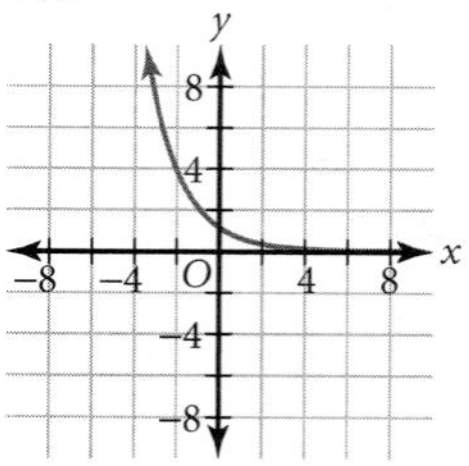

**49.**

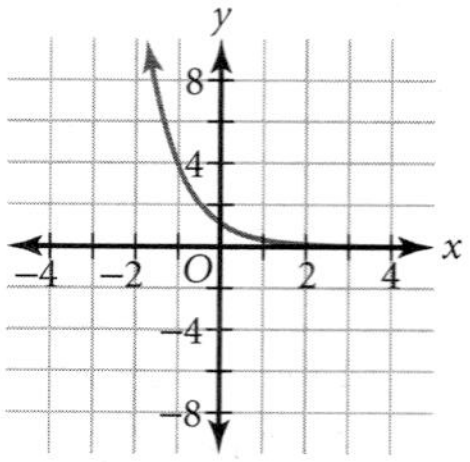

## CHAPTER REVIEW AND ASSESSMENT

**1.** $20a^5$ **3.** $28s^4t^4$ **5.** $12m^4n^2$ **7.** $24x^3y^3z^3$ **9.** $c^8$
**11.** $64x^6y^3$ **13.** $-p^7$ **15.** $-24m^{10}n^2$ **17.** $\frac{x^4}{y^4}$ **19.** $\frac{8}{27w^6}$
**21.** $\frac{y^6}{125x^3}$ **23.** $-243b^5$ **25.** $\frac{1}{3^2}$ **27.** $\frac{c^3}{b^2}$ **29.** $\frac{3}{p^4}$ **31.** $-5t$
**33.** $5.9 \times 10^6$ **35.** $1.5 \times 10^8$ **37.** $1 \times 10^3$ **39.** $3 \times 10^3$
**41.** ≈3,545,347 **43.** 3,839,406

**45.**

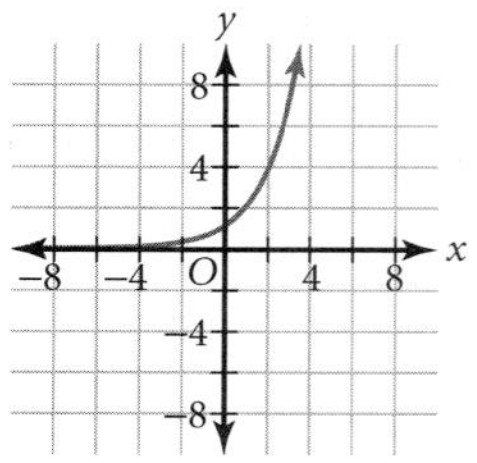

**47.**

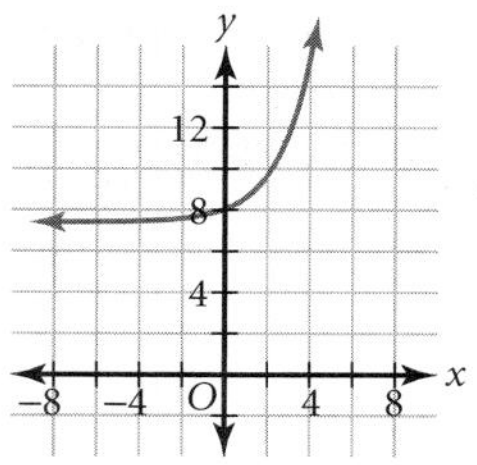

**49.**

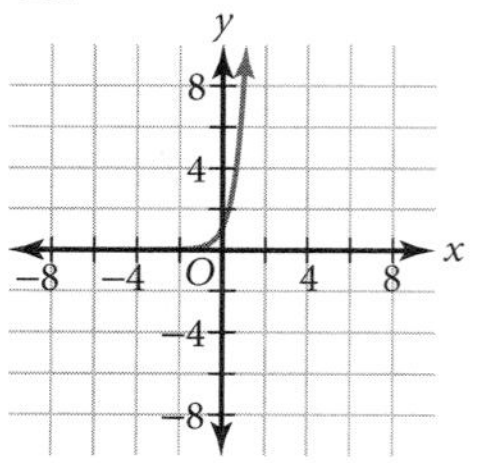

**51.** \$4820.35 **53.** \$3,688.21 **55.** \$1,070,580.46

# Chapter 9

## LESSON 9.1

**TRY THIS** (p. 428, Ex. 2)

$$\begin{array}{r} 3x^2 + 5x + 0 \\ +(-2x^2) - 6x + 4 \\ \hline x^2 - x + 4 \end{array}$$

**TRY THIS** (p. 428, Ex. 3)

$(x^2 + 3x) + (x^2 + 3x)$

$+ (x^3 + 2x^2) + (x^3 + 2x^2)$

$= 2x^3 + 6x^2 + 6x$

**TRY THIS** (p. 429)

$3x^2 - 2x + 3 - (x^2 + 2x - 4) = 2x^2 - 4x + 7$

**Exercises**

**6.** $3x^3 + x - 2$ **7.** $-3x^2 + 2x + 15$
**8.** $3x^4 + 2x^2 - 2x + 23$ **9.** $4x^3 + 2x - 9$
**10.** $18x^2 + 5x - 3$ **11.** $12x^3 + 4x^2 + 10x - 4$
**12.** $10x^4 + 3x^3 - x^2 - 11$ **13.** $8x + 2$ **14.** $14x$
**15.** $2x^2 + 2x + 9$ **16.** $x^4 + 2x^2 - 3$

**17.**

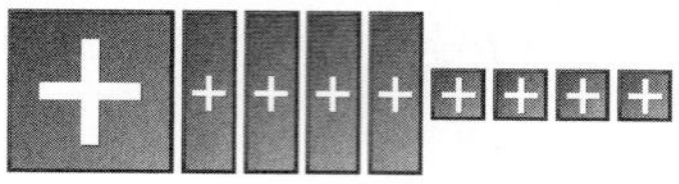

**19.** $c^3 + c + 6$ **21.** $p^7 + 10$ **23.** Polynomial; quadratic **25.** Binomial; linear **27.** Trinomial; quadratic **29.** Answers may vary. Sample answer: $x + 5$ **31.** $7x^4 + 4x^2 - x - 5$ **33.** $4r^4 + 2r^3 + r^2 - 6$ **35.** $x^3 - x + 7$ **37.** $c^3 + c^2 + 10c + 6$ **39.** $y^3 + y^2 - 6$ **41.** $s^3 + 4s^2 + 6s - 7$ **43.** $y^2 + 4$ **45.** $x^2 - 4x + 7$ **47.** $-x^4 - 3x^2 - 3x + 10$ **49.** $4x^5 - 3x^3 + 2x^2 - 7x - 10$ **51.** $10x + 10$ **53.** width: $2x + 3$, length: $3x + 4$ **55.** $6x^2 - 24x$ **57.** $\frac{5}{2} = 2\frac{1}{2}$ **59.** $(1, -1)$ **61.** $8.9 \times 10^9$

## LESSON 9.2

**TRY THIS** (p. 434)

**a.** $10x^2 + 5x$

**b.** $x^2 + 8x - 9$

### Exercises

**5. a.** $2x^2 + 5x + 3$

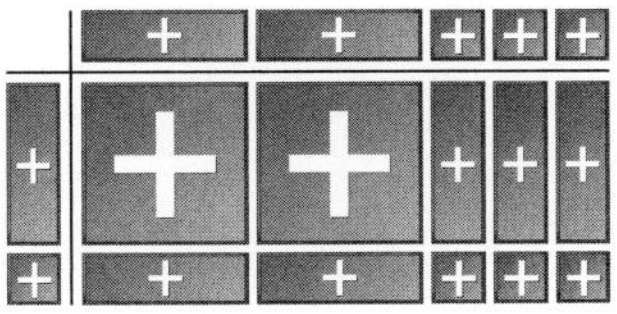

**b.** $6x^2 - 3x - 3$

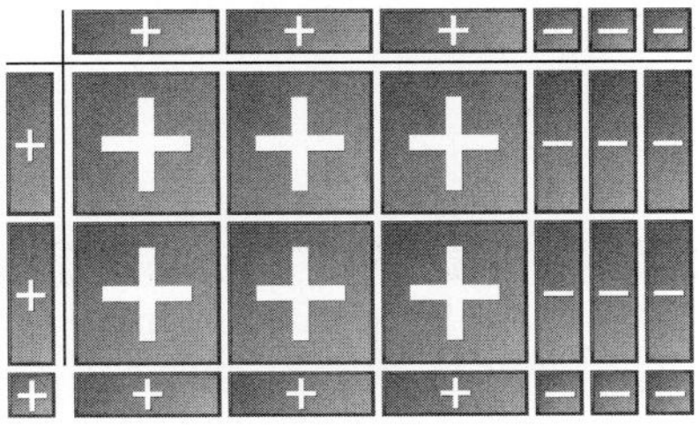

**c.** $x^2 - 2x - 8$

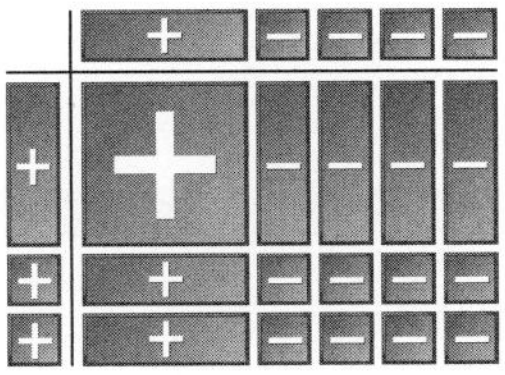

**6. a.** $x^2 + 2x - 15$

**7.** $(x + 2)(x + 1) = x^2 + 3x + 2$

**9.** $(x - 1)(x + 1) = x^2 - 1$

**11.** $(x - 4)(x + 2) = x^2 - 2x - 8$

**13.** $x^2 + 7x + 10$

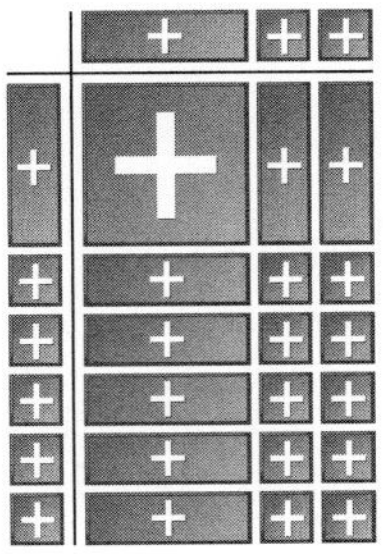

**15.** $2x^2 + 3x + 1$

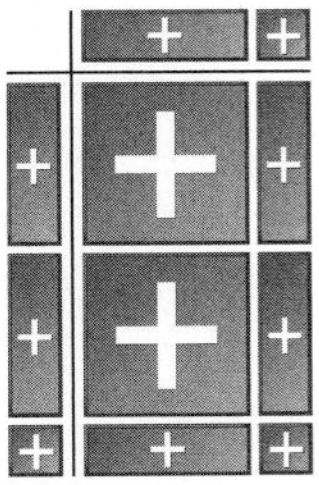

**17.** $x^2 + 3x + 2$

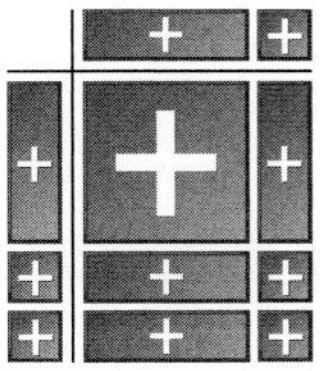

**19.** $4x^2 - 11x + 6$

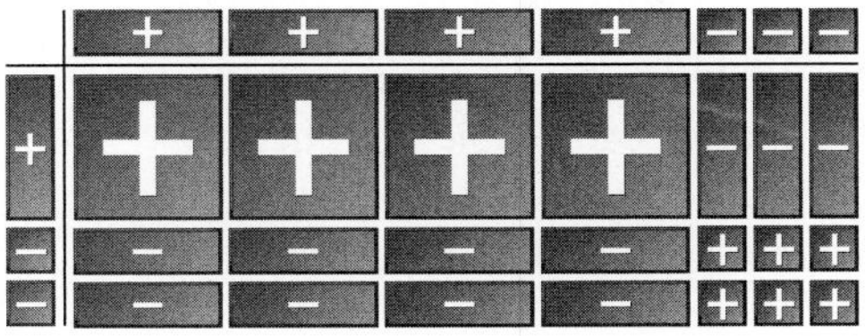

**21.** $2x^2 + 2x - 60$

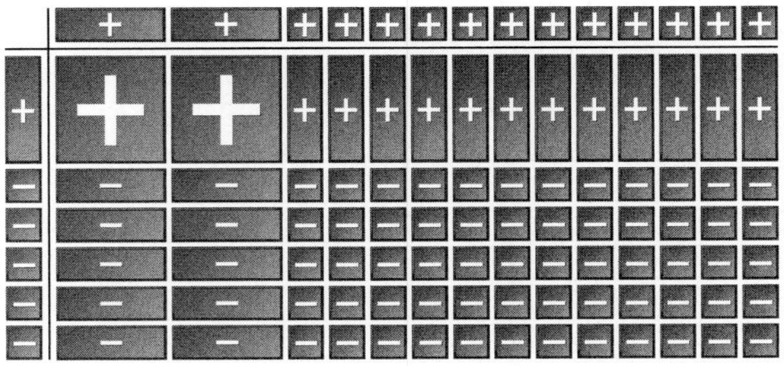

**23.** $12x + 42$ **25.** $4x - 16$ **27.** $-2x + 6$ **29.** $-3x^3 - 12x^2$ **31.** $x^2 - 2x$ **33.** $-4x^2 + 12x$ **35.** $a^2 + a - 12$ **37.** $y^2 + y - 6$ **39.** $d^2 - 25$ **41.** $w^3 + w^2$ **43.** True **45.** True **47.** When $x = 10$, $x - 10 = 0$. Thus the product is zero on both sides. When $x \neq 10$, the products will be different. **49.** $(20)(10)$ yd$^2$ **51.** $0.375x + 0.625y$ **53.** No solution **55.** Solution $\left(\frac{1}{2}, \frac{1}{3}\right)$ **57.** $\left(\frac{3}{4}, \frac{2}{3}\right)$ **59.** $x^2 + x - 1$

## LESSON 9.3

**TRY THIS** (p. 439)

**a.** $2x^2 + 3x - 20$

**b.** $6y^2 - y - \frac{1}{3}$

**TRY THIS** (p. 440, Ex. 2)

**a.** $25x^2 - 4$

**b.** $4x^2 - 16x + 16$

**TRY THIS** (p. 440, Ex. 3)

16ft × 28ft

### Exercises

**5. a.** $x^2 + 9x + 18$ **b.** $a^2 - 13a + 36$ **c.** $m^2 + 6m - 27$ **6. a.** $x^2 - 25$ **b.** $9x^2 + 12x + 4$ **c.** $4x^2 - 32x + 64$ **7.** length: 24; width: 23 **9.** $5p^2 + 8p + 3$ **11.** $4x^2 + 4x - 15$ **13.** $6w^2 - 43w + 72$ **15.** $14s^2 - 17s - 6$ **17.** $y^2 + \frac{2}{9}y - \frac{5}{27}$ **19.** $4a^4 + 12a^2 + 9$ **21.** $p^2 + 2pq + q^2$ **23.** $x^3 + x^2y + xy + y^2$ **25.** $0.96m^2 - 0.8m - 20$ **27.** $25x^2 - 1$ **29.** $x^2 + 2x + 1$ **31.** $9x^2 - 24x + 16$ **33.** $x^2 - 14x + 49$ **35.** $y^2 + 12y + 36$ square units **37.** The area of the square is larger, by 1 square unit **39.** Slope $= -\frac{4}{3}$

**41.** Slope $= \frac{7}{3}$ **43.** Slopes are $\frac{1}{2}, -2$, Lines are perpendicular **45.** $y = -2x - 5$ **47.** No solution **49.** 24 **51.** $c^6b^4$

## LESSON 9.4

**TRY THIS** (p. 443)

$V = 9375$ cubic centimeters

**TRY THIS** (p. 445)

Original volume $= 640\pi$ cubic centimeters
$\approx 2011$ cubic centimeters
New volume $= 800\pi$ cubic centimeters
$\approx 2513$ cubic centimeters

Height of new container $= 12\frac{1}{2}$ centimeters

### Exercises

**4.** 7500 $cm^3$ **5.** original volume $\approx$ 6283 $cm^3$; new volume $\approx$ 7540 $cm^3$; new height $\approx$ 24 cm

**6.**

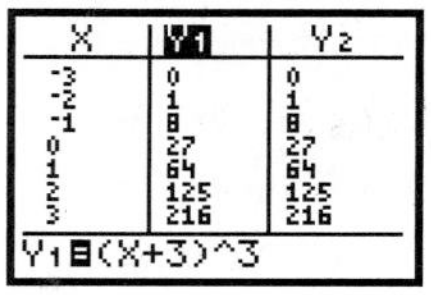

| X | Y1 | Y2 |
|---|---|---|
| -3 | 0 | 0 |
| -2 | 1 | 1 |
| -1 | 8 | 8 |
| 0 | 27 | 27 |
| 1 | 64 | 64 |
| 2 | 125 | 125 |
| 3 | 216 | 216 |

Y1 = (X+3)^3

**7.**

| $x$ | $x^2 + 5x + 6$ | $(x + 2)(x + 3)$ |
|---|---|---|
| −3 | 0 | 0 |
| −2 | 0 | 0 |
| −1 | 2 | 2 |
| 0 | 6 | 6 |
| 1 | 12 | 12 |
| 2 | 20 | 20 |
| 3 | 30 | 30 |

**9.**

| $x$ | $x^2 - 5x + 6$ | $(x - 2)(x - 3)$ |
|---|---|---|
| −3 | 30 | 30 |
| −2 | 20 | 20 |
| −1 | 12 | 12 |
| 0 | 6 | 6 |
| 1 | 2 | 2 |
| 2 | 0 | 0 |
| 3 | 0 | 0 |

**11.**

| $x$ | $x^2 - 9$ | $(x - 3)(x + 3)$ |
|---|---|---|
| −3 | 0 | 0 |
| −2 | −5 | −5 |
| −1 | −8 | −8 |
| 0 | −9 | −9 |
| 1 | −8 | −8 |
| 2 | −5 | −5 |
| 3 | 0 | 0 |

**13.**

| $x$ | $x^3 - 8$ | $(x - 2)(x^2 + 2x + 4)$ |
|---|---|---|
| −3 | −35 | −35 |
| −2 | −16 | −16 |
| −1 | −9 | −9 |
| 0 | −8 | −8 |
| 1 | −7 | −7 |
| 2 | 0 | 0 |
| 3 | 19 | 19 |

**15.**

| $x$ | $x^3 + 12x^2 + 48x + 64$ | $(x + 4)^3$ |
|---|---|---|
| −3 | 1 | 1 |
| −2 | 8 | 8 |
| −1 | 27 | 27 |
| 0 | 64 | 64 |
| 1 | 125 | 125 |
| 2 | 216 | 216 |
| 3 | 343 | 343 |

**17.**

| $x$ | $x^2 - 6x + 9$ | $(x - 3)^2$ |
|---|---|---|
| −3 | 36 | 36 |
| −2 | 25 | 25 |
| −1 | 16 | 16 |
| 0 | 9 | 9 |
| 1 | 4 | 4 |
| 2 | 1 | 1 |
| 3 | 0 | 0 |

**19.** $V = (2x)^3 = 8x^3$ $cm^3$ **21.** $S = (20x + 42)$ $cm^2$ **23.** $C \approx 113.1$ in.; $S \approx 4071.5$ in.$^2$; $V \approx 24{,}429.0$ in.$^3$ **25.** $C \approx 31.4$ in.; $S \approx 314.2$ in.$^2$; $V \approx 523.6$ in.$^3$ **27.** $C \approx 6.3$ in.; $S \approx 12.6$ in.$^2$; $V \approx 4.2$ in.$^3$ **29.** −26 **31.** 19 **33.** $y = x - 1$

## LESSON 9.5

**TRY THIS** (p. 449, Ex. 1)

**a.** $3a(2b + 1)$
**b.** $5x(x^2 + 2x - 4)$
**c.** $2x^2 - 3y^2$ is prime

**TRY THIS** (p. 449, Ex. 2)

$(a + 6)(x - 3)$

**TRY THIS** (p. 449, Ex. 3)

**a.** $ax + bx + ay + by = (ax + bx) + (ay + by)$
$= x(a + b) + y(a + b)$
$= (a + b)(x + y)$

**b.** $ax + bx - ay - by = (ax + bx) - (ay + by)$
$= x(a + b) - y(a + b)$
$= (a + b)(x - y)$

### Exercises

**6.** $5b(2c + 1)$ **7.** Does not factor **8.** $4xy(x + 3y - 5)$ **9.** $(x - 2)(y + 1)$ **10.** $(m + 4)(n^2 + 3)$ **11.** $(m - 4)(m + 1)$ **12.** $(x + 3a)(x - 2a)$

**13.** $4(x-4)$ **15.** $2(n^2+2)$ **17.** $3p(p+4)$
**19.** $2(x^2-2)$ **21.** $3x(x+2)$ **23.** $k^2(k^3+1)$
**25.** $2(2x^2+x-3)$ **27.** $9y(3y^2+2y-9)$
**29.** $25(4+s^5-2s)$ **31.** $3m(m^2-3m+1)$
**33.** $xy(2x^2-8xy+17y^2)$ **35.** $7y(y^2-3y+2)$
**37.** $3x^2(1+7x^2)$ **39.** $2t^3(1-65t^2)$
**41.** $3w^{2n}(3+7w)$ **43.** $(x+1)(x+2)$
**45.** $(x+y)(a+b)$ **47.** $(x-1)(x+2)$
**49.** $(a-3)(5a+4)$ **51.** $(x-2)(2-x)$
**53.** $(r-s)^2(2r-3)$ **55.** $(x+y)(2+a)$
**57.** $15(x+y)$ **59.** $\pi(R^2-r^2)$ **61.** 19
**63.** $-\frac{17}{3}$ **65.** $\left(\frac{1}{2}, -\frac{3}{5}\right)$ **67.** $\approx$ \$4763

## LESSON 9.6

**TRY THIS** (p. 453)
**a.** $(3s+4)^2$
**b.** $(8a^2-b)^2$

**TRY THIS** (p. 454, Ex. 4)
$(5w+9)(5w-9)$

**TRY THIS** (p. 454, Ex. 5)

$$(30+4)(30-4) = 30^2-4^2 = 900-16 = 884$$

### Exercises

**4.** Yes. Explanations may vary. Sample explanation: The first and last terms are perfect squares, and the middle term is twice the product of the square roots of the first and last terms. **5.** Yes. Explanations may vary. Sample explanation: The first and last terms are perfect squares, and the middle term is the opposite of twice the product of the square roots of the first and last terms. **6.** No. Explanations may vary. Sample explanation: The first and last terms are not perfect squares. **7.** $(x+5)^2$ **8.** $(6a-5)^2$ **9.** $(4x+y)^2$ **10.** Yes, both terms are perfect squares. **11.** Yes, both terms are perfect squares. **12.** No, $5y^2$ is not a perfect square. **13.** $(a+b)(a-b)$ **14.** $(6x+7y)(6x-7y)$ **15.** $(2m+12n)(2m-12n)$ **16.** $(50-3)(50+3)=2491$ **17.** $(100-3)(100+3)=9991$ **19.** $(x+2)^2$ **21.** $(x+2)(x-2)$ **23.** $(2t+1)(2t-1)$ **25.** $(5x+3)(5x-3)$ **27.** $(2x-5)^2$ **29.** $(4c+5)(4c-5)$ **31.** $(3c+2d)(3c-2d)$ **33.** $(3a-2)^2$ **35.** $(ax+b)^2$ **37.** $(9a^2+3b)(9a^2-3b)$ **39.** $(x-4)(25-x)=(x+2)(x-2)(5+x)(5-x)$ **41.** $(3x+5)(x+2)(x-2)$ **43.** 49 **45.** 36 **47.** 36 **49.** Answers may vary. Sample answer: The rectangle formed has dimensions $(a+b)(a-b)$, therefore, by comparing areas, $a^2-b^2=(a+b)(a-b)$. **51.** $4n-24$ **53.** $x^2+8x+16$ **55.** $10x+15$

**57.**

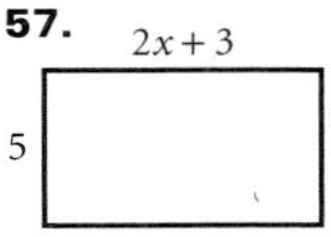


**59.** 95 **61.** Slope $=\frac{3}{4}$

**63.**

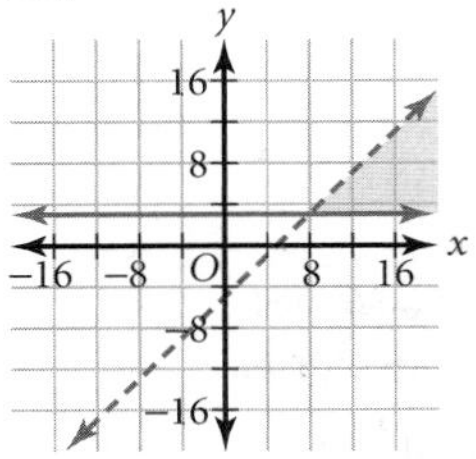


## LESSON 9.7

**TRY THIS** (p. 459)
**a.** $(x+4)(x+10)$
**b.** $(x-2)(x-8)$

**TRY THIS** (p. 460)
**a.** $(x+3)(x+6)$
**b.** $(x-3)(x-7)$
**c.** $(x-3)(x+7)$
**d.** $(x+4)(x-5)$

**TRY THIS** (p. 461, Ex. 3)
**a.** $(x+3)(x+4)$
**b.** $(x+9)(x-2)$

**TRY THIS** (p. 461, Ex. 4)
**a.** $(3x-1)(x-3)$
**b.** $(3x-1)(2x+3)$

**TRY THIS** (p. 461, Ex. 5)
$x(x+3)(2x+3)$

### Exercises

**5. a.** $(x+3)(x+4)$ **b.** $(x-3)(x-4)$ **6. a.** $(x-3)(x+4)$ **b.** $(x+3)(x-4)$ **7.** $(x-8)(x-5)$ **8.** $(2x-7)(2x+1)$ **9.** $(3x)(x+3)^2$ **11.** $(x+3)(x-4)$ **13.** $(x-3)(x-3)$ or $(x-3)^2$ **15.** $(p-2)(p+6)$ **17.** $(b+3)(b-8)$ **19.** $(z-4)(z+5)$ **21.** Prime **23.** $(s-3)(2s-21)$ **25.** Prime **27.** $6y(2y-1)(y-2)$ **29.** $(a+3)(a-1)(a+6)$ **31.** $(x+6)(x-6)(x+2)$ **33.** $16(2p^2+1)(2p^2-1)$ **35.** $(x-1)^2$ **37.** $(a+b)(2x+y)$ **39.** $\pm 11, \pm 4, \pm 1$ **41.** $217\frac{1}{3}$ **43.** $93\frac{7}{9}$ **45.** $y=\frac{1}{2}x-3$

## LESSON 9.8

**TRY THIS** (p. 466)
$-10x^2+100x=-10x(x-10)=0, x=0$ or $x=10$

**TRY THIS** (p. 467)
**a.** $x^2+6x-16=(x+8)(x-2)=0, x=-8$ or $x=2$
**b.** $2x^2+5x-3=(2x-1)(x+3)=0, x=\frac{1}{2}$ or $x=-3$

### Exercises

**6.** $x=-5$ and $x=-2$ **7.** $x=-2$ and $x=10$ **8.** $x=-7$ and $x=6$ **9.** $x=0$ or $x=\frac{1}{5}$

**10.** $x = 0$ or $x = \frac{1}{4}$ **11.** $x = 0$ or $x = 2$ **12.** $x = -2$ or $x = -5$ **13.** $x = 2$ or $x = 5$ **14.** $x = 2$ or $x = -\frac{5}{2}$ **15.** $x = -5$ and $x = 2$ **17.** $x = -6$ **19.** $x = -10$ and $x = 3$ **21.** $x = 3$ **23.** $x = 1$ or $x = 9$ **25.** $x = 3$ or $x = -6$ **27.** $x = -\frac{1}{2}$ or $x = 1$ **29.** $x = 3$ or $x = 12$ **31.** $x = -3$ or $x = 12$ **33.** $x = 3$ or $x = 4$ **35.** $x = -3$ or $x = 4$ **37.** 3750 $ft^2$ **39.** $x = 0$ and $x = 125$. These are the values of the length that will make the area equal zero. **41.** $x = 25$ or $x = 100$, $x$ represents the width, $(125 - x)$ represents the length, and 2500 represents the area. **43.** $r = \frac{m}{4t}$ **45.** $x = 1.2$ **47.** $x \geq -6$ **49.** $\frac{21}{50}$

## CHAPTER REVIEW AND ASSESSMENT

**1.** $5x^2 - x$ **3.** $6d^2 - 5d + 5$ **5.** $8m^2 - 2m + 4$ **7.** $6x^2 + 4x$ **9.** $4t^5 + 4t^4 + 2t^3 - 4t^2$ **11.** $4y^2 + 8y$ **13.** $y^2 + 7y - 18$ **15.** $x^2 - x - 12$ **17.** $4w^2 + 7wz + 3z^2$ **19.** $x^2 - 49$ **21.** $x^2 + 2x + 1$ **23.** $9x^2 - 64$

**25.**

| $x$ | $(x+2)(x-2)$ | $x^2 - 4$ |
|---|---|---|
| −3 | 5 | 5 |
| −2 | 0 | 0 |
| −1 | −3 | −3 |
| 0 | −4 | −4 |
| 1 | −3 | −3 |
| 2 | 0 | 0 |
| 3 | 5 | 5 |

**27.** $b(b^3 + 15b^2 + 5)$ **29.** $2a^2y(a^2 - 2ay + 4)$ **31.** $(x + 4)(x + 2)$ **33.** $(m - 5)(2m + 3)$ **35.** $(2x + 1)^2$ **37.** $3(s - 1)^2$ **39.** $(m + 2)(m - 4)$ **41.** $(r - 7)(r + 8)$ **43.** $(b + 4)(b + 7)$ **45.** $x = 3$ or $x = -5$ **47.** $x = -3$ or $x = -4$ **49.** 13 cm, 9 cm

# Chapter 10

## LESSON 10.1

**TRY THIS** (p. 482)

**a.** Vertex: $(1, -2)$, Axis of symmetry: $x = 1$

**b.** Vertex: $(-3, 1)$, Axis of symmetry: $x = -3$

**TRY THIS** (p. 483)

The zeros of the function are 5 and 2. The vertex is $(3.5, -2.25)$. $y = (x - 3.5)^2 - 2.25$

### Exercises

**5.** $(-6, -4)$, $x = -6$ **6.** $(3, -8)$, $x = 3$ **7.** $(-3, 6)$, $x = -3$ **8.** vertex: $(6, -16)$, $y = (x - 6)^2 - 16$ **9.** vertex: $\left(\frac{3}{2}, -\frac{49}{4}\right)$, $y = \left(x - \frac{3}{2}\right)^2 - \frac{49}{4}$ **10.** vertex: $\left(-\frac{11}{2}, -\frac{49}{4}\right)$, $y = \left(x + \frac{11}{2}\right)^2 - \frac{49}{4}$ **11.** The graph is shifted two units to the right and three units up. **13.** The graph is shifted two units to the right, reflected over the $x$-axis, and shifted one unit up. **15.** The graph is shifted three units to the right, reflected over the $x$-axis, and shifted two units down. **17.** The graph is shifted four units to the left and five units down. **19.** The graph is shifted two units to the left and eight units down.

**21.** vertex: $(2, 3)$; axis of symmetry: $x = 2$

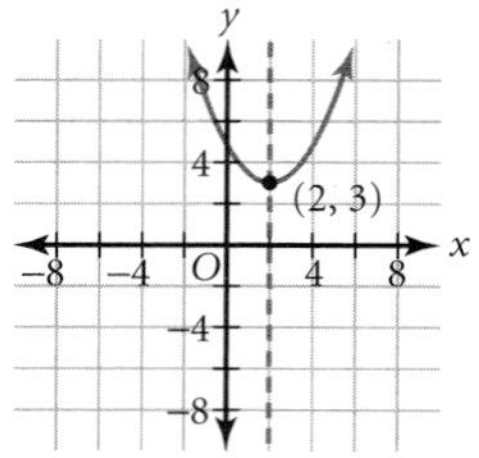


**23.** vertex: $(-2, -4)$; axis of symmetry: $x = -2$

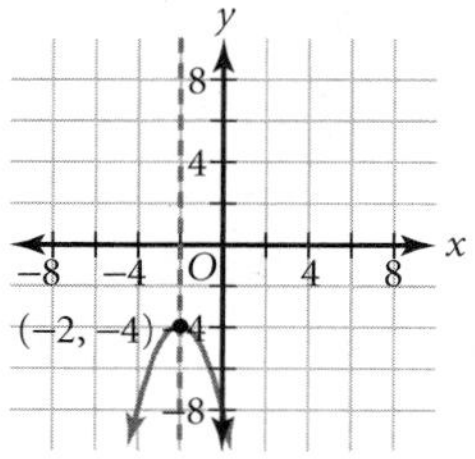


**25.** vertex: $(4, 5)$; axis of symmetry: $x = 4$

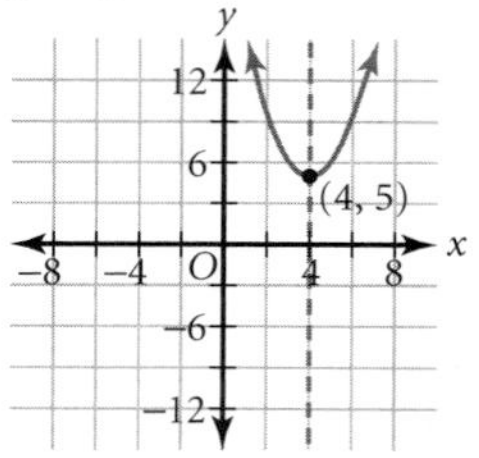


**27.** vertex: $(-5, -7)$; axis of symmetry: $x = -5$

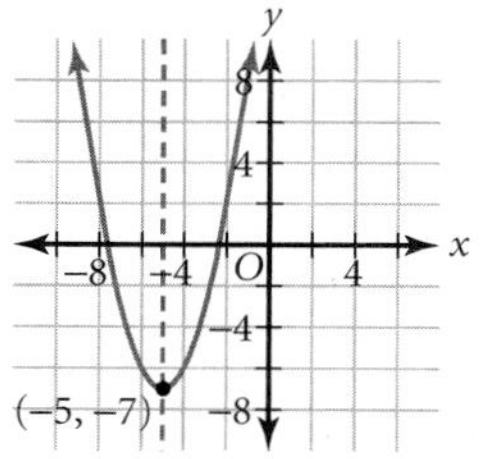


**29.** vertex: $(5, 2)$; axis of symmetry: $x = 5$

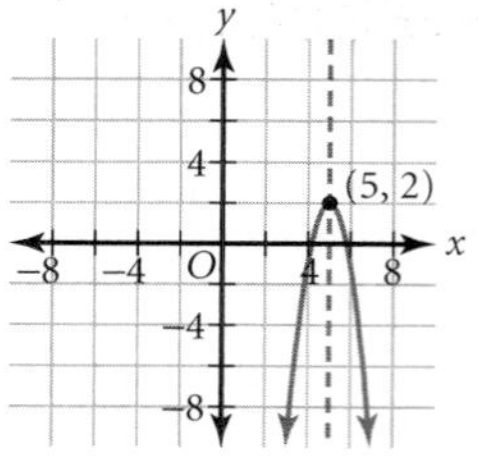


**31.** vertex: $(-5, 7)$; axis of symmetry: $x = -5$

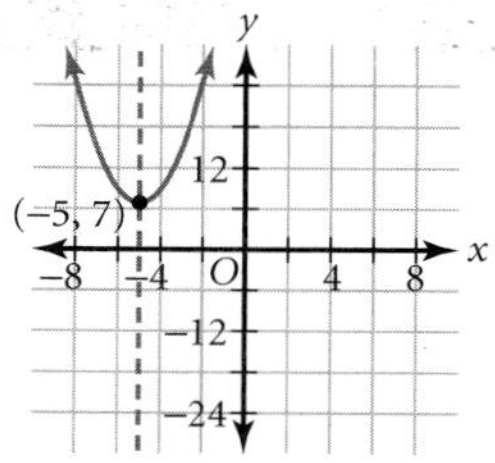


**33.** zero: 10 **35.** zeros: 1, −7 **37.** zeros: 4, −6 **39.** zeros: 7, −9 **41.** −3 and −8, or 3 and 8 **43.** 5 seconds **45.** $x = 5$ **47.** Subtraction inside parentheses, $(x - 3)$ **49.** Multiplication, $2(x - 3)^2$ **51.** $(6, 4)$ **53.** $(x + 5)^2$

## LESSON 10.2

**TRY THIS** (p. 487)

**a.** $x = \pm 10$

**b.** $x = \pm\frac{5}{2}$

**c.** $x = \pm\frac{4}{7}$

**TRY THIS** (p. 488)

**a.** $t = \frac{14}{5}$ or $-2$

**b.** $x = 13$ or $-5$

### Exercises

**6.** $\pm 8$ **7.** $\pm\frac{4}{5}$ **8.** $\approx \pm 4.12$ **9.** about 6.12 seconds **10.** about 7.07 seconds **11.** $x = 10$ or $-14$ **12.** $x = 23$ or $-17$ **13.** 11 **15.** 25 **17.** 6.63 **19.** 9.38 **21.** $x = \pm 5$ **23.** $x = \pm 9$ **25.** $x = \pm 3.46$ **27.** $x = \pm 4.24$ **29.** $x = \pm\frac{5}{9}$

**31.** $x = \pm\frac{6}{7}$ **33.** $x = \pm 5.66$ **35.** $x = \pm 5.29$ **37.** $x = 1$ or $x = -9$ **39.** $x = 0$ or $x = -2$ **41.** $x = -4.76$ or $x = -9.24$ **43.** $x = 3$ or $x = -9$ **45.** $x = 17$ or $x = -1$ **47.** $x = -1.84$ or $x = -8.16$

**49.** vertex: $(4, -3)$; axis of symmetry: $x = 4$ zeros: 5.73 and 2.27

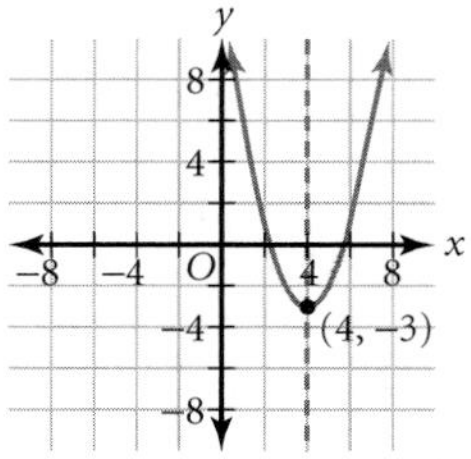


**51.** vertex: $(4, -2)$; axis of symmetry: $x = 4$ zeros: 2.59 and 5.41

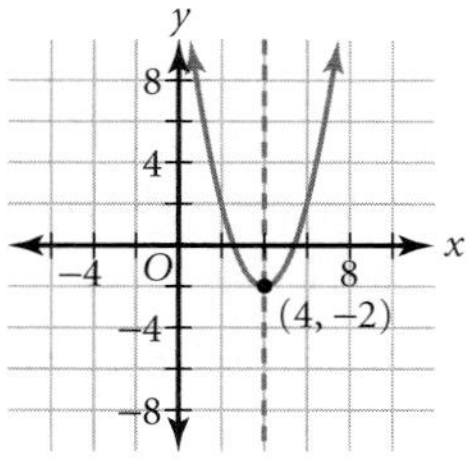


**53.** $x = -4$

**55.**

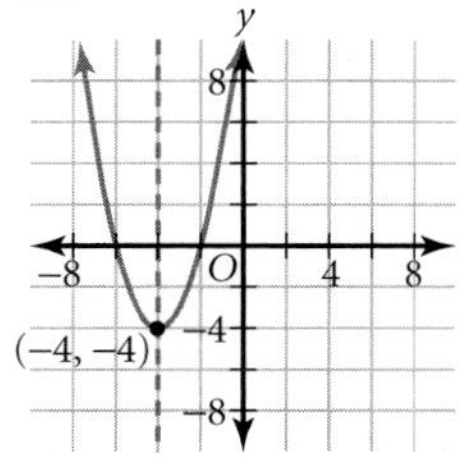


**57.** $r \approx 2.68$ meters **59.** $t = \sqrt{37.5} \approx 6.12$ seconds **61.** $-\frac{2}{3}$ **63.** $\frac{4}{5}$ **65.** 0 **67.** $-3xy^2$ **69.** $4a^4b^6p^8$ **71.** $-x^2 + 7x - 8$

## LESSON 10.3

**TRY THIS** (p. 494)

$\left(\frac{20}{2}\right)^2 = 100$

### Exercises

**6.** $90\frac{1}{4}$ **7.** $2\frac{1}{4}$ **8.** $y = (x + 3)^2 + 9$ **9.** $x^2 + 14x + 49$ **11.** $x^2 + 8x + 16$ **13.** $x^2 + 4x + 4$ **15.** $x^2 + 8x + 16$ **17.** $y = (x + 4)^2 - 16$ **19.** $y = (x - 5)^2 - 25$ **21.** $y = (x - 8)^2 - 64$ **23.** $x^2 + 6x + 9 = (x + 3)^2$ **25.** $x^2 + 12x + 36 = (x + 6)^2$

**27.** $x^2 + 7x + \frac{49}{4} = \left(x + \frac{7}{2}\right)^2$

**29.** $x^2 + 15x + \frac{225}{4} = \left(x + \frac{15}{2}\right)^2$

**31.** $x^2 + 16x + 64 = (x + 8)^2$

**33.** $x^2 - 9x + \frac{81}{4} = \left(x - \frac{9}{2}\right)^2$

**35.** $x^2 - 36x + 324 = (x - 18)^2$

**37.** $x^2 - 17x + \frac{289}{4} = \left(x - \frac{17}{2}\right)^2$

**39.** $x^2 - 13x + \frac{169}{4} = \left(x - \frac{13}{2}\right)^2$

**41.** $x^2 + 22x + 121 = (x + 11)^2$ **43.** $y = (x + 5)^2 - 25$ **45.** $-25$ **47.** $(0, -4)$ **49.** $(0, -1)$ **51.** $(0, -5)$ **53.** $y = (x + 3)^2 - 4$, $(-3, -4)$ **55.** $y = (x - 6)^2 - 1$, $(6, -1)$ **57.** $y = (x - 2)^2 - 5$, $(2, -5)$ **59.** $y = (x - 1)^2 - 4$, $(1, -4)$ **61.** $y = \left(x - \frac{1}{2}\right)^2 + \frac{7}{4}$, $\left(\frac{1}{2}, \frac{7}{4}\right)$ **63.** $y = \left(x + \frac{1}{6}\right)^2 - 3\frac{1}{36}$, $\left(-\frac{1}{6}, -3\frac{1}{36}\right)$

**65.**

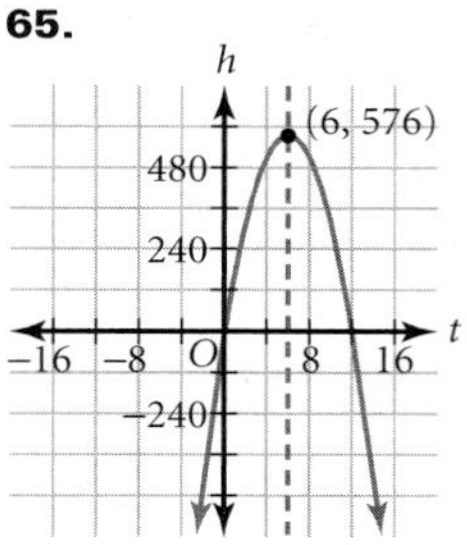


**67.** 576 feet

**69.**

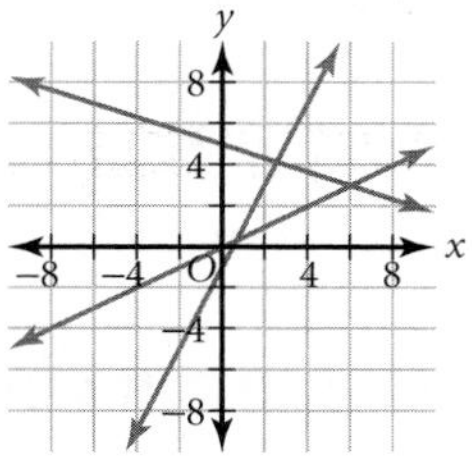


**71.** $(2, 1)$ **73.** $m^7$

## LESSON 10.4

**TRY THIS** (p. 499)

The zeros are $1 \pm \sqrt{3}$ and 5.

**TRY THIS** (p. 500)

$-3 + \sqrt{5} \approx -0.76$
$-3 - \sqrt{5} \approx -5.24$

**TRY THIS** (p. 501, Ex. 4)

The values of $x$ are 2 and 6.

**TRY THIS** (p. 501, Ex. 5)

The graphs intersect at $(-1, 6)$ and $(3, -2)$.

### Exercises

**6.** $-1$ and 8 **7.** $-2 \pm \sqrt{2}$ **8.** $4 \pm 2\sqrt{6}$ **9.** $-3$ or $-9$ **10.** $\approx 4.65$ or $\approx -0.65$ **11.** 6 or 2 **12.** 4 **13.** 7 or 1 **14.** $(6, 7)$ **15.** $(-2, -16)$ or $(7, 65)$ **17.** $-1$ or $-5$ **19.** $-1$ or 3 **21.** $-4$ or $-3$ **23.** $-2$ or 5 **25.** 3 **27.** 1 **29.** 1 or $-5$ **31.** 1.65 or $-3.65$ **33.** 2 or $-3$ **35.** 0.24 or $-4.24$ **37.** 0 or $-10$ **39.** 2.30 or $-1.30$ **41.** 1.19 or $-4.19$ **43.** 5.54 or $-0.54$ **45.** 2.58 or $-6.58$ **47.** $(-1, 4)$ and $(3, 4)$ **49.** $(0, 3)$ and $(5, 8)$ **51.** width = 10 yds, length = 14 yds **53.** width = 18 cm, length = 22 cm **55.** $\left(0, -\frac{1}{2}\right)$ **57.** 40,000 **59.** 0.000096 **61.** $6w^2(w - 1)$

## LESSON 10.5

**TRY THIS** (p. 507)

The solutions are $-6$ and 2.

### Exercises

**4.** $x = 8$ or $x = -1$ **5.** $x = -1$ **6.** $x \approx 0.72$ or $x \approx -1.72$ **7.** $x \approx 4.83$ or $x \approx -0.83$ **8.** no real solutions **9.** 0.41 or $-7.41$ **10.** 1 real solution **11.** no real solutions **12.** 2 real solutions

**13.** $2(3x-7)(x+1)$ **14.** $-3(x+1)(5x+2)$ **15.** $(7x-6)(2x+7)$ **17.** 36; 2 **19.** 100; 2 **21.** 64; 2 **23.** 2 or $-8$ **25.** 4 or 0 **27.** 3 or $-3$ **29.** 1 or $-\frac{1}{3}$ **31.** $2(x-5)(x-2)$ **33.** $4(x-1)(x+3)$ **35.** $(x-1)(2x-1)$ **37.** 2 or $-6$ **39.** $\frac{1}{2}$ or $-3$ **41.** 4 or $-\frac{3}{2}$ **43.** 27 **45.** $3.2 \times 10^7$ **47.** $6.54 \times 10^{-3}$

**49.**

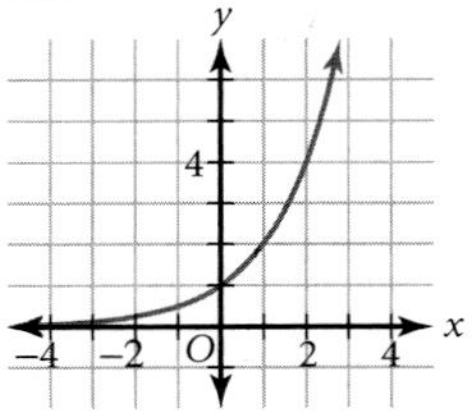

**51.**

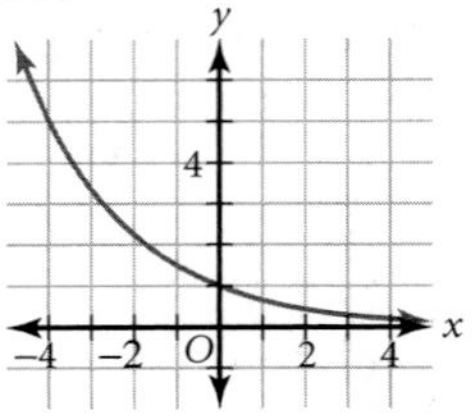

**53.** $4x^2 + 4x - 8$

## LESSON 10.6

**TRY THIS** (p. 512)
$x \leq -4$ or $x \geq 3$

**TRY THIS** (p. 513)

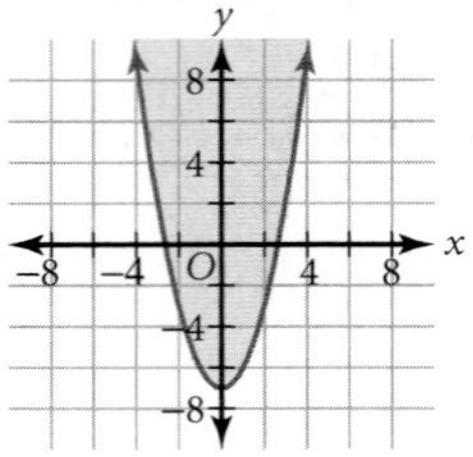

**Exercises**

**5.** $x > -3$ or $x < -3$ **6.** $-2 < x < 2$ **7.** $x < -4$ or $x > -3$

**8.**

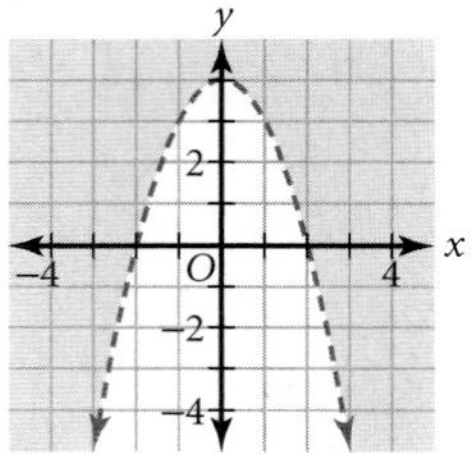

**9.**

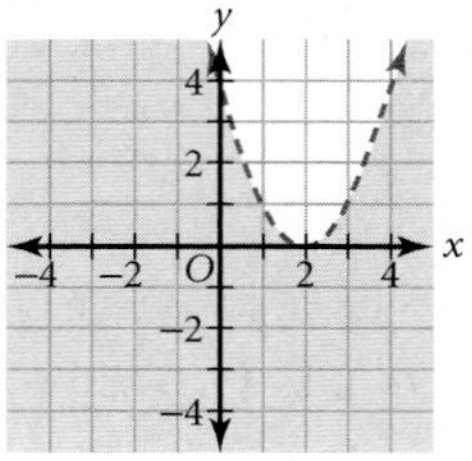

**10.**

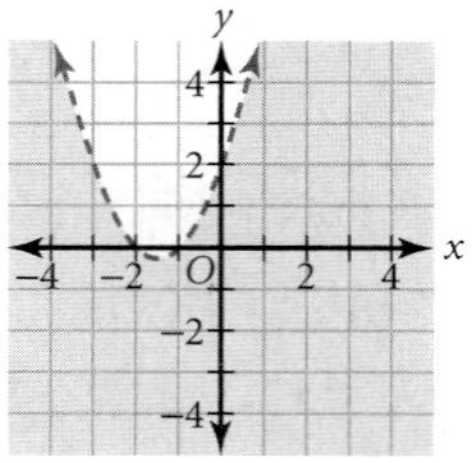

**11.** $x < 2$ or $x > 4$ **13.** $2 \leq x \leq 7$ **15.** $1 < x < 6$ **17.** $x < 3$ or $x > 5$ **19.** $x \leq 5$ or $x \geq 6$ **21.** $x < -7$ or $x > 2$

**23.**

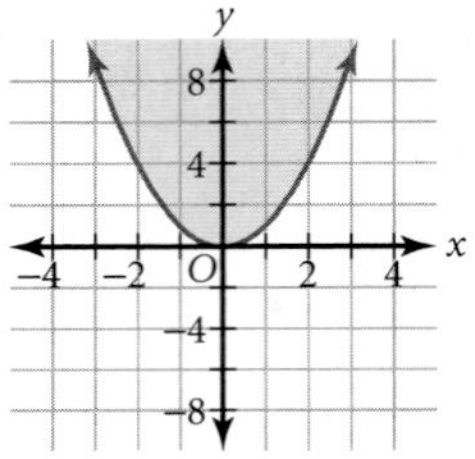

**25.**

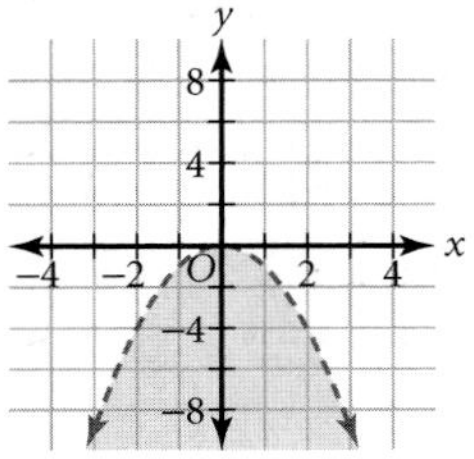

**27.**

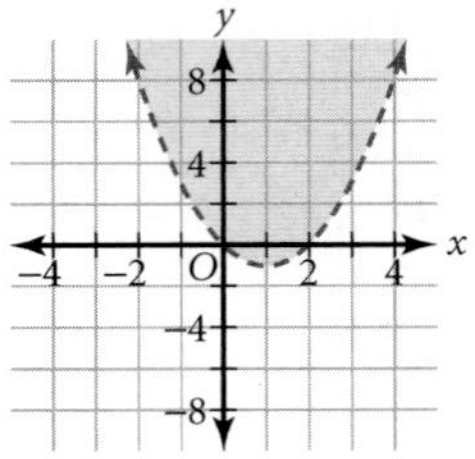

**29.**

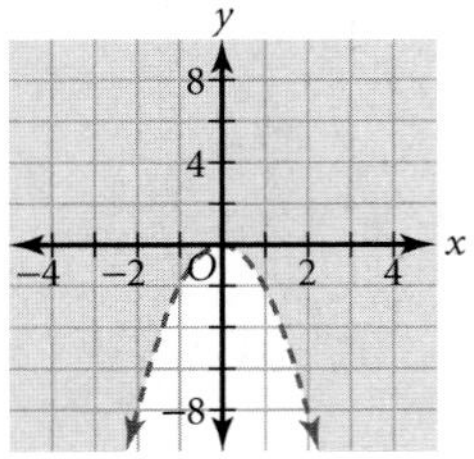

**31.**

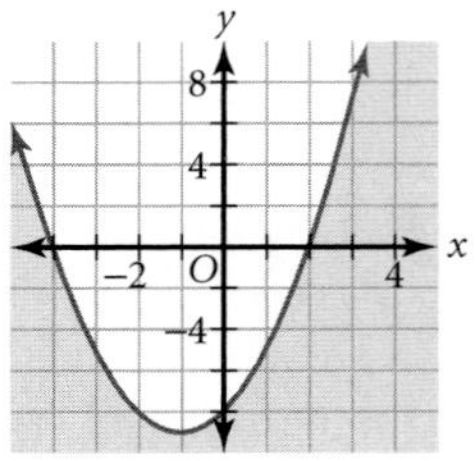

**33.**

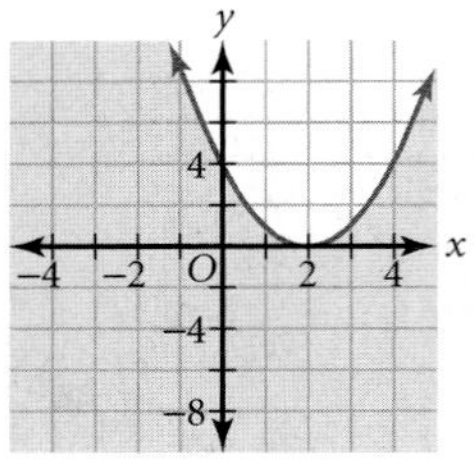

**35.**

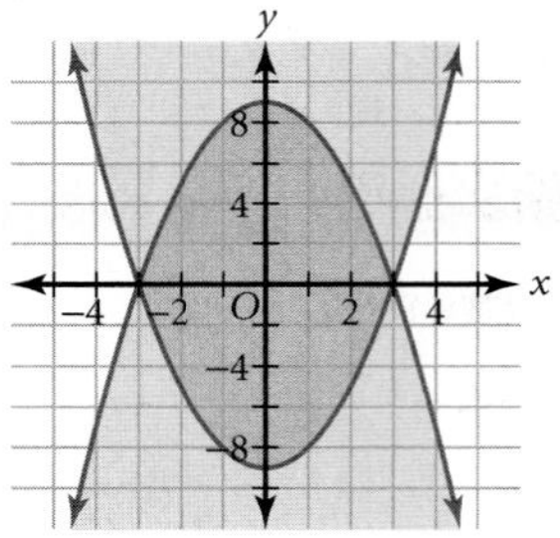

**37.** $4 < t < 16$ **39.** $-a^3b^3c^4$ **41.** $8.25 \times 10^{-1}$ **43.** $7.4 \times 10^{-6}$ **45.** $(2x+b)(a+3)$

## CHAPTER REVIEW AND ASSESSMENT

**1.** vertex: $(-1, -4)$; axis of symmetry: $x = -1$

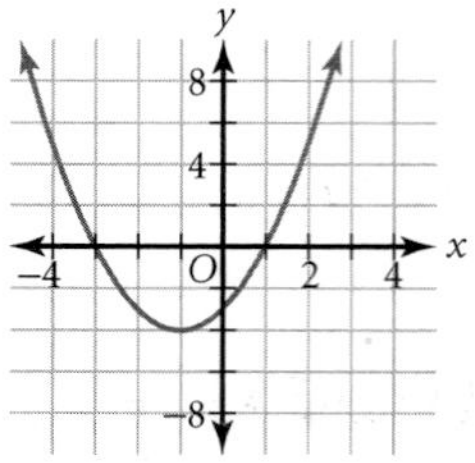

**3.** $x = 2$ **5.** $x = -4$ **7.** $x = -3$ or $x = 4$ **9.** $x = \pm 5$ **11.** $x = \pm 2$ **13.** $x = 1$ or $x = -3$ **15.** $x = 4.73$ or $x = 1.27$ **17.** 16 **19.** $y = (x-4)^2 + 2$, vertex: $(4, 2)$ **21.** $y = (x+6)^2 - 20$, vertex: $(-6, -20)$ **23.** $y = \left(x - \frac{5}{2}\right)^2 - \frac{17}{4}$, vertex: $\left(\frac{5}{2}, -\frac{17}{4}\right)$ **25.** $x = -2$ or $x = -3$ **27.** $x = -5$ or $x = 7$ **29.** 33; 2 solutions

**31.** $x \approx -2.16$ or $x \approx 4.16$ **33.** $\frac{1}{2}$ **35.** $x = \frac{1}{4}$ or $x = -\frac{1}{2}$

**37.** $-2 < x < 5$

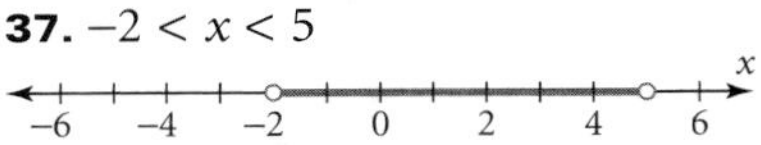

**39.** $-5 \leq x \leq 3$

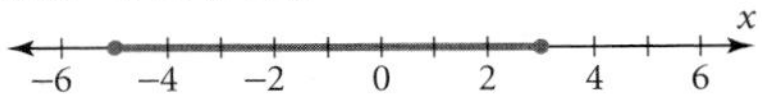

**41.**

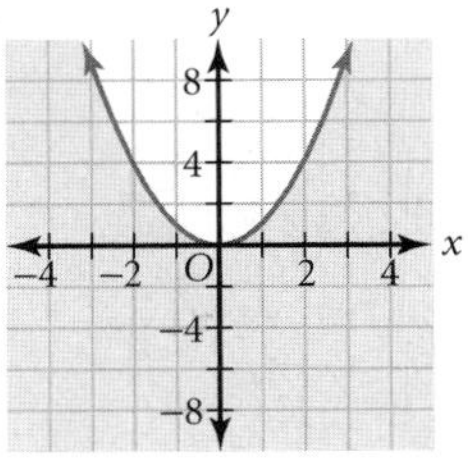

**43.**

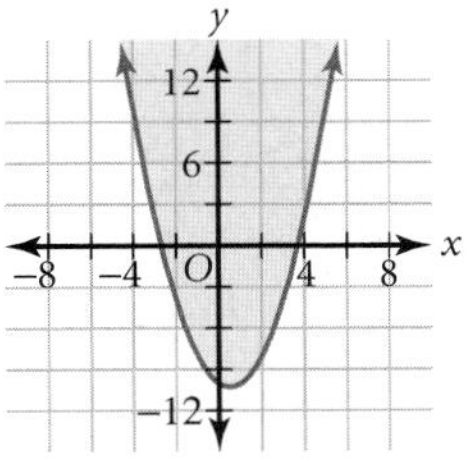

**45**

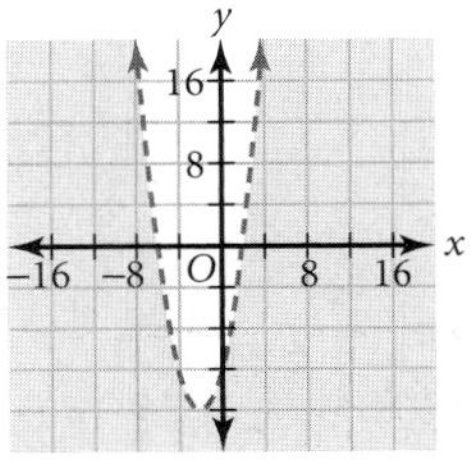

**47.** $\approx$ 15 yards **49.** 5 in.

# Chapter 11

## LESSON 11.1

**TRY THIS** (p. 527)

$y = \frac{48}{x}$; $k = 48$

**TRY THIS** (p. 528)

$pq = 60$

**TRY THIS** (p. 529)

It takes 7.2 years at 10% interest

### Exercises

**5.** $y = \frac{720}{x}$ **6.** $y = \frac{12}{x}$ **7.** $y = \frac{32}{x}$ **8.** $p \cdot q = 24$ is an inverse variation. **9.** $p \cdot q = 120$ is an inverse variation. **10.** $p + q = 12$ is not an inverse variation. **11.** $p \cdot q = 225$ is an inverse variation. **12.** 9 years **13.** 18 years **14.** 6 years **15.** Yes **17.** No **19.** No **21.** Yes **23.** 4 **25.** 24 **27.** −18 **29.** 33 cm **31.** 400 rpm **33.** 5 hr 20 min **35.** 300 m **37.** 22, 35, and 51 **39.** 70 discs

## LESSON 11.2

**TRY THIS** (p. 534)

**a.** All real numbers except −1

**b.** All real numbers except −1 and −2

**TRY THIS** (p. 535)

**a.** $y = \frac{83}{x} - 44.3$

**b.** $x \approx 19.15$

### Exercise

**4.** Domain: all real numbers except 3 **5.** Domain: all real numbers except 0 and 5 **6.** Domain: all real numbers except −4 **7.** $y = \frac{2}{x} - 3$ **8.** $y = \frac{5}{x} + 7$ **9.** $y = \frac{3}{x} - 5$ **11.** $y = 2$ **13.** $x = 1$ or $x = -5$ **15.** $x = 0$ **17.** $l = -\frac{5}{2}$ **19.** 2; −1 **21.** $\frac{1}{5}$; −1 **23.** Undefined; $4\frac{2}{3}$ **25.** $-3\frac{6}{7}$; −9 **27.** 4; 5.5 **29.** 0; undefined **31.** $-\frac{2}{5}$; $\frac{1}{8}$ **33.** $\frac{4}{3}$; undefined **35.** Undefined; $\frac{11}{9}$

**37.**

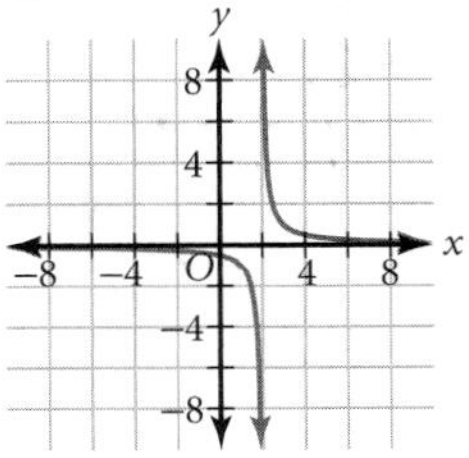

$x \neq 2$

**39.**

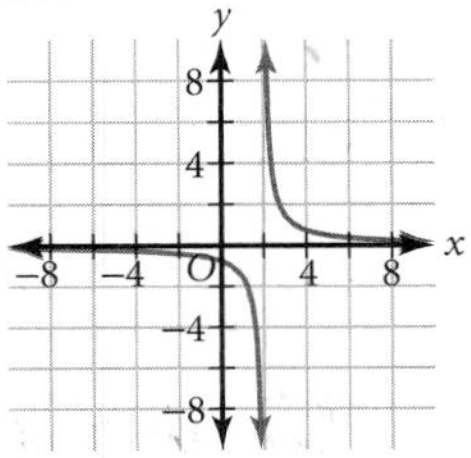

$x \neq 2$

**41.**

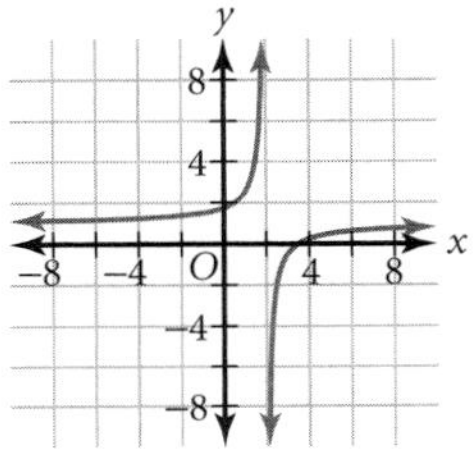

$x \neq 2$

**43.**

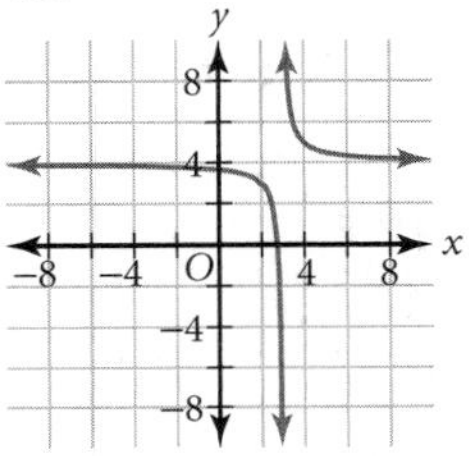

$x \neq 3$

**45.** $y = \frac{4}{x} + 6$ **47.** $y = \frac{1}{2x} - 5$ **49.** $y = \frac{9}{x} - 2$

**51.**

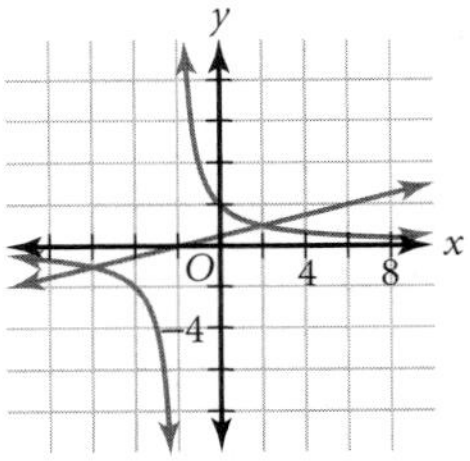

$y = \frac{4}{x+2}$ is undefined at $x = -2$.
Points of intersection: (−6, −1) and (2, 1).

**53.** 98 months **55.** $y - 1 = -\frac{4}{5}(x - 2)$ or $y - 5 = -\frac{4}{5}(x + 3)$ **57.** $y = (x + 7)^2 - 10$ **59.** $y = (x + 2)^2 + 1$ **61.** $y = (x + 3)^2 - 17$

## LESSON 11.3

**TRY THIS** (p. 539)

$\frac{2 - x}{x}$, $x \neq 0$

**TRY THIS** (p. 540, Ex. 2)

$\frac{1}{x + 3}$, $x \neq -2, -3$

**TRY THIS** (p. 540, Ex. 3)

$\frac{x - 5}{-x + 6}$, $x \neq 6, -6$

**TRY THIS** (p. 541)

$\frac{6}{s}$

**Exercises**

**6.** $2(b-3)$ **7.** $\frac{x+4}{2x}, x \neq 0$ **8.** $\frac{5n-3}{2}, n \neq 0$
**9.** $\frac{3}{2(x+2)}, x \neq -2$ **10.** $\frac{2a}{7(a+2)}, a \neq -2$ and $b \neq 0$
**11.** $2, x \neq 5$ **12.** $\frac{1}{a}, a \neq 0, 6$ **13.** $\frac{x-y}{x+y}, x \neq -y$
**14.** $\frac{x-1}{2}, x \neq 1$ **15.** $\frac{1}{x-3}, x \neq -4, 3$
**16.** $\frac{y-7}{y-3}, y \neq -7, 3$ **17.** $\frac{m}{m-3}, m \neq -4, 3$
**19.** $y = 5$ **21.** $p = 4$ **23.** $a = 3, -4$ **25.** $y = 0, 3, 2$
**27.** 3 **29.** $r + 3$ **31.** $\frac{a+b}{2(a-b)}, a \neq b$ **33.** $\frac{x-y}{2}, x \neq -y$
**35.** $7, t \neq -3$ **37.** $d, d \neq -\frac{1}{3}$ **39.** $\frac{1}{x+4}, x \neq 2, -4$
**41.** $\frac{-1}{k+3}, k \neq 4, -3$ **43.** $\frac{3}{n-3}, n \neq 3, 4$
**45.** $\frac{a-b}{a+b}, a \neq -b$ **47.** $\frac{3}{8}$ **49.** $y = -\frac{32}{9}x + \frac{10}{27}$
**51.** $y = -(x-2)^2 + 3$

## LESSON 11.4

**TRY THIS** (p. 548, Ex. 2)
$\frac{5}{2}, x \neq 0$

**TRY THIS** (p. 548, Ex. 3)
**a.** $\frac{x+5}{x-3}, x \neq 2, 3$
**b.** $\frac{1}{x+7}, x \neq -7, 2$

**TRY THIS** (p. 549, Ex. 4)
$\frac{3(x^2-2x-1)}{(x+3)(x-4)}, x \neq -3, 4$

**TRY THIS** (p. 549, Ex. 5)
**a.** $\frac{3x-2}{x^2-1}, x \neq -1, 1$
**b.** $\frac{4x-17}{8x}, x \neq 0$

**Exercises**

**6.** $\frac{x+3}{x-4}, x \neq 4, 5$ **7.** $\frac{k-4}{k+8}, k \neq -8, -7$
**8.** $\frac{3(k-4)}{(k+2)}, k \neq -2, 2$ **9.** $\frac{4(x+4)}{3(x+1)}, x \neq -1, 2$
**10.** $\frac{2(k+1)}{3(2k+1)}, k \neq -6, -\frac{1}{2}$ **11.** $\frac{y+3}{2y}, y \neq 0, 9$
**12.** $\frac{3x-2}{x^2-1}, x \neq -1, 1$ **13.** $\frac{5x+12}{x^2-9}, x \neq -3, 3$
**14.** $\frac{6k-10}{k^2-4k+4}, k \neq 2$ **15.** $\frac{4z+9}{z^2+6z+9}, z \neq -3$
**16.** $\frac{4y-12}{y^2-8y+16}, y \neq 4$ **17.** $\frac{7}{x-1}, x \neq 1$
**18.** $\frac{-5x}{x^2+x-6}, x \neq -3, 2$ **19.** $\frac{-9x}{x^2+x-20}, x \neq -5, 4$
**20.** $\frac{2a}{a^2-1}, a \neq -1, 1$ **21.** $\frac{7}{3x}, x \neq 0$ **23.** $\frac{2}{5}, y \neq -4$
**25.** $\frac{5a+2}{a^2+a}, a \neq -1, 0$ **27.** $\frac{11}{2r}, r \neq 0$ **29.** $\frac{-1}{2(x+1)}, x \neq -1$
**31.** $\frac{d-5}{d-1}, d \neq 1$ **33.** $\frac{5}{2+n}, n \neq -3, -2$
**35.** $\frac{q-1}{q}, q \neq 0, -1$ **37.** $\frac{15}{x^2-x}, x \neq 0, 1$
**39.** $\frac{x+3}{x^2+x}, x \neq 0, -1$ **41.** $\frac{m^2+2m-15}{m^2+m-2}, m \neq 1, -2$
**43.** $\frac{-5b-12}{b^2+b-6}, b \neq 2, -3$ **45.** $\frac{-3}{y+6}, y \neq 5, -6$
**47.** $\frac{7-x}{2x^2-18}, x \neq 3, -3$ **49.** $\frac{x+5}{3x}, x \neq 0, 7$

**51.**

| $x$ | $y = x$ | $y = (x-4)(x+4)$ |
|---|---|---|
| −2 | −2 | −12 |
| −1 | −1 | −15 |
| 0 | 0 | −16 |
| 1 | 1 | −15 |
| 2 | 2 | −12 |

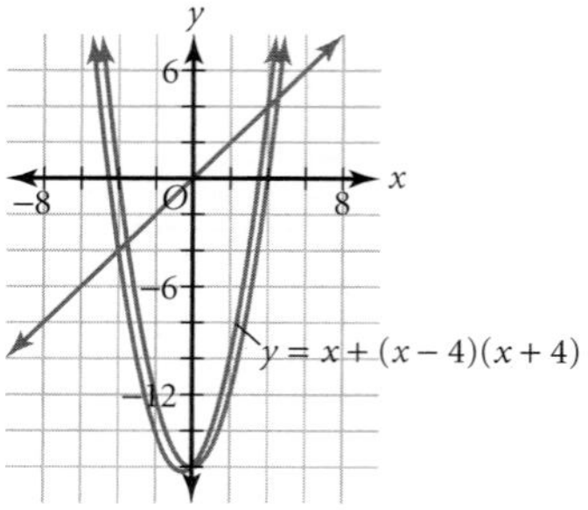


**53.**

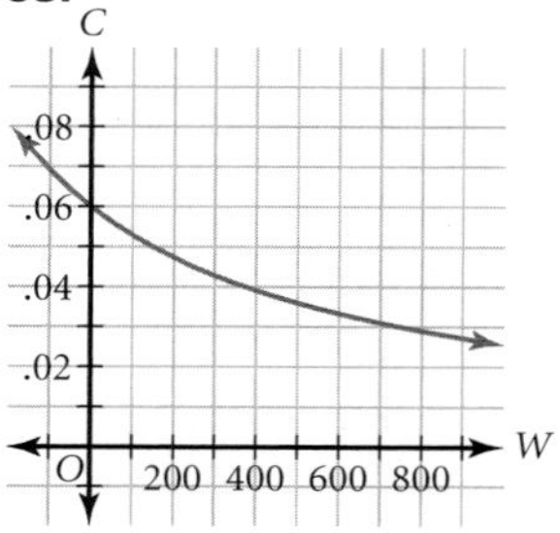


**55.** 39, 54, and 71

**57.**

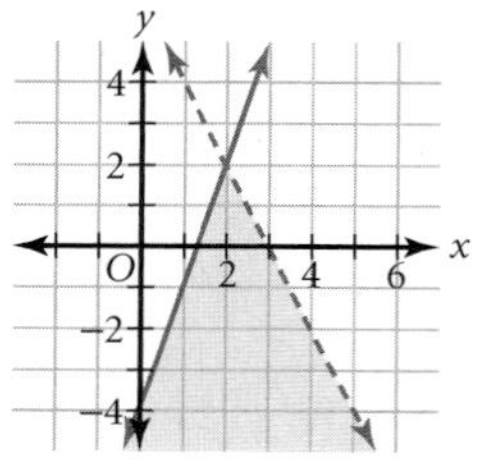


## LESSON 11.5

**TRY THIS** (p. 554)
$x \approx -5.34, 0.84$

**TRY THIS** (p. 555)
72 minutes

**Exercises**

**6.** $x = 5$ **7.** $x \approx 8.62$ or $x \approx -1.62$ **8.** There are no real solutions. **9.** 3.6 days **10.** $13\frac{1}{8}$ hours
**11.** $x = \frac{11}{12}$ **13.** $x = 3.2$ **15.** No solution
**17.** $y = \frac{1}{2}$ **19.** $x = 3$ **21.** $x = -1$ **23.** $x = 2$
**25.** No solution **27.** $x = -1, -\frac{1}{3}$ **29.** 20 and 36
**31.** downstream $= \frac{500}{20+c}$, upstream $= \frac{300}{20-c}$
**33.** Kayla's part is $\frac{x}{5}$ and Emily's part is $\frac{x}{7}$ **35.** $\frac{10}{7}$
**37.** (4, 1) **39.** (−1, 7)

## LESSON 11.6

**TRY THIS** (p. 561)

Since the product of two odds is odd, the product of three odds would be odd also. Since 100 is even, there is no combination of 3 odd integers whose product is 100.

**TRY THIS** (p. 562)

**a.** The statement is true only when $a = 0$ and false for all other values of $a$ and $b$.
**b.** The statement is sometimes true. It is true when $a \geq 0$ and $b \geq 0$. The statement is false when $a < 0$ and $b > 0$ or when $a > 0$ and $b < 0$.
**c.** This statement is always true.

**TRY THIS** (p. 563)

The integers are not closed under division. For example, $1 \div 2 = 0.5$, which is not an integer.

### Exercises

**4.**

| | |
|---|---|
| $\frac{2}{3} = \frac{x}{9}$ | Given |
| $(3 \cdot 9)\frac{2}{3} = (3 \cdot 9)\frac{x}{9}$ | Multiplicative Property of Equality |
| $18 = 3x$ | Simplified form |
| $\frac{18}{3} = \frac{3x}{3}$ | Division Property of Equality |
| $6 = x$ | Simplified form |

Therefore, if $\frac{2}{3} = \frac{x}{9}$, then $x = 6$.

**5.**

| | |
|---|---|
| $a$ is an even number | Given |
| $b$ is an odd number | Given |
| $a = 2k$ | Definition of even numbers |
| $b = 2p + 1$ | Definition of odd numbers |
| $a + b = 2k + 2p + 1$ | Addition of real numbers |
| $= 2(k + p) + 1$ | Distributive property |

$2(k + p) + 1$ is in the form $2n + 1$ where $n$ is equal to $k + p$, and $k + p$ is an integer. Thus $2(k + p) + 1$ is an odd number.

**6.**

| | |
|---|---|
| $k$ is an integer | Given |
| $k + 1$ is an integer | Definition of integer |
| $2(k + 1)$ is even | Definition of even numbers |
| $2k + 2$ is even | Distributive Property |
| $2(2k + 2)$ is even | Definition of even numbers |
| $4k + 4$ is even | Distributive Property |

Therefore, if $k$ is an integer, $4k + 4$ is an even integer.

**7.** Impossible because 100 is even. The product of two odds is odd. The product of four odds would be odd also. **8.** Sometimes true

**9.** Yes, since the product of two integers is always an integer, the set of integers is closed under multiplication. **11.** Inverse Property of Addition
**13.** Division Property of Equality
**15.** Identity Property of Multiplication
**17.** Commutative Property of Addition
**19.** Inverse Property of Addition

**21.**

| | |
|---|---|
| $y = 5x - 3x$ | Given |
| $= x(5 - 3)$ | Distributive Property |
| $= x(2)$ | Simplified form |
| $= 2x$ | Commutative Property of Multiplication |

**23.**

| | |
|---|---|
| $a = b$ | Given |
| $a + c = b + c$ | Addition Property of Equality |

**25.** $(a - b) + (b - a)$

| | |
|---|---|
| $= a + (-b + b) + (-a)$ | Associative Property for Addition |
| $= a + 0 + (-a)$ | Inverse Prop. Addition |
| $= a + (-a)$ | Identity for Addition |
| $= 0$ | Inverse Prop. Addition |

Therefore, by the definition of opposites, $a - b$ and $b - a$ are opposites.

**27.** The sum of 2 odd integers is even. The sum of that even integer and an odd integer is odd. Therefore, the sum of three odd integers must be odd, and cannot be 50. **29.** Sometimes true
**31.** Always true **33.** The set $\{-2, -1, 0, 1, 2\}$ is not closed under subtraction. For example, $(-2) - (2) = -4$, which is not in the set. **35.** $x = -3$
**37.** $y = -\frac{3}{2}x + 3$

## CHAPTER REVIEW AND ASSESSMENT

**1.** 60 **3.** 0.05 **5.** –4; 2 **7.** 2; undefined
**9.** $y = \frac{-2}{x} + 4, x \neq 0$

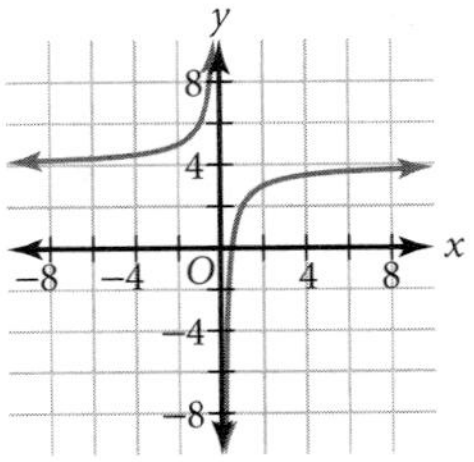


**11.** $y = \frac{2}{x-2} + 5, x \neq 2$

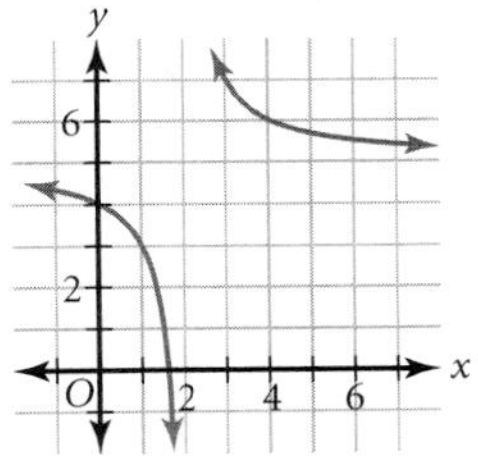

**13.** $\frac{4}{5(x+1)}, x \neq -1$ **15.** $\frac{2x+1}{-1-x}, x \neq -1, 1$ **17.** $\frac{4y+5x}{10xy}$, $x \neq 0, y \neq 0$ **19.** $\frac{-3t-9}{2}, t \neq 3$ **21.** $\frac{2(a-5)}{9}, a \neq -3$
**23.** $\frac{5r^2q^2}{18}, q \neq 0, r \neq 0$ **25.** $x+5, x \neq -4, 3, 5$
**27.** $z = \frac{8}{7}$ **29.** $w = -3$

**31.**

| | |
|---|---|
| $x = (a-b) - (b-a)$ | Given |
| $= a - b - b + a$ | Distributive Property |
| $= a + [-b + (-b)] + a$ | Definition of Subtraction |
| $= a + (-2b) + a$ | Simplified form |
| $= a + a + (-2b)$ | Commutative Property of Addition |
| $= 2a + (-2b)$ | Simplified form |
| $= 2a - 2b$ | Definition of Subtraction |

**33.**

| | |
|---|---|
| $n = (2x+y) - (x-3y)$ | Given |
| $= 2x + y - x + 3y$ | Distributive Property |
| $= 2x - x + y + 3y$ | Commutative Property of Addition |
| $= x + 4y$ | Simplified form |

**35.** $\frac{300+20x}{x}, x \neq 0$ **37.** 42 mph

# Chapter 12

## LESSON 12.1

**TRY THIS** (p. 577)
**a.** 14
**b.** −15

**TRY THIS** (p. 578)
**a.** $3\sqrt{6}$
**b.** $-a - \sqrt{b}$

**TRY THIS** (p. 579, Ex. 3)
$6|m|n^2\sqrt{2n}$

**TRY THIS** (p. 579, Ex. 4)
**a.** 28
**b.** $4\sqrt{2} - 4$

**TRY THIS** (p. 580)
**a.** $\frac{2}{5}$
**b.** $\frac{2\sqrt{3}}{3}$
**c.** $\frac{|b|\sqrt{ac}}{c}$

**Exercises**

**6.** 6 **7.** −8 **8.** ±9 **9.** −11 **10.** $2\sqrt{3}$ **11.** $13 - 2\sqrt{7}$
**12.** $4\sqrt{2}$ **13.** $|x|y^3\sqrt{y}$ **14.** $3|x^3|\sqrt{3}$ **15.** $a^3b\sqrt{19ab}$
**16.** 539 **17.** $2\sqrt{5}$ **18.** $7 - 3\sqrt{3}$ **19.** $\frac{3}{2}$ **20.** $\frac{\sqrt{6}}{7}$
**21.** $\frac{5\sqrt{2}}{2}$ **22.** $\frac{x^3|y^7|\sqrt{xz}}{z^2}$ **23.** 15 **25.** ±3.32 **27.** −6.32
**29.** 31.62 **31.** −0.2 **33.** 7 **35.** 24 **37.** $8\sqrt{3}$ **39.** $7\sqrt{2}$
**41.** $2\sqrt{66}$ **43.** false **45.** true **47.** 12 **49.** 18 **51.** $4\sqrt{3}$
**53.** 5 **55.** $14\sqrt{7}$ **57.** $2\sqrt{3}$ **59.** $x^4y^4\sqrt{y}$ **61.** $\frac{x\sqrt{x}}{|y^3|}$
**63.** $\sqrt{5}$ **65.** $5 + \sqrt{3} - \sqrt{2}$ **67.** $\sqrt{3} + \sqrt{2}$ **69.** 400
**71.** 48 **73.** $6\sqrt{5} - 5\sqrt{3}$ **75.** 1 **77.** $7\sqrt{3} + 12$
**79.** $21\sqrt{5} - 40$ **81.** 12 in. **83.** 26 ft **85.** $x^3y^4$
**87.** $4x^2 - 16x + 16$ **89.** $18b^2 - 3b - 1$

## LESSON 12.2

**TRY THIS** (p. 585, Ex. 1)
**a.** About 1.42 seconds
**b.** About 620.60 centimeters

**TRY THIS** (p. 585, Ex. 2)
$x = 9.5$

**TRY THIS** (p. 586)
$x = 4$

**TRY THIS** (p. 587, Ex. 4)
**a.** $x = \pm 10\sqrt{2}$
**b.** $x = \pm 13$
**c.** $x = \pm\sqrt{y^2 + z^3}$

**TRY THIS** (p. 587, Ex. 6)
$x = 2$ or $x = -12$

**Exercises**

**5.** About 1.55 seconds **6.** About 2.46 seconds **7.** About 2.01 seconds **8.** About 1.35 seconds **9.** About 99.29 centimeters **10.** About 155.15 centimeters **11.** About 304.09 centimeters **12.** About 397.18 centimeters **13.** $x = 11$ **14.** $x = -2$ **15.** $x = 5$ **16.** $x = 4$ **17.** No real number solutions **18.** 2 **19.** $x = 1$ or $x = 2$ **20.** $x = 6$ **21.** $x = \pm 20\sqrt{2}$ **22.** $x = \pm 5\sqrt{10}$ **23.** $x = \pm 10$ **24.** $x = \pm\sqrt{v^2 - t^2}$ **25.** About 2.39 yards **26.** $x = 9$ or $x = -1$ **27.** $x = -2$ or $x = -6$ **28.** $x = -6$ or $x = 2$ **29.** $x = 9$ **31.** $x = 18$ **33.** $x = 20$ **35.** $x = 2$ **37.** $x = 3$ or $x = 2$ **39.** $x = 1$ **41.** $x = 2$ **43.** $x = 2$ **45.** $x = \pm\frac{\sqrt{7}}{2}$ **47.** $x = \pm 2\sqrt{6}$ **49.** $x = 6$ **51.** $x = 81$ **53.** Cannot solve: it is not possible to square a number and get a negative result. **55.** No solution

**57. a.**

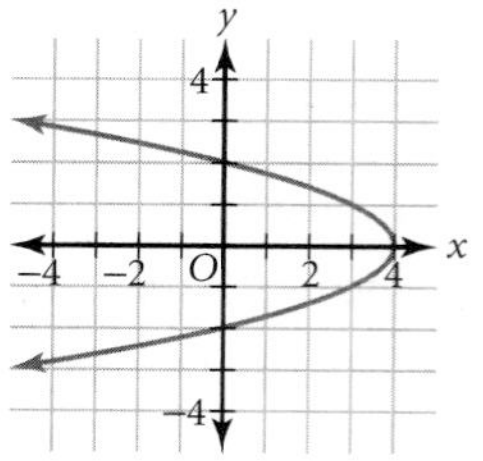


Domain: $x \leq 4$
Range: all real numbers

**b.**

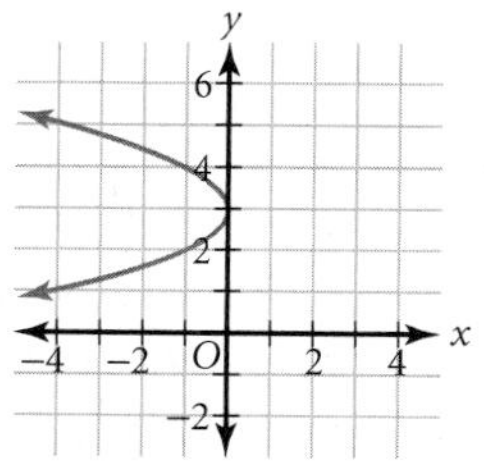


Domain: $x \leq 0$
Range: all real numbers

**59.** If the time required for 1 complete swing is doubled, the length of the pendulum is multiplied by 4. **61. a.** $\frac{\sqrt{2Em}}{m}$ **b.** about 26.73 meters per second
**63.** $4(x^2 - 2y^2)(x^2 + 2y^2)$
**65.** $x > 4$ or $x < -3$

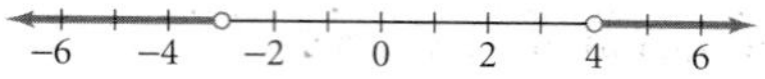


**67.** $-2 < a < 8$

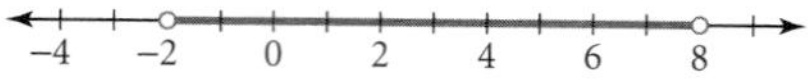

## LESSON 12.3

**TRY THIS** (p. 593)

**a.** $x = 13$
**b.** $y = 25$
**c.** $z = 2\sqrt{2}$

**TRY THIS** (p. 594, Ex. 2)

60 units

**TRY THIS** (p. 594, Ex. 3)

about 57.2 meters

### Exercises

**6.** $x = 5$ **7.** $y = 20$ **8.** $z = 6\sqrt{2}$ **9.** $2\sqrt{34}$ units **10.** 20 meters **11.** hypotenuse $= c = \overline{AB}$, leg $= a = \overline{CB}$, leg $= b = \overline{AC}$ **13.** hypotenuse $= r = \overline{ST}$, leg $= s = \overline{RT}$, leg $= t = \overline{RS}$ **15.** 24 **17.** 10.3 **19.** 24.3 **21.** 1 **23.** yes **25.** yes **27.** no **29.** About 9.8 feet **31.** \$169.32

**33.**

| Lamps Sold | 0 | 10 | 20 | 30 | 40 | 50 |
|---|---|---|---|---|---|---|
| Week's Pay | \$20 | \$45 | \$70 | \$95 | \$120 | \$145 |

**35.** \$92.50 **37.** neither **39.** parallel

## LESSON 12.4

**TRY THIS** (p. 599)

$d \approx 9.22$

**TRY THIS** (p. 600)

yes: $(\sqrt{52})^2 + (\sqrt{13})^2 = (\sqrt{65})^2$

**TRY THIS** (p. 602)

$B(-6, -7)$

### Exercises

**6.** 5 **7.** 13 **8.** 10 **9.** The triangle is a right triangle. **10.** The triangle is a right triangle. **11.** The triangle is not a right triangle. **12.** The triangle is a right triangle. **13.** $(1, 3)$ **14.** $\left(\frac{1}{2}, -4\right)$ **15.** $(13, 0)$ **16.** $\left(15\frac{1}{2}, -\frac{1}{2}\right)$ **17.** $(1, 1)$ **18.** $(-16, -12)$ **19.** $(-1, 5)$ **20.** $(13, -2)$ **21.** 5 **23.** 6.40 **25.** 7 **27.** 8.54 **29.** 9.06 **31.** 4.47 **33.** 16 **35.** 11.70 **37.** $\triangle PQR$ is a scalene right triangle. **39.** $\triangle PQR$ is an isosceles right triangle. $\overline{PQ}$ and $\overline{PR}$ are $\sqrt{17}$ units long. **41.** $\triangle PQR$ is scalene. **43.** $(-3, -4)$ **45.** $(6.5, -0.5)$ **47.** $\left(-1, -\frac{5}{2}\right)$ **49.** $\left(\frac{a+c}{2}, \frac{b+d}{2}\right)$ **51.** $(16, 6)$ **53.** $P(3, 11)$ and $Q(1, 5)$ **55.** Answers may vary. **57.** $Q(7, 2)$; 7.81 **59.** About 12.04 mi **61.** $(x + 12)(x + 18)$ **63.** $(-3, -1)$ **65.** $(-4, 1)$ **67.** $-9$ **69.** 5

## LESSON 12.5

**TRY THIS** (p. 607)

**a.** $x^2 + y^2 = 1$
**b.** $(x + 2)^2 + (y - 5)^2 = 25$

**TRY THIS** (p. 609)

$y = \frac{1}{6}x^2 + \frac{3}{2}$

### Exercises

**5. a.** $x^2 + y^2 = 9$ **b.** $(x + 3)^2 + (y - 4)^2 = 16$ **6.** midpoint of hypotenuse: $C(1.5, 2)$; midpoint of vertical leg: $D(3, 2)$; slope of midsegment $\overline{CD} = 0$, slope of base $\overline{OB} = 0$. The slope is 0 in each case, therefore the lines are parallel. length of midsegment $\overline{CD} = 1.5$, length of base $\overline{OB} = 3$ The length of the midsegment is one-half the length of the third side. **7.** $y = \frac{1}{6}x^2 - \frac{2}{3}x + \frac{13}{6}$ **9.** $x^2 + y^2 = 1.44$ **11.** $x^2 + y^2 = 75$ **13.** $(x + 1)^2 + (y - 4)^2 = 16$ **15.** $(x + 3)^2 + (y + 6)^2 = 16$ **17.** center: $(-1, -3)$, radius: 3 **19.** center: $(-6, -4)$, radius: 10 **21.** center: $(-1, -4)$, radius: $\frac{2}{5}$ **23.** $x^2 + y^2 = 25$ **25.** $(x - 2)^2 + (y - 2)^2 = 4$ **27.** $(x - 5)^2 + (y - 4)^2 = 29$ **29.** $M = (3, 4)$ **31.** $MN = 2\sqrt{5}$ **33.** $MN = \frac{1}{2}QR$ or $QR = 2MN$. **35.** $\overline{MN}$ is parallel to $\overline{QR}$. **37.** $\sqrt{x^2 + (y + 4)^2} + \sqrt{x^2 + (y - 4)^2} = 10$ **39.** The diagonals of a rectangle bisect each other at their midpoints. **41.** The two diagonals of a square are perpendicular. **43.** no **45.** yes **47.** yes **49.** yes **51.** $1.25 \times 10^{-6}$ **53.** $6b^2 - b - 16$ **55.** 25

**57.**

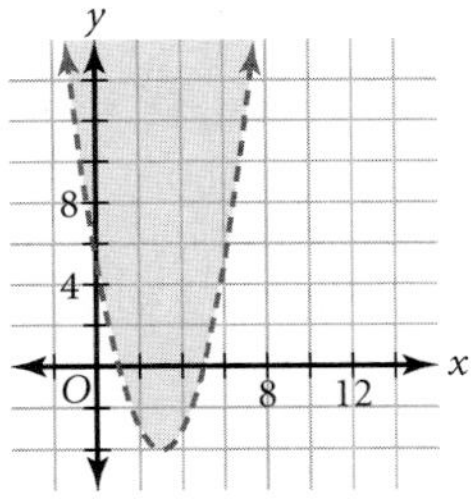


## LESSON 12.6

**TRY THIS** (p. 616)

**a.** $\tan N = \frac{24}{7}$, $\tan O = \frac{7}{24}$
$N \approx 74°$ $O \approx 16°$

**b.** $\tan P = \frac{16}{12}$, $\tan R = \frac{12}{16}$
$P \approx 53°$ $R \approx 37°$

**TRY THIS** (p. 617, Ex. 2)

The angle of the ray of sunlight with the ground is about 37°.

**TRY THIS** (p. 617, Ex. 3)

The wolf is about 2860 feet from the tower.

### Exercises

**4. a.** $\tan A = \frac{11}{60}$; $\tan B = \frac{60}{11}$, $A \approx 10°$, $B \approx 80°$
**b.** $\tan D = \frac{48}{20}$; $\tan E = \frac{20}{48}$, $D \approx 67°$, $E \approx 23°$

**5.** The angle of the ray of sunlight with the ground is about 68°. **6.** The wolf is about 2854 feet from the tower. **7.** 1 **9.** 1.7321 **11.** 0.7813 **13.** 57.2900 **15.** 60° **17.** 15° **19.** $a$ is about 6.93 m **21.** $b$ is 4 cm **23.** $b$ is about 9.99 in. **25.** ≈ 5.1 m **27.** ≈ 18°, ≈ 28°, ≈ 63°, ≈ 73°, ≈ 86°, ≈ 81°29. The angle that the road inclines is about 4°. **31.** $\frac{14}{3} = x$ **33.** $x \geq 3$ or $x \leq 2$ **35.** (7, 7) **37.** $PM = \sqrt{13}$

## LESSON 12.7

**TRY THIS** (p. 622)

$\cos 65° \approx 0.423$, $\sin 25° \approx 0.423$

**TRY THIS** (p. 623, Ex. 1)

57.9

**TRY THIS** (p. 623, Ex. 2)

41.40°

### Exercises

**5.** $s \approx 37.6$ **6.** $X \approx 36.9°$ **7.** The angle of the ride is about 51° which is greater than the maximum allowable angle of 45°. **9.** 0.8480 **11.** 0.3090 **13.** 0.1011 **15.** 0.7501 **17.** 63.1° **19.** 49.4° **21.** 45.0° **23.** 26.4° **25.** 76.4° **27.** 90° **29.** 60.0° **31.** 45.0° **33.** 35.9° **35.** 38.7° **37.** 50.0° **39.** 0.0° **41.** sin; $x \approx 41.8°$ **43.** cos; $x \approx 14.7$ **45.** sin; $x \approx 66.3$ **47.** $A \approx 62.5°$ **49.** $a \approx 13.8$ **51.** $B \approx 41.4°$ **53.** $c = 22$ **55.** $b \approx 2.8$ **57.** The ramp should be about 229.4 inches long. **59.** The base of the ladder is about 3.1 feet from the building. **61.** About 4.34 minutes or 260 seconds **63.** $x = -12$ or $x = 12$ **65.** $x \approx 4.32$ or $x \approx -2.32$ **67.** $x^2 - 6x + 9 = (x - 3)^2$ **69.** $x^2 - 7x + \frac{49}{4} = \left(x - \frac{7}{2}\right)^2$ **71.** $h = -5$ or $h = -1$ **73.** $x \approx 6.5$ **75.** $x \approx 63.5°$

## LESSON 12.8

**TRY THIS** (p. 630)

$$\begin{bmatrix} 8.82 \\ 14.05 \end{bmatrix}$$

**TRY THIS** (p. 632, Ex. 5)

$AB$; not possible; $A$ has 2 columns and $B$ has 1 row. $B \times A = [6\ \ 14]$

**TRY THIS** (p. 632, Ex. 6)

$$\begin{bmatrix} 1 & 0 & 0 & 0 \\ 0 & 1 & 0 & 0 \\ 0 & 0 & 1 & 0 \\ 0 & 0 & 0 & 1 \end{bmatrix}$$

### Exercises

**4.** $H$ and $J$ **5.** $\begin{bmatrix} 25 & 13 \\ -7 & -11 \end{bmatrix}$ **6.** $\begin{bmatrix} -9 & 25 \\ -5 & 1 \end{bmatrix}$

**7.** $\begin{bmatrix} 3 & -4 & 9 \\ -30 & 1 & 1 \\ -3 & 19 & -17 \end{bmatrix}$ **8.** $\begin{bmatrix} 30 & -20 \\ -15 & 25 \end{bmatrix}$

**9.** $\begin{bmatrix} -6 & -18 \\ 12 & 8 \end{bmatrix}$ **10.** $\begin{bmatrix} 6 & -15 & 3 \\ 36 & -9 & -24 \\ 42 & 30 & 12 \end{bmatrix}$

**11.** $\begin{bmatrix} -30 & 32 \\ -189 & 98 \end{bmatrix}$ **12.** $\begin{bmatrix} -33 & 69 \\ -30 & -48 \end{bmatrix}$

**13.** does not exist **14.** does not exist

**15.** $\begin{bmatrix} 1 & 0 \\ 0 & 1 \end{bmatrix}$ **16.** $\begin{bmatrix} 1 & 0 & 0 \\ 0 & 1 & 0 \\ 0 & 0 & 1 \end{bmatrix}$ **17.** $\begin{bmatrix} -3 & 3 \\ -10 & 9 \\ -5 & 0 \end{bmatrix}$

**19.** $\begin{bmatrix} 16 & 145 & 133 \\ -23 & -75 & -115 \\ -84 & 220 & 99 \end{bmatrix}$ **21.** $\begin{bmatrix} -2 \\ 0 \\ 2 \end{bmatrix}$

**23.** $\begin{bmatrix} -6 & -3 \\ -1 & -8 \end{bmatrix}$ **25.** $\begin{bmatrix} 8 & -1 \\ 4 & 6 \\ 10 & -1 \end{bmatrix}$

**27.** $\begin{bmatrix} -15 & -51 & 9 \\ -36 & 24 & -6 \\ 21 & -42 & -36 \end{bmatrix}$ **29.** $\begin{bmatrix} -12 & -86 \\ 68 & -24 \end{bmatrix}$

**31.** $\begin{bmatrix} -72 & -50 & 43 \\ -28 & -18 & 15 \\ 224 & 40 & 4 \end{bmatrix}$ **33.** $\begin{bmatrix} 102 \\ 87 \\ -53 \end{bmatrix}$

**35.** $a = -10$, $b = 7$, $c = 15$, $d = \frac{1}{3}$, $g = 75$, $k = \frac{12}{5}$

**37.**

| | Mu | Ar | Te | He |
|---|---|---|---|---|
| Boys | 834 | 475 | 1003 | 756 |
| Girls | 1003 | 744 | 819 | 836 |

| | | | | |
|---|---|---|---|---|
| Total Enrollment | 1837 | 1219 | 1822 | 1592 |

**39.**

| | |
|---|---|
| Alesia | 51 |
| Jennel | 59 |
| Betsy | 56 |
| Katie | 33 |
| Nancy | 46 |

| | |
|---|---|
| Alesia | 29 |
| Jennel | 11 |
| Betsy | 14 |
| Katie | 18 |
| Nancy | 10 |

**41.** 327 points **43.** −4 **45.** −11 **47.** a **49.** $x = -7$ **51.** $m = 78$ **53.** $\frac{1}{19}$ **55.** $x = 7$ or $x = -5$ **57.** $x = -3 \pm \sqrt{2}$

## CHAPTER REVIEW AND ASSESSMENT

**1.** 4.47 **3.** $|a|b^3\sqrt{b}$ **5.** 12 **7.** −31 **9.** $x = 11$ **11.** no solution **13.** $x = \pm 2\sqrt{10}$ **15.** $x = \pm 5$ **17.** $c = \sqrt{218}$ **19.** $c = 13$ **21.** A triangle with side lengths $\sqrt{5}$ is a right triangle because it satisfies the Pythagorean Theorem: $(2)^2 + (\sqrt{5})^2 = (3)^2$. **23.** $XY \approx 13.60$ **25.** $ST = 5$ **27.** (−5, 1.5) **29.** $\left(\frac{1}{2}, \frac{1}{2}\right)$ **31.** $\left(-\frac{5}{2}, -\frac{1}{2}\right)$ **33.** $\left(2a, \frac{3}{2}b\right)$ **35.** $(x + 2)^2 + (y - 5)^2 = 25$

**37.** $(x+3)^2+(y+2)^2=36$ **39.** Center: $(-0.25, 0)$, radius: 2.5 **41.** $\frac{5}{12}$ **43.** $\frac{8}{6}$ (or $\frac{4}{3}$) **45.** $\frac{6}{8}$ (or $\frac{3}{4}$) **47.** $\frac{5}{5}$ (or 1) **49.** $\frac{1}{2}=0.5$ **51.** $\frac{8}{15}\approx 0.53$ **53.** $\frac{3}{5}=0.6$ **55.** $\frac{8}{15}\approx 0.53$ **57.** $\approx 58.77°$ **59.** $= 26.74°$

**61.** $\begin{bmatrix}-2 & 15\\ 0 & -2\end{bmatrix}$ **63.** $\begin{bmatrix}7 & 11\\ 4 & 30\end{bmatrix}$ **65.** $\begin{bmatrix}-12 & 48\\ 96 & -32\end{bmatrix}$

**67.** $\begin{bmatrix}322 & 46\\ 299 & -138\end{bmatrix}$ **69.** $\begin{bmatrix}30 & -24\\ 49 & 70\end{bmatrix}$

**71.** $\begin{bmatrix}8 & 9\\ 12 & 356\end{bmatrix}$ **73.** $\begin{bmatrix}845 & -765\\ -156 & -943\end{bmatrix}$

**75. a.** The cable is about 514.2 ft long **b.** The cable makes an angle of about 13.5° with the ground.

# Chapter 13

## LESSON 13.1

**TRY THIS** (p. 649, Ex. 1)

There are eight different possible results from tossing a coin three times.

| First Toss | Second Toss | Third Toss |
|---|---|---|
| H | H | H |
| H | H | T |
| H | T | H |
| H | T | T |
| T | H | H |
| T | H | T |
| T | T | H |
| T | T | T |

You can write these in set notation as {HHH, HHT, HTH, HTT, THH, THT, TTH, TTT}.

**TRY THIS** (p. 649, Ex. 2)

The sample space for the experiment is {HHH, HHT, HTH, HTT, THH, THT, TTH, TTT}. The outcomes in which two or more of the tosses are heads are the favorable outcomes. They are {HHH, HHT, HTH, THH}. There are four favorable outcomes.

**TRY THIS** (p. 650, Ex. 3)

The sample space is {HHH, HHT, HTH, HTT, THH, THT, TTH, TTT}. The outcomes in which one or two of the tosses will be heads are {HHT, HTH, HTT, THH, THT TTH}.
$P(1 \text{ or } 2 \text{ heads}) = \frac{6}{8} = \frac{3}{4}$.

**TRY THIS** (p. 650, Ex. 4)

$\frac{17}{71} > \frac{19}{91}$, so 17 chances out of 71 has a greater probability.

**Exercises**

**4.** $\{1-1, 1-2, 1-3, 1-4, 1-5, 1-6, 2-1, 2-2, 2-3, 2-4, 2-5, 2-6, 3-1, 3-2, 3-3, 3-4, 3-5, 3-6, 4-1, 4-2, 4-3, 4-4, 4-5, 4-6, 5-1, 5-2, 5-3, 5-4, 5-5, 5-6, 6-1, 6-2, 6-3, 6-4, 6-5, 6-6\}$

**5.**

| First Toss | Second Toss | Third Toss | Fourth Toss |
|---|---|---|---|
| H | H | H | H |
| H | H | H | T |
| H | H | T | H |
| H | H | T | T |
| H | T | H | H |
| H | T | H | T |
| H | T | T | H |
| H | T | T | T |
| T | H | H | H |
| T | H | H | T |
| T | H | T | H |
| T | H | T | T |
| T | T | H | H |
| T | T | H | T |
| T | T | T | H |
| T | T | T | T |

**6.** 6 favorable outcomes **7.** 1 favorable outcome **8.** $\frac{1}{6}$ **9.** $\frac{1}{8}$ **10.** 18 chances out of 74 **11.** 34 chances out of 71 **13.** $\frac{13}{25}$ **15.** $\frac{1}{5}$ **17.** $\frac{1}{2}$ **19.** $\frac{1}{2}$ **21.** $\frac{5}{26}$ **23.** $\frac{4}{13}$ **25.** $\frac{4}{25}$ **27.** $\frac{1}{2}$ **29.** $\frac{1}{16}$ **31.** $\frac{3.18}{53.00}=\frac{x}{100}$; 6% **33.** 21

**35.**

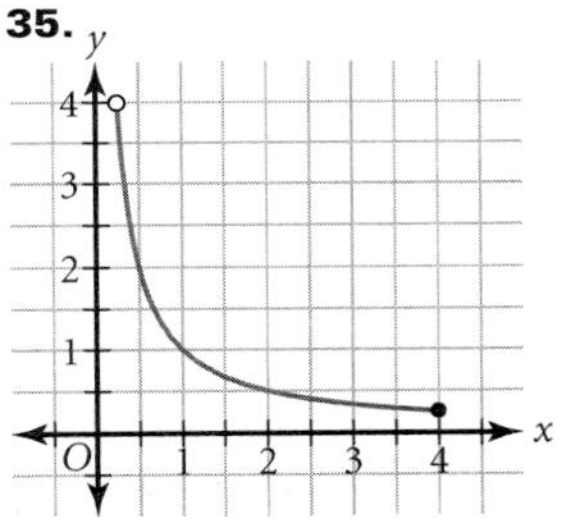


$\frac{1}{4} \le y < 4$

## LESSON 13.2

**TRY THIS** (p. 657, Ex. 1)

13 cards, 4 are aces, but 1 is both and mustn't be counted twice. 16 cards are spades or aces.

**TRY THIS** (p. 657, Ex. 2)

$P(\text{spade OR club}) = \frac{26}{52} = \frac{1}{2}$

**TRY THIS** (p. 658)

$P(\text{odd-numbered spade OR ace}) = \frac{5+3}{52} \approx 0.15$

**Exercises**

**8.** 16 **9.** 26 **10.** 8 **11.** $\frac{4}{13}$ **12.** $\frac{7}{13}$ **13.** $\frac{2}{13}$ **14.** $\frac{9}{13}$ **15.** $\frac{8}{13}$ **16.** $\frac{4}{13}$ **17.** {2, 4, 6, 8, 10} **19.** {6} **21.** 13 **23.** 26 **25.** 20 **27.** {3, 6, 9, 12, 15, 18} **29.** {3, 5, 6, 9, 10, 12, 15, 18, 20}

**31.**

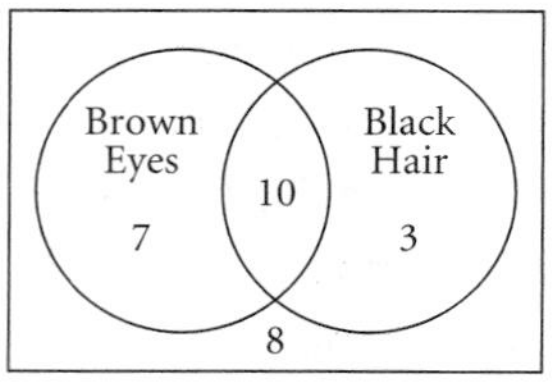


**33.** 111 **35.** $22\frac{1}{2}$ min **37.** 26 **39.** 29 **41.** $\frac{3V}{\pi r^2} = h$
**43.** $x = \frac{3}{2} = 1\frac{1}{2}$

## LESSON 13.3

**TRY THIS** (p. 662, Ex. 1)

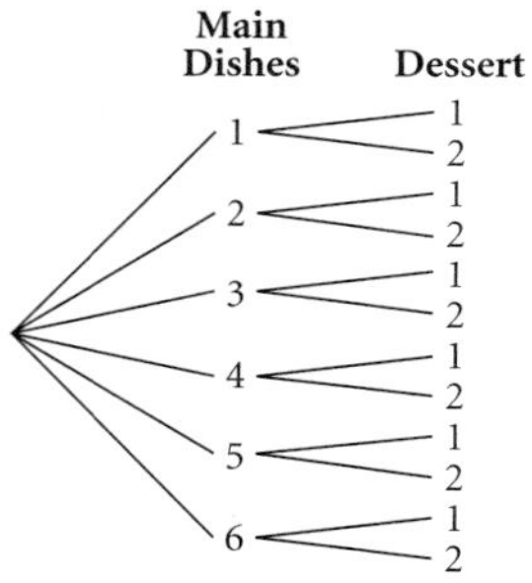


12 sets of choices are possible

**TRY THIS** (p. 662, Ex. 2)

There are 3 • 5 = 15 ways to choose a car.

**TRY THIS** (p. 663)

There are 2 • 4 • 2 = 16 possible selections.

**TRY THIS** (p. 664)

There are 26 • 12 = 312 ways.

### Exercises

**4.** 24 possible choices **5.** 12 possible choices **6.** 8 possible choices **7.** 18 possible choices **8.** 144 choices **9.** 300 pairs of shorts **10.** 52 ways **11.** 312 ways **13.** 6 possibilities **15.** 6 ways **17.** 400 possibilities **19.** 9,000,000 possible phone numbers **21.** 6 ways **23.** 720 ways **25.** 120 and 153 **27.** $x = 2\frac{1}{2}$ **29.** −1 **31.** 382.5 milliliters

## LESSON 13.4

**TRY THIS** (p. 669, Ex. 1)

The probability is $\frac{1}{9}$.

**TRY THIS** (p. 669, Ex. 2)

The probability is 28%.

**TRY THIS** (p. 670, Ex. 3)

The probability is $\frac{1}{169}$.

**TRY THIS** (p. 670, Ex. 4)

The probability is $\frac{35}{36}$.

### Exercises

**5.** $\frac{1}{4}$ **6.** $\frac{1}{4}$ **7.** $\frac{1}{4}$ **8.** 63% **9.** 72% **10.** $\frac{1}{169}$ **11.** $\frac{1}{16}$ **12.** $\frac{1}{676}$ **13.** $\frac{168}{169}$ **14.** $\frac{15}{16}$ **15.** $\frac{675}{676}$ **17.** Independent **19.** 80% **21.** 56% **23.** 0 **25.** $\frac{2}{3}$ **27.** $\frac{4}{25}$ **29.** $\frac{6}{25}$ **31.** $\frac{1}{3}$ **33.** $\frac{1}{12}$ **35.** 20 **37.** 4 **39.** $\frac{1}{4}$ **41.** 59 students

## LESSON 13.5

**TRY THIS** (p. 678)

Students should model their answers on the Example. Results should be close to (0.6)(0.6) = 0.36 or 36%.

### Exercises

**7.** Sample answer: The class should combine their results to calculate the percent of trials that had rain on all three days. **8.** Sample answer: The class should combine their results to calculate the percent of trials that had no rain. **9.** 2 successful shots out of 10 **10.** Out of the 10 trials, 5 had results of more than 50% success. **11.** INT(RAND()*2) **12.** 0 = did not make the shot; 1 = made the shot. **For exercises 13–21 answers may vary.** **23.** $t = 4n$, where $n = 1, 2, 3, 4...$ **25.** 12 cm **27.** $\begin{bmatrix} 1 & -5 \\ 12 & -26 \end{bmatrix}$; $\begin{bmatrix} -22 & -2 \\ -16 & -3 \end{bmatrix}$ No

## CHAPTER REVIEW AND ASSESSMENT

**1.** $\frac{10}{21}$ **3.** $\frac{1}{21}$ **5.** $\frac{1}{2}$ **7.** $\frac{2}{3}$ **9.** 18 **11.** 32 **13.** $\frac{1}{2}$ **15.** 24 **17.** 729 **19.** 48 **21.** $\frac{1}{100}$ **23.** $\frac{9}{100}$ **25.** $\frac{15}{64}$ **27.** $\frac{25}{64}$ **29.** Sample answer: Let heads represent a correct response and tails represent an incorrect response. Flip a coin 10 times; record the results. This represents 1 trial. Repeat for 20 trials. Divided the number of trials where all 10 responses were correct by 20. The quotient represents the experimental probability. **31.** $\frac{3}{4}$ **33.** $\frac{3}{40}$ **35.** 7.5%

# Chapter 14

## LESSON 14.1

**TRY THIS** (p. 695)

7

**TRY THIS** (p. 696)

$\{(-3, 11), (-2, 6), (-1, 3), (0, 2), (1, 3), (2, 6), (3, 11)\}$

**Exercises**

**3.** Yes, because there is only one range value associated with each domain value. **4.** Yes, because there is only one range value associated with each domain value. **5.** No, because there is more than one range value associated with one of the domain values. **6.** No, because there is more than one range value associated with one of the domain values.
**7.** function; Each domain value is associated with only one range value. **8.** not a function; Two of the dots have the same domain value. **9.** 12; 3; 48 **10.** 1; 7; 3 **11.** $\{(-3, 3), (-2, 2), (-1, 1), (0, 0), (1, 1), (2, 2), (3, 3)\}$; **12.** $f(x) = x^2$ **13.** $f(x) = x$ **14.** $f(x) = |x|$ **15.** $f(x)a^x$ **16.** $f(x) = \frac{1}{x}$ **17.** $f(x) = \sqrt{x}$ **19.** Yes, there is only one range value associated with each domain value. **21.** Yes, there is only one range value associated with each domain value. **23.** No, the graph does not represent a function because it has more than one range value for at least one of the domain values. (It would fail the vertical line test.)
**25.** domain: $\{0, 1, 2, 3\}$ range: $\{1, 3, 5, 7\}$
**27.** domain: $\{0.1, 0.2, 0.3\}$ range: $\{1, 2, 3\}$
**29.** domain: $\{2.2, 4.1, 5\}$ range: $\{3, 5.3, 9\}$
**31.** domain: $\{-11, 20, 70\}$ range: $\{13, 53, 91)$ **33.** 10 **35.** −5 **37.** 15 **39.** −10 **41.** 9 **43.** 3 **45.** domain: all real numbers; range: all real numbers **47.** domain: all real numbers; range: all non-negative numbers **49.** domain: all real numbers; range: all non-positive numbers **51.** domain: all real numbers; range: all non-negative numbers **53.** domain: all real numbers except 0; range: all real numbers except 0
**55.** $(-9, 80)$ **57.** $g(x) = x^2$ **59.** $h(x) = \frac{1}{x}$ **61.** yes; Yes, because there is only one Celsius temperature for every Fahrenheit temperature. **63.** Each speed value in the function between time and speed would have to be unique (i.e., no duplications) in order for the speed values to become the domain and the time values to become the range of a new function.
**65.** $r = 2$ **67.** $B = \frac{2A - hb}{h}$, or $\frac{2A}{h} - b$

## LESSON 14.2

**TRY THIS** (p. 701)

Move the graph of $f(x) = |x|$ up by 4 units.

**TRY THIS** (p. 703)

Shift the graph of $f(x) = |x|$ to the right 6 units.

**Exercises**

**4.** The graph of $y = x^2 - 2$ is a translation of the graph of $y = x^2$ 2 units down. **5.** The graph of $y = x^2 + 3$ is a translation of the graph of $y = x^2$ 3 units up. **6.** The graph of $y = 10^x + 1$ is a translation of the graph of $y = 10^x$ 1 unit up.

**7.** The graph of $y = \frac{1}{x} - 3$ is a translation of the graph of $y = \frac{1}{x}$ 3 units down. **8.** The graph of $y = (x - 2)^2$ is a translation of the graph of the parent function $y = x^2$ 2 units right. **9.** The graph of $y = 3x^2$ is not a translation of the graph of the parent function $y = x^2$ because 3 is multiplying the parent function, not being added or subtracted. **10.** The graph of $y = 10^{x+1}$ is a translation of the graph of the parent function $y = 10^x$ 1 unit left. **11.** The graph of $y = \frac{1}{x-3}$ is a translation of the graph of the parent function $y = \frac{1}{x}$ 3 units right.

**12. a.**

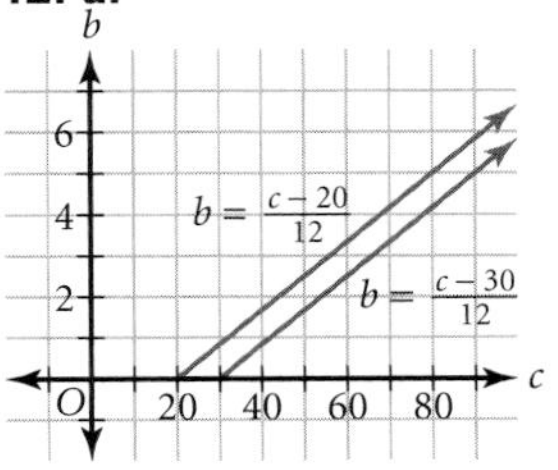


**b.** The graph of $b = \frac{c - 30}{12}$ is a horizontal translation of the graph of $b = \frac{c - 20}{12}$ 10 units to the right.

**13. a.** $y = x^2$ **b.** The graph of $y = (x - 6)^2 - 1$ is a translation 6 units right and 1 unit down

**c.**

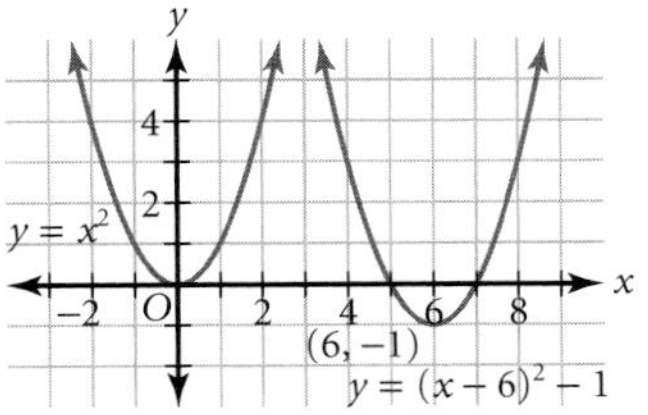


**15. a.** $y = |x|$ **b.** The graph of $y = |x| - 1$ is a translation 1 unit down.

**c.**

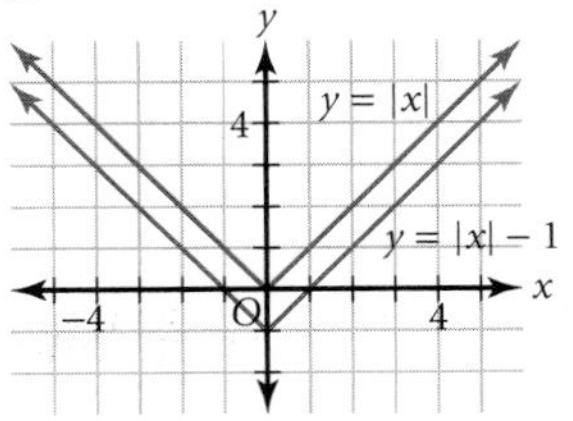


**17.** Because the parent graph is translated 5 units up, the $y$-value of the new coordinate, 14, is 5 units larger than that of the old coordinate, 9. **19.** Because the parent graph is translated 5 units left, the $x$-value of the new coordinate, −2, is 5 units less than that of the old coordinate, 3. **21.** (5, 14) **23.** (5, −2) **25.** (8, 8) **27.** (−7, 8) **29.** no **31.** $y = -\left(\frac{x}{40}\right)^2 + 208$ The graph is shifted down 22 feet.
**33. a.** mean: \$602.50; median: \$537.50; mode: none **b.** The mean is higher; the highest rent is so high that it raises the mean higher than the median.

## LESSON 14.3

**TRY THIS** (p. 708)

**a.** vertical stretch by a factor of 3;

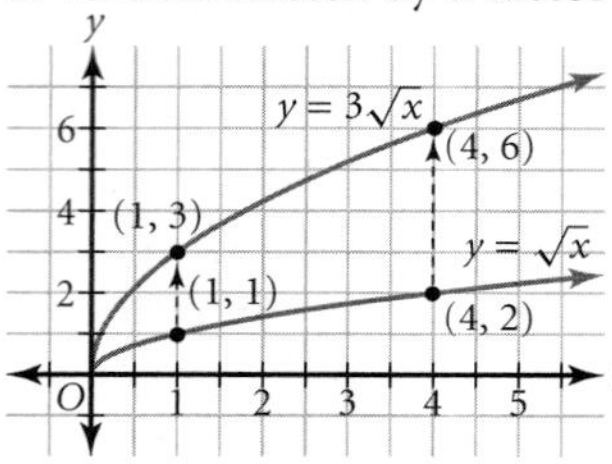


**b.** vertical compression by a factor of $\frac{1}{3}$;

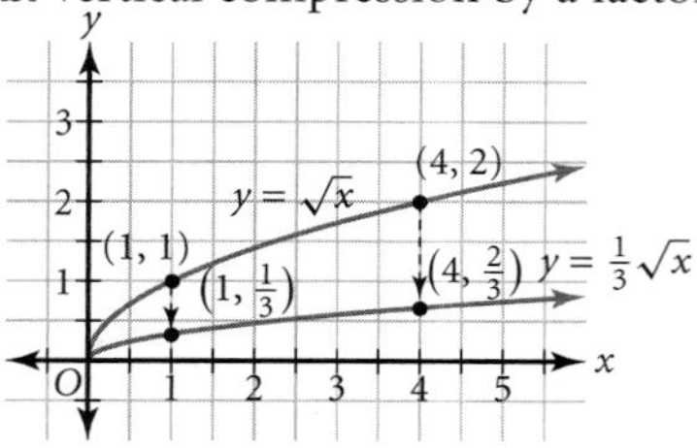


**TRY THIS** (p. 709)

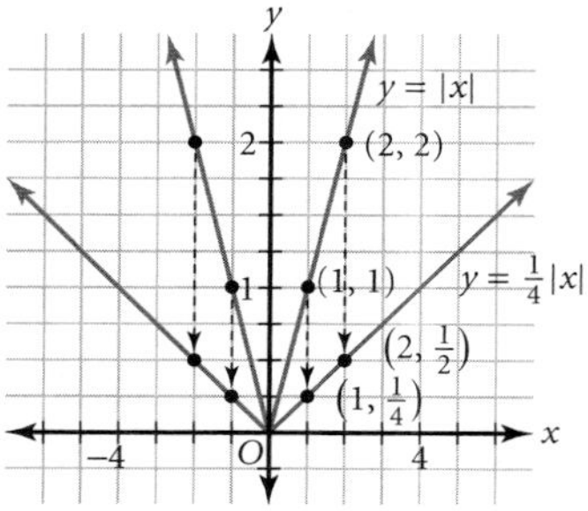


**TRY THIS** (p. 710)

**a.** horizontal compression by a factor of $\frac{1}{3}$

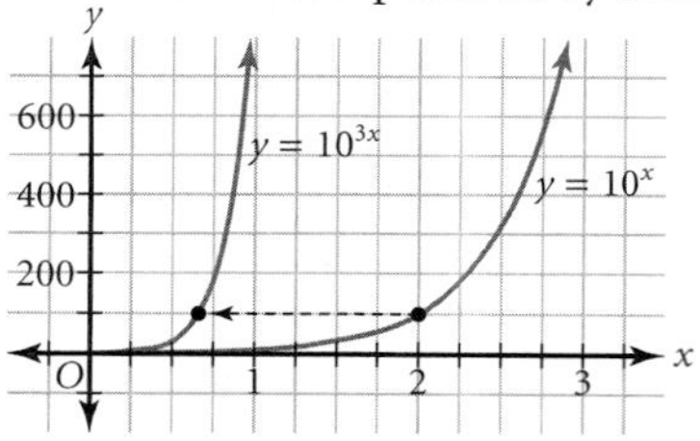


**b.** horizontal stretch by a factor of 3

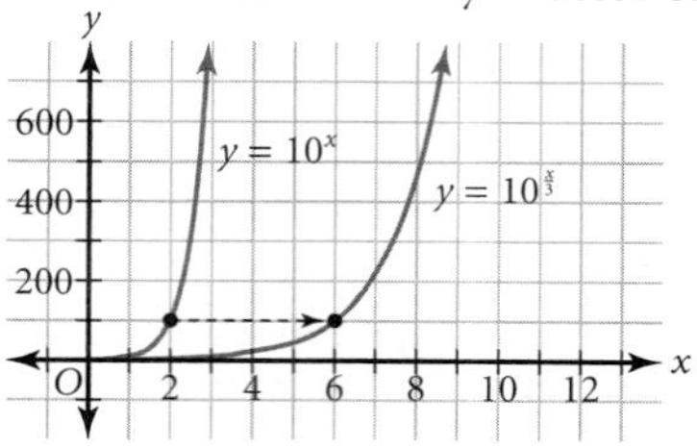


### Exercises

**5.**

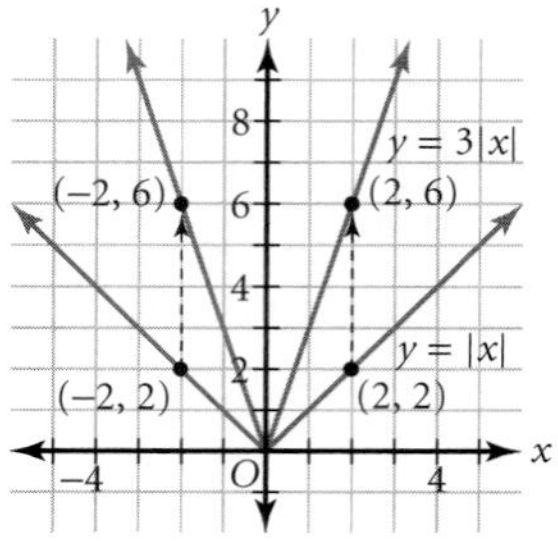


**6.**

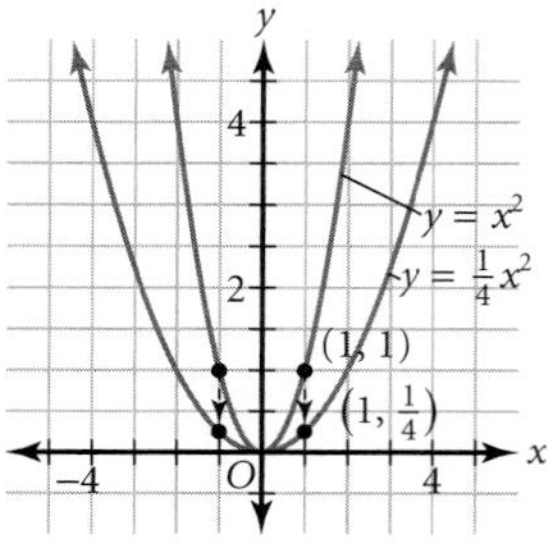


**7.**

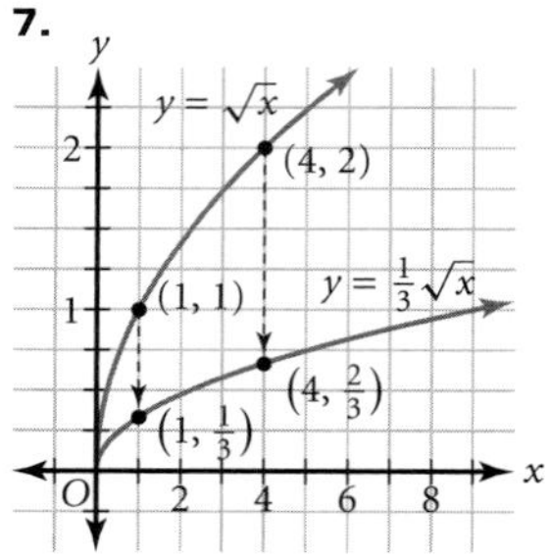


**8.**

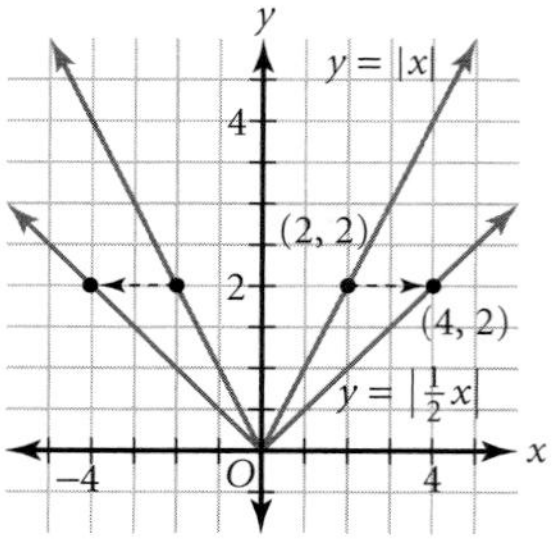


**9.**

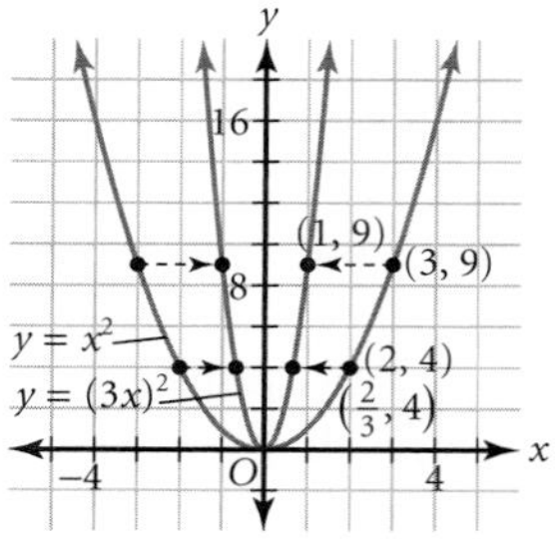


**10.**

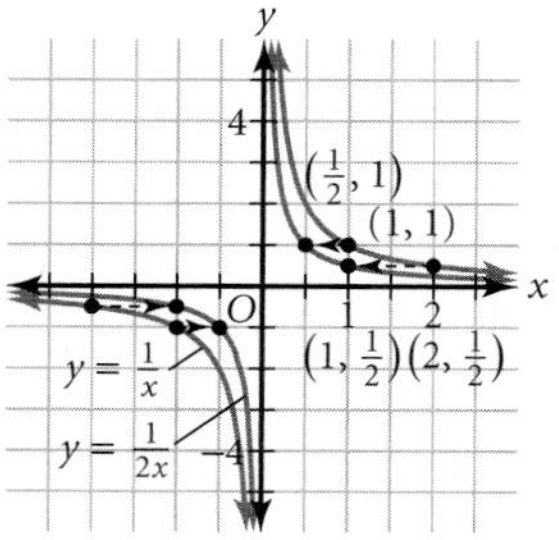


**11.** vertical compression **12.** neither **13.** horizontal compression **14.** neither **15.** neither **16.** horizontal compression

**17.**

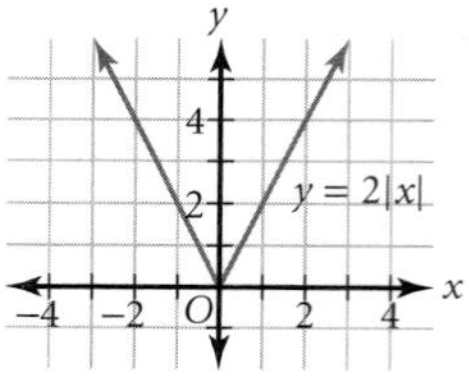


**19.**

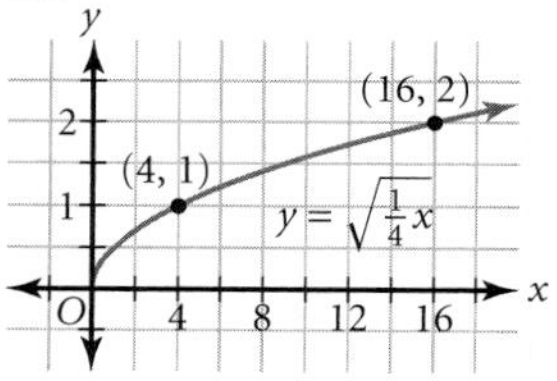


**21.** neither

**23. a.** $y = x^2$

**b.**

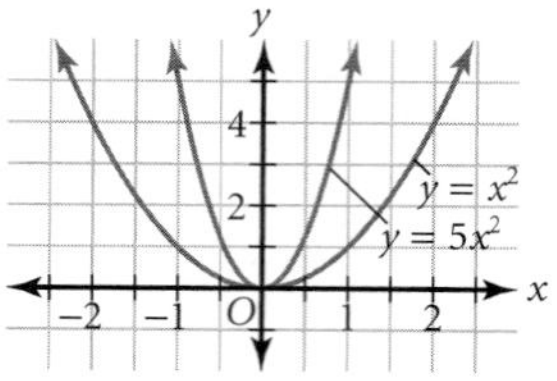


**c.** $(x, 5y)$

**25.** true **27.** $y = 2x^2$, $y = (\sqrt{2}x)^2$

**29.** $y = |x|$;

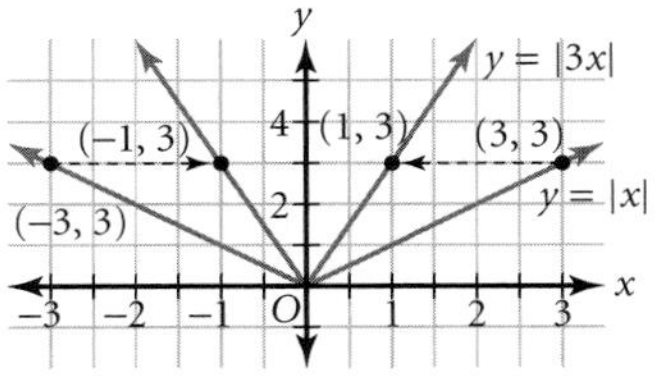


horizontal compression

**31.** $y = x^2$;

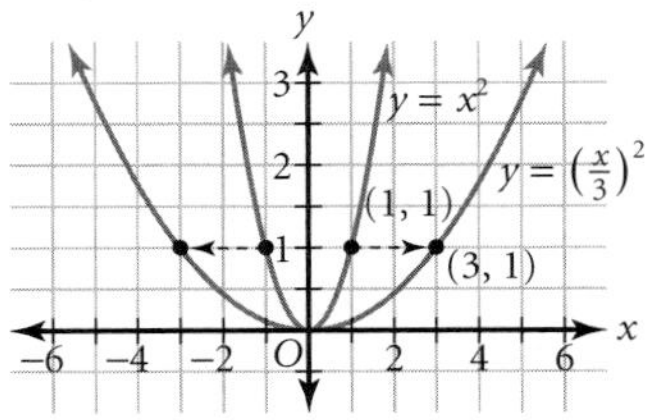


horizontal stretch

**33.**

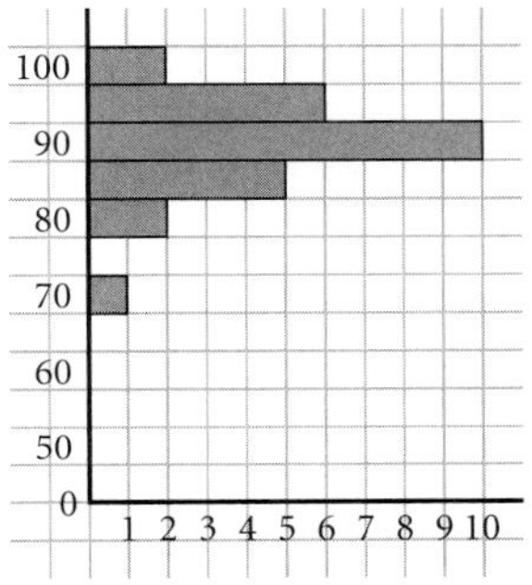

Each bar moves up 20 points.

**35. a.–b.**

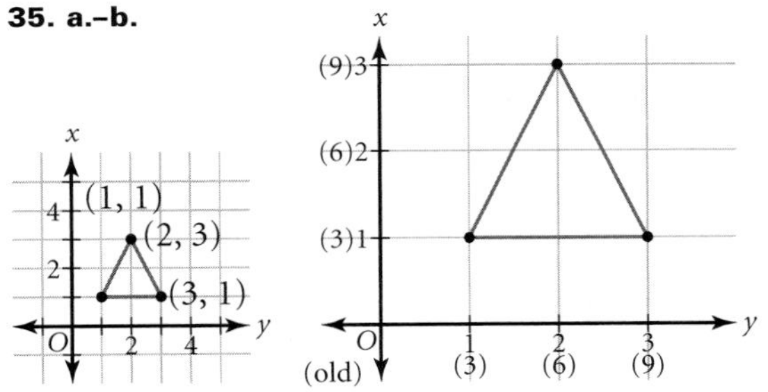


The graph sizes will vary, but the second triangle will have sides 3 times as long as the sides of the first triangle. **c.** The sides of the new grid squares are vertically and horizontally stretched by a factor of 3. Therefore, if the original grid were placed over the enlarged grid, matching the origins, the large triangle would have vertices of (3, 3), (9, 3), and (6, 9).

**37.** $x = 13$ **39.** $x = 17, y = 4$ **41.** 3 girls, 4 boys

## LESSON 14.4

**TRY THIS** (p. 715)

**a.** reflection across the $x$-axis;

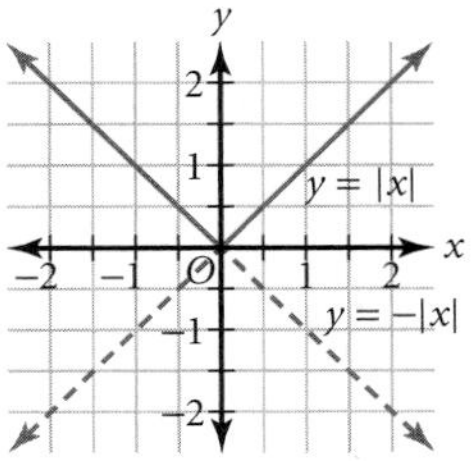


**b.** reflection across the $y$-axis;

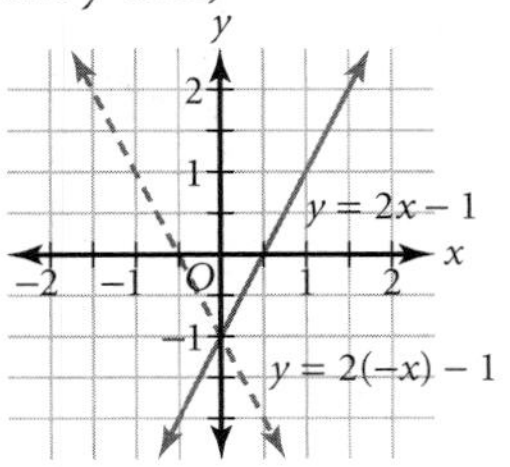


**TRY THIS** (p. 716)

**a.**

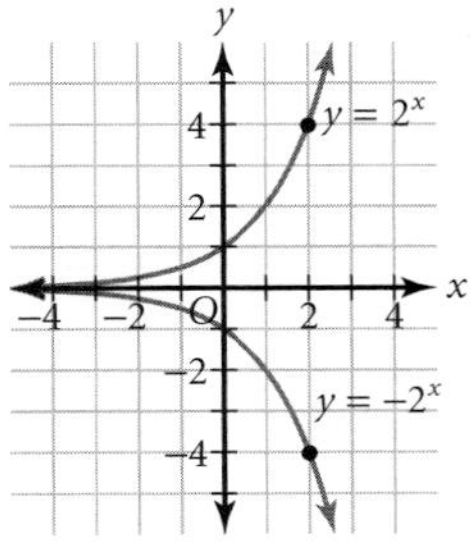


The parent function is reflected across the $x$-axis.

**b.**

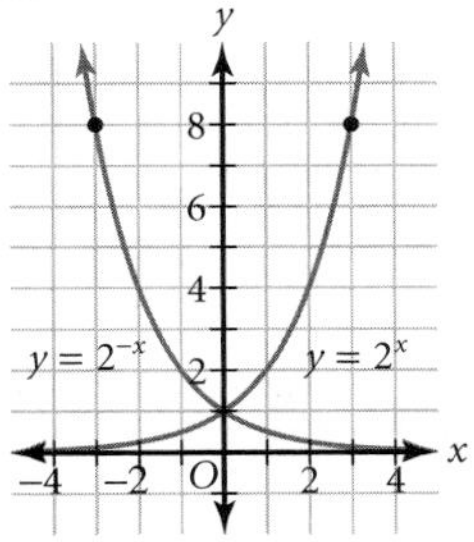


The parent function is reflected across the $y$-axis.

**Exercises**

**5.** reflection across the $x$-axis;

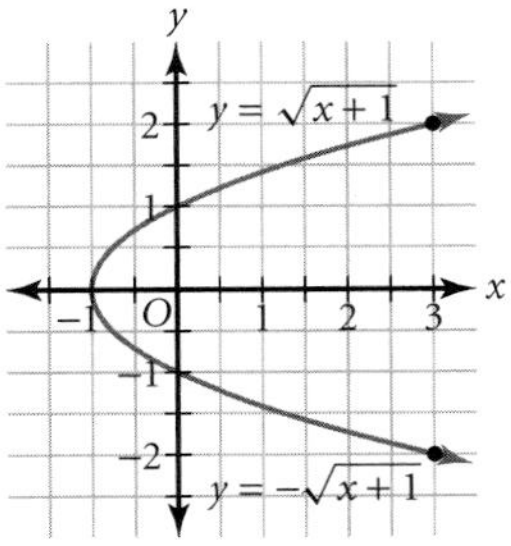


**6.** reflection across the $x$-axis;

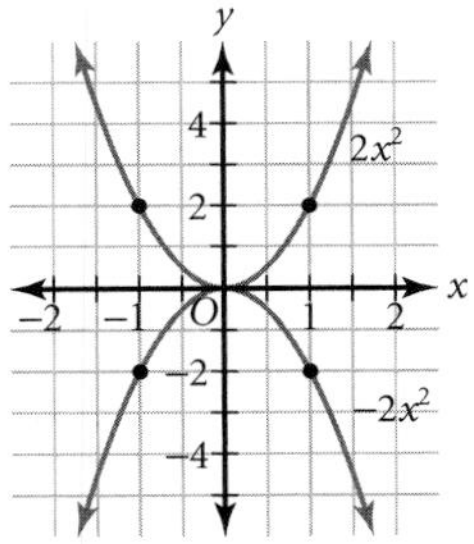


**7.** reflection across the $y$-axis;

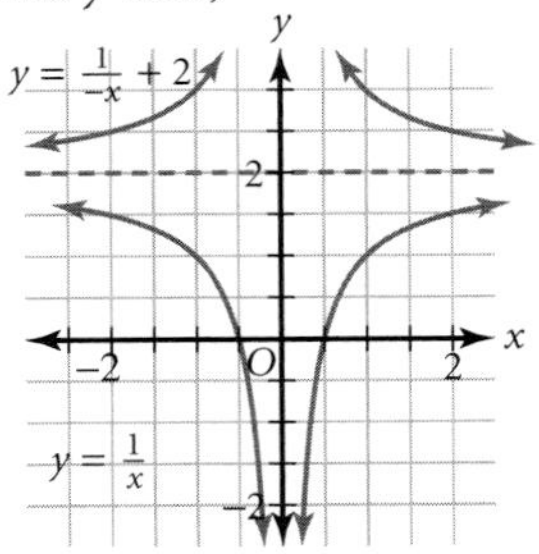


**8.** reflection across the $y$-axis;

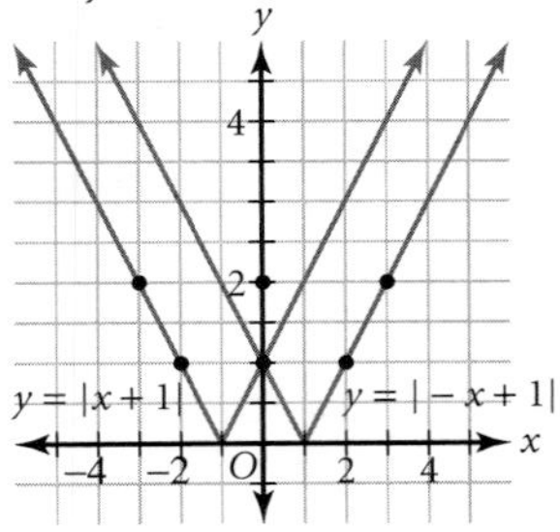

**9.** reflection across the $x$-axis;

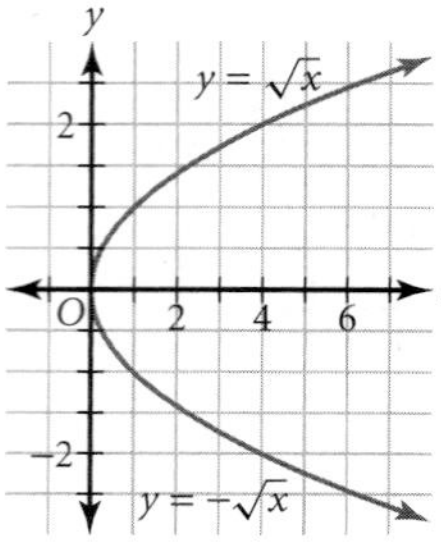


**10.** reflection across the $x$-axis;

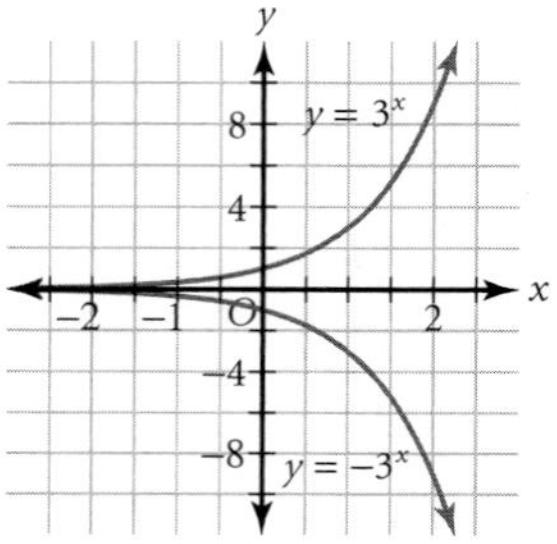


**11.** reflection across the $y$-axis (graphs are identical);

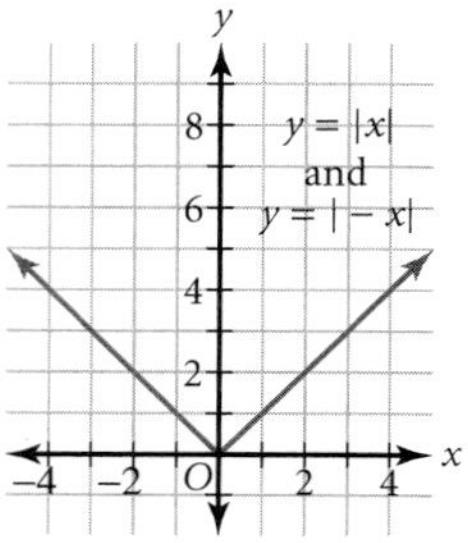


**12.** reflection across the $y$-axis;

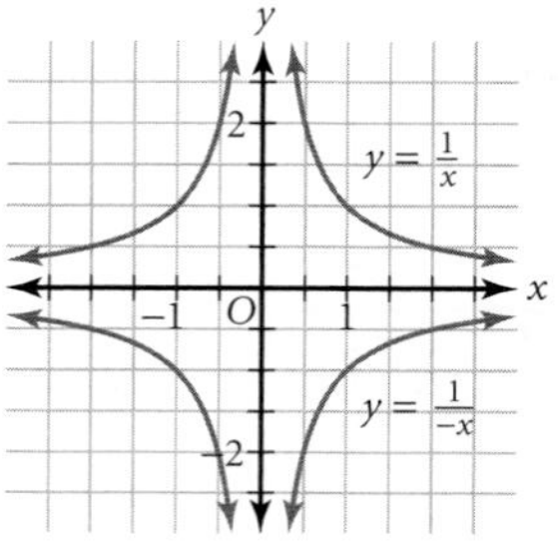


**13.** reflection across the $y$-axis;

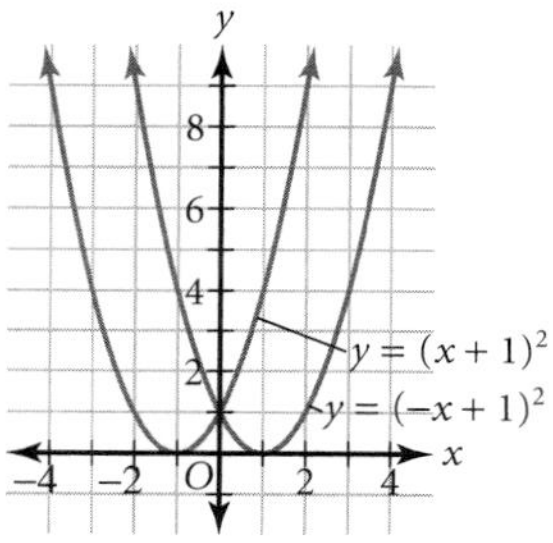


**15.** reflection across the $x$-axis;

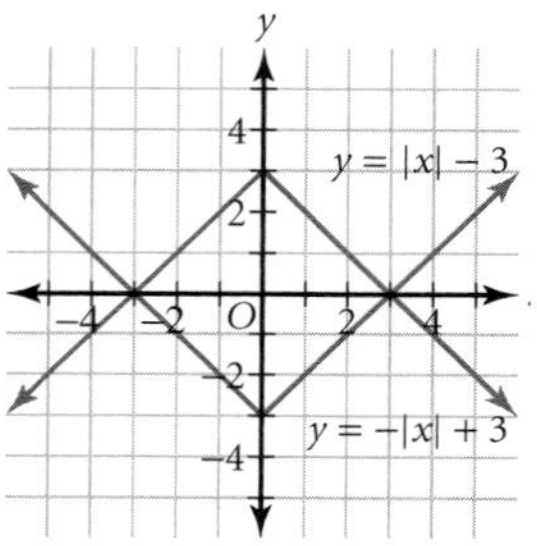


**17.** $f(x) = 5x + 2$ reflected across the $x$-axis: $g(x) = -(5x + 2) = -5x - 2$; $f(x) = 5x + 2$ reflected across the $y$-axis: $h(x) = 5(-x) + 2 = -5x + 2$
**19.** 2310 meters away **21.** $x = 22$
**23.** 9 meters × 27 meters

**25.**

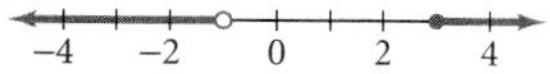


## LESSON 14.5

**TRY THIS** (p. 724)

Rewrite the equation as $y = -\frac{1}{3}\left(2^{-2(x+\frac{1}{2})}\right) + 1$. Translate $\frac{1}{2}$ unit left; compress horizontally by a factor of $\frac{1}{2}$; reflect across the $y$-axis; compress vertically by a factor of $\frac{1}{3}$; reflect across the $x$-axis; translate 1 unit up.

**Exercises**

**5.** Rewrite as $f(x) = 2[-(x+3)]^2 + 1$. Parent function: $f(x) = x^2$. Translate 3 units left; reflect across the $y$-axis; vertically stretch by a factor of 2; vertically translate 1 unit up.

**6.** Rewrite as $f(x) = -2[-3(x-1)]^2 + 1$. Parent function: $f(x) = x^2$. Translate 1 unit right; horizontally compress by a factor of $\frac{1}{3}$; vertically stretch by a factor of 2; reflect over the $x$-axis; translate 1 unit up.

**7.** Rewrite as $f(t) = \frac{1}{2}\left|\frac{1}{2}(t+2)\right| + 3$. Parent function: $f(t) = |t|$. Translate 2 units left; horizontally stretch by a factor of 2; vertically compress by a factor of $\frac{1}{2}$; translate 3 units up.

**8.** Rewrite as $f(t) = -\frac{1}{2}\left|\frac{1}{4}(t+12)\right| - 2$. Parent function: $f(t) = |t|$. Translate 12 units left; horizontally stretch by a factor of 4; vertically compress by a factor of $\frac{1}{2}$; reflect across the $x$-axis; translate 2 units down.

**9.** Rewrite as $f(g) = -3\left[\frac{1}{2\left(g+\frac{1}{2}\right)}\right] + 2$. Parent function: $f(g) = \frac{1}{g}$. Translate $\frac{1}{2}$ unit left; horizontally compress by a factor of $\frac{1}{2}$; vertically stretch by a factor of 3; reflect across the $x$-axis; translate 2 units up.

**10.** Rewrite as $f(x) = 2\left[\frac{1}{2(x-2)}\right] - 6$. Parent function: $f(x) = \frac{1}{x}$. Translate 2 units right; horizontally compress by a factor of $\frac{1}{2}$; vertically stretch by a factor of 2; translate 6 units down.

**11.**

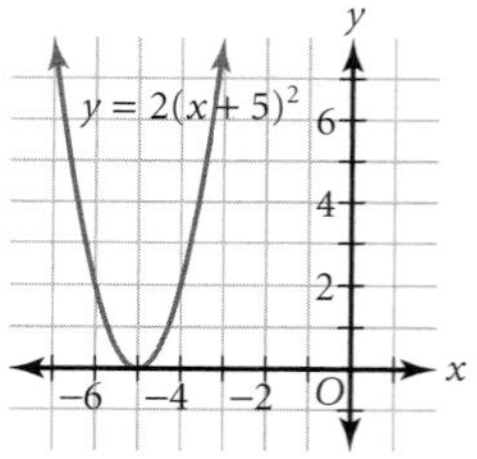


**13.**

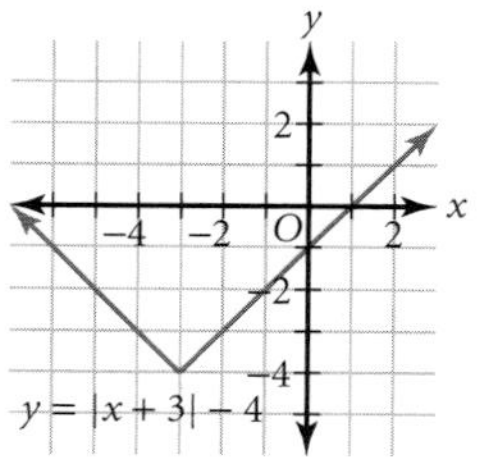


**15.**

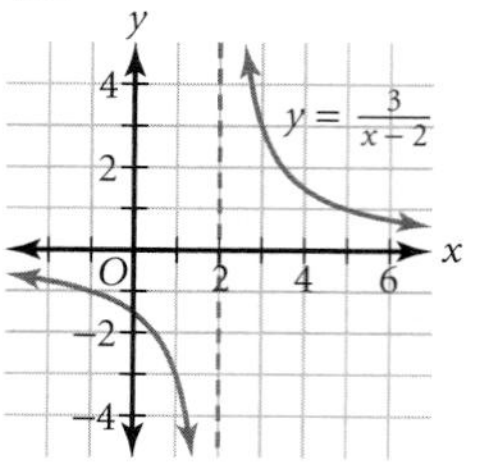


**17.**

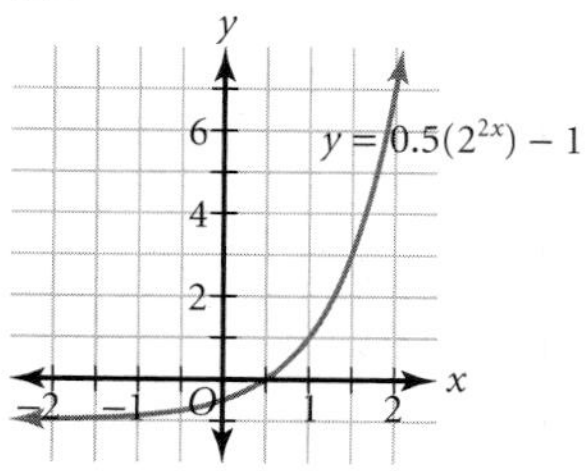

**19.**

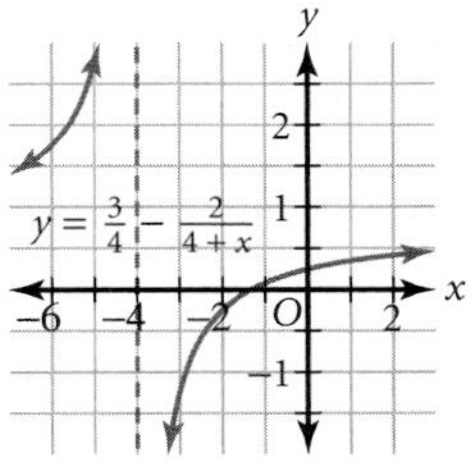


**21.** (4, −20) **23.** (12, 15) **25.** $y = -5x + 6$; parent function: $y = x$ **27.** $A' : (a, -b)$ $A''(a - 18, -b)$; Yes, the other vertices change in the same pattern.

**29.** 42 boxes

**31.** $y = \frac{5}{16}x + 7\frac{15}{16}$, or $\frac{5}{16}x - y = -7\frac{15}{16}$

**33.**

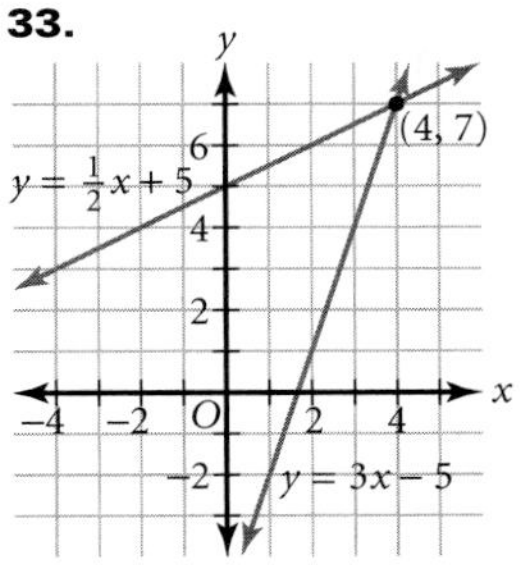


**25.**

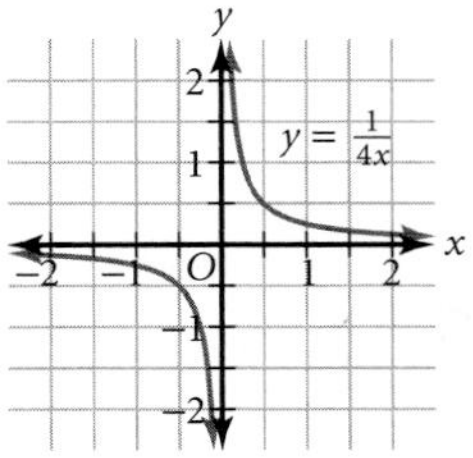


**27.**

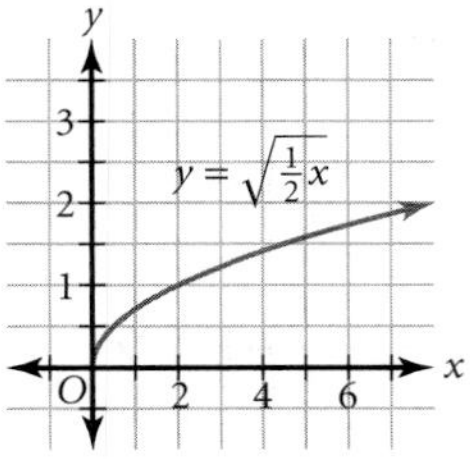


**29.**

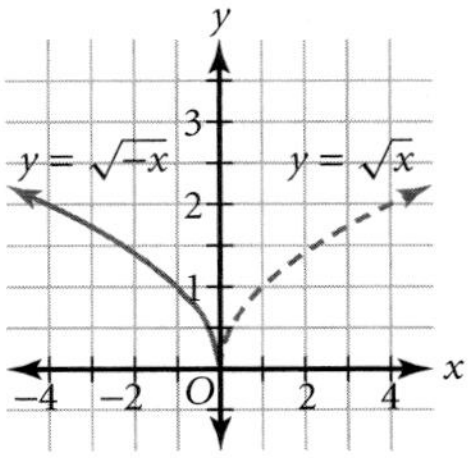


**31.**

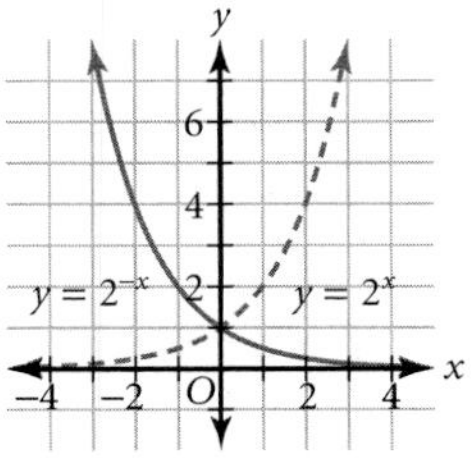


**33.**

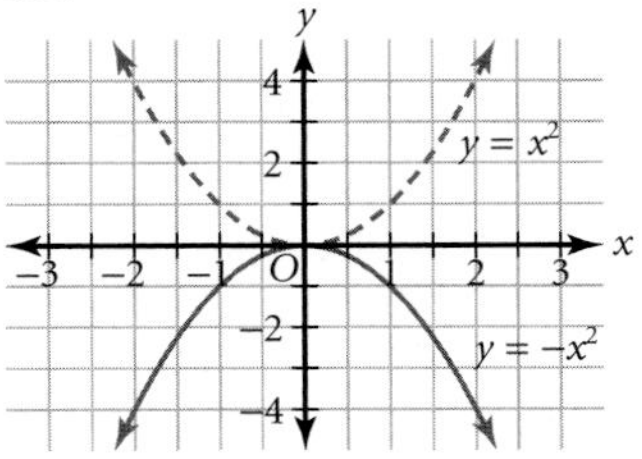


**35.** horizontal translation 6 units right, vertical translation 2 units up

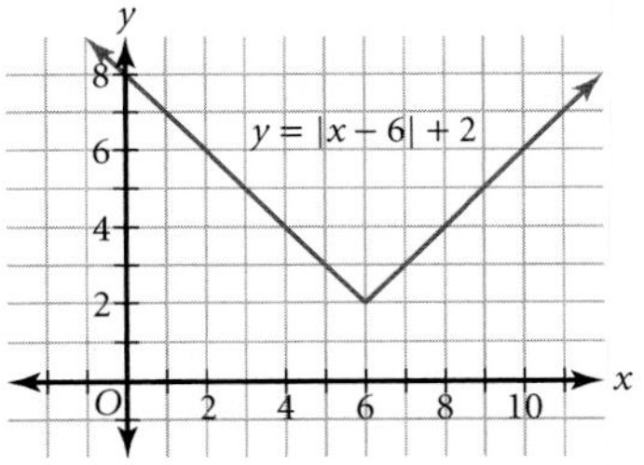


**37.** horizontal translation 3 units right, vertical translation 4 units up

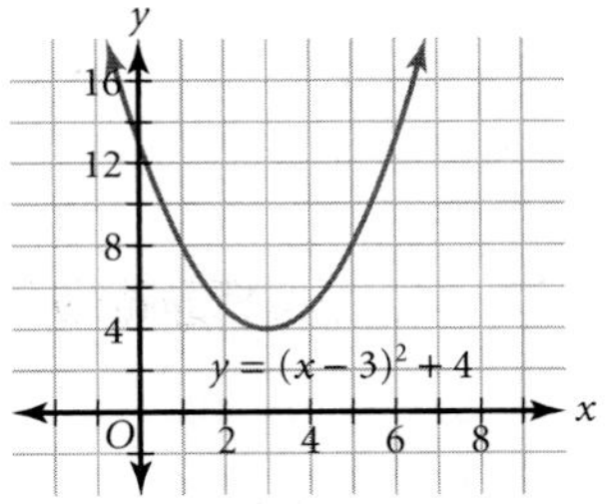


## CHAPTER REVIEW AND ASSESSMENT

**1.** not a function **3.** yes **5.** $f(4) = -4$ **7.** $h(4) = 81$ **9.** $k(4) = -5$ **11.** $f(x) = \frac{1}{x}$

**13.** $y = x^2$, vertical translation

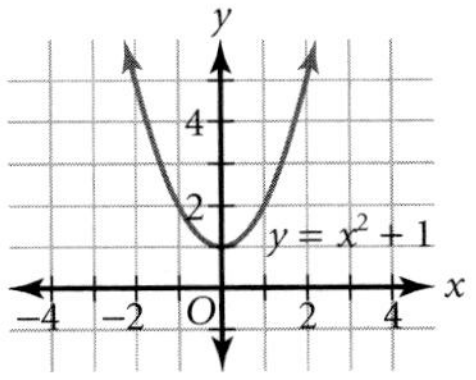


**15.** $y = |x|$, vertical translation

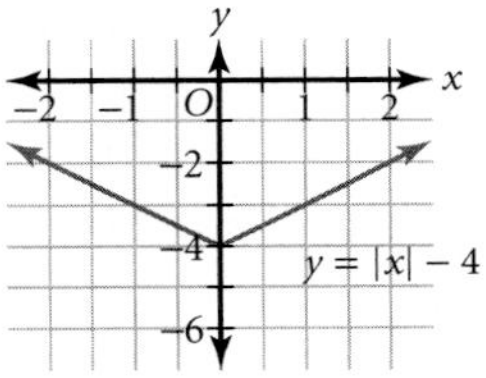


**17.** $y = \frac{1}{x}$, vertical translation

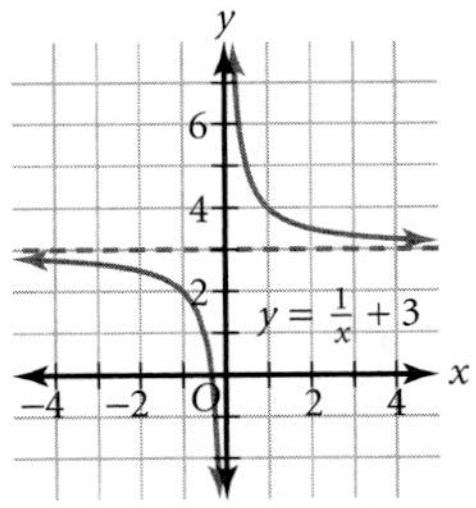


**19.** $y = 2^x$, horizontal translation

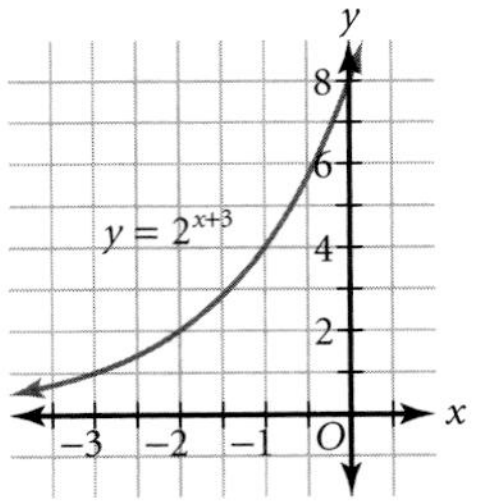


**21.**

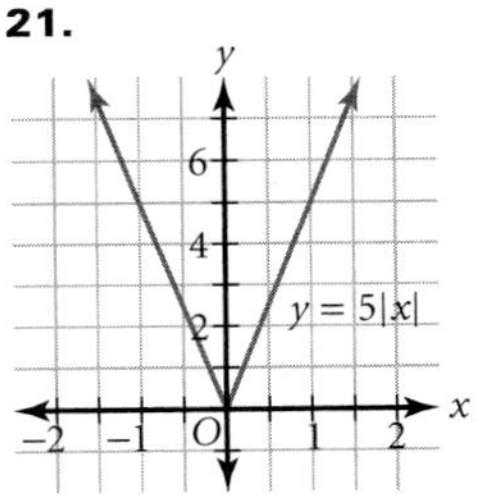


**23.**

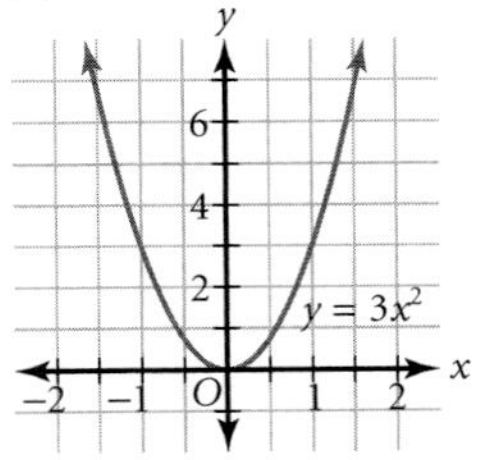

**39.** horizontal translation 4 units left, vertical translation 2 units down

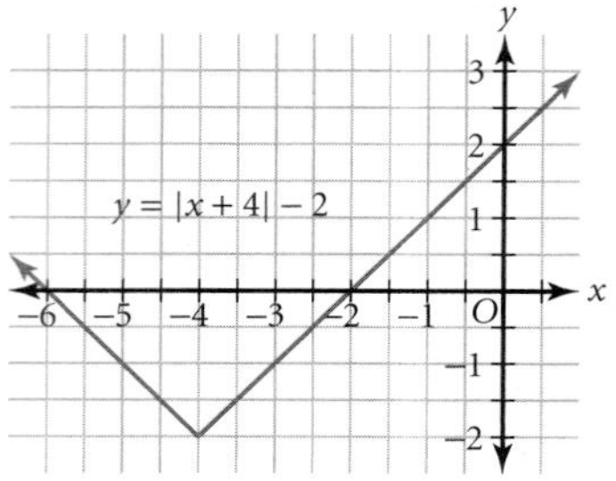


**41.** vertical translation

# Extra Practice Answers

## CHAPTER 1

### Lesson 1.1

**1.** 24, 27, 30 **3.** 28, 33, 38 **5.** 42, 46, 50 **7.** 58, 62, 66 **9.** 43, 49, 55 **11.** 98, 116, 135 **13.** 33, 39, 45 **15.** 75, 90, 105 **17.** 10 **19.** 4095 **21.** 6105 **23.** 3, 6, 11, 18, 27

### Lesson 1.2

**1.** Expression **3.** Expression **5.** Expression **7.** Expression **9.** 3, 6, 9, 12, 15 **11.** 6, 8, 10, 12, 14 **13.** 12, 21, 30, 39, 48 **15.** −24, −15, −6, 3, 12 **17.** 2 **19.** 5 **21.** 8 **23.** 8 **25.** 5

### Lesson 1.3

**1.** 24 **3.** 80 **5.** 68 **7.** 25 **9.** 1050 **11.** 138 **13.** 2 **15.** $\frac{1}{4}$ **17.** 19 **19.** $(24 - 6) \cdot 2 + 2^2 = 4$ **21.** $6 \cdot (8 + 1) \div 3 = 18$ **23.** $(15 + 5) \div 5 - 2 = 2$ **25.** $(39 - 7) \div 4 + 3 = 11$ **27.** 15 **29.** −19 **31.** 44 **33.** 17 **35.** −45

### Lesson 1.4

**1.** yes **3.** yes **5.** no **7.** no **9.** yes **11.** 6, 7, 8, 9, 10 **13.** −3, −2, −1, 0, 1 **15.** −4, −8, −12, −16, −20 **17.** 1, 1, 3, 5, 7 **19.** −5, −9, −13, −17, −21 **21.** 1.5, 3.5, 5.5, 7.5, 9.5 **23.** −2, −9, −16, −23, −30 **25.** 12, 24, 36, 48, 60

### Lesson 1.5

**1.** 7; $y = 7x - 8$ **3.** 11, 9, 7, 5, 3 **5.** −2, −6, −10, −14, −18 **7.** $0 \cdot 0.69$; \$0 **9.** $10 \cdot 0.69$; \$6.90 **11.** $5 \cdot 2.5$; \$12.50 **13.** $8 \cdot 2.5$; \$20 **15.** 4 tickets

### Lesson 1.6

**1.** Strong negative **3.** Strong positive **5.** Plots will vary. **7.** For this data, the correlation is strong negative because winning times decrease as the years increase. In other words, the runners are getting faster every year.

## CHAPTER 2

### Lesson 2.1

**1.** < **3.** < **5.** −13 **7.** 2.36 **9.** −18 **11.** $4\frac{5}{6}$ **13.** 15 **15.** 3.47 **17.** $7\frac{2}{3}$ **19.** 842

### Lesson 2.2

**1.** −2 **3.** −13 **5.** −2 **7.** −7 **9.** −20 **11.** −14 **13.** −26 **15.** −50 **17.** −13 **19.** −13 **21.** −125 **23.** −34 **25.** −1 **27.** 13 **29.** 61 **31.** −545

### Lesson 2.3

**1.** −4 **3.** −9 **5.** −4 **7.** 60 **9.** −71 **11.** 106 **13.** −69 **15.** −58 **17.** −24 **19.** −14 **21.** −73 **23.** 108 **25.** −60 **27.** −79 **29.** −61 **31.** −87 **33.** −221 **35.** 20 **37.** 3 **39.** 20 **41.** 28 **43.** 6 **45.** 63

### Lesson 2.4

**1.** −12 **3.** −4 **5.** 86 **7.** −4 **9.** −4 **11.** −14 **13.** 28 **15.** −70 **17.** −225 **19.** 5 **21.** 105 **23.** −12 **25.** 4 **27.** 2

### Lesson 2.5

**1.** −16 **3.** $-5z$ **5.** $5r$ **7.** $-a + 7$ **9.** $-5x + 7y$ **11.** $-8a + c$ **13.** $10m - 4n$ **15.** $15 + v$ **17.** $2c - 4d + 5$ **19.** $-8a - 4b + c$ **21.** 126 **23.** 360 **25.** 1380 **27.** 229 **29.** 0 **31.** 150 **33.** Commutative **35.** Distributive **37.** Associative **39.** Associative **41.** Associative

### Lesson 2.6

**1.** $12x + 3x = (12 + 3)x = 15x$
**3.** $-5x + 4x = (-5 + 4)x = -x$
**5.** $7x - 12x = (7 - 12)x = -5x$
**7.** $8a - 2$ **9.** $5c - 3$ **11.** $z + 3$ **13.** $-r - 2$ **15.** $a$ **17.** $4w - 8$ **19.** $y + 7$ **21.** $5r - 2s - 3t$ **23.** $4.1x + 5.2y$ **25.** $9w - 1$ **27.** $14q - 2$ **29.** $-y + 2$ **31.** $6h + 12j$ **33.** 3 **35.** $4a + 4b + 2c$ **37.** $-3p - 5q + 11$

### Lesson 2.7

**1.** $12m$ **3.** $10x^2$ **5.** $-440x^2$ **7.** $14x^2 + 21x$ **9.** $7x$ **10.** $-16x$ **13.** $9n - 3$ **15.** $7h - 13$ **17.** $4e + 2$ **19.** $-11t - 9e$ **21.** $-3g + 1$ **23.** $x + 5$ **25.** $2 - 8u$ **27.** $-3x + 2$ **29.** $24ex - 48e$ **31.** $-5t - 5r$ **33.** $-30c^2 + 75c$ **35.** $y^2 + y$ **36.** $30c + 27$

## CHAPTER 3

### Lesson 3.1

**1.** $x = -1$ **3.** $w = 1$ **5.** $y = -6$ **7.** $h = 11$ **9.** $z = 21$ **11.** $t = 6$ **13.** $n = -17$ **15.** $c = 15$ **17.** $b = -8$ **19.** $x = 8$ **21.** $j = -17$ **23.** $w = -7$ **25.** $x = -10$ **27.** $w = 2$ **29.** $y = 4$ **31.** $x = 35$ **33.** $n = 124$ **35.** $w = 7$ **37.** $l = 2$ **39.** $x = -48$ **41.** $u = -44$ **43.** $q = 1094$ **45.** $s = 61$ **47.** $p = 60$

### Lesson 3.2

**1.** Multiplication Property of Equality; $x = -24$ **3.** Multiplication Property of Equality; $x = 4$ **5.** Division Property of Equality; $y = -\frac{1}{3}$ **7.** Division Property of Equality; $x = \frac{1}{7}$ **9.** Division Property of Equality; $y = -12$ **11.** Multiplication Property of Equality; $a = 17.5$ **13.** $x = 1\frac{1}{4}$ **15.** $r = -54$ **17.** $c = 60$ **19.** $x = 34$ **21.** $d = 0$ **23.** $h = 99$ **25.** $m = 4.2$ **27.** $y = 17.62$ **29.** $z = 25$ **31.** $m = 1\frac{2}{3}$ **33.** $w = 2$ **35.** $v = 97$

### Lesson 3.3

**1.** $y = 4$ **3.** $y = 3$ **5.** $d = 7$ **7.** $c = -8\frac{1}{3}$ **9.** $m = 10$ **11.** $n = 6$ **13.** $p = 10$ **15.** $g = -12$ **17.** $x = -7$ **19.** $j = -\frac{1}{45}$ **21.** $x = -1$ **23.** $x = 60$

### Lesson 3.4

**1.** $x = 3$ **3.** $m = 7$ **5.** $c = -2$ **7.** $y = 8$ **9.** $s = 3$ **11.** $r = 6$ **13.** $b = \frac{4}{3}$ **15.** $x = 6$ **17.** $a = -\frac{5}{7}$ **19.** $c = -5$

### Lesson 3.5

**1.** $g = 7$ **3.** $z = 24$ **5.** $d = 6$ **7.** $f = 4$ **9.** $r = 4$ **11.** $v = \frac{1}{2}$ **13.** $x = -4$ **15.** $u = 6$ **17.** $m = 5$ **19.** $p = 12$

### Lesson 3.6

**1.** $s = P + c$ **3.** $a = c - b$ **5.** $s = \frac{P}{4}$ **7.** $t = \frac{d}{r}$ **9.** $m = sh$ **11.** $r = \frac{I}{pt}$ **13.** $y = \frac{r-x}{4}$ **15.** $D = 2P - 30$ **17.** $f = 18 - d - e$ **19.** $n = \frac{s}{180} + 2$

## CHAPTER 4

### Lesson 4.1

**1.** True **3.** True **5.** False **7.** True **9.** False **11.** False **13.** $x = 54$ **15.** $r = 20.25$ **17.** $a = 45$ **19.** $c = 24.2$ **21.** $t = 1.75$ **23.** $r = 64.5$ **25.** $b = 23.04$ **27.** $a = 3.27$ **29.** $x = 1$ **31.** $q = 5.6$ **33.** $x = 11.5$ **35.** $z = 4.1$ **37.** $f = 5$ **39.** $b = 19.25$

### Lesson 4.2

**1.** 0.65 **3.** 0.09 **5.** 0.024 **7.** 2.25 **9.** 87% **11.** 6% **13.** 2.56% **15.** 7.3% **17.** $\frac{1}{4}$ **19.** $\frac{13}{200}$ **21.** $3\frac{1}{20}$ **23.** $\frac{11}{25}$ **25.** 18 **27.** 50% **29.** 15 **31.** 18% **33.** 220 **35.** 133% **37.** 0.5 **39.** 60% **41.** 31.9 **43.** 7.2%

### Lesson 4.3

**1.** no **3.** 49% **5.** Yes; the combined areas are almost the same. **7.** 37%

### Lesson 4.4

| | Mean | Median | Mode | Range |
|---|---|---|---|---|
| **1.** | 13.6 | 12 | 12 | 12 |
| **3.** | 4.09 | 4 | 1, 2, 4 | 8 |
| **5.** | 133 | 133.5 | none | 25 |
| **7.** | 62.42 | 60.5 | none | 38 |

**9.**

| Runs | 0 | 1 | 2 | 3 | 4 | 5 | 6 | 7 | 8 | 9 |
|---|---|---|---|---|---|---|---|---|---|---|
| Freq. | 3 | 2 | 4 | 2 | 2 | 3 | 2 | 2 | 2 | 2 |

**11.** 4.125 **13.** The mean gives the best impression because it shows the greatest number of runs.

### Lesson 4.5

**1.** 28,614 **3.** 42,168 **5.** 66,500 **7.** 34,200

### Lesson 4.6

**1.** key: 5|8 = 58

| | |
|---|---|
| 5 | 8, 9 |
| 6 | 0, 7, 8, 8, 9 |
| 7 | 3, 5, 5, 6, 7 |
| 8 | 0, 0, 0, 3, 7 |
| 9 | 0, 1, 2, 3, 5, 5, 7 |

**3.** 68.5

## CHAPTER 5

### Lesson 5.1

**1.** domain = {1, 2, 3, 4}; range = {2, 4, 6, 8}; function **3.** domain = {–200, 100, 150, 200}; range = {3, 4, 5}; function **5.** domain = {–5, –2, 1, 4}; range = {6, 7, 8, 9, 10}; not a function **7.** domain = {2, 3, 4, 5}; range = {20, 25, 30}; function **9.** $y = 6\frac{1}{2}$ **11.** $x = 6$ **13.** $y = -2$ **15.** $x = 8$ **17.** $y = -26$ **19.** $y = 2\frac{1}{2}$ **21.** $y = 7$ **23.** $x = 8\frac{2}{3}$

### Lesson 5.2

**1.** 2 **3.** 4 **5.** 0 **7.** $-\frac{4}{5}$ **9.** $\frac{9}{4}$ **11.** $\frac{2}{3}$ **13.** $-\frac{1}{2}$ **15.** $-\frac{3}{2}$ **17.** 2 **19.** 1 **21.** $-\frac{1}{3}$

### Lesson 5.3

**1.** $2.00 per hour **3.** Scott's Skates **5.** 2; $y = 2x$ **7.** 3; $y = 3x$ **9.** 1.33; $y = 1.33x$ **11.** 0.875; $y = 0.875x$ **13.** 0.5; $y = 0.5x$ **15.** 1.5; $y = 1.5x$ **17.** 4.88; $y = 4.88x$ **19.** 0.636; $y = 0.636x$

### Lesson 5.4

**1.** 3; (0, 1); $\left(-\frac{1}{3}, 0\right)$ **3.** 4; (0, 0) **5.** 5; (0, 4); $\left(-\frac{4}{5}, 0\right)$ **7.** –11; (0, 10); $\left(\frac{10}{11}, 0\right)$ **9.** 1; (0, 9), (-9, 0) **11.** 7; (0, –12); $\left(1\frac{5}{7}, 0\right)$ **13.** $y = \frac{2}{3}x + 3$ **15.** $y = -2x + 1$ **17.** $y = x + 2$ **19.** $y = x - 1$ **21.** $y = \frac{1}{4}x + 2$ **23.** $y = \frac{1}{2}x - \frac{5}{2}$ **25.** $y = -\frac{3}{4}x + 3$ **27.** $y = -\frac{4}{3}x$

### Lesson 5.5

**1.** $3x - 6y = 18$ **3.** $2x - 5y = -1$ **5.** $5x - y = 4$ **7.** $3x - 7y = -12$ **9.** $4x - 8y = 0$ **11.** $2x - 3y = -1$ **13.** $2x + y = 4$ **15.** $x - 2y = -12$ **17.** $y - 5 = 2(x - 4)$ **19.** $(2, 0), (0, 2)$ **21.** $(8, 0), (0, 4)$ **23.** $(5, 0), (0, -4)$ **25.** $(2, 0), (0, 18)$ **27.** $(0, 0)$ **29.** $(0, 0)$ **31.** $y = x - 1$ **33.** $y = \frac{1}{4}x + 2$ **35.** $y = -x + 5$ **37.** $y = -5$ **39.** $y = -x + 2$

### Lesson 5.6

**1.** 4 **3.** $\frac{5}{2}$ **5.** 1 **7.** $-\frac{1}{3}$ **9.** $\frac{1}{2}$ **11.** $\frac{1}{9}$ **13.** $-\frac{1}{4}$ **15.** $-\frac{1}{5}$ **17.** $-7$ **19.** $\frac{4}{3}$ **21.** $-\frac{1}{2}$ **23.** $-\frac{1}{2}$ **25.** perpendicular **27.** perpendicular **29.** $y = -\frac{1}{3}x - 3\frac{1}{3}$ **31.** $y = -\frac{1}{2}x + 3\frac{1}{2}$ **33.** $y = -2x - 7$ **35.** $y = \frac{1}{2}x + 5$

## CHAPTER 6

### Lesson 6.1

**1.** $x < 3$ **3.** $w \leq 8$ **5.** $m > -1$ **7.** $x \leq -13$ **9.** $b \geq \frac{3}{4}$ **11.** $t > 8.75$ **13.** $w > 2\frac{2}{3}$ **15.** $d \geq 16.25$ **17.** $t \geq \frac{3}{8}$ **19.** $c < \frac{7}{12}$ **21.** $x \geq -13$ **23.** $r \geq 11$ **25.** $m < 21$ **27.** $f \leq -9$ **29.** $k < -11.4$

### Lesson 6.2

**1.** $x \geq 3$ **3.** $y \leq 5$ **5.** $h < 12$ **7.** $c < -50$ **9.** $x \geq \frac{1}{2}$ **11.** $m \leq 2\frac{1}{2}$ **13.** $b > 2\frac{1}{3}$ **15.** $w \leq -32$ **17.** $n > -1\frac{4}{5}$ **19.** $t < \frac{1}{2}$ **21.** $m \geq 36$ **23.** $z \leq 6$ **25.** $W < L$ **27.** $v \leq 4.6$ **29.** $x \leq -2$ **31.** $d > 0$

### Lesson 6.3

**1.** $-5 < x \leq 3$

**3.** $4 \leq s \leq 11$

**5.** $5 \leq r < 13$

**7.** $z > 0$ or $z < -2$

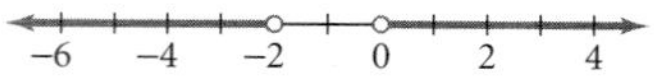


**9.** $b \geq 5\frac{2}{3}$ or $b < 2$

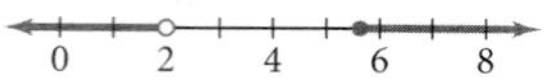


**11.** $4 \leq x < 9$

**13.** $h > 5.3$ or $h < -2$

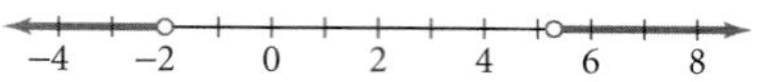


**15.** $-3 \leq t < -1$

**17.** $-5 < s \leq 4$ **19.** $1 \leq t < 2\frac{2}{5}$ **21.** $-3 < g \leq 10$ **23.** $1 \leq h < 5$ **25.** $t < 5$ or $t > 6$ **27.** $-4 < t \leq 5\frac{1}{2}$ **29.** $-\frac{1}{2} < b \leq 3.2$ **31.** $d > 5$ or $d \leq 2$

### Lesson 6.4

**1.** 16 **3.** $-4.62$ **5.** 1.03 **7.** $1\frac{3}{10}$ **9.** $\frac{7}{8}$ **11.** 0.064 **13.** $-1$ **15.** 15 **17.** 1 **19.** 3 **21.** 30 **23.** domain all reals, range: all non negatives **25.** domain all reals, range: all non positive **27.** domain all reals, range: all non positive **29.** domain all reals, range: reals $\leq 6$

### Lesson 6.5

**1.** $x = 2$ or $x = -6$ **3.** $x = 7$ or $x = -5$ **5.** $x = -2$ or $x = -18$ **7.** $x = 8$ or $x = 12$ **9.** $x = 4$ or $x = -4\frac{2}{3}$ **11.** $x = 2$ or $x = -1\frac{1}{2}$ **13.** $x = -1$ or $x = -\frac{1}{2}$ **15.** $x = 2$ or $x = -4\frac{2}{5}$ **17.** $x = 3\frac{1}{2}$ **19.** $x = \frac{1}{2}$ or $x = 4\frac{1}{2}$ **21.** $x > -1$ or $x < -3$ **23.** $x \geq 11$ or $x \leq 3$ **25.** $-2 < x < 8$ **27.** $x = 2$ **29.** $x \leq -1\frac{1}{3}$ or $x \geq 4\frac{2}{3}$ **31.** $x \geq 4$ or $x \leq -4$ **33.** $2 < x < 8$ **35.** all real numbers **37.** $-4 \leq x \leq 2$ **39.** $1 < x < 3$

## CHAPTER 7

### Lesson 7.1

**1.** $(0.9, -0.1)$ **3.** $(0, 2)$ **5.** $(3.9, 0.8)$ **7.** $(2.5, 4)$ **9.** $(-2.5, -2.5)$ **11.** $(1.8, 23)$ **13.** yes **15.** no

### Lesson 7.2

**1.** $(1, 3)$ **3.** $(-2, 4)$ **5.** $(4, -5)$ **7.** $(-3, 10)$ **9.** $(0.5, -0.25)$ **11.** $(3, 0.5)$ **13.** $(0.75, -0.75)$ **15.** $(0.4, 1)$ **17.** $\left(-2\frac{2}{3}, \frac{1}{3}\right)$

### Lesson 7.3

**1.** $(8, -1)$ **3.** $(2, 0.5)$ **5.** $(1, 1)$ **7.** $\left(1\frac{2}{9}, \frac{8}{9}\right)$ **9.** $(7, -2)$ **11.** $\left(1\frac{1}{3}, 3\right)$ **13.** $(6, 9)$ **15.** $(4, -28)$ **17.** $(3, 3)$ **19.** $(-12.8, 6.2)$ **21.** $(-3, 2)$

### Lesson 7.4

**1.** $\left(\frac{16}{29}, -\frac{7}{29}\right)$ **3.** $(36, -26)$ **5.** No solution **7.** $\left(3\frac{4}{7}, \frac{5}{7}\right)$ **9.** $(1, 1)$ **11.** all real numbers **13.** inconsistent **15.** independent **17.** independent **19.** dependent **21.** inconsistent

### Lesson 7.5

**1.**

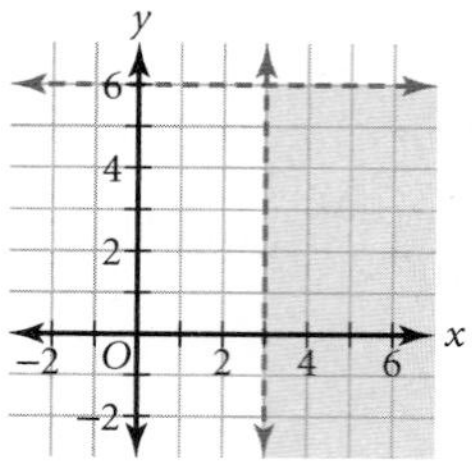


**3.**

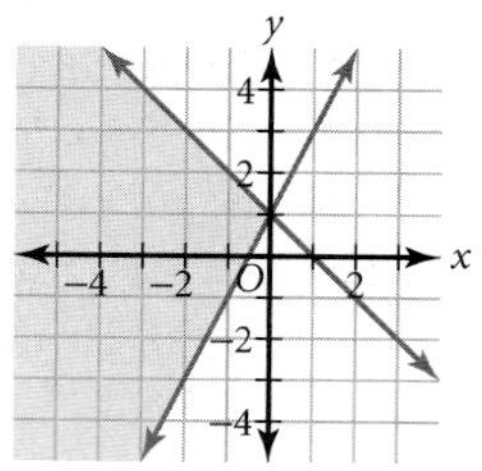

**5.**

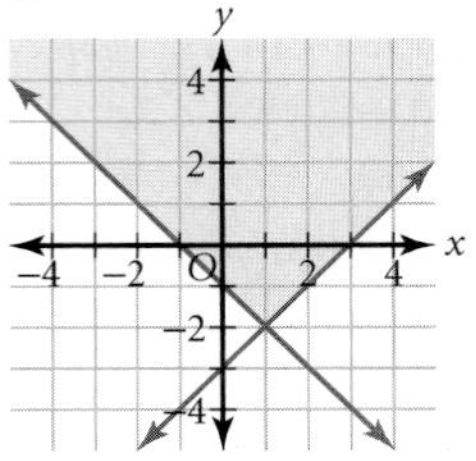


**7.** no common solution **9.** no common solution

**Lesson 7.6**

**1.** 56 dimes and 27 nickels **3.** speed of wind: 25 miles per hour speed of plane with no wind: 425 miles per hour **5.** Tonia is 25years old. Her father is 50 years old. **7.** 20 quarters and 7 nickels

## CHAPTER 8

**Lesson 8.1**

**1.** 64 **3.** 4096 **5.** 49 **7.** 6561 **9.** 810,000 **11.** 10,000 **13.** 512 **15.** 216 **17.** 100,000 **19.** 2187 **21.** 78,125 **23.** 6561 **25.** 343 **27.** $32d^7e^{11}$ **29.** $-16q^8p^{17}$ **31.** $-16w^{10}x^9y^{11}$ **33.** $32c^4d^8e^4$

**Lesson 8.2**

**1.** $\frac{1}{16}$ **3.** $-16$ **5.** $\frac{1}{16}$ **7.** $\frac{1}{1000}$ **9.** $-36$ **11.** $\frac{1}{25}$ **13.** $\frac{r^2}{s^3}$ **15.** $d^4$ **17.** $\frac{a}{c^4}$ **19.** $\frac{256z^{32}}{y^{24}}$ **21.** $w^{18}z^{12}$ **23.** $j^{20}k^{30}$ **25.** $-c^{15}v^5$ **27.** $a^{21}b^5$ **29.** $b^{24}c^{51}d^3$ **31.** $512n^{15}p^{36}$ **33.** $5625a^4b^{14}$

**Lesson 8.3**

**1.** 100, **3.** 343 **5.** $f^3$ **7.** $v^{x-y}$ **9.** $k^5$ **11.** $4hr^3$ **13.** $\frac{s^3}{t^3}$ **15.** $\frac{161{,}051}{x^{15}}$ **17.** $\frac{h^{32}}{g^{48}}$ **19.** $\frac{256z^{32}}{y^{24}}$ **21.** $\frac{-r^{15}}{27q^6}$ **23.** $\frac{1}{f^2}$ **25.** $0.5a^4y^3$ **27.** $a^{22}g^{45}y^{63}$

**Lesson 8.4**

**1.** $\frac{1}{16}$ **3.** $-16$ **5.** $\frac{1}{16}$ **7.** $\frac{1}{1000}$ **9.** $-36$ **11.** $\frac{1}{25}$ **13.** $\frac{r^2}{s^3}$ **15.** $d^4$ **17.** $\frac{h^{32}}{g^{48}}$ **19.** $-\frac{y^3}{z^5}$ **21.** $-\frac{1}{w^5}$ **23.** $\frac{1}{f^2}$ **25.** $n^6$ **27.** $a^{22}g^{45}y^{63}$ **29.** $\frac{4}{d^9}$ **31.** $\frac{2r^7}{s}$ **33.** $15a^6$ **35.** $4b$ **37.** 1 **39.** $-\frac{r^3}{t^5}$

**Lesson 8.5**

**1.** $4 \times 10^6$ **3.** $2.5 \times 10^6$ **5.** $4.4 \times 10^{-5}$ **7.** $5.6 \times 10^{-4}$ **9.** $3.45 \times 10^{-4}$ **11.** $7.95 \times 10^5$ **13.** 4000 **15.** 70,000,000 **17.** 0.0017 **19.** −970,000 **21.** $1.2 \times 10^{10}$ **23.** $1.4 \times 10^{17}$ **25.** $10 \times 10^4$ **27.** $1 \times 10^5$ **29.** $3.9 \times 10^9$ **31.** $3.6 \times 10^{17}$ **33.** $2 \times 10^2$ **35.** $3.5 \times 10^7$ **37.** $6 \times 10^3$

**Lesson 8.6**

**1.**

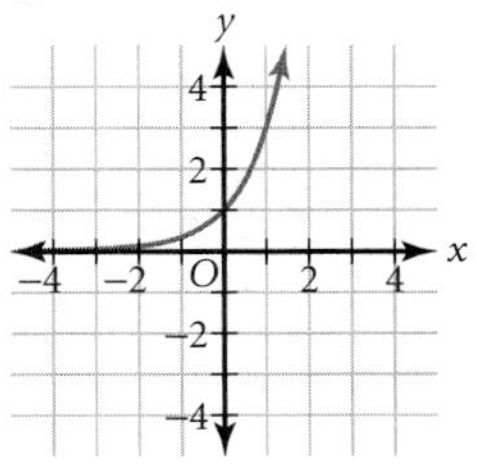


**3.**

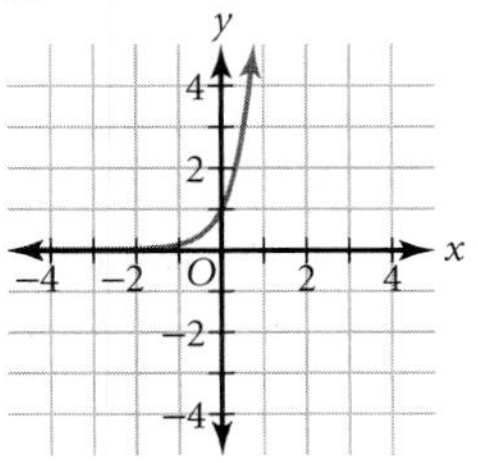


**9.** 9 **11.** 3 **13.** $\frac{1}{9}$ **15.** $\frac{1}{3}$ **17.** 81 **19.** 307,272

**Lesson 8.7**

**1.** about 12,000 years **3.** about 5700 years **5.** about 2900 years **7.** $8701.24 **9.** $13,058.21 **11.** $6638.13 **13.** $5,418.69 **15.** $47,044.80 **17.** $44,291.75 **19.** $48,974.59 **21.** $50,473.71 **23.** about 3 years **25.** about 20 years **27.** about 34 years **29.** about 51 years

## CHAPTER 9

**Lesson 9.1**

**1.** $x^2 + x + 7$ **3.** $-3w^2 + w + 8 + y + 3$ **5.** $-r3 + r$ **7.** 2 **9.** 3 **11.** $10x^4 + 5x^2 + 2$ **13.** $8z^4 - 3z^2 - 2z + 4$ **15.** $x^3 + 2x^2 - 1$ **17.** $x^4 - 5x^2 + 5$ **19.** $w^4 + 4w^2 - 5w - 5$ **21.** $-x^2 + 9x + 3$ **23.** $2x^2 - 10x + 10$ **25.** $3.5x^4 + 7$

**Lesson 9.2**

**1.** $5x + 20$ **3.** $7x + 77$ **5.** $6x - 18$ **7.** $-3x + 6$ **9.** $2x - x^2$ **11.** $-4x - x^2$ **13.** $5x^2 + 15x$ **15.** $-x^2 + 6x$ **17.** $x^2 + 5x + 6$ **19.** $a^2 - 2a - 8$ **21.** $d^2 + 7d + 12$ **23.** $c^2 + 12c + 36$ **25.** $m^2 - 6m + 8$ **27.** $s^2 + s - 12$ **29.** $s^2 + 5s - 14$ **31.** $w^2 + 5w - 6$ **33.** $w^2 - 64$ **35.** $m^2 - m - 100$ **37.** $a^2 - 11a + 10$

**Lesson 9.3**

**1.** $w^2 + 12w + 20$ **3.** $y^2 - 12y + 27$ **5.** $w^2 + 5w - 24$ **7.** $4x^2 - 5x - 6$ **9.** $15x^2 + 4x - 4$ **11.** $12w^2 - 19w + 5$ **13.** $r^2 - s^2$ **15.** $a^2 + ab - 2b^2$ **17.** $p^2 + 2pq + q^2$ **19.** $m^4 - n^2$ **21.** $x^2 + 9x + 18$ **23.** $x^2 - 121$ **25.** $4x^2 - 24x + 36$ **27.** $x^2 - 36$ **29.** $4 - x^2$ **31.** $64 - 16x + x^2$ **33.** $49x^2 + 42x + 9$ **35.** $25x^2 - 60x + 36$ **37.** $x^2 + 8x + 16$

**Lesson 9.4**

**1.** $S = 6x^2$; $V = x^3$ **3.** $S = 24x + 72$; $V = 36x$ **5.** $S = 27.4x + 79.8$; $V = 39.9x$ **7.** True **9.** True **11.** True **13.** True **15.** True **17.** False **19.** True **21.** False

### Lesson 9.5

**1.** can be factored **3.** can be factored **5.** can be factored **7.** can be factored **9.** cannot be factored **11.** can be factored **13.** $4(x^2 + 4)$ **15.** $4a(2a - 1)$ **17.** $25d^2(1 - 3d)$ **19.** $3st^3(s + 5)$ **21.** $5ab^2(5b^2 + 4a^2)$ **23.** $(7 - x)(y + 4)$ **25.** $(4 - v)(w - 3)$ **27.** $(s - t)(x + 5)$ **29.** $(3w - 4)(x + y)$ **31.** $(5 - y)(3a + 4)$ **33.** $(m + 2)(r + 3)$ **35.** $(z + 3)(r + 4)$ **37.** $(c + 7)(l + 2)$ **39.** $(r - 5)(s - 6)$ **41.** $(m - 2)(a - 6)$

### Lesson 9.6

**1.** $x^2 + 4x + 4$ **3.** $m^2 - 8m + 16$ **5.** $16y^2 - 40y + 25$ **7.** $x^2 - 9y^2$ **9.** $25a^2 - 9$ **11.** $20z$ **13.** $4y^2$ **15.** $9x^2$ **17.** $(3t + 1)(3t - 1)$ **19.** $(x - 2)^2$ **21.** $(5w - 4)(5w + 4)$ **23.** $(x + y)(x - y)$ **25.** $(q - 10)^2$ **27.** $(5r + 2)^2$ **29.** $(5b - 1)^2$ **31.** $2(x + 2)(x - 2)$ **33.** $(x - 5)^2$ **35.** $(x + y)^2$

### Lesson 9.7

**1.** $(x + 2)(x - 5)$ **3.** $(x + 5)(x + 4)$ **5.** $(x + 10)(x - 2)$ **7.** $(x - 8)(x - 1)$ **9.** $(x + 8)(x + 2)$ **11.** $(x - 3)(x - 5)$ **13.** $(s + 21)(s + 1)$ **15.** $(h + 13)(h + 2)$ **17.** $(x - 7)(x - 2)$ **19.** $(y + 7)(y - 7)$ **21.** $(z + 12)(z + 2)$ **23.** $2n(14n - 9)$ **25.** prime **27.** $(b + 4m)(y + 3x)$ **29.** prime **31.** $(y + 3)(y + 4)$ **33.** $2(y - 1)(2y + 3)$ **35.** $6(n^2 - 4n + 1)$

### Lesson 9.8

**1.** 5 and −4 **3.** −11 and 10 **5.** −2 **7.** −9 and −6 **9.** 4 **11.** 2.9 and −3.2 **13.** 5 **15.** 13 **17.** 0 and 5 **19.** −1 and −5 **21.** 2 and −2 **23.** −3 **25.** 8 and −8 **27.** −11 and −12 **29.** −8 and −2 **31.** −5 **33.** −10 **35.** 1 **37.** −2 and −11 **39.** 6 and 3 **41.** −6 and 3 **43.** −9 and −4 **45.** 9 and −4

## CHAPTER 10

### Lesson 10.1

**1.** Horizontal shift by 3 units to the right, vertical shift up 1 unit. **3.** Horizontal shift by 1 unit to the left, vertical shift 2 units down. **5.** Horizontal shift by 4 units to the right, opens downward. **7.** Horizontal shift by 1 unit to the right, vertical shift 1 unit down. **9.** Horizontal shift by 2 units to the left, vertical shift by 5 units down, opens downward. **11.** Horizontal shift by 9 units to the left, vertical shift by 3 units down, opens downward. **13.** $(-1, -2)$, $x = -1$ **15.** $(3, 4)$, $x = 3$ **17.** $(-3, 0)$, $x = -3$ **19.** −4, −3 **21.** −5, 6 **23.** −5, 4 **25.** 1 **27.** −9, −1 **29.** −3, −1; $x = -2$; $(-2, -1)$ **31.** −5, −3; $x = -4$; $(-4, -1)$ **33.** −4, 6; $x = 1$; $(1, -25)$

### Lesson 10.2

**1.** 10 **3.** 22 **5.** 7.07 **7.** 5.29 **9.** 9.95 **11.** 14.14 **13.** 9.75 **15.** 11.62 **17.** 6, −6 **19.** 8, −8 **21.** $\frac{4}{5}$, $-\frac{4}{5}$ **23.** −8, 0 **25.** −3, 9 **27.** −2.46, 4.46 **29.** $(9, -16)$; $x = 9$; 5, 13 **31.** $(4, -2)$; $x = 4$; 2.59, 5.41

### Lesson 10.3

**1.** +36 **3.** + 4 **5.** +225 **7.** +225 **9.** $(0, 5)$ **11.** $(0, 15)$ **13.** $(0, 16)$ **15.** $y = (x - 3)^2 - 9$; $(3, -9)$ **17.** $y = (x + 1)^2 - 1$; $(-1, -1)$ **19.** $y = (x - 1)^2 - 4$; $(1, -4)$ **21.** $y = (x - 5)^2 - 4$; $(5, -4)$ **23.** $y = (x + 2.5)^2 - 2.25$; $(-2.5, -2.25)$ **25.** $y = \left(x - \frac{1}{8}\right)^2 + 5\frac{63}{64}$; $\left(\frac{1}{8}, 5\frac{63}{64}\right)$

### Lesson 10.4

**1.** −1, 3 **3.** 2, 6 **5.** 2, 4 **7.** −1, 2 **9.** −2, 3 **11.** −7, 4 **13.** −7, 1 **15.** −9, 5 **17.** −3, 4 **19.** 2, 10 **21.** −4, −3 **23.** 0, 4 **25.** −1, 2 **27.** −6, 1 **29.** $(3, 1)$ $(2, 0)$

### Lesson 10.5

**1.** 1, −4, 5 **3.** 8, −5, 2 **5.** −1, 7, −6 **7.** 25, 2 **9.** 44, 2 **11.** −11, 0 **13.** 16, 2 **15.** 49, 2 **17.** $x = -5, -3$ **19.** $x = 0, 6$ **21.** $x = -10, 0$ **23.** $x = -\frac{3}{4}, -\frac{1}{2}$ **25.** $x = 4, 5$ **27.** $x = -7, -6$ **29.** $x = -1.85, 1.35$ **31.** $x = -0.44, -1.36$ **33.** $x = 1.43, 0.23$ **35.** $x = 6.14, -1.14$

### Lesson 10.6

**1.** $x > 4$ or $x < -3$ **3.** $x \geq 2$ or $x \leq -2$ **5.** $3 < x < 7$ **7.** $-3 < x < 3$ **9.** $x > 3$ or $x < 1$ **11.** $-3 \leq x \leq 1$

**13.**

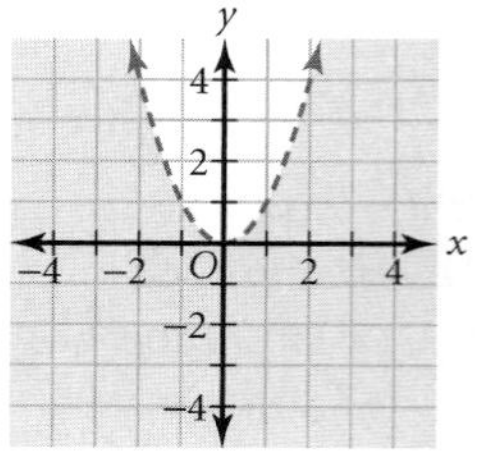


**15.**

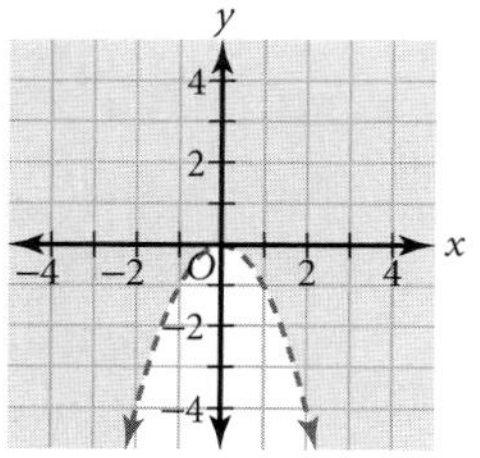


**17.**

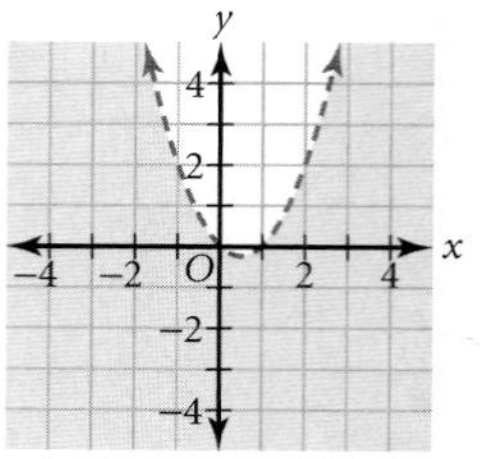


**19.**

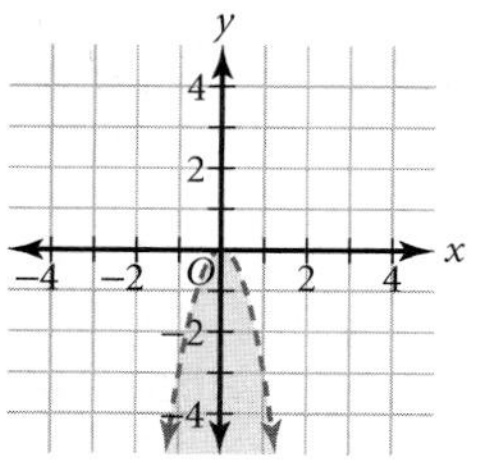


**21.**

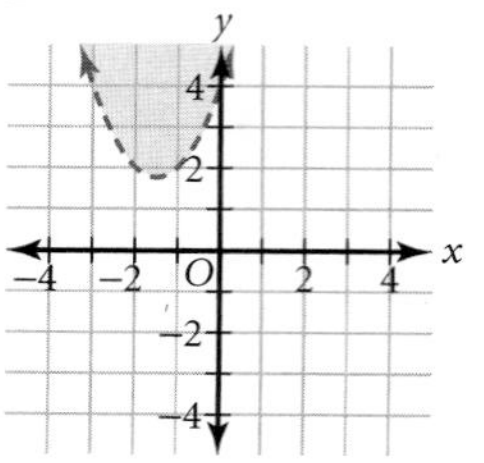


**23.**

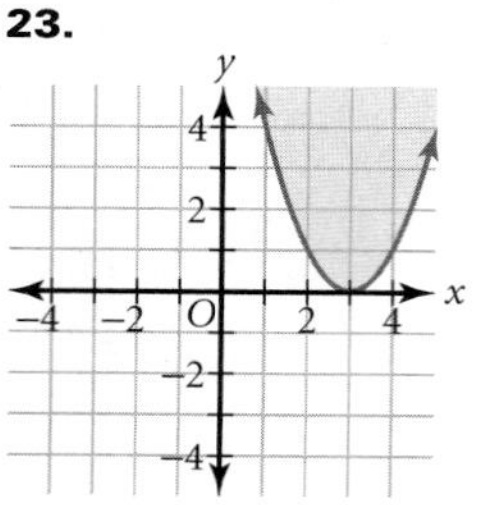

## CHAPTER 11

**Lesson 11.1**

**1.** yes **3.** no **5.** no **7.** no **9.** yes **11.** yes **13.** 48 **15.** 3 **17.** $11\frac{2}{3}$ **19.** $-2\frac{2}{7}$ **21.** 0.5 **23.** $\frac{1}{4}$ **25.** 3.5 **27.** 4.2 **29.** 12.15

**Lesson 11.2**

**1.** $c = 0$ **3.** $b = -3$ **5.** $c = -\frac{1}{2}$ **7.** $y = 2$ **9.** $g = 0$ **11.** $z = -2, -3$ **13.** $\frac{2}{7}, -1$ **15.** $\frac{5}{6}$, undefined **17.** undefined, $-\frac{10}{9}$ **19.** $-\frac{17}{3}$, undefined **21.** $-\frac{1}{2}, -\frac{1}{5}$ **25.** $x \neq 0$ **27.** $x \neq 3$

**Lesson 11.3**

**1.** $x = -15$ **3.** $w = 0$ **5.** $t = 6$ **7.** $s = 2$ **9.** 4 **11.** $x + 3$ **13.** $\frac{x+3}{2(x-1)}$ **15.** $\frac{2}{7}$ **17.** $w - 2$ **19.** $\frac{2a+3b}{a}$ **21.** $w - 5$, $w \neq -5$ **23.** $\frac{g+3}{g+9}$, $g \neq -9, 3$

**Lesson 11.4**

**1.** $\frac{3}{y}$, $y \neq 0$ **3.** $\frac{-5m}{n+3}$, $n \neq -3$ **5.** $1$, $x \neq -17$ **7.** $\frac{6b-c}{ab^2}$, $a \neq 0, b \neq 0$ **9.** $\frac{2n}{m-2}$, $m \neq 2$ **11.** $\frac{2g-7}{g-1}$, $g \neq 1$ **13.** $\frac{7-x}{m+n}$, $m \neq -n, y \neq -4$ **15.** $\frac{x-3}{x}$, $x \neq -3, 0$ **17.** $\frac{12}{r^2-3r}$, $r \neq 3, 0$ **19.** $\frac{4x+13}{x^2+x-2}$, $x \neq -2, 1$ **21.** $\frac{v^2-2v+12}{v^2-v-6}$, $v \neq -2, 3$ **23.** $\frac{h+17}{h^2+6h+8}$, $h \neq -4, -2$

**Lesson 11.5**

**1.** $x = 29$ **3.** no solution **5.** $w = 5$ **7.** $z = 10$ **9.** $y = -2$ **11.** $x = -3$ **13.** $q = -4$ **15.** $a = 3$ **17.** $x = -4$ **19.** $x = 28$ **21.** $x = 6.2$

**Lesson 11.6**

**1.** Subtraction Property of Equality **5.** Distributive **7.** Associative **9.** Addition and Division Properties of Equality **11.** Addition and Multiplication Properties of Equality **13.** Addition and Multiplication Properties of Equality

## CHAPTER 12

**Lesson 12.1**

**1.** 6 **3.** 18 **5.** $6\sqrt{3}$ **7.** $8\sqrt{3}$ **9.** $11\sqrt{5}$ **11.** $18\sqrt{6}$ **13.** $7\sqrt{10}$ **15.** $\sqrt{133}$ **17.** 9 **19.** $\sqrt{67}$ **21.** 40 **23.** $2\sqrt{11}$ **25.** 10 **27.** $7\sqrt{42}$ **29.** 2 **31.** $\sqrt{10}$ **33.** 48 **35.** 128 **37.** $3\sqrt{6} + 12$ **39.** $3\sqrt{6} - 6\sqrt{3}$ **41.** 2 **43.** $3\sqrt{3} - 7$

**Lesson 12.2**

**1.** $x = 4$ **3.** $x = 73$ **5.** $x = 20$ **7.** $x = 4$ **9.** $x = 8$ **11.** $x = 0$ **13.** $x = 6$ **15.** $x = 5$ **17.** $x = 2.4$ **19.** $x = \pm 4\sqrt{5}$ **21.** $x = \pm 7\sqrt{2}$ **23.** $x = \pm\frac{\sqrt{5}}{4}$ **25.** $x = \pm 3$ **27.** $x = \pm 2\sqrt{3}$ **29.** $x = -2 \pm \sqrt{2}$ **31.** $x = \pm 4$ **33.** $x = \pm 16$ **35.** Not possible; a principle square root cannot be negative.

**Lesson 12.3**

**1.** 6.3 **3.** 8 **5.** 16.6 **7.** 17.9 **9.** 23.4 **11.** 26 **13.** 9 **15.** 9.4 1 **17.** 9.8

**Lesson 12.4**

**1.** $Q(1, 5)$; 2, 3, $\sqrt{13}$ **3.** $Q(1, 1)$; 5, 5, $5\sqrt{2}$ **5.** 15 **7.** 9.2 **9.** 8.6 **11.** $(-2, -2)$ **13.** $(5.5, 4.5)$ **15.** $(10, 10)$

**Lesson 12.5**

**1.** $x^2 + y^2 = 9$ **3.** $x^2 + y^2 = 100$ **5.** $x^2 + y^2 = 1$ **7.** $x^2 + y^2 = 121$ **9.** $x^2 + y^2 = 400$ **11.** $x^2 + y^2 = 1024$ **13.** $(x-1)^2 + (y-2)^2 = 4$ **15.** $(x-1)^2 + (y+2)^2 = 4$ **17.** $(x-2)^2 + (y-2)^2 = 9$ **19.** $(x+3)^2 + y^2 = 1$ **21.** center $(1, 2)$; radius 4 **23.** center $(-3, 4)$; radius 5 **25.** center $(-7, 1)$; radius $\sqrt{3}$ **27.** center $(3, 9)$; radius $\sqrt{11}$ **29.** center $(0, 1)$; radius $\sqrt{13}$ **31.** center $(3, 0)$; radius 11 **32.** center $(5, 10)$; radius 10

**Lesson 12.6**

**1.** $\tan Y = \frac{7}{24}$; $\tan Z = \frac{24}{7}$ **3.** 1.7321 **5.** 0.0349 **7.** 0.9965 **9.** 25.0° **11.** 7.0° **13.** 34.0° **15.** 52.0° **17.** 10.8 **19.** 13.3 **21.** 35.8°

**Lesson 12.7**

**1.** 0.7071 **3.** 0.9925 **5.** 0.0349 **7.** 0.9820 **9.** 0.7986 **10.** 0.7278 **11.** 0.9659 **13.** 17.0° **15.** 3.0° **17.** 36.0° **19.** 46.4° **21.** 41.8° **23.** 41.4°

**Lesson 12.8**

**1.** $\begin{bmatrix} -6 & 27 \\ -1 & -20 \end{bmatrix}$ **3.** $\begin{bmatrix} 9 & -27 & -55 \\ 72 & -19 & 8 \end{bmatrix}$

**5.** $\begin{bmatrix} 6.3 & 7.3 \\ 9.1 & 3.1 \\ 7.9 & 10.2 \end{bmatrix}$ **7.** $\begin{bmatrix} 13 & 3.5 & 7.5 \\ -1 & 5 & 10 \\ 21 & 9 & 8 \end{bmatrix}$

**9.** not possible because the number of columns in the first matrix does not equal the number of rows in the second matrix.

## CHAPTER 13

**Lesson 13.1**

**1.** $\frac{26}{51}$ **3.** $\frac{6}{51}$ **5.** $\frac{17}{51}$ **7.** $\frac{1}{51}$ **9.** $\frac{1}{26}$ **11.** $\frac{3}{26}$ **13.** $\frac{1}{2}$

**Lesson 13.2**

**1.** none **3.** 24, 30, 36 **5.** 21, 23, 24, 25, 27, 29, 30, 31, 33, 35, 36, 37, 39 **7.** 20, 22, 24, 25, 26, 28, 30, 32, 34, 35, 36, 38, 40 **9.** 20, 21, 23, 25, 27, 29, 30, 31, 33, 35, 37, 39, 40 **11.** 285 **13.** 136 **15.** 91 **17.** 277 **19.** 573

**Lesson 13.3**

**1.** 360 **3.** 120 **5.** 60 **7.** 4 **9.** 26 **11.** 104 **13.** 676 **15.** 24 **17.** 40320

**Lesson 13.4**

**1.** $\frac{16}{81}$ **3.** $\frac{25}{81}$ **5.** $\frac{40}{81}$ **7.** $\frac{20}{81}$ **9.** $\frac{25}{441}$ **11.** $\frac{4}{441}$ **13.** $\frac{60}{44}$ **15.** $\frac{10}{441}$

**Lesson 13.5**

**Answers for Exercises 1–6 may vary.**

## CHAPTER 14

**Lesson 14.1**

**1.** No, the $x$-value of 2 is paired with two $y$-values, 3 and 8. **3.** Yes, each $x$-value is paired with exactly one $y$-value. **5.** No, there are $x$-values paired with more than one $y$-value. **7.** 12 **9.** 24 **11.** −48 **13.** $\frac{1}{2}$ **15.** −4 **17.** 21 **19.** −25 **21.** 2.5

**Lesson 14.2**

**1.** $y = x^2$; A vertical reflection over the $x$-axis and a stretch by a factor of 2 **3.** $y = x^2$; A shift 1 unit to the right **5.** $y = |x|$; A vertical reflection over the $x$-axis and a stretch by a factor of 3 **7.** $y = |x|$; A shift to the left by 2 units and shift up 2 units **9.** $y = |x|$; A vertical reflection over the $x$-axis and a stretch by a factor of 2 **11.** $y = |x|$; A shift to the left by 1 unit **13.** $y = x$; A shift up 6 units **15.** $y = x^2$; A shift 2 units to the right **17.** Shift 2 units up **19.** Shift 2 units to the left **21.** (6, 9) is translated to (6, 14) **23.** (6, 9) is translated to (6, 0) **25.** (6, 9) is translated to (8, 9) **27.** (6, 9) is translated to (−4, 9) **29.** stretch **31.** stretch

**Lesson 14.3**

**1.** $y = x^2$ **3.** The graph of $y = x^2$ is stretched by a factor of 3 to become the graph of $y = 3x^2$.

**5.**

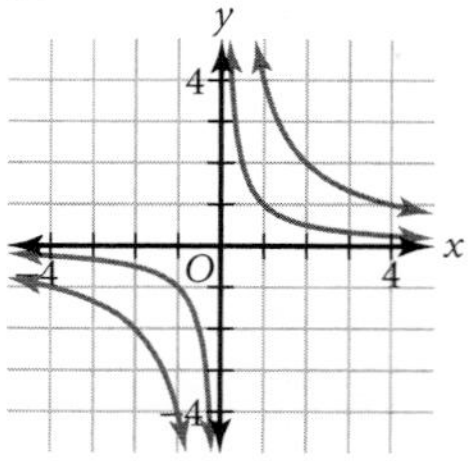


**7.** $y = |x|$ **9.** The graph of $y = |x|$ is stretched by a factor of 5 to become the graph of $y = 5|x|$. **10.** no **11.** yes **12.** yes **13.** no

**Lesson 14.4**

**1.**

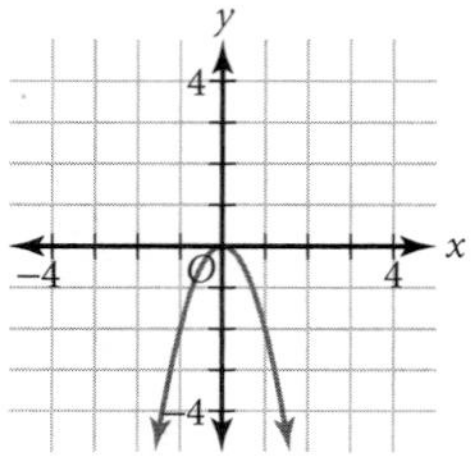


**3.**

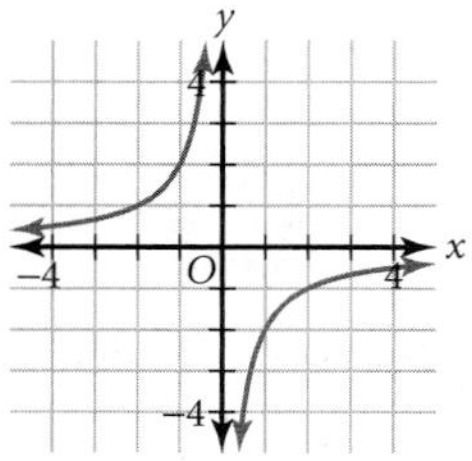


**5.**

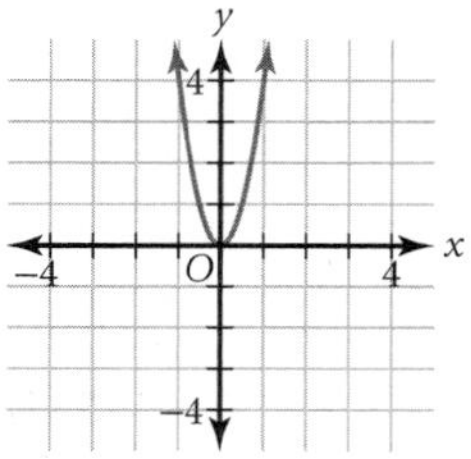


**7.** not a vertical reflection **9.** vertical reflection **11.** vertical reflection **13.** vertical reflection **15.** not a vertical reflection **17.** $f(3) = 3$; not a vertical reflection

**Lesson 14.5**

**1.**

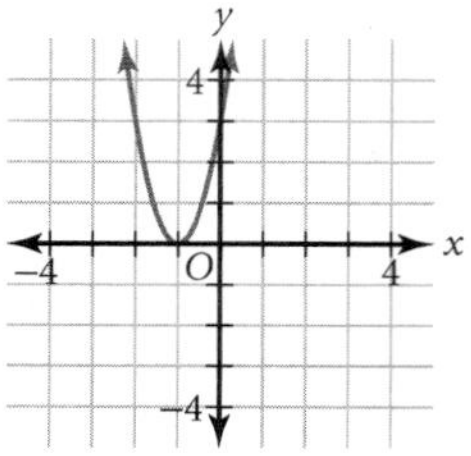


**3.**

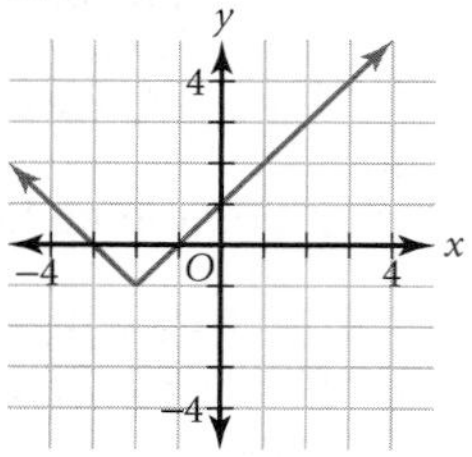


**5.**

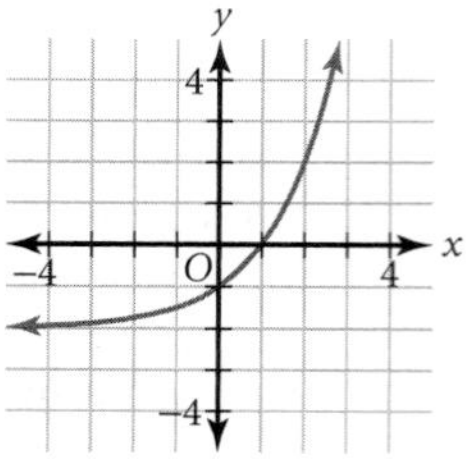


**7.** (4, 6) is translated to (4, 14) **9.** (4, 6) is translated to (4, −8) **11.** (4, 8) is translated to (4, 18) **13.** (4, 8) is translated to (4, −10)

# INDEX



## A

## N



## O



## P

# Credits



## PHOTOS

Abbreviations used: (t) top, (c) center, (b) bottom, (l) left, (r) right, (bckgd) background, (bdr) border.
**FRONT COVER:** (bckgd), Index Stock Photography Inc./Ron Russell; (b), Jean Miele MCMXCII/The Stock Market. **TABLE OF CONTENTS:** Page vi(tl), John Langford/HRW Photo; vi(cl), C. C. Lockwood/Bruce Coleman Inc.; vii(tl), Don Couch/HRW Photo; vii(cl), Stuart Westmorland/Tony Stone Images; viii(tl), Peter Van Steen/HRW Photo; viii(cl), Bob Anderson/Masterfile; ix(tl), David Ducros/Science Photo Library/Photo Researchers Inc.; ix(cl), Comstock; x(tl), Randal Alhadeff/HRW Photo; x(cl), Sarah Stone/Tony Stone Images; xi(tl), SuperStock; xi(cl), Brett Froomer/The Image Bank © 2000; xii(tl), Randal Alhadeff/HRW Photo, props courtesy Recreational Equipment Inc.; xii(cl), David Phillips/HRW Photo. **CHAPTER ONE:** Page 2(br),(bc), Chuck Place/Place Stock Photo; 2(bl), Stephen Trimble, pottery courtesy Richard M. Howard Collection; 2(cr), Tim Davis/Tony Stone Images; 2(tr), Color Box/Masterfile; 2(cr), Comstock/D. Lada/H. Armstrong Roberts; 2(cr), Planet Earth Pictures/FPG International; 3(tr), Artbase Inc.; 3(t), Russell D. Curtis/Photo Researchers Inc.; 3(cl), Andrew Rafkind/Tony Stone Images; 3(bl), James Robinson/Masterfile; 4(t), Warren Faidley/First Light; 7(b), John Langford/HRW Photo; 9(bl), Randal Alhadeff/HRW Photo(teapot); 9(bl), Comnet/First Light(steam); 10(c), eStudios/HRW Photo; 11(t), CORBIS/Joel W. Rodgers; 12(bl), C. Eugene Gebhardt/Masterfile; 13(bl), Sam Dudgeon/HRW Photo; 14(c), Map Art Publishing Corporation; 16(bl), Don Couch/HRW Photo; 17(bl),(br), eStudios/HRW Photo; 18(t), Peter Van Steen/HRW Photo; 20(tl), Artbase Inc.; 23(tr), Sam Dudgeon/HRW Photo; 24(t), Frog Map Co.; 26(cr), Stephen Marks/The Image Bank © 2000; 28(br), Randal Alhadeff/HRW Photo; 29(tl), Corbis-Bettmann; 29(br), James Robinson/Masterfile; 30(t), Peter Van Steen/HRW Photo; 31(br), Jeff Barber/Infocus/The Victoria Clipper; 32(bl), Mark Wagner/Tony Stone Images; 32(bl), Adam Jones/Photo Researchers Inc.; 33(bl), Allsport USA/Gerard Planchenault; 33(bl), Lester Lefkowitz/Masterfile; 34(cl), Hoverspeed; 35(br), Bill Stoughton/Austin Convention & Visitors Bureau; 37(t), Eric Meola/The Image Bank © 2000; 38(cl), Randal Alhadeff/HRW Photo; 39(tl), Edgar T. Jones/Bruce Coleman Inc.; 39(br), A. T. Willett/The Image Bank © 2000; 42(bl), Don Couch/HRW Photo; 44(cl), Kathleen Campbell/Allstock/PNI; 45(tr), eStudios/HRW Photo; 45(b), John Langford/HRW Photo.
**CHAPTER TWO:** Page 52(all), The Granger Collection, New York; 53(tl),(tc),(tr), al-Khwarizmi; 53(c), Artbase Inc.; 54(cr), Marian Bacon/Animals Animals; 54(tl), Hans Reinhard/Bruce Coleman Inc.; 54(cl), Gail Shumway/FPG International; 59(bl), Peter Van Steen/HRW Photo; 60(tr), John Langford/HRW Photo; 64(bl), Sam Dudgeon/HRW Photo; 66(br), MTPA Stock/Masterfile; 67(tr),(tc), John Langford/HRW Photo; 71(bc), C. C. Lockwood/Bruce Coleman Inc.; 72(tr), Blair Seitz/Photo Researchers Inc.; 73(tr),(tc), Sam Dudgeon/HRW Photo; 78(br), David Madison/Tony Stone Images; 79(cr), Tom Kitchin/First Light; 80(t), Peter Van Steen/HRW Photo; 82(tr), Peter Van Steen/HRW Photo; 83(br), John Langford/HRW Photo; 87(b), Michelle Bridwell/HRW Photo; 88(bl), Everett Johnson/Tony Stone Images; 94(l), Kim Westerskov/Tony Stone Images; 94(bl),95(cr), NYT Graphics © *The New York Times*; 96(tr), Robert Holmes; 99(br), D. Mason/First Light; 99(cr), eStudios/HRW Photo; 100(tc), Scott Markewitz/Masterfile; 100(tr), Chris Cole/The Image Bank © 2000; 103(cr), Comstock; 103(cr), Artbase Inc.; 104(bl), Sam Dudgeon/HRW Photo; 105(tr), Spaceshots; 105(b), Stephen Simpson/FPG International. **CHAPTER THREE:** Page 112(c), Comstock; 112(t), R. Essel/First Light; 112(b), Jon Feingersh/First Light; 113(tr), Allsport USA/Otto Creule; 113(tl), David Oliver/Tony Stone Images; 114(tr), Michelle Bridwell/PhotoEdit; 117(br), © Stockbyte; 119(tr), Mark E. Gibson; 120(cr), Don Smetzer/Tony Stone Images; 121(br), Richard Ashworth/Robert Harding; 122(t), Michelle Bridwell/HRW Photo; 124(tl), Bob Daemmrich; 126(tl), Image Copyright ©2001 PhotoDisc, Inc.; 127(bl), Don Couch/HRW Photo; 128(cl),(c),(cr),129(tr), John Langford/HRW Photo; 131(cr), Michelle Bridwell/Frontera Fotos; 134(tr), Stan Goldberg/SuperStock; 135(tr), Sam Dudgeon/HRW Photo; 136(cr), Christine Galida/HRW Photo; 139(bl), Allsport USA/Todd Warshaw; 139(br), Michael Freeman/Bruce Coleman Inc.; 140(tr), David Boyle/Earth Scenes; 141(tr), Michelle Bridwell/HRW Photo; 146(tr), Christine Galida/HRW Photo; 147(tc),(tl), Gay Bumgarner/Tony Stone Images; 147(tr), Artbase Inc.; 149(cr), 152(tr), Peter Van Steen/HRW Photo; 152(cr), Corbis/Joseph Sohm/ChromoSohm Inc.; 152(br), Greg Pease/Tony Stone Images; 153(all), David R. Frazier Photolibrary; 154(t), Peter Van Steen/HRW Photo; 155(b), John Langford/HRW Photo. **CHAPTER FOUR:** Page 162(cr), John Langford/HRW Photo; 163(cl), Joseph Pobereskin/Tony Stone Images; 163(tr), Corbis/Kevin R. Morris; 163(tl), Don Couch/HRW Photo; 164(tr), Michelle Bridwell/HRW Photo; 167(br), Don Couch/HRW Photo; 168(cr), Robert Lubeck/Animals Animals; 169(b), Sam Dudgeon/HRW Photo; 171(tc), Randy Faris/First Light; 171(c), Dietrich Stock Photos, Inc.; 171(tr), Paul Harris/Tony Stone Images; 173(tl), Scala/Art Resource, NY; 174(bl), Jon Riley/Tony Stone Images; 175(tl), Randal Alhadeff/HRW Photo, location courtesy Bill Mundy South; 176(tl), Peter Van Steen/HRW Photo; 177(cl), Peter Van Steen/HRW Photo, location courtesy Recreational Equipment Inc.; 178(cl), Jim Zuckerman/First Light; 179(tr), John Langford/HRW Photo; 179(br), John Warden/Tony Stone Images; 180(tr), Don Couch/HRW Photo; 181(cr), Sam Dudgeon/HRW Photo; 181(bc),182(cr), eStudios/HRW Photo; 182(br), Sam Dudgeon/HRW Photo; 183(tl), Ron Tanaka/HRW Photo; 184(bl),186(tr), Sam Dudgeon/HRW Photo; 186(tc), Allsport USA/Brian Bahr; 186(c), Allsport USA/Matthew Stockman; 189(br), Lori Hdamski Peek/Tony Stone Images; 190(bl), David Madison/Bruce Coleman Inc.; 191(br), Laura Riley/Bruce Coleman Inc.; 191(bc), Charles Palek/Animals Animals; 192(tc), Allsport USA/Shaun Botterill; 192(tr), John Warden/Tony Stone Images; 192(tl), Steve Niedorf/The Image Bank © 2000; 196(bc), Mark Stephenson/First Light; 196(bl), Peter Gridley/Masterfile; 196(br), Rivera Collection/SuperStock; 196(bc), SuperStock; 199(bl), Lee Foster/Bruce Coleman Inc.; 201(tr), Marilyn Kazmers/Peter Arnold, Inc.; 204(tl), C. Allan Morgan/Peter Arnold, Inc.; 206(bl), Stuart Westmorland/Tony Stone Images; 207(c), Rob Gage/Masterfile; 207(cr), Corbis-Bettmann; 208(c), Michelle Bridwell/HRW Photo; 209(tl), Michael Young/HRW Photo; 209(bl), Michelle Bridwell/PhotoEdit. **CHAPTER FIVE:** Page 216(b), Jean-Marc Truchet/Tony Stone Images; 217(c), Sylvain Grandadam/Tony Stone Images; 217(tr), Peter Griffith/Masterfile; 217(cl), Telegraph Colour Library/FPG International; 217(tl), Ron Watts/First Light; 218(tr), John Langford/HRW Photo; 221(cr), John Langford/HRW Photo, props courtesy Gruene River Co.; 222(cl), Mark Burnham/First Light; 222(cl), Hubert Kanus/Photo Researchers Inc.; 223(bl), Peter Van Steen/HRW Photo; 224(br), Brian Bailey/Tony Stone Images; 225(tl), Don Couch/HRW Photo; 226(tl) Image Copyright © 2001 PhotoDisc, Inc.; 228(tr), Randal Alhadeff/HRW Photo; 232(l), Tom Bean; 232(br), Jack Olson; 233(tl), Tony Stone Images; 233(bc), eStudios/HRW Photo; 234(l), Eva Rubenstein/Photonica; 235(cr), Lisa Stancati /Photonica; 236(tr), Bill Brooks/Masterfile; 237(bl), Randal Alhadeff/HRW Photo; 238(tl), Y.C.L.-TCL/Masterfile; 239(tr), Don Couch/HRW Photo; 240(tl), Lori Adamski Peek/Tony Stone Images; 242(tr), John Langford/HRW Photo; 243(cr), Mike McQueen/Tony Stone Images; 243(c), Garry D. McMichael/Photo Researchers Inc.; 244(tc),(tr), John Langford/HRW Photo, location courtsey Chaparral Ice Center at Northcross Mall; 248(bl), Harvey Kennan/Tony Stone Images; 248(b), Ken Straiton/First Light; 251(b), Albert Normandin/Masterfile; 252(tc), Kasala/Gamma Liaison; 252(tr), Joe Larusso/Gamma Liaison; 257(cl), Corbis-Bettmann/John Springer; 258(c), George Hunter/Tony Stone Images; 258(tr), Alec Pytlowany/Masterfile; 259(br), Randal Alhadeff/HRW Photo; 261(tr), Peter Van Steen/HRW Photo; 263(c), Frank Whitney/The Image Bank © 2000; 263(cr), David R. Frazier Photolibrary; 264(tr), M. H. Dunn/First Light; 264(b), John Zimmerman/FPG International; 270(br), P.R. Production/SuperStock. **CHAPTER SIX:** Page 274(br),(tr),275(tc), Artbase Inc.; 275(tl), Index Stock Photography Inc./Lonnie Duka; 275(tr), Andy Snow/Science Photo Library/Photo Researchers Inc.; 276(t), Burke/Triolo Productions/FoodPix; 277(tr), Randal Alhadeff/HRW Photo; 280(br), Alan Sirulnikoff/First Light; 281(tl), Randal Alhadeff/HRW Photo; 281(tl), Artbase Inc.; 282(tc),(tr), JPL/NASA/Gamma Liaison; 286(bl), Bob Anderson/Masterfile; 287(br), Noriyuki Yoshida/SuperStock; 289(tr), John Langford/HRW Photo; 290(tr), Artbase Inc.; 294(cr), Randal Alhadeff/HRW Photo; 294(tr), Michelle Bridwell/HRW Photo; 299(tl), Christine Galida/HRW Photo; 300(tc), Darius Koehli/First Light; 300(tr), Comstock; 302(cr), J. Boutin/Publiphoto; 306(br), Paul Silver/Bruce Coleman Inc. ; 307(tr), Space Frontiers/Masterfile; 308(tc), Index Stock Photography Inc./Benelux Press; 308(bl), Allsport USA/Shaun Botterill; 309(t),(c), Mark E. Gibson; 309(c), Allsport USA/Brian Bahr. **CHAPTER SEVEN:** Page 316(tr), Chris Harvey/Tony Stone Images; 316(b), John Cancalos/Valan Photos; 316(cr), Corbis/Raymond Gehman; 317(c), NASA Space Photo/The Image Bank © 2000; 317(tr), NASA/Science Photo Libary/Photo Researchers Inc.; 318(tr), Ian Crysler; 322(t), Artbase Inc.; 324(b), Ron Sherman/Tony Stone Images; 326(t), Richard Dole; 331(t), Map ©1998 by Rand McNally #98-S-158; 331(c), Chris Speedie/Tony Stone Images; 331(tc),333(bl), David Starrett; 337(tl), Tomb 181 - Bebamun & Ipuky; 343(br), Don Couch/HRW Photo; 344(b), Randal Alhadeff/HRW Photo (driver); 344(b), Peter Van Steen/HRW Photo (car); 345(c), Stratford Festival production of The Tempest, 1976, directed by Robin Phillips and William Hutt, designed by John Ferguson, basic set design by Daphne Dare, music by Berthold Carriere, lighting by Gil Wechsler. William Hutt (left) as Prospero with members of Festival Acting Company. Courtesy of the Stratford Festival Archives; 349(tl), Randal Alhadeff/HRW Photo; 350(tl), Ian Crysler; 351(br), Michelle Bridwell/HRW Photo; 352(bl), Sam Dudgeon/HRW Photo; 353(tr), David Starrett; 354(bl), Comstock; 354(bc), Erik Simonsen/The Image Bank © 2000; 354(br), Artbase Inc.; 356(br), Dennis Fagan/HRW Photo; 357(bl), Bill Lishman/Joseph Duff - Operation Migration; 358(c), John Langford /HRW Photo, props courtesy Gruene River Co.; 359(cl), J. Towers/First Light; 360(bl), Joe Cornish/Tony Stone Images. **CHAPTER EIGHT:** Page 361(cr), John Langford/HRW Photo; 368(br), John Reader/Science Photo Library/Photo Researchers Inc.; 369(tl), Chris Harvey/Tony Stone Images; 369(tr), S. Lowry/Univ. Ulster/Tony Stone Images; 369(cr), Gary Buss/FPG International; 370(tc), Randal Alhadeff/HRW Photo; 374(cr), S. Lowry/Univ.Ulster/Tony Stone Images; 375(br), Victoria Hurst/First Light; 375(bl), Leonard Lee Rue III/Animals Animals; 377(tr), Heatons/First Light; 377(tr), Artbase Inc.; 380(tr),382(cr), John Langford/HRW Photo; 383(tr), Patti Murray/Earth Scenes; 383(cr), Bill Ross/First Light; 385(br), Reza Estakhrian/Tony Stone Images; 386(tl), Natural Selection/First Light; 386(tr), Michael Dick/Animals Animals; 389(c), Don Couch/HRW Photo, props courtesy The Hat Box; 390(tr), Telegraph Colour Library/Masterfile; 392(c),(cr), Randal Alhadeff/HRW Photo; 394(cl), Color Day/The Image Bank © 2000; 395(cl), Corbis-Bettmann; 395(cl), Artbase Inc.; 396(tl), © 1998 Rick Friedman/Black Star; 396(cl), Sam Dudgeon/HRW Photo; 397(tc),(c), Scott Van Osdol/HRW Photo; 398(bckgd), Luke Dodd/Science Photo Library/Photo Researchers Inc.; 398(c), John Sanford/Science Photo Library/Photo Researchers Inc.; 398(tr), Scott Van Osdol/HRW Photo; 402(br), Paul Morrell/Tony Stone Images; 402(bl), Artbase Inc.; 404(tr), Paolo Negri/Tony Stone Images; 408(bl), Corbis-Bettmann; 408(bl), Artbase Inc.; 409(all), Kenneth Garrett/National Geographic Society; 410(br), Sam Dudgeon/HRW Photo; 411(tr), M. Bradley/Bruce Coleman Inc.; 412(c), Douglas E. Walker/Masterfile; 413(cl), Comstock/Colin Quirk; 414(cr), Philip & Karen Smith/Tony Stone Images; 415(cr), Bernd Kappelmeyer/Masterfile; 416(b), Gail Shumway/FPG International; 416(t), Peter McLeod/First Light; 417(cr), Joel Sartorengs/National Geographic Society; 417(cr), T. Kitchin/First Light; 417(br), K&K Ammann/Bruce Coleman Inc.; 417(br), Don Enger/Animals Animals; 417(tr), Tom Brakefield/Bruce Coleman Inc. **CHAPTER NINE:** Page 424(all),425(all), Ron Behrmann/HRW Photo; 426(tr), Randal Alhadeff/HRW Photo; 428(br), Phil Jason/Tony Stone Images; 431(tr),432(c), Peter Van Steen/HRW Photo; 437(br), Corbis-Bettmann; 438(tr), Sam Dudgeon/HRW Photo; 440(cl),442(cr), Peter Van Steen/HRW Photo; 443(tr), Sam Dudgeon/HRW Photo; 445(tl), Peter Van Steen/HRW Photo; 451(tr), Don Couch/HRW Photo; 458(tr), John Langford/HRW Photo; 464(tr), Allsport USA/Mark Thompson; 471(cr), Index Stock Photography Inc./Shaffer-Smith. **CHAPTER TEN:** Page 478(c), David Madison/Tony Stone Images; 478(bl),(br), Randal Alhadeff/HRW Photo; 479(t), Corbis/NASA; 479(c), Henry Groskinsky/Peter Arnold, Inc.; 480(tr),483(bl), Jeff Hunter/The Image Bank © 2000; 486(tc), Corbis-Bettmann; 486(cr), SEF/Art Resource, NY; 488(cr), Sarah Stone/Tony Stone Images; 491(cr), Peter Van Steen/HRW Photo; 492(all), John Langford/HRW Photo; 497(tr), Ezio Geneletti/The Image Bank © 2000; 498(tr), Peter Van Steen/HRW Photo; 503(tr), National Museum of American History Smithsonian Institution; 504(tr), Nancy Engebretson/Phoenix Gazette/Used with permisson. Permission does not imply endorsement.; 504(tl), (cl), Jump Run Productions/The Image Bank © 2000; 504(bl), Jason Hawkes/Tony Stone Images; 505(tr), Kris Coppieters/SuperStock; 506(tr), Allsport USA/Mike Powell; 511(tr), Roy Ooms/Masterfile; 513(bl), Peter Van Steen/HRW Photo; 515(bl), Trevor Bonderud/First Light. **CHAPTER ELEVEN:** Page 524(bc), NSP/PP/Nawrocki Stock Photo Inc.; 524(cr), Dale Sanders/Masterfile; 525(tr), Scott Barrow/International Stock; 525(tr), Corbis/Bettmann; 525(tl), Cirque du Soleil Mystère, Photo: Al Seib, Costumes: Dominique Lemieux; 526(t), David Phillips/HRW Photo; 528(tr), Randal Alhadeff/HRW Photo; 532(tr), Michelle Bridwell/HRW Photo; Michelle Bridwell/HRW Photo; 538(tr), Dennis Fagan/HRW Photo; 541(tr), John Woods/Index Stock Photography Inc.; 542(br), Robert Lubeck/Animals Animals; 543(br), Randal Alhadeff/HRW Photo; 544(tl), Index Stock Photography Inc./Matthew Borkoski; 544(bl), Index Stock Photography Inc./Stewart Cohen; 545(cr), Andrew Syred/Tony Stone Images; 546(cr), Index Stock Photography Inc./The Sports File; 551(tr), Michelle Bridwell/PhotoEdit; 552(tr), Dennis Fagan/

HRW Photo; 555(cr), David Phillips/HRW Photo; 556(br), Comstock/George Hunter; 557(cr), Leonard Lessin/Peter Arnold, Inc.; 565(tr), Randal Alhadeff/HRW Photo; 566(tc), Michelle Bridwell/HRW Photo; 566(tr), Michael Tamborrino/FPG International; 566(tl), Benjamin Rondel/First Light; 567(tl), Jake Rajs/Tony Stone Images; 567(tr), David Noton/Masterfile. **CHAPTER TWELVE:** Page 574(tr), John Mead/Science Photo Library/Photo Researchers Inc.; 574(bc), Hilarie Kavanagh/Tony Stone Images; 575(tr), D. E. Cox/Tony Stone Images; 575(cr), Vladimir Pcholkin/FPG International; 576(tr), Michelle Bridwell/HRW Photo; 583(tr), Dennis Fagan/HRW Photo; 584(tr),585(cr), John Langford Photography; 587(cr), Gary Withey/Bruce Coleman Inc.; 588(cl), Ken Reid/Masterfile; 590(c), Peter Van Steen/HRW Photo; 591(tc), Vic Thomasson/Tony Stone Images; 591(tr), G.A. Plimpton Collection, Rare Book & Manuscript Library, Columbia University; 593(cr), Jan Becker; 595(tl), Corbis-Bettmann; 596(br), Jake Rajs/Tony Stone Images; 597(tr), Randal Alhadeff/HRW Photo, props courtesy Lights Fantastic; 598(tr), Matthew Neal McVay/Tony Stone Images; 599(cr), Nawrocki Stock Photo Inc.; 600(br), Peter Van Steen/HRW Photo; 604(br),(bc), United States Coast Guard; 606(tr), The SKYSCAN Photolibrary; 613(c),(tr), Photo courtesy of the Boeing Company; 615(bl), Mark Wagner/Tony Stone Images; 617(b), Jack Olson; 617(bl), Roy Corral/Tony Stone Images; 617(tr), Brett Froomer/The Image Bank © 2000; 619(br), Peter Van Steen/HRW Photo; 620(tr), Robert P. Carr/Bruce Coleman Inc. ; 621(tr), John Langford/HRW Photo, location courtesy Sea World of Texas; 623(tr), David Lawrence/First Light; 623(bl), Frederick McKinney/FPG International; 624(bl), First Light; 627(cr), Comstock/H. Armstrong Roberts; 627(cr), Toshiba of Canada Limited, Consumer Electronics Group; 628(tc), Aldo Torelli/Tony Stone Images; 635(tl), William Sallaz/The Image Bank © 2000; 636(bl), Peter Van Steen/HRW Photo. **CHAPTER THIRTEEN:** Page 646(tr), Peter Van Steen/HRW Photo(coins); 646(c), Norman Owen Tomalin/Bruce Coleman Inc.; 646(bc), Stephen St. Johnngs/National Geographic; 646(br), Steve Taylor/Tony Stone Images; 647(tl), Peter Cawson/Masterfile; 647(cr), Jerry Cooke/Animals Animals; 647(tr), Leonard Lee Rue III/Photo Researchers Inc.; 648(tr), John Langford/HRW Photo; 649(tr), Peter Van Steen/HRW Photo; 652(br), U.S. Department of the Interior, Indian Arts and Crafts Board, Museum of the Plains Indian; 654(tr), John Langford/HRW Photo; 657(tr), Sam Dudgeon/HRW Photo; 660(tr), Bob Torrez/Tony Stone Images; 661(tr), David Phillips/HRW Photo; 662(tr), John Langford/HRW Photo; 663(cr), Randal Alhadeff/HRW Photo, props courtesy Recreational Equipment Inc.; 665(cr), Michelle Bridwell/HRW Photo; 665(br), Sam Dudgeon/HRW Photo; 666(c),(cr), Peter Van Steen/HRW Photo; 667(tr), David Phillips/HRW Photo; 669(tl), Michelle Bridwell/HRW Photo; 671(br), Leland Bobbe/Tony Stone Images; 673(br), Cyberimage/Tony Stone Images; 674(cl), Allsport USA/Jonathan Daniel; 675(cr), Allsport USA/Vincent Laforet; 676(t), CORBIS/Duomo; 680(br), John Langford/HRW Photo, location courtesy Sea World of Texas; 680(tr), CORBIS/Duomo; 681(bl), John Langford/HRW Photo; 682(c), 683(cl), 683(cr), Peter Van Steen/HRW Photo. **CHAPTER FOURTEEN:** Page 690(b), John Elk III/ELKJO/Bruce Coleman Inc.; 691(tc), Rex Ziak/Tony Stone Images; 691(tl), Jason Hawkes/Tony Stone Images; 691(tr), Comstock/George Gerster; 691(cr), Harald Sund/Image Bank; 692(tr), John Kieffer/Peter Arnold, Inc; 696(c), Comstock; 696(cl), The New York Public Library, Astor, Lenox and Tilden Foundations from Photographs and Prints Division Schomburg Center for Research in Black Culture; 699(tr), I.T.P./International Stock; 700(tr), *Pegasus* (Tiled facade for the hall of a school in the Hague) by M.C. Escher ©1998 Cordon, Art-Baarn-Holland. All rights reserved; 702(t), Sam Dudgeon/HRW Photo; 706(cr), Seattle Mariners/NBBJ Architects; 707(tc), Ron Thomas/FPG International; 713(tr), Chuck Close, *John*, 1992 (portrait of artist with work in progress), oil on canvas, 100" x 84", Photograph by Bill Jacobson, Courtesy of PaceWildenstein; 714(tc),(tr), Peter Van Steen/HRW Photo; 717(br), Raymond G. Barnes/Tony Stone Images; 718(br), 720(bl), 720(br), Sam Dudgeon/HRW Photo; 720(cr), George Lepp/Tony Stone Images; 721(cr), Ron Sanford/Tony Stone Images; 722(tr), Lori Adamski Peek/Tony Stone Images; 725(tl), Dugald Bremmer/Tony Stone Images; 729(tr),(tc), Scott Van Osdol/HRW Photo; 732(br), Mark E. Gibson

## ILLUSTRATIONS

*All art, unless otherwise noted, by Holt, Rinehart & Winston.*
**Chapter One:** Page 14, Mountain High Maps® ©2001; 22, Michael Morrow; 42, Leslie Kell. **Chapter Two:** Page 60, Michael Morrow; 89, Uhl Studios. **Chapter Three:** Page 120, Boston Graphics; 128, Lori Osiecki; 133, Boston Graphics; 150, David Fischer; 154, Lori Osiecki. **Chapter Four:** Page 162, Annie Bissett; 163, Annie Bissett; 165, Michael Morrow; 167, Michael Morrow; 170, Stephen Durke; 171, Michael Morrow; 173, Michael Morrow; 180, Michael Morrow; 186, Michael Morrow; 192, Elizabeth Brandt; 193, Leslie Kell/HRW; 194, Leslie Kell/HRW; 195, Stephen Durke; 196 (c), Michael Morrow; 200, Michael Morrow; 202, Stephen Durke. **Chapter Five:** Page 245, Boston Graphics; 253, Stephen Durke; 257, Michael Morrow. **Chapter Six:** Page 302, Michael Morrow. **Chapter Seven:** Page 324, Michael Morrow; 325, Boston Graphics; 336, Michael Morrow; 337, Stephen Durke; 338, Ortelius Designs; 345, Nenad Jakesevic; 354, Stephen Durke. **Chapter Eight:** Page 372, Christy Krames; 376, Michael Morrow; 390, David Fischer; 407, MapQuest.com. **Chapter Nine:** Page 436, Brent Spraggins; 439 (b), Michael Morrow; 442, Brent Spraggins, 448, Stephen Durke; 452, David Fischer; 456 (t), Stephen Durke; 456 (b), Uhl Studios; 457, Stephen Durke; 463, MapQuest.com. **Chapter Ten:** Page 485, Nenad Jakesevic; 511, Stephen Durke/Nishi Kumar; 513 (cr), Michael Morrow; 517, Uhl Studios. **Chapter Eleven:** Page 527, Boston Graphics; 529, Michael Morrow; 537, David Fischer; 567, Brent Spraggins. **Chapter Twelve:** Page 594, Michael Morrow; 598, Stephen Durke; 609, Stephen Durke; 626, Michael Morrow. **Chapter Thirteen:** Page 653, Michael Morrow; 661, Lori Osiecki; 667, Nishi Kumar; 672, Stephen Durke.

## PERMISSIONS

For permission to reprint copyrighted material, grateful acknowledgment is made to the following sources:

***The Associated Press:*** From "Atlanta Architect Steps Up Quest for Safe Staircases" by Lauran Neergaard from the *Albuquerque Journal*, 1993. Copyright © 1993 by The Associated Press.

***Hill and Wang, a division of Farrar, Straus & Giroux, Inc.:*** From *Innumeracy* by John Allen Paulos. Copyright © 1988 by John Allen Paulos.

***Karol V. Menzie and Randy Johnson:*** From "Count Your Way to Stair Success" by Karol V. Menzie and Randy Johnson from "Homework" column from *The Baltimore Sun*, 1993. Copyright © 1993 by Karol V. Menzie and Randy Johnson.

***The National Council of Teachers of Mathematics:*** From page 84 from *The Rhind Mathematical Papyrus*, translations by Arnold Buffum Chace. Published by The National Council of Teachers of Mathematics, 1979.

***The New York Times Company:*** From "Hot Hands' Phenomenon: A Myth?" from *The New York Times*, April 19, 1988. Copyright © 1988 by The New York Times Company. From graph, "Getting Lost in the Shuffle," and from "In Shuffling Cards, 7 Is Winning Number" by Gina Kolata from *The New York Times*, January 9, 1990. Copyright © 1990 by The New York Times Company. From "Faulty Rivets Emerge as Clues To Titanic Disaster" by William J. Broad from *The New York Times*, January 27, 1998. Copyright © 1998 by The New York Times Company.

***Newsweek, Inc.:*** From "Finding Order in Disorder" by Sharon Begley from *Newsweek*, December 21, 1987. Copyright © 1987 by Newsweek, Inc. All rights reserved.

***Springer-Verlag New York, Inc.:*** Excerpt (retitled "Researchers' Survey of 100 Basketball Fans") from "The Cold Facts About the 'Hot Hand' in Basketball" by Tversky, Amos, and Gilovich from *Change: New Directions for Statistics and Computing*, vol. 2, no. 1, 1989. Copyright © 1989 by Springer-Verlag New York, Inc.

***Time Inc.:*** From "A Miraculous Sky Rescue" from *Time*, May 4, 1987. Copyright © 1987 by Time Inc.

***U.S. News & World Report:*** From "Now, the testing question" from *U.S. News & World Report*, September 21, 1992. Copyright © 1992 by U.S. News & World Report, Inc.

***The Wall Street Journal:*** From "Counting Big Bears" by Marj Charlier from *The Wall Street Journal*, May 2, 1990. Copyright © 1990 by Dow Jones & Company, Inc. All Rights Reserved Worldwide.